“十三五”国家重点出版物
出版规划项目

Series on Advanced Electronic Packaging Technology and Key Materials

先进电子封装技术与关键材料丛书

汪正平（C.P. Wong） 刘胜（Sheng Liu） 朱文辉（Wenhui Zhu） 主编

Modeling and Simulation for Microelectronic Packaging and Integration

Manufacturing, Reliability and Testing

Second Edition

微电子封装和集成的建模与仿真

——制造、可靠性和测试

（第二版）

刘 胜（Sheng Liu） 刘 勇（Yong Liu） 著

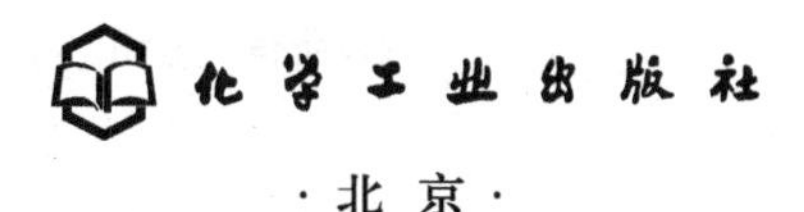

·北 京·

内 容 简 介

随着电子封装的发展，电子封装已从传统的四个主要功能（电源系统、信号分布及传递、散热与机械保护）扩展为六个功能，即增加了DFX及系统测试两个新的功能。其中DFX是为“X”而设计，X包括：可制造性、可靠性、可维护性、成本，甚至六西格玛。DFX有望在产品设计阶段实现工艺窗口的确定、可靠性评估和测试结构及参数的设计等功能，真正做到“第一次就能成功”，从而将计算机辅助工程（CAE）变为计算机主导工程（CE），以大大加速产品的上市速度。本书是全面介绍DFX在封装中应用的图书。作为封装工艺过程和快速可靠性评估及测试建模仿真的第一本专著，书中包含两位作者在工业界二十多年的丰富经验，以及在MEMS、IC和LED封装部分成功的实例，希望能给国内同行起到抛砖引玉的作用。同时，读者将会从书中的先进工程设计和微电子产品的并行工程和协同设计方法中受益。

本书第2版新增了两位作者在电子制造和封装领域新的成果与经验，例如电力电子模块的建模和仿真、电子封装耐热性的分析模型、3D TSV封装等内容。

本书主要读者对象为学习DFX（制造工艺设计、测试设计、可靠性设计等）的研究人员、工程师和学生等。

图书在版编目（CIP）数据

微电子封装和集成的建模与仿真：制造、可靠性和测试 = Modeling and Simulation for Microelectronic Packaging and Integration: Manufacturing, Reliability and Testing：英文 / 刘胜，刘勇著. —2版. —北京：化学工业出版社，2021.8

（先进电子封装技术与关键材料丛书 / 汪正平，刘胜，朱文辉主编）

ISBN 978-7-122-39227-5

Ⅰ. ①微… Ⅱ. ①刘…②刘… Ⅲ. ①微电子技术-封装工艺-系统建模-英文②微电子技术-封装工艺-系统仿真-英文 Ⅳ. ①TN405.94

中国版本图书馆CIP数据核字（2021）第096992号

责任编辑：吴 刚 毛振威　　封面设计：关 飞

责任校对：王素芹

出版发行：化学工业出版社（北京市东城区青年湖南街13号 邮政编码100011）

印　　装：北京建宏印刷有限公司

710mm×1000mm 1/16 印张45¼ 字数1442千字 2021年12月北京第2版第1次印刷

购书咨询：010-64518888　　售后服务：010-64518899

网　　址：http://www.cip.com.cn

凡购买本书，如有缺损质量问题，本社销售中心负责调换。

定　价：498.00元

Preface to the Series

The technical level and development scale of the integrated circuit (IC) industry is one of the important indicators to measure a country's industrial competitiveness and comprehensive national strength, and is the source of modern economic development. The application of IC has already become routine in various industries, such as military satellites, radar, civilian automotive electronics, smart equipment, and consumer electronics, etc. At present, the IC industry has formed three major industrial chains of design, manufacturing and packaging testing, which have become the indispensable pillar in the IC industry.

IC packaging is an indispensable process in the IC industry, which is the bridge from chip to device and device to system. It is a key fundamental manufacturing part of the IC industry and a competitive commanding height for the core device manufacturing of the IC industry.

With the rapid development of IC technology, higher and higher requirements for miniaturization, multi-function, high reliability and low cost of electronic products are put forward. Facing this situation, the electronic packaging materials and technologies are undergoing rapid development, promoting lots of advanced packaging materials. Advanced electronic packaging materials and technologies are the core of IC packaging.

In order to promote the development of China's advanced electronic packaging industry and meet the urgent needs of researchers ranged from teaching and scientific study to engineering developing in the field of electronic packaging, the editorial committee has invited famous specialists to write the *Series on Advanced Electronic Packaging Technology and Key Materials* in recent years (English version). The series includes: "*Advanced Polyimide Materials*", "*From LED to Solid State Lighting* ", "*Freeform Optics for LED Packages and Applications*", "*Modeling, Analysis, Design and Tests for Electronics Packaging beyond Moore*", "*TSV(through-silicon via technology) Package*", etc.

This series of books systematically describes the advanced electronic packaging from three aspects: advanced packaging materials, advanced packaging technologies and advanced packaging simulation design methods. This series covers the most advanced packaging materials such as polyimide materials and packaging technologies such as freeform optical technology, TSV (through-silicon via technology) packaging, and advanced packaging simulation design methods such as multi-physics analysis and applications. In addition, this series also makes a planning outlook and forecast for the development trend of advanced electronic packaging.

This series of books is of great worth for workers engaged in scientific research, production and application in electronic packaging and related industries, and also has great reference significance to teachers and students of related majors in higher education institutions.

We believe that the publication of this series of books will play a positive role in promoting the development of China's IC industry and advanced electronic packaging industry.

Finally, we would like to express our sincere gratitude to our colleagues who have worked hard in the preparation of this series. We also express our heartfelt thanks to those who participated in organizing the publication of this series!

C.P. Wong
IEEE Fellow
Member of Academy of Engineering of the USA
Member of Chinese Academy of Engineering
Former Bell Labs Fellow

Dean of Engineering, The Chinese University of Hong Kong
Regents' Professor, Georgia Institute of Technology, Atlanta, GA 30332, USA

Sheng Liu, Ph.D
IEEE Fellow, ASME Fellow
Chang Jiang Scholar Professor
Dean, School of Power and Mechanical Engineering,
Founding Executive Director, Institute of Technology Sciences
Associate Dean of School of Microelectronics, Wuhan University
Professor of School of Mechanical Science and Engineering
Huazhong University of Science and Technology
Wuhan, Hubei, China

Wenhui Zhu, Ph.D.
National Invited Professor
College of Mechanical and Electrical Engineering,
Central South University
Changsha, Hunan, China

Contents

Foreword

By Jianbin Luo

Over the past three decades, the semiconductor technology has been advancing rapidly and has impacted almost every industry sector including those emerging sectors such as internet of things (IoT), autonomous driving, 5G/6G and internet security. On one hand, increasing fine feature towards deep sub-micron and nano domains follows Moore's law, forming system on chip (SoC) integration. With the increasing cost in investing equipment required in SoC and strong needs for more functions such as the RF, power electronics, microsensors, MEMS, solid state lighting, solar cells, often called the More than Moore, demand a technology, widely called system in packaging (SiP). The real drive behind Moore's law and More than Moore is the increasing business demand on increasing cost reduction, shorter time-to-market and time-to-profit, making it impossible for the whole industry to use trial and error approach to the technology development and product design. Design of these sophisticated products needs expertise in different disciplines. What is more, the design of these products needs to consider the various manufacturing processes that can induce defects in some steps, which may either degrade the performance of the device and system, or these defects may grow to some non-tolerable sizes in those subsequent steps. It is often found that the defect in some particular assemble step, when generated, can result in either yield problem in processing and testing, or catastrophic failures, or causing returns in the field use. Mechanics is so important in microelectronic packaging assemblies and mechanics-based modeling and simulation has been important for aerospace industry and automobile industry and now it is finding more and more importance in microelectronics and nanoelectronics, as the capital investment is so huge for the equipment and any mistake made in the design of product, and design of process window can cause huge penalty. It is often challenging to realize a commercial product to be the first-time pass, which requires robust model-based design, many model-based accelerated reliability tests, and real time in-line testing. The strong industrial needs have attracted many outstanding researchers to overcome barriers in the development of the microelectronic and nanoelectronics products and it has been the dream of mechanics researchers to conduct work in depth towards final goal of virtual manufacturing, virtual reliability and virtual testing. Although, the dream may not fully come true in the near future, as a lot of hard work must be done, but a framework, based on a lot of dedicated efforts, will help significantly reduce the amount of testing, amount of trial and error manufacturing, and amount of uncontrolled testing, which will make our manufacturing and products greener, our society less polluted, more resources reserved, and the life for us and our kids better.

The authors Prof. Sheng Liu and Dr. Yong Liu, very dedicated scientists and outstanding practicing engineers at the same time, have pioneered efforts in almost all the major manufacturing processes since early 1990s for plastic packaging to the 3D TSV to nano-welding and to nano-film deposition. I have known them for more than 20 years and I am impressed that they have been so dedicated in building the needed material database and constitutive models, building robust models, finding ways to validate models, keeping refining them, use the models to solve numerous issues in manufacturing, reliability and testing, and help the product developments in various industry and business sectors such as commercial electronics, portable devices, automobile electronics, powering electronics, CPUs, micro sensors, solid state lighting, and modern power electronics. I am also happy to see that they have applied modern molecular dynamics to various nano-level processes such as nano film depositions and characterization of lower dimensional materials and systems. It is hoped that this book will help both researchers and engineers and encourage them to do more work and solve many more problems that still need fundamental study and to challenge those problems which have not yet appeared. I am very happy to see the second edition of this book completed and feel that it will be an important addition to mechanical science and engineering, mechanics, microelectronics, nanoelectronics and other industries that have manufacturing processes, reliability and testing involved.

Jianbin Luo, PhD
Chang Jiang Scholar Professor of State Key Lab. Of Tribology
Member of Chinese Academy of Sciences
Tsinghua University
Beijing 100084, China

Foreword

By C. P. Wong

Modeling and simulation of microelectronic packaging and assembly is a multi-disciplinary activity that relies on the expertise of sequence dependant complex processes, almost all the material types, and detailed process windows; a very challenging task for both academic people and practicing engineers. Modeling and simulation has been classified in the ITRS Roadmap in the past years as being one of the cross cutting technologies that must be mastered to enable rapid progress in this first industry. The importance of modeling and simulation has been witnessed by the increasing number of design engineers in each corporation from 20% in the early 1980s to 80% in the late 2000s in terms of recruited engineers, as it is essential to design and make the product the first time right.

The most popular methodology of design and manufacturing is called Design for X (DFX, here X refers to manufacturing, assembly, testing, reliability, maintenance, environment, and even cost), which has been widely adopted by those multinational and small high tech start-up companies. The design methodology is being adjusted to meet the requirements of a full-life cycle, the so called "concept/cradle-to-grave" product responsibility, coined by Dr. Walter L. Winterbottom of Ford Science Lab.

A packaging module and related application systems, like any other electronic systems, involve a lot of manufacturing processes from film deposition, etching, chip to wafer and wafer to wafer bonding, dicing/singulation, and extensive reliability testing for extended-life goals of many critical products such as those used for automobile electronics, avionics, portable electronics, and so on. The defects in terms of voids, cracks, delaminations, and microstructure changes, can be induced in any step and may interact and grow in subsequent steps, imposing extreme demands on the fundamental understanding of stressing and physics of failures. Currently, the testing programs have been extensive to assure reliability during product development. An iterative, build-test-fix-later process has long been used in new product development, significant concerns are being addressed as cost effective and fast time-to-market needs may not be achievable with such an approach. In the sense of high reliability, system hardware design and manufacturing and testing are costly and time consuming, and severely limit the number of design choices within the short time frame, and do not allow enough time to explore the optimal design. With the current situation of three to six months for each generation of IC chip, it is challenging to achieve truly optimal and innovative products with so many constraints in design. Design procedure must be modified and DFX must be used so as to achieve prevention with integrated consideration of manufacturing processes, testing, and operation.

Although there is still a long way to go to enable virtual manufacturing, virtual reliability and virtual testing due to the many difficulties involved, pioneering efforts have been made since the early 1990s by outstanding professionals to shorten the gap significantly, and Professor Sheng Liu and Dr. Yong Liu are two of those brave individuals. Professor Sheng Liu and Dr. Yong Liu have been promoting the new design method in the past years to help assist engineers in material selection, manufacturing yield enhancement, appropriate rapid reliability assessment, and testing when the packaging module and system are subjected to uncertainties of material selection, process windows, and various service loadings. All these issues must be addressed prior to hardware build-up and test.

A lot of processes were first time modeled and simulated by the two authors, addressing the critical importance of modeling in manufacturing, reliability, and testing. Many books about packaging have been written. Some books have been written in which mechanics has been applied to reliability issues only. This is the first book focusing on the many detailed processes in front end, back end, even probing, wire bonding, bonding, and so on. It is the first book to cover the broad aspects from manufacturing to reliability, and to testing, with many examples of their pioneering efforts. The authors describe their contributions in detail and provide guidance to those in the field and present a design approach that must ultimately replace the build-test-fix-later process if the efficiencies and potential cost benefits of the microelectronic packaging systems are to be fully realized.

C.P. Wong
IEEE Fellow
Member of Academy of Engineering of the USA
Member of Chinese Academy of Engineering
Former Bell Labs Fellow
Dean of Engineering, The chinese University of Hong Kong
Regents' Professor, Georgia Institute of Technology, Atlanta, GA 30332

Foreword

By Zhigang Suo

The dramatic rise of the semiconductor industry is fundamentally changing our lives. Memory devices store all human experience, or at least the part that can be digitized. Solid-state lighting saves energy. Biochips detect cancers early and at low cost. Our brains are being scanned to unravel the molecular basis of happiness and despair. The world has become a giant computer, connecting people to people, and to a multitude of devices that extend our senses. The distinction between the computer and the human may one day become a matter of choice.

The exciting future aside, the semiconductor industry is facing major challenges. As the features on chips shrink, approaching the nanoscale, the cost for chip-making equipment has been increasing. Diverse functions are being packaged, as represented by cell phones, power electronics, MEMS, solar cells, and deformable electronics. Complex manufacturing processes inevitably generate defects, which may lower the yield or result in low returns. For an industry particularly sensitive to cost and time-to-market, it has become increasingly difficult to create technologies and design products by using the trial-and-error approach.

Simulation based on mechanics has long been important for the aerospace and automobile industries, and is becoming more and more important in the semiconductor industry. The strong industrial needs have attracted many researchers to develop methods to simulate manufacturing processes and testing methods. While the dream of virtual manufacturing and virtual testing may not fully come true in the near future, simulation has already begun to reduce the amount of trial-and-error manufacturing and uncontrolled testing. The sustained effort will make the technology greener.

The authors of this book, Professor Sheng Liu and Dr. Yong Liu, are outstanding researchers and engineers. Since early 1990s, they have been intimately involved in pioneering efforts in simulating almost all the major manufacturing processes, ranging from plastic packaging to nano welding. They have devoted their careers to building material databases, developing constitutive models, finding ways to validate the models, constantly refining them, and using the models to solve practical problems. They have applied mechanics to various industry sectors, including commercial electronics, automobile electronics, powering electronics, CPUs, micro sensors, and solid-state lighting. Their experience is distilled in this book. The book is an important addition to mechanics and microelectronics, as well as to other industries that involve sophisticated manufacturing processes and testing methods.

Zhigang Suo, PhD
Allen E. and Marilyn M. Puckett Professor of Mechanics and Materials
School of Engineering and Applied Sciences, Kavli Institute for Bionano Science and Technology
Member of the U.S. National Academy of Engineering
ASME Fellow
Harvard University
Cambridge, Massachusetts 02138, USA

Preface to Second Edition

The development of the integrated circuit (IC) is probably the most important indictor to measure a modern country's industrial competitiveness and comprehensive national strength, and the source of modern economic development. As described in the preface of the first edition of this book, IC industry is shifting from being powered by Moore's law which has been focusing on IC miniaturization down to nano dimensions and system on chip (SoC) integration to more on system in packaging (SiP), which are coined by "More than Moore", which will be based on not only by silicon technologies but also broad-band semiconductors, flexible electronics and even possibly low dimensional materials. It is estimated that the integration density from 7nm to 5nm may bring less than 20% enhancement, and the most advanced packaging technology may bring 10 × to more than 1000 × enhancement. With appropriate advances of architectures in chips, packaging and system, it is likely that the performance of a system with mixed technology nodes (42nm, 22nm, 14nm, 10nm, 7nm, 5nm, etc.) cannot be worse than the most advanced node, which will depend on the co-design in electrical, thermal, stress, and reliability, etc. It is estimated the SiP research will be very active in our community, in particular in those countries and areas where advanced chip equipment cannot be available. The applications will cover probably every sector of chip application, which forms the core part of the 3rd and 4th industrial revolutions.

With so many processes involved in micro- and nano-electronics related chip and packaging developments, contaminants must be involved and therefore, 40%-50% processes are involved in cleaning which may also involve various physical and chemical methods, with the purpose to enhance the yield of each process, as the total yield of the product is the product of yields for all the major processes. In the framework of virtual manufacturing and even in the digital twin-based intelligent manufacturing, models are needed to address the effects of environments and contaminants on the yield and reliability.

With the trend of 3D integration in both chip and packaging, vertical vias with high aspect of ratio are involved and etching and filling of the vias may involve a lot of challenges, such as the pre-wetting process of TSV electroplating. Micro and nano scale defects may be involved and molecular dynamics modeling must be conducted for the thin film deposition and various substrates.

The second edition of this book is further concerned with the various aspects of the electronic manufacturing and packaging. In particular, we have three new chapters, Chapter 19 on modeling and simulation of power electronic modules, Chapter 20 on analytical models for thermal resistances in electronics packaging, Chapter 21 on 3D through silicon via (TSV) packaging. We also update Chapter 23 with in-depth discussions on the film deposition of AlN, effects of process parameters, AlN on GaN, AlGaN on AlN, mechanical properties of AlN with Raman verification, and the molecular distinctive nanofriction of graphene coated copper foil.

Sheng Liu, Ph.D.
IEEE Fellow, ASME Fellow
Chang Jiang Scholar Professor
Dean, School of Power and Mechanical Engineering
Founding Executive Director, Institute of Technology Sciences
Associate Dean of School of Microelectronics, Wuhan University
Professor of School of Mechanical Science and Engineering
Huazhong University of Science and Technology
Wuhan, Hubei, China

Yong Liu, PhD, IEEE Fellow
On Semiconductor
South Portland, USA

Preface to First Edition

Over the past two decades semiconductor technology has made impressive progress, particularly in electronics, opto electronics, communications, health, automotive applications, computing, consumer electronics, security, and industrial electronics. These progresses are powered by Moore's law which is focusing on IC miniaturization down to nano dimensions and system on chip (SoC) integration. However, there are technologies based on or derived from silicon technologies but which do not simply scale with Moore's law such as the RF, power electronics, sensors, MEMS, opto/lighting and other systems in package; these technologies are called "More than Moore". Along with the technology development trends characterized by Moore's law and "More than Moore", the business trends are mainly characterized by cost reduction, shortertime-to-market, and outsourcing. The combination of these technologies and business trends leads to increased design complexity, decreased design margins, increased chances and consequences of failure in reliability, decreased product development in R&D, and difficulties in assembly manufacture and qualification times. Especially in the assembly manufacturing process, quality and reliability are key technologies to ensure a successful product. In addition to the forward looking trends of technology it is important to recognize that the methods for concurrent engineering of these solutions (both the semiconductor content and high performance package capability) are becoming increasingly dependent on rigorous use of proven multi-physics/finite element analysis (FEA) tools and techniques for both new product development and its assembly processes. The correct use of the modeling tool can definitely save design time and shorten the design cycle. The challenges are along with the development of new package technology: can the modeling tools and methodologies be ready to support the new trends? Examples are various designs, reliability and assembly manufacture modeling which include electro migration simulation; diffusion along the interface of two metal materials; contamination at the interface between leadframe, multiple chips and EMC; thermal resistance definition in system in package (SiP); 3D copper stud bumping, wire bonding simulation, and so on.

Most challenges in modeling for electronic packaging today are the fundamental multiple-physics simulation which couples the electrical, thermal, and mechanical fields for various assembly manufacturing processes and for various reliability tests. Development of a highly efficient modeling algorithm for such a SiP system is critical for the virtual prototyping of the new product. In some cases, the SiP might have strong thermal mechanical performance but is weak in the electrical area, or the SiP has very good electrical performance but is weak in thermal-mechanical design. Therefore, it is necessary to establish the best solution using the modeling design of experiment (DoE) while the actual tests or actual assembly manufacturing will cost more and take much longer.

In industry, we all understand the importance of assembly manufacturing, reliability, and testing, because a lot of packaging failures are related to the assembly process. Examples include the wire bonding process, which would induce silicon cratering, bond pad peel off, and interlayer dielectric (ILD) layer under bond pad cracking; die attaching process for multiple die; the order of the die attaching process will have a big impact on the residual stress after the die bonding. When thin die (today the thinnest die thickness would be below 20 microns) is picked up from the tape, the pick-up process could crack the silicon. The molding process is also a key process which could induce later failure, such as delamination due to the voids tracked in the interface between the leadframe and encapsulate molding compound (EMC). Leadframe forming, punch/simulation, and trim may result in the die and package cracking as well. A lot of initial tiny defects are induced in first assembly manufacture process; later in the further assembly process and reliability test, they become potential product quality and reliability concerns. An example is wafer sorting, which will induce the cratering/marks on the bond pad. When the wire bonding process is applied to the cratered/marked bond pad, it will definitely impact the adhesion strength at the interface between the wire bond and the bond pad. Product line engineers are always interested in knowing wire bonding versus wafer probing: which makes things worse? The study of assembly manufacture processes through modeling and simulation started from the beginning of the IC packaging, and today the need for modeling and simulation in this area has increased much more. However, at the present time, there is no book that has systematically described the modeling methodologies for assembly manufacturing, reliability, and testing, as well as discussing the above challenges and giving the readers some unresolved space for further exploring. This is our goal in writing this book: to share our modeling experiences and systematically introduce our modeling methodologies for electronic packaging assembly manufacturing, reliability and testing.

This book is primarily concerned with studies of electronic packaging in assembly manufacture processes and failure mechanisms in assembly manufacture processes and tests through modeling and simulation. However, the fundamental studies regarding the advanced modeling methodologies including molecular dynamics, state of the art simulation algorithms, material constitutive relations, material behavior testing, and various semiconductor reliability tests are also discussed and presented in the book. Various package layouts including the 3D/TSV/Stacking/SiP, carbon nano-tube and interconnects technology, opto packaging, and MEMS are discussed and presented in the book as well. A lot of case studies provide the most advanced research progress and the topics that industry is interested in. The basic framework of this book is arranged in four parts.The first part (Part I) includes eight chapters that introduce the fundamental mechanics concepts, material constitutive models, basic modeling methodologies such as finite element, and the concurrent engineering background for microelectronics. This part is to provide readers with the background knowledge for modeling and simulation, mechanics, material, and the engineering. The next part (Part II) is concerned with the modeling in microelectronic packaging and the assembly manufacturing process which includes five chapters. Major topics include the electronic packaging front of line (which involves key processes such as wafer sorting, die picking up, die attach process, and wire bonding), end of line (which includes molding, leadframe clamping, forming, and singulation), Opto packaging assembly, MEMS packaging assembly, system in packaging stack/3D assembly and the Nano interconnects and packaging assembly. Part III describes the modeling in reliability and testing which includes the wafer probing test and typical package reliability tests such as the temperature cycle test, powercycle test, drop test, electro-migration test, andprecondition test with reflow which includes moisture sensitivity test, vapor pressure at reflow with the popcorning failure mechanisms. The advanced simulation algorithm will be presented especially in the wafer level CSP solder joint reliability in drop test, temperature cycling, and the mass migration induced failure which includes electro migration, thermal migration, stress migration, and the atomic density gradient induced migration. The final part will introduce the modern modeling and simulation methodologies which include the classical molecular dynamics, advanced molecular dynamics, and coupling with continuum modeling methods. This part will disclose how the modern modeling and simulation methodologies will impact the assembly manufacturing, reliability and the testing for today and future advanced package development. An appendix is available on the book companion web site for those who like to know more details about finite element.

The present edition is co-published with John Wiley & Sons, which follows the typesetting of Wiley's edition, including, but not limited to, the fonts, sizes, subscripts, superscripts, normal or italic letters as a courtesy. Metric units have been used throughout this book , though a few common imperial units appear only on very limited occasions. The readers can refer to Appendix for the conversion factors between imperial units and metric units.

Sheng Liu, Ph.D.
IEEE Fellow, ASME Fellow
Chang Jiang Scholar Professor
Dean, School of Power and Mechanical Engineering
Founding Executive Director, Institute of Technology Sciences
Associate Dean of School of Microelectronics, Wuhan University
Professor of School of Mechanical Science and Engineering
Huazhong University of Science and Technology
Wuhan, Hubei, China

Yong Liu PhD, IEEE Senior Member
Fairchild Senior Member of Tech Staff
Fairchild Semiconductor Corporation
South Portland, USA

Acknowledgments

Development and preparation of *Modeling and Simulation for Microelectronic Packaging Assembly: Manufacturing, Reliability and Testing* was facilitated by a number of dedicated people at John Wiley & Sons, Chemical Industry Press, and Huazhong University of Science and Technology. We would like to thank all of them, with special mentions for Gang Wu of Chemical Industry Press, James W. Murphy of John Wiley & Sons, and Bin Song, the research assistant of the first author, at Huazhong University of Science and Technology. Without them, our dream of this book would not have come true, as they have worked so hard on the preparation of the book. It has been a great pleasure and fruitful experience to work with them in transferring our manuscript into a very attractive printed book.

The materials in this book have clearly been derived mainly from previous papers and research notes by the two authors and some works of their friends who are mainly from both the research community and industry, such as Dr. Gary Li of Freescale, Dr. Yifan Guo of Skyworks, and Professor Zhigang Suo from Harvard University. It would be quite impossible for us to express our appreciation to everyone concerned for their collaboration in the production of this book, but we would like to extend our gratitude. In particular, we would like to thank several professional societies in which we have previously published some of our materials included in this book. They are the American Society of American Engineers (ASME) and the Institute of Electrical and Electronic Engineers (IEEE) for their conferences, proceedings, and journals, including *ASME Transactions on Journal of Electronic Packaging, IEEE Transactions on Advanced Packaging, IEEE Transactions on Components and Packaging Technology* and *IEEE Transactions on Electronics Packaging Manufacturing*. Many important conferences such as the Electronic Components and Technology Conference (ECTC), and International Conference on Electronic Packaging Technology and High Density Packaging (ICEPT–HDP), and EuroSime are also appreciated for the reproduction of some of their publication materials.

We would like to acknowledge those colleagues who helped review some chapters and manuscript. They are Dr. Yi-Hsin Pao of Foxcom, Professor Biao Wang of Sun Yat-sen University, Professor Kikuo Kishimoto at Tokyo Institute of Technology, and Professor Ricky Lee at Hong Kong University of Science and Technology. We would like to thank them for many suggestions and comments which added a lot to this book. Their depth of knowledge and their dedication have been demonstrated throughout the process of reviewing this book and its chapters.

We would like to thank Huazhong University of Science and Technology (HUST), Wuhan National Laboratory for Optoelectronics, the School of Mechanical Science and Engineering, and Fairchild Semiconductor Corporation for providing us with excellent working environment to make this book possible. We also appreciate those places where we have previously studied and worked, such as Stanford University, Florida Institute of Technology, and Wayne State University, where we conducted quite a significant part of our research results in the book. We would like to express our appreciation to those who have worked in microelectronic packaging and assembly for many years, such as those SRC mentors in the first author's research. The first author felt honored to have worked with Dr. William Chen and Dr. T. Y. Wu at ASE (formerly at IBM), Dr. YiFan Guo at Skyworks (formerly at IBM and Freescale), Dr. J. Benson at Harris, Dr. T.D. Dudderar at Bell Labs, Dr. L. Li at Cisco (formerly at IBM), Dr. M.J. Lii at Intel, Dr. Z. Mei at HP, Dr. S. Sidharth at AMD, Dr. L. Stark at TI, and Dr. J. Xie at Cisco (formerly at Motorola), Dr.S.Wu at Motorola, and Dr.Luu T.Nguyen at National.We would also like to thank several outstanding scholars and leaders who once worked or who are still working at HUST, who have been providing us with immense help in many aspects and treating us like good old friends. They are Professor Ji Zhou, now the President of Chinese Academy of Engineering, Professor Peigen Li, President of HUST and a member of Chinese Academy of Engineering, Professor Xinyu Shao, VP of HUST, Professor Youlen Xiong, a member of the Chinese Academy of Sciences, and Professor Han Ding, current Dean of School of Mechanical Science and Engineering of HUST.

We would like to show our thanks to our outstanding students, visiting scientists and colleagues. For the research with the first author in the USA and in China, they include Dr. Yuhai Mei, Dr. Jiansen Zhu, Dr. Jianjun Wang, Dr. Zhengfang Qian, Dr. Daqing Zou, Dr. Wei Ren, Mr. Jian Yang, Ms. Changrong Ji, Dr. Xiaohui Song, Mr. Zhaohui Chen, Professor Fulong Dai at Tsinghua University, Professor Xiaoyuan He at Southeast University, Professor Mingfu Lu, Professor Yuwen Qin at Tianjin University, Professor Wei Yang at Zhejiang University (formerly at Tsinghua University), Professors Yilong Bai, Haiying Wang, and Yapu Zhao at the Institute of Mechanics of Chinese Academy of Sciences, and Dr. Jimmy Hu at Microsoft (formerly at FOKD). For the research with the second author in the USA and China, they include Dr. Timwah Luk, Mr. Scott Irving (formerly at Fairchild), Mr. Richard Qian and Dr. Yumin Liu at Fairchild, Professor Zhigang Suo at Harvard University, Professor Antoinette Maniatty at RPI, Dr. Zhen Zhang at Microsoft, Dr. Lihua, Liang and Ms Shinan Wang at Fairchild-ZJUT Joint Lab, Zhejiang University of Technology.

We would like to express our appreciation to the Chinese Electronics Society and its Electronic Manufacturing and Packaging Branch led by Professor Keyun Bi for providing us with many technical and academic exchange opportunities. We particularly appreciate the excellent platform initiated jointly by us in early 1990s, the International Conference on Electronic Packaging Technology (ICEPT, now ICEPT–HDP), now an important IEEE CPMT conference in the world. We are very happy to have worked with outstanding fellows such as Professor C. P. Wong, Professor Ricky Lee, Professor Y. C. Lee, Professor Shangtong Gao, Professor Jushen Ma, Professor Shouwen Yu, Professor Xiangfu Zong, Professor Ming Li, Professor Qingchun Wang, Professor Johan Liu, Professor Xiaoyuan He, Professor Lixi Wan, Dr. Tom Chang, Dr. Daniel Shi, Dr. Guoqi Zhang, Dr. Dongkai Shanguang, Professor Mike Pecht, Professor Kuoning Chuang, Professor Xuejun Fan, Professor James Morris, and many others. Working and socializing with them has been a privilege and a pleasant experience.

We would also like to acknowledge the support of many funding agencies in the past many years such as the USA National Science Foundation, the USA SRC (Semiconductor Research Corporation), the National Natural Science Foundation of China, the Ministry of Science and Technology of China, the National S&T Major Project, the Hubei Department of Science and Technology, the Wuhan Science and Technology Bureau, the Guangdong Department of Science and Technology, the Foshan Bureau of Science and Technology, and the Nanhai Bureau of Science and Technology.

Lastly the first author would like to thank his parents, Mr. Jixian Liu and Ms. Yanrong Shen, his wife, Bin Chen, and his daughter Amy Liu, his son Aaron Liu, and the second author would like to thank both his parents and parents-in-law, his wife, Jane Chen, and his great sons Junyang Liu and Alexander Liu for their love, consideration, and patience in allowing them to work on many weekends and late nights for this book. The author's simple belief is that the contribution of this book to themicroelectronic packaging industry is worthwhile, and will continue to be worthwhile to our civilization for many years to come. The authors would like to dedicate this book to their families.

Sheng Liu, Ph.D.
IEEE Fellow, ASME Fellow
Chang Jiang Scholar Professor
Dean, School of Power and Mechanical Engineering
Founding Executive Director, Institute of Technology Sciences
Associate Dean of School of Microelectronics, Wuhan University
Professor of School of Mechanical Science and Engineering
Huazhong University of Science and Technology
Wuhan, Hubei, China

Yong Liu, PhD, IEEE Senior Member
Fairchild Senior Member of Tech Staff
Fairchild Semiconductor Corporation
South Portland, USA

About the Authors

Sheng Liu is a Chair Professor of Wuhan University and Huazhong University of Science and Technology. He is the dean of the School of Power and Mechanical Engineering, executive director of the Institute of Technological Sciences and associate dean of the School of Microelec tronics at Wuhan University. He is the committee member of the advanced manufacturing of the 8th Science and Technology Commission. He once was a tenured faculty at Wayne State University. He has over 29 years' experience in LED/MEMS/ NEMS/IC packaging. He once won prestigious White House/NSF Presidential Faculty Fellow Award in 1995, ASME Young Engineer Award in 1996, NSFC Overseas Young Scientist Award in 1999 in China, IEEE CPMT Exceptional Technical Achievement Award in 2009, and Chinese Electronic Manufacturing and Packaging Technology Society Special Achievement Award in 2009, the first prize of the technological invention award of the Ministry of Education in 2012, the second prize of the National Technological Invention Award in 2016, the first prize of the technological invention award of China Electronics Society in 2018, the first prize of the National Science and Technology Progress Award in 2020, etc . He has been an associate editor of *IEEE Transaction on Electronic Packaging Manufacturing* since 1999 and an associate editor of *Journal of Frontiers of Optoelectronics in China* since 2007, an editor of *Microsystems and Nanoengineering*, associate editor of *Science Bulletin* (*Engineering*). He obtained his PhD from Stanford University in 1992, his MS and BS degrees from Nanjing University of Aeronautics and Astronautics in 1986 and 1983 respectively. He was an aircraft designer at Chengdu Aircraft Company for two years. He is ASME Fellow and IEEE Fellow. He has filed and owed more than 165 patents in China and USA, has published more than 700 technical articles, has given more than 100 keynotes and invited talks, has edited more than 9 proceedings in English for ASME and IEEE, and has authored / co-authored 5 books.

Yong Liu has been with ON Semiconductor Corp in South Portland, Maine since September 2016 as a Principal Member of Tech Staff. Before Fairchild was acquired by ON Semiconductor, he was a Distinguished Member of the Technical Staff since 2015, a Senior Member of the Technical Staff from 2008 to 2014, a Member of the Technical Staff from 2004 to 2007, and a Principal Engineer from 2001 to 2004. His main interest areas are advanced analog and power electronic packaging, modeling and simulation, reliability and material characterization. He has been invited to give numerous keynotes talks, presentations and professional short courses. He has authored and co-authored 4 books, 3 book chapters and over 180 papers in journals and conferences. He has been granted 47 US patents and over 10 US Patents pending. Dr. Liu has won numerous awards including Alexander von Humboldt Fellowship in 1994, the Fairchild award for Power of Pen first place in 2004, Fairchild Built in Quality Award in product innovation in 2005, Fairchild Key Technologist in 2006 and 2009, the first Fairchild President Award in 2008, numerous IEEE best papers, the IEEE CPMT Exceptional Technical Achievement Award in 2013, the IEEE EPS Electronic Manufacturing Technology Award, 2021. He was elevated as IEEE Fellow in 2015.

Part I

Mechanics and Modeling

Microelectronic devices, optoelectronic devices, and MEMS (micro-electrical-mechanical-systems) devices and their systems are mainly being manufactured with a silicon chip, or chips, in various compact package configurations to satisfy cost and performance requirements. A typical package and assembly is manufactured through many processes (both front-end and back-end), followed by many severe reliability qualification tests. A typical packaging assembly consists of components of different materials with different mechanical and thermal properties, and geometric discontinuities exist in terms of vias, comers, free edges, interfaces, surfaces, and composites. Non-uniform temperature and moisture fields during each process of packaging and assembly, testing, storage, and operation often subject these materials and components to various failure modes, for example, metal line voiding, passivation cracking, die cracking, fine metal line smearing, package swelling, warpage, and delamination. If failure occurs in a process step of manufacturing, and results in the component not being processed, yield can be a big issue, and popcorning during the reflow of plastic packaging onto the board is such an example. System functionality tests performed on these packages and assemblies, in general, will not detect the majority of these failures. This may not cause catastrophic failure but may cause performance degradation and failure before the designed life cycle. Due to the rapid advancement of the related industries, information on detailed failure initiation, growth and performance degradation is not easily available and it is a challenging task for the modeling and simulation community. Closed-form solutions are mostly unavailable due to the many nonlinearities such as material nonlinearities for many polymer and solder materials, geometrical nonlinearity such as membrane deformation in sensors, force nonlinearity such as contact and debonding, contour change for sequential process steps, micro-structural changes such as for annealing, damage development, and mass transport for electromigration and wetting, just to list a very few examples. Serious study is essential for modeling and simulation. Constitutive models for engineering materials are a must to avoid a wrong analysis which can easily result in engineers arriving at wrong conclusions or no conclusions at all. In the first part of the book, we will introduce constitutive models and leave the definition of the basic concepts such as stress, strain, and so on, to Appendix A. Then we will briefly present the finite element methods and some of the advanced features such as sub-modeling, sub-structure, and element birth and death. Material testing for small samples is presented to show the uniqueness of our samples as compared to bulk structures for traditional industries. User-supplied subroutines are also presented with solders as an example. Multi-physics and multi-scale modeling is presented by briefly introducing molecular dynamics (MD), while leaving most of MD to Part IV. Validation tools and their applications for both on-line testing and off-line quality analysis are presented. Fracture mechanics with the focus on the interfacial fracture mechanics is presented and the concurrent engineering approach is presented as a platform for leading research laboratories which are product oriented and have a strong desire to shorten the time-to-market and time-to-profit.

1

Constitutive Models and Finite Element Method

In this chapter, the basic equations of continuum mechanics are introduced. It is assumed that the readers already have a basic prior knowledge of the subject. Therefore, lengthy derivations are omitted, such information being provided in the Appendix. Only equations that are needed later for deriving the mechanics theories are outlined. Most equations are adopted from several famous references listed in the literature of this chapter, including ABAQUS theory manual (Belytschko *et al.*, 2000; Hibbit *et al.*, 2008).

1.1 Constitutive Models for Typical Materials

1.1.1 Linear Elasticity

In many engineering applications involving small strains and rotations, the response of the material may be considered to be linearly elastic. The most general way to represent elastic tensor $\boldsymbol{C}$ relation between the stress and strain tensors is given by:

$$\sigma_{ij} = C_{ijkl}\varepsilon_{kl} \quad \boldsymbol{\sigma} = \boldsymbol{C} : \boldsymbol{\varepsilon} \tag{1.1}$$

where C_{ijkl} are components of the 4th-order tensor of elastic moduli. This equation is the generalized Hooke's law which incorporates a fully anisotropic material response.

The strain energy per unit volume, often called the elastic potential, W, is generalized to multiaxial states by:

$$W = \int \sigma_{ij} \mathrm{d}\varepsilon_{ij} = \frac{1}{2} C_{ijkl}\varepsilon_{ij}\varepsilon_{kl} = \frac{1}{2}\boldsymbol{\varepsilon} : \boldsymbol{C} : \boldsymbol{\varepsilon} \tag{1.2}$$

The stress is then given by:

$$\boldsymbol{\sigma} = \frac{\partial W}{\partial \boldsymbol{\varepsilon}} \tag{1.3}$$

Modeling and Simulation for Microelectronic Packaging and Integration: Manufacturing, Reliability and Testing, Second Edition. Sheng Liu and Yong Liu.

which is the tensor equivalent of Equation 1.2. The strain energy is assumed to be positive definite:

$$W = \frac{1}{2}C_{ijkl}\varepsilon_{ij}\varepsilon_{kl} = \frac{1}{2}\boldsymbol{\varepsilon} : \boldsymbol{C} : \boldsymbol{\varepsilon} \geqslant 0 \tag{1.4}$$

with equality if and only if $\varepsilon_{ij} = 0$ which implies that $\boldsymbol{C}$ is a positive-definite 4th-order tensor. From the symmetries of the stress and strain tensors, the material coefficients have the so called minor symmetries:

$$C_{ijkl} = C_{jikl} = C_{ijlk} \tag{1.5}$$

and from the existence of a strain energy potential Equation 1.2 it follows that:

$$C_{ijkl} = \frac{\partial^2 W}{\partial \varepsilon_{ij} \partial \varepsilon_{kl}} \quad \boldsymbol{C} = \frac{\partial^2 W}{\partial \boldsymbol{\varepsilon} \partial \boldsymbol{\varepsilon}} \tag{1.6}$$

If W is a smooth function of ε, Equation 1.6 implies a property called major symmetry:

$$C_{ijkl} = C_{klij} \tag{1.7}$$

since smoothness implies:

$$\frac{\partial^2 W}{\partial \varepsilon_{ij} \partial \varepsilon_{kl}} = \frac{\partial^2 W}{\partial \varepsilon_{kl} \partial \varepsilon_{ij}} \tag{1.8}$$

The general 4th-order tensor C_{ijkl} has $3^4 = 81$ independent constants. These 81 constants may also be interpreted as arising from the necessity to relate nine components of the complete stress tensor to nine components of the complete strain tensor, that is, $9 \times 9 = 81$. The symmetries of the stress and strain tensors require only that six independent components of stress be related to six independent components of strain. The resulting minor symmetries of the elastic moduli therefore reduce the number of independent constants to $6 \times 6 = 36$. Major symmetry of the moduli, expressed through Equation 1.7 reduces the number of independent elastic constants to $n(n+1)/2 = 21$, for $n = 6$, that is, the number of independent components of a 6×6 matrix.

Considerations of material symmetry further reduce the number of independent material constants. An isotropic material is one which has no preferred orientations or directions in its properties, so that the stress-strain relation is identical when expressed in component form in any rectangular Cartesian coordinate system. The most general constant isotropic 4th-order tensor can be shown to be a linear combination of terms comprised of Kronecker deltas, that is, for an isotropic linearly elastic material:

$$C_{ijkl} = \lambda\delta_{ij}\delta_{kl} + \mu\left(\delta_{ik}\delta_{jl} + \delta_{il}\delta_{jk}\right) + \mu'\left(\delta_{ik}\delta_{jl} + \delta_{il}\delta_{jk}\right) \tag{1.9}$$

Because of the symmetry of the strain and the associated minor symmetry $C_{ijkl} = C_{ijlk}$ it follows that $\mu' = 0$. Thus Equation 1.9 is written as:

$$C_{ijkl} = \lambda\delta_{ij}\delta_{kl} + \mu\left(\delta_{ik}\delta_{jl} + \delta_{il}\delta_{jk}\right) \quad \boldsymbol{C} = \lambda\boldsymbol{I} \otimes \boldsymbol{I} + 2\mu\boldsymbol{I} \tag{1.10}$$

and the two independent material constants λ and μ are called the Lamé constants. The stress strain relation for an isotropic linear elastic material may therefore be written as:

$$\sigma_{ij} = \lambda\varepsilon_{kk}\delta_{ij} + 2\mu\varepsilon_{ij} = C_{jikl}\varepsilon_{kl} \quad \boldsymbol{\sigma} = \lambda trace(\boldsymbol{\varepsilon})\boldsymbol{I} + 2\mu\boldsymbol{\varepsilon} \tag{1.11}$$

1.1.2 *Elastic-Visco-Plasticity*

In classical formulations of elastic-visco-plasticity, the yield criterion is defined through a loading function $F \equiv F(\boldsymbol{\sigma}, \boldsymbol{q})$, where $\boldsymbol{\sigma}$ denotes the stress state and $\boldsymbol{q}$ denotes the internal variables. As elastic-visco-plastic deformation appears, the stress is permissible outside the closure of the loading surface, that is $F(\boldsymbol{\sigma}, \boldsymbol{q}) > 0$. However, in rate-independent plasticity, $F(\boldsymbol{\sigma}, \boldsymbol{q}) \leqslant 0$, which is the basic difference between viscoplasticity and rate-independent plasticity.

For the classic elastic-viscoplastic constitutive model (Figure 1.1, where σ_s is the yield stress), the total strain rate is the sum of its elastic and viscoplastic components:

$$\dot{\boldsymbol{\varepsilon}} = \dot{\boldsymbol{\varepsilon}}^{\mathrm{e}} + \dot{\boldsymbol{\varepsilon}}^{\mathrm{vp}} \tag{1.12}$$

where the superscript "e" indicates the elastic component, and the superscript "vp" indicates the viscoplastic component.

According to a study by Perzyna (Perzyna, 1971), the viscoplastic strain rate $\dot{\boldsymbol{\varepsilon}}^{\mathrm{vp}}$ is defined by a flow rule:

$$\dot{\boldsymbol{\varepsilon}}^{\mathrm{vp}} = \gamma \frac{\partial F}{\partial \boldsymbol{\sigma}} = \frac{\langle \boldsymbol{\Phi}(F) \rangle}{\eta} \frac{\partial F}{\partial \boldsymbol{\sigma}} \tag{1.13}$$

where F is a yield function of the material; $\boldsymbol{\Phi}(F)$ is a flow function; $\langle \cdot \rangle$ are MacCauley's brackets, $\langle x \rangle = (x + |x|)/2$; $\dot{\gamma}$ is the viscoplastic flow rate parameter; and η is a given viscoplastic material fluidity parameter.

Considering a mixed strain-hardening model which includes both isotropic hardening and kinematic hardening, the loading yield function is:

$$F(\boldsymbol{\xi}, \bar{e}^{\mathrm{vp}}) = \frac{\|\boldsymbol{\xi}\|}{\sqrt{\frac{2}{3}}\kappa(\bar{e}^{\mathrm{vp}})} \tag{1.14}$$

where $\kappa(\bar{e}^{\mathrm{vp}})$ is a function governing the isotropic expansion of the yield surface. $\boldsymbol{\xi} = \boldsymbol{S} - \boldsymbol{\alpha}$, $\boldsymbol{S}$ is the deviatoric stress tensor, and $\boldsymbol{\alpha}$ refers to a back stress tensor (an internal variable), which defines the translation of the center of the yield surface. The evolution of the back stress tensor is usually defined as $\dot{\boldsymbol{\alpha}} = \frac{2}{3} H' \dot{\gamma} \frac{\partial F}{\partial \boldsymbol{\sigma}}$, where H' is called the kinematic hardening modulus. If $H' = 0$, one gets the isotropic hardening. $\|\boldsymbol{\xi}\| = \sqrt{\boldsymbol{\xi} : \boldsymbol{\xi}}$. $\bar{e}^{\mathrm{vp}}$ is the cumulated equivalent viscoplastic strain:

$$\bar{e}^{\mathrm{vp}} = \int_0^t \sqrt{\frac{2}{3}} (\dot{\boldsymbol{\varepsilon}}^{\mathrm{vp}}(\tau) : \dot{\boldsymbol{\varepsilon}}^{\mathrm{vp}}(\tau))^{1/2} \, \mathrm{d}\tau$$

Common choices for the flow function $\boldsymbol{\Phi}(F)$ are exponent-type and power-type, which can be written as:

$$\boldsymbol{\Phi}(F) = e^{cF} - 1 \tag{1.15a}$$

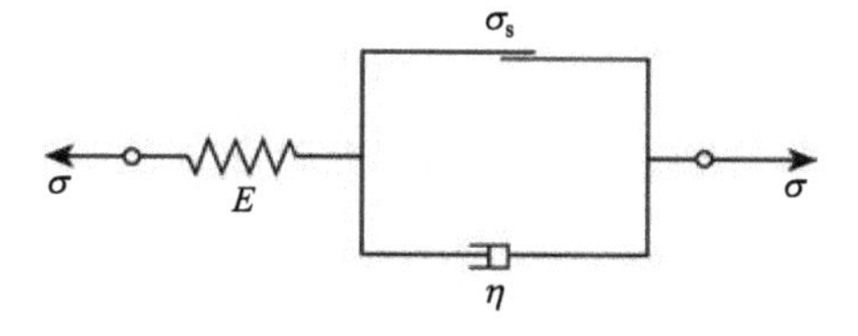

Figure 1.1 One dimensional rheological model

$$\boldsymbol{\Phi}(F) = F^d \tag{1.15b}$$

in which c and d are prescribed constants. In general, the flow function $\boldsymbol{\Phi}(F)$ is the combination of both basic types above.

On the other hand, from Equation 1.12 and Hooke's law, it can be obtained:

$$\dot{\boldsymbol{\sigma}} = \boldsymbol{C} : (\dot{\boldsymbol{\varepsilon}} - \dot{\boldsymbol{\varepsilon}}^{\text{vp}}) \tag{1.16}$$

where $\boldsymbol{C}$ is the 4th-order elastic tensor given by $\boldsymbol{C} = \lambda \boldsymbol{l} \otimes \boldsymbol{l} + 2\mu \boldsymbol{I}$, $\boldsymbol{l}$ is the 2nd-order unit tensor given by $\boldsymbol{l} = \delta_{ij} \boldsymbol{e}_i \otimes \boldsymbol{e}_j$; $\boldsymbol{I} = (1/2)\left[\delta_{ik}\delta_{jl} + \delta_{il}\delta_{jk}\right] \boldsymbol{e}_i \otimes \boldsymbol{e}_j \otimes \boldsymbol{e}_k \otimes \boldsymbol{e}_l$ is 4th-order unit tensor; where $\boldsymbol{e}$ represents basis vector and $\otimes$ denotes the tensor product. λ and μ are the Lame' constants, and μ is the shear modulus. In fact, as the fluidity parameter $\eta \rightarrow 0$, stress state outside of the loading surface becomes increasingly penalized and thus $F \rightarrow 0$, the viscoplastic Equations 1.13 and 1.16 reduce to the rate-independent plasticity problem. As the fluidity parameter $\eta \rightarrow \infty$, $\dot{\gamma} \rightarrow 0$, $\dot{\boldsymbol{\alpha}} \rightarrow \boldsymbol{0}$, $\dot{\boldsymbol{\varepsilon}}^{\text{vp}} \rightarrow \boldsymbol{0}$, Equations 1.13 and 1.16 collapse to the rate form of linear elasticity.

(i) Elastic-Visco-Plastic Radial Return Evolution (RRE)

A trial (Tr) deviatoric stress is introduced as:

$$\boldsymbol{\xi}_{n+1}^{\text{Tr}} = \boldsymbol{\xi}_n + 2\mu \Delta \boldsymbol{e}_n \tag{1.17}$$

where $\boldsymbol{e}$ is the deviatoric strain tensors.

Considering the viscoplastic evolution problem for strain increment in any given finite time step Δt, the elastic-visco-plastic constitutive law reduces to giving a RRE rule that makes $\boldsymbol{\sigma}_{n+1} \equiv \bar{\boldsymbol{\sigma}}(\boldsymbol{\varepsilon}_n, \boldsymbol{\sigma}_n, \boldsymbol{q}_n, \Delta\boldsymbol{\varepsilon}_n) = \boldsymbol{\sigma}_n + \Delta\boldsymbol{\sigma}_n$ finally be consistent with the loading surface $f = F(\boldsymbol{\xi}, \bar{e}^{\text{vp}})$.

Figure 1.2 shows the evolution actions that take place on the π-plane which illustrates a general viscoplastic case with isotropic and kinematic hardening. Note that as the parameter η approaches zero, the yield surface $f_{n+1} = F$ approaches zero, it actually returns to elastic-plastic case. When the kinematic hardening parameter $H' = 0$, this returns to the pure isotropic hardening case (Figure 1.3).

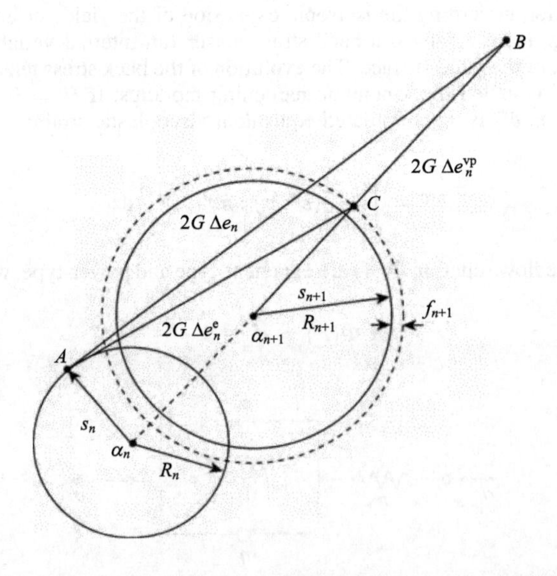

Figure 1.2 Illustration of radial return evolution with both kinematic and isotropic hardening with the radius $R = \sqrt{\frac{2}{3}}\kappa(\bar{e}^{\text{vp}})$

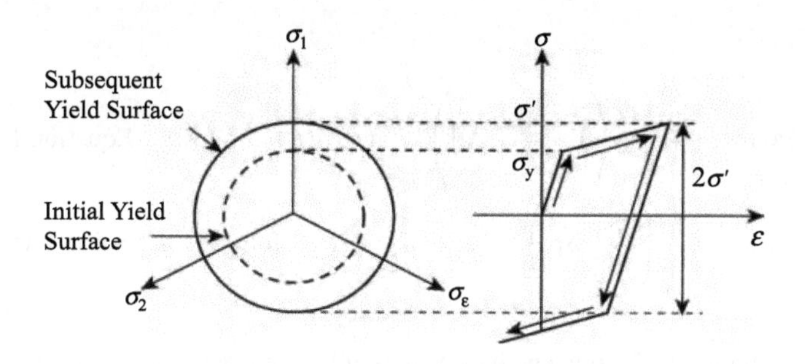

Figure 1.3 Isotropic hardening plasticity

If $F\left(\xi_{n+1}^{\mathrm{Tr}}, \kappa_n\right) \leqslant 0$, that is the elastic deformation, one has:

$$\boldsymbol{\sigma}_{n+1} = \boldsymbol{\sigma} = K\Delta\boldsymbol{\varepsilon}_n : (\boldsymbol{I}\otimes\boldsymbol{I}) + 2\mu\Delta\boldsymbol{e}_n + \boldsymbol{\sigma}_n \tag{1.18}$$

where K is the bulk modulus. This is the elastic constitutive equation in incremental form.
If $F\left(\boldsymbol{\xi}_{n+1}^{\mathrm{Tr}}, \kappa_n\right) > 0$, that is the viscoplastic deformation, one has:

$$\boldsymbol{\sigma}_{n+1} = \bar{\boldsymbol{\sigma}} = K\boldsymbol{\varepsilon}_{n+1} : (\boldsymbol{I}\otimes\boldsymbol{I}) + \boldsymbol{S}_{n+1} \tag{1.19}$$

On the other hand, one can obtain the following counterpart of the implicit backward-Euler difference scheme and viscoplastic RRE consistency equations:

$$\begin{cases} \dot{\boldsymbol{\varepsilon}}^{\mathrm{vp}} = \dot{\gamma}\hat{\boldsymbol{n}} = \dfrac{\langle \boldsymbol{\Phi}(F)\rangle}{\eta}\hat{\boldsymbol{n}} \\ \boldsymbol{\varepsilon}_{n+1}^{\mathrm{vp}} = \boldsymbol{\varepsilon}_n^{\mathrm{vp}} + [\dot{\gamma}\Delta t]\hat{\boldsymbol{n}}_{n+1} \\ \boldsymbol{\alpha}_{n+1} = \boldsymbol{\alpha}_n + \dfrac{2}{3}H'[\dot{\gamma}\Delta t]\hat{\boldsymbol{n}}_{n+1} \\ \bar{e}_{n+1}^{\mathrm{vp}} = \bar{e}_n^{\mathrm{vp}} + \sqrt{\dfrac{2}{3}}[\dot{\gamma}\Delta t] \\ \hat{\boldsymbol{n}}_{n+1} = \dfrac{1}{\left\|\xi_{n+1}^{\mathrm{Tr}}\right\|}\xi_{n+1}^{\mathrm{Tr}} \end{cases} \tag{1.20}$$

along with the condition:

$$\left\|\boldsymbol{\xi}_{n+1}\right\| = \left\|\boldsymbol{\xi}_{n+1}^{\mathrm{Tr}}\right\| - 2\mu[\dot{\gamma}\Delta t]\left(1+\frac{H'}{3\mu}\right) \tag{1.21}$$

In addition, note that:

$$\boldsymbol{S}_{n+1} = \left\|\boldsymbol{\xi}_{n+1}\right\|\hat{\boldsymbol{n}}_{n+1} + \boldsymbol{\alpha}_{n+1} \tag{1.22}$$

Equations 1.14, 1.15, 1.17 and 1.22 solve the viscoplastic RRE consistency equation:

$$\Re(\dot{\gamma}\Delta t) \equiv \left\|\xi_{n+1}^{\mathrm{Tr}}\right\| - \sqrt{\frac{2}{3}}\kappa\left(\bar{e}_{n+1}^{\mathrm{vp}}\right)\cdot\hbar(\dot{\gamma}\Delta t) - 2\mu[\dot{\gamma}\Delta t]\left(1+\frac{H'}{3\mu}\right) = 0 \tag{1.23}$$

where:

$$\begin{cases} \hbar(\dot{\gamma}\Delta t) = 1 + \dfrac{1}{c}\ln\left(1 + \dfrac{\eta}{\Delta t}[\dot{\gamma}\Delta t]\right), & \text{if } \Phi(F) \text{ follows Equation 1.15a} \\ \hbar(\dot{\gamma}\Delta t) = 1 + \left(\dfrac{\eta}{\Delta t}[\dot{\gamma}\Delta t]\right)^{\frac{1}{d}}, & \text{if } \Phi(F) \text{ follows Equation 1.15b} \end{cases}$$

The solution of Equation 1.23 from which the values of $|\dot{\gamma}\Delta t|$ are determined, can be effectively solved by the local Newton iteration procedure.

(ii) Elastic-Visco-Plastic Consistent Tangent Operator (CTO)

The CTO is defined in a 4th-order tensor (Simo and Taylor, 1985):

$$\boldsymbol{C}_{n+1} = \frac{\partial \bar{\boldsymbol{\sigma}}}{\partial \Delta \boldsymbol{\varepsilon}_n} = \frac{\partial \boldsymbol{\sigma}_{n+1}}{\partial \Delta \boldsymbol{\varepsilon}_n} \tag{1.24}$$

which depends on the particular algorithm $\delta \boldsymbol{\varepsilon}_n \rightarrow \boldsymbol{\sigma}_{n+1}(\boldsymbol{\varepsilon}_n, \boldsymbol{\sigma}_n, \boldsymbol{q}_n, \Delta \boldsymbol{\varepsilon}_n)$ chosen. Neglecting the deriving process, the CTO takes the form:

$$\boldsymbol{C}_{n+1} = \boldsymbol{C}^{\text{vp}}_{n+1} = K\boldsymbol{l} \otimes \boldsymbol{l} + 2\mu\beta\left(\boldsymbol{I} - \frac{1}{3}\boldsymbol{l} \otimes \boldsymbol{l}\right) - 2\mu(\beta - \vartheta)\hat{\boldsymbol{n}}_{n+1} \otimes \hat{\boldsymbol{n}}_{n+1} \tag{1.25}$$

where:

$$\beta = 1 - \frac{2\mu[\dot{\gamma}\Delta t]}{\left\|\xi^{\text{Tr}}_{n+1}\right\|}, \quad \vartheta = 1 - \frac{1}{\frac{1}{3\mu}\kappa'\left(\bar{e}^{\text{vp}}_{n+1}\right) \cdot \hbar(\dot{\gamma}\Delta t) + \frac{1}{2\mu}\sqrt{\frac{2}{3}}\kappa\left(\bar{e}^{\text{vp}}_{n+1}\right) \cdot \hbar'(\dot{\gamma}\Delta t) + 1 + \frac{H'}{3\mu}}$$

where:

$$\begin{cases} \hbar'(\dot{\gamma}\Delta t) = \dfrac{\dfrac{h}{\Delta t}}{c\left(1 + \dfrac{\eta}{\Delta t}[\dot{\gamma}\Delta t]\right)}, & \text{if } \Phi(F) \text{ follows Equation 1.15a} \\ \hbar'(\dot{\gamma}\Delta t) = \dfrac{\dfrac{\eta}{\Delta t}}{d}\left(\dfrac{\eta}{\Delta t}[\dot{\gamma}\Delta t]\right)^{\frac{1}{d}-1}, & \text{if } \Phi(F) \text{ follows Equation 1.15b} \end{cases}$$

It should be indicated that when $\boldsymbol{\sigma}_{n+1} = \bar{\boldsymbol{\sigma}}(\boldsymbol{\varepsilon}_n, \boldsymbol{\sigma}_n, \boldsymbol{q}_n, \Delta \boldsymbol{\varepsilon}_n)$ is elastic, one has $\boldsymbol{C}_{n+1} = \boldsymbol{C}^{\text{vp}}_{n+1} = \boldsymbol{C}$.

The above consistent tangent operator based constitutive relation is a unified material constitutive model that has been applied to various nonlinear problems in computational mechanics. It has been proven to be very powerful in dealing with various highly nonlinear and large deformation problems in engineering.

1.2 Finite Element Method

The finite element method is a numerical analysis technique for obtaining approximate solutions to a wide variety of engineering problems.

1.2.1 Basic Finite Element Equations

We begin with the equilibrium statement, written as the virtual work principle:

$$\int_V \boldsymbol{\sigma} : \delta \boldsymbol{D} \mathrm{d}V = \int_S \boldsymbol{t}^{\mathrm{T}} \cdot \delta \boldsymbol{v} \mathrm{d}S + \int_V \boldsymbol{f}^{\mathrm{T}} \cdot \delta \boldsymbol{v} \mathrm{d}V \tag{1.26}$$

the left-hand side of this equation, the internal virtual work rate term is replaced with the integral over the reference volume of the virtual work rate per reference volume defined by any conjugate pairing of stress and strain:

$$\int_{V^0} \boldsymbol{\tau}^{\mathrm{c}} : \delta \boldsymbol{\varepsilon} \mathrm{d}V^0 = \int_S \boldsymbol{t}^{\mathrm{T}} \cdot \delta \boldsymbol{v} \mathrm{d}S + \int_V \boldsymbol{f}^{\mathrm{T}} \cdot \delta \boldsymbol{v} \mathrm{d}V \tag{1.27}$$

where $\boldsymbol{\tau}^{\mathrm{c}}$ and ε are any conjugate pairing of material stress and strain measures. The particular choice of ε depends on the individual element.

The finite element interpolator can be written in general as:

$$\boldsymbol{u} = N_{\mathrm{N}} u^{\mathrm{N}} \tag{1.28}$$

where N_{N} are interpolation functions that depend on some material coordinate system, u^{N} are nodal variables, and the summation convention is adopted for the upper case subscripts and superscripts that indicate nodal variables.

The virtual field, $\delta \boldsymbol{v}$, must be compatible with all kinematic constraints. Introducing the above interpolation constrains the displacement to have a certain spatial variation, so $\delta \boldsymbol{v}$ must also have the same spatial form:

$$\delta \boldsymbol{v} = N_{\mathrm{N}} \delta v^{\mathrm{N}} \tag{1.29}$$

The continuum variational statement Equation 1.27 is, thus, approximated by a variation over the finite set δv.

Now $\delta\varepsilon$ is the virtual rate of material strain associated with $\delta \boldsymbol{v}$, and because it is a rate form, it must be linear in $\delta \boldsymbol{v}$. Hence, the interpolation assumption gives:

$$\delta \boldsymbol{\varepsilon} = \boldsymbol{\beta}_{\mathrm{N}} \delta v^{\mathrm{N}} \tag{1.30}$$

where β_{N} is a matrix that depends, in general, on the current position, $\boldsymbol{x}$, of the material point being considered. The matrix β_{N} that defines the strain variation from the variations of the kinematic variables is derivable immediately from the interpolation functions once the particular strain measure to be used is defined.

Without loss of generality we can write $\boldsymbol{\beta}_{\mathrm{N}} = \boldsymbol{\beta}_{\mathrm{N}}(\boldsymbol{x}, N_{\mathrm{N}})$, and, with this notation, the equilibrium equation is approximated as:

$$\delta v^{\mathrm{N}} \int_{V^0} \boldsymbol{\beta}_{\mathrm{N}} : \boldsymbol{\tau}^{\mathrm{c}} \mathrm{d}V^0 = \delta v^{\mathrm{N}} \left[\int_S N_{\mathrm{N}}^{\mathrm{T}} \cdot \boldsymbol{t} \mathrm{d}S + \int_V N_{\mathrm{N}}^{\mathrm{T}} \cdot \boldsymbol{f} \mathrm{d}V \right] \tag{1.31}$$

Since the δv^{N} are independent variables, we can choose each one to be nonzero and all others zero in turn, to arrive at a system of nonlinear equilibrium equations:

$$\int_{V^0} \boldsymbol{\beta}_{\mathrm{N}} : \boldsymbol{\tau}^{\mathrm{c}} \mathrm{d}V^0 = \int_S \boldsymbol{N}_{\mathrm{N}}^{\mathrm{T}} \cdot \boldsymbol{t} \mathrm{d}S + \int_V \boldsymbol{N}_{\mathrm{N}}^{\mathrm{T}} \cdot \boldsymbol{f} \mathrm{d}V \tag{1.32}$$

This system of equations forms the basis for the standard assumed displacement finite element analysis procedure and is of the form:

$$F^{\mathrm{N}}(u^{\mathrm{M}}) = 0 \tag{1.33}$$

as discussed above. The above equations are valid for static and dynamic analysis if the body force is assumed to contain the inertia contribution. In dynamic analysis, however, the inertia contribution is more commonly considered separately, leading to the equation:

$$M^{\mathrm{NM}} \ddot{u}^{\mathrm{M}} + F^{\mathrm{N}}(u^{\mathrm{M}}) = 0 \tag{1.34}$$

For the Newton algorithm, or for the linear perturbation procedure we need the Jacobian of the finite element equilibrium equations. To develop the Jacobian, we begin by taking the variation of Equation 1.26, giving:

$$\begin{aligned} \int_{V^0} (\mathrm{d}\boldsymbol{\tau}^{\mathrm{c}} : \delta\boldsymbol{\varepsilon} + \boldsymbol{\tau}^{\mathrm{c}} : \mathrm{d}\delta\boldsymbol{\varepsilon}) \mathrm{d}V^0 - \int_S \mathrm{d}\boldsymbol{t}^{\mathrm{T}} \cdot \delta\boldsymbol{v} \mathrm{d}S - \int_S \boldsymbol{t}^{\mathrm{T}} \cdot \delta\boldsymbol{v} \mathrm{d}A_{\mathrm{r}} \frac{1}{A_{\mathrm{r}}} \mathrm{d}S \\ - \int_V \mathrm{d}\boldsymbol{f}^{\mathrm{T}} \cdot \delta\boldsymbol{v} \mathrm{d}V - \int_V \boldsymbol{f}^{\mathrm{T}} \cdot \delta\boldsymbol{v} J \frac{1}{J} \mathrm{d}V = 0 \end{aligned} \tag{1.35}$$

where d() represents the linear variation of the quantity () with respect to the basic variables which are the degrees of freedom of the finite element model. In the above expression, $J = \left|\mathrm{d}V/\mathrm{d}V^0\right|$ is the volume change between the reference and the current volume occupied by a piece of the structure and, likewise, $A_{\mathrm{r}} = \left|\mathrm{d}S/\mathrm{d}S^0\right|$ is the surface area ratio between the reference and the current configuration. The Jacobian matrix is obtained by restricting the above variation, allowing variations in the nodal variables, u^{N}, only. Let such a restricted variation be indicated by $\partial_{\mathrm{N}} = \partial/\partial u^{\mathrm{N}}$. Examining Equation 1.35 term by term with this in mind, we proceed as follows. The first term contains $\mathrm{d}\boldsymbol{\tau}^{\mathrm{c}}$. We now assume that the constitutive theory allows us to write:

$$\mathrm{d}\boldsymbol{\tau}^{\mathrm{c}} = \boldsymbol{H} : \mathrm{d}\boldsymbol{\varepsilon} + \boldsymbol{g} \tag{1.36}$$

where $\boldsymbol{H}$ and $\boldsymbol{g}$ are defined in terms of the current state, direction of straining, and so on, and on the kinematic assumptions used to form the generalized strains. From the choice of generalized strain measure and interpolation function:

$$\partial_{\mathrm{N}}\boldsymbol{\varepsilon} = \frac{\partial\boldsymbol{\varepsilon}}{\partial u^{\mathrm{N}}} = \boldsymbol{\beta}_{\mathrm{N}} \tag{1.37}$$

from the above constitutive assumption:

$$\partial_{\mathrm{N}}\boldsymbol{\tau}^{\mathrm{c}} = \boldsymbol{H} : \boldsymbol{\beta}_{\mathrm{N}} \tag{1.38}$$

Now, since $\delta\boldsymbol{\varepsilon}$ is the first variation of ε with respect to nodal variables:

$$\delta\boldsymbol{\varepsilon} = \partial_{\mathrm{M}}\boldsymbol{\varepsilon} \delta u^{\mathrm{M}} = \boldsymbol{\beta}_{\mathrm{M}} \delta u^{\mathrm{M}} \tag{1.39}$$

thus, the first term in Equation 1.35 of the Jacobian matrix can be written as:

$$\int_{V^0} \boldsymbol{\beta}_{\mathrm{M}} : \boldsymbol{H} : \boldsymbol{\beta}_{\mathrm{N}} \mathrm{d}V^0 \tag{1.40}$$

which is the usual "small-displacement stiffness matrix", except that, since the strain measure ε is always nonlinear in displacement, the β_{N} in this term is a function of displacement.

The second term in Equation 1.35 is:

$$\int_{V^0} \boldsymbol{\tau}^{\mathrm{c}} : \mathrm{d}\delta\boldsymbol{\varepsilon}\mathrm{d}V^0$$

which can be rewritten as:

$$\int_{V^0} \boldsymbol{\tau}^{\mathrm{c}} : \partial_{\mathrm{N}}\delta\boldsymbol{\varepsilon}\mathrm{d}V^0$$

and further leads to:

$$\int_{V^0} \boldsymbol{\tau}^{\mathrm{c}} : \partial_{\mathrm{N}}\boldsymbol{\beta}_{\mathrm{M}}\mathrm{d}V^0$$

where this term contributes to the Jacobian and is the "initial stress matrix".

The external load rate terms in Equation 1.35 are considered next. In general, these load vectors can be written as:

$$\boldsymbol{t} = \boldsymbol{t}(\lambda, \boldsymbol{x}) \text{ and } \boldsymbol{f} = \boldsymbol{f}(\lambda, \boldsymbol{x}) \tag{1.41}$$

where λ represents the externally prescribed loading parameters. Whether the load depends on position or not, it is up to the particular load type, but common types of loading such as pressure, or centrifugal load do depend on position—for example, if $\boldsymbol{t}$ is caused by pressure on the surface, $\boldsymbol{t}$ depends on the pressure magnitude, on the direction of the normal to the surface, and on the current surface area: the latter two are functions of the current position of points on the surface. The variation of the load vector with nodal variables can then be written symbolically as:

$$\partial_{\mathrm{N}}\boldsymbol{t} + \boldsymbol{t}\frac{1}{A_{\mathrm{r}}}\partial_{\mathrm{N}}A_{\mathrm{r}} = \boldsymbol{Q}_{\mathrm{N}}^{\mathrm{S}} \tag{1.42}$$

$$\partial_{\mathrm{N}}\boldsymbol{f} + \boldsymbol{f}\frac{1}{J}\partial_{\mathrm{N}}J = \boldsymbol{Q}_{\mathrm{N}}^{\mathrm{V}} \tag{1.43}$$

and then:

$$\delta\boldsymbol{v} = \boldsymbol{N}_{\mathrm{M}}\delta v^{\mathrm{M}} \tag{1.44}$$

where $\boldsymbol{N}_{\mathrm{M}}$ is obtained directly from the interpolation functions, we can write the Jacobian terms pertaining to the last four terms of Equation 1.35 as:

$$-\int_S \boldsymbol{N}_{\mathrm{M}}^{\mathrm{T}} \cdot \boldsymbol{Q}_{\mathrm{N}}^{\mathrm{S}}\mathrm{d}S - \int_V \boldsymbol{N}_{\mathrm{M}}^{\mathrm{T}} \cdot \boldsymbol{Q}_{\mathrm{N}}^{\mathrm{V}}\mathrm{d}V$$

these are commonly called the "load stiffness matrix". The actual form of the load stiffness is very much dependent on the type of load being considered.

The complete Jacobian matrix is then:

$$K_{\mathrm{MN}} = \int_{V^0} \boldsymbol{\beta}_{\mathrm{M}} : \boldsymbol{H} : \boldsymbol{\beta}_{\mathrm{N}} \mathrm{d}V^0 + \int_{V^0} \boldsymbol{\tau}^{\mathrm{c}} : \partial_{\mathrm{N}} \boldsymbol{\beta}_F \mathrm{d}V^0 - \int_S N_{\mathrm{M}}^{\mathrm{T}} : Q_{\mathrm{N}}^{\mathrm{S}} \mathrm{d}S - \int_V N_{\mathrm{M}}^{\mathrm{T}} \cdot Q_{\mathrm{N}}^{\mathrm{V}} \mathrm{d}V \tag{1.45}$$

With the advances of various kinds of commercial finite element software, most of procedures described above have been automated. The challenges arise when the general purpose finite element software is used to solve specific problems such as in microelectronics with advanced analysis techniques and methods.

1.2.2 Nonlinear Solution Methods

The finite element modeling usually relates to nonlinear problems and can involve from a few to thousands of variables. In terms of these variables the equilibrium equations obtained by discretizing the virtual work equation can be written symbolically as:

$$F^{\mathrm{N}}(u^{\mathrm{M}}) = 0 \tag{1.46}$$

where F^{N} is the force component conjugate to the Nth variable in the problem and u^{M} is the value of the Mth variable. The basic problem is to solve Equation 1.46 for the u^{M} throughout the history of interest.

Many of the problems are nonlinear and history-dependent, so the solution must be developed by a series of "small" increments. Two issues arise: how the discrete equilibrium statement Equation 1.46 is to be solved at each increment, and how the increment size is chosen.

Newton's method is generally used as a numerical technique for solving the nonlinear equilibrium equations. The motivation for this choice is primarily the convergence rate obtained by using Newton's method compared to the convergence rates exhibited by alternate methods, usually modified Newton or quasi-Newton methods for the types of nonlinear problems. The basic algorithm of Newton's method is as follows. Assume that, after an iteration i, an approximation u_i^{M}, to the solution has been obtained. Let c_{i+1}^{M} be the difference between this solution and the exact solution to the discrete equilibrium Equation 1.46. This means that:

$$F^{\mathrm{N}}\left(u_i^{\mathrm{M}} + c_{i+1}^{\mathrm{M}}\right) = 0 \tag{1.47}$$

Expanding the left-hand side of this equation in a Taylor series about the approximate solution then gives:

$$F^{\mathrm{N}}(u_i^{\mathrm{M}}) + \frac{\partial F^{\mathrm{N}}}{\partial u^{\mathrm{P}}}(u_i^{\mathrm{M}}) c_{i+1}^{\mathrm{P}} + \frac{\partial^2 F^{\mathrm{N}}}{\partial u^{\mathrm{P}} \partial u^{\mathrm{Q}}}(u_i^{\mathrm{M}}) c_{i+1}^{\mathrm{P}} c_{i+1}^{\mathrm{Q}} + \cdots = 0 \tag{1.48}$$

If u_i^{M} is a close approximation to the solution, the magnitude of each c_{i+1}^{M} will be small, and so all but the first two terms above can be neglected giving a linear system of equations:

$$K_i^{\mathrm{NP}} c_{i+1}^{\mathrm{P}} = -F_i^{\mathrm{N}} \tag{1.49}$$

where:

$$K_i^{\mathrm{NP}} = \frac{\partial F^{\mathrm{N}}}{\partial u^{\mathrm{P}}}(u_i^{\mathrm{M}}) \tag{1.50}$$

is the Jacobian matrix and:

$$F_i^{\mathrm{N}} = F^{\mathrm{N}}(u_i^{\mathrm{M}}) \tag{1.51}$$

the next approximation to the solution is then:

$$u_{i+1}^{\mathrm{M}} = u_i^{\mathrm{M}} + c_{i+1}^{\mathrm{M}} \tag{1.52}$$

and the iteration continues.

Convergence of Newton's method is best measured by ensuring that all entries in F_i^{N} and all entries in c_{i+1}^{N} are sufficiently small. Both these criteria are checked by default in a solution.

Newton's method is usually avoided in large finite element codes, apparently for two reasons. First, the complete Jacobian matrix is sometimes difficult to formulate; and for some problems it can be impossible to obtain this matrix in closed form, so it must be calculated numerically—an expensive (and not always reliable) process. Secondly, the method is expensive per iteration, because the Jacobian must be formed and solved in each iteration. The most commonly used alternative to Newton is the modified Newton method, in which the Jacobian in Equation 1.50 is recalculated only occasionally (or not at all, as in the initial strain method of simple contained plasticity problems). This method is attractive for mildly nonlinear problems involving softening behavior (such as contained plasticity with monotonic straining) but is not suitable for severely nonlinear cases.

Another alternative is the quasi-Newton method, in which Equation 1.50 is symbolically rewritten:

$$c_{i+1}^{\mathrm{P}} = -\left[K_i^{\mathrm{NP}}\right]^{-1} F_i^{\mathrm{N}} \tag{1.53}$$

and the inverse Jacobian is obtained by an iteration process.

There are a wide range of quasi-Newton methods. The more appropriate methods for structural applications appear to be reasonably well behaved in all but the most extremely nonlinear cases—the trade-off is that more iterations are required to converge, compared to Newton. While the savings in forming and solving the Jacobian might seem large, the savings might be offset by the additional arithmetic involved in the residual evaluations (that is, in calculating the F_i), and in the cascading vector transformations associated with the quasi-Newton iterations. Thus, for some practical cases quasi-Newton methods are more economic than full Newton, but in other cases they are more expensive.

When any iterative algorithm is applied to a history-dependent problem, the intermediate, non-converged solutions obtained during the iteration process are usually not on the actual solution path; thus, the integration of history-dependent variables must be performed completely over the increment at each iteration, and not obtained as the sum of integrations associated with each Newton iteration, c_i. This is done by assuming that the basic nodal variables, u, vary linearly over the increment, so that:

$$u(\tau) = \left(1 - \frac{\tau}{\Delta t}\right) u(t) + \frac{\tau}{\Delta t} u(t + \Delta t) \tag{1.54}$$

where $0 \leqslant \tau \leqslant \Delta t$ represents "time" during the increment. Then, for any history-dependent variable, $g(t)$, we compute:

$$g(t + \Delta t) = g(t) + \int_t^{t+\Delta t} \frac{\mathrm{d}g}{\mathrm{d}\tau}(\tau)\mathrm{d}\tau \tag{1.55}$$

at each iteration.

The issue of choosing suitable time steps is a difficult problem to resolve. First of all, the considerations are quite different in static, dynamic, or diffusion cases. It is always necessary to model the response as a function of time to some acceptable level of accuracy. In the case of dynamic or diffusion problems, time is a physical dimension for the problem and the time stepping scheme must provide suitable steps to allow accurate modeling in this dimension. Even if the problem is linear, this accuracy requirement imposes restrictions on the choice of time step. In contrast, most static problems have no imposed time scale, and the only criterion involved in time step choice is accuracy in modeling nonlinear effects. In dynamic and diffusion problems it is exceptional to encounter discontinuities in the time history, because inertia or viscous effects provide smoothing in the solution. However, in static cases sharp discontinuities (such as bifurcations caused by buckling) are common. Softening systems, or unconstrained systems, require special consideration in static cases but are handled naturally in dynamic or diffusion cases. Thus, the considerations upon which time step choice is made are quite different for the three different problem classes.

Both "automatic" time step choice and direct user control for all classes of problems are provided. Direct user control can be useful in cases where the problem behavior is well understood (as might occur when the user is carrying out a series of parameter studies) or in cases where the automatic algorithms do not handle the problem well. However, the automatic schemes are based on extensive experience with a wide range of problems. Therefore, it generally can provide a reliable approach.

One other ingredient in this algorithm is that a minimum increment size is specified, which prevents excessive computation in cases where buckling, limit load, or some modeling error causes the solution to stall. This control is handled internally, with user override if needed. Several other controls are built into the algorithm; for example, it will cut back the increment size if an element inverts due to excessively large geometry changes. These detailed controls are based on empirical testing.

In dynamic analysis when implicit integration is used, the automatic time stepping is based on the concept of half-step residuals. The basic idea is that the time stepping operator defines the velocities and accelerations at the end of the step $(t+\Delta t)$ in terms of displacement at the end of the step and conditions at the beginning of the step. Equilibrium is then established at $(t+\Delta t)$ which ensures an equilibrium solution at the end of each time step and, thus, at the beginning and end of any individual time step. However, these equilibrium solutions do not guarantee equilibrium throughout the step. The time step control is based on measuring the equilibrium error (the force residuals) at some point during the time step, by using the integration operator, together with the solution obtained at $(t+\Delta t)$, to interpolate within the time step. The evaluation is performed at the half step $(t+\Delta t/2)$. If the maximum entry in this residual vector—the maximum "half-step residual"—is greater than a user-specified tolerance, the time step is considered to be too big and is reduced by an appropriate factor. If the maximum half-step residual is sufficiently below the user-specified tolerance, the time step can be increased by an appropriate factor for the next increment. Otherwise, the time step is deemed adequate. The algorithm is somewhat more complicated at traumatic events such as impact. Here, the time step can also be adjusted based on the magnitude of residuals in the first or second iteration following such events. Clearly, if these residuals are several orders of magnitude greater than those permitted, convergence is unlikely and the time step is altered immediately to avoid unproductive iteration.

1.2.3 Advanced Modeling Techniques in Finite Element Analysis

(i) Sub-Modeling

The submodeling technique is used to study a local part of a model with a refined mesh based on interpolation of the solution from an initial, relatively coarse, global model; it is most useful when it is necessary to obtain an accurate, detailed solution in a local region and the detailed modeling of that local region has negligible effect on the overall solution and can use a combination of linear and nonlinear procedures.

The submodel is run as a separate analysis from the global analysis. The only link between the submodel and the global model is the transfer of the time-dependent values of variables saved in the global analysis to the relevant boundary nodes of the submodel (Figure 1.4). This transfer is accomplished by saving the results from the global model in the results, then reading these results into the submodel analysis. Since the submodel is a separate analysis, submodeling can be used to any number of levels; a submodel can be used as the global model for a subsequent submodel. The results from the global model are interpolated onto the nodes on the appropriate parts of the boundary of the submodel. Thus, the response at the boundary of the local region is defined by the solution for the global model. The driven nodes and any loads applied to the local region determine the solution in the submodel.

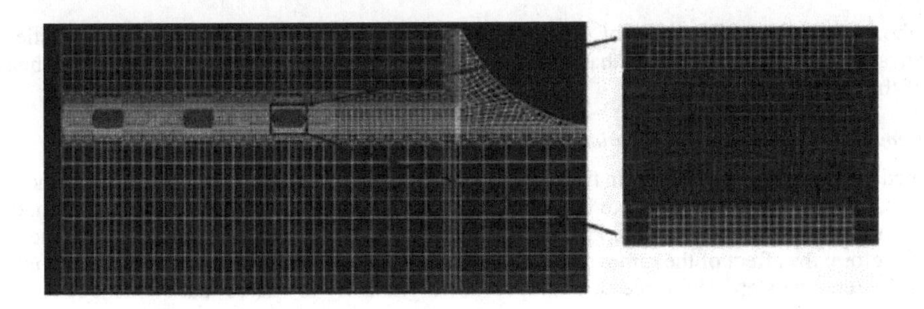

Figure 1.4 Relevant boundary nodes of the submodel

The global model in a submodeling analysis must define the submodel boundary response with sufficient accuracy. It is the user's responsibility to ensure that any particular use of the submodeling technique provides physically meaningful results. In general, the solution at the boundary of the submodel must not be altered significantly by the different local modeling.

(ii) Sub-Structure Modeling

Substructures are collections of elements from which the internal degrees of freedom have been eliminated. Retained nodes and degrees of freedom are those that will be recognized externally at the usage level (when the substructure is used in an analysis), and they are defined during generation of the substructure. System matrices (stiffness, mass) are small as a result of substructuring. Subsequent to the creation of the substructure, only the retained degrees of freedom and the associated reduced stiffness (and mass) matrix are used in the analysis until it is necessary to recover the solution internal to the substructure.

Substructuring can isolate possible changes outside substructures to save time during re-analysis. During the design process large portions of the structure will often remain unchanged; these portions can be isolated in a substructure to save the computational effort involved in forming the stiffness of that part of the structure.

In a problem with local nonlinearities, such as a model that includes interfaces with possible separation or contact, the iterations to resolve these local nonlinearities can be made on a very much reduced number of degrees of freedom if the substructure capability is used to condense the model down to just those degrees of freedom involved in the local nonlinearity.

Substructuring provides a systematic approach to complex analyses. The design process often begins with independent analyses of naturally occurring substructures. Therefore, it is efficient to perform the final design analysis with the use of substructure data obtained during these independent analyses.

Many practical structures are so large and complex that a finite element model of the complete structure places excessive demands on available computational resources. Such a large linear problem can be solved by building the model, substructure by substructure, and stacking these level by level until the whole structure is complete and then recovering the displacements and stresses locally, as required.

(iii) Adaptive Mesh Generation

Adaptive meshing is a tool that makes it possible to maintain a high-quality mesh throughout an analysis, even when large deformation or loss of material occurs, by allowing the mesh to move independently of the material. Adaptive meshing does not alter the topology (elements and connectivity) of the mesh, which implies some limitations on the ability of this method to maintain a high-quality mesh upon extreme deformation. Adaptive re-meshing is typically used for accuracy control, although it can also be used for distortion control in some situations. The adaptive re-meshing process involves the iterative generation of multiple dissimilar meshes to determine a single, optimized mesh that is used throughout an analysis. The goal of adaptive remeshing is to obtain a solution that satisfies mesh discretization error indicator targets that you set, while minimizing the number of elements and, hence, the cost of the analysis.

(iv) Element Removal and Reactivation

Specified elements are removed from the model in a general analysis step. Just prior to the removal step, the forces/fluxes on that the region to be removed is exerting on the remaining part of the model at the nodes on the boundary between them, are stored. These forces are ramped down to zero during the removal step; therefore, the effect of the removed region on the rest of the model is completely absent only at the end of the removal step. The forces are ramped down gradually to ensure that element removal has a smooth effect on the model.

Care must be taken in removing elements in transient procedures. The nodal flux that the removed elements apply at the boundary with the rest of the model is ramped down over the step. In transient heat transfer or fully coupled temperature-displacement analysis if the fluxes are high and the step is long, this ramping down may have the effect of cooling down or heating up the rest of the body. In dynamic analysis if the forces are high and the step is long, kinetic energy can be imparted to the remaining portion of the model. This problem can be avoided by removing the elements in a very short transient step prior to the rest of the analysis. This step can be done in a single increment.

Two distinct types of reactivation are provided for stress/displacement elements: strain-free reactivation and reactivation with strain. Strain-free reactivation resets the initial configuration; reactivation with strain does not.

Although elements cannot be created within an analysis, a similar effect can be achieved by creating elements in the model definition, removing them in the first step, and subsequently reactivating them.

When stress/displacement elements are reactivated in a strain-free state, they become fully active immediately at the moment of reactivation. They are reset to an "annealed" state in the configuration in which they lie at the start of the reactivation step. This configuration depends on whether a small- or large-displacement analysis is being conducted. Alternatively, reactivation in a non virgin state can be specified, as described below.

Since these elements are reactivated in a virgin state that is, with zero stress, they exert zero nodal forces on the rest of the model. This result allows reactivation to be done immediately, without an adverse effect on the smoothness of the solution.

After reactivation the strains and the deformation gradients are based on the displacements subsequent to the moment of reactivation, rather than on their total displacements. Thus, the current configuration at the start of the reactivation step is the new initial configuration for the element.

This kind of reactivation usually is used to model the creation of an undeformed and unstrained region of the model that is sharing a boundary with another, possibly stressed, deformed region. For example, in tunnel excavation an unstressed tunnel liner is added to line the walls of an already deformed tunnel.

(v) Multi-physics Coupling Analysis

A coupled-field analysis is a combination of analyses from different engineering disciplines (physics fields) that interact to solve a global engineering problem. Hence, we often refer to a coupled-field analysis as a multi-physics analysis. When the input of one field analysis depends on the results from another analysis, the analyses are coupled.

Some analyses can have one-way coupling. For example, in a thermal stress problem, the temperature field introduces thermal strains in the structural field, but the structural strains generally do not affect the temperature distribution. Thus, there is no need to iterate between the two field solutions. More complicated cases involve two-way coupling. A piezoelectric analysis in a MEMS structure, for example, handles the interaction between the structural and electric fields: it solves the voltage distribution due to applied displacements, or vice versa. In a fluid-structure interaction problem, the fluid pressure causes the structure to deform, which in turn causes the fluid solution to change. This problem requires iterations between the two physics fields for convergence.

The coupling between the fields can be accomplished by either direct or indirect (load transfer) coupling. Coupling across fields can be complicated because different fields may be the solver for different types of analyses during a simulation. For example, in an induction heating problem,

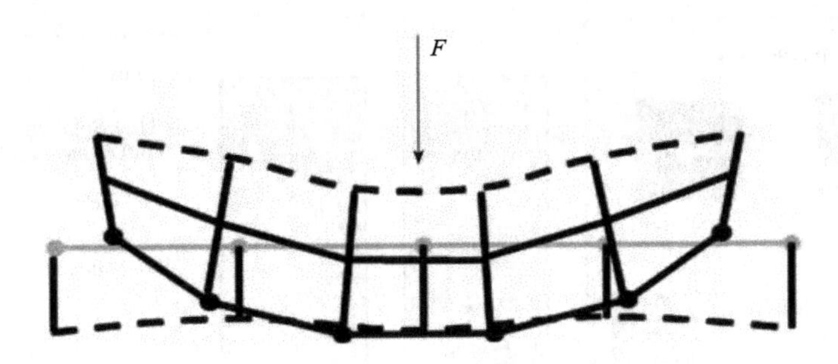

Figure 1.5 Contact pair with interpenetration

a harmonic electromagnetic analysis calculates Joule heating, which is used in a transient thermal analysis to predict a time-dependent temperature solution. The induction heating problem is complicated further by the fact that the material properties in both physics simulations depend highly on temperature.

(vi) Contact Mechanics

When two separate surfaces touch each other such that they become mutually tangent, they are said to be in contact. In the common physical sense, surfaces that are in contact have these characteristics:

- They do not interpenetrate.
- They can transmit compressive normal forces and tangential friction forces.
- They often do not transmit tensile normal forces. They are therefore free to separate and move away from each other.

Contact is a changing-status nonlinearity. That is, the stiffness of the system depends on the contact status, whether parts are touching or separated. Physical contacting bodies do not interpenetrate. Therefore, the finite element program must establish a relationship between the two surfaces to prevent them from passing through each other in the analysis. Figure 1.5 shows the bad contact model with interpenetration. When the program prevents interpenetration, we say that it enforces contact compatibility. Normally, good finite element software should offer several different contact algorithms to enforce compatibility at the contact interface. There are typically three types of algorithms: Augmented Lagrangian, Pure Penalty, and Normal Lagrange Multiplier. Penalty-based methods formulate contact as $[K]\{x\}$, so that there is a concept of contact stiffness[K] and some allowable penetration$\{x\}$. Normal Lagrange solves contact pressure as a DOF directly, so that there is no contact stiffness or penetration, although the solver selection becomes limited because of the unique formulation. Friction describes the tangential behavior between two moving parts. In addition, contact with friction for the tangential or sliding behavior of two contacting bodies is very important in engineering. With friction defined, parts can only slide relative to one another if the tangential force exceeds the product of the normal force and coefficient of friction. Therefore, contact with friction is a function that the advanced finite element should include and carefully consider in the modeling and simulation.

1.2.4 Finite Element Applications in Semiconductor Packaging Modeling

Figure 1.6 shows the flow chart of the solving process of the finite element in the fundamental problems in semiconductor packaging.

Figure 1.7 shows the modeling and simulation mapping for the application of finite elements to various areas of the semiconductor industry. This mapping figure has presented the major functions and roles of modeling and simulation in semiconductor engineering.

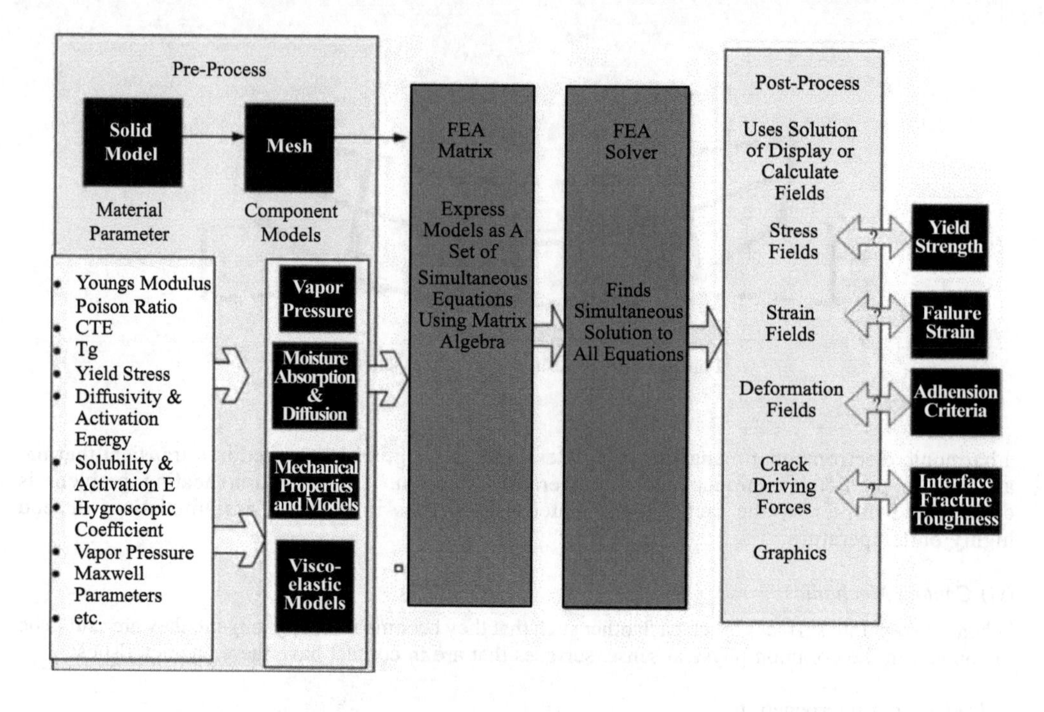

Figure 1.6 Finite element flow chart for the semiconductor packaging modeling

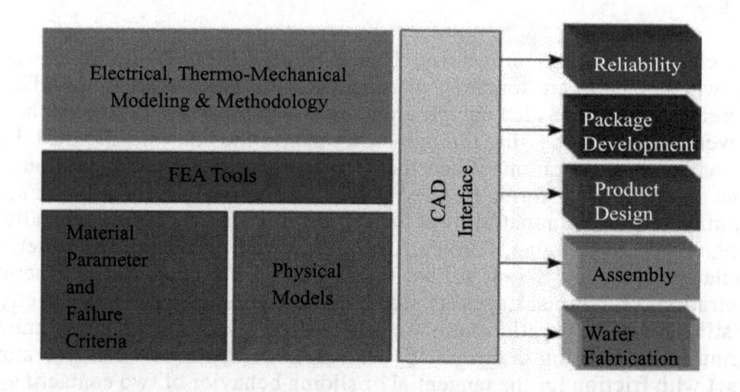

Figure 1.7 Basic mapping of the modeling and simulation in semiconductor industry

1.3 Chapter Summary

In this chapter, we summarized the basic constitutive equations for typical materials which include the linear elasticity, the elastic-visco-plasticity, and the related radial return evolution as well as the consistent tangent operator based on the theory of continuum mechanics. The elastic-visco-plastic consistent tangent operator based constitutive relation is a unified material constitutive model that has been applied to various nonlinear problems in computational mechanics. Regular elasticity and plasticity are its special limit cases. It has been proven to be very powerful in dealing with various highly nonlinear and large deformation problems in engineering. Then the finite element method was briefly presented starting from the virtual work principle and shape functions. Nonlinear iteration solutions based on Newton's theory were also discussed. Advanced modeling techniques such as sub-modeling, sub-structure modeling, adaptive mesh generation, element removal and reactivation, multi-physics coupling analysis, and contact mechanics are presented. Finally, the flow chart of the solving process of the finite element for semiconductor packaging was presented and discussed. Modeling and simulation mapping for the application of finite elements to various areas of the semiconductor industry was also provided.

References

Belytschko, T., Liu, W.K. and Moran, B. (2000) *Nonlinear Finite Elements for Continua and Structures*, John Wiley and Sons Ltd, New York.

Fung, Y.C. (1965) *Foundation of Solid Mechanics*, Prentice-Hall, New Jersey.

Gittus, J. (1975) *Viscoelasticity and Creep Fracture in Solids*, John Wiley-Halsted Press, New York.

Hibbit, H.D., Karlsson B.I. and Sorensen (2008). *ABAQUS Theory Manual*, Version 6.8.

Hill, R. (1982) *Mathematical Theory of Plasticity*, Pergamon Press, Oxford.

Khan, A. and Huang, S. (1995) *Contimuum Theory of Plasticity*, Wiley, New York.

Liu, S. (1992) Damage Mechanics of Cross-ply Laminates Resulting from Transverse Concentrated Loads, Ph.D. dissertation, Mechanical Engineering Department, Stanford University, CA, Aug. 1992.

Liu, Y. and Antes, H. (1999) An improved implicit algorithm for elastic viscoplastic boundary element method, *Zeitschrift für Angewandte Mathematik und Mechanik* **79**:317–333.

Maniatty, A.M. and Liu, Y. (2003) Stabilized finite element method for viscoplastic flow: formulation with state variable evolution, *International Journal of Numerical Methods in Engineering* **56**:185–209.

Paulino, G.H. and Liu, Y. (2000) Implicit consistent and continuum tangent operators in elastoplastic boundary element formulations, *Computational Methods in Applied Mechanics and Engineering* **190**:2157–2179.

Perzyna, P. (1971) Thermodynamic theory of viscoplasticity, *Advances in Applied Mechanics* **11**:313–354. Academic Press, New York.

Simo, J.C. and Taylor, R.L. (1985) Consistent tangent operators for rate-independent elastoplasticity, *Computer Methods in Applied Mechanics and Engineering* **48**:101–118.

Simo, J.C. and Hughes, T.J.R. (1997) *Computational Inelasticity*, Springer, New York.

Timoshenko, S.P. and Goodier, J.N. (1970) *Theory of Elasticity*, McGraw-Hill, New York.

Zhang, G.Q., van Driel, W.D. and Fan, X.J. (2006) *Mechanics of Microelectronics*, Springer, New York.

2

Material and Structural Testing for Small Samples

The reliability of a high performance microelectronic packaging assembly is closely governed by the thermo-mechanical properties of its constituent materials. Therefore, in order to conduct a reliability assessment for advanced electronic packages, it is necessary to understand the behaviors of the materials in the microelectronic package. Due to the extreme complexity of an electronic package system, it is always natural to conduct testing of alternative testing vehicles or sub-assemblies which consist of some key processing steps or key material pairs or combinations, so as to have a fundamental understanding of the behaviors of the whole package. Among all packaging materials, solder alloys play a key role in terms of electronic package reliability. Thus, to experimentally obtain a reliable and consistent property database for solder alloys becomes the first priority. Because underfilled epoxy encapsulant can tremendously extend the fatigue life of solder joints, it is now widely used in advanced electronic packages. In this chapter, solder alloys, underfills, some simplified structures or test vehicles are also used as examples to show the necessity and capability of testing for small samples in terms of materials and structures. Other materials such as thin metal films, thin polymer films, thin flexible substrates epoxy mold compound, solder masks, and thin silicon wafers can also be tested.

2.1 Material Testing for Solder Joints

2.1.1 Specimens

In order to obtain reliable and consistent experimental test data, the first challenge is to develop proper test specimens specialized for the investigation of the solder joints which are very tiny, soft and have low melting temperature and unstable microstructures. There are three major specimens commonly used in the study of solder mechanical behaviors as shown in Figure 2.1. They are dogbone bulk specimens, solder joint-like specimens such as single lap, double laps (Frear, 1989; Xie *et al.*, 1996), and thin strip specimen developed at first author's group (Ren *et al.*, 1997c, 1999a). The advantages of thermo-mechanical fatigue testing of bulk solder specimens are that the accuracy and standard testing systems such as MTS and Instron are available and data is more repeatable due to massive samples. One of drawbacks for this test specimen is that no correlation has been made between large bulk sample and small solder joint. It has been found that there are close similarities between the real solder joints in electronic packages and the solder joint-like specimens (Zhang *et al.*, 1996). The testing data can be directly used for reliability evaluation. But the finite element results show that there generally exists a highly mixed tensile and shear stress of single lap specimen along the interface between solder and copper strip (Figure 2.2). The specimen dimension is shown in Figure 2.3. Thus, the joint is a mini structure that is subjected to

Modeling and Simulation for Microelectronic Packaging and Integration: Manufacturing, Reliability and Testing, Second Edition. Sheng Liu and Yong Liu.

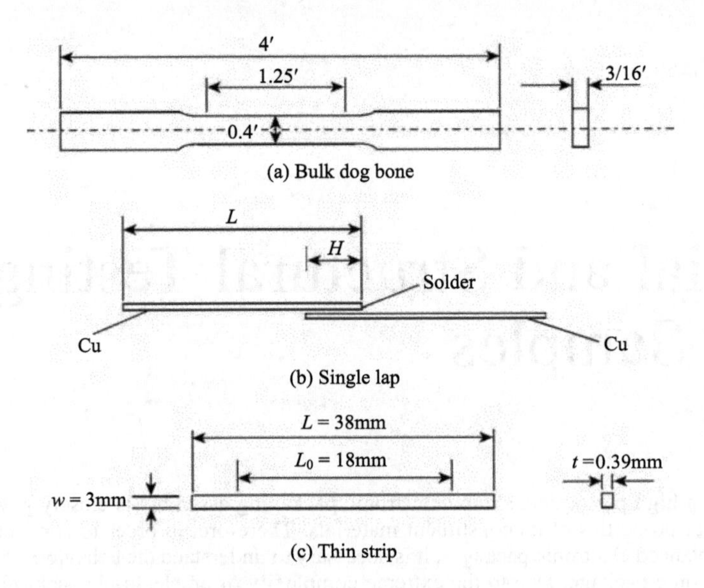

Figure 2.1 Types of specimens

non-uniform strain and stress during testing. This makes the joint specimen be not proper for material property research. It is also not desirable for fatigue tests because fatigue life is very sensitive to local geometric defects. The material properties of solder alloys cannot be isolated from the structure effect of fatigue tests by using solder joint specimens. Therefore, it is essential to develop an alternative thin strip specimen with thickness that is close to the joint size and can be tested under uniform deformation condition. Also, consistent data from the thin strip specimens are more suitable for fundamental understanding of solder fatigue behaviors under different test conditions.

A thin strip specimen is specially designed as a potential candidate of standard specimens for solder alloys. Significant efforts were made to control void level and the uniformity of specimen thickness (Ren *et al.*, 1997c, 1999a). A metal alloy mold was first designed to prepare a thin sheet of a solder alloy. The stiff shim with different thicknesses was used to control the sheet thickness. A careful procedure was

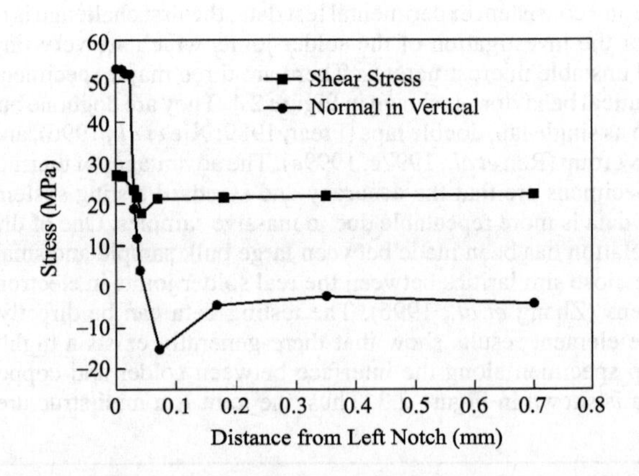

Figure 2.2 Variation of normal stress and shear stress of single lap specimen along interface between solder and copper strip

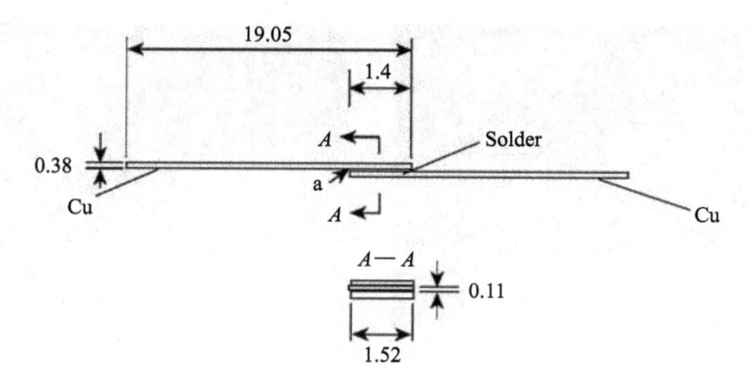

Figure 2.3 Dimension of single lap specimen used in FEA

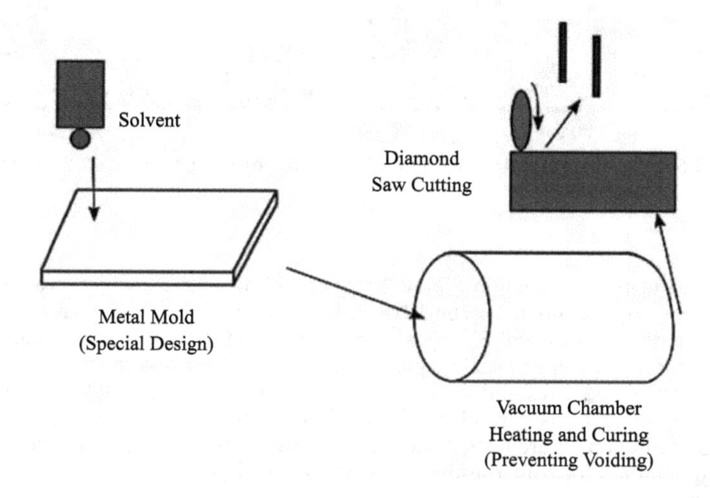

Figure 2.4 Specimen preparation

followed during the reflowing and cooling process of solder alloys in order to minimize void formation and control the quality of solder sheet. The thickness of the solder sheet was checked after it was made. And the variation of thickness was found to be within ±1%. The solder sheet was then cut into thin strip specimens using a diamond saw. The specimen preparation process is schematically shown in Figure 2.4. The specimen used in the study of solder mechanical behaviors has gage length $L = 18\,\mathrm{mm}$, width $w = 3\,\mathrm{mm}$, and thickness $t = 0.39\,\mathrm{mm}$, as shown in Figure 2.1(c). It has the thickness close to the size of the solder joints currently used in electronic industry when the research was done in 1900s.

Prior to testing, all specimens were annealed under vacuum chamber at 100 °C for one hour and then cooled down to room temperature in the chamber to stabilize microstructure and remove the residual stresses produced during the specimen preparation.

2.1.2 A Thermo-Mechanical Fatigue Tester

It is necessary to develop a special tester with high precision and high resolution of load and displacement to investigate the mechanical behaviors of small/mini specimens. A general purpose

(a) (b)

Figure 2.5 (a) 6-axis mini tester. (b) A close-up view of 6-axis mini tester using thin strip specimen

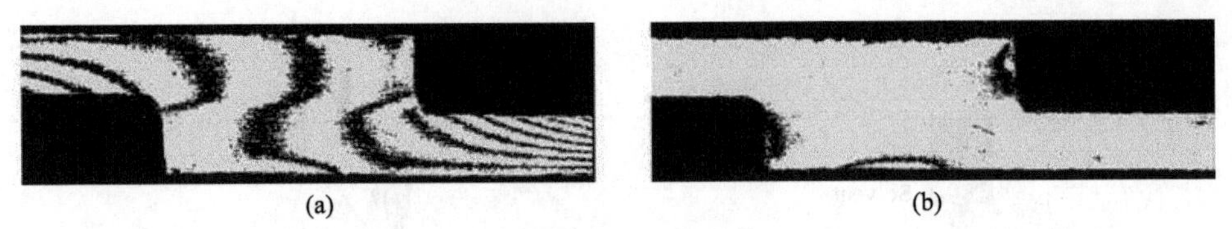

(a) (b)

Figure 2.6 (a) Fringe pattern without alignment. (b) Fringe pattern with alignment

6-axis mini fatigue tester, as shown in Figure 2.5(a) and Figure 2.5(b), which is a close-up view of the tester using thin strip specimen, was built by the first author's group and it has been commercially available by WTI of the USA and Shanghai FineMEMS Inc. It is a reliable, accurate, and well controlled by a microcomputer and easily operated under a window system. The resolution of load is one gram. The displacement resolution is 0.1 micron in translation and 0.001 degree in rotation. Before testing, the machine is carefully calibrated. The mini tester is equipped with a thermal chamber for elevated temperature tests. With this powerful tester, a reliable and consistent material database can be obtained for solder alloys. Most mechanical property tests are carried out on this mini fatigue tester.

In mounting a thin strip specimen, an alignment technique was applied to release preload produced by clamping the specimen. Figure 2.6(a) and (b) show the fringe patterns obtained by a high resolution, non-contact and real time interferometry moiré technique, to be presented in chapter 6. It can be seen that by applying the alignment technique, the preload can be reduced to be nearly zero. This is very important in the sense of the mini specimen bending or buckling due to overloading by improper clamping. This technique also makes testing data more consistent, accurate, and repeatable.

With the thin strip specimen and the computer controlled 6-axis mini fatigue tester a series of systematical experiments for solder materials under various testing conditions were conducted. The details are presented as follows.

2.1.3 Tensile Test

Thin strip specimens were tested on the 6-axis mini tester with strain rates ranging from 5.56×10^{-5}/s to 5.56×10^{-2}/s at 25 °C, 75 °C, and 100 °C, respectively. A thermal couple was kept in contact with the specimens to monitor the testing temperature. All tests were under displacement control. The force and displacement were recorded by computer and then converted to corresponding stress and strain. Typical tensile stress-strain curves at room temperature are shown in Figure 2.7. The tensile strength can be obtained from the curves, which is defined as the maximum engineering stress in the curve. Figure 2.8

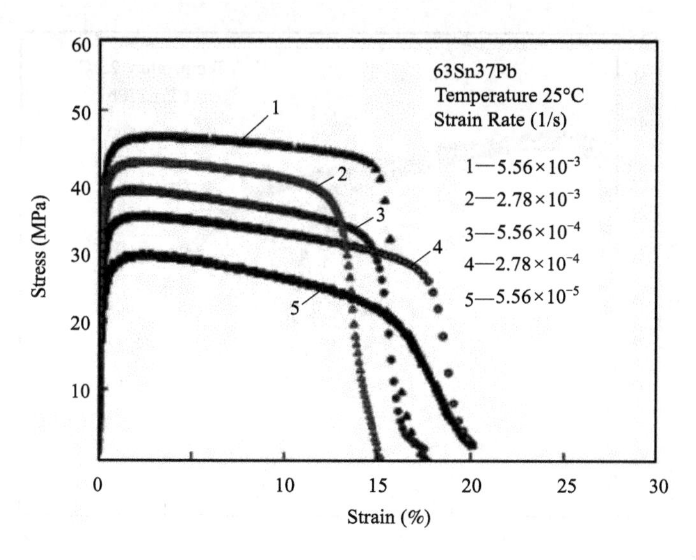

Figure 2.7 Tensile stress-strain curves with different strain rates at room temperature

shows the results of the tests for two solder alloys—eutectic 63Sn37Pb solder alloy and lead-free 80Sn10In9.5Bi0.5Ag solder alloy.

Figures 2.7 and 2.8 indicate the strong viscoplastic behaviors of two solder alloys. Tensile strengths significantly increase with the increase of tensile strain rates. The higher the temperature, the lower the strength. Figure 2.8 also shows that the tensile strength of new lead-free solder is over twice of that of 63Sn37Pb. This is one of the advantages of the new lead-free solder.

In order to investigate the effect of gage length on mechanical properties, the specimens of different gage lengths were tested. The displacement rate was controlled at 0.02mm/s, 0.01mm/s

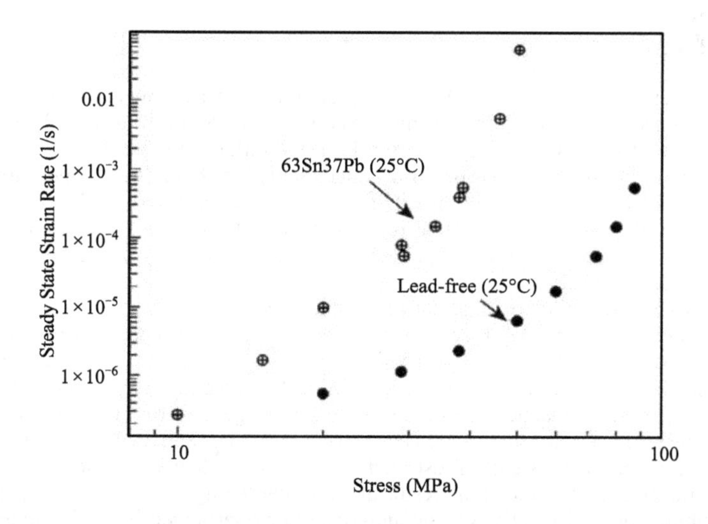

Figure 2.8 Tensile strength for 63Sn37Pb and a new lead-free solder alloy with different strain rates at the temperature investigated

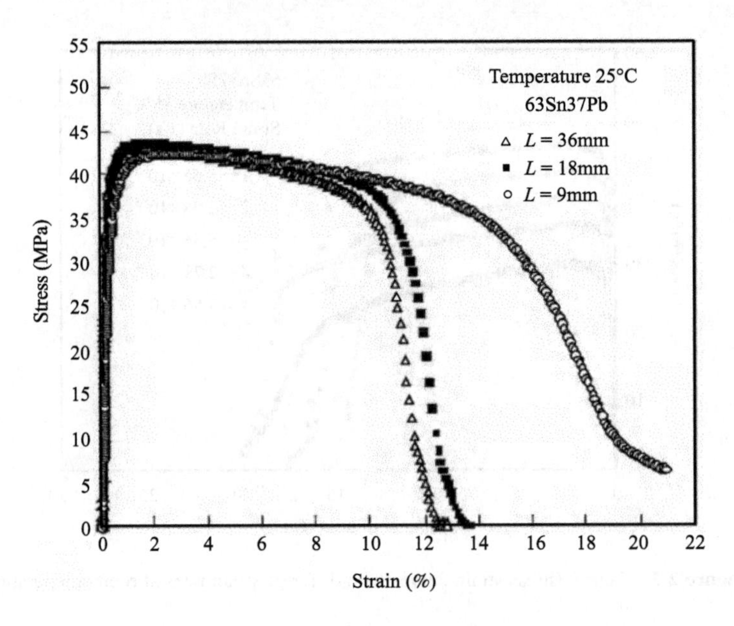

Figure 2.9 Stress-strain curves for 63Sn37Pb with different gage lengths of specimens

($L = 18$ mm), and 0.005 mm/s ($L = 9$ mm), respectively. Therefore, the same strain rate, 5.56/s, was applied for all the tests. The result is shown in Figure 2.9; it can be seen that the gage length of the specimens has no significant effects on the mechanical properties. The gage length that is independent of the test results also clearly demonstrates the reliability of the test, showing the powerfulness of the developed tester.

2.1.4 Creep Test

Power cycling is very important and is the primary source of thermal-mechanical fatigue by turning electronic devices on and off. The creep deformation constitutes most of the accumulated strain in power cycling (Liang *et al.*, 1995). It also plays a key role in the failure of creep-fatigue interaction. Under normal service conditions, solder joints operate at considerably high homologous operation temperatures. For instance, the solder alloy 63Sn37Pb has its melting temperature $T_m = 456$ K, at room temperature around $T = 298$ K, $T/T_m = 0.65$. Thus even at room temperature, creep deformation and creep fracture generally become important. Therefore, it is necessary to fundamentally understand the creep and its damage behaviors of solder alloys.

The creep tests were controlled by load with different stress levels and different temperatures for two solder alloys (Ren *et al.*, 1998a). All three stages of creep deformation, namely primary, secondary, and tertiary creep, were observed in all tests. Typical creep curves are illustrated in Figure 2.10(a) and (b) and Figure 2.11(a) and (b). The pictures show that after loading, the creep rate decays during the primary stage until a constant creep rate is reached during the secondary stage. After secondary stage the creep rate increases again during the tertiary stage which leads to creep damage and final rupture. On the other hand, three stages of creep can be physically illustrated by dislocation theory. As shown in Figure 2.12, the generation and movement of dislocations result in strain hardening, while recovery processes such as dislocation climbing allow dislocations to be annihilated or re-arranged into a low-energy dislocation pattern. In the first stage of creep the rate of work hardening is bigger than that of recovery, resulting in the decrease in the creep strain rate. In the secondary stage of creep, the two rates reach a balance.

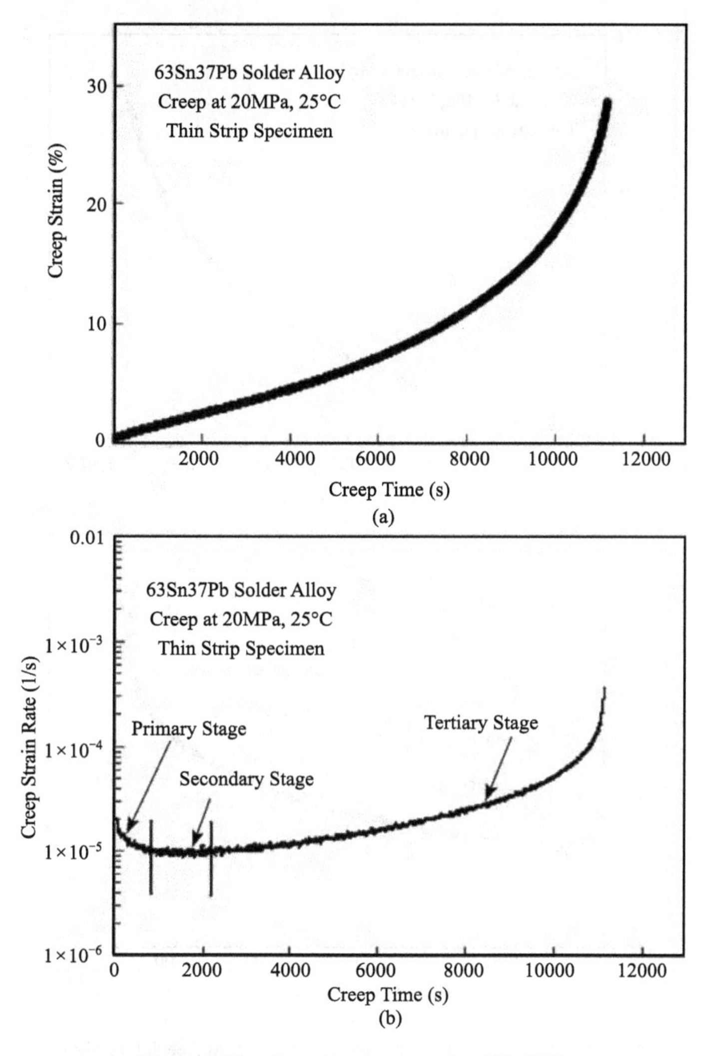

Figure 2.10 Typical creep curve for 63Sn37Pb

Therefore, the creep rate remains a constant. Then, in the tertiary stage, the creep rate increases because the rate of work hardening becomes smaller than that of recovery due to creep damage. For both solder alloys the secondary-stage creep, which is described by Norton's power-law, is relatively short comparing to the tertiary stage of creep. Therefore, for the life prediction of solder alloy it is not sufficient to consider only the second stage of creep, which will significantly over predict creep rupture time. It is necessary to consider creep damage evolution of tertiary creep for life prediction.

The secondary creep under medium stress level is well described by power-law creep, that is, Norton's law, as shown in Figures 2.13 and 2.14. By fitting the power-law creep equation between creep rates in the secondary stage and creep stresses, the stress exponent n for eutectic solder 63Sn37Pb is $n=5.5$. It is much higher than the stress exponent of new lead-free 80Sn10In9.5Bi0.5Ag solder ($n=2.25$). Furthermore, the creep rupture time of the lead-free solder is tremendously longer than that of eutectic solder at the same creep stress, as shown in Figure 2.15. More results are given in Figure 2.16, it shows that the creep

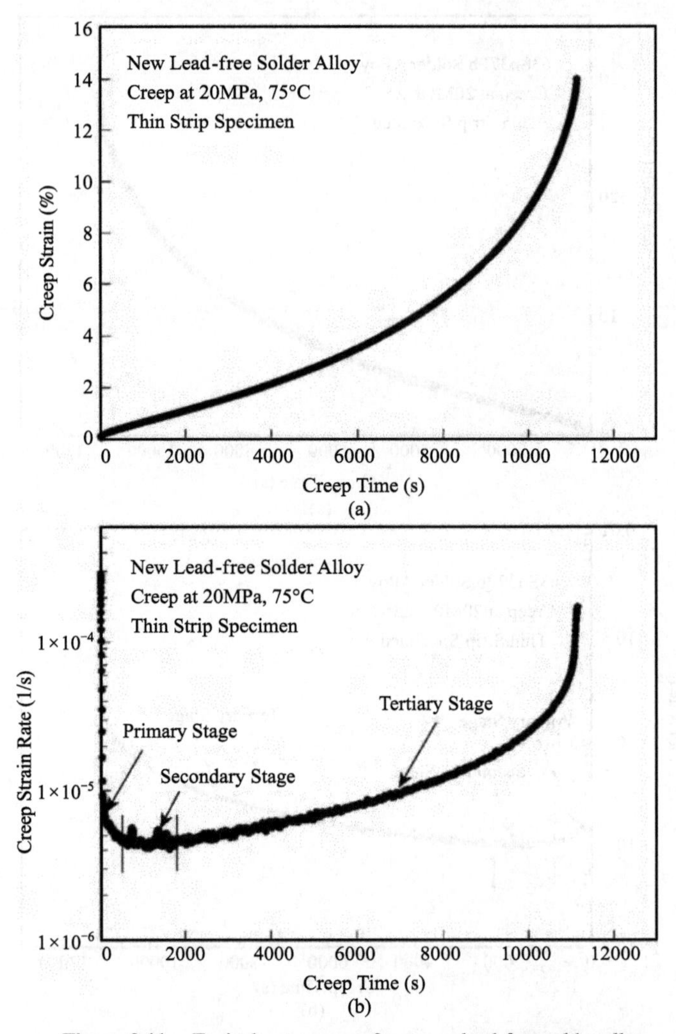

Figure 2.11 Typical creep curve for a new lead-free solder alloy

rupture time of the lead-free solder is one hundred times longer than that of eutectic solder under creep stress 38 MPa.

From the point of view of creep resistance, the new lead-free solder seems to be a good candidate of no-lead solder alloy. Figure 2.17 shows tensile and creep test results at room temperature for two solder alloys.

The concept of steady-state stress and strain rate is convenient to compare testing results from different research papers. A steady-state strain rate is defined as the creep rate at secondary creep stage and as the tensile strain rate at a tensile test. A steady-state stress is defined as creep stress and the tensile strength at engineering stress-strain curve of a tension test correspondingly. The von-Mises equivalent stress and strain rate under these steady states are used to convert testing data from shear or torsion tests to tension. In such a way, three typical data of 60–40 eutectic solder alloy from different research groups are collected and constructed in Figure 2.18. Tensile data of dogbone bulk specimens from (Adams, 1981), Motorola test data of double lap shear joints (Darveaux *et al.*, 1992) are compared with the tensile test data of thin

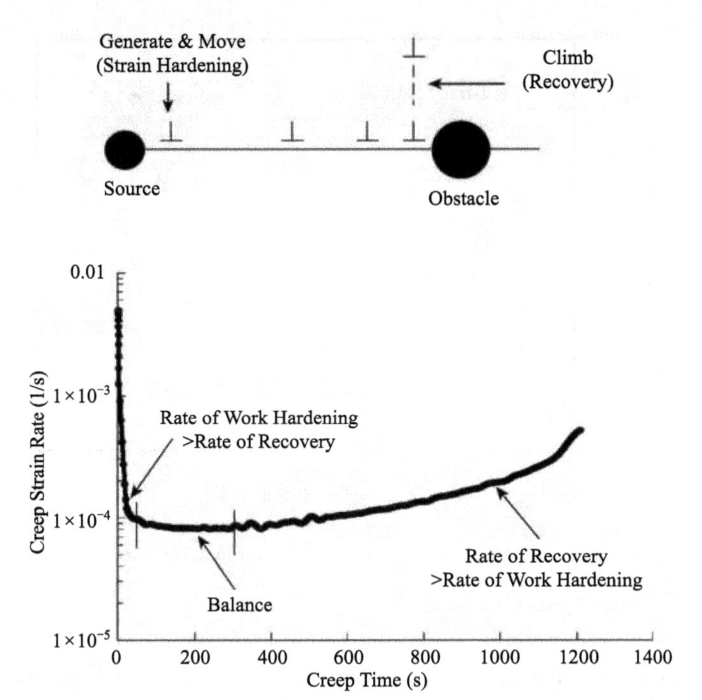

Figure 2.12 Schematic illustration of dislocation creep

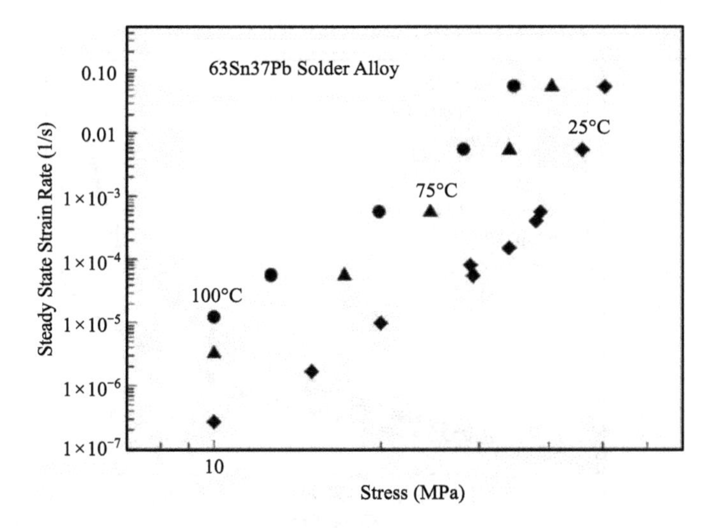

Figure 2.13 Results of tensile and creep tests for 63Sn37Pb solder alloy

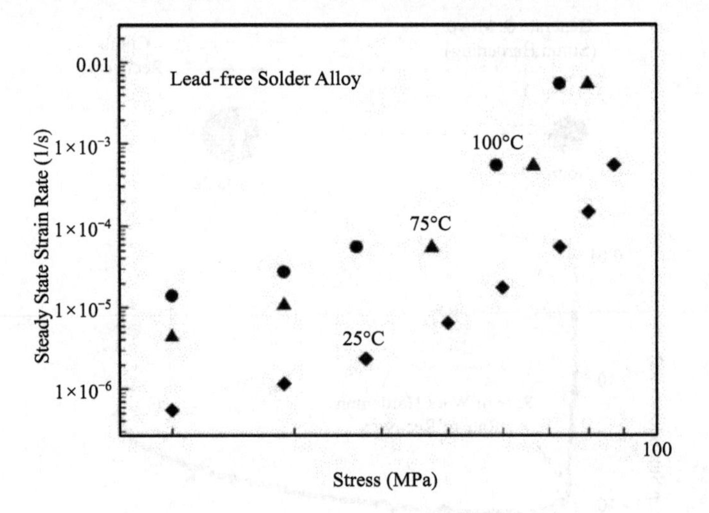

Figure 2.14 Results of tensile and creep tests for new lead-free solder alloy

strip specimens, as shown in Figure 2.18. It is found that the test data of thin strip specimens are close to the test results from double lap joint specimens at 100 °C. The data of thin strip specimens are close to those of bulk specimen (Adams, 1981) at room temperature. The strong deformation resistance of solder joint specimens comes from the effect of intermetallic layers (Darveaux *et al.*, 1992). This effect seems to be more significant at room temperature than in high temperature according to Figure 2.18. It also can be seen from Figures 2.18, 2.13 and 2.14 that both eutectic solder and new lead-free solder obey the power-law creep at low steady-stare stresses, and are subjected to power-law breakdown at high steady-state stresses. The data from thin strip specimens are more consistent and reasonably compatible to data from the literature with different specimen geometry.

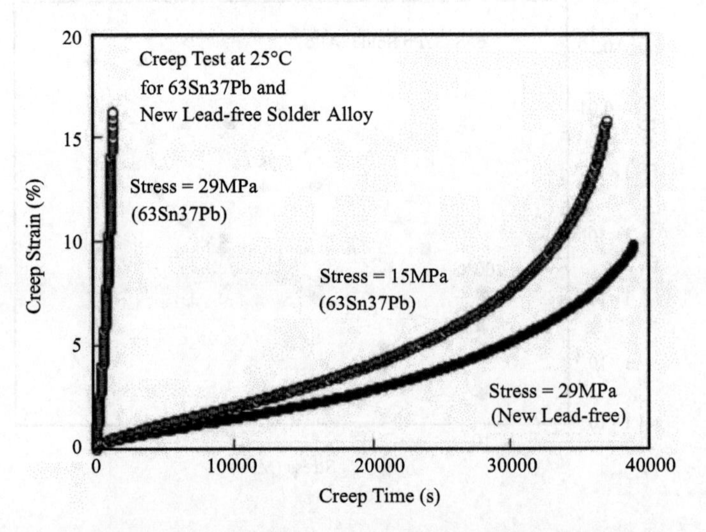

Figure 2.15 Typical creep curves for two solder alloys

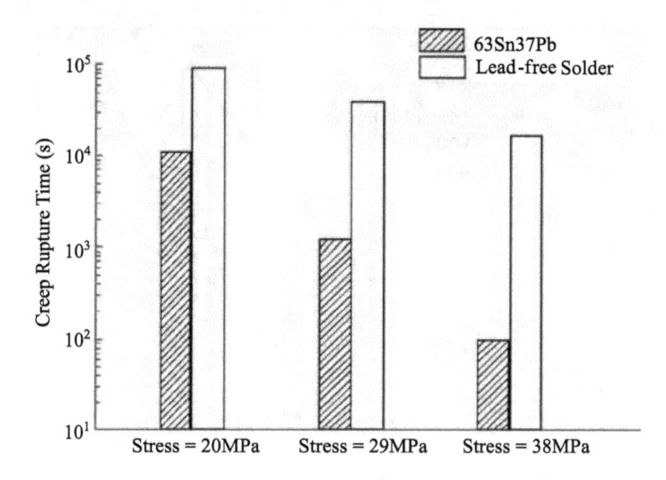

Figure 2.16 Comparison of creep rupture time (temperature 25 °C)

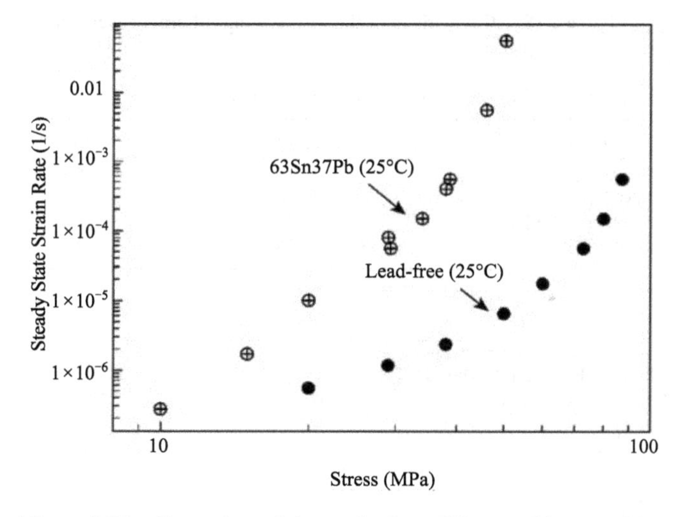

Figure 2.17 Comparison of the results from different solder materials

2.1.5 Fatigue Test

The tensile-tensile fatigue tests under load control were performed with a triangular waveform at different maximum stresses. The strain rate was held constant, 1×10^{-4}/s, for each test. The fatigue life versus maximum stress is listed in Table 2.1.

Figure 2.19 shows the typical stress-strain and time-strain relations during a fatigue test of the lead-free 80Sn10In9.5Bi0.5Ag solder under a stress of 56 MPa. Cyclic ratcheting is observed is Figure 2.19. The ratcheting rate then reaches a constant. The acceleration of ratcheting is because of fatigue-induced material damage. Therefore, the strain softening is observed at the last stage of ratcheting. It is very interesting that the three stages of the ratcheting are similar to three stages of creep. Actually, the ratcheting is a kind of important deformation failure mode that is found in the chemical and nuclear vessels served at high temperatures. The failure phenomenon could be important in electronic package structures subjected to thermal cyclic loading.

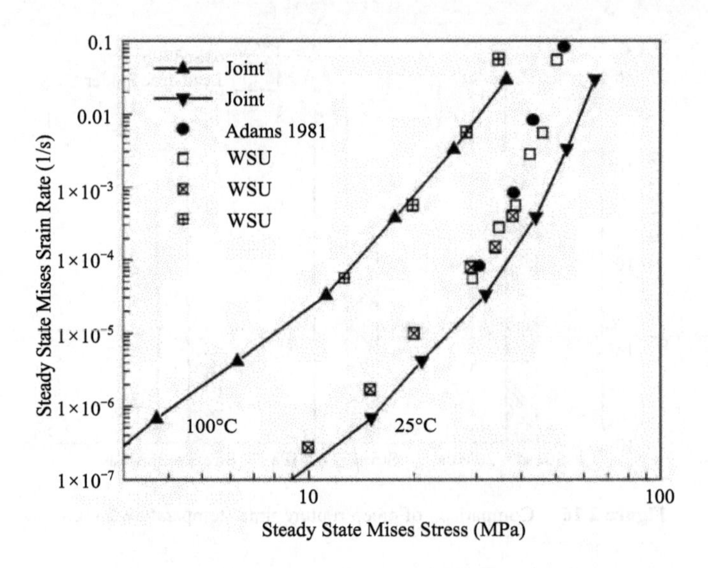

Figure 2.18 Comparison of the results from different specimens

Table 2.1 Fatigue life versus maximum stress

Material	σ_{max}(MPa)	Total Cycle
New Lead-Free Solder	45	205
80Sn10In9.5Bi0.5Ag	56	75
Eutectic Solder	25	208
63Sn37Pb	28	110

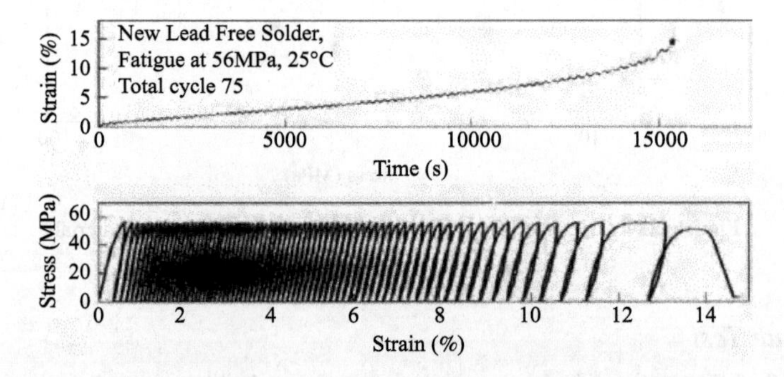

Figure 2.19 Stress-strain and time-strain relations recorded during the fatigue test

2.2 Scale Effect of Packaging Materials

The electronic packaging trend is moving towards further miniaturization. Flip-chip technology with high-density interconnections provides a solution to the increasing demands on miniaturization and weight reduction of portable electronics products. The technology enables the direct chip attachment of silicon chips to substrates, also called flip-chip-on-board (FCOB) assembly

(Lau, 1996). The space between the chip and board is often encapsulated by underfilled epoxy encapsulant to reinforce the mechanical performance of the solder joints. Then the fatigue life of solder joints can be extended tremendously (Suryanarayana *et al.*, 1991; Zhang *et al.*, 1998). Therefore, quite a few researchers have been studying new and reworkable underfills (Wang *et al.*, 1998; Wang and Wong, 1998), wafer-level flip-chip underfilling (Nguyen, 1999), and the thermo-mechanical nonlinear behaviors (Qian *et al.*, 1998).

While the electronic packages become smaller and smaller, the scale effect in terms of package materials and structures might become one issue. For those materials like underfilled epoxy encapsulant that are highly filled with silica particles, this concern becomes even more serious. On the other hand, it is also a challenge to obtain thermo-mechanical properties of package materials when the specimen scale tends to be sub-micron. The related technique is specimen alignment and failure mode control for testing thin/mini specimens. Currently available single-axis tester is, in general, not adequate.

In this research, underfill HYSOL® FP4526 and FP4511 (Product Bulletin of HYSOL, 1997) and eutectic solder alloy 63Sn37Pb are selected to investigate the scale effect of materials by combining specially-designed and well-manufactured thin strip specimens with a computer-controlled 6-axis mini fatigue tester (Ren *et al.*, 1999b). The specimen thickness investigated in this research ranges from 80 microns to 600 microns. This is also a critical way to test the consistency of test data from both precious tester and well-made specimens.

2.2.1 Specimens

The similar procedure was followed to make the thin strip specimens for underfill HYSOL® FP4526 and underfill HYSOL® FP4511 as mentioned in the previous section for solder alloys. Significant efforts were made to obtain void-free and uniform specimens with the thickness down to 80 microns. A metal alloy mold was used to prepare a thin sheet of underfill encapsulants. The stiff shim with different sizes was used to control the sheet thickness. The curing processes recommended by the Dexter product bulletin (Product Bulletin of HYSOL®, 1997) were followed. For underfill FP4526® specimen, the frozen underfill stored at $-40\,°C$ was allowed to reach room temperature before use. The solvent was then cast into the mold that had been preheated at 85 °C for a long time and reached a uniform temperature distribution. The curing was finished at 165 °C for 15 minutes and then cooled down to room temperature in the oven. The fine surface and uniform thickness sheet of the underfill was successfully fabricated in this way. The range of the thickness was from 0.08 mm to 0.60 mm. The thickness of the underfill sheet was checked after the sheet had been made. It was found that the thickness variation of the underfill sheet was within a very small range. For example, the thickness variation of the 0.24 mm thickness underfill sheet was found to be less than $\pm 1\%$. The underfill sheet was then cut into thin strip specimens by a diamond saw. The similar procedure was followed to make the specimens for underfill HYSOL® FP4511.

The underfill specimen used in all tests has gage length $L_0 = 16$ mm, width $w = 2.2$ mm, and thickness $t = 0.08$ mm–0.60 mm. The specimen thickness of 80 microns is very close to the typical solder joint size. Therefore, the test results are realistic and can be directly used to guide the design of advanced packages involved underfill encapsulants. Figure 2.20 shows the typical thin strip specimens of underfill HYSOL®

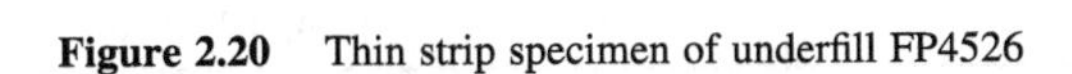

Figure 2.20 Thin strip specimen of underfill FP4526

Figure 2.21 SEM microstructure of underfill FP4526 specimen

FP4526. One of the SEM pictures is presented in Figure 2.21 to show the microstructure of the underfill. The material is highly filled with silica particles. No settling of particles is found.

2.2.2 *Experimental Results and Discussions*

The tensile test for underfills was carried out on the 6-axis mini tester with the constant strain rate 6.25×10^{-4}/s at 25 °C, 75 °C, and 100 °C, respectively. The thickness of specimens was 0.08, 0.10, 0.14, 0.20, 0.24, and 0.60 mm, respectively. The test details of combination of thickness and temperatures are listed in Table 2.2. Two to five specimens were tested for each combination to check the repeatability and consistency of test data.

Figure 2.22 shows room temperature test data from six different values of thickness. In this case, there is no temperature control variation. The local view of the stress/strain curves of Figure 2.22 at linear elastic stage is shown in Figure 2.23(a) that indicates all curves nearly have the same slop with very narrow scatter band. Therefore, the Young's modulus can be obtained with high accuracy. This demonstrates that the 6-axis mini tester performs a precious control of test conditions such as strain rate and data acquisition with high accuracy. The excellent quality of void-free specimens with uniform thickness also leads to the consistency and repeatability of test data at linear stage. Figure 2.23(b) shows a magnified view of the curves. The curves tend to be apart for larger strain. The scatter band increases with the increase of strain, in particular, when the strain is larger than 1.2%, as shown in Figures 2.22 and 2.23(b). There is a certain extension of scatter at strength, that is, the maximum stress at rupture strain. However, no certain correlation

Table 2.2 Specimen thickness and test temperature for underfill HYSOL® FP4526

Test Temperature (°C)	Specimen Thickness (mm)
0	0.24, 0.14
25	0.60, 0.24, 0.20, 0.14, 0.10, 0.08
50	0.24, 0.20, 0.10, 0.08
75	0.24, 0.20, 0.10, 0.08
100	0.24, 0.20, 0.10

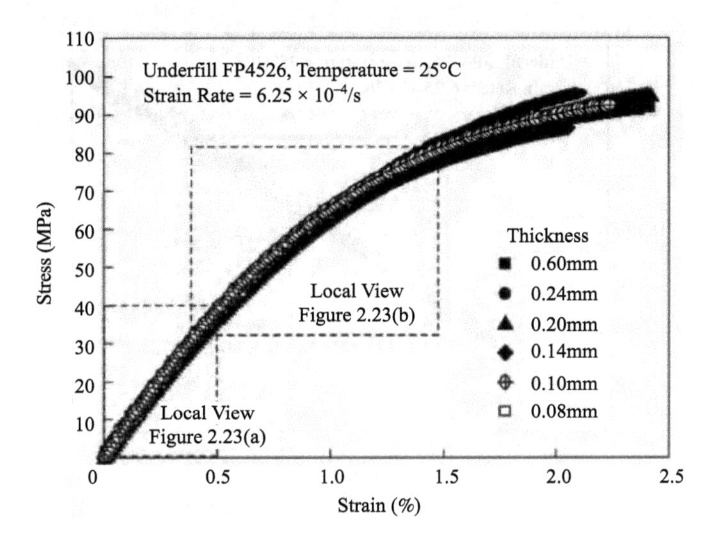

Figure 2.22 Stress-strain curves for underfill FP4526 at 25 °C

with thickness is identified. For instance, the stress curve of 0.60 mm thickness specimen is in the middle where is very close to the curves of specimens with the thickness of 0.10 mm and 0.20 mm. The larger scatter of test data at the nonlinear stage of the curves is because the stress and strain are more sensitive to the minor defects of the material and specimens. The rupture strain of specimen thickness of 0.08 mm is found to be quite small at around 1%, comparing with larger thickness specimens. The nonlinear part of the stress strain curve of the thinnest specimen is short, as shown in Figure 2.22. The test was repeated several times. And the same results were obtained. A possible reason is that the rupture strain of such a thin specimen (80 microns in thickness) is quite sensitive to the edge defect of the specimen. The tiny crack, once generated, is easy to propagate along silica interfaces through the specimen. More investigation of the microstructure of the rupture section is on-going to identify the phenomena. The maximum silica filler size of underfill HYSOL® FP4526 is found to be around 27microns. Therefore, when the thickness of specimen approaches a certain critical value, the filler size effect could be significant to the rupture strain. The surface roughness of underfill interfaces with silicon chip and FR4 substrate could be also critical for underfill delamination when very thin underfill layer is used for high density interconnects. The smaller filler will benefit the increase of the ductility. On the other hand, the nonlinear properties are found to be important for the failure mode and mechanism of flip-chip packages at corners and interfaces. Therefore, it seems that there is a critical thickness for underfilling. Further investigation is needed to justify these points.

Moreover, the repeatability of test data from specimens with the same thickness is shown in Figure 2.24 at room temperature 25 °C and 50 °C, respectively. The data shows a very narrow scatter band. Again, there is a little larger scatter near rupture strains. Figures 2.25, 2.26, and 2.27 show the test results of specimen thickness ranging from 0.08 mm to 0.24 mm at 50 °C, 75 °C, and 100 °C, respectively. The larger scatter at later nonlinear stage at 100 °C is possibly caused by the temperature which is close to the T_g (133 °C) of the underfill. Figure 2.28 shows test data of a specimen with two thicknesses at the lower temperature of 0 °C. The linear stage is significantly longer than that at higher temperatures. On the other hand, the strength is higher. The underfill FP4511 is found to be a little stronger than the FP4526, with higher Young's modulus and strength, as shown in Figure 2.29. Two underfills show very similar nonlinear characterizations.

The Young's module of FP4526 at different temperatures and the same strain rate 6.25×10^{-4}/s are plotted in Figure 2.30, together with the data reported before (Qian *et al.*, 1998). The Young's modulus is

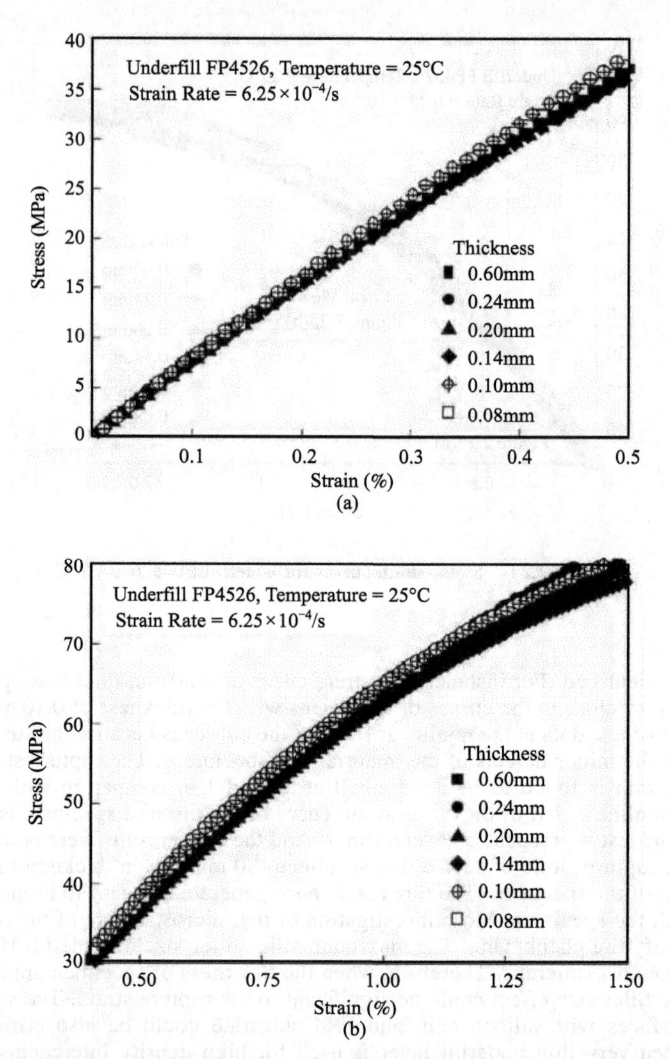

Figure 2.23 (a) Local view of the stress-strain curves for underfill FP4526 at 25 °C. (b) Local view of the stress-strain curves for underfill FP4526 at 25 °C

decreased quickly at a lower temperature range. The data is fitted to an exponential decay relationship perfectly. The equation is:

$$E(T) = 5.275 + 4.0\exp(-T/39.8) \quad \text{(GPa)} \tag{2.1}$$

where the unit of temperature T is °C.

Table 2.3 lists the Young's Moduli E_t obtained from the specimens with different thicknesses at room temperature. The E_t is, in fact, the average value of Young's moduli obtained from samples with the same thickness. The difference ΔE_t defined by Equation 2.2 and 2.3 is also

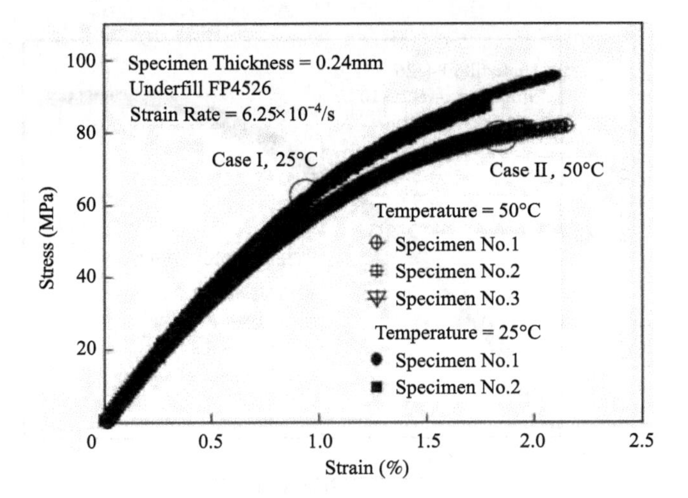

Figure 2.24 Typical stress-strain curves obtained from the same specimen thickness and the same test condition

presented in the Table 2.3.

$$E = \frac{\sum E_t}{N} \tag{2.2}$$

where E is the average Young's Modulus obtained from tests with different thickness, N is total number of the tests.

$$\Delta E_t = \frac{E_t - E}{E} \tag{2.3}$$

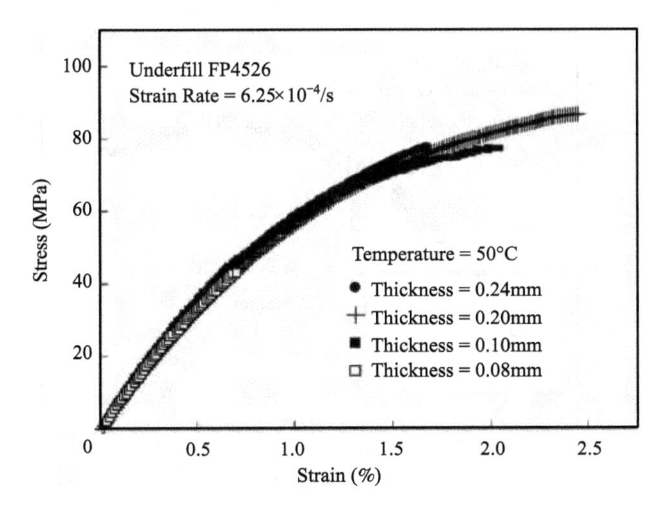

Figure 2.25 Stress-strain curves for underfill FP4526 at 50 °C

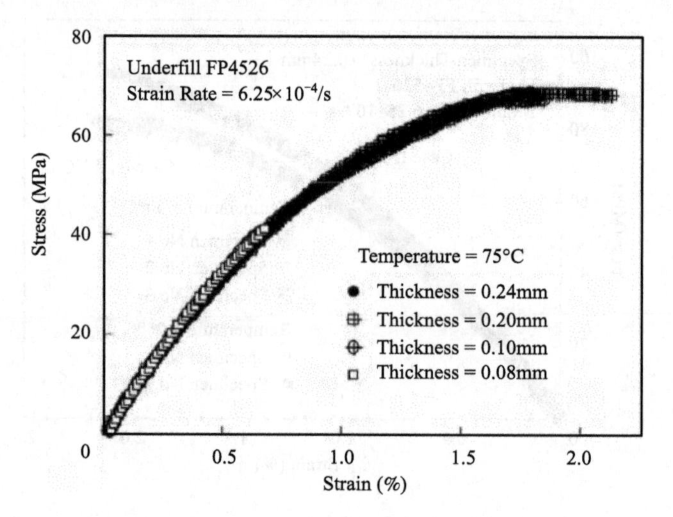

Figure 2.26 Stress-strain curves for underfill FP4526 at 75 °C

The maximum difference ΔE_t at room temperature is found to be only 2.82%, which indicates excellent repeatability and consistency of all tests.

The tensile tests were also conducted for eutectic solder alloy 63Sn37Pb at room temperature. Test data from different thickness show no significant difference, as illustrated in Figure 2.31.

Test data show excellent repeatability and consistency. The thermo-mechanical properties of underfills are temperature dependent. The Young's Modulus decreases with the increase of the temperature. The comprehensive investigation shows that the specimen thickness has no significant effect on the test results of underfills and solders within the scale studied in this research. However, the scale effect needs further investigation when the scale is down to a critical size.

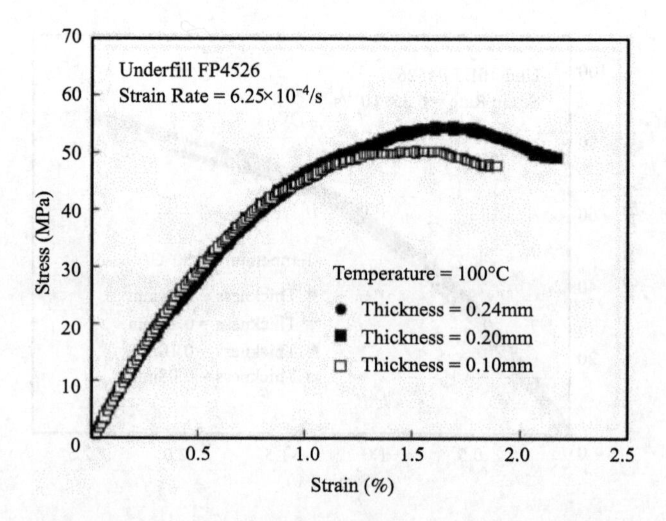

Figure 2.27 Stress-strain curves for underfill FP4526 at 100 °C

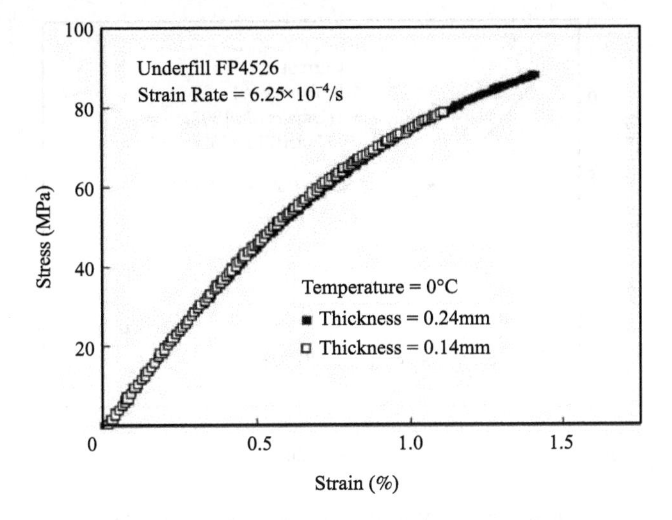

Figure 2.28 Stress-strain curves for underfill FP4526 at 0 °C

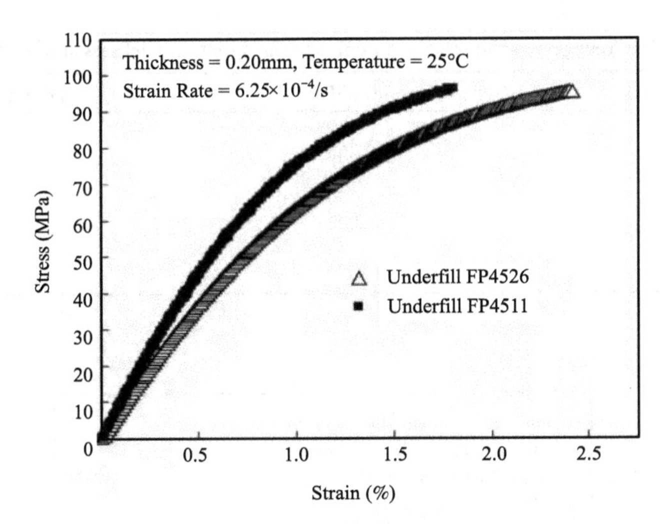

Figure 2.29 Stress-strain curves for underfill FP4526 and FP4511 at room temperature

2.2.3 *Thin Film Scale Dependence for Polymer Thin Films*

The scale effect may become more important with the decrease of ball pitch sizes and the development of high density interconnects. Unlike metal films, the thinner the polymer films, the stronger the deformation hardening. The phenomenon demonstrates the significant difference of mechanical properties between bulk polymers and thin films. An example is shown in Figure 2.32 for the test of Kapton films with different thicknesses at 23 °C and 100 °C, respectively. The micro-mechanism of the phenomena relies on a molecular chain interaction and orientation hardening. More fundamental research is needed for future research in the community. The thickness effect is believed to be an important aspect for polymer adhesives and interlayers.

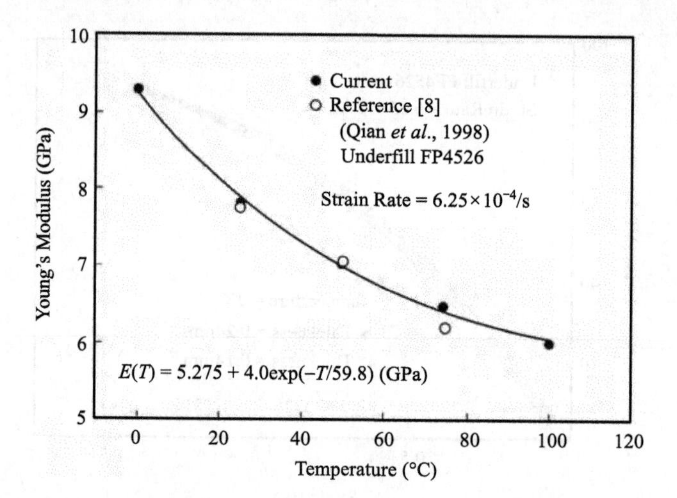

Figure 2.30 Young's modulus of underfill FP4526

Table 2.3 The Young's modulus obtained from different specimen thickness at room temperature

Thickness (mm)	E_t (GPa)	Difference ΔE_t (%)
0.60	7.70	−1.41
0.24	7.68	1.66
0.20	7.83	0.26
0.14	7.59	−2.82
0.10	8.02	2.69
0.08	8.03	2.82
Average	7.81	0.53

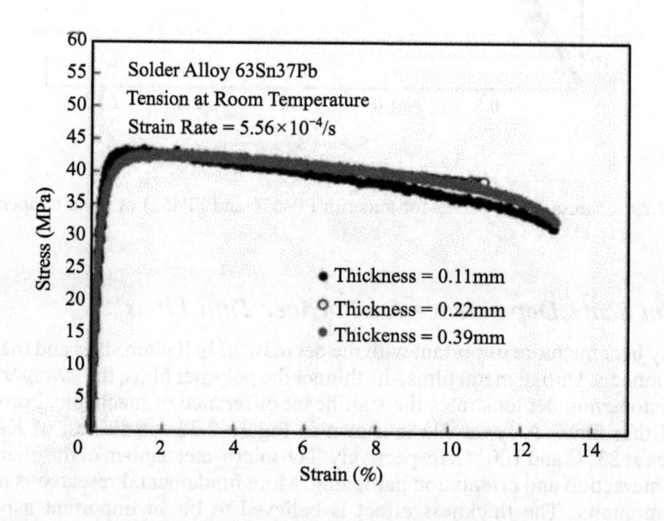

Figure 2.31 Stress-strain curves for solder alloy 63Sn37Pb at room temperature

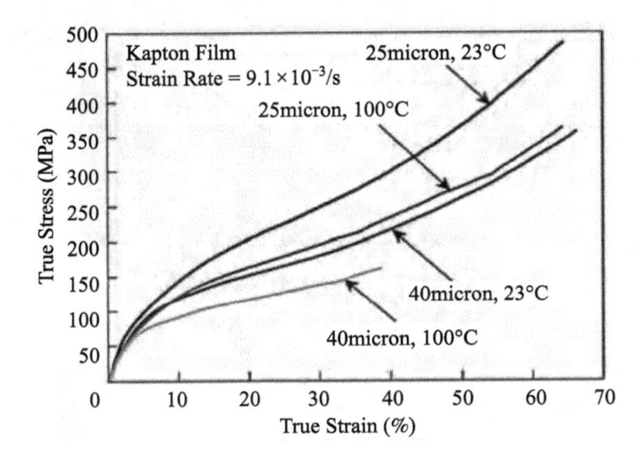

Figure 2.32 Strain hardening and scale effect of Kapton films

2.3 Two-Ball Joint Specimen Fatigue Testing

In general, it is desirable to test packages or package sub-assemblies in an isothermal condition with mechanical loading. However, there exist differences between isothermal mechanical testing and thermal cycling for full packages due to the many solder balls with each ball experiencing difference shear strain. A two-ball solder joint sample would be a unique one, as each ball will experience the same shear when the simplified structure is subjected to either isothermal mechanical fatigue or thermal cycling. We were motivated to build this two-ball sample by removing other balls and cutting upper BT board and low FR4 board.

After the sample was made, we make a fixture which can hold the two boards and the sample is then subjected to tension-compression fatigue. Figures 2.33(a) and (b) show load versus shear displacement curves showing hardening, cracking, and softening behaviors of 63Sn37Pb at room temperature. From Figure 2.33(a), very nice loading and unloading curves are presented and we clearly observed the ball cracking and cracking propagation. From Figure 2.33(b), we found the symmetric locations of crack initiation, corresponding to the cracking initiation, and damage growth. This example will provide an interesting alternative test vehicle for research in accelerated testing.

2.4 Chapter Summary

The thin strip specimen used in this research serves as an ideal specimen for the study of thermo-mechanical properties of solder alloys underfills, and other packaging materials. The thickness is very close to the package solder joint size and underfill layer thickness. The solder alloy test data are comparable to solder joint test results. All underfill test data show excellent repeatability and consistency. Therefore, this kind of specimen has the potential to be used as a standard specimen in the further study.

By using the newly developed 6-axis mini tester, a series of consistent data of two types of solder alloys were obtained. The main conclusions from the test results are as follows:

(1) The two solder alloys—eutectic 63Sn37Pb solder alloy and lead-free 80Sn10In9.5Bi0.5Ag solder alloy present strong viscoplastic behaviors, that is, temperature and strain rate dependence, while the gage length has no significant effect on mechanical properties;
(2) The new lead-free solder alloy 80Sn10In9.5Bi0.5Ag has high tensile strength. The tensile strengths for different temperatures and strain rates investigated in this research are over twice of that of 63Sn37Pb;

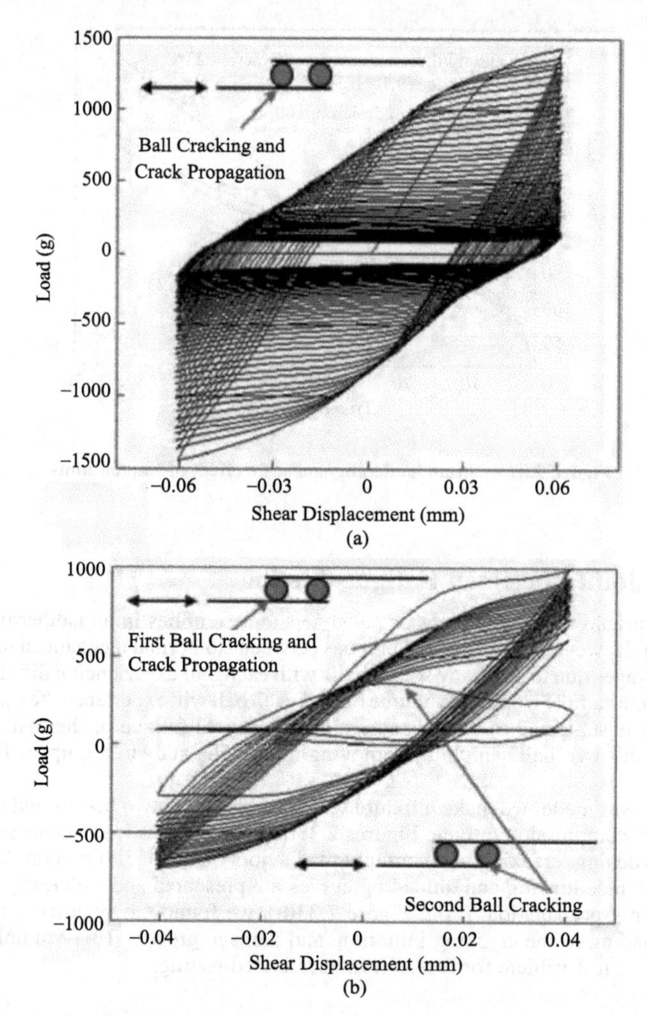

Figure 2.33 Load versus shear displacement for a two-ball solder (63Sn37Pb at room temperature, showing hardening, cracking, softening behaviors). (a) First curves showing hardening; (b) last curves showing cracking and softening

(3) The new lead-free solder 80Sn10In9.5Bi0.5Ag displays much stronger creep resistance than that of eutectic solder 63Sn37Pb. From this point of the view, the new lead-free solder alloy seems to be a good candidate of no lead solder alloy;
(4) The new lead-free solder has a longer fatigue lifetime corresponding to stress controlled fatigue tests.

Unique material characteristics combined with the mechanical properties make the new lead-free solder alloy 80Sn10In9.5Bi0.5Ag a suitable candidate for electronic packaging applications.

The comprehensive investigation shows that the specimen thickness has no significant effect on the test results of underfills and solders within the scale studied in this research when the thickness is close to the underfill thickness. However, the scale effect needs further investigation when the scale is down to a critical size.

The scale effect of Kapton film has been identified, as the thickness is in the range of 20 to 50 microns, which is typical of most polymer materials.

A two-ball solder sample is used as a typical simplified test vehicle to study the damage initiation, growth, and collapse.

References

Adams, P.J. (1981) Thermal fatigue of solder joints in micro-electronic devices, M.S. thesis, Department of Mechanical Engineering, MIT, Cambridge, MA.

Darveaux, R. and Banerji, K. (1992) Constitutive relations for tin-based solder joints, *IEEE Transactions on Components, Hybrids, and Manufacturing Technology* **15**(6):1013–1024.

Engelmaier, W. (1983) Fatigue life of leadless chip carrier solder joints during power cycling, *IEEE Transactions on Parts, Hybrids and Packaging*, **CHMT-6**(3):232–237.

Ewell, G.J. and Robertson, S.R. (1996) Development of Pb-free solder: an overview, *The Second International Symposium on Electronic Packaging Technology, Shanghai, China*, 73–78.

Ferry, J.D. (1980) *Viscoelastic Properties of Polymers*, 3rd Edition, John Wiley and Sons Ltd, Chichester.

Fox, L.R., Sofia, J.W. and Shine, M.C. (1985) Investigation of solder fatigue acceleration factors, *IEEE Transactions on Components, Hybrids, and Manufacturing Technology*, **CHMT-4**(2):275–285.

Frear, D., Grivas, D. and Morris, J.W. (1988) A microstructural study of the thermal fatigue failures of 60Sn-40Pb solder joints, *Journal of Electronic Materials* **17**(2):171–180.

Frear, D.R. (1989) Thermo-mechanical fatigue of solder joints: a new comprehensive test method, *IEEE Transactions on Components, Hybrids, and Manufacturing Technology* **12**:492–501.

Frear, D.R., Jones, W.B. and Kinsman, K.R. (1990) *Solder Mechanics,* A State of the Art Assessment, TMS, New Mexico.

Frear, D.R., Burchett, S.N. and Neilsen, M.K. (1997) Life prediction modeling of solder interconnects for electronic systems, *Advances in Electronic Packaging*, ASME EEP **19**(2):1515–1522.

Freed, D. and Walker, K.P. (1990) Steady-state and transient Zener parameter in viscoplasticity: drag strength versus yield strength, *Applied Mechanics Reviews* **43**:328–436.

Gektin, V., Bar-Cohen, A. and Ames, J. (1997) Coffin-manson fatigue model of underfilled flip-chips, *IEEE Transaction.* CPMT Part A **20**(3):317–326.

Glazer, J. (1994) Microstructure and mechanical properties of Pb-Free solder alloys for low-cost electronic Assembly: A Review, *Journal of Electronic Materials* **23**:693–700.

Gurson, A.L. (1977) Continuum theory of ductile rupture by void nucleation and growth: part I - yield criteria and flow rules for porous ductile media, *Journal of Engineering Materials and Technology,* ASME **99**:2–15.

Lau, J.H. (1996) *Flip-Chip Technologies*, McGraw-Hill, New York.

Lu, M. and Liu, S. (1996) A multi-axial thermo-mechanical fatigue tester for electronic packaging materials in sensing, *Modeling and Simulation in Emerging Electronic Packaging*, ASME, EEEP **17**:87–92.

Liang, J., Gollhardt, N., Lee, P.S., Schroeder, S.A. and Morris, W.L. (1995) Fatigue and cyclic deformation behavior of high temperature solder alloys, *ASME International Mechanical Engineering Congress and Exposition*, EEP **13**, AMD **214**:1–9, San Francisco.

Mavoori, H. (1996) Mechanical properties and fatigue lifetime prediction of solders for electronic applications: tin-silver and tin-zinc eutectics, Ph.D Dissertation, Northwestern University.

McDowell, D.L. (1994) Multiaxial effects in metallic materials, *Durability and Damage Tolerance* ASME, AD **43**:213–267.

Nguyen, L. (1999) Rationale for wafer level flip-chip underfilling, *IMAPS, International Advanced Technology Workshop on Flip-Chip Technology*, Braselton, GA, March 12–14.

Qian, Z. and Liu, S. (1997) A unified viscoplastic constitutive model for tin-lead solder joints, *Advances in Electronic Packaging*, ASME EEP **19**(2):1599–1604.

Qian, Z., Lu, M., Wang, J. and Liu, S. (1997) Testing and constitutive modeling of thin polymer films and underfills by a 6-axis submicron tester, *International Mechanical Engineering Congress and Exposition*, Dallas ASME EEP **22**, AMD **226**:105–112.

Qian, Z. and Liu, S. (1998) On the life prediction and accelerated testing of solder joints, in *Thermo-mechanical Characterization of Evolving Packaging Materials and Structures*, ASME EEP **7**:1–11.

Ren, W., Lu, M.F., Liu, S. and Shangguan, D.K. (1997a) Thermal mechanical property testing of new lead free solder joints, *Journal of Soldering and Surface Mount Technology* **9**, No. 3:37–40.

Ren, W., Lu, M.F., Liu, S. and Shangguan, D.K. (1997b) A study on a lead-free solder joint, *SEM Spring Conference on Experimental Mechanics and Experimental/Numerical Mechanics in Electronic Packaging*, Bellevue, Washington.

Ren, W., Qian, Z.F., Lu, M.F., Liu, S. and Shangguan, D.K. (1997c) Thermal mechanical properties of two solder alloys, *International Mechanical Engineering Congress and Exposition*, Dallas, ASME EEP **22**, AMD **226**:125–130.

Ren, W., Qian, Z.F. and Liu, S. (1998a) Thermo-mechanical creep of two solder alloys, *48th Electronic Components and Technology Conference*, Seattle, Washington, 1431–1437.

Ren, W., Wang, J.J. and Liu, S. (1998b) Investigation of creep behavior for a flip-chip package, *TECHCON'98, Semiconductor Research Corporation Fifth National Conference*, Las Vegas, Nevada.

Ren, W., Qian, Z.F., Liu, S. and Shangguan, D.K. (1999a) Investigation of a new lead free solder alloy using thin strip specimens, *ASME Transactions, Journal of Electronic Packaging* **121**:271–274.

Ren, W., Qian, Z.F. and Liu, S. (1999b) Scale effect on packaging materials, *49th Electronic Components and Technology Conference*, San Diego, California, 1229–1234.

Suryanarayana, D., Hsiao, R., Gall, T.P. and McCreary, J.M. (1991) Enhancement of flip-chip fatigue life by encapsulation, *IEEE CHMP* **14**:218–223.

Wang, J., Qian, Z., Zou, D. and Liu, S. (1998) Creep behavior of a flip-chip package by both FEM modeling and real time moiré interferometry, *Transactions of the ASME, Journal of Electronic Packaging* **120**(2):179–185.

Wang, L. and Wong, C.P. (1998) Novel thermally reworkable underfill encapsulants for flip-chip applications, *48th Electronic Components and Technology Conference*, Seattle, 92–100.

Xie, D.J., Chan, Y.C., Lai, J.K.L. and Hui, I.K. (1996) Fatigue life estimation of surface mount solder joints, *IEEE Transactions on Components, Packaging, and Manufacturing Technology–Part B* **19**(3):669–677.

Zhang, Z., Yao, D.P. and Shang, J.K. (1996) Fatigue crack initiation in solder joints, *Journal of Electronic Package* **118**:41–44.

Zhang, W., Wu, D., Su, B., Hareb, S.A., Lee, Y.C. and Masterson, B.P. (1998) The effect of underfill epoxy on warpage in flip-chip assemblies, *IEEE Transaction on Components, Packaging, and Manufacturing Technology-Part A* **21**(2):323–329.

3

Constitutive and User-Supplied Subroutines for Solders Considering Damage Evolution

Solder joint reliability is an important issue for electronic packaging. Solder joints provide thermal and electronic connections in addition to the traditionally mechanical function. The solder alloys used for the joints are subjected to thermal and mechanical loading during assembly processes, testing, and service operation. Therefore, it is highly desirable to develop a sophisticated solder constitutive model for the prediction of its service life. A damage coupling framework is proposed with a unified viscoplastic constitutive model for the fatigue life prediction of solder alloys, including a consistent and systematic procedure for the determination of material parameters. The model also incorporates the effects of grain size and cyclic hardening/softening. The phenomenological framework of continuum damage mechanics (CDM) is modified to embed intensive research on the ductile damage fracture by void nucleation, growth, and coalescence, based on a lower bound for the viscoplastic potential of voided materials. The damage coupling with elasticity and viscoplasticity is formulated by an irreversible thermodynamix approach.

3.1 Constitutive Model for Tin-Lead Solder Joint

3.1.1 Model Formulation

The creep deformation processes are expected to dominate the deformation kinetics of solder alloys due to their low melting points according to the study of deformation mechanisms of metal alloy. Steady-state creep is generally expressed by a relationship of the form:

$$\dot{\gamma}_{ss} = \frac{CGbD_0}{kT}\left(\frac{b}{d}\right)^p\left(\frac{\tau}{G}\right)^n \exp\left(-\frac{Q}{kT}\right) \tag{3.1}$$

where $\dot{\gamma}_{ss}$ is the shear creep rate under the steady-state, τ is the shear creep stress, G is the shear modulus, b the magnitude of Burgers vector, k the Boltzmann constant, T the absolute temperature, d the grain size, D_0 the diffusion coefficient, Q the activation energy for the creep processes, n the stress exponent, p the grain size exponent, and C a material constant.

Modeling and Simulation for Microelectronic Packaging and Integration: Manufacturing, Reliability and Testing, Second Edition. Sheng Liu and Yong Liu.

For the generalization and the case of no detail information of grain sizes, the general expression for steady states of creep and plasticity can be written as:

$$\dot{\varepsilon}^{\mathrm{I}} = \Theta(T) Z_{ss}(\sigma_{ss}) \tag{3.2}$$

where the strain rates of creep and plasticity are unified as one inelastic strain rate $\dot{\varepsilon}^{\mathrm{I}}$, and moreover, the temperature dependence $\Theta(T)$ is separated from stress dependence, the so called Zener parameter. From Equation 3.1, the expression of $\Theta(T)$ can be:

$$\Theta(T) = \dot{\varepsilon}_0 \frac{G}{T} \exp\left(-\frac{Q}{kT}\right) \tag{3.3}$$

where the reference strain rate $\dot{\varepsilon}_0$ includes the grain size dependence. There are various viscosity functions of Z_{ss} which are used in literature (Chaboche *et al.*, 1989). The hyperbolic function of Zener and Hollomon (Freed *et al.*, 1993) type is believed to be able to cover a wider range of strain rate, including power-law breakdown region of creep deformation. The function is given by:

$$Z_{ss}(\sigma_{ss}) = \left[\sinh\left(\frac{\sigma_{ss}}{C}\right)\right]^n \tag{3.4}$$

where, again, n is the stress exponent. The constant C in the above equation characterizes the critical stress in the power-low breakdown region, where dislocation gliding is more dominant than dislocation climbing in the region. When applied stress is lower than C, the Equation 3.4 reduces to power-law form. Equation 3.4 transfers smoothly to the exponential behavior of the power-law breakdown region when applied stress is larger than C. Therefore, the characteristic stress will be used to determine material constants in the following section.

The description of transient behaviors of materials is to introduce the evolution equation of internal variables. The back stress or internal stress is a most useful one which is used to model kinematical hardening and cyclic hardening/softening due to dislocation movements. In fact, its evolution equation is formulated as three parts, including strain hardening, strain (dynamic) recovery, and thermal (static) recovery (Chaboche *et al.*, 1989; Freed *et al.*, 1990; Qian *et al.*, 1991; Lowe *et al.*, 1986). One conventional assumption is that the steady state behaviors of creep and plasticity act as the bonds of transient behaviors (Qian *et al.*, 1991; Lowe *et al.*, 1986; Freed *et al.*, 1993). The physical base of the assumption as dislocation pattern formation had been proposed in reference (Qian *et al.*, 1991). This assumption imposes the restraint for the evolution equation of back stress. The proposed constitutive model incorporating the above ideas into back stress evolution, we call it SodMod, and it includes following differential equations in one-dimensional form:

$$\dot{\varepsilon}^{\mathrm{I}} = \Theta(T)\left[\sinh\left(\frac{\sigma - X_1 - X_2}{D}\right)\right]^n \frac{\sigma - X_1 - X_2}{|\sigma - X_1 - X_2|} \tag{3.5}$$

$$\dot{X}_1 = H_1 \dot{\varepsilon}^{\mathrm{I}}\left[1 - \frac{A_1 X_1 \dot{\varepsilon}^{\mathrm{I}}}{A\sigma_{ss}|\dot{\varepsilon}^{\mathrm{I}}|}\right] + H_1 C_1\left[\dot{\varepsilon}^{\mathrm{I}} - \Theta(T)\sinh\left(A_1 X_1 \frac{X_1}{|X_1|}\right)\right] \tag{3.6a}$$

$$\dot{X}_2 = H_1 \dot{\varepsilon}^{\mathrm{I}}\left[1 - \frac{A_2 X_2 \dot{\varepsilon}^{\mathrm{I}}}{A\sigma_{ss}|\dot{\varepsilon}^{\mathrm{I}}|}\right] + H_2 C_2\left[\dot{\varepsilon}^{\mathrm{I}} - \Theta(T)\sinh^n\left(A_2 X_2 \frac{X_2}{|X_2|}\right)\right] \tag{3.6b}$$

where $H_1, H_2, C_1, C_2, A_1, A_2$, and A are the material parameters, the D is drag stress and taken as a constant for simplification. The components X_1 and X_2 of the back stress in Equation 3.6 are introduced to cover short range and long range behaviors of dislocation internal stresses (Lowe *et al.*, 1986). The first term in Equations 3.6a and 3.6b is to describe the additional strain hardening and thermal induced recovery. The characteristics of these variables and the parameter relationship under steady states of deformation are discussed below.

At the steady states of creep tests, the following relationship is kept:

$$A_1 X_{1\mathrm{ss}} = A_2 X_{2\mathrm{ss}} = \frac{\sigma_{\mathrm{ss}} - X_{1\mathrm{ss}} - X_{2\mathrm{ss}}}{D} = A\sigma_{\mathrm{ss}} \tag{3.7a}$$

$$A = \frac{A_1 A_2}{A_1 + A_2 + A_1 A_2 D} \tag{3.7b}$$

On the other hand, in the case of quick loading for a very short time, traditionally called plastic stress, this can be approximated by dropping a thermal recovery term in Equation 3.6 as follows:

$$\begin{aligned} A_1 X_{1\mathrm{ss}} &= (1 + C_1) A\sigma_{\mathrm{ss}} \\ A_2 X_{2\mathrm{ss}} &= (1 + C_2) A\sigma_{\mathrm{ss}} \end{aligned} \tag{3.8a}$$

$$\begin{aligned} \frac{\sigma_{\mathrm{ss}} - X_{1\mathrm{ss}} - X_{2\mathrm{ss}}}{D} &= A_{\mathrm{p}}\sigma_{\mathrm{ss}} \\ A_{\mathrm{p}} &= \frac{A_1 A_2 D - C_1 A_2 - C_2 A_1}{D(A_1 + A_2 + A_1 A_2 D)} \end{aligned} \tag{3.8b}$$

Obviously, the constants C_1 and C_2 account for the unequal hardening between creep and plasticity (Qian *et al.*, 1991, 1993) by comparing Equation 3.7 and Equation 3.8, which is designed as an optional ability of the SodMod. When the back stress is equal for the same quantity of plastic strain and creep strain, the parameters of C_1 and C_2 should be taken as zero. The creep-plasticity unequal hardening is still not confirmed for solder alloys although there are some experimental results (Busso *et al.*, 1992). The $C_1 = 0$, $C_2 = 0$, therefore, $A_{\mathrm{p}} = A$, will be used in this investigation. In this case, the Equation 3.6 can be integrated under the tests of creep or constant strain rate tension. We obtain:

$$X_1 = \frac{A\sigma_{\mathrm{ss}}}{A_1}\left[1 - \exp\left(-\frac{A_1 H_1}{A\sigma_{\mathrm{ss}}}\varepsilon^{\mathrm{I}}\right)\right] \tag{3.9a}$$

$$X_2 = \frac{A\sigma_{\mathrm{ss}}}{A_2}\left[1 - \exp\left(-\frac{A_2 H_2}{A\sigma_{\mathrm{ss}}}\varepsilon^{\mathrm{I}}\right)\right] \tag{3.9b}$$

The expression is convenient for the determination of material constants involved in back stress evolution equation.

3.1.2 *Determination of Material Constants*

(i) Determination of n Q A $\dot{\varepsilon}_0$ by Steady State Feature

At the steady states of creep and plasticity, combining Equations 3.2, 3.3, and 3.4 gives:

$$\dot{\varepsilon}^{\mathrm{I}}_{\mathrm{ss}} = \dot{\varepsilon}_0 \frac{G}{T}\left[\sinh\left(\frac{\sigma_{\mathrm{ss}}}{C}\right)\right]^n \exp\left(-\frac{Q}{kT}\right) \tag{3.10}$$

where the $\dot{\varepsilon}^{\mathrm{I}}_{\mathrm{ss}}$ is the steady creep rate at constant creep stress $\sigma_0 = \sigma_{\mathrm{ss}}$ or the plastic strain rate at the saturated stress σ_{ss} for the test of a constant strain rate, and the constant $C = 1/A$. The temperature dependence of elastic shear modulus G can be usually expressed by a linear relationship:

$$G(T) = G_0 - G_1 T \tag{3.11}$$

where the constant G_0 and G_1 can be found, for instance, in reference (Darveaus *et al.*, 1995) for solder joint alloys. As discussed in the above section, the hyperbolic sine in Equation 3.11 will reduce to power-law function at power-law creep region where $\sigma_{ss} = \sigma_0 < C$. Therefore, in a double logarithmic coordinate system, the Equation 3.10 can be written as:

$$y = a(T) + nx, x = \ln(\sigma_{ss}), y = \ln(\dot{\varepsilon}^{\mathrm{I}}_{ss}) \tag{3.12a}$$

$$a(T) = \ln(\dot{\varepsilon}_0) + n \ln A + \left[\ln G(T) - \ln T - \frac{Q}{k}\frac{1}{T}\right] \tag{3.12b}$$

$$\Delta a(T) = a(T_1) - a(T_2) = \ln\left[\frac{G(T_1)}{G(T_2)}\right] - \ln\left(\frac{T_1}{T_2}\right) - \frac{Q}{k}\left(\frac{1}{T_1} - \frac{1}{T_2}\right) \tag{3.12c}$$

The stress component n can be uniquely determined by the slope of linear fitting to power-law creep region according to Equation 3.12a. The activation energy Q is also conveniently obtained for Equation 3.12c by using the linear offsetting at two different temperatures T_1 and T_2 because the G_0, G_1, and k are known. If the experimental data are available in more than two different temperatures, the averaging values of n and Q determined by the above procedure are recommended for use. The determination of the constant $A = 1/C$ is to use the characteristic stress C at the power-law breakdown point. Once the A is determined, the constant $\dot{\varepsilon}_0$ can be easily given by Equation 3.12b where n, Q, A have been determined by the above procedure. However, as the power-law breakdown points on the stress via strain rate curves under steady states at different temperatures lie within a narrow range, a little adjustment is necessary to best fit all curves. The constants n Q A and $\dot{\varepsilon}_0$ are listed in Table 3.1 for one commonly used solder alloy 62Sn36Pb2Ag. They were determined by the proposed procedure. The Boltzmann constant k is 8.63×10^{-5}. The raw data was taken from the reference (Darveaus *et al.*, 1995) for the shearing of solder joints. The comparison of experimental data with model prediction from the above parameters is shown in Figure 3.1 for the solder alloy. The agreement is excellent. The reasonable value of Q is also obtained by the proposed procedure.

Table 3.1 Material constants of a 60–40 solder alloy at steady states

Solder	G_0(psi)	G_1(psi)	n	Q(eV)	A(1/psi)	$\dot{\varepsilon}_0$ (0K/s/psi)
62Sn36Pb2Ag	4.111×10^6	8.10×10^3	3.525	0.855	7.340×10^{-4}	1.967×10^4

(ii) Determination of H_1, H_2, A_1, A_2, D by Transient Feature

The transient behavior of plastic or creep deformation can be used to obtain constants H_1, H_2, A_1, A_2, and D since we have taken the simplification that $C_1 = C_2 = 0$. The determination of constant D can be worked out by the information at initial yielding or creep points. At these points, the back stress is zero. From Equation 3.5, for initial yielding we have:

$$\dot{\varepsilon}^{\mathrm{I}}_y = \Theta(T)\left[\sinh\left(\frac{\sigma_y}{D}\right)\right]^n$$
$$\dot{\varepsilon}^{\mathrm{I}}_y = \dot{\varepsilon}^{\mathrm{t}}\left(1 - \frac{G_y}{G}\right) \tag{3.13a}$$

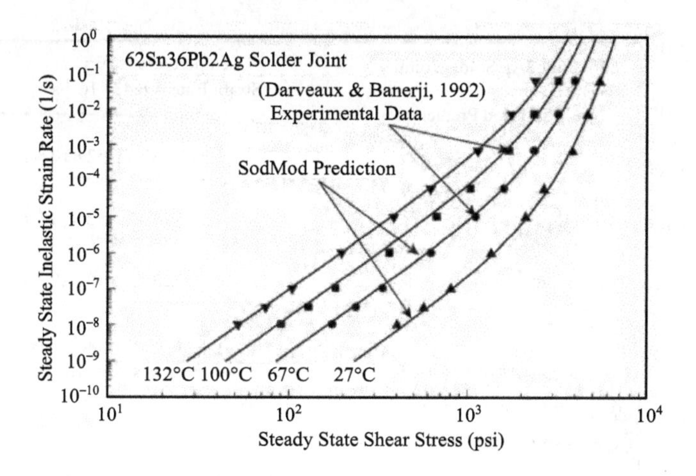

Figure 3.1 Model prediction and experimental data for steady-state behavior

and for initial creep we have:

$$\dot{\varepsilon}_c^I = \Theta(T)\left[\sinh\left(\frac{\sigma_0}{D}\right)\right]^n \tag{3.13b}$$

The alternative of Equation 3.13a is:

$$D = \frac{\sigma_y}{\ln(x + \sqrt{1 + x^2})}$$
$$x = \left[\dot{\varepsilon}_y^I / \Theta(T)\right]^{\frac{1}{n}} \tag{3.14}$$

where the $\dot{\varepsilon}^t$ is known as the constant strain rate of a tension test, and the G_y is the tangent modulus of a yield point with the yielding stress σ_y. The constant D was determined by Equation 3.14. The D determined is 660.0 psi for available data of 62Sn36Pb2Ag alloy. If the initial creep rates $\dot{\varepsilon}_c^I$ at different creep stresses σ_0 are available, the D is easier to obtain by using Equation 3.13b. The fitting of constant H_1, H_2, A_1, A_2 can be worked out by a stress strain curve with a highest constant tensile rate. The test at a tensile rate of 6.45×10^{-2}/s was employed here. The analytical solution Equation 3.9a, 3.9b of back stress is also used. Actually, only three of those constants are independent since the relationship in Equation 3.7 must be satisfied. A least-square minimization program was utilized to obtain those constants. The results are listed in Table 3.2 for 62Sn36Pb2Ag.

Table 3.2 Material constants of a 60–40 solder alloy at steady states

Solder	H_1(psi)	H_2(psi)	A_1(1/psi)	A_2(1/psi)	D(psi)
62Sn36Pb2Ag	3.291×10^4	5.068×10^3	1.648×10^{-3}	1.046×10^{-2}	660.0

3.1.3 Model Prediction

The model prediction by using the material constants listed in Tables 3.1 and 3.2 is shown in Figure 3.1 for 62Sn36Pb2Ag solder alloy. The model accurately predicts almost all curves for four different strain rates

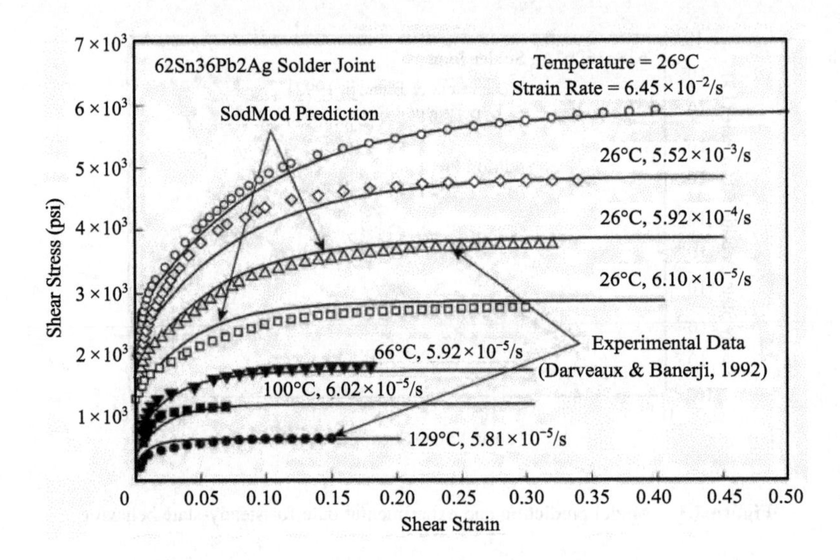

Figure 3.2 Model prediction and experimental data for constant strain rate tension

at four different temperatures. The agreement between model prediction and experimental data for constant stress creep is reasonably good, as shown in Figure 3.2. The proposed model, therefore, is verified by the available experimental data, and is found to be able to accurately predict the mechanical behaviors of solder alloys.

3.2 Visco-Elastic-Plastic Properties and Constitutive Modeling of Underfills

3.2.1 Constitutive Modeling of Underfills

This section focuses on modeling mechanical properties of underfills under 110 °C because the mechanical behaviors of HYSOL FP4526 dramatically degrade above 110 °C (Qian *et al.*, 1999). Although the thermo-mechanical properties around T_g (glass transition temperature) need further investigation, the modeling work can be used to investigate the thermo-mechanical response of underfills under service conditions of packaging devices where the temperature is normally lower than 100 °C (Merton *et al.*, 1997).

Figure 3.3 shows the strong strain rate-dependence of Young's modulus. The logarithmic-linear relationship of strain rate with Young's modulus was found from experimental data taken carefully from the initial part of tensile tests at different strain rates. Experimental data of Young's modulus can be fitted in the following equation in terms of Mises strain rate and temperature (degrees in Kelvin), as shown in Figure 3.3:

$$E = 17.3145 + 0.187 \ln \dot{\bar{\varepsilon}} - 0.027595T(\text{GPa}) \tag{3.15a}$$

with the definition of Mises total strain rate:

$$\dot{\bar{\varepsilon}} = \sqrt{\frac{2}{3}\dot{\varepsilon}_{ij}\dot{\varepsilon}_{ij}} \tag{3.15b}$$

The property is, in fact, the characterization of viscoelasticity due to polymer matrix. Although traditional viscoelastic models can be formulated, the direct implementation of Equation 3.15 in commercial

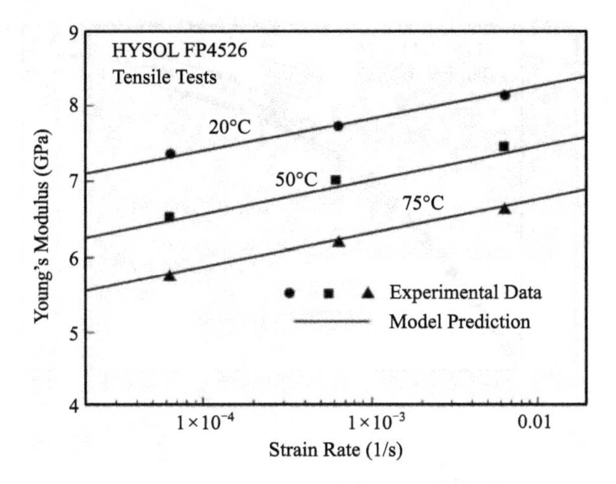

Figure 3.3 Strain rate and temperature-dependence of the Young's modulus of the FP4526 underfill

software such as ABAQUS is convenient. More test data at a wide range of strain rates are needed to identify the upper and lower bounds of the Young's modulus.

It is also found that the viscoplastic constitutive model developed for solder alloys (Qian *et al.*, 1997) is suitable to describe the viscoplastic properties of underfills. There is a very similar characterization of steady-state properties between solder alloys (Ren *et al.*, 1998) and the underfill, as shown in Figure 3.4. The steady-state inelastic strain rate is defined as a creep rate at second-stage creep or a tensile strain rate at the maximum stress point. Its counterpart, the steady-state stress is defined as the creep stress for a creep test or the maximum stress for a tensile test. In particular, the HYSOL FP4526 underfill shows power-law behavior at low stress levels and elevated temperatures.

There is also a regime of power-law breakdown at room temperature, as shown in Figure 3.4. Although the deformation mechanism of underfills remains to be investigated, the phenomenological similarity of deformation characterization to metals (Frost *et al.*, 1982) shows that many sophisticated models for solder alloys are available and underfill modeling could be under a similar constitutive framework. This will definitely benefit the finite element simulation and parameter study of flip-chip packages where underfills and solder alloys are two essential component materials.

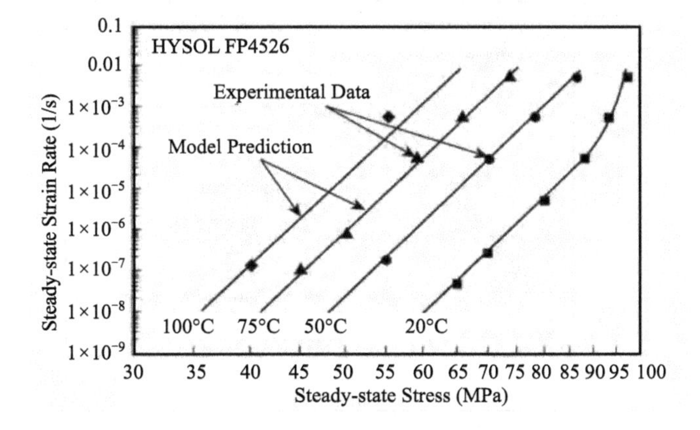

Figure 3.4 Relationship between stress and inelastic strain rate at steady state

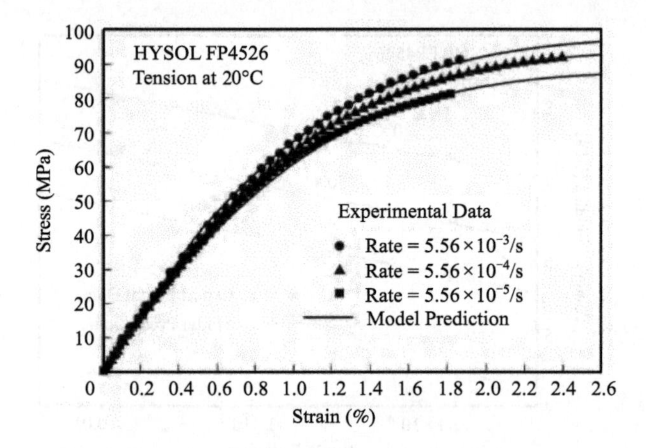

Figure 3.5 Comparison of model prediction with experimental data of viscoplastic properties of the FP4526 at 20 °C

Moreover, the underfill exhibits strong metal-like deformation hardening during tensile tests at the strain rates ranging from 5.56×10^{-5}/s to 5.56×10^{-3}/s, as shown in Figures 3.5, 3.6 and 3.7. The characterization also indicates that transient creep dominates the creep deformation, as shown in Figures 3.8 and 3.9. Therefore, a back stress $X_{ij} = X_{ij}^{(1)} + X_{ij}^{(2)}$ with a fast evolution component and a slow one is introduced to describe the transient property of the metal-like deformation hardening. The evolution equations of the back stress components are obtained from the constitutive model proposed for solder alloys (Qian *et al.*, 1997). Based on experimental results, the small deformation assumption is sufficient for the highly filled polymers such as underfills. Underfills are usually subjected to deformation of no more than 4%, as reported here.

In summary, the set of three-dimensional constitutive equations is listed below (details can be found in Qian and Liu (Qian *et al.*, 1997)). Total strain rate tensor $\dot{\varepsilon}_{ij}$ is composed of an elastic strain rate $\dot{\varepsilon}_{ij}^{\varepsilon}$ and an inelastic strain rate $\dot{\varepsilon}_{ij}^{\mathrm{I}}$:

$$\dot{\varepsilon}_{ij} = \dot{\varepsilon}_{ij}^{e} + \dot{\varepsilon}_{ij}^{\mathrm{I}} = \frac{1+\nu}{E}\dot{\sigma}_{ij} - \frac{\nu}{E}\dot{\sigma}_{kk}\delta_{ij} + \dot{\varepsilon}_{ij}^{\mathrm{I}} \tag{3.16}$$

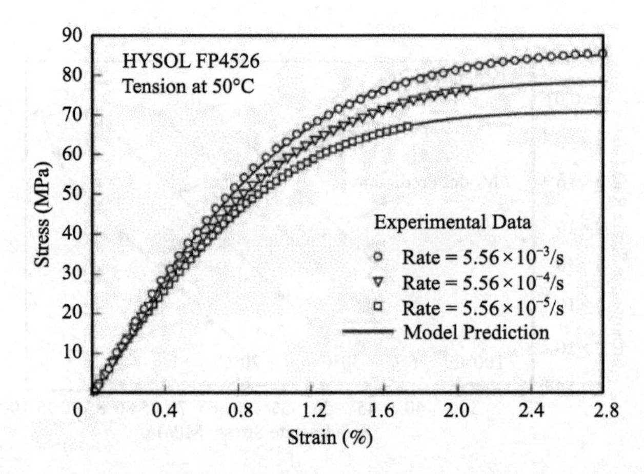

Figure 3.6 Comparison of model prediction with experimental data of viscoplastic properties of the FP4526 at 50 °C

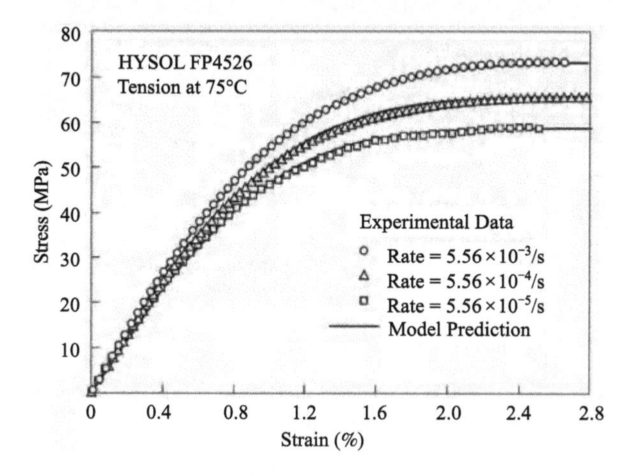

Figure 3.7 Comparison of model prediction with experimental data of viscoplastic properties of the FP4526 at 75 °C

where E is the Young's modulus, and ν the Poisson's ratio. And isotropic elasticity is assumed. The inelastic strain rate, $\dot{\varepsilon}_{ij}^{\mathrm{I}}$, is constructed as:

$$\dot{\varepsilon}_{ij}^{\mathrm{I}} = \dot{p}\frac{\frac{3}{2}\left(S_{ij} - X_{ij}^{\prime(1)} - X_{ij}^{\prime(2)}\right)}{\sigma_{\mathrm{e}}} \tag{3.17a}$$

with the definition of Mises inelastic strain rate $\dot{p}$

$$\dot{p} = \sqrt{\frac{2}{3}\dot{\varepsilon}_{ij}^{\mathrm{I}}\dot{\varepsilon}_{ij}^{\mathrm{I}}} \tag{3.17b}$$

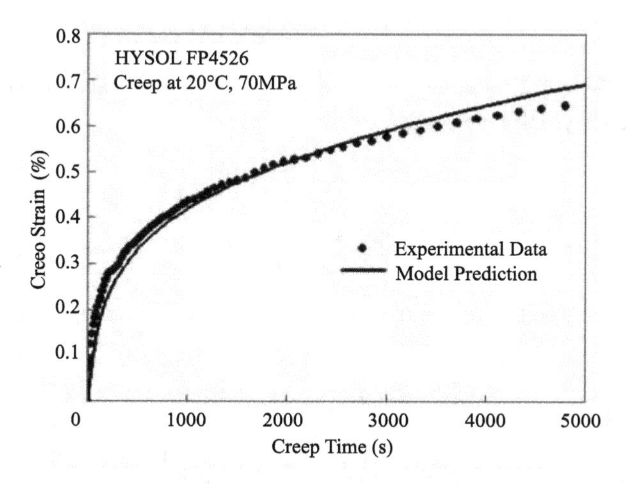

Figure 3.8 Comparison of model prediction with experimental data of a creep test of the FP4526 at 20 °C

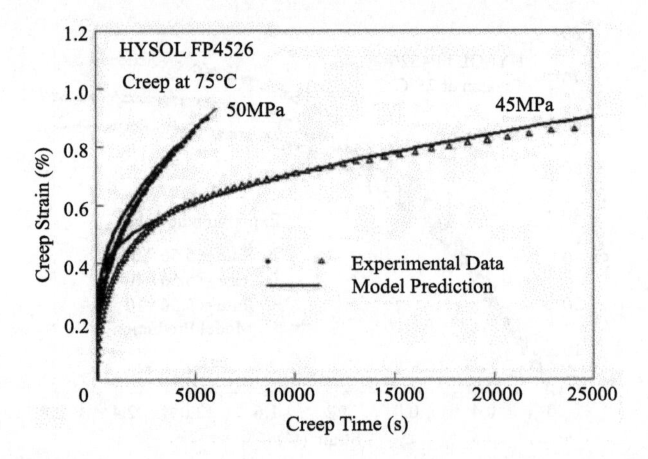

Figure 3.9 Comparison of model prediction with experimental data of a creep test of the FP4526 at 75 °C

and its hyperbolic-sine stress-dependence:

$$\dot{p} = \Theta(T)\sinh\left[\left(\frac{\sigma_e}{D}\right)^n\right] \tag{3.17c}$$

as well as the Arrhenius relationship of temperature dependence:

$$\Theta(T) = \dot{\varepsilon}_0 \exp\left(-\frac{Q}{kT}\right) \tag{3.17d}$$

where n is the stress exponent, D the drag stress, $\dot{\varepsilon}_0$ the reference strain rate, k the Boltzmann constant, Q the activation energy for inelastic deformation processes, and T the absolute temperature, respectively. In above equations, σ_e the Mises equivalent stress which is defined by:

$$\sigma_e^2 = \frac{3}{2}\left(S_{ij} - X_{ij}'^{(1)} - X_{ij}'^{(2)}\right)\left(S_{ij} - X_{ij}'^{(1)} - X_{ij}'^{(2)}\right) \tag{3.17e}$$

where S_{ij} and $X_{ij}'^{(k)}$ $(k = 1, 2)$, are the deviatoric tensors of Cauchy stress σ_{ij} and back stress components $X_{ij}^{(k)}$, respectively. They are defined by:

$$S_{ij} = \sigma_{ij} - \frac{1}{3}\sigma_{kk}\delta_{ij} \tag{3.17f}$$

$$X_{ij}' = X_{ij} - \frac{1}{3}X_{kk}\delta_{ij} \tag{3.17g}$$

where δ_{ij} is the Kronecker delta. The evolution equations of back stress components $X_{ij}'^{(k)}$ are derived as:

$$\dot{X}_{ij}'^{(1)} = \frac{2}{3}H_1\left[\dot{\varepsilon}_{ij}^{\mathrm{I}} - \frac{\dot{p}}{X_{ss}^{(1)}}\frac{3}{2}X_{ij}'^{(1)}\right] \tag{3.18a}$$

$$\dot{X}_{ij}'^{(2)} = \frac{2}{3}H_2\left[\dot{\varepsilon}_{ij}^{\mathrm{I}} - \frac{\dot{p}}{X_{ss}^{(2)}}\frac{3}{2}X_{ij}'^{(2)}\right] \tag{3.18b}$$

with a relationship between saturated back stress and steady-state stress σ_{ss}

$$X_{ss}^{(1)} = \frac{A\sigma_{ss}}{A_1}, \quad X_{ss}^{(2)} = \frac{A\sigma_{ss}}{A_2} \tag{3.18c}$$

and the definition of the steady-state stress σ_{ss}

$$\sigma_{ss} = \sqrt{(\sigma_{ij}\sigma_{ij})_{ss}} \tag{3.18d}$$

where H_1 and H_2 are the coefficients of strain hardening, A, A_1, and A_2 the material constants.

The analytical forms of its components of back stress are obtained under uniaxial loading, which is convenient for the determination of material constants

$$X_{ss}^{(1)} = \frac{A}{A_1}\sigma_{ss}\left[1 - \exp\left(-\frac{H_1 A_1}{A\sigma_{ss}}\varepsilon_{11}^{I}\right)\right] \tag{3.19a}$$

$$X_{ss}^{(2)} = \frac{A}{A_2}\sigma_{ss}\left[1 - \exp\left(-\frac{H_2 A_2}{A\sigma_{ss}}\varepsilon_{11}^{I}\right)\right] \tag{3.19b}$$

3.2.2 Identification of Material Constants

The elastic modulus E of HYSOL FP4526 is found as 3.15a. The consistent procedure of material constants determination developed before can be directly used for the current framework, which is summarized as follows. In general, constants n, Q, A, $\dot{\varepsilon}_0$ can be determined by a steady-state feature, as shown in Figure 3.4, and H_1, H_2, A_1, A_2, D by a transient feature of deformation from tensile tests. Particularly, the stress exponent n is uniquely determined by the power-law creep. The constant A is determined from the region of power-law breakdown. The activation energy Q is conveniently obtained by using the data of the steady state of power-law creep at two different temperatures. The constant $\dot{\varepsilon}_0$ can be given by a derived parameter relation when it has been determined. Drag stress D is determined by initial yielding points or initial creep rates. H_1, H_2, A_1, and A_2 are fitted to the tensile curve at a relatively high strain rate by a least square program. A relation among A_1, A_2, D, and A is found. Therefore, only one of A_1, and A_2 needs to be determined. The constants determined for HYSOL FP4526 underfill are listed in Table 3.3. The above procedure without trial and error has also been verified for the Motorola test data of a 60 : 40 solder-joint at different temperatures and tensile rates (Qian *et al.*, 1997) and for the benchmark tests of the eutectic solder 63Sn37Pb conducted by using a thin-strip specimen and six-axis mini fatigue tester (Ren *et al.*, 1998).

Table 3.3 indicates that high stress exponent and activation energy Q have been obtained for the underfill, comparing with solder alloys (Qian *et al.*, 1997; Ren *et al.*, 1998).

3.2.3 Model Verification and Prediction

The model prediction is presented by using the parameters listed in Equation 3.15 and Table 3.3. Figure 3.4 shows the excellent agreement between experimental data and model prediction for the steady-state properties at different temperatures. Figure 3.5 presents the comparison of model prediction with the

Table 3.3 Material constants for 63Sn37Pb solder

n	Q(eV)	$\dot{\varepsilon}_0$(K/s/MPa)
23.0	1.40	1.31×10^{20}
A(1/MPa)	D_0(MPa)	H_1(MPa)
1.111×10^{-2}	38.0	1.780×10^{4}
H_2(MPa)	A_1(1/MPa)	A_2(1/MPa)
9.031×10^{3}	6.565×10^{-2}	2.720×10^{-2}

tensile tests at three strain rates and room temperature. The comparison results for tensile tests at 50 °C and 75 °C are shown in Figures 3.6 and 3.7, respectively. The model predicts the viscoplastic behavior well at all three temperatures. Among the stress strain curves, only the curve at room temperature at the strain rate of 5.56×10^{-4}/s was used to fit material constants, others are pure prediction of the constitutive model, which demonstrates the predictive power of the proposed model. The model can predict creep deformation as well, as shown in Figure 3.8 for room temperature and in Figure 3.9 for elevated temperature with different creep stresses. Transient (first-stage) creep dominates the creep behavior of the HYSOL FP4526. No tertiary-stage creep was observed until the rupture of a specimen. The creep property is significantly different from that of solder alloys in which a long tertiary-stage creep is observed. Therefore, the back stress is necessary to describe the transient deformation property and kinematical hardening for cyclic loading. The hyperbolic-sine law of creep without back stress evolution will significantly under predict the creep deformation of the underfill due to the transient creep-dominated characterization.

The model, furthermore, is able to be applied to finite element analysis of advanced packages with underfills and to clarify the role of underfills and interaction with solder joints in the potential FCOB technology.

3.3 A Damage Coupling Framework of Unified Viscoplasticity for the Fatigue of Solder Alloys

The very ductile materials such as solder alloys are damaged by the micro-processes of void nucleation on phase and/or grain boundaries (Frear *et al.*, 1990; Lee, 1995), growth, and their coalescence to result in microcrack initiation (Cocks *et al.*, 1980). The McClintock (McClintock, 1968) and Rice and Tracey (Rice and Tracy, 1969) have done pioneering work to obtain the solutions for the growth of voids of cylindrical and spherical shapes, respectively. Analysis of the effect of microvoids on plastic flow has been carried out by Gurson (Gurson, 1977). Specially, an analytical plastic potential has been obtained based on an upper bound solution for a porous material of perfectly plastic matrix. An extension of the basic Gurson model has been proposed to account for final material failure (Tvergaard *et al.*, 1984), kinematic hardening and void shape effects (Becker *et al.*, 1986). The extensive investigation of void growth in an infinite power-law material has been carried out by Budiansky (1982). Duva and Hutchinson (1984) have obtained constitutive potentials for the same material. However, the Gurson-like models do not easily comply with the mathematical structure of constitutive frameworks which are proposed in unified viscoplastic models and the CDM. Alternatively, lower bounds and self-consistent estimates of the constitutive potentials have been obtained by Ponte, Castaneda, and Willins (Ponte *et al.*, 1988). Following this direction, Cocks (1989) derived an analytical strain rate potential which can be compared with some CDM constitutive models like Leckie and Hayhurst (Leckie and Hayhurst, 1977). This is the start point of this research.

The CDM is based on a strain equivalence principle (Lemaitre, 1996) which, originally from the geometric point of view of damage degradation, neglects the effects of strain and stress concentration around a void. This oversimplification results in the damage evolution and its dependence on stress triaxility built in a pure phenomenological manner. This may limit the CDM application to real micromechanical deformation and failure processes.

A new constitutive framework is proposed to integrate CDM with the Cocks' lower bound of strain rate potential, which is extended to kinematic hardening in this chapter. The damage coupling with elasticity and viscoplasticity are formulated by thermo-dynamics, including a unified constitutive model for solder alloys. The model is extended, furthermore, to three-dimensional and large deformation cases. The damage evolution of voids due to viscoplastic deformation is derived analytically. The framework is then applied to creep damage.

3.3.1 Damage Coupling Thermodynamic Framework

The viscoplastic deformation process with damage evolution is an irreversible thermodynamic process. The methodology of thermodynamics is to postulate the existence of energy potentials from which one can derive the constitutive equations in three dimensions. The two thermodynamic potentials, Helmholz free energy and dissipation potential, have been introduced to the current constitutive framework of CDM (Lemaitre, 1996). Within the assumption of small strains and small displacements, as the first step, the total strain rate $\dot{\varepsilon}_{ij}$ is composed of an elastic part $\dot{\varepsilon}^{e}_{ij}$ and an inelastic part $\dot{\varepsilon}^{I}_{ij}$

$$\dot{\varepsilon}_{ij} = \dot{\varepsilon}^{e}_{ij} + \dot{\varepsilon}^{I}_{ij} \tag{3.20}$$

The isotropic elasticity is assumed, and the elastic modulus tensor C is written as:

$$C_{ijkl} = \lambda\delta_{ij}\delta_{kl} + \mu(\delta_{ik}\delta_{jl} + \delta_{il}\delta_{jk}) \tag{3.21}$$

where λ and μ are Lame's coefficients of elasticity. And δ_{ij} is Kronecker delta. Taking the Helmholtz specific free energy (Lemaitre, 1996)

$$\psi = \psi(\varepsilon^{e}_{ij}, \alpha^{(1)}_{ij}, \alpha^{(2)}_{ij}; T, f) \tag{3.22}$$

where $\alpha^{(1)}_{ij}$ and $\alpha^{(2)}_{ij}$ are back strain tensors associated with back stress tensors $X^{(1)}_{ij}$ and $X^{(2)}_{ij}$, which are defined later and used to describe both short term and long term kinematical hardening, respectively. T is the absolute temperature, and f is the damage variable, that is, the volume fraction of voids. In particular, the f in the free energy ψ is a parameter variable, and not considered as an independent state one. Its evolution equation will be given by a well-known relation with a volume plastic strain rate.

For simplification, an isothermal process is considered here. The Clausius-Duhern inequality reads (Lemaitre, 1996), in the isothermal case,

$$\sigma_{ij}\dot{\varepsilon}_{ij} - \rho\psi \geqslant 0 \tag{3.23}$$

where ρ is the mass density of a material. The free energy ψ is the function of the state variables ε^{e}_{ij}, $\alpha^{(1)}_{ij}$, and $\alpha^{(2)}_{ij}$. Its rate can be written as:

$$\psi = \frac{\partial\psi}{\partial\varepsilon^{e}_{ij}}\dot{\varepsilon}^{e}_{ij} + \frac{\partial\psi}{\partial\alpha^{(1)}_{ij}}\dot{\alpha}^{(1)}_{ij} + \frac{\partial\psi}{\partial\alpha^{(2)}_{ij}}\dot{\alpha}^{(2)}_{ij} \tag{3.24}$$

substituting into Equations 3.20 and 3.24, the inequality (3.23) becomes:

$$\left(\sigma_{ij} - \rho\frac{\partial\psi}{\partial\varepsilon^{e}_{ij}}\right)\dot{\varepsilon}^{e}_{ij} + \sigma_{ij}\dot{\varepsilon}^{I}_{ij} - \rho\frac{\partial\psi}{\partial\alpha^{(1)}_{ij}}\dot{\alpha}^{(1)}_{ij} - \rho\frac{\partial\psi}{\partial\alpha^{(2)}_{ij}}\dot{\alpha}^{(2)}_{ij} \geqslant 0 \tag{3.25}$$

for any elastic strain rate $\dot{\varepsilon}^{e}_{ij}$, one obtains:

$$\sigma_{ij} = \rho\frac{\partial\psi}{\partial\varepsilon^{e}_{ij}} \tag{3.26}$$

$$\sigma_{ij}\dot{\varepsilon}^{I}_{ij} - X^{(1)}_{ij}\dot{\alpha}^{(1)}_{ij} - X^{(2)}_{ij}\dot{\alpha}^{(2)}_{ij} \geqslant 0 \tag{3.27}$$

with the definition of a back stress X_{ij} associated with a back strain:

$$X^{(1)}_{ij} = \rho\frac{\partial\psi}{\partial\alpha^{(1)}_{ij}}, X^{(2)}_{ij} = \rho\frac{\partial\psi}{\partial\alpha^{(2)}_{ij}} \tag{3.28}$$

Obviously, the dissipation inequality in Equation 3.27 consists of the sum of the products of thermodynamic fluxes (inelastic strain rate, negative back strain rates) multiplied by their dual thermodynamic forces (applied stress and back stress). Therefore, a continuous and convex function F, so-called dissipation potential, is postulated, from which the flow rule of an inelastic strain rate and kinetic evolution laws of back strains can be derived. With the so-called "normality rule of generalized standard materials," one has (Lemaitre, 1996):

$$\dot{\varepsilon}_{ij}^{\mathrm{I}} = \dot{\lambda}\frac{\partial F}{\partial \sigma_{ij}} \tag{3.29a}$$

$$\dot{\alpha}_{ij}^{(1)} = -\dot{\lambda}\frac{\partial F}{\partial X_{ij}^{(1)}} \tag{3.29b}$$

$$\dot{\alpha}_{ij}^{(2)} = -\dot{\lambda}\frac{\partial F}{\partial X_{ij}^{(2)}} \tag{3.29c}$$

The inequality in Equation 3.27 requires that the scalar multiplier $\dot{\lambda}$ s always positive. This ensures the normality condition of the inelastic strain rate to the surface of a dissipation potential, including both plasticity and viscoplasticity.

For the elastic damage coupling with viscoplasticity, a specific Helmholtz free energy is taken as (Lemaitre, 1996):

$$\psi = \frac{1}{\rho}\left[\frac{1}{2}C_{ijkl}\varepsilon_{ij}^{e}\varepsilon_{kl}^{e}(1-D_f) + \frac{1}{3}H_1\alpha_{ij}^{(1)}\alpha_{ij}^{(1)} + \frac{1}{3}H_2\alpha_{ij}^{(2)}\alpha_{ij}^{(2)}\right] \tag{3.30}$$

where H_1 and H_2 are the strain hardening coefficients of a material, From Equations 3.26, 3.30, and 3.31, the elastic law is obtained as:

$$\begin{aligned}\varepsilon_{ij}^{e} &= \frac{1+\nu}{E}\frac{\sigma_{ij}}{(1-D_f)} - \frac{\nu}{E}\frac{\sigma_{kk}}{(1-D_f)}\delta_{ij} \\ &= \frac{1+\nu}{E}\sigma_{ij} - \frac{\nu}{E}\sigma_{kk}\delta_{ij}\end{aligned} \tag{3.31}$$

where ν is the Poisson's ratio. The elastic modulus E of a virgin material is degraded to E_D due to the damage D:

$$E_D = (1-D_f)E \tag{3.32}$$

The damage variable D_f for elastic degradation due to viscoplastic deformation induced voids is related to variable f by a well-established estimate approach. The self-consistent estimate is also available. Specifically, for a dilute void distribution, a simple solution of the D_f is given by (Hill, 1965):

$$D_f = a(\nu)f, a(\nu) = \frac{15(1-\nu)}{7-5\nu} \tag{3.33}$$

Clearly, the damage D_f coupling with the Poisson's ratio ν comes from the elastic behavior. However, the evolution equation of D_f via f is to be derived from void growth micromechanics, where the stress triaxility is incorporated by coupling with viscoplastic behavior. This is quite different from the CDM model, where the evolution of D_f is assumed in a phenomenological way. The viscoplastic coupling damage is also not well established (Lemaitre, 1996)

With the Helmholtz free energy Equation 3.30 and the definition of back stress Equation 3.28, the following relation are derived:

$$X_{ij}^{(1)} = \frac{2}{3}H_1\alpha_{ij}^{(1)}, Y_{ij}^{(1)} = \frac{2}{3}H_1\alpha_{ij}^{\prime(1)} \tag{3.34a}$$

$$\dot{X}_{ij}^{(1)} = \frac{2}{3}H_1\dot{\alpha}_{ij}^{(1)}, \dot{Y}_{ij}^{(1)} = \frac{2}{3}H_1\dot{\alpha}_{ij}^{\prime(1)} \tag{3.34b}$$

$$X_{ij}^{(2)} = \frac{2}{3}H_1\alpha_{ij}^{(2)}, Y_{ij}^{(2)} = \frac{2}{3}H_1\alpha_{ij}^{\prime(2)} \tag{3.35a}$$

$$\dot{X}_{ij}^{(2)} = \frac{2}{3}H_1\dot{\alpha}_{ij}^{(2)}, \dot{Y}_{ij}^{(2)} = \frac{2}{3}H_1\dot{\alpha}_{ij}^{\prime(2)} \tag{3.35b}$$

where $Y_{ij}^{(k)}$ and $\alpha'^{(k)}_{ij}$ are the deviatoric part of $X_{ij}^{(k)}$ and $\alpha_{ij}^{(k)}$ respectively. The superscript k takes 1 and 2. The form of dissipation potential needs more fundamental study. We start here from Cocks' work on porous material with a distribution of spherical voids. A basic result of Cocks' lower bound solution of strain rate potential is that a transition between Mises stresses of the damaged material with voids and the undamaged material (matrix) can be reached, keeping the same mathematical structure of strain rate potential in matrix and in damaged porous material. That is (Cocks, 1989):

$$\bar{\sigma}_e^c = \frac{\sigma_e^c}{(1-f)}g\left(f, n, \frac{\sigma_{kk}}{\sigma_e^c}\right) \tag{3.36a}$$

$$g = \sqrt{1 + \frac{2nf}{(n+1)(f+1)}\left(\frac{1}{2}\frac{\sigma_{kk}}{\sigma_e^c}\right)^2} \tag{3.36b}$$

where:

$$\left(\sigma_e^c\right)^2 = \frac{3}{2}S_{ij}S_{ij}, S_{ij} = \sigma_{ij} - \frac{1}{3}\sigma_{kk}\delta_{ij} \tag{3.36c}$$

The superscript c in Equation 3.36 denotes the Cocks' solution and the solution for power-law creep materials. It is shown that Equation 3.36 includes the pure geometric factor $(1-f)$ like the CDM and an important function g that accounts for a stress triaxility $\frac{\sigma_{kk}}{\sigma_e^c}$ a strain rate exponent n, and a damage parameter (current void volume fraction) f. Therefore, the strain equivalence principle (Lemaitre, 1996) should be modified and stated as Equation 3.36, to incorporate naturally into the effects of stress triaxilities and strain rates. Since the Cocks' solution for a power-law creep material is a lower bound, it can be extended directly to unified viscoplasticity to include kinematical hardening characterized by a back stress with two components (Lowe *et al.*, 1986; Qian *et al.*, 1991). This modification is similar to the extension of Gurson's model to include kinematical hardening (Becker *et al.*, 1986). Further theoretical study and comparison among the different bounds (Duva *et al.*, 1984; Ponte *et al.*, 1988) will be given elsewhere. The extension of the Equation 3.36 can be given elsewhere. The extension of the Equation 3.36 can be:

$$\bar{\sigma}_e = \frac{\sigma_e}{(1-f)}g\left(f, n, \frac{\sigma_{kk} - X_{ij}^{(1)} - X_{ij}^{(2)}}{\sigma_e}\right)$$

$$g = \sqrt{1 + \frac{2nf}{(n+1)(f+1)}\left(\frac{1}{2}\frac{\sigma_{kk} - X_{ij}^{(1)} - X_{ij}^{(2)}}{\sigma_e}\right)^2} \tag{3.37a}$$

With the new definition of Mises equivalent stress:

$$\sigma_e^2 = \frac{3}{2}\left(S_{ij} - Y_{ij}^{(1)} - Y_{ij}^{(2)}\right)\left(S_{ij} - Y_{ij}^{(1)} - Y_{ij}^{(2)}\right) \tag{3.37b}$$

where:

$$\begin{aligned} S_{ij} &= \sigma_{ij} - \frac{1}{3}\sigma_{kk}\delta_{ij} \\ Y_{ij}^{(1)} &= X_{ij}^{(1)} - \frac{1}{3}X_{kk}^{(1)}\delta_{ij} \\ Y_{ij}^{(2)} &= X_{ij}^{(2)} - \frac{1}{3}X_{kk}^{(2)}\delta_{ij} \end{aligned} \tag{3.37c}$$

The dissipation potential is then proposed based on the new strain equivalent principle, that is, Equation 3.37. The proposed F is of the form:

$$F = \bar{\sigma}_e + \frac{3}{4}\frac{1}{(1-f)^2}\left(\frac{X_{ij}^{(1)}X_{ij}^{(1)}}{X_{ss}^{(1)}} + \frac{X_{ij}^{(2)}X_{ij}^{(2)}}{X_{ss}^{(2)}}\right) \tag{3.38a}$$

with the definition of saturated back stresses:

$$\begin{aligned} X_{ss}^{(1)} &= \sqrt{\frac{3}{2}\left(X_{ij}^{(1)}X_{ij}^{(1)}\right)_{ss}} \\ X_{ss}^{(2)} &= \sqrt{\frac{3}{2}\left(X_{ij}^{(2)}X_{ij}^{(2)}\right)_{ss}} \end{aligned} \tag{3.38b}$$

The form of the potential F is similar to that adopted by Lemaitre (Lemaitre, 1996). The factor $1/(1-f)^2$ is added to couple the viscoplastic damage with back stress evolution. The saturated back stresses $X_{ss}^{(1)}$, $X_{ss}^{(2)}$ serve as upper bounds of back stress evolution (Qian *et al.*, 1997; Freed *et al.*, 1993). The equations of inelastic strain rate and back stress evolution can therefore be derived by combining Equations 3.29, 3.38, and 3.37 with Equations 3.34 and 3.35. The derivation is straightforward, but is omitted due to the length limit of this section. The following are the results. The inelastic strain rate is composed of its deviatoric part $\dot{\varepsilon}_{ij}^{\prime I}$ and a dilational part $\dot{\varepsilon}_{kk}^{I}$, which is:

$$\dot{\varepsilon}_{ij}^{\prime I} = \dot{p}\frac{\frac{3}{2}\left(S_{ij} - Y_{ij}^{(1)} - Y_{ij}^{(2)}\right)}{\sigma_e} \tag{3.39a}$$

$$\dot{\varepsilon}_{kk}^{I} = 3\dot{p}\frac{fn\left(\sigma_{kk} - X_{kk}^{(1)} - X_{kk}^{(2)}\right)}{2(n+1)(f+1)\sigma_e} \tag{3.39b}$$

with the definition of Mises equivalent strain rate $\dot{p}$:

$$\dot{p} = \frac{\dot{\lambda}}{(1-f)g} = \sqrt{\frac{2}{3}\dot{\varepsilon}_{ij}^{\prime I}\dot{\varepsilon}_{ij}^{\prime I}} \tag{3.39c}$$

The relation of Mises equivalent strain rate $\dot{p}$ with modified Mises equivalent stress $\bar{\sigma}_e$ was proposed to cover the region of both power-law creep and power-law breakdown, as a viscoplastic Garofalo's form:

$$\dot{p} = \Theta(T)\left[\sinh\left(\frac{\bar{\sigma}_e}{D}\right)\right]^n \tag{3.40}$$

with the temperature dependence:

$$\Theta(T) = \dot{\varepsilon}_0 \frac{E}{T} \exp\left(-\frac{Q}{kT}\right) \tag{3.41}$$

where D is a drag stress, $\dot{\varepsilon}_0$ is a reference strain rate, k the Boltzmann constant, Q the activation energy for inelastic deformation processes, and T the absolute temperature, respectively. Equation 3.41 is similar to the Arrhenius relation. The evolution equations of back stress are derived as:

$$\dot{Y}_{ij}^{(1)} = \frac{2}{3}H_1\left[\dot{\varepsilon}_{ij}^{\prime\mathrm{I}} - \frac{g\dot{p}}{(1-f)X_{\mathrm{ss}}^{(1)}}\frac{3}{2}Y_{ij}^{(1)}\right] \tag{3.42a}$$

$$\dot{Y}_{ij}^{(2)} = \frac{2}{3}H_2\left[\dot{\varepsilon}_{ij}^{\prime\mathrm{I}} - \frac{g\dot{p}}{(1-f)X_{\mathrm{ss}}^{(2)}}\frac{3}{2}Y_{ij}^{(2)}\right] \tag{3.42b}$$

$$\dot{X}_{kk}^{(1)} = \frac{2}{3}H_1\left[\dot{\varepsilon}_{kk}^{\mathrm{I}} - \frac{g\dot{p}}{(1-f)X_{\mathrm{ss}}^{(1)}}\frac{3}{2}X_{kk}^{(1)}\right] \tag{3.43a}$$

$$\dot{X}_{kk}^{(2)} = \frac{2}{3}H_2\left[\dot{\varepsilon}_{kk}^{\mathrm{I}} - \frac{g\dot{p}}{(1-f)X_{\mathrm{ss}}^{(1)}}\frac{3}{2}X_{kk}^{(2)}\right] \tag{3.43b}$$

Equation 3.43 shows that the dilation of the back stresses increases linearly with the dilatational strain. A combination of upper bounds of back stresses with the cyclic softening/hardening function $h(p)$ can be specified by:

$$X_{\mathrm{ss}}^{(1)} = \frac{A\sigma_{\mathrm{ss}}}{A_1}h(p) \tag{3.44a}$$

$$X_{\mathrm{ss}}^{(2)} = \frac{A\sigma_{\mathrm{ss}}}{A_2}h(p) \tag{3.44b}$$

where the definition of saturated applied stress σ_{ss} is defined by:

$$\sigma_{\mathrm{ss}} = \sqrt{\left(\sigma_{ij}\sigma_{ij}\right)_{\mathrm{ss}}} \tag{3.45}$$

This $h(p)$ is taken to evolve according to (Chaboche, 1989):

$$\dot{h} = b(h_\infty - h)\dot{p} \tag{3.46a}$$

Integrating Equation 3.46a gives:

$$h = h_\infty + (h_0 - h_\infty)\exp(-bp) \tag{3.46b}$$

where A, A_1, A_2, b, h_0, and h_∞ are material constants. In the case of cyclic softening of a material, for instance, the h_∞ is less than the h_0. The softening/hardening function $h(p)$ is constructed to distinguish cyclic loading from uniaxial loading. It is also used to separate the cyclic softening of solder alloys (Knecht *et al.*, 1990) with damage induced softening. Another characterization in the evolution Equation 3.42) of back stresses is that the coupling of void growth induces the release of back stresses. One of the mechanisms of void growth comes from the annihilation of dislocations and the cluster of micro vacancies by grain boundary diffusion (Vitek, 1980), which results in the relaxation of the dislocation back stresses (Qian *et al.*, 1991, 1993). Furthermore, the grain size effect of solder alloys can also be incorporated into this framework. A way is to introduce the grain size d into the drag stress via a Hall-Petch relation, that means:

$$D = D_0 + \frac{D_1}{\sqrt{d}} \tag{3.47}$$

where D_0 and D_1 are two material constants. The evolution equation of d can be carried out (Frear *et al.*, 1997). No further attempt was made here due to the lack of detailed microstructure information of grain size evolution in the most of technical reports (CINDAS, 1998). Finally, the evolution equation of void volume fracture f is obtained by a well-known relation (Gurson, 1977):

$$f = (1-f)\dot{\varepsilon}^{\mathrm{I}}_{kk} \tag{3.48}$$

Substituting Equation 3.39b into Equation 3.48 gives:

$$\frac{\dot{f}}{(1-f)} = 3\dot{p}\frac{fn(\sigma_{kk} - X^{(1)}_{kk} - X^{(2)}_{kk})}{2(n+1)(f+1)\sigma_{\mathrm{e}}} \tag{3.49}$$

In summary, the damage coupling framework of unified viscoplasticity consists of the elastic Hook's law Equation 3.31, the inelastic strain rate Equations 3.39, 3.40 and 3.41, the evolution equation of back stresses, Equations 3.42, 3.43 and 3.45, and the evolution Equation 3.49 of voids volume fraction, including the damage coupling with elasticity Equations 3.32 and 3.33, as well as the effects of cyclic softening/hardening Equation 3.46a and grain sizes Equation 3.47.

3.3.2 Large Deformation Formulation

It is necessary to extend the proposed framework to large deformation kinematics for the local damage approach. The deformation is transformed between an initial configuration and a current configuration and a current configuration by a deformation gradient tensor *F*. The decomposition of the velocity gradient tensor $\boldsymbol{L}$ gives:

$$\boldsymbol{L} = \dot{\boldsymbol{F}}\boldsymbol{F}^{-1} = \dot{\varepsilon} + \boldsymbol{W} = \dot{\varepsilon}^{\mathrm{e}} + \dot{\varepsilon}^{\mathrm{I}} + \boldsymbol{W} \tag{3.50}$$

where $\boldsymbol{W}$ is the material spin tensor. The material time derivations of stress tensors such as Cauchy stress σ, back stress and should be replaced by their Jaumann derivative. This replacement is the same as the work of Becker and Needleman (Becker and Needleman, 1986). The definition of the Jaumann derivative of a tensor $\boldsymbol{R}$ is of the form:

$$\hat{\boldsymbol{R}} = \dot{\boldsymbol{R}} - \boldsymbol{WR} + \boldsymbol{RW} \tag{3.51}$$

where hat denotes the Jaumann derivative, and dot the material time derivative. Therefore, the elastic law of Equation 3.31 is rewritten as:

$$\dot{\varepsilon}_{ij}^{e} = \frac{1+v}{E_D}\hat{\sigma}_{ij} - \frac{v}{E_D}\hat{\sigma}_{kk}\delta_{ij} + \left[\frac{1+v}{E_D}\sigma_{ij} - \frac{v}{E_D}\sigma_{kk}\delta_{ij}\right]\frac{D}{(1-D_f)^2} \tag{3.52}$$

The evolution equations of back stresses are changed from the material time derivative of a back stress to its Jaumann derivative, that is:

$$\hat{Y}_{ij}^{(1)} = \frac{2}{3}H_1\left[\dot{\varepsilon}_{ij}^{\prime I} - \frac{g\dot{p}}{(1-f)X_{ss}^{(1)}}\frac{3}{2}Y_{ij}^{(1)}\right] \tag{3.53a}$$

$$\hat{Y}_{ij}^{(2)} = \frac{2}{3}H_2\left[\dot{\varepsilon}_{ij}^{\prime I} - \frac{g\dot{p}}{(1-f)X_{ss}^{(2)}}\frac{3}{2}Y_{ij}^{(2)}\right] \tag{3.53b}$$

$$\hat{X}_{kk}^{(1)} = \frac{2}{3}H_1\left[\dot{\varepsilon}_{kk}^{I} - \frac{g\dot{p}}{(1-f)X_{ss}^{(1)}}\frac{3}{2}X_{kk}^{(1)}\right] \tag{3.54a}$$

$$\hat{X}_{kk}^{(2)} = \frac{2}{3}H_2\left[\dot{\varepsilon}_{kk}^{I} - \frac{g\dot{p}}{(1-f)X_{ss}^{(2)}}\frac{3}{2}X_{kk}^{(2)}\right] \tag{3.54b}$$

and the rest of equations keep their forms.

3.3.3 Identification of the Material Parameters

The material parameters involved in above constitutive framework can be identified from uniform deformation properties of one dimensional experiments such as tension/compression, torsion, or shearing. In the case, damage does not exist, therefore:

$$f = D_f = 0,\ \dot{f} = 0, \dot{\varepsilon}_{kk}^{I} = 0, \dot{X}_{kk}^{(1)} = \dot{X}_{kk}^{(2)} = 0 \tag{3.55}$$

The conditions of uniaxial axisymmetric loading can be written as:

$$\begin{aligned} S_{11} &= \frac{2}{3}\sigma, S_{22} = S_{33} = -\frac{1}{3}\sigma \\ Y_{11}^{(1)} &= \frac{2}{3}X_1, Y_{22}^{(1)} = Y_{33}^{(1)} = -\frac{1}{3}X_1 \\ Y_{11}^{(2)} &= \frac{2}{3}X_2, Y_{22}^{(2)} = Y_{33}^{(2)} = -\frac{1}{3}X_2 \\ &= W_{12} = W_{23} = W_{31} = 0 \end{aligned} \tag{3.56}$$

where σ is the tensile stress and is the tensile viscoplastic strain rate. The constant D_1 is set to zero and $h(p)$ is set to 1.0 to simplify the identification. Actually, the constants b, h_0, and h can be determined by a cyclic stress-strain curve. By using the conditions of Equations 3.55 and 3.56, constitutive equations for uniaxial loading are obtained from Equations 3.20, 3.31 3.37, 3.39, 3.40, 3.41, 3.42 3.44 and 3.45 as follows:

$$\dot{\varepsilon} = \dot{\varepsilon}_p + \frac{\sigma}{E} \tag{3.57a}$$

$$\dot{\varepsilon}_p = \Theta(T)\left[\sinh\left(\frac{|\sigma - X_1 - X_2|}{D_0}\right)\right] \tag{3.57b}$$

$$\dot{X}_1 = H_1\dot{\varepsilon}_p\left[1 - \frac{A_1X_1}{A\sigma_{ss}}\frac{\dot{\varepsilon}_p}{|\dot{\varepsilon}_p|}\right] \tag{3.57c}$$

$$\dot{X}_2 = H_2\dot{\varepsilon}_p\left[1 - \frac{A_2X_2}{A\sigma_{ss}}\frac{\dot{\varepsilon}_p}{|\dot{\varepsilon}_p|}\right] \tag{3.57d}$$

$$\Theta(T) = \dot{\varepsilon}_0\frac{E}{T}\exp\left(-\frac{Q}{kT}\right) \tag{3.57e}$$

where $\dot{\varepsilon}$ is the total applied strain rate.

Above equations are the same as the unified model proposed by authors (Qian *et al.*, 1997). The constitutive framework proposed here includes, furthermore, the extension of that model to three-dimensional form. Therefore, the consistent procedure of material constants determination developed before can be directly used with the current framework. The points are summarized as follows, and the details refer to Qian and Liu (Qian and Liu, 1998). In general, constants n, Q, A, and $\dot{\varepsilon}_0$ can be determined by the steady-state feature and H_1, H_2, A_1, A_2, and D_O by the transient feature of deformation. Particularly, the stress exponent n is uniquely determined by the power-law creep on the curve of the steady-state creep rates via creep stresses. The constant A is determined from the power-law breakdown region. The activation energy Q is conveniently obtained by using the linear offsetting of power-law creep at two different temperatures. The constants can be given by a derived parameter relation when n, Q, and A have been determined. D_O is determined by initial yielding or initial creep rates. H_1, H_2, A_1, and A_2 are fitted to the tensile curve at a relatively high strain rate by a least square program. A relation among A_1, A_2, D_O, and A is found. Therefore, only one of A_1, and A_2 needs to be determined. The approach has been verified for the Motorola data of 60–40 solder-joint tests at different temperatures and tensile rates (Qian *et al.*, 1997). For the benchmark tests of the eutectic solder 63Sn37Pb conducted by using a thin-strip specimen and six-axis mini fatigue tester, the constants are determined and listed in Table 3.3.

By using the above model constants, the excellent agreement of experiment data with model prediction without damage at small and uniform deformation stages is achieved, as shown in Figures 3.10, 3.11 and 3.12. Figure 3.10 shows the comparison of model prediction and the data at steady states. The typical

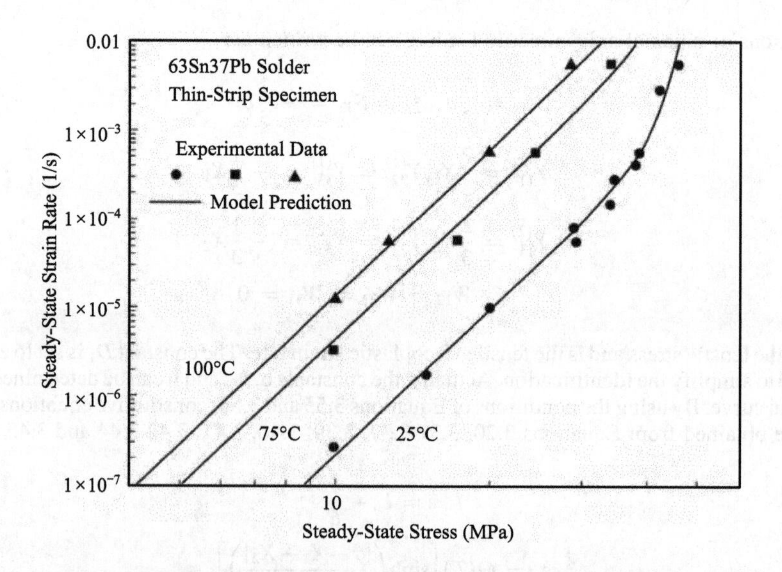

Figure 3.10 Steady-state properties from creep and tensile tests

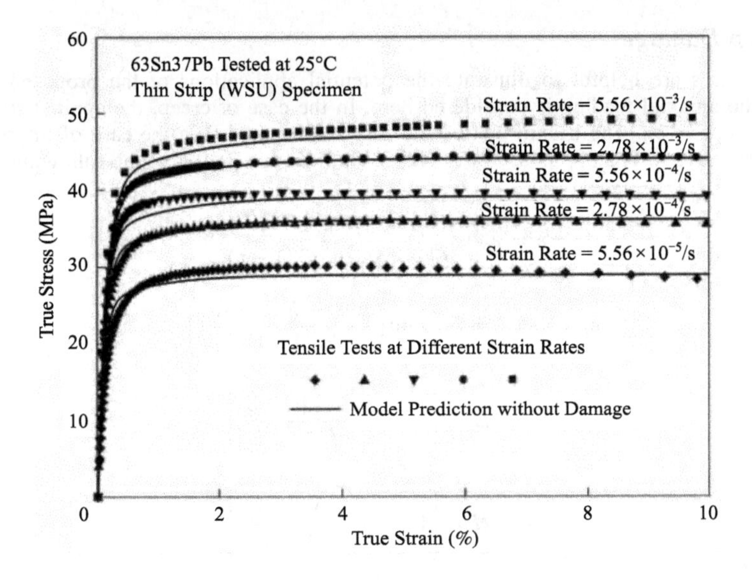

Figure 3.11 Tensile tests at different stain rates at 25 °C

comparison of model prediction and experimental curves is shown in Figure 3.11 and Figure 3.12 for the tensile tests at different strain rates and the temperatures 25 °C and 100 °C respectively.

Damage is initial by specimen necking or localization. The coupling evolution equations, Equations 3.39, 3.42, 3.43, and 3.49, for damage stages can be numerically integrated after the material constants are determined by the above procedure. Only the initial void volume fraction is unknown. This can be carried out by best fitting to experimental data such as creep damage curves, as illustrated in the next section.

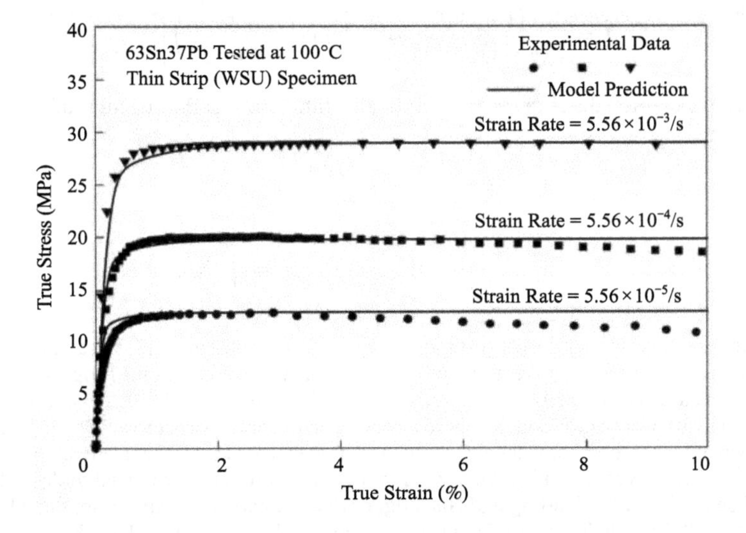

Figure 3.12 Tensile tests at different stain rates at 100 °C

3.3.4 Creep Damage

Some special cases are helpful to illustrate the potential applications of the proposed constitutive framework. The uniaxial loading is considered here. In the case of creep, the applied stress remains constant, that is, $\sigma = \sigma_0$. The triaxility stress ratio is a constant 1.0 in the case of uniaxial loading. Equations 3.20, 3.31, 3.37, 3.39, 3.40, 3.41, 3.42, 3.44, 3.45, and 3.49 give the following results:

$$\dot{\varepsilon} = \dot{\varepsilon}_p + \frac{a\sigma_0 \dot{f}}{E(1-af)^2} + \frac{1}{3}\frac{\dot{f}}{(1-f)} \tag{3.58a}$$

$$\dot{\varepsilon}_p = \dot{p} = \Theta(T)\left[\sinh\left(\frac{g(\sigma_0 - X_1 - X_2)}{D_0(1-f)}\right)\right]^n \tag{3.58b}$$

$$\dot{X}_1 = H_1\dot{\varepsilon}_p\left[1 - \frac{gA_1X_1}{(1-f)A\sigma_{ss}}\frac{\dot{\varepsilon}_p}{|\dot{\varepsilon}_p|}\right] \tag{3.58c}$$

$$\dot{X}_2 = H_2\dot{\varepsilon}_p\left[1 - \frac{gA_2X_2}{(1-f)A\sigma_{ss}}\frac{\dot{\varepsilon}_p}{|\dot{\varepsilon}_p|}\right] \tag{3.58d}$$

$$g = \sqrt{1 + \frac{nf}{2(n+1)(f+1)}} \tag{3.58e}$$

The damage evolution equation is then obtained from Equation 3.49:

$$\frac{\dot{f}}{(1-f)} = \frac{3}{2}\frac{nf}{(n+1)(f+1)}\dot{p} \tag{3.58f}$$

Or:

$$\dot{f} = \frac{3}{2}\frac{(1-f)nf}{(n+1)(f+1)}\Theta(T)\left[\sinh\left(\frac{g\sigma_0}{D_0(1-f)}\right)\right]^n \tag{3.58g}$$

The integration of Equation 3.58f gives an analytical solution of f as the function of p, which is:

$$f = \frac{2w + 1 - \sqrt{4w+1}}{2w} \tag{3.59a}$$

$$w = \frac{f_0}{(1-f_0)^2}\exp\left(\frac{3}{2}\frac{n}{n+1}p\right), p = \int_0^t |\dot{\varepsilon}_p| dt \tag{3.59b}$$

where f is the initial value of void volume fraction. The mathematical structure of constitutive Equations 3.58b and 3.58g are similar to creep damage constitutive equations proposed by Leckie and Hayhurst (Leckie and Hayhurst, 1977), as pointed out by Cocks (Cocks, 1989). This, in fact, is followed the original works of Kachanov and Rabotnov (Kachanov, 1986; Rabotnov, 1969). In addition, the temperature dependence, both power-law creep and power-law breakdown are also included here. The effect of stress rate exponent n is involved. And, the back stress is introduced to describe kinematic hardening and deal with cyclic loading and Bausinger effect. Creep curve can be obtained by integrating Equations 3.58 and 3.59 with the material constants listed in Table 3.4. Figure 3.13 shows the comparison of a creep curve with and without damage. The initial value of void volume fraction, f_0, is then taken as

Table 3.4 Material constants for 63Sn37Pb solder

n	Q(eV)	$\dot{\varepsilon}_0$(K/s/MPa)
5.5	0.633	1.359×10^5
A(1/MPa)	D_0(MPa)	H_1(MPa)
2.857×10^{-2}	20.0	1.360×10^4
H_2(MPa)	A_1(1/MPa)	A_2(1/MPa)
4.319×10^2	9.062×10^{-2}	0.252

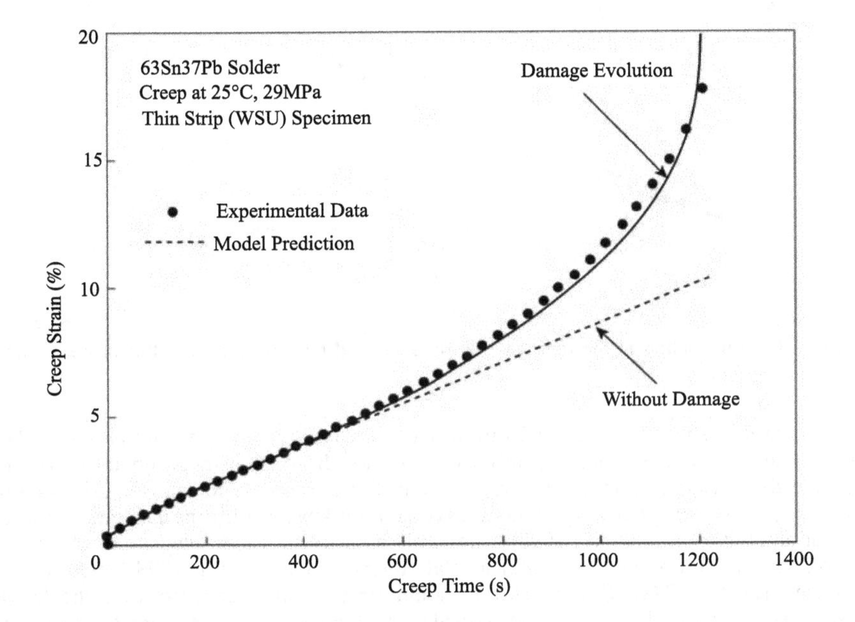

Figure 3.13 Damage softening of the tertiary creep

0.01 by matching this creep curve. The model satisfactorily predicts the experimental data of creep tests at different creep stresses and temperatures, as shown in Figure 3.14. The model prediction of creep damage evolution is shown in Figure 3.15. The creep damage maintains a small value from transient to steady-state (second-stage creep). The accuracy of the model prediction of second-stage creep will affect the accuracy of tertiary-stage creep, as shown in Figure 3.14 for the case of the creep test at 10MPa and 75MPa.

3.4 User-Supplied Subroutines for Solders Considering Damage Evolution

3.4.1 Return-Mapping Algorithm and FEA Implementation

The implementation of advanced and elaborated constitutive models of viscoplasticity with and without damage-evolution has been a challenge. Generally speaking, the reason is that the equations involved in the models are stiff differential equations with possible softening regime in the case of damage evolution. The nonlinear stability of numerical algorithms is quite important for such differential equation system (Dekker *et al.*, 1984). Convergence requires deriving an analytically consistent tangent modulus, that is, a Jacobian matrix, which is not an easy task even for experts. This can explain why those advanced models are not available in current commercial FEA codes.

Figure 3.15 shows schematically the outline of numerical finite element methods, which includes two major steps: spatial and time discretization. The spatial discretization is almost a standard technique

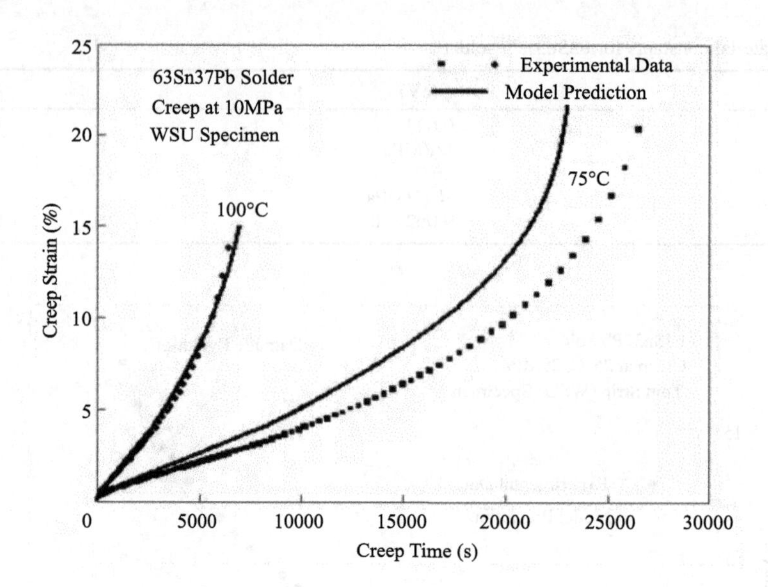

Figure 3.14 Model prediction of the transient, steady state, and tertiary creep at different creep stresses and temperatures

whereas the time discretization keeps updating when new constitutive models are proposed due to the requirement of new material development and characterization. In addition, the Jacobian matrix is required to transfer the time discretization procedure to spatial discretization. The analytical derivation of the Jacobian matrix, that is, consistent tangent modulus (CM) is the key to ensure the quadratic convergence of the nonlinear algorithm (Simo *et al.*, 1998). The implicit backward Euler integration scheme has been approved to be both A- (linear) and B- (nonlinear) stability method (Simo *et al.*, 1998; Dekker *et al.*, 1998), which has been used by ABAQUS for most of its nonlinear material models. However, the derivation of required stress computation algorithm and analytical tangent modulus is complicated for internal variable based models (Dekker *et al.*, 1984; Chaboche *et al.*, 1989, 1996).

The research has derived the consistent tangent modulus and the generalized return-mapping algorithm of stress computation for a class of viscoplastic models with linear/nonlinear isotropic/kinematical hardening and elastic/viscoplastic damage evolution. The implicit backward Euler integration scheme has been used to ensure the unconditional stability of the developed algorithm. The algorithm finally yields only a scalar nonlinear algebraic equation for the increment of inelastic von-Mises strain that can be solved locally. Rate-independent plasticity can be considered as the limit of viscoplasticity and programmed in a unified manner. The derivation was quite general and straightforward. Figure 3.16 shows the key

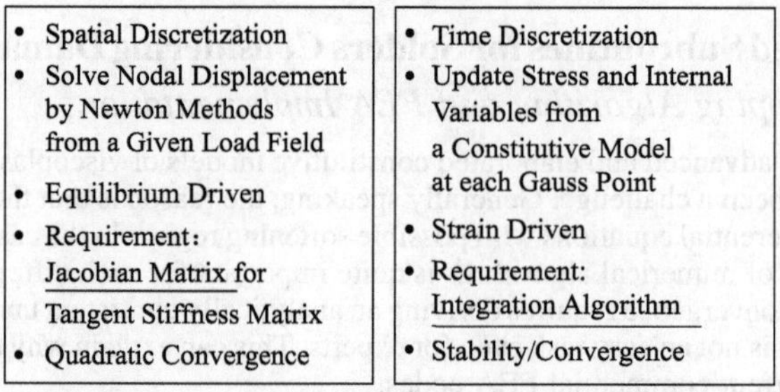

• Spatial Discretization • Solve Nodal Displacement by Newton Methods from a Given Load Field • Equilibrium Driven • Requirement: Jacobian Matrix for Tangent Stiffness Matrix • Quadratic Convergence	• Time Discretization • Update Stress and Internal Variables from a Constitutive Model at each Gauss Point • Strain Driven • Requirement: Integration Algorithm • Stability/Convergence

Figure 3.15 Outline of numerical finite element methods

• Stress Updating • Return-Mapping • Backward-Euler Integration Method • Elastic/Viscoplastic Damage Evolution • A- and B-Stability • Unified Framework	• Jacobian Matrix Derivation • Algorithm Consistent Tangent Modulus (CM) • Coupling between Elastic Damage and Nonlinear Hardening • CM Symmetry-Broken

Figure 3.16 Key developments of the algorithm

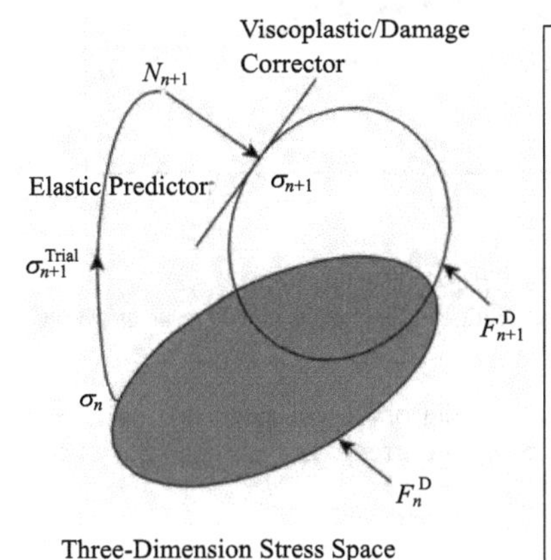

Step 1: Compute Trial Elastic State-Elastic Predictor

Step 2: Check Yielding Condition
If Satisfied: Exit otherwise to Step 3

Step 3: Solve Mises Strain Increment and Viscoplastic/Damage Corrector

Step 4: Update Stress and Strain

Step 5: Update Internal Variables

Step 6: Compute Algorithm Modulus

Figure 3.17 Generalized return-mapping algorithm

development of the generalized return-mapping algorithm. Details can be found in reference (Qian *et al.*, 1999).

The return-mapping algorithm for stress updating with damage evolution is illustrated in Figure 3.17. The algorithm can be divided as elastic predictor and viscoplastic/damage corrector in order to update stress-state from the increment time step n to $n+1$. Trial elastic state is very effective computationally to judge viscoplastic loading and elastic loading/unloading conditions.

3.4.2 Advanced Features of the Implementation

(i) Rate-dependent Tensile Deformation

The implementation of the generalized return-mapping algorithm is able to simulate solder deformation with a long range of softening caused by damage evolution. Figure 3.17 shows such a simulation of whole deformation stage to failure of eutectic solder samples that were subjected to tensile tests at different strain rates at 100 °C.

(ii) Creep Deformation

With the implementation of damage evolution and kinematical hardening, the unified constitutive model can describe three stages of creep deformation, whereas, most of current FEA models are able to simulate only second stage creep. A typical creep test is shown in Figure 3.18, which demonstrates

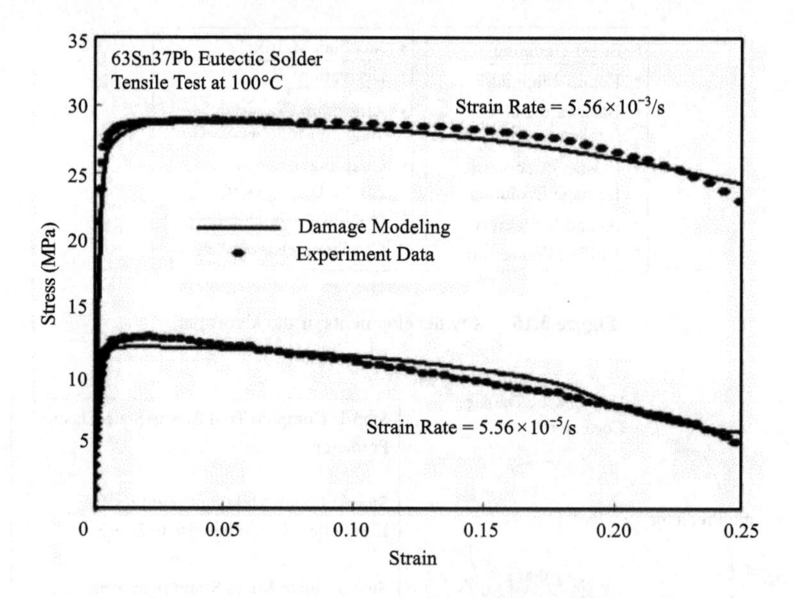

Figure 3.18 Whole deformation stage of eutectic solder during tensile tests at 100 °C

the effectiveness with and without creep damage simulation. Moreover, the implementation also successfully simulated the deformation behaviors of cyclic creep and mechanical cycling of solder materials (Qian *et al.*, 2000).

(iii) Cyclic Characteristics of the Algorithm

As an illusion of damage evolution-induced degradation of material deformation resistance, Figure 3.19 shows the back stress relaxation during cyclic loading.

Algorithmic modulus must keep its positiveness to ensure the nonlinear convergence due to damage-induced softening (Qian *et al.*, 1999). Figure 3.20 demonstrates that the algorithmic modulus always remains positive. The positive definitive property of the algorithmic modulus is rooted in the constitutive framework of rate-dependent viscoplasticity. This fact indicates one of the advantages of viscoplasticity over rate-independent plasticity in the case of damage-induced material softening. An interesting feature is that the modulus jumps from viscoplastic loading to elastic unloading or to elastic reloading.

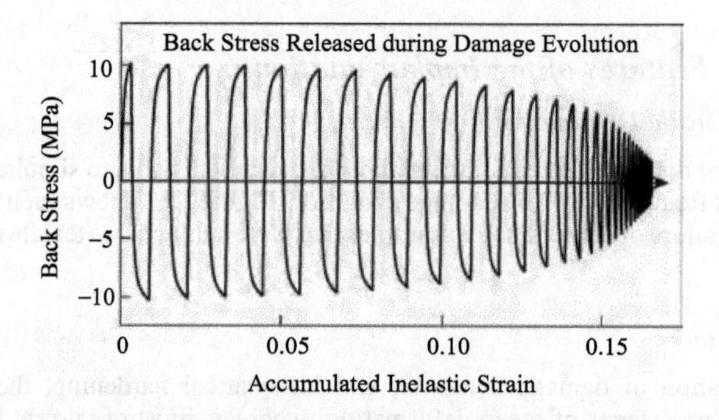

Figure 3.19 Back stress relaxation during cyclic loading

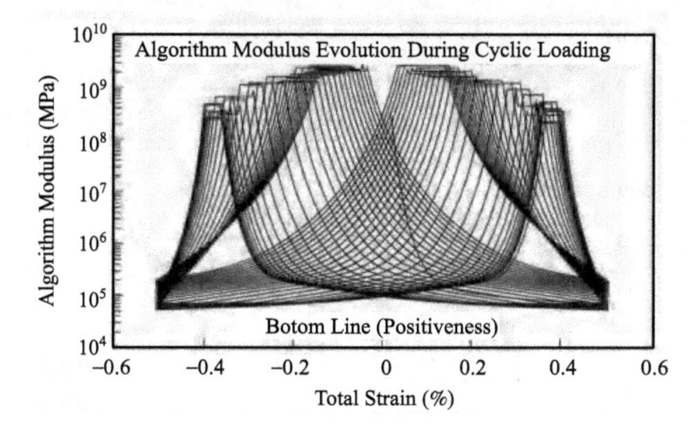

Figure 3.20 Algorithmic modulus evolution during cyclic loading and its positiveness

3.4.3 Applications of the Methodology

Virtual simulation of failure processes under cyclic loading can be explored by damage mechanics-based methodology. Nonlinear algorithm development and the implementation of the unified constitutive framework make the simulation possible. The advanced approach, in fact, has much more prediction power than traditional approaches.

(i) Solder Joint Fatigue

The first illusion example is the fatigue of solder joint specimens. Figure 3.21 shows the test configuration and the FEA model of a specimen of two solder balls under cyclic mechanical loading. The specimen was subjected to cyclic tensile test by a mini fatigue tester (Lu *et al.*, 1998). Typical load-displacement curves are shown in Figure 3.22. Due to crack initiation and propagation, the measured load was decreasing during displacement controlled cycling.

Figure 3.23 shows the contour plot of von-Mises stress of the solder balls. It indicates that the maximum stress zones are near the central areas on the top and bottom of each ball. However, the crack initiation did not locate in the areas. Instead, crack initiated at one of corner edges on the top and bottom of each ball, as shown in Figure 3.24. The reason is that the bending was concentrated at the corner edges. Damage mechanics is able to consider the effect in terms of tri-axial stress ratio (Qian *et al.*, 1999). In fact, due to the bending and peeling effect, the stress components at the corner edges are much larger than von-Mises stress.

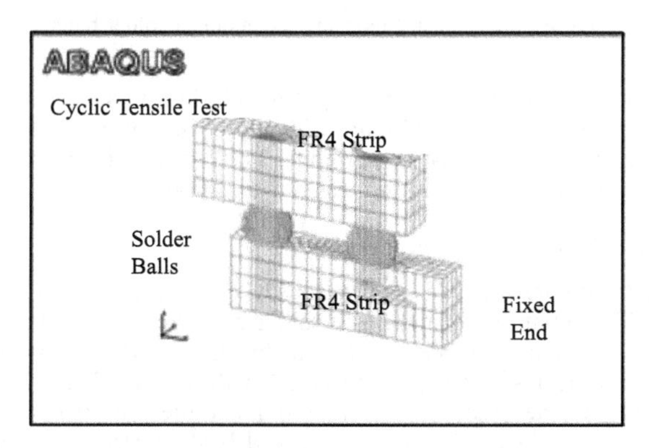

Figure 3.21 FEA model of the solder joint specimen

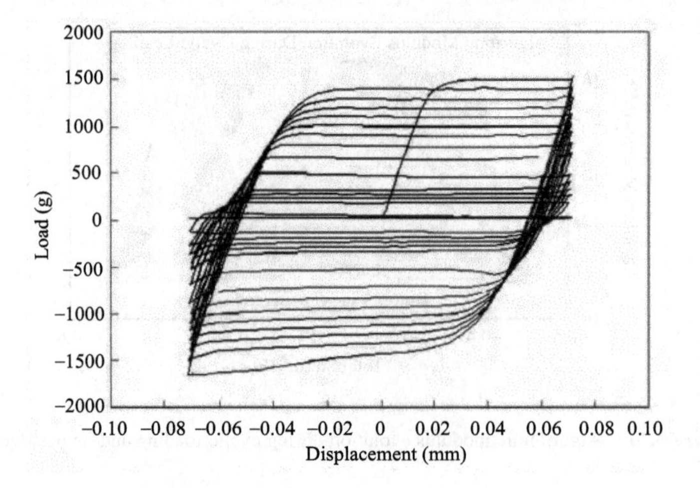

Figure 3.22 Cyclic load-displacement curves

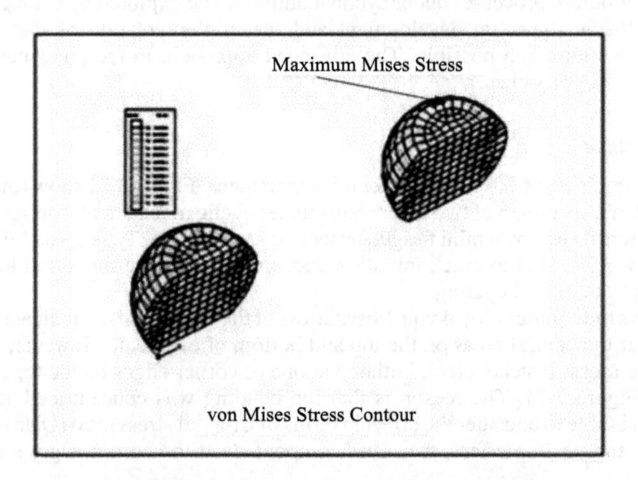

Figure 3.23 von-Mises stress contour of solder balls

Performed SEM (Scanning Electronic Microscopy) picture indeed supports the simulation result of crack initiation (Lu *et al.*, 1998), as shown in Figure 3.25. Moreover, the evidence of the SEM observation shows that the crack propagated along the corner edge. The characteristics of crack propagation were successfully reproduced by the FEA simulation, as shown in Figure 3.26.

(ii) BGA Package Under Thermal Cycling

BGAs, micro BGAs, and CSPs are widely used in various IC products due to their high density, small size, and high performance. The reliability of their tiny solder balls (interconnects) has been an issue. The example is to illustrate the damage mode and failure location of the BGAs under thermal cycling. A typical thermal cyclic profile has been applied to a BGA with 30 solder balls. Only a quarter-sized FEA model of the BGA structure is necessary for simulation due to the symmetry, as shown in Figure 3.27.

As expected, the effect of solder balls on the chip and FR4 substrate were localized, as shown in Figures 3.28 and 3.29 in terms of von-Mises stress contour.

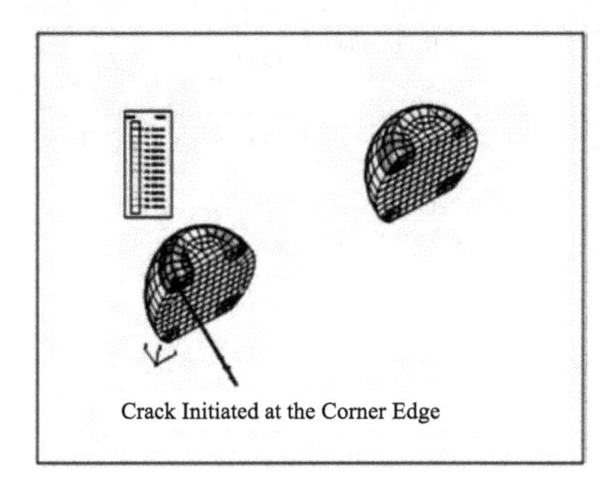

Figure 3.24 Location of crack initiation

Figure 3.25 SEM observation of failed ball after test

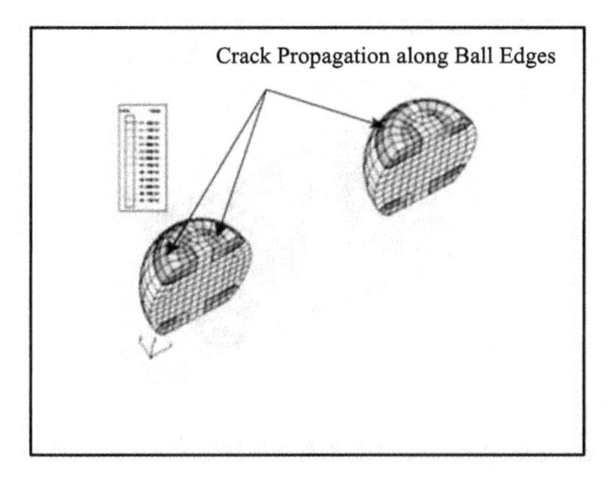

Figure 3.26 Characteristics of crack propagation

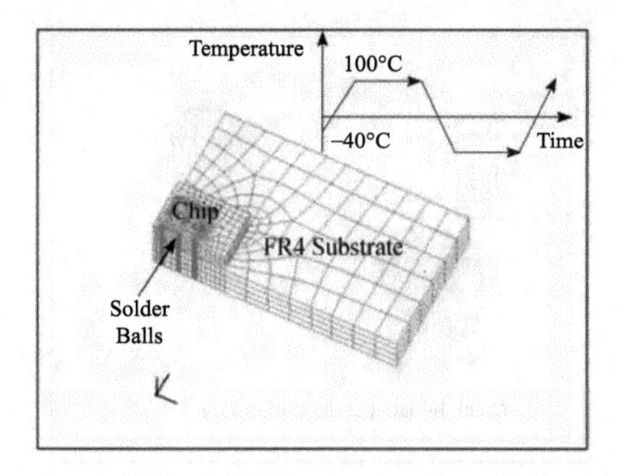

Figure 3.27 BGA model and thermal profile

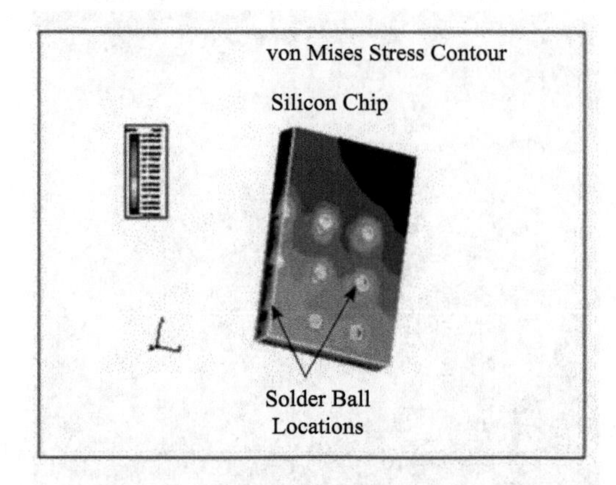

Figure 3.28 Local effect of solder balls on silicon chip

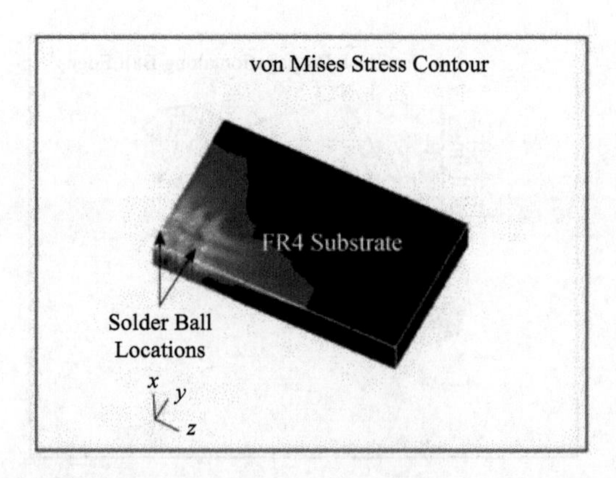

Figure 3.29 Local effect of solder balls on FR4 substrate

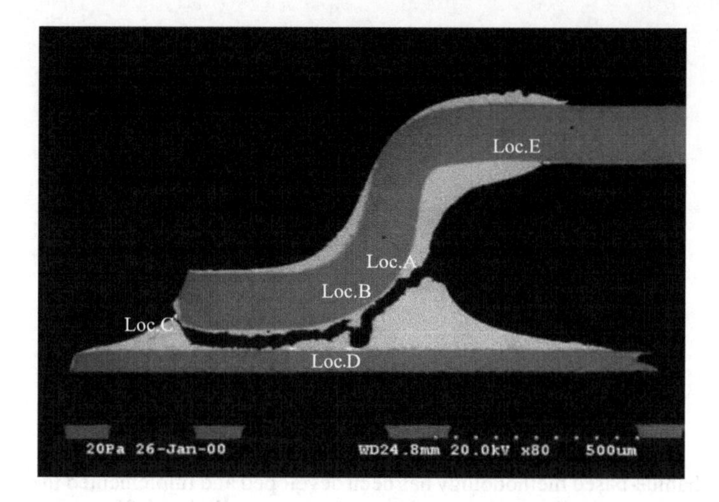

Figure 3.30 Damage location of the BGA under thermal cyclic loading

The damage location of the BGA under thermal cycling is shown in Figure 3.30. As indicated, the most damaging solder ball is the corner one. The crack initiation is located on the edges of the corner balls. Crack propagation and damage will spread from the outside of the balls towards the inside of the balls. All those characteristics have been reported in literature (Lau *et al.*, 1997).

(iii) TSOP Package Under Keypad Actuation

Figure 3.31 shows the FEA simulation of a failed solder joint of a Gull-Wing leaded TSOP package under keypad actuation. The comparison with the SEM picture is shown in Figure 3.32. Details were reported by Qian *et al.* (2000).

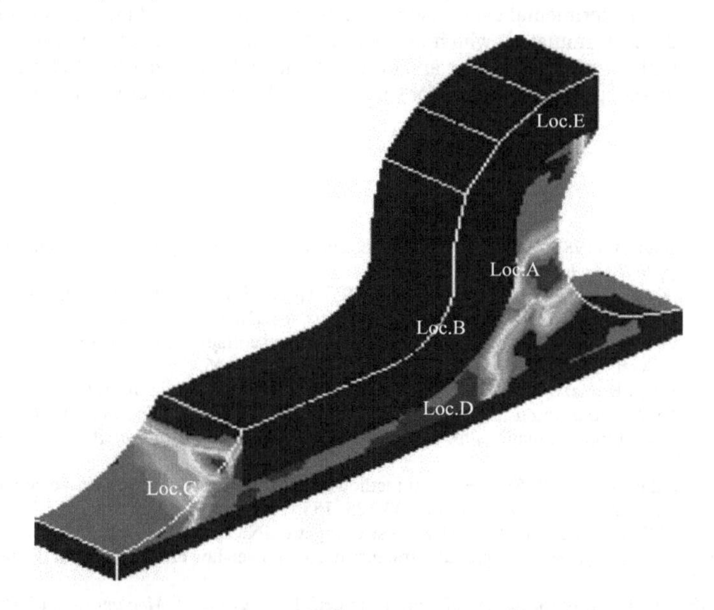

Figure 3.31 FEA simulation of solder joint failure of a Gull-Wing leaded TSOP under keypad actuation

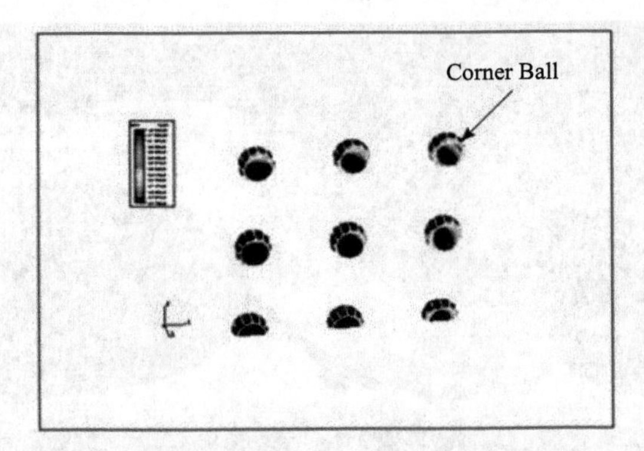

Figure 3.32 SEM analysis of the failed solder joint

A damage mechanics-based methodology has been developed and implemented in commercial FEA codes. Its prediction power has been illustrated by the virtual simulations of failure processes of solder joint fatigue under mechanical and/or thermal cyclic loading. The characteristics of crack initiation and propagation, observed by performed SEM analyses, have been successfully reproduced in detail. The methodology can be used for virtual qualification to predict potential damage mode and locations for product design. Further work will focus on evaluating the effect of defects and residual stress-strain induced by manufacturing/assembly processes. The further implementation of polymer-based materials is on-going. The implementations will significantly enhance the FEA tool power for virtual product prototyping.

3.5 Chapter Summary

In this chapter, constitutive models for several key packaging materials such as solders, and underfills are described mathematically in detail, along with the detailed procedures to derive material constants and the model validation by experimental data. Visco-elastic-plastic models and models considering damage coupling are studied and formulated, which are believed to be essential for the realistic modeling of accelerated life prediction for creep damage and fatigue damage. Finally, in order to facilitate the ease of use for the users, a user-supplied subroutines for solders considering damage evolution have been derived and applied.

References

Becker, R. and Needleman, A. (1986) Effect of yield surface curvature on necking and failure in porous plastic solids, *ASME Journal of Applied Mechanics* **53**:491–499.

Bhatti, P., Gschwend, K., Hwang, A.Y. and Syed, A.R. (1995) Three-dimensional creep analysis of solder joints in surface mount devices, *ASME Transactions, Journal of Electronic Packaging* **117**:20–25.

Budiansky, B., Hutchinson, J.W. and Slutsky, S. (1982) Void growth and collapse in viscous solids, *Mechanics of Solids*, Hopkins, H.G. and Sewell, M.J., eds., Pergamon Press, New York, 13–45.

Busso, E.P., Kitano, M. and Kumazawa, T. (1992) A viscoplastic constitutive model for 60/40 tin-lead solder used in IC package joints, *ASME Transactions, Journal of Engineering Materials and Technology* **114**:331–337.

Chaboche, J.L. (1989) Constitutive equations for cyclic plasticity and cyclic viscoplasticity, *International Journal of Plasticity* **5**:247–277.

Chaboche, J.L. and Cailletaud, G. (1996) Integration methods for complex plastic constitutive equations, *Computer Methods in Applied Mechanics and Engineering* **133**:125–155.

CINDAS Data Base, SRC Doc., Purdue University, West Lafayette, IN, 1998.

Cocks, A.C.F. and Ashby, M.F. (1980) Intergranular fracture during power-law creep under multiaxial stresses, *Metal Science* **14**:395–402.

Cocks, A.C.F. (1989) Inelastic deformation of porous materials, *Journal of Mechanics and Physics of Solids* **37**:693–715.

Darveaux, R. and Banerji, K. (1992) Constitutive relations for tin-based solder joints, *IEEE Transactions, CHMT* **15**:1013–1024.
Darveaus, R., Banerji, K., Mawer, A. and Dody, G. (1995) Reliability of plastic ball grid array assembly, in *Ball Grid Array Technology*, Lau J.H., ed, McGraw-Hill, New York.
Dasgupta, A., Oyan, C., Barker, D. and Pecht, M. (1992) Solder creep-fatigue analysis by an energy-partitioning approach, *ASME Journal of Electronic Packaging* **114**:152–160.
Dekker, K. and Verwer, J.G. (1984) *Stability of Runge-Kutta Methods for Stiff Nonlinear Differential Equations*, North-Holland, New York, 1–100.
Doi, H., Kawano, K., Yasukawa, A. and Sato, T. (1998) Reliability of underfill encapsulated flip-chips with heat spreaders, *ASME Transactions, Journal of Electronic Packaging* **120**:322–327.
Duva, J.M. and Hutchinson, J.W. (1984) Constitutive potentials for dilutely voided nonlinear materials, *Mechanics of Materials* **3**:41–54.
Frear, D.R., Jones, W.B. and Kinsman, K.R. (1990) Solder mechanics, a state of the art assessment, TMS, New Mexico.
Frear, D.R., Burchett, S.N. and Neilsen, M.K. (1997) Life prediction modeling of solder inter connects for electronic systems, *Advances in Electronic Packaging*, ASME EEP **19-2**:1512–1522.
Freed, A.D. and Walker, K.P. (1990) Steady-state and transient Zener parameter in viscoplasticity: drag strength versus yield strength, *Applied Mechanics Review* **43**:328–436.
Freed, A.D. and Walker, K.P. (1993) Viscoplasticity with creep and plasticity bounds, *International Journal of Plasticity* **9**:213–242.
Frost, H.J. and Ashby, M.F. (1982) *Deformation Mechanism Maps*, Oxford Press, London, 1–12, 1982.
Gurson, A.L. (1977) Continuum theory of ductile rupture by void nucleation and growth: Part I - yield criteria and flow rules for porous ductile media, *ASME Journal of Engineering Materials and Technology* **99**:2–15.
Harper, B.D., Lu, L. and Kenner, V.H. (1997) Effects of temperature and moisture upon the mechanical behavior of an epoxy molding compound, *Advances in Electronic Packaging* **19-1**:1207–1212.
Hartman, S., Luhrs, G. and Haupt, P. (1997) An efficient stress algorithm with applications in viscoplasticity and plasticity, *International Journal of Numerical Methods in Engineering* **40**:991–1013.
Hill, R. (1965) A self-consistent mechanics of composite materials, *Journal of Mechanics and Physics of Solids* **13**:213–222.
Hong, B.Z., Yuan, T.D. and Burrell, L.G. (1996) Anisothermal fatigue analysis of solder joints in a convective CBGA package under power cycling, *Sensing Modeling and Simulation in Emerging Electronic Packaging*, ASME EEP **17**:39–46.
Kachanov, L.M. (1986) *Introduction to Continuum Damage Mechanics*, Martinus Nijhoff, The Netherlands.
Knecht, S. and Fox, L.R. (1990) Constitutive relation and creep-fatigue life model for eutectic tin-lead solder, *IEEE Transactions on Components Hybrids and Manufacturing Technology* **13**:424–433.
Lau, J.H. (1991) *Solder Joint Reliability-Theory and Applications*, Van Nostrand Reinhold, New York.
Lau, J.H. (1995) *Ball Grid Array Technology*, McGraw-Hill, New York.
Lau, J.H. (1996) *Flip-Chip Technologies*, McGraw-Hill, New York.
Lau, J.H. and Pao, Y. (1997) *Solder Joint Reliability of BGA, CSP, Flip-Chip, and Fine Pitch SMT Assemblies*, McGraw-Hill, New York, 1–408.
Leckie, F.A. and Hayhurst, D.R. (1977) Constitutive equations for creep rupture, *Acta Metallurgica* **25**:1059–1070.
Lee, S.M. (1995) Cavitational failure phenomenon in Pb-Sn eutectic solders, *Journal of Applied Physics* **34**: L1475–1477.
Liu, S. and Mei, Y. (1994) Effects of voids and cracks on SMT solder joint by a nonlinear and time dependent finite element under thermal cycling, *Journal of Solder and Surface Mount Technology*, October, **21–28**.
Lemaitre, J. (1996) *A Course on Damage Mechanics*, Springer, New York.
Lowe, T.C. and Miller, A.K. (1986) Modeling internal stresses in the non-elastic deformation of metals, *ASME Transactions, Journal of Engineering Materials and Technology* **108**:365–373.
Lu, M. and Liu, S. (1996) A multi-axial thermo-mechanical fatigue tester for electronic packaging materials, *Sensing Modeling and Simulation in Emerging Electronic Packaging* **17**:87–92.
Lu, M., Qian, Z., Liu, S. and Wu, J. (1998) Linear motor driven mini Tester and its application to thermal/mechanical fatigue testing of solder joint specimens, in *Thermal-Mechanical Characterization of Evolving Packaging Materials and Structures*, S. Liu, Z. Qian, and C.P. Yeh, eds, *ASME EEP* **7**:40–46.
Lu, M., Qian, Z., Ren, W., Liu, S. and Dongkai, S. (1999) Investigation of electronic packaging materials by using a 6-axis mini thermo-mechanical tester, *International Journal of Solids Structure* **36**:65–78.
McClintock, F.A. (1968) A criterion for ductile fracture by the growth of holes, *ASME Journal of Applied Mechanics* **35**:363–371.
Merton, M.M. Mahajan, R.L. and Nikmanesh, N. (1997) Alternative curing methods for FCOB underfill, *Advances in Electronic Packaging* **19–1**:291–300.
Product Bulletin of HYSOL FP4526, Electronic Materials Division, Dexter Corporation, 1997, pp. 1–2.

Pao, Y.H., Jih, E., Siddapureddy, V., Song, X., Liu, R., McMillan, R. and Hu, J.M. (1996) A thermal fatigue model for surface mount leadless chip resistor (LCR) solder joints, *Sensing, Modeling and Simulation in Emerging Electronic Packaging*, ASME EEP **17**:1–12.

Ponte Castaneda, P. and Willis, J.R. (1988) On the overall properties of nonlinearly viscous composites, *Proceedings of the Royal Society (London)* **A416**:217–244.

Qian, Z. and Fan, J. (1991) Constitutive model for creep-plasticity interaction based on continuous kinetics of dislocations, in *High Temperature Constitutive Modeling Theory and Application*, Freed, A.D. and Walker, K.P., eds, ASME MD-26/AMD-121, 49–63.

Qian, Z., Duan, Z. and Fan, J. (1993) A constitutive model for creep-plasticity interaction based on dislocation dynamics, *Acta Mechanica Sinica* **9**:261–268.

Qian, Z., Lu, M., Wang, J. and Liu, S. (1997) Testing and constitutive modeling of thin polymer films and underfills by a six-axis submicron tester, *Application of Experimental Mechanics to Electronic Packaging* **22/AMD–226**:105–112.

Qian, Z. and Liu, S. (1997) A unified viscoplastic constitutive model for tin-lead solder joints, *Advances in Electronic Packaging*, ASME EEP **19-2**:1599–1604.

Qian, Z., Lu, M. and Liu, S. (1998) Constitutive modeling of polymer films from viscoelasticity to viscoplasticity, *ASME Transactions, Journal of Electronic Packaging* **120**:145–149.

Qian Z. and Liu, S. (1998) On the life prediction and accelerated testing of solder joints, *Thermo-Mechanical Characterization Evolving Packaging, Material, Structure* **7**:1–11.

Qian, Z. and Liu, S. (1999) A damage coupling framework of unified viscoplasticity for the fatigue of solder alloys, *ASME Transactions, Journal of Electronic Packaging* **121**:162–168.

Qian, Z. and Liu, S. (1999) Implementation of unified viscoplasticity with damage evolution and its application to electronic packaging, in *Proceedings of the InterPack'99*, 1–8, Hawaii.

Qian, Z., Wang, J., Yang, J. and Liu, S. (1999) Visco-elastic-plastic properties and constitutive modeling of underfills, *IEEE Transactions on Components and Packaging Technologies* **22**:152–157.

Qian, Z., Shi, L. and Liu S. (2000) A unified approach to solder joint life prediction, *SAE Technical Paper 2000-01-0454, SAE 2000 World Congress*, Detroit, Michigan, 1–8.

Qian, Z. and Wang, H. (2000) Root cause identification of failed IC components of StarTAC phones by keypad actuation in terms of component/system level solder joint reliability, in *2000 HEMES Symposium*, Ft. Lauderdale, FL.

Rabotnov, Y.N. (1969) *Creep Problems in Structural Members*, North-Holland, Amsterdam, The Netherlands.

Ren, W., Qian, Z., Lu, M., Liu, S. and Shangguan, D. (1997) Thermal mechanical property of two solder alloys, *Applications of Experimental Mechanics to Electronic Packaging*, ASME EEP **22**/AMD **226**:125–130.

Ren, W., Qian, Z. and Liu, S. (1998) Thermal mechanical creep of two solder alloys, in *Proceedings of the 48th ECTC*, Seattle, WA, 1431–1437.

Ren, W., Qian, Z. and Liu, S. (1999) Scale effect on packaging materials, in *Proceedings of the 49th ECTC*, San Diego, CA, June 1–4, 1–5.

Rice, J.R. and Tracey, D.M. (1969) On the ductile enlargement of voids in triaxial stress fields, *Journal of the Mechanics and Physics of Solids* **17**:201–217.

Rzepka, S., Korhonen, M.A., Meusel, E. and Li, C.Y. (1998) The effect of underfill and underfill delamination on the thermal stress in flip-chip solder joints, *ASME Trans. Journal of Electronic Packaging* **120**:342–348.

Sarihan, V. (1994) Energy based methodology for damage and life prediction of solder joints under thermal cycling, *IEEE Transactions, CPMT*, B17, 626–631.

Schubert, A., Dudek, R., Auersperg, J., Vogel, D., Michel, B. and Reichl, H. (1997) Thermo-mechanical reliability analysis of flip-chip assemblies by combined microdac and the finite element method, *Advances in Electronic Packaging* **19**(2):1647–1654.

Simo, J.C. and Hughes T.J.R. (1998) *Computational Inelasticity*, Springer, New York, 71–153, 1998.

Skipor, A.F., Harren, S.V. and Botsis, J. (1996) On the constitutive response of 63/37 Sn/Pb Eutectic solder, *ASME Transactions, Journal of Engineering Materials and Technology* **118**:1–11.

Solomon, H.D. (1991) Predicting thermal and mechanical fatigue lives from isothermal low cycle data, *Solder Joint Reliability-Theory and Applications*, Lau, J., ed., Van Nostrand Reinhold, New York: 406–454.

Suryanarayana, D., Hsiao, R., Gall, T.P. and McCreary, J.M. (1991) Enhancement of flip-chip fatigue life by encapsulation, *IEEE Transactions on Components, Hybrids, Manufacturing Technology* **14**:218–223.

Tvergaard, V. and Needleman, A. (1984) Analysis of cup-cone fracture in a round tensile bar, *Acta Metallurgica* **32**:157–169.

Viteck, V. (1980) Diffusional growth of intergranular cavities in uniform stress field and ahead of crack-like stress concentrator, *Metal Science* **14**:403–407.

Wang, L. and Wong, C.P. (1998) Novel thermally reworkable underfill encapsulants for flip-chip applications, in *Proceedings of the 48th ECTC*, Seattle, WA, 92–100.

Wang, J., Qian, Z. and Liu, S. (1998) Process induced stresses of a flip-chip packaging by sequential processing modeling technique, *ASME Transactions, Journal of Electronic Packaging* **120**:309–313.

4

Accelerated Fatigue Life Assessment Approaches for Solders in Packages

The solder joints play a key role in mechanical and electrical interconnects of advanced packaging technologies such as Ball Grid Array (BGA), Flip-Chip Package, Chip Scale Package (CSP), Plastic Quad Flat Pack (PQFP), Direct Chip Attach (DCA), fine pitch SMT assemblies (Lau, 1991), and those emerging packaging types such as packaging in packaging (PoP), system in packaging (SiP), power electronics packaging, and three-dimensional packaging. The fatigue failure of solder joints and their life prediction are two of the most important issues of solder joint reliability, in particular, for military, aerospace, medical, and automotive electronics. Design-for-reliability (DfR) is driven by high-density interconnection, high performance of electronic products, and packaging miniaturization. Electronic industries typically subject new product design to a series of reliability qualification tests to prove that the product they are planning to introduce will be sufficiently robust to function in various environments. Product design currently determines 75% of manufacturing costs. Additionally, companies in electronic industry have to pay for product and pre-production prototype failures. Therefore, increasing numbers of manufacturers now turn to solid mechanics based computer-aided design and engineering to address escalating electronic packaging challenges. In terms of fatigue failure of solder joints, the current paradigm for assessing the in-service reliability of electronic packages is based on thermal/mechanical cycling and thermal shock tests with humidity, which is a time-consuming practice. Therefore, accelerated testing becomes more important and is the focus of recent intensive research, driven by short-time-to-market, short-time-to-profit, and low-cost. Rapid reliability assessment is highly desirable for electronic manufacturing industries. Concurrent engineering approaches, in particular, for solder joint reliability assessment, include the methodologies of life prediction, computer modeling, and accelerated testing of solder joints under various accelerated hygro-thermal-mechanical testing conditions.

4.1 Life Prediction Methodology

Approaches of fatigue life prediction are developed based on experimental stress/strain/energy data from thermal/mechanical cycling tests. A number of life prediction approaches have been proposed for solder joint fatigue during the past few years. These approaches can be classified into four categories: (1) stain-based approach; (2) energy-based approach; (3) fracture mechanics-based approach; and (4) damage evolution-based approach. A stress-based approach is less used in practice although it can be useful for vibration or physically shocked or stressed components (Lee *et al.*, 1997; Frear *et al.*, 1991).

Modeling and Simulation for Microelectronic Packaging and Integration: Manufacturing, Reliability and Testing, Second Edition. Sheng Liu and Yong Liu.

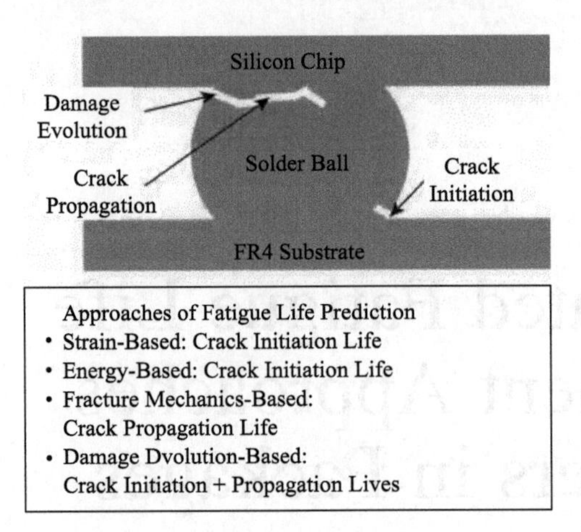

Figure 4.1 Illustration of life prediction methodology

The definition of fatigue failure of a solder joint plays a critical role in understanding the above approaches and the validity of each approach. Basically, fatigue failure comprises both the initiation of a macroscopic crack and the propagation of the crack. The initiation of a macroscopic crack occurs at a microscopic material level. The very ductile materials such as solder alloys are damaged by the micro-processes of void nucleation on phase and/or grain boundaries, void growth, and their coalescence to result in macroscopic crack initiation (Frear *et al.*, 1991; Lee, 1995; Cocks *et al.*, 1980; Solomon, 1993). It is believed that the irreversible inelastic deformation, including both plastic and creep strains, is the driving force of micro damage processes. This is the origin of the strain-based approach. On the other hand, the energy-based approach presents a combination of strain-based and stress-based approaches. Clearly, the approaches only predict the fatigue life of a solder joint until the appearance of a macroscopic crack.

On the other hand, the propagating mechanism of the crack inside a solder joint under cyclic loading is controlled by facture quantities. Therefore, the fracture mechanics-based approach predicts only the part of fatigue life when a macroscopic crack propagates through the solder joint to cause its electric and/or mechanical failure.

There are, in fact, wide discrepancies on the failure definition due to fatigue crack initiation and propagation in literature. For instance, expedient fatigue failure criteria include complete electrical circuit opens, a 50% reduction in the measured stress amplitude on the solder joint, and a 20% crack propagation across the solder joint. The predicted fatigue life, therefore, also depends on the definition. When comparing the fatigue life data from various sources, it is important to take into account the definition of the solder joint fatigue failure employed (Lee *et al.*, 1997; Solomon, 1993).

The damage evolution-based approach, combined with the finite element technique, is able to predict both parts of fatigue lives due to crack initiation and propagation. Figure 4.1 illustrates the scopes of the different approaches for fatigue life prediction. Details of the methods and their mathematical formula are described in following sections.

4.1.1 Strain-Based Approach

The well-known Coffin–Manson equation is an inelastic strain ranged-based method for low cyclic fatigue (Coffin, 1954; Manson, 1965). The method has been widely used for fatigue life prediction of many solder alloys subjected to shear strain-dominated deformation (Solomon, 1993; Hong *et al.*, 1996). There is a survey of many fatigue models available in literature (Lee, 1997). The models are, in fact, based on the Coffin–Manson equation. Extending the original one-dimensional form Equation 4.1 to three-dimensional form Equation 4.2 for solder alloys, the Coffin–Manson equation is given by:

$$(N_{\rm f})^{\beta}\Delta\gamma^{p} = C^{p} \tag{4.1}$$

$$N_{\rm f} = C^{\rm I}\left(\Delta\varepsilon^{\rm I}_{\rm Mises}\right)^{-\alpha} \tag{4.2}$$

With:

$$\alpha = \frac{1}{\beta}, \quad C^{\rm I} = \left(\frac{C^{p}}{\sqrt{3}}\right)^{1/\beta} \tag{4.3}$$

where β is called the fatigue ductility exponent, and C^{p} the fatigue ductility coefficient, $\Delta\gamma^{p}$ and $\Delta\varepsilon^{\rm I}_{\rm Mises}$ are the ranges of plastic shear strain and viscoplastic Mises strain, respectively. $N_{\rm f}$ is the fatigue life that is usually taken as the life of 50% load drop of a test specimen. In order to account for frequency effect, a frequency-modified version of the Coffin–Manson equation was also proposed. It was found that between $-55\,^{\circ}$C and 125 °C, the life of isothermal fatigue tests of 60/40 solder can be approximated by a single Coffin–Manson expression with material constants $\beta = 0.51$, $C^{p} = 1.14$, at a constant frequency 0.3 Hz. Therefore, $\alpha = 1.96$, $C^{\rm I} = 0.4405$ are obtained from Equation 4.3. Material constants used in this chapter for eutectic solder joints subjected to thermal fatigue between $-40\,^{\circ}$C and 125 °C are listed in Table 4.1.

Table 4.1 Material constants for 60/40 solder fatigue

α	$C^{\rm I}$	δ	$C^{\rm E}$
1.96	0.4405	1.818	347.55

It should be pointed out the Equation 4.2 must be used for the life prediction of solder joints in 3D deformation states. The correct identification of the $\Delta\varepsilon^{\rm I}_{\rm Mises}$ from 3D finite element calculation will be shown in later section for thermal cyclic loading. There is, however, some confusion in the literature of the identification. Moreover, inelastic Mises strain range is the only input quantity of the Coffin–Manson equation, which actually separates the life calculation with the deformation computation that is the task of constitutive modeling.

On the other hand, a more sophisticated method for creep-plasticity interaction is the Strain Range Partitioning (SRP) technique. Rate-independent plastic and rate-dependent creep strains are separated by a special procedure for each cycle. Some applications of the SRP method of solder alloys and a modified version of the SRP technique can be found in references (Solomon, 1993).

It is well known that the nonlinear viscoplastic property of solder alloys is highly rate/temperature-dependent (Lau, 1991; Frear *et al.*, 1991; Solomon, 1993; Qian *et al.*, 1999). The critical question is then how to define the rate-independent plastic curves from the experimental data. In literature, however, quite different curves have been used for rate-independent plasticity for eutectic solder. The real rate-independent curve at a specific temperature might not be reached actually by any test machine, which will be required to run a tensile test at a strain rate as low as 1.0×10^{-10}/s to find out the true rate-independent property for a solder alloy. Another expedient definition of plasticity can be the tensile strain/stress curve at a relatively high strain rate, about, 1.0×10^{-1}/s. On the other hand, solder joints of an electronic package during temperature ramping stage, in fact, are subjected to a continuous change of strain rates. Therefore, any particular choice for the plasticity curve at a specific tensile strain rate cannot be suitable for and consistent with the whole ramping stage of thermal cyclic loading. There is always the time-dependent deformation, creep strain, during thermal cyclic loading.

Considering the difficulty and confusion of obtaining pure rate-independent plastic deformation from a high rate-dependent viscoplastic strain of a solder alloy, the SRP method is not recommended in this chapter. Using the Coffin–Mason equation by the ranges of plastic strain and creep strain, respectively, for life prediction, is not recommended either. Instead, the range of time rate/temperature dependent viscoplastic Mises strain that unifies plastic strain with creep strain is strongly recommended to be used for life prediction in Equation 4.2.

4.1.2 Energy-Based Approach

Recently, strain energy-based methods have been applied to the fatigue of solder joints (Solomon, 1993; Engelmaier, 1989) to name a few. Actually, the method was used very early (Morrow, 1964). A similar power-law relationship of the Coffin–Mason equation in terms of inelastic hysteresis energy density, W^{I}, is written as follows:

$$N_{\mathrm{f}} = C^{\mathrm{E}}\left(W^{\mathrm{I}}\right)^{-\delta} \tag{4.4}$$

where δ and C^{E} are the material constants. It was reported that $\delta = 1.818$ and $C^{\mathrm{E}} = 347.55$ for the 60/40 solder isothermal fatigue between $-50\,°\mathrm{C}$ and $125\,°\mathrm{C}$. These constants are also listed in Table 4.1 the unit of W^{I} has been transferred to mm·N/mm^3 for convenience. It should be pointed out that some authors in literature have interpreted the term of inelastic strain energy density W^{I} in their papers instead of inelastic or plastic strain energy. This can result in some confusion in using Equation 4.4, and calibrating material parameters.

Similar to the SRP method, the energy density partitioning approach has been proposed (Dasgupta *et al.*, 1992). Total strain energy density can be divided into elastic, plastic, and creep strain energy density in each cycle. The relationship of each component of the energy density with fatigue life has been assumed to follow a Coffin–Manson power-law relationship such as Equation 4.24. Total creep-fatigue damage then follows the Miner's linear superposition rule. However, the same argument about the difficult separation of creep and plastic strains for the SRP method could be true for the application of the strain energy density partitioning approach. From this point of view, the total inelastic energy density-based approach seems favored for the life prediction of thermal fatigue.

The energy approach does have its technical merit for three-dimensional deformation states (Dasgupta *et al.*, 1992). The extended Coffin–Mason equation, is also convenient for application to three-dimensional deformation states. In fact, the ranges of inelastic Mises strain and inelastic energy density are only two suitable quantities for thermal fatigue life prediction of solder joints, based on finite element analysis.

There is, however, a complication that must be addressed when utilizing W^{I} for life prediction (Solomon *et al.*, 1995, 1996). Moreover, fatigue life was not found to be a single valued function of the hysteresis energy density for changing frequency tests (Solomon *et al.*, 1995, 1996). We have also identified the questionable tendency of predicted fatigue lives by the energy-based approach for accelerated testing via frequency change, as shown in later sections.

4.1.3 Fracture Mechanics-Based Approach

For the large solder joint structures such as SMT (Surface Mounting Technology) LCR (Leadless Chip Resistor) solder joints, the fatigue life of crack propagation from a tiny crack to final joint failure electrically and/or mechanically can be a significant portion of the whole fatigue life. Therefore, fracture mechanics approaches may play an important role in characterizing the crack behavior and thus can lead to the formulation of life prediction methods for such joints. Details can be found in references (Subrahmanyan *et al.*, 1989; Pao, 1992).

4.2 Accelerated Testing Methodology

Accelerated testing is a very important industrial practice for package design and rapid reliability assessment. The qualification and reliability tests of advanced packages such as BGAs will take 3000 to 15000 cycles to reach failure typically. The time to failure is about 4 to 10 months as 1–2 cycles/hour. It is a time-consuming process in addition to purchasing expensive test equipment. The accelerated testing method is to shorten the time to failure by over-stressing solder joints. The utmost goal is to run qualification and reliability tests developed by computer tools. The test procedure is realized currently in the electronics industry by rising extreme temperatures, shortening dwell time, and increasing temperature ramping rates. The accelerated testing methodology, actually, is closely related with the methodologies of fatigue life prediction and constitutive modeling. Acceleration factors and their bounds can be determined by integrating a fatigue life prediction approach with a powerful viscoplastic model that is able to be used to investigate various accelerated conditions. It is still not certain how to determine the accelerated testing factor and how to correlate the fatigue life of accelerated tests with product/package life during service conditions (Lau *et al.*, 1997; Solomon, 1986). More systematic research is needed.

4.2.1 Failure Modes via Accelerated Testing Bounds

It is very important that proper considerations are given to failure mechanisms and modes when designing an accelerated test for solder joint reliability assessment. A life prediction method can be a very useful tool if it can predict failure modes of solder joints and embed the domination failure mechanisms such as intergranular or transgranular cracking, grain boundary sliding, and grain growth-induced shear banding.

There will be bounds of accelerated factors regarding failure mode change or thermal gradient limitation under extreme accelerated test conditions. For instance, the increase of the ramping rate to that of thermal shock may induce the failure mode change from intergranular fracture of solder joints to underfill delamination of flip-chip package and failure location shifting. Details will be discussed in later sections.

4.2.2 Isothermal Fatigue via Thermal Fatigue

One related issue of accelerated testing is the relationship between isothermal fatigue and thermal fatigue. Isothermal fatigue can be performed much faster than thermal fatigue since it is an easier experiment to operate under well-controlled laboratory testing conditions. It is often used to calibrate constitutive models and the methods of fatigue life prediction. Electronic industries have been accumulating reliability testing data of thermal fatigue of electronic packages (Lee *et al.*, 1997; Dougherty *et al.*, 1998; Popelar, 1997). Isothermal fatigue and thermal fatigue exhibit a number of similarities and differences. However, the mechanisms of deformation and failure include an effect in thermal fatigue that can be absent in isothermal fatigue. Accelerated testing can also create new failure modes, as discussed above. The microstructure evolution occurs in near eutectic solders in thermal fatigue (Frear *et al.*, 1997). The temperature cycle can induce solder recrystalization and then grain growth. In fact, the strain rate keeps changing during thermal cyclic loading due to the temperature dependence of viscoplastic behaviors of solder. The mean strain/stress and energy density generated during the initial ramping of the cooling stage of thermal cycling are also an important factor for thermal fatigue life, which has been ignored during investigation at isothermal fatigue tests. Therefore, strain rate and temperature history, microstructure evolution, and mean strain/energy density are three additional factors that need to be evaluated for constitutive modeling, fatigue life prediction, and accelerated testing for the thermal fatigue of solder joints.

4.3 Constitutive Modeling Methodology

As discussed in the above sections, the life prediction of solder joints under creep-fatigue interaction is still a difficult problem in terms of crack initiation and crack propagation. Constitutive modeling is one of the keys that account for the visoplastic property of solder alloys and compensate for the deficiency of life prediction model by strain and energy-based approaches. The range of inelastic Mises strain or strain energy density must be accurately calculated by finite element analysis, based on a realistic and elaborated nonlinear constitutive model of solder alloys. The effects of extreme temperatures, temperature ramping rates, dwell time and temperature, and solder microstructure on thermal fatigue life under accelerated test conditions can be explored only by powerful constitutive models that incorporate the effects of time/temperature/rate, deformation and temperature history, and microstructure evolution.

4.3.1 Separated Modeling via Unified Modeling

Currently, the constitutive models of solder alloys, used in electronic packaging, are mostly separated as rate-independent plasticity and steady-state creep. Classical plastic flow theories and power-law creep are still widely used. The corners and interfaces of solder joints under thermal shock and accelerated test conditions can reach the regime of power-law breakdown. The major deficiency of separated-models is the lack of predictive power to correctly describe the stress/strain hysteresis curve during thermal cyclic loading. The separated constitutive models also lack the ability to investigate the effects of heating/cooling rate changing. The inaccuracy of the hysteresis curve calculation substantially magnifies the error of fatigue life prediction of solder joints due to the power-law dependence of fatigue life on the ranges of inelastic strain and energy density according to Coffin–Manson equation. On the other hand, unified constitutive modeling considers plastic strain and creep strain as an inelastic strain. The approach has been developed since the 1970's for the design and reliability assessment of structure and parts made from super alloys served at high temperatures. The alloys exhibit the similar deformation behaviors at high temperatures as those of solders even at room temperature. Due to its intrinsic merit of unified constitutive modeling for rate/time/temperature- dependent nonlinear properties of solder alloys, it is strongly recommended in this chapter for solder joint reliability assessment.

4.3.2 Viscoplasticity with Damage Evolution

New advances in this area have been achieved, including the application of unified viscoplastic constitutive models to solder alloys (Busso *et al.*, 1992, 1994; Fu *et al.*, 1996; Qian *et al.*, 1997). The unified visocplasticity with damage evolution is proposed for the fatigue life prediction of solder joints (Qian *et al.*, 1999; Basaran *et al.*, 1998). In addition, cyclic hardening/softening, microstructure evolution, and scale effect have been also incorporated into the constitutive framework (Qian *et al.*, 1999) that also covers both power-law and power law breakdown regimes. The implementation of the framework into commercial code such as ABAQUS has been used to calculate hysteresis curves and predict thermal fatigue life at critical points, interfaces, and corners for flip-chips and BGAs (Qian *et al.*, 1999). The new methodology of fatigue life prediction, based on unified models incorporating damage evolution and crack/failure criteria, can be developed. By the local approach to fracture, the fatigue life of solder joints during both crack initiation and crack propagation can be predicted by finite element techniques.

4.4 Solder Joint Reliability via FEA

Several examples performed by ABAQUS finite element analysis (FEA) are presented as follows, to illustrate the significance of solder joint reliability assessment and highlight the points discussed above.

4.4.1 Life Prediction of Ford Joint Specimen

The Ford joint specimen (Pao *et al.*, 1992) is selected as a typical example to perform parametric studying under accelerated testing conditions since the joint specimen is mainly subjected to shear deformation and well calibrated by experimental data and finite element calculation. The geometry of the specimen and its finite element meshing are shown in Figure 4.2(a) and (b). The specimen is composed of an aluminum beam and an alumina beam which are soldered together by the solder joint. The aluminum and alumina are modeled as thermal elasticity whereas solder alloys are modeled as rate-dependent plasticity. All data of the materials are taken from references (Pao *et al.*, 1992).

Figure 4.3 shows an FEM result that reasonably agrees with experimental data and is totally identical with the FEM analysis data by user-defined subroutine for the same temperature profile (Pao *et al.*, 1992). Therefore, the finite element technique of rate-dependent plasticity in ABAQUS adapted in this chapter is verified to effectively model the viscoplastic behaviors of solder alloys. The details can be found in reference (Qian *et al.*, 1999). The isotropic hardening of solder alloys is assumed in this work.

Although this solder joint is well designed for pure shearing, there is unavoidable bending near the corners of the joint. Therefore, the range of inelastic shear strain can be different at different joint locations.

Corner elements 3, 7, 26, 30 and the middle element 13 on the joint cross section are selected, respectively, to investigate the effect of non-uniformity of the joint deformation, as shown in Figure 4.4(a). The inelastic strain ranges of the elements under the temperature profile [Figure 4.4(b)] is shown in

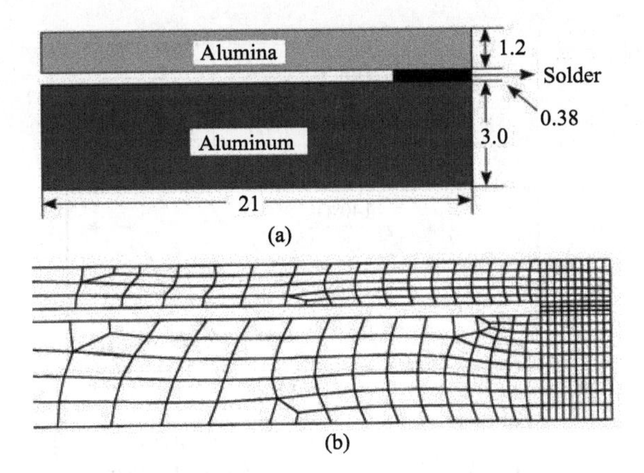

Figure 4.2 (a) Dimensions of Ford joint specimen; (b) 2D finite element meshing of the specimen (Reprinted with permission from Y.H. Pao, E. Jih, S. Badgley and J. Browning, "Thermal cyclic behavior of 97Sn-3Cu solder joints," *ASME 92-WA, EEP*, **21**, 1992. © 1992 ASME.)

Figure 4.4(c), which is calculated as a plane stress analysis. Based on the Coffin–Manson Equation 4.4 and materials constants in Table 4.1, predicted fatigue lives of five elements are shown in Figure 4.5 with different dwell time. The longer the dwell time, the shorter the fatigue life, due to creep damage accumulation during dwell period. Although the bending effect is not significant, the cycles of fatigue life at different locations of the joint are substantially different, see Figure 4.5.

This fact can be used to propose concept of bounds for the fatigue life of a solder joint, depending on the failure criteria and failure locations. Calculated fatigue life at the position where maximum range of inelastic strain is located will be a conservative lower bound that presents the shortest fatigue life of the joint and indicates a macroscopic crack initiation of the solder joint. In the case of the Ford joint specimen, the lower bound of its fatigue life at element 3 and the upper bound of the life at element 7 have been predicted due to the non-uniformity of shear deformation. The strain of middle element 13 gives the

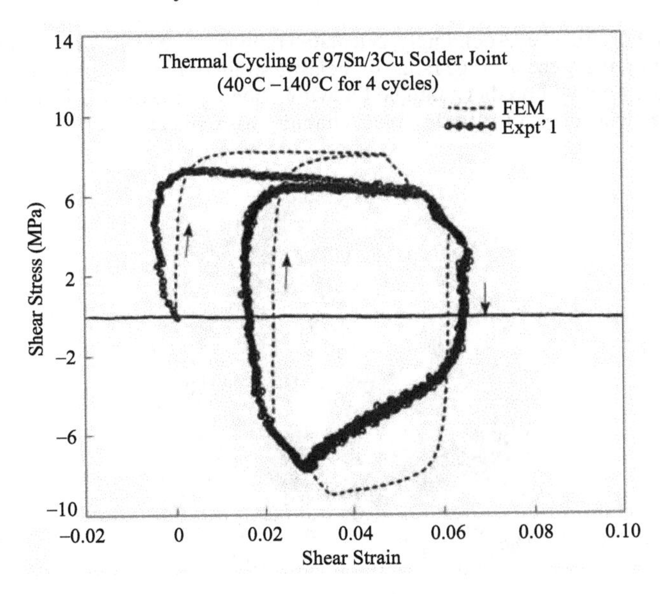

Figure 4.3 Comparison of FEM calculation with experimental data for thermal cyclic loading

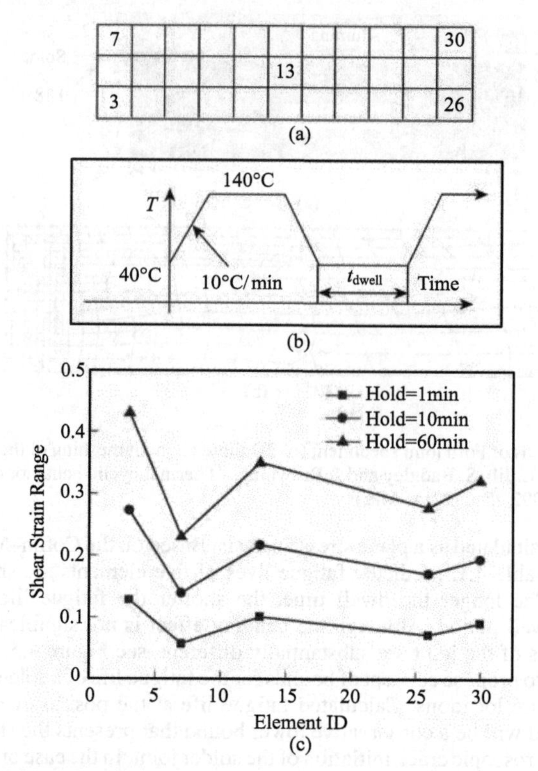

Figure 4.4 (a) Locations of five elements of the joint; (b) Applied temperature profile; (c) Ranges of inelastic shear strain of five elements with different hold time periods

average value over the joint. Correspondingly, calculated fatigue life at element 13 is indeed between two bounds. Alternatively, using the average range of inelastic strain or the average of inelastic strain energy density over a solder joint will predict a fatigue life within bounds. Therefore, the concept of bounds for fatigue life gives a clear margin for solder joint reliability assessment.

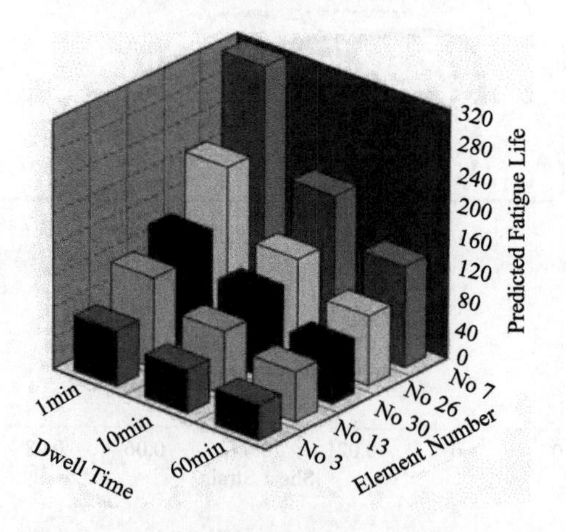

Figure 4.5 Predicted fatigue lives of five elements

4.4.2 Accelerated Testing: Insights from Life Prediction

In principle, the Coffin–Manson equation predicts the fatigue life of the solder joint on the material level, which indicates the material element located at maximum inelastic strain range or energy density reaches the failure criterion. In terms of damage mechanics (Qian *et al.*, 1999; Lemaitre, 1996), it is known that so-called tri-axial stress ratio, that is, the average stress of three principal stress components over the Mises stress, accelerates damage evolution and crack initiation. The effect thus shortens the fatigue life at the corners of the solder join where the solder material element is always subjected to three dimensional stress states. Therefore, it is necessary to distinguish the physical accelerated factor at the material level from the geometrical accelerated factor at the structural level for joint reliability assessment. Accelerated testing actually accelerates both factors simultaneously. Moreover, the physical accelerated factor can be determined by testing solder material samples. The geometrical accelerated factor must be determined by performing a finite element analysis, depending on the shape and size of solder joints.

Figure 4.5 shows that the average accelerated failure factor at each element is around 2.0 for a dwell time from 1minute to 60 minutes for the joint specimen. The geometry-induced local failure has an accelerated factor around 3 to 4.5. It is defined by the ratio of N_f at element 3. The factor decreases with the increase of dwell time since creep during dwell time causes the uniformity of deformation within a solder joint.

In order to investigate how the accelerated failure factor is amplified by accelerated test conditions, a set of special virtual tests via FE analysis was designed under different ramping rates without dwell time. Extreme temperatures were selected as 125 °C for maximum and − 40 °C for minimum. The ramping rate, that is both heating and cooling rate in this case, was changed from 0.5 °C to 600 °C per minute, as shown in Figure 4.5. The ramping frequency is defined as the inverse of one cycle time taken at heating and cooling. Figure 4.6 shows shear stress/strain curves of middle element 13 at three typical ramping rates of 1 °C/min, 10 °C/min, and 100 °C/min, respectively. Maximum magnitudes of shear stress increase with the increase of ramping rates, indicating viscoplastic deformation characterization. On the contrary, the ranges of inelastic shear strain significantly decrease with the increase of ramping rates. It is pointed out that only rate-dependent constitutive modeling can predict the stress and the strain characterization under different ramping rates. The rate-dependent plasticity or creep laws are not able to obtain this kind of mechanical response.

Figures 4.7 and 4.8 show how to calculate the ranges of inelastic Mises strain and inelastic strain energy density from the accumulated curves of the quantities with temperature cycles. It should be noted that the inelastic strain is accumulated twice per cycle. Moreover, the inelastic strain or energy generated

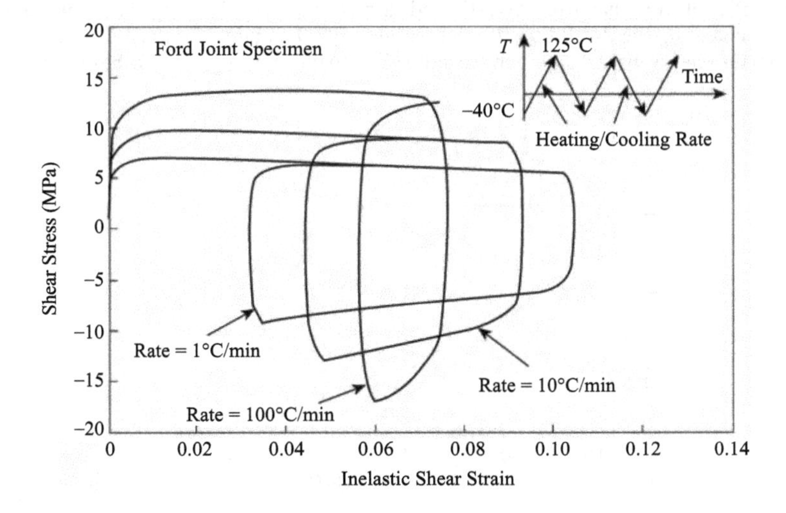

Figure 4.6 Shear stress/strain curves under accelerated tests with different ramping rates

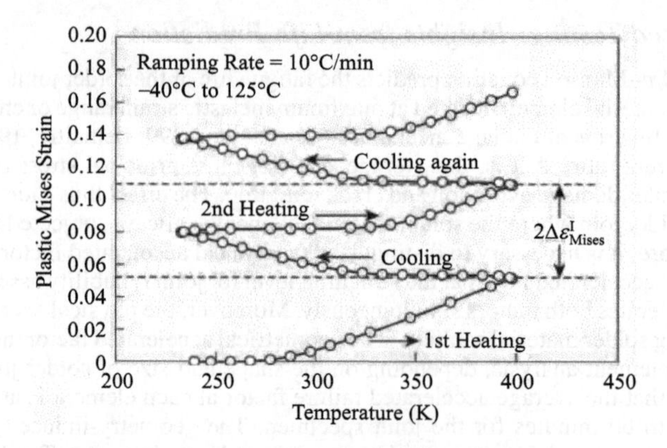

Figure 4.7 Calculation of inelastic strain range

during the first heating or cooling stage is relatively large but it does not contribute to the range calculation whereas it directly contributes to the calculation of the mean strain or energy, see Figure 4.6. From Equations 4.1 to 4.4, the fatigue life can be predicted based on the calculated ranges of inelastic Mises strain and strain energy density. The results are presented in Figure 4.9 illustrating the fact that the faster the ramping frequencies, the longer the fatigue life for strain-based approach.

However, the energy-based approach predicted a questionable tendency at low ramping frequencies and slow ramping rates. At very slow ramping rates, the energy-based approach predicts almost the same fatigue cycles. The reason of this phenomenon comes from the rate-dependent behavior of solder materials. When the ramping rate is decreasing, the stress decreases but the strain range increases. Inelastic strain energy density at low range of frequencies, therefore, is not changed significantly. If the energy density is divided by both the maximum stress and a factor 2, an effective strain range thus is obtained. Predicted fatigue life based on the effective strain range approach is almost the same as that from strain-based Coffin–Mason equation, as shown in Figure 4.9. The example shows that the energy density might not be a proper quantity for fatigue life prediction under some accelerated test conditions such as different ramping rates. In fact, it was found the energy density predicted a wrong tendency of fatigue life for fatigue tests under different frequencies.

The effect of dwell time has been investigated in a similar way, keeping the same ramping rate of 10 °C/min and changing the dwell time from 1 minute to 600 minutes. Figure 4.10 shows that the lives predicted from the energy-based approach are quite close to those from the strain-based approach for such

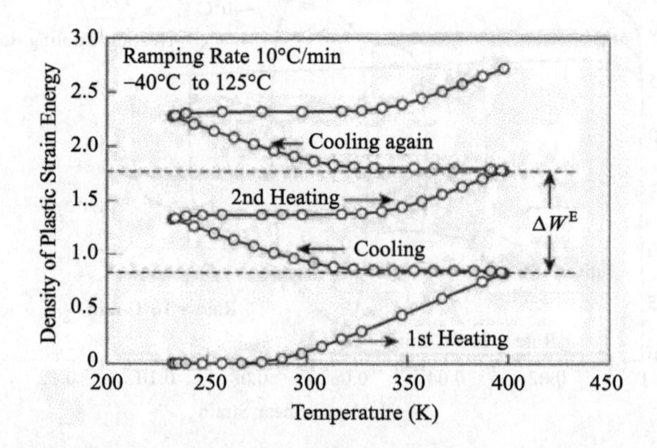

Figure 4.8 Calculation of the range of inelastic strain energy density

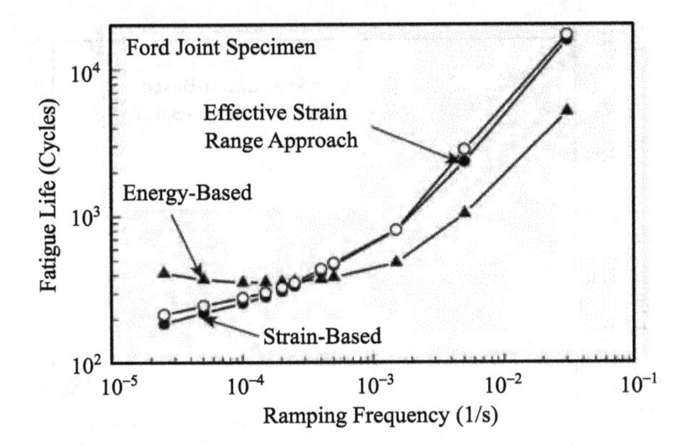

Figure 4.9 Fatigue life prediction via different approaches

type of accelerated testing. Furthermore, Figure 4.11 shows that life cycles decreases exponentially with the increase of the time of one cycle. Therefore, the time to failure, that is, the total test time of one sample, can be a proper parameter to measure accelerated factor.

The time to failure can be calculated by multiplying fatigue cycles by one cycle time. When the data of the time to failure were plotted against ramping frequencies, a surprising curve was obtained, as shown in Figure 4.12, for changing the ramping rate, and Figure 4.13 for changing the dwell time. There is an upper bound for accelerated testing, in addition to a linear relationship in a double logarithmic coordinate system.

The significance of this result indicates that it is not necessary to accelerate the fatigue test to equipment limitation. The fastest test will take about one week for both accelerated methods. For the joint specimen and extreme temperatures chosen, the total time of accelerated tests could be shortened from three months to one week; that is the shortest period for a thermal fatigue test. Maximum ramping rate is about 30 °C/min in this case whereas the shortest dwell time is about 15 minutes. These results are really meaningful to industry practices for the design of accelerated testing. If one wants to run an accelerated thermal cyclic test of an electronic package of less than one week, the alternative method must be used. For instance, isothermal mechanical fatigue with meaningful test procedure can be a fast process for solder joint reliability assessment of BGAs.

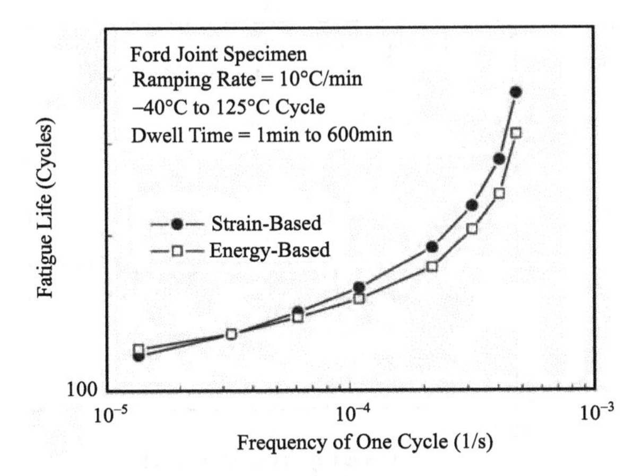

Figure 4.10 Fatigue life with different dwell time

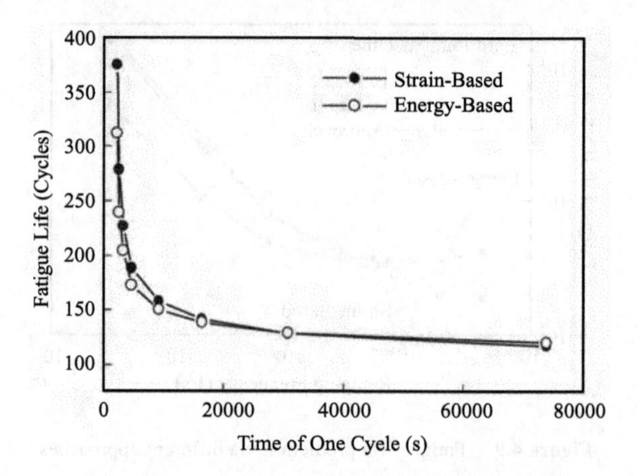

Figure 4.11 Fatigue life with the time of one cycle

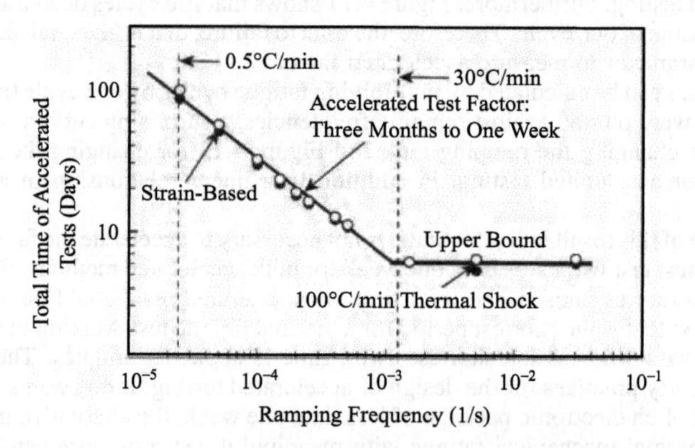

Figure 4.12 There is an upper bound for accelerated testing

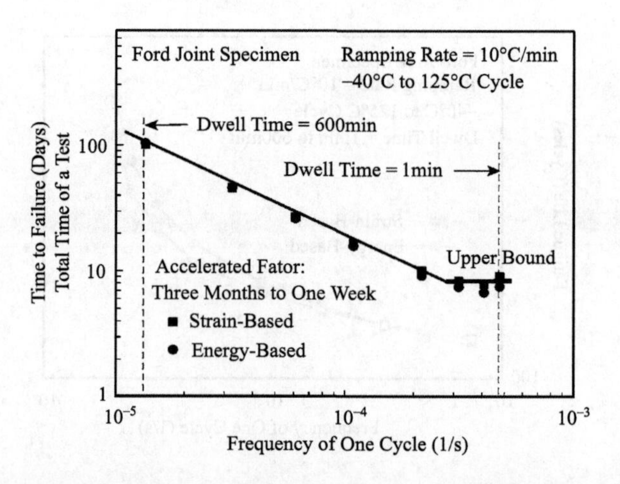

Figure 4.13 Accelerated testing factor via reducing dwell time

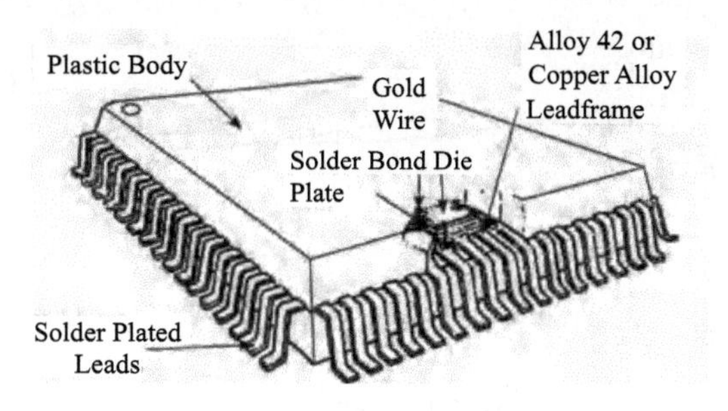

Figure 4.14 A typical PQFP package

4.4.3 Fatigue Life Prediction of a PQFP Package

A PQFP package was chosen to illustrate the importance of using unified or rate-dependent constitutive model for the methods of strain-based and energy-based life prediction. A typical PQFP package and its 2D finite element model are shown in Figure 4.14, and Figure 4.15(a) with local amplification of a solder joint, respectively.

The size of the package is the same as that used in reference (Syed, 1997). The solder material data were taken from experimental tests (Lau *et al.*, 1997). The thermo-mechanical material data of chip, FR-4, copper lead, and molding compound were taken from the CINDAS database. The shear stress/strain curve at point A, the toe of a solder joint, under the thermal cyclic loading with applied temperature profile in Figure 4.15(b) is shown in Figure 4.16(a). Moreover, the cyclic ratcheting was also predicted. This shows that there is a relatively large amount of mean shear strain that increases step-by-step with the increase of thermal cycles. Its effect on fatigue life needs to be explored for both experimental and theoretical studies in the future.

Figure 4.16(b) shows the result of equivalent stress/strain curve calculated by separated constitutive modeling, which is taken from reference (Gupta *et al.*, 1993) for a similar PQFP package at the same location of the solder joint. Although slight different solder data were used to calculate Figures 4.16(a)

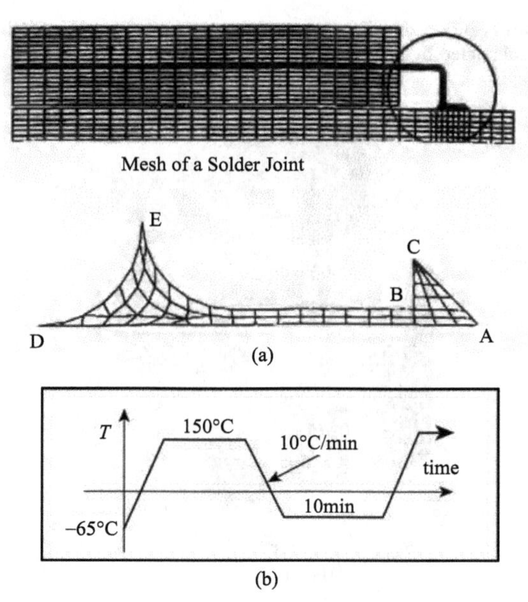

Figure 4.15 (a) 2D finite element model of the PQFP; (b) Applied temperature profile

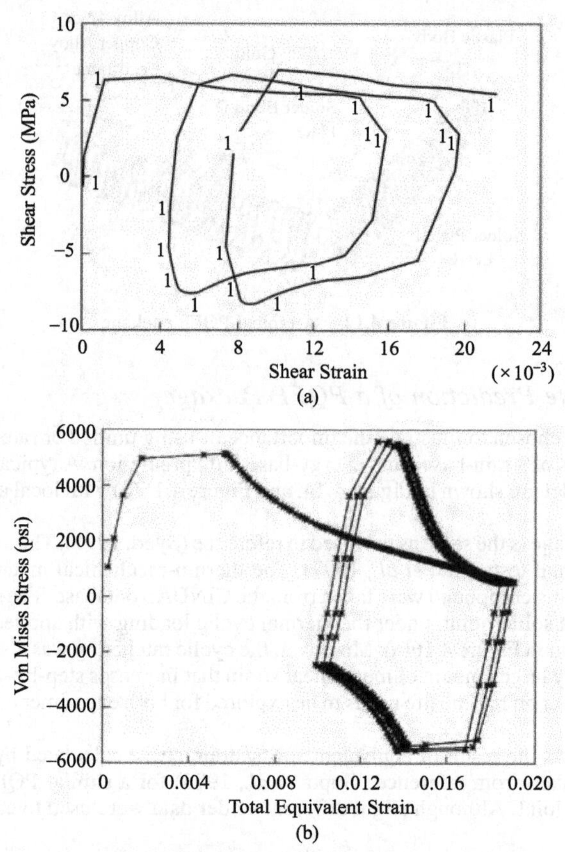

Figure 4.16 (a) Shear stress and shear stain at toe as calculated by unified constitutive modelings; (b) Shear stress and shear strain at toe as calculated by separated constitutive modeling

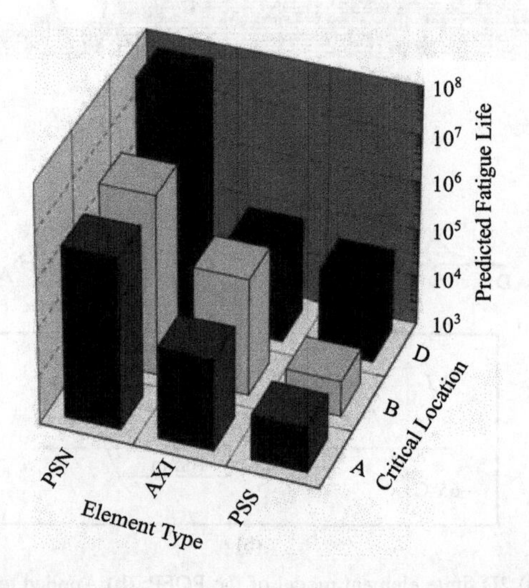

Figure 4.17 Fatigue life prediction via different 2D FEM models

and (b), the qualitative difference in terms of the shape of two hysteresis curves is significant due to totally different constitutive modeling methods. This comparison illustrates the importance of correctly using the rate-dependent constitutive modeling to calculate the hysteresis curve. The significant difference in the calculation of strain energy density and inelastic strain during thermal cycling could produce a substantial difference in the fatigue life prediction due to the power-law characterization of Coffin–Manson Equations 4.1 to 4.4.

Another important issue has been ignored in literature concerning different 2D finite element modeling for the reliability assessment of solder joints. Figure 4.17 shows enormous differences in fatigue life prediction where different 2D finite element models were used. Actually, this situation exists in many reports of finite element modeling. The plane stress (PSS) model predicts the shortest life, a lower bound. On the other hand, the plane strain (PSN) model predicts the longest life, an upper bound. The life prediction based on axisymmetrical modeling (AXI) is, as expected, between the two bounds. Therefore, considering the time and cost saving of running finite element modeling, 2D axisymmetric modeling is recommended, at least, for rapid reliability assessment.

4.5 Life Prediction of Flip-Chip Packages

Flip-chip packages become more important due to the development of potential chip-on-board technology and the successful utilization of underfill encapsulant. The fatigue life of solder joints (Lau *et al.*, 1997; Yeh *et al.*, 1996; Qian *et al.*, 1999; Popelar, 1997; Suryanarayana *et al.*, 1991; Schubert *et al.*, 1997) of a flip-chip can be extended tremendously.

Figure 4.18(a) and (b) show schematically flip-chip packages with a ball pitch size from 250 microns to 50 microns. The basic geometric dimensions of the chip and FR-4 are the same as in Reference. Instead of plane stress/plane model, axis-symmetrical 2D finite element has been selected as the geometric model. Over 4000 elements were used with the local fine meshing of solder balls, underfill corners, and interfaces. Currently, most of the analyses still use thermal elastic properties of underfill encapsulants to perform finite element simulation of flip-chip packages. The reason can be attributed to the lack of experimental data of nonlinear properties of underfills. Consistent and systematic experimental data of underfill FP4526 have been generated by combining a 6-axis mini tester with a well-manufactured thin strip specimen (Qian *et al.*, 1997, 1999). The nonlinear behaviors of the underfill have been explored in terms of temperature dependence of Young's modulus and CTE (Coefficient of Thermal Expansion), viscoelastic and viscoplastic properties (Qian *et al.*, 1999). The thermal-mechanical data of chip and FR-4 are taken from references (Wang *et al.*, 1998). The database is then incorporated into the ABAQUS material model library with the nonlinear viscoplastic behaviors of eutectic solder and underfill FP4526 for rate-dependent constitutive modeling (Qian *et al.*, 1999). All material constants for fatigue life prediction were taken from Table 4.1. With the implementation, the nonlinear finite element analysis of the flip-chip package with ball pith of 250-microns has been verified by the moiré measurement for simple creep tests.

4.5.1 Fatigue Life Prediction with and without Underfill

Deformation over a solder ball becomes more uniform due to the enhancement of the underfill encapsulant. The deformation mode of solder balls of a flip-chip with underfill also changes from shear-dominated mode for a flip-chip without underfill to more uniform and 3D deformation states due to different stress/strain boundary conditions around a solder ball. As a result, the range of inelastic Mises strain of the outmost solder ball under thermal cyclic loading has been substantially reduced due to the underfill strengthening mechanism. Therefore, the thermal fatigue life of flip-chip packages has been increased tremendously.

Nonlinear finite element analysis showed inelastic strain localization inside the outmost solder ball of the flip-chip package without underfill encapsulant. Therefore, early failure can be induced by the localized deformation as the initiation of cracks. Predicted failure location is very close to the corner B, as shown in Figure 4.18(b). Therefore, short fatigue life has been predicted. Calculations also indicated that inelastic strain distribution becomes much more uniform in the underfilled flip-chip inside every solder ball. The underfill indeed transfers the shear deformation from one ball to another. Moreover, the

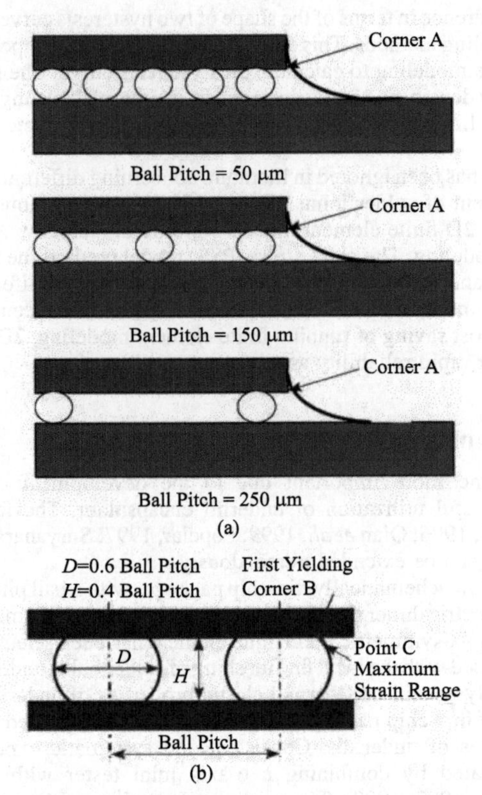

Figure 4.18 (a) Schematic flip-chips with different ball pitch sizes; (b) Ball size and critical point at outmost solder ball

critical point of the flip-chip package with the underfill is also shifted from the corner B of the outmost solder ball to the interface C between the solder ball and the underfill as shown in Figure 4.18(b), with matches qualitatively the experimental observation of crack initiation.

The flip-chip was subjected to thermal cyclic loading from −40 °C to 100 °C with the ramping rate of 10 °C/min and the dwell time of 10 minutes. For the flip-chip without underfill, predicted life cycles, based on inelastic shear strain range and Mises strain range, respectively, are very close to each other due to shear dominated deformation. In the case of flip-chip with underfill, the range of inelastic Mises strain must be used for life prediction due to 3D-deformation states. The fatigue life due to an underfill strengthening mechanism increases roughly ten times for the flip-chip package, as summarized in Table 4.2.

Table 4.2 Fatigue life of flip-chip with/without underfill

Flip-Chip	Without Underfill		With Underfill
Inelastic Strain Range	$\Delta\gamma^{I}$ 0.0429	$\Delta\varepsilon^{I}_{Mises}$ 0.02293	$\Delta\varepsilon^{I}_{Mises}$ 0.00732
Fatigue Life (cycles)	618	720	6758

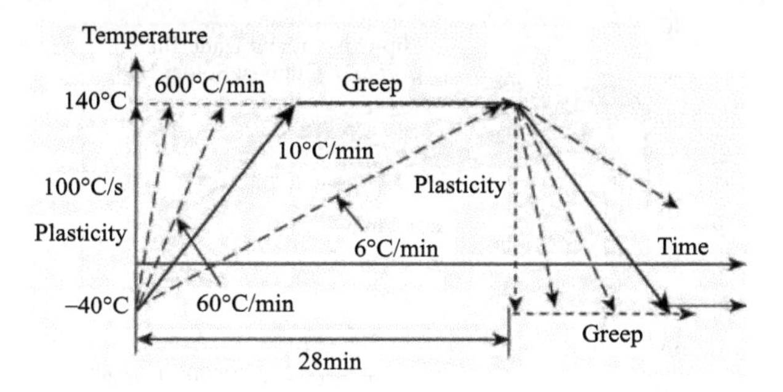

Figure 4.19 Path-dependence of shear stress/strain curves via unified constitutive modeling

4.5.2 Life Prediction of Flip-Chips without Underfill via Unified and Separated Constitutive Modeling

In this case, the flip-chip without underfill is equivalent to a type of BGA package. In order to compare separated with unified constitutive modeling, a special design loading paths is shown in Figure 4.19. During separated modeling, approximate temperature profile (Dasgupta *et al.*, 1992) has to be adopted to match industrial thermal cyclic profile. Due to the difficulty of calculating the rate/time-independent plastic strain during temperature ramping, the approximation procedure of applying temperature was taken as jumping to extreme temperatures as plasticity and followed by creep during a dwell time with is equal to the temperature ramping period (Dasgupta *et al.*, 1992; Gupta *et al.*, 1993), as shown in Figure 4.19. Many authors used a similar procedure to calculate the stress/strain hysteresis curves during a temperature ramping stage, based on separated constitutive modeling. Therefore, five paths were selected for unified constitutive modeling by changing temperature ramping rates from 6 °C/min to 600 °C/min (as virtual plasticity), whereas the time of one-cycle and extreme temperatures are kept unchanged, as shown in Figure 4.19. As expected, the separated modeling will predict the same results for all loading paths.

On the other hand, since the nonlinear property of solder alloys is highly rate/time/temperature-dependent, in literature, authors have used quite different curves as plasticity for eutectic solder, lacking clearly consistency and uniqueness. Three tensile curves (Busso *et al.*, 1992) at the strain rates of 1.67×10^{-2}/s, 1.67×10^{-4}/s were taken as rate-independent plastic definition at the temperatures of −55 °C, –15 °C, 22 °C, 50 °C, 75 °C, and 125 °C, respectively. The hyperbolic law of creep with the same material constants used in reference (Wang *et al.*, 1992) was applied to calculate the creep deformation. All stress/strain curves are plotted at the corner B of outmost solder ball. Figure 4.20 shows that the strain

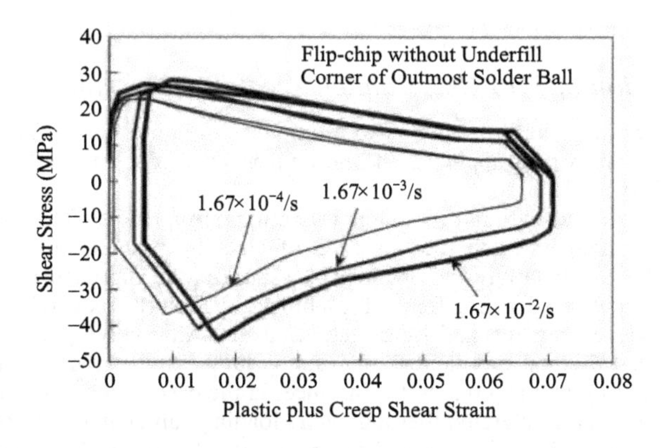

Figure 4.20 Shear stress/strain curves via separated constitutive modeling

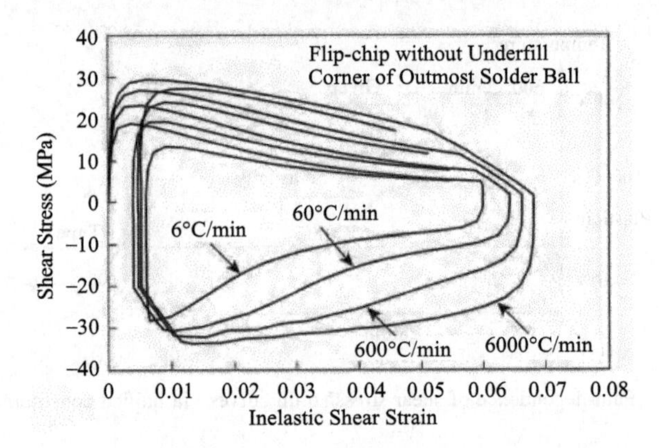

Figure 4.21 Path-dependence of shear stress/strain curves via unified constitutive modeling

range of plastic plus creep deformation is shifted with the increase of mean strain when the tensile stress/strain curve at a higher strain rate was used. Therefore, the predicted fatigue life is almost the same if a strain-based approach is used. However, the density of strain energy significantly increased for higher rate case. The 40% difference of predicted fatigue lives by the energy-based approach was found among the three cases.

Figure 4.21 shows shear stress/strain curves at the corner B of the outmost solder ball, obtained by unified constitutive modeling for planned four loading paths. Inelastic shear strain range increases with the increase of ramping rates. The qualitative difference of the hysteresis curve shape can be found by comparing Figure 4.21 with the curve of 600 °C/min (100 °C/s, virtual plasticity) path in Figure 4.21 by unified modeling. The curves in Figure 4.21 match the qualitatively nonlinear characterization measured experimentally by solder joint specimens subjected to thermal cycling. Clearly, inelastic shear strain was generated mainly during heating and cooling stages. The inelastic energy density, however, is very sensitive to temperature ramping rates. In terms of fatigue life prediction, the 400% difference of fatigue lives among the four cases has been predicted based on energy approach. Although these findings need to be verified by well-designed accelerated tests, the conclusion is that a strain-based life prediction approach is most likely combined with separated constitutive modeling to surpass its intrinsic deficiency. In contrast, an energy based life prediction approach could mislead the results obtained by unified constitutive modeling. The combination of an energy-based approach with separated constitutive modeling could result in significant scatter of life prediction data, therefore, is not recommended in this chapter.

4.5.3 Life Prediction of Flip-Chips under Accelerated Testing

Accelerated testing conditions with the same dwell time of 10 minutes and different ramps rates from 0.3 °C/min to 600 °C/min have been applied to the flip-chip package of 50-micron ball pitch under thermal cycling from −40 °C to 125 °C.

Figure 4.22 shows the predicted fatigue lives against ramping frequencies. The energy-based approach produced questionable life prediction for this kind of tests, as pointed out earlier. A logarithmic linear relationship of the time to failure was obtained by a strain-based approach, as shown in Figure 4.23. The time-to-failure of accelerated testing for the flip-chip package can be reduced from three years to one week. An upper bound was not reached by current analysis. Alternatively, the upper bound can be considered from coupled thermal-mechanical analysis by temperature gradient inside a real flip-chip package. When the ramping rate is too fast, for instance, in the case of thermal shock, the temperature gradient can be so large that accelerated thermal cyclic loading can change failure modes totally.

The details of characterization of stress and strain at critical location C are shown in Figure 4.24 for accumulated Mises strain and Figure 4.25 for the normal stress/strain curve under the thermal

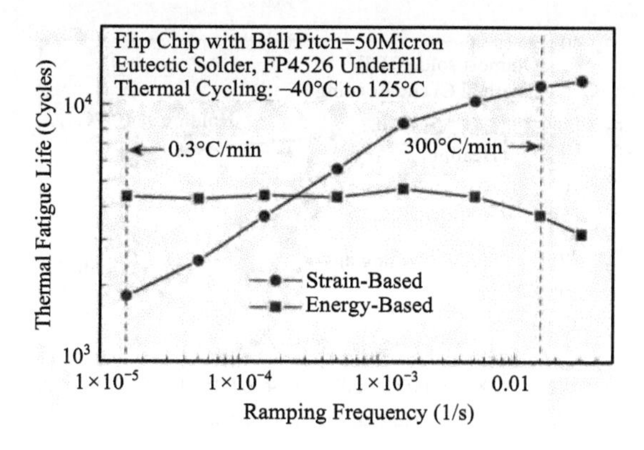

Figure 4.22 Fatigue life prediction of the flip-chip under different ramping rates of thermal cyclic loading

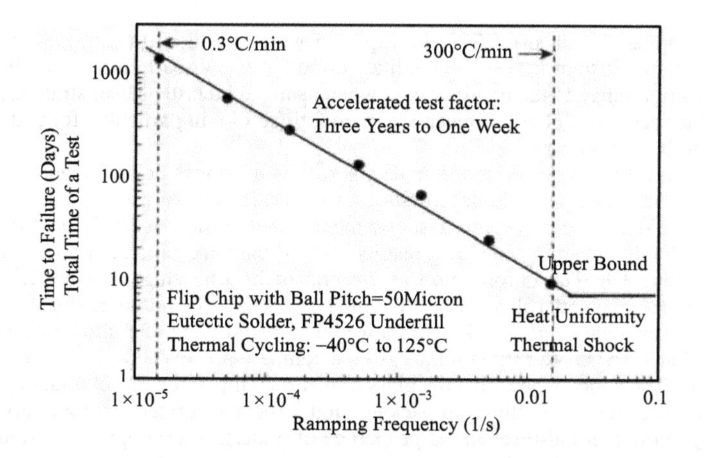

Figure 4.23 Accelerated testing factor by increasing ramping rates

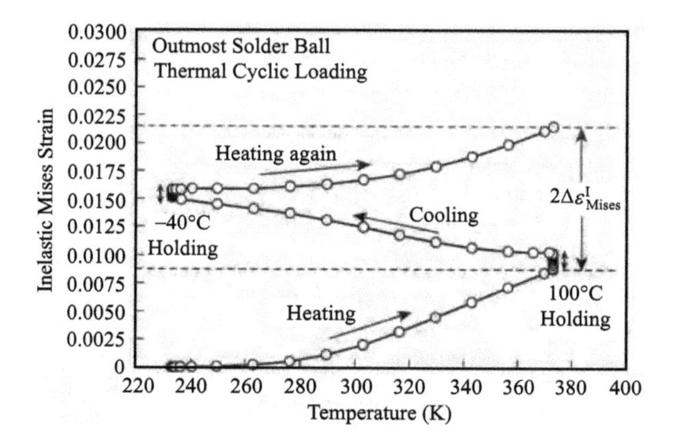

Figure 4.24 Accumulated inelastic Mises strain with temperature cycling

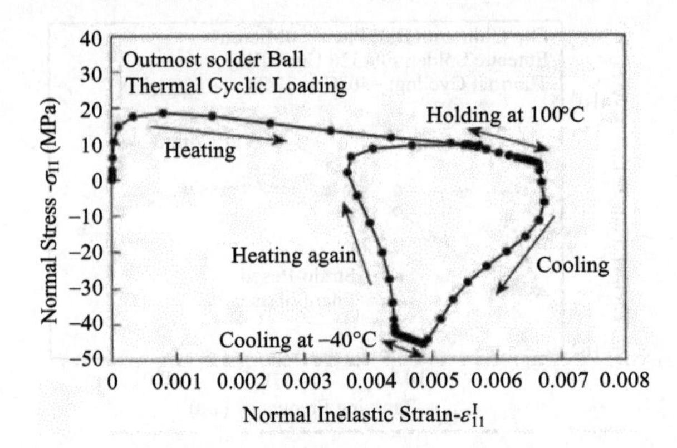

Figure 4.25 Curve of normal stress/strain at point C

cyclic loading. Compared with the BGA-like flip-chip without underfill (Figures 4.20 and 4.21), the characterization of stress/strain curves are qualitatively different. In particular, mean inelastic strain is about half of the strain range in the previous case whereas mean inelastic Mises strain is greater than the strain range in later case. The effect of mean strain on fatigue life, in particular, for underfilled flip-chip packages needs to be investigated.

On the other hand, the corner A of the underfill side is a critical point for underfill delamination (Madenci *et al.*, 1998). The inelastic deformation of underfill FP4526 accumulated with temperature cycling is shown in Figure 4.26. The picture shows the fact that most of underfill inelastic deformation is generated during the first heating state. Test results showed the maximum strain of underfill FP4526 at 100 °C is only about 2%. The deformation at the end of heating stage reaches 1.5% according to Figure 4.25. Therefore, the underfill at corner A could be broken at the initial thermal loading. Nonlinear finite element analysis (Qian *et al.*, 1999) also showed that the fatigue life of flip-chips with the underfill was reduced seriously when the maximum extreme temperature exceeds 100 °C due to significant material softening. The conclusion is that the glass transition temperature T_g of underfill encapsulants is critical for the fatigue life of the flip-chip package under the accelerated testing temperatures near or above the T_g since the thermal-mechanical properties of underfills are degraded seriously around T_g. Therefore, an over-accelerated testing temperature is not necessary, which can result in an unrealistic requirement for underfill material selection.

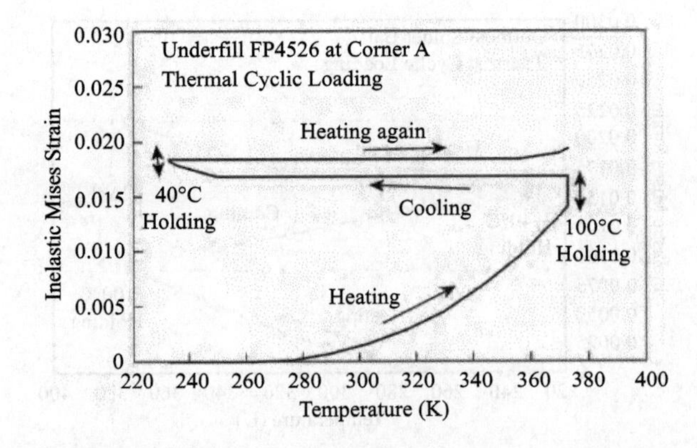

Figure 4.26 Accumulation of inelastic Mises strain of underfill

4.6 Chapter Summary

The life prediction methodology of solder joints under thermal cyclic loading has been summarized and critically reviewed in this chapter. An energy-based approach could mislead the life prediction of solder joints under some accelerated conditions such as different ramping rates. An energy-based approach might not be suitable for life prediction of accelerated testing conditions, as shown in this chapter and literature (Solomon *et al.*, 1995, 1996). A strain-based approach is strongly recommended to combine with unified constitutive modeling for the reliability assessment of solder joints. The constitutive modeling plays an important role in the life prediction in terms of correct calculation of inelastic strain ranges, inelastic strain energy density of hysteresis loop, and creep-plasticity interaction, under various accelerated test conditions. The separated constitutive modeling of rate-independent plasticity and creep should be used cautiously, in particular, considering the inconsistency and difficulty for the calibration of model parameters and the finite element analysis of solder joints subjected to thermal cyclic loading. The proposed bound concept of life prediction and the upper bound of accelerated testing factors obtained in this chapter are very meaningful for industry practice. However, further investigation is definitely needed. Finally, the investigations of underfill strengthening mechanisms and life prediction of flip-chip packages give valuable insights for advanced package design and reliability assessment.

References

Basaran, C., Desai, C.S. and Kundu, T. (1998) Thermo-mechanical finite element analysis of problems in electronic packaging using the disturbed state concept: part I – theory and formulation, part II – verification and application, *Transaction Journal of Electronic Packaging*, ASME **120**:41–53.

Bolton, S.C., Mawer, A.J. and Mammo, E. (1995) Influence of plastic ball grid array design/materials upon solder joint reliability, *International Journal of Microcircuits & Electronic Packaging* **18**(2):109–121.

Busso, E.P., Kitano, M. and Kumazawa, T. (1992) A visco-plastic constitutive model for 60/40 tin-lead solder used in IC package joints, *ASME Transaction Journal of Engineering Materials & Technology* **114**:331–337.

Busso, E.P., Kitano, M. and Kumazawa, T. (1994) Modeling complex inelastic deformation processes in IC packages' solder joints, *ASME Transaction Journal of Electronic Packaging* **116**:6–15.

Chaboche, J.L. (1989) Constitutive equations for cyclic plasticity and cyclic viscoplasticity, *International Journal of Plasticity* **5**:247–277.

CINDAS Database, *Semiconductor Research Corporation*, Purdue University, 1998.

Cocks, A.C.F. and Ashby, M.F. (1980) Intergranular fracture during power-law creep under multiaxial stresses, *Metal Science* **14**:395–402.

Coffin, L.F., Jr. (1954) A study of the effects of cyclic thermal stresses on a ductile metal, *ASME Transactions* **76**:931–950.

Darveaux, R., Banerji, K., Mawer, A., and Dody, G., Reliability of plastic ball grid array assembly, *Ball Grid Array Technology*, Lau J.H. ed, McGraw-Hill, New York, 1995.

Dasgupta, A., Oyan, C., Barker, D. and Pecht, M. (1992) Solder creep-fatigue analysis by an energy-partitioning approach, *ASME Transaction Journal of Electronic Packaging* **114**:152–160.

Dougherty, D.J., Culbertson, D.M. and Fusaro, J.M. (1998) Life prediction methodology for a discrete transistor package, *International Journal of Microcircuits & Electronic Packaging* **21**(1):59–66.

Engelmaier, W. (1989) Surface mount solder joint long-term reliability: design, testing, and prediction, *Soldering and Surface Mount Technology* **1**:12–22.

Frear, D.R., Jones W.B. and Kinsman, K.R. (1991) *Solder Mechanics, A State of the Art Assessment*, TMS, New Mexico, 1991.

Frear, D.R., Burchett, N.B. and Neilsen, M.K. (1997) Life prediction modeling of solder interconnects for electronic systems, *Advances in Electronic Packaging*, ASME EEP **19-2**:1515–1522.

Freed, A.D. and Walker, K.P. (1990) Steady-state and transient zener parameter in viscoplasticity: drag strength versus yield strength, *Applied Mechanics Review* **43**:328–436.

Fu, C, Ume, I.C. and McDowell, D.L. (1996) Thermo-mechanical stress analysis of plated-through holes in PWB using internal state variable constitutive models, *Sensing, Modeling and Simulation in Emerging Electronic Packaging*, ASME, EEP **17**:65–72.

Gektin, V. and Bar-Cohen, A. (1998) Coffin-Manson based fatigue analysis of underfilled DCAs, *IEEE Transactions*, CPMT **A21**:577–584.

Gupta, V.K. and Barker, D.B. (1993) Influence of surface mount lead end geometry on fatigue life, *ASME 93* WA, EEP-5, 1–7.

Halford, G.R., Hirschberg, M.H. and Manson, S.S. (1973) Temperature effects on the strain-range partitioning approach for creep-fatigue analysis, *ASTM STP 520*, PA, 658–667.

Halpern, M. (1988) Pushing the design envelope with CAE, *ASME Mechanical Engineering*, November, 66–71.

He, J., Morris, W.L., Shaw, M.C., Mather, J.C. and Sridhar, N. (1998) Reliability in large area solder joint assemblies and effects of thermal expansion mismatch and die size, *International Journal of Microcircuits & Electronic Packaging* **21**(3):297–304.

Hong, B.Z., Yuan, T.D. and Burrell, L. (1996) An isothermal fatigue analysis of solder joints in a convective CBGA package under power cycling, *Sensing, Modeling and Simulation in Emerging Electronic Packaging*, ASME, EEP **17**:39–46.

Hong, B.Z. and Yuan, T.D. (1998) Integrated flow-thermo-mechanical analysis of solder joints fatigue in a low air flow C4/CBGA package, *International Journal of Microcircuits & Electronic Packaging* **21**(2):137–144.

Lau, J.H. (1991) *Solder Joint Reliability - Theory and Applications*, Van Nostrand Reinhold, New York.

Lau, J.H. and Pao Y.-H. (1997) *Solder Joint Reliability of BGA, CSP, Flip-Chip, and Fine Pitch SMT Assemblies*, McGraw-Hill, New York.

Lead-free Solder Project, NCMS report, Ann Arbor, Michigan, pp. 1–80, (1997).

Lee, W.W., Nguyen, L.T. and Selvaduray, G.S. (2000) Solder joint fatigue models – critical review and applicability to chip scale packages, *IEEE Transaction, Microelectronics. Reliability* **40**(2):231–244.

Lee, S.M. (1995) Cavitational failure phenomenon in Pb-Sn eutectic solders, *Japanese Journal of Applied Physics* **34**: L1475–L1477.

Leicht, L. and Skipor, A. (1999) Mechanical cycling fatigue of PBGA package interconnects, *International Journal of Microcircuits and Electronic Packaging* **22**(1):57–61.

Lemaitre, J. (1996) *A Course on Damage Mechanics*, Springer, New York, 1–100.

Madenci, E., Shkarayev, S. and Mahajan, R. (1998) Potential failure sites in a flip-chip package with and without underfill, *ASME Transaction Journal of Electronic Packaging* **120**:336–341.

Manson, S.S. (1965) Fatigue: a complex subject – some SPL approximations, *Experimental Mechanics* **5**:193–226.

Mei, Z. (1996) TSOP Solder Joint Reliability, *Proceedings of 46th Electronic Components and Technology*, 1232–1238.

Mei, Z. (1999) Hewlett-Packard, Private Communication.

Morrow, J.D. (1964) Cyclic plastic strain energy and fatigue of metals, *ASTM STP* **378**:45–87.

National Technology Roadmap for Semiconductors, Semiconductor Industry Association, San Jose, California, 1997.

Pao, Y.H. (1992) A fracture mechanics approach to thermal fatigue life prediction of solder joints, *IEEE Transaction on CHMT* **15**:559–570.

Pao, Y.H., Jih E., Badgley S. and Browning J. (1992) Thermal cyclic behavior of 97Sn-3Cu solder joints, *ASME 92-WA*, EEP **21**:1–6.

Popelar, S.F. (1997) A parametric study of flip chip reliability based on solder fatigue modeling, *IEEE/CPMT International Electronic Manufacturing Technology Symposium*, 299–307.

Qian, Z. (1991) Non-classic plasticity: a micro-mechanical based constitutive model for creep-plasticity iteraction, Ph. D. Dissertation, Chongqing University, 1–90.

Qian, Z. and Liu, S. (1997) A unified viscoplastic constitutive model for tin-lead solder joints, *Advances in Electronic Packaging*, ASME EEP **19-2**:1599–1604.

Qian, Z, Lu, M., Wang, J. and Liu, S. (1997) Testing and constitutive modeling of thin polymer films and underfills by a 6-axis submicron tester, *Applications of Experimental Mechanics to Electronic Packaging*, ASME EEP **22**/AMD **226**:105–112.

Qian, Z., Ren, W. and Liu, S. (1999) A damage coupling framework of unified viscoplasticity for the fatigue of solder alloys, *ASME Transaction Journal of Electronic Packaging* **121**, September.

Qian, Z. and Liu, S. (1999) Implementation of unified viscoplasticity with damage evolution and its application to electronic packaging, *InterPack'99*, 1–8, Hawaii, June.

Qian, Z., Wang J., Yang, J. and Liu, S. (1999) Visco-Elastic-Plastic properties and constitutive modeling of underfills, *IEEE Transactions on Components and Packaging Technology* **22**:152–157.

Qian, Z., Lu, M., Ren, W., and Liu, S. (1999) Fatigue life prediction of flip-chips in terms of nonlinear behaviors of solder and underfill, *Proceeding of 49th Electronic Component and Technology Conference*, 141–148.

Ren, W., Qian, Z., Lu, M., Liu, S. and Shangguan, D. (1997) Thermal mechanical properties of two solder alloys, *Applications of Experimental Mechanics to electronic Packaging*, ASME EEP **22**/AMD **226**:125–130.

Schubert, A., Dudek, R., Auersperg, J., Vogel, D., Michel, B. and Reichl, H. (1997) Thermo-mechanical reliability analysis of flip-chip assemblies by combined microdac and the finite element method, *Advances in Electronic Packaging*, ASME EEP **19-2**:1647–1654.

Schubert, A., Michel, B. *et al.* (1999) Do chip size limits exist for DCA? *Proceedings of International Symposium on Advanced Packaging Materials*, March 14–17, Chateau Elan, GA, 150–157.

Solomon, H.D. (1986) Fatigue of 60/40 Solder, *IEEE Transaction CHMT* **9**:423–432.

Solomon, H.D. (1993) Life prediction and accelerated testing, *The Mechanics of Solder Alloy Interconnects*, S.N. Burchett, *et al.* eds, Van Nostrand Reinhold, 199–313.

Solomon, H.D. and Tolksdorf, E.D. (1995) Energy approach to the fatigue of 60/40 solder: Part I – influence of temperature and cycle frequency, *ASME Transaction Journal of Electronic Packaging* **117**:130–135.
Solomon, H.D. and Tolksdorf, E.D. (1996) Energy approach to the fatigue of 60/40 solder: Part II—influence of hold time and asymmetric loading, *ASME Transaction Journal of Electronic Packaging* **118**:67–71.
Subrahmanyan, R., Wilcox, J.R. and Li, C.Y. (1989) A damage integral approach to thermal fatigue of solder joints, *IEEE Transaction on CHMT* **12**:480–491.
Suryanarayana, D., Hsiao, R., Gall, T.P. and McCreary, J.M. (1991) Enhancement of flip-chip fatigue life by encapsulation, *IEEE CHMP* **14**:218–223.
Syed, A.R. (1997) Factors affecting creep-fatigue interaction in eutectic Sn/Pb solder joints, *Advances in Electronic Packaging*, ASME EEP **19-2**:1535–1542.
Wang, J., Qian, Z., Zou, D. and Liu, S. (1998) Creep behavior of a flip-chip package by both FEM modeling and real time moiré interferometry, *ASME Transaction Journal of Electronic Packaging* **120**:179–185.
Yang, L.Y. and Mui, Y.C. (1998) Solder joint reliability study for plastic ball grid array packages, *International Journal of Microcircuits & Electronic Packaging* **21**(1):100–108.
Yeh, C.P., Zhou, W.X. and Wyatt, K. (1996) Parametric finite element analysis of flip chip reliability, *International Journal of Microcircuits & Electronic Packaging* **19**(2):120–127.
Zhang, X.W., Lee, S. and Ricky, W. (1998) Critical issues in computational modeling and fatigue life analysis for PBGA solder joints, *International Journal of Microcircuits & Electronic Packaging* **21**(3):253–261.

5

Multi-Physics and Multi-Scale Modeling

During the manufacturing and testing of microsystems, issues of multi-physics and multi-scale modeling are involved due to the intrinsic nature of nano-micro scales, and multi-physics behaviors occurring in those processes. Despite the advances made in the modeling of the structural, thermal, mechanical, and transport properties of materials at the macroscopic level (finite element analysis of complicated structures), there remains tremendous uncertainty about how to predict many properties of interest. For instance, exploiting the tremendous physical and mechanical properties of new nano-materials by understanding materials at atomic, molecular, and supramolecular levels is useful for designing devices including nanoscale materials and structures (Gates *et al.*, 2005). In addition, optical, thermal, electrical, and deformation/stress occur concurrently, which requires a coupled-field analysis to determine the combined effects of multiple physical phenomena (fields) on a design.

5.1 Multi-Physics Modeling

In the development of microsystems such as microelectromechanical systems (MEMS), multi-physics effects are inherent and absolutely they must be addressed. Microsystems have components with micrometer dimensions. Extensive applications for these devices exist in both commercial and industrial systems. Well-known components such as integrated silicon pressure sensors, accelerometers, and motion detectors have been used for several years in automotive and industrial applications. Research activity in microfluidics is changing medical-diagnostic processes such as DNA analysis, and it is spurring the development of successful commercial products. These microminiature systems (some even thinner than a human hair) derive their function from the interrelated effects of stress, temperature, electrostatic, piezoelectric, and electromagnetic effects. Two numerical techniques are available for combining physics fields: direct-coupled and sequential-coupled fields (Haiping *et al.*, 2009).

5.1.1 Direct-Coupled Analysis

Direct-coupled analysis assembles all the physics fields as finite-element equations in one matrix and solves the matrix as a whole. One example of a direct-coupled field is the combination of thermal and electrical effects to study Joule heating produced by electromagnetic energy passing through a resistive or dielectric material. In certain transducers, piezoelectric direct coupling of electrical and mechanical physics is useful in determining both the amount of deformation resulting from an applied voltage and

Modeling and Simulation for Microelectronic Packaging and Integration: Manufacturing, Reliability and Testing, Second Edition. Sheng Liu and Yong Liu.

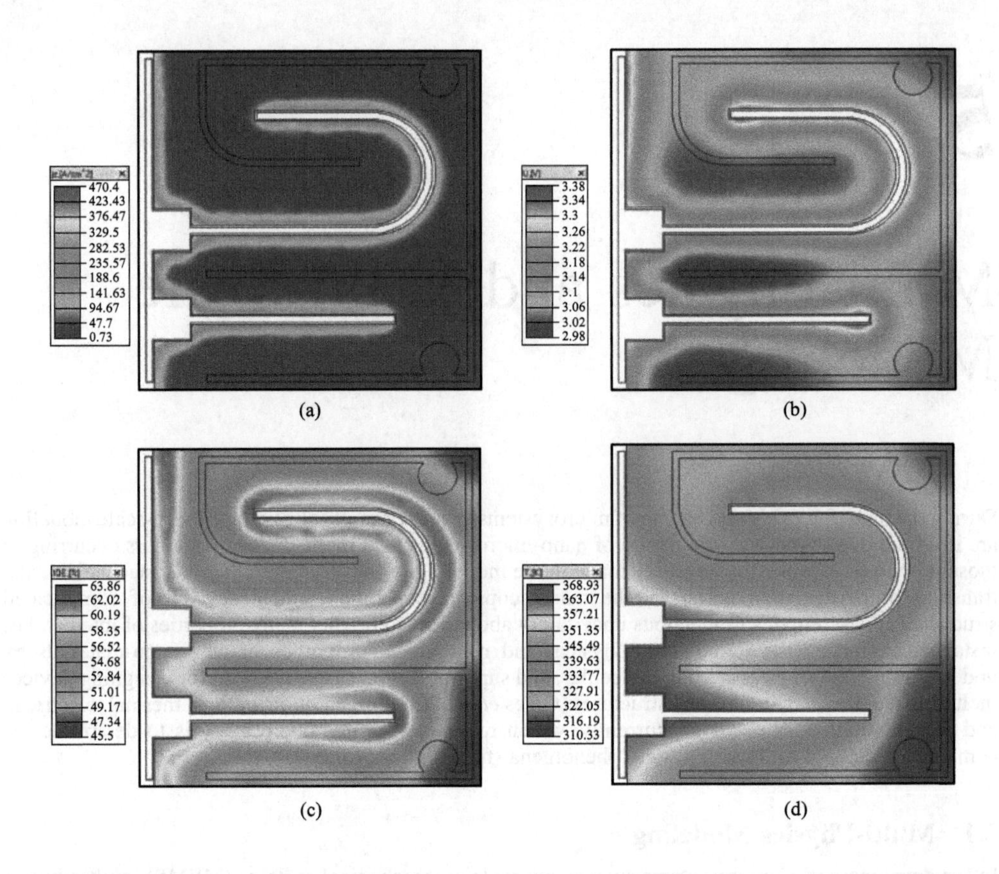

Figure 5.1 Simulation results of LED chip, all results are displayed on active layer: (a) current density distribution; (b) p–n junction bias distribution; (c) IQE distribution; and (d) temperature distribution (*Color version of this figure is available online*)

vice versa, as in piezoelectric ignition devices. In these types of analyses, all physics fields can be accounted for in a single solution.

For example, an LED heterostructure (Liu *et al.*, 2008; Alan *et al.*, 2003; Wang *et al.*, 2008) is considered as a set of planar semiconductor layers and electrodes. Electrode pads are considered as a fragment of metal electrodes where the electric potential has a given constant value. The considerable difference in the scales of thickness of the active region and contact layers allows us to model injection of carriers into the active region and their recombination as a separate one-dimensional problem. Therefore, a one-dimensional model is used to simulate the LED band diagrams as a function of bias, electron, and hole transport inside a heterostructure, no-radiative and radiative carrier recombination that provides light emission. As a result we can obtain the relationship between Internal Quantum Efficiency (IQE), current density and emission spectra versus p–n junction bias and temperature (Liu *et al.*, 2008) shown in Figure 5.1.

5.1.2 Sequential Coupling

In sequential coupling (often referred to as load-vector or staggered coupling), the equation for one field is partially solved and the results passed as loads (the results of one physics field interacting with another) to the next physics field to drive its partial solution. The analysis software then passes this iteration to the next

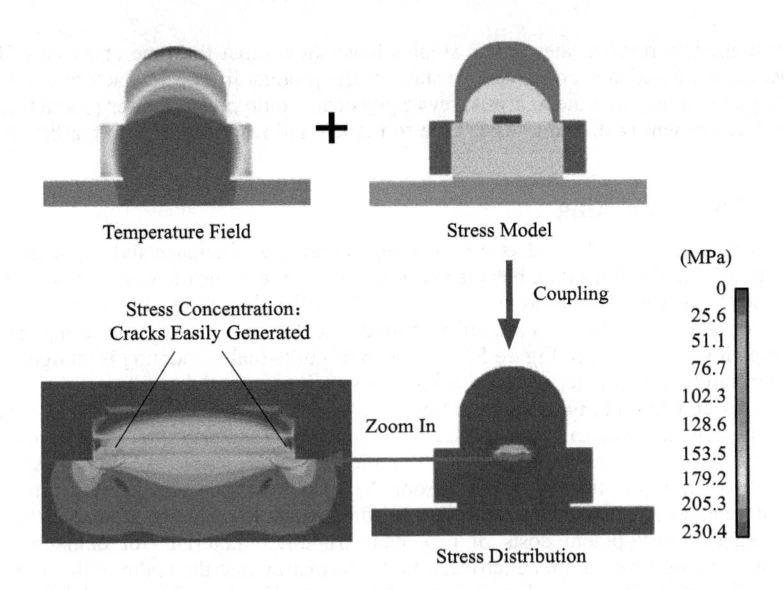

Figure 5.2 Thermal stress analysis of high power LED (*Color version of this figure is available online*)

physics field, and so on, down to the last field. Then the sequential iteration process begins all over and continues until a final solution is achieved.

For example, defects in terms of voids, cracks, and delaminations are often generated in LED devices and modules during various manufacturing processes, accelerated testing, inappropriate handling and field applications. Defects are mostly frequently induced in the early stage of process development. One loading is due to the non-uniform loads caused by temperature, moisture, and their gradients. As shown in Figure 5.2, defects in various cases are modeled by a nonlinear finite element method (FEM) to investigate the existence of interfaces, interfacial open and contacts in terms of thermal contact resistance, stress nonlinearity, and optical interfaces so as to analyze their effects on LED's thermal and optical performance. In a sequential-coupled field analysis, the initial temperature field is computed and then passed to the structural analysis.

The simulation results show that voids and delaminations in the die attachment would enhance the thermal resistance greatly and decrease LED's light extraction efficiency, depending on the defects' sizes and locations generated in our considered package. The modeling result for the thermal behavior is significant as it shows that, without an appropriate thermal management scheme, the time to reach the steady state is slow, and there will be a significant difference for the results by different vendors and there will also be a significant difference between the transient thermal resistance and steady state thermal resistance. This could be significant for the testing method and facility development. It is also found that both kinds of stresses could be high enough so as to generate cracking at the interface of the die attachment and Cu base when the hard die attach material is used. Effects of defects with different kinds and sizes on thermal resistance were also studied, and the maximum thermal resistance of the die attachment can be enhanced by six times as compared to the die attachment without defects (Liu *et al.*, 1995; Tan *et al.*, 2007; Chen *et al.*, 2008).

Typically, with multiphysics analysis, the exchange of data between the physics fields requires careful coordination, and the different mesh requirements for the various fields, loads, and boundary conditions must be correlated. For all these to function correctly requires a complex feedback loop between the various fields so that the coupled analysis converges to an accurate solution. The predictive FEA simulations more closely represent the real world than the simplified assumptions that often neglect accounting for issues that turn out to be critical. Performing multiphysics analysis early in product development enables companies to easily and inexpensively spot and fix problems that can become costly and time consuming to resolve later. Some studies have shown that the time and cost of fixing such problems increases 10-fold at each successive stage of the product life cycle. Factoring in the effect of more physics yields more accurate analysis, fewer physical prototypes, a shorter product-development cycle, lower development cost, and a faster time to market and response time to market changes.

5.2 Multi-Scale Modeling

Multi-scale simulation can be defined as the enabling technology of science and engineering that links phenomena, models, and information between various scales of complex systems. It is recognized that within the scope of materials and structures research, the breadth of length and time scales may range more than 12 orders of magnitude (Gates *et al.*, 2005), and different scientific and engineering disciplines are involved at each level as shown in Figure 5.3. The idea of multi-scale modeling is straightforward: one computes information at a smaller (finer) scale and passes it to a model at a larger (coarser) scale by leaving out, that is, coarse-graining, degrees of freedom. From a "bottom-up" perspective, the multi-scale approach should consider the intrinsic attributes of the constituent materials for the system of study. The ultimate goal of multi-scale modeling is then to predict the macroscopic behavior of an engineering process from first principle, that is, starting from the quantum scale and passing information into molecular scales and eventually to process scales. In addition, multi-scale modeling has the potential to significantly reduce development costs of new nanostructured materials for demanding structural applications by bringing physical and microstructural information into the realm of the design engineer. The intent is to assist the material developer by providing a rational approach to material development and concurrently assist the structural designer by providing an integrated analysis tool that incorporates fundamental material behavior.

From a "bottom-up" perspective, the multi-scale approach should consider the intrinsic attributes of the constituent materials for the system of study. Much of the current work focuses on the use of nanostructured materials

The nanostructured materials based on carbon nanotubes and related carbon structures are of current interest for much of the materials community. Although at the time of their discoveries, other materials with well-defined nanoscopic structure were known, investigators were intrigued to find that these new forms of carbon could be viewed as either individual molecules or as potential structural materials. This realization in turn energized a whole new culture of nanotechnology research accompanied by worldwide efforts to synthesize nano-materials and to use them to create multifunctional composite materials. More broadly then, nanotechnology presents the vision of working at the molecular level, atom by atom, to create large structures with fundamentally new molecular organization.

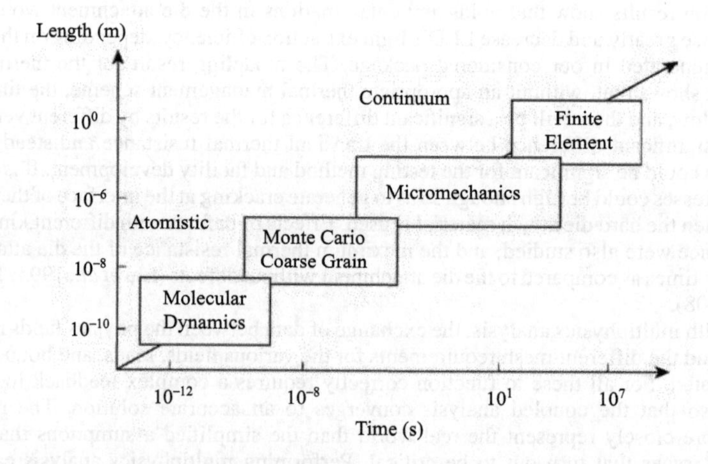

Figure 5.3 Range of length and time scales of the key simulation methods

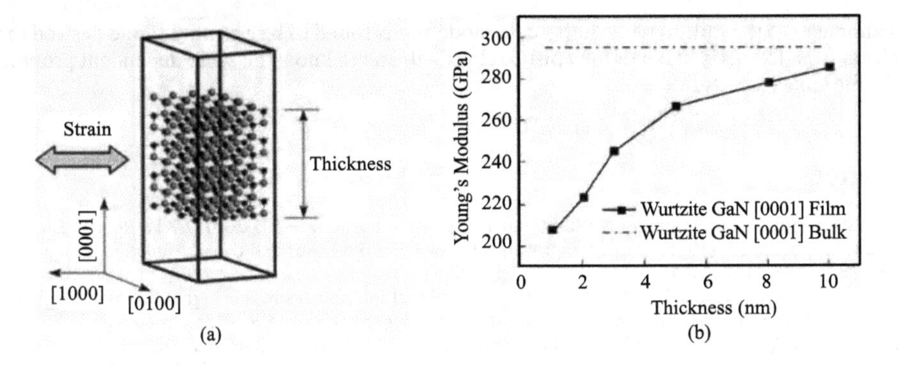

Figure 5.4 (a) Molecular models for elastic modulus of wurtzite gallium nitride nanofilm structures; and (b) Elastic modulus with respect to thickness for wurtzite gallium nitride nanofilms

The effective continuum approach for connecting atomistic models to continuum models uses relevant input from the atomistic simulations and carries forward the critical information to represent the continuum with the intrinsic nano-scale features incorporated into the model. The design of large-scale engineering structures requires a complete knowledge of the bulk-level behavior and properties of a material. For structural analysis, the bulk-level material behavior is described or predicted using continuum-based approaches, such as the micromechanical and finite element methods described above. Continuum mechanical parameters, such as Young's modulus or stress, are classically defined with the assumption that the material is a mathematical continuum. However, a set of atoms in a molecular modeling simulation, which possess a structure that is in thermodynamic equilibrium, clearly does not resemble a mathematical continuum, but a discrete lattice structure. Therefore, the direct application of continuum-mechanics analyses for molecular models is problematic unless steps are taken to secure their equivalency.

For example, Young's modulus for various thicknesses of gallium nitride nanofilms is obtained by Car-Parrinello molecular dynamics (CPMD) simulations (Car *et al.*, 1985; Izvekov *et al.*, 2005; McCullocha *et al.*, 2000; Moram *et al.*, 2007). The elastic properties can be obtained by static simulation runs in which an external tensile strain is imposed in the [1000] crystallographic direction. The Poisson ratio is considered as 0.205. The Young's modulus can be obtained by applying a generalized Hooke's law of the form $\sigma = Y\varepsilon$, where σ, Y and ε denote the stress, Young's modulus and strain, respectively. As compared with the bulk gallium nitride, the Young's modulus decreases monotonically by as much as 29.8% as the thickness of the gallium nitride film decreases as shown in Figure 5.4.

5.3 Chapter Summary

In this chapter, concepts of multi-physics and multi-scale modeling were presented in the framework of manufacturing and testing of micro-systems and their packaging. Direct-coupled analysis or two-way coupling was first introduced, which assembles all the physics fields as finite-element equations in one matrix and solves the matrix as a whole and all physics fields can be accounted for in a single solution. The LED heterostructure was considered as a set of planar semiconductor layers and electrodes, with the concurrent consideration of thermal, optical, and electrical aspects. Then the sequential coupling methodology was also introduced, in which the equation for one field is partially solved and the results passed as loads (the results of one physics field interacting with another) to the next physics field to drive its partial solution. The analysis then passes this iteration to the next physics field until a final solution is achieved. The hygro-thermal-stress analysis of an LED packaging is an example solved by the sequential method due to the rapid temperature dependent process and slow moisture absorption process. Multi-scale simulation can be defined as the enabling technology of science and engineering that links phenomena, models, and information between various scales of complex systems. Particularly in the next generation semiconductor IC packaging, the size scale effect is so large a difference from nanometer to the millimeter. Mutli-scale modeling is found to be essential if one desired to study the defects from the early stage of the epi-taxial growth and to know the scale dependent properties of the thin film materials formed.

References

Alan, M. (2003) Solid state lighting-a world of expanding opportunities at LED 2002, *III-Vs Review* **16**(1):30–33.

Car, R. and Parrinello, M. (1985) Unified approach for molecular dynamics and density-functional theory, *Physics Review Letters* **55**:2471–2474.

Chen, Z.H., Li, J., Wang, K. and Liu, S. (2008) The simulation of interfacial stress in higher brightness light emitting diodes, *5th China International Forum on Solid State Lighting*, Shenzhen, P.R. China, 349–353.

Gates, T.S., Odegard, G.M., Frankland, S.J.V., and Clancy, T.C. (2005) Computational materials: multi-scale modeling and simulation of nanostructured materials, *Composites Science and Technology* **65**:2416–2434.

Haiping, Y., Chunfeng, L., Jianghua, D. (2009) Sequential coupling simulation for electromagnetic-mechanical tube compression by finite element analysis, *Journal of Materials Processing Technology* **209**:707–713.

Izvekov, S. and Votha, G.A. (2005) Ab initio molecular-dynamics simulation of aqueous proton solvation and transport revisited, *The Journal of Chemical Physics* **123**:044505.

Liu, S., and Mei, Y.H. (1995) Behavior of delaminated plastic IC packages subjected to encapsulation cooling, moisture absorption, and wave soldering, *IEEE Transactions on Components, Packaging, and Manufacturing Technology–Part A* **18**(3):634–645.

Liu, S., Gan, Z., Luo X., Wang, K., Song, X., Chen, Z., Yan, H., Liu, Z., Wang, P. and Wei, W. (2008) Multi-physics multi-scale modeling issues in LED, *Proceedings of the International Society for Optical Engineering*, 7375, 737507.

Liu, S., Wang, K., Liu, Z.Y., Chen, M.X. and Luo, X.B. (2008) Roadmap for LED packaging development: a personal view, *5th China International Forum on Solid State Lighting,* Shenzhen, P.R. China, 396–400.

McCullocha, D.G., McKenzie, D.R. and Goringe, C.M. (2000) Ab initio study of structure in boron nitride, aluminum nitride and mixed aluminum boron nitride amorphous alloys, *Journal of Applied Physics* **88**:5028–5033.

Moram, M.A., Barber, Z.H. and Humphreys, C.J. (2007) Accurate experimental determination of the poisson's ratio of GaN using high-resolution X-ray diffraction, *Journal of Applied Physics* **102**:023505.

Tan, L.X., Li, J., Liu, S., Gan, Z.Y. and Wang, K. (2007) Effect of defects on thermal and optical performance of high power LEDs, *4th China International Forum on Solid State Lighting*, Shanghai, P.R. China, 304–309.

Wang, K., Liu, S., Chen, F., Liu, Z.Y. and Luo, X.B. (2008) Optical design for LED lighting, *5th China International Forum on Solid State Lighting*, Shenzhen, P.R. China, 343–348.

6

Modeling Validation Tools

Electronic packages are complicated material systems operating under electrical, thermal, and mechanical loading conditions. The package design and reliability analysis requires knowledge in many technical disciplines and specialized experimental methods and tools. Many advanced techniques have been adopted from other technical areas to be applied to the packaging applications and many experimental tools have been developed correspondingly. In this chapter, advanced experimental techniques in electronic packaging are reviewed. The techniques and methods for mechanical analysis are emphasized as well as the new developments in equipment and tools. The applications of these techniques are introduced by numerous examples and reference publications. Most of the applications are directly related to new product development and have made a significant impact on the electronics industries by assisting product designs, qualifications, reliability predictions, and cycle time reductions.

6.1 Structural Mechanics Analysis

Excluding the techniques used in electrical designs and thermal management, the reliability analysis in electronic packaging are conducted in three major areas: structural mechanics analysis, process characterization, and material characterization.Table 6.1 shows the existing techniques and tools used in these three areas. In the area of structural mechanics analysis, the techniques and methods are used to measure the structural deformations, strain, stress, and to verify the package designs. An important application of those techniques is to assist numerical and analytical methods to conduct design optimizations.

In the design and process characterization area (area B), destructive testing methods are usually used to determine the process quality and to verify the package designs. The strength and the failure molds of the components, interconnections, as well as the whole assemblies are studied. In electronic packaging, the strength and the failure mode of a package is closely related to the physical designs and process conditions, those testing methods can be used to define the design weakness and characterize each process step used in the manufacture.

The functions of the techniques used in area C are material, surface, and chemical property characterizations. In this area, image and substance studies are conducted. It provides the understanding of material properties (such as the CTE, T_g, modulus, and thermal properties), the surface properties (such as the surface roughness and textures) and the chemical properties (such as the chemical contents and surface contamination).

Some analyses require tools from more than one area such as the determinations of material properties, mechanical, thermal, and chemical testing that may be conducted in parallel. For testing the adhesion property, an interfacial adhesion test, which is a destructive test, and a surface analysis may be conducted. On the other hand, some of the tools can perform functions in different areas. For example, the X-ray

Modeling and Simulation for Microelectronic Packaging and Integration: Manufacturing, Reliability and Testing, Second Edition. Sheng Liu and Yong Liu.

Table 6.1 Methods and tools used for the reliability analysis in electronic packaging

A. Structural Mechanics Analysis (deformation study)	B. Design/Process Characterization (destructive testing)	C. Material, Surface, Chemical Property Characterization (image and substance study)
Strain Gages and Extensometers Moiré Interferometry Speckle, Holography Interferometry	Die Fracture Strength Test Bump/Ball Shear and Pull Test Leads/Wire Pull Test	Microscopic Analysis Scanning Electron Microscope (SEM) Atomic Force Microscope (AFM)
Speckle Correlation Methods ESPI	Die(module) Shear and Pull Test Underfill Adhesion Test	Dielectric Analyzer (DEA) Dynamic Mechanical Analyzer (DMA)
Twyman–Green Interferometer	Mechanical Fatigue Test	Differential Scanning Calorimeter (DSC)
Shadow Moiré, Projecting Moiré Test Chip Technology Micromachining Technology	Vibration Test Impact Test	Differential Thermal Analyzer (DTA) Thermogravimetric Analyzer (TGA) Thermomechanical Analyzer (TMA)

system can be used to characterize process parameters as well as the residual stresses. The Wyko measurements system can be used to characterize certain process parameters and the surface roughness. The testing machines are used to perform structural tests, determine material strength, and measure material properties.

All the techniques and methods directly impact the reliability of the final product of an electronic package. They are critical in package design optimizations, material selections, and process analysis. In this chapter, the experimental methods and tools used in the structural mechanics analysis are emphasized. The methods in the other two areas are more established and the tools are mostly commercialized. The equipment manufacturers usually provide detailed information on the capabilities of the techniques and the instructions on how to operate the equipment. In contrary, the structural analysis techniques and tools, except the strain gauges and extensometers, are comparatively new and most of them are advanced optical techniques. Few techniques have been commercialized. New applications in electronic packaging are still under development.

In recent years, many major computer and electronic companies have established special groups to focus their efforts on the structural mechanics analysis. Those efforts include the development of new analytical and experimental techniques such as advanced numerical simulations and the in-situ measurement tools for local strain/stress measurements. Using these state-of-the-art techniques, strain/stress concentrations in very localized areas can be determined and failure mechanisms can be better understood. The knowledge has been used to assess the dependence of reliability on various structural and material configurations. Together with other analytical tools, the structural mechanics analysis is conducted at the design and qualification stage, such that the reliability of the system can be estimated and optimized before the packages are put into production. The implementations of those advanced techniques have produced significant impact on packaging development such as increased efficiency, improved reliability, and reduced design cycle time, which are extremely important for the competitiveness of all electronic products. In practice, the task of structural mechanics analysis is conducted by a mechanical group, design/process characterization is conducted by a process group, and the material, surface, and chemical property characterization is conducted by a materials group. There are always interactions and collaborations in these three areas. The ultimate goal is achieving a reliable package, which is usually determined by the accelerated thermal cycles. The accelerated testing methods and techniques are not included in this chapter because they are more a qualification tool than a testing tool.

The main tasks of the mechanical analysis in electronic packaging are the failure analysis, design optimizations, and life predictions. The current prediction theories and simulation models use thermal-mechanical strains and stresses as parameters for estimating damages and failures. In order to obtain those critical strains and stresses in the packages, advanced techniques in mechanical and material characterizations are needed. As electronic technology advances, more novel materials are used in packages. Most of the materials are non-homogeneous and have strong nonlinear material properties.

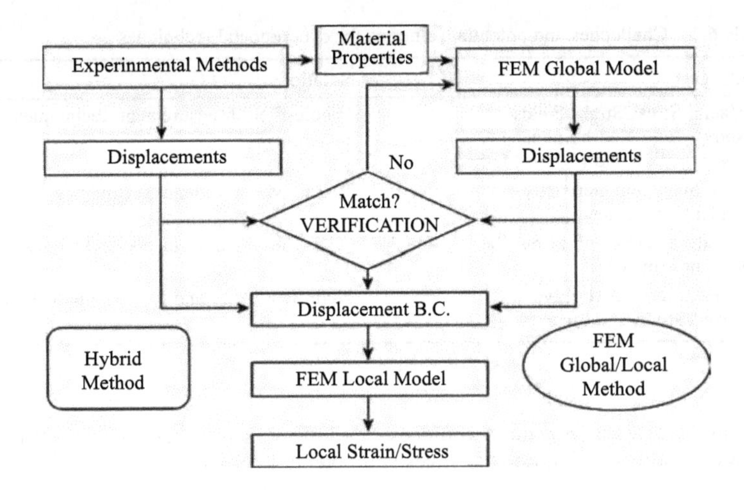

Figure 6.1 Current methodology used in the electronic industries for the mechanical analysis

Electronic components usually consist of multi-materials and interfaces with highly compact and integrated features. Those conditions generated demand for special experimental techniques in the failure and reliability analysis. For the strain/stress analysis and the determinations of the failure mechanism, in-situ and whole-field measurements are crucial for local strain/stress distributions in very localized regions. These are serious challenges for experimental methods and equipment.

As shown in Figure 6.1, the current methodology used in the electronic industries for the structural mechanical analysis can be summarized as the numerical, experimental and hybrid approaches. Numerical methods have been used extensively to calculate strain/stress in the package structures. It is a very important tool for design optimization and failure predictions in electronic package. An important issue in the numerical analysis is that the accuracy is dependent on the material property and the geometry. The model verification with experimental results can help to improve the accuracy and confidence. The experimental methods in many cases can be very effective if the specimens are available. However, as the components and interconnections used in electronic packages are becoming smaller and smaller, the strain concentrations are frequently localized in a very tiny area with high magnitudes. Those situations make the experimental determinations of strain/stress distributions extremely difficult.

Electronic packaging is a relatively new and multi-discipline field where many new and exciting developments have happened in last few years. The development in new experimental techniques and methodologies for structural mechanics analysis is one of them.

6.2 Requirements of Experimental Methods for Structural Mechanics Analysis

Electronic packages are complicated material systems operating under thermal and mechanical loading conditions. Many materials used are organic materials which have highly non-linear properties and are very process and scale dependent. Some materials are also temperature and strain rate dependent, even with the evolution of microstructures. Because of the developments in the advance processing technologies, the dimensions of the devices and interconnections are becoming smaller and smaller. It requires experimental techniques with high sensitivity and resolution to characterize the material properties and measure the local strains and stresses. Conventional experimental tools such as strain gauges and extensometers can only measure point displacements and are not capable of determining deformations and strain distributions at the interfaces and in the non-uniform media. The conventional techniques are also very sensitive to temperature changes and difficult to practice in thermal loading conditions.

Table 6.2 Challenges and potential solutions for experimental techniques

Challenges	Solutions
Interfacial Stress/Strain Fields Anisotropy and Non-homogeneity FEA Verification and Hybrid Method	Whole-Field Measurement Techniques
Process, Scale and Geometry Dependent Properties	In-Situ Measurements Techniques
Temperature, Time and Strain Rate Dependent Properties	Real-Time Measurements Techniques
Small Features and Highly Localized Strain and Stress	High Sensitivity and Resolution Techniques

The requirements for new experimental techniques are whole field, in-situ, real-time, and high resolution. Table 6.2 shows a summary of those challenges and potential solutions.

6.3 Whole Field Optical Techniques

The optical techniques are usually whole field measurements which provide displacement contour maps over a testing area. They are well suited for in-situ measurements with various displacement sensitivities and resolutions. They are also insensitive to temperature and can be used easily for thermal deformation measurements. The whole field and high resolution properties fit the application requirements of packaging systems very well. The optical techniques, such as moiré interferometry, holographic interferometry, Twyman–Green interferometry, and shadow moiré method are becoming more and more important tools in characterizations of electronic materials, components, and assemblies (Guo *et al.*, 1997; Pappalettere *et al.*, 1997; Dai *et al.*, 1997; Teng *et al.*, 1997; Han *et al.*, 1996; Dadkhah *et al.*, 1996). These optical methods produce whole field displacement contour maps which are ideal for FEM verifications and hybrid analysis (Guo *et al.*, 1992, 1994, 1995, 1996, 1997). The hybrid method, such as the FEM-moiré interferometry method, taking advantages of both numerical and experimental techniques, has great potential for micro-mechanical analysis in electronic packaging (Guo *et al.*, 1992, 1994, 1995).

In recent years, many advanced experimental tools using optical principles have been developed for the applications in electronic packaging. Table 6.3 lists some of the techniques. For the in-plane displacement measurements, moiré interferometry is widely used. The key element in moiré interferometry applications is the specimen grating. Moiré interferometry was first applied in packaging analysis since late 80s (Guo *et al.*, 1997). The technique was not quickly adopted because of the difficulties of producing specimen gratings under the thermal conditions. The electronic components have interconnections. As a cross-section is made through these interconnections, many small internal edges are formed. Special replication techniques were developed to ensure clean internal edges and easy separations with a simple process in producing specimen gratings (Wang *et al.*, 1997). New grating techniques and processes are still under development. It is a very challenging task to produce high efficiency and clean specimen gratings when specimen features become smaller and smaller (Zhu *et al.*, 1997). Using the speckle

Table 6.3 Whole-field optical techniques used in electronic packaging analysis

Optical Techniques	In-plane Deformation	Out-of-plane Deformation
Low Sensitivity	Geometric Moiré Speckle Correlation	Shadow Moiré Projecting Moiré
High Sensitivity	Moiré Interferometer Speckle Interferometer	Twyman–Green Interferometer Holographic Interferometer
Very High Sensitivity	Speckle Correlation in SEM Micro Moiré Interferometer Electron-beam Moiré	ESPI Wyko Interferometer

technique, specimen grating is eliminated, which is a great advantage of the technique. In comparison with the moiré technique, it is difficult for the speckle technique to obtain a large field measurement with a wide dynamic range of strain distributions. However, with the improvement in the resolution of the CCD camera and computation power of computers, the speckle technique has great potential in the near future.

For the high sensitivity and resolution deformation measurement, techniques such as the speckle correlation in SEM (Jones *et al.*, 1988; Moore *et al.*, 1990), the micro moiré interferometer (Guo *et al.*, 1996, 1997) and the electron-beam moiré (Dally *et al.*, 1993; Drexler *et al.*, 1997) are developed for packaging applications. Using these advanced techniques, small deformations in a very localized area such as the area in solder joint or near a two material interface, can be resolved and determined.

6.4 Thermal Strains Measurements Using Moiré Interferometry

This method is a whole-field in-plane displacement measurement technique and schematically shown in Figure 6.2. The technique provides half-wavelength resolution which is depending on the type of laser used in the system. The fringe patterns are contours of in-plane displacements which are calculated as: $W = GN$, where N is the fringe order and G is the resolution factor of the optical system. In our photomechanics lab, a HeNe laser with a wavelength of 633 nm is used, which results in a displacement resolution of 316 nm per fringe order.

6.4.1 *Thermal Strains in a Plastic Ball Grid Array (PBGA) Interconnection*

The application of the moiré interferometry in the thermal strain measurements of a PBGA package is a typical example of using the optical method for reliability analysis. The mechanical reliability of the solder joints in the Ball Grid Array (BGA) technology has significant impact on the system reliability of an electronic device and, therefore, is critical in package designs and qualifications. The major concern of solder joint reliability is the solder fatigue during system operation. Since the solder fatigue life is related to the thermal-mechanical strains, the mechanical reliability analysis of the solder joints can be conducted through the analysis of the thermal-mechanical strains.

The CTE mismatch in the package induces thermal strain and stress in the solder joints during thermal cycles and causes solder fatigue failures. The typical dimension of the solder ball in a PBGA package is 25 mil (0.63 mm). Due to the small scale of the local ball joint structure, it is extremely difficult to measure accurate strain values directly using an experimental tool. Moiré interferometry can provide accurate relative displacements in the vicinity of a solder joint by measuring the global deformations of the whole package assembly (Guo *et al.*, 1995, 1996). However, the local strain distribution in a particular solder joint, which determines the maximum strain, cannot be resolved accurately. A local FEM model, which precisely models the local joint geometry with fine meshes, can offer much better resolution and accuracy in the local strain concentration of a solder joint.

Moiré interferometry has suitable sensitivity and resolution for the global displacement measurements in the PBGA structures. In the moiré experiments, cross-sections of specimens were prepared. The cross-sections contained all the solder balls in the symmetrical plane of the PBGA module (Figure 6.3). The thermal loading was applied by replicating the specimen grating when the specimen was held at an elevated temperature T_1 (102 °C), and the measurements were conducted at the room temperature T_2 (22 °C). The deformations of the specimen grating reflect the deformations experienced by the specimen under the thermal loading ΔT ($T_2 - T_1$). In this work, the ΔT is -80 °C. In order to place a uniform and known-frequency grating (unreformed grating) on the specimen at an elevated temperature, a ULE (Ultra Low Expansion) glass grating mold was first produced; then specimen gratings were replicated from the ULE grating mold at an elevated temperature.

Figure 6.3 shows the fringe patterns obtained by moiré interferometry, which are the U and V displacement fields of the PBGA module. It is clear that by this resolution, the strain distributions affected by SMD or NSMD joint configurations cannot be detected from the fringe patterns. However, the relative displacements between the module and the PCB at the location of each solder ball can be determined. In Figure 6.4, the relative displacements between the module and PCB are plotted against the locations of the

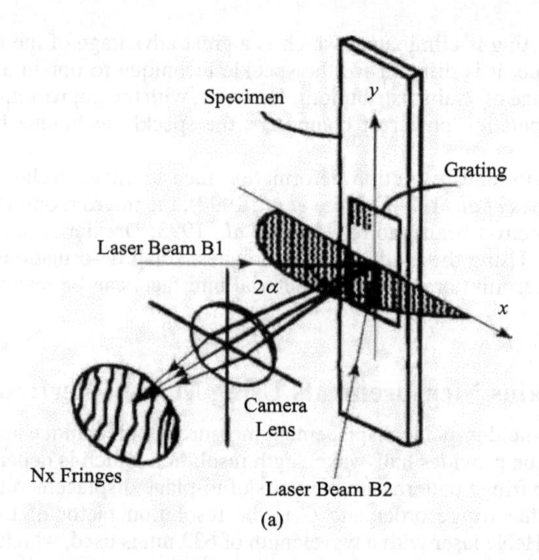

(a)

(b)

Figure 6.2 (a) Schematic of 4M interferometry. (b) Photograph of macro-micro moiré measurement interferometer developed by WTI USA/FineMEMS

nine solder balls at the right-hand side of the cross-section. The maximum relative displacement is at the location of the end solder ball which has the largest Distance to Neutral Point (DNP) in the cross-section. The fringe patterns provide the total displacements and the mechanical part shown in Figure 6.4 that is obtained by subtracting the free thermal expansion from the total displacements. In order to determine the accurate relative displacements at the end solder ball, carrier frequencies were used to modulate the fringe patterns by adding a rigid body rotation to the specimen during the moiré experiments. The carrier

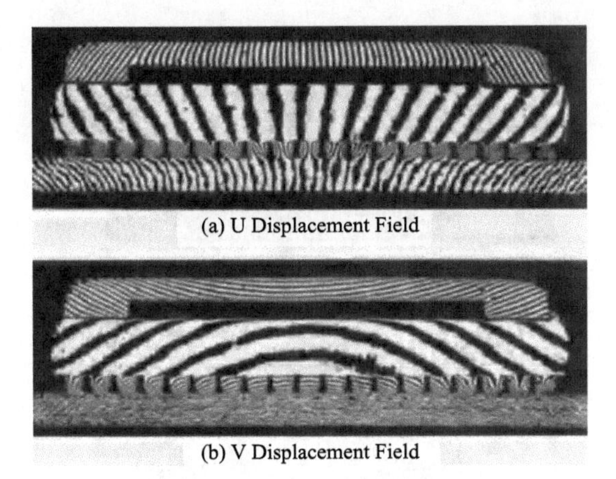

(a) U Displacement Field

(b) V Displacement Field

Figure 6.3 Fringe pattern in symmetrical plane of the PBGA module

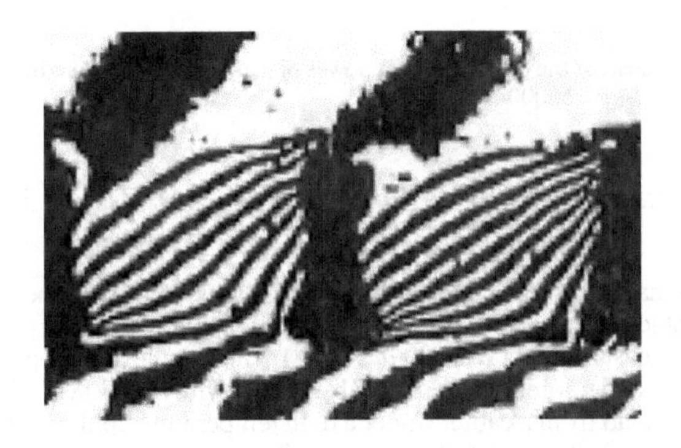

Figure 6.4 Fringe pattern of the nine solder balls

frequency cancels the rigid body rotation at the end solder joint, such that the x and y displacements can be measured independently from the two displacement fields (U and V fields).

Because the experimental data is obtained from the cross-section parallel to the longer side of the module, the end solder joint represents the maximum deformation in the symmetrical plane of the assembly. In order to determine the boundary conditions of the solder ball with maximum deformations in the package, which is the corner solder ball, linear extrapolation using the DNP ratio is applied. The DNP ratio between the corner solder ball and the end solder ball is 1.08. A factor of 1.08 is applied to calculate the displacement boundary conditions for the corner solder ball from the end solder ball. Because of the high length to width ratio in this particular BGA interconnection, using linear extrapolation to determine the displacements in the corner solder ball will not result in significant error.

The global displacement field and the critical solder ball with the highest thermal deformations are determined from the experimental results. The relative displacements for end solder ball are also obtained from the experimental data. At this stage, PBGA assemblies with different modules and PCBs are measured. The parameters which affect the global deformations are determined. The boundary conditions at the corner solder ball which has the highest DNP in the assembly was determined from the experimental results of the end solder joint and a linear extrapolation using the DNP ratio of 1.08. The displacement boundary conditions are prescribed at the two parallel surfaces, plane I and plane II which also define the boundaries of the local FEM model.

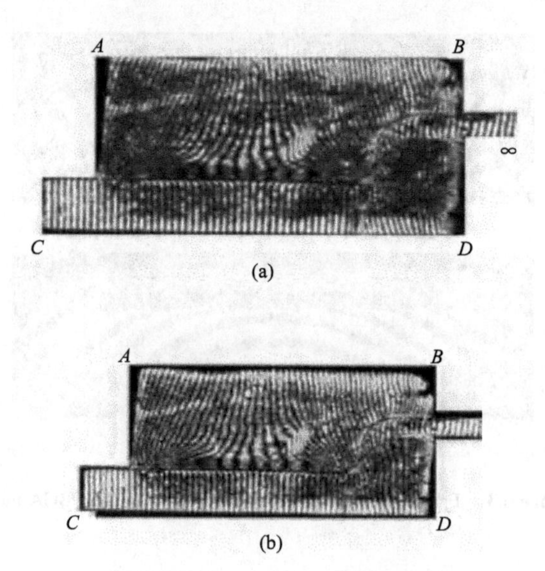

Figure 6.5 (a) *U* displacement of the creep test on a power plastic package $t = 0$ s. (b) *U* displacement of the creep test on a power plastic package $t = 50000$ s

6.4.2 Real-Time Thermal Deformation Measurements Using Moiré Interferometry

The optical techniques have the advantage in the applications of in-situ and real-time measurements because of the non-contact, high sensitivity, and thermally stable natures. Figure 6.5 shows a time and temperature dependent measurement (Wang *et al.*, 1997). The sample is a copper connector soldered to a printed circuit board. The solder joint area is 12 mm by 12 mm with a thickness of 0.11 mm. The connector is subjected to a constant load from a cable. Under a high temperature application condition (100 °C), the solder layer creeps and eventually crack failure occurs. The creep strain rate of the solder was determined using moiré interferometry under simulated temperature and load. The fringe patterns, which are the U displacement fields (normal to the solder layer), provide the solder creep strains as a function of time.

6.5 In-Situ Measurements on Micro-Machined Sensors

In-situ measurement capabilities are extremely useful in characterizing deformations in accelerated reliability tests and under real operating conditions. Electronic components are very often operated under power cycles which induce cyclic strains and stresses to the structures and interconnections. These cyclic loads can cause fatigue and fracture failures if the strain and stress reach a certain level. The service life is usually a function of the strain and stress. Because of the temperature insensitivity and non-contact properties of the optical techniques, they are ideal for the applications of in-situ measurements and monitoring of the deformations in the electronic components which operate under power cycles.

6.5.1 Micro-Machined Membrane Structure in a Chemical Sensor

The new generation of sensors are using the MEMS (Micro Electron Mechanical Sensor) technology where the sensors are built on a thin silicon membrane. The carbon mono-oxide sensor is an example (Bosc *et al.*, 1998). The thin membrane structure is used to minimize the thermal mass, increase the speed of temperature response and reduce power consumption during operation. As shown in Figure 6.6, the silicon wafer is etched to form a 1.8 mm by 1.8 mm membrane with a thickness of 2.5 micrometers.

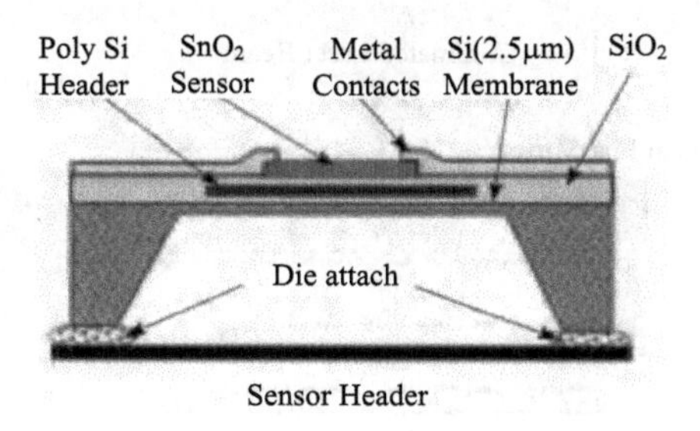

Figure 6.6 Physical structure of a CO sensor

Various materials are deposited and patterned on the membrane to form the heating and sensing elements and interconnections. A photograph of the membrane and the sensor structure is shown in Figure 6.7. In order to detect CO level and eliminate humidity effect in the environment, CO sensors are operating in a pulsed mode which regulates the membrane temperature from 100 °C to 500 °C using a power cycle of 15 s at 5 V across the heater for 5 s and 1 V for 10 s (Figure 6.8).

In the qualification process of chemical sensors, the reliability data on the membrane structures under operational pulsed conditions has to be obtained. It is also important to determine the relative reliability when material or design changes are made. Under a pulsing operation condition, sensor failures are usually caused by the high deformations in the membrane which produce the local stresses in the sensor structure and interconnections. It is important to determine the membrane deformation during the power cycles.

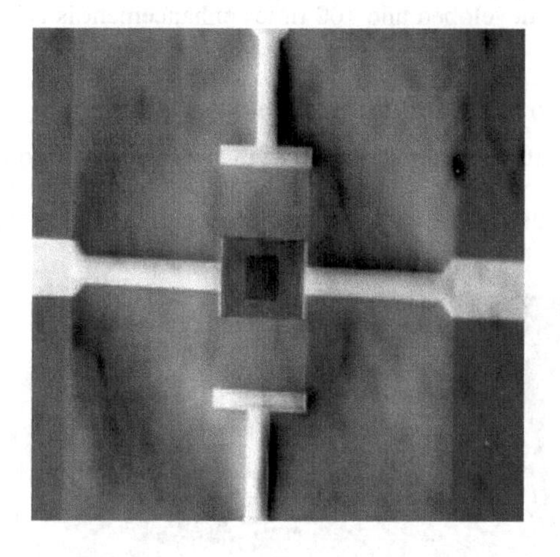

Figure 6.7 A photograph of the silicon membrane

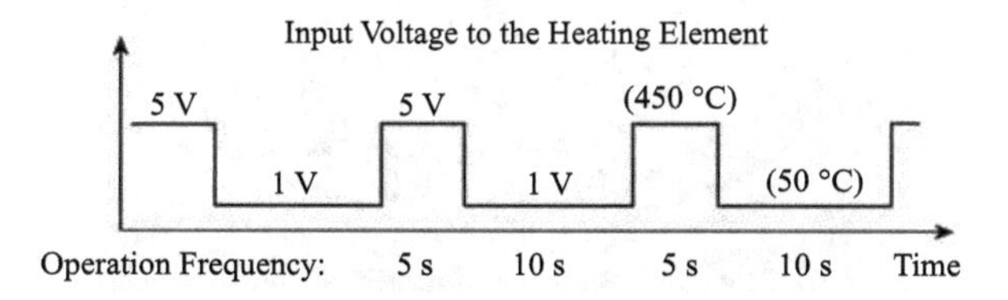

Figure 6.8 CO sensors are operating in a pulsed mode

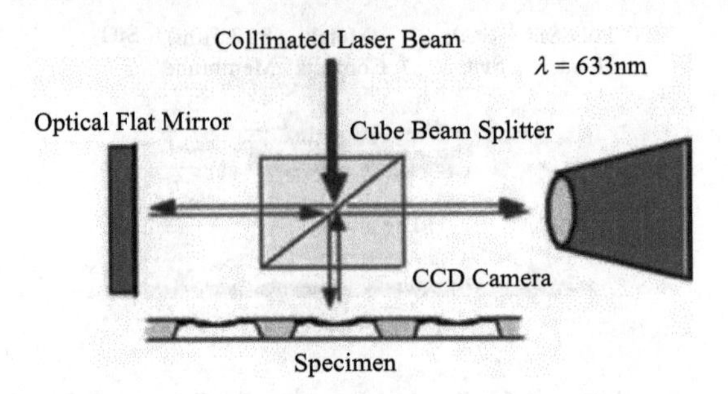

Figure 6.9 A schematic diagram of the Twyman–Green interferometry (Reproduced with permission from V. Sarihan, Y.F. Guo, T. Lee and S. Teng, "Impacting electronic package design by validated simulations," *48th IEEE Electronic Components & Technology Conference*, 330–335, 1998. © 1998 IEEE.)

6.5.2 In-Situ Measurement Using Twyman–Green Interferometry

The deformation behavior of the membrane was studied by the Twyman–Green interferometry, which is a whole-field out-of-plane displacement measurement technique, schematically shown in Figure 6.9. This technique provides half wavelength displacement sensitivity, which depends on the type of laser used in the system. The fringe patterns are contours of out-of-plane displacements which are calculated as: $W = G \times N$, where N is the fringe order and G is the resolution factor of the optical system. Here, a HeNe laser with a wavelength of 633 nm is used, which results in a displacement sensitivity of 316 nm per fringe order. By collaborating with the first author's former lab at Wayne State University, a closed-loop controlled phase shifter was developed and 100 times enhancement is achieved.

6.5.3 Membrane Deformations due to Power Cycles

Figure 6.10 shows a fringe pattern obtained by the Twyman–Green interferometry, which is a contour map of out-of-plane displacement of a membrane. The measurement results are plotted in the Figure 6.11. The plot shows the steady state deformation of the membrane at room temperature and at 45 °C with two

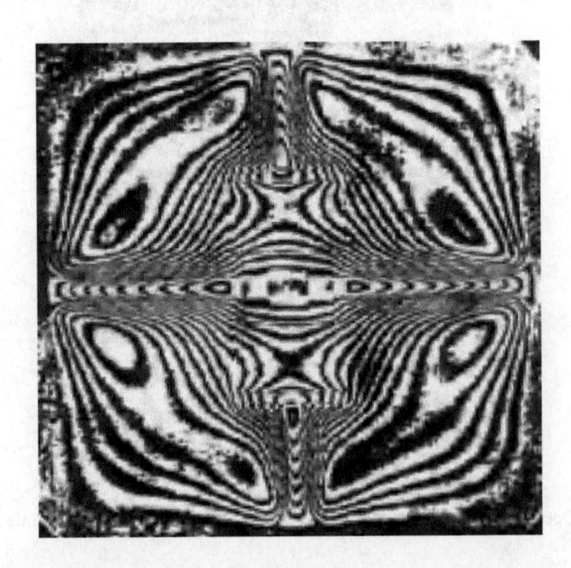

Figure 6.10 Fringe patterns obtained by the Twyman–Green interferometry

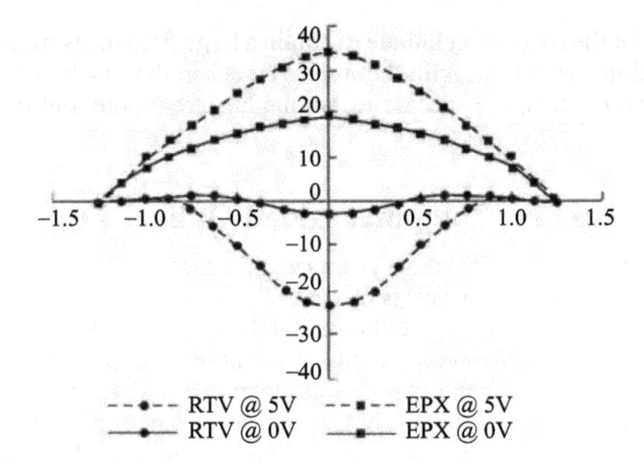

Figure 6.11 Plot of the membrane deformation at 50 °C and 450 °C

deflection modes. The membrane deformations were determined at both steady state and during a step function power input using 5 V and 1 V. For the steady state, the measurements were conducted after the thermal equilibrium was reached (2 minutes after the power input). For the measurements during the step function power input, a video tape was used to record the membrane deformation, and the extreme deformations and the corresponding times were measured from the video tape. The frame speed of the video tape is 30 ms (0.03 s), which is the time resolution of the measurement.

Figure 6.11 shows the membrane deformations at 0 V and 5 V power input. Under the step function power input (5 V and then 0 V), the video tape showed that the membrane reaches the same deformations (steady state positions) in the time of two video frames which was 60 ms (0.06 s) for both 5 V and 0 V. It indicates that the time needed for the mechanical deformation to reach to the steady state position is less than 100 ms (0.1 s). Because the membrane deformation is the result of the temperature gradient in the membrane, the deformation response must follow the temperature response. When the deformation reaches the extreme positions as in the steady state, the temperature distribution should reach the same distribution as in the steady state. In addition, the stress in the membrane is determined by the membrane deformation. The stress in the membrane should also reach the same level as in the steady state when the deformation reaches the same level as in the steady state. The stress in relation with the membrane deformation was analyzed using a finite element simulation. Simulations and experiments have shown that the membrane deforms severely during the power cycles due to the local thermal gradient of the silicon membrane. The repetitive buckling is the major cause of the material and interface fatigue failure in the sensor structures.

Die and membrane deformations were measured using the Twyman–Green interferometry technique. The fringe patterns provide the whole-field membrane shapes and deflection magnitudes before and after the die attachment and also during the power cycles. The experimental results provide the thermal deformations which are important for obtaining an optimum design of the membrane and its package. Different layouts and materials were also evaluated by this testing method and the testing data was used to give a parametric description of device failure which was then used for design optimization and for reliability prediction under differing operational conditions.

6.6 Real-Time Measurements Using Speckle Interferometry*

In the applications of moiré interfrometry, it is a very challenging task to produce high efficiency and clean specimen gratings when specimen features become smaller and smaller. Using the speckle technique, the specimen grating is eliminated, which is a great advantage of the technique. In comparison with the moiré

*Reproduced with permission from J.M. Bosc, Y.F. Guo, V. Sarihan and T. Lee, "Accelerated life testing for micromachined chemical sensors," *IEEE Transactions on Reliability*, **47**, 2, 135–141, 1998. © 1998 IEEE.

technique, it is difficult for the speckle technique to obtain a large field measurement with a wide dynamic range of strain distributions. However, with the improvement in the resolution of the CCD camera and computation power of computers, the speckle technique has great potential in the near future.

6.7 Image Processing and Computer Aided Optical Techniques

In combining with the optical methods, image processing programs are developed to interpret the fringe patterns and convert them to 3D contour plots or strain fields. With computer aided analysis and phase shifting technique, higher sensitivity and resolution can be achieved. There are great potentials in the image processing and computer aided computational methods to expand the capability of the optical techniques. Recently, the speckle image correlation technique and ESPI (Electronic Speckle Pattern Interferometer) method have been successfully transferred to electronic packaging applications. The speckle interferometry and speckle correlation under SEM (Scanning Electron Microscope) were also developed for high special resolution applications, such as the studies of interface stress and deformations of small solder joints.

6.7.1 Image Processing for Fringe Analysis

Fringe patterns obtained from whole-field optical methods, such as the moiré interferometry or shadow moiré method, are displacement contours where displacement data is usually represented by fringe center lines. From the fringes, 3D surface shape and strain contour maps can be obtained from image processing software. In the applications of electronic packaging, Choi *et al.* have developed a technique using Fourier Transformation to calculate strain fields from the moiré interferometry fringe patterns (Choi *et al.*, 1993). The algorithm is shown in Figure 6.12. Other researchers have also developed image processing software with different algorithms for other applications (Reid, 1986/1987; Robinsong *et al.*, 1983; Morimoto *et al.*, 1988; Voloshing, 1986).

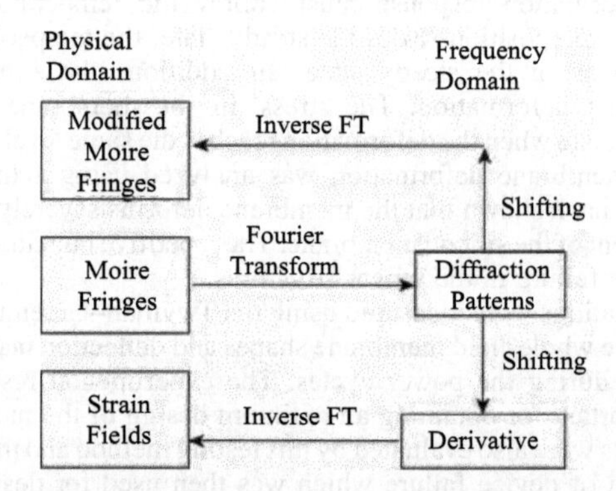

Figure 6.12 Algorithm of the image processing program

6.7.2 Phase Shifting Technique for Increasing Displacement Resolution

A typical example of the computer aided experimental technique is the Phase Shifting Shadow Moiré technique developed by Y. Wang *et al.* (1997). As shown in Figure 6.13 a phase shifting system is implemented to translate the shadow moiré grating in the vertical direction by fractions of the grating pitch. For example, if three-step shifting images are used, the distance of the translation of the gratings will be 0, $1/4\pi$, and $4/3\pi$ which are produced by the grating translations of 0, one third, and two thirds of the grating pitch. Computer software is used to capture three frames of the images with different phase. Using these three images the phase image can be computed. The equations used to calculate the phase image are:

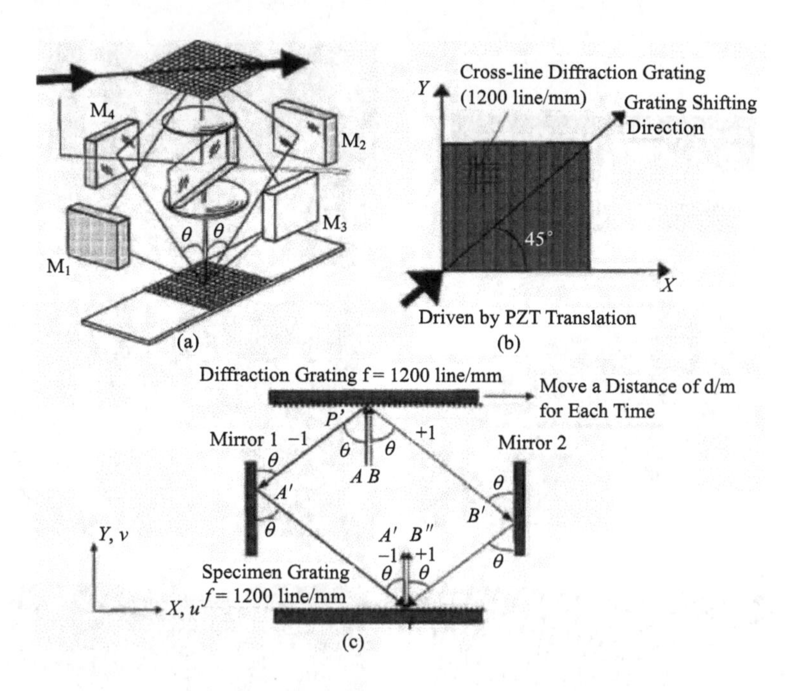

Figure 6.13 Schematics of the optical arrangement and the principle of phase shifting

$$\left.\begin{aligned} I_1 &= I_0 + A\cos[\varphi(x,y)+\alpha_1] \\ I_2 &= I_0 + A\cos[\varphi(x,y)+\alpha_2] \\ I_3 &= I_0 + A\cos[\varphi(x,y)+\alpha_3] \end{aligned}\right\} \tag{6.1}$$

$$w(x,y) = p\frac{\varphi(x,y)}{2\pi} \tag{6.2}$$

where I is the intensity distribution of the fringe images, φ is the phase distribution over the fringe patterns, and α is the phase angle which is related to the amount of the phase shifting applied. Using the three images with known phase angles (controlled by the amount of phase shifting), the phase distribution of the fringe patterns can be calculated.

When the phase information is obtained, the displacement can be calculated from the phase distribution:

$$w(x,y) = p\frac{\varphi(x,y)}{2\pi} \tag{6.3}$$

where w is the out-of-plane displacement, φ is the phase distribution, and p the unrapping factor. As shown in Figure 6.14, contour maps can be generated from the phase map with different contour intervals. In this study, the original grating pitch is 300 lines per inch. The displacement sensitivity of the original fringe patterns is 3.3 mil per fringe order. Using the computer program in combination with the phase shifting technique, a 0.5 mil contour interval was obtained, which increased the displacement sensitivity by six times. The same principle has been implemented in the micro moiré systems introduced in the reference (He *et al.*, 1997).

Micro moiré interferometry systems using phase shifting techniques are also developed together with image processing software. By using the phase shifting technique and advanced computer software,

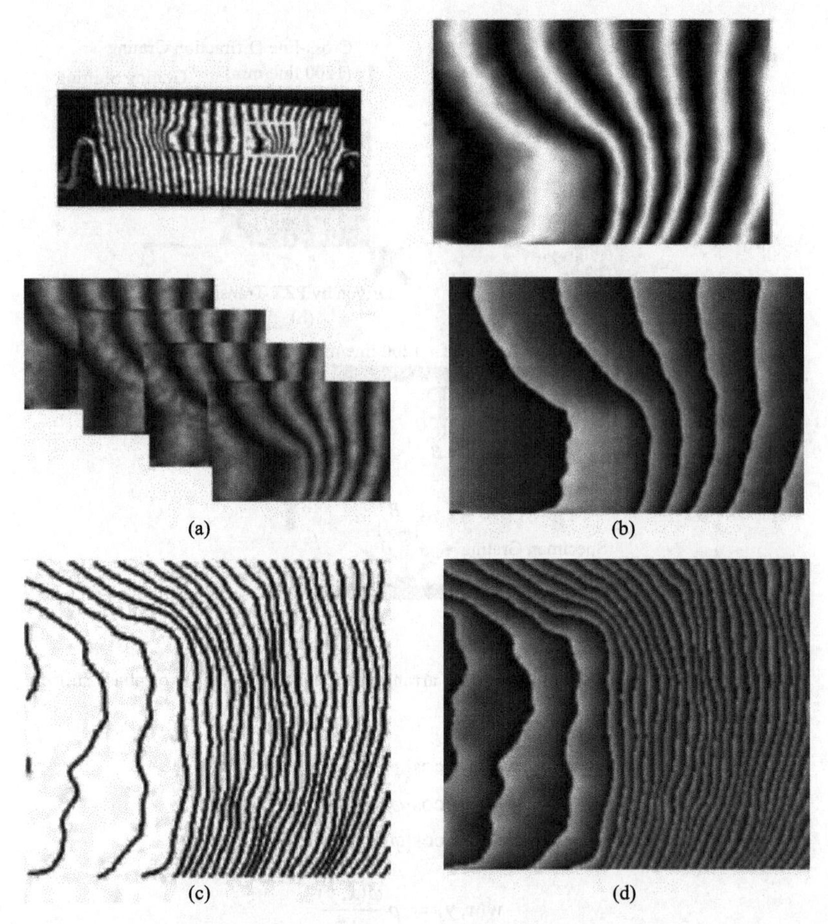

Figure 6.14 Fringe patterns of the *u* field on the cross section of a power plastic package (PFOP-3) parallel to the leads (80 to 24 °C.). (a) Phase shifting fringe pattern (U-field). (b) Phase diagram of fringe pattern (U-field). (c) Refined phase diagram of fringe pattern (U-field). (d) Refined displacement contour (52 nm/per fringe contour)

displacement sensitivity was pushed beyond the limit of the wavelength of the visible laser source used in the interferometric techniques. The displacement sensitivity is 0.4 μm per fringe order in a standard moiré interferometer. The micro moiré interferometer can have displacement sensitivity of 0.02 μm, twenty times of the standard sensitivity which is limited by the light wavelength. To increase the spatial resolution, a microscope system is usually used in the micro moiré systems. It is capable of studying a small solder joint with a dimension around 100 μm or even smaller.

The specimen grating was replicated at elevated temperature of 80 °C. The measurement is carried out at room temperature. Figure 6.14 shows the fringe patterns for both macro and micro thermomechanical measurement on the power plastic package. Thermal strain concentration and delamination caused by processing were found at the corner of the die. The delamination between the die and die attach can be seen clearly. Figure 6.1 shows the procedures of the phase-shifting method for moiré interferometry. Four fringe patterns with constant phase difference ($\pi/2$) were used for calculation of the phase diagram of fringe patterns. From that phase diagram, refined displacement contours were obtained, from which resolutions of 52 nm per fringe contour and even higher can be achieved.

The specimen grating for the flip-chip package was replicated onto its cross section at 100 °C. The measurement was performed at room temperature 25 °C. Figure 6.15(a) and (b) show the local

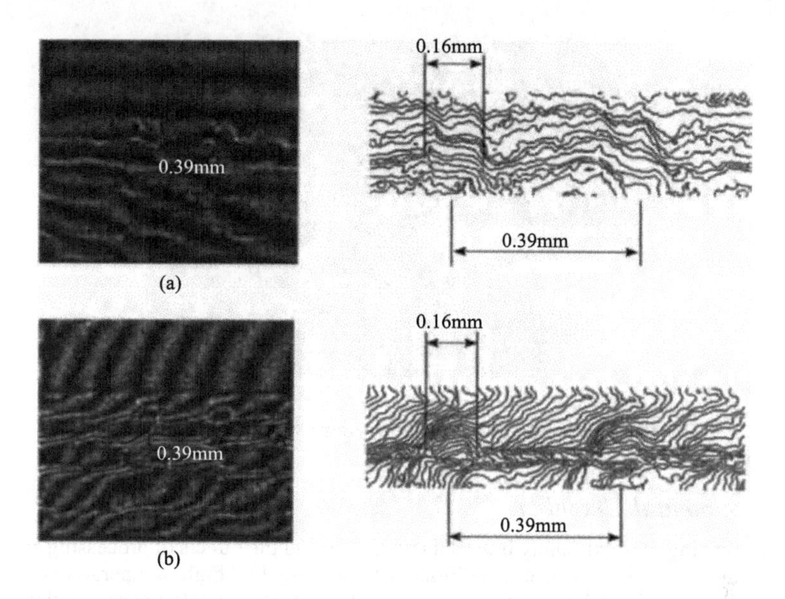

Figure 6.15 (a) Local fringe pattern of U-field and refined displacement contour (52 nm/per fringe contour). (b) Local fringe pattern of V-field and refined displacement contour (52 nm/per fringe contour)

fringe patterns and refined displacement contour about the underfill with two and a half solder balls. Because of thermal mismatch between the materials in the package such as those of the die, underfill, solder, and FR-4 board, there are bending and shearing effects on the layer between the die and FR-4, as can be seen in the fringe patterns. Solder balls and underfill are also included in this area. Large shear strains can be found for both solder and underfill material. The deformation of the solder can be resolved from the refined displacement contour, even though it is very tiny. From this refined u-displacement contour, a large shear deformation can be seen between the die and solder. From the refined v-displacement contour, a large shear deformation caused by bending can be seen between the solder and the underfill.

6.8 Real-Time Thermal-Mechanical Loading Tools

6.8.1 Micro-Mechanical Testing

Because the material systems used in electronic packages are in very small scale, as described in chapter 2, the material properties may be different from their bulk material counter part. In order to determine the material properties in special forms such as thin films and small volumes, high displacement and load resolution are required for loading tools. Micro testing machines have been developed to conduct material tests with submicron displacement resolution and sub gram load resolution. Improvements are also implemented in sample alignment, load capacity, dynamic response, stability, and versatility. The developments in the micro-testers include highly automated control software, excellent graphic capability, and high degree user friendly interface (Liu 1992; Lu *et al.*, 1996).

Those micro-testers are ideal for real-time mechanical loading in combination with optical measurement techniques. The 6-axis force sensing and displacement adjustability offers easy sample alignment and provides precise and desired loading statues. Extremely small electronic components and assemblies can be precisely loaded under an optical system and the whole-field displacement can be obtained. Because of the high resolution nature of the optical methods, highly concentrated deformations and the precise stress-strain relations can be obtained.

Figure 6.16 Thermal vacuum chamber with optical windows

6.8.2 *Environmental Chamber*

In electronic packaging, it is extremely important to understand the effects of processing and application conditions on the reliability. Those conditions usually involve high temperatures and humidity experimental techniques have been developed to conduct measurements under simulated processing conditions (Zou *et al.*, 1997). Those loading devices are combined with high sensitivity whole field measurement techniques to determine displacement fields on real electronic components under various environmental conditions (Figure 6.16). The information is directly applicable to the FEM verifications and hybrid methods.

6.9 Warpage Measurement Using PM-SM System

In an environment marked by the higher-speed and miniaturizing trend of the electronic products, BGA packages have been used in most of the electronic products thanks to the advantages of higher pin count and compact body size features. Along with high volume usage of the BGA packages, analysis data of soldering failures have accumulated and the package warpage at elevated temperature has been gaining attention as a cause of these failures, which is shown in Figure 6.17. This phenomenon means that the package warps during the rising temperature of the reflow process and solder joints fail in an open or short mode, even if the package meets the coplanarity requirement at room temperature. Migration to thinner package body, finer pitch balls, and lead-free material has increased package warpage during the reflow process and raised problems of open solder joints or solder bridges between balls. It is known that the more a package is moisturized, the more likely the package warps during the reflow process, with one example shown in Figure 6.18. This example is an unpublished result we did in house for a BGA package, with and without moisture saturation for one week. The BGA with moisture experienced popcorning, even though

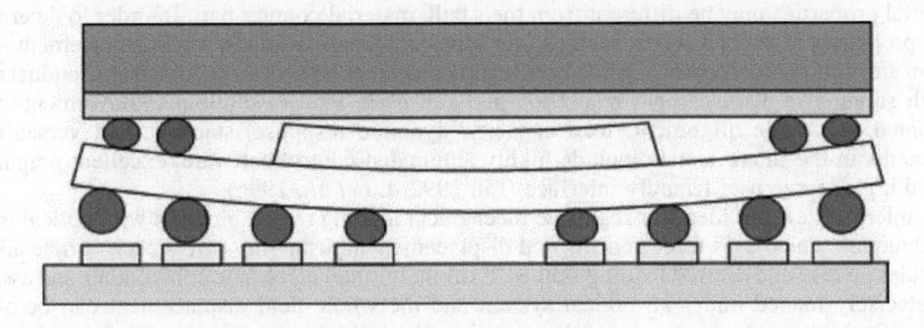

Figure 6.17 Schematic of soldering failure due to warpage

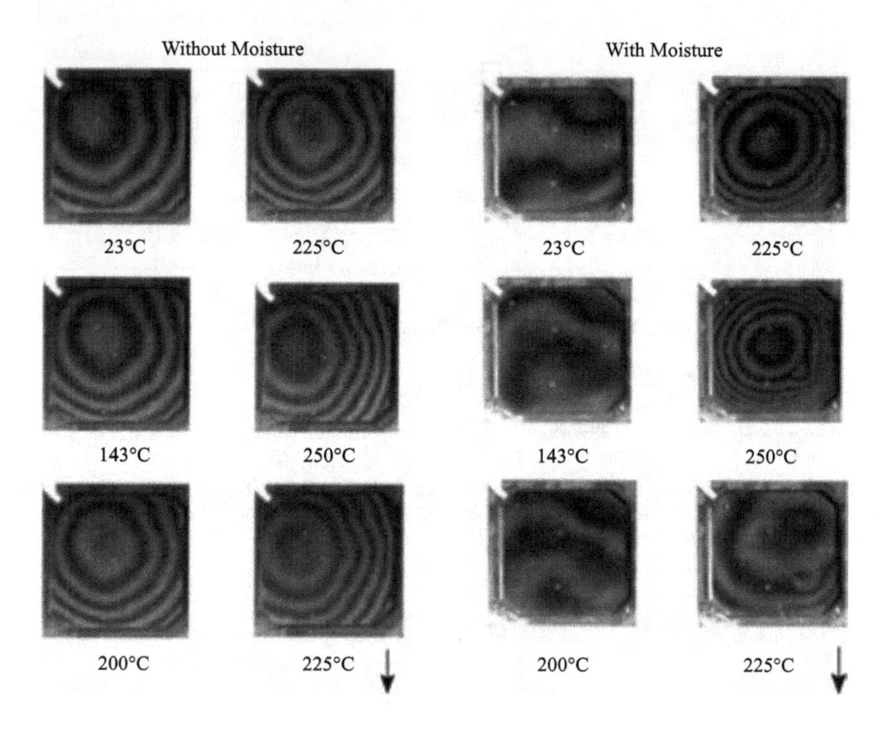

Figure 6.18 Fringe pattern of solder warpage at different temperature with/without moisture

the package was initially flat after the moisture was saturated. The community reached an agreement that the semiconductor suppliers shall specify the maximum permissible package warpages at elevated temperature. It is similar to the package delamination specification at reflow stress. This specification is aimed at agreement of the common terms, unification of measurement methods, and establishment of the criteria.

6.9.1 Shadow Moiré and Project Moiré Setup

As stated earlier, superposing two periodic images generates moiré patterns. In the shadow moiré setup, as shown in Figure 6.19 (Stiteler *et al.*, 1997), one of the two periodic images comes from a glass grating. The other is the shadow of the grating lines on a surface being measured, as the name suggested. The glass grating is a duplicate of a precision master grating, and is used as the reference. Small variations from the reference grating are magnified by moiré fringes and give a quantitative measure of surface topology. A CCD camera captures the moiré patterns and sends them to a computer to interpret the out-of-plane displacement with respect to the reference grating. Diverging white light is used as an illuminating source. Precision motors connected to the sample holder provide better resolution and accuracy capability, which will be discussed later in the phase-stepping technique. Since the setup is mainly used for measuring the warpage of electronic products under thermal processes, thermocouples are used to measure the temperature of the sample at different locations.

Figure 6.20 shows one of several possible projection-moiré setups (Ding *et al.*, 2002). The grating is projected onto the sample surface from faraway. Instead of a glass grating as in the shadow moiré setup, the periodic structure is the interference pattern of laser light. Two projected grating images captured by a CCD camera at different times will be superposed by a computer, and the resulting moiré image will correspond to the surface topology change during that time interval. Again, collimated beams are preferred, but they are not practical in this case, since the aperture of the interferometer is usually much

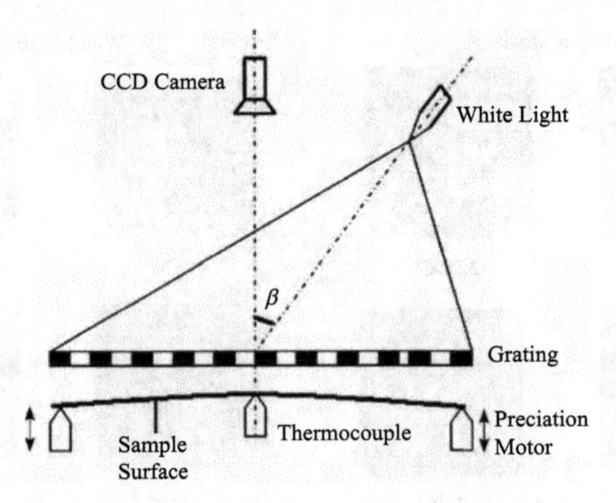

Figure 6.19 Shadow moiré set up (Reproduced with permission from H. Ding, R.E. Powell and I.C. Ume, "Warpage measurement comparison using shadow Moiré and projection Moiré methods," *IEEE Transactions on Components and Packaging Technologies*, **25**, 4, 714–721, 2002. © 2002 IEEE.)

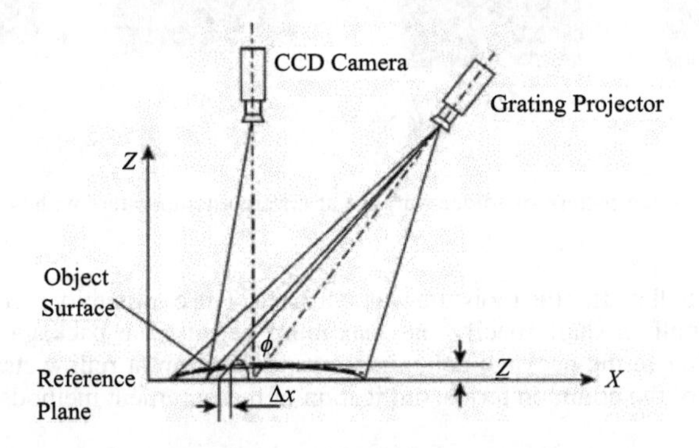

Figure 6.20 Project moiré set up

smaller than the sample surface size. A piezoelectric transducer (PZT), on which the reference mirror is mounted, enables phase stepping, as do the precision motors in the shadow moiré setup.

Supporting mechanisms, control circuits, and computer software, have to be designed and built around the moiré setups to fully exploit their capabilities. An integrated system is shown in Figure 6.21. As a stand-alone subsystem, the oven was designed to simulate thermal processes, for example, solder reflow and composite material curing. Combining the logic of a blowing fan and an exhaust valve, the oven can heat a sample up by infrared and/or by convective heating, or cool it down to room temperature. Thermocouples attached to the sample provide feedback to the oven temperature controllers. A pre-specified temperature profile is stored in the computer and automatically followed, thus allowing warpage study under various thermal processes. When using the shadow moiré method, four precision motors controlled by the central computer enable phase-stepping implementation. Moiré fringes are sensitive to vibration. To get rid of low frequency disturbances from the building floor, two sets of dampers are used to support both the oven and the optical workbench.

A calibration block containing steps of different heights is used to demonstrate the resolution and accuracy of the developed systems. The out-of-plane heights of five stairs are measured by both moiré systems, PM-SM system calibration is shown in Figure 6.22. The results of two systems are shown in Figure 6.23.

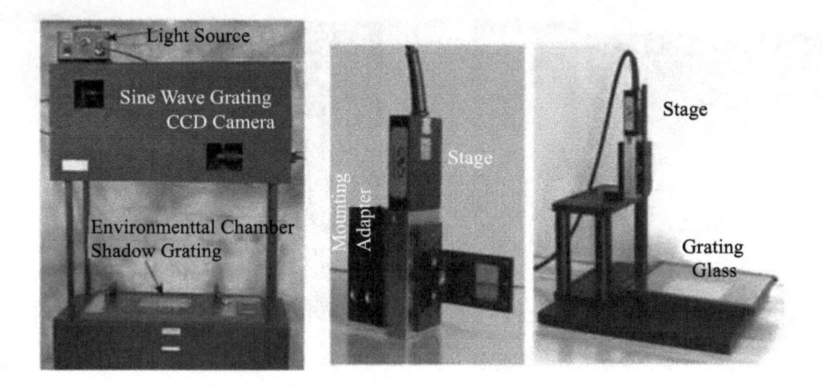

Figure 6.21 Photograph of the integrated system

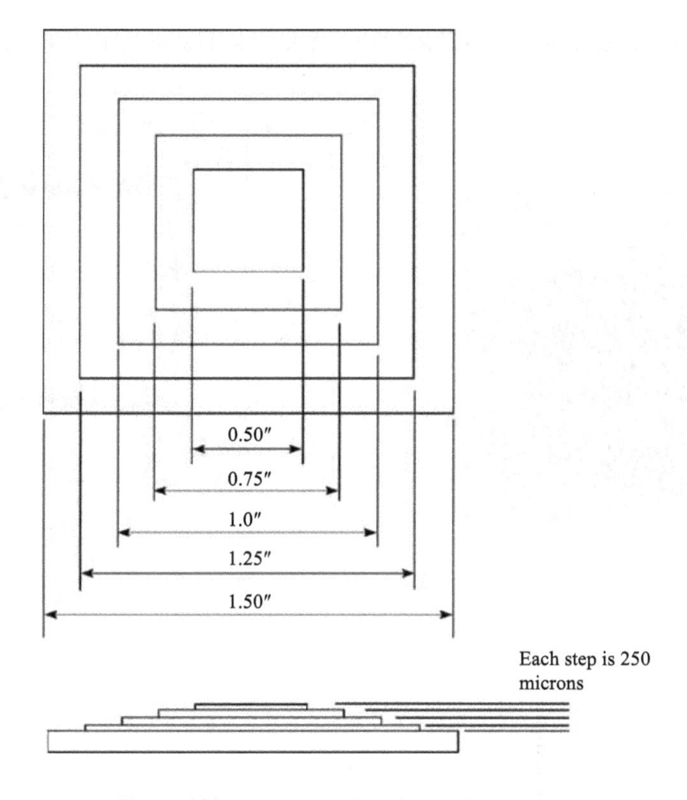

Figure 6.22 A square calibration device (step-step)

6.9.2 Warpage Measurement of a BGA, Two Crowded PCBs

BGA deforms due to many different materials with CTE and Young's modulus and different processing windows (temperature, pressure, and so on). The warpage deformation is important as it will determine if this BGA will experience deformation when it is surface reflowed onto the next level board. One critical issue for the crowded board, for instance, a PCB with many active and passive components, is that it needs

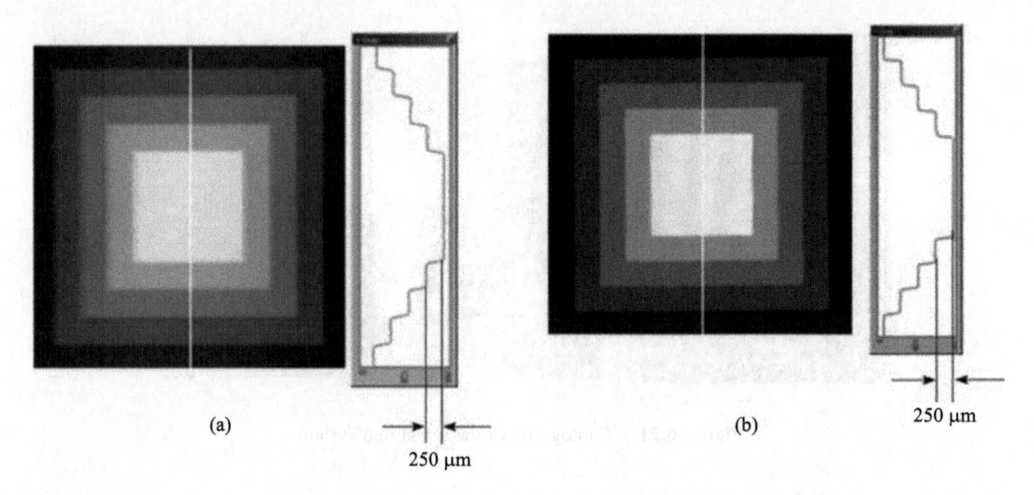

Figure 6.23 (a) Measure result of the project moiré system. (b) Measure result of the shadow moiré system

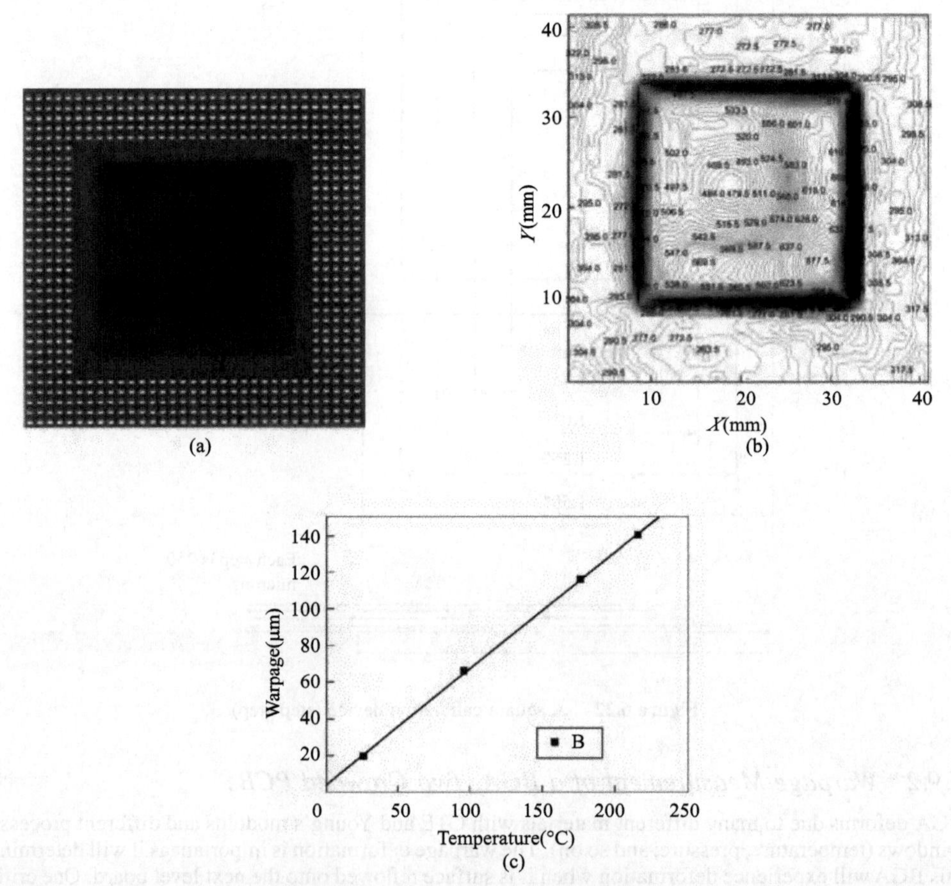

Figure 6.24 (a) A plastic BGA from a commercial vendor. (b) Out of deformation measurement in contour of a plastic BGA. (c) Warpage measurement of a plastic BGA as a function of the temperature

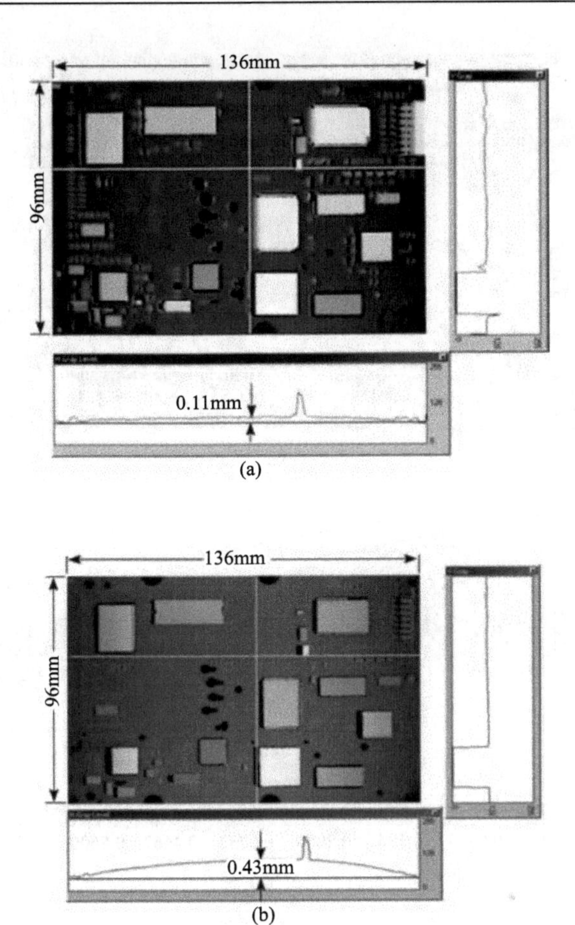

Figure 6.25 (a) Out of deformation in one horizontal line and one vertical line of a crowded board at a room temperature (25 °C). (b) Out of deformation in one horizontal line and one vertical line of a crowded board at a temperature (130 °C)

to have the last component surface mounted onto it. Therefore, the warpage deformation of those crowded boards during one reflow is then critical. In the following, we demonstrate the effectiveness of the combined SM-PM method to the components, modules, and two crowded boards (Figures 6.24, 6.25 and 6.26). Figure 6.24 is a typical plastic ball grid array packaging provided by a commercial vendor. By using SM-PM, the information on the molding compound and BT boards can be easily obtained. The representation can be in terms of the contour of the out of plane deformation and the maximum warpage as a function of reflow temperature, as shown in Figure 6.24(a) and (b), respectively.

Figure 6.25 is a plane crowded board provided by a system vendor in the automotive industry, in which yield and low return rate in terms of parts per million (ppm, 1ppm=10^{-6})are two key issues for a success business. During the manufacturing, it has been found that the warpage in the last process step can create yield problem for the next process step. For instance, when many flip-chip packages need to be surface mounted to the board, the local warpage and co-planarity will be important for the success of mounting of a particular package onto the board. In this case, the deformation or warpage information is critical for the yield of next surface mounting process and this information is important knowledge for process planning. By using SM-PM, the information on the crowded board, the local deformation on the components, and

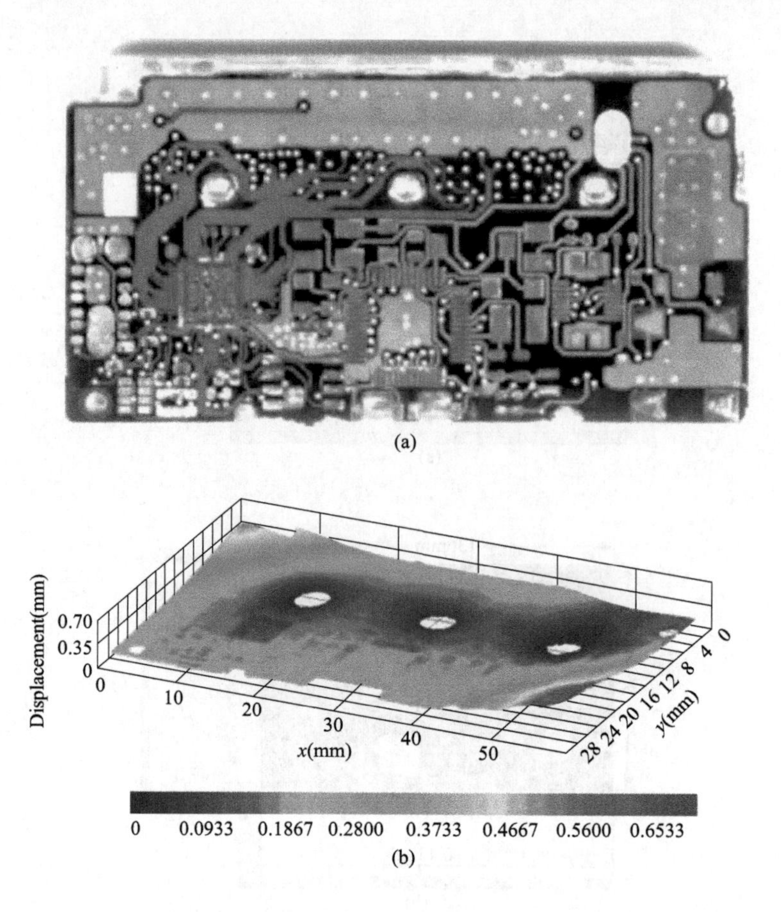

Figure 6.26 (a) Second board for a cell phone with many components. (b) Three dimensional deformation for the second board for a cell phone with many components (*Color version of this figure is available online*)

the deformation on that specific location to mount that particular package can be determined from the room temperature, to the reflow temperature, and to the complete reflow cycle. We can combine this with virtual manufacturing processes and help the information to select various materials and components from different vendors, so as to make the design and processes right first time. The representation can be in terms of the contour of the out of plane deformation and the maximum warpage in different temperatures, along some particular lines, as shown in Figure 6.25(a) and (b), respectively. It is clearly shown that the particular lines do show the existence of the components in both horizontal and vertical directions.

Figure 6.26 presents the second board from a cell phone vendor in the telecommunication industry, in which many active and passive components are focused on integration, such as packaging on packaging (PoP), CSP, and the more complex system in packaging (SiP). By using SM-PM, three dimensional global and local deformation on both the components and the board can be easily presented with given reflow conditions, as shown in Figure 6.26(a) and (b), respectively. The information has been successfully used for the virtual design and virtual manufacturing of the wireless systems for many vendors.

6.10 Chapter Summary

The developments in experimental techniques in structural mechanics analysis have added a new dimension in the testing and measurements in electronic packaging. With these advanced experimental techniques, better understandings of phenomenological behaviors and more accurate reliability predictions can be achieved. It has great potential in the applications of validations of numerical simulations to assist failure analysis and design optimizations. The challenges are in-situ measurements with high displacement and spatial resolutions and testing under precisely simulated environment conditions. Because of the lack of experimental methods which can measure strain and stress inside a component or an assembly, experimental studies on an exact 3D body is difficult. The new experimental techniques should be developed in considerations of hybrid approaches in complementing with numerical analysis. With help from numerical simulations, surface deformations measured by experimental techniques can be much more useful than simply interpreting the measurement data.

References

Bastawros, A. and Voloshin, A. (1990) Thermal strain measurements in electronic packages through fractional fringe moiré interferometry, *ASME Journal of Electronic Packaging*, **112**, December.

Bosc, J.M., Guo. Y., Sarihan, Y.V. and Lee, T. (1998) Reliability evaluation of a silicon based micromachined sensor, to be published in the IEEE Transactions on Reliability, *Optical Engineering* **47**(2):135–141.

Brownell, J. and Parker, R. (1991) Automatic fringe analysis for moiré interferometry, *International Society for Optical Engineering* **1554**, Fringe Pattern Analysis.

Choi, H.C., Guo. Y., Lafontaine, W. and Lim, C.K. (1993) Solder ball connect (SBC) assemblies under thermal loading: II. Strain analysis via image processing, and reliability consideration, *IBM Journal of Research and Development*, **37**, No. 5, September.

Dadkhah, M. and McKie, A. (1996) Real-Time holographic interferometry of power modules, *SEM International Congress on Experimental Mechanics*, Nashville.

Dai, X., Kim, C., Willecke, R., Poon, T. and Ho, P. (1997) Temperature dependent mapping and modeling verification of thermomechanical deformation in plastic packaging structure using moiré interferometry, *SEM Expermental-Numerical Mechanics in Electronic Packaging*, Volume. **1**.

Dally, J. and Read, D. (1993) Electron beam moiré, *Experimental Mechanics* **33**:270–277.

Ding, H., Powell, R.E. and Ume, I.C. (2002) A projection moiré system for measuring warpage with case studies, *Proceeding of International Conference Advances Packaging System*, Reno, NV, Mar.

Drexler, E.S. and Berger, J.R. (1997) Mechanical behavior of conductive adhesives using electron-beam moiré, *Proceedings of 1997 SEM Spring Conference on Experimental and Applied Mechanics*, Bellevue, WA, June.

Guo, Y., Chen, W. and Lim, C.K. (1992) Experimental determinations of thermal strains in semiconductor packaging using moiré interferometry, *Proceeding of the ASME/JSME Electronic Packaging Conference*, San Jose, April.

Guo, Y., Chio, H.C. and LaFontaine, W. (1992) A modified moiré interferometry system for thermal strain measurements, *Proceedings of VII SEM Internationsl Congress on Experimental Mechanics*, Las Vegas, June.

Guo, Y. and Woychik, C.G. (1992) Thermal strain measurements of solder joints in second level interconnections using moiré interferometry, *ASME Journal of Electronic Packaging*, March.

Guo, Y., Lim, C.K., Chen, T. and Woychik, C.G. (1993) Solder ball connect (SBC) assemblies under thermal loading: I. deformation measurement via moiré interferometry, and its interpretation, *IBM Journal of Research and Development*, **37**, No. 5, September.

Guo, Y. and Lim, C.K. (1994) Hybrid method for strain/stress analysis in electronic packaging using moiré interferometry and FEM, *Proceedings of 1994 SEM Spring Conference*, Baltimore, Maryland, June.

Guo, Y. (1995) Determinations of coefficient of thermal expansion in electronic packaging using moiré interferometry, *Proceeding of 1995 International Electronic Packaging Conference*, Lahina, Hawaii, March 26–30.

Guo, Y. (1995) Applications of shadow moiré method in determinations of thermal deformations in electronic packaging, *SEM Spring Conference on Experimental Mechanics*, Grand Rapids, MI, June.

Guo, Y. and Li, L. (1996) Hybrid method for local strain determinations in PBGA Solder Joints, *SEM Experimental/ Numerical Methods on Electronic Packaging*, Vol. **1**.

Guo, Y. and Sarihan, V. (1996) Characterizations of thermal deformation and effective CTE of PBGA Modules, *Proceedings of the Second International Symposium on Electronic Packaging Technology*, Shanghai, China, December.

Guo, Y. and Liu, S. (1997) Development in optical methods for reliability analysis in electronic packaging applications, *Experiment/Numerical Methods in Electronic Packaging SEM*, Vol. **2**.

Guo, Y., Sarihan, V. and Lee, T. (1997) Accelerated reliability testing method for sensors with a pulsing membrane structure, *Applications of Experiment Mechanics to Electronic Packaging*, ASME. EEP **22**/AMD **226**.

Guo, Y. and Sarihan, V. (1997) Residual strain/stress in electronic packages due to processing conditions, *Post Conference Proceedings of 1997 SEM Spring Conference on Experimental and Applied Mechanics*, Bellevne, WA, June.

Guo, Y. and Sarihan, V. (1997) Testing and measurement techniques applied to electronic packaging development, *Applications of Experimental Mechanics to Electronic Packaging, ASME* EEP **22**/AMD **226**.

Guo, Y. and Boeman, P., Ifju, R. and Dai, F. (1997) Formation of specimen gratings for moiré interferometry applications, *Post Conference Proceedings of 1997 SEM Spring Conference on Experimental and Applied Mechanics*, Bellevue, WA, June.

Guo, Y., Yeung, B. and Sarihan, V. (1998) Effect of substrate thickness on reliability of flip-chip PBGA packages, to be published in the *Proceedings of 1998 SEM Spring Conference on Experimental Mechanics*, Houston TX, June.

Guo, Y. and Corbin, J. (1995) Reliability of ceramic ball grid array assembly, *Handbook of Ball Grid Array Technologies*, J.H. Lau, ed., McGraw-Hill, Inc., New York.

Han, B. (1992) Higher sensitivity moiré interferometry for micro-mechanical studies, *Optical Engineering*, Vol. **31**.

Han, B. and Guo, Y. (1995) Thermal deformation analysis of various electronic packaging products by moiré and microscopic moiré interferometry, *Journal of Electronic Packaging, Transaction of ASME*, **117**.

Han, B., Guo, Y., Lim, C.K. and Caletka, D. (1996) Verification of numerical models used in microelectronics packaging design by interferometric displacement measurement method, *ASME Journal of Electronic Packaging*.

Han, B. and Guo, Y. (1996) Photomechanics tools as applied to electronic packaging product development, *SEM Experimental/Numerical Methods on Electronic Packaging*, Vol. **1**.

Han, B. and Guo, Y. (1996) Determination of an effective coefficient of thermal expansion of electronic packaging components: a whole-field approach, *IEEE Transactions on Components, Packaging and Manufacturing Technology, Part A*, Vol. **19**, No, 2.

He, X., Zou, D., Liu, S. and Guo, Y., (1997) A window based graphics interface for phase shifting analysis in moiré interferometry, *Application of Experimental Mechanics to Electronic Packaging*, ASME EEP **22**/AMD **226**.

He, X., Zou, D., Liu, S. and Guo, Y. (1998) Phase shifting analysis in moiré interferometry and its applications in electronic packaging, *Optical Engineering* **37**(5):1410–1419.

Jones, R. and Wykes, C. (1988) *Holographic and Speckle Interferometry*, 2nd edn., Chapter 4, Cambridge University Press, Cambridge.

Lau, J. (1991) *Solder Joint Reliability, Theory and Applications*, Van Nostrand Reinhold, New York.

Liao, J. and Voloshin, A. (1993) Enhancement of the shadow-moiré method through digital image processing, *Experimental Mechanics*, March.

Liu, S. (1997), SRC Annual Report, Contact: 97-PJ-457, December.

Lu, M. and Liu, S., (1996) A multi-axis thermal-mechanical fatigue tester for electronic packaging materials, *Sensing, Modeling and Simulation in Emerging Electronic Packaging*, ASME, EEP **17**.

Masy, W. (1983) Two-dimensional fringe pattern analysis, *Applied Optics* **22**(23).

Michel, B., Vogel, D., Schubert, A., Auersperg, J. and Reichi, H. (1997) The microDAC method; a powerful means to microdeformation analysis in electronic packaging, *Applications of Experimental Mechanics to Electronic Packaging*, ASME, EEP **22**/AMD **226**.

Moore, A. and Tyrer, J. (1990) An electronic speckle pattern interferometer for complete in-plane displacement measurement, *Measurement Science and Techniques* **1**(10).

Morimoto, Y., Seguchi, Y. and Higashi, T, (1988) Application of moiré analysis of strain using Fourier Transform, *Optical Engineering* **27**(8), August.

Niu, T. (1992) 6-axis sub-micron fatigue tester, *Advanced in Electronic Packaging, Proceedings of the Joint ASME/JSME Conference in Electronic Packaging*, Vol. **2**.

Pappalettere, C., Rizzo, G., Sun, W. and Trentadue, B. (1997) Thermal stress determination in a transistor by speckle interferometry, *SEM Spring Conference on Experimental and Applied Mechanics*, Bellevue, WA, June.

Park, S., Dadkhah, M.S. and Motarmedi, E. (1996) Characterization of silicon nitrides thin film by interferometry and FEA, *Proceedings of SEM International Congress on Experimental Mechanics*, Nashville.

Petersen, K. (1982) Silicon as a mechanical material, *Proceeding of the IEEE* **70**(5) May.

Post, D., Han, B. and Ifju, P. (1994) *High Sensitivity Moiré*, Springer-Verlag, New York.

Reid, G. (1986/1987) Automatic fringe analysis- a revise, *Optical and Laser Engineering*, Vol. **7**.

Robinsong, D. (1983) Automatic fringe analysis with computer image processing system, *Applied Optics* **22**(14).

Stiteler, M. and Ume, I.C. (1997) System for real-time measurements of thermally induced warpage in a simulated infrared soldering environment, *ASME Journal of Electronics Packaging* **119**:1–7.

Teng, S., Guo, Y. and Sarihan, V. (1997) A comparative analysis of chip-scale package designs, *Structural Analysis in Microelectronics and Fiber Optics, ASME*, EEP **21**.

Voloshing, A. (1986) Fractional moiré strain analysis using digital imaging techniques, *Experimental Mechanics* **26**.

Wang, Y.P., Prakash, V. and Guo, Y. (1995) Strain sensitivity to geometry and material property in CGA solder connect using hybrid analysis, *Proceedings of the 1995 International Mechanical Engineering Congress and Exhibition, the ASME Winter Annual Meeting*, San Francisco, November.

Wang, Y. and Hassel, P. (1997) Measurement of thermal deformation of BGA using phase shifting shadow moiré, *Post Conference Proceedings of 1997 SEM Spring Conference on Experimental and Applied Mechanics*, Bellevue, WA, June.

Wang, J., Zou, D. and Liu, S. (1997) Creep behavior of a power packaging by both FEM modeling and real time moiré interferometry, *ASME WAM Symposium on Processing Sensing, Monitoring, Modeling, and Validation of Electronic Packaging*, Dallas, November.

Wang, Q., Guo, Y., Chang, S. and Chiang, F.P. (1997) Measurement of local thermal-mechanical deformation in a solder joint using speckle interferometry with electron microscoopy, *Proceeding of 1997 SEM Spring Conference on Experimental and Applied Mechanics*, Bellwvue, WA, June.

Wang, J., Lu, M., Zou, D. and Liu, S. (1997) Investigation of interfacial fracture behavior of a flip-chip package under a constant concentrated load, *Application of Fracture Mechanics in Electronic Packaging*, ASME AMD **222**/EEP **20**: 103–113.

Wang, Y., Prakash, M. and Guo, Y. (1997) Three dimensional optical interferometry/finite element hybrid analysis of a PBGA package, *Advances in Electronic Packaging-1997* Volume 1, ASME, EEP **19-2**.

Wu, T., Tsukada, Y. and Chen, W. (1996) Material and mechanics issues in flip-chip organic packaging, *46th Electronic Components and Technology Conference*, San Jose, CA, May.

Zhu, J., Zou, D., Dai, F., Yang, W. and Liu, S. (1997) High temperature deformation of area array packages by moiré interferometry/FEM hybrid method, *47th Electronic Components and Technology Conference*, San Jose, CA, May 18–21.

Zou, D., He, X., Liu, S. and Guo, Y. and Dai, F. (1997) Resolving deformation field and interfaces by a phase-shifting moiré interferometry, *Applications of Experimental Mechanics to Electronic Packaging*, ASME EEP **22**/AMD **226**.

Zou, D., Lu, M. and Liu, S. (1997) Thermo-mechanical behaviors of electronic packaging material and structures by high temperature moiré interferometry, *Advances in Electronic Packaging-1997, ASME*, EEP **19-1**:1191–1196.

Zou, D., Wang, J., Lu, M. and Liu, S. (1997) Creep behavior study of plastic power package by real time moiré interferometry and FEM modeling, *ASME Symposium on Application Mechanics to Electronic Packaging*, Dallas, November 16–21.

Zou, D., Zhu, J., Dai, F., Liu, S. and Guo, Y. (1997) Global/Local thermal deformation study of power plastic package by a multi-purpose macro/micro moiré interferometer, *Proceedings of 1997 SEM Spring Conference on Experimental and Applied Mechanics*, Bellevue, WA, June.

7

Application of Fracture Mechanics

Fracture mechanics has a distinguished history in the fields of mechanics and materials science from the times of Griffith and Irwin (Griffith, 1921; Irwin, 1960). Fracture theories have found many useful applications in engineering. Many of the materials used in electronic packaging are brittle in nature. Silicon and ceramics are two prime examples. Many of the polymer materials used in packaging are highly filled with inorganic fillers, rendering them brittle also. Perhaps most importantly, with the trend in electronic products towards mass production of cost sensitive, small, light products, the trend for physical design of electronic components is moving towards highly complex composites processed from many different materials with correspondingly many different interfaces. The integrity of materials and their interfaces from chip attach, component fabrication, assembly, test, rework, final assembly, shipment, and field operation are of high concern to the industry.

7.1 Fundamental of Fracture Mechanics

The concepts of fracture mechanics are basic ideas for developing methods of predicting the loading-carrying capacities of structures and components containing cracks. The concepts deal with basic quantities and the parameters of fracture mechanics. The basic quantities in fracture mechanics can be discussed in relation to a simple example: a center crack in a plate remotely loaded by a uniform tensile stress shown in Figure 7.1. When the half-crack length, a, is less than 1/10 of the total plate width, the relationship among stress-intensity factor, K, applied stress, σ, and half crack-length, a, is very close to the relationship for a crack in an infinitely wide plate, which is:

$$K = \sigma\sqrt{\pi a} \tag{7.1}$$

The stress applied to the component, the length of crack, and the stress-intensity factor in the loaded component with a crack are the basic quantities of fracture mechanics. The example in Figure 7.1 also provides a simple explanation for the units of stress-intensity factor, that is, the product of stress and square root of length. But more important is the concept that the stress-intensity factor, K, is a single parameter which includes both the effect of the stress applied to a specimen and the effect of a crack of a given size in a specimen. Still using the example of Figure 7.1, if the combination of σ and a in Equation 7.1 were to exceed a critical value of K, then the fracture strength of the plate would be exceeded and the crack would be expected to propagate.

Modeling and Simulation for Microelectronic Packaging and Integration: Manufacturing, Reliability and Testing, Second Edition. Sheng Liu and Yong Liu.

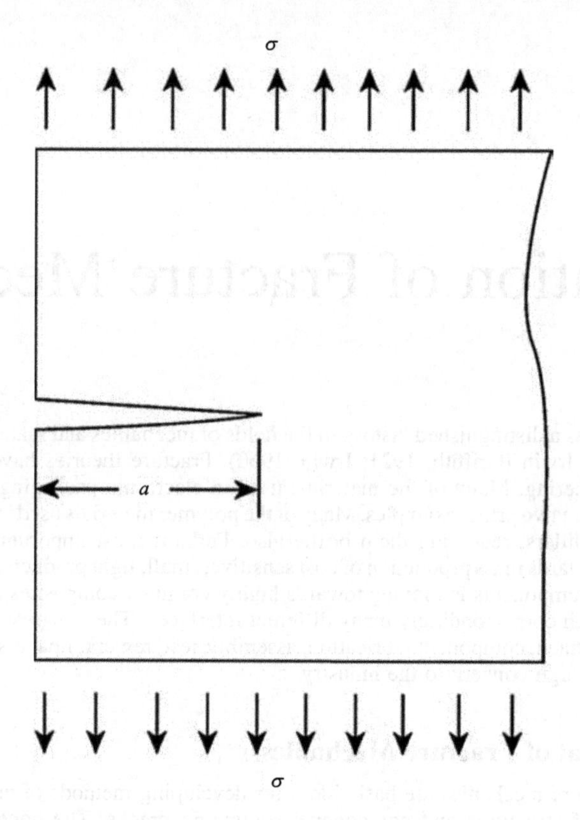

Figure 7.1 Schematic illustration of a center crack in a wide plate

7.1.1 Energy Release Rate

The origins of modern-day fracture mechanics may be traced to Griffith (Griffith, 1921), who established an energy release rate criterion for brittle materials. Observations of the fracture strength of glass rods had shown the longer the rod, the lower the strength. Thus the idea of a distribution of flaw sizes evolved, and it was discovered that the longer the rod, the larger the chance of finding a large natural flaw. This physical insight led to an instability criterion which involved the elastic energy released in a solid at the time a flaw grew catastrophically under an applied stress.

From the theory of elasticity comes the concept that the strain energy contained in an elastic body per unit volume is simply the area under the stress-strain curve, or:

$$U_0 = \frac{\sigma^2}{2E} \tag{7.2}$$

where σ is the applied stress and E is Young's modulus. However, there is a reduction (that is a release) of energy in an elastic body containing a flaw or crack because of the inability of the unloaded crack surface to support a load. We assume that the volume of material whose energy is released is the area of an elliptical region around the crack (as shown in Figure 7.2) multiplied by the plate thickness B; the volume is $\pi(2a)(a)B$. This is based on the area of an ellipse being $\pi r_a r_b$, where r_a and r_b are the

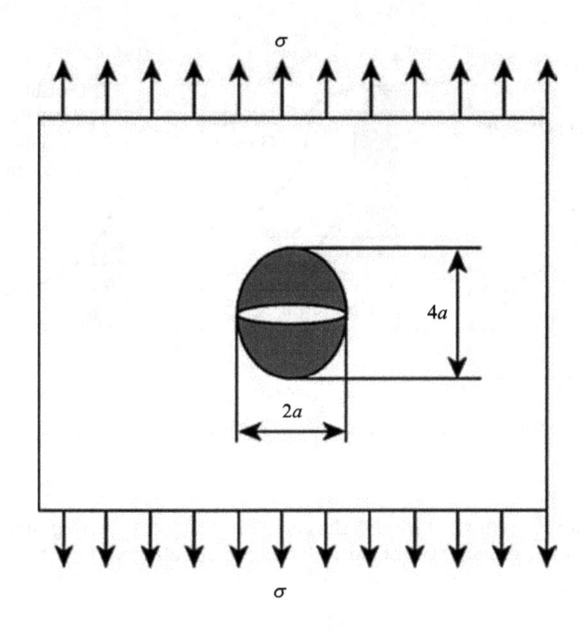

Figure 7.2 Schematical of the concept of energy release around a center crack in a loaded plate

major and minor radii of the ellipse. Then, the total energy released from the body due to the crack is the energy per unit volume multiplied by the volume, which is:

$$U = \pi(2a)(a)B\frac{\sigma^2}{2E} = \frac{\pi\sigma^2 a^2 B}{2E} \tag{7.3}$$

In ideally brittle solids, the released energy can be offset only when the surface energy is absorbed, which is:

$$W = (2aB)(2\gamma_s) = 4aB\gamma_s \tag{7.4}$$

where $2aB$ is the area of crack and $2\gamma_s$ is twice the surface energy per unit area (because there are two crack surfaces).

Griffith's energy-balance criterion, in the simplest sense, is that crack growth will occur when the amount of energy released due to an increment of crack advance is larger than the amount of energy release rate absorbed:

$$\frac{dU}{da} \geqslant \frac{dW}{da} \tag{7.5}$$

Performing the derivatives indicated in Equation 7.5 and rearranging it give the Griffith criterion for crack growth:

$$\sigma\sqrt{\pi a} = \sqrt{2E\gamma_s} \tag{7.6}$$

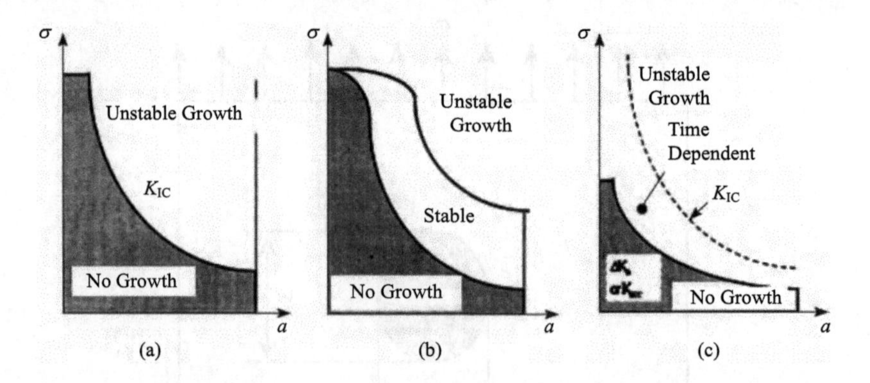

Figure 7.3 Relationship between stress and crack length, showing regions and types of crack growth

Fracture theory was built upon this criterion in the early 1940's by considering that critical energy release rate, G_c, required for crack growth was equal to twice the effective surface energy, γ_{eff}:

$$G_c = 2\gamma_{eff} \tag{7.7}$$

This γ_{eff} is predominantly the plastic energy absorption around the crack tip, with only a small part due to the surface energy of the crack surface. Then, with the development of complex variable and numerical techniques to define the stress fields near cracks, this energy view is supplemented by stress concepts—that is, the stress-intensity factor, K, and a critical value of K for crack growth, K_c. Replacing γ_s with γ_{eff} in Equation 7.6 and noting that the energy and stress concepts are essentially identical (that is, $K = \sqrt{EG}$) gives:

$$K_c = \sqrt{EG_c} = \sigma\sqrt{\pi a} \tag{7.8}$$

which is the crack growth criterion equivalent of Equation 7.2. Thus, K_c is the critical value of K which, when it is exceeded by a combination of applied stress and crack length, will lead to crack growth. For the thick-plate plane-strain conditions, this critical value became known as the plane-strain fracture toughness, K_{IC} (or G_{IC}), and any combination of applied stress and crack length that exceed this value could produce unstable crack growth, as indicated schematically in Figure 7.3(a) (linear-elastic).

In work with tougher, lower-strength materials, it was later noted that stable slow crack growth could occur, though accompanied by considerable plastic deformation. Such phenomena led to the nonlinear J-integral and R-curve concepts which could be used to predict the onset of stable slow crack growth and final instability under elastic-plastic conditions, as noted in Figure 7.3(b). Finally, the fracture mechanics approach was applied to characterize subcritical crack growth phenomena where time-dependent slow crack growth, da/dt, or cyclic crack growth, da/dN, may be induced by special environmental threshold, K_{ISCC}, or fatigue threshold, ΔK_{th}, subcritical growth occur, as indicated in Figure 7.3(c).

7.1.2 J Integral

The J-Integral, proposed by Rice (Rice, 1967; Rice, 1968), characterizes the stress-strain field at a crack tip by integrating a path far from the crack tip and substituting this path for another close to the crack tip (Figure 7.4). The integral path taken, the J-Integral, is path independent

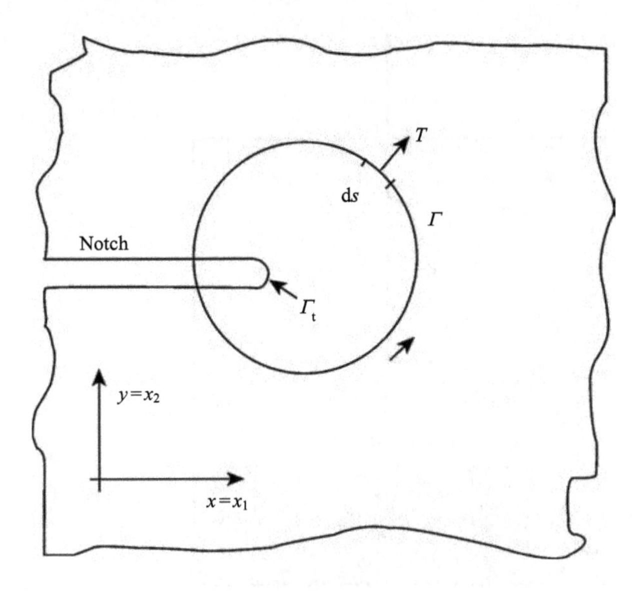

Figure 7.4 Contour of the J-Integral

(Barson *et al.*, 1987). Rice defines a path independent integral for elastic-plastic deformation, in two dimensions, as follows:

$$J = \int_{\Gamma} \left(w\mathrm{d}y - T \cdot \left[\frac{\partial u}{\partial x} \right] \mathrm{d}s \right) \tag{7.9}$$

where x and y are rectangular coordinates normal to the crack front; ds is an increment of an arc length along any contour Γ; T is the traction vector exerted on the material along the contour; u is the displacement and w is the strain energy density.

7.1.3 *Interfacial Crack*

An interfacial crack which lies along the interface of two dissimilar elastic materials is shown in Figure 7.5. Williams was the first researcher to study the interface crack in an isotropic dissimilar material (Williams, 1959). After the complex singularity at the interface crack-tip was discovered by Williams, the full asymptotic elastic field in the vicinity of a crack-tip under the completely open crack surface condition was obtained in (Erdogan, 1963; Erdogan, 1965; England, 1965; Sih, 1964; Rice, 1965). In these works, a complex interface stress intensity factor does not possess the same physical meaning as in a homogeneous crack. Rice (Rice, 1988) restudied the complex stress intensity factor and defined a new stress intensity factor such that the crack-tip field normalized at a characteristic length was given by:

where the vector K:

$$K^{\mathrm{T}} = \{K_{\mathrm{I}}, K_{\mathrm{II}}, K_{\mathrm{III}}\} \tag{7.10}$$

defines the interfacial mode "I, II, III" stress intensity factors. These factors have the classical units of FL^2

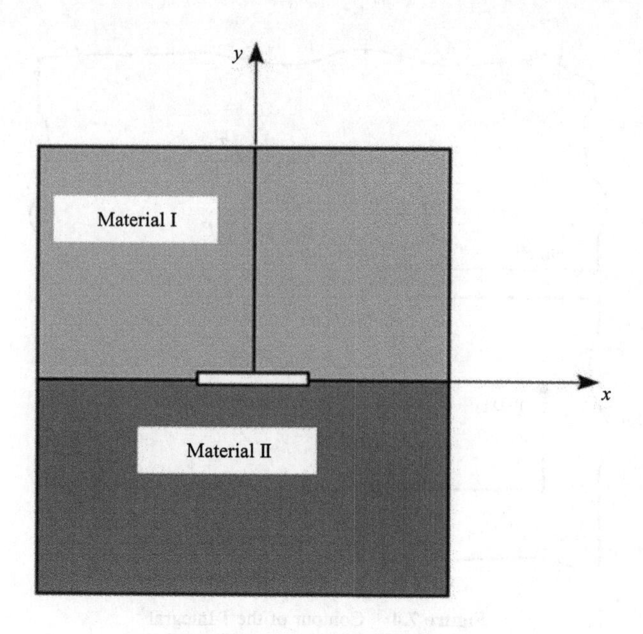

Figure 7.5 An interfacial crack

For a bimaterial system with a crack along interface (the boundary of two materials), the fracture conditions of interest at the bond interface are mixed mode in nature, that is, conditions with combined *K* and *K* stress intensity factors. In order to evaluate the mode mixity of the crack tip, the term phase angle which is related to the ratio of *K* to *K* is introduced. The crack tip mode mixity can be expressed as:

$$\varphi = \tan^{-1}\left(\frac{K_{\mathrm{II}}}{K_{\mathrm{I}}}\right) \tag{7.11}$$

Based on the crack tip displacement field (Sun *et al.*, 1996), the energy release rate and phase angle for a bimaterial system with a crack along the interface can be written as (Hutchinson *et al.*, 1992; Wang *et al.*, 1998):

$$G = \frac{1}{4}\left[\frac{D_{11}}{\cosh^2 \pi\varepsilon} K_{\mathrm{I}}^2 + \left(D_{11} - \frac{W_{21}^2}{D_{11}}\right) K_{\mathrm{II}}^2\right] \tag{7.12}$$

$$\varphi = \tan^{-1}\left(\frac{K_{\mathrm{II}}}{K_{\mathrm{I}}}\right) \tag{7.13}$$

The above parameters D_{11}, K_{I}, K_{II}, W_{21} and ε are defined as follows:

$$D_{11} = \frac{1-\nu_1}{G_1} + \frac{1-\nu_2}{G_2} \tag{7.14}$$

$$K_{\mathrm{I}} = \frac{\Delta u_1 \mathrm{Im}(\zeta_2) + \Delta u_2 \mathrm{Re}(\zeta_2)}{\mathrm{Im}^2(\zeta_2) + \mathrm{Re}^2(\zeta_2)} \frac{1}{D_{11}} \sqrt{\frac{\pi}{2r}} \tag{7.15}$$

$$K_{\mathrm{II}} = \frac{\Delta u_1 \mathrm{Re}(\zeta_2) + \Delta u_2 \mathrm{Im}(\zeta_2)}{\mathrm{Im}^2(\zeta_2) + \mathrm{Re}^2(\zeta_2)} \frac{1}{D_{11}} \sqrt{\frac{\pi}{2r}} \tag{7.16}$$

$$W_{21} = \frac{1-2v_1}{2G_1} - \frac{1-2v_2}{2G_2} \tag{7.17}$$

$$\varepsilon = \frac{1}{\pi} \ln \frac{G_1 + G(3-4v_1)}{G_2 + G_1(3-4v_2)} \tag{7.18}$$

Parameter ζ_2 is of the form:

$$\zeta_2 = \frac{1}{(1+4\varepsilon^2)\cosh \pi\varepsilon}(1-2i\varepsilon)\left\{\cos\left[\varepsilon \ln\left(\frac{r}{2a}\right)\right] + i \sin\left[\varepsilon \ln\left(\frac{r}{2a}\right)\right]\right\} \tag{7.19}$$

where: Δu_1 and Δu_2 are the near-tip displacements in both opening and shearing directions respectively; G_1 and v_1, the shear modulus, and Poison ratio of material I respectively; and G_2 and v_2, the shear modulus, and Poisson ratio of material II respectively. It can be seen from Equation 7.12 to Equation 7.19 that the energy release rate and phase angle are only the near crack tip displacement-based functions. That is, the energy release rate and phase angle can easily be calculated through Equation 7.12 to Equation 7.19 if the near tip displacement fields are previously determined. The near tip displacement fields can be obtained through either experimental or numerical work such as finite element analysis. A high density laser moiré interferometry technique will be used to determine the near crack tip displacement fields of the three-point bending specimen.

7.2 Bulk Material Cracks in Electronic Packages

7.2.1 Background

In microelectronic packaging, various types of cracks could occur during manufacturing, testing, and operation. Besides delaminations along interfaces between dissimilar materials in electronic packages, chip cracking, molding compound cracking, or any other cracking located inside bulk materials in electronic packages are also the main damage modes which may cause fatigue problems. These cracks in bulk materials may propagate and sooner or later reach the life of the devices.

It has been shown (Hutchinson, 1968; Rice *et al.*, 1968; Begley *et al.*, 1972) that the plastic stress and strain singularities at the crack tip are related to the J-integral. Thus, the J-Integral is a parameter that can be used as a failure criterion in the case of large-scale plasticity at the crack tip.

In general, we are given two group values for the delaminated bimaterial systems, such as energy release rates. One is for the interfaces and the other for the bulk materials. In order to judge if the crack propagates along the interface or into the bulk material, these two groups of values will be evaluated. For instance, given two ratios $J^{\mathrm{B}}/J^{\mathrm{B}}_{\mathrm{C}}$ and $J^{\mathrm{I}}/J^{\mathrm{I}}_{\mathrm{C}}$

If:

$$\frac{J^{\mathrm{B}}}{J^{\mathrm{B}}_{\mathrm{C}}} \geqslant \frac{J^{\mathrm{I}}}{J^{\mathrm{I}}_{\mathrm{C}}} \tag{7.20}$$

the crack will branch into the bulk material (Hutchinson *et al.*, 1968). Otherwise, the crack will propagate along the interface. In the above equation, superscript I refers to the interfaces; superscript B to the bulk material, and the subscript C to the critical data. Figure 7.6 shows a schematic of interfacial cracking and branching for a bimaterial system, and applied loads. The growth of interfacial crack branching into brittle materials has been discussed previously (Liu *et al.*, 1993; Liu *et al.*, 1994; Liu *et al.*, 1995).

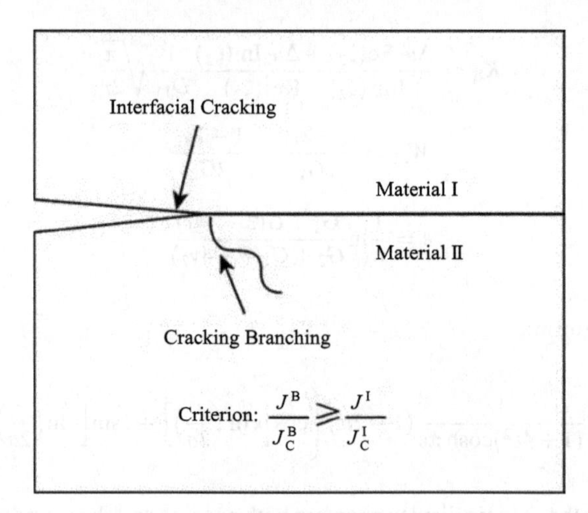

Figure 7.6 Interfacial cracking and branching for a bimaterial system

If the crack branches into the bulk material, the direction at which the total energy release rate reaches the maximum value can be used to predict the direction of the crack initiation. During the crack growing phase, the direction of the crack propagation needs to be adjusted to fit the curve of the crack growth correctly.

The hoop stress around the crack tip is generally the function of the angle ψ and there exists a maximum hoop stress around the crack tip (Deng, 1994). In this chapter it is argued that, for a steadily growing interface crack in an isotropic bimaterial, the singular part of the circumferential stress at the crack tip is always positive in one solid and negative in the other, which implies that if the crack is curved out of the interface, it is likely that it is going to branch into the solid with positive hoop stress. Furthermore, the crack branching or kinking is more likely to occur towards the softer material because the hoop stress is found to be mostly positive values of the fracture toughness of the interface and the component materials. The following parts will present some numerical analysis results.

7.2.2 *Crack Propagation in Ceramic/Adhesive/Glass System*

Die attach materials, such as filled epoxy, are important for various electronic packages. Some studies (Levis *et al.*, 1994; Chen *et al.*, 1993; Wakizaka *et al.*, 1993) have indicated that the system reliability depends on the failure of die attach materials and their associated interfaces.

Alumina/glass system bonded by conductive adhesive is a typical material combination widely used for electronic packaging. Its thermal-mechanical performance and fracture behaviors at different temperatures are critically important for reliability of electronic devices. Therefore an alumina/ conductive adhesive/glass system was selected as a model system to address several issues in terms of crack initiation and growth. A combined experimental and analytical study was employed to investigate the mechanical behaviors of the material system with a surface notch and predict the initiation and the growth of cracking.

(i) Experiments

The experimental work was conducted to investigate the strength of material systems including glass, Al_2O_3, and adhesive, as well as their thermal-mechanical properties. Three-point-bending tests were undertaken at different temperatures. The single-notched bending (three-point loading) specimen is shown in Figure 7.7. The specimens were made by bonding glass (borosilicate) with Al_2O_3. The thickness of the glass and Al_2O_3 are $0.7621 + 0.01$ mm and $0.787 + 0.01$ mm, respectively. The widths of both the glass and Al_2O_3 are $4.0 + 0.01$ mm, and the lengths of both materials are 25 mm (not critical). The two

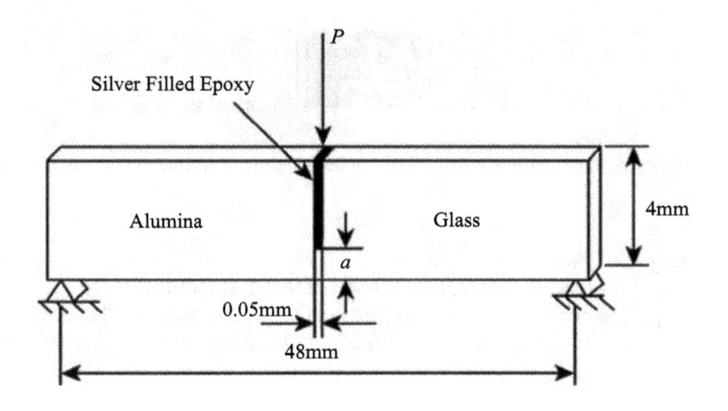

Figure 7.7 Three-point bending specimen

materials were glued together by JM7000 silver filled epoxy. The notch was made by controlling the bonding line manually. The depth of the notch was controlled by inserting a Teflon layer (0.035 mm) and adjusting the claps before curing. According to the specification of the bonding material, the specimens were put in an oven at 200 °C for 7 minutes or 15 minutes.

Tests were performed on a Rheometrics solid analyzer (RSA-2). As shown in Figure 7.8, the specimens were gripped by a bending fixture between a linear servo motor and a machine transducer. The motor applied a strain to the specimen and the transducer measured the corresponding force.

In the tests, the inner volume of an oven chamber and the fixtures in contact with the specimen were cooled down to $-55\,°C$ by boiling liquid nitrogen or heated to $+140\,°C$ by a controlled ac voltage applied to the convection gun heater. The test temperature was controlled to be within $+2\,°C$. The changes of temperature and loading were monitored by a microcomputer.

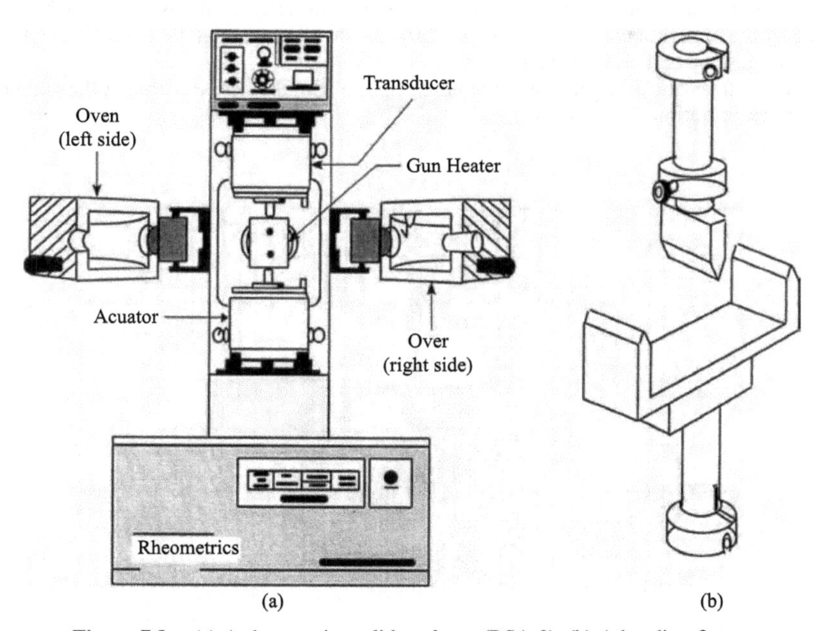

Figure 7.8 (a) A rheometrics solid analyzer (RSA-2). (b) A bending fixture

Figure 7.9 A finite element mesh for a three-point bending alumina/adhensive/glass specimen

Three-point bending specimens were loaded gradually all the way to the fracture point. The load rate was 10^{-6} m/s. The applied load and load line displacement (deflection) were recorded by computer automatically. After specimen breaking, the fracture surfaces were examined to determine the fracture pattern and to measure the initial crack length and branching angle under optical microscope.

(ii) Stress and Fracture Analysis

In the stress and fracture analysis, a constant temperature drop has been considered and therefore residual stresses due to conductive epoxy curing have been taken into consideration. The specimen is idealized by a finite element mesh shown in Figure 7.9. It is noted that a very fine local mesh is used in order to handle the thin conductive epoxy and predict the local crack initiation and growth. A close-up view of the mesh is shown in Figure 7.10, which covers the area of the adhesive, to appear experimentally. The J-Integral is implemented. Although the 'true' stress and strain relationship needs to be implemented in a fully nonlinear finite element code, only engineering stress and strain curves are given here, as shown in Figure 7.11. It is noted that, except for the high temperature case, no significant necking was observed in specimens for low temperatures, and room temperatures should provide a good approximation as the input for a nonlinear finite element code. However, certain errors may occur for specimens at high temperature and the true stress and strain curve for high temperature is not available due to the difficulty in measuring the true necking area. The coefficients of thermal expansions (CTE's) are also temperature dependent and are implemented in the nonlinear finite element code.

It is found in the preliminary stress analysis that there exists a stress and strain concentration around the notch tip near the interface of glass and epoxy because of the combination effect of temperature drop and bending moment. It can be fairly assumed that the crack initiation begins at this point and this assumption was verified by the experiments, even though the cracks may propagate along the interface between the epoxy and the glass or branch into the glass.

Both the hoop stress and J-integral near the notch tip are adopted here to verify if they can be used to predict the crack initiation and growth.

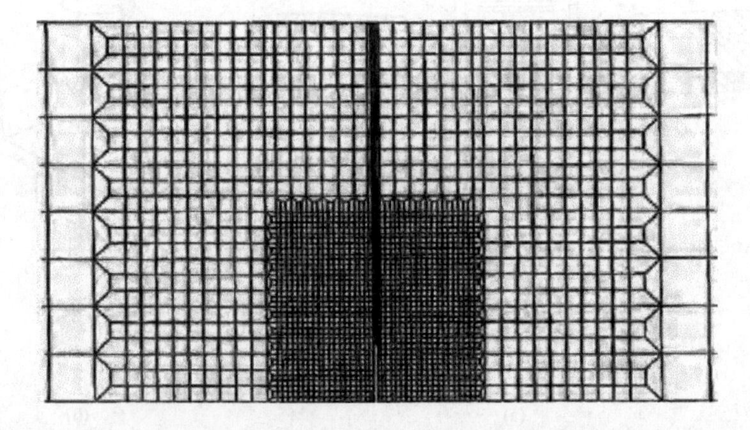

Figure 7.10 A local general finite element fine mesh for a three-point bending alumina/adhensive/glass specimen

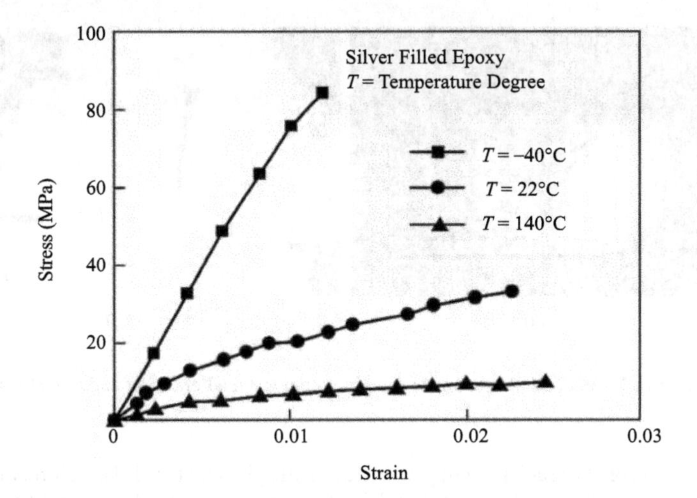

Figure 7.11 Engineering stress and strain curves for conductive epoxy at three different temperatures

When the crack propagates, there are two possibilities for the crack propagation direction. One is the crack along the interface (Figure 7.12(b)). The other is the crack branching into the bulk material (Figure 7.12(c)). These were confirmed by the experiments (Figure 7.13). The Equation 7.20 is used to judge the direction of crack propagation.

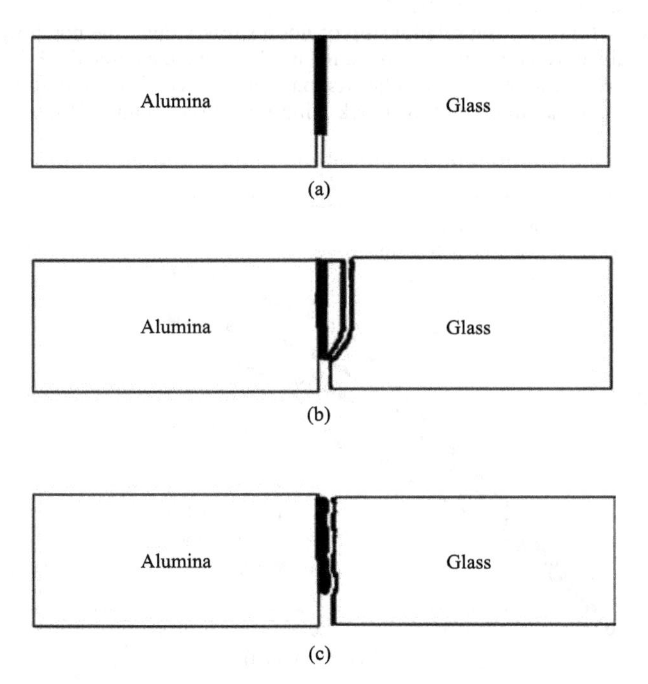

Figure 7.12 A schematic of interfacial cracking and branching into the glass

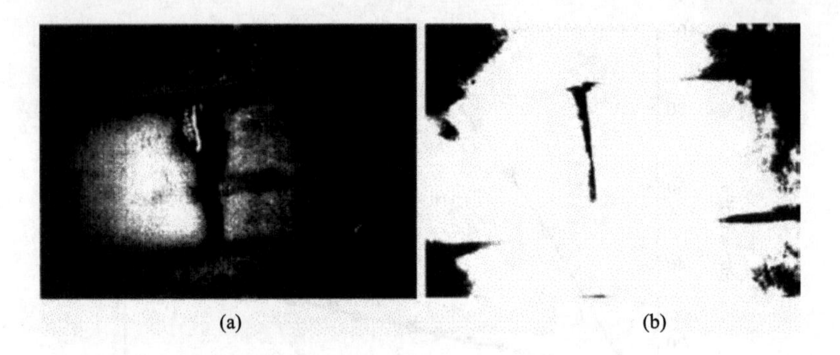

Figure 7.13 Failed specimens showing the interfacial cracking and branching

Here the attention is on the crack growth path inside brittle glass material. Therefore, the emphasis is on the temperature $-40\,^{\circ}C$ and $125\,^{\circ}C$ as it is found that the crack grows inside the glass for the case studied. The growth of interfacial crack branching in brittle materials has also been discussed in (Nguyen *et al.*, 1997; Jiao *et al.*, 1997; Yan *et al.*, 1997).

7.2.3 Results

Figure 7.14 shows the load deflection responses for specimens with different notch depths. It can be clearly observed that nonlinear behavior exists in room temperature due to the nonlinear constitutive relationship for conductive epoxy at room temperature. It is found that the deeper the notch depth, the softer the loads deflection response is.

Figures 7.15 and 7.16 present the relationship of hoop stress around the notch tip with the angle θ. These results are obtained by a locally very fine radial mesh around the notch tip. It is observed that there exists an angle at which the hoop stress reaches its maximum value. Thus, it is supposed that the hoop stress can be used as an index to predict the crack initiation and branching. However, it is necessary to

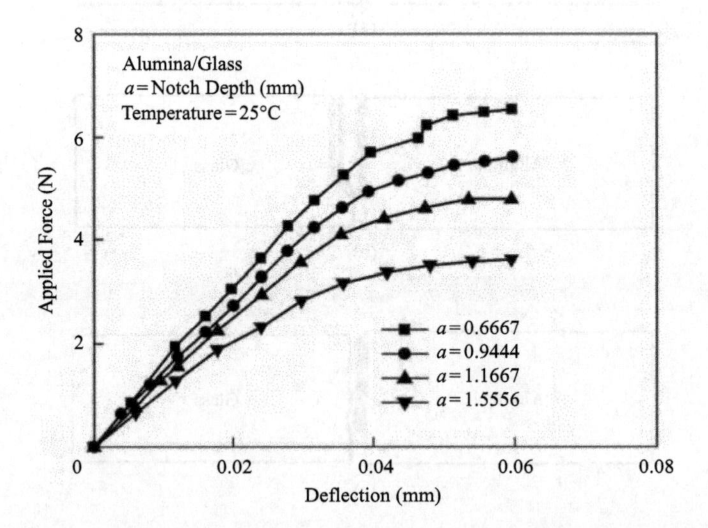

Figure 7.14 Load-deflection curves for specimens with different notch depths

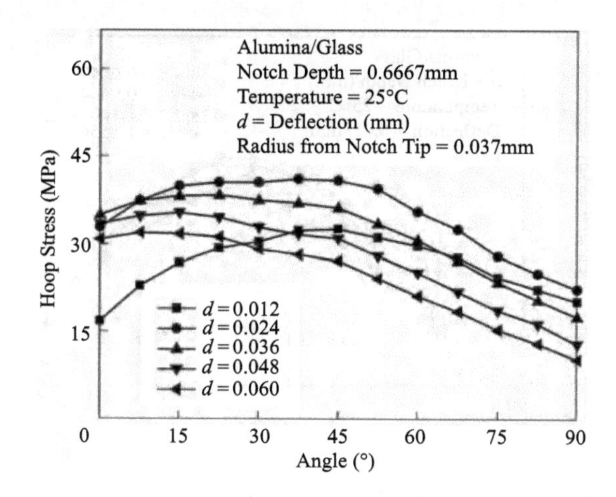

Figure 7.15 Hoop stress for different displacements on the radius 0.037 mm

generate a particular mesh to obtain the results. In the following it will be shown that the J-integral can be used to predict the crack initiation and growth even by regular mesh.

Figure 7.17 shows the values of J-integral with respect to angle θ corresponding to different notch depths and an applied displacement 0.018 mm.

In terms of the location and direction of the crack initiation in the glass, it is also noted from the above figures that the maximum J-integral value falls in an angle between 11.25° and 25° depending on the notch depth. In experiments, four specimens failed due to glass cracking and two failed along the epoxy/glass interface. For the specimens that failed in glass, the cracking angle is relatively small and falls in the predicted range as shown above. For the specimens which failed in the interface, it is found that the failure loads are relatively low compared with the loads for glass failure. It is supposed that this is caused by the weak interface bonding. Due to the availability of the fracture toughness data for both the interface and the bulk glass at room temperature, it is believed that the prediction is fairly good as compared with the experiments.

Figure 7.18 present one of the deformed local mesh where the crack surface is open. It is observed that the peak value of J-integral changes toward the interface (that is, decreasing the angle θ) with the increase

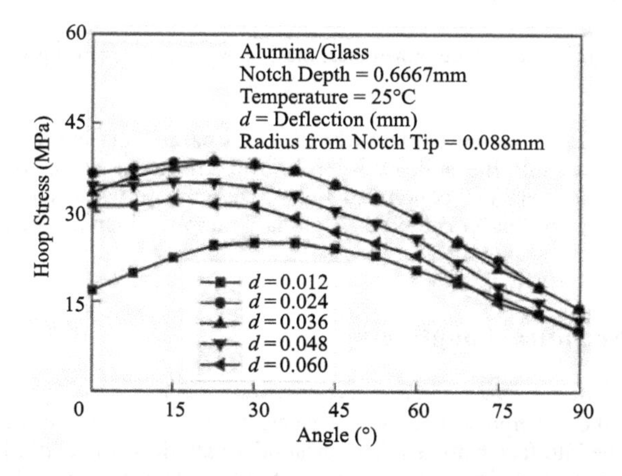

Figure 7.16 Hoop stress for different displacements on the radius 0.088 mm

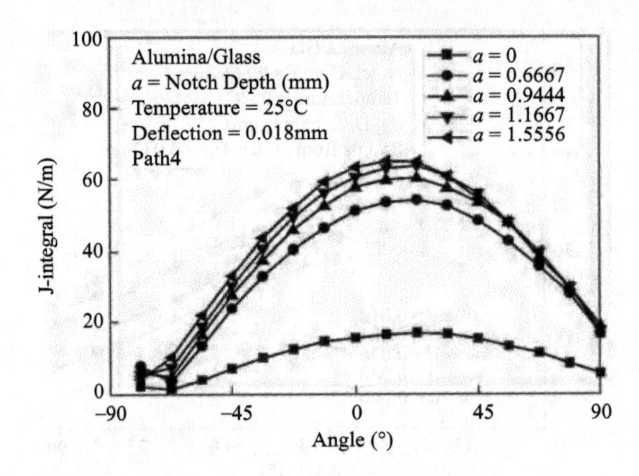

Figure 7.17 J-integral value for various notch depths with respect to the angle θ corresponding to an indentation depth (0.018 mm)

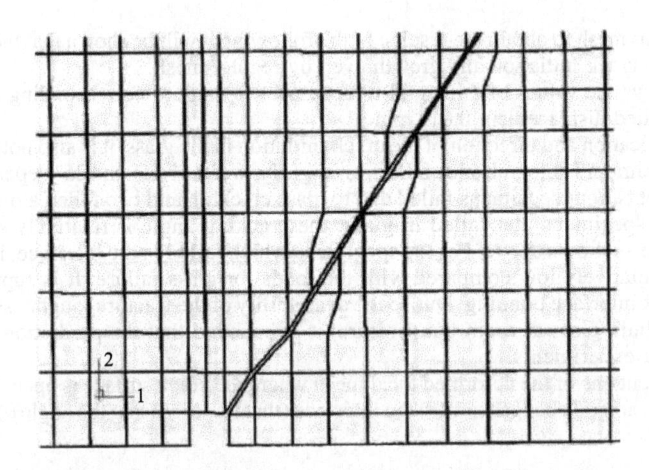

Figure 7.18 A deformed local mesh showing crack growth path

of the arc length. This observation may explain why the crack path is curved towards the direction of the interface after it penetrates into the bulk glass. If the fracture toughness is assumed to be constant (brittle system), it can be therefore concluded that the cracks grows unstably, which results in a catastrophic failure of the system considered. For a large applied displacement, Figure 7.19 shows a similar phenomena.

7.3 Interfacial Fracture Toughness

7.3.1 Background

The mechanical integral of layered structures such as electronic packages is directly determined by the adhensive strength of the interface between these dissimilar materials (Liu *et al.*, 1993; Liu *et al.*, 1994; Liu *et al.*, 1995). Therefore, experimental measurement of the adhensive strength between interfaces in plastic electronic packages for integrated circuits is of critical importance for the determination of

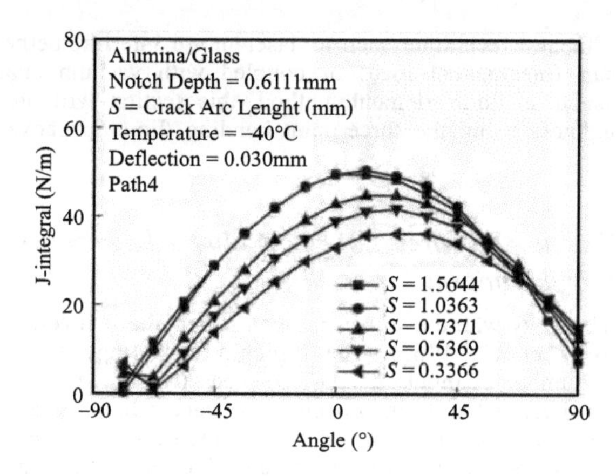

Figure 7.19 J-integral values with respect to the angle θ for five different crack arc lengths and applied displacement of 0.030 mm

package reliability. In particular, increasing the adhensive strength of the interfaces may significantly help improve the life of electronic devices. Therefore, precisely measuring and evaluating the adhensive strength of the electronic packages has become a major concern.

It is common knowledge that one of the most usually encountered failure modes in electronic packages is delamination. Unlike crack problems in homogeneous media, delamination cracks found in electronic packages are often subjected to mixed-mode loading situations. Application of fracture mechanics concepts in the measurement of interface bond strength is now commonly recognized as the preferred methodology for characterizing interface integrity. It is also commonly recognized that the fracture conditions of interest at the bond interface are mixed mode in nature, that is, conditions with combined K_I and K_{II} stress intensity factors. However, only a qualitative assessment of crack resistance can be measured using Mode I test and there definitely exists a level of uncertainty as to how to apply such data in the design of new packages.

A failure envelope containing interfacial fracture toughness as a function of mode-mixity would allow a more quantitative assessment of the interfacial crack resistance and facilitate the design of new microelectronic package. It is common practice to use a number of different specimen geometrics such as the double cantilever beam and the edge notched flexure. It is obvious that such a strategy can be time-consuming and expensive. Moreover, it is often difficult to produce the different specimen shapes using identical processing conditions. Therefore, it would be useful to evaluate test methods that use the same specimen geometry, but vary the geometry of the test fixture or change the ratio of the displacement-controlled multi-axial concentrated line loads to control the mode-mixity at the crack tip.

In an effort to perform fracture tests under mixed-mode conditions, Reeder and Crews (Reeder *et al.*, 1990) have developed a mixed-mode bending test to measure interlaminar fracture toughness as a function of mode-mixity for carbon reinforced epoxy composites. Such a test method used the same specimen geometry and varies the mode-mixity at the crack-tip by changing the fixture settings. Other investigators (Suo, 1992; Goodle *et al.*, 1995; Brandenburger *et al.*, 1995; Lu *et al.*, 1996; Guo *et al.*, 1993; Guo *et al.*, 1994; Guo *et al.*, 1995) have also developed various test fixtures that permit control of the mixed-mode conditions, through control of the ratio of applied shear stress to normal stress. However, the above mentioned test methods only provide one value of interfacial fracture toughness with one specimen during each test. Since the interfacial fracture toughness of the interface between dissimilar materials is related to the phase angle, a number of tests need to be conducted in order to acquire the relationship between the interfacial fracture toughness and the phase angle for a bimaterial system. Such a strategy can also be time-consuming and expensive. Therefore, in order to shorten time and reduce the cost of the test, a new facility and methodology need to be developed. They should be capable of providing a series of data of interfacial fracture toughness which are related to different phase angles with one specimen during each test.

By combining experimental technique such as laser moiré interferometry coupled with 6-axis submicron fatigue tester (mechanical load) or coupled with vacuum chamber (thermal load) with numerical work such as finite element method, this section will present some results of interfacial fracture toughness using the three-point bending flip-chip package specimen and the bimaterial specimen.

7.3.2 *Interfacial Fracture Toughness of Flip-Chip Package between Passivated Silicon Chip and Underfill*

Flip-chip attach is a technology where a solder-bumped silicon die is inverted and attached directly to a chip carrier (Norris *et al.*, 1969). The flip-chip technology that has been applied in interconnecting a large number of input/output (I/O) leads to an area array solder bumps on the silicon chips to chip carriers. It has been proven to be very reliable with ceramic as substrates since the thermal mismatch between the chip and the ceramic substrate is not high. In particular, the introduction of underfill encapsulation led to improved reliability of the flip-chip interconnection. Liquid encapsulant is dispensed and subsequently drawn by capillary action into the space between the chip and the substrate. The encapsulant then solidifies upon oven curing and reinforces all the solder bumps, resulting in an enhancement of solder bump fatigue lifetime when compared to an unencapsulated package (Clementi *et al.*, 1993; Suryanarayana *et al.*, 1993; Guo *et al.*, 1992). Recent development of high performance filled epoxy based material suitable for use as underfill encapsulant provides dramatic fatigue life enhancement with minimal impact on the manufacturing flow. Therefore, the plastic flip-chip packaging (that is, flip-chip mounted on organic chip carriers) is being evaluated as a promising packaging technology for the next generation of electronic devices. Since interfaces are numerous in a flip-chip package (that is, die/passivation, passivation/underfill, underfill/solder mask, and solder mask/circuit board) and susceptible to decohension, interfacial strength always plays a key role in determining the reliability of the flip-chip structures in electronic packaging. The interfaces in the flip-chip structures are becoming the important loci of failure initiation during testing, storage, and operation. Increasing the fracture toughness may significantly help improve the life of the electronic devices. It has been observed that the interfacial fracture toughness correlates well with the thermal cycling life (Yan *et al.*, 1997; LeGall *et al.*, 1997). Therefore, how to precisely measure the interfacial fracture toughness of the flip-chip package has become a major concern.

The objective of the current study is to select a real-scale three-point bending flip-chip specimen as a model system to evaluate the interfacial fracture toughness of the flip-chip package.

The seven major tasks of this study include: (a) measurement of the variation of the load line deflection with time under a given constant concentrated load applied to the center point of the flip-chip specimen; (b) real time monitoring of the crack propagation and crack length; (c) calibration of the relationship between crack length and load line deflection; (d) computation of the load line speed and crack growth rate; (e) calculation of the near tip displacement fields; (f) prediction of the energy release rate and the phase angle corresponding to various crack lengths; and (g) determination of the interfacial fracture toughness G_c.

In order to fulfill these objectives, a unique 6-axis submicron tester coupled with a high density laser moiré interferometry was used. The real-scale three-point bending flip-chip specimen, capable of measuring the crack growth rate (along the interface) and the interfacial fracture toughness was developed. The moiré interferometry technique was used to monitor and measure the crack length during the test. The crack growth rate along the interface of the passivated silicon chip/underfill was calculated in terms of the load line deflection versus time curve obtained from the test. In addition, the relationship between the crack length and the load line deflection was calibrated by using finite element analysis. The near tip displacement fields of the flip-chip package were also determined by the same method. The energy release rate was computed by using these near tip displacement variables through an analytical expression (see Equation 7.12 to Equation 7.19). The interfacial fracture toughness G_c was determined by calculating the energy release rate corresponding to the crack length at the quasi-crack arrest stage measure in the test.

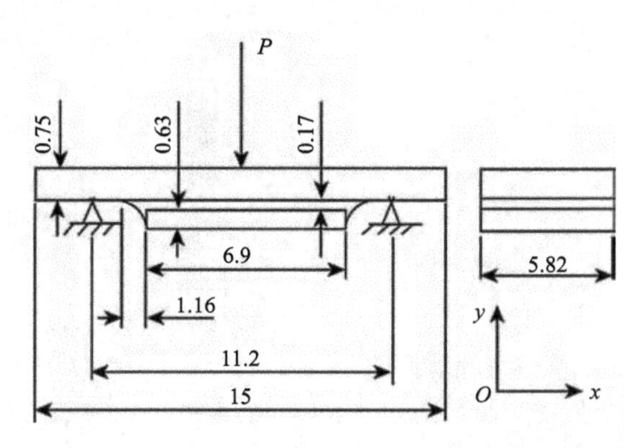

Figure 7.20 Dimensions, loading, and boundary conditions of three-point bending specimen (the unit of dimensions in this figure is mm)

(i) Experiments

The experimental work was conducted to investigate the interfacial fracture behavior of the flip-chip package. The real-scale three-point bending flip-chip specimen was used. The specimen consists of a chip (with passivation,) underfill, and circuit board (with solder mask) layers without an initial crack along the interface between the silicon chip and the underfill. The dimensions, loading, and boundary condition of the specimen considered are shown in Figure 7.20 schematically.

The test was undertaken on the 6-axis submicron tester which was shown in Figure 2.5. The high density laser moiré interferometry (1200 lines/mm) is a whole-field in-plane displacement measurement technique, featuring both high displacement sensitivity and high spatial resolution. It is especially effective for the non-uniform in-plane deformation measurements and has been used in the research and development for microelectronic packages (Guo *et al.*, 1993; Zuo *et al.*, 1997; He, 1997). It was, therefore, adopted to monitor the crack propagation and measure the crack length in the test.

The test setup of a real-scale three-point bending flip-chip specimen in conjunction with the high density laser moiré interferometry measurement system is presented in Figure 7.21, while the schematic drawing of the test fixtures are illustrated in Figure 7.22. A close-up view of a real-scale three-point bending flip-chip specimen test via 6-axis high tester is shown in Figure 7.23.

In the test, a three-point bending specimen was gradually loaded all the way to a given load (about 29 nm), then this concentrated load was constantly controlled until the test was finished. The stage speed with respect to the loading rate is 0.001 mm/s. The load line displacement (deflection) with time was automatically recorded by a computer when the crack propagated. It was observed that the crack first initiated downward from the edge of vertical interface between the underfill and chip, then branched to the horizontal interface between the underfill and the passivated chip. The constant concentrated load was held for 14 hours until the crack arrested.

(ii) Finite Element Modeling

In order to evaluate the energy release rate G and the phase angle φ corresponding to various crack lengths and to determine the critical energy release rate G_c caused by the given constant concentrated load, the same three-point bending flip-chip specimen was taken into consideration. However, the underfill fillet were assumed to be triangle fillets. The problem is idealized by a finite element mesh shown in Figure 7.24. It is noted that a fine local mesh is arranged along the interface of a silicon chip/underfill to handle crack initiation and crack growth and to accurately calculate the near tip displacement fields. A fine mesh is plotted in Figure 7.25.

Due to the fact that the flip-chip package specimen is a segment cross-sectioned and subjected to a line load across the width of the sample, plane strain conditions were assumed. Two types of boundary conditions were considered. These include: (1) the nodes at left and right supported points were set to

Figure 7.21 Test setup of real-scale three-point bending flip-chip specimen in conjunction with the high density laser moiré interferometry measurement system

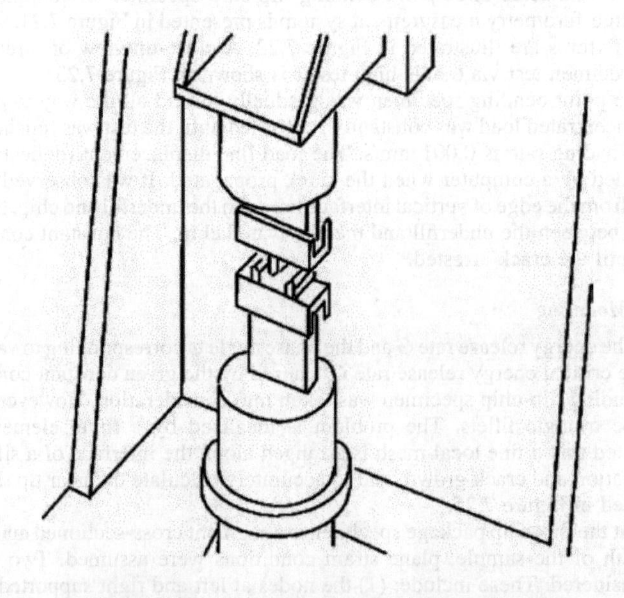

Figure 7.22 Schematic drawing of the test fixture

Figure 7.23 Real-scale three-point bending flip-chip specimen test via 6-axis high precision tester

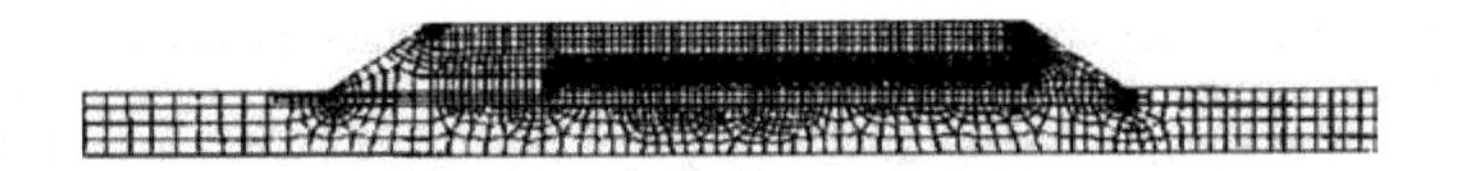

Figure 7.24 Finite element mesh for three-point bending flip-chip specimen

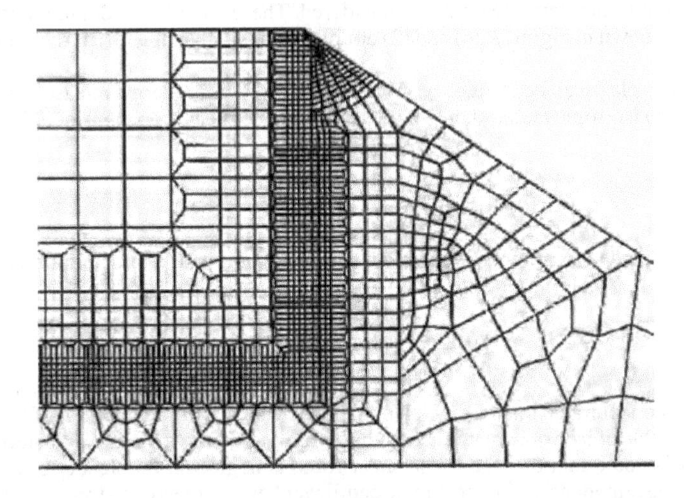

Figure 7.25 Local general finite element fine mesh for three-point bending flip-chip specimen

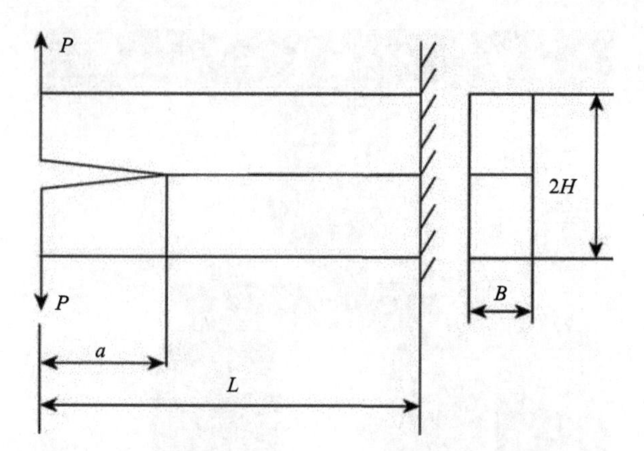

Figure 7.26 Double cantilever beam with isotropic material

$u = v = 0$; (2) the node at left supported point was set to $u = v = 0$, while the node at right supported point was set to $v = 0$.

Since the creep behavior of the viscoelastic property of the underfill does not have a significant effect on the warpage of the flip-chip package due to the constrained small volume of the underfill involved (Wang *et al.*, 1997), the properties of the materials involved were assumed to be linear elastic. The Young's modulus and Poisson ratio of the silicon chip are 169.5 GPa and 0.278 respectively (CINDAS, 1997). The Young's modulus and Poisson ratio corresponding to the underfill material are 2.0 GPa and 0.4 respectively, which is from the sample donated company. The substrate is composed of FR-4 material which was assumed to have an orthotropic property. The Poisson ratios v_{xz}, v_{xy} and v_{yz} are 0.02, 0.143 and 0.143 respectively. The Young's moduli E_x, E_z and E_y are 22.4 GPa, 22.4 GPa and 1.6 GPa respectively. The x and y direction are defined in Figure 7.20. The z direction follows the right hand spiral rule.

(iii) Verification

In order to verify Equation 7.12 to Equation 7.19, numerical calculations based on these equations were compared with the available analytical results. Numerical calculations were performed to determine the strain energy release rate of a double cantilever beam subjected to a pair of loads at the free ends. Two properties of the double cantilever beam were considered. These include: (1) double cantilever beam with isotropic material (shown in Figure 7.26) and (2) double cantilever beam with a bimaterial system (shown in Figure 7.27).

According to the classical beam theory (Williama, 1988), the energy release rate for the double cantilever beam with isotropic material, in which the transverse shear is not considered, can be derived as:

$$G = \frac{12P^2a^2}{EB^2H^3} \tag{7.21}$$

Similarly, the energy release rate for the double cantilever beam with a bimaterial system is defined as:

$$G = \frac{6P^2a^2}{B^2H^3}\left(\frac{1}{E_1} + \frac{1}{E_2}\right) \tag{7.22}$$

where P stands for concentrated load; a crack length; H the width of the beam; the half height of the beam; E the Young's modulus of isotropic material; E_1 and E_2 are the Young's moduli of material I and material II respectively. It is noted that the transverse shear force is neglected for slender beams.

The near tip displacement fields for the double cantilever beams considered were calculated using finite element analysis. In the calculations, B, H, and P were chosen as one unit. The crack length, a, was

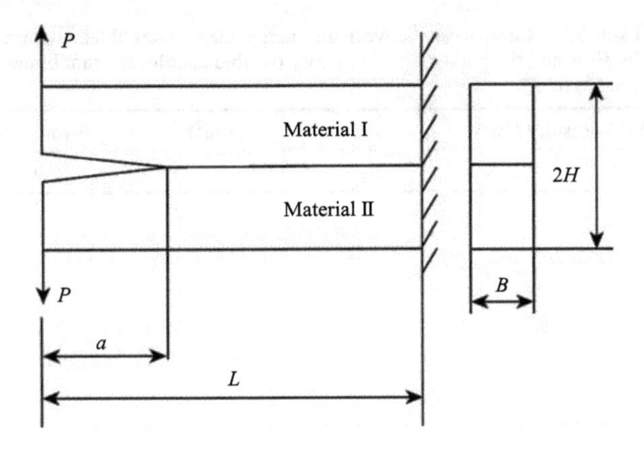

Figure 7.27 Double cantilever beam with bimaterial system

assumed to be 20 units long. The Young's modulus of isotropic material, E, was assumed to be 11900 units. The Young's moduli of material I and material II, E_1 and E_2, were assumed to be 11900 unit and 169540 units respectively. In addition, the beam length, L, was assumed to be 40 units long. Meanwhile, the unit system selected in the numerical calculations was assumed to be compatible.

The results of the energy release rates calculated from Equation 7.12 to Equation 7.19 and the classical beam theory (Equation 7.21 to Equation 7.22) are listed in Table 7.1 and Table 7.2. It is found that the relative error between the energy release rates obtained from both Equation 7.12 to Equation 7.19 and Equation 7.21 for the double cantilever beam with isotropic material is only 1.6%, while the relative error between the energy release rates computed from both Equation 7.12 to Equation 7.19 and Equation 7.22 for the double cantilever beamwith a bimaterial system is less than 5%. It is obvious that the energy release rates predicted from Equation 7.12 to Equation 7.19 agree with the classical beam theory.

(iv) Comparison of Results between Numerical Work and Test

Since the near tip displacement field were calculated using finite element analysis to evaluate the energy release rate and phase angle, it is necessary that the FE results be validated by the test. In order to prove the test, two deformed configurations of the three-point bending flip-chip specimens were selected during the test as the benchmark. These include: (1) the initial deformed configuration indicating the crack initiation after the load reached the prescribed value (shown in Figure 7.28); (2) the final deformed configuration indicating the crack arrest after the constant concentrated load was held for 14 hours (shown in Figure 7.29). The load line displacements obtained from the test and finite element analysis for the two deformed configurations of the three-point bending flip-chip specimen are tabulated in Table 7.3 and Table 7.4. It is found that the relative error between the load line deflections obtained from the element finite analysis and the test is less than 5.2%. Therefore, it is shown that the load line deflections obtained from the finite element analysis are in agreement with those obtained from the test.

(v) Experimental Results

Figure 7.30 shows the real-time measurement of the load line deflection against time under a constant concentrated loading due to the crack propagation in the three-point bending flip-chip specimen using the

Table 7.1 Comparison between the energy release rates obtained form the FEA and the classical beam theory (double cantilever beam with isotropic material)

FEA Result (J/m^2)	Beam Theory (J/m^2)	Error (%)
37.84	37.24	1.6

Table 7.2 Comparison between the energy release rates obtained form the FEA and the classical beam theory (double cantilever beam bimaterial system)

FEA Result (J/m^2)	Beam Theory (J/m^2)	Error (%)
205.82	215.83	4.6

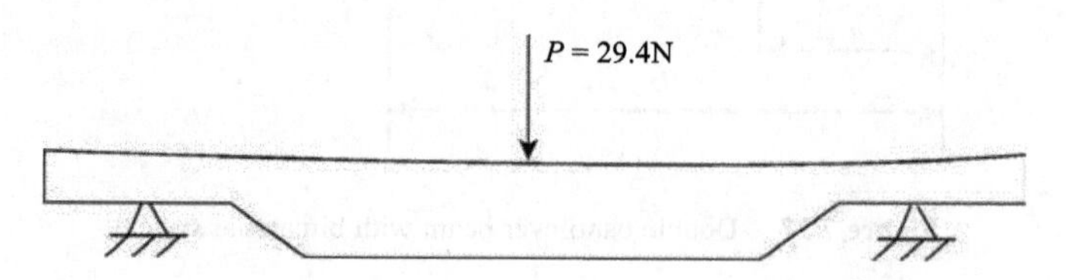

Figure 7.28 Initial deformed configuration of three-point bending flip-chip specimen (without crack)

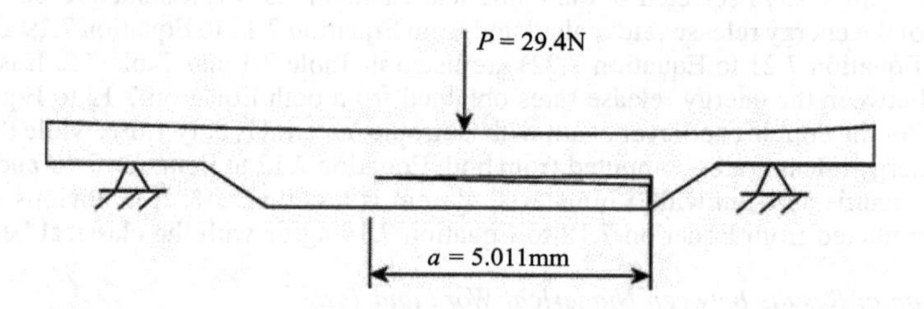

Figure 7.29 Final deformed configuration of three-point bending flip-chip specimen. (with crack length 5.011 mm)

Table 7.3 Comparison between the load line deflections obtained from the test and the FEA (initial deflection configuration)

FEA Result (mm)	Test Result (mm)	Error (%)
0.0578	0.0591	2.2

Table 7.4 Comparison between the load line deflections obtained from the test and the FEA (final deflection configuration)

FEA Result (mm)	Test Result (mm)	Error (%)
0.264	0.251	5.2

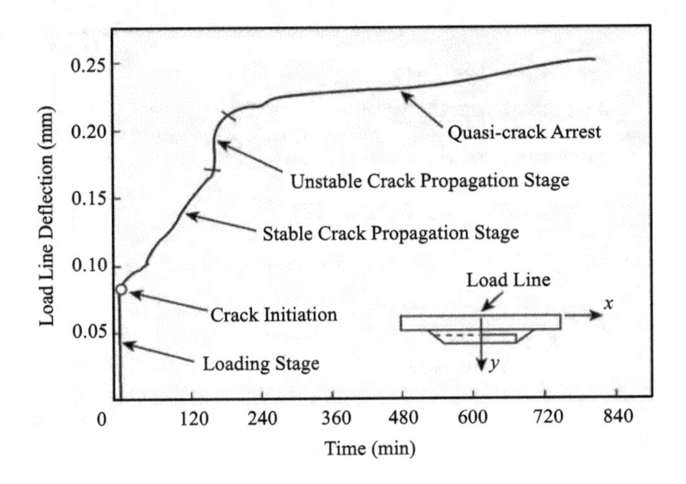

Figure 7.30 Variation of load line deflection in y direction against time

6-axis submicron tester. It is clearly observed that the variation of the load line deflection as a function of time is composed of several stages. These include the uniform change, the nonuniform change, and finally the invariant (stationary state). Since the applied load on the three-point bending flip-chip specimen was constantly controlled during the test, the variation of the load line deflection versus time shows the alternation of the compliance of the three-point bending flip-chip specimen against time. This change is actually caused by crack propagation. Therefore, the characteristics of the variation of the load line deflection with time directly reflects the several stages of crack propagation. They can be classified as the crack initiation stage, the stable crack propagation stage, the unstable crack propagation stage, and the quasi-crack arrest stage. Such kinds of cracking phenomena mainly depend on the form of the applied loading condition (constant concentrated load) and the form of the selected specimen (three-point bending flip-chip specimen). Since the crack arrest stage under the constant concentrated load, occurs during the test, the interfacial fracture toughness can be determined by the energy release rate corresponding to the crack length at the quasi static crack arrest stage measured during the test.

The displacement fields of the three-point bending flip-chip specimen by 1200 lines/mm grating moiré measurement in both directions accompanied by the deformation simulated with the FEA technique are presented in Figure 7.31. It is noted that there clearly exists a discontinuity along the interface between

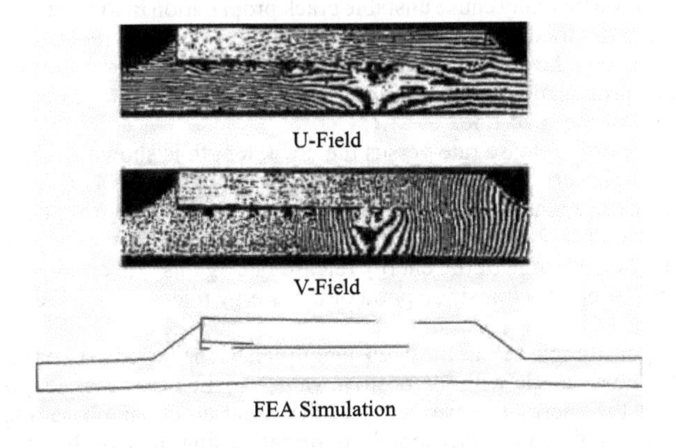

Figure 7.31 Fringe pattern and FEA simulation of specimen during test

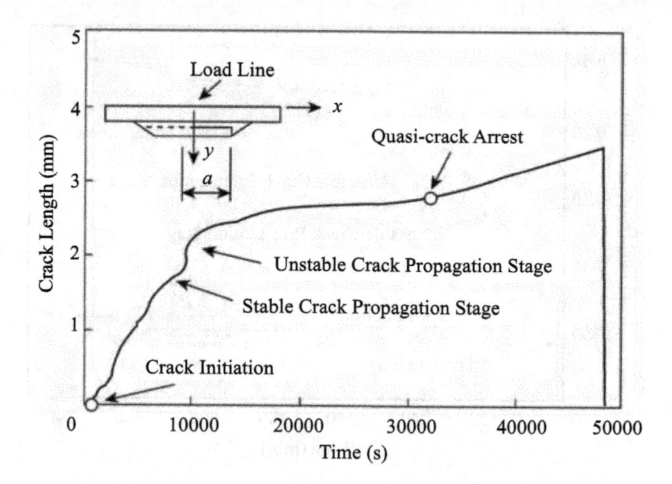

Figure 7.32 Variation of crack length against time

the die and the underfill in the u and v displacement fringe patterns. Therefore, the crack length can be more precisely measured from the discontinuity of the fringe patterns. It is also found that the configurations of the three-point bending flip-chip specimen captured by the moiré interferometry technique is the same as that modeled by the finite element method.

(vi) Modeling Results

Through the calibration of the relationship between the load line deflection and the crack length, using finite element analysis, a real-time measurement of the load line deflection against time can be easily transformed into that of the crack length versus time.

Figure 7.32 shows the variation of the crack length against time. It is obvious that the variation of the crack length against time is quite similar to that of the load line deflection against time. The several stages of the crack propagation are clearly marked on the crack propagation curve. These are the crack initiation stage, the stable crack propagation stage, the unstable crack propagation stage, and the quasi-crack arrest stage. It is noted that the range of the crack length in the unstable crack propagation stage is in the vicinity of the load line, but does not surpass the load line. Since the bending moment has a larger change in this area during the crack propagation, a larger variation of the stress distribution near the crack tip may be produced. This is the reason that may cause unstable crack propagation in this area. It is observed that after the crack length is over the load line, the crack growth rate is rapidly reduced by up to nearly zero. The crack propagation enters the range of the quasi-crack arrest. It is shown that there is a sharp change from the unstable crack propagation stage to the quasi static crack arrest stage in the area where the crack length is in the vicinity of the load line.

The variation of the energy release rate versus the crack length is shown in Figure 7.33, whereas the variation of the phase angle versus the crack length is drawn in Figure 7.34. The FE results indicate that the energy release rate increases as the crack length increases, then decreases with a further increase of the crack length. Like the trend of the variation of the crack growth rate in the range of the unstable crack propagation stage, the sharp change of the energy release rate against the crack length also occurs in the vicinity of the load line. There is a transition point of the energy release rate which is near the position of the load line.

The variation of the phase angle versus the crack length can be identified in two stages. These consist of: (1) the variation of the phase angle with the positive values against the crack length which corresponds to the increase stage of the energy release rate; (2) the variation of the phase angle with the negative values against the crack length which corresponds to the decreasing stage of the energy release rate. It is, however, found that there is no significant change for the phase angle during the crack propagation,

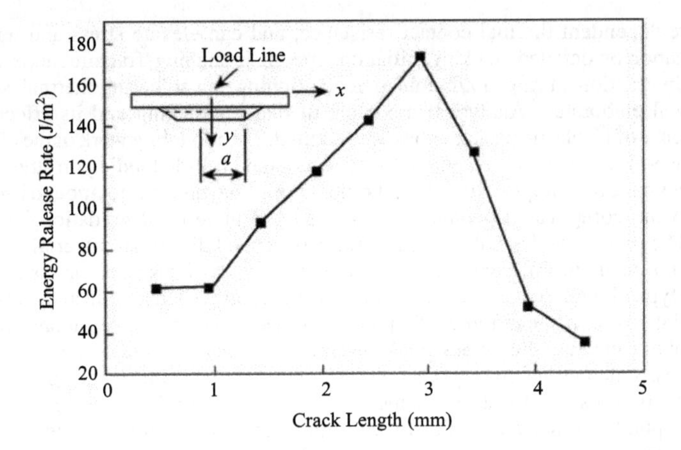

Figure 7.33 Variation of energy release rate against crack length

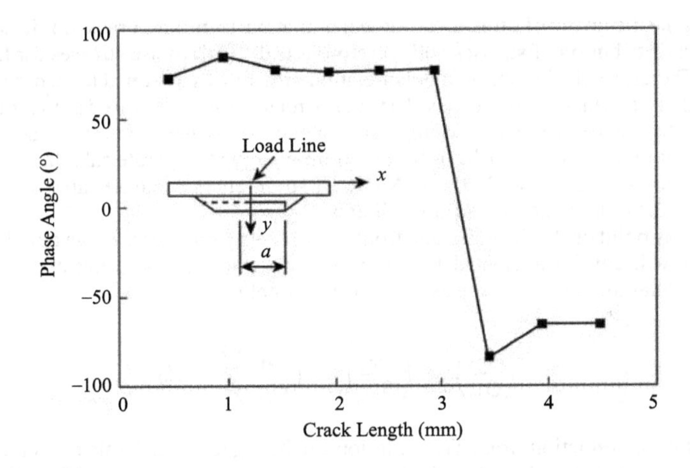

Figure 7.34 Variation of phase angle against crack length

whether in the range of the phase angle with the positive values or in the range of the phase angle with the negative values except for the sign change of the phase angle in the vicinity of the load line.

The interfacial fracture toughness and the phase angle are determined by the values of the energy release rate and the phase angle corresponding to the crack length of 4.5 mm. They are about 35 J/m and 65°, respectively.

7.4 Three-Dimensional Energy Release Rate Calculation

Cracks extending from the surface of the package to the die provide a path allowing flux and other contaminants to reach the die pad. The device may eventually fail due to wire bond pad corrosion, which is considered a long-term reliability failure. In general, system functionality tests performed on cracked plastic packages will not detect the majority of these failures, which will further degrade the thermal and electrical integrity of the package. For example, debonding between the leadframe and the plastic can enhance pressure-dependent thermal contact resistance, and can elevate stress and temperature levels. However, information on detailed cracking initiation, growth, branching, fracture modes, and the cracking and dedonding interaction during hot/humid preconditioning, or soldering/thermal shock, or thermal cycling is not well elaborated. Analytical modeling of plastic cracking and its effect on the thermal-mechanical response of IC plastic packages has been limited. The modeling work of the effect of debonding has been either two-dimensional or axisymmetric, and assumed full-dedonding on stress distribution. The fracture mechanics have been applied mainly to the cracking of the molding compound itself or solder itself. However, due to the complicated geometries, loading, and material degradation involved, the actual debonding could be partial and irregular at one or more of several dissimilar material interfaces inside the package and could occur in different locations such as corners, free edges, or inside the IC package. For instance, the typical interfaces associated with plastic packages such as chip and encapsulant, chip and die-attach material, chip and leadframe (die-pad), and leadframe and encapsulant could more or less experience partial or full debonding/cracking. When the stress singular areas are concerned and due to the limitations of stress calculations by regular finite elements, fracture mechanics provides an alternative approach to more rigorous interfacial mechanics evaluation for the reliability concern.

This section applied fracture mechanics to determine the delamination propagation in three dimensions. The strain energy release rates were calculated by a crack closure technique.

7.4.1 Fracture Analysis

Due to the singular behavior of stresses at the delamination front and crack tip, it is very difficult to calculate correct distribution of stresses and therefore it is difficult to use stresses for failure criteria for crack devices. Therefore, the initiation of delamination growth was assumed to be predictable by energy concept in linear elastic fracture mechanics. The energy release rates were used as the fracture quantity, as they are not sensitive to the mesh size and singular elements are not needed for the crack front. The virtual crack closure technique served as the basis for the strain-energy release rate calculation (Liu, 1992). This procedure determines Mode I, Mode II and Mode III strain energy release rates (G_{I}) from the energy required to close the delamination over a small area.

Therefore, at any point on the delamination front, strain energy release rates can be calculated. Since the delamination growth can be attributed to a mixed mode fracture, the criterion for determining the initiation of the delamination growth was selected for simplicity as:

$$\left(\frac{G_{\mathrm{I}}}{G_{\mathrm{Ic}}}\right)^{\alpha} + \left(\frac{G_{\mathrm{II}}}{G_{\mathrm{IIc}}}\right)^{\beta} + \left(\frac{G_{\mathrm{II}}}{G_{\mathrm{IIIc}}}\right)^{\gamma} = E_d \tag{7.23}$$

for any pint on the delamination front. This criterion can be regarded as the first order approximation for the two phase angles extension based on the two dimensional case (Suo *et al.*, 1992; Wang *et al.*, 1997). Here, G_{Ic}, G_{IIc}, and G_{IIIc} are the critical strain energy release rates corresponding to Mode I, Mode II, and Mode III fracture, respectively. It was assumed that G_{Ic}, G_{IIc}, and G_{IIIc} did not change with delamination size. $\alpha = 1$, $\beta = 1$, and $\gamma = 1$ were on the delamination front. Unfortunately, these values of fracture toughness have not been reported in the literature for the concerned material interface, but some of the data were available for some laminated composite interfaces (Liu, 1992). Therefore, the current analysis was mainly restricted to calculating the distribution of these quantities along the delamination front. Further work is needed for the fracture toughness measurement for typical bi-material debonding interfaces.

7.4.2 Results and Comparison

Consider a simplified packaging microlaminate subjected to a hydrothermal and/or mechanical load as shown in Figure 7.35. Each layer may have arbitrary material embedded in the edges, the structure may be free on four edges but clamped in the bottom surface, and subjected to a quisi-static load. The structure may contain a pre-existing delamination located at a chosen interface. The delamination can be located in the center of the plate, or at the free-edge of the plate, or at the corner of the plate, or through the width of the plate. One surface layer cracking in conjunction with the delamination can be considered in the study.

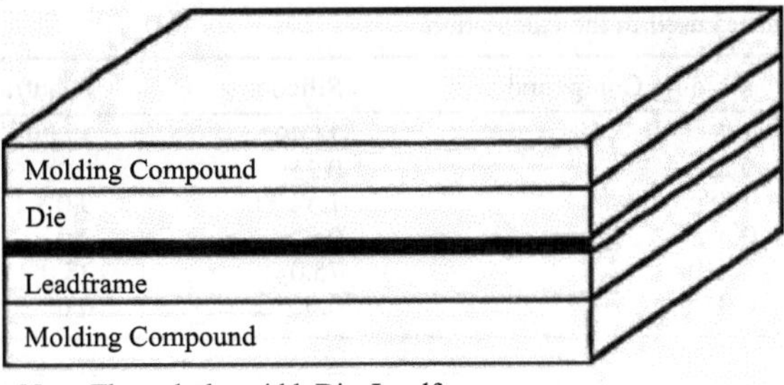

Figure 7.35 Simplified plastic packaging

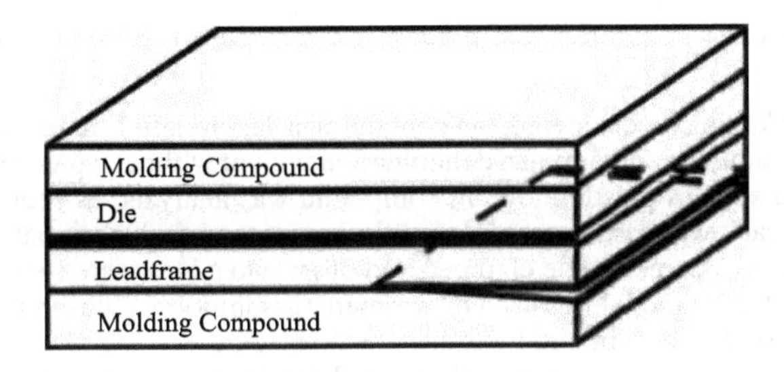

Figure 7.36 Simplified plastic packaging with a through-the-width delamination

Accordingly, for a given set of loading and boundary conditions, models accounting for assumed damage modes are listed and shown in Figures 7.36 and 7.37. In this chapter only results of thermal-moisture loading are shown as examples.

In order to verify the model, numerical solutions were generated and compared with existing analytical and numerical solutions. Overall, the agreements were very good (Liu, 1992). In the following, numerical simulation for plastic packaging specimens are shown and discussed. The properties of materials involved are listed in Table 7.5. The stress free temperature for molding compound and adhesive was assumed to be 300 °F. The stress free temperature for silicon and leadframe was assumed to be 75 °F. The moisture expansion coefficient for molding compounds and adhesives was assumed to be 0.6, while for leadframe and silicon, it was assumed to be zero. As the first step in the analysis, uniform temperature

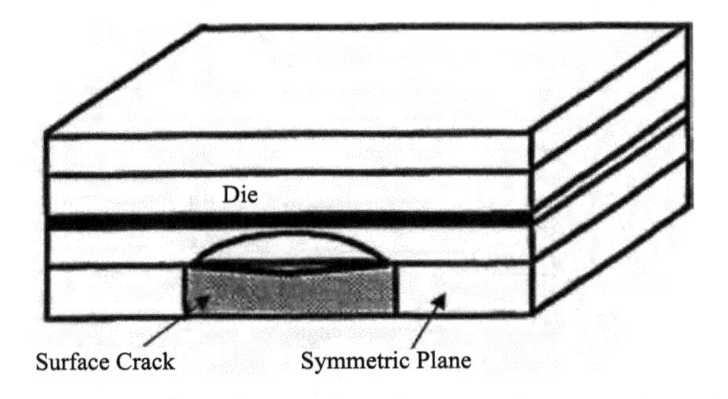

Figure 7.37 A Simplified plastic package with an elliptical delamination

Table 7.5 Material properties used in the calculation

	Molding Compound	Silicon	Leadframe	Adhesive
E (msi)	1.7	24.49	17.49	1.28
θ	0.25	0.25	0.25	0.25
α (1/°F)(1.0×10^{-6})	7.78	1.44	9.0	4.44
β	0.6	0.0	0.0	0.6
T_R	300.0	75.0	75.0	300.0
C_R	0.0	0.0	0.0	0.0

and moisture distributions were assumed and material properties were assumed to be temperature independent. However, due to the temperature and moisture gradients involved in the practical packages and temperature dependent material properties, certain caution should be exercised when applying the conclusions to the practical packaging.

Specimens with a Center Delamination and a Surface Crack Subjected to Different Loading and Boundary Conditions

Now the simplified packaging is considered, but containing a delamination and a surface crack attached to the delamination. The laminate containing a delamination with or without a pre-introduced surface crack in the surface material such as plastic molding compound was analyzed as shown in Figure 7.38. The length of the surface crack was assumed to be slightly longer than the delamination size in X direction. Two opposite edges were assumed to be clamped and other two edges were free. The surface crack ran along the X direction. Quarter model was used to symmetry assumption. Two circular delaminations were considered, on being $a=b=0.05$ in and the other $a=b=0.3$ in.

Figure 7.39 shows a deformed configuration with a delamination located at the molding compound/leadframe interface and a surface cracking subjected to heating only. It was also clearly shown that the interface was fully open and significant bending and warping occurred. By examining the energy release rates (Figure 7.39(a)), Mode I and Mode II controlled the total energy release rate with Mode I dominating the delamination growth. It was also seen that the energy release rates were several order of magnitude higher than the case without any surface cracking, indicating that the interaction between the delamination and surface (transverse) cracking was significant. In terms of full discrete modeling, both surfaces cracking and debonding have to be modeled. Also it was predicted that the interface was fully open.

Now, the loading was changed to a room temperature of ($T=75$ °F) and 0.5% moisture content while everything else was kept unchanged. Figure 7.40(b) shows that mixed mode exists along the delamination

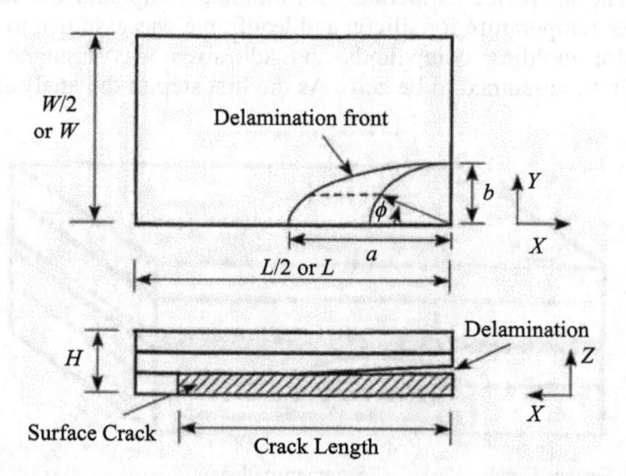

Figure 7.38 A delaminated laminate with or without a surface crack and the coordinate system in its finite element modeling

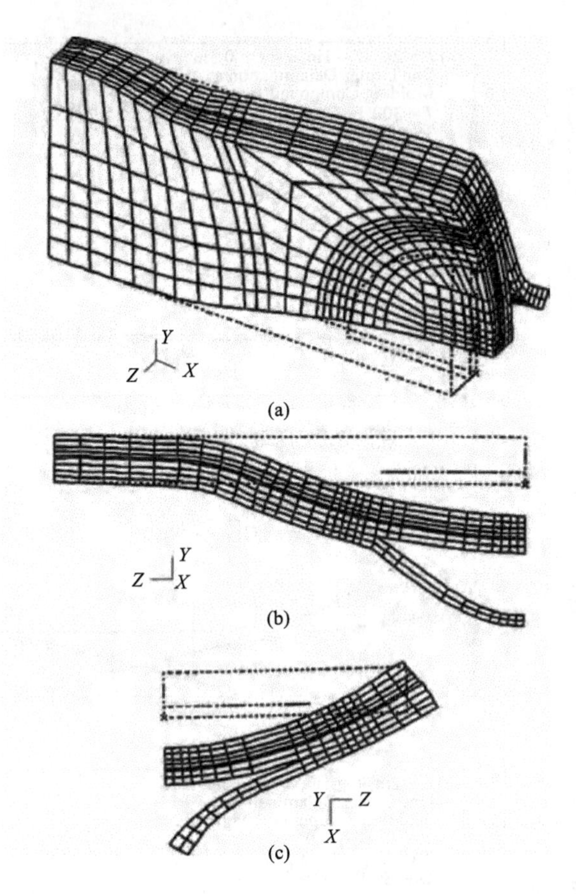

Figure 7.39 Center delaminated specimen with surface crack attached to delamination subjected to; (a) a view of a three-dimensional deformation for with respect to its undeformed boundary in dashed lines; (b) a view of deformation of a cross-section of X-Z, $Y=0$; (c) a view of deformation of a cross-section of Y–Z, $X=0$

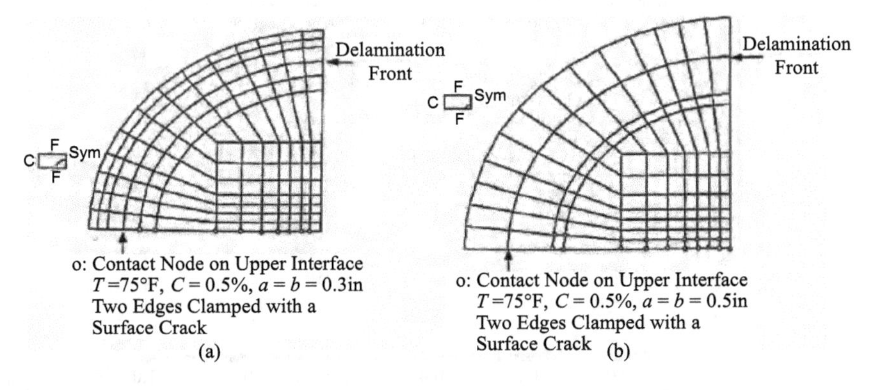

Figure 7.40 Contact distribution for center delaminated specimens with a surface crack subjected to various loading and boundary conditions

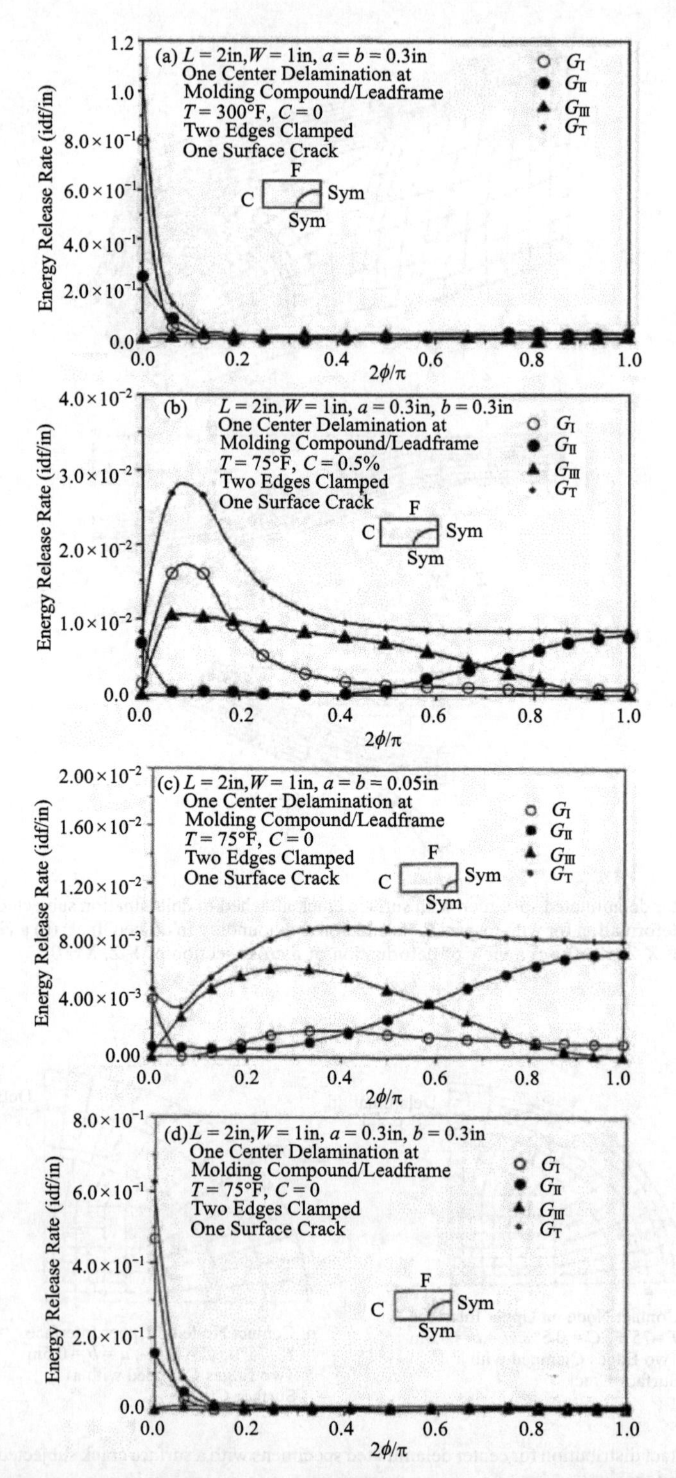

Figure 7.41 Energy release rates distribution along the delamination front for a center delaminated specimen with a surface crack subject to various loading and boundary conditions

front and all three modes could play important roles. At the peak, Mode I and Mode II were more important. By examining the contact information (Figure 7.40(b)), only few edge nodes were found in contact, which was not surprising by examining the $G_i s'$ distribution.

A small delamination ($a = b = 0.05$ in) was studied while keeping everything unchanged for the above example (Figure 7.41). Similar response was observed except that the peak in $G_i s$ disappeared. The Mode I component seemed less important at the intersection point between the surface crack and delamination. Mixed mode was still observed to control the fracture growth.

7.5 Chapter Summary

In this chapter, some basic concepts of fracture mechanics were described. As applications of fracture mechanics in electronic packages, several typical examples were presented and studied. Many important issues such as the crack propagation of solder connectors and fatigue life prediction based on fracture mechanics have not been addressed in this chapter due to the space limitation. In addition, although a lot of researchers have also addressed this topic and have already obtained many valuable results, their work cannot be summarized in this chapter due to the space limitation.

References

Barson, J.M. and Rolfe, S.T. (1987) *Fracture and Fatigue Control in Structure*, 2nd edn., Prentic-Hall Inc., Englewood Cliffs, NJ, 547–575.

Begley, J.A. and Landes, J.D. (1972) The J-integral as a fracture criterion, *Fracture Toughness, ASTM STP 514, American Society for Testing and Materials, Philadephia, PA*: 24.

Brandenburger, P.D. and Pearson, R.A. (1995) Mixed-mode fracture of organic chip attachment adhensives, *Application of Fracture Mechanics in Electronic Packaging and Materials*, ASME EEP **11**:179–185.

Chen, A.S. *et al.* (1993) Effect of material interacions during thermal shock testing on IC package reliability, *IEEE Transaction on Components Packaging Manufacturing Technology* **16**(8), December, 932–939.

CINDAS SRC Document, Purdue Universty, 1997.

Clementi, J., McCreary, J., Niu, T.M., Palomaki, J., Varcoe, J. and G. Hill. (1993) Flip-chip encapsulation on ceramic substrates, in *Proceedings of the 43rd Electronic Components and Technology Conference, 1993* pp. 175–181.

Dai, X., Brillhart, Mark V. and Paul S.H. (1997) Investigation of underfill adhension in plastic flip-chip packages, *Application of Fracture Mechanics in Electronic Packaging*, ASME AMD **222**, EEP **20**:115–124.

Deng, X.M. (1994) An asymptotic analysis of stationary and moving cracks with frictional contact along bimaterial interfaces and in homogeneous solids, *International Journal of Solids & Structures.*

Erdogan, F. (1963) Stress distribution in a nonhomogeneous elastic plane with crack, *Journal of Applied Mechanics* **30**:232–236.

Erdogan, F. (1965) Stress distribution in bonded dissimilar materials with crack, *Journal of Applied Mechanics* **32**:403–410.

England, A.H. (1965) A crack between dissimilar media, *Journal of Applied Mechanics* **32**:400–402.

LeGall, C.A., Qu, J. and McDowell, D.L. (1997) Some mechanic issues related to the thermomechanical reliability of flip-chip DCA with underfill encapsulation, *Application of Fracture Mechanics in Electronic Packaging*, ASME AMD 222, EEP **20**:85–95.

Goodle, J.P., Pearson, R.A. and Wu, T.Y. (1995) Interlaminar fracture toughness of a glass-filled FR-4 epoxy composite as a function of mode-mixity, *Application of Fracture Mechanics in Electronic Packaging and Materials*, ASME EEP **11**:163–169.

Griffith, A.A. (1921) The phenomena of rupture and flow in solids, *Philosophical Transactions of the Royal Society A* **221**:163–197.

Guo, Y., Chen, W.T. and Lim, C.K. (1992) Experiment determinations of thermal strains in semiconductor packaging using moiré interferometry, in *Proceedings of the ASME Conference on Electronic Packaging, 1992*, 779–783.

Guo, Y., Lim, C.K., Chen, W.T. and Woychik, C.G. (1993) Solder ball connect (SBC) assemblies under thermal loading: I. deformation measurement via moiré interferometry and its interpretation, *IBM Journal of Research and Development* **37**(5), September, 635–647.

Guo, Y. and Lim, C.K. (1994) Hybrid method for strain/stress analysis in electronic packaging using moiré interferometry and FEM, *SEM Spring Conference*, Baltimore, Maryland, June 6–8.

Guo, Y. (1995) Experimental determination of effective coefficients of thermal expansion in electronic packaging, *ASME International, Intersociety Electronic Packaging Conference & Exhibition*, Lahaina, Hawaii, March 26–30.
Hutchinson, J.W. (1968) Singular behavior at the end of a tensile crack in a hardening material, *Journal of the Mechanics and Physics of Solids* **16**:16.
Hutchinson, J.W. and Suo, Z. (1992) Mixed mode cracking in layered materials, *Advances in Applied Mechanics* **29**:63–191.
He, X., Zou, D., Liu, S. and Guo, Y. (1997) A window-based graphics interface for phase shifting analysis in moiré interferometry, *Applications of Experimental Mechanics to Electronic Packaging*, ASME EEP **22**/AMD **226**:97–103.
Irwin, G.R. (1960) Plastic zone near a crack tip and fracture toughness, *Proceeding of the Sagamore Conference*, IV-63.
Jiao, J., Kramer, E.J., *et al.* (1997) Effect of thermal residual stress on the measurement of the adhesion between polyimide and underfill using an asymmetric double cantilever beam specimen, *Application of Fracture Mechanics in Electronic Packaging*, ASME AMD **222**, EEP **20**:97–102.
Levis, G.L. *et al.* (1994) Role of materials evolution in VLSI plastic packages in improving reflow soldering performance, in *Proceedings of the 44th Electronic Components and Technology Conference*, 177–185.
Liu, S. (1992) Damage mechanics of cross-ply laminates resulting from transverse concentrated loads, Ph.D. Thesis, Mechanical Engineering Department, Stanford University, August.
Liu, S. (1993) Matrix cracking and delaminations in laminated composite due to transverse concentrated Loads, *Journal of Composite Structures*, Special Issue, July.
Liu, S. (1993) Debonding and cracking of microlaminates due to mechanical and hydo-thermal loads for plastic packaging, *6th Symposium on Structural Analysis in Microelectronics and Fiber Optics*. EEP **7**, ASME WAM, Nov. 28-Dec. **3**:1–11.
Liu, S., Kutlu, Z. and Chang, F. (1993) Matrix cracking and delamination in laminated composite beams subjected to a transverse concentrated line load, *Journal of Composite Materials* **27**(5):436–470.
Liu, S. (1993) Delaminations and matrix cracking of cross-ply laminates due to spherical indenter, *Journal of Composite Structure*, **25**.
Liu, S. and Chang, F. (1994) Matrix cracking effect on delamination growth in laminated composites induced by a spherical indenter, *Journal of Composite Materials* **28**(10):940–977.
Liu, S. (1994) Quasi-impact damage initiation and growth of thick-section and toughened composite materials, *International Journal of Damage Mechanics* **31**(22):3079–3098.
Liu, S. and Mei, Y. (1994) Effect of voids and their interactions on SMT solder joint reliability, *Journal of Surface Mount & Related Technologies* **18**:21–28.
Liu S., Mei Y. *et al.* (1995) Bimaterial interfacial crack growth as a function of mode-mixity, *IEEE Transaction on Components, Packaging, and Manufacturing Technology –Part A* **18**:618–626.
Liu, S. and Zhu, J. (1995) Micromechanics of high temperature composites containing cracking and debonding, *International Journal of Damage Mechanics* **4**(4):380–401.
Liu, S., Zhu, J., Hu J.M. and Pao, Y.-H. (1995) Investigation of crack propagation in ceramic/adhensive/glass system, *IEEE Transaction on Component, Packaging, Manufacturing Technology, Part A* **18**(3):627–633.
Liu, S. and Mei, Y. (1995) Behavior of delaminated plastic IC packages subjected to encapsulation cooling, moisture absorption and wave soldering, *IEEE Transaction on Component, Packaging, Manufacturing Technology, Part A* **18**(3):634–645.
Lu, M., Ren, W. and Liu, S. (1996) A multi-axial thermo-mechanical fatigue tester for electronic packaging material, *ASME Symposium of Sensoring, Monitoring, Modeling and Verification of Electronic Packaging*, November 17–21, Atlanta, 87–92.
Nguyen, L.T. *et al.* (1997) Interfacial fracture toughness in plastic packaging, *Application of Fracture Mechanics in Electronic Packaging*, ASME AMD **222**, EEP **20**:15–24.
Norris, K.C. and Landzberg, A.H. (1969) *IBM Journal of Research and Development* **13**(3):266–271, May.
Reeder, J.R. and Crew, J.H. (1990) Mixed-mode bending method for delamination, *The American Institute of Aeronautics and Astronautics Journal* **28**:1270–1276.
Rice, J.R. and Sih, G.C. (1965) Plane problems of cracks in dissimilar media, *Journal of Applied Mechanics* **32**:418–423.
Rice, J.R. (1967) A path independent integral and the approximate analysis of strain concentration by notches and cracks, Brown University ARPA SD-86, Report E39, May.
Rice, J.R. and Rosengren, G.F. (1968) Plane-strain deformation near a crack tip in a power law hardening material, *Journal of the Mechanics and Physics of Solids* **16**:1.
Rice, J.R. (1968) A path independent integral and the approximate analysis of strain concentration by notches and cracks, *Transaction of the ASME Journal of Applied Mechanics*, 379 June.
Rice, J.R. (1988) Elastic fracture mechanics concept for interfacial cracks, *Journal of Applied Mechanics* **55**:98–103.
Sih, G.C. and Rice, J.R. (1964) The bending of plates of dissimilar materials with cracks, *Journal of Applied Mechanics* **31**:447–482.

Sun, C.T. and Qian, W. (1997) The use of finite extension strain energy release rates for fracture interface crack, *International Journal of Solids and Structure* **34**:2595–2609.

Suryanarayana, D., Hsiso, R., Gall, T.P. and McCreary, J.M. (1993) Flip-chip solder bump fatigue enhancement by polymer encapsulation, *IEEE Transaction on Components, Hybrids, and Manufacturing Technology* **16**:858–862.

Wakizaka, S. *et al.* (1993) Solder crack caused by die attach paste, in *Proceeding 44th Electronic Components and Technology Conference*, 336–340.

Wang, J., Qian, Z. and Liu, S. (1997) Process induced stresses of flip-chip packaging by sequential processing modeling technique, *ASME Transactions Journal of Electronic Packaging*.

Wang, J., Lu, M., Zou D. and Liu, S. (1998) Investigation of interfacial fracture behavior of a flip-chip package under a constant concentrated load, *IEEE Transaction on Component, Packaging, Manufacturing Technology-Part B: Advanced Packaging*.

Williama, J.G. (1988) On the calculations of energy release rates for cracked laminates, *International Journal of Fracture* **36**:101–119.

Williams, M.L. (1959) The stress around a fault or cracks in dissimilar media, *Bulletin of the Seismological Society of American* **49**:199–204.

Yan, X. and Agarwall, R.K. (1997) Two test specimen for determining the interfacial fracture toughness in flip-chip assemblies, *Manufacturing Science and Engineering*, ASME MED **6-1**:383–390.

Zou, D., He, X., Liu, S., Guo, Y. and Dai, F. (1997) Resolving deformation field near corners and interface by phase shifting moiré interferometry, *Applications of Experimental Mechanics to Electronic Packaging*, ASME EEP **22**/AMD **226**:69–76.

8

Concurrent Engineering for Microelectronics

The electronic design process is a very complex dynamic activity encompassing numerous interdisciplinary interactions, voluminous quantities of data, frequent iteration, and many design tools. Fierce market competition continues to propel electronics companies to shrink the product development cycle time. In the meanwhile, these companies need to accommodate emerging manufacturing technologies to develop ultra large scale integrated circuits (ULSIC), multi-chip module (MCM), and very high speed integrated circuits (VHSIC). This results in a concurrent reduction in connector trace width and an increase in conductor layers, leading to a dramatic increase in capacity for circuit density and complexity. The current electronic design process has many drawbacks that hinder its ability to meet these challenges. For example, the current design process lacks two-way interdisciplinary communications. These drawbacks result from an insufficient understanding of the electronic design process and the lack of a powerful design process modeling methodology to allow one to flexibly characterize the ever changing nature of the modern electronic design process.

8.1 Design Optimization

Due to its multidisciplinary complex nature, the electronic design process is therefore carried out using hierarchical decomposition techniques. In this section, a decomposition design process model methodology suitable for modeling the electronic design process is proposed and formalized. Some concerns regarding structural optimization will also be discussed.

There are three ways to represent the design process: the graph–algebra (G–A), the matrix, and the design object. The matrix is the best for design process optimization. The G–A representation provides sophisticated graph symbols that are capable of capturing the semantics of the design process, however, it is not suitable for a design process trade-off study; for example, if one would like to explore different design alternatives or alter the sequence of some design sequences to streamline the design process. It is awkward to use the G–A representation to model such alternations especially when the system is complex and the number of design tasks is large. The matrix representation can provide the flexibility to allow one to easily modify the electronic design process structure.

After the design tasks have been identified from the G–A representation, a design task precedence matrix, an $N \times N$ matrix, where N is the total number of design tasks, can be constructed. With matrix representation, each design task is represented by a mark in the diagonal and is typically assigned to a unique sequential integer number. The matrix element e_{ij} contains a different mark if and only if there is an

Modeling and Simulation for Microelectronic Packaging and Integration: Manufacturing, Reliability and Testing, Second Edition. Sheng Liu and Yong Liu.

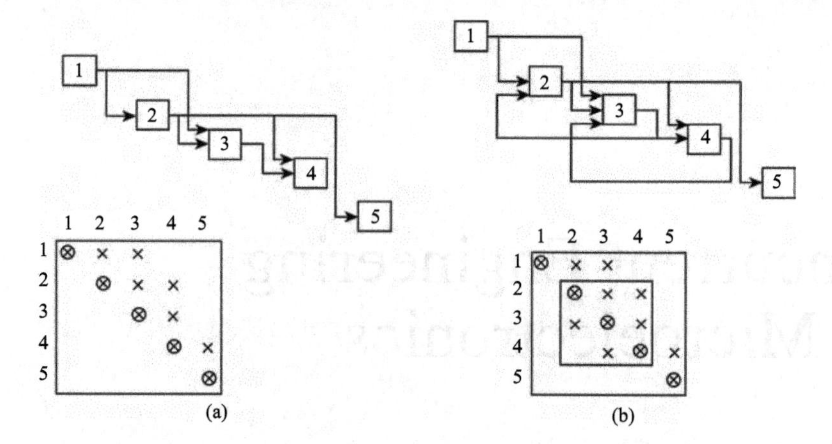

Figure 8.1 (a) Design process with feed-forward paths; (b) Design process with both feed-forward and feedback paths

interaction from design task *i* to task j. Task *i* is called the predecessor, and Task *j* is called the successor. The task whose behavior is affected by the behavior of the other tasks is given a higher number. This ordering scheme is to enforce a sequence that the low-numbered tasks precede the high-number tasks. This allows a natural way to analyze the design process by following the order of the task numbers.

The interaction from a lower-numbered task to a higher-number task is called feed-forward interaction and is designated in the upper triangular areas above the diagonal of the matrix. The design process with only feed-forward iteration is the most desirable since it is straight forward to carry out and easy to quantify and schedule (Figure 8.1(a)). In real-life electronic design, however, such a feed-forward-only matrix rarely exists. Feedbacks each corresponding to an interaction from a higher-numbered task to lower-numbered task often exist to form clusters. A feedback represents where the behavior of a higher-number task is needed to determine the behavior of a lower-number task. Yet the behavior of a higher-numbered task is governed by the lower-number task. Feedbacks are located in the lower triangular area (below the diagonal) of the matrix. A cluster is defined as a block on the diagonal of the matrix which contains both feed-forward and feedback interactions (Figure 8.1(b)). A cluster corresponds to a simultaneous subsystem, that is, within a cluster there is a path from every task in the cluster to very other task in the same cluster. Simultaneous subsystems can only be solved by iteration. In such a case the feedbacks represent where guesses are used to initiate the iteration. Feedbacks normally increase the cost and the time for solving the problem.

With the matrix representation, the design information flow involved in the design process and the interactions/dependencies between design tasks can be easily denoted understood and analyzed. A design structure matrix also consists of vital information in the design decision making process. For instance it often occurs that a modification is made in some variables during the design process due to a change in design specification a new design idea or an error. By following down columns the designers can trace the successors of these modified design variables until all the affected tasks are found.

One of the most important assets of the matrix representation is that it can be processed and analyzed using a computer and advanced algorithms such as the artificial intelligence (AI) techniques in an effective, rapid manner as such the matrix representation is well-suited for automating and managing the electronic design process. It can be used as a management and documentation tool for design change and design verification. Given appropriate estimates for the design task duration number of iterations, and learning factor, the matrix also can be used to predict the design schedule for the entire design process or individual clusters.

In the context of the matrix representation, the optimal design process often corresponds to the matrix with the least number of feedbacks. The matrix representation clustering of design tasks allows one to determine a potential group of activities that might be scheduled simultaneously or identify activities that might cause bottleneck effects in the design process. For example if the clusters of design tasks in the design structure matrix are mutually exclusive in the absence of precedence constraints the design activities within that cluster could be scheduled in parallel.

In structural optimization, depending on the efforts one would like to spend, when the packaging type and processing condition are fixed, many variables in geometry and design can be changed to obtain the best combination of geometry/materials. Actually, as long as you have the confidence in your modeling in materials and device, the package type can be a variable. That is to say, you can also compare two or choose among several package types.

A model mentioned above, once verified, can be used to conduct design optimization. Mathematically, a design optimization is to find the maximum or minimum solution for an objective function with certain constraints (Yeh, 1997; Guo *et al.*, 1997). The objective function can be the weight/mass for weight sensitive systems, cost, stress, strain, strain energy density, plastic work, inelasticity energy, warpage, resonant frequency or frequencies, buckling load, fatigue life, and so on. The design variables can be the thickness of the substrate, the number of conductors/power planes, solder height, bonding layer thickness, corner shapes, solder shapes, die thickness and size, and so on. Constraints can be the minimum or maximum thickness for a layer, size for a substrate, and so on. Usually, the sensitivity information of the objective function with respective to each design variable is needed, which is important for the design improvement even without conducting the full optimization study. There is no intent here to present theories and algorithms in this book and the readers can be referred to extensive literature on optimization. Optimization, as a mathematical tool, has been applied to structural topological and dimension design for thirty years in aerospace, machinery, and civil structures such as composite wing structure or bridges. Optimization design usually is not a global optimal one, but a local one, depending on how many initial solutions you have for your iterations in your optimization process.

In particular, for electronic packaging assembly, the fully automatic optimization practices have been rare (Yeh, 1997; Guo *et al.*, 1997). Instead, a parametric finite element technique has been used. That is to say, judging by the experience the designer accumulates in terms of, the matrix containing typical variables and failure modes, a series of modelings are conducted to obtain the so-called acceptable or best solutions (Mertol, 1997). For instance, encapsulant Young's Modulus multiplier, encapsulant thermal coefficient of expansion, package substrate thickness, encapsulant thickness, die attach Young's Modulus, die attach TCE, die attach thickness, power/ground plane thickness are chosen to be variables for a plastic BGA. Die cracking, various delaminations, solder ball cracking, PWB warpage are failure criteria for reliability design judgment.

In the following, a design optimization for flip-chip reliability is used to illustrate how the parametric finite element analysis and structural optimization can be used for preliminary design including providing guides for various geometric sizes, desired material properties, which could satisfy the design criteria.

Case Study: Design Optimization for Flip-Chip Reliability

Flip-chip on board (FCOB) technology, also called direct chip attach (DCA), or flip-chip attach (FCA), has recently received increasing attention as a way to improve package density and electrical performance. Compared to the similar but more mature C4 (Controlled Collapse Chip Connection) technology that requires reflow temperature up to 320 °C for lead-rich tin/lead (3/97) solder bumps, the FCOB process involves a lower temperature reflow temperature by using eutectic solder (63/37 tin/lead alloy) to join the chip solder bumps to the substrate. For this reason, the FCOB process can be used for the traditional low cost organic, epoxy-based printed wiring board (PWB) assembly process. Although the FCOB technology provides definite advantages over the C4 and traditional SMT packaged component assemblies, reliability concerns have been raised due to a higher coefficient of thermal expansion (CTE) mismatch between the silicon die (around 3–4 ppm/°C) and the organic substrate (16–26 ppm/°C)

A significant number of reliability studies for FCOB assemblies have been performed over the past few years. These studies however are based on a design of experiment (DOE) approach. Based upon many companies' experience, during the early process development stage, item 1, 2, 3, and 5 are the most likely dominating failure mechanisms/modes. As the FCOB process becomes more stable and mature, items 2 and 4 are more prominent.

Although the DOE-based approach has been proven to be a useful vehicle for process characterization and optimization, it involves construction of many prototypes. This is a very time consuming and expensive process. In addition, testing alone sometimes results in "trial-and-error" and "ad-doc" experimental procedures. This, in combination with inherent testing noise, in many cases, may cause one to reach erroneous conclusions when extrapolating experimental data for new designs/processes that are different from the specimens tested. Most importantly, testing without analytical simulation would not provide sufficient understanding and insight of the physical behaviors of the structures. Experimental data inevitably reflects the combined effects of a wide variety of factors, while it is the knowledge of the role of each particular design/process parameter that is needed for reliability prediction. For these reasons, a comprehensive parametric finite element analysis has been proposed and implemented. The subsequent sections describe the modeling approach and results.

Parametric Finite Element Analysis(FEA)

Approach: The objective of this parametric FEA study is to: (1) investigate the reliability impact due to various geometric design parameters, material properties, process conditions, and modeling techniques, and (2) construct a sufficient knowledge base in an attempt to develop comprehensive FCOB design guidelines for reliability enhancements. From this study, qualitative comparisons between different parameters will be performed and critical parameters will be identified. To obtain optimal reliability, design optimization has also been carried out. The ANSYS finite element code was selected and used for this analysis because of its advanced parametric design language and design optimization capabilities.

Modeling Assumptions:

1. Most efforts involved in this parametric study have been based on 2D linear elastic plane-strain analysis. To validate this assumption, both 3D linear and 2D nonlinear models were analyzed. In addition, experiments conducted using laser moiré interferometry have also confirmed that this assumption is valid over the temperature range studied.
2. To ensure that both the analyses and material properties stay in the linear region, an isothermal temperature loading swing from 100 to 25 °C was used.
3. All material properties including coefficients of thermal expansion (CTE), Young's Moduli (E), and Poisson's ratio (n) are temperature independent, homogeneous, and isotropic except for the PWB, which is treated as an orthotropic material (different in-plane and out-of-plane properties).
4. Perfect bonding is assumed at all interfaces between heterogeneous materials.
5. The maximum effective elastic strain (EEP) or von-Mises strain was calculated and used as the indicator for determining the reliability of the structures.
6. No initial residual stress is considered in the model and the structure is at a zero stress stage. As indicated earlier, the purpose of this study is not to determine the absolute true stress in the FCOB structure but rather to conduct a qualitative-based comparative analysis for the various design concepts selected.

Finite Element Models: A baseline finite element model has been established based on a typical FCOB reliability test vehicle. The geometry of the FCOB structure is shown in Figure 8.2. The size of the die Si is 340mil×340 mil.The PWB is made of FR-4 laminated layers. To minimize the CTE mismatch between the PWB and the silicon die, the underfill materials is a composite material that consists of thermoset polymer mixed with silica fillers. A net temperature swing of 75 °C (from 25 to 100 °C) was used for all analyses unless otherwise denoted.

Material Properties: The mechanical material properties of the PWB and underfill have been measured using a Rheometrics Dynamic Spectrometer or 6-axis micro tester at different temperatures. The solder joint and silicon material properties have been extracted from public material database. Because the applied temperature of 100 °C is below the glass transition temperature (T_g) of the PWB and the underfill (approximately 130 °C and 150 °C respectively), the material properties at room temperature were used with good justification.

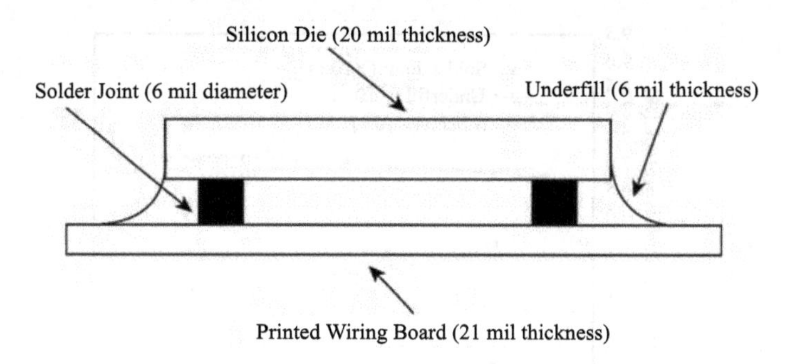

Figure 8.2 Schematic view of a typical FCOB structure

Parametric FEA Study Results

The key parameters investigated in this study include temperature shock effects, underfill thickness, die thickness, PWB thickness, solder joint location, die size, underfill fillet profile, initial PWB warpage, material properties, and 2D versus 3D models. The detailed results are described as follows.

Temperature Cycling Effects: The FCOB test vehicle normally undergoes a liquid-to-liquid temperature shock cycling testing. During the test, the structure incurs a large temperature swing (from 125 °C to − 55 °C) in a very short period of time. It is of interest to understand the induced EEP resulting from such a temperature gradient during the cool-down and the heat-up stages respectively. A coupled transient heat transfer/structural analysis has been carried out to study the maximum EEP in the underfill and the corresponding temperature at the location where the maximum EEP occurs are calculated. As shown in Figure 8.3, the results show that there is a significant EEP variation (0.0052 to 0.0006) during the cool down cycle while the EEP remains stable at 0.0043 throughout the heat-up stage. Therefore, the cool down stage is more responsible for material fatigue than the heat up stage.

Underfill Thickness: As shown in Figure 8.4, a reduction in underfill thickness can increase the maximum EEP considerably in the underfill (42%), especially for the thinner underfill layers, but decrease the EEP only slightly in the die (6%).

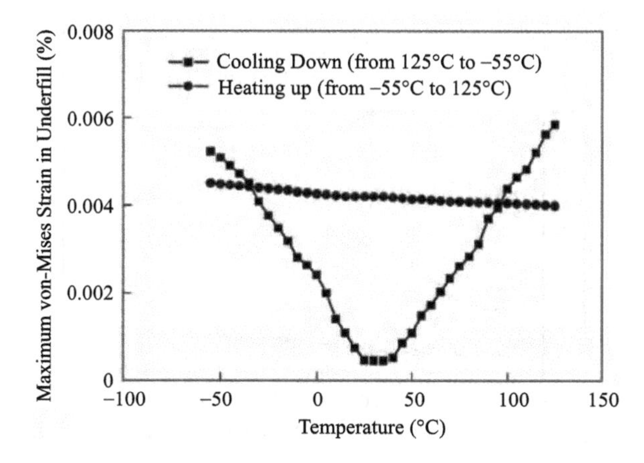

Figure 8.3 Maximum EEP in underfill (baseline configuration) during both cool-down and heat-up stage

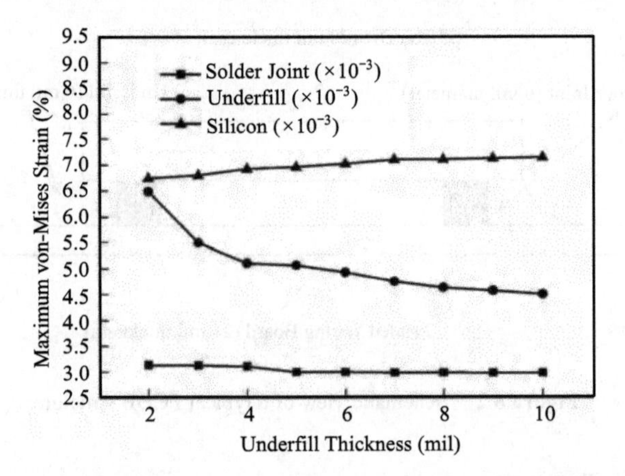

Figure 8.4 Parametric study results for underfill thicknesses

Die Thickness: As shown in Figure 8.5, the maximum EEPs in the die, underfill, and the PWB increases as the die thickness becomes larger. However, it should be noted that there exists a local maximum EEP in the die even when the die thickness remains relatively small (approximately 15 mil).

PWB Thickness: The parametric study result for the board thickness (5–50 mil) is shown in Figure 8.6. The maximum EEP in the die peaks when the board thickness is around 25 mil. The ratio of the largest and the smallest EEPs in the board thickness range studies is 33%. The maximum EEP of the underfill increases somewhat (12%) as the board thickness increases. The maximum EEP in the solder joint is insensitive to the board thickness variation (1% difference).

Solder Joint Location: The maximum EEP has been calculated for various distances (10, 15, and 20 mil) between the center of the solder joint and the die edges. It was found that the EEPs in the underfill and the die are insensitive to this distance. There is a very small EEP decrease (2%) in the solder joint as the distance become larger.

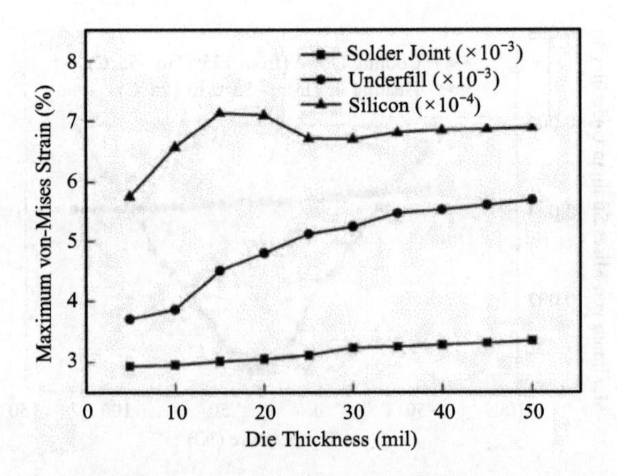

Figure 8.5 Parametric study results for die thicknesses (baseline configuration)

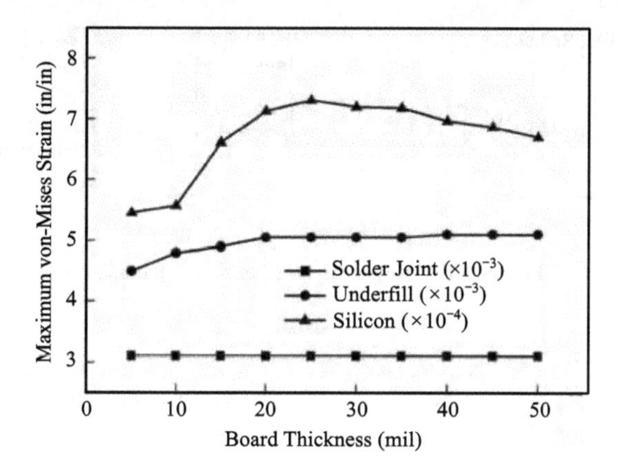

Figure 8.6 Parametric study results for board thicknesses

Die Size: The maximum strain in the underfill is often located at the interface near the lower corner of the die (Figure 8.7). In general, without underfill, the EEP in the solder joint for the larger die is much larger than that of the smaller die due to the larger DNP. Because of the laminated structure, the maximum EEPs in the underfill, the die, and the board remain the same regardless of the size and the DNP. It is also evident that the use of underfill can effectively reduce the solder joint strain by a factor of 2 to 2.5.

Underfill Fillet Profile: Three key parameters governing the profile geometry were investigated: fillet radius, fillet base length, and fillet height (Figure 8.8). The results show that the fillet height is more critical than the other two parameters. As shown in Figure 8.9, the maximum die EEP increases significantly (55%) as the fillet height decreases. The underfill EEP, however, does not follow the same trend. It decreases by 21% over the height range studied. The solder joint EEP shows little movement. This prediction has been observed through an independent DOE study. It has been found that the fillet base length and the fillet radius have minimum impact on the EEP (only approximately 3%–5% strain variation) over the sufficiently large parameter ranges.

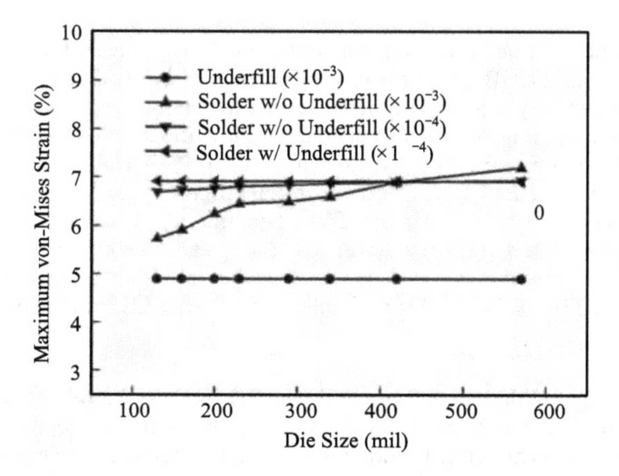

Figure 8.7 Parametric study results for die sizes. (baseline configuration)

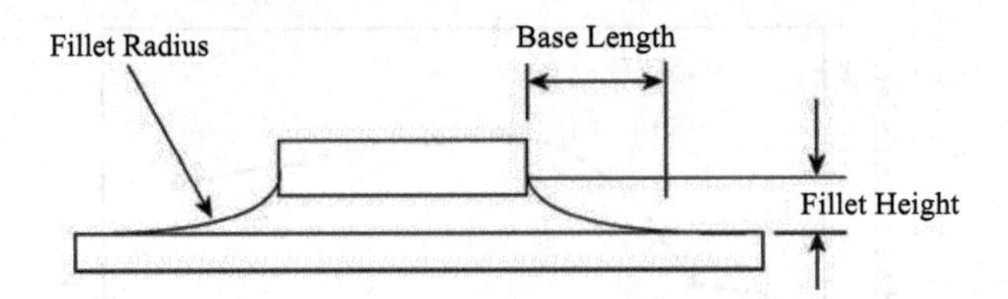

Figure 8.8 Key underfill fillet profile parameters

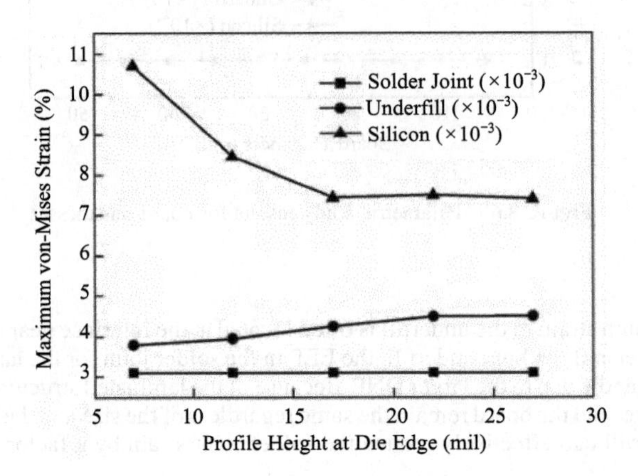

Figure 8.9 Parametric study results for underfill fillet heights (baseline configuration)

Initial PWB Warpage: The initial bare PWB warpage, combined with various elevated temperature assembly processes, may often produce warped FCOB structures. This warpage is typically created by process induced residual stress. It will be of great interest to understand the reliability impact due to such warpage. A positive 75 °C temperature was applied to the models. As shown in Figure 8.10, the positive warpage (concave shape) will further worsen the reliability by adding as much as 25% additional strain to that of the flat PWB case. In contrast, the convex initial warpage will alleviate the maximum EEP at a lesser degree (8%).

Material Properties: The CTE and the Young's modulus are the two most influential material properties in dictating the reliability of FCOB assemblies. Evidence also shows that these material properties of underfill and the PWB can vary significantly from vendor to vendor depending on the composition of the materials, the process conditions, or even storage environment. It is very useful to understand how the changes in material properties can affect the aforementioned failure mechanisms. Figures 8.11 to 8.15 show the changes in maximum EEPs as different material properties vary between 80% and 120% of their baseline values. It is apparent that maximum EEPs become larger when the underfill CTE increases, although this trend for the underfill is more dramatic (32%) than those for the solder (1%) and the underfill (10%) dwindle and the EEP in the die (3%) increases as the Young's modulus of the underfill increases. It seems that the PWB CTE has the most significant impact on the EEPs in the die and the underfill (37% and 18% respectively).

2D versus 3D Models: To validate the aforementioned 2D models, 3D models were developed with an emphasis on geometric details especially for the solder joints. Only a one-eighth of the assembly was modeled due to geometric symmetry. Figure 8.16 shows the parametric FEA results for the underfill thickness study for both the 2D and the 3D models. The results show that the 2D model underestimates the underfill and the die strains by as high as 73% and 23% respectively, but overestimates the solder EEP by

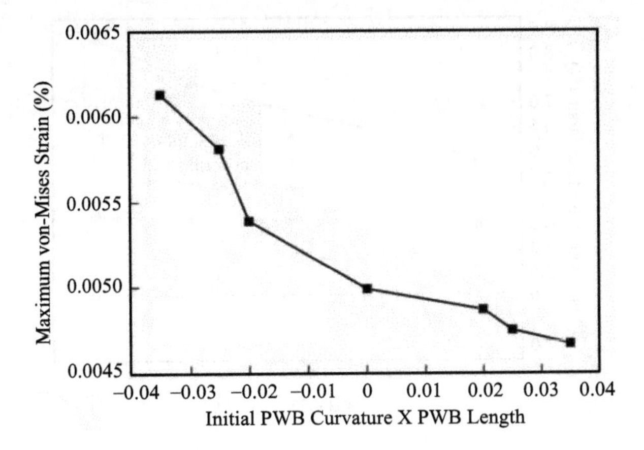

Figure 8.10 Parametric FEA results for underfill fillet heights (baseline configuration)

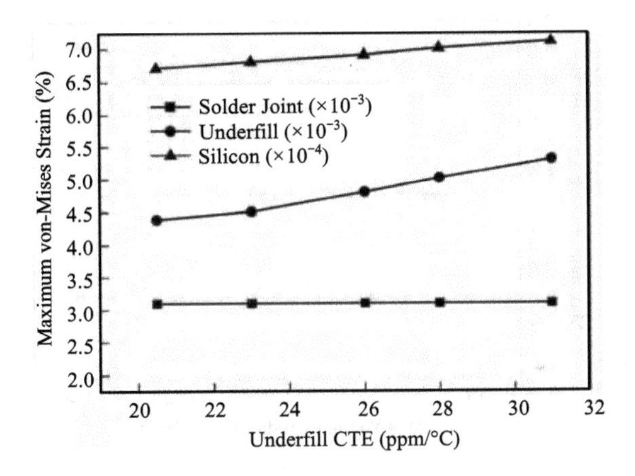

Figure 8.11 Parametric FEA results for underfill CTEs (baseline configuration)

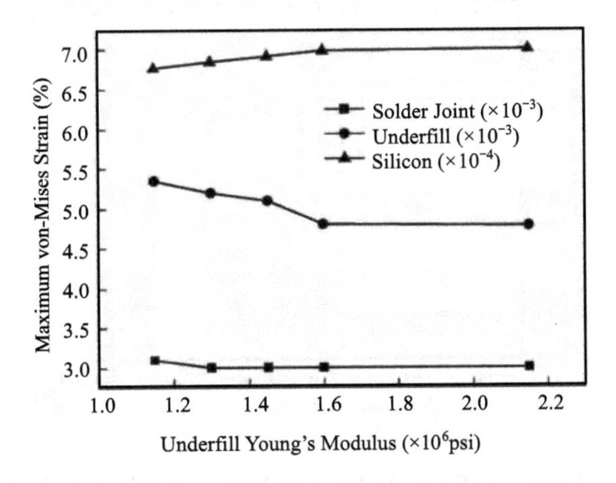

Figure 8.12 Parametric FEA results for underfill Young's modulus (baseline configuration)

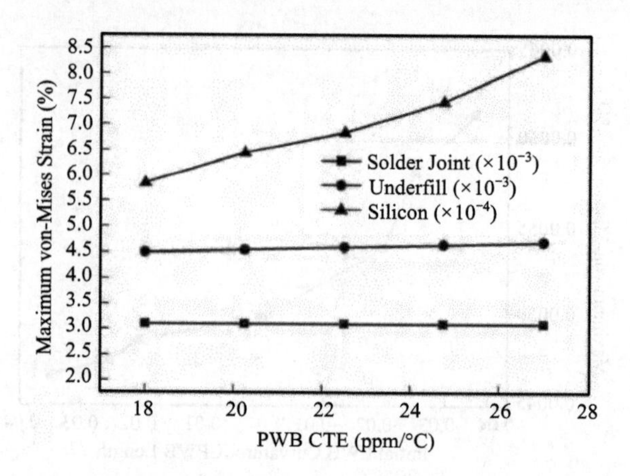

Figure 8.13 Parametric FEA results for PWB CTEs (baseline configuration)

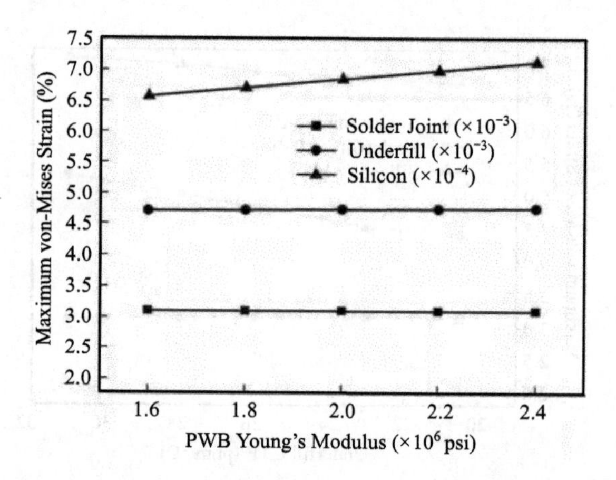

Figure 8.14 Parametric FEA results for PWB Young's modulus (baseline configuration)

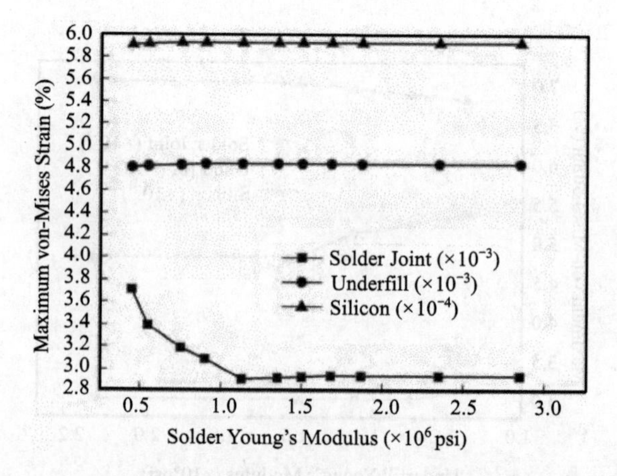

Figure 8.15 Parametric FEA results for solder Young's modulus (baseline configuration)

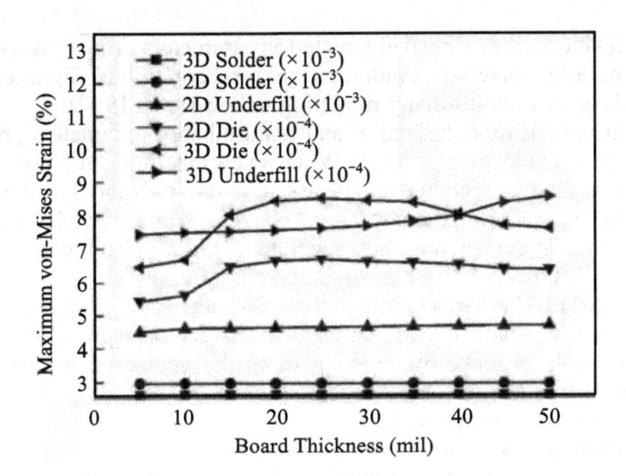

Figure 8.16 Parametric FEA results comparisons for 2D and 3D models

6%. It is suspected that these discrepancies are due to the 2D plane-strain assumption and the 3D corner effects. However, these over-/under- estimations are found to be very consistent for all thicknesses studied. For the purpose of comparing various design options, 2D results are still qualitatively trustworthy and useful.

Design Optimization: Design optimization using ANSYS has been performed for minimizing the EEP in the die, the solder joint, and the PWB. ANSYS uses the Sequential Unconstrained Minimization Technique (SUMT). It took ANSYS 15 tries to reach the goal. Depending on the criticality of different failure mechanism, the values of these design variables can vary. For instance, thinner PWB and die but thicker underfill can effectively minimize the maximum EEP in the underfill and the solder joint. On the other hand, thinner die and underfill, coupled with medium thick PWB, are more desirable for reducing the EEP in the die.

8.2 New Developments and Trends in Integrated Design Tools

As mentioned in the introduction of this chapter, a virtual development environment is essentially needed for multi-disciplinary design and prototyping integration. Although significant efforts are being paid in the single disciplinary modeling and the validation, as described in the last section, significant progress has also been made in the integrated tools.

A virtual development environment can realize design and prototyping virtually to create a model to analyze the device characteristics prior to building the actual device. There are several major elements in creating this model: a user-friendly interface, device library, scalable component library, material library, environment, feature-based parametric FEM, interactive FEM, various modelers including thermomechanical, electromagnetic, CFD, and reliability capabilities.

Currently, integrated design and prototyping have not yet been included into commercial tools, which should have the complete capabilities described above. Some tools target integrating reliability analysis tools in CAD/CAE, such as (Zhou *et al*., 1997; Pao *et al*., 1995; Bot, 1997; Osterman *et al*., 1997), or integrating CFD and thermal modeling (Kjellberg *et al*., 1997).

To reduce the modeling repetition, concepts of taxonomy and thermomechanical component library were introduced (Zhou *et al*., 1997). A thermomechanical component can be generated as a shared, manageable FE model. It can be handled as an icon (modularized FE model) for computational simulation of the movement/replacement of an electronic component. Since a geometric sensitivity and uncertainty study is one of the key parts in thermomechanical design and optimization, flexible modeling capability is greatly needed. Pure traditional geometry parameter modeling and/or FEM no longer satisfy these needs. To overcome the deficiency, the parametric finite element modeling in general includes parameters both for geometry creation and for mesh control in the modeling process. In addition, relations, such as "with/

without" and "if-then" relations, are extensively treated as parameters in creating a model. A single model can be used for different tasks, materials, configurations, and conditions. Significant time savings have been found as the result of this one-to-many modeling (Zhou *et al.*, 1997).

Utilization of the concepts of modularization and feature based parameters provide for coupling of thermomechanical design with electronic design. When a new electronic component is added to existing design, a corresponding mechanical module can be assembled to the model, and be ready for a solution. "What-if" scenarios can be performed with design changes. Failure modes can be recognized and eliminated before final product design, and thus generate a "failure-free" product, which is "correct by design". This single pass design will result in time and cost savings.

Compared to a traditional FEM approach, interactive FEM integrates pre-processor, solution, and post-processer into a single user interface. It allows a user (designer) to graphically modify the loading and boundary conditions, as well as make minor changes to the geometry and elements. By adding an interactive FEM capability into a tradition FE modeling package, a developed FE model can be used as both as an analysis tool and as a design tool.

The mechanical complexity of advanced electronic systems and the throughput limitations of existing computing platforms preclude the use of numerical models for detailed design. It is believed that a fully developed integrated modularized FEM method is a vital tool to create the bridge to a virtual design environment for electronic products.

Various materials with different properties exist in an electronic device. Depending on the modeling requirements, physical, chemical, thermal, or mechanical properties may be needed. These properties can be represented in single numbers, tables, curves, or mathematical equations. Depending on the needs, the properties can be in simple linear range or significant nonlinear range. The properties can be temperature, time, rate, or history dependent. CINDAS is the SRC funded material database for electronic packaging materials. However, caution has to be exercised as most data collected are from the literature and quite often no rigorous conditions are recorded. For instance, solder is an excellent example.

In order to be compatible with the models, all the above properties have been in terms of mathematical formulations, which we call constitutive laws. Quite a few commercial codes contain extensive constitutive laws, such as in ABAQUS and ANSYS. For those new constitutive laws, users have to write their own subroutines to describe complicated material responses, for instance, in cyclic loading for solder alloys and polymers films.

The loading environment is very critical for reliability prediction in both manufacturing and operation conditions. Aerospace and automobile industries have made significant efforts to provide loading spectrums for actual loading environments to be able to predict the true response and life for vehicles. Similarly, an electronic device has to go through complicated loading conditions for both its processing and operation. During the manufacturing, the oven/chamber thermal conditions, temperature and pressure profiles, cleaning and no-cleaning conditions, are all important for the yield and reliability during manufacturing and their impact on the reliability of made products. For instance, the underhood conditions for automobile electronics are important for an engine controller if it is placed near the engine. On the other hand, if the control is placed near the rear seat, the loading/temperature profile will be different.

Currently, there is no way to exactly provide the loading conditions for one practical device. Actually, the stress accelerating tests are used to evaluate the performance of the product. There is still a way to actually collaborate the accelerating life and the true device life.

Thermomechanical, electromagnetic, CFD, reliability capabilities are actually solvers for the packaging performance. Currently, there are limited capabilities in the coupling between them. In general, a multi-level global/local approach discussed in section 6.3 is used to predict the fatigue life and failure modes for interconnects and interfaces.

In the following, a practice by Ford Motor Company (Pao *et al.*, 1995) is used to demonstrate the basic concepts and applications of integrated CAD/CAE analysis for a printed wiring board assembly and the fatigue life for the solders.

CAIR System for Thermal Reliability Prediction

In order to reduce the product development time and accelerate the time to market, the number of iterations in the current prototype build-and-fixed mode needs to be minimized, and the time for each iteration has to be shortened. With the advancement in computer modeling techniques for electronic packaging design and analysis during the past ten years and an increasing understanding of thermal fatigue failure of all packaging levels, it has become an emerging trend to replace testing with modeling in product development using integrated computer tools with physics of failure reliability approaches. The following section gives an introduction to such a system, Computer Aided Interconnect Reliability (CAIR), developed for thermal fatigue life prediction of solder interconnects at Ford (Pao *et al.*, 1994).

Architecture

The CAIR system is developed on a UNIX based workstation. It uses PCB/Explorer (Pacific Numerix Corp.) as a front end for PWB/component layout and board-level thermal/vibration analysis. In addition to a variety of software tools and language, for example, LISP, C++, FORTRAN, and PATRAN Command language, CAIR uses SUIT (Simple User Interface Toolkit) as a user interface to model the data structure. Two other commercial programs, PATRAN (PDA Engineering) and ABAQUS (SIMULIA Inc), are incorporated for automated finite element analysis of component lead frame stiffness. The results of global thermal analysis from PCB/Explorer are transported in neutral file for detailed local interconnect reliability analysis, Moreover, these functions are supported by a built-in material library/database. The system allows for a modular assembly of heterogeneous programs for solving various reliability problems.

Capability and Applications

The system currently provides six major functions:

1. PWB layout and component placement,
2. Heat transfer analysis (transient and steady state),
3. Thermal fatigue life prediction of leaded, for example, gull-wing, J-bend, and butt, and leadless solder joint, for example, chip capacitor, resistor, carrier or flip-chip bump,
4. Interfacial thermal stress analysis of multilayered transistor stacks, including solder interlayer,
5. Automated 3D finite element modeling of lead frame, and
6. A material database of physical, mechanical, and thermal properties for 300 electronic materials.

Figure 8.17 shows the flow chart of how to apply this system to reliability prediction in electronic packaging design. Three types of solder interconnects are considered in CAIR: Leaded surface mount solder joints and solder interlayer in multilayered stacks. Both leaded and leadless solder joints are considered as inelastic materials. Different failure criteria have been developed and implemented in the database for life prediction. The failure criteria for the leadless solder joints are based upon the elastic and viscoplastic material, as discussed in the previous section.

The material database is divided into four major categories: Die, Interconnect, Substrate, and Packaging materials. For each material 43 physical, thermal and mechanical properties are identified, for example, melting temperature, density, elastic modulus, Poisson's ratio, strength, hardness, fracture toughness, creep and plasticity parameters, coefficient of thermal expansion (CTE), thermal conductivity, dielectric constant. A unique feature of this database is that any property can be specified as a function of other variables, for example, temperature or time. This feature is of particular importance to electronic material since most properties are temperature dependent. Screen plots of Young's modulus and CTE versus temperature for some commonly used materials that can also been shown.

Furthermore, a reference ID is assigned to each property in the database to document the source. The ID and the associated reference are also displayed on the interface. The database currently incorporates 300 materials commonly used in electronic packaging, with over 4000 properties. The database can either be used in conjunction with CAIR or as a separate source for material selection.

Future Packaging Trends and CAIR Development

New electronic packaging technologies need to be continuously developed not only to meet the increasing demands in low and high reliability, but to address more stringent governmental regulations and environmental requirements. The following list enumerates several new packaging technologies,

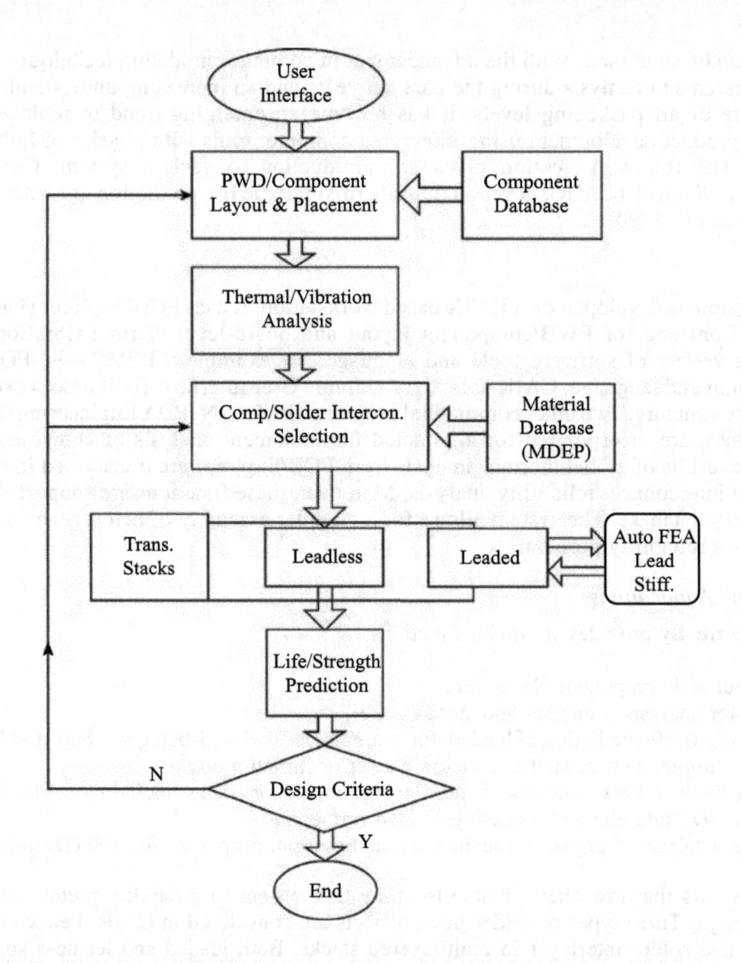

Figure 8.17 Flow chart of CAIR system capabilities for solder interconnect reliability prediction

particularly of levels 2 and 3, that may be considered for future automotive applications:

- Low cost, high T_g , and low dielectric multilayer PWB,
- Thinner and smaller substrate and component,
- High aspect ratio, low temperature via technology (vertical interconnect),
- Multichip modules (MCM),
- Ball grid array (BGA) for coarser pitch and higher I/Os, flip-chip, chip scale packaging,
- Low temperature processes, for example, solders or conductive adhesives,
- Pb-free solders, no-clean or water-clean fluxes, and associated soldering processes,
- Optoelectronics, and
- Silicon micromaching (MEMS devices).

The future development of CAIR will be driven by user's needs for specific packaging designs, the trend of packaging technology, reliability requirements, and availability of CAE tools for thermal and mechanical analysis of packaging. Examples of future CAIR capabilities include:

- Automated, parametric, and database using neural net work technology,
- Wirebond design,

- Life prediction of BGA joints,
- Thin and thick film reliability,
- Interface delamination (particularly the role of moisure),
- Vibration/thermal coupled effect on solder joint fatigue life,
- Solder reflow simulation,
- Prediction of solder joint shape,
- Design of accelerated thermal cycling tests, and
- Incorporation of more process-dependent thermal and mechanical properties of organic materials into the database.

In addition, the current capabilities will be continuously improved, and the material database and failure criteria will be updated on a regular basis. Other methodologies, such as DOE (design of experiment), fracture mechanics based fatigue models, and improved material modeling capabilities of PWB in the current vibrations analysis will also be incorporated. The system is initially designed for more general purpose applications, but its modular structure of knowledge sources is flexible enough to be tailored for analyzing any specific interconnect configurations under prescribed environmental and loading conditions.

8.3 Chapter Summary

The modern electronic design and product prototyping process are highly complex, sophisticated, and multidisciplinary processes. Ever-stringent electronic product specifications and reduced design cycle time have imposed even bigger challenges on electronic designers. With the continued emphasis on current engineering, it has been found that the current design process has many drawbacks in meeting these challenges. This chapter reviewed practice in integrated design and prototyping and optimization. In order to fully develop a virtual environment for integrated framework for electronic packaging design and prototyping, significant modeling work needs to be done. In particular, modularized and parameterized modeling systems, modeling validation techniques, a consistent and reliable material database or library, and coupled solvers need to be refined or developed.

References

Baehmann, P.L., Sham, T.L., Song, LY. and Shephard, M.S. (1992) Thermal and thermomechanical analysis of multichip modules using adaptive finite element techniques, *Computer Aided Design in Electronic Packaging*, ASME **3**:57–63.

Bar-Cohen, A. (1993) Physical design of electronic systems-methodology, technology trends, and future challenges, Bar-Cohen, A., Kraus, A.D., eds., *Advances in Thermal Modeling of Electronic Components and Systems* **3**:1–60, ASME, New York.

Bhatti, P.K. Gschwend, K., Kwang, A.Y. and Syed, A.R. (1993) Computer simulation of leaded surface mount devices and solder joint in a thermal cycling environment, *Advances in Electronic Packaging*, ASME **4-2**:1047–1054.

Bot, Y. (1997) Integrating reliability analysis tool in CAD. CAE, *ASME EEP-Vol. 19-1, Advances in Electronic Packaging-1997*, Volume 1, Hawaii, Big Island, June: 885–890.

Classon, F. (1993) *Surface Mount Technology for Concurrent Engineering and Manufacturing*, McGraw-Hill, New York.

Cox, F.L. and Jazbutis, G.B. (1997) Models of electronic package engineering, *ASME EEP-Vol. 19-1, Advance in Electronic Packaging – 1997*, Volume 1. Hawaii, Big Island, June, 901–910.

Dally, J.W. *et al.* (1992) Electronic packaging education for mechanical engineer: a panel response, *Advances in Electronic Packaging*, ASME. 1099–1012.

Demmin, J.C. (1994) Choosing an MCM technology for thermal performance, *MCM 94 Proceedings*, 549–554.

Engelmaier, W. (1993) Long-term surface mount solder joint reliability in electronic systems with multiple use environments and multiplicity of component, *Advances in Electronic Packaging*, ASME **4-1**:479–486.

The Institute for Interconnection and Packaging Electronic Circuits (IPC). Electronic Packaging Handlebook:

Funk, J.M., Bomholt, L.H., Paganini, R.P., Lien, H.P. and Fichtner, W. (1997) Solids: a TCAD environment for packaging simulation, *ASME EEP-Vol. 19-1, Advance in Electronic Packaging – 1997*, Volume 1. Hawaii, Big Island, June: 919–910.

Godfrey, W.M. *et al.* (1994) Interactive thermal modeling of electric circuit boards, *ASME Journal of Electronic Packaging*, 65–71.

Grosses, I.R., Schaubert, D.H., Chio, T., Corkill, D.D. and Stoklosa, M. (1997) A framework for integration of analysis tools for microelectronic devices, *ASME EEP-Vol. 19-1*, Advances in Electronic Packaging – 1997, Volume 1. Hawaii, Big Island, June, 891–900.

Guo, Y.F. and Sarihan, V. (1997) *Symposium on Applications of Experiment Mechanics to Electronic Packaging*, 1997.

Heinrich, S.M., Liedtke, P.E. *et al.* (1993) Effect of chip pad geometry on solder joint formation in SMT, *Advances in Electronic Packaging*, ASME, 603–609.

Ikegami, K. (1993) Some topics of mechanical problems in electronic packaging, *Advances in Electronic Packaging*, ASME: 567–573.

Keegan, K.K. Kearney, D. and Monghan, P.F. (1993) Thermal fatigue life prediction for a middle gull-wing joint using finite element analysis, *Advances in Electronic Packaging*, ASME **4-1**:499–507.

Keister, G. (1990) Surface-mounted components, *Electronic Component New*, February 25–39.

Kjellberg, S., McGuckin, D., Kisielewicz, T. and Ando, K. (1997) An integrated software tool for the thermomecanical analysis of electronic chips, *ASME EEP-Vol. 19-1, Advance in Electronic Packaging – 1997*, Volume 1. Hawaii, Big Island, June: 957–962.

Ladd, S.K. and Mandry, J.E. (1993) Solving real world MCM design problems, *MCM 94 proceeding*, 567–573.

Lau, J.H. (1991) *Chip on board technology for multichip modules*, Van Nostrand Reinhold, New York.

Ling, S.X., and Dasgupta, A. (1993) A design model for through-hole component based on lead-fatigue consideration, *Advances in Electronic Packaging, ASME* **4-1**:217–225.

Mcbridge, R.R., Zarnow, D.F. and Brown, R.L. (1994) The essential role of custom design kits for cost-effective access to the MCM foundry, *MCM proceeding*, 535–539.

Mertol, A. (1997) Optimization of high pin count cavity-up enhanced plastic ball grid array (EPBGA) packages for robust design, *ASME EEP-Vol. 19-1*, Advance in Electronic Packaging – 1997, Vol 1, Hawaii, Big Island, June, pp. 1079–1088.

Nolan, L. (1992) Mechanical CAD/CAM/CAE contribution to qualify in concurrent engineering, *Manufacturing Aspects in Electronic Packaging*, ASME **2**:1–6.

Osterman, M., Stadterman, T. and Wheeler, R. (1997) CAD/CAE requirements and usage for reliability assessment of electronic products, *ASME EEP-Vol. 19-1*, Advance in Electronic Packaging – 1997, Volume 1. Hawaii, Big Island, June: 927–938.

Pao, Y-H, Jih, E. and Reddy, V. (1994) Thermal fatigue life prediction of surface mount solder interconnects: a design integrated system approach, *ASME AMD-Vol. 195*, Mechanics and Materials for Electronic Packaging, Volume 1- *Design and Process Issues in Electronic Packaging*, 181–191.

Pao, Y-H., Jih, E. and Reddy, V. (1995) Thermal reliability prediction of automotive electronic packaging, SAE Technical Paper Series, No. 950991, International Congress and Exposition, Detroit, Michigan, February 27 - March 2.

Peak, R.S., Fulton, R.E. and Sitaraman, S.K. (1997) Thermomechanical CAD/CAE integration in the Tiger PWA toolse, *ASME EEP-Vol. 19-1, Advance in Electronic Packaging – 1997*, Volume 1. Hawaii, Big Island, June: 963–970.

Rassaian, M., Lee, J.C. and Twigg, D.W. (1997) Integrated durability design tool for electronic packaging, *ASME EEP-Vol. 19-1*, Advance in Electronic Packaging-1997, Volume 1. Hawaii, Big Island, June: 911–910.

Shah, M.K. (1990) Analysis of parameters influencing stress in the solder joints of leadless chip capacitors, *ASME Journal of Electronic Packaging* **112(2)**:147–153.

Shephard, M.S. (1988) Approaches to the automatic generation and control of finite element meshes, *Applied Mechanics*, ASME **41**(4), April.

Tummala, R. and Rymaszewski, E.J. (1989) *Microelectronics Packaging Handbook*, Van Nostrand Reinhold, 1989, Cadigan, Mike, Trends in Electronic Packaging, Circuits.

Zhou, W.X., Hsiung, C.H., Fulton, R.E., Yin, X.F., Yeh, C.P. and Waytt, K. (1997) CAD-Based analysis tools for electronic packaging design: a new modeling methodology for a virtual development environment, *ASME EEP-Vol. 19-1*, Advances in Electronic Packaging-1997, Volume 1. Hawaii, Big Island, June, 971–979.

Part II

Modeling in Microelectronic Packaging and Assembly

9

Typical IC Packaging and Assembly Processes

Assembly manufacturing processes can be broken into front of line (FOL) and end of line (EOL) components, each having several process steps.

In FOL, typical processes include the wafer process and thinning, die pick up process, die attach process and the wire bonding process. Die pick up from tape that becomes more of a problem as die thinning becomes more extreme. FOL die attach processes can induce residual stresses due to process shrinkage and CTE mismatch. Wire bonding processes can induce bond pad cratering or failure in the device under the bond pad. The major FOL simulations include die pick up simulation, multiple die attach simulation, and wire bonding simulation. Multiple 3D contact pairs are set up between collet and die, die and tape, tape and die holder, and tape and eject pins for the die pick up simulation. A debonding criterion for die surface and tape is developed to show the die separation process from tape. Transient dynamics modeling shows the stress distribution in the die so that the process can be optimized and we can verify that the die strength is strong to withstand the dynamic pick up and ejection pin stresses. A nonlinear (creep and elastic plastic) material constitutive model is applied to the solder paste for the multiple die attach process simulation. The die attach process order is modeled by element death and birth technique to show the impact of the multiple die attach process. Transient dynamics modeling method will be used to simulate the wire bonding process with different bonding forces, frequency, and friction between the bond pad and bond wire free air ball (FAB).

At EOL, typical processes include molding process with curing stress and residual stress, molding ejection, and clamping process that can potentially cause cracking in the molding compound, trim and form, and the singulation process that would induce the stress wave to crack the die or the EMC. For the molding process simulation, material properties are the function of both time and temperature, also the shrinkage during curing will impact the residual stress for the molding process. A user supplied FORTRAN subroutine code is developed in ANSYS in both time and temperature of the epoxy mold material. The simulation shows the mold injection, the curing process, and the residual stress build-up during molding. For the molding ejection and clamping simulation, a 3D model is developed to show the stress at the interface between the leadframe and mold compound that would induce the delamination. Prior to trim and form, a clamping process is needed to flatten the leadframe due to warpage after molding. The simulation is applied to this process to make sure the clamping process will not damage the package and die inside. For the package Trim/Form/Singulation simulation, a 3D transient dynamic large deformation finite element method will be used to determine what stresses are going to be induced while singulating the package, to check if the package can withstand punch stress waves and to determine potential design weaknesses.

Modeling and Simulation for Microelectronic Packaging and Integration: Manufacturing, Reliability and Testing, Second Edition. Sheng Liu and Yong Liu.

9.1 Wafer Process and Thinning

During back end processing a sequence of depositions and etches are used to build up the interconnect layers and the final passivation layer. At minimum this involves one metal layer and one dielectric layer per level of interconnect, with one to six levels of interconnect being common in the industry.

The dielectric and metal layers are thin films deposited by vapor deposition or sputtering techniques. The conductive metal layers may consist of one or more thin films, typically TiW, TiN, Al(Cu), W, Cu, and so on. The insulating dieletric layer typically consists of oxides (TEOS, BPTEOS, PSG, SOG, and so on), nitride (Si_3N_4), oxynitride (SiON), and so on. Each thin film has unique material properties associated with it. The intrinsic stress of each material will vary with the deposition process, and thermal history that occurs after deposition.

Typically a large portion(30%–70%) of the deposited metal layers is removed during patterning. For the dielectric layers the proportion is much smaller. For the dielectric layers the patterning is for the contact, via or bond pad openings.

With the exception of passivation, the dielectric layers will often have the in-die areas refilled with tungsten plugs, or the metal from the next conductive layer. All films above the silicon are normally removed from the saw street region of the wafers. Given the above, the dielectric layers can usually be assumed to be continuous, except for the saw street. For the conductive layers others have estimated that the stress relief due to patterning is proportional to the area removed (Shen *et al.*, 1996; van Silfhout *et al.*, 2002).

The backside grinding of wafers is a common operation for thinning. The final wafer thickness is set either by our ability to mechanically handle the wafer without breakage or for high voltage vertical devices by the need to support deep depletion regions. In vertical devices, thinning the wafer improves the Rdson that can be achieved. For all types of devices wafer thinning improves the transfer of heat from the front device side of the wafer to the backside for dissipation by the paddle, the package substrate, or other portions of the package capable of efficient conduction. The drawback of wafer thinning is an increase in wafer warpage and fragility.

9.1.1 *Wafer Process Stress Models*

The stress and strain in a thin film layer results either from the deposition process or is due to CTE mismatch. The stress from deposition, known as the intrinsic stress, is due to the non-equilibrium growth of the film microstructure (van Silfhout *et al.*, 2002). Intrinsic stress begins at the start of the deposition, at temperature and will vary as a result of changing process conditions or subsequent process steps including anneals.

The stress due to CTE mismatch is the well known thermo-mechanical phenomenon. The total stress in the film is given by:

$$\sigma_t = \sigma_{int} + \sigma_{th} \tag{9.1}$$

where σ_t is the total stress, σ_{int} is the intrinsic stress, and σ_{th} is the thermo-mechanical stress due to CTE mismatch. In this study we did not include the intrinsic stress; earlier internal studies showed the intrinsic stress to be small compared to other sources.

Further we can relate the stress in a film to the wafer warpage using Stoney's formula (Stoney, 1909), knowing only the materials properties of the substrate and the thickness of the deposited film.

$$\sigma_t = \frac{E_s t_s^2}{6(1-\upsilon_s)t_f}(\kappa_i - \kappa_e) \tag{9.2}$$

where σ_t is the biaxial film stress, t_s and t_f are the thickness of the substrate and film respectively, υ_s is Poisson's ratio for the substrate, E_s is Young's modulus, and κ_i and κ_o are the final and initial values of the wafer curvature (1/bending radius).

Euler beam theory gives the load limit of a sample as:

$$P_{\text{lim}} = \frac{\sigma_p b h^2}{1.5L} \tag{9.3}$$

where σ_p is the chip strength, $L, b,$ and h are the span, width, and thickness of the sample respectively, and P_{lim} is the load limit.

Modeling wafer warpage and backside grind is technically challenging because of the scale differences between the wafer size and the thinness of the deposited films. Backside grind presents similar issues because of the relative thinness of the layers removed.

In order to handle the wide variety of materials and conditions easily the models presented were built using the parameterization features of the FEM software (ANSYS). With the parameterized code available any number of variations can be run easily. This includes the ability to change the number of deposited films and the associated materials parameters, the wafer diameter, and backside grind thickness.

9.1.2 Thin Film Deposition

We start the modeling at the first deposited dielectric layer. The grown oxides and other films have a relatively insignificant effect on wafer warpage and stress. Table 9.1 below lists the material parameters used in this study. Ideally the study would use parameters extracted from the wafer fabrication process itself but these were unavailable so values from the vendors were used.

The system simulated in this study was a single conductor made up of a TiW barrier layer and Al(Cu). The layer thicknesses are given in Table 9.2 below. We could not account for the wafer patterning in this study as it increases the complexity of the simulation beyond normal computational limits.

Figure 9.1 below shows the temperature history used in the simulation. Dwell times at temperature were not included as no time dependent material properties or effects of anneals were simulated.

The deposition of layers is done using the element birth mechanism within ANSYS.

Figure 9.2 shows the incremental warpage due to each sequential deposition. The extremely thick aluminum deposition is responsible for the majority of the deformation as can be seen in the increase from step 2 to step 3. The final warpage is displayed in Figure 9.3.

The black circle shows the measured final wafer warpage. Given the amount of patterning for these wafers the difference is reasonable. If a patterning step was included we would expect to see a decrease in warpage at the time of the etch.

Table 9.1 Deposited materials parameters

Material	Modulus (MPa)	CTE (ppm/°C)	Poisson Ratio
Silicon	169.5×10^3	3.2	0.23
ILD 1	70.0×10^3	4	0.25
TiW	117.0×10^3	10.2	0.25
Al(Cu)	70.0×10^3	10	0.35
SiON	115.0×10^3	2.92	0.25

Table 9.2 Layer thicknesses

Layer	Material	Thickness
Substrate	Si	670 μm
ILD 1	LTO/BPSG	0.5 μm
Metal 1	TiW	0.18 μm
Metal 2	Al(Cu)	5.0 μm
Passivation	SiON	1.2 μm

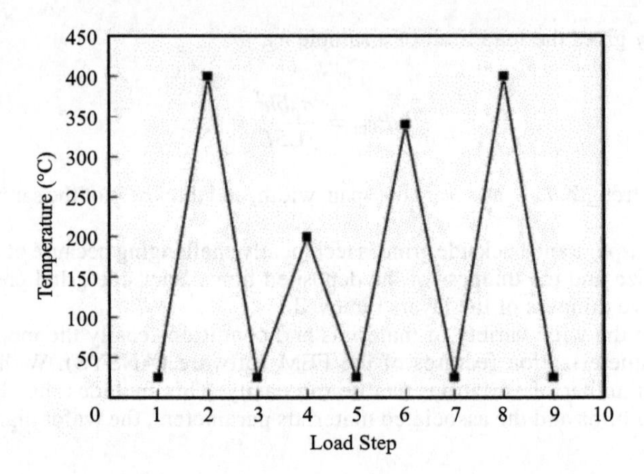

Figure 9.1 Deposition temperature of load history

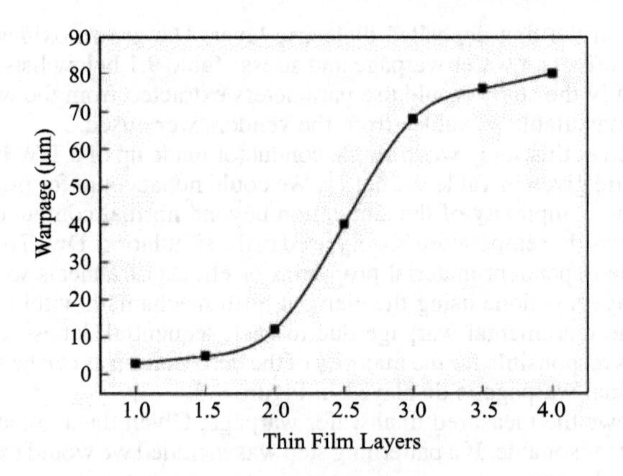

Figure 9.2 Warpage versus deposition step

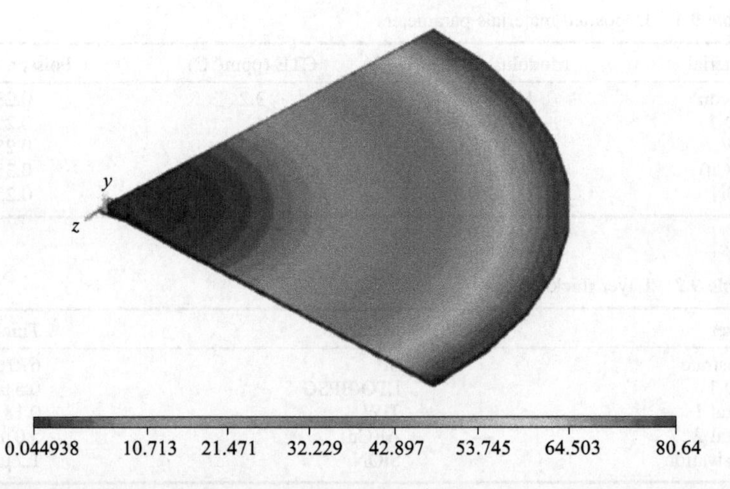

Figure 9.3 Wafer warpage from deposition, unit in microns (*Color version of this figure is available online*)

9.1.3 Backside Grind for Thinning

The backside grind process uses the initial wafer stresses from the previous steps. During the backside grind process the frontside of the wafer is protected with a film, the wafer is loaded onto a vacuum chuck, and a two-step grind process is used to thin the wafer. We examined two cases with the simulation, with final wafer thicknesses of 113 μm and 226 μm.

The steps in the backside grind process model are:

1. Apply vacuum chuck
2. Heat to process temperature
3. Coarse grind
4. Fine grind
5. Cool down
6. Remove wafer from chuck

Table 9.3 shows the process parameters used for simulation in this research. The same is shown schematically in Figure 9.4.

Due to the geometries involved the elements used have a large aspect ratio. The results converged, however with a negative Jacobian during the solution process. As a result the values in the above table should not be taken as being predictive, they are valid for a relative comparison between different films or backside grind thicknesses. Table 9.4 and Figure 9.5 show the wafer warpage and the radial stress of 4 mil

Table 9.3 Backside grind process parameters

Parameter	Thin	Thick
Initial Thickness	670 μm	670 μm
Grind Force	6 lbs	6 lbs
Grind Temperature	60 °C	60 °C
Coarse Grind Thickness	537 μm	424 μm
Fine Grind	20 μm	20 μm
Final Thickness	4 mil (113 μm)	8 mil (226 μm)

Table 9.4 Uncalibrated backside grind results

	4 mil	8 mil
Warpage	2639 μm	687 μm
Maximum Stress	239.0 MPa	243.6 MPa

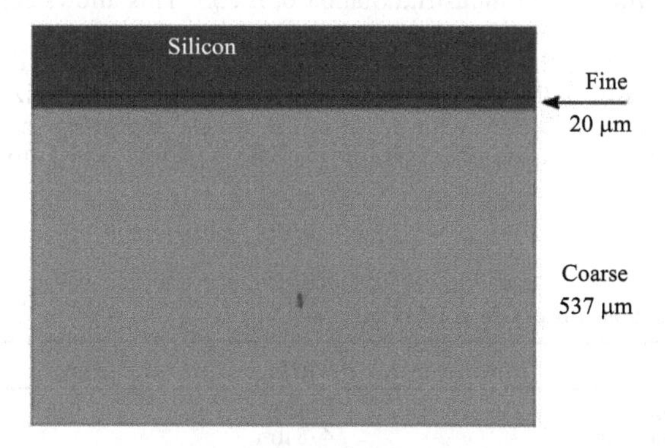

Figure 9.4 Grind scheme for 4 mil wafer

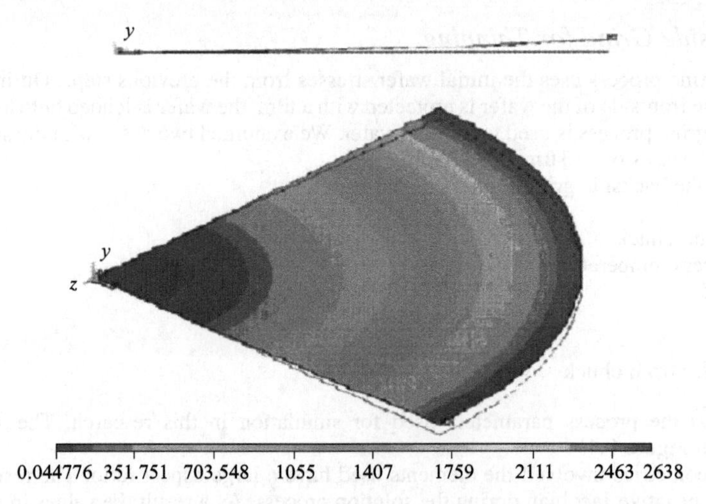

Figure 9.5 Wafer after grind to 4 mil (*Color version of this figure is available online*)

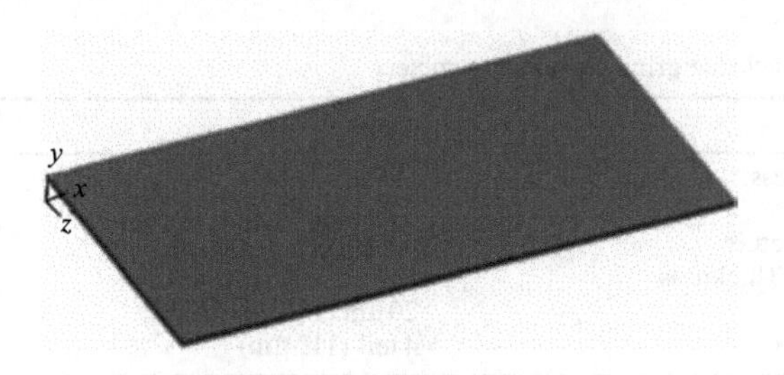

Figure 9.6 Reduced model

and 8 mil wafer. The result shows the stress to be independent of backside grind thickness and the ratio between the grind thicknesses to be 3.84.

In order to derive a predictive model, a smaller geometry was constructed that allows a reduced aspect ratio. The new model used a 13.6 mm × 6.3 mm sample size as shown below in Figure 9.6, constructed to comply with ASTM Industrial Standard, E 855. This allows comparison to the results of Wu *et al*. (2002).

The reduced model results in the same maximum stresses as were found in the quarter model above. Table 9.5 shows the results of the new model that uses the smaller sample size. This new model does not have the same problem as before with negative Jacobians appearing in the solution. The ratio between the 4 mil and 8 mil backside grind samples is 3.85 which is similar to the result in the larger uncalibrated model.

Table 9.5 Calibrated backside grind results

	4 mil	8 mil
Sample Warpage	87 μm	63.4 μm
Normalized Warpage	478 μm	124 μm
Maximum Stress	239.4 MPa	243.8 MPa

It confirms the experimental results of Wu *et al.* (2002), that the stress is independent of the size of the wafer or sample. This result makes sense, that the films and grind processes include a given amount of stress that is independent of the sample size. The stress in turn induces a warpage that could be expressed as a radius of curvature that is independent of sample size. This is important for future work that will try to model the effects of backside grind and stress on die damage, but will need to be modeled as smaller samples to be practical.

Wu *et al.* (2002) show that average tensile yield strength for a wafer is approximately 240 MPa. They also showed a variability in the post backside grind die strength of about 40%, with the weakest areas being 30% weaker than the average. This weakness occurred on machines produced by several vendors, although in differing patterns. Therefore in the weak areas of the wafer the yield strength will be reduced to be about 161 MPa.

Using Equation 9.3 above allows us to calculate the load limits that can be placed on a wafer of different backside grind thicknesses.

$$P_{4\text{mil}} = 0.635\text{N}$$
$$P_{8\text{mil}} = 2.54\text{N}$$

Based on the load limits an 8 mil thick wafer can withstand four times the load of a 4 mil wafer before it is damaged.

9.2 Die Pick Up

For the manufacturing process of package assembly, the wafer is first adhered to tape and is sawn into the dies as shown in Figure 9.7. After the wafer is sawed, the die will be debonded and placed to the bond position on the leadframe.

Figure 9.8 shows that a vacuum pick-up tool commonly known as a 'collet' mounted on a bond head grabs the aligned die from the sawn wafer. The vacuum collet will push the die with a force to prevent air moving into the vacuum pipe. The die for mounting is then ejected from the wafer by one or more ejector needles under the tape. The vacuum collet then picks the die up from the sawn wafer and sticks it to the leadframe.

Figure 9.9 shows the structure of die holder. The ejection needles eject die through the holes in the die holder. The die holder is fixed during the whole die pick up process. The die holder fixes the tape through vacuum air in the holes.

Thin die (50–100 μm) is more and more widely used in small packages of high-power applications as it gives excellent electrical resistance performance, reduced space, and weight (Reiche *et al.*, 2003; Peter *et al.*, 2002). For thin wafer, the die pick up operation is a critical aspect of the manufacturing process. The use of inappropriate tape stickiness, inappropriate ejector needles, and improper ejection parameter settings can cause die backside tool marks or microcracks that can eventually lead to die cracking. This work shows the impact of tape stickiness and die thickness on die stress during the separation process from tape.

Figures 9.10 and 9.11 show the mesh for modeling work. Multiple 3D contact pairs are set up between the collet and die, die and tape, tape and die holder, and tape and eject pins for the simulation. A debonding criterion for die surface and tape is developed to show the die separation process from tape.

Figure 9.7 Sawn wafer stuck to tape

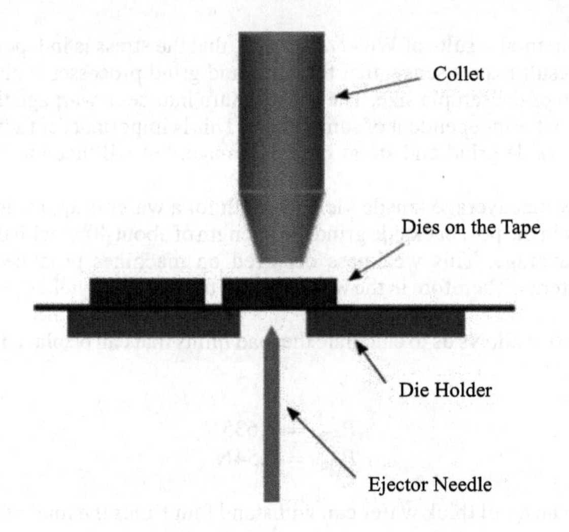

Figure 9.8 Die pick up process

Transient dynamics modeling shows the stress distribution in the die so that the process can be optimized and we can verify that if the die strength is strong to withstand the dynamic pick up and ejection pin stresses.

Tie break (Tied with Failure) contact (ANSYS12.0 Theory Manual 2009) actually 'glue' the contact surfaces to the target surfaces. The contact and target surfaces are initially coplanar so that during initialization, an isoparametric position of the contact surface within the target segment is calculated. Thereafter, upon application of loads or initial velocities, the contact surfaces are forced to maintain their

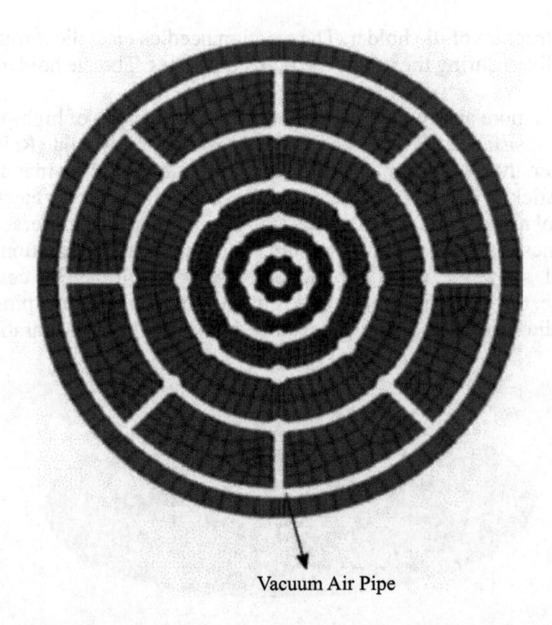

Figure 9.9 Die holder with vacuum air holes

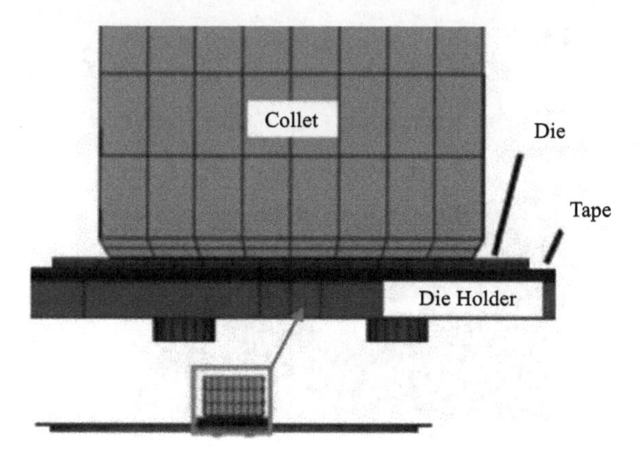

Figure 9.10 Die, tape, collet, tape, and die holder

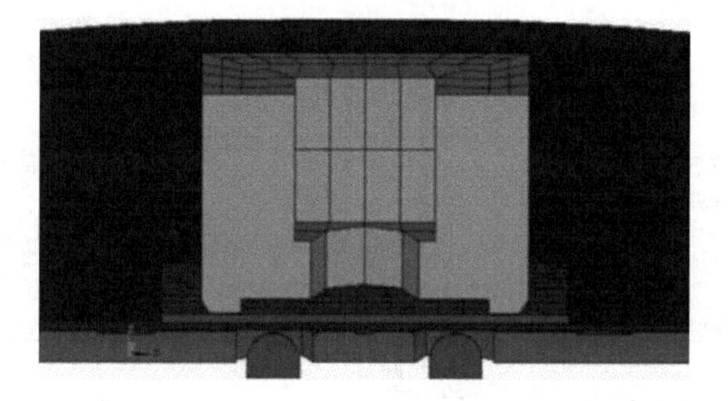

Figure 9.11 Cross-section of die, tape, collet, tape, and die holder

isoparametric position within the target surface. The effect of tied break contact is that the target surfaces can deform and the slave nodes are forced to follow that deformation. The contact surfaces are tied to the target surfaces only until a failure criterion is reached. This is done by 'pinning' the contact surfaces to the target using a penalty stiffness; after the failure criterion is exceeded, the contact surfaces are allowed to slide relative to, or separate from, the target surface (Figure 9.12). Tied break contact failure is based on the normal failure stress and the shear failure stresses. Failure of the connection will occur when:

$$\left(\frac{|f_{\mathrm{n}}|}{f_{\mathrm{n,fail}}}\right)^{m1}+\left(\frac{|f_{\mathrm{s}}|}{f_{\mathrm{s,fail}}}\right)^{m2}\geqslant 1 \tag{9.4}$$

Figure 9.13 shows die stress sequence when die is separated from the tape. The sequence shows the stress starts when the collet pushes the die and prevents vacuum leakage (phase 1). As the ejection pin pushes the die up, the force from the tape changes as die separates from the tape. There is a peak stress (phase 2) before the die is separated from the tape (phase 3). At the time the die is separating from the tape, the stress changes from the die's edge to the die's center. That reduces the die deformation caused by tape. Die stress is only caused by the vacuum after the die is separated from the tape (phase 4).

Figure 9.14 shows the tape's first principal stress and von-Mises stress at phase 2. Figure 9.15 shows the tape's first principal stress and von-Mises stress at phase 3. The tape's stress will increase as the

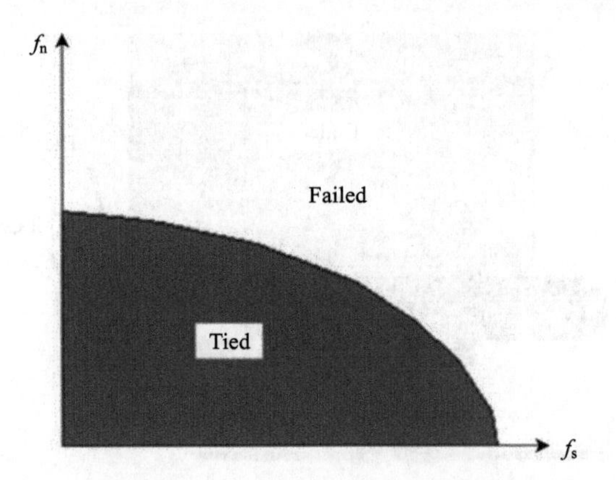

Figure 9.12 Tied break contact failure criteria (upper cases for the first Alph)

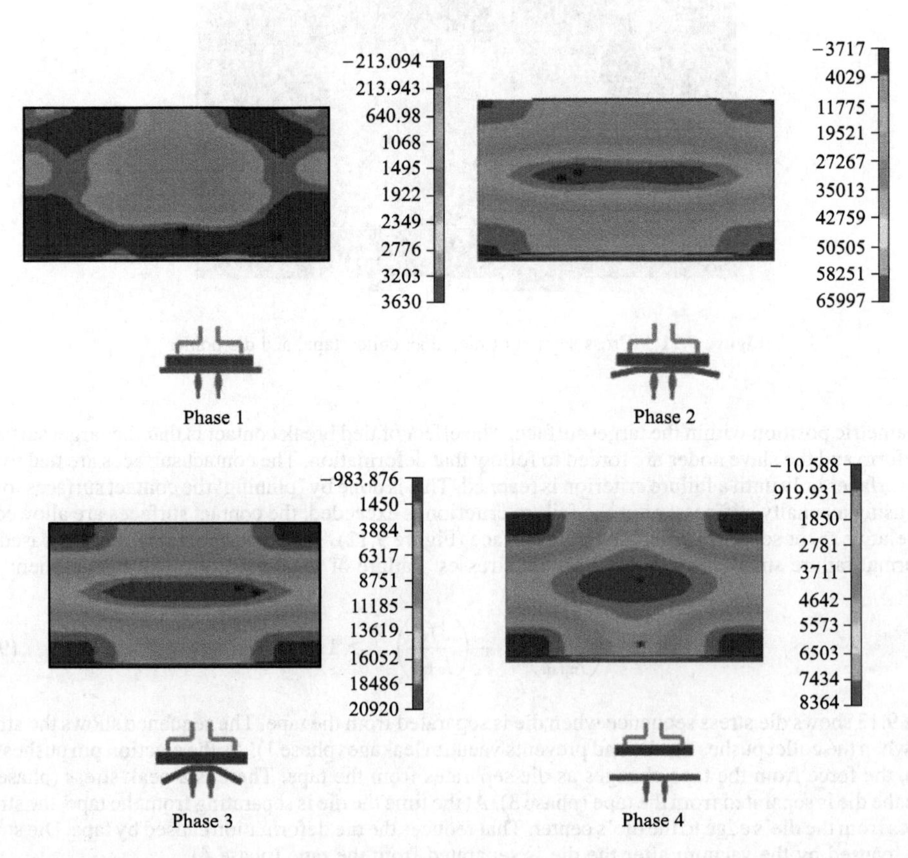

Figure 9.13 First principal stress in die during debonding process from the tape *(Color version of this figure is available online)*

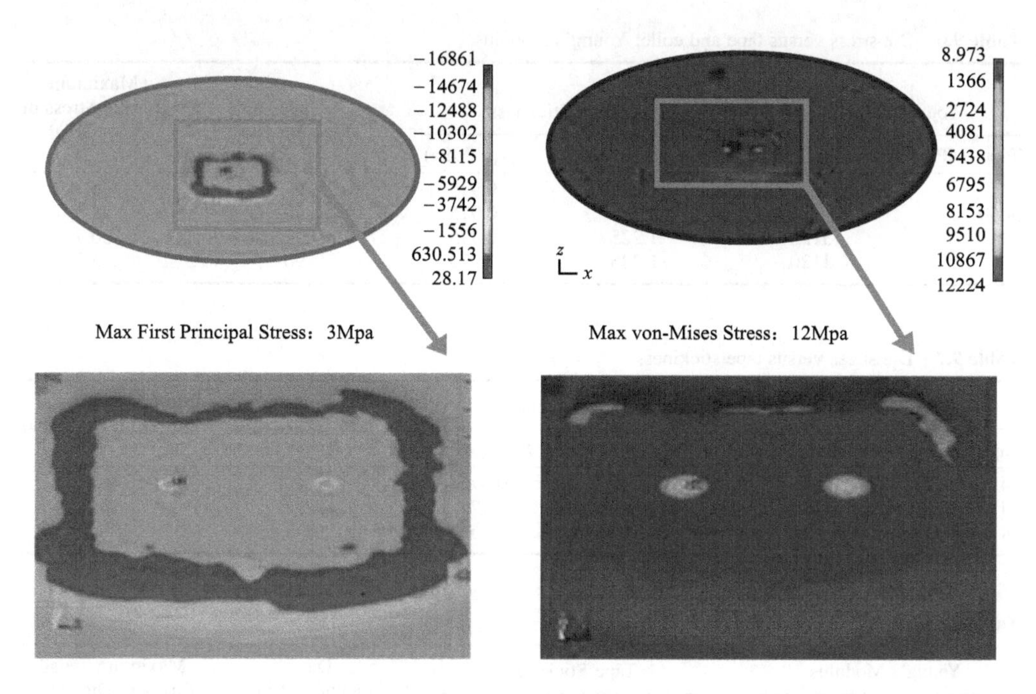

Figure 9.14 Tape's stress when die begins to separate from the tape (location of max stress is die's edge) *(Color version of this figure is available online)*

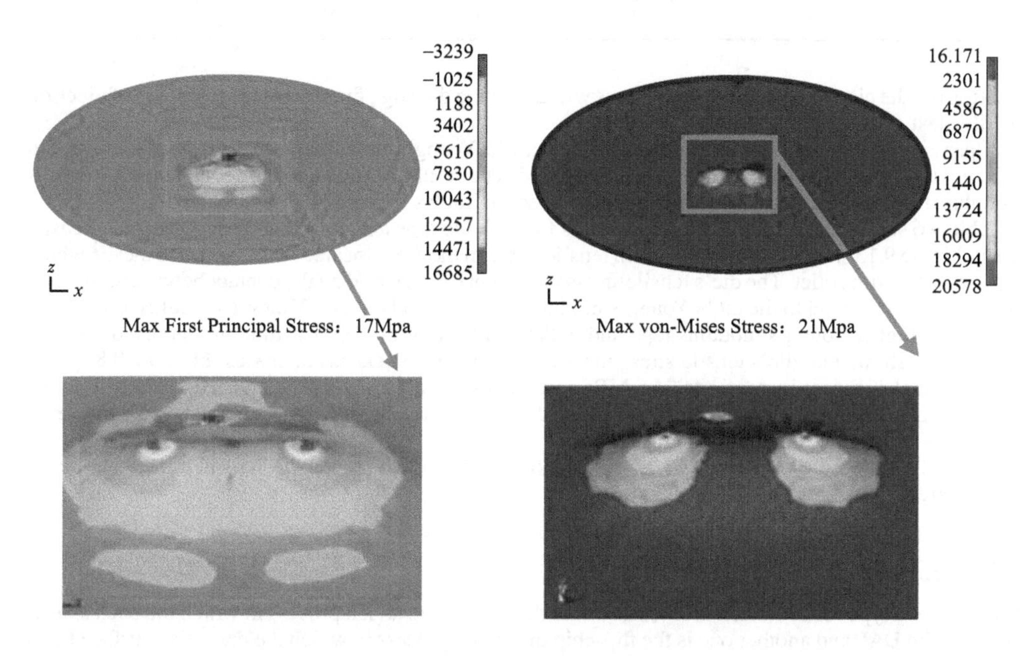

Figure 9.15 Tape's stress when die is fully separated from the tape *(Color version of this figure is available online)*

Table 9.6 Die stress versus tape and collet Young's modulus

Young's Modulus		Tape Stickness		Die Thickness (μm)	Maximum Tensile Stress in Die (MPa)
Tape (MPa)	Collet (MPa)	$f_{n,fail}$ (MPa)	$f_{s,fail}$ (MPa)		
54	18.4	1.225	0.975	90	66
3059	18.4	1.225	0.975	90	28
54	3120	1.225	0.975	90	50.7
3059	3120	1.225	0.975	90	15.3

Table 9.7 Die stress versus tape stickiness

Young's Modulus		Tape Stickiness		Die Thickness (μm)	Maximum Tensile Stress in Die (MPa)
Tape (MPa)	Collet (MPa)	$f_{n,fail}$ (MPa)	$f_{s,fail}$ (MPa)		
54	18.4	1.225	0.975	90	66
54	18.4	12.25	0.975	90	78.7
54	18.4	24.5	1.95	90	129.5

Table 9.8 Die stress versus die thickness

Young's Modulus		Tape Stickiness		Die Thickness (μm)	Maximum Tensile Stress in Die (MPa)
Tape (MPa)	Collet (MPa)	$f_{n,fail}$ (MPa)	$f_{s,fail}$ (MPa)		
54	18.4	1.225	0.975	90	66
54	18.4	1.225	0.975	350	11.6
3059	3120	1.225	0.975	90	15.3
3059	3120	1.225	0.975	350	5.2

ejection needle ejects the die. The tape stress will act on the die edge first and move to the top of ejection needles last.

A parametric study about different collet and tape's Young's modulus, different die thickness and different tape stickiness are conducted and discussed. The results of impact of collet and tape's Young's modulus, tape stickiness, and die thickness are show in Tables 9.6 and 9.7.

Table 9.6 shows that higher Young's modulus for both the tape and collet can reduce the die's tensile stress. Figure 9.13 shows that the maximum tensile stress in the die is located in the center of the die where not in contact with collet. The die's tensile stress seems not to be caused by the contact between collet and die directly. Compared to the die's Young's modulus, the tape and collet's Young's modulus are relative small. The higher Young's modulus tape and collet can prevent die deformation and reduce die stress.

Table 9.7 shows that die's tensile stress increases as the tape stickness increases. In Table 9.8, as the tape's normal stickness increases to 24.5 MPa and shear stickiness increase to 1.95 MPa, die will not be picked up from the tape. Then the tensile stress on die is very large because tape's impact on die locates at die's edge which maximum die's deformation.

Table 9.8 shows die tensile stress as die thickness changes. Thinner die will increase maximum tensile stress.

9.3 Die Attach

There are two types of die attach processes, one is the regular die attach process in which the back side of die faces the DAP and another one is the flip-chip die attach process in which the die active surface faces the DAP.

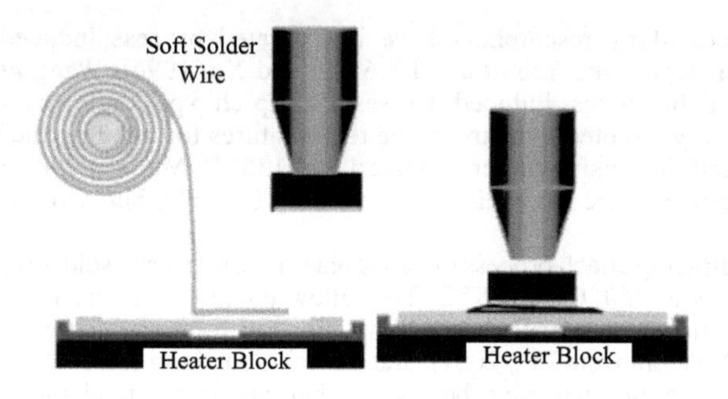

Figure 9.16 Eutectic solder die attach process

A typical soft solder die attach process is shown in Figure 9.16. As the vacuum collect head picks up the die with metallization from the wafer tape, a leadframe is held on the surface of heater block. The soft solder wire is then placed on the leadframe. The temperature of the heater block ramps up to the melting temperature of the solder wire that makes the wire fuse. Finally the cavity collet head moves the die to the melt solder in the leadframe and as soon as the solder cools down, a solid connection is established. A controlled temperature profile is required to define the liquidus/solidus transition. The die attach process is a very nonlinear process that needs to develop a strong finite element method to obtain the converged solution.

Figure 9.17 shows a typical flip-chip die attach process. A lot of efforts have been made to minimize and optimize the manufacturing steps for flip-chip die attach processing with low cost. However, the quality and reliability have to be assured. Particularly, the assembly process induced stresses have to be

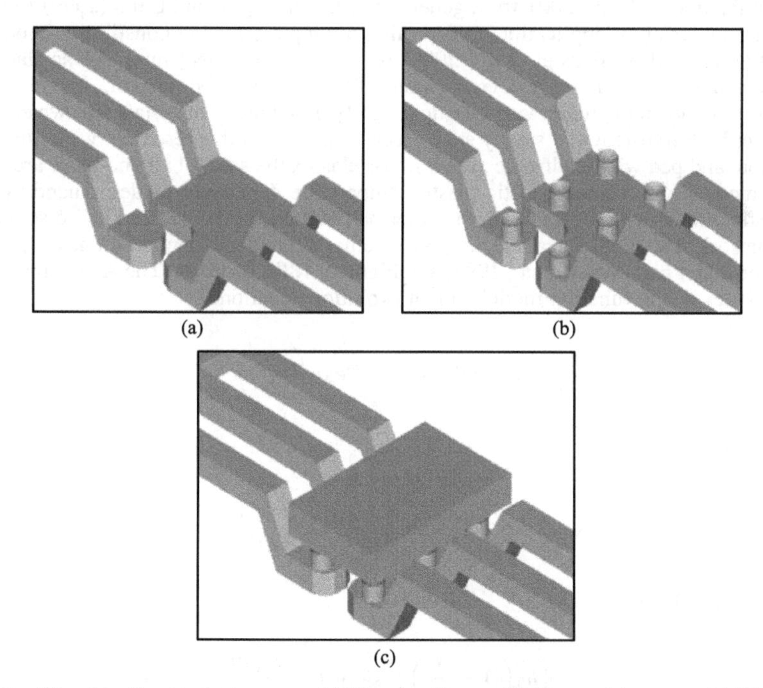

Figure 9.17 Flip-chip die attach processes. (a) Place leadframe. (b) Place solder paste. (c) Flip-chip die attach and reflow *(Color version of this figure is available online)*

reasonablely modeled. Many researchers have investigated process induced stress analysis for various plastic IC packages (see Yeh *et al.*, 1996; Yu and Xu, 1996). Wang *et al.* (1998) were the first to have developed the process induced stresses of a flip-chip packaging by sequential processing technique, which mainly targeted two stress free temperatures for bonding the silicon chip onto the substrate at 180 °C and dispensing underfill material at 135 °C. Mannan *et al.* (2000) simulated the solder paste reflow coalescence for a flip-chip assembly by CFD, but without the results of stress induced after reflow.

Typical assembly flip-chip attach process includes place or screen print solder paste on leadframe, flip-chip attach and reflows at 260 °C or 285 °C. The reflow process is a very nonlinear process both in materials and in geometry. As it is well known, simulation results are influenced by the material models (Liu and Antes, 1999; Paulino and Liu, 2001), and the solution algorithm strategies (Manniaty and Liu *et al.*, 2001). From a material viewpoint, both solder ball and paste are characterized by deformation behavior which is dependent on temperature and time. A correct solder joint material model is necessary to obtain reasonable results. On the other hand, highly nonlinear system equations are used to obtain results, which must be solved by a highly efficient solution algorithm. Normally, the algorithm based on Newton's method with a tangent operator converges quickly if a trial solution inside the radius of convergence is obtainable and used as an initial guess. However, a reference (Eggert *et al.*, 1991) indicates that for viscoplastic problems, obtaining a trial solution within the radius of convergence for Newton's method has been problematic. Therefore, the goal for die attach or for flip-chip attach simulation is development of a robust finite element framework with both an effective material model and solution algorithm strategies to target the assembly process induced stress and to assure the reliability of our products.

9.3.1 Material Constitutive Relations

There are a lot of research efforts for the plastic and viscoplastic constitutive models (see Liu and Antes, 1999 and Paulino and Liu, 2001 for a general relation and Qian and Liu's (1999) formulations in semiconductor solder alloy applications). Typically, a simple effective constitutive model, which shows good fitting capability (Feustel *et al.*, 2000) is a time-independent multi-linear elastic-plastic model which includes a time dependent creep model for solder joint's behavor. Anand's evolution model is also a useful model to predict viscoplastic property of hot metal materials (Brown *et al.*, 1989), which may reflect both transient and steady state creep behavors. In this research, we assume the die, UBM, passivation, and pad with leadframe are all linear elastic, the solder ball and paste are nonlinear. Three material models (multi-linear elastic-plastic temperature dependent model, Anand viscoplastic model, and plastic with creep model) for both solder ball and paste will be used and discussed in the simulation framework. The multi-linear elastic-plastic temperature dependent relations are presented in Figure 9.18(a) for solder ball (Wang *et al.*, 1998) and Figure 9.18(b) for paste. The Anand model (Brown *et al.*, 1989) includes a sinh function model and an evolution equation:

$$\frac{d\varepsilon_{vp}}{dt} = A[\sinh(\xi\sigma/s)]^{1/m}\exp\left(\frac{-Q}{kT}\right) \tag{9.5}$$

or, it can be written as:

$$f_{ANAND} = \bar{\sigma} = \frac{s}{\xi}\sinh^{-1}\left(\frac{\dot{\varepsilon}_{vp}}{A}\exp\left(\frac{Q}{kT}\right)\right)^m \tag{9.6}$$

where the evolution equation:

$$\frac{ds}{dt} = \left\{h_0\left(\left|1-\frac{s}{s^*}\right|\right)\right\}^a \mathrm{sign}\left(1-\frac{s}{s^*}\right)\frac{d\varepsilon_{vp}}{dt} \tag{9.7}$$

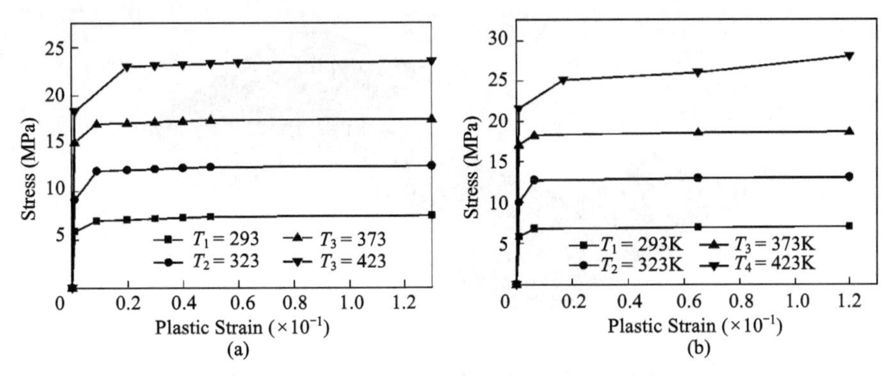

Figure 9.18 Elastic-plastic stress-strain relations for solder joint. (a) Solder ball. (b) Solder paste

and:

$$s^* = \hat{s}\left[\frac{\mathrm{d}\varepsilon_{\mathrm{vp}}/\mathrm{d}t}{A}\exp\left(\frac{Q}{kT}\right)\right]^n \tag{9.8}$$

where the constants $Q/k = 11935$ (K) for ball and 9400 (K) for paste, which stands for the activation energy/Boltzmann's constant; $A = 3.785\text{e}7(1/\text{s})$ for ball and 4.0e6 (1/s) for paste, which is a pre-exponential factor; $\xi = 5.91$ for ball and 1.5 for paste, means a multiplier of stress; $m = 0.143$ for ball and 0.3 for paste, which is the strain rate sensitivity index of stress; h_0 (unit: MPa) is the hardening/softening constant, its number is 1579 MPa for both ball and paste; $\hat{s} = 13.79$ MPa is the coefficient for deformation resistance saturation value for both ball and paste; $n = 0.07$, is the strain rate sensitivity index of deformation resistance; a is the strain rate sensitivity of hardening/softening, $a = 1.3$ for both ball and paste.

For the elastic-plastic and creep model, the elastic-plastic relation is selected as in Figure 9.18, the creep model is chosen to be a Norton's law as following:

$$\varepsilon_{\mathrm{creep}} = c\exp\left(\frac{-Q}{kT}\right)\sigma^n \tag{9.9}$$

The parameter c [1/(s ·MPa)] is 1.936 for solder ball and 100.6 for paste respectively; n is 11.03 for solder ball and 4.434 at 25 °C and 1.262 at 285 °C for paste; Q/k (K) is 1.1935×10^4 for solder ball and 7.45×10^3 for paste.

The coefficient of thermal expansion (CTE) of the solder ball is presented as in (Wang *et al.*, 1998). The Poisson's ratios are selected as 0.4 for solder alloy and 0.35 for paste. The silicon is assumed to be linear elastic and its Young's modulus and Poisson's ratios are 169.5 GPa and 0.278, respectively. The CTE of the silicon chip is presented in (Wang *et al.*, 1998). The passivation is composed of elastic material, the Young's modulus is 4.2 GPa, Poisson's ratio is 0.34, and the CTE is 3.59 ppm/°C. The Young's modulus of UBM is 186.5 GPa, Poisson's ratio is 0.3, and the CTE is 12.4 ppm/°C. Leadframe is copper based material, its Young's modulus is 120.5 GPa, Poisson's ratio is 0.3, and the CTE is 18.3 ppm/°C.

9.3.2 Modeling and Numerical Strategies

A typical flip-chip attach on pad and leadframe is shown in Figure 9.19, only one package with related uncut leadframe model is taken into consideration for simplicity.

In the die attach processing the package components include die, passivation, UBM, solder ball and paste, and leadframe. The die attach process is a very nonlinear process, its reflow curve is applied as in

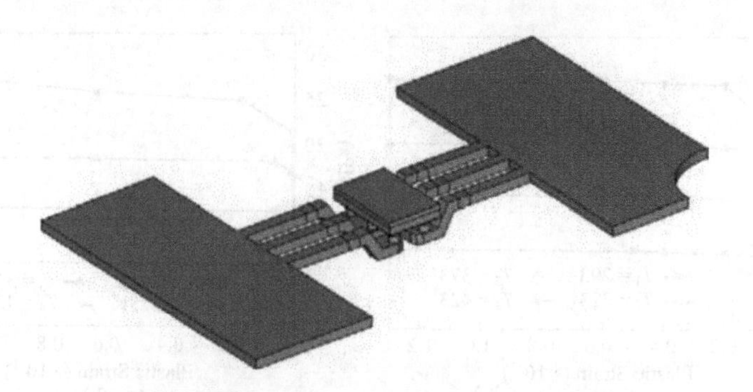

Figure 9.19 A flip-chip package with leadframe

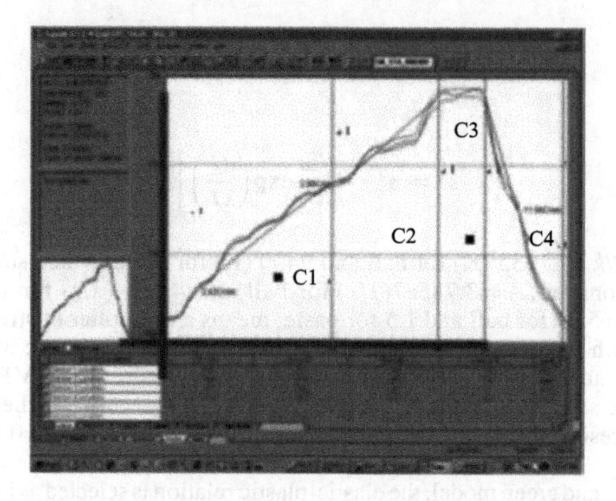

Figure 9.20 Reflow profile of flip-chip attach process

Figure 9.20 the melting temperature for the solder ball is 312 °C, and the solder paste is 285 °C. In order to simplify the problem, we assume: (1) Compression flow placement is perfectly done by the machine, no chip float before reflow process; (2) the initial stresses induced before reflow temperature are too small to be considered; and (3) the stress in the solder paste is free at reflow temp 285 °C (it almost fully melts at this temperature), and it starts to be generated from 285 °C cooling down. Based on these assumptions, the die attach simulation is presented through a 3D model (Figure 9.21).

Figure 9.21 3D model of the flip-chip attach processing

Loading for the flip-chip attach process is defined such that during the flip-chip processing, the thermal loading is based on the reflow profile. The reflow profile has four phases: phase one (C1) is the preheating and ramp up with about rate 2.62 °C/s for about 47 s ; and phase two (C2) is ramp up to 285 °C with 2.8 °C/s for about 30 s; phase three (C3) dwells at 285 °C for 13 s, solder paste melts at this phase, and then phase four (C4) cools down to a room temperature of 25 °C with rate 11.36 °C/s (Figure 9.20).

Numerical Strategies

(i) Simple Progressive Solution Strategy

It is known that for a nonlinear problem, linearized finite element analysis (FEA) formulation is developed to allow an implementation in a Newton–Raphson solution procedure. Iterative schemes based on the Newton–Raphson method with a tangent operator, which is consistent with the numerical method employed, converge rapidly as long as the initial trial solution is within the radius of convergence. Because of the high degree of nonlinearity typically arising in viscoplastic constitutive models, obtaining a trial solution within the radius of convergence for Newton–Raphson's method has been problematic because the equations have a very narrow radius of convergence. Therefore in this research, a simple progressive solution method is used according to our previous research results (Maniatty *et al.*, 2001). Based on this method, the system equations are solved progressively in a step-wise manner and the solution from the previous step is used as the trial solution in the next step: two constitutive material models, a simple easy converged trial function model, and the targeting material model are considered, the progressive formulation can be expressed as:

$$F(t) = \left(1 - \frac{t}{t_L}\right) \exp\left(\beta \frac{t}{t_L}\right) f_{\text{trial}} + \left[1 - \left(1 - \frac{t}{t_L}\right) \exp\left(\beta \frac{t}{t_L}\right)\right] f_{\text{ANAND}} \qquad (9.10)$$

where t varies from 0 to t_L, f_{trial} is a trial function, in a special case, it may be selected as the linear elastic function. Parameter β controls the rate at which the trial function approachs the targeting material model, an example is shown in Figure 9.22. t_L is the local time within which the trial function may make the system solution inside the radius of convergence. Equation 9.9 may be written as a user supplied subroutine code in any commerical code, like ANSYS or ABAQUS. If one just simply takes an elastic trial solution in a very short time as the initial guess, one may obtain very good convergence in all the remaining step simulations, particularly for viscoplastic models.

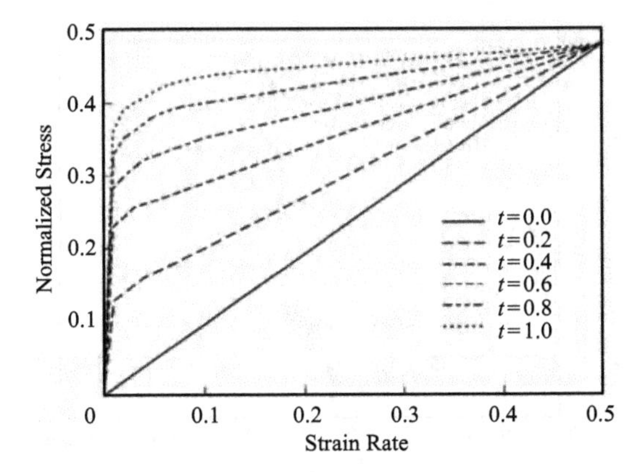

Figure 9.22 Progressive approach process for a viscoplastic Anand model with $\beta = -1$, initial solution is a linear function

(ii) Implicit Algorithm

The implicit scheme is a robust, fast, and accurate algorithm which may allow you to use a larger time step (Liu and Antes, 1999):

$$\Delta\varepsilon_{\mathrm{vp}}^{n} = \Delta t_n\left[\left(1-\alpha^*\right)\dot{\varepsilon}_{\mathrm{vp}}^{n} + \alpha^*\dot{\varepsilon}_{\mathrm{vp}}^{n+1}\right] \tag{9.11}$$

where the parameter α^* in the range of [0,1] defines various explicit and implicit schemes.

(iii) Solver Technology

Solvers currently have direct solver and iterative solver.

Normally, for the degree of freedoms (DOFs) within 500 K using sparse solver might get speed and robustness solution. If the multiple processors are used, an algebraic multigrid iterative solver can increase speed for large problems (DOFs of more than 0.5 M), and is also very strong in targeting the ill-conditioned problems compared to the conjugate class gradient solvers.

9.3.3 FEA Simulation Result of Flip-Chip Attach

The system is meshed with 15353 elements. The goal of this simulation is to simulate the stress distribution during the flip-chip attach. The simulation results are shown in Figures 9.23 to 9.26. Implicit and sparse solvers are used in this simulation.

Figure 9.23 shows the history of maximum von-Mises stress in solder ball versus time in three material models (elastic-plastic, Anand viscoplastic, and elastic-plastic with creep models). From the figure that it may be seen that the maximum von-Mises stresses in three cases increase when the temperature ramps up, the stress value is slightly higher than the yield strength at about 100 °C. After that when the temperature continues to ramp the stress decreases, this is because the yield stress of solder material becomes lower as the temperature increases.

The stress in solder material will not be higher than its yield strength because of the plastic flow properties. At the end of dwelling reflow temperature 285 °C, all the stresses reach their lowest numbers. After the reflow temperature 285 °C, the temperature cools down, all the stresses increase, and reach the highest values at the end of room temperature. In the pure elastic-plastic and Anand viscoplastic models, the stress reaches the yield stress while the plastic and creep model does not. It seems that the stress value of Anand model is the highest at the end of reflow. The results further show that after the end of reflow,

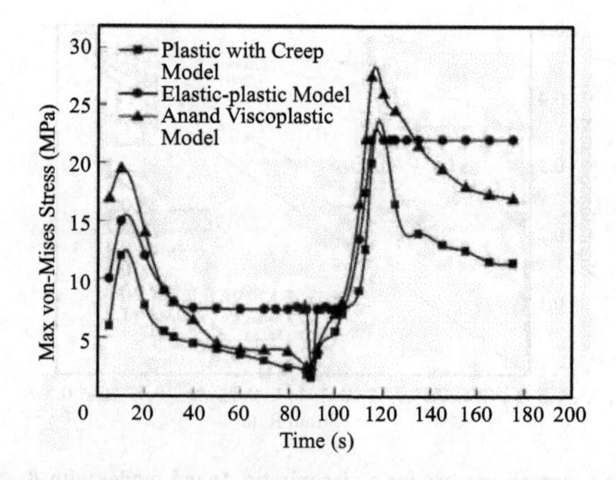

Figure 9.23 Maximum von-Mises stresses in solder ball with three material models

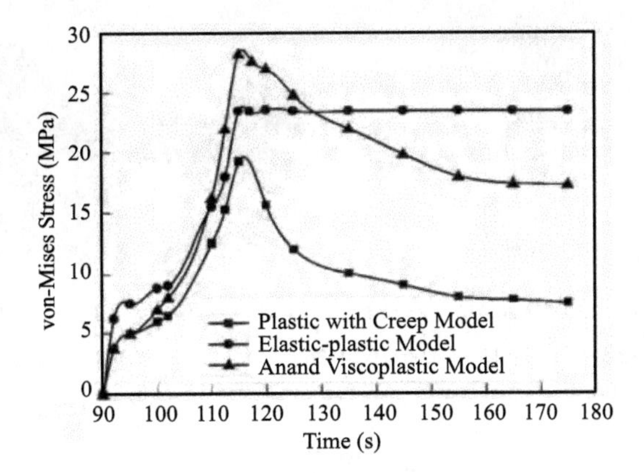

Figure 9.24 Maximum von-Mises stresses in paste with three material models

if the package dwells at room temperature for 60 s, the stresses are released for both Anand viscoplastic and elastic-plastic with creep model.

Figure 9.24 shows the history of maximum von-Mises stress in solder-paste versus time in three material models, the time range is from the melting point to the room temperature point, and then dwell 60 s. Basically during room temperature, the stress is very small due to the attach placement. When the temperature reaches reflow temperature of 285 °C, which is the melting point of the paste, the stress is free. Therefore, for the paste, we use two models, one is the element birth and death model, in which the paste elements are set to be dead from room temperature to 285 °C, active right after 285 °C and starting to cool point. This method makes it difficult to obtain a converged solution. Another method is the continuum model, which assumes the paste elements are active during the whole reflow process. Figure 9.25 gives the comparison of elastic-plastic von-Mises stresses with both element birth-death model and the continuum model. The results show that the maximum von-Mises stress in the continuum model have similar behavior as in the solder ball before melting point, while from the melting point cooling down to room temperature, the stresses obtained by the element birth and death model agree with the results of the continuum model. The prediction of the continuum model is slightly higher than that of the element birth and death model. Again, the Anand material model predicts a higher stress value (Figure 9.24 at the end of reflow).

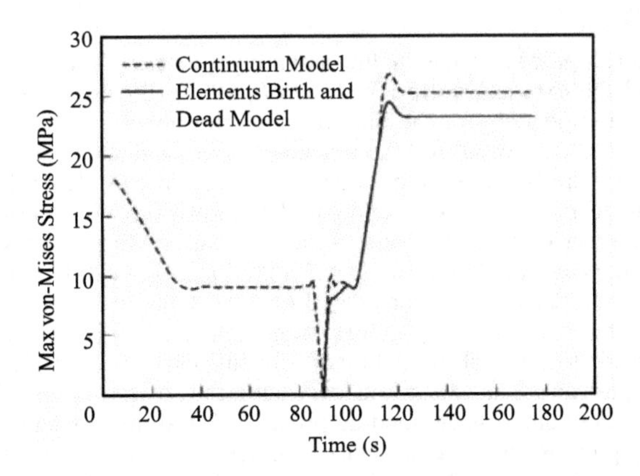

Figure 9.25 Comparison of elastic-plastic von-Mises stresses in paste with element birth and dead model and continuum model

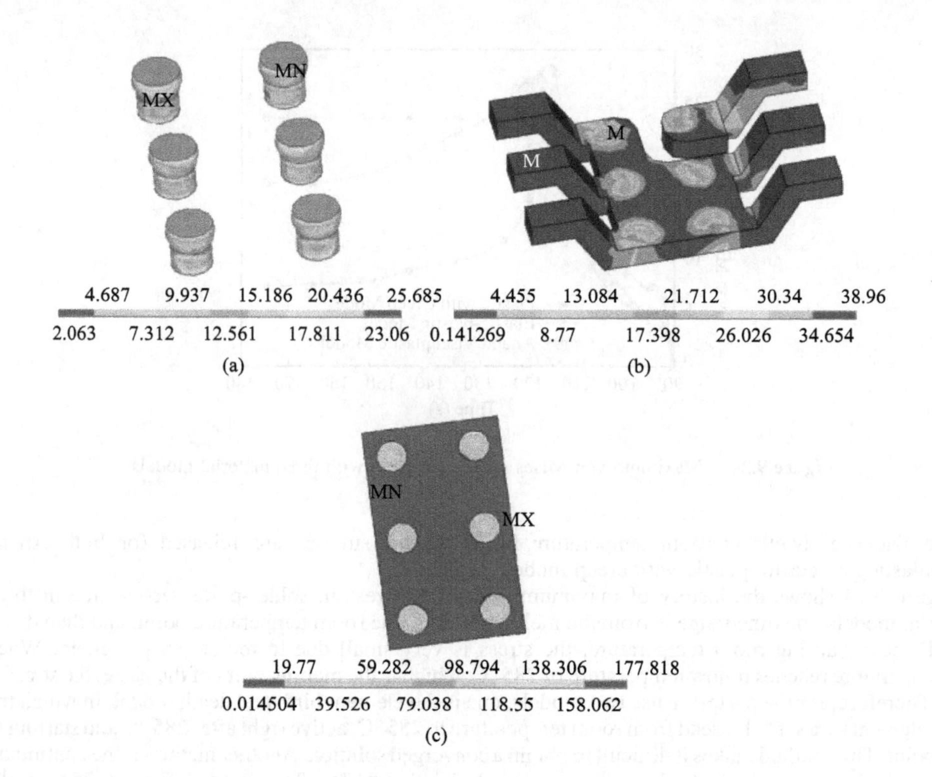

Figure 9.26 von-Mises stress in solder joint, leadframe, and UBM and passivation. (a) Solder joint. (b) Leadframe. (c) UBM/passivation *(Color version of this figure is available online)*

Figure 9.26 shows the von-Mises stresses distribution in the solder joint, leadframe, UBM, and passivation after reflow. The maximum stress in the solder joint appears in the gate solder joint. The stress value of the UBM edge in connection with the solder ball is very high, which could be the potential factor for delamination.

9.4 Wire Bonding

Wire bonding is a critical process stage in the assembly process for the connection between the semiconductor chip and the external world (Ikeda *et al.*, 1999). By this stage, most of the device's costs have been absorbed, especially for the high density wire bonding and the bond pad over active (BPOA) design (Hess *et al.*, 2003; Awad, 2004). To reduce the cost and obtain the optimized wire bonding solution, modeling of the wire bonding process has been used to help determine the optimized wire bonding parameters and to help identify the potential failure mechanisms (Takahashi *et al.*, 2002). Currently, there are a number of modeling studies on the wire bonding process (Gillotti *et al.*, 2002; Van Driel *et al.*, 2004; Degryse *et al.*, 2004; Liu *et al.*, 2006; Fiori *et al.*, 2007; Liu *et al.*, 2008). Degryse *et al.* (2004) and Fiori *et al.* (2007) have considered Cu/low K interconnects under the bonding pad; Liu *et al.* (2004) studied the wire bonding loop formation in the electronic packaging assembly process. Yeh *et al.* (2006) conducted the transient simulation of wire pull test on Cu/Low-K wafers. However, most authors only considered pure mechanical bonding loads with static or quasi-dynamic methods to simulate the free air ball (FAB) under compressive bonding process. Few simulations include both the dynamic nonlinear wire bonding process and silicon device stresses in the same model due to difficulties in achieving convergence. In reality, wire bonding is a complicated, multi-physics, transient dynamic process, which is completed within a very short period of time. The dynamic impact to both the wire bond and devices on silicon is critical and significant. Therefore, previous modeling methods seem to be insufficient (Liu *et al.*, 2008).

It is known that the common failure modes of wire bonding are bond pad cratering, peeling, and cracking below the bond pad. There are five major factors that relate to the failure modes and affect the quality of bonding process and bond pad devices (Liu *et al.*, 2008; Harman, 1997): bonding force or deformation, ultrasonic amplitude and frequency, friction and intermetallic compounds between FAB and bond pad, substrate temperature, and time duration. The challenges here are how to describe the ultrasonic dynamic effects. If we change the bond pad design, how does it affect the bonding process and stress distribution? When the system cools down to room temperature, how does the residual stress impact the bond pad device? In this chapter, the methodology of wire bonding modeling and simulation is presented and the finite element framework for both static and transient nonlinear dynamic wire bonding analysis are developed, which integrates both wire bonding process and the interconnects/silicon under the bond pad. Dynamic simulation focuses on the ultrasonic transient dynamic bonding process and the stress wave transferred to the bond pad device and silicon. The bonder capillary is considered as a rigid body due to high hardness. This results in a rigid and elastic-plastic contact pair between the capillary and the FAB. While the contact surfaces between the FAB and the bond pad are a nonlinear contact pair with consideration of the dynamic friction. The Pierce strain rate dependent model is introduced to model the impact stain hardening effect.

Four topics of interest will be presented in this section: (1) Wire bonding process with different parameters which includes ultrasonic amplitude, frequency, friction between the FAB and the bond pad, the bond pad and the below device, residual stress after substrate cooling down; (2) Comparison of the impacts between the wire bonding and the wafer probing for a bond pad over active device (BPOA); (3) Wire bonding above a laminate substrate; (4) Impact of the wedge bonding versus the thermal-mechanical stress.

9.4.1 Assumption, Material Properties and Method of Analysis

Simulation may help us to understand the stress impact and to examine the relative effects of elements within the bond pad structure over the active device on the stresses developed during wire bonding. However, modeling cannot solve every part of the bonding process. To conduct an effective simulation, the following assumptions are made:

Assume that the temperature of the FAB is the same as substrate (in reality, there is some difference due to the transient temperature cooling from FAB forming and moving to contact bond pad).

Assume that the FAB is rate dependent elastic-plastic material during the bonding process. Bond pad and rest metal layers are elastic-plastic materials. All the other materials are linear elastic.

The contact intermetallic effect and diffusion in the bond formation due to ultrasonic energy will not be considered in this chapter. It will require further work to determine an equivalent way to model the intermetallic impact, such as the equivalent material parameters over certain local ranges and the coefficient of friction between FAB and bond pad.

Assume the capillary is a rigid body due to a much higher Young's modulus and hardness. The inertia force from the capillary transferred to the FAB is not considered here.

The heat and temperature induced by the friction between FAB and bond pad are not included.

When ultrasonic energy is applied to the FAB by capillary, it causes a reduction in yield strength and increases the mobility and density of dislocations after some dwell time (Harman, 1997). The strain rate is in the "slip by dislocation shifting region", as the deformation occurs, the material strain hardens. When the hardening material transmits energy to the ball-pad interface, slip planes shift at the interface, opening up new metal surfaces. Contact diffusion bonding (with intermetallic effects at certain temperature rises by dynamic friction), enhanced by ultrasonics, occurs at the newly exposed metal surfaces; As the frequency increases, after some point (for example, 120 kHz or above), the FAB material may not be significantly softened at the beginning, while the strain rate is in the "simultaneous several lattice slip" region, the material behaves as a hard material transmitting energy to the ball-pad interface (Langenecker, 1966; Shirai *et al.*, 1993). Ikeda *et al.* indicated (1999): a gold ball is impacted by a capillary at the loading speed of 0.98 N/s, which may result in the strain rate of the gold ball being more than 1000 1/s locally. Based on the Hopkinson impact bar tests by Ikeda, the yield stress of FAB with strain rate hardening may be approximated by:

$$\sigma_s = \sigma_0 + H'\dot{\varepsilon}^{pl} \tag{9.12}$$

where $\sigma_0 = 0.0327\text{GPa}$, $H' = 0.00057\text{GPa} \cdot \text{s}$.

Table 9.9 Materials parameters

Material	Modulus (GPa)	CTE (ppm/°C)	Poisson Ratio	Yield Stress (GPa)
Silicon	169.5	3.2	0.23	
ILD	70.0	4	0.25	
TiW	117.0	10.2	0.25	
Al(Cu)	70.0	10	0.35	0.2 (25 °C) 0.05 (450 °C)
Au(FAB)	60.0	14	0.44	0.0327 (200 °C)
W (plugs)	409.6	4.5	0.28	

Equation 9.12 can be further expressed as the rate dependent Peirce Model:

$$\sigma_s = \left[1+\frac{\dot{\varepsilon}^{pl}}{\gamma}\right]^m \sigma_0 \tag{9.13}$$

where $m = 1$ and $\gamma = 561.4$ (1/*s*).

The material parameters are listed in Table 9.9, FAB, bond pad, and metal layers are nonlinear (bi-linear) materials, all of the rest of the materials are considered to be linear elastic.

A general finite element code, ANSYS®, is used in the modeling. A nonlinear large deformation and transient dynamics implicit algorithm with the above rate dependent Peirce model is selected. Since the bonder capillary is considered as a rigid body due to high hardness, this leads to the rigid and elastic-plastic contact pair between capillary and FAB. While the contact surfaces between FAB and bond pad are a nonlinear contact pair with consideration of the dynamic friction.

9.4.2 Wire Bonding Process with Different Parameters

A conceptual 2D model is shown in Figure 9.27, which is a cut from a typical die with three layer metallization and three dielectric (ILD) layers above the silicon. The typical diameter of a FAB is 70 µm, the bond pad length is 90 µm. The bottom of silicon is fixed and two sides are constrained in horizontal direction. Figure 9.28 gives the meshes of the FAB and bond pad system. The capillary moves down a certain height (bonding height) to press the FAB with a high speed and different frequency. The ultrasonic horizontal motion cycle of the capillary with a typical amplitude of 1 µm and a typical frequency of 100 kHz is shown in Figure 9.29.

Figure 9.30 shows the typical bonding force versus time. Two phases are defined in Figure 9.30. Phase 1 includes the contact impact with strain hardening and after about 100–150 ultrasonic cycles it becomes softened (similar to Levine, 1995); and it goes into the second phase with a lower constant bonding force.

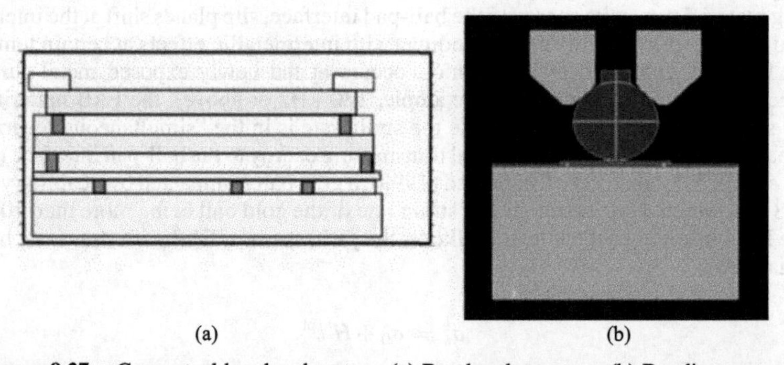

(a) (b)

Figure 9.27 Conceptual bond pad system. (a) Bond pad structure. (b) Bonding system

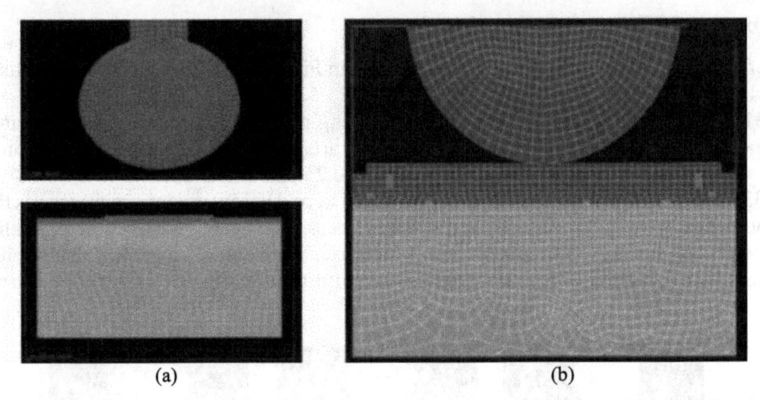

Figure 9.28 Meshes of FAB and bond pad model. (a) FAB and substrate. (b) FAB bond pad

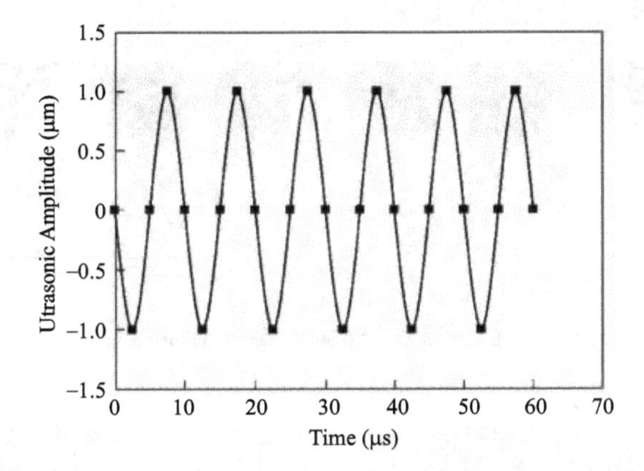

Figure 9.29 Ultrasonic cycle movement versus time (μs) with a amplitude of 1 μm and 100 kHz

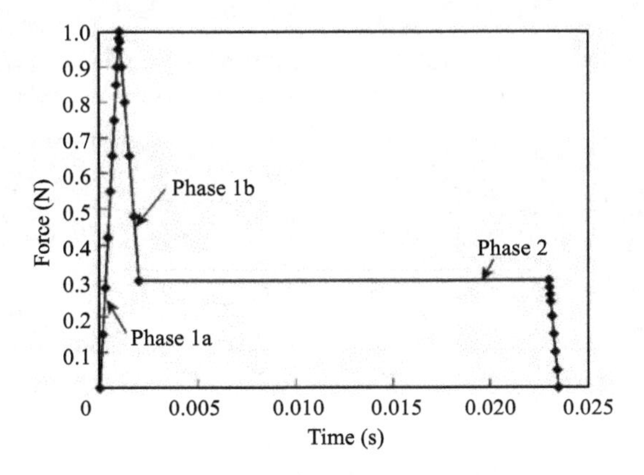

Figure 9.30 Typical bonding force versus time

9.4.3 Impact of Ultrasonic Amplitude

The results of impact of ultrasonic amplitude are shown in Figures 9.31 to 9.36. These results are obtained under a fixed ultrasonic frequency 138 kHz.

Figure 9.31 shows that the stress in bonding processing moves as the capillary moves and symmetric case only appears when the capillary moves to the center area. Figure 9.32 shows that the von-Mises stress at amplitude 1.0 μm is about 37% greater than that at 0.25 μm amplitude.

Figure 9.33 shows that the maximum principal stress in pad below devices increases as the ultrasonic amplitude increases. However, maximum von-Mises stress and shear stresses decrease at the beginning, after amplitude is larger than 0.5 μm, their values increase. Figure 9.34 gives similar situation in stresses transferred to silicon. Figure 9.35 shows that as the amplitude increases, cratering deformation in the

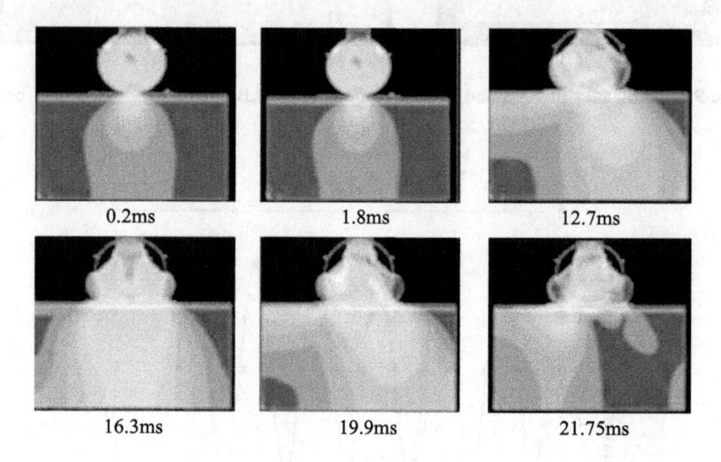

Figure 9.31 Evolution of von-Mises stress distribution at different time

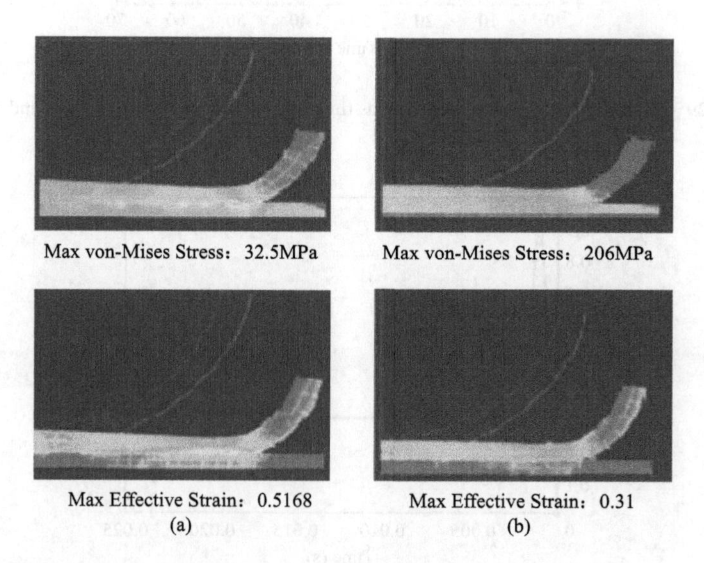

Figure 9.32 Contact layers between pad and ball with ultrasonic amplitude 1.0 μm versus 0.25 μm. (a) Amplitude 1.0 μm. (b) Amplitude 0.25 μm *(Color version of this figure is available online)*

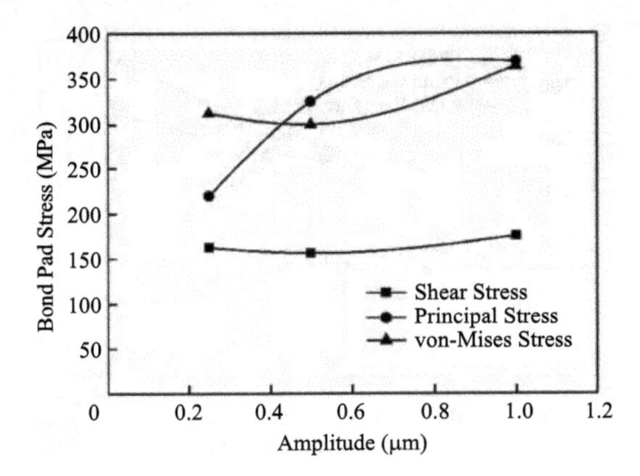

Figure 9.33 Stresses in pad below device versus ultrasonic amplitude

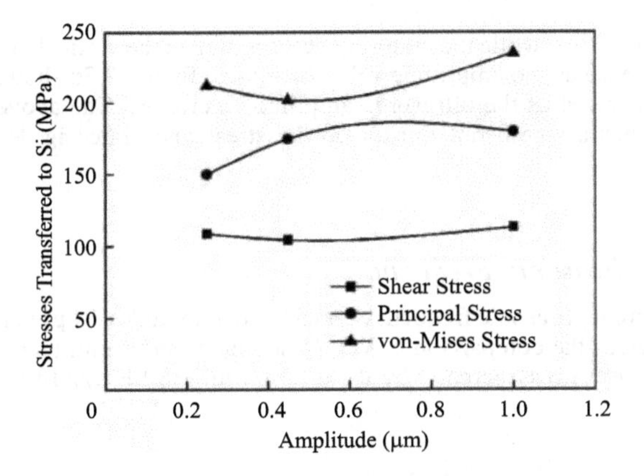

Figure 9.34 Stresses transferred to silicon

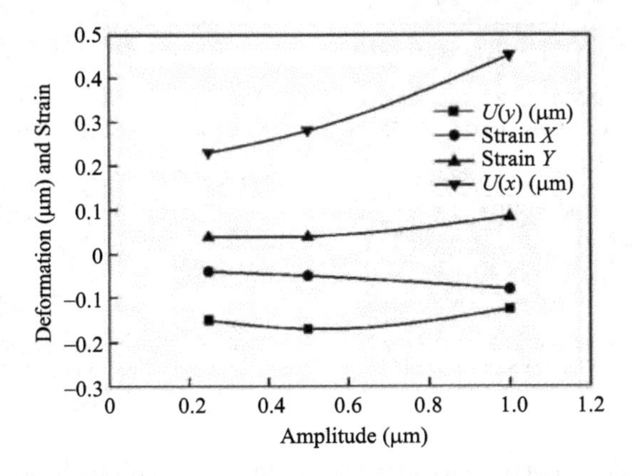

Figure 9.35 Bond pad cratering deformation and strains

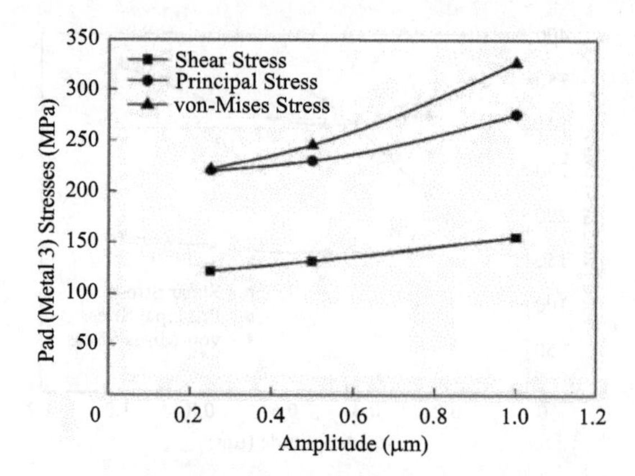

Figure 9.36 Bond pad stresses versus ultrasonic amplitude

horizontal direction increases, while the cratering deformation in the vertical direction reduces. Strains in both directions increase as the amplitude value increases. Figure 9.36 shows all the stresses in the bond pad (Metal 3) increase as the ultrasonic amplitude increases. The above results show that the ultrasonic amplitude has a significant impact on the stress and cratering deformation during wire bonding process.

9.4.4 Impact of Ultrasonic Frequency

The results under different ultrasonic frequency and a fixed amplitude of 1 μm are listed in Figures 9.37 to 9.41. Figure 9.37 gives the comparison of von-Mises stress and strain under 60 kHz and 138 kHz frequencies. The von-Mises stress increases by about 7.5% with 138 kHz, and effective strain increases by about 8.5%.

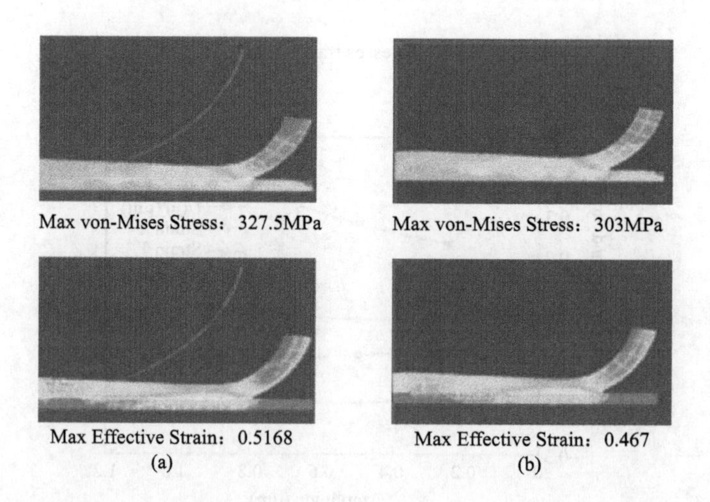

Figure 9.37 Contact ball and pad layers with 138 kHz and 60 kHz. (a) 138 kHz. (b) 60 kHz *(Color version of this figure is available online)*

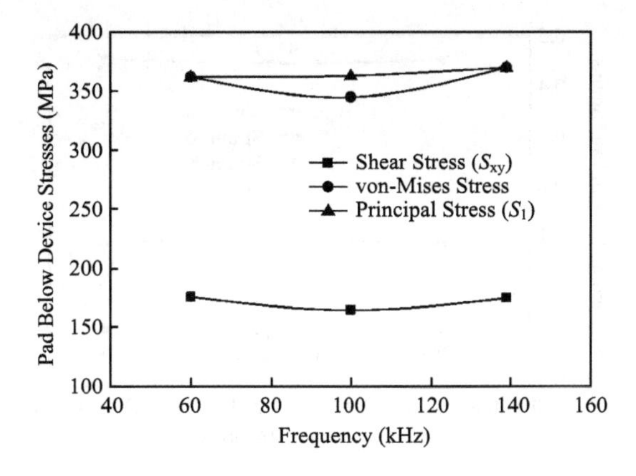

Figure 9.38 Stresses of pad below device under different frequencies

Figure 9.38 shows the maximum principal stress in the pad below the device increases as the ultrasonic frequency increases. However, the maximum von-Mises stress and shear stress decrease at the beginning, and their value increases when the frequency is larger than the 100 kHz. This may explain the effects of frequency. At the beginning, the FAB becomes softened at a lower frequency, and the strain hardening properties due to rate dependence become dominant and make the stresses increase after 100 kHz. However, the changes are not so significant. Figure 9.39 shows that all the stresses transferred to silicon have similar properties, that is at beginning the stresses decrease after the frequency is larger than 100 kHz, their values increase. Figure 9.40 shows there is no significant difference in the bond pad cratering strain and deformation as the frequency goes up, though vertical cratering deformation has the same properties. Figure 9.41 shows that stresses in the bond pad increase as the ultrasonic frequency increases. Overall, the impact of ultrasonic frequency is not as significant as the impact of ultrasonic amplitude. Perhaps when in our modeling assumptions, the inertia force from the capillary is not considered, that could induce an error.

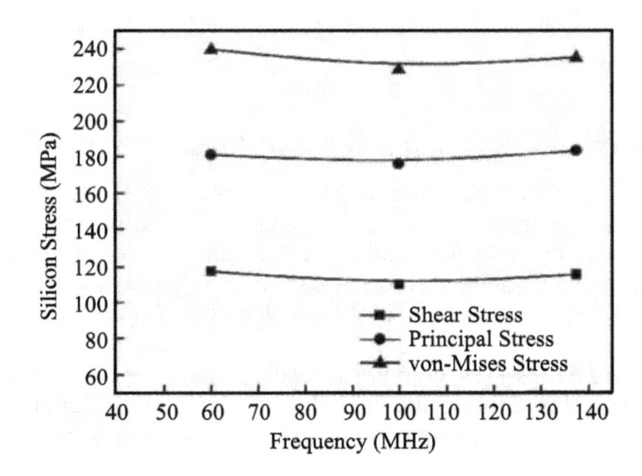

Figure 9.39 Stresses transferred to silicon under different frequencies

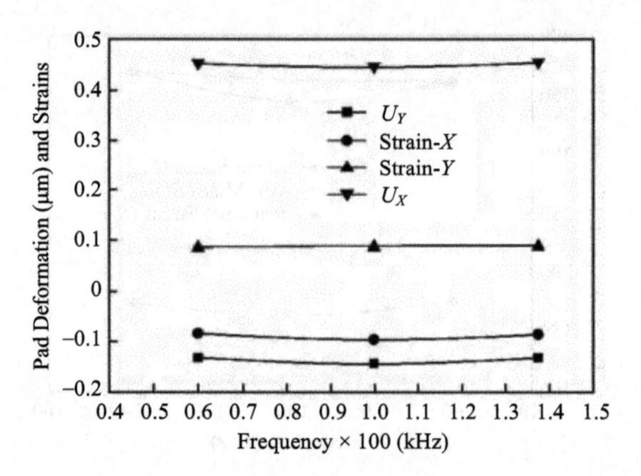

Figure 9.40 Bond pad cratering deformation and strain versus frequency

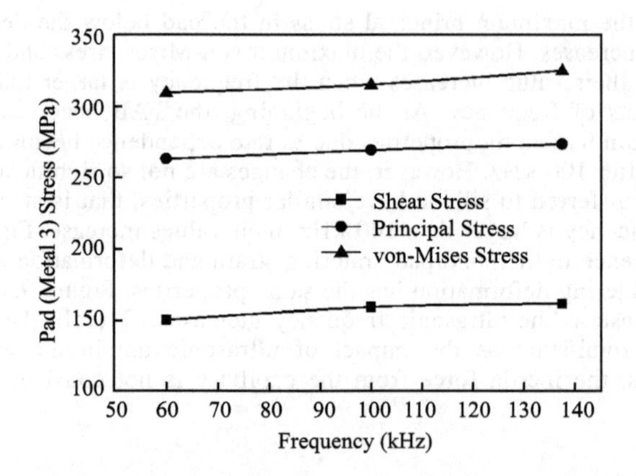

Figure 9.41 Bond pad stresses versus frequency

9.4.5 Impact of Friction Coefficients between Bond Pad and FAB

Bonding friction is a complicated multiple physics process at the interface between the ball and pad. Bonding occurs when ample energy is available to overcome the active energy of barrier and surface oxidation, and the relative motion at the interface of ball and pad is zero. Based on this assumption, the heat induced by friction and thereby induced by the intermetallic diffusion problem is not considered. The results are listed in Figures 9.42 to 9.45 with different friction coefficients.

Figure 9.42 shows the von-Mises stress comparison at different phases with a higher friction and a lower friction. The results have disclosed that at the final phase, although the stress in a higher friction case is about 1.18 times greater than the lower friction case, the radius of ball/pad bonding is about 10% greater than that of lower friction.

Figures 9.44 and 9.45 show that the pad deformation is very non-uniform in the horizontal (radial) direction. Higher friction makes a larger ball/pad contact area and co-deformed bonding. However, this may result in higher stress and greater cratering. If the stress and deformation induced in the wire bonding

von-Mises Stress Distribution of Lower Friction

Max：65.06MPa Max：119.04MPa Max：178.67MPa

von-Mises Stress Distribution of Higher Friction

Max：77.94MPa Max：132.91MPa Max：212.53MPa

Figure 9.42 von-Mises stress comparison of higher friction (friction coefficient is 1.5) and lower friction (friction coefficient is 0.2) *(Color version of this figure is available online)*

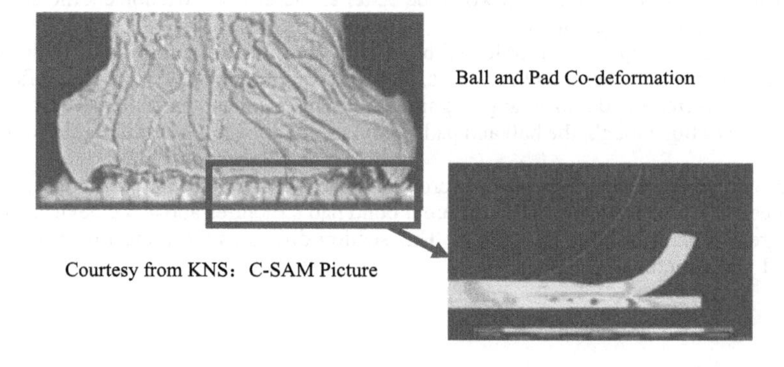

Figure 9.43 Co-deformed ball/pad with friction *(Color version of this figure is available online)*

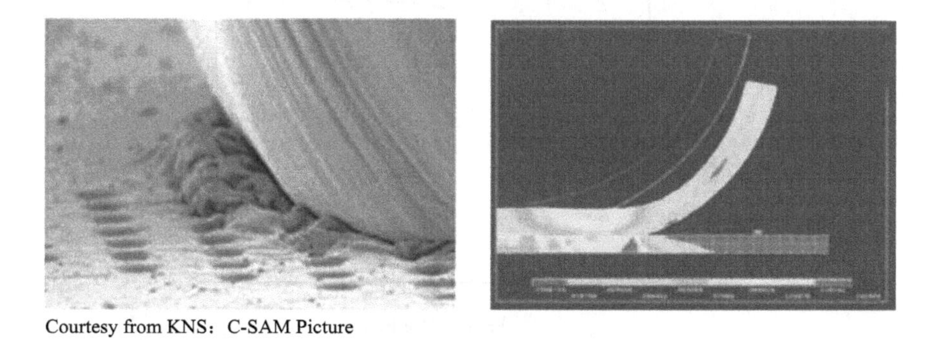

Figure 9.44 Deformed ball/pad during bonding process (maximum stress appeared at the contact edge interface) *(Color version of this figure is available online)*

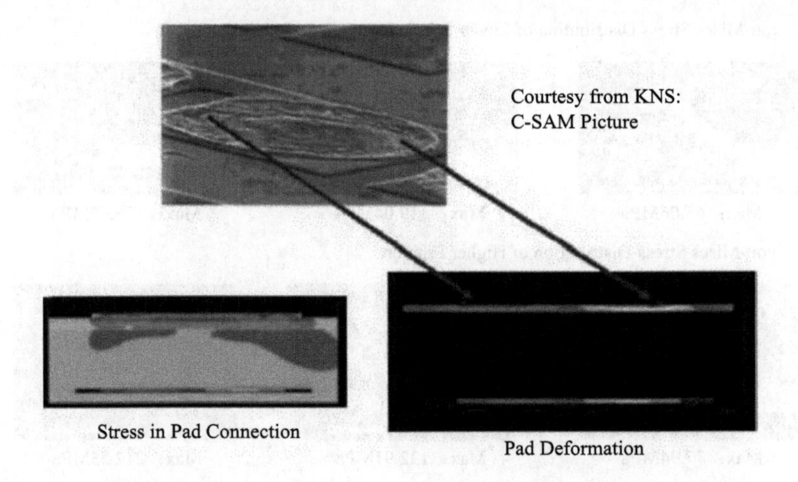

Figure 9.45 Pad cratering and 2D section modeling results

process are within the failure criterion, it would be better to increase the friction coefficient. Figure 9.45 shows the damage mark in the silicon.

Figures 9.46 and 9.47 give the profiles of bond pad stresses and deformation versus the friction coefficients. As the friction coefficient increases, the stress and deformation increases. After some point (1.5), stress and deformation do not change significantly. This is understandable, because as the friction coefficient becomes big enough, the ball and pad become stuck together and initial relative movement is not possible.

Optimization of the wire bonding assembly process is one way to reduce the cratering and crack failure of a BPOA design. Another way is to test different bond pad structures above the device, and determine how much stress is transferred to the silicon. This section discusses the impact of different bond pad thickness and pad structure.

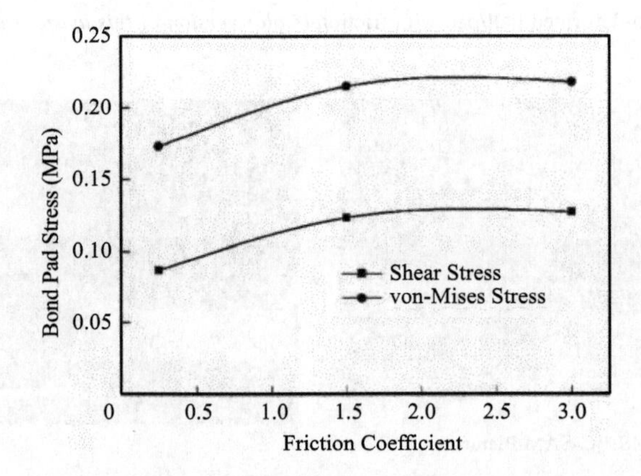

Figure 9.46 Bond pad stresses versus friction coefficient

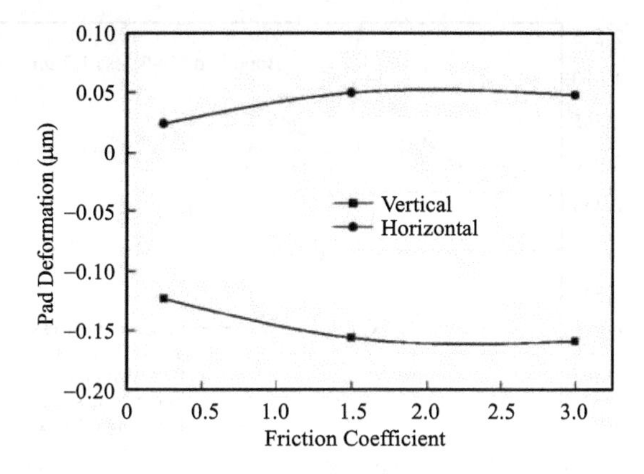

Figure 9.47 Bond pad deformation versus friction coefficient

9.4.6 Impact of Different Bond Pad Thickness

Increasing bond pad thickness is an easy and low cost way to reduce cratering. The simulation results for different bond pad thicknesses are listed in Figures 9.48 to 9.51.

Figure 9.48 gives the maximum principal stresses comparison for die under three different bond pad thicknesses. Figure 9.49 gives the profiles comparison of stresses transferred to silicon interface. Both have shown that as the bond pad thickness increases, the stresses transferred to silicon see no significant reduction. However, the plastic effective strain and plastic strain density in the bond pad reduces rapidly (Figures 9.50 and 9.51). It is this property of the bond pad that reduces the cratering during wire bonding.

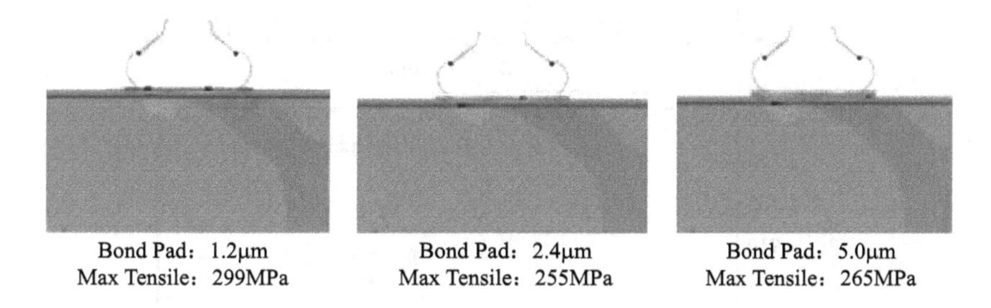

Figure 9.48 Maximum principal stresses in die versus bond pad thickness at the first step of cycle 6 (capillary towards left) *(Color version of this figure is available online)*

9.4.7 Impact of Different Bond Pad Structures

Here we discuss two different layouts below the bond pad; one layout adds a thin TiW layer under the bond pad (Figure 9.52) and another layout enables a higher uniform density of plugs in the three ILD layers (Figure 9.53).

Table 9.10 gives the von-Mises stress comparison of two structures with and without the TiW thin layer. These data show that adding a TiW thin layer could reduce the stress transferred to the silicon by a small amount. However, this induces greater stress in ILD and metal layers.

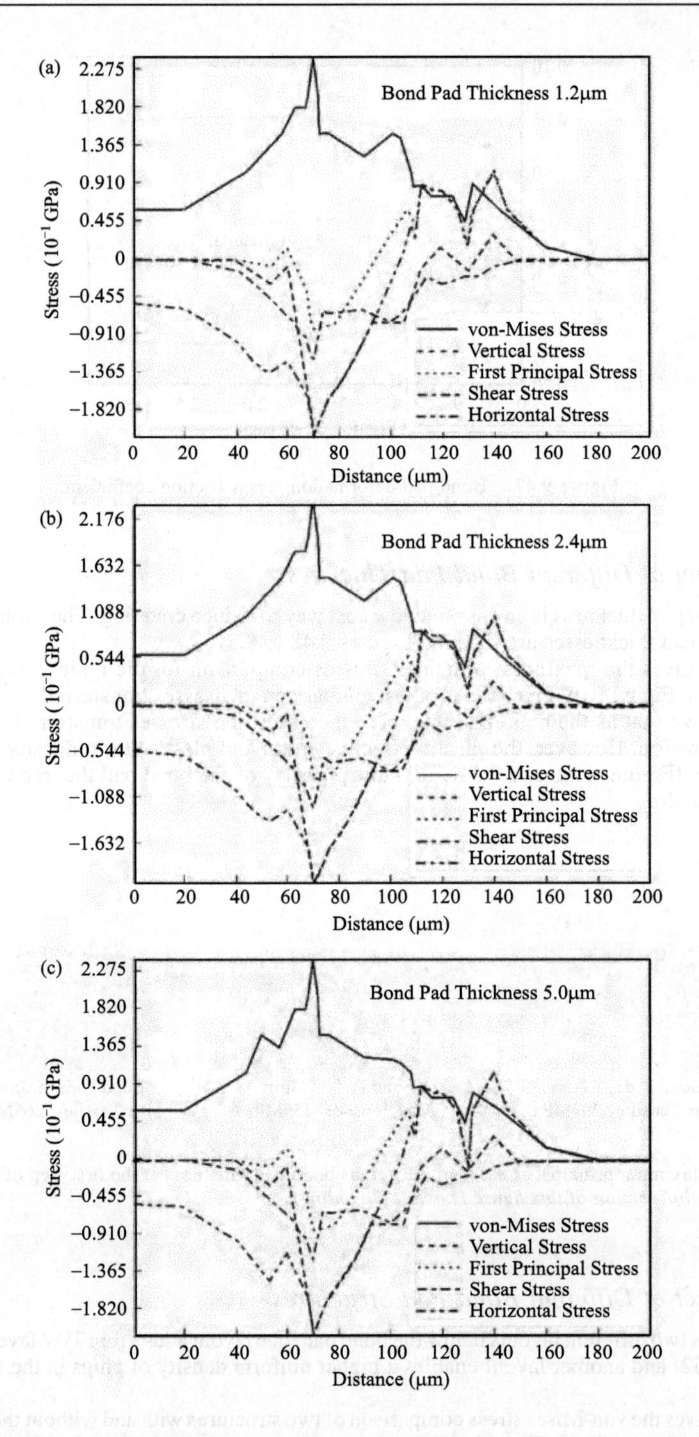

Figure 9.49 Stresses transferred to the interface between ILD and silicon at the first step of cycle 6 (capillary toward left and 35–135 is the pad area)

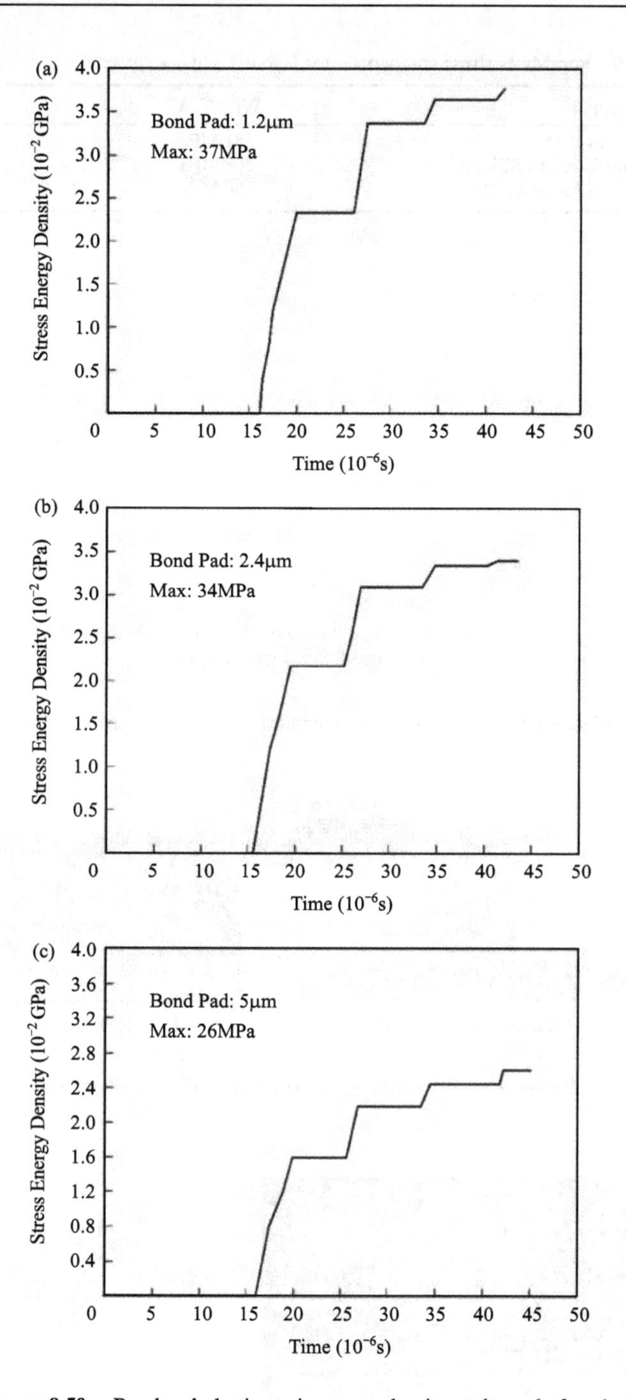

Figure 9.50 Bond pad plastic strain energy density at the end of cycle 6

Table 9.10 von-Mises stress comparison for layout 1 without TiW

Quasi-dynamics	With TiW	No TiW
Max Stress in ILD	281 MPa	276 MPa
Stress Transferred to Silicon	268 MPa	272 MPa
Max Stress in Metal Layers	180 MPa	171 MPa

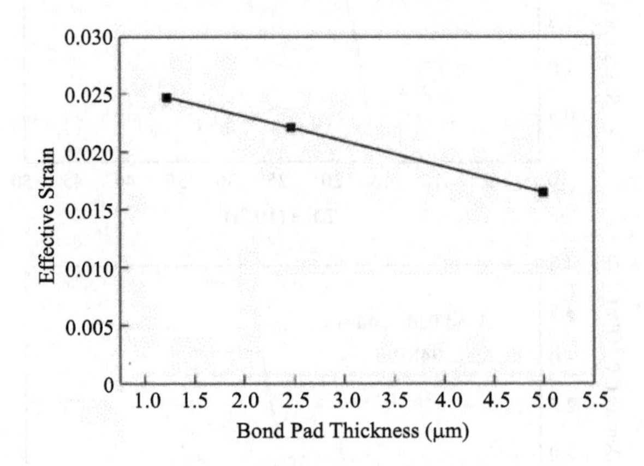

Figure 9.51 Bond pad maximum plastic strain versus pad thickness

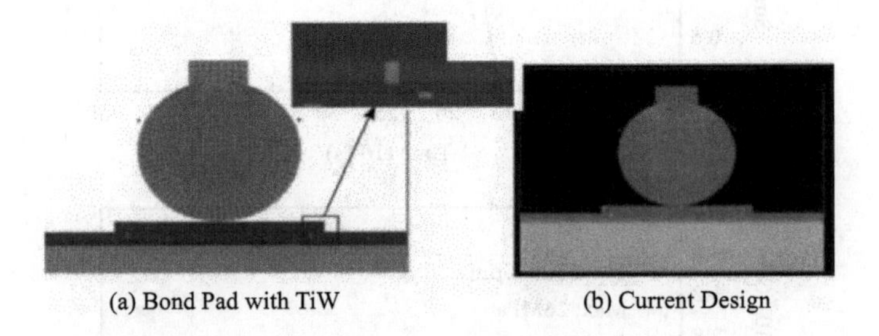

Figure 9.52 A TiW layer (0.3 μm) is added under bond pad

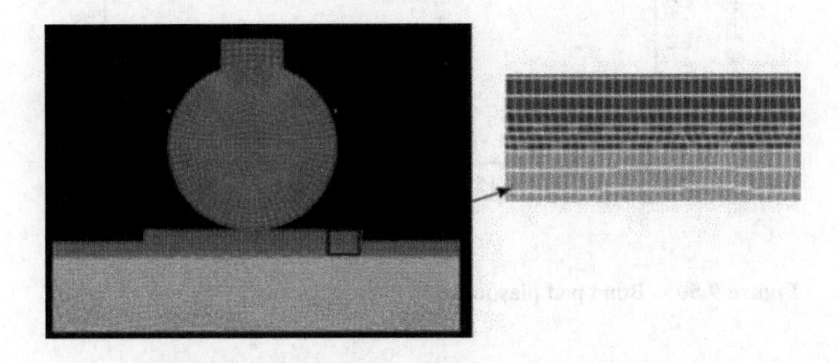

Figure 9.53 A higher density plugs in ILD layers

Table 9.11 and Figure 9.54 show the von-Mises stress comparison with and without higher density plugs in ILD. The results show that with a uniform high density of plugs, it reduces the stress transferred to silicon. However, the stress in ILD layers increases.

From the above it can be seen that there is a trade-off in changing the bond pad structure. One needs to do reliability testing to make sure that the new device can withstand the wire bonding process.

Table 9.11 von-Mises stress comparison for layout 2

Quasi-dynamics	With Higher Density Uniform Plugs	Current Design
Max Stress in ILD	294 MPa	276 MPa
Stress Transferred to Silicon	260 MPa	272 MPa
Max Stress in Metal Layers	173 MPa	171 MPa

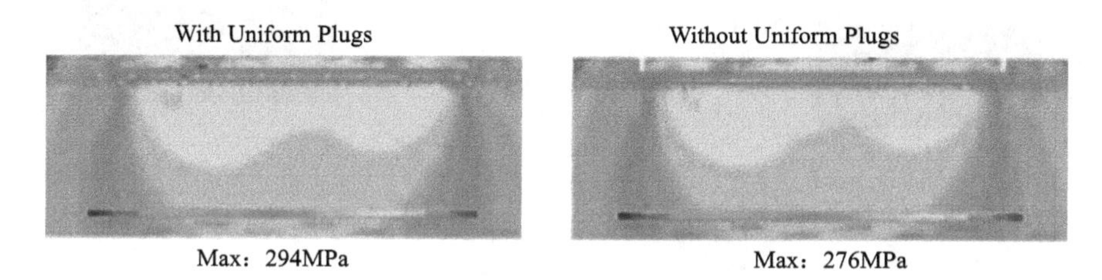

Figure 9.54 von-Mises stress comparison with and without higher density uniform plugs *(Color version of this figure is available online)*

9.4.8 *Modeling Results and Discussion for Cooling Substrate Temperature after Wire Bonding*

The residual stress after wire bonding is an interesting topic, which relates to the substrate temperature and bond pad peeling failure (Mayer *et al.*, 1999; Tan *et al.*, 2003). Mayer *et al.* (1999) studied the optimization of the thermosonic ball bonding process with different substrate temperatures. Here, ultrasonic wire bonding with amplitude 0.25 μm and 138 kHz, its substrate temperature is 240 °C. After wire bonding, the system cools down to 50 °C. The simulation of a fully transient dynamic wire bonding modeling at substrate temperature 240 °C is done first, then remove the capillary and cool down to 50 °C. The results show that most of the stresses in the ball decrease when the temperature cools down to 50 °C. Figure 9.55 shows an example of shear stress distribution before and after cooling. However, the stresses below the ball increases, the maximum von-Mises stress appears at the interface between plug and ILD (Figure 9.56).

Figure 9.57 shows the shear stress distribution at the contact interface between ball and pad, the position value stands for the shear stress towards the counter clockwise direction, it reduces after cooling to 50 °C, however, the shear stress along the clockwise direction increases significantly. Figure 9.58 shows the comparison of stresses transferred to the interface between the ILD and silicon, after cooling down, the stresses increase due to CTE mismatch. However, since the capillary is removed, stress jumping at the ball and pad contact edge relaxes.

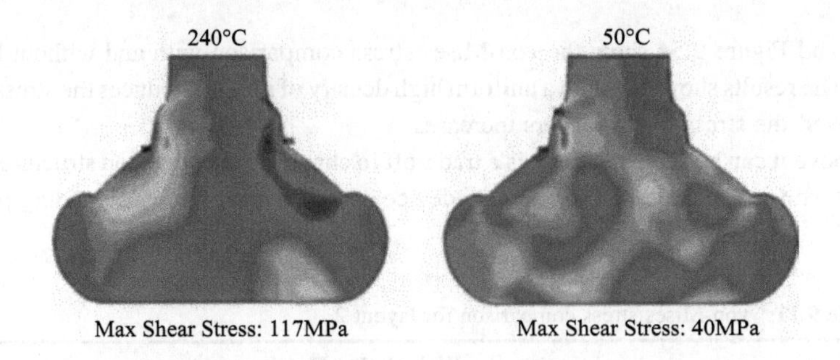

Figure 9.55 Shear stress before and after ball cools to 50 °C

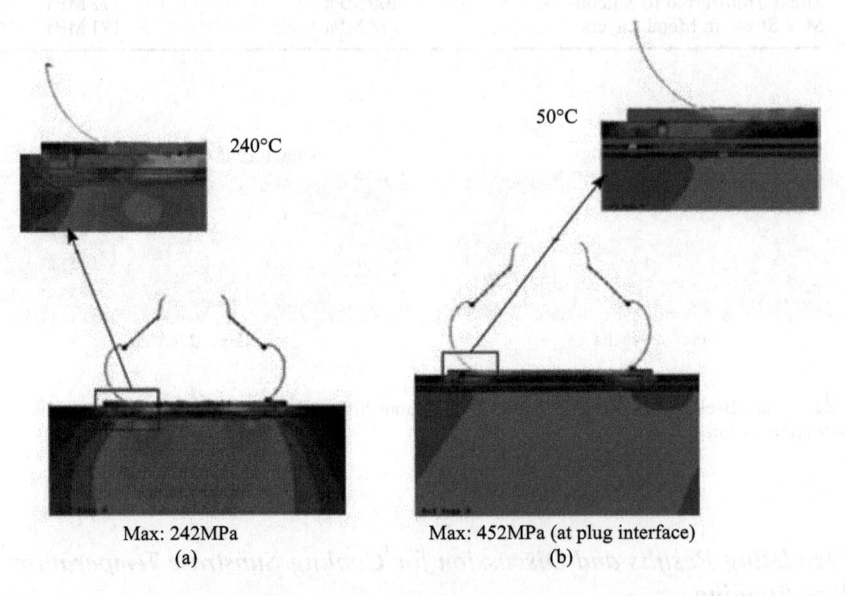

Figure 9.56 Maximum stress appearing at the plugs interface after cooling to 50 °C. (a) Before cooling; (b) after cooling

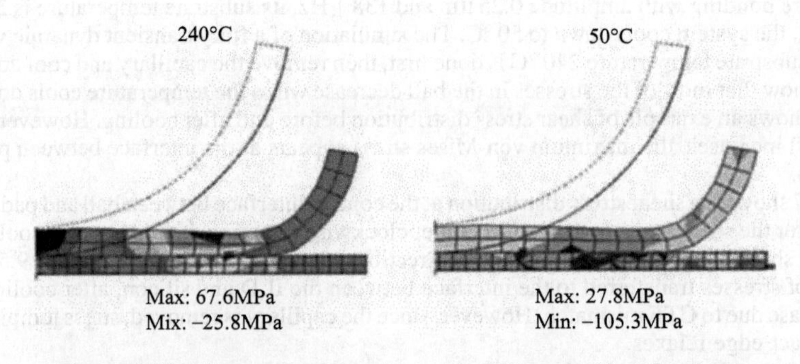

Figure 9.57 Shear stress comparison at the ball and pad interface layers

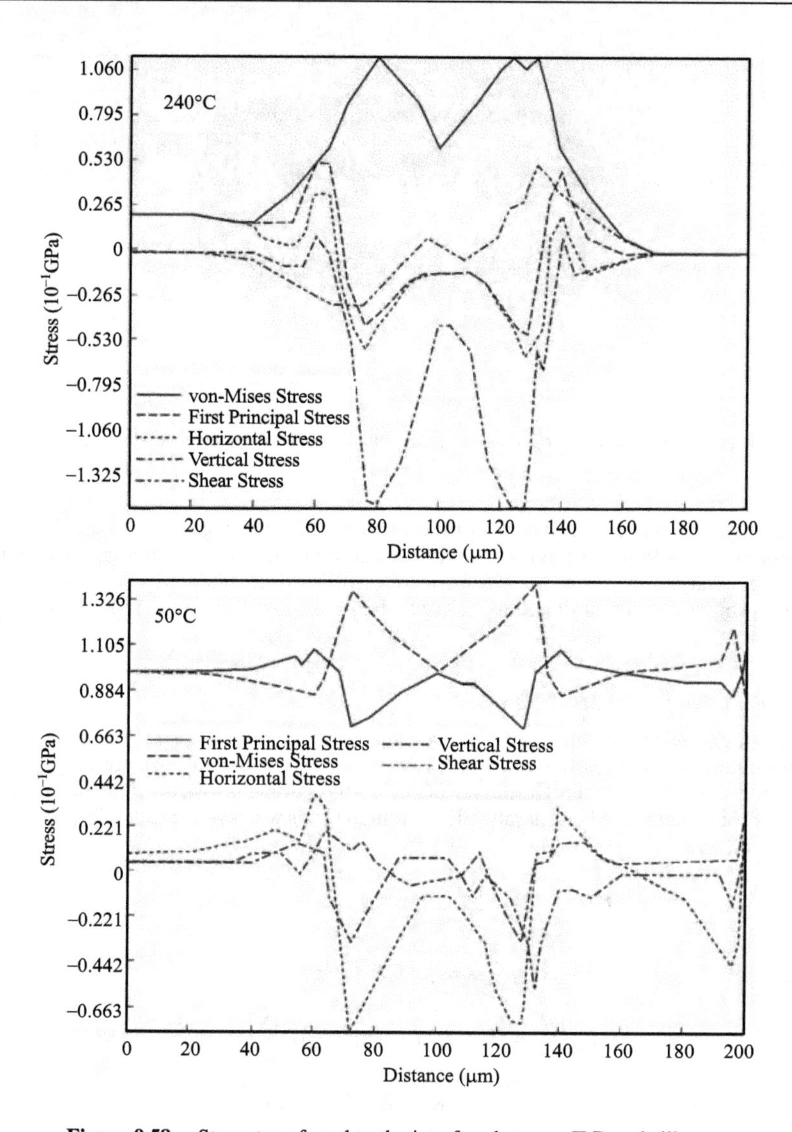

Figure 9.58 Stress transferred to the interface between ILD and silicon

9.5 Molding

In the injection molding process, most of the molding compounds are initially liquid at high temperature. It changes to solid as the temperature decreases after some time and never changes back to liquid (for thermoset materials). The material properties vary over temperature and time. Cure shrinkage strain is the sum of the thermal strain and the chemical strain. It is caused by the thermal expansion and chemically-induced cure shrinkage, which are both modeled.

9.5.1 Molding Flow Simulation

Semiconductor chips are often encapsulated with an encapsulant molding compound (EMC) for protection. Transfer molding technology has advantages for IC packages with small size, low cost,

Figure 9.59 Size comparison of general packages versus power module package

high productivity, and liability. It has been used in various IC packages as well as the various power packages. For better mold flow design and optimization of processing, the molding flow simulation and analysis are necessary.

Figure 9.59 shows the power discrete packages, IC Packages, and power module package after the molding.

(i) Basic Formulation for 3D Mold Flow

Typically the thickness of the chip cavity is much smaller than its width, so the generalized Hele-Shaw approximation may be used (Bird *et al.*, 1987; Han *et al.*, 2000). Because we are only interested in the filling stage of encapsulation where the degree of cure of the sample is low, the fluid elasticity is neglected and thus the flow is assumed to be generalized Newtonian. Thus we have the following equations for mold flow (Bird *et al.*, 1987):

$$\frac{\partial \rho}{\partial t} + \nabla \cdot (\rho \boldsymbol{v}) = 0 \tag{9.14}$$

$$\rho \frac{\mathrm{D}\boldsymbol{v}}{\mathrm{D}t} = -\nabla P + \nabla \boldsymbol{\tau} + \rho g \tag{9.15}$$

$$\rho C_{\mathrm{p}} \frac{\mathrm{D}T}{\mathrm{D}t} = \nabla \cdot (k \nabla T) + \eta \dot{\gamma}^2 + \beta T \frac{\mathrm{D}P}{\mathrm{D}t} + H \frac{\mathrm{D}\alpha}{\mathrm{D}t} \tag{9.16}$$

$$\frac{\mathrm{D}\alpha}{\mathrm{D}t} = (K_1 + K_2 \alpha^{\mu})(1-\alpha)^{\nu} \tag{9.17}$$

where:

$$K_1 = A_1 \mathrm{e}^{(-(\frac{E_1}{T}))}, \quad K_2 = A_2 \mathrm{e}^{(-(\frac{E_2}{T}))} \tag{9.18}$$

The first three Equations 9.14 to 9.18 describe conservation of mass, momentum, and energy respectively. The third term in Equation 9.16 is the irreversible internal energy loss due to viscous dissipation. The last term in Equation 9.16 represents the heat generated due to curing. The stress tensor $\tau (= \sigma + PI)$ is expressed explicitly as a function of the deformation rate tensor $\dot{\gamma}$, (that is, generalized Newtonian fluid model). Equation 9.17 and Equation 9.18 are Kamal's curing equations.

Equations 9.14 to 9.16 are solved for an incompressible fluid by both a 3D Navier–Stokes and a 3D fast solution method. The fast solution method uses an approximation of the full Navier–Stokes equations and is generally sufficiently accurate for many cases of molding simulation.

For the viscosity, we use the cross-exponential-Macosko viscosity model. This model can be represented by the following equations:

$$\eta(\alpha, T, \dot{\gamma}) = \frac{\eta_0(T)}{1 + \left(\frac{\eta_0(T)\dot{\gamma}}{\tau^*}\right)^{1-n}} \left(\frac{\alpha_g}{\alpha_g - \alpha}\right)^{(C_1 + C_2\alpha)} \tag{9.19}$$

$$\eta_0(T) = Be^{(T_b/T)} \tag{9.20}$$

In molding flow simulation, the venting analysis is used to calculate the pressure drop of the air that passes through the vent and to include the effects of air pressure on the flow simulation of the microchip encapsulation process.

Pressure drop through the vent can be calculated using the viscous flow analysis. Each vent can have different properties (dimensions, exit pressure). The volume of air connected to a vent can be calculated by adding the volume of all the nodes that are identified as air nodes connected to a vent. The viscosity of air is assumed to be constant (Lee *et al.*, 2008). The pressure of the air at the vent exit can be specified. There can be several vents connected to a cavity.

Neglecting the side wall effect, the formula of the pressure drop in a vent can be obtained as following (Lee *et al.*, 2008):

$$\Delta p \frac{12\eta QL}{h^3 w} \tag{9.21}$$

When side wall effect is considered, the pressure drop formula can be given by:

$$\Delta p = \frac{12\eta QL}{h^3 w} \times \frac{16}{3} \frac{l}{s} \tag{9.22}$$

$$s = \frac{16}{3} - \frac{1024}{\pi^5} \frac{h}{w} \sum_{n=0}^{\infty} \frac{\tanh\left[\frac{(2n+1)\pi w}{2h}\right]}{(2n+1)^5} \tag{9.23}$$

In the above equations, Δp is pressure drop at the vent, η is the viscosity, Q is the flow rate, l is the length of the vent, h is the thickness of the vent, and w is the width of the vent. When the pressure drop at the vent is small, the simulation value and the value obtained from the above analytical equations match well.

(ii) Molding Flow Simulation Results

In this chapter, ANSYS–FluentTM with user defined curing function, E–MoldTM and MoldflowTM software are used for saving time and money during the development stage. The material properties of encapsulant mold compound are listed as follows:

1. For reaction kinetics properties
   $H = 37693$ J/Kg, $B_1 = 0$, $B_2 = 0$, $m = 0.5955$, $n = 1.437$, $A_1 = 0.1/s$, $A_2 = 1.9381/s$, $E_1 = 3670.9$K, $E_2 = 8886.54$K
2. Reactive viscosity model coefficients
   $n = 0.55$, $Tau^* = 0.0001$Pa, $B = 1.8242\text{e}^{-014}$Pa·s, $T_b = 22549.7$ K, $c_1 = 5.456$, $c_2 = 0.4491$

Three simulation examples are presented for the molding flow and curing processes.

Example A. A BGA molding and curing case by ANSYS FLUENT software

Figure 9.60 shows the process of the molding flow filling the cavity of BGA. The simulation shows the flow front situation and the air track during the molding filling. Figure 9.61 shows the curing after molding injection. This contour figure gives the curing process in the BGA cavity which clearly shows the non-uniform curing distribution.

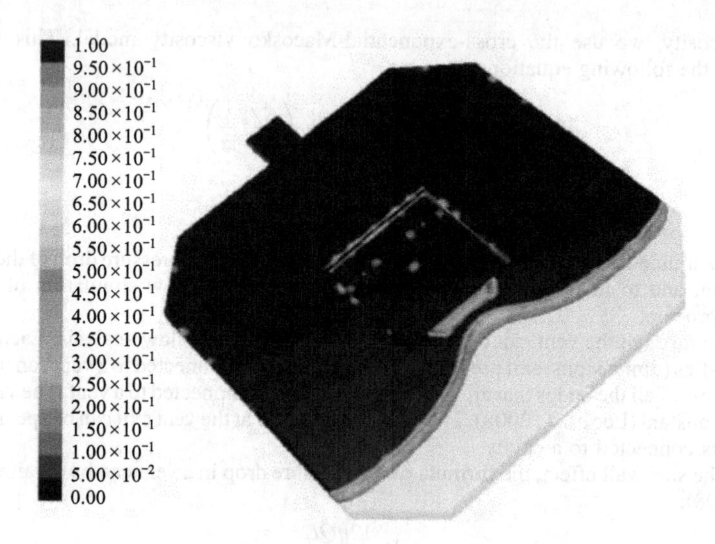

Figure 9.60 BGA molding filling process

Example B. Molding flow front simulation for a group of cavities

Figure 9.62 shows the molding flow front simulation for a group of cavities in which the die and leadframe are placed. Figure 9.62(a) shows the molding injection at the beginning when the mold just starts to fill the cavities. Figure 9.62(b) shows that the molding filling has been completed for most of the cavities. The mold flow front is clearly seen at the unfinished area. The red and yellow area stands for the flow front with injection time (s). The advantage of the molding flow simulation is to watch the cavity design and layout for IC package as well as to improve the mold runner system. The simulation of mold front was done by E-Mold.

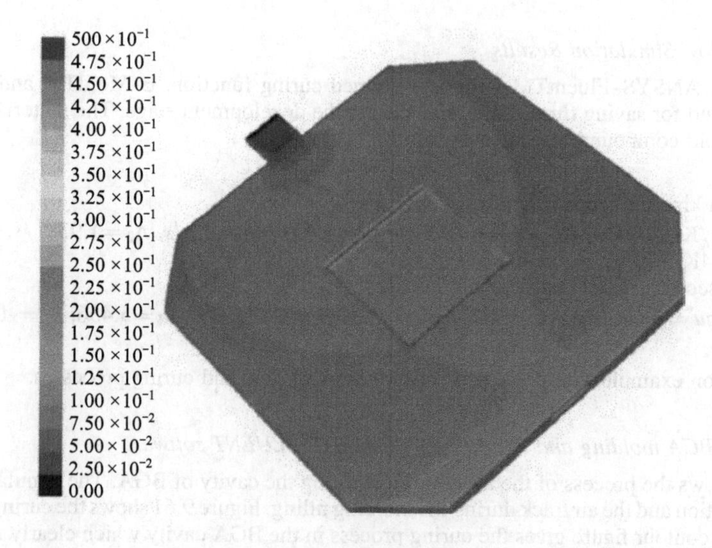

Figure 9.61 Curing process at 50 s

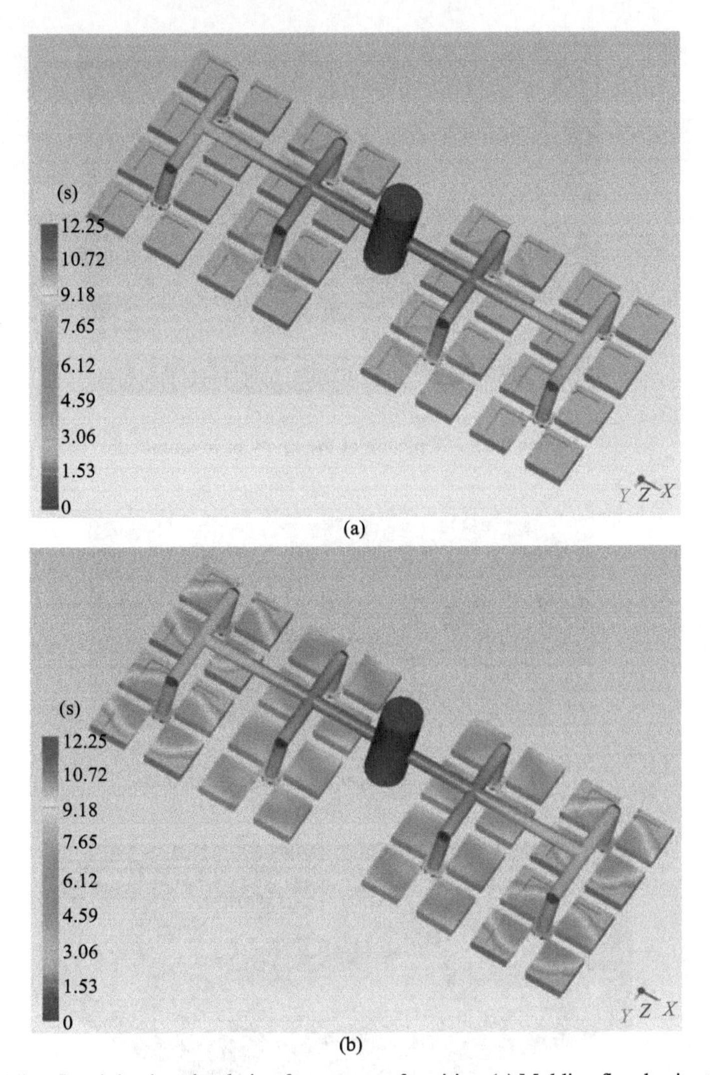

Figure 9.62 Molding flow injection simulation for a group of cavities. (a) Molding flow begins to fill the cavity. (b) Molding flow has filled most of the cavities *(Color version of this figure is available online)*

Example C. Molding simulation with consideration of vent effects

This example is selected from reference (Lee *et al.*, 2008). Figure 9.63 shows one cavity of an actual mold die, and Figure 9.64 shows a magnified image of the region indicated by a red ellipse in Figure 9.63.

After making a finite element model including package and gate, air vents are added onto the model. Figure 9.65 is one sample of a finite element model with air vent using MoldflowTM. Air vent model in black circle is assigned to have thickness, width, and length.

Figure 9.66 is a full power package type and has very small gap (less than 1 mm) between the die attach pad and the package bottom surface. It might have incomplete molding on the back side of package, so this problem can cause some yield loss. Therefore perfect mold ability is strongly needed to improve the design.

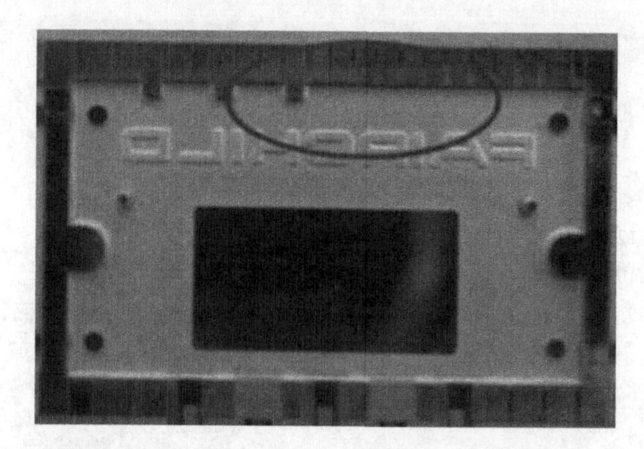

Figure 9.63 A picture of the cavity of in a mold die

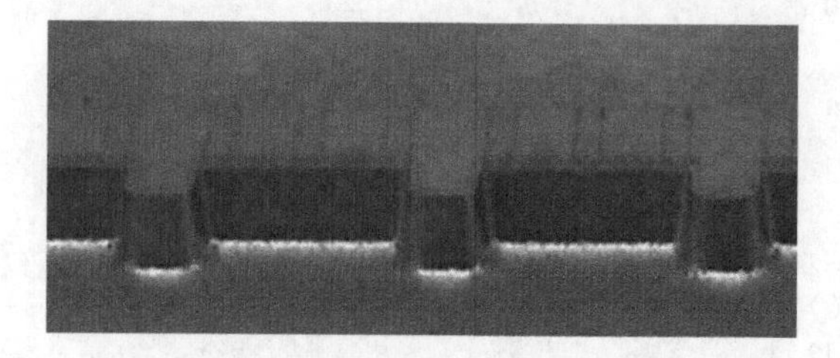

Figure 9.64 A magnified picture of air vents

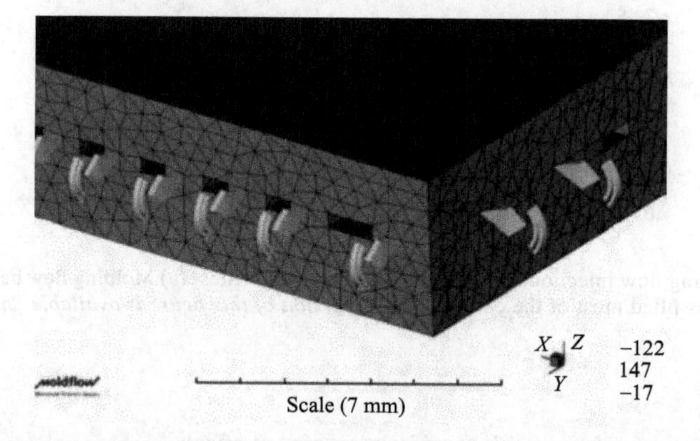

Figure 9.65 An example of finite elements and air vents

The result of air venting analysis shows that the package has some air trap on the bottom side in the center area when flow front reaches the end of the package; that is the root cause of the incomplete molding seen in Figure 9.67. Figure 9.67 by simulation and Figure 9.66 by actual molding correlate very well in mold flow front with air vent analysis. But non-air-venting analysis case shows no air trap at this time (Figure 9.66). Figure 9.68 shows the simulation result without consider the effect of air vent. As compared to the actual molding process, there is larger difference.

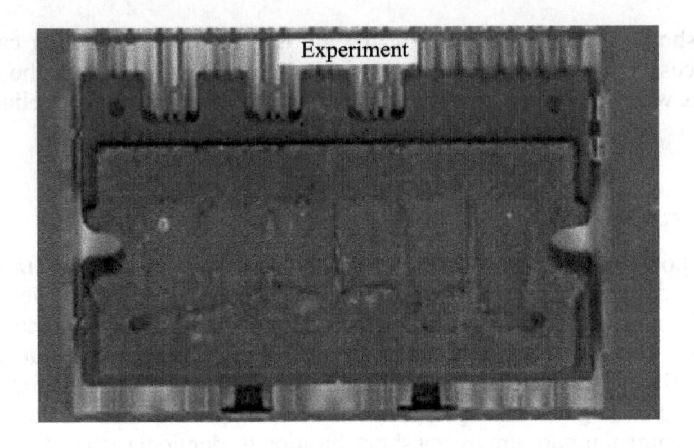

Figure 9.66 Actual incomplete molding in package

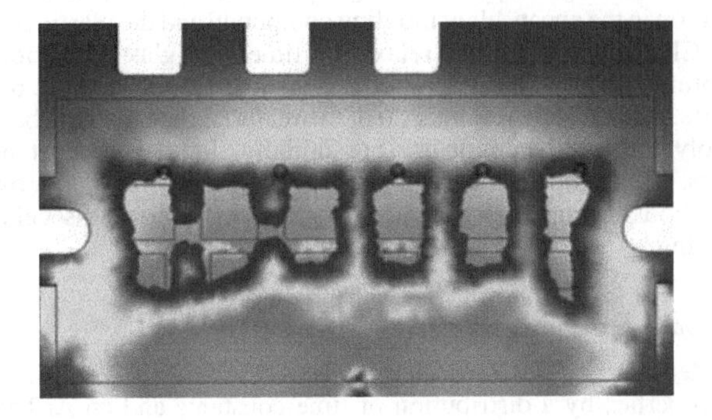

Figure 9.67 Molding simulation result of package with air vent option

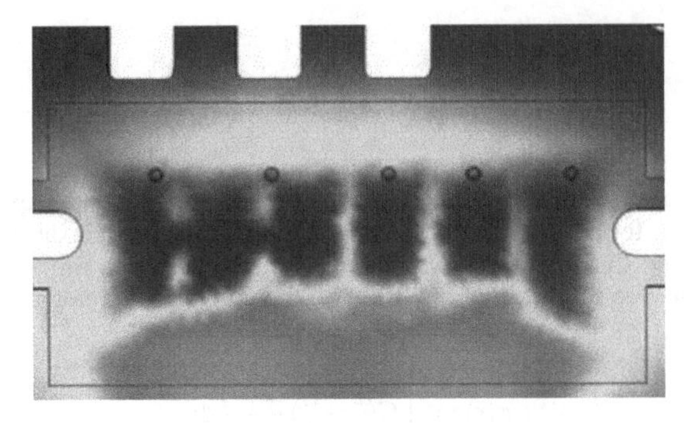

Figure 9.68 Molding simulation result of package without air vent option

Molding flow simulation is important for the design and assembly process of IC packages with encapsulant mold compound. It may help to improve the product quality and reliability so as to control and reduce the air trap in the molding assembly process. This simulation can help the IC package design engineers and assembly manufacturing engineers to understand the mold flow front situation in a single cavity and in a group of cavities with runner and sub-runner system. In the molding process, the simulation has shown that consideration of the air vent option in modeling can indeed reflect the actual molding process with air trap. The curing simulation has shown that the curing is actually a non-uniform process which may induce the curing stress that may impact the reliability of IC package as well.

9.5.2 *Curing Stress Model*

Many polymers or epoxy materials used in the molding process are viscoelastic in nature. A material is said to be viscoelastic, if the material has an elastic (recoverable) part as well as a viscous (non-recoverable) part. Upon application of a load, the elastic deformation is instantaneous while the viscous part occurs over time. The viscoelastic model incorporated in ANSYS usually depicts the deformation behavior of glass or glassy materials and may simulate cooling and heating sequences. The material is restricted to be thermorheologically simple (TRS). The concept of TRS materials implies the material response to a load at a high temperature over a short duration is identical to that at a lower temperature but over a longer duration.

The purpose of this section is to determine the curing stress and related mechanical behavior of the epoxy (die attach) and molding compound during cure. This work provides a method of calculating the residual stress distribution in the encapsulant molding compound and the plastic strain in the solder during cure and cool down. Changes in modulus, relaxation time, and glass transition temperatures can be changed to find an optimal set of material properties. Standard ANSYS input of material data may be based upon temperature. As currently implemented, the material data may not be a function of any other parameters. Many polymers used in industry are initially liquid and must be "cured" during the manufacturing process. This process means that the properties will change over time. This section demonstrates one method of implementing a curing process into a viscoelastic polymer used in semiconductor packaging.

(i) Conventional Approach for Curing Stress

The polymer is modeled with a VISCO88 element in ANSYS. This element characterizes the time-dependent material properties by a distribution of time constants and an Arrhenius activation energy driven temperature dependence. A pseudo time (ζ), based on the time-temperature superposition principle, is used. The equation representing the effective shear modulus (G) is depicted in following viscoelastic equation:

$$G(\zeta) = \sum_{i=1}^{I} C_i[G(0) - G(\infty)]\exp\left(\frac{-\zeta}{\lambda_i}\right) + G(\infty) \tag{9.24}$$

where G for shear modulus at specific times, $G(0)$ for immediately after the applied load, $G(t)$ for any arbitrary time, and $G(\infty)$ for a long time after the applied load. λ_I for relaxation times of the Maxwell viscoelastic elements, and C_i for weights associated with the relaxation times.

The stresses are related to the strain at any time by the convolution integral:

$$\sigma(t) = \int_0^t G(\zeta - \zeta')\frac{\mathrm{d}\varepsilon(t')}{\mathrm{d}t'}\mathrm{d}t' \tag{9.25}$$

where σ for polymeric viscoelastic stress, and ε for polymeric viscoelastic strain.

The integration of time zero allows the representation of the Boltzmann superposition principle. The Boltzmann principle states that the creep at any time is a function of the entire loading history and that each load step makes an independent contribution to the final deformation. For the VISCO88 element, the fictive temperature that represents the temperature hysteresis was chosen to be the glass transition temperature (T_g) of the underfill material.

Since the completeness of a polymeric material cure is judged by its mechanical integrity, the mechanical degree of cure (DOC) is calculated by:

$$p_{m}(t)=\frac{G'(t)}{G'_{max}-G'_{min}} \tag{9.26}$$

where $p_m(t)$ for mechanical degree of cure (DOC) as a function of time, G'_{max} for maximum storage modulus achieved at the end of cure, and G'_{min} for minimum storage modulus of the uncured material.

For epoxy molding compound or underfill materials used in semiconductor packaging processes, the viscosity in the uncured liquid state is small, so that the minimum modulus is effectively zero.

For this model, cure shrinkage is included into the thermal expansion by calculating an effective coefficient of thermal expansion (α_{eff}). The strains caused by the thermal expansion (ε_T) and chemically-induced cure shrinkage (ε_C) are modeled respectively as:

$$\varepsilon_{T}(t)=\alpha[T(t)-T_{0}] \tag{9.27}$$

And:

$$\varepsilon_{C}(t)=bp_{m}(t) \tag{9.28}$$

where α for coefficient of thermal expansion (underfill), $T(t)$ for temperature as a function of time, T_0 for reference temperature (underfill), and b for total chemical strain at the end of cure (negative for shrinkage).

A typical value for volumetric cure shrinkage in unfilled polymers is about 5%. When the effect of the filler particles is accounted for, the resulting linear chemical strain (b) is -0.005 for a typical commercial underfill material. The total strain (ε_{total}) is the sum of the thermal strain and the chemical strain. After some algebraic manipulation, the expression for the total strain is:

$$\varepsilon_{total}(t)=\left[\alpha+\frac{bp_{m}(t)}{T(t)-T_{0}}\right](T(t)-T_{0}) \tag{9.29}$$

This equation shows that the effective CTE is a function of time and is expressed as:

$$\alpha_{eff}(t)=\alpha+\frac{bp_{m}(t)}{T(t)-T_{0}} \tag{9.30}$$

With the fact that the degree of cure (DOC) and the temperature are known in every time step, the effective coefficient of thermal expansion (α_{eff}) can be determined in every time step. Each time, viscoelastic material parameters need to be changed due to the curing property, which may be completed by changing the related material by TB between the load steps. However, it works fine for a small deformation of viscoelasticity but may not be good for non-linear large deformation analysis especially in the solder joint, and the viscoplastic material model is used. Therefore, it is necessary to develop a user defined programmable code approach to target the curing induced stress problem.

(ii) User-Defined Programmable Code UsrFictive. F in ANSYS

Using a combination of the conventional approach and a UPF approach provides a more user-friendly definition of a polymer curing simulation. The UsrFictive subroutine in ANSYS is particularly well suited for this type of application (for elements VISCO88 or VISCO89), since the entire viscoelastic material table is passed when the subroutine is called using TBDATA, 5,11 in the TB, EVISC command set. Subsequently, the data that enters the subroutine may be replaced with the "appropriate" data based upon time and returned to the calling subroutine where the viscoelastic calculations are performed. Sample FORTRAN code for this application is shown in the file USRFICTIVE.F for ANSYS70. The subroutine has been modified in the following manner.

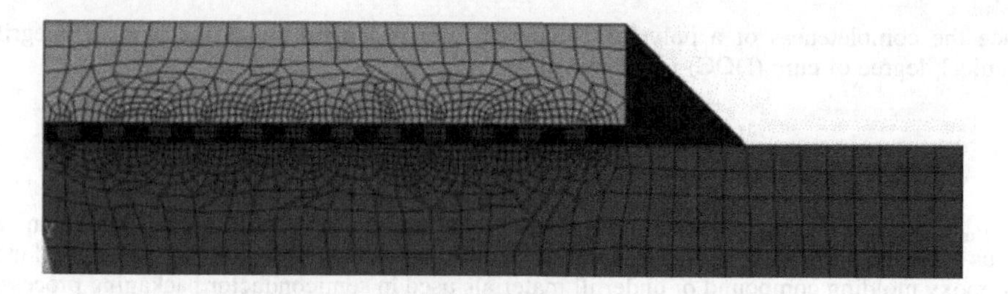

Figure 9.69 A 2D flip-chip model with underfill curing

As originally coded, TIME is not included in the subroutine either as an argument or in one of the COMMON blocks. But in ANSYS, time is included in the STEPCM.INC include file and can be made available to the subroutine in this manner.

Two arrays are defined to hold the material data for interpolation. Array CURTIM is used for the curing time and array VJMT is used for the viscoelastic data. Array VJMT is made row major to facilitate the interpolation of the data. As coded here, up to 10 times may be input but this may be changed by the user.

The material data is made available to the subroutine as an ANSYS parameter. This will permit the user to change the data table without re-linking the ANSYS code.

The user defined program needs to be re-compiled and re-linked at the user directory (refer: ANSYS Guide to UPF).

(iii) Curing Stress in a Flip-Chip

Using a direct call to UsrFictive subroutine, a 2D flip-chip with underfill curing processing is modeled. The solder joint is considered to be the viscoplastic materials with Anand model. The results are shown in Figures 9.69 to 9.74.

(iv) Curing Stress in a Tssop Package

The model is a 2D Tssop20 package with an initial void/crack inside the die attach (Figure 9.75), the simulation tries to see the curing stress from starting to 100% curing process after the mold injection. The results are listed in Figures 9.76 to 9.79.

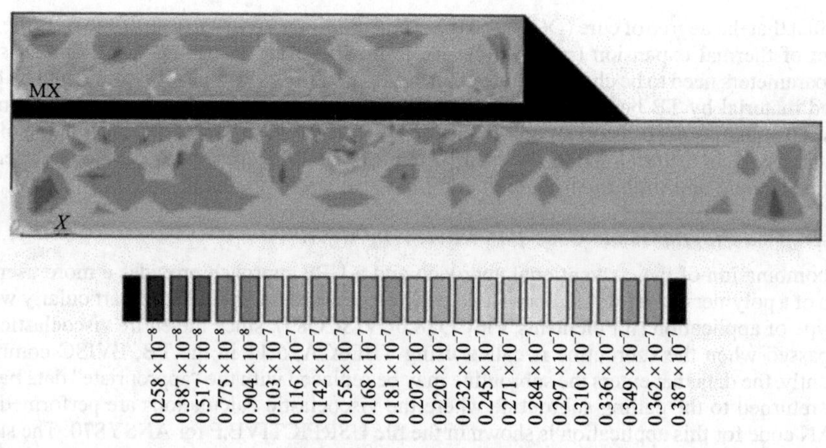

Figure 9.70 Starting curing

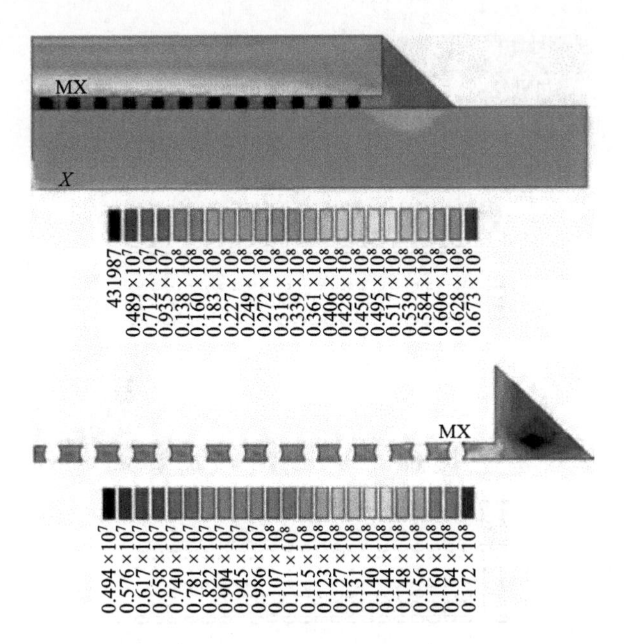

Figure 9.71 5% curing degree property used

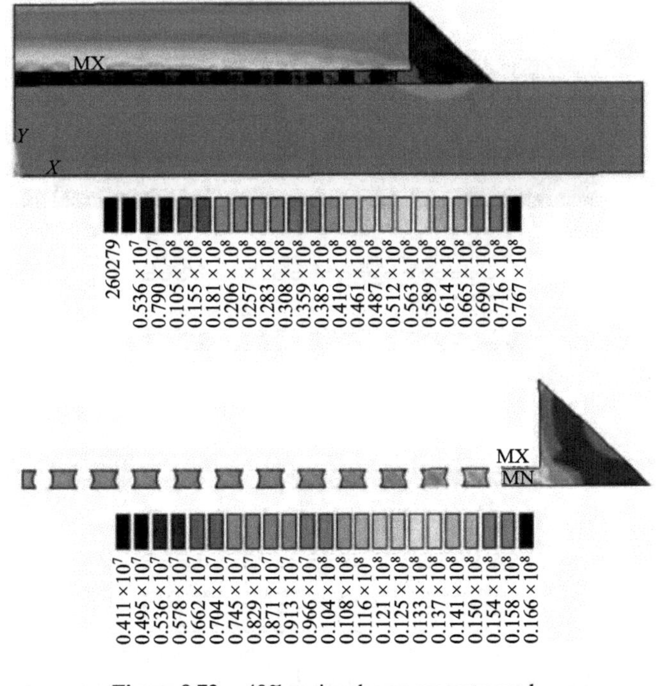

Figure 9.72 40% curing degree property used

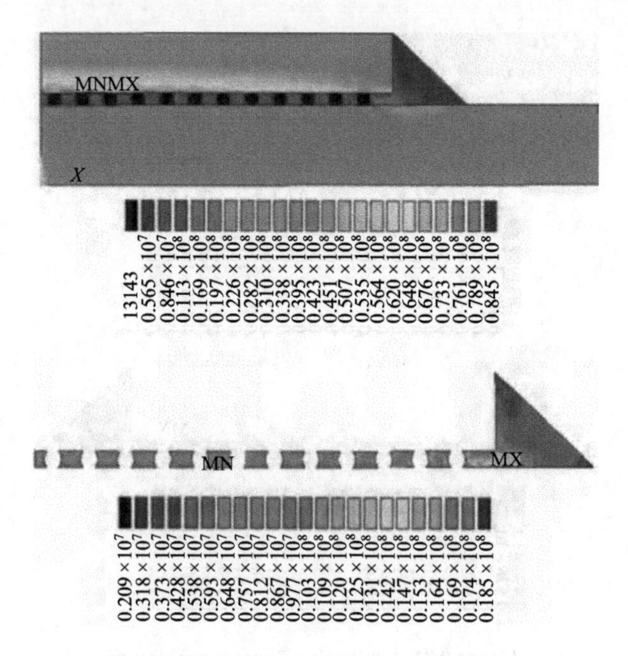

Figure 9.73 100% curing degree property used

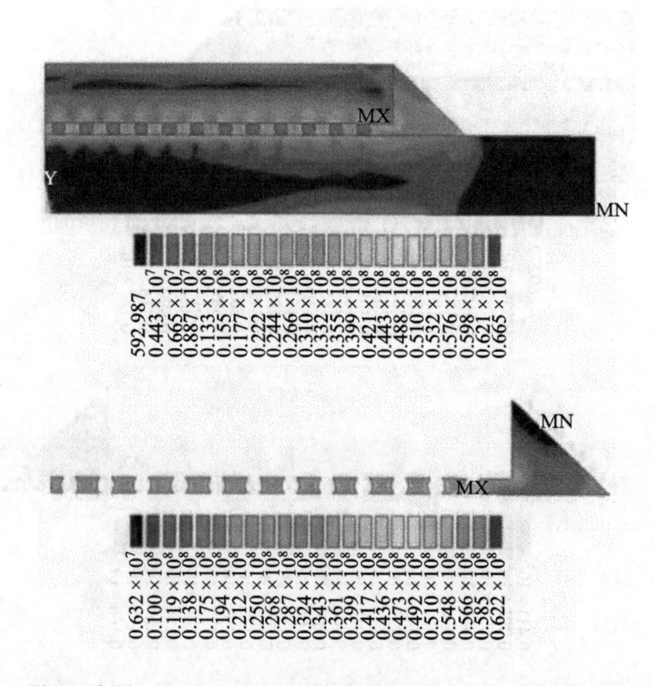

Figure 9.74 Cools down and shows the residual stress distribution

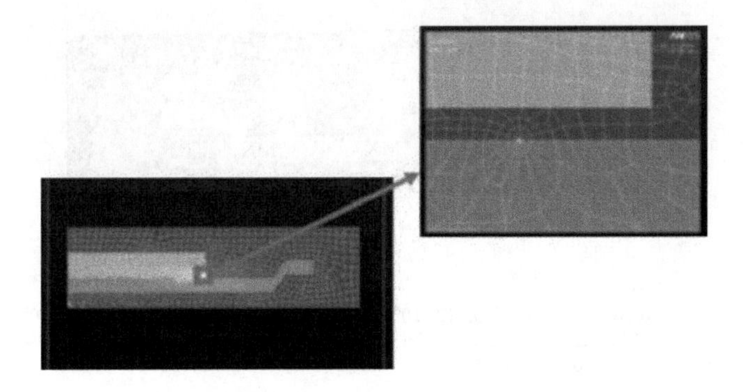

Figure 9.75 2D Tssop20 model with initial crack at die attach

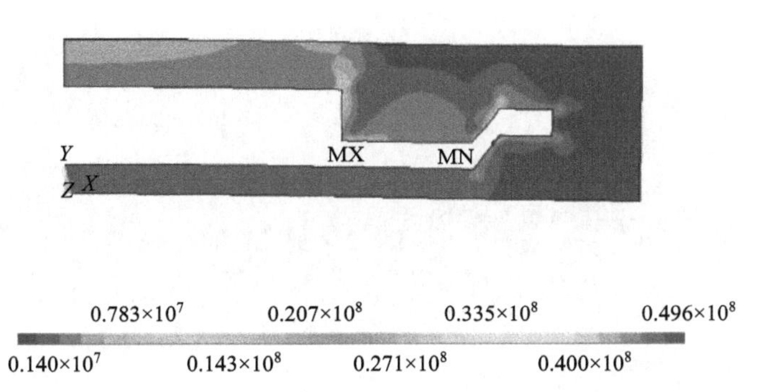

Figure 9.76 Curing at 5% *(Color version of this figure is available online)*

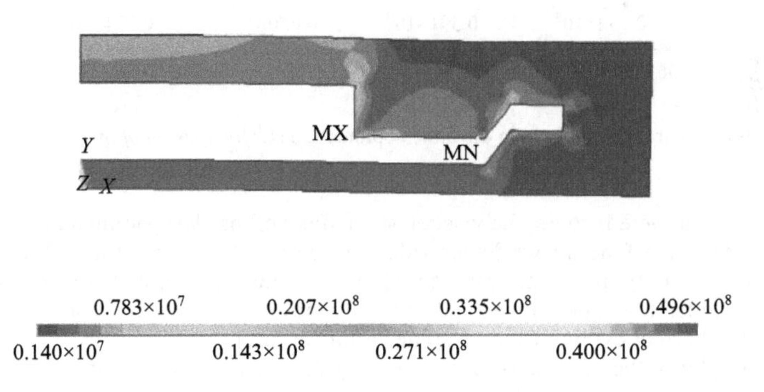

Figure 9.77 Curing at 40% *(Color version of this figure is available online)*

(v) Summary of Curing Stress Simulation

The change of material properties with time could be done with a simple shift in temperature, taking advantage of the time-temperature superposition principle. A shift function could be defined that would permit the material properties to change over time by changing the temperature. An alternative is to specify the material properties as a simple lookup table whose entry value is based upon time.

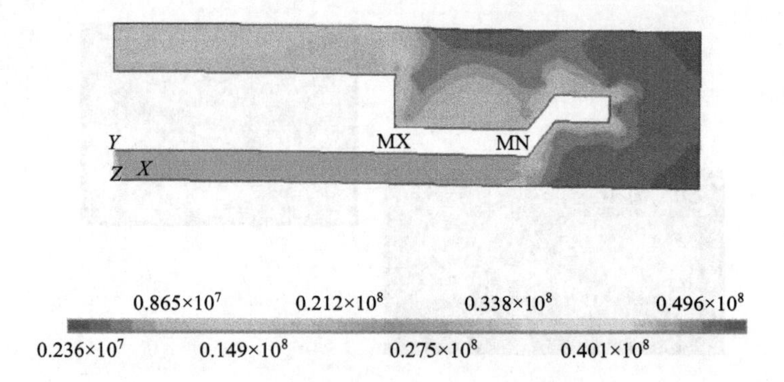

Figure 9.78 Curing at 100% *(Color version of this figure is available online)*

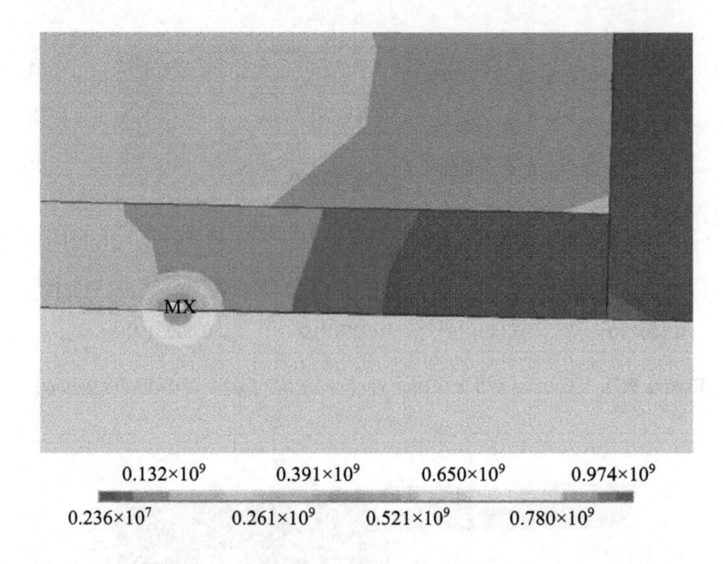

Figure 9.79 Stress field around crack tip in die attach connection *(Color version of this figure is available online)*

Therefore, the key point here is to get the viscoelastic (Maxwell model) parameters at each curing time, which may validate the new function we developed for the curing stress simulation. Other material models (for example, viscoplasticity in solders) can be included for more complex, realistic simulations. Stress simulations can be added together to investigate not only the cure process, but also subsequent reliability test conditions like thermal cycling (that is, solder joint fatigue).

In summary, supplying the viscoelasticity material properties via UsrFictive can allow for property definition as a function of both time and temperature independent variables. Since viscoelasticity is a time dependent phenomenon, the coupling time, temperature, and material property changes are essential. The material property parameters may be obtained by the test.

9.5.3 Molding Ejection and Clamping Simulation

(i) Molding Ejection Process Simulation:

The molding ejection process starts after the molding injection and curing is complete. A location pin in the mold will fix the leadframe position before the molding process. It insures the leadframe is located at

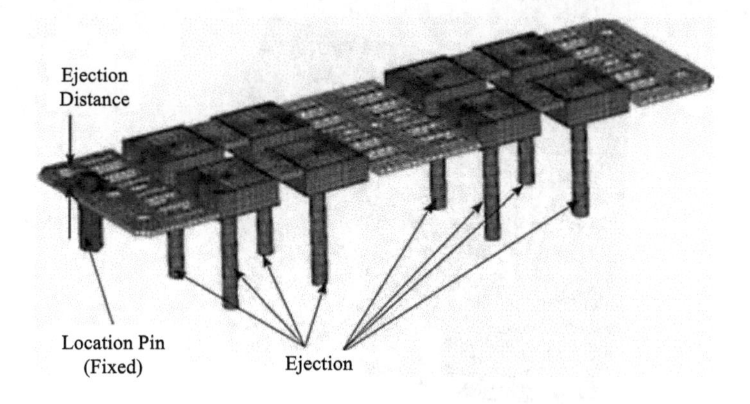

Figure 9.80 Location pin, ejection pins, and leadframe

the right place during the molding process. But it may also prevent the leadframe from being ejected from the mold because of the friction between the location pin and leadframe. There is a tradeoff; small clearances between the location pin and location hole in the leadframe can fix the leadframe accurately, but induce large friction stress during mold eject. Large clearance can reduce the friction stress but induce the large positioning tolerance of molding process. There is an optimized size of the clearance. The purpose of molding eject process simulation is to reduce the package failure during molding ejection process. A 3D solid model with contact between leadframe and eject pin, contact between eject pins and molding compound is developed to simulate the process (Figure 9.80). A parametric study of friction coefficient and leadframe ejection distance is studied.

Figure 9.81 shows stress of mold compound with 0.1 friction coefficient and 1.3 mm eject distance. There is large stress in the bottom package's molding compound which is near the location pin. The molding ejection does not induce large stresses to the packages far away from the location pin. Only the package near the location pin will be discussed in the following studies.

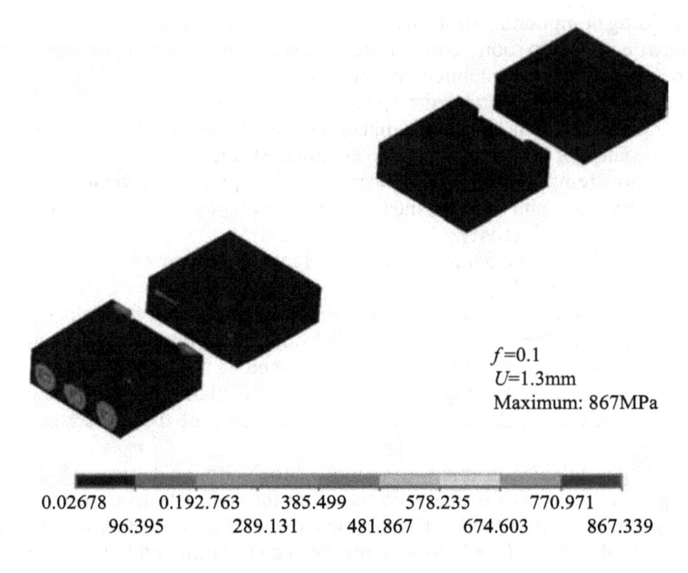

Figure 9.81 Stress in molding compound with $f = 0.1$

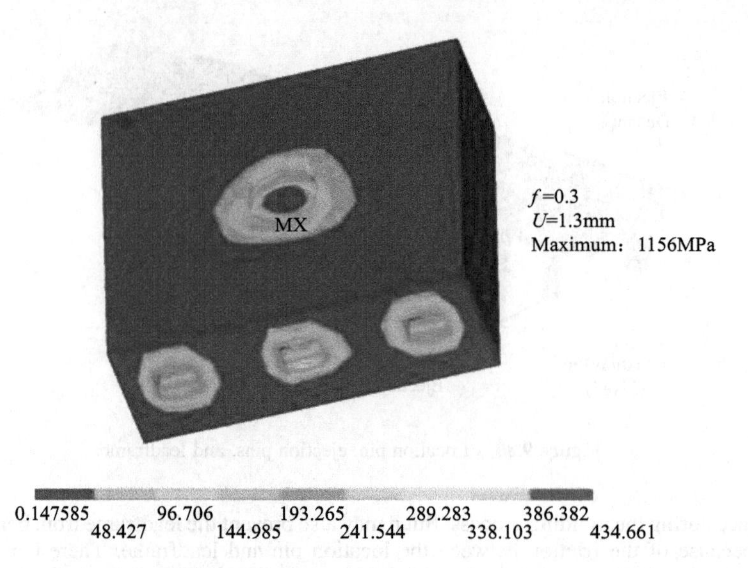

Figure 9.82 Stress in mold compound with $f = 0.3$ *(Color version of this figure is available online)*

Figures 9.81 and 9.82 show that larger friction coefficient between the location pin and leadframe will induce higher stress in the molding compound. The friction coefficient of 0.3 is preferred to make sure the leadframe is fixed at the right place during molding. The results of impact of ejection distance are shown in Figure 9.83.

Figure 9.83 shows that reducing the ejection distance from 1.1 mm to 0.5 mm can reduce stress in the molding compound by about 62%. It can reduce the risk of delamination and package cracking.

(ii) Clamping Process Simulation:

As we know the molding compound will shrink after it changes from initially liquid at high temperature to solid as temperature decreases to room temperature. The shrinkage strain of the molding compound is the sum of the thermal strain and the chemical strain. It is caused by thermal expansion and chemically-induced cure shrinkage which are modeled respectively. Normally the coefficient of thermal expansion (CTE) of the molding compound is larger than that of the leadframe. Molding compound will shrink more than the leadframe because of CTE mismatch and chemical strain of molding compound. This will cause package warpage at room temperature. Before the package is separated from the leadframe, the leadframe will be clamped to be flattened and fixed for the later singulation process. This process is normally done at high speed to improve the productivity.

Figure 9.84 shows the leadframe without warpage. Figures 9.85 and 9.86 show the photo and mesh of package with warpage after molding curing.

An explicit scheme based the software LS-DYNA is used because of its high speed to simulate large structures and high impact problems (Liu *et al.*, 2006). Contact between the clamps and the leadframe and contact between the guide rail and the leadframe are defined to model the process.

In this study a parametric study of package warpage and clamping parameters is conducted and discussed. From this study, we will have a better understanding of the impact on the package that is induced by clamping.

The package warpage of the leadframe is described as the warpage degree as in Figure 9.87. Package warpage from two to eight degrees is analyzed based on the real data from test.

The package is singulated from the leadframe one by one. Figure 9.88 shows that the guide rail can locate the leadframe in the right place before clamping. The 0.2 mm and 0.7 mm clearance between the leadframe and the guide rail are analyzed.

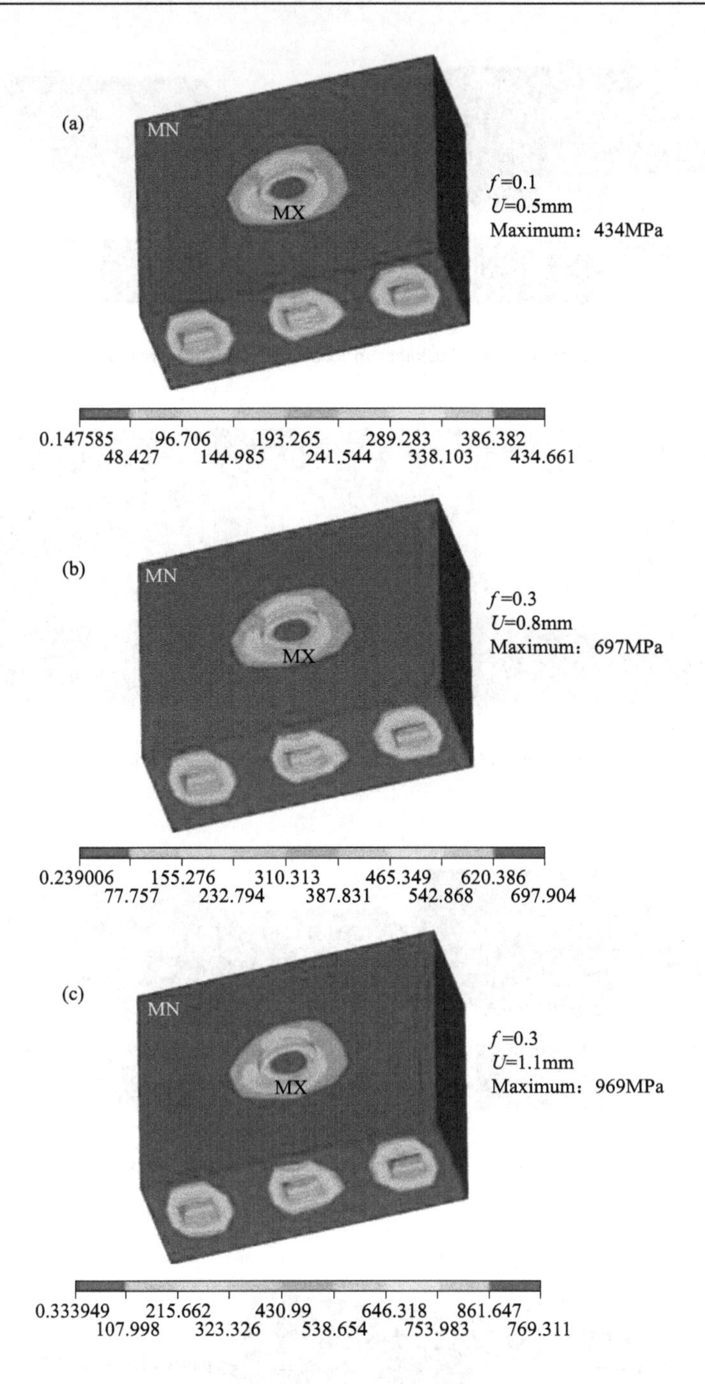

Figure 9.83 Stress in mold compound versus ejection distance *(Color version of this figure is available online)*

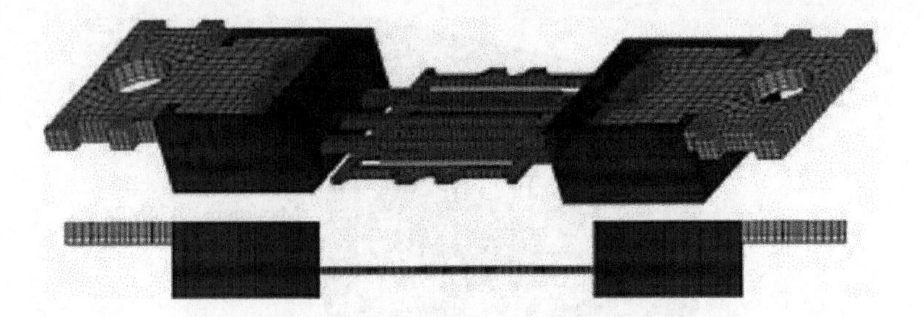

Figure 9.84 Packages in leadframe without warpage

Figure 9.85 Photo of warpage after molding curing

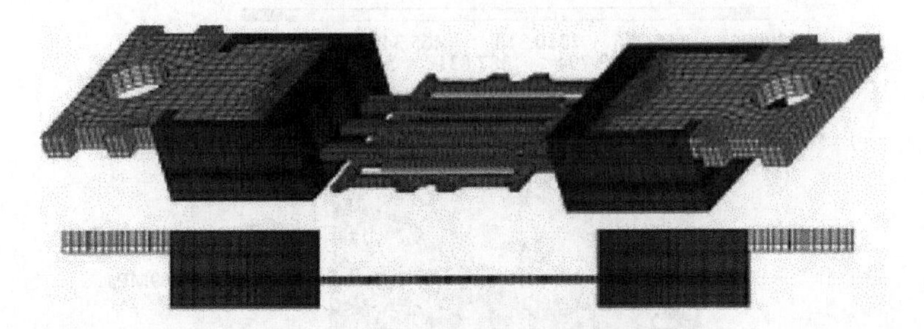

Figure 9.86 Package mesh in leadframe with warpage

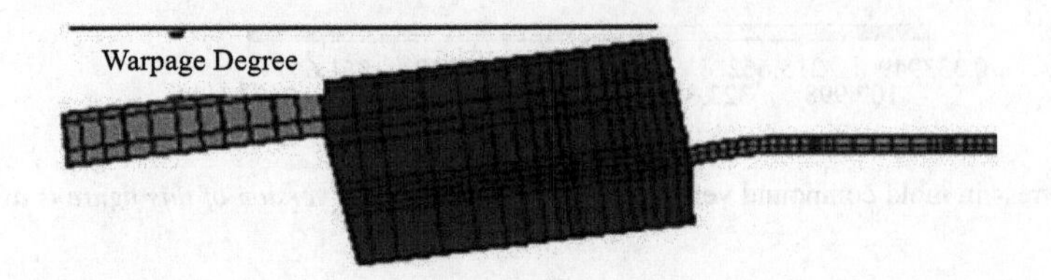

Figure 9.87 Warpage degree

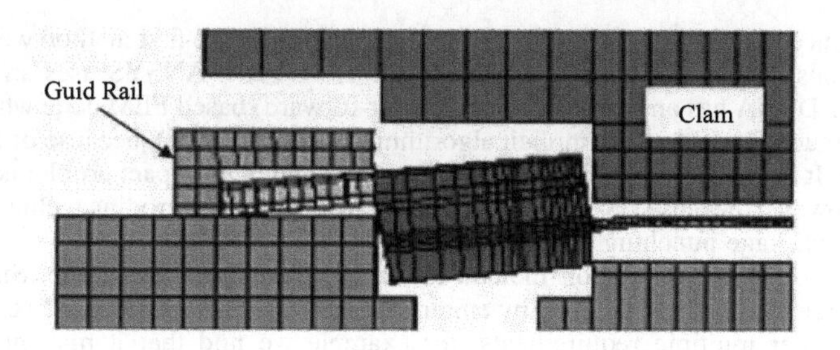

Figure 9.88 Symmetry model of leadframe and clamp

Table 9.12 Package stress versus package warpage

Warpage (°C)	Stress in Package		
	Max Seqv (MPa)	Max S1 (MPa)	Max S3 (MPa)
2	25	30	−26
5	52	59	−57
8	57	62	−68

Table 9.13 Package stress versus clearance between guide rail and leadframe

Warpage Clearance (mm)	Stress in Package		
	Max Seqv (MPa)	Max S1 (MPa)	Max S3 (MPa)
0.2	83	94	−89
0.7	57	62	−68

Table 9.12 shows the impact of the degree of warpage. The maximum stress in the package is located in the interface between the mold compound and the leadframe. The stress in the die is much lower. The stress in the molding compound shows that the maximum stress in molding compound will increase dramatically as package warpage increase from 2 ° to 5 °, it does not change a lot as package warpage increases from 5 ° to 8 °. The leadframe with 5 ° or larger package warpage should be flattened before clamping and trim form process.

Table 9.13 shows the impact of clearance between guide rail and leadframe thickness. Reducing the clearance between the guide rail and the leadframe can increase the stress in the package. During the clamping, some areas of the leadframe are impacted at high speed. The other part of the leadframe will vibrate dramatically because of warpage and hit the guide rail, which induce the damage of package.

9.6 Leadframe Forming/Singulation

The use of leadframes during assembly processing is important in the case of handling and economics of scale. These leadframes can contain anywhere from 2 to thousands of individual products that must be separated prior to completion. There are several methods by which singulation can be achieved, including saw or punch. The use of a punch is preferred for its speed. In this section we will look at the use of a punch process to separate assembled product, and in particular the use of FEA to model the stresses in the process. This allows us to maximize final product yield, and quality, by minimizing the possibility of die cracking.

We have examined die cracking using two modeling methods. In the first method we utilized standard ANSYS simulations, and in a second method we used LS-DYNA. ANSYS uses an implicit solution method, while LS-DYNA has an explicit scheme (Euler-forward) based FEM code which has very high efficiency and is much faster than an implicit algorithm (Euler-backward), because of the lack of a local implicit-iteration. It is therefore suitable for large structures and high impact problems like automobiles and aerospace. However, because LS-DYNA has the function of element eroding technique, it is good tool to be used for IC package punching process

The advantages of using simulation include lower cost and faster turn time. Simulation enables experiments that are too costly to be done by empirical methods. This includes the ability to optimize a package design under multiple requirements, for example we find that it may not be practical to empirically look at multiple combinations of materials, or leadframe geometry. Using simulation we can quickly and affordably find the conditions that optimize cost, performance, and reliability under different sets of conditions.

In the studies presented here we examine the stresses on the die as we vary the shape and depth of grooves that have been designed into the leadframe.

9.6.1 Euler Forward versus Backward Solution Method

During the simulation of the punching process, the mesh used becomes highly deformed. This is a dynamical problem with a very small time step required during the solution to resolve the desired detail. In this work we use standard ANSYS to obtain an implicit solution, and the ANSYS LS-DYNA program for an explicit solution.

The Euler backward method (implicit) computes as:

$$y_{n+1} = y_n + hf(x_{n+1}, y_{n+1}) \tag{9.31}$$

This solution method is unconditionally stable. The stability allows us to specify larger time steps. However in practice this is limited because if the time step is too large we will miss the effects such as the stress wave, which we are looking for in our study.

The Euler forward method (explicit) computes as:

$$y_{n+1} = y_n + hf(n_n, t_n) \tag{9.32}$$

and it is only stable if the Courant-Friedrich-Lévy condition is met.

In both equations above, the step size h is computed as:

$$h = t_n - t_{n-1} \tag{9.33}$$

9.6.2 Punch Process Setup

A simplified illustration of the tooling setup for the punch process is shown in Figure 9.89. The punch tooling consists of three pieces, the punch, its body guide, and the anvil. For the simulation we will assume the punch, guide, and anvil to be rigid bodies.

A more detailed view of the primary section of the solid model is shown in Figure 9.90. This illustration also shows the guide and punch relationship. Note the punch used in these studies is hollow in the region over the die. The action of the punch is such that when the punch is forced downwards, it cuts the leadframe and the freed leadframe falls through the anvil to a receptacle for collecting the units.

Simulations were set up using the following material parameters.

The material properties are listed in Table 9.14. For the copper alloy an elastic-plastic bi-linear stress-strain relation is used. The yield strength of copper alloy yield stress is 68.97e3 MPa,

Table 9.14 Materials parameters

	Leadframe	Silicon Die	Die Attach	Solder Ball
Young's Modulus (GPa)	117.72	169.5	41.6	23.5
Poisson's Ratio	0.35	0.278	0.4	0.4
Density (kg/mm^3)	8.9×10^{-6}	2.39×10^{-6}	7.39×10^{-6}	1.13×10^{-5}
Failure Strain	5%–10%		20%	
Yield Stress (MPa)	68.97		31	

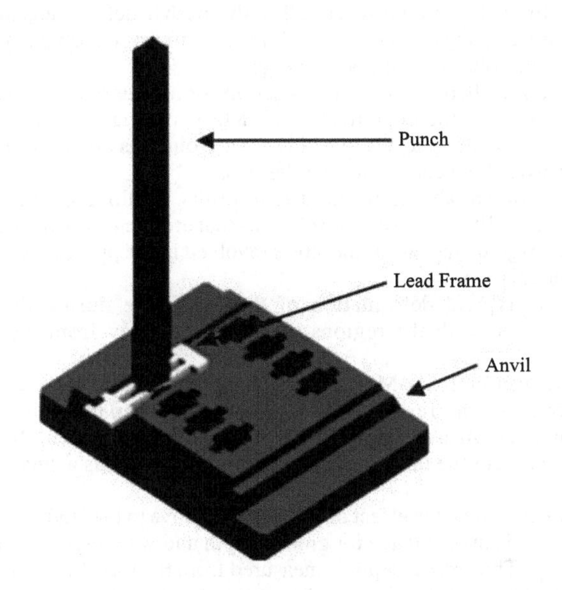

Figure 9.89 Typical punch and leadframe

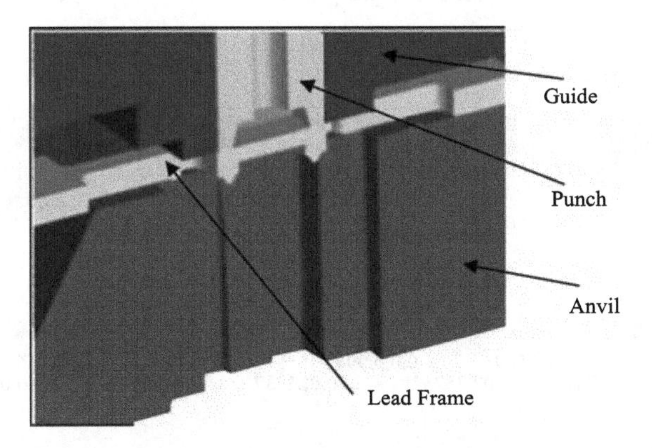

Figure 9.90 Punch and leadframe

the tangent modulus is 6.897e6 kPa. For the purposes of the study we assumed a strain failure criteria of 0.05-0.1 in the copper during the LS-DYNA simulations, which needs to be experimentally measured.

9.6.3 *Punch Simulation by ANSYS Implicit*

The first method we discuss is implicit modeling by the use of ANSYS Structural, Mechanical, or Multiphysics. This mode of modeling the punch process is advantageous since all the capabilities of ANSYS are available to the analyst, including a large array of element types with which we mesh the geometries. However, due to the use of implicit solution techniques in ANSYS, there is no element eroding function. We are limited in how far we can allow the mesh to deform, and the technique is likely to over predict the stresses since there is no mechanism for the leadframe material to yield. We find that if the deformation is too great the solution will not converge.

In the simulations described here the punch was set up for a penetration of approximately 0.05 mm during the punch process. These simulations treat the die attach material as a linear elastic material, and as a result we expect the modeled stress levels to be high. The solutions are quasi static and the results are used for relative comparisons between leadframe designs.

Despite the restrictions we are able to obtain useful results. Minimizing the stress problems in this portion of the punch process will lead to similar solutions that are found when modeling the entire punch process. This is because a significant part of the stress involved in the process occurs in the initial phases that can be implicitly modeled.

Figure 9.91 shows the typical deformation of the leadframe during the implicit simulation. The features are as expected with the regions of leadframe away from the punch edge showing no deformation.

Figure 9.92 shows the resultant von-Mises stress on the part with die, die attach, and solder balls, while Table 9.15 summarizes the results of die stress for a comparison between punches of two widths. The narrow punch has an area that is 36% less than the wide punch. Both leadframes have no pre-groove. It can be seen from the table that narrowing the punch has a small effect on the stress in the die.

As part of this work we examined the effect of adding a V groove to the leadframe during the leadframe stamping operation. Comparisons are made for groove depth and wall angle, as well to as the no-groove leadframe discussed above. The groove depth is measured from the top of the leadframe and the angle is measured from the vertical wall of the V as can be seen in Figure 9.93.

Table 9.16 shows the effect of the groove angle experiment on stresses in the die, at a fixed 3 mil groove depth. We can see from the data presented and post processing that there is a significant decrease in the stress levels in die. The groove acts to concentrate the stress at the vertex of the groove, and the stress transmitted to the die is less than without the pre-groove.

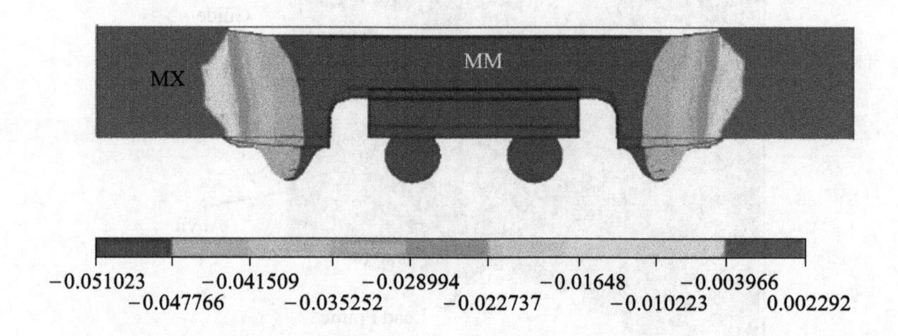

Figure 9.91 Deformation of the frame *(Color version of this figure is available online)*

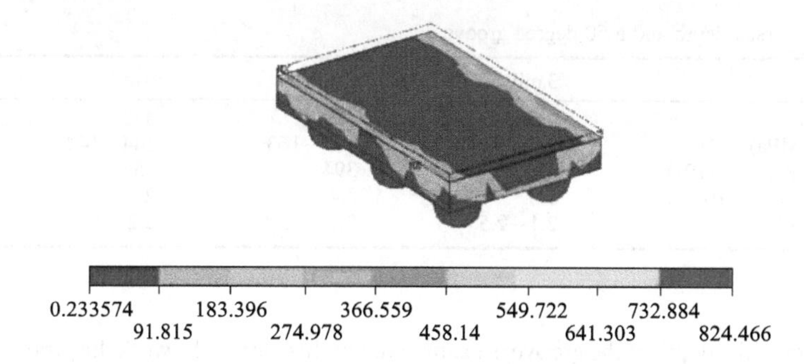

Figure 9.92 von-Mises stress on die *(Color version of this figure is available online)*

Table 9.15 Die stress results for no-groove leadframes

Property	Wide Punch	Narrow Punch
von-Mises (MPa)	542	502
Shear Stress (MPa)	212–222	209–191
Max Principle Stress S1 (MPa)	469	428
Total Effective Strain ($\times 10^{-3}$)	2.9	2.7
Shear Strain ($\times 10^{-3}$)	3.2–3.3	3.2–2.9

Figure 9.93 3 mil groove depth with 67° angle

Table 9.16 Results for the die with different groove angles

Property	50°	67°	74°
von-Mises (MPa)	417	431	367
Shear Stress (MPa)	142–168	175–161	151–157
Max Principle Stress S1 (MPa)	373	410	330
Total Effective Strain ($\times 10^{-3}$)	2.3	2.5	2.1
Shear Strain ($\times 10^{-3}$)	2.1–2.5	2.6–2.1	2.3–2.4

Table 9.17 Die versus depth and a 50 degree groove

Property	3 mil	6 mil	9 mil	12 mil
von-Mises (MPa)	417	429	372	274
Max/Min Shear (MPa)	142–168	183–163	144–142	120–114
Max Principle Stress S1 (MPa)	373	394	355	246
Total Effective Strain ($\times 10^{-3}$)	2.3	2.4	2.1	1.5
Shear Strain ($\times 10^{-3}$)	2.1–2.5	2.7–2.5	2.2–2.1	1.8–1.7

Next the effect of the depth of the groove is examined. For this part of the work the groove angle is held fixed at 50 °. Compared to the existence of the groove, the actual groove depth used has a lesser effect on the leadframe stress levels.

Table 9.17 shows the effect of varying the depth on the die stress levels. Increasing the groove depth has a beneficial effect, however this needs to be balanced as a tradeoff with any degradations it may induce in handling, due to increased flexibility of the overall leadframe.

Ideally the above work would be extended to include simulations for all depths and angles, or would be extended as a second order central composite design.

9.6.4 Punch Simulation by LS-DYNA

Next the LS-DYNA is used to simulate the punch process. LS-DYNA is capable of modeling the shearing action of the punch through the copper leadframe during the punch process.

In setting up the analysis problem we assume that the punch is a rigid body due to its extreme hardness and young's modulus. The punch has a velocity of 120 mm/s and no acceleration during the actual punch operation. The model is symmetric so a half model can be used for computational efficiency, and it is meshed with brick elements. The half model is shown in Figure 9.94 below.

After a short time after the initial contact there is a noticeable difference between the leadframe with and without a built in groove. Figure 9.95(a) and (b) show the simulation at $t = 0.0002$ s after the initial contact. It is clear that with the groove the strain is more confined to the desired region.

Further examination of the stress in the die bears out the initial view. Figure 9.96 shows a comparison of the maximum von-Mises stress that occurs in the die, for leadframes with and without grooves. The stress in the die attached to the leadframe without the groove is 1.54 times greater than the leadframe with the groove. So similar to the results of the quasi static analysis, the groove is found to have a very significant and beneficial effect.

Figure 9.94 Meshed LS-DYNA solid model

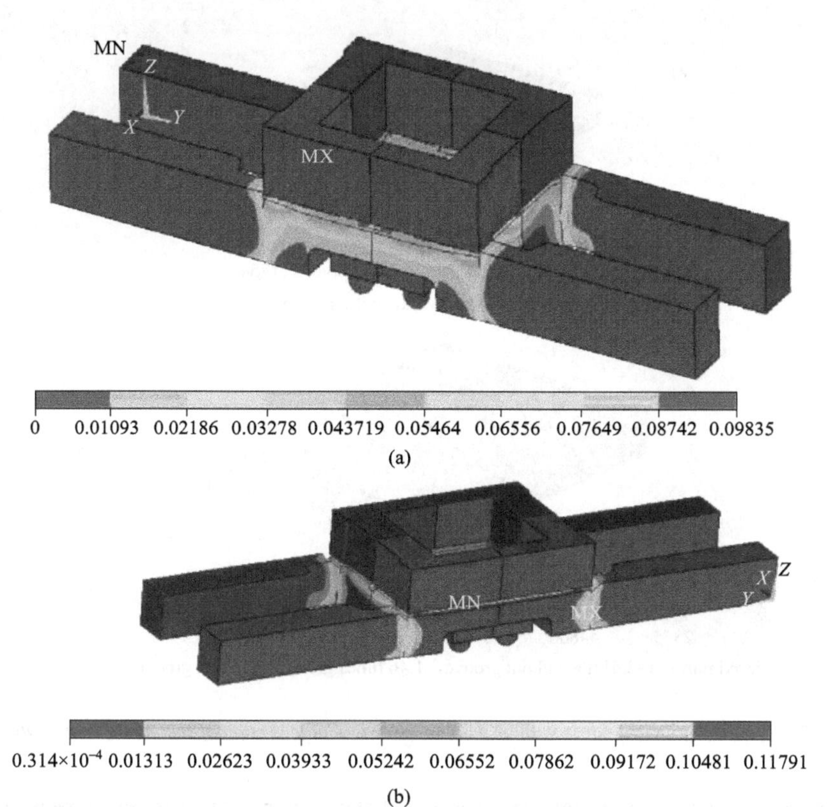

Figure 9.95 (a) Plastic strain without grooved frame. (b) Plastic strain with grooved frame *(Color version of this figure is available online)*

Figure 9.97(a) and (b) show the first principal stress in die, S1, with and without the groove. The stress without the groove is higher by a factor of 1.46. Both simulation techniques show a substantial beneficial effect of adding a groove to the leadframe.

Figure 9.98 shows a sequence of the die level stress as a function of time. The stress wave sequence was taken from a study of a package that is smaller than the package of Figure 9.95, by a fact of 2.8. As a result the stress values should not be directly compared to previous figures.

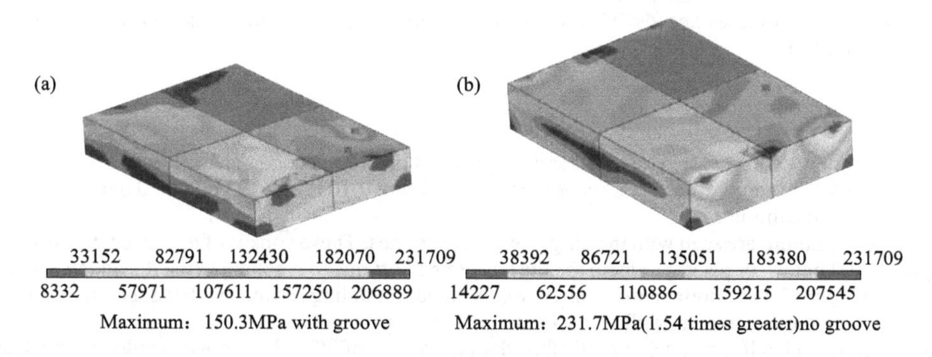

Figure 9.96 Die von-Mises stress *(Color version of this figure is available online)*

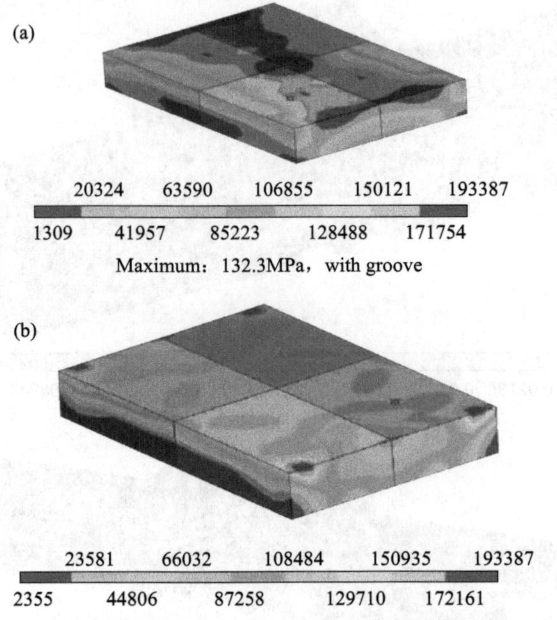

Figure 9.97 Die first principal stress with or without groove *(Color version of this figure is available online)*

The sequence shows the principal stress wave as it starts, propagates, and finally peaks at the die corners. This sequence provides a clearer understanding of the stress mechanisms within the device.

Similarly we are able to map out the stress wave propagation in the die attach. Figure 9.99 shows a snapshot of the von-Mises stress in the die attach.

The von-Mises stress in the die attach propagates directly into the die itself as can be seen in Figure 9.100.

Figure 9.101 plots the time evolution of the peak first principal stress in the die during the punch process. The model uses a leadframe failure strain of 5%.

During the studies the effect of the failure strain rate of the leadframe, and the effect of die size is also examined. Figure 9.102 shows the maximum first principal stress, S1, in the die for two different die sizes in two different leadframe designs.

For the smaller of the two die the data shows that we will exceed the die failure criteria if the failure strain rate of the leadframe exceeds 7%. For larger die there is little concern for likely leadframe design/material to be used.

9.6.5 Experimental Data

In line experiments comparing a grooved and un-grooved die were run to determine the efficacy of the grooved design. The sample size consists of 1280 units. Figure 9.103 shows a good final unit. This unit is from the grooved sample set.

Two types of failure are seen with the un-grooved leadframes. These consist of a cracked die, as is seen in Figure 9.104 below, or package where the die has popped off due to cracking in the die attach material, as is seen in Figure 9.105. Both pictures agree well with our modeling results that have shown the highest stress at the die corner area, as well as at the die attach corner area.

The un-grooved leadframe packages displayed a failure rate of 3%. The grooved leadframe packages show no failures.

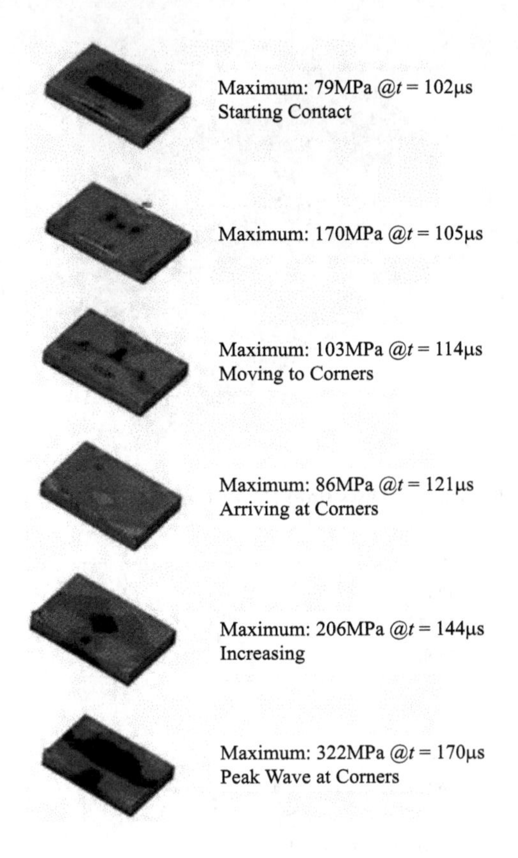

Figure 9.98 Principal stress S1 wave in die

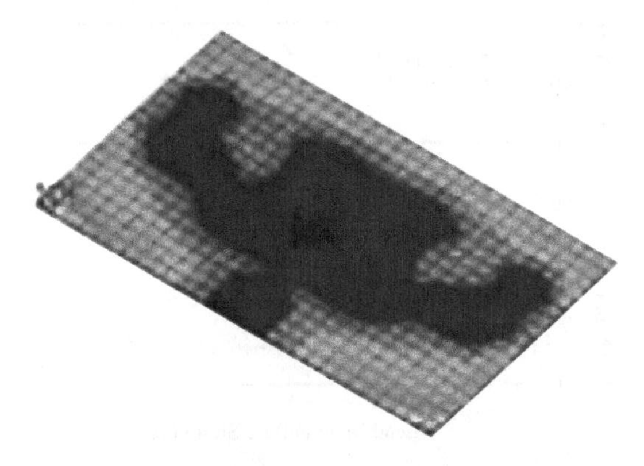

Figure 9.99 Die attach von-Mises stress *(Color version of this figure is available online)*

Figure 9.100 Die von-Mises stress *(Color version of this figure is available online)*

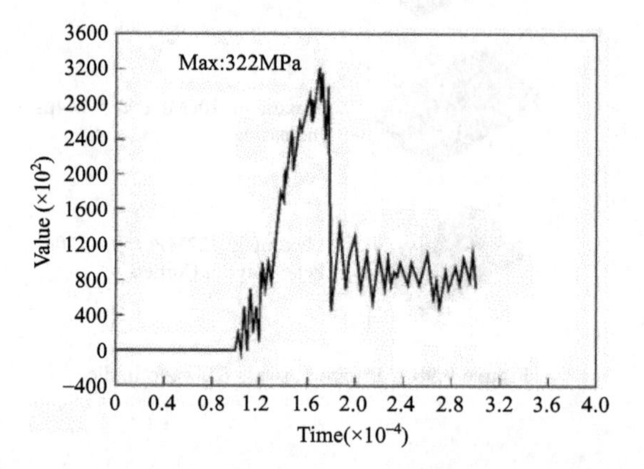

Figure 9.101 Die first principal stress S1 versus time comparison on die. (L.F.S. = 5%)

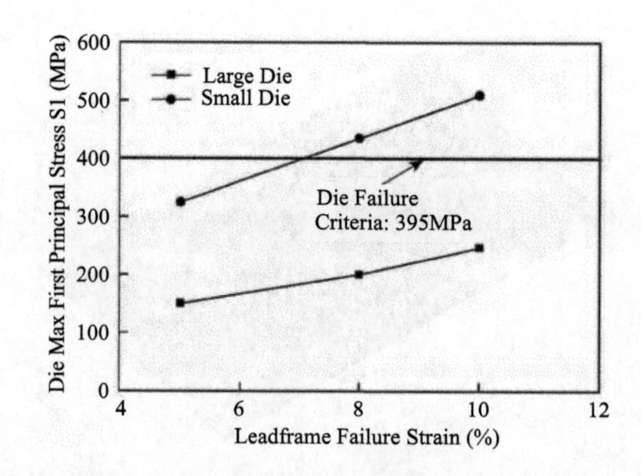

Figure 9.102 First principal S1 in die versus leadframe strain failure

Figure 9.103 Good die

Figure 9.104 Crack at die corner

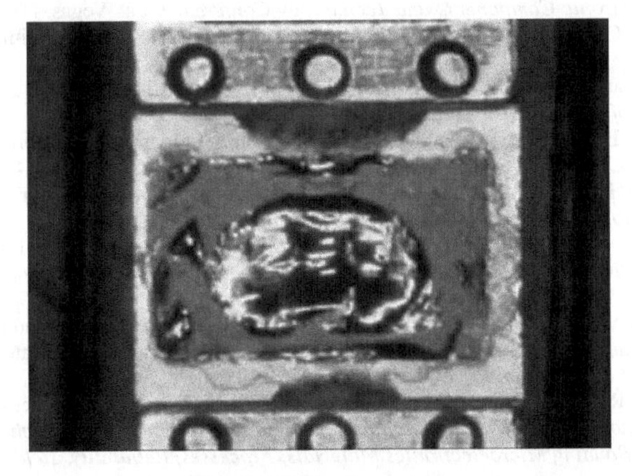

Figure 9.105 Popped die

9.7 Chapter Summary

This chapter presents the modeling and simulation methodologies for major IC packaging assembly manufacturing processes in FOL and EOL.

The major FOL modeling and simulation methodologies include wafer thinning process, die pick up simulation, multiple die attach simulation, and wire bonding simulation. Multiple 3D contact pairs in the simulation platform are set up between the die pick up collet and die, die and tape, tape and die holder, and tape and eject pins for the die pick up simulation. A debonding criterion for the die surface and tape is developed to show the die separation process from tape. Transient dynamics modeling shows the stress distribution in the die so that the process can be optimized and we can verify that if the die strength is strong enough to withstand the dynamic pick up and ejection pin stresses. A nonlinear (creep and elastic plastic) material constitutive model is applied to the solder paste for the multiple die attach process simulation. The attach process order is modeled by an element death and birth technique to show the impact of multiple die attach process. Transient dynamics modeling method is used to simulate the wire bonding process with different bonding forces, frequency, and friction between the bond pad and bond wire free air ball (FAB). From the model we can check the stress between FAB and bond pad, cratering in the bond pad as well as the failure of interconnect device under bond pad.

In EOL, typical processes include molding process with curing stress and residual stress, molding ejection, and clamping process that can potentially cause cracking in molding compound, trim and form and separation process that would induce the stress wave to crack the die or the EMC. For the molding process simulation, material properties are the function of both time and temperature, also the shrinkage during curing will impact the residual stress for the molding process. A user supplied FORTRAN subroutine code is developed in ANSYS in both the time and temperature of the epoxy mold material. The simulation shows the mold injection, the curing process and the residual stress build-up during molding. For the molding ejection and clamping simulation, a 3D model is developed to show the stress at the interface between the leadframe and mold compound that would induce the delamination. Prior to trim and form, a clamping process is needed to flatten the leadframe due to warpage after molding. The simulation is applied to this process to make sure the clamping process will not damage the package and die inside. For the package Trim/Form/Singulation simulation, a 3D transient dynamic large deformation finite element method is used to determine what stresses are going to be induced while singulating the package, to check if the package can withstand punch stress waves and to determine potential design weaknesses. A nonlinear finite element contact eroding model is developed and used for simulation.

References

ANSYS12.0 Theory manual, (2009).

Awad, E. (2004) Active devices and wiring under chip bond pads: Stress simulations and modeling methodology, *Proceedings of 54 Electronic Components and Technology Conference*, Las Vegas, NV, May 1784–1787.

Bird, R.B., Armstrong, R.C. and Hassager, O. (1987) *Dynamics of Polymeric Liquids*, Volume 1, John Wiley & Sons, New York.

Brown, S.B., Kim, K.H. and Anand, L. (1989) An internal variable constitutive model for hot working of metals, *International Journal of Plasticity* **5**:95–130.

Degryse, D., Vandevelde, B., Beyne, E. (2004) Mechanical FEM simulation of bonding process on Cu LowK wafers, *IEEE Transactions on Components and Packaging Technologies*, **27**(4), December, 643–650.

Eggert, G.M., Dawson, P.R. and Mathur, K.K. (1991) An adaptive descent method for non-linear viscoplasticity, *International Journal for Numerical Methods in Engineering* **31**:1031–1054.

Feustel, F. and Wiese, S. and Meusel, E. (2000) Time-dependent material modeling for finite element analyses of flip-chips, *Proceedings of 50th Electronic Components and Technology Conference*, Las Vegas, NV, May:1548–1553.

Fiori, V., Beng, L.T., Downey, S. (2007) 3D multi scale modeling of wire bonding induced peeling in Cu/low-k interconnects: application of an energy based criteria and correlations with experiments, *Proceedings of 57th Electronic Components and Technology Conference 2007*, ECTC '07, 256–263.

Gillotti, G. and Cathcart, R. (2002) Optimizing wire bonding processes for maximum factory portability, *SEMICON West 2002 SEMI Tecgnology Symposium: International Electronics Manfacturing Technology Symposium.*

Harman, G. (1997) *Wire Bonding Microelectronics Materials, Processes, Reliability, Yield*, Volume 1, McGraw-Hill, New York.

Han, S. and Wang, K.K. (2000) Flow analysis in a chip cavity during semiconductor encapsulation, *Transaction ASME Journal of Electronic Packaging* **122**:160–167.
Hess, K.J., Downey, S.H. and Hall, G.B. *et al.* (2003) Reliability of bond over active pad structures for 0.13-um CMOS technology, *Proceedings of 53th Electronic Components and Technology Conference*, New Orleans, May.
Ikeda, T., Miyazaki, N., Kudo, K., *et al.* (1999) Failure estimation of semiconductor chip during wire bonding process, *ASME Journal of Electronic Packaging* **121**:85–91.
Irving, S. and Liu, Y. (2005) An effective method for improving IC package die failure during assembly punch process, *EuroSimE* 2005.
Langenecker, B. (1966) Effects of ultrasound on deformation characteristics of metals, *IEEE Transactions on Sonics and Ultrasonics*, **SU-13, 1.**
Levine, L. (1995) The ultrasonic wedge bonding mechanism: two theories converge, *ISHM 1995, Proc*, Los Angeles, California, October 24–26, 242–246.
Lee, T.K. *et al.* (2008) Some case studies on air venting analysis of semiconductor package using mold flow, *Electronic Packaging Technology Conference 2008 Singapore*, 444–449.
Liu, Y. and Antes, H. (1999) An improved impicit algorithm for the elastic viscoplastic boundary element method, *ZAMM-Zeitschrift für Angewandte Mathematik und Mechanik* **79**:317–333.
Liu, D.S., Chao, Y.C., Wang, C.H. (2004) Study of wire bonding looping formation in the electronic packaging process using the three-dimensional finite element method, *Finite Elements in Analysis and Design* **40**(3) January, 263–286.
Liu, Y., *et al.* (2006) Simulation and analysis for typical package assembly manufacture, *EuroSimE* 2006.
Liu, Y., Allen, H. and Luk, T. (2006) Simulation and experimental analysis for a ball stitch on bump wire bonding process above a laminate substrate, *Proceedings of 56th Electronic Components and Technology Conference*, 1918–1923.
Liu, Y., Irving, S. and Luk, T. (2008) Thermosonic wire bonding process simulation and bond pad over active stress analysis, *IEEE Transactions on Electronics Packaging Manufacturing* **31**:61–71.
Maniatty, A.M., Liu, Y., Ottmar, K. *et al.* (2001) Stabilized finite element method for viscoplastic flow: formulation and a simple progressive solution strategy, *Computer Methods in Applied Mechanics and Engineering* **190**:4609–4625.
Mannan, S.H. *et al.* (2000) Solder paste reflow modeling for flip-chip assembly, *3rd Electronics Packaging Technology Conference*, Singapore, December, 103–109.
Mayer, M., Paul, O., Bolliger, D. and Baltes, H. (1999) Integrated temperature microsensores for characterization and optimization of thermosonic ball bonding process, *Proceedings of 49 th Electronic Components and Technology Conference*, San Diego, California, May, 463–468.
Paulino, G.H. and Liu, Y. (2001) Implicit consistent and continuum tangent operators in elastoplastic boundary element formulations, *Computer Methods in Applied Mechanics and Engineering* **190**:2157–2179.
Peter, B. and Domenico, T. (2002) The back-end process: step 4-die attach today's challenges, *Advanced Packaging*, April.
Qian, Z. and Liu, S. (1999) A damage coupling framework of unified viscoplasticity for the fatigue of solder alloys, *ASME, Journal of Electronic Packaging* **121**:162–168.
Reiche, M. and Wagner, G. (2003) Wafer thinning: techniques for ultra-thin wafers, *Advanced Packaging*, March.
Shen, Y.L. *et al.* (1996) Stresses, warpages, and shape changes arising from patterned lineson silicon wafers, *Journal of Applied Physics* **80**:1388–1398.
Shirai, Y., Otsuka, K., Araki, T. *et al.* (1993) High reliability wire bonding by the 120 kHz frequency of ultrasonic, *ICEMM Proceedings*, 366–375.
Simo, J.C. and Hughes, T.J.R. (1997), *Computational Inelasticity*, Springer, New York, 114–151.
Stoney, G.G. (1909) The tension of metallic films deposited by electrolysis, *Proceedings of the Royal Society A: Mathematical* **82**:172–175.
Takahashi, Y. and Inoue, M. (2002) Numerical study of wire bonding - analysis of interfacial deformation between wire and pad, *ASME Journal of Electronic Packaging* **121**:27–36.
Tan, C.T. and Gan, Z.H. (2003) Failure mechanisms of aluminum bond pad peeling during thermosonic bonding, *IEEE Transactions on Device and Materials Reliability* **3**(2):44–50.
Van Driel, W.D., Janssen, J.H.J. and Van Silfhout, R.B.R. (2004) On wire failures in micro-electronic packages, *EuroSimE2004*, 53–57.
van Silfhout, R.B.R. *et al.* (2002) Prediction of back-end process-induced wafer warpage and experimental verification, *Proceedings of the 2002 Electronic Components and Technology Conference*, San Diego, CA, May, 1182–1187.
Wang, J., Qian, Z. and Liu, S. (1998) Process induced stresses of a flip-chip packaging by sequential processing modeling technique, *ASME Journal of Electronic Packaging* **120**:309–313.
Wu, E. *et al.* (2002) Influence of grinding process on semiconductor chip strength, *Proceedings of the 2002 Electronic Components and Technology Conference*, San Diego, CA, May, 1617–1621.
Yeh, C. and Zhou, W. (1996) Parametric finite element analysis of flip-chip reliability, *International Journal of Microcircuits and Electronic Packaging* **19**(2):120–127.
Yeh, C.L., Lai, Y.-S., Kao, C.-L. (2006) Transient simulation of wire pull test on Cu/Low-K wafers, *IEEE Transactions on Advanced Packaging* **29**(3):631–638.
Yu, S. and Xu, Q. (1996) Structural mechanics and manufacturing mechanics in electronic packaging, *Proceedings of the Second Intnetnational Symposium on Electronic Packaging Technology*, Shanghai, December,120–124.

10

Opto Packaging and Assembly

Micro technology is generating a completely novel basis for many industries including optical communication, where future innovation will be dominated by miniaturization. Manufacturing these devices involves complex work of design for device, systems, and packaging. As potential new products and markets emerge for these miniature devices, designers face continuous pressure to get the design done right and fast the first time. Engineers have identified the challenging components in the progress of micro systems as the need for a multidisciplinary design team, development of a device-specific process flow, and the need for multi-physics analysis and development of custom packages. Currently a huge portion of cost is involved for the packaging and testing of these devices. It is well known that the fiber optics industry once sufferred from the burst of the internet bubble in late 1990s and early 2000s, which was believed to be caused by the broken chain between the carriers, equipment OEMs, module and subsystems suppliers, components, and chip sets suppliers. However, the demand for more bandwidth is still growing due to the need for transmission of data for both business and personal uses driven by on-line services, communications, media, and gaming. While we suffered badly in the telecomm sector, the data communication sector is growing rapidly. There is still good demand for data communication modules. With the continuing demand of reducing the prices for those modules, non-hermetic modules with passive coupling is a good direction to go in the data communication sector. silicon optical bench (SOB) technology is a technology that can meet the needs of the data communication sector. Analysis was performed for sequential modeling for a laser module based on hermetic packaging. The approach proposed in the chapter is expected to help designing the process flow for general opto packaging modules.

10.1 Silicon Substrate Based Opto Package Assembly

10.1.1 State of the Technology

It is well known that InGaAsP diode lasers are the fundamental building blocks of optical transmission systems. Significant efforts have been made in the optimization of design and manufacturing, and enhancement of yield and long term reliability. Laser dimensions are between 0.3 mm × 0.25 mm × 0.11 mm with some variations in thickness and length, and have a high beam divergence with far field angles of 15–35 degrees. This requires maintaining a very tight tolerance of the elements in the optical path to even a fraction of microns to minimize the losses. For instance, a 980 nm pump laser needs an active alignment either manually or by an expensive and automotive aligner, with each taking 15 minutes. Optical alignment makes the optical, packaging different from conventional IC packaging, where self alignment of the solders will easily satisfy the alignment requirements for IC packaging.

Therefore, packaging has been a troublesome issue that lowers the yield and it is slow and therefore costly, as it takes significant efforts to couple the lasing beam into a single mode fiber core. When hermetic

Modeling and Simulation for Microelectronic Packaging and Integration: Manufacturing, Reliability and Testing, Second Edition. Sheng Liu and Yong Liu.

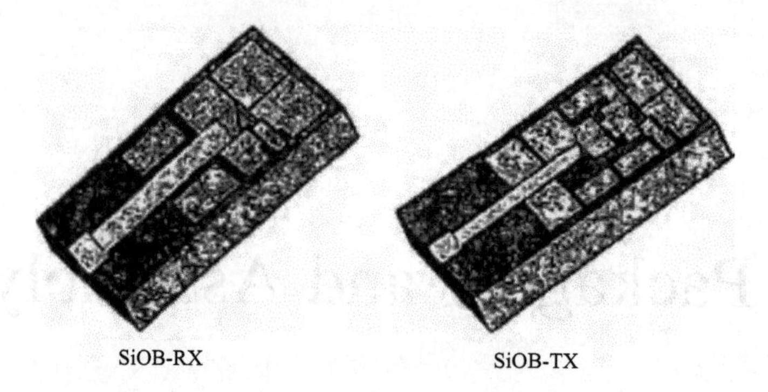

Figure 10.1 Silicon benches for receiving (RX) and transmitting (TX)

packaging is used, various processes such as seam welding, bonding, and other processes may change the coupling efficiency during the manufacturing. In passive coupling, a V-groove is produced with about one micron accuracy so that the fiber and the laser can be aligned without searching in a six-degrees-of-freedom space. A typical structure of the silicon bench made in house is shown in Figure 10.1.

10.1.2 Monte Carlo Simulation of Bonding/Soldering Process

Due to the fact that there are various sources of errors that may contribute to the misalignment of the laser diode to the core of the fiber, critical issues must be identified. Variables that may affect the alignment are the post-bonding accuracy in the transverse direction to the laser in the horizontal plane, the solder laser thickness, laser die active up or down, fiber bowing in V-groove, and variations due to the V-groove. Monte Carlo simulation is a powerful tool to identify the key parameters that will affect the coupling efficiency to be within the acceptable level.

The simple principal model is shown in Figure 10.2.

The optical performance can be simplified into LD to Fiber calculated by optical modeling software CODE V by changing the LD and fiber relative position. Detailed modeling is omitted here. The following conclusions are made out of modeling and experimental validation:

1. Post-bonding accuracy must be less than two to three microns in the *x-y* plane, therefore requiring a one-micron directly coupling model. The coupling efficiency can be assured by a one-micron bonding machine.
2. Thickness of the solder alloy must be within three to five microns therefore a flip-chip type bonding with active area is favored.
3. Posting bonding accuracy in the *z*-axis may not be more than one micron.
4. LD must be controlled to be parallel to the surface of SOB, or the bonding accuracy in the horizontal plane will be required; fortunately it can be easily achieved.
5. Both force and displacement must be exercised in the *z*-axis so that the accurate control of the lasing light point and the center of the core are assured.
6. The bow angle of the fiber on the V-groove is critical and V-cure fixing is essential.

10.1.3 Effect of Matching Fluid

Match fluid must have a refractive index of approximately 1.5. It has to be pointed out that the match fluid must maintain a certain viscosity, bonding strength to silicon substrate, fiber, and diode. One has to make sure that the optical reflection has to be minimum in the interfaces between fiber and LD. Another matching fluid must be dispensed as a good round shape. A schematic of use of the matching fluid is shown in Figure 10.3.

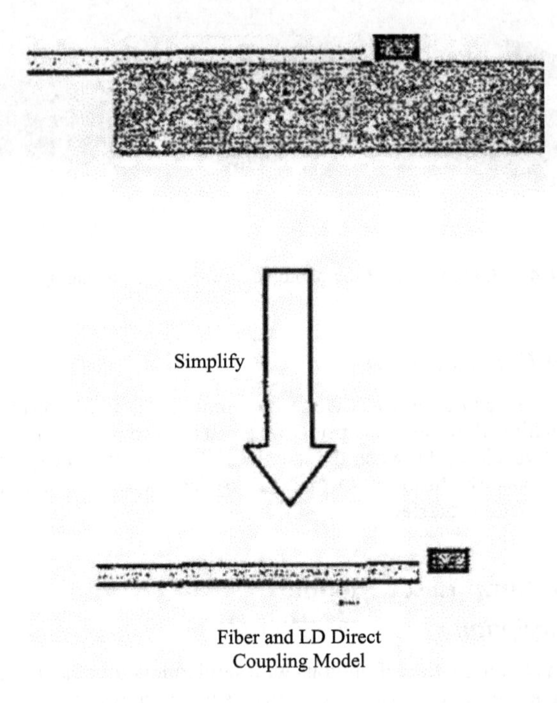

Figure 10.2 A fiber and LD direct coupling model for Monte Carlo simulation

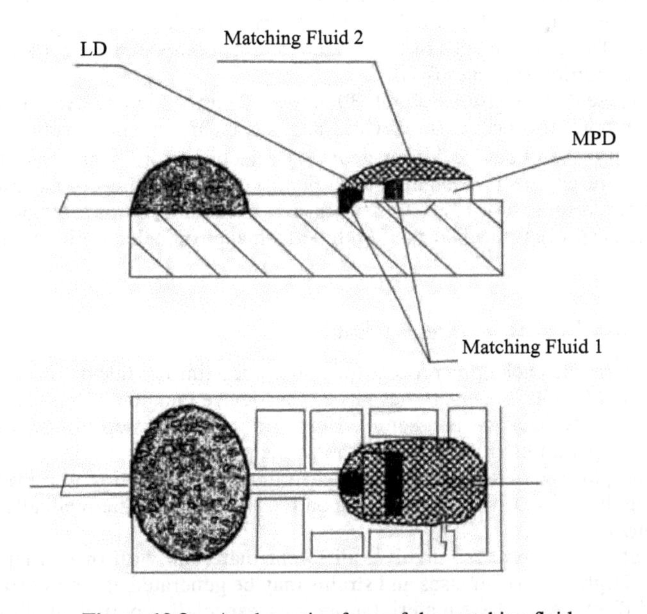

Figure 10.3 A schematic of use of the matching fluid

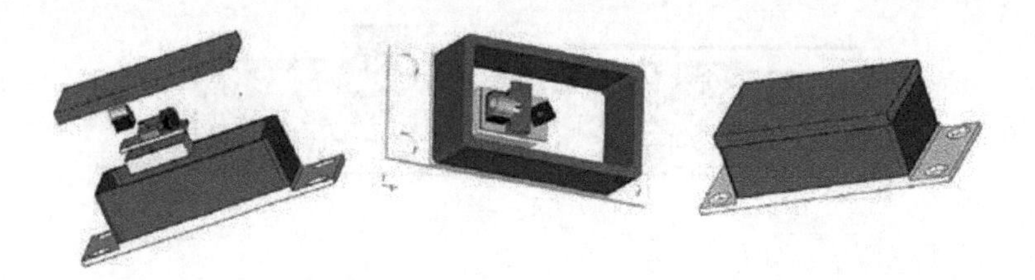

Figure 10.4 3D rendition of the photonic mini-module packaging assembly

10.1.4 Effect of the Encapsulation

For a non-hermetic case, encapsulation by silicone type material is used. Perfect bonding need to be assured between the matching fluid and encapsulation, as the thermal cycling in both the testing and operation may cause relative sliding between the matching fluid and the encapsulation. The dark color (see Figure 10.3) helps to absorb the light so that the optical reflection can be minimized along the matching fluid/encapsulation interface.

10.2 Welding of a Pump Laser Module

10.2.1 Module Description

Optoelectronics devices have interaction of photons with semiconductors that use light energy known as photons for operation. The devices provide the optical sources and detectors that allow broadband telecommunications and data transmission over optical fibers. Conversion of electronic signals to optical signals and vice-versa is based on optoelectronic principles. Photonic devices are also called "optoelectronics" devices. For these kinds of devices, the emitters of light energy/photon are coherent sources in the form of laser (light amplification by stimulated emission of radiation) diodes (LD) and converter of photon into electrical energy are photodiodes (PD). LD is a semiconductor device, which works as the source of highly directional, monochromatic, coherent light (Streetman *et al.*, 2001). The other major components of this module are the TEC (Thermo Electric Cooler), optical bench, micro lens, and the module case. The CAD images of the module and its components are shown in exploded and hermetically sealed condition (Figure 10.4).

The module being sealed measures about 30 mm × 10 mm × 13 mm. Hermeticity of the device packaging is very important since these devices are desired in adverse conditions communication. This involves a smart process of device design, geometry selection, manufacturing process, and above all packaging solutions. The engineers, designers, and scientists are addressing critical issues for this type of packaging process for a long time. For this work, we have chosen finite element based numerical approach to simulate the process mechanics of high performance optoelectronics communication module packaging.

10.2.2 Module Packaging Process Flow

We have designed a hermetic packaging process for a photonic Mini-module of 10Gbps type. Emergence of the 10 Gbps Ethernet (10GbE) standard has placed this device manufacturing volume in higher scale (Verdiell *et al.*, 2002). The package is assembled in various step by step flux-less bonding/soldering processes at higher temperatures (Figure 10.5). The whole packaging assembly involves many steps of components pick and place with solder layers between them and reflowing in a thermal chamber and subsequently cooling them. Finally, a cap is placed on the assembly and hermetically sealed around the corner of the cap and case.

This multi step packaging process involves different materials with dissimilar thermal and mechanical properties. High level of stresses and strains may be generated at the interfaces of the various component due to differences in these material properties, especially coefficient of thermal expansions (CTE). We have used other solder material with higher melting points (80/20 AuSn) since the most

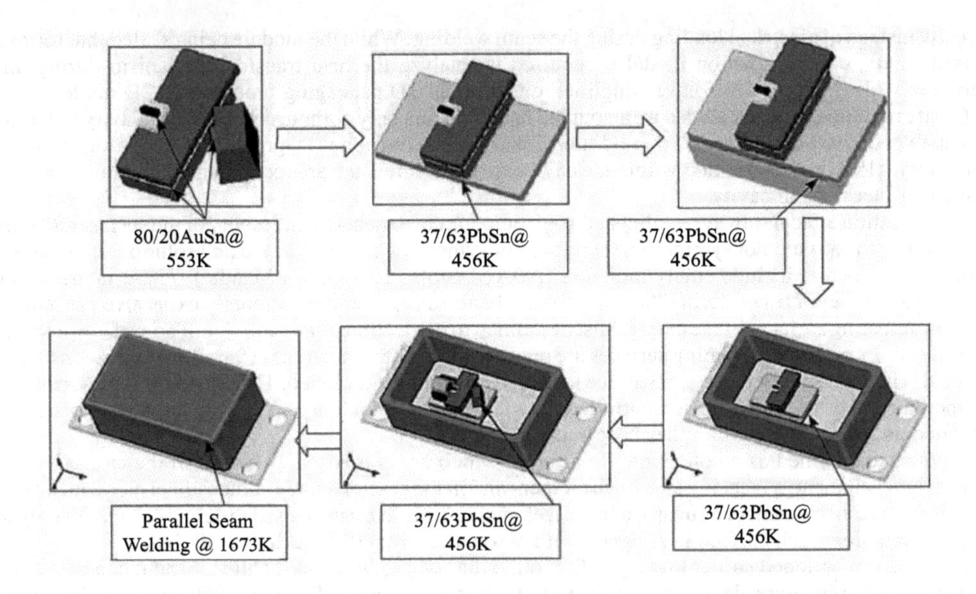

Figure 10.5 Photonic mini-module packaging process flow using different solder layers at different temperatures

commonly used solder (PbSn) is very poor structural material (Liu *et al.*, 1992; Liu *et al.*,1995; Cao, 1997; Zhu, 1996; Guo *et al.*, 1994). Beside the mechanical integrity from the structural point of view there is another concern of optical coupling between the laser diode (LD) and fiber for this kind of packaging. Misalignment of the optical axes of the LD and the fiber lens has a most significant effect on the performance and stability of this device. Hence, the most crucial part of the whole packaging process is fiber-to-laser diode coupling and attachment. Some active as well as passive manual and automated alignment with laser welding is being developed to get the best possible coupling. One of the most promising methods is welding the fiber sub-mount by laser and applying lateral adjustment after determining post-weld shift (Kuang *et al.*, 2001). The simulated packaging process consists of a total of 15 steps (benchmark) including five soldering and five cooling steps, although lot of efforts have been made to minimize the process steps for producing an efficient optoelectronics device without compromising the reliability (Zhou *et al.*, 1992). The main purpose of this work is to demonstrate the capability and effectiveness of the finite element method to simulate the manufacturing of complex MEMS/photonic packaging. However, in this section we are only going to describe the radiation heat transfer simulation of the final packaging assembly process of parallel sealing (a type of spot welding) at a very high temperature. Radiation heat from the weld spots to the LD at this high temperature welding process is a great concern for this kind of hermetic packaging. It is important to know the contribution of radiation towards heat transfer during the parallel sealing process for hermetic sealing of the package. During sealing at 1400 °C/1673 K, if there is considerable amount of heat transferred from the module top to the LD, which is susceptible to high temperature, the LD might get damaged.

10.2.3 Radiation Heat Transfer Modeling for Hermetic Sealing Process

All the components and interconnects are assembled in the module package by the multiple steps flux less solder reflow process. At the end, the module needs to be hermetically sealed to protect the components from adverse environmental conditions. The sealing process is a solder-less metal to metal seam welding process at a very high temperature. Although this is a type of resistance welding that is accomplished at a high voltage, it is the heat generation at the weld spot that needs to be predicted to monitor the temperature rise of the delicate components, such as the LD. Therefore, we have considered using thermal loading directly instead of electrical loading during the seam welding. While the module being sealed that forms a closed cavity, cavity radiation model is required to analyze the heat transfer mechanism during this hermetic sealing process. We have simplified our original 3D packaging process to a 2D model.

Cavity radiation can be included in uncoupled heat transfers only without deformation. Cavity radiation is capable of two-dimensional, three-dimensional, and axisymmetric problems. Radiation surfaces symmetry, blocking, and motion within the cavity are considered and are recommended for considerably distant surfaces of the cavity.

The radiation effect is always nonlinear due to temperature dependent emissivity of the surfaces. Cavity radiation equations are not symmetric, therefore nonsymmetric matrix storage and solution schemes are used in models that include cavity radiation (ABAQUS/Standard User's Manual). These matrices are updated a number of times during the analysis and hence this is a computationally expensive procedure. The participating surfaces of the cavity must be defined to model the cavity radiation. The surface property and emissivity of the participating surfaces are defined and constant such as Stefan–Boltzmann constant is defined. The radiation symmetry or surface motion if any must be included. The boundary is prescribed as temperature and the radiation is controlled by view factor calculation. View factor matrices can be supplied as output along with other heat transfer outputs.

Cavities are formed as a collection of surfaces, which are composed of facets. In the case of a two dimensional element, a facet is a side of an element and in the case of a three-dimensional element, a facet is a face of a solid element or a surface of a shell element. A surface can be used to define one cavity only to avoid ambiguity. Each surface is assigned with associated properties such as emissivity.

It has been mentioned earlier that, by definition, radiation problems are nonlinear and it can be further nonlinear if temperature dependent surface emissivity is assigned. The emissivity used in the cavity radiation formulation is associated with radiation flux per unit area into a cavity facet as

$$q_i^c = \frac{\sigma\varepsilon_i}{A_i}\sum_j \varepsilon_j \sum_j F_{ij}C_{ij}^{-1}\left(\left(\theta_j-\theta^Z\right)^4-\left(\theta_i-\theta^Z\right)^4\right) \tag{10.1}$$

where A_i is the area of facet i seeing all cavity facets $j = 1, n$, ε_i, ε_j are the emissivity of the facets i, j, σ is the Stefan–Boltzmann constant; F_{ij} is the geometrical view factor matrix, C_{ij} is the reflection matrix, θ_i, θ_j are the temperature of facets i, j and θ_Z is the absolute zero on the temperature scale used (ABAQUS/Standard User's Manual; Glass, 1999). Emissivity is a non-dimensional quantity with a value between zero to one. A value of zero means a perfect reflective surface where all the radiation is being reflected by the surface and a value of one means a perfect black body where all the radiation is absorbed by the surface. In case of blackbody radiation in a cavity, the reflection is ignored but in a normal radiation case the reflection is always present.

Cavity radiation occurs when the surfaces of the model can see each other and, thus, exchange heat with each other by radiation. Such exchange depends on view factors that measure the relative interaction between the composing surfaces of the cavity. View factor calculation is rather complicated for anything but the most trivial geometries. ABAQUS offers an automatic view factor calculation capability for two- and three-dimensional cases as well as for axisymmetric situations. This capability can take into account general surface blocking (or shadowing) as well as the most common forms of radiation symmetry. The view factor calculation can also be automatically repeated a number of times throughout the analysis history (this is user-controlled) if cavity surfaces are moved in space causing the view factors to change (Hibbit *et al.*, 2002). This can be used in applications such as the simulation of manufacturing sequences where radiation view factors change during the simulation. The technology adopted in ABAQUS for the radiation view factor calculation is originally developed by Atomic Energy Authority of the United Kingdom (Glass, 1988; Glass, 1999; Johnson, 1987; Hibbit *et al.*, 2002).

10.2.4 Two-Dimensional FEA Modeling for Hermetic Sealing

We have simplified our original 3D module model to a 2D model (Figure 10.6). After all the components are soldered together by reflow soldering, the package is hermetically sealed by placing a cap on top and welded by a resistance seam welding process. During this welding process the temperature of the weld

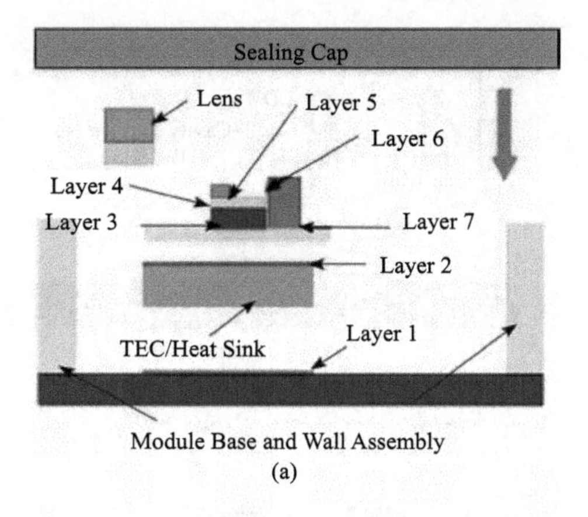

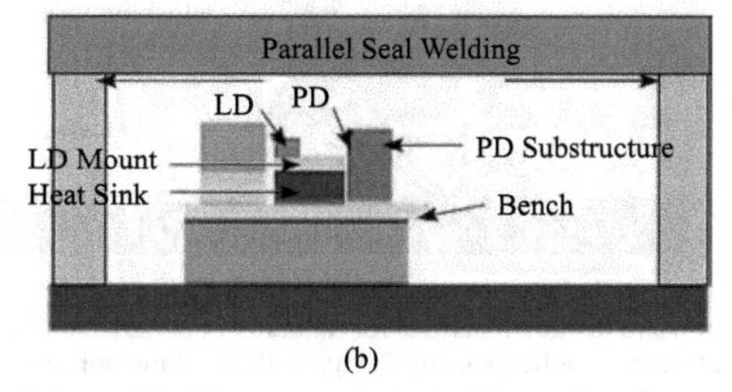

Figure 10.6 Simplified 2D module model with components and solder perform layers. (a) Layer for solder perform (thickness 10–100 micron). (b) Module is hermetically sealed

spots can reach around 1400 °C/1673.16 K. To simulate the radiation mechanism of heat transfer at this high temperature, we have modeled the closed cavity radiation problem. Since the package is sealed and no opening is allowed for maintaining hermeticity, it is a perfectly closed cavity problem without any consideration of radiation in the outside environment. The cavity is formed with a collection of surfaces of module wall, cap, and LD element facets, since we are interested in the radiation between these components (Figure 10.7). The module cap and wall is welded at a very high temperature (1673.16 K) and they become the source of radiation and the LD is the receiver of radiating heat energy. Different situations with different parameters such as heating area, cavity surface emissivity, and view factor blocking effect are simulated. 2D diffusive heat transfer elements DC2D3 and DC2D4 are used for cavity radiation simulations. The surface emissivity for cavity forming surfaces is assigned along with other material thermal properties.

(*i*) *Cavity Radiation Simulation Methodology–1*

At the first simulation, the module cap and wall corners were welded together by ramping up the temperature to 1673.16 K at 100 ms. The simulation was continued until the steady state condition. The steady state condition was defined by a temperature change rate of 2 °/s. The emissivity for wall-cap surface and LD surfaces are 0.5 and 1.0 respectively and the blocking surfaces are removed.

(*ii*) *Cavity Radiation Simulation Methodology–2*

For the second simulation, the whole cap surface was heated at the same magnitude of the first simulation. All other parameters were the same as the first simulation.

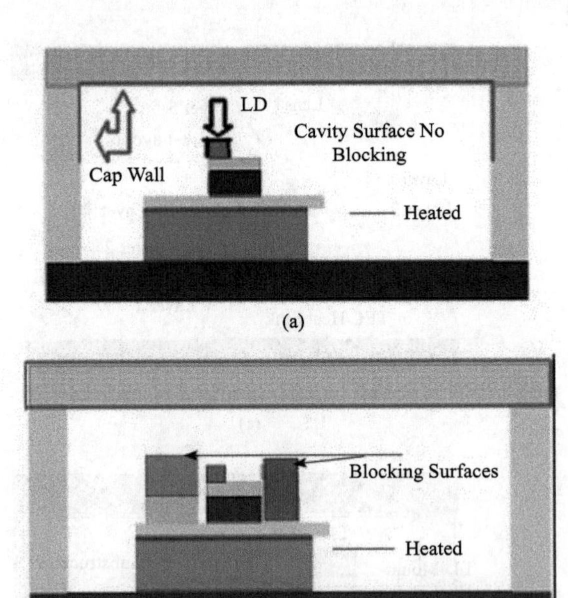

Figure 10.7 Cavity formation and heating locations for radiation heat transfer analysis. (a) Cavity with no blocking surfaces and only the cap wall joints are heated. (b) Cavity with blocking surfaces and whole cap is heated

(iii) Cavity Radiation Simulation Methodology–3

In the third case, the same parameters were taken as the second simulation except for the presence of the two blocking surfaces (Figure 10.8).

(iv) Cavity Radiation Simulation Methodology–4

The fourth simulation had the emissivity of the wall-cap surface changed to 0.8 from 0.5 and all other parameters are the same as the third simulation.

(v) Cavity Radiation Simulation Methodology–5

One cavity radiation process was simulated with a very high temperature (2800 °C/3073 K) to examine an extreme case. All other parameters were kept the same as the fourth simulation. In all simulations the initial condition was defined as the room temperature that is 298 K. In all of the methodologies conduction heat transfer mechanism was implemented along with radiation.

10.2.5 Cavity Radiation Analyses Results and Discussions

Our main objective of these simulations was to predict the temperature rise of the LD due to the radiation mechanism of heat transfer during the hermetic sealing process. The five simulation results in terms of an LD nodal temperature rise along with time were plotted (Figure 10.8). In the first cavity radiation simulation, the LD node was not heated at all. This confirmed that during the sealing at a high temperature spot at the cap and wall corner, the LD would have no temperature rise due to radiation even if the LD was a perfect black body. In the second simulation, at 100 ms the LD node

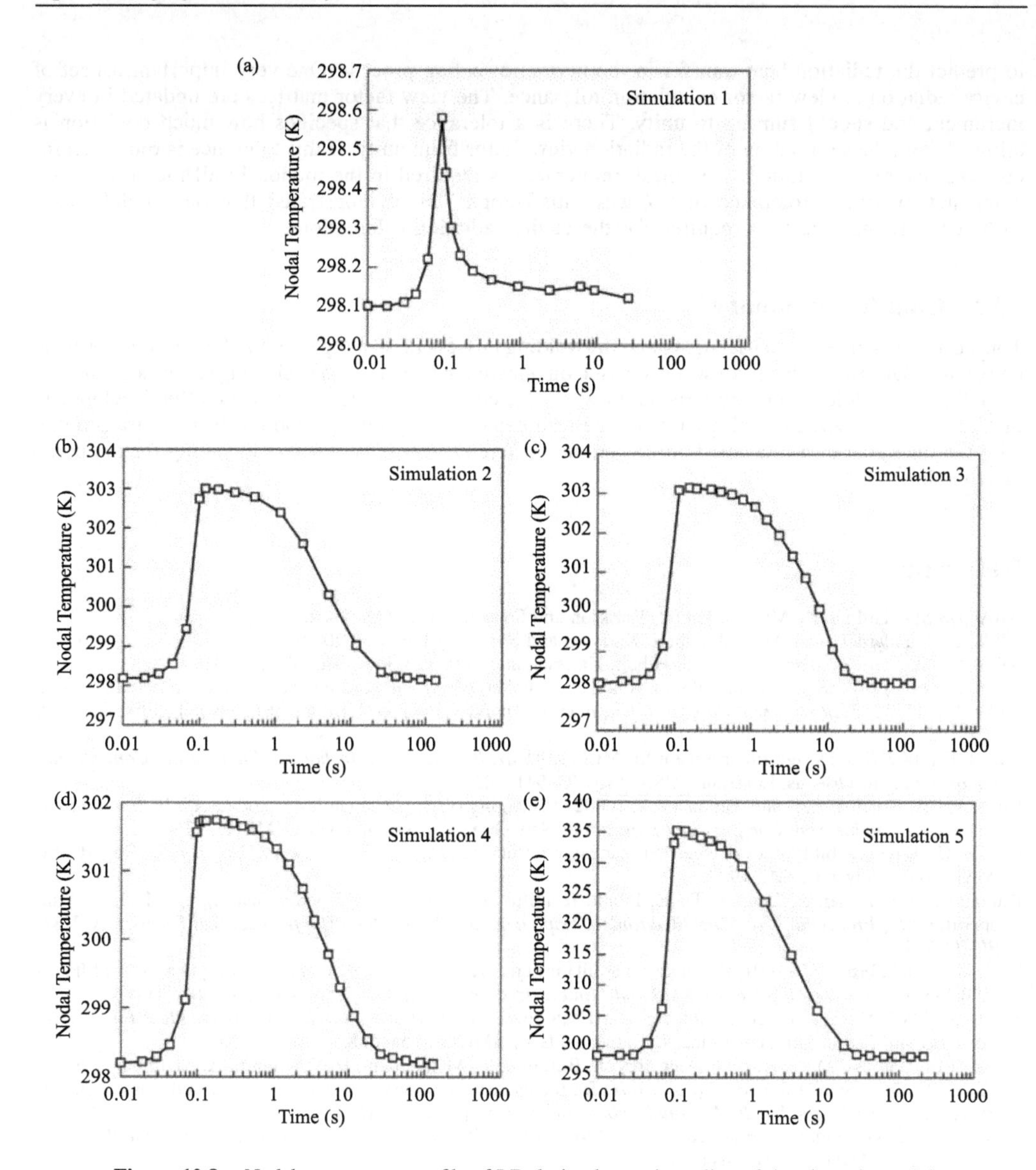

Figure 10.8 Nodal temperature profile of LD during hermetic sealing of the photonic module

temperature rose by almost 5 degrees. This proved the efficacy of our cavity radiation model. In the third simulation, the blocking surfaces are present. The LD is partially blocked by lens and PD substructure (Figure 10.8). The result indicated that even if the blocking surfaces were present, the LD nodal temperature did not change. This indicated that in micro scale, tiny surfaces like these did not block the radiation heat transfer. The effect of surface emissivity on cavity radiation analysis was shown in the fourth simulation. It was quite obvious that surface emissivity did affect the radiation heat transfer. The result indicated that the LD gets a temperature rise of around 3 K. This was considerably less than the second and third simulation. In the fifth simulation, which was an extreme case, the LD nodal temperature did rise almost by 37 K. That reiterated the effectiveness of our model to predict the radiation heat transfer in the hermetic sealing process. One very important aspect of cavity radiation is view factor calculation tolerance. The view factor matrices are updated in every increment and should sum up to unity. There is a tolerance that specifies how much deviation is allowed from the summation of the radiation view factor from unity. If this tolerance is too much for convergence of the solution, then mesh refinement is required in the model. In all the simulations from first to fifth, a tolerance of 0.7 was satisfactory. This demonstrated that our model was a sufficiently refined mesh as required for the cavity radiation calculations.

10.3 Chapter Summary

This chapter discusses opto packaging and the welding process for a pump laser module. A non-hermetic packaging platform has been constructed based on our silicon bench based packaging technology and the modeling methodology for cavity radiation was selected. Further work is focused on the development of a ceramic MiniDIL packaging case and their implementation on a high end receiver and transmitter, with the final goal of a low cost transceiver, and different process parameters modeling for the pump laser module.

References

ABAQUS/Standard User's Manual, Hibbit, Karlsson and Sorensen, Inc., CA, 2002.
ABAQUS/Standard Theory Manual, Hibbit, Karlsson and Sorensen, Inc., CA, 2002.
ABAQUS/Standard Benchmark Manual, Hibbit, Karlsson and Sorensen, Inc., CA, 2002.
Allan, R. (2000) MEMS packaging solutions open new market, *Electronic Packaging and Production* **40**(6):49–58.
Bathe, K.J. (1982) *Finite Element Process in Engineering Analysis*, Prentice-Hall, Inc., Englewood Cliffs, NJ, USA, 114–145.
Bart, S.F. (2002) The Design Environment for MEMS, *ASME International Mechanical Engineering Congress and Exposition*, New Orleans, Louisiana, USA, IMECE, 541–545.
Borgesen, P., Kondos, P.A. and Yamunan, V. (2002) Packaging of single mode laser diodes, *ASME International Mechanical Engineering Congress and Exposition*, New Orleans, Louisiana, USA, IMECE, 471–475.
Cao, W. (1997) A reliability study of electronic packaging structures, MS Thesis, Mechanical Engineering Department, Wayne State University, MI, 1997.
Dautartas, M.F., Fisher, J., Lou, H., Datta, P. and Jeantilus, A. (2002) Hybrid optical packaging, challenges and opportunities, *Proceedings of 52nd Electronic Components and Technology Conference*, San Diego, CA, USA, ECTC, 787–793.
Ehrfeld, W. and Ehrfeld, U. (2001) Progress and profit through micro technologies: commercial applications of MEMS/MOEMS, *Proceedings of SPIE MOEMS and Miniaturized Systems II*, San Francisco, CA, USA, xi–xix.
Finch, N. (2002) Design to succeed: integrated MEMS development, *ASME International Mechanical Engineering Congress and Exposition*, New Orleans, Louisiana, USA, IMECE, 555–558.
Glass, R.E., Burgess, M., Livesey, E., Geoffrey, J., Bourdon, S., Mennerdahl, D., Cherubini, A., Giambuzzi, S. and Nagel, P. (1988) Standard thermal problem set for the evaluation of heat transfer codes used in the assessment of transportation packages, *Sandia National Laboratories*, Albuquerque, NM.
Glass, R.E. (1999) Standard thermal problem set, *Proceedings of the Ninth International Symposium on the Packaging of Radioactive Materials*, 275–282.
Guo, Y. and Lim, C.K. (1994) Hybrid Method for strain/stress analysis in electronic packaging using moiré interferometer and FEM, *Proceeding of 1994 SEM Spring Conference and Exhibits*, 321–327.
Johnson, D. (1987) Surface to surface radiation in the program tau, taking account of multiple reflection, *United Kingdom Atomic Energy Authority Report* ND-R-1444 (R).
Kopola, H., Lenkkeri, J., Kautio, K., Trokkeli, A., Rusanen, O. and Jaakola, T. (2001) MEMS sensor packaging using LTCC substrate technology, *Proceedings of SPIE, Device and Process Technology for MEMS and Micro-technology*, Adelaide, SA, Australia 148–158.
Kuang, J.H., Sheen, M.T., Wang, S.C., Wang, G.L. and Cheng, W.H. (2001) Post weld shift in dual in line laser package, *IEEE Transaction on Advanced Packaging* **24**(1):81–85.
Liu, S. (1992) *Damage Mechanics of Cross-Ply Laminates Resulting from Transverse Concentrated Loads*, Ph.D. Thesis, Mechanical Engineering Department, Stanford University, CA, 1992.
Liu, S., Zhu, J.S., Hu, J.M. and Pao, Y.H. (1995) Investigation of crack propagation in ceramic/adhesive/glass system, *IEEE Transactions on Components, Packaging, and Manufacturing Technology, Part A* **18**(3):627–633.

Liu, S., Mei, Y. and Wu, T.Y. (1995) Bimaterial interfacial crack growth as a function of mode-mixity, *IEEE Transactions on Components, Packaging, and Manufacturing Technology, Part A* **18**(3):618–626.
Miller, M.P. (1993) *Getting Started with MSC/NASTRAN User's Guide*, The MacNeal-Schwendler Corporation, LA, CA, USA, 2–3.
Streetman, B.G. and Banerjee, S. (2001) *Solid State Electronic Devices*, 5th edn., Prentice-Hall, Englewood Cliffs, NJ, USA, 379–396.
Verdiell, J.-M., Kohler, R., Epitaux, M., Finto, M., Kirkpatrick, P., Lake, R., Colin, S., Mader, T., Bennett, J., Yao, J., Zbinden, E., Buchheit, S. and Walker, J. (2002) Automated opto-electronic packaging for 10 Gb/s transponders, *Proceedings of 52nd Electronic Components and Technology Conference*, San Diego, CA, USA, ECTC, 808–810.
Zhu, J.S. (1996) *Modeling and Validation of Thermally-Induced Failure Mechanisms in Microelectronic Packaging*, Ph.D. Thesis, Mechanical Engineering Department, Wayne State University, MI, 1996.
Zhou, W.X., Hsuing, C.H., Fulton, R.E., Yin, X.F., Yeh, C.P. and Wyatt, K. (1997) CAD-based analysis tools for electronic packaging design, *ASME Advances in Electronic Packaging*, Kohala Coast, Hawaii, 917–979.

11

MEMS and MEMS Package Assembly

Micro-electrical-mechanical- systems (MEMS) or microsystems have been widely used. However, the development is uneven in different parts of the world due to the complexity in chip manufacturing processes, application specific integrated circuits (ASIC), chip and packaging design, packaging processes, application specific testing, and reliability, and durability testing. Sustained development has been a challenging even for those multinational companies, which has been shown by limited products for each company, as compared to those regular IC products. With the development trend of system-in-packaging technology and more than Moore technology, it is highly desirable that major processes and reliability must be modeled in MEMS and MEMS packaging so that a modulized development can be realized in the future. With this goal in mind, in this chapter, efforts are made to choose some typical examples that our group and some colleagues in the community have researched to show those important issues in MEMS and MEMS packaging. Piezoresistive pressure sensors, thermo-fluid based accelerometer, the capacitance based accelerometer, and TPMS sensor and its application to vehicles, thermo-fluid gyroscope, thermo based air flow sensor, microjet device and associated cooling system, micro-channel based DEP separation device, and capsule endoscope microsystem are examples to be presented. Packaging and testing have been regarded as being important bottlenecks for the commercialization of MEMS and model based chip and packaging design may help overcome the difficulties sooner or later, accelerating the wide spread applications of MEMS devices in our daily.

11.1 A Pressure Sensor Packaging (Deformation and Stress)

Piezoresistive transducer (PRT) pressure sensor utilizing piezoresistance of semiconductor and micro-machined diaphragm structure to sensor pressure level has been successfully applied in automotive, process control, aerospace, medical, and consumer sectors for decades. Recently, a low pressure sensor application in the automotive industry, for instance fuel tank pressure transducer (FTPT) sensor, has grown quickly. Low pressure sensing level, wide operating temperature range (−40 to 125°C) and chemically harsh environments require sensors to be precise and reliable. The thermal shock durability and chemical resistance are considerable important reliability issues to the packaging process and die-bonding material. The sensing precision, including temperature zero shift, pressure sensitivity, temperature and pressure hysteresis, as well as thermal drift is controlled by die design, micromachining processes, wafer bonding/packaging process, and sensor packaging process. Since the Wheatstone bridge circuit, commonly used in PRT design, can compensate for thermal resistance change bonding and die attachment become major issues in temperature zero shift, hysteresis, and drift. The packaging-induced

Modeling and Simulation for Microelectronic Packaging and Integration: Manufacturing, Reliability and Testing, Second Edition. Sheng Liu and Yong Liu.

stress and thermal zero shift can be determined experimentally and calculated by the finite element method (FEM) (Nysæther *et al.*, 1998; Kelly *et al.*, 1996; Jaeger *et al.*, 1994). This study applies FEM coupled with experimental data to predict PRT output performance.

11.1.1 Piezoresistance in Silicon

For a p-type silicon crystal in the <110> direction, the piezoresistance change with the applied stress and temperature is expressed as:

$$\frac{\Delta R}{R} = \pi_l \sigma_l + \pi_t \sigma_t + \alpha \Delta T \tag{11.1}$$

where σ_l and σ_t are longitudinal and transversal stress components with respect to the current, α is the temperature coefficient of resistivity, and ΔT is the temperature change, π_l and π_t are the longitudinal and transversal piezoresistive coefficients given by:

$$\pi_l = 0.5\,(\pi_{11} + \pi_{12} + \pi_{44}) \tag{11.2}$$

$$\pi_t = 0.5\,(\pi_{11} + \pi_{12} - \pi_{44}) \tag{11.3}$$

The piezoresistive coefficients are sensitive to temperature, doping level, conductivity type, and orientation of resistors. A commonly used value for π_{11} is 3×10^{-12} cm^2/dyn and π_{12} is -1×10^{-12} cm^2/dyn (Clark *et al.*, 1979). Temperature and doping level dependent π_{44} was reported by Tufte and Stelzer (Tufte *et al.*, 1963), and are expressed as following for the lowest doping level:

$$\pi_{44} = (118.93 - 0.321T + 0.00097T^2) \times 10^{-12}\text{cm}^2/\text{dyn} \tag{11.4}$$

Figure 11.1 illustrates the schematic cross section of PRT sensor. Figure 11.2(a) shows the internal Wheatstone bridge structure. Referring to Figure 11.2(b), the voltage output of the Wheatstone bridge normalized by excitation input voltage is defined as following:

$$\frac{\Delta V}{V_C} = \frac{R_1 R_4 - R_2 R_3}{(R_1 + R_2)(R_3 + R_4)} \tag{11.5}$$

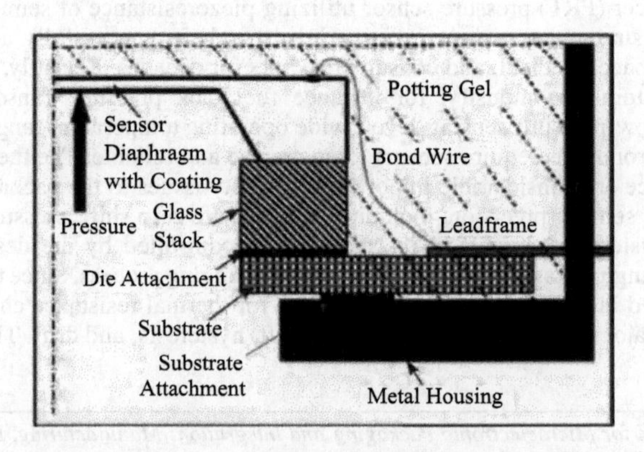

Figure 11.1 Schematic of pressure sensor assembly

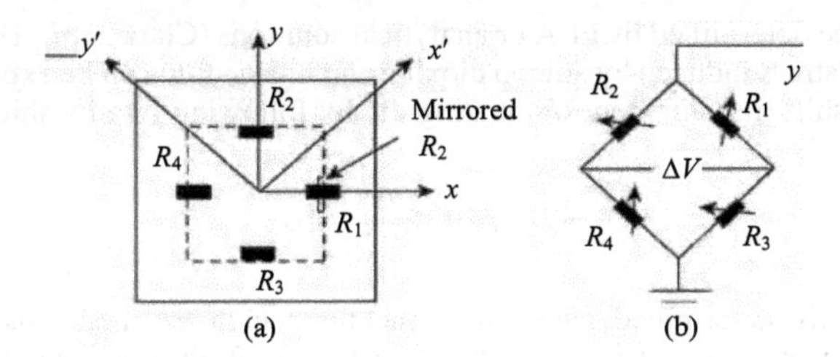

Figure 11.2 P-type piezoresistor configuration on a membrane. (a) Resistors R1 and R4 are stressed longitudinally while R2 and R3 are transversely. R2 is mirrored to R1 location when a 1/8 model is defined. (b) Wheatstone bridge configuration

where:

$$R_i = R_{i0}\left[1+\frac{\Delta R}{R_{i0}}\right], \quad i = 1,2,3,4 \tag{11.6}$$

Assume R_{i0} is equal, $\Delta R_1 = \Delta R_4$, $\Delta R_2 = \Delta R_3$ then Equation 11.5 becomes:

$$\frac{\Delta V}{V_C} = \frac{(\Delta R_1/R_1)-(\Delta R_1/R_2)}{2+(\Delta R_1/R_1)+(\Delta R_1/R_2)} \tag{11.7}$$

Considering axially (x-axis) and diagonally (x'-axial) symmetrical, a square diaphragm model can be simplified to be a one-eighth model as shown in Figure 11.2(a). Submitting Equations 11.1 to 11.6 into Equation 11.7 and converting all stresses to x and y components, one could obtain:

$$\frac{\Delta V}{V_C} = \frac{\pi_{44}\left(\sigma_x-\sigma_y+\sigma_\perp^{\mathrm{p}}-\sigma_{11}^{\mathrm{p}}\right)}{2+(\pi_{11}+\pi_{12})\left(\sigma_x+\sigma_y+\sigma_\perp^{\mathrm{p}}+\sigma_{11}^{\mathrm{p}}\right)} \tag{11.8}$$

where σ_x and σ_y are thermal stress components induced by material thermal mismatch respectively. Equation 11.8 suggests that the bridge output $\frac{\Delta V}{V_C}$ is much more sensitive to stress term $(\sigma_x-\sigma_y)$ than $(\sigma_x+\sigma_y)$. Since both π_{44} and $(\sigma_x-\sigma_y)$ are temperature dependent, the bridge output $\frac{\Delta V}{V_C}$ highly relies on temperature change. Expressing $\frac{\Delta V}{V_C}$ as a two-order polynomial function about temperature variable, the second-order term of this function is the second-order temperature coefficient of offset voltage (ZeroTc^2). The value of ZeroTc^2 indicates the nonlinearity level of voltage offset. The terms of $\sigma_\perp^{\mathrm{p}}$ and σ_{11}^{p} are pressure-contributed stresses perpendicular and parallel to the diaphragm edges respectively.

It is noted that when the thermal stresses and pressure stresses are not extremely large. Equation 11.8 could be simplified as:

$$\frac{\Delta V}{V_C} = \frac{\pi_{44}\left(\sigma_x-\sigma_y+\sigma_\perp^{\mathrm{p}}-\sigma_{11}^{\mathrm{p}}\right)}{2} \tag{11.9}$$

if measuring $\frac{\Delta V}{V_C}$ at elevated pressure load, under unchanged temperature, π_{44} would be estimated at certain temperature levels from following equation:

$$\pi_{44} = 2\frac{\Delta V}{V_C}/\left(\sigma_\perp^{\mathrm{p}}-\sigma_{11}^{\mathrm{p}}\right) \tag{11.10}$$

where $(\sigma_{\perp}^{p} - \sigma_{11}^{p})$ can be determined by FEA or analytical solutions (Clark *et al.*, 1979). In a similar way, the overall differential stress induced by silicon dioxide and nitride film can be experimentally estimated through thermal zero shift measurement on floating die by following relationship:

$$\sigma_x - \sigma_y = 2\frac{\Delta V}{V_C}/\pi_{44} \tag{11.11}$$

Based on experimentally determined parameters, the finite element model usually predicts sensor packaging performance precisely. It is noted that Equation 11.9 only presents the external influence on bridge output. Therefore, all internal factors, such as resistance variation of piezoresistors, non-symmetrical piezoresistance change, and internal circuit error, are neglected from the equation.

11.1.2 Finite Element Modeling and Geometry

Microfabrication processes, including IC fabrication, wafer bonding/packaging processes, and sensor packaging process combine several materials into one unit (Madou, 1997). Because different thermal expansion capabilities of these materials, thermal stresses are induced in the system. The finite element model is an efficient method to calculate the average thermal stresses at the piezoresistor area. To do so, a one-eighth x-axial and diagonal symmetric three-dimensional finite element model was developed, shown in Figure 11.3. The model consists of an 8-node incompatible quadratic brick element to provide reasonable accurate average stresses.

In the finite element analysis, a 1000 Å silicon dioxide film and 500 Å silicon nitride film are added into the model at 1000 °C and 450 °C respectively. Since silicon material is an elastic material, silicon, dioxide, and silicon nitride forming processes and the film thermal stress are pre-analyzed. After performing thermal stress analyses due to Pyrex glass stack and die-attachment, the film stresses are included into the final results during the post-process. This technique reduces computation time significantly.

11.1.3 Material Properties

The temperature-independent material properties used in the analysis are listed in Table 11.1. The temperature-dependent thermal expansion coefficients for silicon and Pyrex 7740 are plotted in Figure 11.4. The viscoplasticity and hyperelasticity are assumed for epoxy and silicone adhesive, respectively.

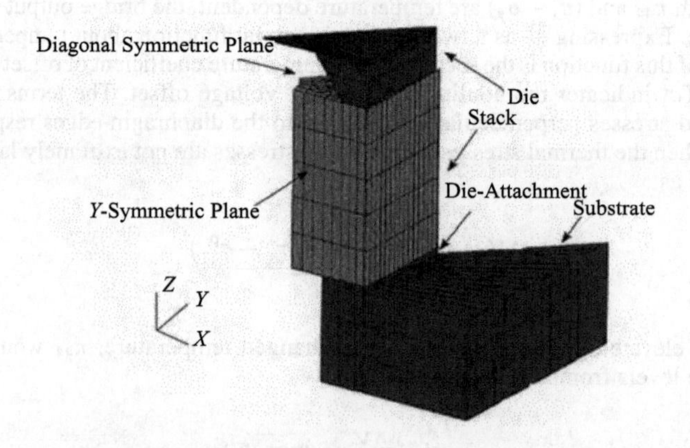

Figure 11.3 3D FEM of pressure sensor assembly

Table 11.1 Material properties for finite element analysis

Material	Young's Modulus (GPa)	Poisson's Ratio	CTE (ppm/°C)	Reference
Silicon	$C_{11} = 165.7$ $C_{12} = 63.9$ $C_{44} = 79.56$		Figure 11.4	(Brantley, 1973)
Pyrex 7740	62.0	0.2	Figure 11.4	(Corning Glass, 1988)
SiO_2	66.0	0.18	0.28	(Jaccodine *et al.*, 1966)
Si_3N_4	270.0	0.27	1.6	(Sinha *et al.*, 19789; Retajczyk *et al.*, 1980)
Al_2O_3 (96%)	310.0	0.25	6.3	
Epoxy #	3.9	0.3	52	
Adhesive	0.001	0.49	3000	

11.1.4 Results and Discussion

Figure 11.5 presents a comparison of thermal stresses induced by silicon dioxide/nitride film and Pyrex glass stack, as well as the total stresses. Although the film-induced thermal stresses are not very high, they still influence nonlinearity. Case studies suggest that the thickness variation could change the total stresses and bridge output.

Figure 11.6 illustrates the comparison of FEA results and experimental data for various sensor element design groups. In the first group, bare die is directly bonded to substrate by silicone adhesive. Another method is for a die wafer to be anodically bonded to the Pyrex glass stack which is then bonded to the substrate by the same bonding material. One can find that the finite element models are comparable to experimental results. It is noted that the experimental data shown in Figure 11.6 are averaged values. The variations of the experimental data may be caused by many factors, such as film thickness variation, doping level variation, packaging, process variation, measurement errors, and so on.

Figure 11.7 shows the effects of stack thickness, stack materials, and bonding materials on the output nonlinearity. It is a typical modeling application in sensor element design. In this study, two different stack material (silicon and Pyrex 7740), two different die-attachment (silicone adhesive and epoxy), as well as

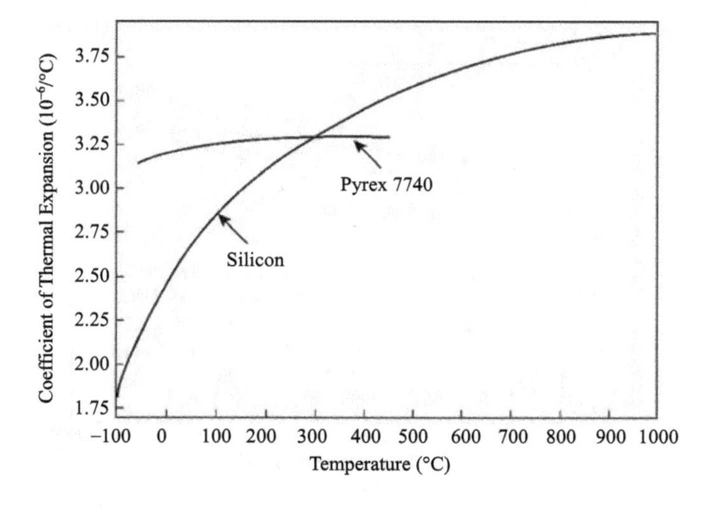

Figure 11.4 Total thermal expansion coefficients for silicon and Pyrex 7740. The reference temperature is 20 °C for silicon and 25 °C for Pyrex 7740 (Lyon, 1977; Okada, 1984; Roberts, 1981; Corning Glass, 1988)

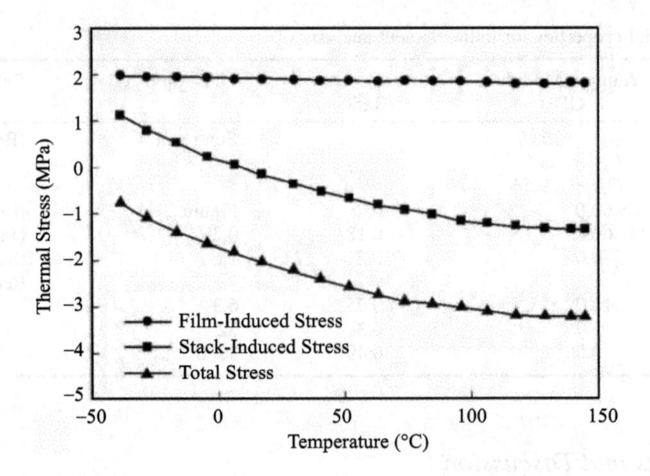

Figure 11.5 Film induced thermal stresses and Pyrex stack induced thermal stresses are calculated separately and then combined into total thermal stresses to reduce computation time

various stack thicknesses (0.05 mm through 3.0 mm) are considered. Since silicone adhesive makes a slight contribution to nonlinearity, the nonlinearity curve for a silicon stack with soft die-attachment indicates that most thermal stresses come from film coating and 0.004 mm sputtering glass bonding line, while another one for Pyrex stack with the same die-attachment presents the thermal stresses resulting from film coating and glass anodic bonding. Therefore, Pyrex stack affects nonlinearity significantly as its thickness increases. However, when the thickness of the Pyrex stack is over 1.8 mm, the Pyrex stack induced nonlinearity becomes a constant. On the other hand, epoxy also dramatically influences nonlinearity if the thickness of stack is less than 2.0 mm. It must be noted that the effect of hard die-attachment relies on the material stiffness. The softer the die-attachment is, the lower the stack height is required.

Figures 11.8 and 11.9 illustrate PRT sensor output drift history and hysteresis loop over a temperature range of −40 to 125 °C under different bonding conditions. The viscoelasticity and viscoplasticity of the

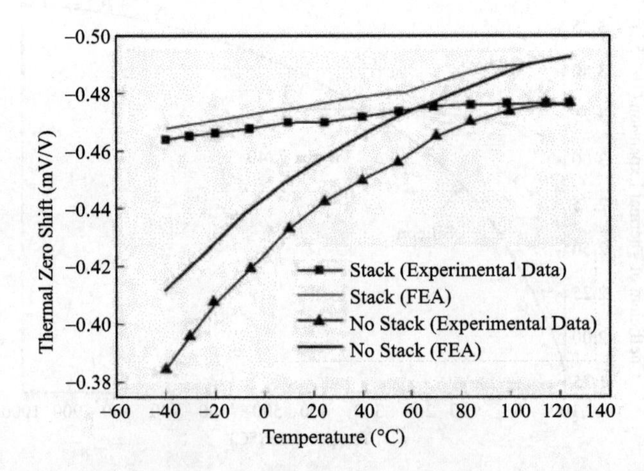

Figure 11.6 Comparison of thermal zero shift between experimental data and FEA for sensing element with of without Pyrex 7740 stack when soft die-attachment is applied. FEA results are comparable to experimental data

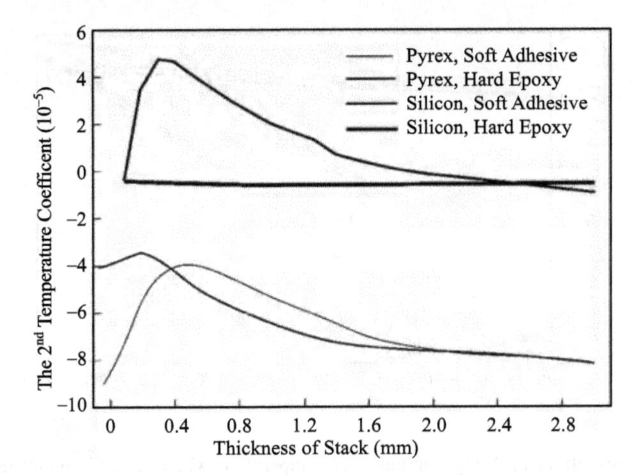

Figure 11.7 Second temperature coefficient of offset voltage presents nonlinear level than silicon stack. Hard epoxy nonlinearity effect decreases with increasing stack thickness. Since film coating and sputter glass bonding line induce thermal stresses, second temperature coefficient will not reach zero at non-stack points

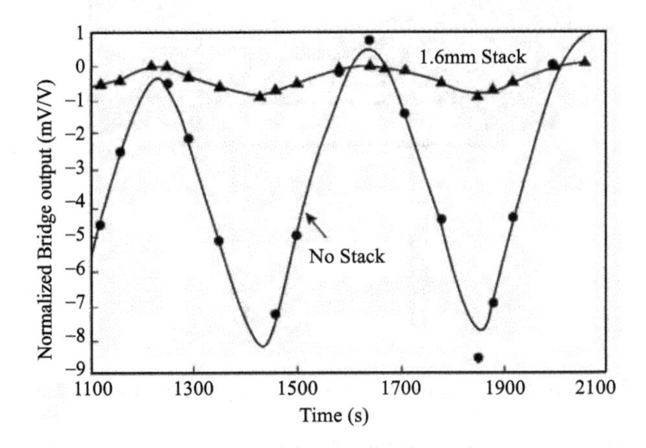

Figure 11.8 Bridge outputs drift over time during thermal cycling test when epoxy bonding materials are applied. However, increasing stack height can reduce drifting effect. The stack material is Pyrex 7740

bonding materials results in hysteresis and drift errors of sensor output. In general, when bonding material becomes flexible or Pyrex stack height increases, the hysteresis and drift efforts would decrease. The critical Pyrex stack height is about 1.6 mm for most epoxy bonding materials. Under a certain sensor element design configuration, this output prediction model is helpful to select bonding materials.

11.2 Mounting of Pressure Sensor

11.2.1 Mounting Process

Because of the non-corrosive environment which the pressure sensor works in, a metal package with silicon gel coating on the chip and wire is employed. Figure 11.10 illustrates the configuration of the packaged pressure sensor.

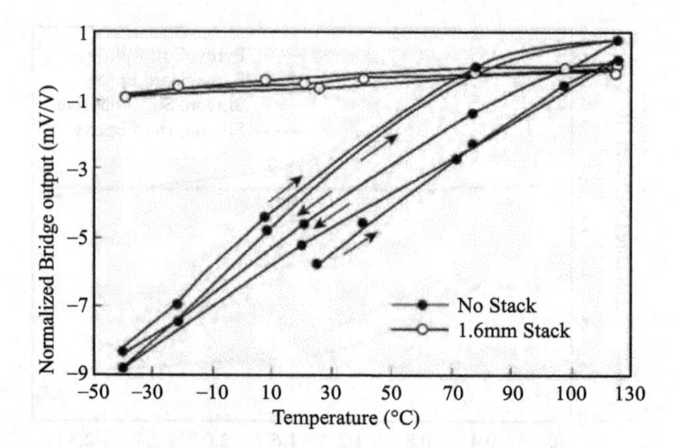

Figure 11.9 Hysteresis effect on bridge output over temperature when epoxy bonding materials are applied. However, increasing stack height can reduce hysteresis effect. The stack material is Pyrex 7740

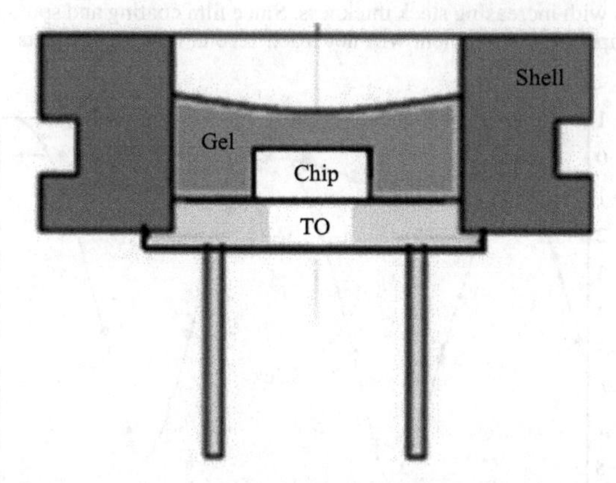

Figure 11.10 Cross section of the packaged pressure sensor

In order to obtain the correct pressure value of the test position, the pressure transducer must be mounted in the right position by a certain method. Figure 11.11 shows a cross-section of the mounted transducer. When being mounted, the pressure sensor is packed between the gasket and the press ring, and the press ring is an interference fit with the mounting shell, thus the pressure sensor is pressed tightly onto the gasket and the hermetic requirement is satisfied. Therefore, the pressure sensor can be connected well with the medium to be tested by the mounting shell through its external thread connector.

The press ring applies a great force on the pressure sensor shown in Figure 11.12, and this force is transferred onto the diaphragm of the pressure chip through a certain path, the internal stress varies, hence the zero point output changes. Different sensors suffer from different mounting stresses when being mounted. In addition, all the signal conditions are completed before being mounted, so that the zero point shifts could not be solved by signal conditioning any more.

11.2.2 Modeling

The coating material on the chip and wire is selected to be a very soft gel whose modulus is only about 110 MPa. It has little effect on the performance of pressure sensor. Thus the coating material is ignored for simplicity when modeled in this research.

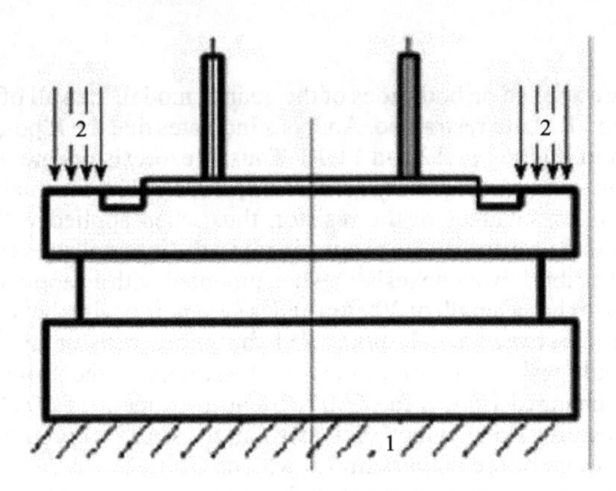

Figure 11.11 Force analysis of pressure sensor when being mounted

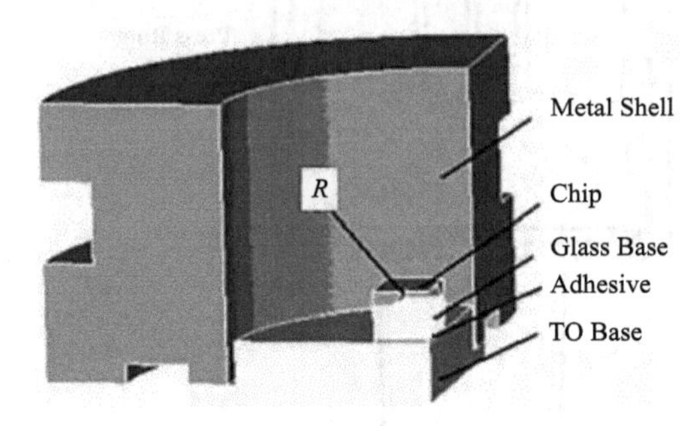

Figure 11.12 1/4 model of pressure sensor

As shown in Figure 11.12, the sensor chip, which consists of a sensor die and glass base, has dimensions of 3.45 mm × 2.7 mm × 0.4 mm with a 1.2 mm × 2.2 mm × 0.03 mm diaphragm. The thickness of the adhesive is 0.1 mm. The inner diameter of the metal shell is 10 mm. The material properties used in the modeling are listed in Table 11.2.

Table 11.2 The material properties

Part	Material	Modulus (GPa)	Poisson Ratio
Metal Shell	304	194	0.28
TO Base	Kovar	138	0.37
Adhesive	Epoxy	3.56	0.33
Glass Base	Pyrex 7740	62	0.2
Chip	Silicon	127	0.28
Ceramic Base	Alumina	310	0.25

11.2.3 Results

Symmetric boundaries are applied on both sides of the quarter model, then all of the degrees of freedom of surface 1 shown in Figure 11.11 are restrained. Analysis indicates that the R point has the maximum stress of the diaphragm shown in Figures 11.12 and 11.13. Thus, piezoresistors are usually positioned thereby in order to obtain the maximum sensitivity (Ah *et al.*, 2001). The internal stress of the diaphragm is proportional to the relative changes of the resistor, thus when applied with a certain excitation the output voltage of the pressure sensor is also proportional to the internal stress (Song *et al.*, 2003; Zarnik *et al.*, 2004). The stress distribution of one resistor when mounted with an applied force of 3000 N is shown in Figure 11.14. The resistor has a length of 300 μm and is symmetric with the R point. Because of the little difference of stress value between the R point and the entire resistor, the R point is taken as the representation of the whole resistor for simplification. FEA obtains the mounting stress of the R point when the applied force changed from 0 to 3000 N by an increment of 500 N, shown by curve S1 in Figure 11.15. S1 has a negative slope, which indicates that the stress is compressive stress at the R point, and the deformation direction of the diaphragm is opposite from that of the sensor chip under practical working condition. As a result, mounting stress makes the zero point output decrease.

The finite element analysis results shown as Figure 11.16 indicate that the mounting stress is transferred by the following path: metal shell–TO base adhesive-glass base-diaphragm of chip.

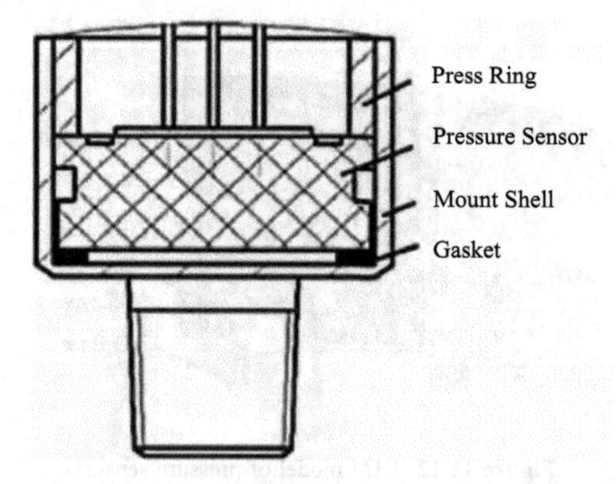

Figure 11.13 Local cross-section of pressure transducer

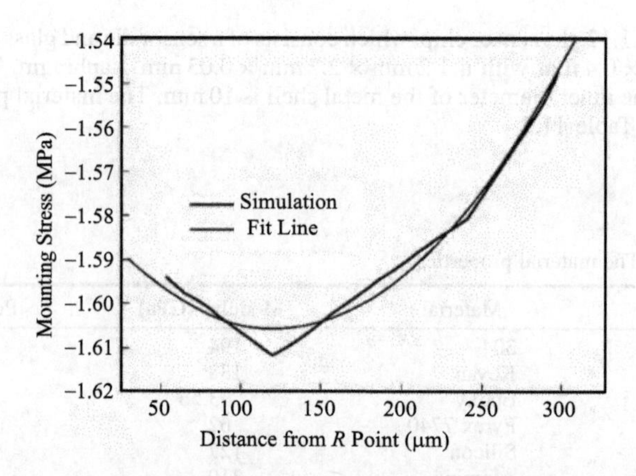

Figure 11.14 Simulation stress of one resistor

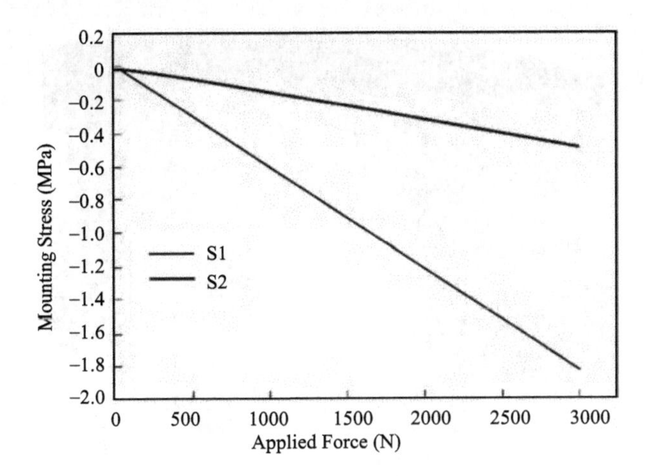

Figure 11.15 Simulation results of mounting stress

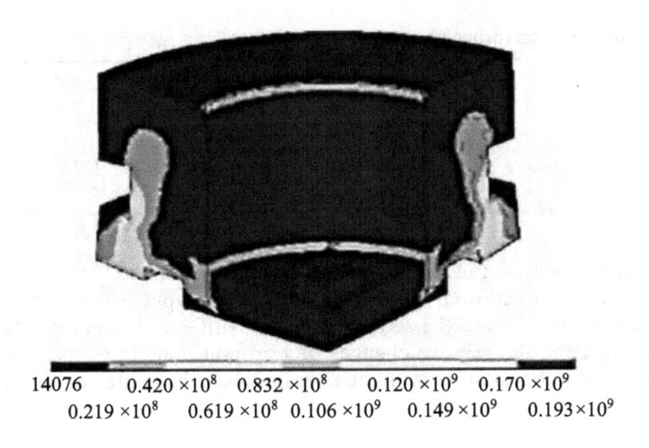

Figure 11.16 Mounting stress distribution

A new package configuration is adopted to minimize the effect of the mounting stress on the diaphragm of pressure sensor. A ceramic base is employed for the desired improvement based on both the demands of customers and reliabilities of pressure transducer. A ceramic base is adhered on the TO base, and then the sensor is adhered in the groove of the ceramic base. The FEA results of the new configuration (Figure 11.17) are shown by curve S2 in Figure 11.15. S1 and S2 indicate that the mounting stress of the new configuration becomes only a quarter of the former one. The reason is that the transfer path becomes longer, and in addition, the ceramic base has a good effect of stress isolation due to its high rigidity and smaller material mismatch to the chip.

11.2.4 Experiments and Discussions

Mounting experiments are conducted to verify the FEA results. A force of 3000 N is applied to these sensors with increments of 500 N by a programmed high-frequency fatigue testing machine, and the zero outputs are tested at the same time. Table 11.3 lists the experiment results.

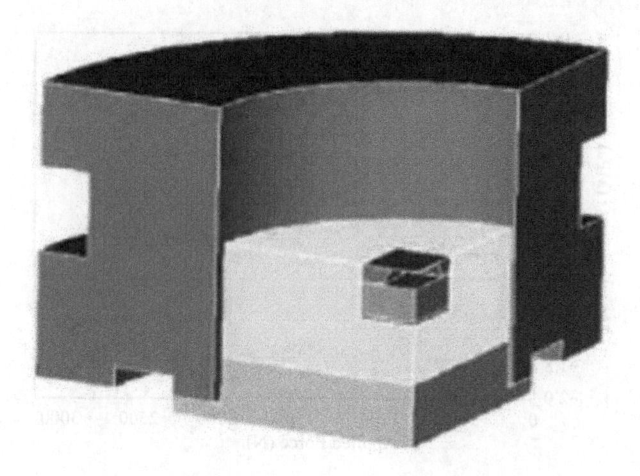

Figure 11.17 Quarter model of amended sensor

Table 11.3 Zero output of samples when being mounted (mV)

Force (N)		0	500	1000	1500	2000	2500	3000
Zero Output	1	3.82	3.73	3.68	3.65	3.64	3.63	3.62
	2	8.00	7.95	7.91	7.89	7.86	7.84	7.83
	3	9.22	9.20	9.19	9.18	9.18	9.18	9.18
	4	6.74	6.73	6.72	6.72	6.72	6.72	6.72

Because of the mismatched parameters of the fabrication process, the four bridge resistors are not consistent with each other. In addition, the test of zero point output is conducted in the atmosphere environment. The above reasons make the zero point outputs different from each other. In order to make the differences more evident, the relative changes of zero point output of these samples are studied. Figure 11.18 illustrates that the samples without a ceramic base have greater zero point shifts with an

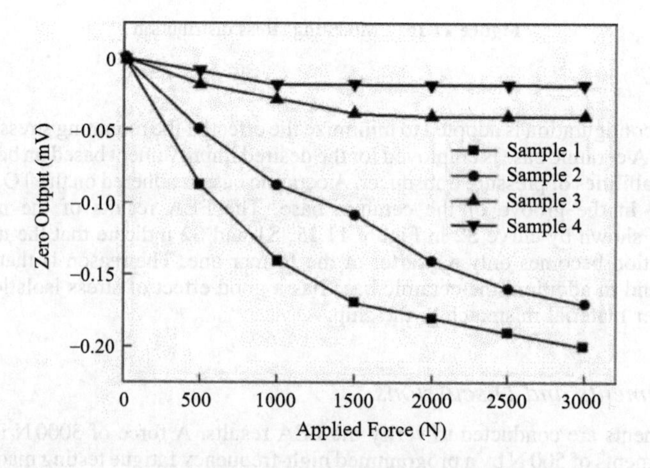

Figure 11.18 Results of mount experiment

average of 0.18 mV, but sample three and sample four which have ceramic bases only have a shift with 0.03 mV, which decrease by five-sixths as compared from former ones.

Compared with the simulation results shown as Figure 11.15, experiment results curve shown as Figure 11.18 seem to be nonlinear. Refined models will need to be established. However, the trends of both can be found to be consistent with each other.

11.3 Thermo-Fluid Based Accelerometer Packaging*

Demands for low-cost and high-performance accelerometers have been increasing in many fields including the automobile industry, navigation systems, the military industry, robotics systems, consumer electronics, and toys (Song *et al.*, 1998). Many efforts have been made in developing micromachined accelerometers in recent years to meet the cost and performance requirements (Yazdi *et al.*, 1998; Kim *et al.*, 1995; Nemirovsky *et al.*, 1996; Kubena *et al.*, 1996; Dauderstadt *et al.*, 1995; Bochobza-Degani *et al.*, 2000).

So far, the common designs of micromachined accelerometers involve solid proof mass, which is allowed to move under accelerating conditions. The existence of the proof mass brings some disadvantages to the accelerometers. Firstly, the ability to resist shock declines, the overload range cannot be wide. Secondly, the fabrication is complex and not suitable for IC techniques, consequently the size cannot be very small and the start-up cost cannot be reduced. Thirdly, the motion sensor of solid mass proof suffers from some problems, for example capacitive sensing, the most common sensing method, suffers from the electromagnetic interference and the influence of parasitic electrostatic force (Bochobza-Degani *et al.*, 2000).

Recently, a novel concept and device structure for acceleration measuring were developed by Dao *et al.* The Operation of the accelerometer is based on free convection of a tiny hot air bubble in an enclosed chamber (Dao *et al.*, 1996). It does not require solid proof mass and is compact, lightweight, inexpensive to manufacture, and sensitive to small acceleration. Leung *et al.* reported the implementation of the device structure by bulk silicon fabrication, the test for the device under natural gravity demonstrates that its sensitivity can reach 0.6 mg (Leung *et al.*, 1997). Milanovic *et al.* fabricated two kinds of the convective accelerometers, thermopile and thermistor types in standard IC techniques. Their accelerometers exhibit some significant advantages such as low cost, miniaturization, integration, and good frequency response (Milanovic *et al.*, 2000).

Linearity is an important index for accelerometers. Good linearity will offer much convenience to applications. Since the convective accelerometer is convection-based, the heat and mass transfer in the device are complicated. The following problems will come forth before us, whether the output of the convective accelerometer can be linear with acceleration/Grash of number or not, under what conditions can a good linearity be obtained?

In this section, numerical simulations are conducted for studying the factors affecting the linearity of convective accelerometer and optimizing the working parameters. According to the analyses, a kind of micromachined convective accelerometer is fabricated. It is noted that materials in this section are the courtesy of Dr. Luo *et al.* (2003).

11.3.1 Device Structure and Operation Principle

Figure 11.19 shows the device structure of a micromachined convective accelerometer. It includes an electric heater and two temperature sensors mounted within a sealed enclosure containing a kind of gas. The sealed enclosure aims to prevent environmental air flow from disturbing the device's operation. The temperature sensors are positioned symmetrically on both sides of the microheater. The microheater is used to heat the gas and therefore to create a free convection when acceleration is applied to the gas. The heater temperature and gas temperature distributions along the X axis are illustrated in Figure 11.20. In the

*Reprinted from *Microelectronic Engineering*, **65**, X.B. Luo, Z.X. Li, Z.Y. Guo and Y.J. Yang, "Study on linearity of a micromachined convective accelerometer," 87–101, 2003, with permission from Elsevier.

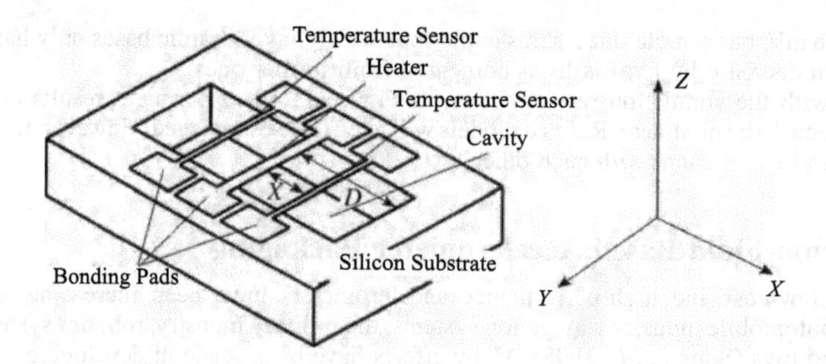

Figure 11.19 Device structure of the micromachined convective accelerometer

case of no acceleration or the acceleration normal to the *X* axis of the accelerometer, the flow pattern and gas temperature distribution are symmetrical as the dash lines show due to the symmetric locations of two temperature sensors. As a result, no temperature difference exists between two temperature sensors no matter how high the heater temperature is. As long as the acceleration along the *X* axis appears, the free convection along the *X* axis is produced or the gravity induced convection is skewed. This must lead to the asymmetric distributions of the gas temperature shown by the solid lines in Figure 11.20. Consequently, the temperature difference between two sensor positions becomes non-zero. Apparently, the temperature difference between two sensors increases with increasing the acceleration along the *X* axis. By measuring the temperature difference, the acceleration information can be acquired.

11.3.2 Linearity Analysis

For the linearity of a convective accelerometer, as illustrated in the above section, it requires that the voltage engendered between two temperature sensors is linear with the exerted acceleration. To actualize this, it is necessary to assure the following two points for the device. Firstly, the temperature difference between the two temperature sensors linearly relates with acceleration. Secondly, the output voltage is linear with the temperature difference. The former is the key factor to obtain good linearity for the convective accelerometer, which will be mainly focused on in this section. As for the latter, it is mainly decided by the material property of the temperature sensor. Linear dependence of resistance on the temperature of the material is appreciated for the linearity. Usually, the temperature sensors are made of polysilicon in micromachined process. Polysilicon produced in different processes is of different variances of resistance with temperature, in order to make the output of the convective accelerometer

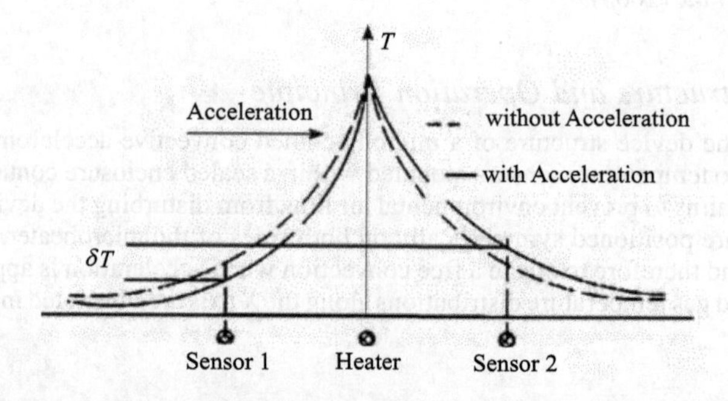

Figure 11.20 Symmetric temperature distribution generated by the central heater is disturbed by applied acceleration

being linear with acceleration, a suitable fabrication process should be chosen so as to provide a linear relation between the sensor temperature and its resistance. In the following analysis, we will focus on the relation of the temperature difference of two sensors with acceleration.

(i) Governing Equations and Numerical Simulation

Since the dimensions of the micromachined convective accelerometer and the gas velocity are small, the gas flow is usually laminar. The free convection is caused by the body force due to acceleration and the spatial density variation resulted from the spatial temperature change in the enclosure. The governing equations are as follows:

Equation of continuity:

$$\Delta \cdot u = 0 \tag{11.12}$$

Equation of momentum:

$$\rho u \cdot \nabla u = -\nabla p + \mu \nabla^2 u + \rho a \tag{11.13}$$

Equation of energy:

$$\rho c_p u \cdot \nabla T = k \nabla^2 T \tag{11.14}$$

Equation of state:

$$\rho = \frac{p}{RT} \tag{11.15}$$

where u, ρ, p, a, T are velocity, density, pressure, acceleration, and temperature respectively, μ, c_p, k are dynamical viscosity, specific heat, and thermal conductivity respectively, and R denotes the ideal gas constant.

Obviously, it is very difficult to have analytical solution for the above-mentioned equations, thus we conduct numerical simulation. In the computations, μ, c_p, k are given the constant values at the average temperature of the heater and environment.

For the device shown in Figure 11.19, because the ratio of the lengths of heater and temperature sensors to their widths and heights is very large, a two dimensional model can be adopted. Figure 11.21 shows the convective accelerometer model we simulate. L denotes the heater width, D denotes half of the cavity length at the bottom part, d represents the cavity depth, and x represents the distance between sensor and heater. For simplicity, both heater surface and cavity wall are regarded as isothermal. In the computations,

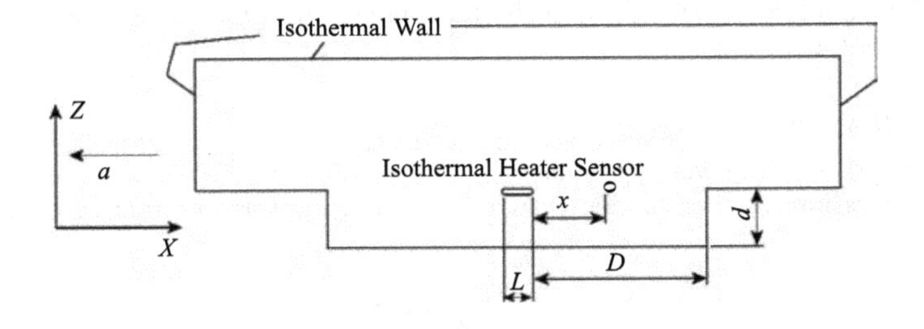

Figure 11.21 Schematic of simulated convective accelerometer

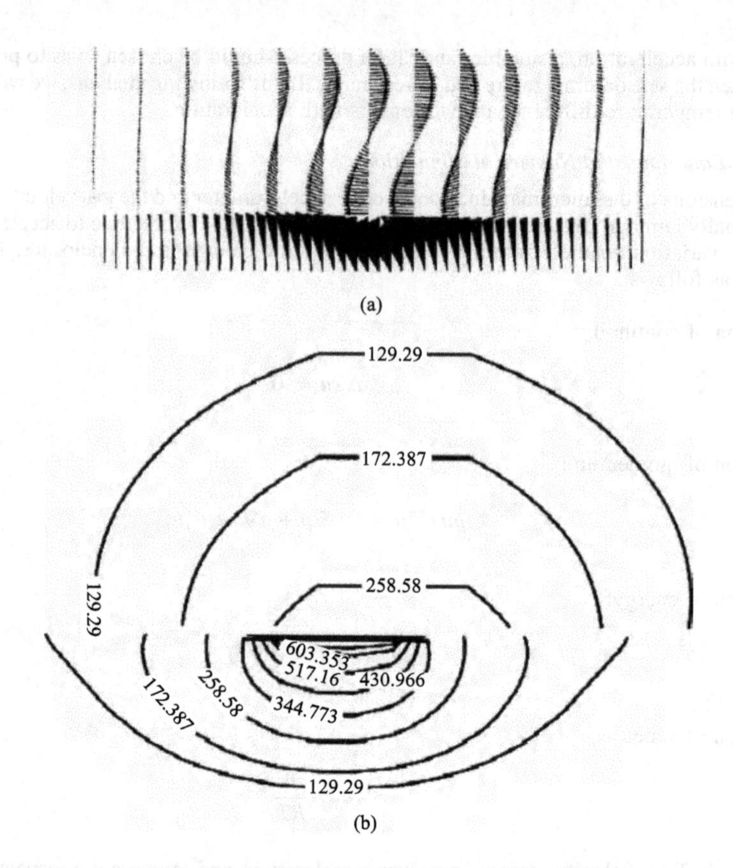

Figure 11.22 Simulation results at heater temperature of 973 K. (a) Velocity distribution, (b) temperature distribution

different grid sizes are adopted, fine grid is adopted in the region near heater, direction of acceleration a is towards the left.

The following numerical simulations and analyses are made starting from a realized prototype having such characteristics: the temperature of heater surface is 973 K, the cavity size is 960 × 100 mm, and the gas media is air. The simulation results of flow and temperature distributions for one case that the acceleration along sensitive axis is 10 g ($g = 9.81\text{m/s}^2$) are shown in Figure 11.22. Figure 11.22(a) shows the velocity distribution in the convective accelerometer. It can be seen that the flow direction of the fluid is towards the left near the heater when subjected to acceleration, asymmetrical flow forms in the cavity. Figure 11.22(b) demonstrates the temperature distribution in the convective accelerometer. It is obvious from Figure 11.22(b) that the isothermal lines in the cavity are not symmetrical to the heater. The temperature of the fluid in the left side of the heater is higher than that in the right side because of the asymmetrical flow in the cavity.

(ii) Results and Discussion

Dimensional analysis on the governing Equations 11.12, 11.13 and 11.14 shows that two dimensionless parameters, the Grashof number Gr, the Prandtl number Pr, govern the free convection (Retajczyk *et al.* 1980):

$$Gr = \frac{\alpha\beta\Delta TL^3}{\upsilon^2} \tag{11.16}$$

$$Pr = \frac{\upsilon}{\alpha} \tag{11.17}$$

where β, υ, α are the bulk expansion coefficient, kinetic viscosity, and thermal diffusivity, respectively, and ΔT is the temperature difference between the heater and cavity wall.

It is obvious from Equation 11.17 that the Prandtl number is decided by the working gas filled in the convective accelerometer. Thus for an accelerometer, if gas media is given, the flow and heat transfer in it are determined only by the Grashof number. It implies that the temperature difference between the two sensors is also decided by the Grashof number. In fact, the convective accelerometer is given in a certain application. The parameters in the definition of the Grashof number such as β, ΔT, L, υ will be given, thereby the Grashof number is just proportional to acceleration based on Equation 11.16. Therefore, the dependence of the temperature difference on the Grashof number represents the relation of the temperature difference with acceleration. This will be the base of the following linearity discussion. Figure 11.23 depicts the dependence of the temperature difference on the Grashof number.

In Figure 11.23, both abscissa and ordinate adopt logarithmic coordinates. The ordinate δT represents the temperature difference between two sensors. It can be seen from Figure 11.23 that for the Grashof number, in the range of larger than 10^{-2} and smaller than 10^3, the temperature difference is almost linear with the Grashof number. That is to say, in the above range for *Gr*, the output of the convective accelerometer is linear with acceleration. Why does the linear relation exist in this *Gr* region? It can be explained as follows. There are three forces, inertial force, viscous force, and buoyancy force, which govern the natural convection. For the natural convection with a large Grashof number resulted from large scale, large temperature, or large acceleration and the inertial force term is not too small to be neglected compared to viscous force. Thus the governing equations show nonlinearity characteristics, consequently, the flow and heat transfer in the accelerometer exhibit nonlinearity. For the natural convection with a very small Grashof number, heat conduction in the convective accelerometer overwhelmingly prevails over the convection, thereby the nonlinearity in small *Gr* condition may be contributed to the increasing influence of the heat conduction where the inertial force is very small compared to the viscous force. Only in the range of *Gr* larger than 10^{-2} and smaller than 10^3, the convective accelerometer exhibits a performance of output linear with acceleration. The reason is that the natural convection in the accelerometer is governed by both viscous force and buoyancy force, and the inside heat conduction is not too strong compared to the natural convection.

Because the temperature difference is linear with the Grashof number, obviously, larger Grashof number will create larger temperature difference, that is to say, the accelerometer design of a large Grashof number will result in high sensitivity for a given accelerometer.

The effect of the sensor position on the linearity of the convective accelerometer is also simulated, and the result is illustrated in Figure 11.24. The abscissa *x*/*D* denotes the non-dimensional position of the

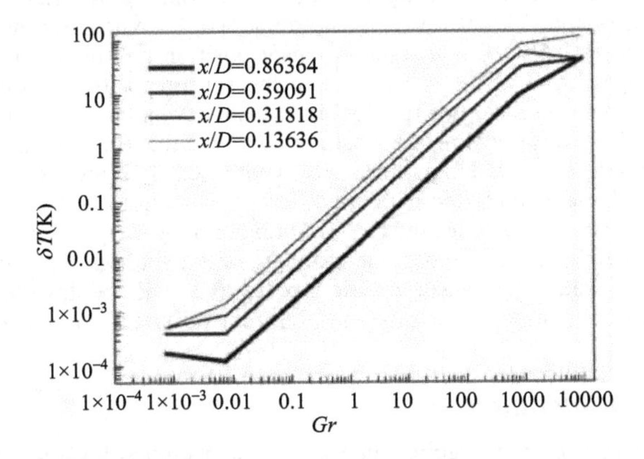

Figure 11.23 Temperature difference versus *Gr*

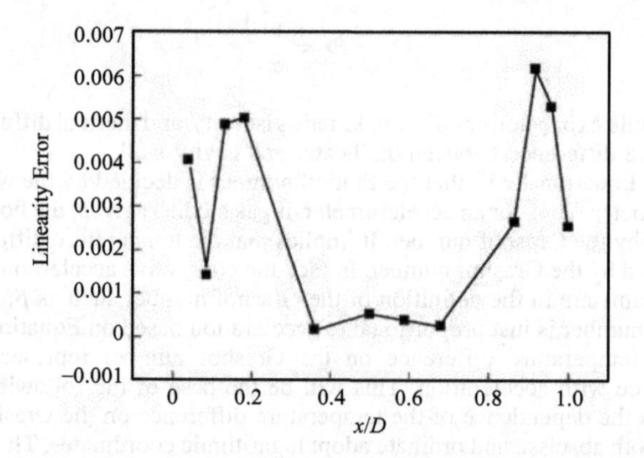

Figure 11.24 Linearity error versus the sensor position

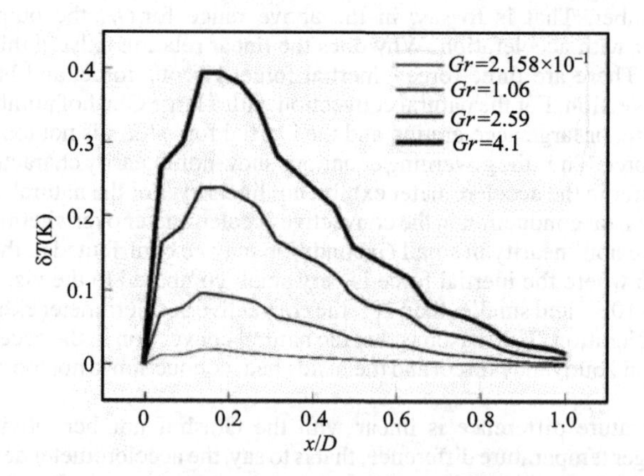

Figure 11.25 Variation temperature difference with sensor position

temperature sensor. The ordinate denotes the linearity error, which is defined as the real numerical results divided by the value obtained from the fitting line. It is noted from Figure 11.24 that the linearity error is smallest when x/D ranges from 0.3 to 0.7, where it is around 0.05%. When the sensor is positioned at other position, it can increase to 0.5%, so that suitable sensor position will be advantageous to the linearity improvement. The dependence of the sensitivity on the sensor position at different Gr is shown in Figure 11.25. It is seen that the sensor position for high sensitivity has an optimum place, which is around $x/D=0.2$. When apart from this position, the sensitivity will decline. Based on the above discussions, the sensor position for good linearity and high sensitivity synchronously is around $x/D=0.3$. It is also noted from Figure 11.25 that the sensitivity (temperature difference) increases with a Grashof number increase.

To sum up, it is necessary to guarantee the Grashof number in the range of $10^{-2}-10^{3}$ to achieve good linearity. The Grashof number should be designed as a large value for high sensitivity providing that good linearity has been achieved. The sensor position, $x/D=0.3$, will be favorable for the convective accelerometer design to obtain good linearity and high sensitivity synchronously in applications.

11.3.3 Design Consideration

Based on Equation 11.16 and the foregoing analysis, to achieve good linearity and high sensitivity, the temperature difference ΔT, heater width L, the filled gas, sensor position x/D should be considered in design.

In addition, compared with the heater width, if the cavity size is not large enough to be regarded as an infinite space, it should also be taken into account because of the boundary suppressing action on fluid flow.

It is obvious from Equation 11.16 that Grashof number Gr increases with increasing ΔT (in other words, increasing the heating power), heater width L and bulk expansion coefficient β of the gas filled in the accelerometer and decreases with increasing the kinetic viscosity y. Especially, increasing the heater width L will lead to a significant increment of Gr owing to the cubic relation of Gr with L. Therefore, to assure the accelerometer obtaining good linearity, the above parameters, especially L should be carefully chosen. At the same time, to increase the sensitivity, we should make the temperature of the heater surface high enough, and the heater width a large size. As for cavity design, it should be designed to be large to reduce the boundary suppressing action. For sensor position, as analyzed in the preceding discussions, the best place is near $x/D = 0.3$.

11.3.4 Fabrication

The device fabrication sequence is illustrated in Figure 11.26. Starting with a <100> silicon substrate, a layer of silicon dioxide is formed using wet thermal oxidation, then a uniform layer of polysilicon is deposited in a low-pressure chemical vapor deposition (LPCVD) reactor and another layer of silicon dioxide is grown on it [Figure 11.26(a)]. The top oxide and polysilicon layers are patterned to engender heater, temperature sensor, and polysilicon contact. After patterning, the polysilicon sidewalls are exposed [Figure 11.26(b)]. Then an oxide layer is produced on the polysilicon sidewalls by another oxidation step, which protects polysilicon from anisotropic etching using ethylene diamine-pyrocatechol (EDP). On the oxide layer, the bonding pad and cavity windows are opened [Figure 11.26(c)]. A boron layer is then used to dope the silicon regions for making electrical connections to the heater and sensors, afterward by the patterned photolithography, a layer of Ti/Pt/Au metal is sputtered onto the top surface [Figure 11.26(d)]. When the metal is stripped, bonding pads and the cavity windows form [Figure 11.26(e)]. Finally, the wafer is wet-etched in EDP until the cavity's creation [Figure 11.26(f)]. Figure 11.27 is a scanning electron microscopy (SEM) photograph of the device that is sealed in an enclosure.

11.3.5 Experiment

One micromachined convective accelerometer is fabricated according to the above process. The heater width is 80 mm, the height is 2 μm, the cavity size is 3000μm (length)×2000μm (width)×250μm (depth), the sensor position is near $x/D = 0.34$.

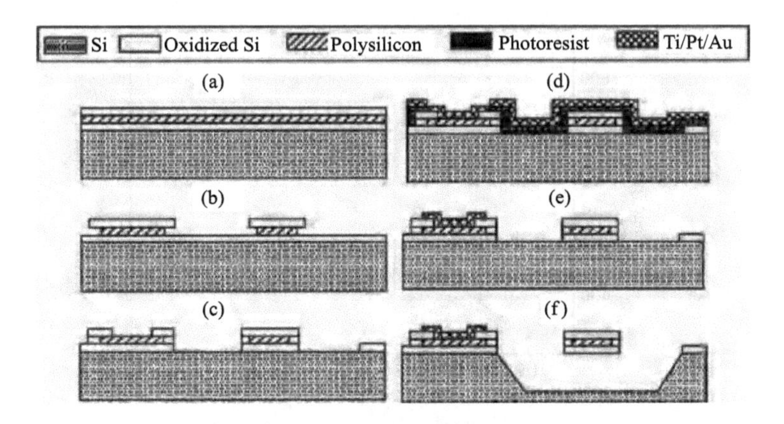

Figure 11.26 Device fabrication sequence

Figure 11.27 SEM photography of the device sealed in an enclosure

It is measured under two kinds of application conditions. Firstly, it is tested from -*g* to *g* under gravitation by rotating of the sensitivity axis with respect to the earth gravitational field. Figure 11.28 illustrates the linearity of the accelerometer under gravitation. It shows a very good fit with the expected linear trend. The linearity error is smaller For a comparison, the nonlinear coefficient of the accelerometer in reference (Leung *et al.*, 1997) under gravitation is obviously larger than 1% at the 1 *g* point and that of the device in reference (Leung *et al.*, 1997) is 0.5%.

The device is also measured on a vibration shaker with an acceleration range from 0 to 20 *g* and frequency from 0 to 200 Hz. In our experiments, we find that when the acceleration range exceeds 10 *g*, good performance especially for good linearity cannot be achieved. Figure 11.29 shows that the device has good linearity in the range from 0 to 10 *g* at 45 Hz. The linearity error is larger than that measured under gravitation, which reaches 2%. It is noted from Figure 11.29 that the sensitivity of the optimized accelerometer for operating power of 87 mW is 600 μV/*g*, where the sensitivity does not include the circuit amplification. For a comparison, the sensitivity of the accelerometer in reference (Milanovic *et al.*, 2000) is 146 μV/*g* for an operating power of 430 mW, and the linearity error in acceleration range from 0 to 7 *g* is 2%. Owing to the absence of relative comparable data, here the comparison with that in reference (Leung *et al.*, 1997) cannot be carried out.

The frequency response of the device is shown in Figure 11.30. It is noted that the frequency response is flat up to about 75 Hz, where the sensitivity decreases substantially. The response frequency is larger than that in reference (Leung *et al.*, 1997), in which it is 20 Hz. However, compared with the response frequency numbered 300 Hz in reference (Milanovic *et al.*, 2000), it is much smaller. The main reason for this may

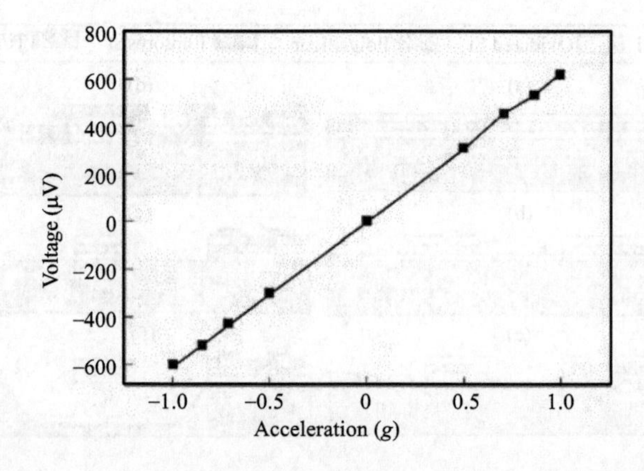

Figure 11.28 Linearity of the optimized accelerometer at 87 mW under gravitation

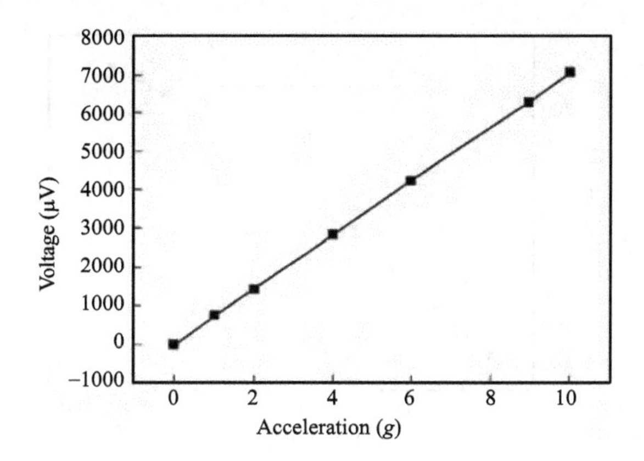

Figure 11.29 Measured performance of the accelerometer at 87 mW/45Hz from 0 to 10*g*

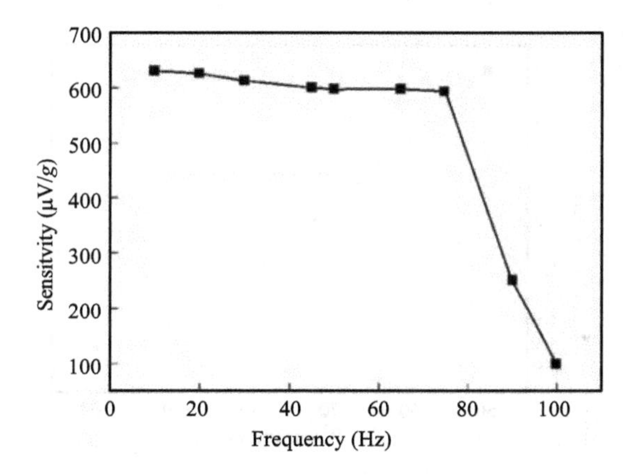

Figure 11.30 Measured frequency responses at 87mW

contributes to the fabrication method and the accelerometer size. In our present work, as in reference (Leung *et al.*, 1997), customary bulk-silicon fabrication is adopted, the accelerometer size including cavity size are larger than that in reference (Milanovic *et al.*, 2000), in which the standard integrated circuits technology was employed. Based on foregoing optimization analysis, the frequency characteristics of the present accelerometer cannot be satisfactory as that in reference (Milanovic *et al.*, 2000).

The noise test is also performed. The 25 Hz output noise at operating power of 87 mW is 0.6 $\mu V\sqrt{Hz}$, so that the noise equivalent acceleration (NEA) of the present convective accelerometer at 25 Hz is approximately 1 $mg/\sqrt{Hz}$.

An investigation into the dependence of the sensitivity on the heating power is also conducted in the test under gravitation. Figure 11.31 shows that the sensitivity of the accelerometer is nearly linear with the heating power. It proves that increasing heating power can increase sensitivity.

The variation of the NEA with the heating power at 25 Hz is demonstrated in Figure 11.32. It is noted that the NEA of the convective accelerometer decreases with the heating power increasing, which can be explained by the fact that when the heating power increases, the sensitivity increases linearly; however, the noise just increases slightly. Therefore, higher heating power will be favorable for gaining better resolution.

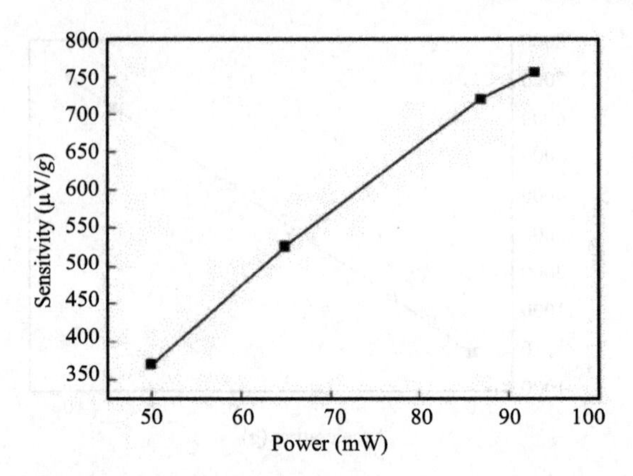

Figure 11.31 Variation of sensitivity with heater power

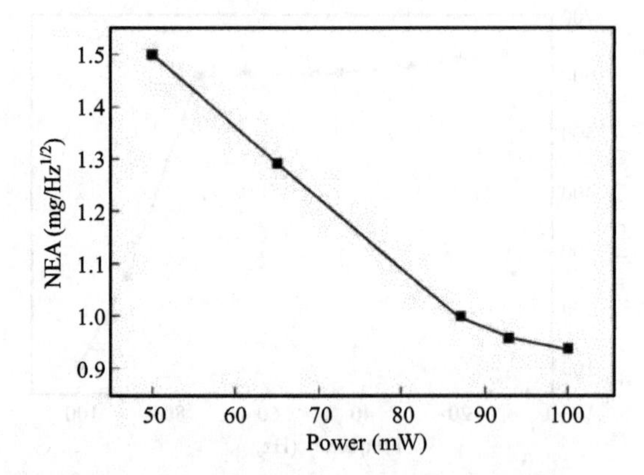

Figure 11.32 Variation of resolution with the heating power

11.4 Plastic Packaging for a Capacitance Based Accelerometer*

In the past decade, tremendous progress has been achieved in designing and manufacturing prototypes of microelectromechanical systems (MEMS). Among the numerous applications, automotive airbags have, by far, the largest market for micromachined accelerometers and rate gyroscopes (Romig *et al.*, 1997). The micromachined devices are batch processed, small in size, and thus result in a low cost relative to traditional mechanical sensing devices. They are also fragile and easily damaged, and therefore generally require two levels of packaging:

(1) Wafer level packaging, which is usually hermetic to provide damping control and to protect the MEMS devices from the subsequent dicing and testing; and
(2) Conventional electronic packaging of die-bonding, wire bonding, and molding to provide housing for handling, mounting, and board level interconnection.

---

*Reproduced with permission from G. Li and A.T. Tseng, "Low stress packaging of a micromachined accelerometer," *IEEE Transaction on Electronics Packaging Manufacturing*, **24**, 1, 18–25, 2001. © 2001 IEEE.

There are many factors, which must be carefully considered when designing a package for MEMS devices. First and foremost, the package must fulfill several basic functions:

(1) To provide electrical connections and isolation;
(2) To dissipate heat through thermal conduction; and
(3) To provide mechanical support and to isolate stress.

Secondly, the packaging process must be stable, robust, and easily automated. Lastly, the design must take testability into account. An efficient means to test and trim an individual part and to isolate the source of defective components is essential in both reducing cost and subsequent quality improvement.

The cost of MEMS packaging is relatively high, and in some cases, can be as high as 80% of the total cost (Beardmore *et al.*, 1997). Reducing packaging costs is therefore a prime concern for both integrated devices, where the transducer and its signal conditioning unit are manufactured on the same silicon chip, and for two-chip (or multichip) designs. The two-chip approach, where the transducer and its signal processing microprocessor control unit (MCU) are fabricated separately and independently, becomes essential when some of the processing of the transducer is not compatible with the processing of the MCU. The acceleration sensor described in this section is a two-chip system. The MCU are fabricated separately and independently, which is essential when some of the processing of the transducer is not compatible with processing of the MCU. The acceleration sensor described in this research is a two-chip system.

It is noted that the materials in this section are taken as courtesy of (Gary *et al.*, 2001).

11.4.1 Micro-Machined Accelerometer

The capacitive transducer considered in the present section is designed in the form of a differential capacitor pair made from three highly doped polysilicon layers using surface micromachining technology. The acceleration sensor is designed to be conveniently integrated into typical board configurations. As an example, Figure 11.33 shows the two-chip accelerometer in an industry standard 16-pins dual-in line package (DIP). The detail of the micromachined transducer is shown in Figure 11.34 a SEM micrograph obtained from (Shemansky *et al.*, 1997). The transducer is designed for 50 *g* full-scale acceleration having a supply voltage of 5 V with a 50 mV/*g* sensitivity. It operates in an open loop mode.

Figure 11.33 Two-chip accelerometer in an industry standard 16-pins dual-in line package (DIP)

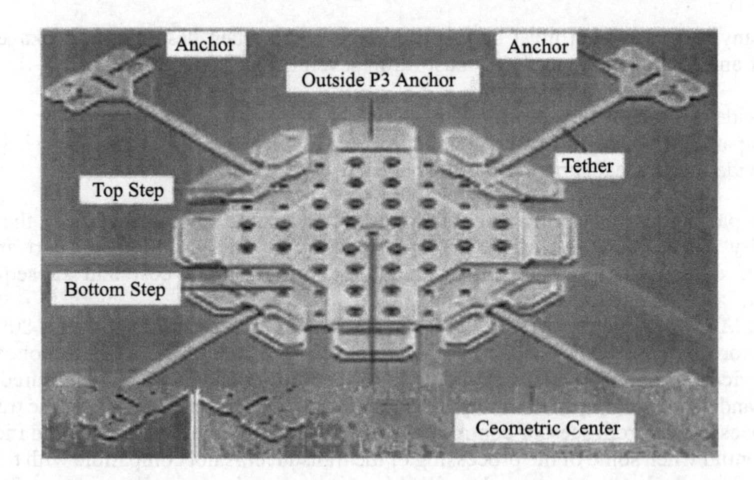

Figure 11.34 SEM graph of a micromachined accelerometer (g-cell)

The three-layer transducer is structured having a bottom polysilicon layer adhering to the substrate, a top polysilicon layer anchored at various points in addition to the substrate, and a middle layer of polysilicon suspended by four tether beams. The middle layer, with the top and bottom layers respectively, forms two capacitors. When subjected to an inertial force, the middle layer is deflected. This motion in turn results in changes in capacitance for both capacitors, generating a differential output signal.

Mechanically, both the bottom and the top polysilicon layers have little motion relative to the substrate. The middle movable layer is an octagonal plate supported by four tethers, which are anchored to the substrate. The plate serves as the "proof mass" for the accelerometer, and contains 52 smaller holes spread over the plate and a larger hole in the center. The four tether suspension beams are made of polysilicon and silicon nitride layers. The combination of both nitride and polysilicon stresses is tensile along the tether, and therefore able to suspend the central plate. This transducer can be modeled as a mass-spring-dashpot oscillator.

11.4.2 Wafer-Level Packaging

Before the accelerometer can be put into a package of conventional leadframe-and-mold assembly, it must be sealed at the wafer level to protect the movable components from being damaged during the die sawing and subsequent molding. The wafer level package also provides the sensor with a controlled ambient to preserve the damping characteristics of the proof mass.

Motorola's wafer-level packaging is achieved through frit-glass bonding (Audet *et al.*, 1997). Sealing glass is first applied to the whole cap wafer in the form of a paste, which consists of a mixture of glass frit and an organic binder. The glass frit is deposited through a standard screen process. Following the screen process the capped wafer is allowed to dry, and then fired at high temperature to burn off the organic binder and sinter the glass. To join the sensor wafer and the glass-coated cap wafer, the two are aligned and placed in close contact with one another. The assembly is then heated to a temperature exceeding the softening point of the glass and thermocompression-bonded to form a hermetic seal. Figure 11.35 shows a schematic of a wafer-level packaged accelerometer. The micromachined accelerometer is sealed inside the cavity formed by the device, cap wafers, and the glass frit.

(i) Experiment and Simulation

This section presents the study of thermal hysteresis for a prototype z-axis acceleration transducer (Figure 11.36). A prototype z-axis accelerometer was experiencing thermal hysteresis of offset of

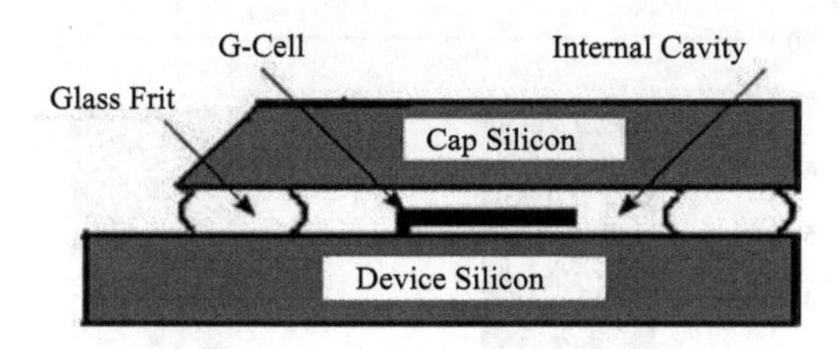

Figure 11.35 Schematic of a wafer-level package of the micromachined accelerometer

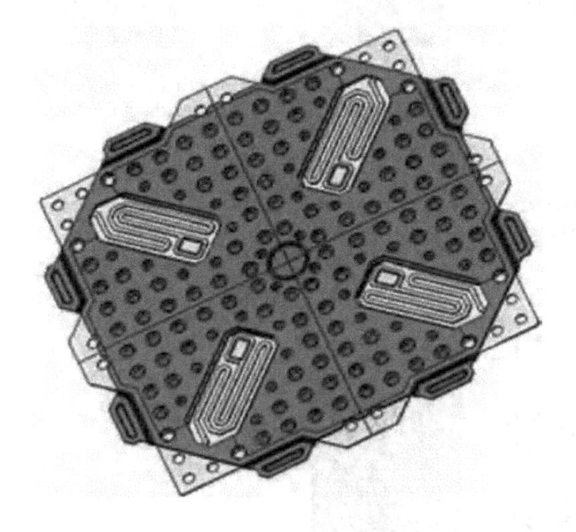

Figure 11.36 Numerical model of a *z*-axis capacitive transducer

approximately 20% of the proposed accuracy requirements. Hysteresis was measured on these parts by measuring the static offset on the part at 25 °C (T_{11}), then cycling the part from 25 °C to − 40 °C (T_c) to 25 °C (T_{12}) up to 125 °C (T_H) and back to 25 °C (T_{13}). Thermal hysteresis was then calculated as the difference in reading at T_{13}, and T_{12}.

When this phenomenon was first observed, the first step taken was to determine if it was caused by the control circuit, the package, or from the transducer itself. The first theory was that the plastic package caused the hysteresis. An experiment was run where the transducer was isolated from the package, and was only connected to the control circuit by wire bonds. It was found that isolating the transducer from the package provided no benefit in terms of thermal hysteresis (Figure 11.37).

Once the package was eliminated as the root cause for this hysteresis, the control circuit was then investigated. Hysteresis on the control circuit was tested by electrically isolating the transducer from the inputs, and using built-in capacitors to simulate the transducer. Experimental results showed that without the transducer, the thermal hysteresis was practically eliminated as illustrated in Figure 11.38. This excluded the control circuit as the root cause of hysteresis.

Upon completion of these two experiments, focus was placed on the transducer itself as the possible root cause of hysteresis. The transducer is hermetically sealed between two silicon wafers (a substrate wafer that the transducer is anchored to and a cap wafer that is tied to a constant voltage for shielding purposes) using a glass-frit bonding process. The cap wafer's top surface has a thin aluminum layer evaporated onto it in order to help bond wires to it. The cap wafer is then tied to the same potential as the substrate to provide shielding to the transducer (Figure 11.39). It was noted that the aluminum thin film on the cap wafer exhibits plastic deformation over temperature (Townsend, 1987), and could thus be a root

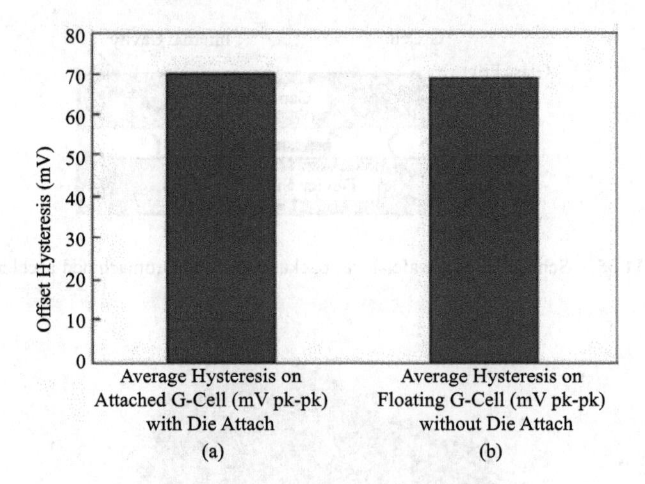

Figure 11.37 (a) Thermal hysteresis on a device with standard packaging. (b) Thermal hysteresis on a device with a mechanically floating transducer

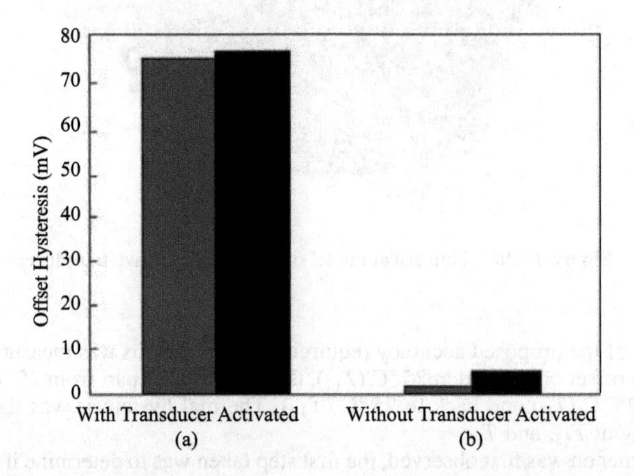

Figure 11.38 (a) Thermal hysteresis on a device with transducer activated. (b) Thermal hysteresis on a device without an electrically activated transducer (2 runs each)

cause for thermal hysteresis. A brief experiment was run where the metal was removed from the cap, and the shield wires were re-attached to the cap with a silver paste. Results showed that the parts without the full metal cap exhibited 68% less hysteresis than those with the full aluminum layer on the cap (Figure 11.40). This provided empirical evidence that the aluminum layer on the cap is the root cause for thermal hysteresis. The next step was to model and understand the phenomenon and fix the issue using these models.

(ii) Numerical Models

The effect of plastic strains in the aluminum layer on the device output was modeled by partitioning the device into two FEA models: a wafer level package model and a transducer model (since the transducer package is much larger than the actual transducer) (Figure 11.41). First, the wafer level package model

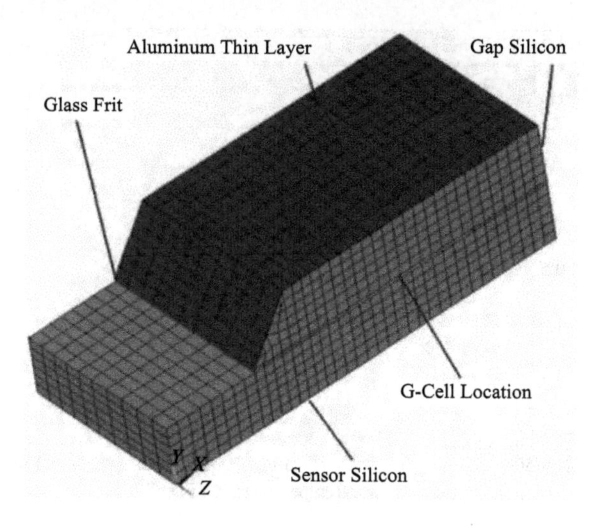

Figure 11.39 Transducer product after wafer cap sealing (half symmetry model)

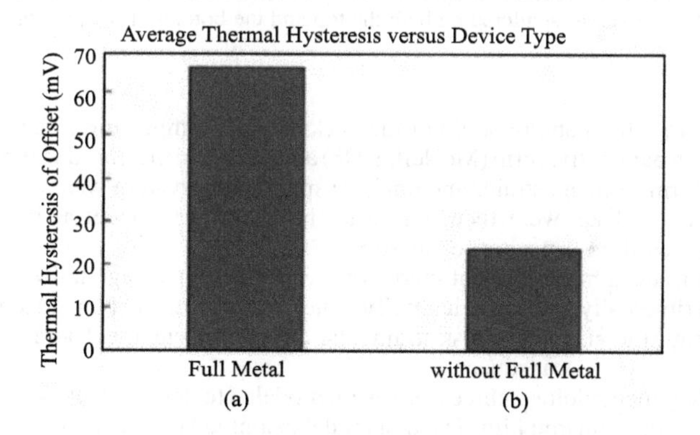

Figure 11.40 (a) Thermal hysteresis on a device with full metal. (b) Thermal hysteresis on a device without full metal

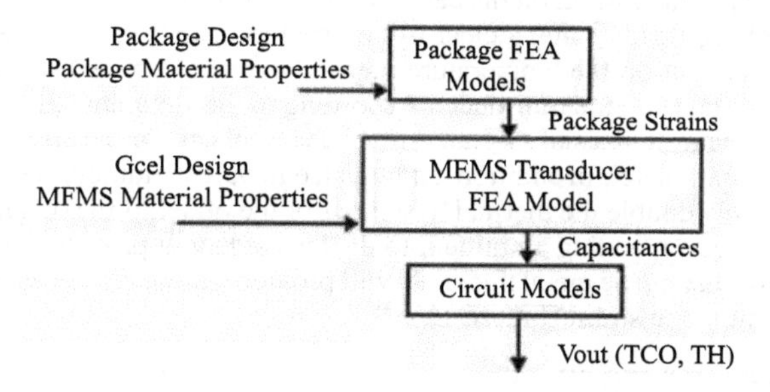

Figure 11.41 Summary of modeling process. Thermal hysteresis was modeled by partitioning sensor into two FEA models: a package model and a transducer model

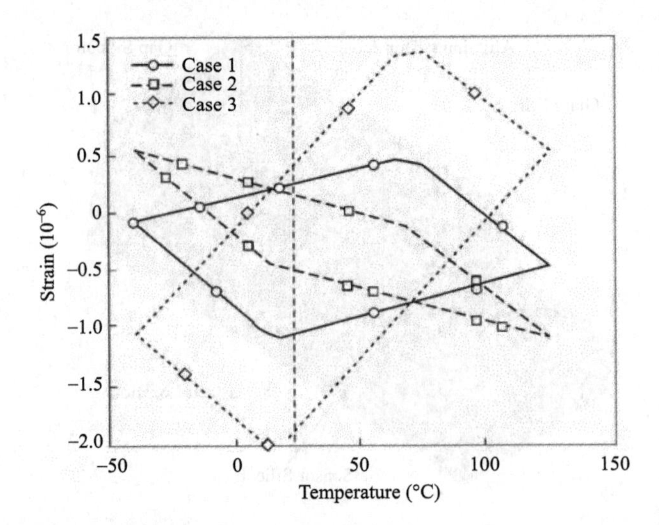

Figure 11.42 Steady state hysteresis loop of strain at the transducer spot. Case 1 is "Standard" where the aluminum metal was sputtered only on the top of the package. Case 2 is "Half Metal" where metal was evaporated on the top with half the usual thickness (this case provides the smallest amount of hysteresis with respect to in plane strain). Case 3 is "Back Metal" where the metal was sputtered on both the top and the bottom of the package

was run to calculate the stress and/or strain of the package over temperature. These stress/strain effects were then expressed in parametric form (McNeil, 1998) and passed to the transducer model (Figure 11.36). In this case the uniform in-plane strain, and uniform spherical curvature were recorded and saved as a function of temperature. They were then passed to the transducer model as boundary conditions to calculate the output based on capacitance variation.

Figure 11.39 illustrates a finite element model of a wafer-level package utilized in our evaluation of hysteresis both experimentally and numerically. Two pieces of silicon are bonded together by glass frit. A thin layer of aluminum silicon, whose stress-strain relation is nonlinear, is evaporated on the top surface to provide EMC shielding to the transducer.

Using this modeling methodology, three cases were modeled to determine how to reduce the hysteresis effect caused by the aluminum thin film. The first model examined the existing transducer package at the wafer level, with the standard metal thickness. A second model was created where the aluminum layer thickness was reduced in half. Lastly, a third model was created in which an aluminum film of nominal thickness was sputtered on to both the bottom and top of the transducer package with intentions of balancing out the curvature.

Figure 11.42 shows the in-plane strain (at the center point of the transducer) versus temperature for the three models. After reaching steady state, a clear hysteresis loop exists such that at room temperature, there are two strains dependent on the temperature path. These two different strains would cause two different readings of capacitance in the transducer. According to the simulation, if the thermal hysteresis of the offset is caused by the hysteresis of in-plane strain, the thickness of the aluminum film will have a direct relationship on this parameter. In addition, adding an equivalent aluminum film on the backside of the wafer will approximately double the thermal hysteresis of the offset. If the offset hysteresis is caused by spherical curvature, however, adding aluminum to the backside will provide the optimal solution for reducing the hysteresis. Reducing the metal thickness will provide similar results whether the hysteresis is caused by curvature, or in plane strain (Figure 11.43).

(iii) Experimental Results

After the simulations were run, each of the three wafer level package conditions was fabricated with the z-axis transducer, shown in Figure 11.36, to get experimental results. To reiterate, the first packaging

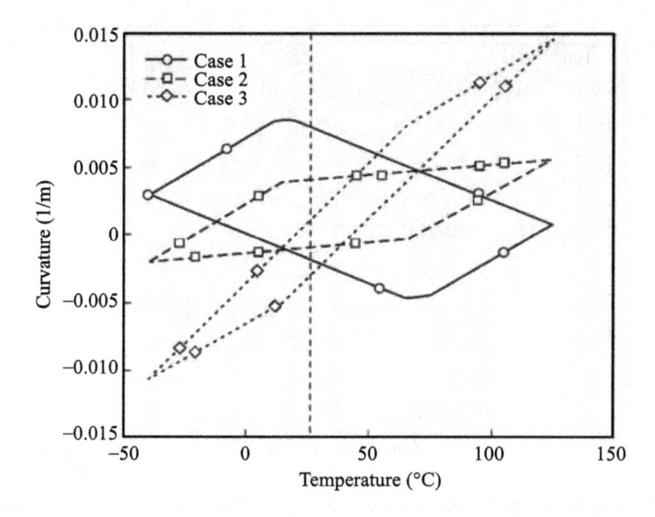

Figure 11.43 Steady state hysteresis loop of curvature at the transducer spot

condition (case 1) was the reference group, which is the existing package, with the standard aluminum thickness on the cap. The second group (case 2) of devices made had only one half the aluminum thickness sputtered on to the top of the g-cell cap. The final group (case 3) had the standard aluminum thickness sputtered on both the top and bottom of the g-cell level package. All three groups of silicon were assembled with a common control circuit in a ceramic package to eliminate device level package stress as a factor.

Figure 11.44 shows a comparison between simulated hysteresis and experimentally measured hysteresis for all three groups. Clearly, the comparison is good both qualitatively and quantitatively with the simulated values slightly smaller. In case 2 where the backside of the package was also coated with metal, the package resulted in more plane strain and less curvature at the transducer spot. Yet, the

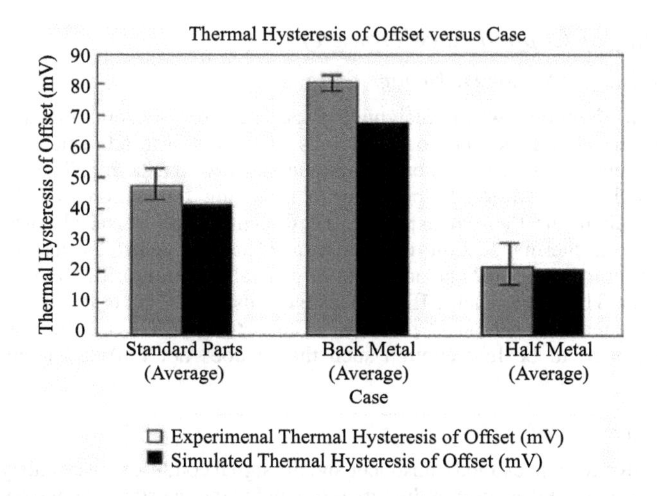

Figure 11.44 Results of average simulated and measured hysteresis for transducers for each of the studied cases

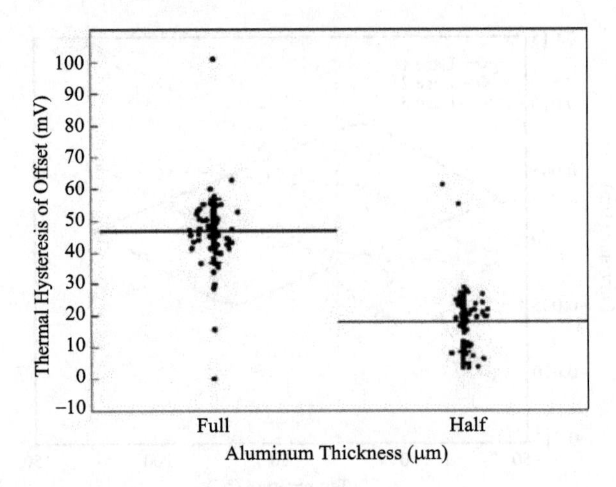

Figure 11.45 Results of larger scale characterization of thermal hysteresis of offset versus aluminum film thickness. In this case, the temperature was cycled from −40 to 105 °C

hysteresis was almost doubled indicating that the plane strain was mainly responsible for the hysteresis. In case 3, with metal thickness reduced by half, the hysteresis was also approximately reduced by half.

Based on these results, the final product was qualified using the half thick metal process. On a sample size of 100 parts taken from three wafer lots, the average hysteresis dropped from 47 mV to 18 mV (Figure 11.45). The lower values in hysteresis in the larger scale characterization relative to the initial experimental data are due to the lowering of T_H from 125 °C to 105 °C in the characterization. Despite this difference, the fundamental theory held true for large quantities of parts, which enabled a solution to the offset hysteresis while maintaining the integrity of the final product. Future work is being done to nullify this thermal hysteresis by modifying and improving the design of the transducer itself.

11.4.3 Packaging of Capped Accelerometer

(i) Basic Considerations for Sensor Packaging

There are three considerations when designing a package for sensors: electrical, thermal, and mechanical. The electrical and thermal considerations are common to all varieties of semiconductor devices, while the mechanical constraints can be unique in sensor packaging. The unique challenge of sensor packaging is that in addition to providing a mounting foundation to a PC board, stresses induced by a mismatch in the thermal expansion coefficients of the materials used to fabricate the package and the external thermal loading of the package must be controlled and kept at a level low enough to avoid impact to the sensor performance. In general, different MEMS devices have different stress tolerance levels. Therefore, each MEMS package must be uniquely designed and evaluated to meet special requirements (Dickerson *et al.*, 1997; Tang *et al.*, 1997). Lastly, the packaging stress should also be controlled to be low enough such that it does not impact the performance of the control IC.

(ii) Assemble Sequence

After capping, the wafer-level packaged transducer and the signal conditioning circuitry are first mounted to a leadframe flag using epoxy as die-bonding material and using the standard die-bonding techniques. The two silicon chips are connected via standard wire bonding. A thin layer of silicone gel is then

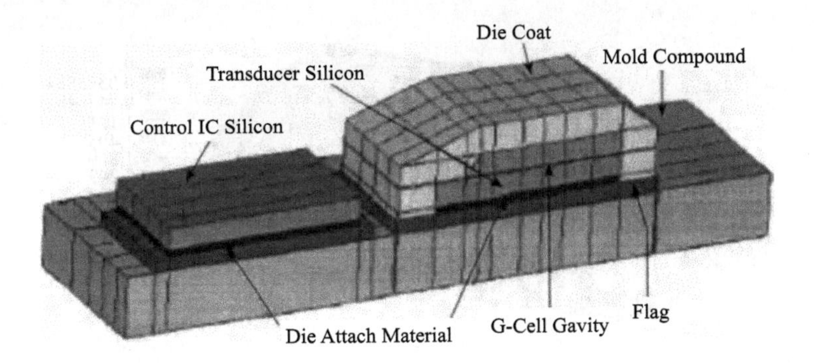

Figure 11.46 Finite element model of a package of the microaccelerometer system. For illustration clarity, only half of the total system and only the partial molding compound (below the flag) are shown

deposited over the transducer silicon die as passivation. In both the die-bonding and die-coating processes, curing is needed to ensure proper orientation after the viscous gel becomes solid. The final step involves the encapsulation of the two silicon chips in an epoxy mold compound, using conventional molding techniques, at a temperature of 165 °C. During the process of cooling (to room temperature), the die-coat shrinks more quickly than the mold compound, and a gap of approximately 25.4 μm (1 mil) is created. This air gap is used to shield stress transfer from the surrounding mold compound to fulfill the purpose of partial stress isolation.

Figure 11.46 shows a partial view of a finite element model package model. For clarity, only the mold compound under the leadframe flag is shown. The front face is actually the symmetric plane and the other half of the package is not shown. On the left is the control IC silicon, and on the right is the capped transducer silicon coated by silicone gel (commonly referred to as the die coat or dome coat). The tiny accelerometer (less than 1 mm in dimension) is hidden inside the g-cell cavity. The overall package footprint is 20 mm×6.4 mm (780 mil × 250 mil). In general, the size of a MEMS die may bear little relation to the package size, which is mainly dictated by compatibility requirements to the custom print circuit board.

(iii) Full Die-Attach Approach

The full die-attach approach refers to a die-bonding process in which the entire transducer silicon die is glued to the leadframe flag without void. It is achieved through dispensing a single drop of die-bond RTV into the flag and the silicon is squeezed into it in a controlled manner. This approach was recently developed by Motorola to appropriately control the amount of die-coat underflow. In die bonding, unnecessary stresses can be generated if too much die-coat gel gets under the silicon die, while not enough die-coat coverage can result in the silicon die to be unprotected. Details of this approach have been reported by Dougherty and Adams (Dougherty, 1997; Adams *et al.*, 1997).

The packaging induced stress or strain at room temperature can be calculated by the finite element model. Due to the symmetric nature of the physical system, with proper boundary conditions, only half of the package needs to be modeled. It is understood that the attachment of electric leads into the board could add some stresses and strains to the silicon transducer. However, based on a separate finite element calculation, the stress induced by the leads is smaller than 2% of the total stress and it has never been found that the induced stress is large enough to cause any problems or failures. Also, our focus is in evaluating the packaging impact to the transducer die. As a result, without complicating the modeling effort, the electric leads are not included in the FE model.

Assuming a stress free state of all components at the elevated temperature of 165 °C at which the molding takes place, both the package and the transducer die deformation at room temperature (25 °C) can be calculated. The resulting deformation contour is shown in Figure 11.47, for the transducer silicon and its surrounding. It is clear that the maximum deformation occurs in the die-coat layer, whereas the transducer die suffers very little deformation and thus stress is indicated by a dark color in the shade spectrum.

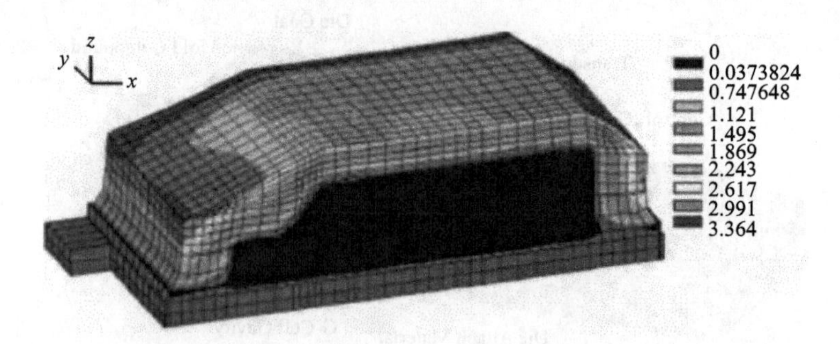

Figure 11.47 Displacement contour for part of the total package at room temperature. Significant deformation occurred in the die coat while the silicon suffered very little deformation as illustrated by the dark color

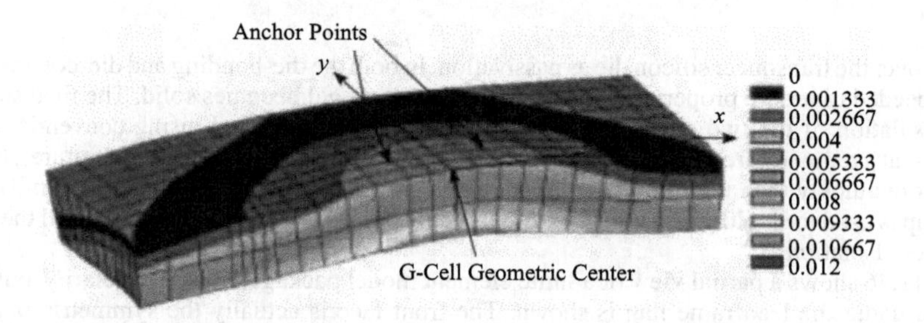

Figure 11.48 Displacement contour of the device silicon (half is shown due to symmetry) at room temperature. Although very small, the silicon was bent by a force transmitted through the die bond

If the transducer silicon is isolated, a closer view of the deformation and stress near the region where the transducer is anchored can be obtained. By modifying the shade index and changing the amplifying factor, the deformed shape of the transducer silicon is shown in Figure 11.48. Note that the maximum displacement in the silicon die is less than 0.305 m (0.012 mil). Considering that the lateral dimension of the silicon die is 3.3 mm (130 mil), a displacement of 0.30 m (0.012 mil) or less is indeed small, and the silicon die remains virtually flat under the packaging stress. After packaging, the accelerometer is trimmed for its sensitivity at room temperature. Therefore, the variation of silicon deformation away from room temperature, to either hot (90 °C) or cold (40 °C), is critical and must be kept below the maximum level that the transducer can tolerate.

To help explain the simulation data, a Cartesian coordinate system is constructed and set on the top surface of the silicon die with the origin coinciding with the material point directly beneath the g-cell geometric center. The x-axis is along the longitudinal direction, and the y-axis is along the lateral direction and is perpendicular to x. Running the simulation for temperatures at -40, 25, and 90 °C, the resulting profiles of the vertical displacement are shown in Figure 11.49. Figure 11.49(a) depicts the displacement variations along the x-axis while Figure 11.49(b) pictures the profiles cut along the y-axis. Clearly, the transducer silicon bends more at a cold (-40 °C) and less at a hot (90 °C) temperature. If the silicon state at room temperature is considered neutral, where trimming is performed, the bending due to a 65 °C temperature drop (to -40 °C) is about twice the bending caused by a 65 °C temperature rise (to 90 °C).

In order to evaluate the adverse impact on the performance characteristics of the transducer itself, in-plane displacements on the silicon surface at multiple locations are necessary. For simplicity and clarity, strain is calculated on the silicon surface rather than displacements. Once strain is known, displacement can be calculated in a straightforward manner. To evaluate the stress impact purely due to packaging, the

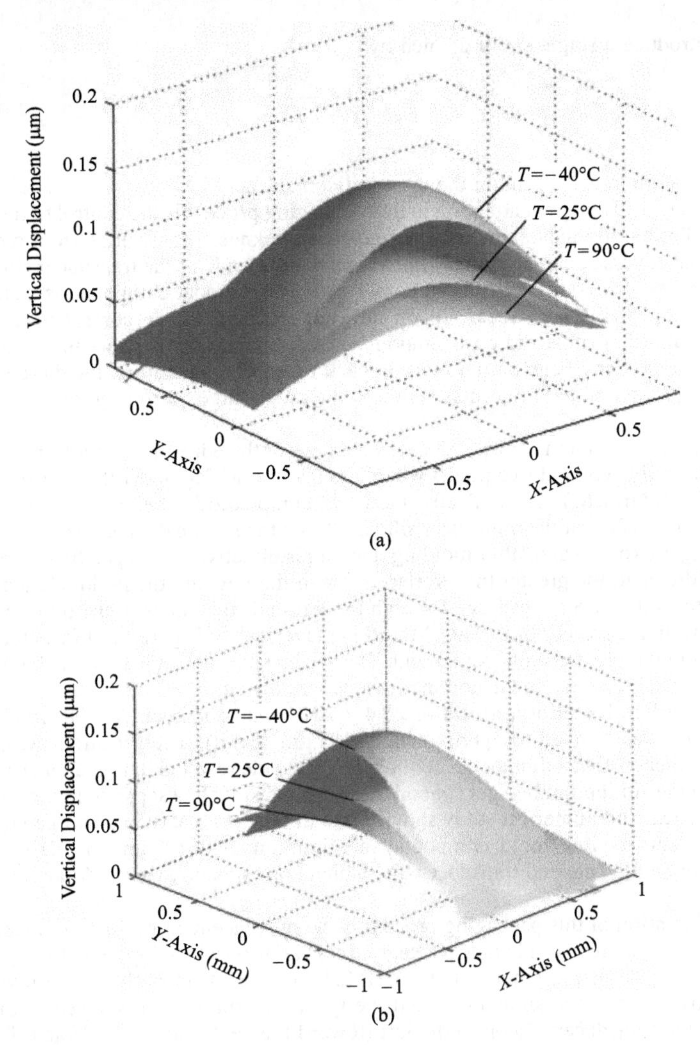

Figure 11.49 Displacement profiles of device silicon bending at three temperatures (−40, 25, and 90 °C). An upward bending of the silicon is clear. (a) Displacement profiles cut along the *x*-axis; (b) displacement profiles cut along the *y*-axis

silicon stress or strain induced by temperature variation must be subtracted from the total stress or strain. Consequently, we define the planar strain in the following manner:

$$\bar{\varepsilon}_x = \varepsilon_x - \varepsilon_0 \tag{11.18a}$$

$$\bar{\varepsilon}_y = \varepsilon_y - \varepsilon_0 \tag{11.18b}$$

where

ε_0 denotes strain of the silicon alone (without the packaging effect);

ε_x and ε_y represent the total strain on the upper silicon surface after packaging;

$\bar{\varepsilon}_x$ and $\bar{\varepsilon}_y$ are used in analyzing the performance of the transducer itself.

We further introduce a single strain defined by:

$$\varepsilon = \sqrt{\varepsilon_x^2 + \varepsilon_y^2} \tag{11.19}$$

which represents a combined strain on the upper silicon surface.

An issue that has received great attention in this packaging process is the control of die-coat thickness and uniformity. Production data show that the die-coat thickness often varies from die to die, and in extreme cases coat coverage can be entirely missed on the backside of the transducer silicon die causing direct contact between silicon and the mold compound. The die-coat compound is a sticky silicon gel dispensed to the top of the transducer silicon die (not the one for the control IC die) as shown in Figure 11.46. The molding of the die-coat compound takes place at 160 °C. Since the coverage compound has a very high thermal coefficient, after cooling back to room temperature, the die-coat coverage can become nonuniform and be partially missed due to the much larger shrinkage of the die coat than the silicon die.

When in full coverage (nominal case), the die-coat ends at the edge of the leadframe flag and the flag is completely covered (except at two points where the leadframe flag under the transducer silicon joins the flag under the control IC). Thus, ideally, the mold compound in the nominal case is not in direct contact with the top surface of the transducer silicon. When the silicon die receives less coat coverage, a portion of the flag is exposed and, after molding, the flag is effectively clamped by the mold compound. The thinner the die-coat, the greater the overlap between the flag and the mold. Numerical simulation was performed for thinned coat coverage for both the left and right sides of the transducer silicon, and the result is shown in Figure 11.50(a). The "Mold/Flag Overlap" of Figure 11.50(a) refers to the overlap between the leadframe flag and mold compound. The vertical axis is strain as defined in Equation 11.19. At zero overlap, which represents the nominal case where the gel coat covers the entire transducer die and the underneath flag footprint, a strain of 28.4×10^{-6} was obtained. At 508 μm (20 mil) overlap, where the gel coat was thinned to approximately 127 μm (5 mil), a lower strain was observed. The simulation data indicates that a thinner gel coat on the right side resulted in a smaller strain as compared to the strain from the similar thinned gel coat on the left-hand side. In a separate report (Vujosevic *et al.*, 1998), both analytical and numerical study showed that in order to maintain an air gap between the die-coat gel and the surrounding mold compound, a minimum of a 127 μm (5 mil) gel coat must be dispensed surrounding the capped transducer die, which corresponds to a mold/flag overlap of 508 μm (20 mil).

Another consideration of this packaging process is the misplacement of the transducer silicon on the leadframe flag. It is of interest to calculate package stress variation when the die is slightly misplaced. For illustration purposes, a misplacement of up to 152.4 μm (6 mil), in both the positive and negative directions, was assumed in the simulation, and the resulting strain ε is displayed in Figure 11.50(b). Clearly, moving the transducer silicon to the left (toward the center of the leadframe flag) results in a lower stress.

Finally, it is of paramount importance to understand packaging stress variation relative to temperature such that a special design scheme can be incorporated into the signal control IC to compensate for the sensitivity gain or loss due to temperature variation, since the sensitivity of the packaged micromachined transducer depends on the temperature induced stress or strain. By varying the temperature from − 40 °C to 90 °C, and calculating the strains on the silicon surface, the results are shown in Figure 11.51. The dashed and dotted curves represent strain as defined in Equation 11.18a in the x-and y-axis, respectively, and the solid curve is the combined strain as defined in Equation 11.19. The packaging strain is high at the cold temperature and is monotonically decreasing as the temperature increases. The sudden change in slope is due to the nonlinear property (with temperature) of the die-bond material.

(iv) Four-Dot Die-Attach Approach

In addition to the full die attach process, an alternative die attach approach (Adams *et al.*, 1997) was also researched and was adopted for some of the product lines. Realizing that the high stress level in the full die attach was primarily caused by silicon bending, it was hypothesized that removing the die-bond material in the central region directly beneath the sensing element should alleviate the bending and thus reduce packaging stress. By re-programming some of the die-bonding machines, the die-bond RTV could be

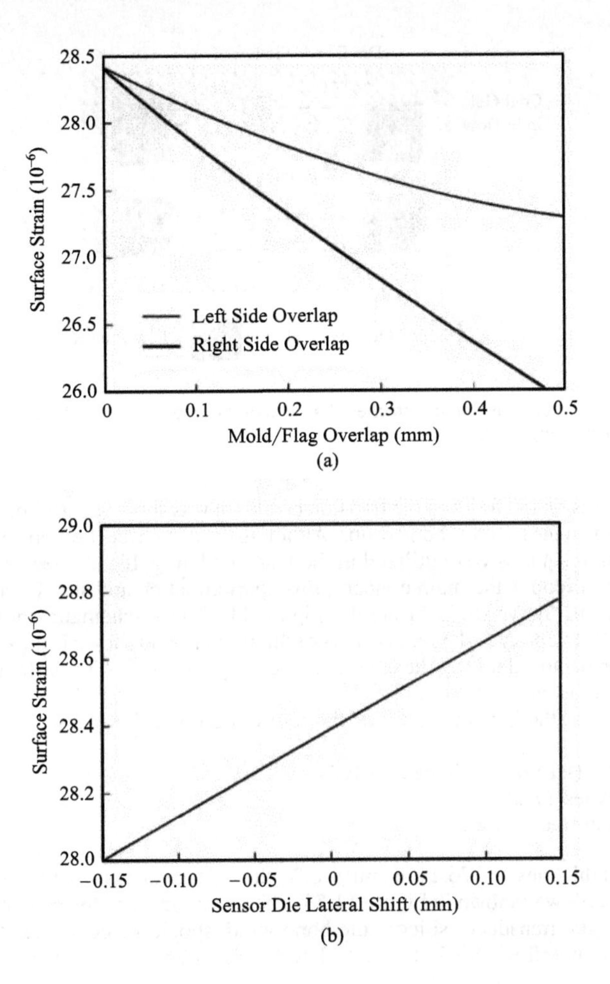

Figure 11.50 Variation of surface strain ε, at room temperature. (a) Mold/flag overlap, (b) sensor die misplacement

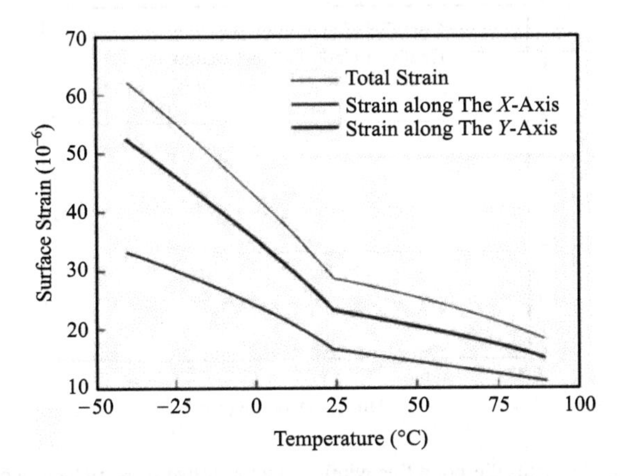

Figure 11.51 Temperature dependence of the surface strain. The sudden change in slopes is caused by a change in the die-bond material property with temperature

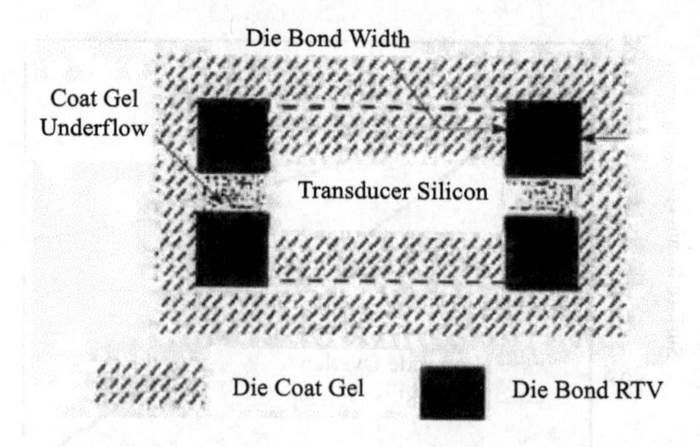

Figure 11.52 Schematic of a four-dot die attaches. The size of die-bond drops can be controlled to ensure bonding reliability and a low packaging stress

dispensed at four points on the flag and the transducer silicon was subsequently squeezed to the flag with its four corners sitting on the four die-bond drops. Although the dispensed die-bond epoxy is usually round or elliptical, four small squares were utilized in the FEA modeling. If the stress and deformation in the central region of the silicon is the main concern, this approximation and simplification should result in little error according to St. Venant's Principle. Figure 11.52 is a schematic showing four dots (four squares) of the die-attach epoxy at the four corners of the silicon. Also shown is the die-coat gel underflow (hatched area) which occurred during the coating and curing processes. The dashed rectangle represents the footprint of the transducer silicon.

There are three controllable parameters for the four-dot die-attach process:

(1) Die-bond width (size of the dispensed RTV);
(2) Die-bond thickness; and
(3) The die-coat gel underflow.

Assuming a nominal thickness of 76.2 μm (mil) RTV, the surface stress as a function of die-bond width and die-coat gel underflow is shown in Figure 11.53. The result indicates that to achieve a lower stress in the central region of the transducer silicon, the bond width should be controlled to less than 1.02 mm (40 mil), and the gel underflow should be controlled to within 762 μm (30 mil).

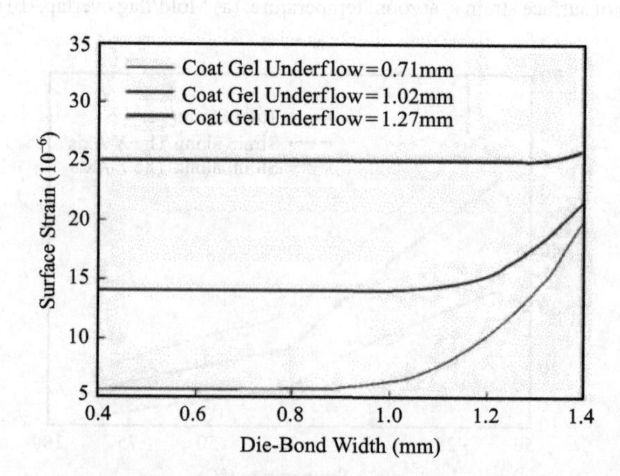

Figure 11.53 Surface strain versus die-bond line width. A factor of five stress reduction can be achieved when the bond line width is controlled to be less than 889 μm (35 mil) and the coat gel underflow near 762 μm (30 mil)

11.5 Tire Pressure Monitoring System (TPMS) Antenna

The accidents resulted from over- or under-inflated tire have attracted more and more attention in the community in recent years. There have been problems such as unnecessary tire wear, reduced gas mileage, and less than optimal vehicle performance when tires are over inflated. In addition, under-inflated tires result in increased tire wear, decreased vehicle performance, and compromise the ability of the tires to maintain a safe interface with the road (Fischer, 2003). Because the TPMS can predict the condition when tires are inflated with abnormal pressure and temperature and warn the driver about the problem, TPMS has attracted public attention to transportation safety. Furthermore, it is required by the US government that a TPMS must be installed on all passenger cars and light trucks sold after September 2007. In recent years, there are mainly two installation methods for TPMS in the market, such as fixing TPMS onto the wheel hub with a steel belt or by an inflating valve. FineMEMS Inc developed its belt-based TPMS and valve-based TPMS as shown in Figure 11.54, by collaborating with major automobile original equipment manufacturers (OEMs).

A typical example of TPMS is a wireless TPMS system, which is shown in Figure 11.55(a) (Leng *et al.*, 2007). The TPMS system consists of four battery-based wireless sensors, a high frequency (HF) antenna, and a central receiver. The wireless sensor is mounted inside the tire to measure pressure and temperature, and send the data to the central receiver via wireless link. The HF antenna is fixed in the underbody. The central receiver consists of a data processing module in the vehicle interior, CAN (Control Area Network) bus, and diagnosis interfaces. The data sent from the wireless sensor are received by the HF antenna mounted in the underbody and are transferred to the central receiver via shielded coaxial cables. Then, the data are relayed to the Instrument Cluster Module (ICM) via a CAN bus interface. Once the tire pressure is abnormal, the system alerts the driver to take appropriate action.

The wireless sensor represents the core component of TPMS. An electronic sensor unit is combined with the aluminum valve to form a compact wireless sensor, which is bolted to the standardized valve bore on the rim inside the wheel, shown in Figure 11.55(b). The wireless sensor mainly consists of three integrated circuits. The first circuit contains micro-mechanical sensors and a signal-conditioning module in order to measure pressure, temperature, acceleration, and voltage. The second circuit contains a 315 MHz transmission power stage, which produces an RF signal and feeds it to the aluminum valve used as an antenna. The third circuit wakes up the wireless sensor when it receives a trigger signal of 125 kHz, thus setting the system to an active mode (Li *et al.*, 2006).

As illustrated in Figure 11.55(c), the wireless sensor is composed of an electronic sensor unit, valve, and washer ring. The electronic sensor unit and the valve are connected by a metallic spherical surface.

Many methods have been applied to improve the vehicle-carrying performance of the TPMS system. But most previous studies of TPMS emphasized the effects of sensor signal conditioning, temperature compensation, software control strategy, communication protocol, and thermal-mechanical characteristics of packaging, without considerations of TPMS antennas in the real application environment (Wang *et al.*, 2005).

As the wireless sensor's antenna is made of metal, the valve is electrically a small antenna because the valve's length is less than 5 cm, much smaller than the working wavelength of TPMS. The wheels

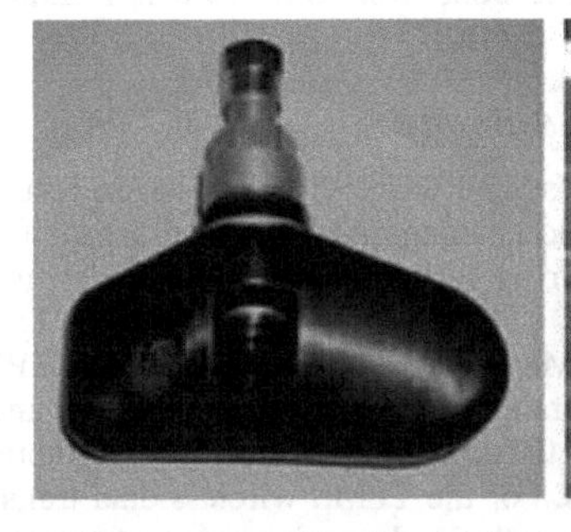

Figure 11.54 Two installation methods of TPMS

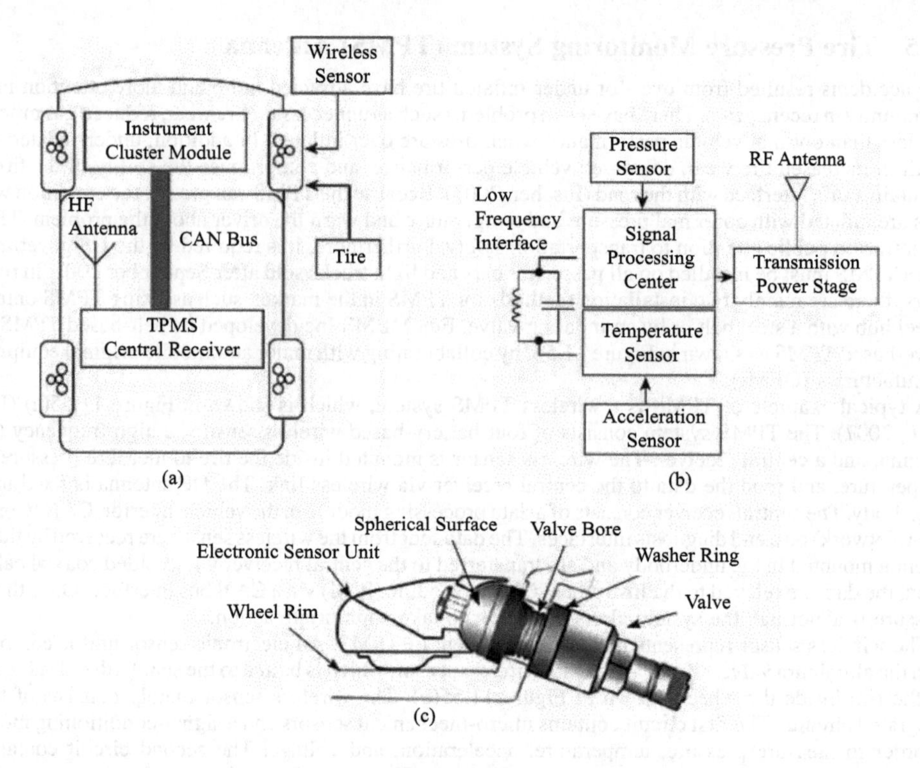

Figure 11.55 (a) TPMS architecture; (b) wireless sensor architecture; (c) illustration of wireless sensor structure on wheel rim

and the vehicle body structure strongly influence the electrical performance of the wireless sensor's antenna, especially when the wheel is rotating (Wiese *et al.*, 2005). Therefore, the valve is called a dynamic antenna due to its dynamic variety with the metallic wheel structure and rotation (Li *et al.*, 2007). Therefore, it is critical to come up with a new antenna so that the performance of the TPMS can be greatly realized.

In this section, the 3D electromagnetic simulation model of the wheel antenna is presented. The simulation results are used for the detailed design of wireless sensor circuit. Some numerical analysis had been done on the TPMS based on the inflating valve in a previously published paper (Wang *et al.*, 2005). The thermo-mechanical and centrifugal force characteristics of the TPMS model are also studied. A three-dimensional model is used to simulate various operation conditions to validate the reliability of the TPMS.

11.5.1 Test of TPMS System with Wheel Antenna

Two measurement methods, power transmission test and wireless transmission test, are used in this section. The wireless sensors with both the dynamic antenna and the wheel antenna were tested. The receiver antenna is installed in the underbody during test. The circuits are designed according to corresponding antenna parameters.

Figure 11.56 shows our specific design of our TPMS wireless test system. For the TPMS transmission power test, the radio frequency signal, which is generated by the RF signal generator and radiated by the valve antenna, is transmitted to the receiver antenna in the underbody. The receiver HF antenna is connected to the spectrum analyzer by shielded coaxial cables. For the TPMS wireless data transmission test, the adopted test samples are the existing TPMS system with the dynamic antenna from an existing vendor and the newly-designed TPMS system with the wheel antenna manufactured by the first author's group.

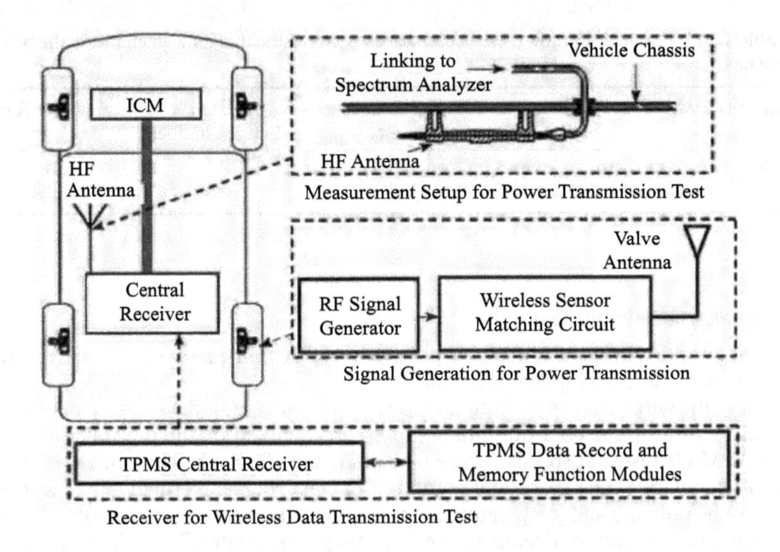

Figure 11.56 TPMS test system setup diagram

The data record and memory function modules are added to the TPMS receivers in order to automatically record wireless data frames from rotating wireless sensors in the course of vehicle road test.

(i) Power Transmission Test

To obtain an accurate test result of the total radiated power from the wireless sensor, the receiver antenna is placed in the fixed location and the wireless sensor on the left rear wheel was at various orientations. The wireless sensor provides a continuous sinusoidal carrier at 433.92 MHz and the input power is +5 dBm. Then, we turned the wheel incrementally in 30° steps and recorded twelve signal power values displayed on the spectrum analyzer respectively. The received power values from the different wireless sensors are compared, as illustrated in Figure 11.57. The signal power values from the wheel antenna shown in Figure 11.57 are 10–20 dB stronger than that from the dynamic antenna.

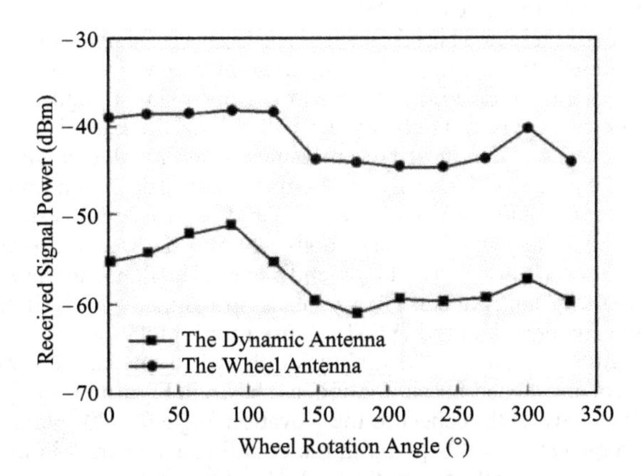

Figure 11.57 Experimental results of the received signal power from the dynamic antenna and the wheel antenna

Table 11.4 Frame error ratio (FER) comparison between the dynamic antenna and the wheel antenna

Antenna Type	Total Frame Number	Erroneous Frame Number	Error Ratio
TPMS with Wheel Antenna	300	12	4%
TPMS with Dynamic Antenna	300	88	29.3%

(ii) Wireless Data Transmission Test

In order to determine the wireless transmission performance, the frame error statistics measurements were carried out in the vehicle-carrying environment.

After installing the wireless sensors in the left rear wheel and fixing the receiver onto the vehicle, we inflate the vehicle's tire to the cold tire inflation pressures marked on the tire inflation pressure label. The vehicle's TPMS is tested at speeds between 50 km/h and 100 km/h. The vehicle is driven for up to 600 minutes of cumulative time along any portion of the test course. Normally, the electronic wireless sensor measures the pressure and temperature of the air in the tire every four seconds, while transmitting one data frame with air pressure and temperature data every two minutes. Thus, the total transmitting data frame number is 300 during testing. Meanwhile, considering the improvement of the wheel antenna parameters such as the radiation resistance, the gain pattern, and the radiation efficiency, the input power is decreased to $+2$ dBm, which is only one-half the power supply of the previous wireless sensor with the dynamic antenna (Wiese *et al.*, 2005).

The vehicle-carrying test results for the TPMS system with wheel antenna are listed in Table 11.4. For comparison, the test results for TPMS with dynamic antenna are also listed in the same table. During the test, the frame length is 10 bytes.

From the table, we can see that the FER of the TPMS system with the wheel antenna is far less than the dynamic antenna's.

11.5.2 3D Electromagnetic Modeling of Wheel Antenna

The wheel antenna presented in this section is composed of a valve and a wheel, as shown in Figure 11.58 (a), and it has already been used on cars in one OEM. The electrical connection between the sensor and the wheel is realized through a metallic washer ring. The metallic washer ring electronically connects the valve with a wheel, providing a means for conducting electric RF signals from a sensor in the wheel to the tire valve and the wheel, which radiates the RF power.

An optimization has been conducted based on the needs from a passenger car body of SQR 7160 series by a local supplier. A feasible design is chosen for modeling the wireless sensor's antenna, which is shown in Figure 11.58. The vehicle dimensions are approximately 4.3 m (length) × 1.7 m (width) × 1.4 m (height). The wheel diameter is approximately 60 cm and the wheel thickness is approximately 18 cm. The wireless sensor model consists of a circuit board submodel and a valve submodel. The circuit board submodel is 5.5 cm × 2.3 cm in size. The valve is 5 cm in length and 1.5 cm in diameter. The wireless sensor is mounted in the wheel and the valve makes an angle of 15°s with respect to the surface of the wheel well. In addition to the wheel and the vehicle body, ground effect cannot be neglected due to the fact that the distance from the sensor's antenna to the ground plane is less than one wavelength and the distance from the wheel bottom to the ground plane is 12 cm. An electromagnetic (EM) simulation software, based on a 3D full-wave Finite Element Method (FEM), is used for simulation.

Assuming that the wireless sensor is mounted inside the wheel of the left rear tire, the four typical locations around the wheel are chosen for simulation, as shown in Figure 11.58(b). Figure 11.59 shows a plot of the azimuth gain pattern with respect to the elevation angle $\theta = 90°$ plane for the left rear wheel antenna at the TPMS frequency. $\theta = 90°$ plane in the coordinate system is in parallel with the ground plane. The azimuth angle at $-90°$ is the front of the vehicle and it rotates in a counter-clockwise direction. The scale is labeled in the figures.

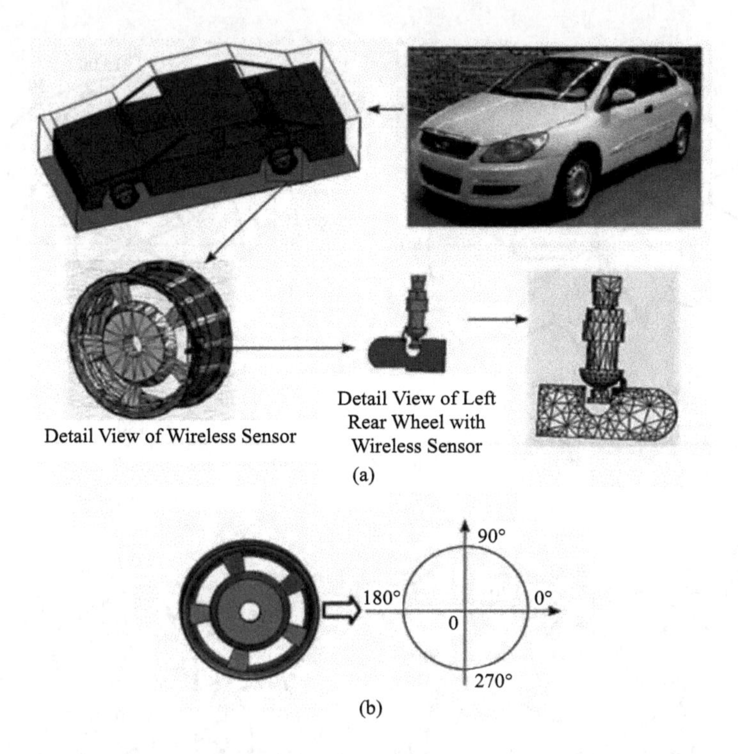

Figure 11.58 (a) 3D EM simulation model for the wheel antenna. (b) Tire valve with orientations at 0°, 90°, 180°, and 270° respectively

Compared with the dynamic antenna, there is bigger power gain and less blind areas (Wiese, 2005). In this case, the effect of the wheel and the vehicle body are to increase the peak gain of the 90 °s orientation from approximately 12 dB to 15 dB (that is an increase of 25%). There are the same gain increases for the other orientations.

As the four locations around the wheel may not sufficiently represent the scenarios for gains and impedance, more simulations for the TPMS wheel antenna rotated incrementally in 30° steps are carried out for the left rear wheel. Hence twelve simulations are run for the wheel antenna. The simulation results are shown in Figure 11.60(a) and (b) respectively. The simulated gain range is from 2.5 to 6 dB. Typical gain values are above 3 dB. The resistance fluctuation range is 0.6 to 0.89 Ω and the reactance fluctuation range is − 100 to 104.41 Ω. Most values of the resistance are above 0.7 Ω.

Compared with the gain and the impedance's sharp fluctuation for the dynamic antenna, the gain value of the wheel antenna fluctuates within a small range, the gain and the resistance part are much bigger than the dynamic antenna's, which makes the wheel antenna become a better radiator of electromagnetic power (Wiese *et al.*,2005). In particular, the small impedance variation range helps to design a fixed antenna matching network.

11.5.3 Stress Modeling of Installed TPMS

TPMS installed on the wheel hub is used for regular tires without inner tube tires. Generally, the whole TPMS body is subjected to centrifugal force and thermal stress during the rolling condition.

The TPMS includes a printed circuit board (PCB) which is mounted with some components, such as a pressure sensor, a crystal, a battery, and an antenna. All these components are encapsulated by a plastic

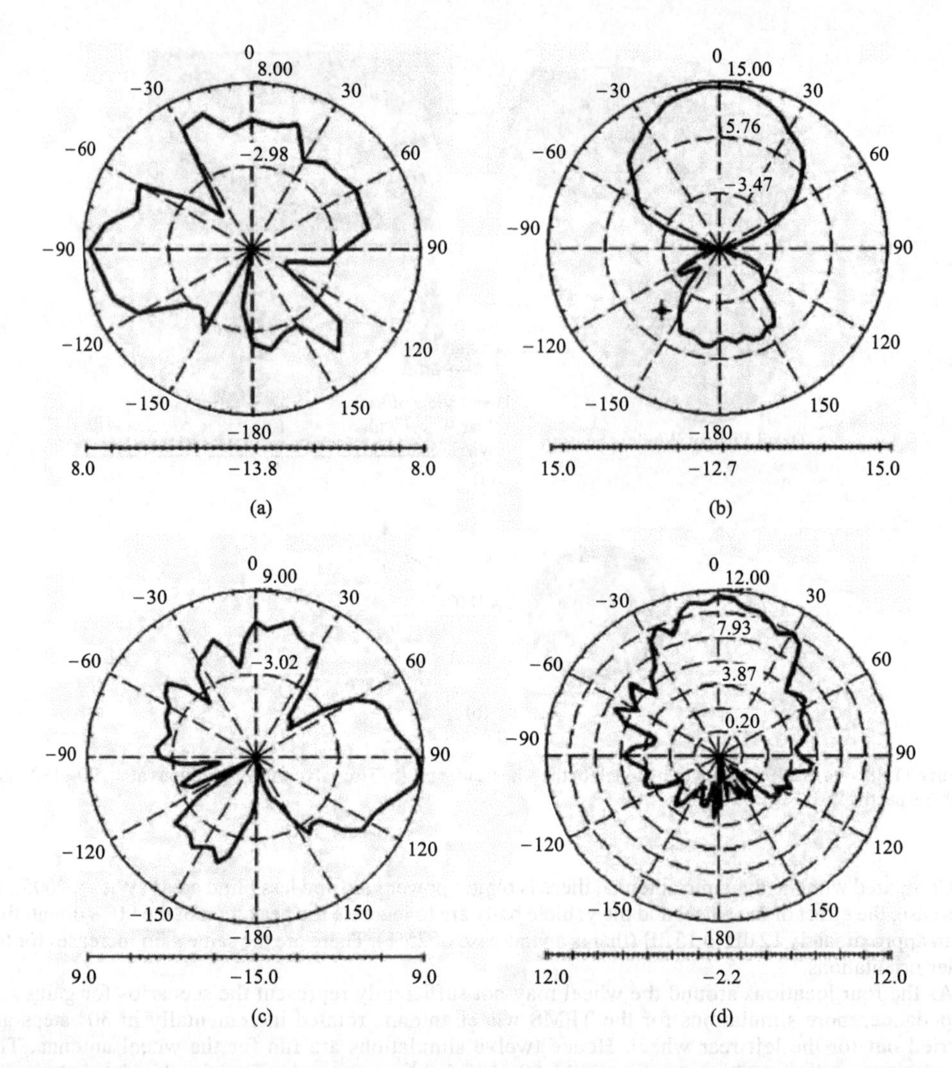

Figure 11.59 Plots of the azimuth gain patterns at four orientations. (a) 0°s; (b) 90°s; (c) 80°s; (d) 270°s

package with some filling glue as shown in Figure 11.61. The whole TPMS is fixed on the wheel hub at the defined location with some fixing glue before the system operation. In this research, the TPMS with the fixing glue of different thickness is analyzed by numerical simulation.

The automobile's driving velocity is considered to vary from 40 km/h to 200 km/h while the diameter of rolling track of the TPMS installed on wheel hub is 598 mm, and the rigorous temperature operation condition of −40 °C to 125 °C is required by the standard in accelerating testing. In order to analyze the reliability and durability of the fixing glue, the thickness of the fixing glue in the TMPS model varies from 0.5 mm to 2.5 mm in different simulation models. Thermo-mechanical material properties of components are shown in Table 11.5.

The TPMS installed on the wheel hub is studied with finite element analysis (FEA). The PCB with chips and fixing glue are the key components for system reliability and durability. Under normal operation condition of 80 °C and velocity of 80 km/h, the result indicates the stress distribution contour of the

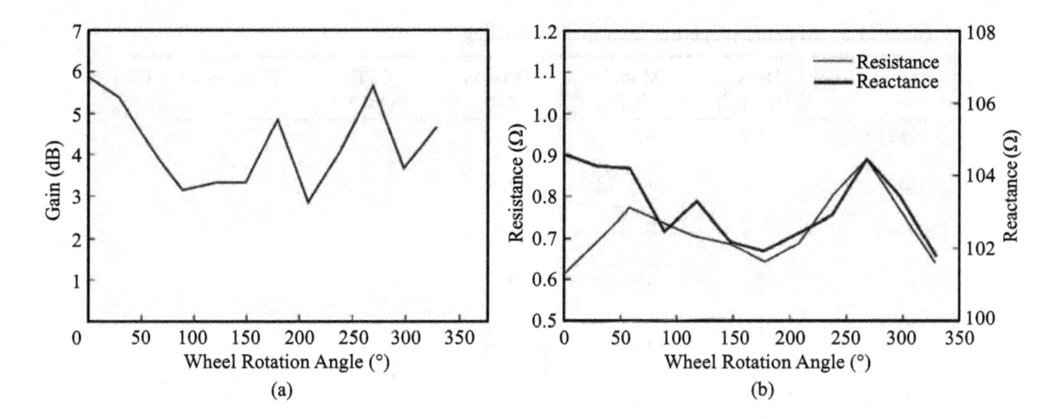

Figure 11.60 (a) Simulated gain of the wheel antenna directed towards HF antenna at twelve locations. (b) Computed impedance for the wheel antenna

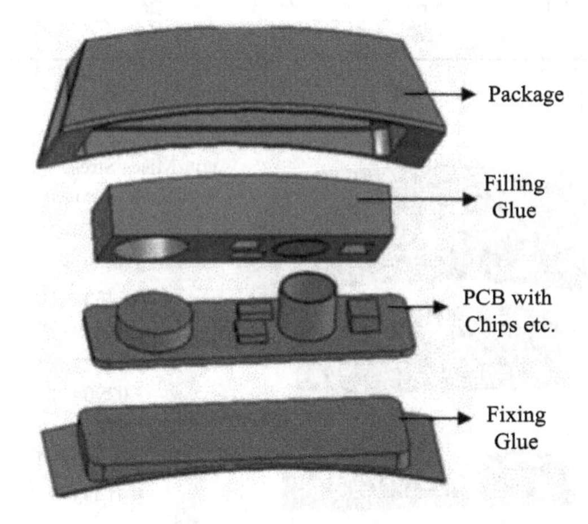

Figure 11.61 Modeling construction of TPMS

internal components in TPMS as shown in Figure 11.62. It can be seen that the maximum stress appears at the linking location between the fixing glue and the wheel hub, but the maximum stress in the fixing glue is low enough for system safe operation. As shown in Figure 11.63, the maximum stress in the fixing glue is only about 1.6 MPa which is lower than the fixing glue's yield strength 7.3 MPa, even when the rolling velocity approaches 200 km/h.

In the course of TPMS installation, the thickness of the fixing glue could be varied easily. But the result shows the thickness of the fixing glue has little effect on the maximum stress in the fixing glue. As shown in Figure 11.64, the maximum stress in the fixing glue is low enough for the TPMS to operate well.

Finally, the study shows that the harshest condition for TPMS operation is the temperature. Results for 120 km/h at operation temperature between −40 °C and 125 °C are all obtained by the finite element method (FEM) as shown in Figure 11.65, and the maximum stress of the TPMS with 1 mm fixing glue is respectively 83.9 MPa in the PCB, 21.7 MPa in the fixing glue, and 11.9 MPa in the package. It can be found that TMPS is still safe under the most possible harsh conditions.

Table 11.5 Material properties used in the modeling

Mat ID	Density (g/cm^3)	Modulus (GPa)	Poisson Ratio	CTE (ppm/°C)	Temperature (°C)
Package	1.2	4.5	0.42	20	25
				60	125
PCB	4.0	*x*, *y* 22.4	0.13	*x*, *y* 16	25
		z 1.6		*z* 65	
Battery	3.5	80	0.3	2.1	25
				2.3	−43
Chip	2.6	196	0.22	2.7	26
				3.0	126
Sensor	1.2			2.3	−43
		100	0.22	2.7	26
				3.0	126
Antenna	8.9	110	0.27	24	25
Crystal	5.9	107	0.16	0.55	25
		3.61	0.42	21	−40
Filling Glue	1.12	2.65	0.42	28	25
		0.04	0.49	63	125
		6.58	0.32	19	−40
Fixing Glue	1.18	5.67	0.35	26	25
		1.71	0.4	56	120

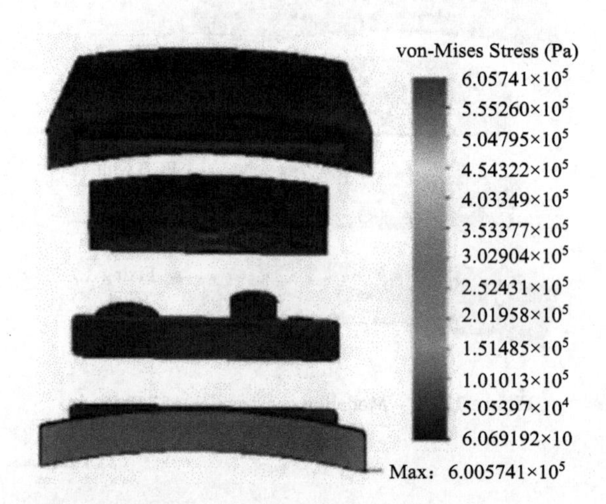

Figure 11.62 Stress distribution under rolling velocity

11.6 Thermo-Fluid Based Gyroscope Packaging

Gyroscopes have been widely used in engineering applications, including missile inertial navigation, automobile industry, robotics systems, and consumer electronics. However, high-performance gyroscopes of smaller size, lower cost, and acceptable reliability are still desirable, which will accelerate the commercialization of micromachined gyroscopes. The existing micromachined gyroscopes mainly utilize Coriolis force induced by external rotation to drive perpendicular vibrating for angular rate detection, which has been based on capacitive, piezoelectric, piezoresistive, and electrical tunneling effects.

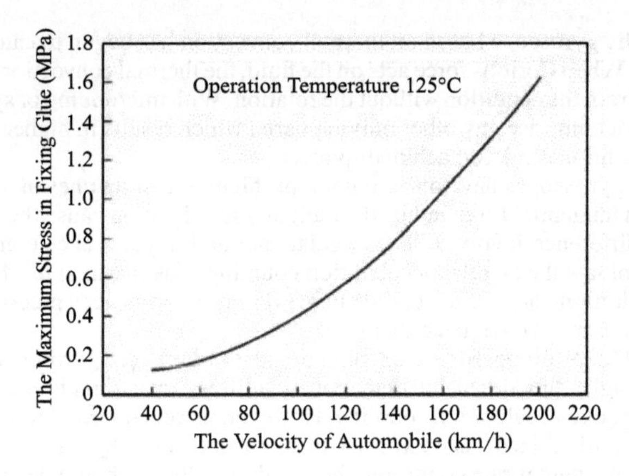

Figure 11.63 Maximum stress of fixing glue under various velocities

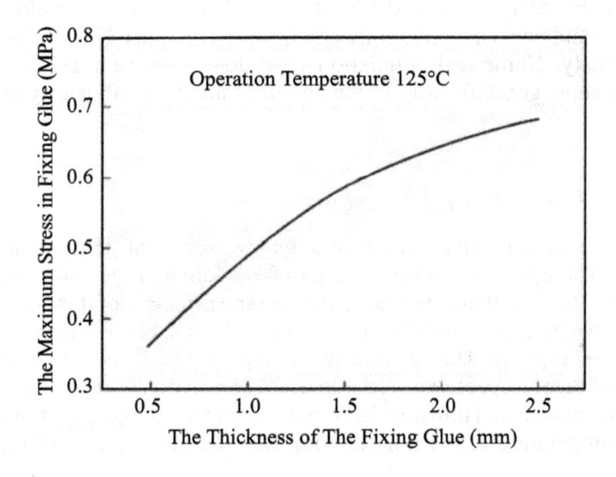

Figure 11.64 Maximum stress of fixing glue under various thicknesses

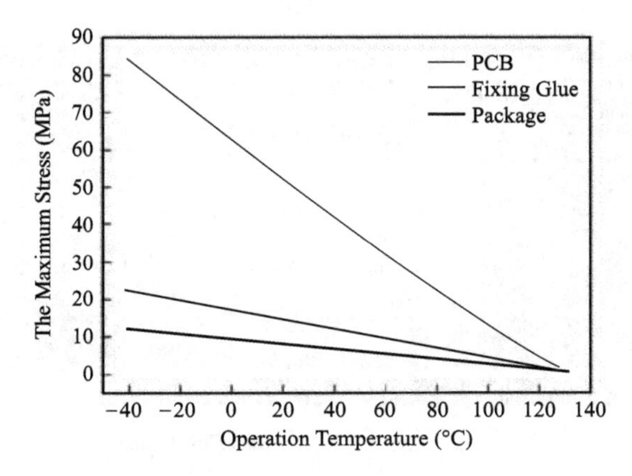

Figure 11.65 Maximum stress under various temperature points

In recent years, fluidic gyroscope based on thermal convection has begun to catch the attention of the gyroscope community. When Coriolis force acts on the fluid, the thermal convection between the fluid and thermistors will differ from the condition without the rotation. With micropump or synthetic jet as the core working part, it does not comprise any other moving parts, which results in higher reliability, durability, and anti-shock than traditional micromachined gyroscopes.

The reported fluidic gyroscopes have one common problem: under an uneven angular rate condition, both Coriolis force and tangential force induced by angular acceleration cause the gas flow to deflect, so that the temperature difference involves both angular rate and angular acceleration, but the reported gyroscopes cannot eliminate the angular acceleration coupling (Takahashi *et al.*, 1991; Dau *et al.*,2006; Yang *et al.*, 2004). In addition, they are not based on MEMS. Some parts need precision machining, which makes assembly difficult and the device expensive.

China patent 01129222.9 proposed a dual directional synthetic jet gyroscope (Luo *et al.*, 2001). Compared with the single directional jet detection, it utilizes dual directional gas flow, which can effectively eliminate angular acceleration coupling. However, no real micro gyroscope was made and the jet deflection method is still utilized. As we know, Coriolis acceleration $a_k = -2\omega \times V$. In order to obtain enough sensitivity in low angular rate condition, the gas flow velocity should be high enough. While the angular rate is also large, the gas flow deflection may exceed the chamber limit. It is clear that the sensitivity and the measurement range cannot be optimal simultaneously.

This section reports the design and analysis of a novel micro thermo-fluidic gyroscope which is applicable for both even and uneven angular rate conditions. The sensitivity and measurement range can be optimal simultaneously. Numerical simulation concludes that the mass flow rate difference and temperature difference show good linearity to the angular rate, the nonlinearity are 0.099% and 0.15%, respectively.

11.6.1 Operating Principle and Design

Schematic view of the proposed micro thermo-fluidic gyroscope utilizing dual directional liquids is shown in Figure 11.66. The gyroscope is symmetric to the midline of the two vibrating membranes. The left half part is introduced to illuminate the basic structure and principle of this novel micro thermo-fluidic gyroscope. The fluid flows from the left outlet to a buffered channel for rectification and the main channel is driven by the central micropump. Then it splits into two parts and flows through split channels with one thermistor inside each of them respectively. After thermal convection with the thermistors, the fluid flows back from the backflow channel and transfers heat to the shell with a high heat transfer coefficient, which is used to ensure the temperature of the fluid flowing into the micro-pump is uniform and repeatable.

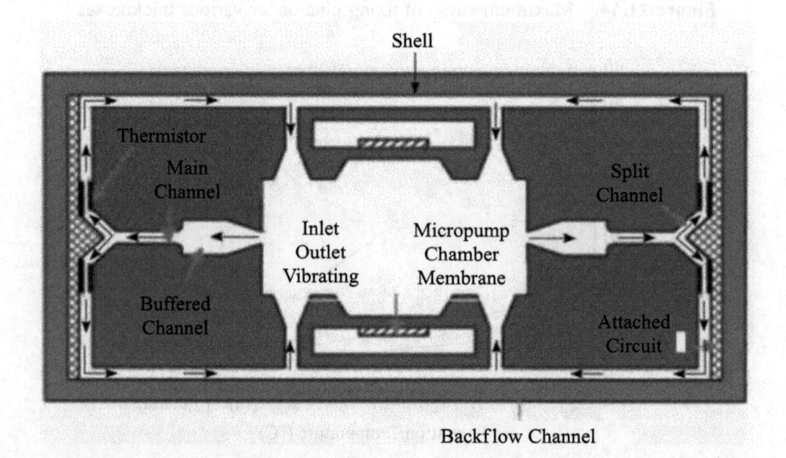

Figure 11.66 Schematic view of the proposed micro thermo-fluidic gyroscope utilizing dual directional liquids

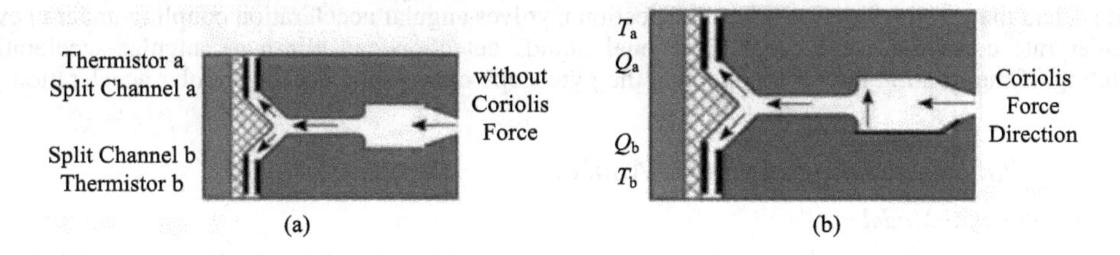

Figure 11.67 Principle of the proposed micro thermo-fluidic gyroscope. (a) No rotation applied: $Q_a = Q_b, T_a = T_b$; (b) rotation applied: $Q_a > Q_b, T_a < T_b$

The basic principle of the gyroscope is shown in Figure 11.67. Q_a, Q_b are the mass flow rates of split channel a, and split channel *b* respectively. T_a, T_b are the temperatures of thermistor *a,* and thermistor *b* respectively. When no rotation is applied (Figure 11.67(a)), no Coriolis force acts on the fluid, the mass flow rates of the two symmetric split channels are equal ($Q_a = Q_b$), so that the temperature difference between the thermistors does not exist ($T_a = T_b$); When the rotation is applied (Figure 11.67(b)), Coriolis force acts on the fluid, the force direction is shown in Figure 11.67(b), the mass flow rate of split channel *a* is larger than that in split channel *b* ($Q_a > Q_b$), so that the temperature of thermistor *a* is lower than that in thermistor *b* ($T_a < T_b$). The angular rate is obtained by simply measuring the temperature difference.

11.6.2 Analysis of Angular Acceleration Coupling

Principle of dual directional liquids detection is shown in Figure 11.68. F_K, F_Q respectively represent Coriolis force and tangential force induced by angular acceleration, while R, β, ω respectively represent rotation radius, angular acceleration, and angular rate. The initial velocity and mass of the concerned fluid are V_0, m respectively.

The rotation direction is shown in Figure 11.68. The Coriolis force and tangential force acting on the fluid is respectively expressed by:

$$F_K = -2m\omega \times V_0 \tag{11.20}$$

$$F_Q = -m\beta \times R \tag{11.21}$$

Under even angular condition, β is zero, so that F_Q is zero. The total force acting on the fluid flowing to the left is $2m\omega V_0$, with the direction upwards; while the total force acting on the fluid flowing to the right is $2m\omega V_0$, with the direction downwards. The output signal of the gyroscope is the function of the summation force acting on the fluid flowing to the left and right which is $4m\omega V_0$ in total.

Under uneven angular condition, β is not zero. We assume β is positive. The total force acting on the fluid flowing to the left is $m(2\omega V_0 - \beta R)$, with the direction upwards; while the total force acting on the fluid flowing to the right is $m(2\omega V_0 + \beta R)$, with the direction downwards. The output signal of the gyroscope is the function of the summation force acting on the fluid flowing to the left and right which is still $4m\omega V_0$ in total.

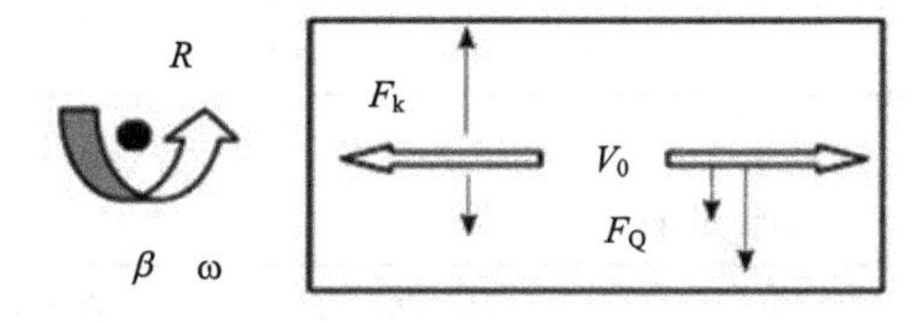

Figure 11.68 Principle of dual directional liquids detection

It is clear that single directional liquid detection involves angular acceleration coupling under uneven angular rate condition, while dual directional liquids detection can eliminate angular acceleration coupling. If the rotation radius R is known, the gyroscope can also detect the angular acceleration β.

11.6.3 Numerical Simulation and Analysis

(i) Computational Model

The governing equations involved in the computational model are mathematically expressed below:

$$\frac{\partial \rho}{\partial t} + \Delta(\rho \vec{u}) = 0 \tag{11.22}$$

$$\frac{\partial(\rho \vec{u})}{\partial t} + \rho \vec{u} \nabla(\vec{u}) = \nabla(\mu \nabla(\vec{u})) + \rho \vec{a} - \nabla \rho \tag{11.23}$$

$$\frac{\partial(\rho T)}{\partial t} + \nabla(\rho \vec{u} T) = \nabla\left(\frac{k}{c_p} \nabla T\right) + S_T \tag{11.24}$$

$$p = \rho R T \tag{11.25}$$

Equations 11.22, 11.23, 11.24 and 11.25 refer to continuity equation, momentum equation (also called Navier–Stokes equation), energy equation, and state equation respectively, where $\vec{u}$, $\vec{a}$, ρ, p, T, c_p, S_T, k, and μ refer to the velocity vector, acceleration vector, fluid density, pressure, temperature, specific heat capacity, heat generation, thermal conductivity, and dynamic viscosity, respectively.

The physical properties of the fluid (in this case water) and channel specifications that are used in the numerical simulation are summarized in Table 11.6 and Table 11.7, respectively.

A three-dimensional model to simulate the incompressible laminar steady flow is shown in Figure 11.69. A sensitive enough mesh with a brick element is applied in the computational model.

The initial velocity V_0 of inlet is 1 m/s, prescribing the Coriolis force direction to be the same as the X axis positive direction. The heat generation of thermistor is 10 mW.

(ii) Mass Flow Rate Simulation Result and Analysis

Figure 11.70 shows the velocity distribution of line I in the main channel with no rotation applied and with rotation applied. Coriolis force induced by the rotation makes the velocity profile differ from that without rotation, which results in the mass flow rate difference of the two symmetric split channels.

Figure 11.71 shows the velocity distribution of face I and face II respectively in split channel I and II when ω is 400 deg/s. It reveals that the velocity profile of face II is higher than that of face I, which means the mass flow rate of split channel II is higher than that of split channel I.

Figure 11.72 shows the relationship between the angular rate ω and the mass flow rate difference of Outlet I and Outlet II. It is clear that the mass flow rate difference shows good linearity to the angular rate. The nonlinearity of mass flow rate difference to the angular rate is 0.099%. When the angular rate ω is

Table 11.6 Physical properties of water at 300 K

Thermal Conductivity	Kinematic Viscosity	Density
0.618 W/(m·K)	0.81×10^{-6} m²/s	997 kg/m³

Table 11.7 Channel specifications

Length	Width	Height	Angle of Split Channels
8 mm	0.4 mm	0.25 mm	90°

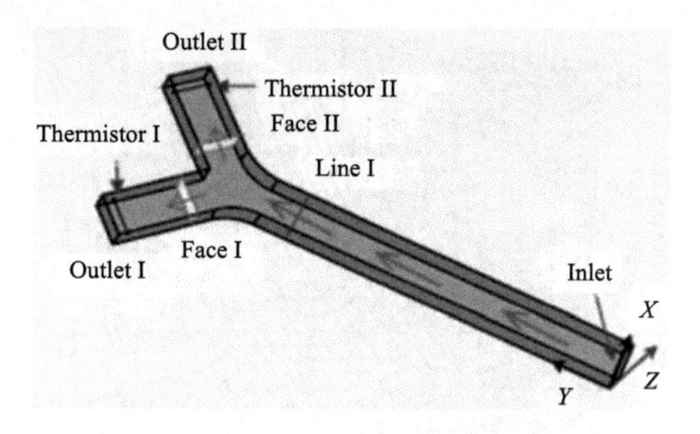

Figure 11.69 Computational model

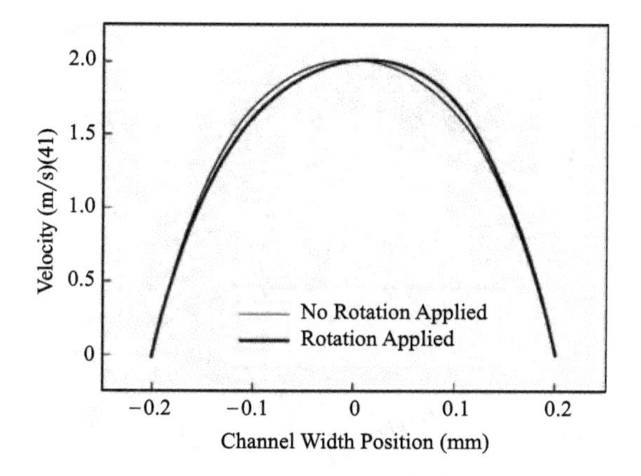

Figure 11.70 Velocity distribution of line I with no rotation applied and with rotation applied

400 deg/s, the mass flow rate difference is 0.00359 g/s, which is 7.2% of 0.0499 g/s, the mass flow rate of one single split channel without rotation.

(iii) Thermal Convection Simulation Result and Analysis

The heat energy transfer from thermistor to fluid through thermal convection is given by Newton's Law of cooling:

$$\Phi = hA(T_{\text{surface}} - T_{\text{fluid}}) \tag{11.26}$$

where h, A, T_{surface}, and T_{fluid} are thermal convection coefficient, contact surface area, temperature of thermistor surface and fluid, respectively. The contact surface area A and temperature difference are the same for the symmetric split channels. The convection coefficient h is a complicated function of fluid flow, thermal properties of the fluid and the geometry of thermistor. The mass flow rate is the only different factor of the symmetric split channels, which causes the temperature difference between the symmetric thermistors inside each split channel. Thermal mass flowmeter is based on the same principle that temperature of thermistor is related to the mass flow rate (Roha *et al.*, 2006).

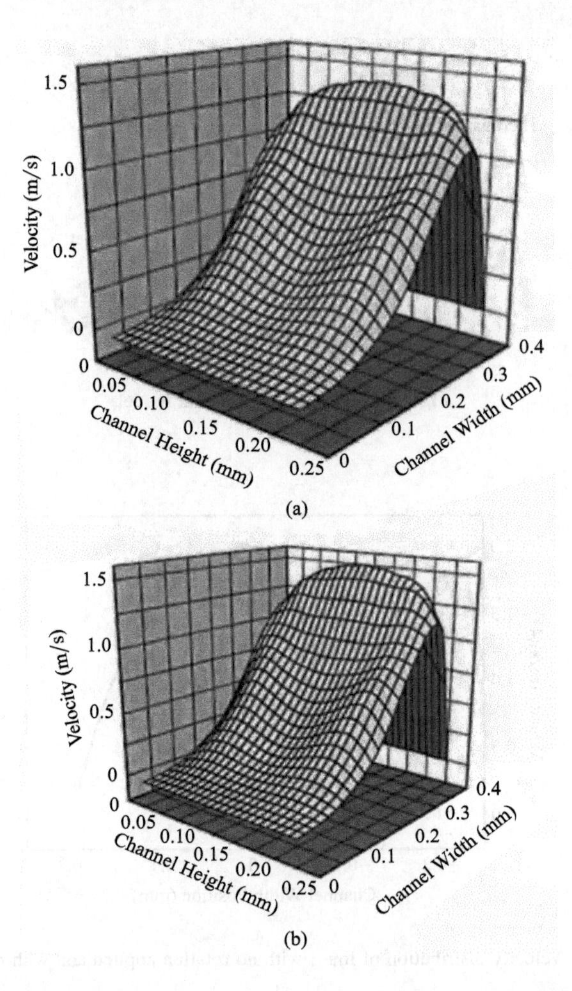

Figure 11.71 Velocity distribution of face I and face II when $\omega = 400$ deg/s. (a) Velocity distribution of face I; (b) velocity distribution of face II

Figure 11.73 shows the relationship between the angular rate ω and the average temperature difference of Thermistor I and II, which shows good linearity to the angular rate. The nonlinearity of the average temperature difference to the angular rate is 0.15%. The average temperature difference is 0.14 K when ω is 400 deg/s with a power consumption of 10 mW.

The parameters of the proposed gyroscope will be optimized to increase sensitivity and frequency response subsequently. Some other fluids such as mercury, various gases such as CO_2 will be considered instead water. In addition, the linearity of the mass flow rate difference and temperature difference to the angular rate will be verified in the prototype experiments.

11.7 Microjets for Radar and LED Cooling

High power light emitting diode (LED) based semiconductor illumination technology, also called solid state lighting (SSL), has become the focus for both research and commercial applications in recent years, because it has a lot of advantages compared with the traditional illumination technologies. Currently, the major general light sources are incandescent lamps, halogen lamps, and fluorescent lamps.

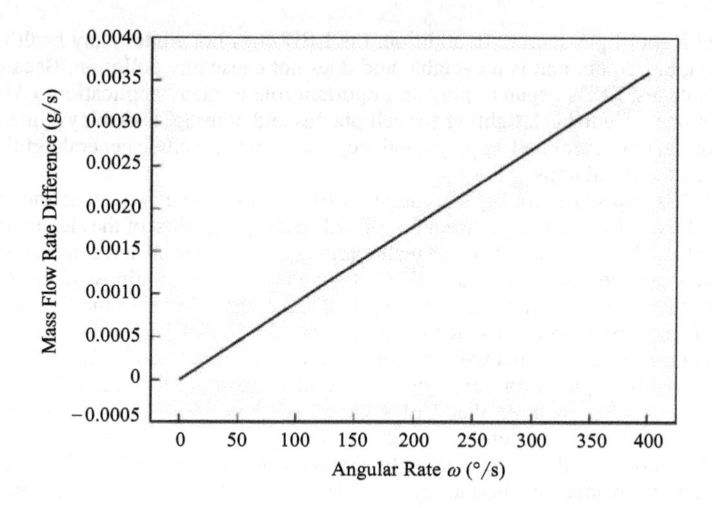

Figure 11.72 Relation between angular rate ω and mass flow rate difference

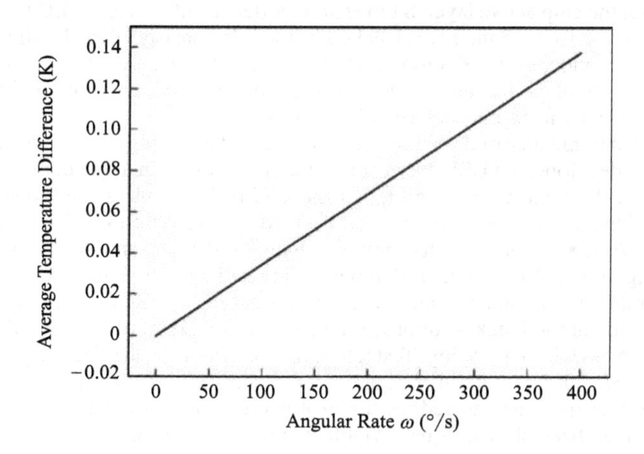

Figure 11.73 Relation between angular rate ω and average temperature difference

Compared with these general light sources, theoretically, LED has several advantages as follows (Zukauskas *et al.*, 2002):

(1) High luminescence efficiency, good optical monochromatic, narrow spectrum, and the ability to emit visible light directly;
(2) Energy saving and low power consumption, as the power consumption of a LED is only one eighth of an incandescent lamp and one half of a fluorescent lamp under the same illumination efficiency and there is still potential to grow significantly in the next years to come;
(3) Long lifetime; and
(4) They are safe and environment-friendly.

In theory, the LED's average lifetime is as long as 100,000 hours, which is ten times that of general lamps. LED is a solid state light source and has low calorific value and no thermal radiation, so that the LED is generally called a cool light source. In addition, the LED does not contain any health hazard such as mercury, sodium, and so on, and is recyclable and does not cause any pollution. Because of the above mentioned advantages, LEDs began to play an important role in many applications (Alan *et al.*, 2003). Typical applications include back lighting for cell phones and other LCD displays, interior and exterior automotive lighting, large signs and displays, and are soon going to include general lighting such as street lighting and even the headlamps.

For present LEDs, especially for high-brightness LEDs, both optical extraction and thermal management are critical factors for the high performance of LED packaging. Most of the electrical power in LEDs will be converted into heat for the state of art technology with low internal and external optical extraction efficiency. In general, the produced heat will greatly reduce device luminosity. In addition, the high junction temperature will shift the peak wavelength of LED, thus changing the light color of the LED. Narendran *et al.* have experimentally demonstrated that the life of LEDs decreases with increasing junction temperature in an exponential manner (Narendran *et al.*, 2005). Therefore, low operation temperature is strongly required for LEDs, which is a distinguishing feature of SSL versus traditional lighting. Since the market requires that LEDs have high power and packaging density, it poses a contradiction between the power density and the operation temperature, especially when applications demand LEDs to operate at full power or more to obtain the desired brightness. The problem leads to the emergence of major advances in thermal management of LED illumination. To address the thermal problem of LEDs, numerous researchers both in China and in other countries have conducted relevant studies. Arik and Weaver carried out a numerical study to understand the chip temperature profile due to the bump defects (Arik *et al.*, 2004). Finite element techniques were utilized to evaluate the effects of localized hot spots at the chip active layer. Sano *et al.* reported an ultra-bright LED module with excellent heat dissipation characteristics (Sano *et al.*,1993). The module consists of an aluminum substrate having outstanding thermal conductivity and a mount for LED chips being formed into fine cavities with high reflectance to improve light recovery efficiency. Furthermore, the condenser of this LED module is filled with high-refraction index resin on the basis of optical simulation.

An improvement in luminance of 25% or more was observed by taking the aforementioned measures in their paper. Petroski developed an LED-based spot module heat sink in a free convective cooling system (Petroski, 2003). A cylindrical tube, longitudinal fin (CTLF) heat sink is used to solve the orientation problem of LED. Chen *et al.* presented a silicon-based thermoelectric (TE) for cooling of high power LEDs (Chen *et al.*, 2005). The test results show that their TE device can effectively reduce the thermal resistance of the high power LED. Hsu *et al.* reported a metallic bonding method for LED packaging to provide good thermal dissipation and ohmic contact (Hsu *et al.*, 2003). Zhang *et al.* used multi-walled carbon nanotubes and carbon black to improve the thermal performance of thermal interface materials (TIM) in high-brightness LED packaging. Tests show that such a TIM can effectively decrease the thermal resistance (Zhang *et al.*, 2005). Acikalin *et al.* used piezoelectric fans to cool LEDs (Acikalin *et al.*, 2004). Their results show that the fans can reduce the heat source temperature by as much as 37.4 °C. The piezoelectric fans have been shown to be a viable solution for thermal management of the electronic component and the LED.

In China, a lot of researchers have also tried to find the solutions to the LED thermal management. Wu *et al.* used a finite element method to analyze the temperature distribution in one watt high power white LEDs (Wu *et al.*, 2005). The simulation results show good agreement with the experimental data. Based on the simulation model, they studied the dependence of chip size on the junction temperature. The results show that with a certain lighting efficiency and packaging structure, the size and maximum input power of the chip should be optimized because of the thermal dissipation limit. Yu and Li analyzed the dependence of bonding materials on the thermal characteristics of high power LEDs (Yu *et al.*, 2005). They established a correlation of bonding material thickness with thermal conductivity and thermal resistance for flip-chip type LEDs. Examples with three typical materials are provided to prove the function. The results show that the bonding material plays an important role in decreasing LED thermal resistance. Chen and Song utilized the thermal resistance model to predict the junction temperature and guide the thermal design of high power LEDs (Chen *et al.*, 2005). By comparing experiments, the feasibility of the model was proven. Ma and Bao introduced a novel electrical measurement method for thermal resistance and junction temperature of the high power LED (Ma *et al.*, 2005). They also investigated the influence factors and constructed an experiment setup, and the experimental results show that the method has the advantages of a simple structure and good stability. It can be used to measure thermal resistance and junction temperature of high power LEDs. Wang *et al.* came up with an equation, which established the relation of LED output power with thermal characteristic parameters under transient condition (Wang *et al.*, 1994). Based on the equation, the dependence of the LED output power on different current pulses was analyzed. The variation of optical output and heat with pulse width and duty ratio was presented.

Because of the high power density of the LED array modules and systems, an active cooling solution, a microjet based cooling system was proposed for LED thermal management. Liu's group had already conducted a series of studies of this cooling system (Luo *et al.*, 2007). They studied its performance for thermal management of high power LEDs which were less than 10 W. In the preliminary experiments, the temperature was measured by an infrared thermometer (Luo *et al.*, 2007). Only the LED surface temperature was obtained, which should be different from the junction temperature of LED chips. In the enhanced investigations, several thermocouples were packaged into the chip substrate to conduct on-line temperature measurement and evaluate the cooling performance of the proposed system (Luo *et al.*, 2007). Numerical simulations were also provided. The experimental and numerical results demonstrated that the microjet-based cooling system has good cooling performance.

In this section, the microjet cooling system for a 220 watts LED light source was investigated. Different from the previous work in which the power of LED packaging was less than 10 watts, the present work studied a microjet cooling system for a 220 watts LED light source. Because of much larger power density, design of the present microjet device was more complicated. Experimental investigation on the thermal characteristics of the 220 watts LED light source was conducted. To understand the internal flow and heat transfer characteristics of its cooling device, circumstantial numerical simulations were also conducted. The numerical model was proven by the comparison with experimental results. After checking the feasibility of the simulation model, it was used to simulate and optimize the microjet device structures. Three kinds of structures were investigated and compared; an optimized microjet structure was found that compared well with the present experiment al one.

11.7.1 Microjet Array Cooling System

Figure 11.74 illustrates the proposed close-looped LED cooling system. It is composed of three parts: a microjet array device, micropump, and a mini fluid container with a heat sink. When the LED needs to be cooled, the system is activated. Water or other fluids or gases in the close-looped system are driven into the microjet array device through an inlet by a micropump. Many microjets will form inside the jet device, which are directly impinged onto the bottom plate of the LED array. Since the impinging jet has a very high heat transfer coefficient, the heat created by the LEDs is easily removed by the recycling fluid in the system. The fluid is heated and its temperature increases after flowing out of the jet device, and then the heated fluid enters into the mini fluid container. The heat sink, which has a fan on the fluid container, will cool the fluid and the heat will be dissipated into the external environment. The cooled fluid is delivered into the jet device to cool the LED array, again driven by the force of the micropump in the system. The above processes constitute one operation cycle of the total system. It should be noted here that the real size of the proposed system can be designed as one small package according to the requirements of the application.

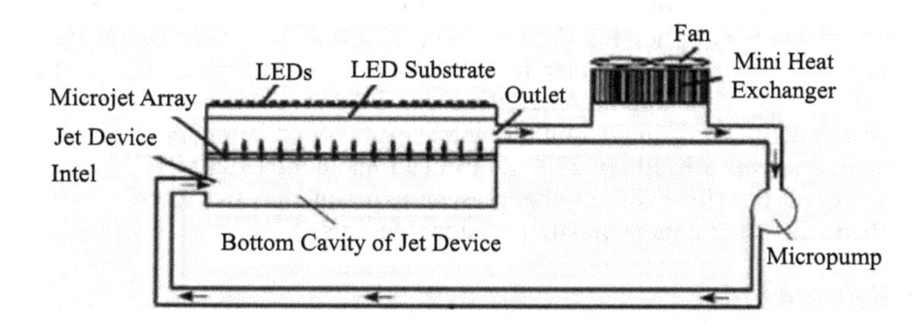

Figure 11.74 Principle schematic of the microjet cooling system

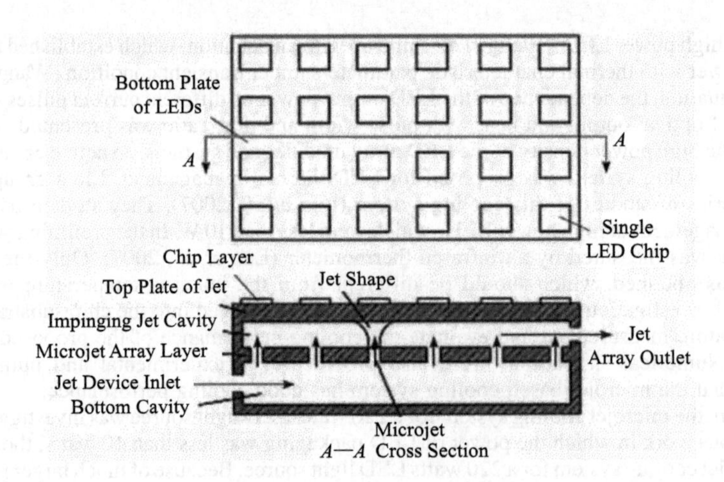

Figure 11.75 Microjet array device

Figure 11.75 shows the structure of the jet device of Figure 11.74 in detail. It consists of several layers, which are (from top to bottom) the chip array layer, the top plate of the jet cavity, the impinging jet cavity, the microjet array layer, and the bottom cavity. Cooled fluid enters into the device through the inlet, which is open at one side of the bottom cavity layer. The fluid flows through the microjet array and forms many microjets, as shown in Figure 11.75.

With sufficient driving force, the jets will impinge onto the top plate of the jet cavity which is bonded with the chip substrate of the LEDs. The heat conducted into the top plate of the jet cavity through the LED chips will be dissipated into the cooled fluid quickly due to the high heat transfer efficiency of the impinging jet. The fluid temperature increases and the heated fluid flows out from the jet array outlet, which is open at one side of the top jet cavity layer. Through this process, the heat from the LED chips will be transferred into the fluid efficiently.

11.7.2 Preliminary Experiments

(i) System and Temperature Measurement

The experimental system was constructed as shown in Figure 11.76. The details of the microjet device are shown in Figure 11.77. To understand the thermal performance of the 220 watts LED light source cooled by the microjet device, the temperatures of the silicon gel which was coated on the chips were measured. The positions of the thermocouples were at the center part of the light source frame, as shown in Figure 11.76. The working media of the cooling system was water. The light source frame was used to support the glass coated with phosphor. It is directly connected with the LED substrate by soldering. Since the thermal resistance between the LED substrate and the frame was very small, the measured temperature can represent the LED substrate temperature. The data obtained by the thermocouples was transferred to the data acquisition system and displayed on the notebook monitor. The model of the data acquisition system in the experiment was a Keithley 2700 multimeter and control unit 7700. It has 20 channels to collect data synchronously. However, in the present experiments, two thermocouples were used. Therefore, two channels of the data acquisition system were used.

(ii) Measurement Accuracy

In the experiments, temperature was the main parameter for system evaluation, and its values at various locations were directly measured by thermocouples. The errors associated with temperature measurement

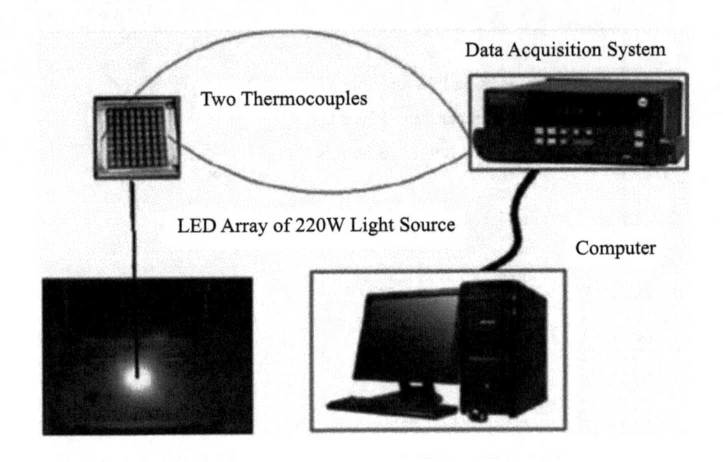

Figure 11.76 Experimental system setup

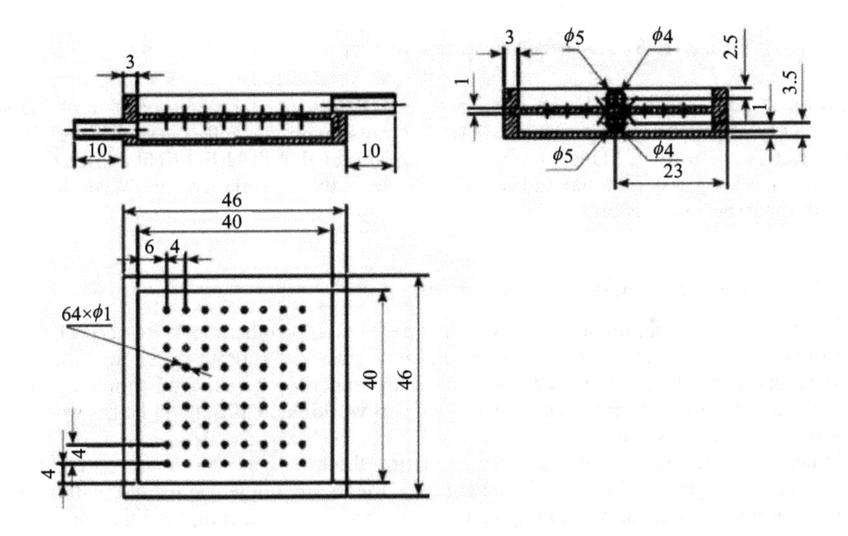

Figure 11.77 Sizes of the microjet array in experiment (all dimensions are in mm)

mainly included measurement error of the thermocouples and reading error of the digital multimeter. Standard T-type thermocouples (Cu–CuNi) were used in the experiments for temperature tests. During the temperature range from −30 °C to 150 °C, their measurement error was about 0.2 °C, its relative measurement error was about 0.75%.

The data acquisition system had a reading error of 1 °C and a 1% relative measurement error since the cold junctions of the thermocouples used the default setup supplied by the system, not the ice bath with constant 0 °C. Therefore, the total error of the temperature measurement for the experiments was about 1.2 °C, and the relative measurement error was about 1.75%.

(iii) Experimental Results

Figure 11.78 shows the experimental results. In the experiments, the environment temperature was about 30.8 °C, the flow rate of the cooling system was about 18.5 mL/s, the working media was water. It can be

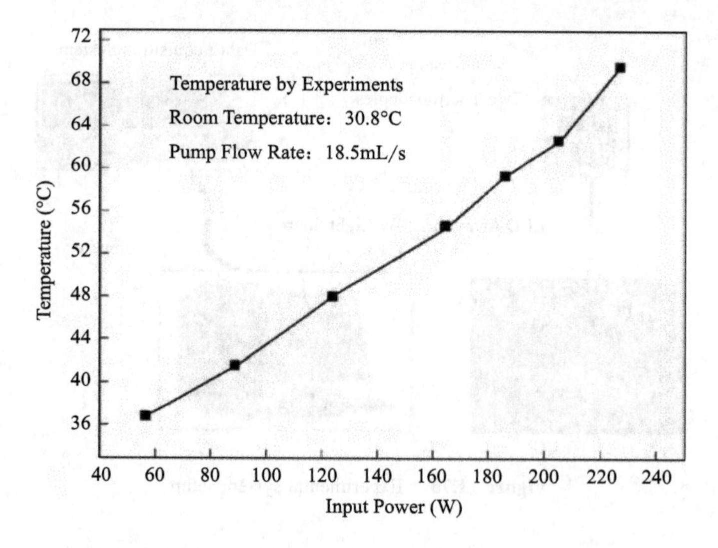

Figure 11.78 Temperature variation with input power of the light source

observed from Figure 11.78 that as the input power increases, the temperature of the LED substrate increases quickly. When the input power is 227 W, its corresponding heat flux is about 14.18 W/cm^2, and the temperature is about 69 °C. This simple test demonstrates that the LED light source cooled by the present system has a low risk for temperature reliability since the maximum endurable temperature of the LED chip in the experiment is about 120 °C.

11.7.3 Simulation and Model Verification

It is necessary to compare simulation results with experimental results in order to verify the simulation model before using the model to conduct optimization. Figure 11.79 shows the simulation model of the microjet device; the structure of the microjet device in the experiment is shown by three figures acquired from three different views. The model parameters such as working conditions and sizes were exactly the same as those in the experiments.

The diameters of the inlet and the outlet were 4 mm, the microjets were eight by eight arrays, the diameter of each microjet was 1 mm, and the heights of the impinging jet cavity and bottom cavity were 5 mm. Other parameters were shown in Figure 11.77. It was a steady, incompressible, and turbulent flow since the Reynolds number based on the inlet was about 7360. A standard k-ε model was used for turbulence simulation. A first-order upwind method was used for the discretization of the governing equations. The Reynold time-averaging governing equations were as follows:

Equation of continuity:

$$\frac{\partial \bar{u}}{\partial x}+\frac{\partial \bar{v}}{\partial y}+\frac{\partial \bar{\omega}}{\partial z} \tag{11.27}$$

Equations of momentum:

$$\begin{aligned}\rho\left(\bar{u}\frac{\partial \bar{u}}{\partial x}+v\frac{\partial \bar{u}}{\partial y}+\omega\frac{\partial \bar{u}}{\partial z}\right) &= -\frac{\partial \bar{p}}{\partial x}+\frac{\partial}{\partial x}\left(\mu\frac{\partial \bar{u}}{\partial x}-\rho\overline{u'^2}\right)+\\ &\quad \frac{\partial}{\partial y}\left(\mu\frac{\partial \bar{v}}{\partial y}-\rho\overline{u'v'}\right)+\frac{\partial}{\partial z}\left(\mu\frac{\partial \omega}{\partial z}-\rho\overline{u'\omega'}\right)\end{aligned} \tag{11.28}$$

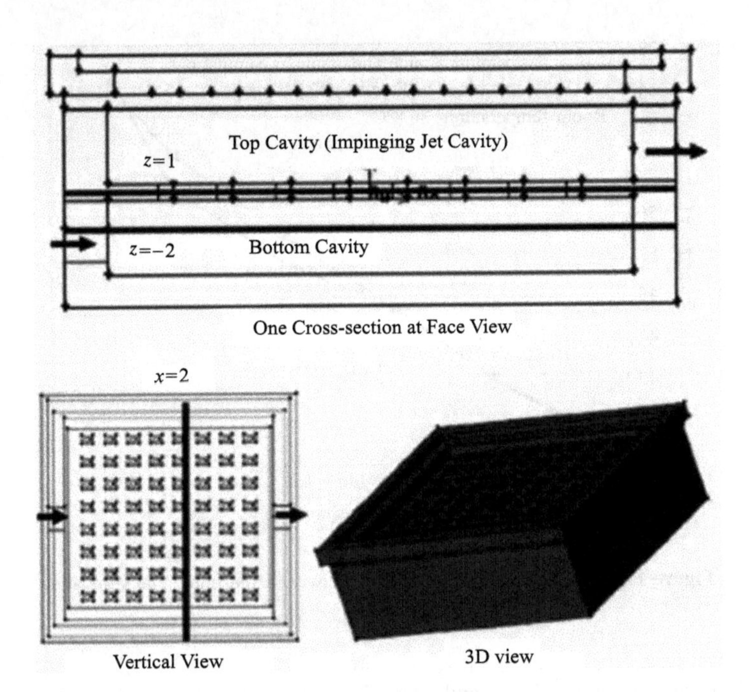

Figure 11.79 Numerical model

$$\rho\left(\bar{u}\frac{\partial\bar{\upsilon}}{\partial x}+\bar{\upsilon}\frac{\partial\bar{\upsilon}}{\partial y}+\bar{\omega}\frac{\partial\bar{\upsilon}}{\partial z}\right) = -\frac{\partial\bar{p}}{\partial y}+\frac{\partial}{\partial x}\left(\mu\frac{\partial\bar{u}}{\partial x}-\rho\overline{u'\upsilon'}\right)+ \frac{\partial}{\partial y}\left(\mu\frac{\partial\bar{\upsilon}}{\partial y}-\rho\overline{\upsilon'^2}\right)+\frac{\partial}{\partial z}\left(\mu\frac{\partial\bar{\omega}}{\partial z}-\rho\overline{\upsilon'\omega'}\right) \quad (11.29)$$

$$\rho\left(\bar{u}\frac{\partial\bar{\omega}}{\partial x}+\bar{\upsilon}\frac{\partial\bar{\omega}}{\partial y}+\bar{\omega}\frac{\partial\bar{\omega}}{\partial z}\right) = -\frac{\partial\bar{p}}{\partial z}+\frac{\partial}{\partial x}\left(\mu\frac{\partial\bar{u}}{\partial x}-\rho\overline{u'\omega'}\right)+ \frac{\partial}{\partial y}\left(\mu\frac{\partial\bar{\upsilon}}{\partial y}-\rho\overline{\upsilon'\omega'}\right)+\frac{\partial}{\partial z}\left(\mu\frac{\partial\bar{\omega}}{\partial z}-\rho\overline{\omega'^2}\right) \quad (11.30)$$

Equation of energy:

$$\rho c_p\left(\bar{u}\frac{\partial\bar{t}}{\partial x}+\bar{\upsilon}\frac{\partial\bar{t}}{\partial y}+\bar{\omega}\frac{\partial\bar{t}}{\partial z}\right) = -\frac{\partial\bar{p}}{\partial z}+\frac{\partial}{\partial x}\left(\mu\frac{\partial\bar{t}}{\partial x}-\rho c_p\overline{u't'}\right)+ \frac{\partial}{\partial y}\left(\mu\frac{\partial\bar{t}}{\partial y}-\rho c_p\overline{\upsilon't'}\right)+\frac{\partial}{\partial z}\left(\mu\frac{\partial\bar{t}}{\partial z}-\rho c_p\overline{\omega't'}\right) \quad (11.31)$$

For the fluid inlet, the boundary condition of constant flow rate was adopted. The heat transfer coefficients of the natural convection boundary were 5 W/(m^2·K), which was used based on the authors' previous experimental tests. The system flow rate was 18.5 mL/s. A study of the convergence of the grids was conducted before the calculation. The grids were refined until the flow field changed by 0.1%. Finally, a

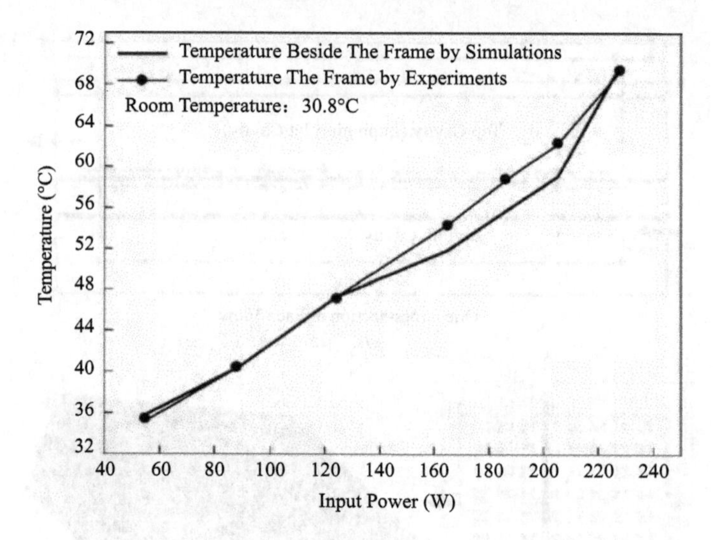

Figure 11.80 Temperature comparison between experiment and simulation

mesh of 1321262 grids was used in the simulation. The grid number of each microjet was about 25. The residual controls of the energy equation and continuity equation were 1×10^{-9} and 1×10^{-3}.

Figure 11.80 shows a comparison between the temperatures in the experiments and simulations. Here the pump flow rate was 18.5 mL/s, and the input power of the LEDs was variable, the working media was water. In this comparison, the temperature sampling positions in experiments and simulations were exactly the same. It can be seen from Figure 11.80 that seven different input powers were used. The simulation temperatures were close to the measured values at the input powers of 55, 88.5, 124, and 227.5 W. However, the temperature differences between the experiments and simulations at the input powers of 165, 186.5, and 205.2 W were much larger than those at the other four input powers. They were 2.2, 3.2, and 3.2 °C respectively.

Correspondingly, the errors of the simulations compared with the experiments at these three input powers were about 3.71%, 5.42%, and 5.1%. Thus, the maximum error between the experiments and the simulations in Figure 11.81 was 5.42%. The difference between the simulation and the experiments can be explained as following. In the experiments, water in the loop was not boiled to remove the gas dissolved inside, but in the simulation, the working media was pure water. The simulation case was an ideal one compared with the experimental conditions. Based on the above comparison and analysis, it was found that this model could be used for simulation and further analysis.

11.7.4 Comparison and Optimization of Three Microjet Devices

Simulation was used to compare the temperature and flow distribution in three kinds of different microjet devices. To compare them effectively, all simulation conditions except the structure sizes were the same.

(i) Flow and Temperature Distributions inside the Microjet Device Used in the Experiments

The simulation model in this case is shown in Figure 11.79. Based on this model, the temperature distribution of the microjet device used in the experiment was achieved and shown in Figure 11.81 when the input power was 220 watts and the corresponding heat flux was 13.75 W/cm^2. From Figure 11.81, it can be seen that the temperatures of the LED substrate at the center zone of the microjet device are the highest, and the maximum temperature is 355 K (81.9 °C).

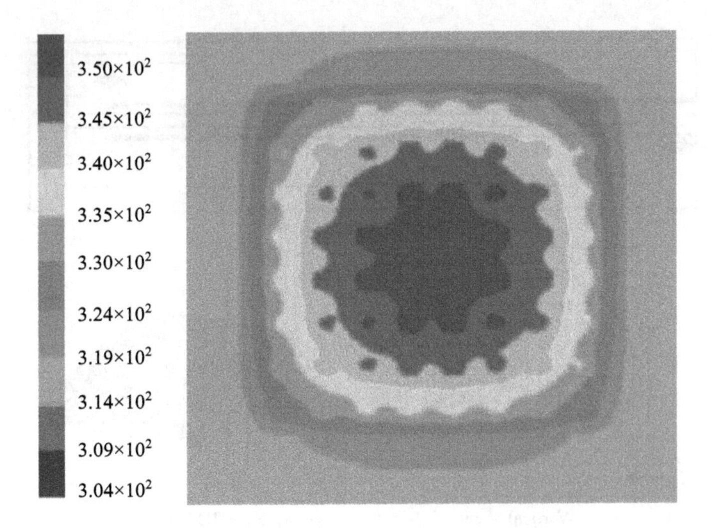

Figure 11.81 Temperature distribution of LED substrate *(Color version of this figure is available online)*

Flow field distributions in the microjet device are shown in Figure 11.82. From Figure 11.82(a), it is clear that there are two vortexes in the bottom cavity when the fluid enters into the bottom cavity through the inlet. The impinging jets are formed as shown in Figure 11.82(b) and (c) as the fluid flows through the microjets. The flow in the top cavity (or impinging jet cavity shown in Figure 11.79) is disturbed strongly. However, because of the vortexes in the bottom cavity, as shown in Figure 11.82(c), the velocities of the impinging jets at the center part of the microjet device are much smaller than those of jets at the edge part of the microjet device, which shows that the heat transfer at the center zone of the microjet device is weaker compared with those at the other parts. Such a flow distribution clearly explains the reason why the chip temperatures at the center zone were the highest.

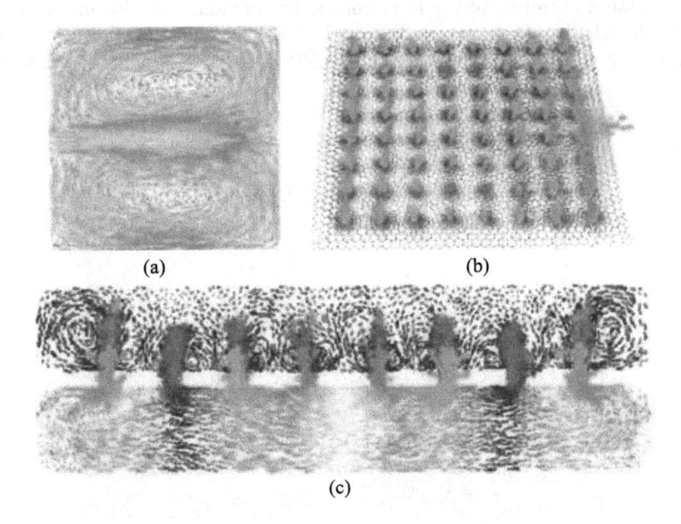

Figure 11.82 (a) Flow distribution at the cross-section of $z = -2$; (b) flow distribution at the cross-section of $z = 1$; (c) flow distribution at the cross-section of $x = 2$ *(Color version of this figure is available online)*

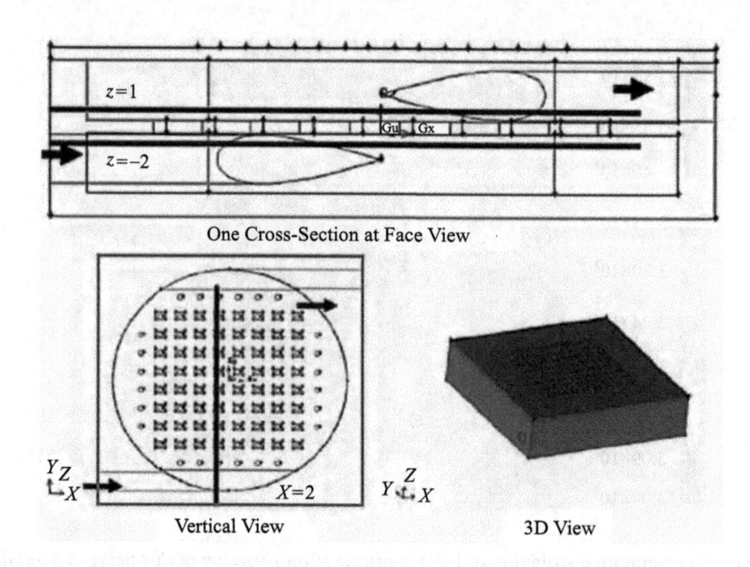

Figure 11.83 Simulation model of an improved microjet device

(ii) Flow and Temperature Distribution for An Improved Microjet Device

The simulation model of an improved microjet device is shown as Figure 11.83. The similarity between the current microjet device and the one in the experiments was that it also had a single inlet and a single outlet. The diameters of the inlet and the outlet channel were 4 mm. What was different from the above-studied microjet device was that the shape of the impinging jet cavity and bottom cavity was a cylinder whose height was 5 mm. Because of the shape change, the distribution of the microjets was different from that of the experimental model, 88 microjets whose diameter was 1 mm were uniformly distributed as shown in Figure 11.83.

In this simulation, the total mesh number was 540338, the turbulence model was standard k-ε model, velocity inlet and pressure outlet were the inlet and outlet boundary conditions respectively.

The temperature distribution is shown as Figure 11.84. It is seen that the temperature of the LED substrate at the center zone is the highest, which is 355 K (81.9 °C). This is the same as that in the

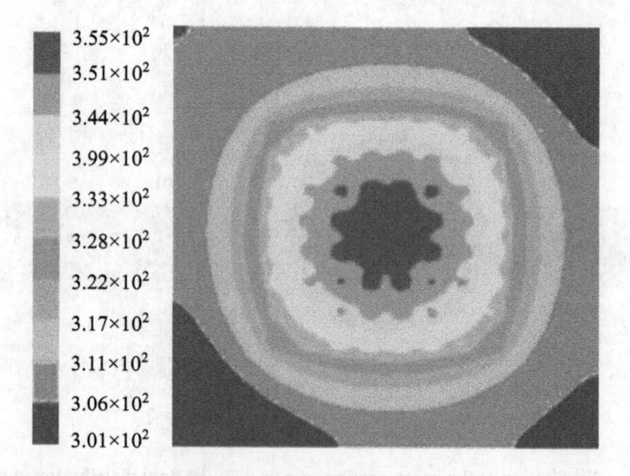

Figure 11.84 Temperature distribution of LED substrate *(Color version of this figure is available online.)*

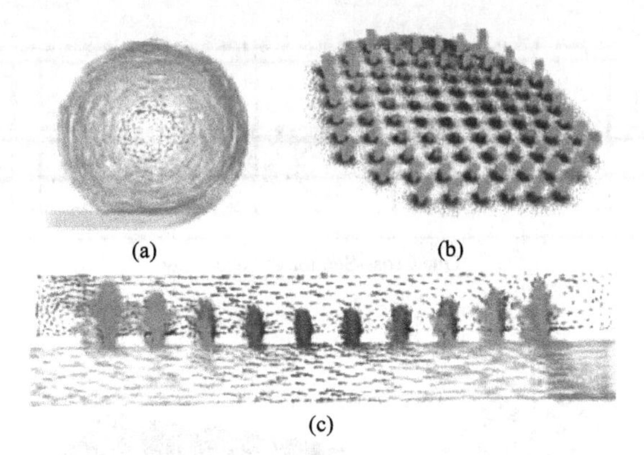

Figure 11.85 (a) Flow distribution at the cross-section of $z = -2$; (b) flow distribution at the cross-section of $z = 1$; (c) flow distribution at the cross-section of $x = 2$ *(Color version of this figure is available online)*

experimental model, and it demonstrates that the cooling performance of the present microjet device does not show much improvement.

Flow field distributions in the cavity are shown in Figure 11.85. It is clear from Figure 11.85(a) that there is one big vortex in the bottom cavity; therefore, the jet velocities in the center part are much smaller than those in the edge part as shown in Figure 11.85(b) and (c). Such a flow distribution will result in the non-uniform temperature distribution as shown in Figure 11.84, where the temperatures at the center zone are much higher than those at the other zones.

Obviously, compared with the experimental model, changing the shape of the cavity like this case has little effect on improving the cooling performance. It is necessary to find other cavity structures to improve the flow distribution and achieve uniform temperature distribution.

(iii) Flow and Temperature Distribution for Another Improved Microjet Device

Another improved microjet system is shown in Figure 11.86, which has one inlet but two outlets. The diameters of the inlet and the outlet were 4 mm. The shape of the impinging jet cavity and the bottom cavity was a cuboids type whose height was 5 mm.

The microjet array was composed of eight by eight microjets, and the diameter of each microjet was 1 mm. In the simulation, the total mesh number was 591841, other boundary and working conditions were the same as those described in the above two models. The steady temperature distribution is shown in Figure 11.87. It is noted that the maximum temperature of the LED substrate is 332 K (58.9 °C), which is 23 K lower than those achieved in the above two models. This demonstrates that the present microjet device has better cooling performance compared with the above two designs.

The flow distributions in the microjet device are shown in Figure 11.88. There are two vortexes in the bottom cavity as shown in Figure 11.88(a). Even though, from Figure 11.88(b) and (c), it can be seen that the impinging jets are much more uniform compared with those in the above two models, which is due to the relatively uniform temperature distribution shown in Figure 11.87.

From Figure 11.88(c), it is also noted that turbulence in the top cavity is much stronger than those in the above two models. Therefore, the maximum temperature of the LED substrate is 23 K lower as described in Figure 11.87.

11.8 Air Flow Sensor

The air intake of the combustion engine is a very important testing parameter. This rate should be known for each intake stroke. Knowledge of this combustion process parameter is essential if one tries to minimize both the engine's fuel consumption and pollution of the environment (Qian, 1998). For the

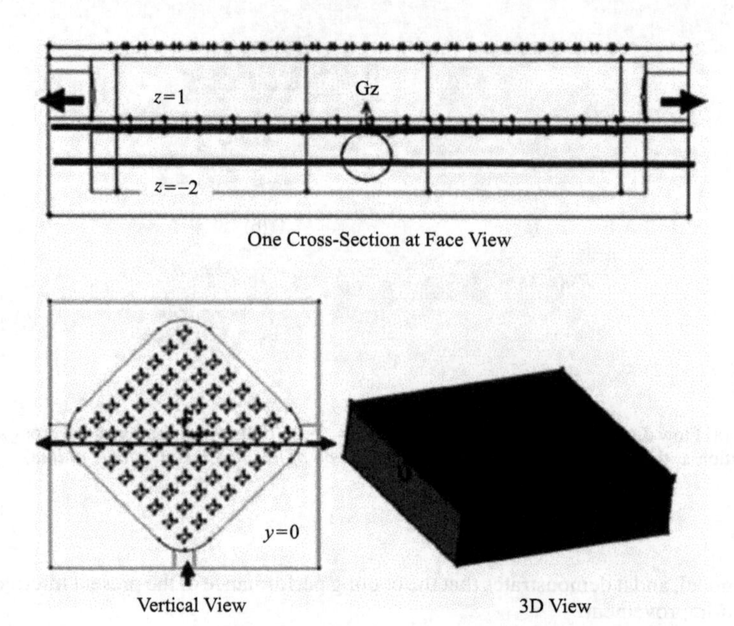

Figure 11.86 Simulated model of the microjet device with one inlet and two outlets

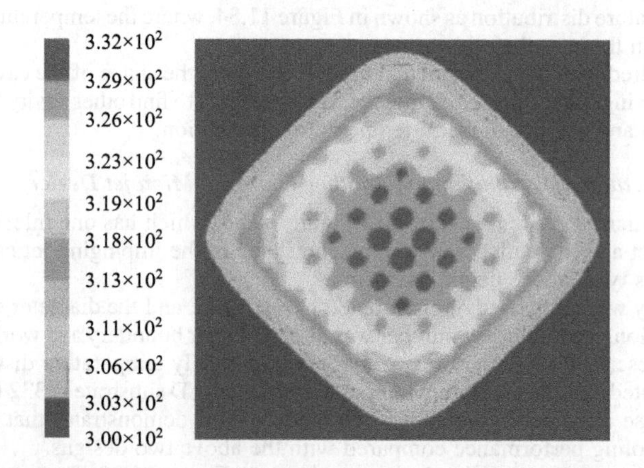

Figure 11.87 Temperature distribution of LED substrate *(Color version of this figure is available online)*

development of such engines, a wide velocity measuring range and high resolution monitoring of the time course of the air velocity during the stroke is desirable. Depending on the number of revolutions per minute and the geometry of the suction pipe, the air flow can change from simple pulsation to an oscillating flow with large amplitudes. The type of electro calorimetric sensor discussed is an attempt to meet the demands of such applications in terms of speed and measurement range.

The thermal flow sensor is based on a heat transfer principle in which a heated body is cooled by a passing flow and the local rate of cooling depends on the flow velocity. The measuring principle can be based on thermistors thermopiles, pyroelectric elements, pn-junctions, resonating microbridges, Prandtl tubes, and several other effects (Nguyen *et al.*, 1997; Nam *et al.*, 2004; Furjes *et al.*, 2004;

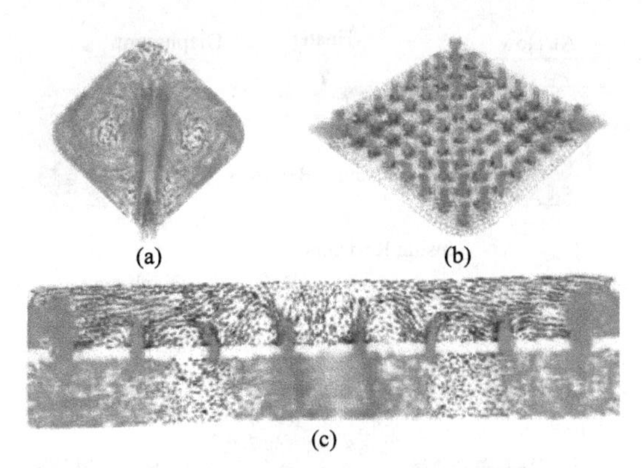

Figure 11.88 (a) Flow distribution at the cross-section of $z = -2$; (b) flow distribution at the cross-section of $z = 1$; (c) flow distribution at the cross-section of $y = 0$ *(Color version of this figure is available online)*

Buchner *et al.*, 2005). Micromachining is adopted to achieve high sensitivity, quick response and low power consumption.

Laminar flow is an indispensable measuring condition of the thermal flow sensor. The range of air intake of combustion engine is about 2 g/s–10 g/s commonly (Sabate *et al.*, 2004; Van Putten *et al.*, 1974; Glaninger *et al.*, 2000; Si *et al.*, 2005; Sun *et al.*, 2005; Mailly *et al.*, 2001; Wang *et al.*, 2003; Li *et al.*, 2004). According to the size of the manifold, the flow in the pipe is turbulence. Therefore, the air intake cannot be measured directly by the thermal flow sensor. In this section, a small channel is designed in the manifold center, and the flow in the small channel is a laminar flow under normal air intake range. Through measuring the flow velocity in the small channel the air intake of the combustion engine is obtained. The flow conditions are also analyzed by using Fluent software, and an ideal Reynolds number Re value is obtained. The temperature of the heated resistor is constant, and temperature conditions around the heater under given flow conditions are simulated by the finite volume method (FVM). The optimal distance between the measuring location and heater location is obtained, and the flow sensor meets the demand of the combustion engine air intake in terms of range and precision.

11.8.1 Operation Principle

A temperature difference based flow sensor is based on the three-element calorimetric measurement principle. The heat transferred from a heater to the fluid is transported by forced convection, causing the temperature levels of two sensing elements placed at both sides to change. The flow response is obtained from the difference in the temperature between the two sensing elements. The response of a calorimetric flow sensor depends basically on the thermal properties of the processing materials and on the geometry of the channel through which the fluid to be measured flows. Many studies on thermal properties of the processing materials have been reported in the past, but study on the channel geometry was rarely reported. In our work, a new flow sensor structure is designed as shown in Figure 11.89. Under the diaphragm of the sensor, a micro channel is structured. There is air flow on both sides of the diaphragm of the flow sensor; therefore, the forced convective thermal transfer between the diaphragm and fluid would exist on both sides of diaphragm. The temperature of the heater is constant which is controlled by a Wheatstone bridge, and an electrical test signal is exported by the other Wheatstone bridge with test resistors Pt as Wheatstone bridge arms.

Figure 11.90 shows the schematic fabrication processes of the micro-channel integrated gas mass flow sensor. The starting material is a 100 mm, 350 μm thick, <100> *p*-type, double sided polished silicon wafer. The supporting dielectric membrane consists of a 500 nm thick LPCVD silicon nitride layer on a thin pad oxide. Over this membrane and for resistor implementation, a 150 nm thick Ti/Ni-bilayer of 0.5 Ω per square is deposited, and the distance of the center of the sensing resistor pair from the edge of the heater is 140 μm. After the patterning of the nickel resistors by means of a liftoff process, a layer of 600 nm of

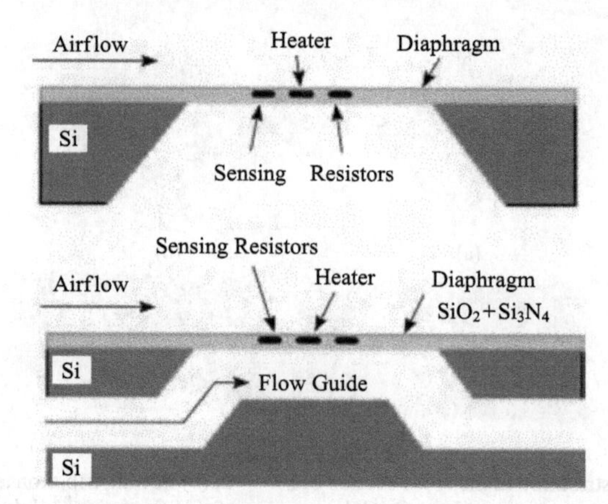

Figure 11.89 Cross sections of the flow sensor structure without and with flow guide

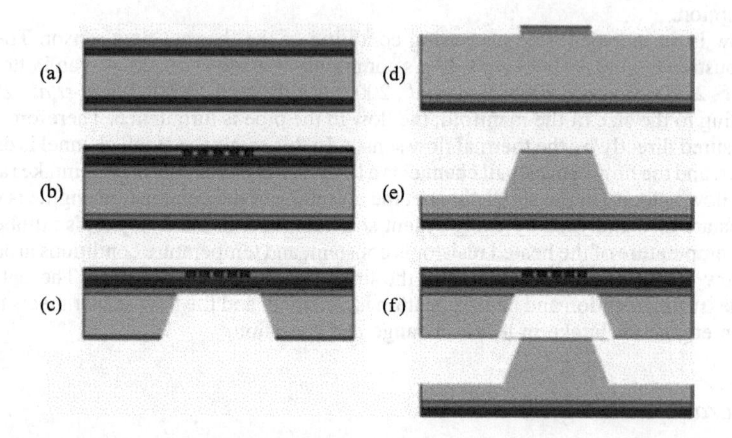

Figure 11.90 Schematic fabrication processes of the microchannel integrated flow sensors. (a) Thermal oxidation and LPCVD nitride deposition on double sides; (b) re-oxidation, sputtering of Pt, PECVD SiO_2, and coating of Si_3N_4; (c) membrane fabrication by KOH etching, RIE Si_3N_4 and SiO_2, and Au coating; (d) window structuring; (e) channel groove fabrication; (f) a micro channel for the flow guide is formed

PECVD oxidized nitride is deposited for passivation. Finally, the back side anisotropic etching of the silicon below the membrane has been made with KOH at 50 °C. The etch stops automatically when it reaches the dielectric membrane.The size of the diaphragm is 1800 μm × 800 μm. The other Si wafer is used as down substrate, and a micro channel groove is structured by etching with SiO_2 as mask, and the etching depth is controlled by etching time. Then an Au layer of 5 μm is coated. By the bonding contacts between the two silicon wafers, a micro channel for the flow guide is formed.

For this flow sensor, laminar flow conditions in the test section are maintained by the flow channel structure design. Under the presence of an incoming flow, the conservation of thermal energy for the heater resistor can be expressed by follows:

$$P_c + Q_{fc} + Q_{nc} + Q_c + Q_{rad} = 0 \tag{11.32}$$

where P_c is the heater input power, Q_{fc} is the force convection heat transfer, Q_{nc} is the nature convection heat transfer, Q_c is the heat conduction by diaphragm, and Q_{rad} is the heat radiation.

Under laminar flow conditions, the coefficient of force convection heat transfer $\bar{h}$ and Q_{fc} are expressed respectively as follows (Furjes *et al.*, 2004):

$$\bar{h} = kC_1 \cdot P_\gamma^m \cdot m_a^{1/2}(A_c \cdot L \cdot \mu)^{-1/2} = cm_a^{1/2} \tag{11.33}$$

$$Q_{fc} = c \cdot m_a^{1/2} \cdot A_h \cdot (T_h - T_\infty) \tag{11.34}$$

where μ is the dynamic viscosity of fluid, P_r is the Prandt contanst, $P_r = 0.71$ when air temperature is − 50 °C–100 °C, L is the character size of flow, A_c is the character area of flow, m_a is the mass flux, and A_h is the surface aera of heater. Comparing with Q_{fc}, nature convection heat exchange Q_{nc} can be ignored.

For the constant temperature of heater, Q_{rad} is constant. Therefore heat radiation has no effect on the temperature difference between upstream and downstream. Therefore, the P_c can be written:

$$P_c + C \cdot m_a^{1/2}(T_h - T_\infty) + \lambda \cdot A_0 \cdot \frac{\partial T}{\partial s} + Q_{rad} = 0 \tag{11.35}$$

where P_c depends on the relation to the mass flux m_a, the thermal conductivity λ, and size A_0 of diaphragm. The temperature of the heater is constant which is controlled by a Wheatstone bridge.

The temperature governing equation on diaphragm can be written as:

$$Q_{rad} + c \cdot m_a^{1/2} \cdot A_s \cdot (T_s - T_\infty) + A_0 \cdot \lambda \cdot \frac{\partial T_s}{\partial r} = \rho \cdot c_p \cdot \frac{\partial T}{\partial \tau} \tag{11.36}$$

where A_0 is the across area of diaphragm, A_s is the surface area of the test temperature elements, and λ is the nominal thermal conductivity. There are film interface scattering and grain interface scattering for films, which decrease the mean free path of phonon and result in a low thermal conductivity of film. When the thickness of Si_3N_4 film is 500 nm, the thermal conductivity decreases to 6 W/(m·K).

11.8.2 Simulation of Flow Conditions

The flow conditions in the manifold depend on the pipe structure and size, except the flow rate. According to the pipe size, the flow rate in the inlet pipe is 0.58–2.97 m/s, and $Re = \frac{\rho \cdot \upsilon^{\circ}}{\mu}$ is 2×10^3–14×10^3, where ρ is density of air, υ is air flow rate, and d is the pipe diameter. Therefore, the flow in the manifold is turbulence. In order to create a laminar flow condition, a small channel is designed in the center of inlet pipe. Figure 11.91 shows the structure of the manifold and the small channel. The air flow direction is from the right to the left. Flow sensor chip is placed in the middle uniform part of the small channel.

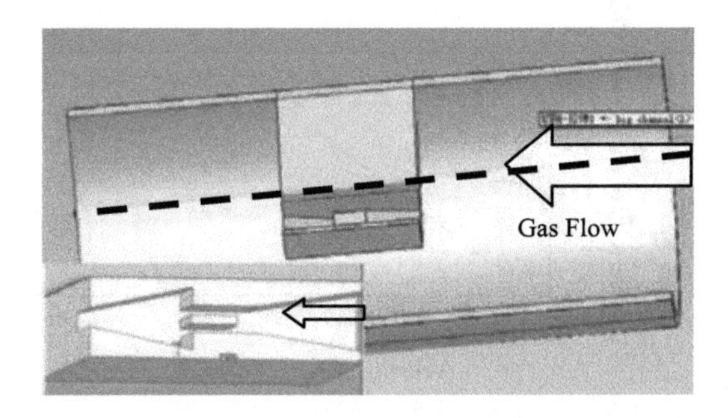

Figure 11.91 Structure of manifold and small channel *(Color version of this figure is available online)*

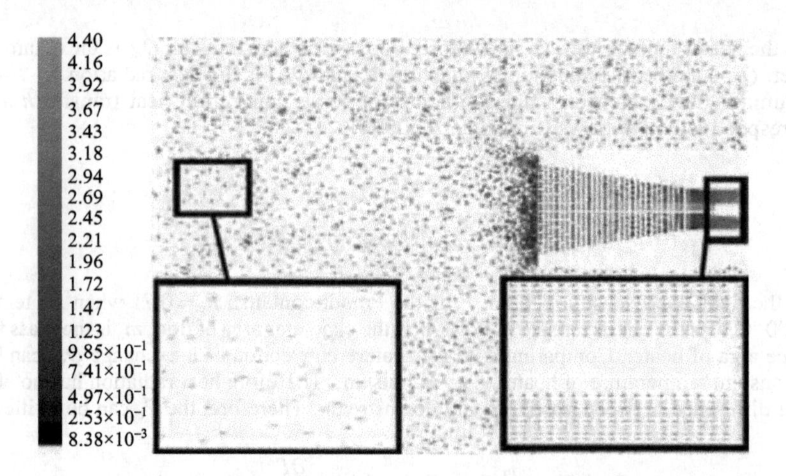

Figure 11.92 Microscopic flow conditions in all the locations

A sharp change of the channel wall along the flow decreases the flow rate in the small channel. The increase of the ratio of the cross-section perimeter to the area results in more flow restriction from the channel wall. The microscopic flow conditions in all the locations were simulated, as shown in Figure 11.92 by streamline vectors. It can be seen, remarkably, that in the manifold the flow conditions are turbulence and streamline vectors are inconsistent. However, in the small channel, streamline vectors are consistent; therefore laminar flow conditions over the sensor chip are maintained by the geometry of the small channel. Re decreases to 12–110. It is concluded that there exists a laminar flow condition in the small channel.

Figure 11.93 shows the relationship between the flow velocity in the manifold and in the small channel above those sensor elements. From this figure, it can be seen that the flow velocity in the small channel is linear to the flow velocity in the manifold, and the proportion coefficient is about 1.03. The flow velocity of the small channel varies from 0.59 m/s to 3.06 m/s. Therefore through measurement of the flow velocity of the small channel above the sensor chip the air intake of combustion engine is obtained.

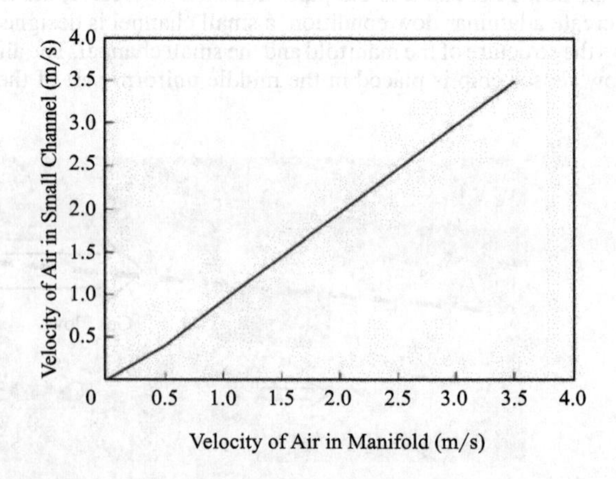

Figure 11.93 Relationship between the flow velocity in the manifold and in the small channel

11.8.3 Simulation of Temperature Field on the Sensor Chip Surface

Figure 11.89 shows the structure sketch of the sensor chip. A heater is placed on a micro machined diaphragm. The fluid temperature, which is normally close to the substrate temperature, can be measured with two additional thermistors arranged along the rim of the silicon chip. Silicon nitride has been applied as the diaphragm material, and the size of diaphragm is 2000 μm × 500 μm. Silicon nitride exhibits a low thermal conductivity resulting in high flow sensitivity. The thermal conductivity of silicon nitride is about 2.3 W/(m·K) as compared to 150 W/(m·K) for silicon. A further advantage of the silicon nitride diaphragm is how thin it is, resulting in a small thermal conduction. Heater and measuring resistor are made of Pt with 150 μm in thickness.

At zero flow, the symmetrical temperature distribution is achieved by heat conduction through the supporting material of the heater, a dielectric membrane, as well as through the surrounding air. However, under the presence of an incoming flow, the effects of forced convection make this temperature distribution change. That is, as the heat dissipated by the heater is transported by the flow, there is a cooling effect in the upstream zone whereas there is an increase in temperature downstream. Temperature variation at each point of the membrane depends on the distance from it to the heating element. Figure 11.94 shows the evolution of the temperature distribution obtained from a two-dimensional FEM simulation.

As can be seen, at low flow rates, the heating effect in the downstream is remarkable, and stronger than the cooling effect in the upstream. Therefore, it is expected that the response of the sensor will be mainly determined by the downstream resistor. However, at high flow rates, the temperature of downstream decreases slightly for increasing flows, especially at the point far from heater downstream.

The cooling effect in the downstream is due to the imbalance between the heat coming from the heater by conduction and the heat loss due to the forced convection. The temperature increase reaches a saturation that depends on the distance to the heating element, thus agreeing with the simulation results of Figure 11.94. On the other hand, the heat carried by the flow makes the downstream temperature rise only at low velocity flows whereas for higher velocity flows a global cooling effect over the whole structure predominates.

Figure 11.95 shows the dt, which expresses the temperature difference between the corresponding points in both the upstream and the downstream on the membrane for different incoming flows. As can be seen, at low flow rates the dt increases rapidly with the increase of the measurement distance; and at high flow rates, dt decreases slowly for increasing measurement distance. There is the highest dt at the some measurement distance for every flow rate, and the distance with the highest dt increases for increasing flow. For instance, when flow rates are 1.5 m/s, and measure distances are 80 μm, 140 μm, and 200 μm, the

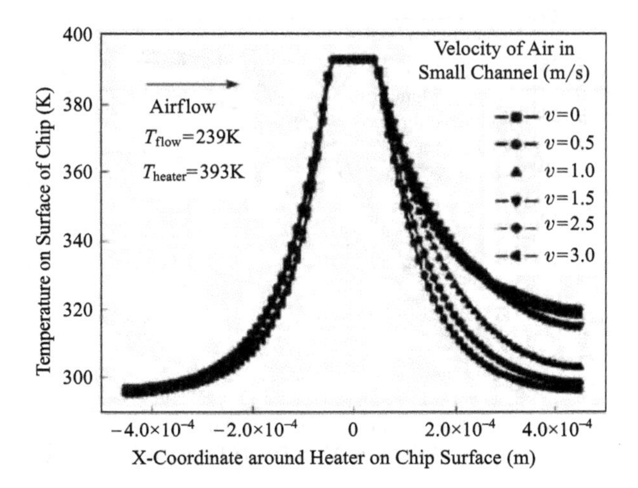

Figure 11.94 Temperature distribution on the sensor chip surface at different incoming flow rates

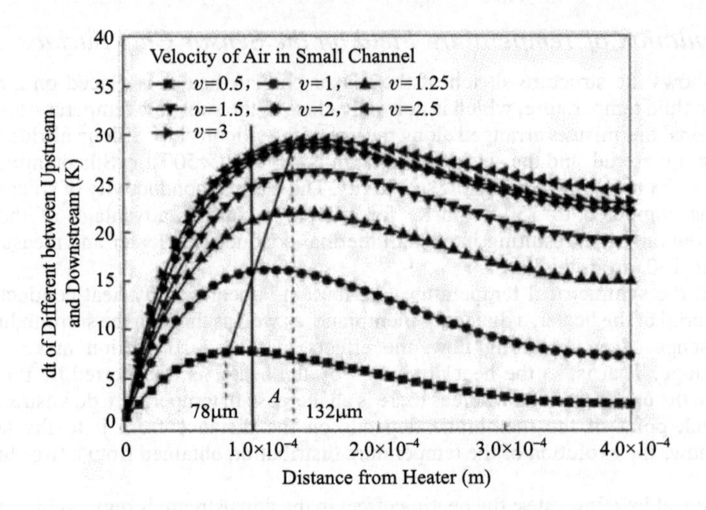

Figure 11.95 dt on the membrane for different incoming flows

dt are 22.7 K, 26.1 K, and 25.1 K respectively. When the range of flow rates is from 0.5 m/s to 3 m/s, the highest dt increases from 7.5 K to 29.5 K, and the corresponding distance varies from 78 μm to 132 μm.

Figure 11.96 shows the relationship of dt and incoming flow rates for different measure distance. As can be seen, at low flow rates, with flow rates increasing dt increases with 1/2 index. However, there is a saturation effect on dt variation at high flow rates, and it becomes more important as the distance from the heater increases. Therefore, the resistors placed far away from the heater saturate at lower flows than the ones placed closer to it.

It can be concluded that there is an optimal position of the sensing resistor pair depending on the flow range and precision to be measured. For distant measurement distance, despite the saturation effect, the best available sensitivity is high. For nearby measurement distance, the measurement range is wide

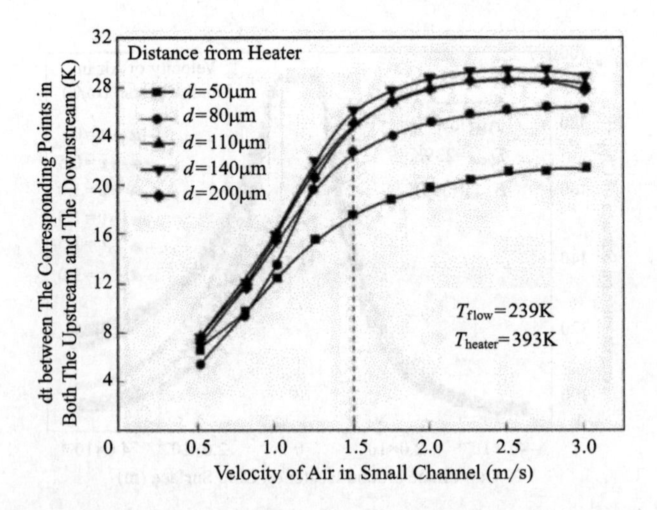

Figure 11.96 Relationship of dt and flow rates for different measure distance

despite the low sensitivity. It can be concluded that for the combustion engine the optimal measure distance is about 80 μm.

11.9 Direct Numerical Simulation of Particle Separation by Direct Current Dielectrophoresis

Analogy to the integrated electronic circuits, lab-on-a-chip systems may bring a revolution of automation to biological and medical applications (Dittrich *et al.*, 2006; Gomez, 2008; Weibel *et al.*, 2006; Su *et al.*, 2006). Manipulation of bio-particles, such as cells, proteins, and bacteria, is of great interest in biomedical analysis and clinical diagnostics (Castillo *et al.*, 2009; Kang *et al.*, 2009). For example, blood is a very complex mixture of various cells, such as red blood cells (RBCs) and white blood cells (WBCs). Therefore, blood samples usually need to be separated prior to further genomic analysis or clinical diagnostics (Toner *et al.*, 2005). Numerous separation techniques have been proposed and accomplished for continuous particle separation in lab-on-a-chip devices, as discussed in two comprehensive reviews by Pamme (2007) and Kersaudy-Kerhoas *et al.* (2008).

Among all the present continuous separation techniques, direct current (DC) dielectrophoresis (DEP), arising from the interaction between dielectric particles and spatially non-uniform electric fields, may be one of the most promising methods. It offers several advantages including easy fabrication, mechanical and chemical stability, high electric field generation, and no bubble evolution due to the absence of embedded metallic electrodes inside the channels. Kang *et al.* (2006) experimentally observed that particles with different sizes experience different trajectory shifts through a non-uniform microchannel due to the particle size dependence of the DEP force. Subsequently, researchers implemented poly (dimethylsiloxane) (PDMS) (Kang *et al.*, 2006; Kang *et al.*, 2009; Kang *et al.*, 2008; Lewpiriyawong *et al.*, 2008), silicon hurdles (Hawkins *et al.*, 2007), and reconfigurable oil droplets (Thwar *et al.*, 2007; Barbulovic-Nad *et al.*, 2006) in microchannels to generate spatially non-uniform electric fields for particle separation.

However, a comprehensive analysis of the particle separation due to the DC DEP effect is still limited so far. Most previous numerical studies of particle electrokinetic transport in non-uniform channels, such as L-shaped (Davison *et al.*, 2008), T-shaped (Ye *et al.*, 2004) microchannels, converging-diverging nanopores (Qian *et al.*, 2006), and microchannel/nanopore junctions (Liu *et al.*, 2007), have neglected the induced DC DEP effect. Our previous study demonstrated that the trajectory shift observed by Kang *et al.* (2006) cannot be correctly predicted from the existing model without considering the DEP effect (Ai *et al.*, 2009). Although a Lagrangian tracking method considering the DEP effect has been developed to understand the DEP particle separation (Kang *et al.*, 2006), the effects of the particle on the flow field and electric field are both neglected. In addition, this method is only appropriate for spherical particles, excluding lots of existing rod-like biological entities.

In this research, we propose a multiphysics model to investigate the DEP separation of particles through a constricted microchannel with a full consideration of the particle-fluid-electric field interactions. The DEP force exerting on the moving particle is obtained by integrating the Maxwell stress tensor over the particle surface, which is the most rigorous method to calculate the DEP force (Wang *et al.*, 1997). The structure of the rest of this research is listed as follows. Section 11.9.2 introduces the mathematical model composed of the Navier–Stokes equations for the flow field and the Laplace equation for electric field defined in an arbitrary Lagrangian-Eulerian (ALE) framework. The numerical results are discussed in Section 11.9.3 with focuses on the main factors of the DEP separation. Concluding remarks are given in the final section.

11.9.1 Mathematical Model and Implementation

A remarkable agreement between the numerical predictions of electrokinetic particle transport in a converging-diverging channel, obtained from a two-dimensional (2D) mathematical model, and the existing experimental data suggests that a 2D model is sufficient to capture the essential physics of the

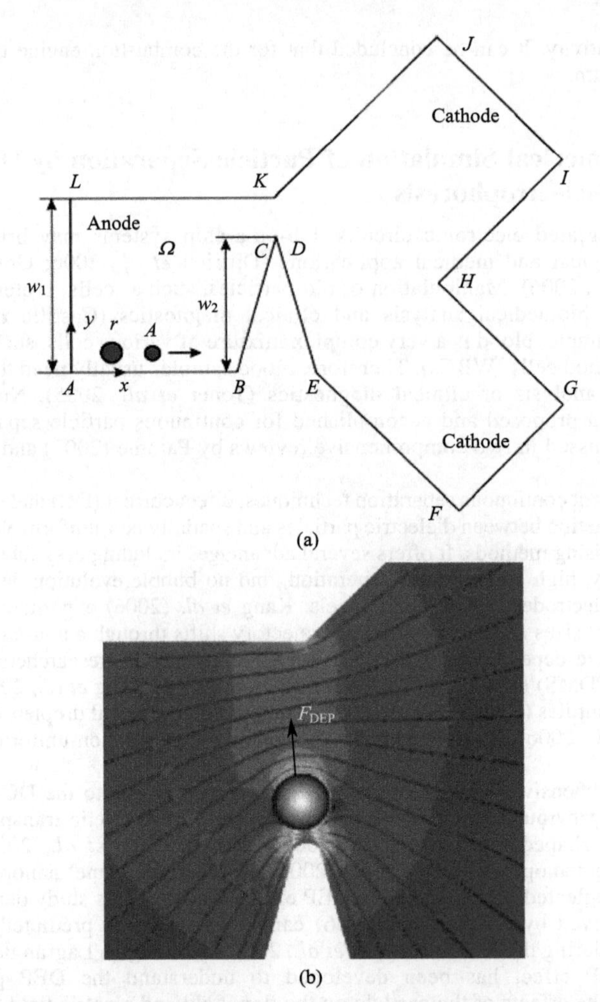

Figure 11.97 (a) A two-dimensional schematic view of two particles moving in a constricted microchannel. AL represents the inlet, while FG and IJ represent the outlets. (b) Distribution and streamlines of the electric field around the constricted section in the presence of a particle. The color levels indicate the electric field intensity, with the red color representing high electric field. The black arrow denotes the direction of the induced DEP force exerting on the particle *(Color version of this figure is available online)*

electrokinetic particle transport process (Ai *et al.*, 2009). Therefore, a 2D mathematical model is adopted in this study.

We consider two circular particles initially located at the upstream of the microchannel moving toward the cathode, as shown in Figure 11.97(a). The constricted microchannel is obtained by adding a triangular hurdle prior to the Y junction, similar to an actual microfluidic device fabricated by Kang *et al.* (2008). A two-dimensional Cartesian coordinate system (x, y) with the origin located at point A is used in the present study. The computational domain Ω, surrounded by the channel boundary ABCDEFGHIJKL and the particle surface Γ and Λ, is filled with an aqueous solution. The segment AL represents the anode and also the inlet of the channel, meanwhile the segments FG and IJ act as the cathode and the outlets of the channel. The particle and channel wall are assumed to be rigid and non-conducting.

The fluid in the computational domain Ω is Newtonian and incompressible. The Brownian motion of micron-sized particles is negligible (Davison *et al.*, 2008).

Electric double layer (EDL) adjacent to the charged surface is commonly approximated to be infinitesimal in micro-scale electrokinetics (Ai *et al.*, 2009), concerned in the present study. Hence, the net charge density in the computational domain Ω is zero and the electric potential, ϕ, satisfies the Laplace equation:

$$\nabla^2\phi = 0 \quad \text{in}\,\Omega \tag{11.37}$$

All rigid surfaces are electrically insulating:

$$\boldsymbol{n}\cdot\nabla\phi = 0 \tag{11.38}$$

and a potential difference between the inlet and the outlet is imposed by:

$$\phi = \phi_0 \text{ on AL} \tag{11.39}$$

and:

$$\phi = 0 \text{ on FG and IJ} \tag{11.40}$$

The particle and its adjacent EDL are regarded as a single entity due to the infinitesimal EDL approximation. Since the Reynolds number of micro-scale electrokinetic flows is usually very small, the inertial term in the Navier–Stokes equations is neglected. The conservation of mass and momentum in the fluid are thus given by:

$$\nabla\cdot\boldsymbol{u} = 0 \text{ in W} \tag{11.41}$$

and:

$$\rho\frac{\partial u}{\partial t} = -\nabla p + \mu\nabla^2\boldsymbol{u} \text{ in W} \tag{11.42}$$

where u, p, ρ, and μ are, respectively, the fluid velocity vector, the pressure, the fluid density and the dynamic viscosity.

Electroosmosis induced by the charged channel wall is approximated by the Smoluckowski slip velocity on the charged surface. Hence, the fluid velocity adjacent to the channel wall is:

$$\boldsymbol{u} = \frac{\varepsilon_f\zeta_w}{\mu}(\boldsymbol{I}-\boldsymbol{nn})\cdot\nabla\phi, \text{ on ABCDEF, GHI, and JKL} \tag{11.43}$$

where ε_f and ζ_w ware, respectively, the fluid permittivity and the zeta potential of the channel wall. The quantity $(\boldsymbol{I}-\boldsymbol{nn})\cdot\nabla\phi$ defines the electric field tangent to the charged channel wall, with I and n representing, respectively, the second-order unit tensor and the unit normal vector pointing from the channel wall to the fluid domain.

Besides the slip velocity, the fluid boundary condition on the particle surface includes the velocity related to the particle motion, expressed as:

$$\boldsymbol{u} = \boldsymbol{U}_{pi} + \boldsymbol{\omega}_{pi}\times\left(\boldsymbol{x}_{si}-\boldsymbol{x}_{pi}\right) + \frac{\varepsilon_f\zeta_{pi}}{\mu}(\boldsymbol{I}-\boldsymbol{nn})\cdot\nabla\phi \quad i = \Gamma \text{ and } \Lambda \tag{11.44}$$

In the above, the first term, U_{pi}, denotes the translational velocity of the ith particle. The second term denotes the ith particle rotation, including the angular velocity, ω_{pi}, the position vector of the particle surface, x_{si}, and the position vector of the particle center, x_{pi}. The last term in Equation 11.44 represents the electroosmotic slip velocity with *pi* representing the zeta potential of the ith particle.

The total force exerting on the ith particle consists of the hydrodynamic force, F_{Hi}, arising from the fluid motion outside of the EDL, and the electrokinetic force, F_{Ei}, which are obtained, respectively, by integrating the hydrodynamic stress tensor T^{H} and the Maxwell stress tensor T^{E} over the ith particle surface:

$$F_{Hi} = \int T^{\text{H}} \cdot n \mathrm{d}s_i = \int \left[-pI + \mu(\nabla u + \nabla u^{\text{T}}) \right] \cdot n \mathrm{d}s_i \tag{11.45}$$

and:

$$F_{Ei} = \int T^{\text{E}} \cdot n \mathrm{d}s_i = \int \left[\varepsilon_{\text{f}} EE - \frac{1}{2} \varepsilon_{\text{f}} (E \cdot E) I \right] \cdot n \mathrm{d}s_i \tag{11.46}$$

where s_i is the ith particle surface, and E is electric field related to the electric potential by $E = -\nabla\phi$. Using Equation 11.38, the first term of the integrand in the RHS of Equation 11.46 vanishes. Therefore, Equation 11.46 represents the pure DEP force acting on the particle.

The translational velocity of the *i*th particle is governed by the Newton's second law:

$$m_{pi} \frac{\mathrm{d}U_{pi}}{\mathrm{d}t} = F_{Hi} + F_{Ei} \tag{11.47}$$

where m_{pi} is the mass of the ith particle.

The angular velocity of the ith particle is determined by:

$$I_{pi} \frac{\mathrm{d}\omega_{pi}}{\mathrm{d}t} = Q_i = \int (x_{si} - x_{pi}) \times (T^{\text{H}} \cdot n) \mathrm{d}S_i + \int (x_{si} - x_{pi}) \times (T^{\text{E}} \cdot n) \mathrm{d}s_i \tag{11.48}$$

where I_{pi} is the moment of inertia of the *i*th particle and Q_i is the torque exerting on the ith particle. When the DEP effect is not considered, the last terms in the RHS of Equation 11.47 and 11.48 are not considered.

The center x_{pi} and the orientation θ_{pi} of the ith particle are expressed by:

$$x_{pi} = x_{pi0} + \int_0^t U_{pi} \mathrm{d}t \tag{11.49}$$

and:

$$\theta_{pi} = \theta_{pi0} + \int_0^t \omega_{pi} \mathrm{d}t \tag{11.50}$$

where x_{pi0} and θ_{pi0} denote, respectively, the initial location and orientation of the particle. t is the time interval.

The aforementioned fluid flow field and electric field are solved in an Eulerian framework, and meanwhile the particle motion is tracked in a Lagrangian manner according to an ALE algorithm, well developed for the simulation of particulate system (Hu *et al.*, 1992; Hu *et al.*, 2001). Based on the algorithm, the mesh is deformed to adapt the new location and orientation of the particle based on Equation 11.49 and Equation 11.50 after each computational time step. When the mesh quality degrades to a designated level as the particle translates and rotates, a new geometry with an undeformed mesh is created based on the preceding deformed mesh. Subsequently, the previous solutions of the fluid flow and the electric field in the deformed mesh are projected onto the undeformed mesh to continue the computation until the next mesh regeneration. Therefore, the ALE algorithm can track a long way particle transport.

The particle-fluid-electric field system is solved in a coupled manner with a commercial finite-element package COMSOL (version 3.4a, www.comsol.com) operating in a high-performance cluster. A higher density of unstructured mesh is generated around the particle and the constricted section to ensure a mesh-independent result. The proposed ALE algorithm has been successfully used to simulate the pressure-driven particle transport through a converging-diverging channel, showing good agreement with the experimental results (Ai *et al.*, 2009). In addition, the particle trajectory shift in a constricted channel due to the DEP effect, predicted by the present particle-fluid-electric field coupled model, shows quantitative agreement with the experimental observation (Ai *et al.*, 2009).

11.9.2 Results and Discussion

The microchannel is assumed to be made of PDMS with zeta potential $\zeta_w = -80\,\text{mV}$. The electric potentials on the two cathodes are fixed to be zero in the following study. The channel width, w1, is 300 micon. The densities of fluid and particle are, respectively, $1.0 \times 10^3\,\text{kg/m}^3$ and $1.05 \times 10^3\,\text{kg/m}^3$. The fluid dynamic viscosity and permittivity are, respectively, $\mu = 1.0 \times 10^{-3}\,\text{kg/(m·s)}$ and $\varepsilon_f = 6.9 \times 10^{-10}$ F/m.

Figure 11.97(b) shows that a significantly non-uniform electric field is induced around the particle due to the constricted structure and the presence of particle. As the negative DEP force always points from a higher electric field to a lower electric field, the particle experiences a net force across the electric field streamlines, as shown in Figure 11.97(b). Hence, the particle trajectory could shift from the electric field streamlines near the tip of the hurdle to that far away from the tip due to the induced DEP force. From Equation 11.46, we can see that the DEP force increases as the particle size and the electric field increase. Thus, particles with different sizes experience different trajectory shifts, which is the basic principle of the DEP particle separation. To gain a comprehensive understanding of the DEP separation, the main factors controlling the particle separation are investigated in the following sections.

(i) Effect of Particle Size

Figure 11.98 illustrates the separation of two particles with diameters of 10 μm and 20 μm through the constricted channel. These trajectories are obtained by superposing sequential particle locations into one figure. The electric potential applied on the anode (inlet AL) is 7 V. The zeta potentials of the two particles are both −32 mV. The constriction ratio, defined as the ratio of the hurdle width to the channel width, w_2/w_1, is 0.753. The 20 μm particle experiences a DEP force which is large enough to shift the particle to the upper outlet IJ. However, the 10 μm particle only experiences a slight trajectory shift from its original streamline, thus, comes out from the lower outlet FG. This kind of particle separation based on their sizes has been successfully implemented in a real microfluidic device, and our numerical predictions are in qualitative agreement with the experimental observations (Kang *et al.*, 2008).

(ii) Effect of Particle Zeta Potential

Figure 11.99 illustrates the trajectories of two 20 μm particles bearing different zeta potentials through the constricted channel. The electric potential applied on the anode is 7 V. The zeta potentials of the two particles are, respectively, 32 mV and −32 mV. The constriction ratio, w_2/w_1, is 0.753. As the particle velocity is linearly proportional to its zeta potential, the velocity of the 20 μm particle bearing a zeta potential of 32 mV (particle A) is almost 2.3 times of that bearing a zeta potential of −32 mV (particle B) considering the electroosmosis. Thus, the effective time of the DEP force acting on the particle B is nearly 2.3 times of that for particle A. Although the two particles experience an identical DEP force, the magnitudes of the two trajectory shifts are quite different, resulting in a separation.

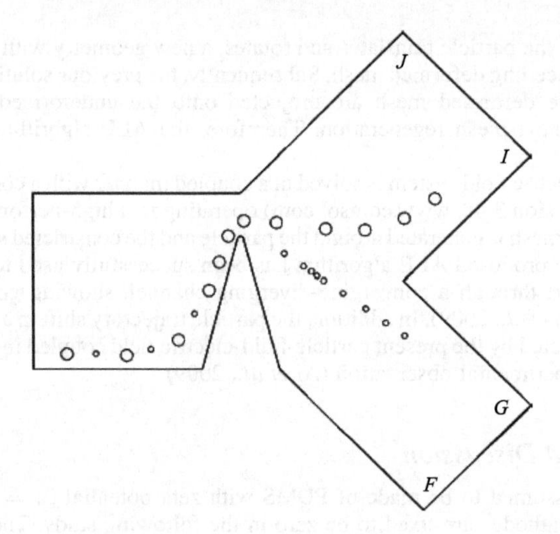

Figure 11.98 Trajectories of 10 μm and 20 μm particles through the constricted microchannel. The electric potential applied on the inlet is 7 V. The zeta potentials of the two particles are both − 32 mV. The constriction ratio w_2/w_1 is 0.753

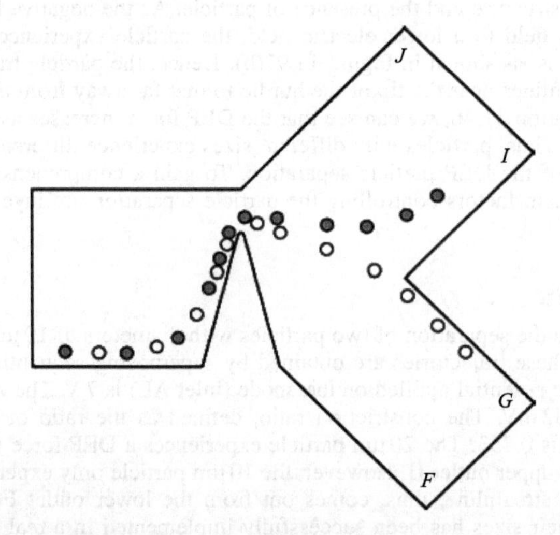

Figure 11.99 Trajectories of two 20 μm particles bearing different zeta potentials through the constricted microchannel. The electric potential applied on the inlet is 7 V. The constriction ratio w_2/w_1 is 0.753. The zeta potentials of the red and black particles are, respectively, 32 mV and − 32 mV

(iii) Effect of Electric Field

Besides the effect of the particle properties including particle size and zeta potential mentioned above, the electric field and channel geometry also affect the DEP separation. Figure 11.100 illustrates the trajectories of 10 μm and 20 μm particles through the constricted channel. The electric potential applied on the anode is 3 V. The zeta potentials of the two particles are both − 32 mV. The constriction ratio, $w2/w1$, is 0.753.

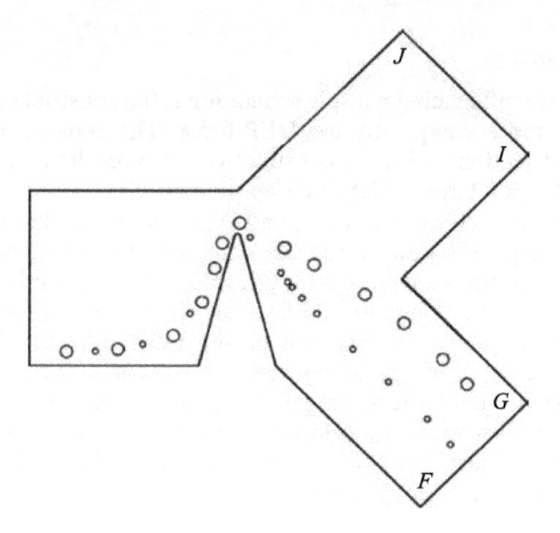

Figure 11.100 Trajectories of 10 μm and 20 μm particles through the constricted microchannel. The electric potential applied on the inlet is 3 V. The zeta potentials of the two particles are both − 32 mV. The constriction ratio w_2/w_1 is 0.753

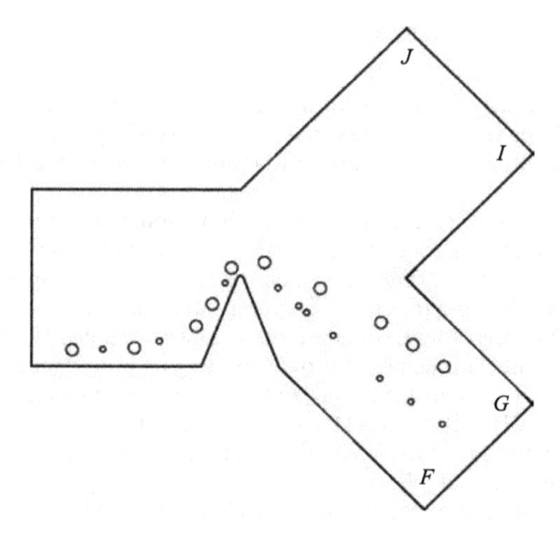

Figure 11.101 Trajectories of 10 μm and 20 μm particles through the constricted microchannel. The electric potential applied on the inlet is 7 V. The zeta potentials of the two particles are both − 32 mV. The constriction ratio w_2/w_1 is 0.513

Except for the decreased electric field, the other parameters in the present study are exactly the same as those in section *(i)*. The two particles experience slighter trajectory shifts than the corresponding particles when 7.0 V is applied to the anode. As the electric field decreases, even 20 μm particles cannot be shifted to the upper outlet. Under the current parameters, the separation threshold of the particle size is larger than 20 μm. Thus, one can adjust the applied electric potential on the anode to control the separation threshold of particle size. Furthermore, a higher electric potential leads to a smaller separation threshold of the particle size.

(iv) Effect of Channel Geometry

The most important geometry influencing particle separation is the constricted section, which induces the non-uniform electric field and consequently the DEP force. The constriction ratio, w_2/w_1, is used to characterize the constricted section. A higher construction ratio results in a more non-uniform electric field and subsequently induces a larger DEP force on the particles.

Figure 11.101 depicts the trajectories of 10 μm and 20 μm particles through the constricted channel when the electric potential applied on the anode is 7 V, the zeta potentials of the two particles are both −32 mV, and the constriction ratio, w_2/w_1, is 0.513. Except for the constriction ratio, all other parameters in Figure 11.101 are the same as those in Figure 11.98. Since a less non-uniform electric field is present in the constricted section, the 20 μm particle cannot be separated from the 10 m particle, as shown in Figure 11.101. Thus, one can also adjust the constriction ratio to control the separation threshold of particle size. In order to get rid of the Joule heating problem in DC electrokinetics induced by high electric fields (Xuan, 2008), several researchers successfully utilized a reconfigurable oil droplet to construct different constriction ratios to control the particle separation (Barbulovic-Nad *et al.*, 2006) and focusing (Thwar *et al.*, 2007) under a lower electric field.

11.10 Modeling of Micro-Machine for Use in Gastrointestinal Endoscopy

Capsule endoscopy is an examination of the gastrointestinal track using a wireless capsule endoscope which contains digital camera, ASIC transmitter, antenna, illuminating LEDs assisted imaging, and battery.

Capsule endoscope with the size of a normal pill can be easily swallowed by patients in various ages. It takes pictures throughout the gastrointestinal tract by sending images to an external recorder and thus provides useful information for clinic diagnosis of gastrointestinal tract diseases (Iddan *et al.*, 2000).

However, recent products of capsule endoscope still have some problems: (1) no external guidance control system; (2) time and money consuming; (3) inability to conduct on-time treatments such as drug delivery and body tissue collection. The movement of the capsule depends solely on the peristalsis system. Therefore, it takes 6–8 hours for an examination and doctors cannot perform a pinpoint analysis once an irregular image has been found. In addition, a camera in a capsule takes thousands of photos in every part of the tract to prevent the possibility of missing an important picture. Therefore, hours of time are needed for doctors to organize and analyze a large number of images, still with a 60%–70% rate of success in diagnostics. All of these drawbacks render capsule endoscopy extremely high in cost but extremely inefficient in terms of time spent and resources used (Sandrasegaran *et al.*, 2008; Hadithi *et al.*, 2006; Christodoulou *et al.*, 2007).

With the development of micro-electro-mechanical systems (MEMS) technology, research and progress on self-propelled capsule endoscope has been the topic of interest for several years (Daveson *et al.*, 2008; Delvaux *et al.*, 2008; Iakovidis *et al.*, 2008). A wide variety of active and wireless control mechanism have been developed for the self-propelled endoscopic capsule such as shape memory alloy (SMA) actuator, spiral type micro machine, magnetic navigation, motor legged robot, and hydraulic manipulator (Kim *et al.*, 2005; Ishiyama *et al.*, 2000; Kassim *et al.*, 2003; Carpi *et al.*, 2007; Quirini *et al.*, 2008; Peirs *et al.*, 1999). K. Ishiyama and his group in Tohoku University proposed one type of micro-machine for capsule endoscopy (Ishiyama *et al.*, 2001). Their micro-machine was composed of a capsule dummy, an internal permanent magnet, and a spiral structure. The micro-machine was rotated and propelled wirelessly by applying an external rotational magnetic field. They used a 3D finite volume method for analyzing the swimming properties of a spiral-type magnetic micro-machine and calculated its swimming velocity, drag, and load torque (Yamazaki *et al.*, 2001).

In this research, the authors theoretically analyzed the swimming properties of a spiral-type magnetic micro-machine using theories of fluid dynamics lubrication and the finite difference methods. We also simulated rotating magnetic field and obtained optimal controlling parameters. We examined this analysis method and compared our calculation result with results from existed publications.

11.10.1 Methods

(i) Fluid Dynamics Model

We use theories of hydrodynamic lubrication to analyze the micro-machine's movement in the intestinal tract. For the model development, the following assumptions are made: (1) the fluids in the tube are incompressible viscous fluids; (2) effects of anterior and posterior end of the machine is neglected; (3) active and passive deformation of the wall of the intestinal tract is not considered; (4) there is no inclination of the machine relative to the tube axis. Based on these assumptions, the fluid dynamics model for the micro-machine's motion in the intestinal tract is established as shown in Figure 11.102. The micro-machine, whose radius is r, rotates with the angular velocity ω, whereas the intestinal tract with a radius of R keeps still relative to selected coordinates. According to the basic theory of lubrication, a thin film will form between the micro-machine's surface and the intestinal tract's inner wall. Pressure will also be established on part of the capsule's surface, whose distributions are determined by the rotating direction and geometric profiles. Due to the pressure difference between the front and back of the spiral blade, the capsule endoscope will travel in a forward or backward direction in terms of the rotating direction as shown in Figure 11.103. But we should know that it is appropriate to use lubrication theory when the distance between the wall and the machine surface is small relative to the diameters of tube. According to

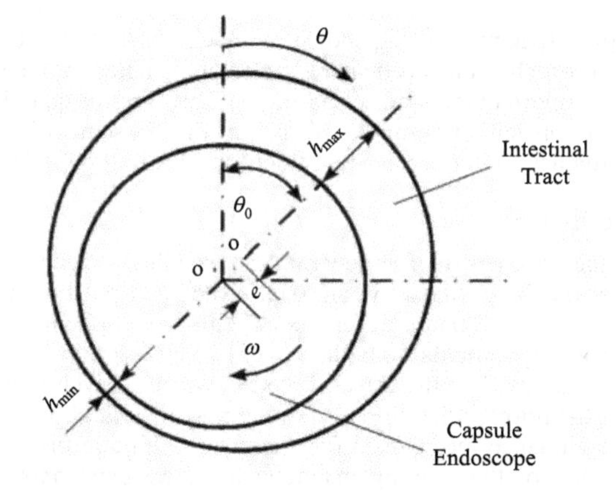

Figure 11.102 Fluid dynamics models of capsule endoscope

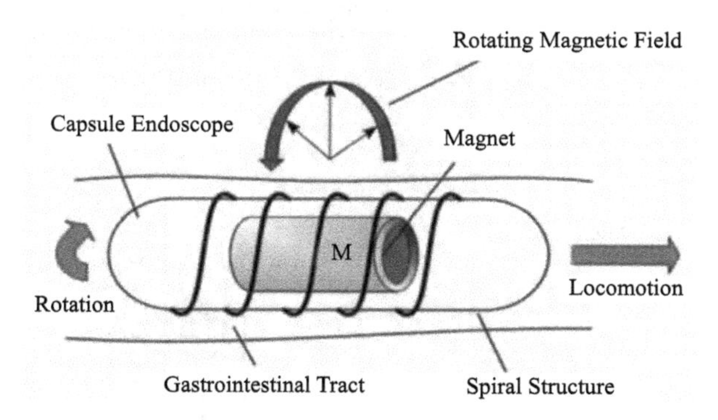

Figure 11.103 Illustration of micro-machine-driven by external magnetic field (redrawn based on Sendoh *et al.*, 2004)

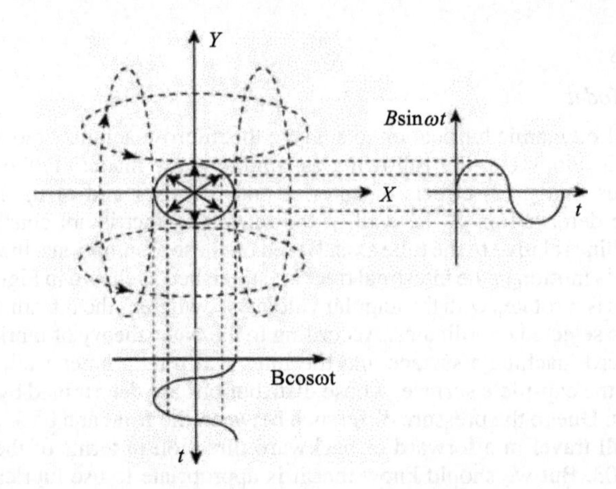

Figure 11.104 Illustration of rotating magnetic field's generation

capsule size and the intestinal specimen of pig shown in (Baek *et al.*, 2004), the actual scales of film thickness is about 0–0.9 mm, which is far less than the scales of diameters of intestinal tract (8–12 mm). Surface roughness is another important factor to be considered; however, the quantitative values of surface roughness of the intestinal tract are still unavailable. We assume that the scales of surface roughness are far less than the scales of the diameters in the intestinal tract for developing the following model.

(ii) Generation of Rotating Magnetic Field

Figure 11.104 illustrates the mechanism for generating a rotating magnetic field. Two groups of Helmholtz coils (circular or quadrate) are set in an axial orthogonal position and are stimulated by a two-way sinusoidal current with a $\pi/2$ phase difference. Then, we can obtain a rotating magnetic field plane, whose normal vector is perpendicular to both axes of two coils groups. A micro-machine with an inside permanent magnet magnetized in the radial direction will be driven under an external rotating magnetic field and move in the normal direction of rotating magnetic field plane (Sendoh *et al.*, 2004).

Figure 11.105 shows the guidance mechanism for a micro-machine in the gastrointestinal tract. By changing the normal direction of the rotating magnetic field, we can control the micro-machine's movement in any direction.

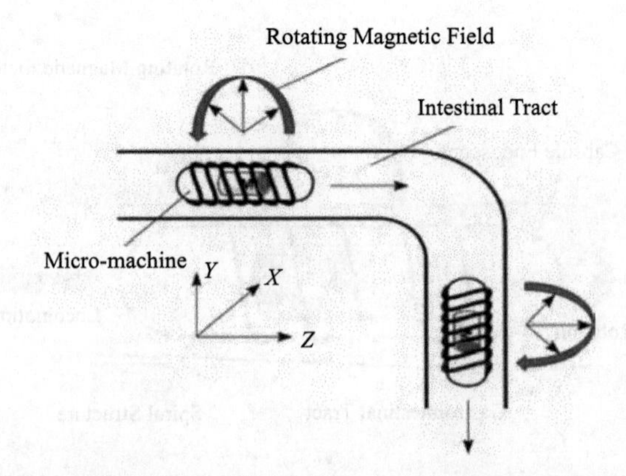

Figure 11.105 Illustration of guidance mechanism for micro-machine in gastrointestinal tract

(iii) Reynolds Equation and Electromagnetic Equation

Compressible fluid's lubrication problem is governed by Reynolds Equations (Hamrock, 1994):

$$\frac{\partial}{\partial x}\left(\frac{\rho h^3}{12\eta}\frac{\partial p}{\partial x}\right)+\frac{\partial}{\partial z}\left(\frac{\rho h^3}{12\eta}\frac{\partial p}{\partial z}\right)=U\frac{\partial(\rho h)}{\partial x}+V\frac{\partial(\rho h)}{\partial z}+\frac{\partial(\rho h)}{\partial t} \tag{11.51}$$

where $U=(u_1+u_2)/2$ is the velocity in x direction, $V=(v_1+v_2)/2$ is the velocity in z direction, p is the pressure, η is the absolute viscosity, ρ is fluid density, and h is the film thickness.

$$h=h_0[1+\varepsilon\cos(\theta-\theta_0)]+h_g \tag{11.52}$$

where h_0 is the radial gap between intestinal tract's inner wall and micro-machine's outside surface, $\varepsilon=e/h_0$ is the relative eccentricity, where e is the eccentric distance, θ_0 is the attitude angle, and h_g is the groove depth.

Vector form of Reynolds Equations can be written as follows:

$$\nabla\cdot\left(\frac{\rho h^3}{12\eta}\nabla p\right)=\nabla\cdot(\rho h\vec{U})+\frac{\partial(\rho h)}{\partial t} \tag{11.53}$$

For incompressible fluid lubrication, the governing equation is:

$$\nabla\cdot\left(\frac{h^3}{12\eta}\nabla p\right)=\nabla\cdot(h\vec{U})+\frac{\partial(h)}{\partial t} \tag{11.54}$$

Based on the Maxwell's equations, we can obtain following equations for transient analysis of rotating magnetic field (Feynman *et al.*, 1964):

$$\sigma\frac{\partial A}{\partial t}+\nabla\times\left(\mu_0^{-1}\mu_r^{-1}\nabla\times A\right)=j \tag{11.55}$$

$$n\times A=0 \tag{11.56}$$

where μ_0 is the permeability of vacuum, μ_r is the relative permeability, σ is the electric conductivity, A is the vector potential, n is the normal vector of solution boundary, and j denotes the externally applied current density.

The Reynolds equation suitable for solving spiral-type micro-machine can be written as:

$$\frac{\partial}{\partial x}\left(\frac{\rho h^3}{\eta}\frac{\partial p}{\partial x}\right)+\frac{\partial}{\partial z}\left(\frac{\rho h^3}{\eta}\frac{\partial p}{\partial z}\right)=6U\cos\phi\frac{\partial(\rho h)}{\partial x}+6U\sin\phi\frac{\partial(\rho h)}{\partial z} \tag{11.57}$$

Reynolds boundary conditions are given as:

$$p=0,\quad \partial p/\partial n=0 \quad p=0,\quad \partial p/\partial n=0 \tag{11.58}$$

Define relevant dimensionless variables as follow, denoting them by an overbar:

$$\bar{h}=\frac{h}{h_0}\left(h_0=H-R-h_g\right) \tag{11.59}$$

$$\theta=\frac{x\cos\phi}{r}\left(r=R\ \text{or}\ r=R+h_g\right) \tag{11.60}$$

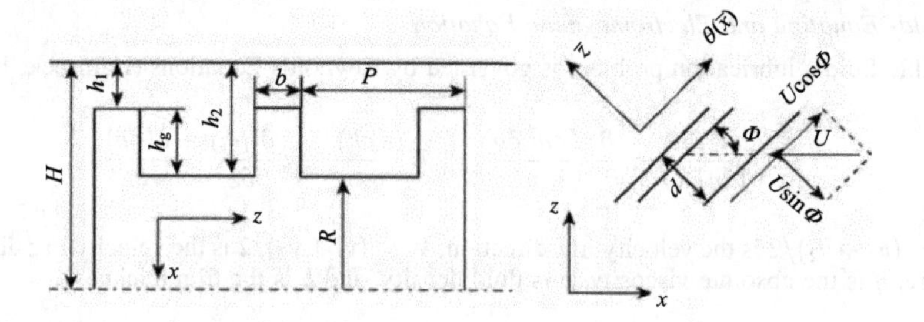

Figure 11.106 Illustration of spiral parameters

$$\bar{z} = \frac{z}{d} \tag{11.61}$$

$$\bar{p} = \frac{h_0^2 \cos\phi}{\eta r U} p \tag{11.62}$$

$$\delta = \frac{2\pi}{n} \sin\phi \cos\phi = \frac{d}{R} \cos\phi \tag{11.63}$$

$$d = P\cos\phi = \frac{S}{n}\cos\phi = \frac{2\pi R}{n}\tan\phi\cos\phi \tag{11.64}$$

where U is the circumferential velocity of micro-machine, d is the vertical distance between two screw threads, h_g is the groove depth, H is the inner radius of intestinal tract, P is the pitch of screw thread, S is the lead of screw thread, n is the number of leads, R is the radius of cylinder, and ϕ is the lead angle. Key parameters for spiral-type micro-machine are shown in Figure 11.106. The parameters used in our calculation are $p = 6.6$ mm, $R = 1$ mm, $b = 0.2$ mm, $h_g = 0.2$ mm, $h_2 = 1$ mm, $n = 1, 2,$ and 4. The length of the capsule is 15 mm.

Substitute these variables into Equation 11.57, the resulting dimensionless Reynolds Equation is:

$$\frac{\partial}{\partial\theta}\left(\frac{\bar{h}^3}{12}\cdot\frac{\partial\bar{p}}{\partial\theta}\right) + \frac{1}{\delta^2}\cdot\frac{\partial}{\partial\bar{z}}\left(\frac{\bar{h}^3}{12}\cdot\frac{\partial\bar{p}}{\partial\bar{z}}\right) = \frac{\cos\phi}{2}\cdot\frac{\partial\bar{h}}{\partial\theta} + \frac{\sin\phi}{2}\cdot\frac{1}{\delta}\cdot\frac{\partial\bar{h}}{\partial\bar{z}} \tag{11.65}$$

We suppose h is independent of z and can obtain the following equation:

$$3h^2\frac{\partial\bar{h}}{\partial\theta}\cdot\frac{\partial p}{\partial\theta} + h^3\frac{\partial^2 p}{\partial\theta^2} + \frac{1}{\delta^2}h^3\frac{\partial^2 p}{\partial z^2} = 6\cos\phi\cdot\frac{\partial\bar{h}}{\partial\theta} \tag{11.66}$$

Friction force in x and z direction can be written as follow (Ma *et al.*, 2004; Wu *et al.*, 1999):

$$F_x = \iint\left(\frac{h}{2}\frac{\partial p}{\partial x} - \frac{\eta\cdot U\cos\phi}{h}\right)\mathrm{d}x\mathrm{d}z \tag{11.67}$$

$$F_z = \iint\left(\frac{h}{2}\frac{\partial p}{\partial z} - \frac{\eta\cdot U\sin\phi}{h}\right)\mathrm{d}x\mathrm{d}z \tag{11.68}$$

Define dimensionless variables as follow, denoting them by an overbar:

$$\overline{F_x} = \frac{h_0}{L\eta r U}F_x \tag{11.69}$$

$$\overline{F_z} = \frac{h_0}{L\eta r U} F_z \tag{11.70}$$

The resulting dimensionless equations of friction force are:

$$\overline{F_x} = \iint \left(\frac{\bar{h}}{2} \cdot \frac{\partial \bar{p}}{\partial \theta} \cdot \frac{1}{\cos \phi} - \frac{1}{\bar{h}} \right) \mathrm{d}\theta \mathrm{d}\bar{z} \tag{11.71}$$

$$\overline{F_z} = \iint \left(\frac{\bar{h}}{2} \cdot \frac{\partial \bar{p}}{\partial z} \cdot \frac{1}{\delta} \cdot \frac{1}{\cos \phi} - 1\bar{h} \frac{\sin \phi}{\cos \phi} \right) \mathrm{d}\theta \mathrm{d}\bar{z} \tag{11.72}$$

(iv) Finite-Difference Method

Our finite-difference technique divides the flow field into equally spaced nodes, as shown in Figure 11.107. To economize on the use of parentheses of functional notation, subscripts i and j denote the position of an arbitrary, equally spaced node, and $p_{i,j}$ denotes the value of the pressure function at that node (White, 1999).

$$p_{i,j} = p(\theta_0 + i\Delta\theta, \quad z_0 + j\Delta z) \tag{11.73}$$

An algebraic approximation for the derivative is:

$$\frac{\partial p}{\partial \theta} \approx \frac{1}{\Delta\theta} \left(p_{i+1,j} - p_{i,j} \right) \tag{11.74}$$

$$\frac{\partial^2 p}{\partial \theta^2} \approx \frac{1}{\Delta\theta^2} \left(p_{i+1,j} - 2p_{i,j} + p_{i-1,j} \right) \tag{11.75}$$

In an exactly similar manner we can derive the equivalent difference expressions for the z direction as follows:

$$\frac{\partial p}{\partial z} \approx \frac{1}{\Delta z} \left(p_{i,j+1} - p_{i,j} \right) \tag{11.76}$$

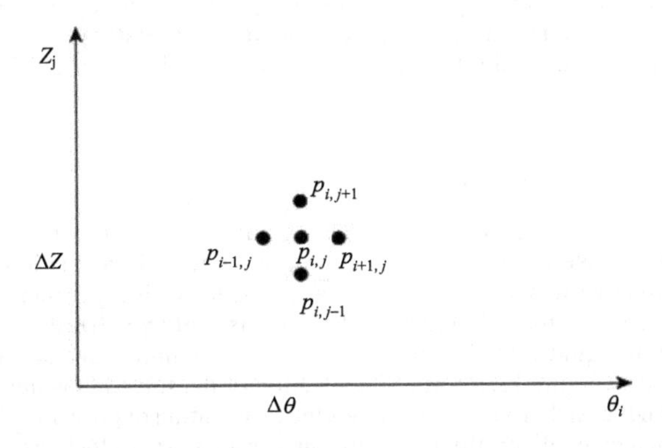

Figure 11.107 Definition sketch for capsule's two-dimensional finite-difference grid

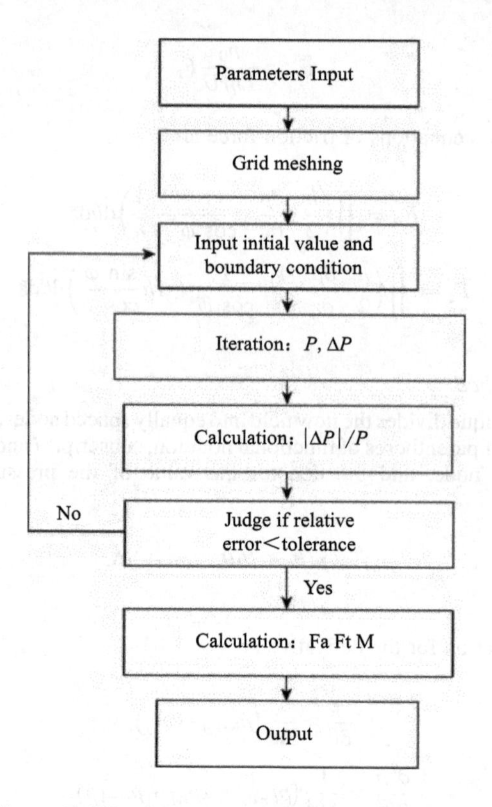

Figure 11.108 Flow chart of fluid dynamics calculation

$$\frac{\partial^2 p}{\partial z^2} \approx \frac{1}{\Delta z^2}\left(p_{i,j+1} - 2p_{i,j} + p_{i,j-1}\right) \tag{11.77}$$

A finite difference algorithm for calculating fluid dynamics properties was programmed. Figure 11.108 shows the flow chart of fluid dynamics calculation. First, we input geometry and physics parameters such as radius, length, pitch, film depth, density, and absolute viscosity. A grid is generated and initial value and boundary condition are input. Furthermore, we use iterative computation methods to calculate a finite-difference equation and then judge the level of relative error. Once the relative error satisfies the tolerance requirement, fluid dynamics parameters such as friction, drag, and load are output for subsequent analysis.

11.10.2 Results and Discussion

The intestinal mucus is known as non–Newtonian fluid; however, previous studies have revealed that assuming the fluid as Newtonian is appropriate for numerical calculation of the intestinal tract (Powell *et al.*, 1974). In order to compare our calculation values with experimental results under similar fluid parameters, we used one type of liquid with absolute viscosity $\eta = 100\,\mathrm{Pa{\cdot}s}$ and density $\rho = 985\,\mathrm{kg/m^3}$, which is analogous to liquid with kinematic viscosity $\nu = 10^5\,\mathrm{mm^2{\cdot}s}$ and density $\rho = 967\,\mathrm{kg/m^3}$ used in (Yamazaki *et al.*, 2003). Figure 11.109 is a visualization of the solved flow field. The arrows show the velocity of the fluid, and the background color shows the distribution of pressure. The inner line shows the profile of the micro-machine. When the micro-machine rotates in a clockwise direction; there was a pressure difference between two sides of the spiral blade. The load can be calculated by integrating

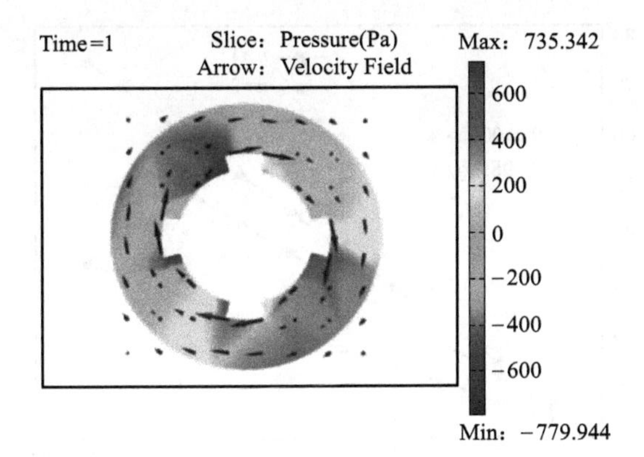

Figure 11.109 Visualization of the solved fluid field *(Color version of this figure is available online)*

pressure on the micro-machine's surface. The drag and the load torque are calculated from the axial and tangential component of the viscous force acting on the micro-machine's surface, respectively.

Figure 11.110 shows the relationship between the frequency and the swimming velocity of the micro-machine. In this figure, the solid line plots the numerical calculation result. We use three types of spiral-blade with a number of spiral lines one, two, and four, respectively. The swimming velocity increases in proportion to the frequency and the four line spiral-blade micro-machine has a greater swimming velocity than those of the one and two line spiral-blade. The order of magnitude of numerical calculation result is similar to experimental results in reference (Yamazaki *et al.*, 2003).

Figure 11.111 shows the relationship between the frequency and the drag. The main contribution to the drag is the viscous force in z direction generated on the micro-machine's surface (including the cylinder and spiral-blade). We use three types of spiral-blade with a number of spiral lines one, two, and four, respectively. As a result of analysis, drag increases in proportion to the frequency and the four line spiral-blade micro-machine has a greater drag than those of the one and two line spiral-blade. We cannot discuss whether the calculated values of drag are exact or not. However, the order of magnitude in our calculation result is the same as that of K. Ishiyama's calculation under similar fluid parameters (Yamazaki *et al.*, 2003).

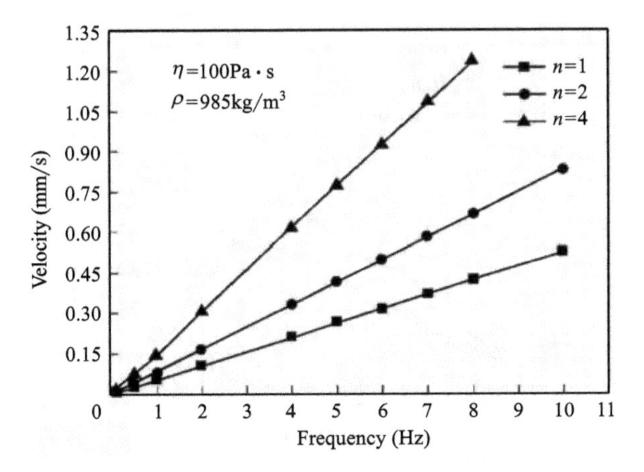

Figure 11.110 Relation between frequency and velocity

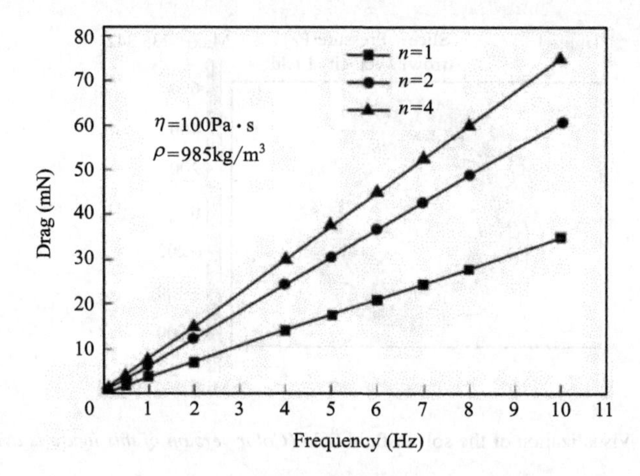

Figure 11.111 Relation between frequency and drag

Figure 11.112 shows the relationship between the frequency and the load torque. We use three types of spiral-blade with a number of spiral lines one, two, and four, respectively. As a result of analysis, load torque increases in proportion to the frequency. It is found that the load torque of the four line spiral blade is larger than those of the one and two line spiral blade. Since the velocity and load torque are much larger for the four line spiral-blade than those of the one and two line spirals, it is clear that in order to swim micro-machine at a faster speed with the same frequency, multiple lines of spiral-blade are needed, and therefore greater driving force needs to be generated to overcome the load torque. We cannot discuss whether the calculated values of load torque are exact or not. However, the order of magnitude in our calculation result is the same as that of K. Ishiyama's calculation under similar fluid parameters (Yamazaki *et al.*, 2003).

However, the value of locomotion velocity does not match with the experimental results very well; they are only similar from the view of order of magnitude. This may be because, effects of the anterior and posterior end of the micro-machine are neglected, which may have certain contributions to reduce the frictional resistance and thus raise the locomotion velocity. In order to estimate the actual locomotion

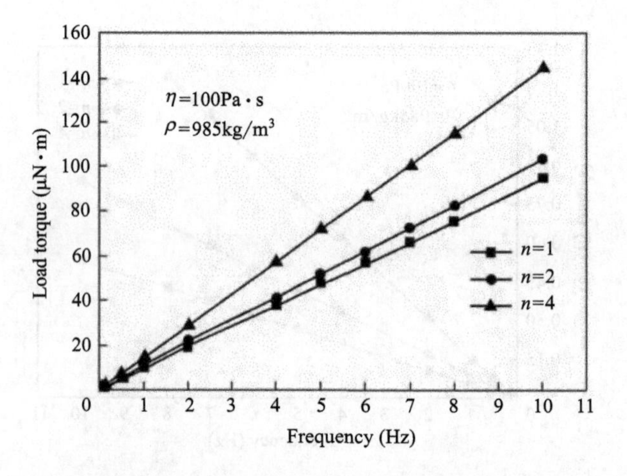

Figure 11.112 Relation between frequency and load torque

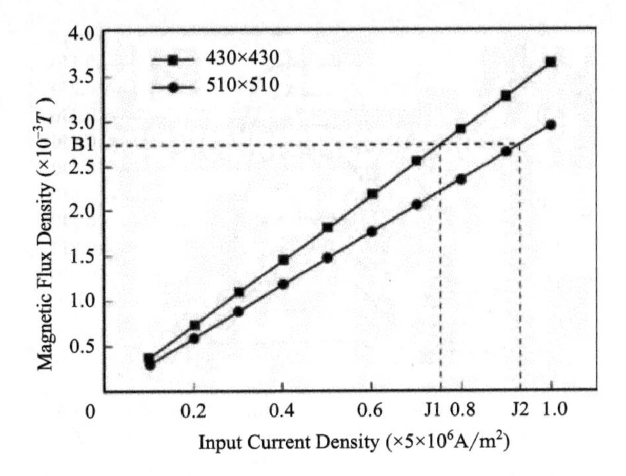

Figure 11.113 Relation between input current density and magnetic flux density

velocity of the micro-machine moving inside the body, the effects of anterior and posterior end of the machine, active and passive deformation of the wall of intestinal tract, and inclination of the machine relative to the tube axis should be considered.

Figure 11.113 shows the relationship between the input current density and corresponding magnetic flux density with different sizes of quadrate Helmholtz coils (510 mm in length, 225 mm^2 in area of cross section, and 430 mm in length, 225 mm^2 in area of cross section, respectively). In order to generate a rotating magnetic field, the magnitude of magnetic flux density for two groups–Helmholtz coils must be equivalent. As the figure above illustrates, when the magnitude of magnetic flux density required for the driving capsule has been determined as B_1, the input current density for two group-Helmholtz coils should be set to J_1 and J_2, respectively.

Figure 11.114 shows the relationship between driving torque and magnetic flux intensity with remanence 10000 G, 5000 G, and 2000 G, respectively. The size of the permanent magnet used in the calculation is $\phi 2$ mm × 20 mm, the driving torque of the micro-machine, depends on the magnetic flux

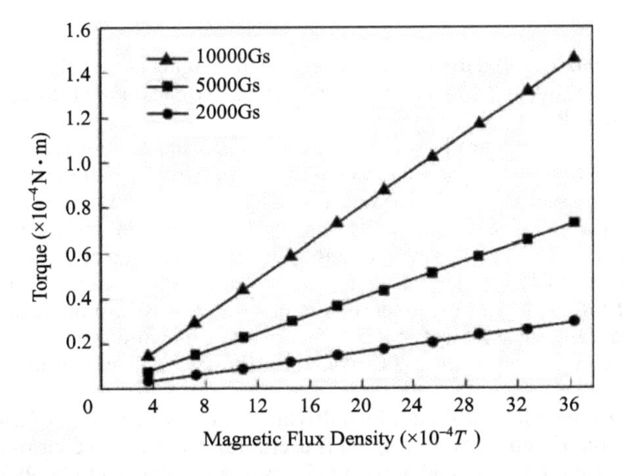

Figure 11.114 Driving torque as a function of magnetic intensity

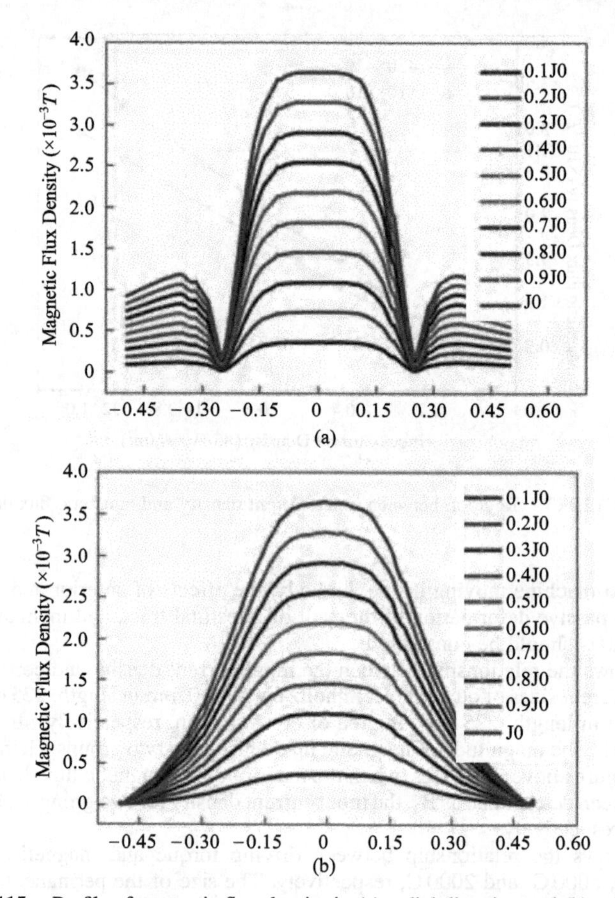

Figure 11.115 Profile of magnetic flux density in (a) radial direction and (b) axial direction

density, the level of remanence, and the volume of permanent magnet. As the figure illustrates, the driving torque increases in proportion to the magnetic intensity, thus, we can obtain the required driving force by increasing magnetic density.

Figure 11.115 illustrates the profile of magnetic flux density in the radial and axial direction, respectively. The magnitude of magnetic flux density varies with the differences of coordinate in the axial and radial direction and a high degree of uniformity is achieved in central region (−0.1 m to 0.1 m). Two group of quadrate–Helmholtz coils are used to generate required magnetic field. The parameters of Group 1 are 510 mm in length and 225 mm^2 in area of cross section. The parameters of Group 2 are 430 mm in length and 225 mm^2 in area of cross section. Figure 11.115 shows the calculation results of Group 2. The magnetic flux density increases in proportion to the input current density J_0. When the value of J_0 is set to 5×10^6 A/m^2, the magnitude of magnetic intensity in the uniform area is approximately equal to 35.2 G, which is sufficient to drive capsule endoscope moving in the intestinal tract.

Figure 11.116 shows the relationship between driving torque as a function of magnetic intensity and load torque as a function of frequency, this figure is useful for determining control parameters of micromachine. The values of fluid density and absolute viscosity are supposed to be 985 kg/m^3 and 100 Pa·s, respectively. Once the frequency f of the capsule is determined by external magnetic equipment, the load

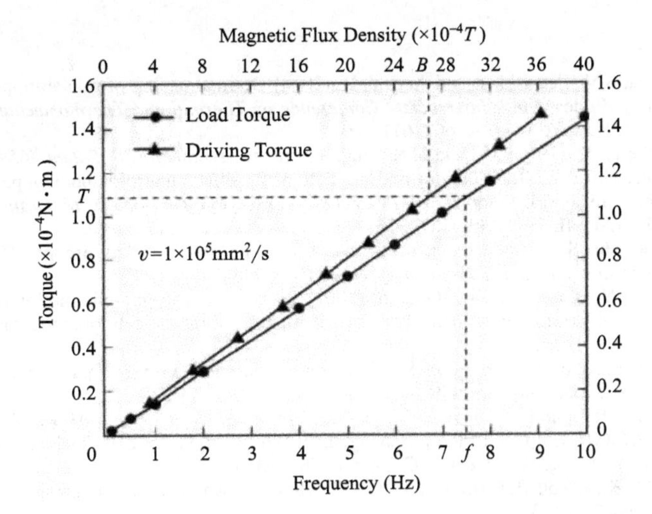

Figure 11.116 Relation between driving torque and load torque

torque can be obtained according to the circle plot line as shown above. In order to overcome the load torque, the magnitude of magnetic intensity must be adjusted to an appropriate value for generating enough driving force torque T, which can be obtained in terms of triangle plot line. In this figure, B is the threshold value of magnetic flux density.

Therefore, this result shows it is suitable to analyze the micro-machine's locomotion in the intestinal tract by using theories of fluid dynamics lubrication.

11.11 Chapter Summary

In this chapter, several main MEMS sensors and systems in industry and research sectors were investigated and different simulation methodologies were presented. Hysteresis and zero shift of the pressure sensor output caused by the viscoelasticity and viscoplasticity of bonding material in the assembly were simulated first, then the design for the thermo-fluid based accelerometer and the capacitance based accelerometer was discussed. A 3D EM simulation model of the wheel antenna was described, along with the test methods applicable for designing the TPMS system and predicting the performance of the installed TPMS system with wheel antenna. The parameters of the gyroscope and air flow sensor such as main channel length and width, angle of two split channels and so on, will be optimized to increase the sensitivity and applicable range through the fluid dynamic simulation.

Three kinds of microjet devices were studied, and their flow and temperature distributions were provided for LED cooling; the comparison results show that under the same conditions, the microjet device design with one inlet and two outlets can achieve a much better cooling performance.

The proposed numerical models have been used to study the effects of the induced DC DEP force, arising from the interactions between the spatially non-uniform electric fields and the dielectric particles, on particle separation through a constricted microchannel. Size-based DEP separation depends largely on the particle size and zeta potential, magnitude of the applied electric field, and the microchannel geometry. A higher electric field and a higher constriction ratio could decrease the separation threshold of the particle size. The proposed numerical model can be exploited further for the design and optimization of DEP separation of particles with arbitrary shapes and sizes through microfluidic channels of complex geometries.

Finally theories of fluid dynamics lubrication and a finite difference method have been used to analyze and calculate the micro-machine's fluid dynamics properties in the gastrointestinal tract.

References

Acikalin, T., Garimella, S.V., Petroski, J. and Arvind, R. (2004) Optimal design of miniature piezoelectric fans for cooling light emitting diodes, *Ninth Intersociety Conference on Thermal and Thermomechanical Phenomena in Electronic Systems*, 1, New York, USA, 663–671.

Adams, V. and Dougherty, D. (1997) Method of forming a package assembly, *U.S. Patent 5659950.*

Ah J.P. and Desmulliez, M.P.Y. (2001) Modeling and simulation of a silicon micro-diaphragm piezoresistive pressure using finite element analysis (FEA) tools, *MEMS Design, Fabrication, Characterization, and Packaging Conference,* Edinburgh, ROYAUME-UNI **4407**:327–336.

Ai, Y., Luo, X.B. and Liu, S. (2006) Design of a novel micro thermo-fluidic gyroscope, *The 7th International Conference on Electronics Packaging Technology*, Shanghai, China.

Ai, Y., Joo, S.W., Jiang, Y., Xuan, X.C. and Qian, S. (2009) Transient electrophoretic motion of a charged particle through a converging-diverging microchannel: Effect of direct current-dielectrophoretic force, *Electrophoresis* **30**(14):2499–2506.

Ai, Y., Joo, S.W., Jiang, Y., Xuan, X.C. and Qian, S. (2009) Pressure-driven transport of particles through a converging-diverging microchannel, *Biomicrofluidics* **3**(2):022404.

Alan, M. (2003) Solid state lighting—a world of expanding opportunities at LED 2002, *III–Vs Review* **16**:30–33.

Arik, M. and Weaver, S. (2004) Chip scale thermal management of high brightness LED Packages, *Fourth International Conference on Solid State Lighting*, Denver, Colorado, USA **5530**:214–223.

Audet, S.A., Edenfeld, K.M. and Bergstrom, P.L. (1997) Motorola wafer-level packaging for integrated sensors, *Micromachine Devices* **2**(1):1–3.

Baek, N.K., Sung, I.H. and Kim, D.E. (2004) Frictional resistance characteristics of a capsule inside the intestine for microendoscope design, *Proceedings of the Institution of Mechanical Engineers Part H-Journal of Engineering in Medicine,* vol. 218, pp. 193–201.

Barbulovic-Nad, I., Xuan, X.C., Lee, J.S.H., and Li D.Q. (2006) DC-dielectrophoretic separation of microparticles using an oil droplet obstacle, *Lab Chip* **6**(2):274–279.

Beardmore, G. (1997) Packaging for microengineered devices. Lessons from the real world, *The Institution of Electrical Engineers on Assembly and Connections in Microsystems*, London, UK, 2/1–2/8.

Bochobza-Degani, O., Socher, E., Nemirovsky, Y. and Seter, D.J. (2000) Comparative study of novel micromachined accelerometers employing MIDOS, *Sensors Actuators A* **80**(2):91–99.

Bosc, J.M., Guo, Y.F., Sarihan, V. and Lee, T. (1998) Accelerated life testing for micro-machined chemical sensors, *IEEE Transactions on Reliability* **47**(2).

Brantley, W.A. (1973) Calculated elastic constants for stress problems associated with semiconductor devices, *Journal of Applied Physics* **44**(1):534–535.

Buchner, R., Rohloff, K., Benecke W. and Lang, W. (2005) A high-temperature thermopile fabrication process for thermal flow sensors, *The 13th international Conference on Solid- State Sensors Actuators and Microsystems Seoul,* Korea: 575–578.

Carpi, F., Galbiati, S. and Carpi, A. (2007) Controlled navigation of endoscopic capsules: Concept and preliminary experimental investigations, *IEEE Transactions on Biomedical Engineering* **54**:2028–2036.

Castillo J., Dimaki M. and Svendsen W.E. (2009) Manipulation of biological samples using micro and nano techniques, *Integrative Biology* **1**(1):30–42.

Chen, J.H., Liu, C.K., Chao, Y.L., and Tain, R.M. (2005) Cooling performance of siliconbased thermoelectric device on high power LED, *24th International Conference on Thermoelectrics,* Clemson, South Carolina, USA, University of Clemson, 53–56.

Chen, Y. and Song, X.J. (2005) Application of thermal resistance model in high power LED, *China Illumination Equipment* **7**:5–7 (in Chinese).

Christodoulou, D., Haber, G., Beejay, U., Tang, S.J., Zanati, S., Petroniene, R., Cirocco, M., Kortan, P., Kandel, G., Tatsioni, A., Tsianos, E. and Marcon, N. (2007) Reproducibility of wireless capsule endoscopy in the investigation of chronic obscure gastrointestinal bleeding, *Canadian Journal of Gastroenterology* **21**:707–714.

Clark, S.K. and Wise, K.D. (1979) Pressure sensitivity in anisotropically etched thin-diaphragm pressure sensor", *IEEE Transactions On Electron Devices*, **26**(12):1887–1895.

Corning Glass material data sheet. (1988) Corning Glass Works, New York.

Dao, R., Morgan, D.E., Kries, H.H. and Bachelder, D.M. (1996) Convective accelerometer and inclinometer, *US Patent 5581034.*

Dau, V.T., Dao, D.V. and Shiozawa, T. Kumagai, H. and Sugiyama, S. (2006) Development of a dual-axis thermal convective gas gyroscope, *Journal of Micromechanics and Microengineering* **16**(7):1301–1306.

Dauderstadt, U.A., Hiratsuka, R., Sarro, P.M. and Devries, P.H.S. (1995) Silicon accelerometer based on thermopiles, *Sensors Actuators A* **46**(1–3):201–204.

Daveson, J. and Appleyard, M. (2008) Future perspectives of small bowel capsule endoscopy, in *10th Endoscopy Forum Japan*, Otaru, JAPAN, pp. 262–270.

Davison, S.M. and Sharp, K.V. (2008) Transient simulations of the electrophoretic motion of a cylindrical particle through a 90 degrees corner, *Microfluid Nanofluid* **4**(5):409–418.
Delvaux, M. and Gay, G. (2008) Capsule endoscopy: technique and indications, *Best Practice and Research in Clinical Gastroenterology* **22**:813–837.
Dickerson, T. and Ward, M. (1997) Low deformation and stress packaging of micromachined devices, *The Institution of Electrical Engineers on Assembly and Connections in Microsystems, London, UK*, 7/1–7/3.
Dittrich, P.S. and Manz, A. (2006) Lab-on-a-chip: microfluidics in drug discovery, *Nature Reviews Drug Discovery* **5**(3):210–218.
Dougherty, D. (1997) Improved die-attach method for reducing package stress on the accelerometer sensor, *1997 Winter Motorola AMT Symposium*, Schaumburg, IL 1211–1219.
Elias, P.M. (1989) The stratum corneum as an organ of protection old and new concepts, in Fritsch P, Schuler G, Hintner H, eds., *Current Problems in Dermatology*, Switzerland, 10–21.
Feynman, R.P., Leighton, R.B. and Sands, M. (1964) *The Feynman Lectures on Physics*, Volume 2, Addison-Wesley, Reading, MA, Chapter 14.
Fischer, M. (2003) Tire pressure monitoring, *Verlag Moderne Industrie*, Landsberg, Germany.
Fu, X.M., Liu, S. and Wang, X.J. (2006) The effect of mounting stress on the zero point output of pressure sensors, *The 7th International Conference on Electronics Packaging Technology*: 1–5.
Furjes, P., Legradi, G., Ducso, C., Aszodi, A. and Barsony, I. (2004) Thermal characterization of a direction dependent flow sensor, *Sensors and Actuators A: Physical* **115**(2–3):417–423.
Gary, L. and Ampere, A.T. (2001) Low stress packaging of a micromachined accelerometer, *IEEE Transaction on Electronics Packaging Manufacturing* **24**(1):18.
Glaninger, A., Jachimowicz, A., Kohl, F., Chabicovsky, R. and Urban, G. (2000) Wide range semiconductor flow sensors, *Sensors and Actuators* **85**(1–3):139.
Gomez, F.A. (2008) *Biological Applications of Microfluidics*, John & Wiley Interscience, New Jersey, 2008.
Gong, Q., Zhou, Z., Yang, Y. and Wang, X. (2000) Design, optimization and simulation microelectro magnetic pump, *Sensor Actuators A* **83**:200–207.
Hadithi, M., Heine, G.D.N., Jacobs, M., von Bodegraven, A.A. and Mulder, C. J. J. (2006) A prospective study comparing video capsule endoscopy with double-balloon enteroscopy in patients with obscure gastrointestinal bleeding, *American Journal of Gastroenterology* **101**:52–57.
Hamrock, B.J. (1994) *Fundamentals of Fluid Film Lubrication*, McGraw-Hill, New York, pp. 141–165.
Hawkins, B.G., Smith, A.E., Syed, Y.A. and Kirby, B.J. (2007) Continuous-flow particle separation by 3D insulative dielectrophoresis using coherently shaped, dc-biased, ac electric fields, *Analytical Chemistry* **79**(19):7291–7300.
Hsu, C.C., Wang, S.J. and Liu, C.Y. (2003) Metallic wafer and chip bonding for LED Packaging, *5th Pacific Rim Conference on Lasers and Electro-Optics*, 1, Taipei, Taiwan, 2003, 26.
Hu, H.H., Joseph, D.D., and Crochet, M.J. (1992) Direct simulation of fluid particle motions, *Theoretical and Computational Fluid Dynamics* **3**(5):285–306.
Hu, H.H., Patankar, N.A. and Zhu, M.Y. (2001) Direct numerical simulations of fluid-solid systems using the arbitrary Lagrangian-Eulerian technique, *Journal of Computational Physics* **169**(2):427–462.
Iakovidis, D.K., Tsevas, S., Maroulis, D., Polydorou, A. and IEEE. (2008) Unsupervised summarization of capsule endoscopy video, in *4th International IEEE Conference Intelligent Systems*, Varna, BULGARIA, pp. 140–145.
Iddan, G., Meron, G., Glukhovsky, A. and Swain, P. (2000) Wireless capsule endoscopy, *Nature* **405**:417.
Ishiyama, K., Arai, K.I., Sendoh, M., Yamazaki, A. and IEEE. (2000) Spiral-type micro-machine for medical applications, in *International Symposium on Micromechatronics and Human Science*, Nagoya, Japan, pp. 65–69.
Ishiyama, K., Sendoh, M., Yamazaki, A. and Aral, K.I. (2001) Swimming micro-machine driven by magnetic torque, *Physical* **A91**(1–2):141–144.
Jaccodine, R.J. and Schlegel, W.A. (1966) Measurement of strains at Si-SiO_2 interface, *Journal of Applied Physics* **37**(6):2429–2434.
Jaeger, R.C., Suhling, J.C., and Ramani, R. (1994) Errors associated with the design, calibration and application of piezoresistive stress sensors in (100) silicon, *IEEE Transactions on Components, Packaging, and Manufacturing Technology, Part B: Advanced Packaging*: 171.
Kang, K.H., Xuan, X.C., Kang, Y. and Li, D. (2006) Effects of dc-dielectrophoretic force on particle trajectories in microchannels, *Journal of Applied Physics* **99**(6):064702–8.
Kang, K.H., Kang, Y.J., Xuan, X.C. and Li, D.Q. (2006) Continuous separation of microparticies by size with direct current-dielectrophoresis, *Electrophoresis* **27**(3):694–702.
Kang, Y.J., Li, D.Q., Kalams, S.A. and Eid, J.E. (2008) DC-Dielectrophoretic separation of biological cells by size, *Biomedical Microdevices* **10**(2):243–249.
Kang, Y. J. and Li, D.Q. (2009) Electrokinetic motion of particles and cells in microchannels, *Microfluid Nanofluid* **6**(4):431–460.
Kang, Y., Cetin, B., Wu, Z., and Li, D.Q. (2009) Continuous particle separation with localized AC-dielectrophoresis using embedded electrodes and an insulating hurdle, *Electrochimica Acta* **54**(6):1715–1720.

Kassim, I., Ng, W.S., Feng, G., Phee, S.J., Dario, P., Mosse, C.A. and IEEE. (2003) Review of locomotion techniques for robotic colonoscopy, in *20th IEEE International Conference on Robotics and Automation (ICRA)*, Taipei, Taiwan, pp. 1086–1091.

Kaushik, S., Hord, A.H., Denson, D.D., McAllister, D.V., Smitra, S., Allen, M.G. and Prausnitz, M. (2001) Lack of pain associated with microfabricated microneedle, *Anesthesia & Analgesia* **92**:502–504.

Kelly, G. (1996) Packaging induced thermo-mechanical stress, Micromechanic Micromech Europe, Barcelona, 297–299.

Kersaudy-Kerhoas, M., Dhariwal, R. and Desmulliez, M.P.Y. (2008) Recent advances in microparticle continuous separation, *IET Nanobiotechnology* **2**(1):1–13.

Kim, K.H., Ko, J.S., Cho, Y.H., Lee, K. and Kwak, B.M. (1995) A skew–symmetric cantilever accelerometer for automotive airbag applications, *Sensors Actuators A* **50**(1–2):121–126.

Kim, B., Lee, S., Park, J.H. and Park, J.O. (2005) Design and fabrication of a locomotive mechanism for capsule-type endoscopes using shape memory alloys (SMAs), *IEEE-ASME Transactions on Mechatronics* **10**:77–86.

Kubena, R.L., Atkinson, G.M., Robinson, W.P. and Stratton, F.P. (1996) A new miniaturized surface micromachined tunneling accelerometer, *IEEE Electron Device Letters* **17**(6):306–308.

Leng, Y., Zhao, G.B., Li, Q.X. and Dong, T.L. (2007) A self-identifying car tire pressure monitoring system based on wireless sensor and CAN bus, *Instrument Technique and Sensor (Chinese)*, 6.

Leung, A.M., Jones, J., Czyzewska, E., Chen, J. and Pascal, M. (1997) Micromachined accelerometer with no proof mass, *Technical Digest of International Electron Device Meeting*, 899–902.

Lewpiriyawong, N., Yang, C. and Lam, Y.C. (2008) Dielectrophoretic manipulation of particles in a modified microfluidic H filter with multi-insulating blocks, *Biomicrofluidics* **2**(3):034105.

Li, C.W. and Liang, G.W. (2004) Research of thermal air mass flow sensor applicated in car engine, *Measure technology* **11**:5–6.

Li, Q.X., He, F.M. and Leng, Y. (2006) Research on the low-frequency communication applied in TPMS, *Computer Engineering and Applications (Chinese)* **42**(17):191–193.

Li, Q.X., Zhang, H.X. and Leng, Y. (2007) Study on dynamic antenna in tire pressure monitor system, *Automotive Engineering (Chinese)*, 6.

Liu, H., Qian, S.Z. and Bau, H.H. (2007) The effect of translocating cylindrical particles on the ionic current through a nanopore, *Biophysical Journal* **92**(4):1164–1177.

Liu, S., Yang, J.Y., Gan, Z.Y. and Luo, X.B. (2008) Structural optimization of a microjet based cooling system for high power LEDs, *International Journal of Thermal Sciences* **47**:1086–1095.

Luo, X.B., Li, Z.X. and Yang, Y.J. (2001) Dual directional synthetic jet gyroscope, *China Patent 01129222.9*.

Luo, X.B., Li, Z.X., Guo, Z.Y. and Yang, Y.J. (2003) Study on linearity of a micromachined convective accerometer, *Microelectronic Engineering* **65**(1–2):87–101.

Luo, X.B. and Liu, S. (2006) A closed micro jet cooling system for high power LEDs, *The 7th International Conference on Electronics Packaging Technology.*

Luo, X.B., Liu, S., Jiang, X.P. and Cheng, T. (2007) Experimental and numerical study on a micro jet cooling solution for high power LEDs, *Sciences in China, Series E* **50**(4):478–489.

Luo, X.B. and Liu, S. (2007) A microjet array cooling system for thermal management of high-brightness LEDs, *IEEE Transactions on Advanced Packaging* **30**(3):475–484.

Lyon, K.G., Silinger, G.L. and Swenson, C.A. (1977) Linear thermal expansion measurement on silicon from 6 to 340 K, *Journal of Applied Physics* **148**(3):865–868.

Ma, X.F., Zhou, Y.S. and Chen, B. (2004) Theoretical analysis and experimental research on a medical micro-robot, *Chinese Journal of Mechanics Engineering* **40**(7):124–127 (in Chinese).

Ma, C.L. and Bao, C. (2005) A novel measuring method of thermal resistance for high power LED, *Optics Instrument* **27**:13–17 (in Chinese).

Ma, B., Liu, S., Gan, Z.Y., Liu, G.J., Cai, X.X., Zhang, H.H. and Yang, Z.G. (2006) A PZT insulin pump integrated with a silicon micro needle array for transdermal drug delivery, *Microfluidics and Nanofluidics* **2**(5):417–423.

Madou, M. (1997) *Fundamentals of Microfabrication*, CRC Press, Boca Raton, New York, 378–399.

Mailly, F., Giani, A., Bonnot, R., Temple-Boyer, P., Pascal-Delannoy, F., Foucaran, A. and Boyer, A. (2001) Anemometer with hot platinum thin film, *Sensors and Actuators A: Physical* **94**(1–2):32.

McNeil, A.C. (1998) A parametric method of linking MEMS package and device models, *Proceedings of The Solid-State Sensor and Actuator Workshop*, Hilton Head IS, SC 166–169.

Milanovic, V., Bowen, E., Zaghloul, M.E., Tea, N.H., Suehle, J.S., Payne, B. and Gaitan, M. (2000) Micromachined convective accelerometers in standard integrated circuits technology, *Applied Physics Letter* **76**(4):508–510.

Nam, T., Kim, S. and Park, S. (2004) The temperature compensation of a thermal flow sensors by changing the slope and the ratio of resistances, *Sensors and Actuators A* **114**:212–218.

Narendran, N. and Gu, Y.M. (2005) Life of LED-based white light sources, *Journal of Display Technology* **1**(1):167–171.

Nemirovsky, Y., Nemirovsky, A., Muralt, P. and Setter, N. (1996) Design of a novel thin film piezoelectric accelerometer, *Sensors Actuators A* **56**(3):239–249.

Nguyen, N.T. (1997) Micromachined flow sensors—a review, *Flow Measurement and Instrument* **8**(1):7–16.

Nguyen, N.T. and Dotzel, W. (1997) Asymmetyrical locations of heaters and sensors relative to each other using heater arrays: a novel method for desiging multi-range electrocaloric mass-flow sensors, *Sensors and Actuators A* **62**:506–512.

Nguyen, N.T. (1997) Micromachined flow sensors-a review, *Flow Measurement and Instrumentation* **8**(1):7.

Nysæther, J.B., Larsen, A., Liverød, B. and Ohlckers, P. (1998) Measurement of package-induced stress and thermal zero shift in transfer mold silicon piezoresirtive pressure sensor, *Journal of Micromechanics and Microengineering* **8**:168–171.

Okada, Y. and Tokymaru, Y. (1984) Precision determination of lattice parameter and thermal expansion coefficient of silicon between 300 and 1500 K, *Journal of Applied Physics* **56**(2):314–320.

Pamme, N. (2007) Continuous flow separations in microfluidic devices, *Lab Chip* **7**(12):1644–1659.

Peirs, J., Reynaerts, D. and Van Brussel, H. (1999) Design of miniature parallel manipulators for integration in a self-propelling endoscope, in *Eurosensors XIII Meeting*, The Hague, Netherlands, pp. 409–417.

Peng, J.G., Ying, Z.Z. and Ye, X.Y. (2005) Progresses on micromachined flow sensors based on MEMS technologies, *Advances and mechanics* **35**:361–376.

Petroski, J. (2003) Understanding longitudinal fin heat sink orientation sensitivity for Light Emitting Diode (LED) lighting applications, *International Electronic Packaging Technical Conference and Exhibition*, Maui, Hawaii, USA, 111–117.

Powell, R.L., Aharonson, E.F. and Schwarz, W.H. (1974) Rheological behavior of normal tracheobronchial mucus of canines, *Journal of Applied Physiology* **37**:447–451.

Qian, R.Y. (1998) Air mass flow sensor and environment temperature compensation of electric-control engine, *Automobile Technology* **6**:13.

Qian, S., Wang, A.H. and Afonien, J.K. (2006) Electrophoretic motion of a spherical particle in a converging-diverging nanotube, *Journal of Colloid Interface Science* **303**(2):579–592.

Quirini, M., Menciassi, A., Scapellato, S., Stefanini, C. and Dario, P. (2008) Design and fabrication of a motor legged capsule for the active exploration of the gastrointestinal tract, *IEEE-ASME Transactions on Mechatronics* **13**:169–179.

Retajczyk, T.F. and Sinha, A.K. (1980) Elastic stiffness and thermal expansion coefficients of various refractory silicides and silicon nitride films, *Thin Solid Film* **70**(2):241–247.

Roberts, R.B. (1981) Thermal expansion reference data: silicon 300–850 K, *Journal of Physics D: Applied Physics* **114**:163.

Roha, S.C., Choi, Y.M. and Kim, S.Y. (2006) Sensitivity enhancement of a silicon micro-machined thermal flow sensor, *Sensors and Actuators A* **128**(1):1–6.

Romig, A.D., Dressendorfer, P.V. and Palmer, D.W. (1997) High performance microsystem packaging: Aperspective, *European Symposium on Reliability of Electron Devices, Failure Physics and Analysis*, Arcachon, France, **37**(10–11):1771–1781.

Sabate, N., Santander, J., Fonseca, L., Gracia, I. and Cane, C. (2004) Multi-range silicon micromachined flow sensor, *Sensors and actuators A* **110**(1–3):282–288.

Sandrasegaran, K., Maglinte, D.D.T., Jennings, S.G. and Chiorean, M.V. (2008) Capsule endoscopy and imaging tests in the elective investigation of small bowel disease, *Clinical Radiology* **63**:712–723.

Sano, S., Murata, H. and Hattori, K. (1993) Development of flat panel LED module with heat sink, *Mitsubishi Cable Industry Review* **86**:112–118.

Sarihan, V., Guo, Y.F., Lee, T. and Teng, S. (1998) Impacting electronic package design by validated simulations, *Electronic Components and Technology Conference*, 330–335.

Sendoh, M., Ishiyama, K. and Arai, K.-I. (2003) Fabrication of magnetic actuator for use in a capsule endoscope, *IEEE Transactions on Magnetics* **39**(5):3232–3234.

Shemansky, F., Ristic, L., Koury, D. and Joseph, E. (1997) A two-chip accelerometer system for automotive applications, *Microsystem Technology* **1**(3):121–125.

Si, D.H. and Li, Y.H. (2005) Thermal mass flow sensor based on technology microelectronic systems, *Sensor Technology* **1**:49–52.

Sinha, A.K. and Smith, T.E. (1978) Thermal stresses and cracking resistance of dielectric films, *Journal of Applied Physics* **49**(4):2424–2426.

Song, C. and Shinn, M. (1998) Commercial vision of silicon-based inertial sensors, *International Conference on Solid-State Sensors and Actuators*, Chicago, IL, **66**(1–3): 231–266.

Song, X. and Liu, S. (2003) A Performance Prediction Model for a Piezoresistive Transducer Pressure Sensor, *The 5th International Conference on Electronics Packaging Technology*, Shanghai, China, 30–35.

Stemme, E. and Stemme, G. (1993) A valve less diffuser/nozzle-based fluid pumps, *Sensor Actuators A* **39**(22):159–167.

Su, F., Chakrabarty, K. and Fair, R.B. (2006) Microfluidics-based biochips: Technology issues, implementation platforms, and design-automation challenges, *IEEE Transactions on Computer-Aided Design of Integrated Circuits and Systems* **25**(2):211–223.

Sun, C.S. and Li, R. (2005) Research and development of thermal-film air-mass sensor, *Sensorworld Technology and Application, China*, **10**:29–32.

Takahashi, T., Nishio, T., Ikegami, M. and Gunji, T. (1991) Gas rate sensor system, *US Patent 5102676.*

Tan, L.X., Liu, S., Zhang, H.H., Gan Z.Y., Chen, C., Hou, B., Wu, Z.G., Wang, X.P., Leng, Y. and Chen, J.J. (2006) Numerical analysis of the reliability of tire pressure monitoring system installed on wheel hub with glue, *The 7th International Conference on Electronics Packaging Technology*, Shanghai, China.

Tang, T.K., Gutierrez, R.C., Stell, C.B., Vorperian, V., Arakaki, G.A., Rice, J.T., Li, W.J., Chakraborty, I., Shcheglov, K., Wilcox, J.Z. and Kaiser, W.J. (1997) A packaged silicon MEMS vibratory gyroscope for microspacecraft, *IEEE 10th Annual International Workshop Micro Electro Mechanical System, Investigation Micro Structures, Sensors, Actuators, Machine, Robots*, New York, NY, 500–505.

Thwar, P.K., Linderman, J.J. and Burns, M.A. (2007) Electrodeless direct current dielectrophoresis using reconfigurable field-shaping oil barriers, *Electrophoresis* **28**(24):4572–4581.

Toner M. and Irimia D. (2005) Blood-on-a-chip, *Annual Review of Biomedical Engineering* **7**:77–103.

Townsend, P. H. (1987) Inelastic strain in thin films, Ph. D thesis, Stanford University.

Tufte, O.N. and Stelzer, E.L. (1963) Piezoresistive properties of silicon diffused layers, *Journal of Applied Physics* **34**(2):313–318.

Van Dde Pol, F.C.M., Van Lintel, H.T.G., Elvenspoek, M. and Fluitman, J.H.J. (1990) A thermopneumatic micropump based on microengineering techniques, *Sensor Actuators A* **21**(1–3):198–202.

Van Lintel, H.T.G., van de Pol, F.C.M. and Bouwstra, S. (1988) A piezoelectric micropump based on micromachining of silicon, *Sensor Actuators A* **15**(2):153–167.

Van Putten, A.F.P. and Middelhoek, S. (1974) Intergrated silicon anemometer, *Electronic Letters* **10**(21):425.

Vujosevic, M. and Shah, M. (1998) Stress isolation of the sensing device in the 40g accelerometer PDIP/SIP: Impact of dome coat coverage, *1998 Winter Motorola AMT Symposium*, IL, 36–43.

Wang, Y.T., Shi, J.S. and Wang, L.T. (1994) Analysis and calculation on junction LED thermal characteristics in engineering, *Semiconductor Optoelectron* **15**:143–146 (in Chinese).

Wang, X.J., Wang, X.B. and Gascoyne, P.R.C. (1997) General expressions for dielectrophoretic force and electrorotational torque derived using the Maxwell stress tensor method, *Journal of Electrostatic* **39**(4): 277–295.

Wang, P., Wang, J.W. and Dong, Y.R. (2000) Appearance and solution measure of the mount stress in the pressure transducer, *Journal of Transducer Technology* **19**(3):49–50.

Wang, B. (2003) The failure diagnosis of MAF sensor on motor, *Journal of Shaoguan University (Natural Science)* **24**:36.

Wang, X.J., Wu, Z.G. and Liu, S. (2005) Modeling and simulation of thermal-mechanical characteristics of the packaging of tire pressure monitoring system, *The 6th International Conference on Electronic Packaging Technology*: 693–695.

Weibel, D.B. and Whitesides, G.M. (2006) Applications of microfluidics in chemical biology, *Current Opinion in Chemical Biology* **10**(6):584–591.

White, F.M. (1999) *Fluid Mechanics*, 4th edn., McGraw-Hill, New York, pp. 540–543.

Wiese, R.W., Song, H.J. and Hsu, H.P. (2005) RF link budget analysis of a 315 MHz wireless link for automotive tire pressure monitoring system, *SAE 2005 World Congress & Exhibition, Session*, In-Vehicle Networks, Detroit, MI, USA.

Williams, A.C. and Barry, B.W. (1992) Skin absorption enhancers, *Critical reviews in therapeutic drug carrier systems* **9**:305–353.

Woias, P. (2001) Micropumps: summarizing the first two decades, *Microfluidics and BioMEMS* **4560**:39–52.

Wu, J.K., Ma, X.N. and Huang, Y.Y. (1999) Parameter comparison calculations for oil film stability of grooved liquid-lubricated journal bearing, *Transaction on Tribology* **19**(1):56–60 (in Chinese).

Wu, K.G. (2002) Temperature-difference type hot-film air mass flow sensor, *Journal of Chang an University (Natural Science Edition)* **22**:86–88.

Wu, H.Y., Qian, K.Y., Hu, F. and Luo, Y. (2005) Study on thermal performances of flip-chip high power LEDs, *Journal of OptoelectronicsLaser* **16**:511–514.

Xuan, X.C. (2008) Joule heating in electrokinetic flow, *Electrophoresis* **29**(1):33–43.

Yamazaki, A., Sendoh, M., Ishiyama, K., Arai, K.I. and Hayase, T. (2003) Three-dimensional analysis of swimming properties of the spiral-type magnetic micro-machine, *Sensors and Actuators A* **105**:103–108.

Yang, K., Zhou, J. and Yan, G.Z. (2004) The research on MEMS fang bar fluidic angular rate sensor, *The 7th International Conference on Solid-State and Integrated Circuits Technology* **3**:1808–1811.

Yazdi, N., Ayazi, F. and Najafi, K. (1998) Micromachined inertial sensors, *Proceedings of the IEEE* **86**(8):1640–1659.

Ye, C.Z. and Li, D.Q. (2004) 3D transient electrophoretic motion of a spherical particle in a T-shaped rectangular microchannel, *Journal of Colloid Interface Science* **272**(2):480–488.

Yu, B.H. and Li, X.F. (2005) Impact of flip-chip substrate adhesive on the thermal characteristic of high power LED, *Semiconductor Technology* **30**:49–51 (in Chinese).

Yu, B.L., Gan, Z.Y., Liu, S., Luo, X.B., Cao, H. and Xu, J.P. (2006) Velocity and temperature simulation of a new air flow sensor, *The 7th International Conference on Electronics Packaging Technology*, Shanghai, China.

Zarnik, M.S., Rocak, D.M. and Macek, S. (2004) Residual stresses in a pressure-sensor package induced by adhesive material during curing: a case study, *Sensors and Actuators A* **116**(3):442–449.

Zhang, K., Xiao, G.W., Wong, C.K.Y., Gu, H.W., Yuen, M.M.F., Chan, P.C.H. and Xu, B. (2005) Study on thermal interface material with carbon nanotubes and carbon black in high-brightness LED packaging with flip-chip technology, *55th Electronic Components and Technology Conference*, Orlando, Florida, USA, 60–65.

Zukauskas, A., Shur, M.S. and Gaska, R. (2002) Introduction to solid-state lighting, *John Wiley and Sons,* New York.

12

System in Package (SIP) Assembly

System in package (SIP) is a trend of electronic package for multiple functions (ITRS 2008). In this section, a leadframe based system in package (SIP) for power management is examined. This package is built using multiple die including power IGBTs (Insulated Gate Bipolar Transistor), diodes, and IC controllers. To maximize product performance the power components use a thin die to minimize on-resistance (RDS (on)), to maximize thermal performance, and to minimize the board standoff height by allowing the package to be thinner. However, the ultra thin die could be a potential risk for die cracking during the assembly process. Therefore, it is critical to understand the impact of a thining die on the reliability of the product in assembly manufactures. Then a flip-chip after the assembly process is studied with six material models and underfill process models. A stacked die flip-chip SIP layout, assembly process and the material selection are investigated for a flip-chip-on-silicon ball-grid-array (FSBGA) package. It is useful to know how the impact of these assembly processes and material selection on the deformation, stress and strain of the SIP.

The simulation of package assembly process of side by side placed SIP consists of three parts: (1) multiple die attach processes with different IGBT die and die attach thickness (by element birth and death techniques) to show its impact on the die stresses; (2) the package warpage induced die stresses after the molding process when the temperature of the package decreases from the molding temperature (around 170 °C) to the room temperature (25 °C); and (3) dynamic trim process (with element eroding technique) to show the impact of stress waves to the thin die. The above assembly processes are simulated by using the finite element code ANSYS®. Next, this chapter discusses the impact of the nonlinear materials behaviors on the flip-chip packaging assembly reliability with comparison of the measured displacement (warpage); Finally, the SIP for a stacked die flip-chip assembly layout, the material selection, and its effects including substrate on the SIP stress are investigated and discussed.

12.1 Assembly Process of Side by Side Placed SIP

12.1.1 Multiple Die Attach Process

Once the thin die are picked up, they can then be placed on a leadframe based SIP (see Figure 12.1 for a smart power module). Because the SIP package has multiple die inside, the impact of different die attach process to die stress will be not be neglected. The multiple die includes power IGBTs, diodes, and IC controllers. The material properties (Liu *et al.*, 2006) of the package are listed in Tables 12.1 to 12.3. The leadframe is considered to be elastic-plastic, and the solder material is elastic-plastic with creep.

The effect of the die thickness and the die attach thickness is investigated by the FEA code ANSYS®. The SIP after, the die attach process on the leadframe, is shown in Figure 12.2. An element birth and death technique is used during the simulations of the die attaching/bonding process. For the die attach process

Modeling and Simulation for Microelectronic Packaging and Integration: Manufacturing, Reliability and Testing, Second Edition. Sheng Liu and Yong Liu.

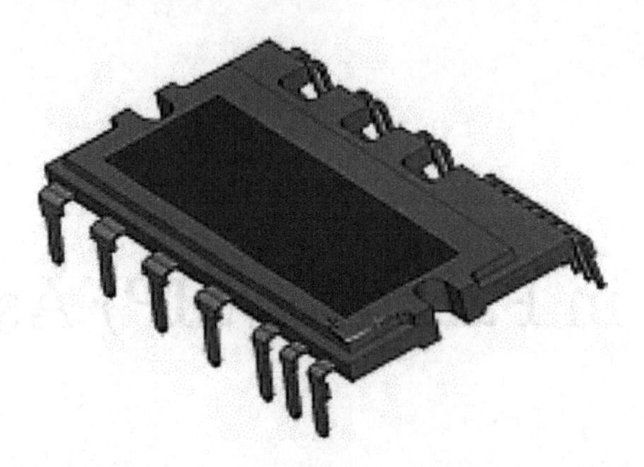

Figure 12.1 A leadframe based system in package

Table 12.1 Material properties of the SIP

	Density (g/cm^3)	Elastic Modulus (GPa)	Poisson's Ratio	Yielding Stress (MPa)	CTE (ppm)
Leadframe	8.96	118	0.3	345	17.7
Ceramic	3.6	340	0.22		6.8
Die	2.3	160	0.23		2.6
EMC	1.9	24 T_g<150 °C 2.4 T_g>150 °C	0.25		10 T_g<150 °C 43 T_g>150 °C
Solder	10.5	Table 12.3	0.4	Table 12.3	30.5
Ag Epoxy	3.5	Table 12.2	0.4		60 T_g<150 °C 13 T_g>150 °C
Silicone Elastomer	2.74	0.085	0.4		130
Punch	14.0	526	0.23		

Table 12.2 Material properties of Ag epoxy

	−65 °C	25 °C	150 °C	250 °C
Elastic Modulus (GPa)	6.33	4.65	1.39	0.517

order simulation for different die types, the procedure of the IGBT die attach first and diode die attach second includes: (1) preheating leadframe; (2) killing both elements for IGBT die and diode die elements, dwelling the temperature in the die attach process; (3) activating IGBT die elements for the die attach process; (4) cooling down to 25 °C; (5) preheating both leadframe and IGBT; (6) activating diode die elements for the die attach process; (7) cooling down to 25 °C. The procedure is similar for the process of diode die attach first and IGBT die attach later.

Table 12.3 Material properties of solder

Elastic Modulus E(GPa)/Temperature(°C)	6.10/25 2.67/150	5.61/50	3.16/100 2.23/200
Plastic Characteristic Stress (MPa)/Plastic Strain	25 °C	14.8/0	22.83/0.552
	50 °C	13.03/0	17.25/0.338
	100 °C	9.21/0	12.15/0.247
	150 °C	7.74/0	8.92/0.31
	200 °C	6.47/0	7.64/0.157

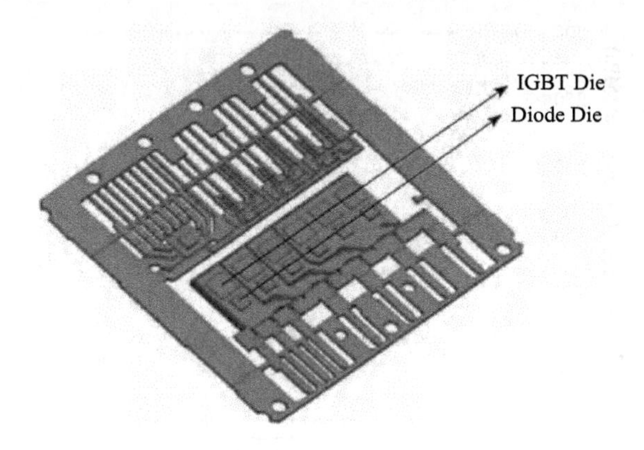

Figure 12.2 SIP after die attach process

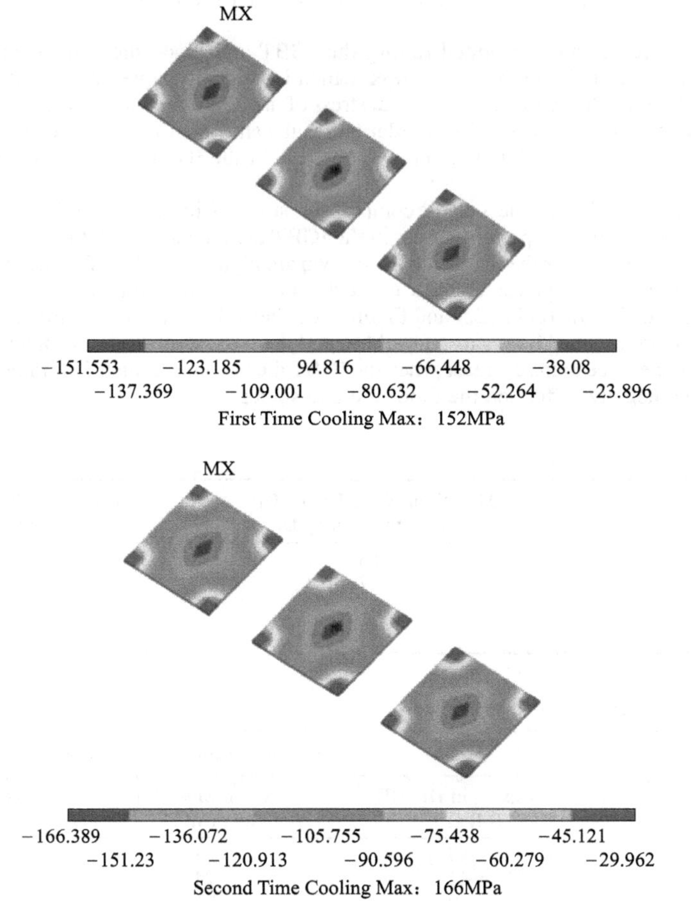

Figure 12.3 Contour of the third principal stress in IGBT die *(Color version of this figure is available online)*

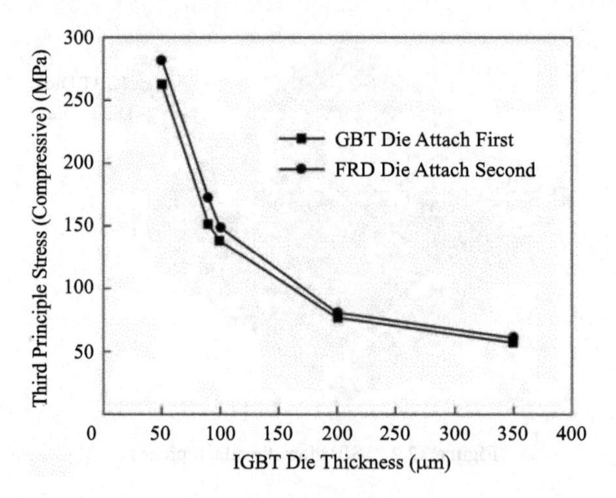

Figure 12.4 Compressive stress of IGBT versus thickness

Two cooling processes are performed during the IGBT and diode die attachment, therefore, the dominated stress in the die is compressive stress, which is generally represented by the third principal stress S_3. Figure 12.3 illustrates the third principal stress of the IGBT die with the thickness of 90 μm after the first and second cooling process. The simulation results show that if the IGBT die are attached first and the diode die later, the maximum compressive stress in the IGBT die increases after the second cooling process.

The effect of the IGBT die thickness on the compressive stress in the IGBT die is listed in Table 12.4 and Figure 12.4. The maximum compressive stress in the IGBT die increases with the decrease of the IGBT die thickness. The stresses after the second cooling down are higher than that after the first cooling down. However, the maximum compressive stress is much lower than the compressive strength of silicon 1100 MPa (Liu *et al.*, 2006). Table 12.5 and Figure 12.5 show the maximum compressive stress in the diode die, the maximum von-Mises stress in solder, and the maximum strain in solder after the second cooling down. All the three items decrease when the IGBT die become thinner. This indicates that the thin IGBT die have no negative effect on the diode die and solder.

Table 12.4 Maximum S_3 in IGBT die

IGBT Die Thickness (μm)	Max S_3 in IGBT Die (MPa) after the First Cooling Down	Max S_3 in IGBT Die (MPa) after the Second Cooling Down
355.6	58	64
203.2	77	84
90	152	166
50	263	282

Table 12.5 Stresses after the second cooling down

IGBT Die Thickness (μm)	After the Second Cooling down		
	Max S_3 in DIODE Die (MPa)	Max von-Mises Stress in Solder (MPa)	Max Nonlinear Strain in Solder (%)
355.6	42	27	45.6
203.2	41	24	33.2
90	40	22	25.8
50	39	21	22.7

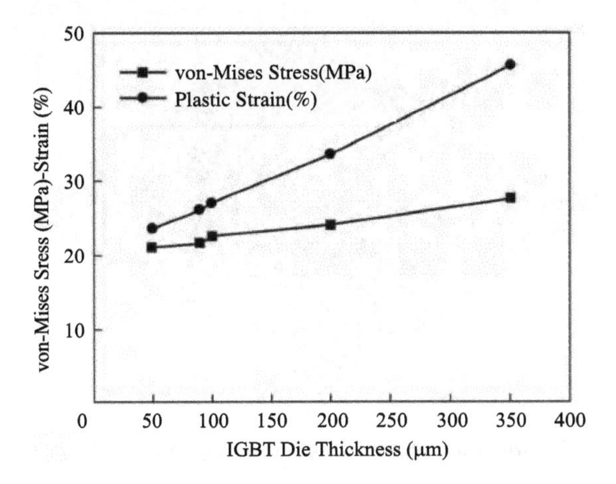

Figure 12.5 Die attach stress-strain versus IGBT thickness

For the IGBT die with a thickness of 90 μm, the effect of the IGBT die attach thickness is examined. The simulation results are listed in Tables 12.6 and 12.7. A thicker IGBT die attach can reduce the stresses in the die and solder. But there should be a trade-off balance during the assembly process.

12.1.2 Cooling Stress and Warpage Simulation after Molding

Molding is the process of encapsulating the device in plastic material, most commonly known as Encapsulate Epoxy Molding Compound (EMC). The molding/curing temperature of the SIP is around 170 °C. After the molding process the package (Figure 12.6) is cooling down to room temperature (25 °C). The material property data in Tables 12.1,12.2 and 12.3 are used for the simulations. The warpage of the package and the thermal stresses in the die are investigated.

For the model with an IGBT die thickness of 90 μm and die attach thickness of 1 mil, the package warpage and stresses in the die are shown in Figures 12.7 to 12.11. The maximum warpage of the package is 0.1 mm. The maximum compressive stress in the die occurs in the IC die with the value of 706 MPa.

Table 12.6 Maximum S_3 in IGBT die

IGBT Die Attach Thickness (mil)	Max S_3 in IGBT Die (MPa) after the First Cooling Down	Max S_3 in IGBT Die (MPa) after the Second Cooling Down
1	152	166
2	140	145
3	132	134

Table 12.7 Stresses after the second cooling down

IGBT Die Attach Thickness (mil)	After the Second Cooling Down		
	Max S_3 in DIODE Die (MPa)	Max von-Mises Stress in Solder (MPa)	Max Nonlinear Strain in Solder (%)
1	40	22	25.8
2	39	18	13
3	39	17	9.4

Figure 12.6 FEA model of the package after meshing

The effect of IGBT die thickness and die attach thickness on the maximum compressive stress in the die is illustrated in Figure 12.11. The maximum compressive stresses in the IGBT and diode die are very low compared with those in the IC die, which are below the compressive strength of silicon (around 1100 MPa). The maximum compressive stress in the IGBT die decreases steadily when the IGBT die become thinner for the same IGBT die attach thickness. The maximum compressive stresses in the IGBT die decrease steadily with the increase of the IGBT die attach thickness for the same IGBT die thickness. The change of the IGBT die thickness and die attach thickness has little impact on the maximum compressive stresses in diode and IC die.

12.1.3 Stress Simulation in Trim Process

Trim is the process of cutting the dam bars that short the leads together. The dam bars to be cut in Figure 12.12 are quite near to the IGBT and diode die, and they are connected to the leadframe substrate which support these die. Therefore, cutting these dam bars is most dangerous during the trim process.

The material property data shown in Tables 12.1 to 12.3 are used for simulations, and the punch is assumed to be rigid. The leadframe is considered to be elastic-plastic. The initial velocity of the punch is 200 mm/s, and is assumed to be constant during the punch process. The nonlinear finite element code

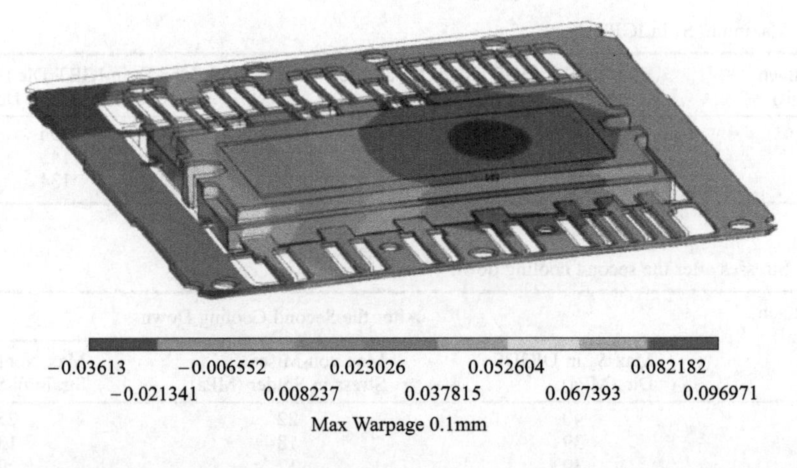

Figure 12.7 Warpage of the SIP *(Color version of this figure is available online)*

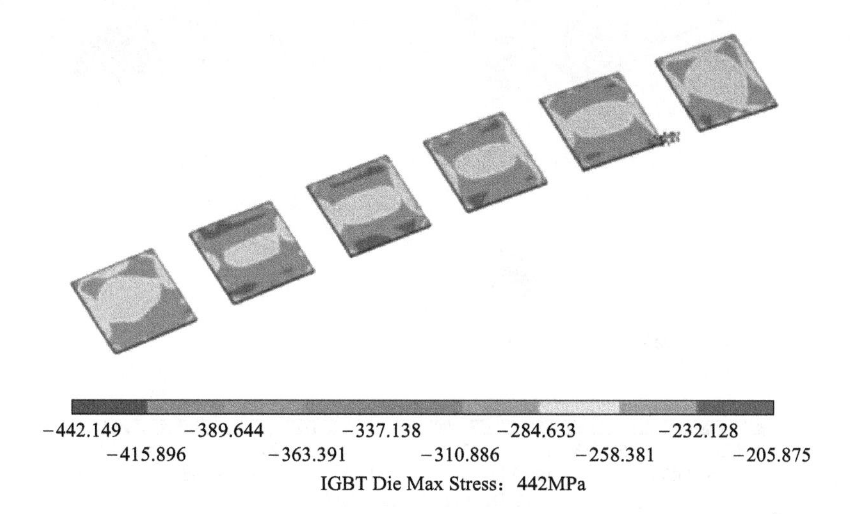

Figure 12.8 S_3 in IGBT die *(Color version of this figure is available online)*

ANSYS®/LS–DYNA is adopted to simulate the dynamic trim process. The contact type between the punch and package is defined as *contact_eroding_single_surface (ANSYS/LS–DYNA manual book). In this contact type a failure strain of the leadframe can be defined. By this technique the element of the leadframe will be eroded if its plastic strain reaches the defined failure strain.

It is shown in Figure 12.13 that the maximum first principal stress in the die with respect to time for the model with 90 μm IGBT die. Table 12.8 illustrates the effect of IGBT die thickness on the maximum first principal stress in die. When the IGBT die thickness decreases from 355.6 μm to 90 μm, the maximum first principal stress in the IGBT die increases from 2.4 MPa to 3.0 MPa, and the maximum first principal

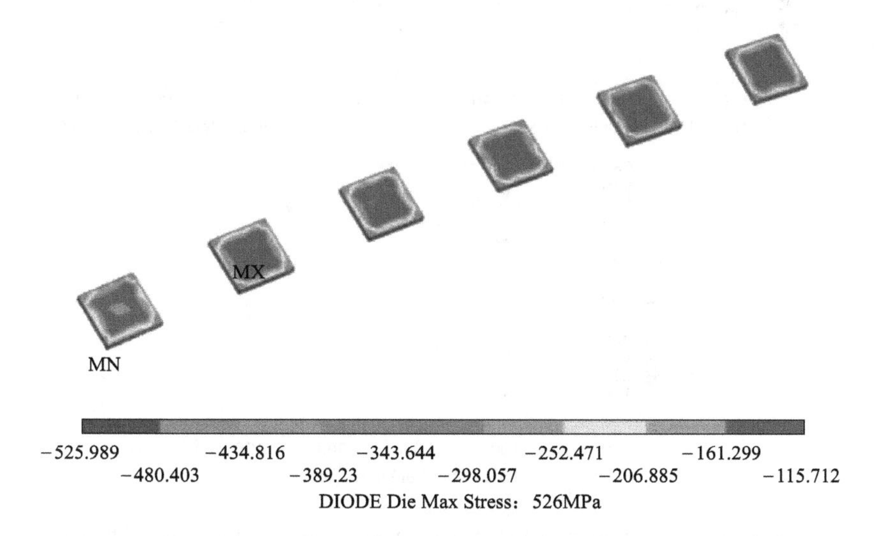

Figure 12.9 S_3 in DIODE die *(Color version of this figure is available online)*

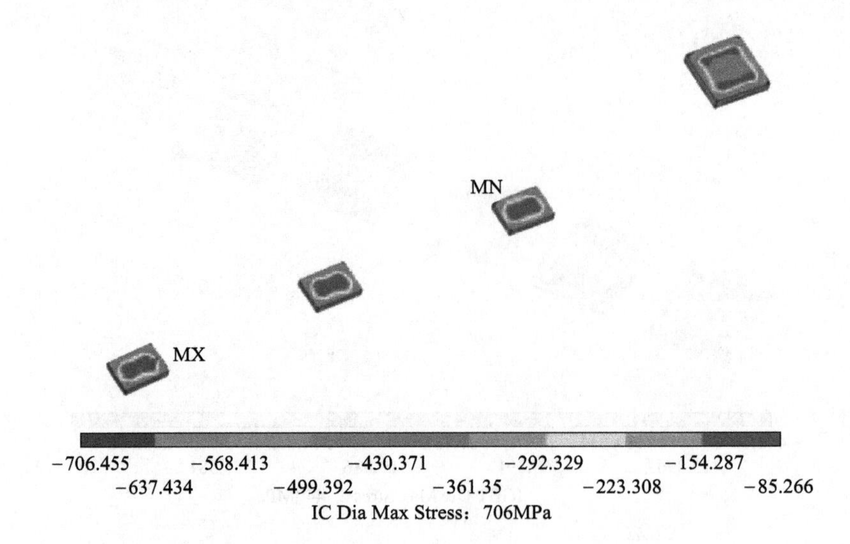

Figure 12.10 S_3 in IC die *(Color version of this figure is available online)*

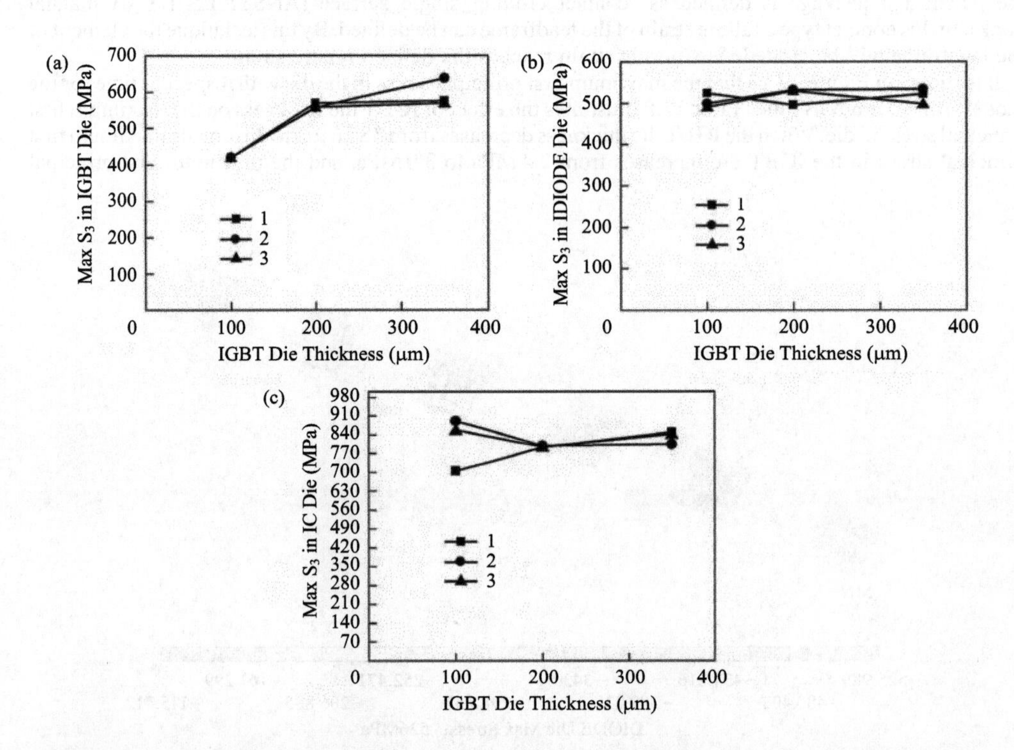

Figure 12.11 The maximum third principal stress in die. 1, 2, and 3 correspond to the IGBT die thickness 1, 2, and 3 mil, respectively

Figure 12.12 FEA model of the package after meshing

stresses in the diode and IC die have almost no change. The maximum first principal stress in the die (3.0 MPa) is far less than the tensile strength of silicon (around 170 MPa). This has shown the robust assembly process that induces very small dynamic stress in trim process.

Table 12.8 Maximum first principal stresses in the die

IGBT Die Thickness (μm)	Maximum First Principal Stress (MPa)		
	IGBT Die	DIODE Die	IC Die
90	3.0	2.3	2.6
203.2	2.4	2.3	2.6
355.6	2.4	2.2	2.6

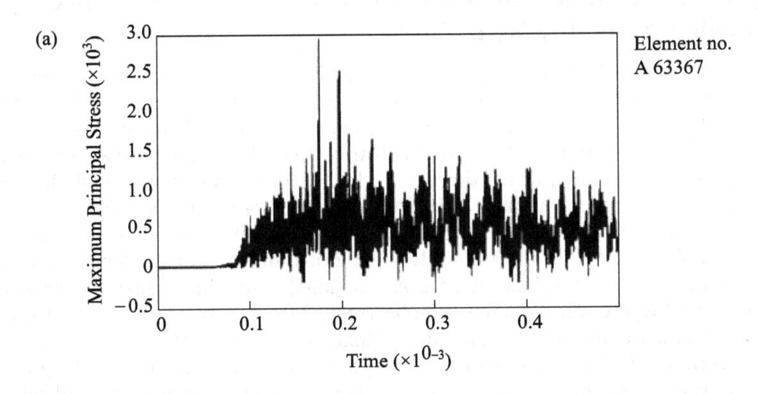

Figure 12.13 The first principal stress in the die with respect to time

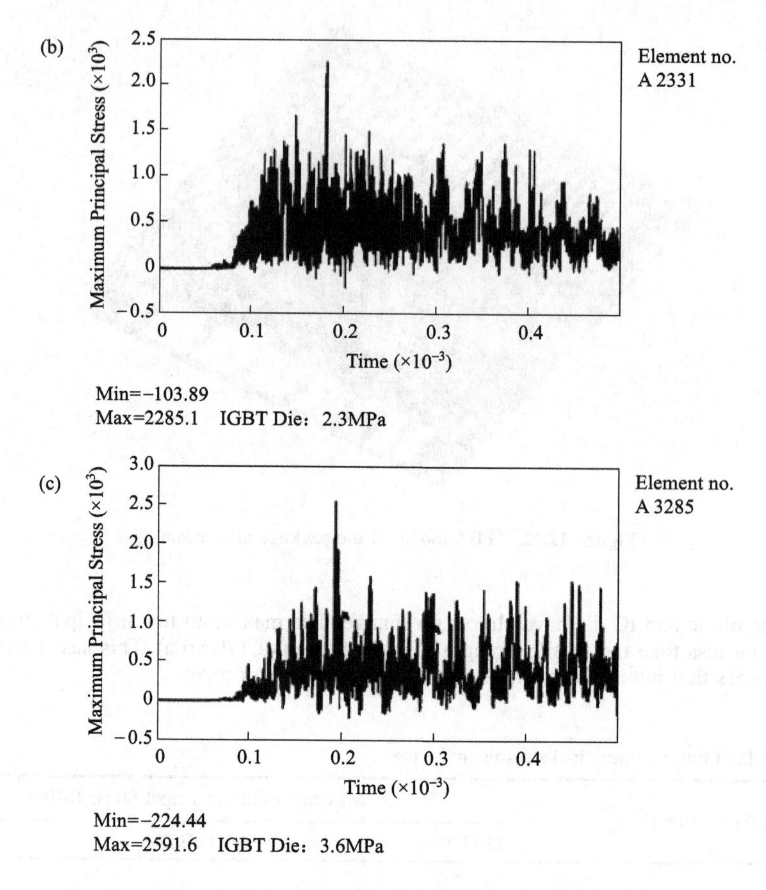

Figure 12.13 (*Continued*)

12.2 Impact of the Nonlinear Materials Behaviors on the Flip-Chip Packaging Assembly Reliability

The reliability evaluation of solder joints is based on the joint deformation and strain. As the joint interconnections become smaller and smaller, direct experimental measurement of the thermal strain and deformation is getting more and more difficult. The experimental tools for deformation measurement can not offer enough resolution for local strain distributions that are essential for the estimation of reliability. On the other hand, measuring time-dependent deformation behaviors of electronic packages is usually expensive and time-consuming. Therefore, recently more and more effects have been made on the reliability study by using numerical method such as finite element method (Wang *et al.*, 1998; Lau and Yi-Hsin, 1997; Bor *et al.*, 1998; to list a few). It is the global developing tendency for future reliability evaluations. However, due to the lack of accurate, detailed viscoplastic material properties for both underfill and solder alloy, the studies mainly emphasize the effect of mismatch of thermal expansion and neglect the viscoplastic nature of the encapsulant (Zhao *et al.*, 1998; Pang *et al.*, 1998) and sometime solder joints (Gektin *et al.*, 1997; Darbha *et al.*, 1997). However, as described in the previous sections, the solder alloy shows a strong viscoplastic nature even at room temperature. And the underfill also exhibits a strong viscolpastic behavior at the elevated temperature environment in which the package is usually subjected in operation. Therefore, it may cause some error in the finite element analysis if the visco nature for both solder and underfill or only for the underfill is neglected. The knowledge of the visco behavior in the advanced packages, such as flip-chip package, is an essential step to approach the precise reliability prediction.

Advanced microelectronic packaging requires higher I/O density and reliable interconnection technology to achieve higher levels of system performance at a competitive cost. To satisfy this demand, the flip-chip has now been widely used in the automotive, avionics, computers, and mobile electronics. However, the difference in the coefficients of thermal expansion (CTE) between the chip and the substrate makes flip-chip configurations vulnerable to thermally-induced strains and can potentially lead to the temperature dependent viscoplastic deformation in and around flip-chip solder joints. It often results in solder joint creep and fatigue (Lau, 1997; Lau and Pao, 1997). Therefore, the ability to further understand creep behavior and to precisely predicting the creep and potentially cyclic behaviors of a flip-chip package are critically important for the reliability of electronic devices.

In this section a flip-chip package which is composed of a silicon chip, underfill, solder balls, and FR-4 substrate is selected. Based on our experimental test, different constitutive laws, including elastic, elastic-plastic, viscoelastic or viscoplastic for both solder balls, and underfill are used to investigate creep behaviors and potential cyclic behaviors of the flip-chip package (Ren *et al.*, 1998b; Ren *et al.*, 1999c).

12.2.1 Finite Element Modeling and Effect of Material Models

The subject under investigation is a flip-chip package specimen. The dimensions and boundary conditions are shown in Figure 12.14. Assume the flip-chip package specimen satisfies the axisymmetric condition. Therefore, it can be simplified to be a 2D axisymmetric FEA model. A finite element mesh is shown in Figure 12.15. It is noted that a fine local mesh is arranged along all interfaces. In the analysis, the commercial code ABAQUS is applied and the eight-node element, CAX8 in ABAQUS, is selected. Two stages of analyses are performed in the current study. First the flip-chip specimen is heated up from room temperature to 115 °C. Subsequently, the specimen undergoes creep at 115 °C for about 10 hours and 40 minutes. The temperature profile is illustrated in Figure 12.16.

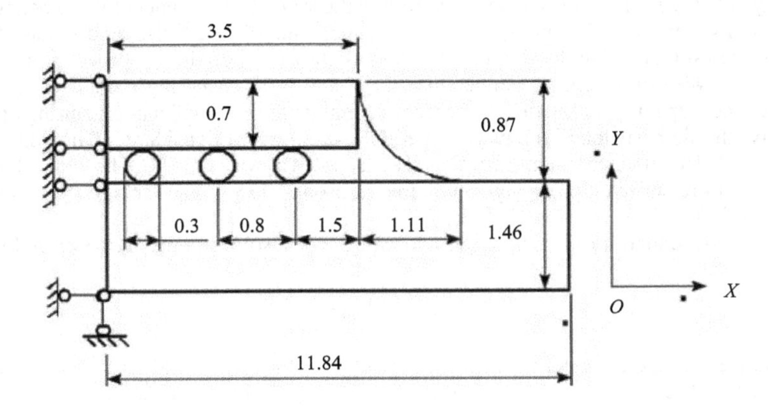

Figure 12.14 Dimensions and boundary conditions of flip-chip package

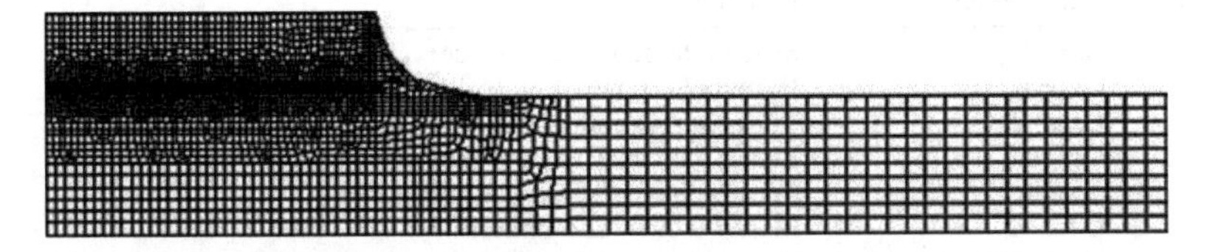

Figure 12.15 Finite element mesh for flip-chip package

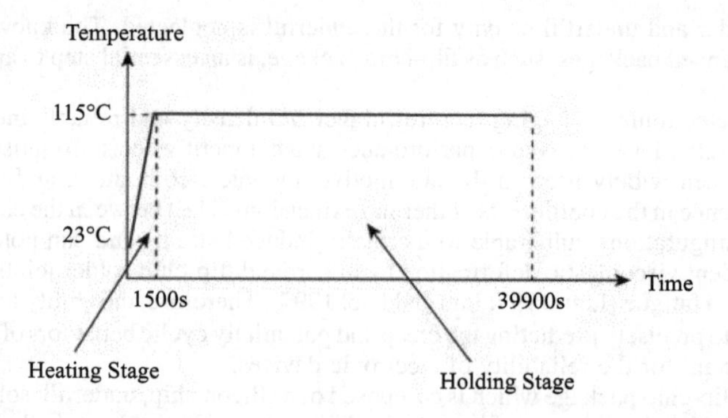

Figure 12.16 Temperature profile for creep load

The properties of the silicon chip and the substrate are assumed to be linear elastic. The Young's modulus and Poisson ratio of the silicon chip are 169.5 GPa and 0.278 respectively. The coefficient of thermal expansion of the silicon chip is 2.8 ppm/°C. The FR-4 substrate is assumed to have an orthotropic property. The Poisson ratios v_{xz}, v_{xy}, and v_{yz} are 0.02, 0.143, and 0.143 respectively. The Young's moduli E_x, E_z and E_y are 22.4 GPa, 22.4 GPa, and 1.6 GPa respectively. The CTEs of FR-4, α_x, α_y and α_z are 16 ppm/°C, 65 ppm/°C, and 16 ppm/°C respectively (Auersperg, 1997). The *x* and *y* directions are defined in Figure 12.14. The z direction follows the right-hand rule.

In order to investigate the creep behaviors of solder joints and underfill for the flip-chip specimen by different constitutives laws and processes, six FEA models are considered as listed in Table 12.9. These include (1) both solder and underfill are considered as strain rate dependent plasticity (viscoplasticity); (2) the creep behaviors of both solder joint and underfill are described by hyperbolic power law; (3) the creep behavior of the solder joint is still described by hyperbolic power law, but the underfill's by viscoelasticity; (4) the solder join is assumed to follow the strain rate-dependent visco behavior, but the underfill is described by elastic behavior, (5) both solder and underfill are considered as elastic and plastic materials; (6) both solder and underfill are considered as elastic materials; (7) processing model, which underfilled model with two steps of temperature drop, and non-processing model, which underfilled model with second step temperature drop. In order to simulate the practical manufacturing process in the chip assembly, the thermal loads for two temperature drops were considered. The silicon chip and substrate are bonded together at around 180 °C, and then cooled down to 135 °C. The underfill is dispensed under and around the entire chip package and cured at around 135 °C, and then cooled down to room temperature (20 °C).

The strain rate dependent properties of the underfill and the solder joints are described by:

$$\bar{\sigma} = \sigma^0\left(\varepsilon^{pl}, T\right)R\left(\dot{\bar{\varepsilon}}^{pl}, T\right) \tag{12.1}$$

where $\bar{\sigma}$ is the yield stress at nonzero plastic strain rate; σ^0 is the static yield stress; $12\bar{\varepsilon}^{pl}$ is the uniaxial equivalent plastic strain; R is the ratio of the yield stress; $\dot{\bar{\varepsilon}}^{pl}$ is the inelastic strain rate; and T is

Table 12.9 FEA model description

FEA Model	Description
Case I	Strain rate dependent (solder and underfill)
Case II	Hyperbolic power law (solder and underfill)
Case III	Viscoelastic (underfill) Hyperbolic power law (solder)
Case IV	Elastic (underfill) Strain rate dependent (solder)
Case V	Elastic-plastic (solder and underfill)
Case VI	Elastic (solder and underfill)
Case VII	Processing model and non-processing model

the temperature. It is a kind of unified constitutive modeling with isotropic hardening and can cover both power-law and power-law breakdown regimes.

Hyperbolic power law, which is commonly used to describe the viscoplastic behavior of material, is described as follows:

$$\dot{\varepsilon}^{\mathrm{cr}} = A(\sinh B\sigma)^{n}\exp(-\Delta H/RT) \tag{12.2}$$

where $\dot{\varepsilon}^{\mathrm{cr}}$ is the uniaxial equivalent creep strain rate; σ the equivalent stress; ΔH the activation energy; R the universal gas constant; T the temperature; A, B and n some constants which are related to temperature and loading and so on.

The viscoelastic model is based on the Maxwell model using a series combination of dashpot and spring element (Ferry, 1980). The model was defined by the Prony series expansion:

$$G(t) = G_{\mathrm{e}} + \sum_{i=1}^{N} G_i e^{-\frac{t}{\tau_i}} \tag{12.3}$$

where G_{e} represents the equilibrium modulus, G_i and τ_i are the modulus and relaxation time for each element respectively. A total of 13 Maxwell elements are used in the study. For thermorheologically simple materials, the temperature effect was introduced through a reduced time concept. The reduced time, $\xi(t)$, was defined by:

$$\xi(t) = \int_0^t \frac{\mathrm{d}\tau}{a_{\mathrm{T}}(T(\tau))} \tag{12.4}$$

where a_{T}, the shift factor at time t, can be approximated by the Williams –Landel–Ferry (WLF) equation:

$$\log a_{\mathrm{T}} = \frac{-C_1(T - T_{\mathrm{r}})}{C_2 + (T - T_{\mathrm{r}})} \tag{11.5}$$

where T_{r} is the reference temperature. C_1 and C_2 are calibration constants obtained at this temperature.

The elastic-plastic model uses the "true stress-strain" (actually true stress-logarithmic strain) relationships at different temperatures. Here the true stress and logarithmic plastic strain follow the definition given in ABAQUS User's Manual. If the properties are isotropic, the nominal (or engineering) stress-strain data obtained in a uniaxial test may be converted to true stress and logarithmic plastic strain by using the relationships:

$$\sigma_{\mathrm{true}} = \sigma_{\mathrm{nom}}(1 + \varepsilon_{\mathrm{nom}}) \tag{11.6}$$

And:

$$\varepsilon_{\mathrm{pl}} = \ln(1 + \varepsilon_{\mathrm{nom}}) - \sigma_{\mathrm{true}} \tag{11.7}$$

where σ_{true} is true stress, σ_{nom} and $\varepsilon_{\mathrm{nom}}$ are engineering stress and strain respectively, and $\varepsilon_{\mathrm{pl}}$ is logarithmic plastic strain.

All constants used in different FEA models can be determined from our test data described in the previous sections. The typical true stress-strain relationships and the creep test results of solder alloy 63 Sn37Pb are illustrated in Figures 2.8 and 2.11. The typical true stress-strain relationships and the creep test results of underfill FP4526 are illustrated in Figures 12.17 and 12.18.

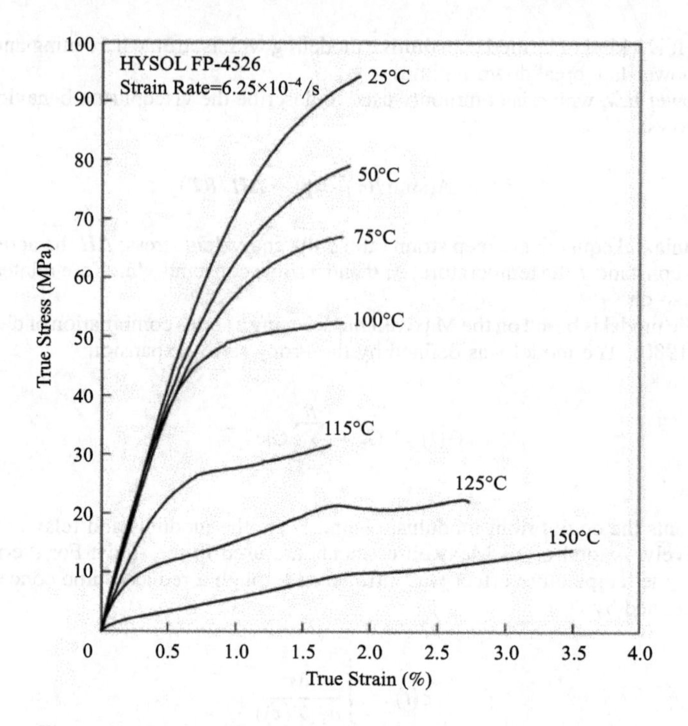

Figure 12.17 Typical stress-strain relationship of underfill FP-4526

12.2.2 Experiment

The high density laser moiré interferometry is a whole-field in-plane displacement measurement technique with both high displacement sensitivity and high spatial resolution. It is especially effective for the measurement of non-uniform in-plane deformation measurements (Post *et al.*, 1993; Zou *et al.*, 1997). In this section, real time moiré interferometry is applied to study the creep behavior of the flip-chip package. Figure 6.8 schematically shows the experiment setup in Chapter 6.

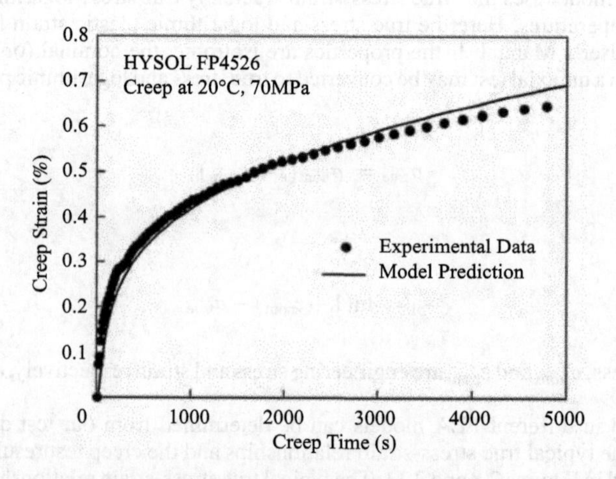

Figure 12.18 Typical creep test result of underfill

The grating was replicated onto the surface of the same flip-chip package sample considered in the FEM analysis at room temperature. Then the flip-chip specimen was subjected to thermal load from room temperature (23 °C) to 115 °C. Isothermal condition was maintained and was controlled at 115 °C for about 10 hours and 40 minutes. The thermal deformations of the flip-chip package specimen in relation to the temperature were measured. The fringe patterns were recorded by a CCD camera and stored into a PC computer in a digital format. In addition, the computer image processing technique was used to analyze the fringe patterns.

12.2.3 Results and Discussions

(i) Comparison of Results between FEA and Test

To compare the results between the finite element analysis and the test conducted by the real-time moiré interferometry, the steady state deformed configuration of the flip-chip package is selected, that is, the deformation state after 10 hours and 40 minutes creep at 115 °C. The displacements at the prescribed location A on the specimen obtained from the test and the finite element analysis are shown in Table 12.10. It is found that the differences are less than 22%. It is shown that the displacements obtained from the finite element analysis for all FEA models are in agreement with those obtained from the test.

The typical displacement contours simulated by the finite element method and the fringe patterns captured by the moiré interferometry technique both in the X and Y directions are shown in Figures 12.19 and 12.20. It is noticed that the configurations of the flip-chip package specimen captured by the moiré interferometry technique show similar patterns compared with those modeled by the finite element method. In particular, the variation of the deformation predicted by the finite element method has the same trend as that obtained by the moiré test. Therefore, the FEA results are comparable to the test results. In another word, the FEA model is verified by the test results.

(ii) Comparison of the Displacements Obtained by Different FEA Models

Figures 12.21 and 12.22 show the variations of X and Y direction displacements at the location A with respect to the location O versus time for six different FEA models. From the results, it can be

Table 12.10 Comparison of steady state displacements at A between FEA and moiré test (location O selected as reference point)

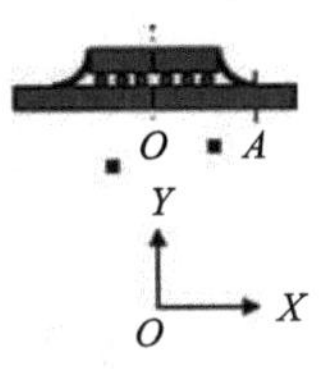

FEA model	Direction	Displacement (μm)	Difference (%)
Case *I*—Strain rate-dependent	X	7.04	20.55
(solder and underfill)	Y	4.28	2.64
Case II—Hyperbolic power law	X	7.05	20.72
(solder and underfill)	Y	4.54	8.87
Case III—Hyperbolic power law	X	7.0	19.86
(solder) Viscoelastic (underfill)	Y	3.25	−22.00
Case IV—Elastic (underfill)	X	6.98	19.52
Strain rate dependent (solder)	Y	4.29	2.88
Case V—Elastic-plastic	X	7.05	20.72
(solder and underfill)	Y	4.6	10.31
Case VI—Elastic	X	7.06	20.89
(solder and underfill)	Y	4.98	19.42
Test Results (μm)	X	5.84	
	Y	4.17	

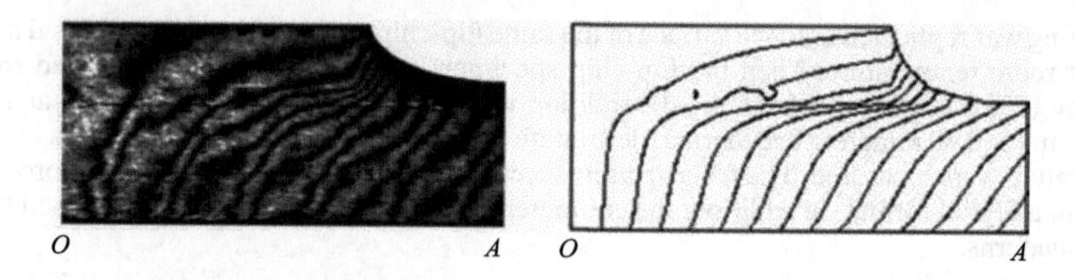

Figure 12.19 Steady state U-field (*X* direction) fringe patterns and FEM *X* direction displacement contours

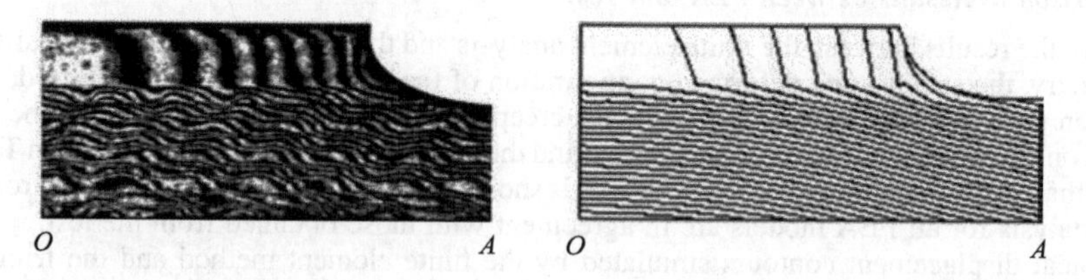

Figure 12.20 Steady state V-field (*Y* direction) fringe patterns and FEM *Y* direction displacement contours

found that there is no obvious difference of displacements obtained from different FEA models in which the material behaviors for solder and underfill are described by different constitutive models due to the constrained small volumes of the solder balls and underfill comparing with the whole flip-chip package specimen.

Furthermore, the displacements both in the *X* and *Y* direction do not change significantly during the temperature holding time. For example, the variation range of the *X* direction displacements is within the range of several manometers. For Case I—strain rate-dependent model (solder and underfill), the *X* direction displacement decreases from 7.0486 μm to 7.0404 μm during the temperature holding time.

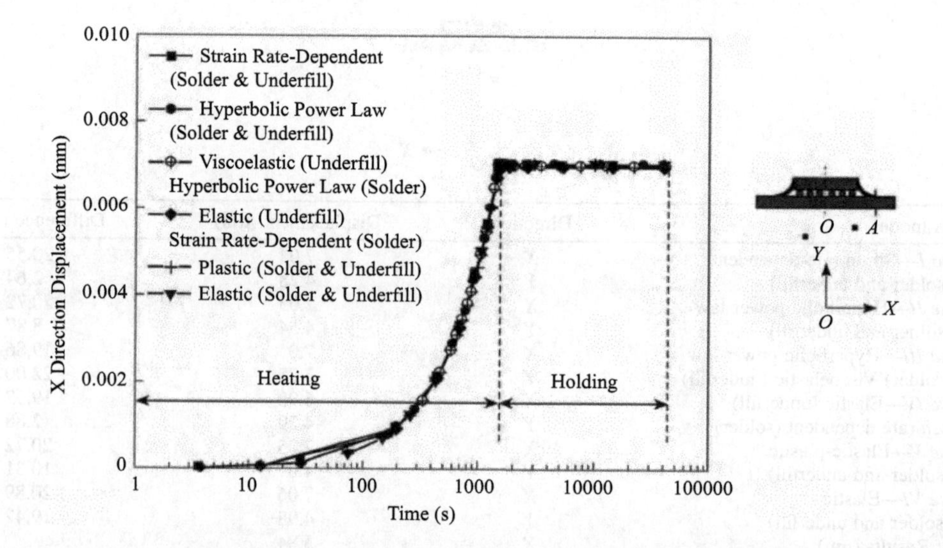

Figure 12.21 Variation of X-displacement at A with respect to O

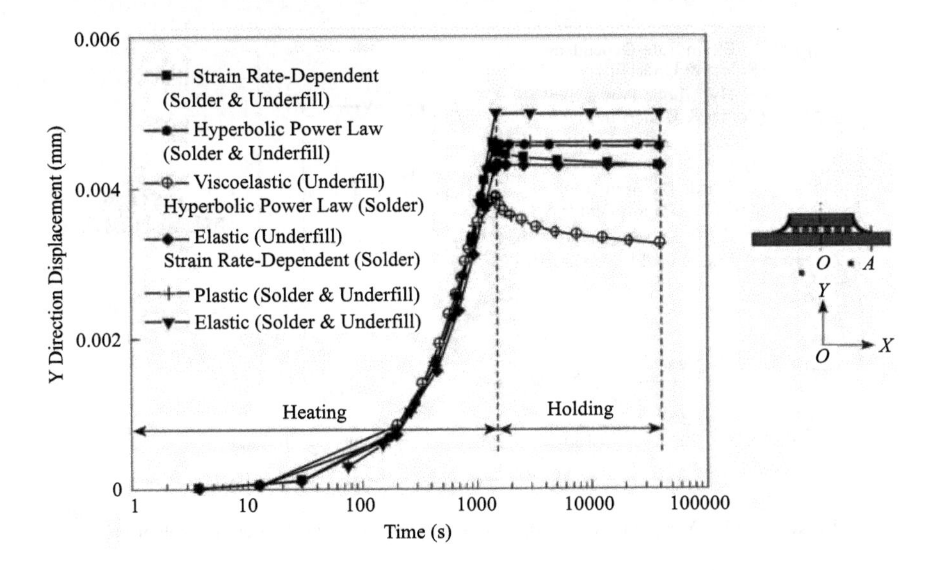

Figure 12.22 Variation of Y-displacement at A with respect to O

This means that the creep behavior of the underfill and the solder balls does not have significant effect on the warpage of the flip-chip under the considered thermal load.

(iii) Comparison of von-Mises Stress in Solder Balls

In order to demonstrate the effect of the creep behavior of the underfill and the solder balls on the stresses in the solder balls, the von-Mises stress in the solder balls is analyzed. The results show that the highest von-Mises stress occurs at the corner of the outmost solder ball. Figure 12.23 shows the variations of von-Mises stress at the corner B of the outmost solder ball versus time predicted by different FEA models.

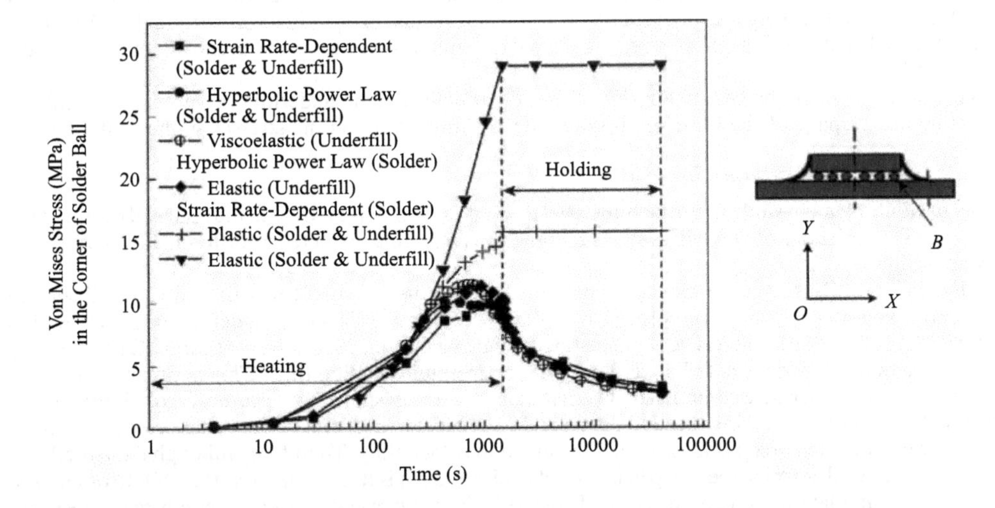

Figure 12.23 Variation of von-Mises stress at the corner B of the outmost solder ball

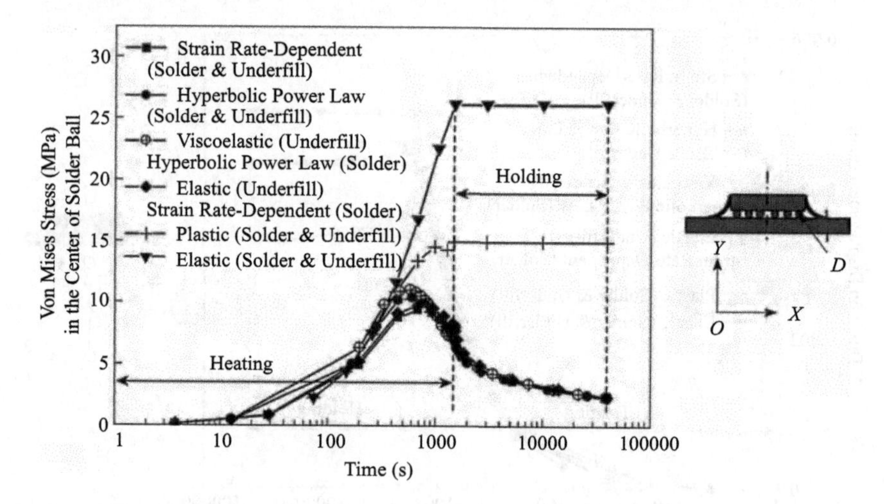

Figure 12.24 Variation of von-Mises stress at the center of the outmost solder ball

The following results can be obtained.

1. As expected the von-Mises stresses predicted by elastic-plastic model-Case V and elastic model-Case VI are much higher than those of predicted by the models in which the visco behavior of the material is considered. At a steady state, as described in the previous text, the values of von-Mises stress in the corner of the outmost solder ball are one order of magnitude higher.
2. The stresses predicted by all visco models (Case I to Case IV as described in Table 12.9) are sharply reduced due to the creep behavior of solder balls. The stress relaxation occurs even in the heating stage.
3. For the prescribed location the stress values predicted by all visco models have no big difference. Furthermore, for all visco models the variations of the stresses versus time have the same trend.

In order to see if the above results are still available for all the locations in the solder ball, the von-Mises stress at the center D of the outmost solder ball is also analyzed. Figure 12.24 shows the variations of von-Mises stress at the center D of the outmost solder ball versus time predicted by different FEA models. It is observed that the von-Mises stress has the same trend as those at the corner B of the outmost solder ball.

Therefore, although the creep behavior of the solder balls and the underfill does not have significant effect on the warpage of the flip-chip, it does have a strong effect on the stresses in the solder balls.

(iv) Comparison of von-Mises Stress in Underfill

From the finite element study, it can be found that the highest von-Mises stress in the underfill occurs at the corner *C*. The variations of von-Mises stress at corner *C* versus time predicted by different FEA models are given in Figure 12.25.

Like the von-Mises stress in the solder ball, the creep behavior of solder balls and underfill sharply reduces the von-Mises stress at the corner C of underfill predicted by the visco models (Case I to Case III as described in Table 12.9). But it is very interesting to find that the von-Mises stress at corner C predicted by hyperbolic power law does not change during the temperature holding time. This is because the creep behavior of the underfill is dominated by first stage creep as shown in Figure 12.18. However, the hyperbolic power law is mainly used to describe the secondary stage and tertiary stage creep, which are usually the dominated creep stages such as the creep in solder alloy. Therefore, although the hyperbolic power law can well describe the creep behavior of solder alloy, it is not the case for underfill. In order to be able to well simulate the creep behavior of underfill, it is suggested to use the strain rate-dependent visco model.

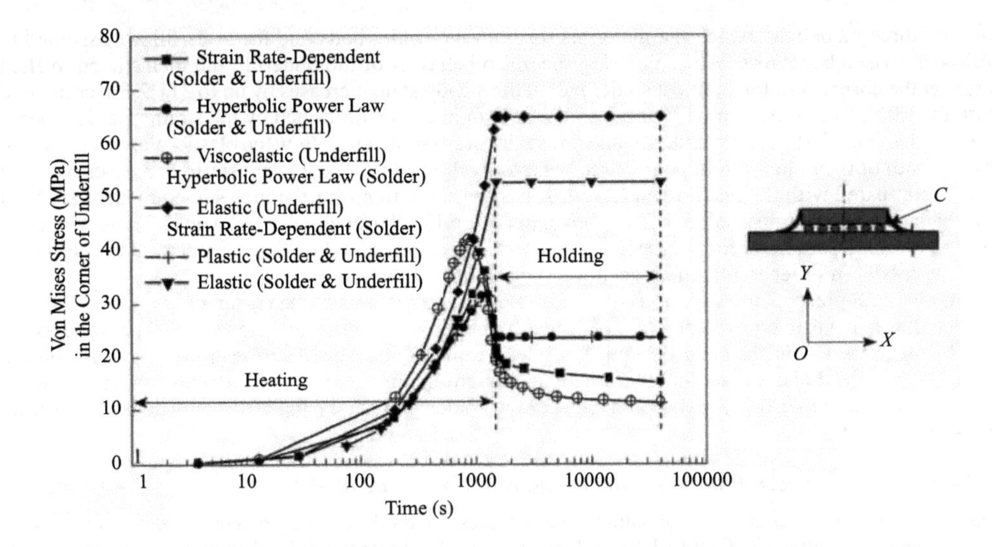

Figure 12.25 Variation of von-Mises stress at the corner C of underfill

(v) Comparison of Inelastic Equivalent Strain in Solder Ball and Underfill

The inelastic equivalent strain $\bar{\varepsilon}$ can be obtained by:

$$\bar{\varepsilon} = \sqrt{\frac{2}{3} e^i_{ij} e^i_{ij}} \tag{12.8}$$

where $e^i_{ij} = \varepsilon^i_{ij} - \varepsilon^i \delta_{ij}$ and $\varepsilon^i = \frac{1}{3}\left(\varepsilon^i_{11} + \varepsilon^i_{22} + \varepsilon^i_{33}\right)$. ε^i_{ij} are the total inelastic components which can be obtained from ABAQUS output files.

Figure 12.26 shows the variations of inelastic equivalent strains at the corner B of the outmost solder ball. It seems that there is no significant difference among the inelastic equivalent strains predicted by

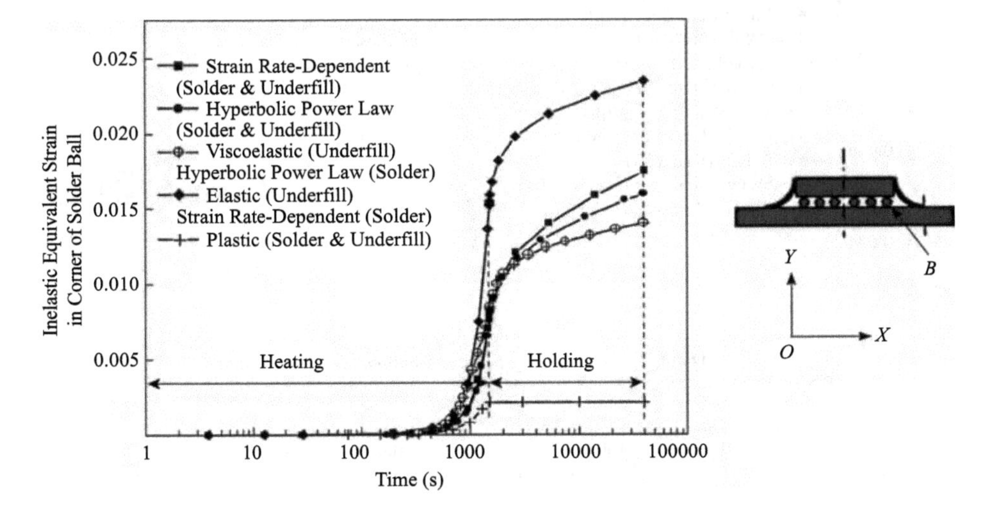

Figure 12.26 Variation of inelastic equivalent strain at the corner B of the outmost solder ball

the first three models, in which the materials for both the solder balls and the underfill are assumed to follow the visco behaviors. While ignoring the visco behavior of the underfill, the inelastic equivalent strain at the corner B of the outmost solder ball at the steady state increases by up to 311.2.6% compared with the result from Case I in which both solder and underfill are described by the strain rate-dependent constitutive model. If both the underfill and the solder are assumed to follow the elastic-plastic behaviors, that is, both of them do not follow the visco behaviors, the inelastic equivalent strain decreases by up to 87.6% compared with the result form Case I. It has the same trend for the inelastic equivalent strain at the center D of the outmost solder ball as shown in Figure 12.27. Usually the inelastic equivalent strain is used as the main parameter for fatigue life prediction. Therefore, one should be very careful to choose the suitable model in order to obtain an accurate fatigue life prediction.

Figure 12.28 presents the variations of inelastic equivalent strains at the corner C of the underfill. As expected, the inelastic equivalent strain obtained by hyperbolic power law shows the same trend as the von-Mises stress, that is, the inelastic equivalent strain remains unchanged during the temperature holding time. On the other hand, for another two visco models, strain rate-dependent (solder and underfill) model and viscoelastic (underfill) + hyperbolic power law (solder) model, the inelastic equivalent strain shows the similar results.

(vi) Comparison between Processing Model and Non-Processing Model

The variations of stresses at the given point between processing model and non-processing model versus time are presented in Figures 12.29 and 12.30(as shown in page 382).From the finite element study, it can be found that theresidual stresses obtained from this model are generally smaller than those obtained from the processing model due to the negligence of the bonding process induced residual stresses in the non-processing model. The values of the stresses at the given point obtained from the processing model are about 20% higher than that obtained from the non-processing model. The variation of deflection at the given points between processing model and non-processing model versus time are drawn in Figures 12.31 and 12.32(as shown in page 382).The FE results also show that there are larger differences between two models.The deflection values at the given points obtained from the processing model are usually 25% higher than those obtained from the non-processing model. It is, therefore, shown that the processing model which is based on the FEM framework set up in this research can more realistically simulate a series of practical manufacturing process in the flip-chip assemblies, where a larger error could be caused by using the non-processing

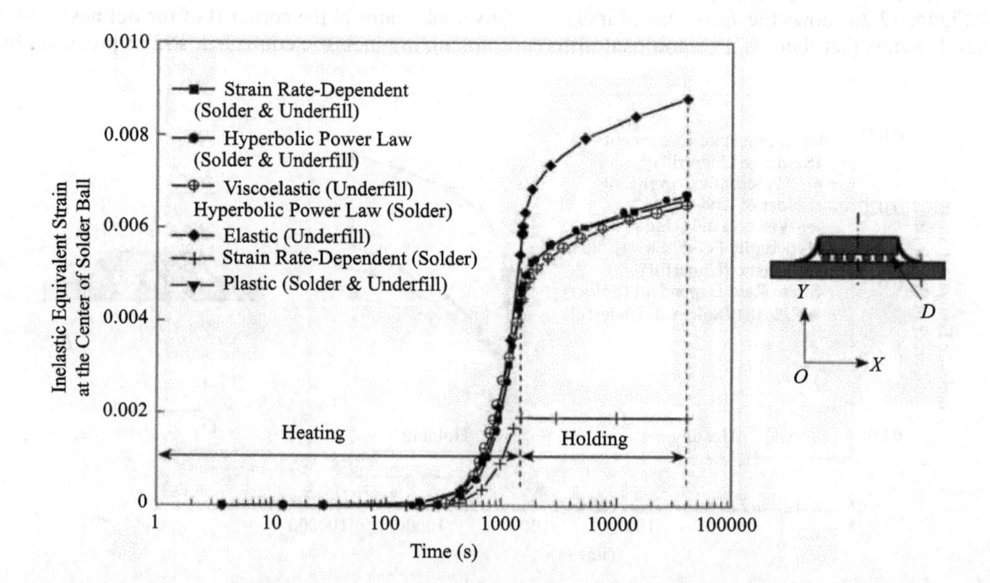

Figure 12.27 Variation of inelastic equivalent strain at the center *D* of the outmost solder ball

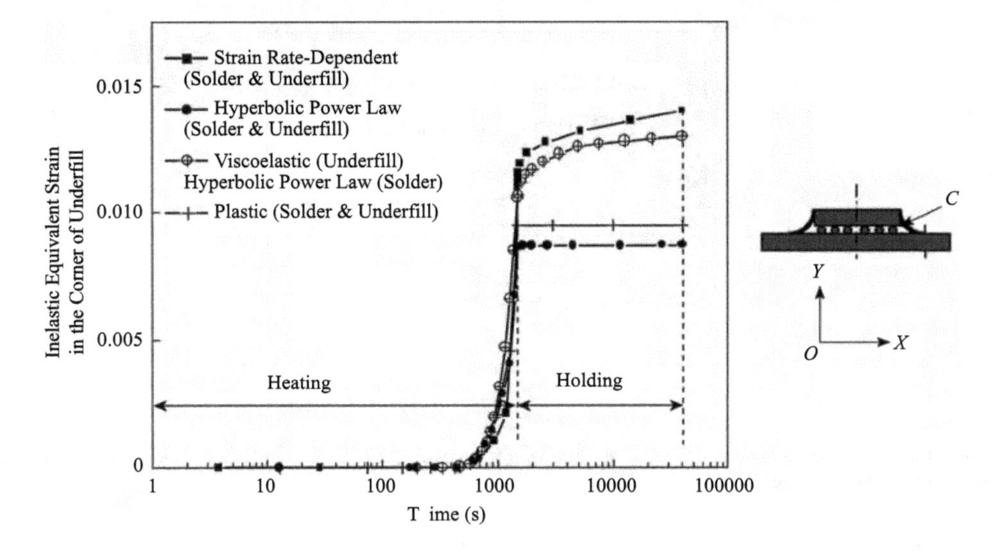

Figure 12.28 Variation of inelastic equivalent strain at the corner C of underfill

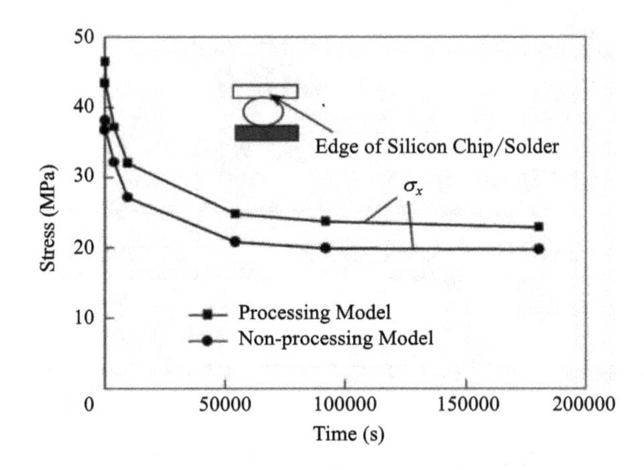

Figure 12.29 Variation of normal stresses in x direction at edge of interface between outmost solder ball and silicon chip against time

model in the analysis of process-induced residual stress field and warpage in the packaging assemblies due to the negligence of the bonding process during cooling from temperature T1 to temperature T2. Therefore, the results obtained from the non-processing model cannot well reflect the manufacturing process-induced thermal residual stresses and warpage in the flip-chip assemblies.

12.3 Stacked Die Flip-Chip Assembly Layout and the Material Selection

Stacked die SIP is a 3D package technology which is different from the die side by side placement. The stacking function can allow a smaller package size with multifunction, which is the trend of product development today. A more important point is its excellent electrical performance (Liu *et al.*, 2008). In this section, a "Flip-chip-on-Silicon BGA" (FSBGA) package (Ren, 2000) is considered, as shown

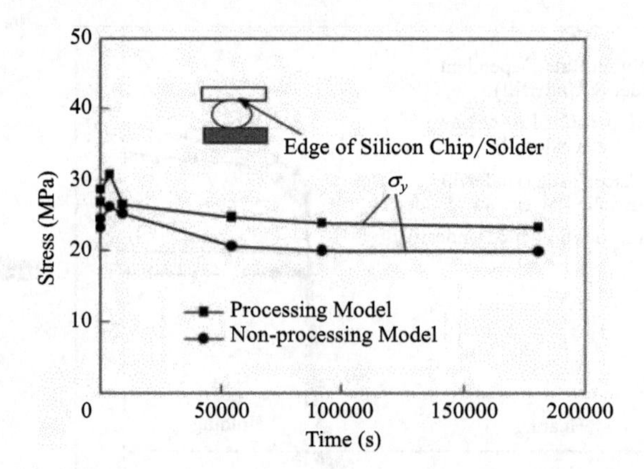

Figure 12.30 Variation of normal stresses in *y* direction at edge of interface between outmost solder ball and silicon chip against time

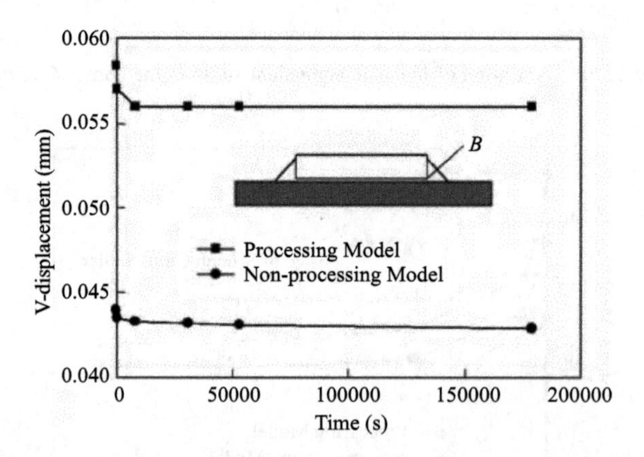

Figure 12.31 Variation of v-displacement at given point of flip-chip versus time

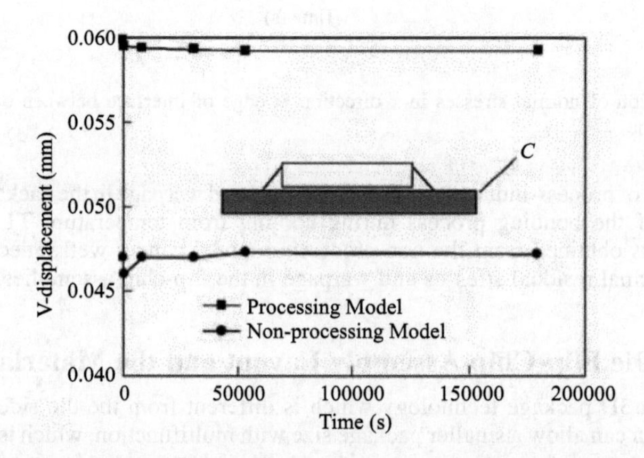

Figure 12.32 Variation of v-displacement at given point of flip-chip versus time

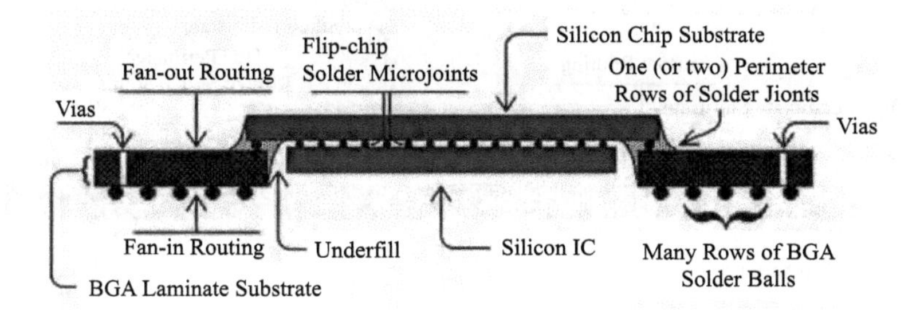

Figure 12.33 Cross sectional diagram of one of the applications of the flip-chip-on-silicon BGA package

in Figure 12.33. This type of package has obtained wide applications. For example, because of the low level of integration in ICs developed for RF applications, most RF products are designed and built using many discrete components, usually inductors and capacitors. These components can be incorporated into the silicon substrate of an FSBGA package.

Nevertheless, FSBGA packages will be used only if they can be fabricated and assembled cost effectively, and can be shown to be reliable; its assembly layout and material selection would be a key in the development.

Consider a conventional flip-chip BGA (FBGA) package design where the use of underfill between the silicon and the BGA laminate substrate is critical to assuring long-term reliability. However, in all FSBGA designs the presence of a flip-soldered chip on the silicon substrate necessitates incorporating a significant assembly opening or a manufactured layout hole in the BGA laminate substrate to accommodate the chip as shown in Figures 12.33. This introduces two important changes in the physical system as compared to the more conventional FBGA architecture. First, depending on its relative size, the hole will increase the compliance of the BGA laminate substrate. In principle, this should be beneficial since a more compliant laminate substrate should be better able to accommodate the various thermal deformation mismatches and, like the perimeter array of compliant leads on a leaded package, thereby reduce the loads developed on the solder joints. On the other hand, by limiting the area between the silicon substrate and the BGA laminate substrate that can be underfilled, the hole may reduce the role of the underfill and thereby increase the thermal mismatch loading on the solder joints.

Obviously, every one of the FSBGA designs is subject to potential failure as a result of thermally-induced mismatch strains, especially if the design fails to take proper account of them. Unfortunately, there are many variables to be considered: material properties, physical dimensions, assembly process, temperature range/history, as well as structural features such as plated vias, solder mask, and multiple layers of PWB circuitry, and so on. At the same time, the available analytical tools, material property data, and time are limited. Consequently, no practical analysis can be expected to consider more than a few, even in simplified terms. Therefore, in this section, first the FSBGA assembly layouts are carefully investigated to insure this unique packaging architecture is reliable. Then material selection is conducted. Since the results raise interesting questions, a study is performed in order to gain a better understanding of its implications.

12.3.1 Finite Element Model for the Stack Die FSBGA

An FSBGA chip-on-chip design, as shown in Figure 12.34, is selected for finite element study, although it does not show the use of underfill between the silicon and the laminate, is analyzed both with and without it.

Because the chip and the silicon substrate have the same thermal properties and are both good conductors (especially when compared to the laminates), the chip and its connections are assumed to be relatively free of thermal stress/strain loading and can be omitted from the finite element model. At the same time, in order to evaluate the behavior of the BGA-to-motherboard solder joints, the finite element model which included the BGA solder balls as joints to the motherboard is developed.

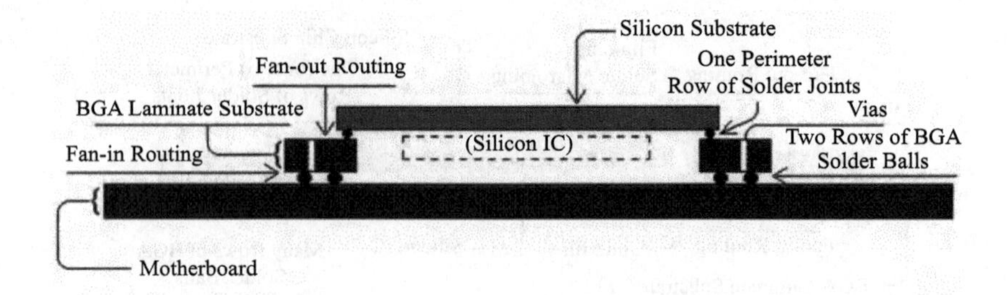

Figure 12.34 Cross sectional diagram of the basic FSBGA FEM model package (not to scale)

Thanks to the four-fold symmetry of the package, it is necessary to analyze only a quarter of the package using a finite element model. While this choice somewhat limits the results by not allowing for anti-symmetric deformations, and so on, it also (1) facilitates the use of a 3D (rather than 2D) analysis and (2) reduces the required programming and computer processing times, thereby permitting more runs on more versions.

Figure 12.35 shows a 3D finite element mesh of the Flip-chip-on-Silicon BGA (FSBGA) package which has both the assembly layout opening hole and an adhesive underfill between the silicon substrate and the BGA substrate (used in subsequent sections as assembly parameter Variation 3). The boundary conditions are: (1) symmetry about $x = 0$ and $y = 0$, that is, the nodes along $x = 0$ have $u_x = 0$ and the nodes along $y = 0$ have $u_y = 0$; (2) to prevent rigid body motion, the node at location O (as shown in Figure 12.35) is fixed along three directions.

The commercial ABAQUS FEM code is utilized to carry out the analysis using eight-node reduced integration element (C3D8R). Initial trials are run on the basic model (but with the mother board omitted) using a twenty-node element (C3D20) and a twenty-node reduced integration element (C3D20R) to determine if the greater processing time they required is justified by the results. The results, which are shown in comparison with the C3D8R results in Table 12.11.

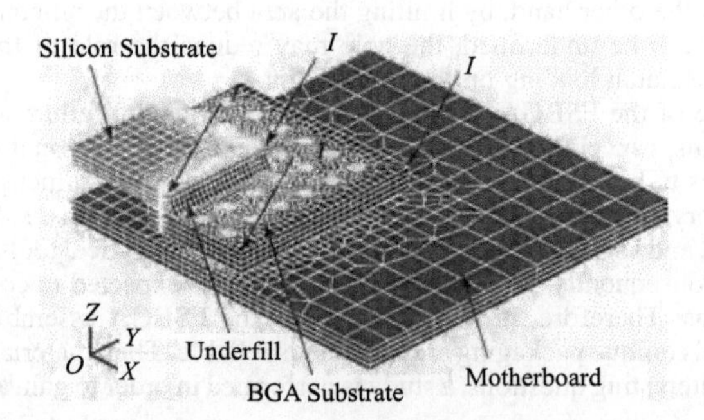

Figure 12.35 A 3D finite element mesh of flip-chip-on-silicon BGA package (Variation 3) which has both the hole and an adhesive underfill between the silicon substrate and the BGA substrate

Table 12.11 Comparison of the results from FEM analysis by different element types

Element type	C3D20	C3D20R	C3D8	C3D8R
U_Z (I) (mm)	1.492×10^{-2}	1.477×10^{-2}	1.486×10^{-2}	1.499×10^{-2}
U_Z (II) (mm)	1.031×110^{-2}	1.015×10^{-2}	1.050×10^{-2}	1.032×10^{-2}
von-Mises (MPa) (Corner Solder Joint)	19.80	19.80	19.80	19.70
Total CPU (s)	37077	35155	3588	2214

Clearly, the variations in von-Mises stresses and in the Z direction displacements at locations I and II are insignificant. However, by using the eight-node reduced integration element (C3D8R) instead of the twenty node element (C3D20), the total CPU time is reduced almost 17 times. Therefore, in order to save processing time, the eight-node reduced integration element (C3D8R) is used in all of the following analysis. However, it should be cautioned that the stability (or node configuration independence) of these results may not strictly apply to other models.

12.3.2 Assembly Layout Investigation

In the FSBGA stack 3D architecture, an opening or manufactured "hole" must be provided in the laminate substrate in order to assembly the stack die. However, this unique novel feature will be used only if its reliability can be proved. Therefore, the present study initially analyses four variations in the finite element model. Two of which are manufactured with open holes, while two of which are without hole. All are based on the elementary package shown in Figure 12.33 and are assumed to be initially stress (and strain) free and then subjected to a temperature change of 23 °C to 125 °C due to a preheating assembly process. Variation 1, as a "basic" model, which is manufactured with a hole in the BGA substrate to accommodate a stacked flip-chip, even though in this case the stacked chip was omitted from the model. Variation 2 is generated to study the effect of the assembly opening, which is based on Variation 1, but has no opening hole in the BGA laminate substrate. The finite element meshes for Variation 1 and Variation 2 are shown in Figures 12.35 and 12.36. To address the effect of underfill on FSBGA architecture, Variation 3 is analyzed, which has both the hole and an adhesive underfill between the silicon substrate and the BGA substrate. The finite element mesh is given in Figure 12.35. In order to represent a conventional underfilled FBGA type application, Variation 4 is adopted, which has no hole in the BGA substrate and underfilled between the silicon substrate and the BGA substrate. Figure 12.36 shows the finite element mesh. The model detail descriptions are also given in Table 12.12.

The properties of the silicon substrate and the FR-4 boards used for the BGA laminate substrate and the laminate motherboard are both assumed to be linear elastic and temperature dependent. The Young's modulus and Poisson ratio of the silicon chip at room temperature are 169.5 GPa and 0.278 respectively. The coefficient of thermal expansion of the silicon chip at room temperature is 2.8 ppm/°C. The FR-4 substrate is assumed to have an orthotropic property with room temperature Poisson's ratios, ν_{xz}, ν_{yz}, and ν_{xy} of 0.143, 0.143, and 0.02, respectively. The Young's moduli, E_x, E_y and E_z, are 22.4 GPa, 22.4 GPa, and 1.6 GPa, and the coefficients of thermal expansion (CTEs), α_x, α_y and α_z, are 16 ppm/°C, 16 ppm/°C, and 65 ppm/°C, also at room temperature. The relevant X, Y, and Z material directions, which follow the right hand spiral rule, are as shown Figure 12.35.

The solder used for both the flip-chip joints connecting the silicon substrate to the BGA laminate substrate (hereafter the flip-solder joints), and the joints connecting the BGA substrate to the motherboard (hereafter the BGA solder joints), is a eutectic Sn/Pb alloy. To save the computer running time, thereby more structures are able to investigate, in this section, structure investigation section, the eutectic Sn/Pb

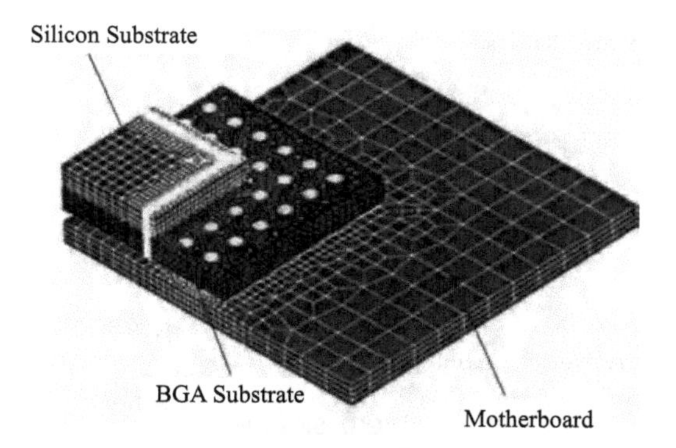

Figure 12.36 Finite element mesh of Variation 1 which had the opening hole in BGA substrate

Table 12.12 Descriptions of the FEM models

Variation	Description	FEM Mesh
Variation 1	Has a hole in the BGA substrate and without underfill between silicon substrate and BGA substrate	Figure 12.36
Variation 2	Has no hole in the BGA substrate and without underfill between silicon substrate and BGA substrate	Figure 12.37
Variation 3	Has a hole in the BGA substrate and with underfill between silicon substrate and BGA substrate	Figure 12.35
Variation 4	Is a conventional flip-chip package (no hole & with underfill between silicon substrate and BGA substrate)	Figure 12.38

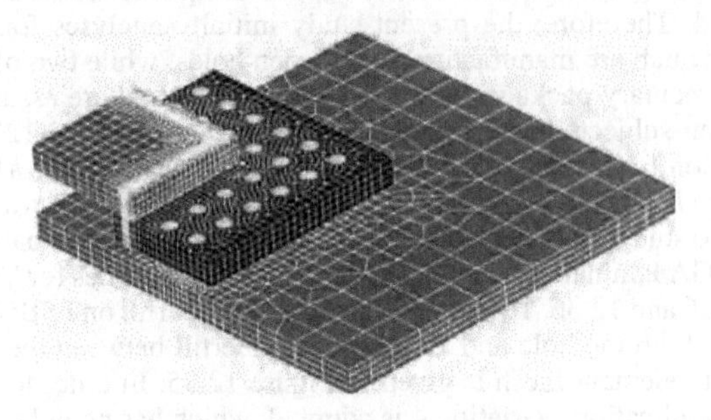

Figure 12.37 Finite element mesh of Variation 2 which had no hole in BGA substrate

solder is assumed to only follow the elastic-plastic and temperature-dependent behavior. The underfill is also assumed to be an elastic-plastic and temperature-dependent material. All material properties for the 63Sn37Pb eutectic solder alloy and underfill FP-4526 are taken from the test database obtained in this research (Ren *et al.*, 2000; Ren *et al.*, 1998).

Throughout this structure study focus has been on the behavior of the solder interconnections which are flip-solder joints and BGA solder joints. The maximum von-Mises stresses as well as the von-Mises stresses at elements A-C in the flip-solder joints as shown in Figure 12.39 and at elements D-F in BGA solder joints as shown Figure 12.40, are listed in Table 12.13. The maximum von-Mises stresses as well as the von-Mises stresses at elements A-C in the flip-solder joints are also plotted in Figure 12.41. Element A-F are chosen here to show the stress distribution along the row of solder joints. From these results, it can be seen that the maximum von-Mises stress levels in the flip-solder joints significantly exceed those in the BGA solder joints, which implies that package is significantly more likely to fail than the package to motherboard connections.

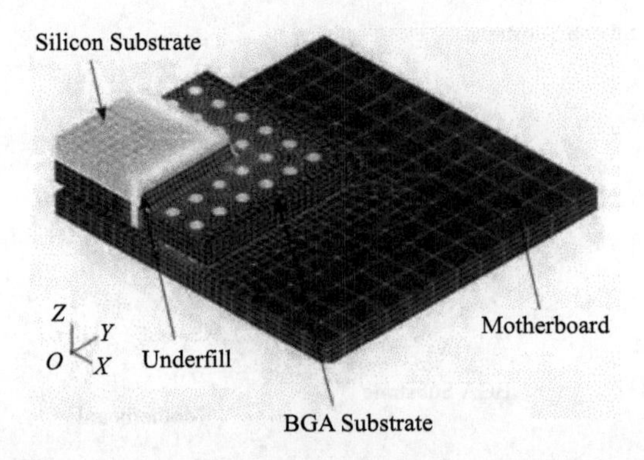

Figure 12.38 Finite element mesh of Variation 4 which is a conventional underfilled FBGA (no hole in the BGA substrate and underfilled between the silicon substrate and the BGA substrate)

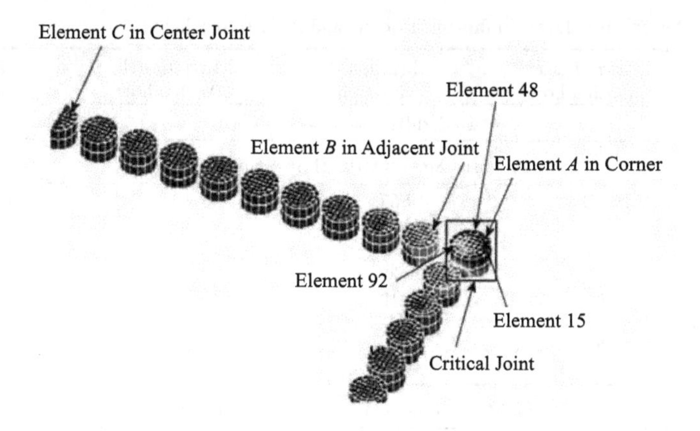

Figure 12.39 von-Mises stress contour in the flip-solder joints for Variation 3

For the flip-solder joints, the maximum von-Mises stress occurs in the joint that is located in the corner of the silicon substrate. The presence or absence of the hole makes no significant difference in von-Mises stress level for the corner joints, but the presence of the hole greatly reduces its level in the center joints, as shown in Figure 12.41. The addition of underfill slightly lowers the maximum von-Mises stress level (in the corner joint), but significantly raises its level in joints nearer to the mid point of the silicon substrate (center joints), altogether bringing forth a more uniform stress distribution along the row of solder joints. Thus the hole or opening in the BGA laminate substrate does not have a significant effect on the maximum stress level. For all variations, the critical solder joint is the same that is in flip-solder joints and located at the corner of the silicon substrate as shown in Figure 12.39.

However, while von-Mises stresses are related to the onset of yield under load, it is the maximum inelastic (plastic) strain amplitude, not the maximum von-Mises stress level, which governs joint failure in fatigue. This is especially true under conditions of low cycle strains that are usually associated with fatigue at the relatively high homologous temperatures experienced with solders. Table 12.14 shows the equivalent plastic strain PEEQ of flip-solder joints. The equivalent plastic strain PEEQ is defined as follows:

$$\mathrm{PEEQ} = \int \sqrt{\frac{2}{3} \mathrm{d}\varepsilon^{\mathrm{pl}} : \mathrm{d}\varepsilon^{\mathrm{pl}}} \tag{12.9}$$

where the strains are tensors.

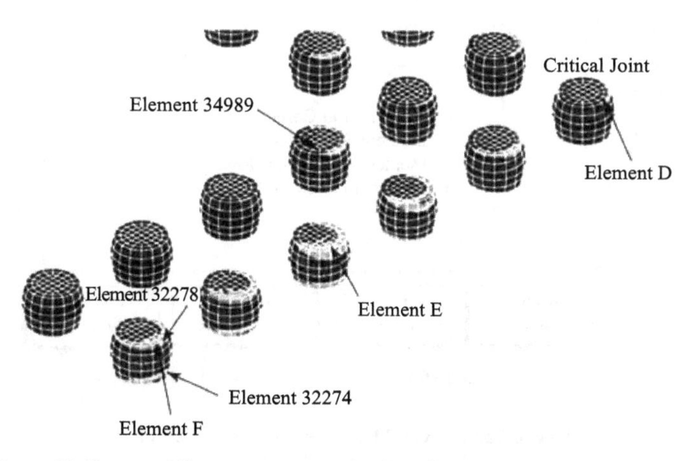

Figure 12.40 von-Mises stress contour in the BGA solder joints for Variation 3

Table 12.13 von-Mises stress levels in flip-solder joints and BGA solder joints

Location	*Variation 1* (with a hole)	*Variation 2* (with no hole)	*Variation 3* (with a hole)	*Variation 4* (with no hole)
	Not Underfilled		Underfilled	
	von-Mises Stress (MPa) (Flip-Solder Joints)			
Element A (in corner joint)	19.67	19.67	18.49	18.46
Element B (in adjacent joint)	19.19	19.53	17.49	17.52
Element C (in center joint)	12.63	19.16	16.93	17.48
Max. Stress Element	19.80 (element 15)	19.80 (element 15)	18.99 (element 48)	18.76 (element 48)
	von-Mises Stress (MPa) (BGA solder joints)			
Element D	2.327	2.307	2.556	2.608
Element E	2.831	3.288	3.506	4.272
Element F	3.462	3.650	2.685	4.731
Max. Stress Element	3.854 (element 32274)	3.679 (element 32278)	3.762 (element 32274)	4.789 (element 32278)

Clearly, the maximum equivalent plastic strain PEEQ for the structures with the underfill is much smaller than that for the structures without underfill. For example, the maximum equivalent plastic strain for the underfilled FSBGA architecture, Variation 3, is 69% lower than that of the no underfilled FSBGA architecture, Variation 1. Therefore, the underfill does help to increase fatigue life by reducing the maximum plastic strain.

Furthermore, comparing the underfilled FSBGA architecture, Variation 3, with the conventional underfilled FBGA, Variation 4, there is no significant difference for the von-Mises stresses and equivalent plastic strains in the flip-solder joints. Therefore, based on the above study, the opening or "hole" in the

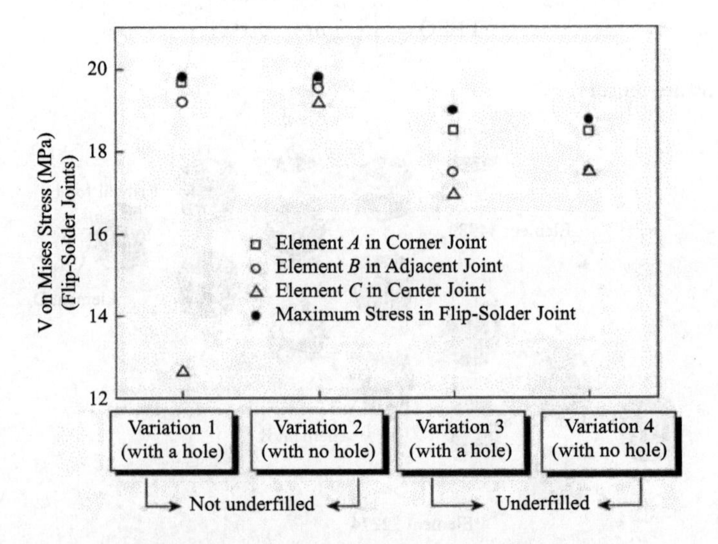

Figure 12.41 von-Mises stress in flip-solder joints

Table 12.14 Equivalent plastic strain PEEQ (flip-solder joints)

Location	*Variation 1* (with a hole)	*Variation 2* (with no hole)	*Variation 3* (with a hole)	*Variation 4* (with no hole)
	Not Underfilled		Underfilled	
Element A	1.1922e-02	1.2175e-02	4.7362e-03	4.6538e-03
Element B	6.8793e-03	9.1328e-03	3.3538e-03	3.3799e-03
Element C	1.1895e-03	6.5914e-03	2.8693e-03	3.3478e-03
Max. Strain Element	1.945e-02 (element 15)	2.513e-02 (element 15)	5.3309e-03 (element 48)	5.1849e-03 (element 48)

FSBGA package won't have big effect on the reliability of the package. In terms of reliability, underfilled FSBGA is comparable to the conventional underfilled FBGA package.

12.3.3 Material Selection

A package design is subject to potential failure as a result of thermally-induced mismatch strains. Many efforts have been made to find a new substrate material that has much lower coefficient of thermal expansion (CTE) that is the CTE is closer to that of silicon components. Thus the thermally-induced mismatch strains can be reduced.

Therefore, another FSBGA model, Variation 5 is also analyzed with the FR-4 laminate replaced by a non-woven substrate material, which has in-plane CTEs closer to the CTE of the silicon components. Variation 5 has the same finite element mesh as that for Variation 3, which has both the hole and adhesive underfill between the silicon substrate and the BGA substrate as shown in Figure 12.35. According to data provided by the manufacturer, the mechanical properties of this non-woven substrate material in the x and y (in-plane) directions are: Young's modulus $E = 13.8$ GPa and Poisson's ratio $\mu = 0.32$. In the absence of additional data, and based on the understanding that this non-woven substrate material is not a laminate, the z-direction mechanical properties are taken to be comparable. In the absence of more detailed material property information, the shear modulus is approximated by using the relationship for an isotropic material, $G = E/2(1+\mu)$. The CTEs of this non-woven substrate material, α_X, α_Y, and α_Z are given as 10.5 ppm/°C, 10.5 ppm/°C, and 108.5 ppm/°C respectively.

Since this non-woven substrate material has an x-y plane CTE half way between that of silicon and FR-4, it is expected to offer some relief from the thermal deformation mismatch related stresses that arise with FR-4, whose x-y plane CTE is approximately six times higher than that of silicon.

The maximum von-Mises stresses as well as the von-Mises stresses at elements A-C in the flip-solder joints as shown in Figure 12.39 and at elements D-F in BGA solder joints as shown Figure 12.40, are listed in Table 12.15. Again Element A-F are chosen here to show the stress distribution along the row

Table 12.15 von-Mises stress levels in flip-solder joints and BGA solder joints

Location	Variation 3	Variation 5
	von-Mises Stress (MPa) (flip-solder joints)	
Element A	18.49	18.97
Element B	17.49	17.89
Element C	16.93	17.46
Max. Stress Element	18.99 (element 48)	19.17 (element 92)
	von-Mises Stress (MPa) (BGA solder joints)	
Element D	2.556	8.221
Element E	3.506	8.395
Element F	2.685	8.358
Max. Stress Element	3.762 (element 32274)	9.722 (element 34989)

Table 12.16 Equivalent plastic strain PEEQ (flip-solder joints)

Location	Variation 3	Variation 5
Element A	4.7362×10^{-3}	5.4431×10^{-3}
Element B	3.3538×10^{-3}	3.7561×10^{-3}
Element C	2.8693×10^{-3}	3.3295×10^{-3}
Max. Strain Element	5.3309×10^{-3} (element 48)	6.6328×10^{-3} (element 92)

of solder joints. Table 12.16 presents the equivalent plastic strain PEEQ of flip-solder joints. The equivalent plastic strain PEEQ is defined as Equation 12.9. As comparison, the results for Variation 3 are also given in Tables 12.15 and 12.16.

It is very interesting to find that the maximum von-Mises stress in the BGA solder joints becomes much higher when the non-woven substrate material is used for the BGA laminate substrate and the laminate motherboard. It is almost two times higher than the maximum von-Mises stress in the BGA solder joints for the Variation 3. In addition, the equivalent plastic strains PEEQ are also larger than those of Variation 3. The results seem unreasonable. Because this non-woven substrate material has lower CTE, it is usually considered that the driving force will be smaller and the stress and strain in the solder joints should decrease.

In order to investigate the issue raised above, another 4 variations, Variation 6-Variation 9, are analyzed to explore the influences of various parameters. In Variation 6 and Variation 7, BGA substrate and motherboard both are assumed to be non-woven substrate material with CTE, $\alpha_X = \alpha_Y =$ 10.5 ppm/°C and, $\alpha_Z =$ 108.5 ppm/°C, which are the same as Variation 5, but with different mechanical parameters. Variation 8 and Variation 9 are different combinations with different thickness of motherboard or BGA substrate for laminate substrate. The details are as listed below:

Variation 6: BGA substrate and motherboard both are assumed to be non-woven substrate material, but with the engineering constants as:

$E_X = 13800$ MPa (the same as Variation 5)
$E_Y = 13800$ MPa (the same as Variation 5)
$E_Z = 1600$ MPa (the same as FR-4's E_Z at 30 °C)
$\mu_X = \mu_Y = \mu_Z = 0.32$ (the same as Variation 5)
$G_{XY} = G_{XZ} = G_{YZ} = 5230$ MPa (the same as Variation 5)

Variation 7: BGA substrate and motherboard both are assumed to be non-woven substrate material, but with the engineering constants as:

$E_X = 13800$ MPa (the same as Variation 5)
$E_Y = 13800$ MPa (the same as Variation 5)
$E_Z = 1600$ MPa (the same as FR-4's E_Z at 30 °C)
$\mu_X = \mu_Y = \mu_Z = 0.32$ (the same as Variation 5)
$G_{XY} = 630$ MPa (the same as FR-4's G_{XY} at 30 °C)
$G_{XZ} = G_{YZ} = 199$ MPa (the same as FR-4's G_{XZ}, G_{YZ} at 30 °C)

Variation 8: Based on Variation 3, both BGA substrate and motherboard are FR-4, but the thickness of motherboard is twice of that of Variation 3.

Variation 9: Based on Variation 3, both BGA substrate and motherboard are FR-4, but the thickness of BGA substrate is twice of that of Variation 3.

The results obtained from Variation 6 and Variation 7 together with the results from Variation 3 and variation 5 are given in Tables 12.17 and 12.18. Table 12.19 presents the engineering constants used for FR-4 taken from the report (Ren, 2000). It can be seen that there is no significant effect on the results when only the Young's modulus E_Z of non-woven substrate material is changed from 13.8 GPa (Variation 5) to 1.6 GPa (Variation 6). However, the maximum von-Mises stress in the BGA solder joints tends to be smaller (4.411 MPa) when the shear moduli of non-woven substrate material are changed

Table 12.17 *Z* direction displacement (UZ) at different locations

Location	*Variation 3*	*Variation 5*	*Variation 6*	*Variation 7*
Location I	7.5677×10^{-3}	3.8096×10^{-3}	4.0869×10^{-3}	3.1336×10^{-3}
Location II	6.7825×10^{-3}	6.6705×10^{-3}	6.7597×10^{-3}	3.5407×10^{-3}
Location III	1.1596×10^{-3}	1.7082×10^{-3}	1.6385×10^{-3}	7.8045×10^{-4}
Location IV	3.7069×10^{-3}	4.1871×10^{-3}	4.2668×10^{-3}	2.3515×10^{-3}

Table 12.18 von-Mises stress in flip-solder joints and BGA solder joints

Location	*Variation 3*	*Variation 5*	*Variation 6*	*Variation 7*
	von-Mises Stress (MPa) (flip-solder joints)			
Element A	18.49	18.97	18.75	17.77
Element B	17.49	17.89	17.77	17.03
Element C	16.93	17.46	17.37	17.03
Max. stress Element	18.99 (element 48)	19.17 (element 92)	19.01 (element 92)	17.99 (element 92)
	von-Mises Stress (MPa) (BGA solder joints)			
Element D	2.556	8.221	7.937	3.924
Element E	3.506	8.395	8.026	4.122
Element F	2.685	8.358	8.024	3.698
Max. stress Element	3.762 (element 32274)	9.722 (element 34989)	8.253 (element 35048)	4.411 (element 32274)

from $G_{XY} = G_{XZ} = G_{YZ} = 5230$ MPa to $G_{XY} = 630$ MPa and $G_{XZ} = G_{YZ} = 199$ MPa (Variation 7). It is close to that in Variation 3 (3.762 MPa). It should be noted that the material properties E_X, E_Y, E_Z, and G_{XY}, G_{XZ}, G_{YZ} assumed for Variation 7 are temperature independent. While in variation 3, they are temperature dependent. So it is reasonable that the maximum von-Mises stress in the BGA solder joints for Variation 7 is still a little bit higher than that for Variation 3. Therefore, it appears that the higher shear moduli assumed for non-woven substrate material is the major reason for the higher maximum von-Mises stress in the BGA solder joints for Variation 5.

Figures 12.42 and 12.43 show the deformed shapes for Variation 3 and Variation 5 and Variation 6 and Variation 7, respectively. It can be seen that the structures for Variation 3 and Variation 7 where the lower G_{XY}, G_{XZ}, G_{YZ} are used are more compliant and are easy to distortion. Therefore, the von-Mises stress levels in Variations 3 and 7 are lower than those in Variations 5 and 6 for both the flip-solder joints and BGA solder joints.

Table 12.19 Engineering constants for FR-4 (Ren, 2000)

Property	Temperature (°C)				
	30	95	125	150	270
E_X (MPa)	22400	20680	19300	17920	16000
E_Y (MPa)	22400	20680	19300	17920	16000
E_Z (MPa)	1600	1200	1000	600	450
μ_X	0.02	0.02	0.02	0.02	0.02
μ_Y	0.1425	0.1425	0.1425	0.1425	0.1425
μ_Z	0.1425	0.1425	0.1425	0.1425	0.1425
G_{XY} (MPa)	630	600	500	450	441
G_{XZ} (MPa)	199	189	167	142	139
G_{YZ} (MPa)	199	189	167	142	139

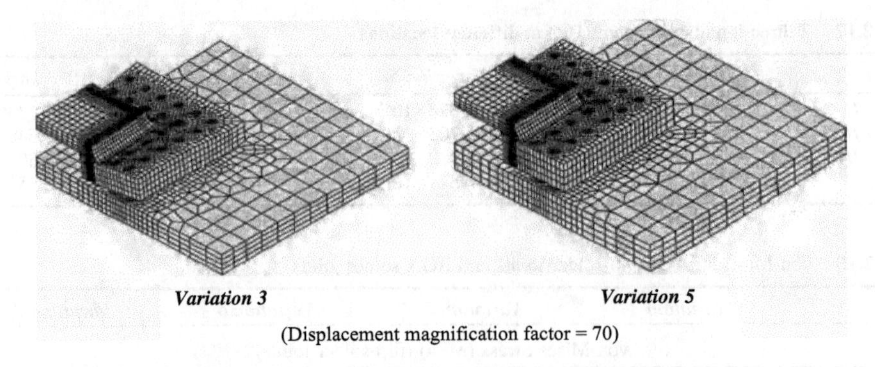

Figure 12.42 The deformed shapes of FSBGA packages

The shear deformations are related to the shear moduli, which can be seen from the stress-strain relationship for the orthotropic elasticity material as defined below:

$$\begin{bmatrix} \varepsilon_X \\ \varepsilon_Y \\ \varepsilon_Z \\ \gamma_{XY} \\ \gamma_{XZ} \\ \gamma_{YZ} \end{bmatrix} = \begin{bmatrix} 1/E_X & -\mu_{YX}/E_Y & \mu_{ZX}/E_Z & 0 & 0 & 0 \\ -\mu_{XY}/E_X & 1/E_Y & -\mu_{ZY}/E_Z & 0 & 0 & 0 \\ -\mu_{XZ}/E_X & -\mu_{YZ}/E_Y & 1/E_{YZ} & 0 & 0 & 0 \\ 0 & 0 & 0 & 1/G_{XY} & 0 & 0 \\ 0 & 0 & 0 & 1/G_{XZ} & 0 & 0 \\ 0 & 0 & 0 & 0 & 0 & 1/G_{YZ} \end{bmatrix} \begin{bmatrix} \sigma_X \\ \sigma_Y \\ \sigma_Z \\ \tau_{XY} \\ \tau_{XZ} \\ \tau_{YZ} \end{bmatrix} \quad (12.10)$$

The higher shear moduli restrict the bending and flexure of laminate substrate and, as a result, the stresses and strains in the solder joints will change because they must, as a consequence, accommodate a greater portion of the mismatch deformation.

Increasing the thickness of the laminate substrates will also increase their values of stiffness and similarly restrict the flexure. As a result, this should also increase the levels of stress and strain in the solder joints. In Variation 8 and Variation 9, either the thickness of motherboard (Variation 8) or the thickness of BGA substrate (Variation 9) is doubled. The maximum von-Mises stresses in the BGA solder joints, as listed in Table 12.20, do increase as expected. It increases from 3.762 MPa to 5.166 MPa (Variation 8) and to 4.350 MPa (Variation 9), respectively.

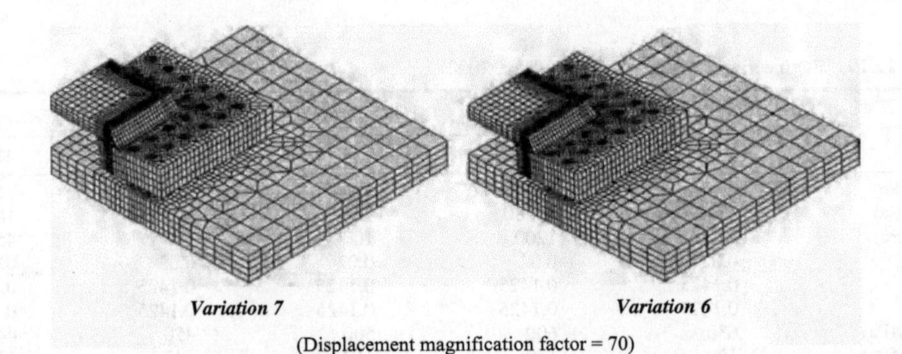

Figure 12.43 The deformed shapes of FSBGA package

Table 12.20 von-Mises stress in flip-solder joints and BGA solder joints

Location	Variation 8	Variation 9
	von-Mises Stress (MPa) (flip-solder joints)	
Element A	18.53	18.58
Element B	17.51	17.56
Element C	16.93	16.97
Max. Stress Element	18.98 (element 48)	18.95 (element 48)
	von-Mises Stress (MPa) (BGA solder joints)	
Element D	2.703	3.218
Element E	3.652	4.301
Element F	2.941	4.263
Max. Stress Element	5.166 (element 32274)	4.350 (element 62760)

Based on the available material properties for the non-woven substrate material, its substitution for FR-4 (even both in the BGA substrate and in the motherboard) cannot be recommended as a means to reduce stress levels in the solder interconnections joining the silicon to the BGA laminate for the FSBGA architecture. The parametric experiments suggest that the much higher (compared to FR-4) shear modulus of the non-woven substrate material limit flexure. It seems to raise the maximum von-Mises stress level in the BGA joints by almost 160%, a factor of almost 2.7. It also raises the maximum equivalent plastic strain in the flip-solder joints.

12.4 Chapter Summary

This chapter discusses the system in package assembly process and the related modeling methodology. The multiple chip power module is investigated first. The modeling focuses the assembly process for die attaching, cooling after molding, and trim/form process. Simulation of the assembly process contains the impact of the die thickness on different assembly processes by using finite element. Then the material behavior of a flip-chip after the assembly process is studied with six material models and underfill process models. It seems that there is no significant difference among the inelastic equivalent strains predicted by the FEA models, in which the materials for both the solder balls and the underfill are assumed to follow the visco behaviors. However, if the visco behaviors of both the solder and the underfill or only the visco behavior of the underfill is neglected, the inelastic equivalent strains in the solder ball have much difference. Therefore, it is suggested that the suitable model be carefully chosen in order to be able to obtain an accurate fatigue life prediction. Then, a stacked die flip-chip assembly layout and the material selection are discussed for a flip-chip-on-silicon ball-grid-array (FSBGA) package. The simulation results show that the assembly opening or the manufactured hole in the FSBGA package won't have big effect on the stress level of the package. The underfill placement in the FSBGA assembly process does help to increase fatigue life by reducing the maximum plastic strain. The substitution for FR-4 (even both in the BGA substrate and in the motherboard) cannot be recommended to reduce the stress levels in the solder interconnections joining the silicon to the BGA laminate for the FSBGA with stacked die in the assembly.

References

ANSYS LS-DYNA Manual Book, Volume 12, 2009.

Auersperg, J. (1997) Fracture and damage evaluation in chip scale packages and flip-chip assemblies by FEA and microdac, *ASME, Symposium on Applications of Fracture Mechanics in Electronic Packaging,* Dallas, November *16–21*:133–138.

Bor, Z.H. (1998) Thermal Fatigue Analysis of a CBGA Package with Lead-free Solder Fillets, *Inter Society Conference on Thermal Phenomena,* 205–211, Seattle, WA, USA, May 6.

Darbha, K., Okura, J.H. and Dasgupta, A. (1997) Impact of underfill filler particles on reliability of flip-chip interconnects, *1st IEEE International Symposium on Polymeric Electronics Packaging,* PEP'97 Norrkoping, Sweden.

Ferry, J.D. (1980) *Viscoelastic Properties of Polymers*, 3rd edn, John Wiley & Sons, New York.

Gektin, V., Bar-Cohen, A. and Ames, J. (1997) Coffin-Manson fatigue model of underfilled flip-chips, *IEEE Transactions CPMT Part A* (**20**)3:317–326.

ITRS (2008), SIP white paper v9.0, The nest step in assembly and packaging: system level integration in the package (SIP), 2008.

Lau, J.H. and Pao, Y.-H. (1997) *Solder Joint Reliability of BGA, CSP, Flip-chip, and Fine Pitch SMT Assemblies,* McGraw-Hill, New York.

Liu, Y., Irving, S. and Luk, T. (2006) Systematic evaluation of die thinning application in a power SIP by simulation, *56th Electronic Components and Technology Conference 2006.*

Liu, Y., Irving, S., Luk, T. and Kinzer, D. (2008) Trends of power electronic packaging and modeling, *Electronic Packaging Technology Conference 2008,* Singapore.

Pang, J.H.L., Tan, T.-I. and Sitaraman, S.K. (1998) Thermo-mechanical analysis of solder joint fatigue and creep in a flip-chip on board package subjected to temperature cycling loading, *48th Electronic Components and Technology Conference,* Seattle, Washington, 878–883.

Post, D., Han, B. and Ifju, P. (1993) *High Sensitivity Moiré: Experimental Analysis for Mechanics and Materials,* Springer-Verlag, NY.

Ren, W., Qian, Z. and Liu, S. (1998) Thermo-mechanical creep of two solder alloys, *48th Electronic Components and Technology Conference,* Seattle, Washington, 1431–1437.

Ren, W., Wang, J., Qian, Z., Zou, D. and Liu, S. (1999c) Investigation of nonlinear behaviors of packaging materials and its application to a flip-chip package, *Proceedings of International Symposium on Advanced Packaging Materials,* Georgia, 31–40.

Ren, W. (2000) Thermo-mechanical Properties of Packaging Materials and Their Applications to Reliability Evaluation for Electronic Packages, PhD Dissertation, Wayne State University.

Wang, J., Qian, Z., Zou, D. and Liu, S. (1998) Creep Behavior of a flip-chip package by both FEM modeling and real time moiré interferometry, *Transactions of the ASME, Journal of Electronic Packaging* **120**(2):179–185.

Zhao, J.-H., Dai, X. and Paul S.H. (1998) Analysis and modeling verification for thermal-mechanical deformation in flip-chip packages, *48th Electronic Components and Technology Conference,* Seattle, Washington, 336–344.

Zou, D. Wang, J., Zhu, J., Lu, M. and Liu, S. (1997) Creep behavior study of plastic power package by real time moiré interferometry and FEM modeling, *ASME, Symposium on Applications of Experimental Mechanics to Electronic Packaging* Dallas, November *16–21*:51–57.

Part III

Modeling in Microelectronic Package and Integration: Reliability and Test

13

Wafer Probing Test

This chapter presents the simulation of parameters for wafer probe test by finite element modeling with consideration of probe over travel (OT) distance, scrub, contact friction coefficient, probe tip shapes, and diameter. The goal is to minimize the stresses in the device under the bond pad and eliminate wafer failure in a probe test. In probe test modeling, a nonlinear finite element contact model is developed for the probe tip and wafer bond pad. Modeling results have shown that the probe test OT, probe tip shape and tip diameters, contact friction between the probe tip and bond pad, as well as the probe scrub of the probe tip on the bond pad are important parameters that impact the failure of interlayer dielectric (ILD) layer under bond pad. Comparison between probe test damage and wire bonding failure shows the degree of damage to both probe test and wire bonding on the same bond pad structures. In addition, a design of experiment (DOE) probe test with different ILD and metal thickness is carried. The correlation between the modeling and the DOE test is studied. The results show that the modeling solution agrees with the DOE probe test data. Modeling results have further revealed that a probe test can induce the local tensile (or bending) first principal stress in ILD layer, which may be a root cause of the ILD failure in probe test.

13.1 Probe Test Model

The probe pin structure is shown in Figure 13.1. Because the geometry size of the probe beam is much larger than the probe tip, in order to conduct an effective simulation and analysis, the following assumptions are made:

(1) The probe beam follows Euler Beam theory (Timoshenko, 1951), therefore the relationship of probe over travel (OT) distance and probe tip force P could be analytically obtained:

$$P = \frac{\pi E \Delta}{64\left\{\frac{d + C_d(L_1 - L)}{3C_d^2[d + C_d(L_1 - L)]^2} + C_1 L + D_1\right\}} \tag{13.1}$$

where Δ is the OT, $C_d = \frac{d - d_t}{L_2}$; $I = \frac{\pi d^4}{64}$

$$C_1 = \frac{(LL_1 - L_1^2)}{EI} - \frac{32[d + 2C_d(L - L_1)]}{3\pi E C_d d_d^2 d^3}$$

$$D_1 = \frac{(LL_1^2/2 - L_1^3/6)}{EI} - C_1 L_1 - \frac{32[2d + C_d(L - L_1)]}{3\pi E C_d^3 d^2}$$

Modeling and Simulation for Microelectronic Packaging and Integration: Manufacturing, Reliability and Testing, Second Edition. Sheng Liu and Yong Liu.

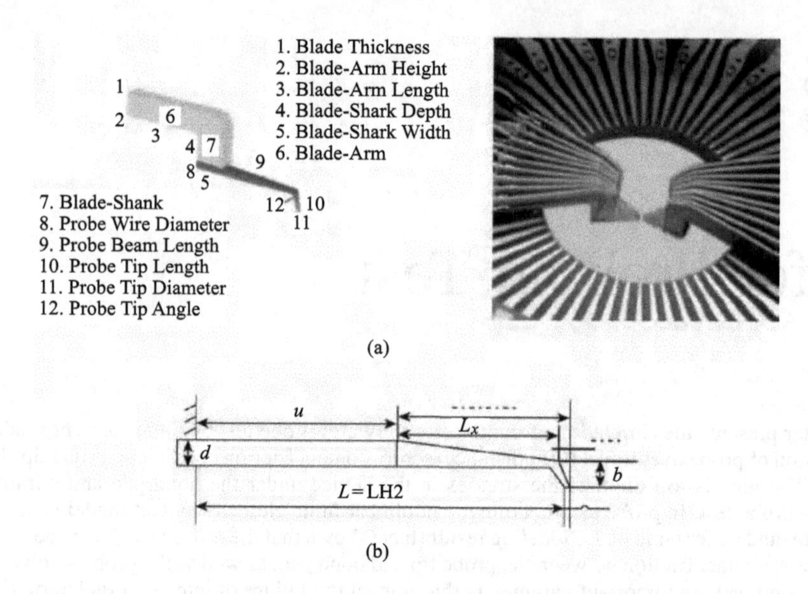

Figure 13.1 Probe pin geometry. (a) Probe test structure; (b) probe beam geometry

(2) Based on assumption (1), the FEA model may be set up for a local probe tip and a bond pad structure (see Figures 13.2 and 13.3), which includes a contact pair for probe tip and bond pad.

For a particular probe test pin used in this study, the relationship between the OT and probe tip force is shown in Table 13.1 with an effective balance contact force (BCF) of 2.15 g/mil. A conceptual bond pad structure is given in Figure 13.3, which includes a SEM probe pin picture from reference (Hartfield *et al.*, 2003) and a conceptual bond pad structure. The conceptual bond pad structure consists of three ILD layers and three metal layers. An FEA model may be set up with a local probe tip on a bond pad structure, which includes a contact pair for the probe tip and bond pad with consideration of contact friction. Figure 13.3 gives three typical probe test FEA models in which the probe tip shapes are considered as flat, flat-round, and round. The material mechanical properties are shown in Table 13.2 (King, 1998), in which the bond pad and metal layers are nonlinear (elastic-plastic) materials; all the rest of the materials, including the probe pin, are considered to be linear elastic.

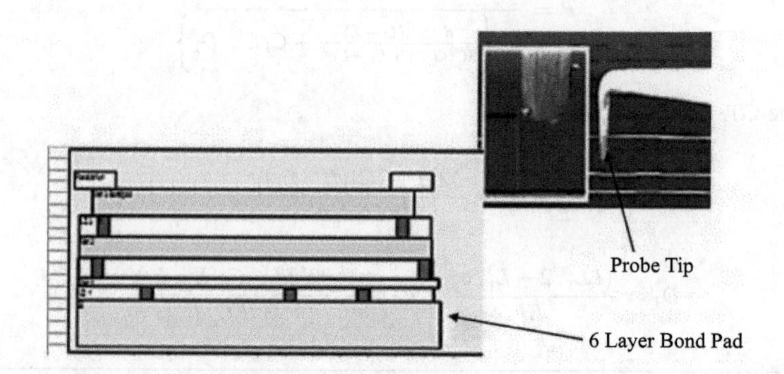

Figure 13.2 Probe tip (Note: the SEM picture is from Hartfield *et al.*, 2003) and a conceptual six-layer bond pad structure

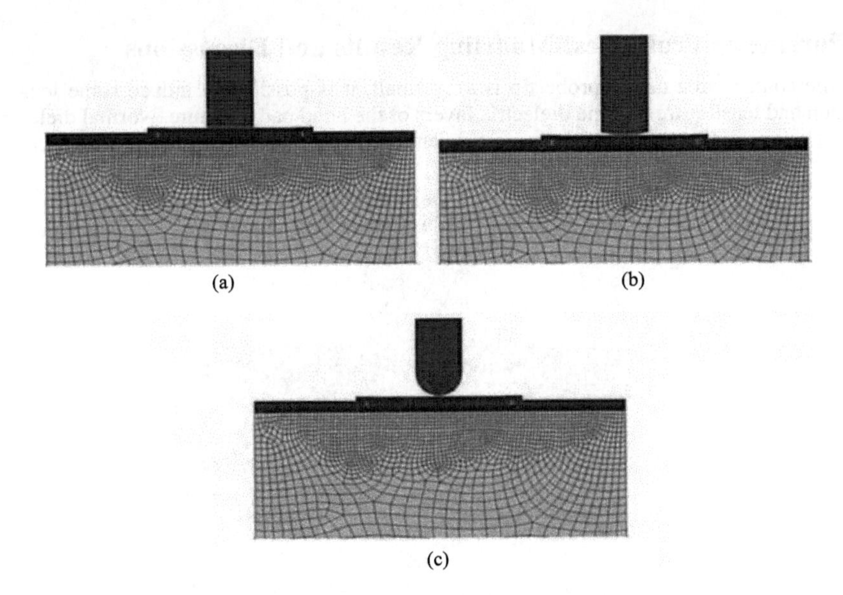

Figure 13.3 Three typical probe test models. (a) Model with flat probe tip; (b) model with flat-round probe tip; (c) model with round probe tip

The friction coefficient in the contact model between the bond pad and probe tip is 0.4. Different contact friction coefficients are chosen to simulate surface roughness. The parameter OT values selected for simulation are 2 mil, 4 mil, 6 mil, and 8 mil. Three types of probes with different tip diameters are considered in the modeling. The horizontal probe scrub during probe touch down is about 15 μm based on testing. To examine the impact of probe scrub, different values of probe scrub are selected during modeling. Normally, a probe tip is subjected to two boundary conditions during probe test. One is the probe OT, which is converted into the probe tip pressure, and the other is the probe scrub along horizontal direction on bond pad.

Table 13.1 OT versus contact force (BCF = 2.15 g/mil)

OT (mil)	Probe Force P (g)	Probe Tip Force P (mN)
2	4.31	42.2
4	8.62	84.4
6	12.93	126.7
8	17.2	168.9
10	21.55	211.1

Table 13.2 Material mechanical properties (King, 1998)

Material	Poisson Ratio	Modulus (GPa)	Yield Stress (GPa)
Silicon	0.23	169.5	
ILD	0.25	70.0	
TiW	0.25	117.0	
Al(Cu)	0.35	70.0	0.2 (25 °C), 0.05 (450 °C)
Au(FAB)	0.44	60.0	0.0327(200 °C)
W	0.28	409.6	
75W/Re25(probe)	0.3	430.3	

13.2 Parameter Probe Test Modeling Results and Discussions

Because the contact area of the probe tip is very small, it is possible to induce some local bending deformation and tensile stress in the dielectric layers of the bond pad structure. Normal dielectrics have very strong compressive strengths, but they are much weaker with regard to their tensile strength. One of

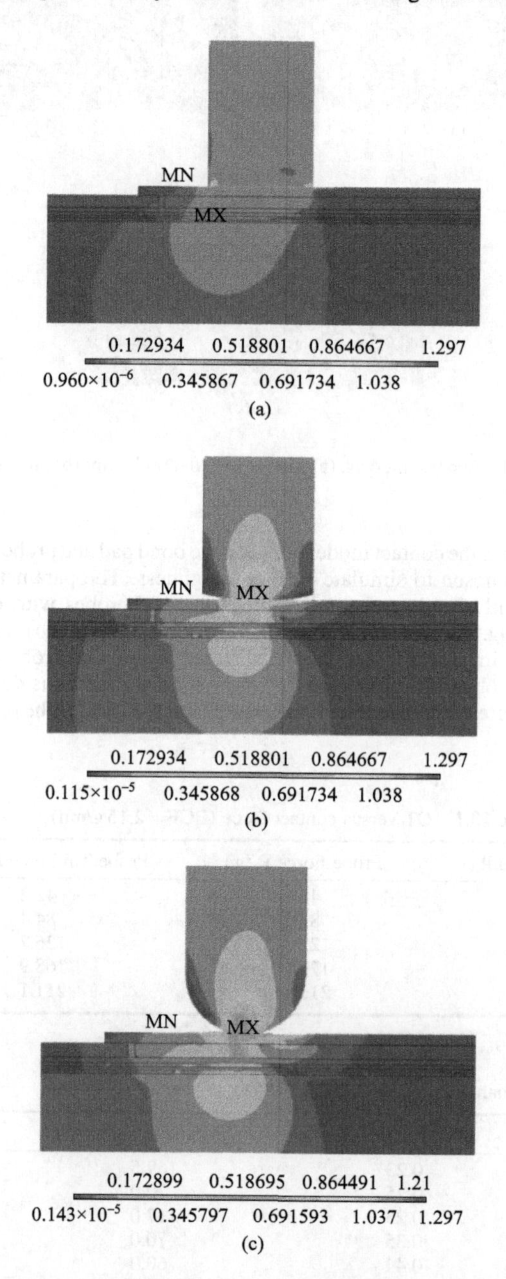

Figure 13.4 von-Mises stress distribution under 4 mil OT with three probe tip shapes. (a) Flat tip, maximum stress: 272 MPa; (b) flat-round tip, maximum stress: 584 MPa; (c) round tip, maximum stress: 1297 MPa *(Color version of this figure is available online)*

the key failure criteria is to examine the dielectric failure as a reliability screen. In this section, the dielectric layer failure criterion is used to judge the modeling and test results. If the first principal stress of the ILD layer induced by a probe test is below the dielectric yield tensile strength, the ILD layer is considered as safe without failure.

13.2.1 Impact of Probe Tip Geometry Shapes

The probe modeling results with different probe tip geometries are shown in Figures 13.4 to 13.7.

Figures 13.4 and 13.5 show the von-Mises stress distribution for the probe test system and the first principal stress distribution in the bond pad structure under an OT of 4 mil with three probe tip shapes, separately. The results show that the round probe tip induces the maximum von-Mises stress in the bond pad and highest first principal stress in the ILD layer. The flat-round probe tip induce the median von-Mises stress in bond pad and median first principal stress in ILD layer while the flat probe induces the minimum von-Mises stress in the bond pad and smallest first principal stress in the ILD layer. This is because the round probe tip has produced the highest stress concentration on the bond pad followed by high stress transferred to the ILD layer below the bond pad. In Figure 13.5, the maximum first principal

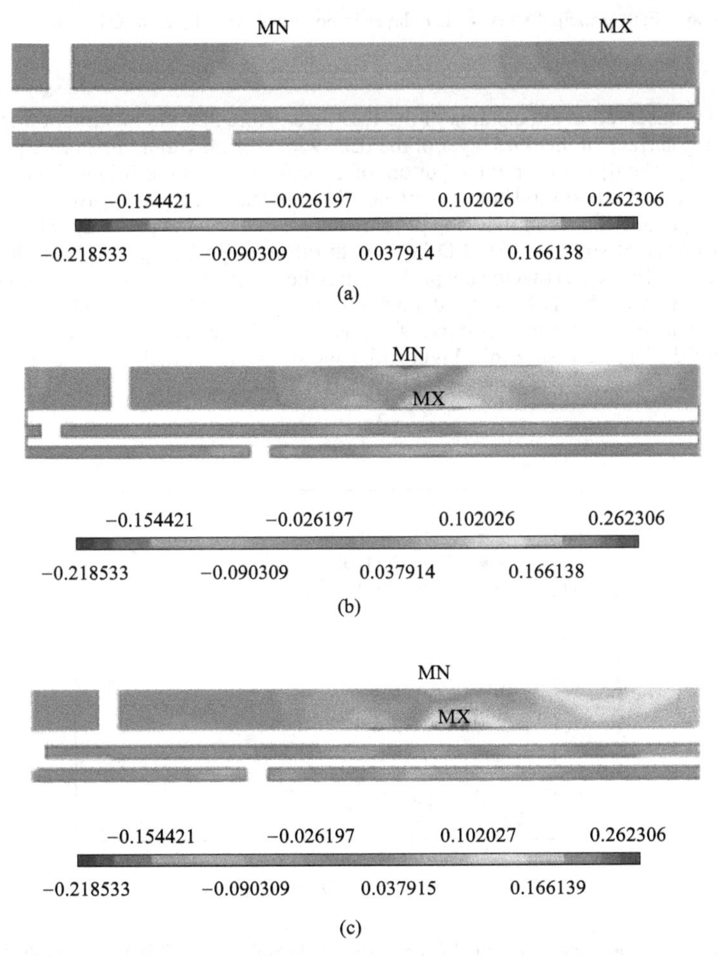

Figure 13.5 First principal stress of ILD layer in bond pad under 4 mil OT and three probe tip shapes. (a) Flat tip, maximum: 47 MPa; (b) flat-round tip, maximum: 186 MPa; (c) round tip, maximum: 262 MPa *(Color version of this figure is available online)*

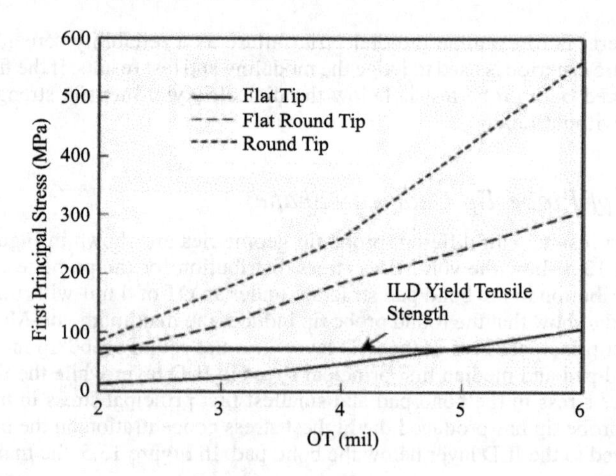

Figure 13.6 First principal stress of ILD layer in bond pad with different OTs and probe tip shapes

stress (tensile) is observed in a local area of the ILD layer. Note that the maximum first principal stress appears at the top surface of the third layer of the ILD, while the maximum first principal stresses of flat-round and round probe tips appear at the bottom of the same layer of the ILD while larger tensile stress have been observed in both top and bottom surfaces. This indicates that the flat-round probe tip and round probe tip have a greater chance of inducing probing failure than the flat probe tip. Figure 13.6 shows the curves of first principal stress of the ILD layer with different probe tip shapes in the same diameter versus OT. The result indicates that the flat probe tip has the smallest stress curve. As compared to the ILD tensile strength criterion, the ILD layer under a flat probe tip can withstand the OT up to 5 mil while the ILD layer with flat-round tip can withstand OT about 2.4 mil and the ILD layer with round probe tip cannot support OT of more than 2 mil. Figure 13.7 shows the first principal stress curves of ILD layers

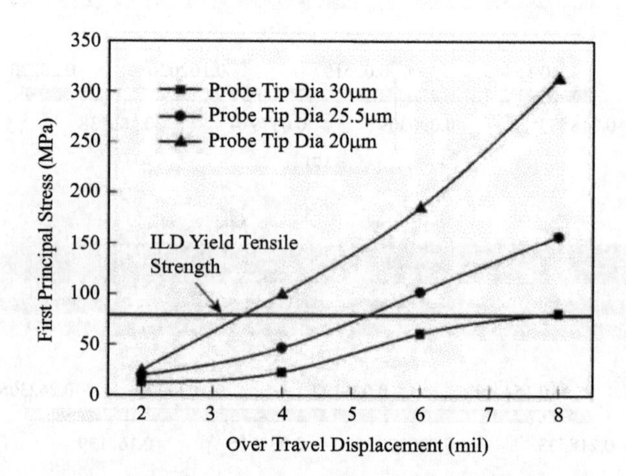

Figure 13.7 First principal stress of ILD layer versus OT (mil) with different flat probe tip diameters

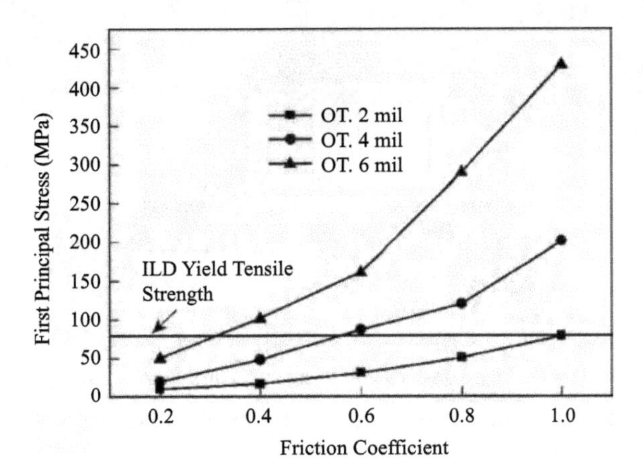

Figure 13.8 First principal stress of ILD layer versus friction coefficient with three OTs in a flat probe tip

versus OT with different probe tip diameters in the same flat probe tip shape. The results disclose that the probe tip shape has a large impact on the probe test damage in the dielectric layer. In the case of the probe diameter of 30 μm, the ILD layers can withstand OT up to 8 mil without the ILD cracking. While in the case of a probe diameter of 25.4 μm (1 mil) the ILD can support OT up to 5 mil. However, the 20 μm diameter probe makes the ILD only able to withstand OT about 3.4 mil. Note that the safe area is under the red line (ILD yield tensile strength) in Figures 13.6 and 13.7. It shows that the flat probe tip is the best case which can reach 5 mil OT in Figure 13.6 and 8 mil OT in Figure 13.7 without damaging the ILD layer.

13.2.2 Impact of Contact Friction

Normally, the friction coefficient between the probe tip and bond pad is about 0.4. However, the dynamic contact processing of the probe tip surface and bond pad surface can change the friction. The results of impact on first principal stress of the ILD layer due to different friction coefficients are given in Figure 13.8.

Figure 13.8 shows the contact friction between the bond pad surface and the probe tip has significant impact on the first principal stress in the ILD layer. In the case of 6 mil OT, the ILD layer will break if the friction coefficient is larger than 0.35. In the case of OT = 4 mil, the ILD layer can withstand tensile stress with a friction coefficient 0.5 without cracking. In the case of OT = 2 mil, the first principal stress of the ILD layer is lower than the ILD tensile yield strength, even though the friction coefficient has reached a large value of 1.0.

13.2.3 Impact of Probe Tip Scrub

As the probe pin touches down and undergoes over travel in the test, the scrub of the probe tip at the bond pad surface occurs. Figures 13.9 and 13.11 give the results of the von-Mises stress in the probe test system and the first principal stress in the ILD layer with different amounts of probe tip scrub.

In Figures 13.9 and 13.10, the impact of probe tip probe scrub on stress distribution is shown. For von-Mises stress in the probe test in Figure 13.9, larger probe scrub may induce greater stress. However, for the first principal stress in the ILD layer; if the probe scrub is larger than 3 μm, there are almost no

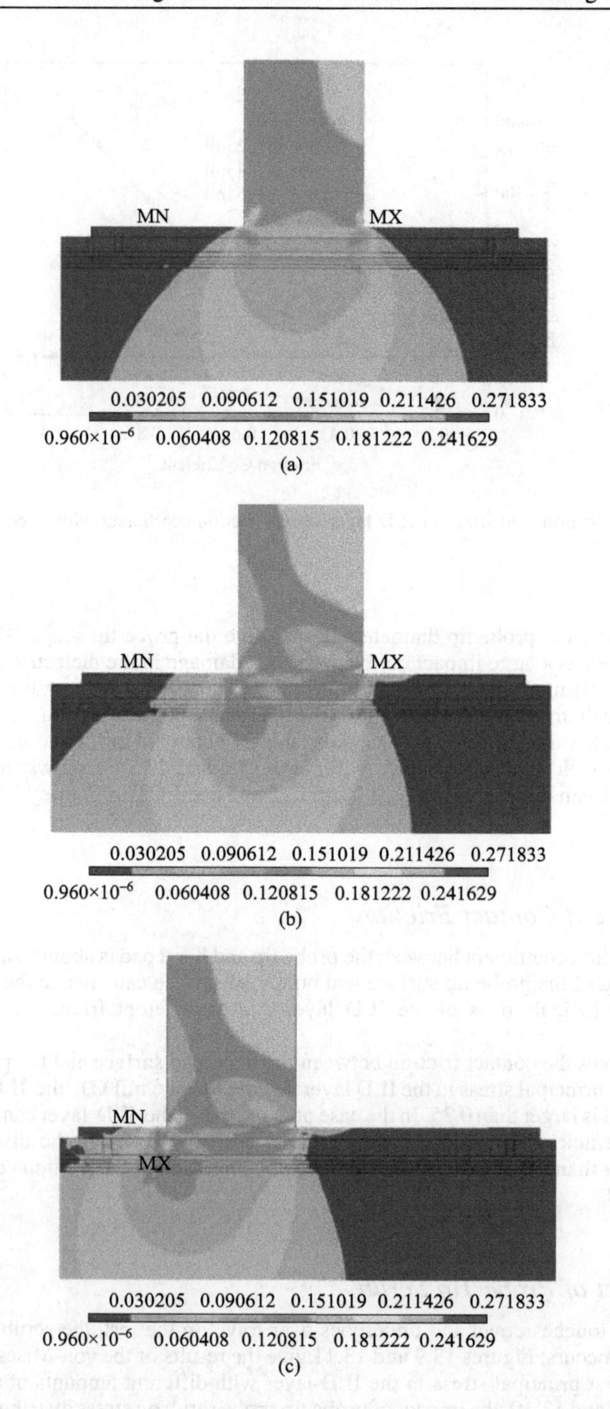

Figure 13.9 von-Mises stress distribution with three probe scrubs under OT = 4 mil. (a) Probe scrub 0 μm, max: 203 MPa; (b) probe scrub 0.5 μm, max: 263 MPa; (c) probe scrub 15 μm, max: 272 MPa *(Color version of this figure is available online)*

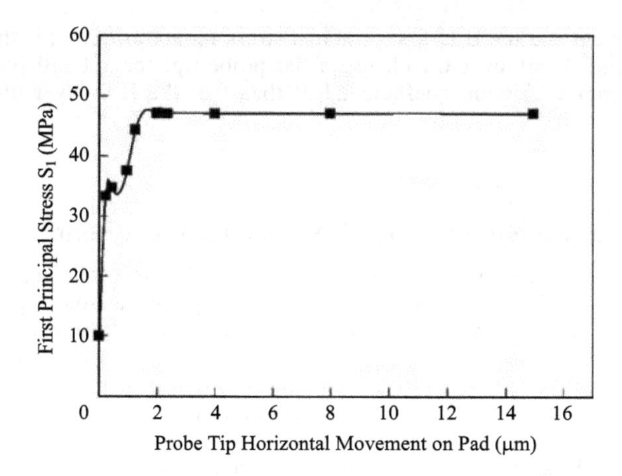

Figure 13.10 First principal stress of ILD layer versus probe tip scrub (horizontal movement) under OT = 4 mil

changes in the stress (Figure 13.10). If there is no probe scrub, both the von-Mises stress and first principal stress are minimized. Figure 13.11 shows that without the horizontal probe tip scrub, the ILD layer is not damaged, even with probe OT up to 8 mil. The first principal stress of the ILD layer however has exceeded the ILD yield tensile strength after the OT is larger than 8 mil, then the ILD layer will break after that.

There is a trade off between the probe scrub and the contact friction coefficient. A larger friction coefficient normally induces greater stress; however, it makes the probe scrub shorter, which may reduce the stress in the ILD layer. The trade-off point and optimization solution are different for different wafer structures and probe test systems. Based on the parameter modeling results

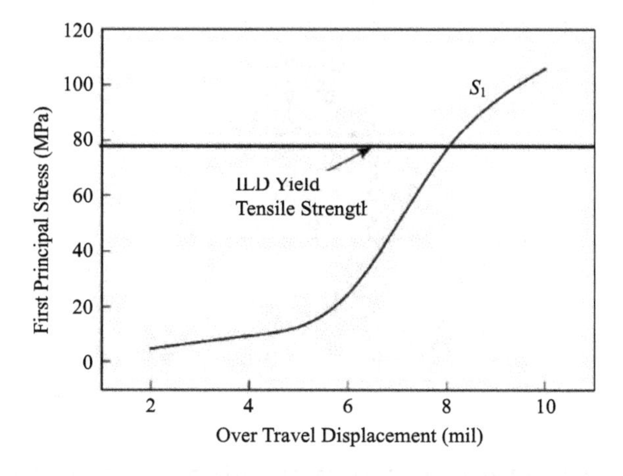

Figure 13.11 First principal stress of ILD without probe scrub

for the probe test system and the ILD first principal stress failure criterion; if the probe parameters can be selected within 4 mil over travel, use a flat probe tip; for a 1 mil probe tip diameter or above with normal contact friction coefficient less than 0.5, the ILD layer under bond pad does not crack.

13.3 Comparison Modeling: Probe Test versus Wire Bonding

Whether probe test or wire bonding has the greatest potential impact to dielectric cracking is always of interest to the industry. This section provides the modeling comparison of probe test versus wire bonding. The bond pad structure in Figure 13.12 is different from the bond pad in Figure 13.2, which has three ILD layers and three metal layers. Figure 13.12 considers a probe test model with a flat probe tip and a two-layer bond pad structure (one layer of ILD and one layer of metal) and a wire bonding model with the same two-layer bond pad structure. The modeling results are shown in Figures 13.13 to 13.16.

Figures 13.13 and 13.14 give the von-Mises stress and first principal stress of the ILD in a probe test with 8 mil OT and wire bonding under 1 N bonding force. Both cases induce equivalent first principal stress in the ILD layers. Both first principal stresses of the ILD layers are higher than dielectric yield

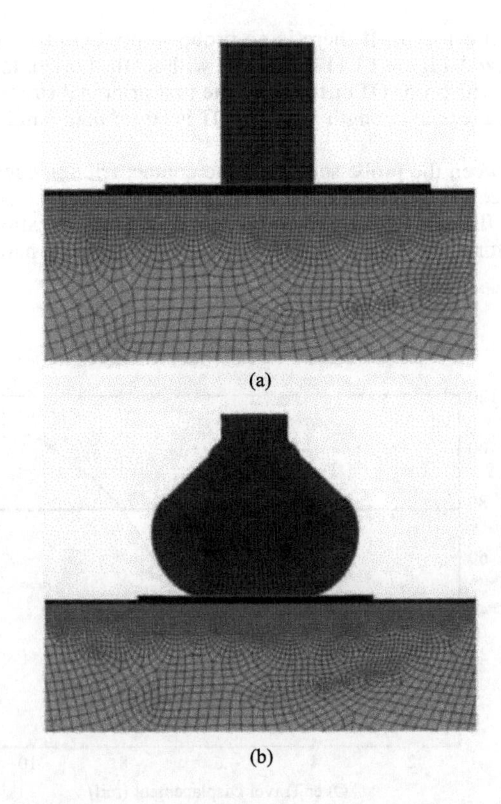

Figure 13.12 Probe test with flat tip and wire bonding models. (a) Probe test model (flat probe tip and two-layer bond pad); (b) deformed wire bonding model

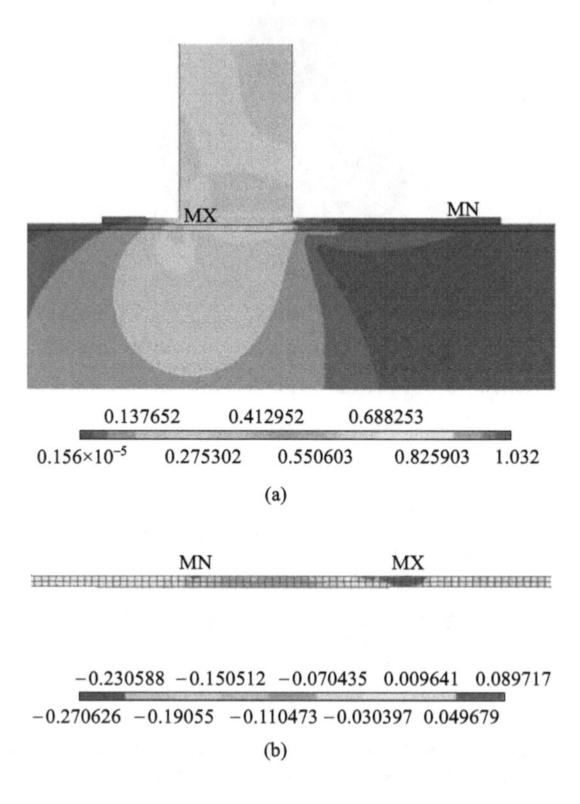

Figure 13.13 Probe test stress distribution under OT 8 mil. (a) von-Mises stress; (b) ILD first principal stress, maximum: 89.7 MPa *(Color version of this figure is available online)*

tensile strength. The only difference is that the maximum tensile first principal stress of the ILD in probing is located below the right side of the bond pad. While the maximum tensile first principal stress of ILD in wire bonding locates with equally chance below the two sides of bond pad but not directly under the wire bond area. This is because the fundamental stress distributions are different due to probing and wire bonding.

Figures 13.15 and 13.16 show the curves of first principal stresses in ILD versus different OT (Figure 13.15) and different wire bonding force (Figure 13.16) respectively. Both curves give the ILD yield tensile strength (red line). When the concentrated stress is below this tensile strength, the ILD will not break, and when it is above the tensile strength, the ILD will be damaged. Comparing the probe test modeling results in Figure 13.15 with wire bonding results in Figure 13.17, it may be seen that the ILD first principal stress in a probe test reaches the dielectric tensile yield strength (76 MPa) with OT above 6 mil, while the ILD first principal stress in wire bonding gets the dielectric tensile yield strength with wire bonding force above 800 mN. Therefore, in the bond pad structure of Figure 13.12, a 6 mil OT probe test is equivalent to 800 mN wire bonding based on dielectric failure criteria. In a normal probe test, the OT is about 2–3 mil. This induces the ILD first principal stress of about 17–35 MPa, while normal wire bonding force is about 650 mN, which induces ILD first principal stress of about 31 MPa. Therefore, the modeling results show that both probe test and wire bonding in the same bond pad structure considered in this section induce the same level first principal stress with a sufficient safety margin.

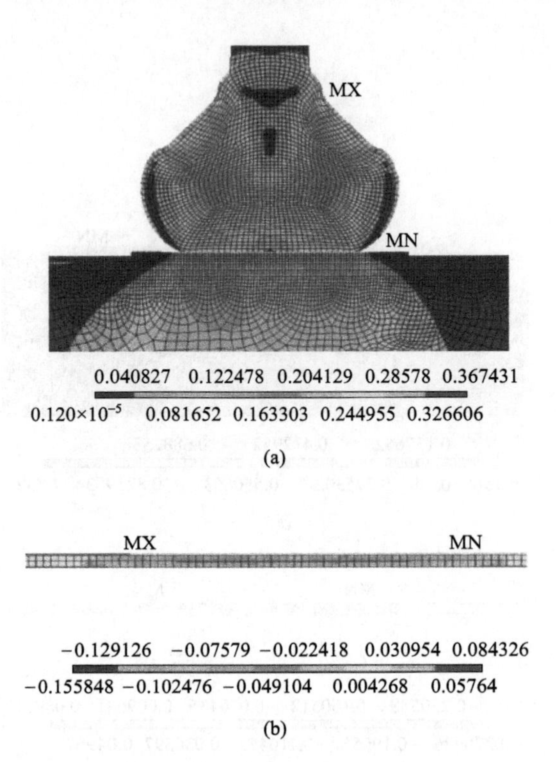

Figure 13.14 Wire bonding stress distribution under bonding force of 1 N. (a) von-Mises stress; (b) ILD first principal stress, maximum: 84 MPa *(Color version of this figure is available online)*

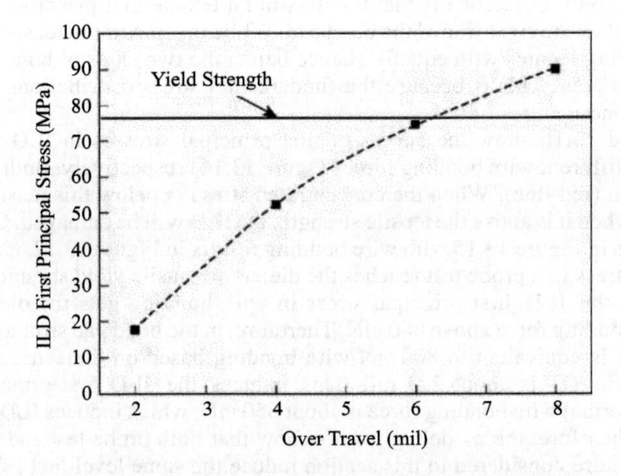

Figure 13.15 First principal stress of ILD versus OT

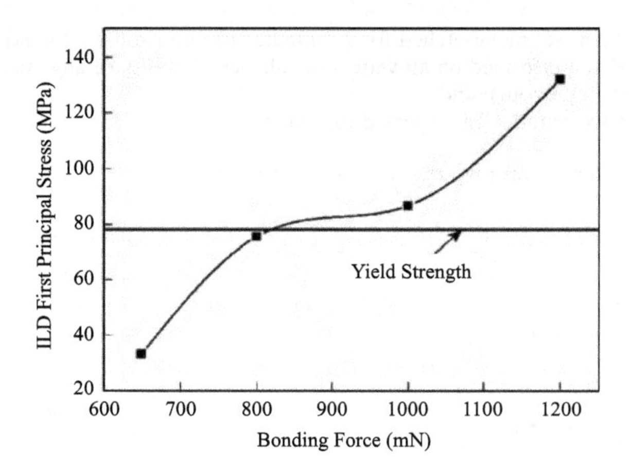

Figure 13.16 First principal stress of ILD in wire bonding versus bonding force

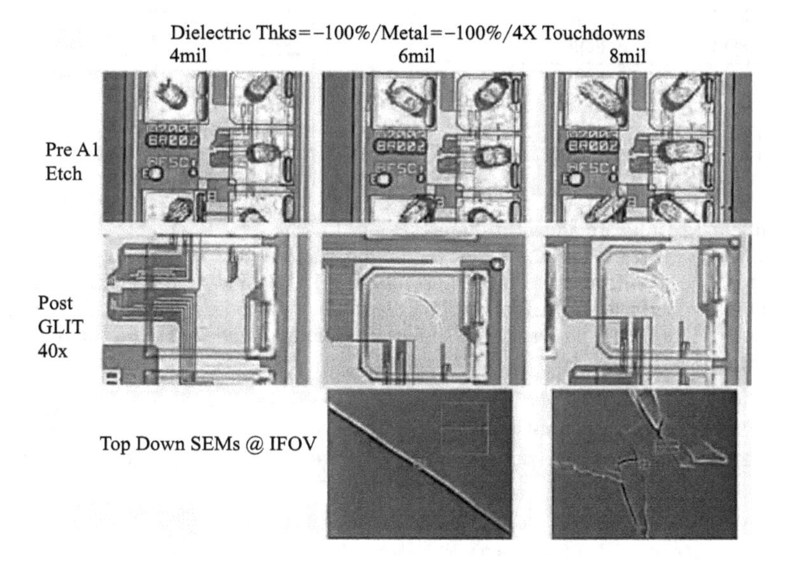

Figure 13.17 Top view of damage under different OT

13.4 Design of Experiment (DOE) Study and Correlation of Probing Experiment and FEA Modeling

The purpose of the DOE study is to simulate potential stresses on a bond pad structure (Figure 13.2) through probing experiments that could induce dielectric crack similar to that observed on the problem sample unit.

1. Full loop wafers were processed in Fab per DOE;
2. Each of the wafer samples were then probed manually with a single site probe card on an EG50 per the sort DOE (20 die/Sort DOE setting);

3. The wafers were then wet metal etched for 25 minutes to remove the Al bond pads;
4. A GLIT test was then performed on all wafers to enhance visibility of any cracks in the passivation (NH4OH/Peroxide Vone hour); and
5. Finally the wafers were optically inspected for cracks.

The experiment DOE for the probing test is shown as following.

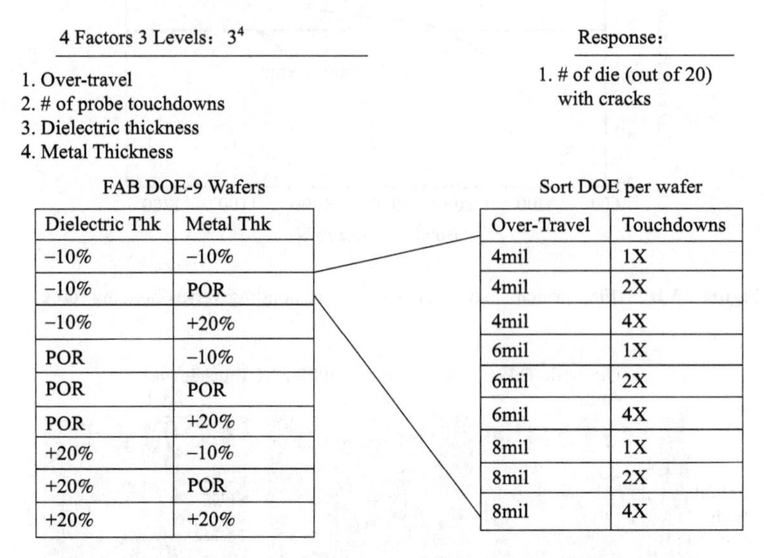

4 Factors 3 Levels: 3^4

1. Over-travel
2. # of probe touchdowns
3. Dielectric thickness
4. Metal Thickness

Response:

1. # of die (out of 20) with cracks

FAB DOE-9 Wafers

Dielectric Thk	Metal Thk
−10%	−10%
−10%	POR
−10%	+20%
POR	−10%
POR	POR
POR	+20%
+20%	−10%
+20%	POR
+20%	+20%

Sort DOE per wafer

Over-Travel	Touchdowns
4mil	1X
4mil	2X
4mil	4X
6mil	1X
6mil	2X
6mil	4X
8mil	1X
8mil	2X
8mil	4X

The experimental results are shown in Figures 13.17 to 13.18.

Figure 13.17 shows that the OT number beyond 6 mil starts to induce cracking, and at OT = 8 mil, cracks may clearly be seen from the SEM pictures. Figure 13.18 shows the statistical data. It shows that the OT is the number one factor for ILD crack. In order to understand the root cause, FEA modeling for the DOE is conducted. The modeling results are shown in Table 13.3 and failure mechanism comparison is given in Figure 13.19.

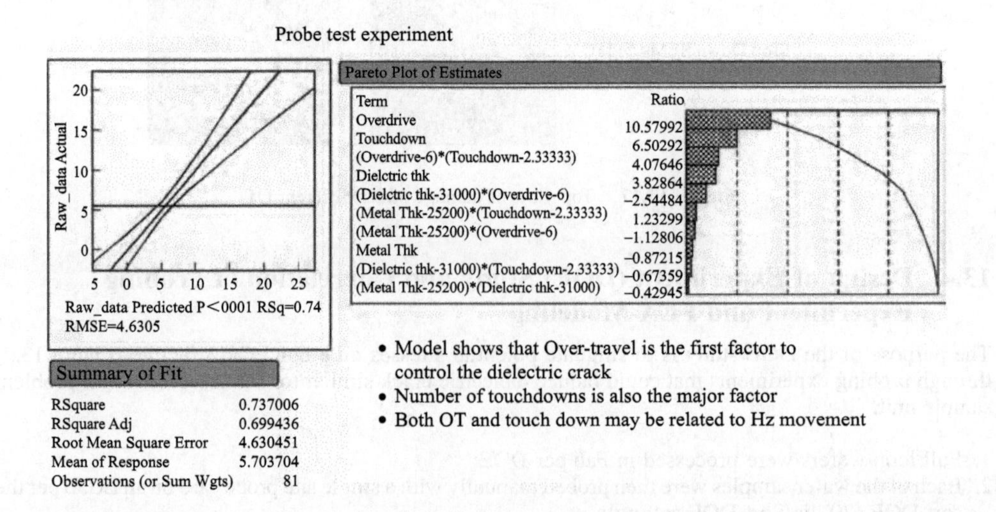

Figure 13.18 DOE results

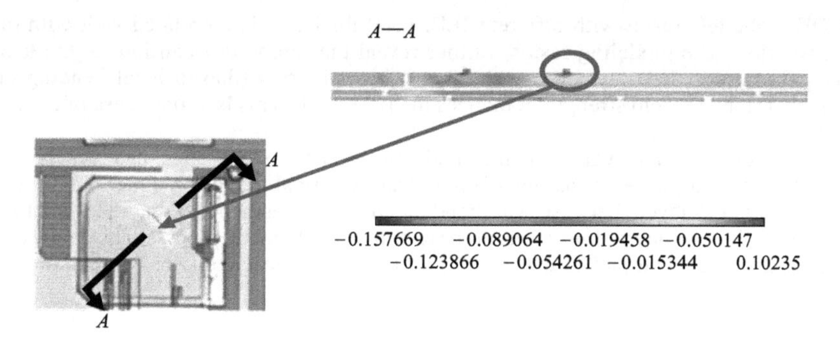

Figure 13.19 Failure comparison of modeling and test with OT = 6 mil, the maximum first principal stress is 102.3 MPa *(Color version of this figure is available online)*

Table 13.3 DOE modeling results for first principal stresses in ILD, the max tensile strength for ILD is 76 MPa (King, 1998)

	2.16 (−10%)μm	2.4 μm	3.0 (+20%)μm
2.7 (−10%)μm	4 mil 51 MPa	6 mil 103.2 MPa	6 mil 91.8 MPa
3 μm	4 mil 52 MPa	6 mil 105 MPa	6 mil 91.7 MPa
3.6 (+20%)μm	6 mil 105 MPa	6 mil 105 MPa	8 mil 129.3 MPa

Table 13.3 gives the DOE modeling results of the first principal stress in ILD with different metal and ILD thickness. From Table 13.3 we may see that as the max tensile yield strength of ILD is 76 MPa, so all the yellow area data beyond this yield strength will induce the crack. The simulation results match the experimental results (Figure 13.17). Figure 13.19 has shown the failure comparison, it may be seen that during the probe touch down, the local tensile stress (first principal stress) is induced. When the local tensile stress is higher than ILD tensile yield strength, the crack appears. That is the root cause of ILD failure in probing.

13.5 Chapter Summary

This chapter develops the FEA simulation framework for probe test parameter modeling. A DOE probe test and the correlation between modeling and the test are investigated and discussed. The work shown in the chapter is part of the modeling efforts to optimize the probe test and to reduce the damage of the bond pad structure. The major results are summarized as follows.

The dielectric failure criterion based on first principal stress law is selected to judge the modeling solution. The probe test modeling results show that the probe test OT, the probe tip shape and tip diameters, contact friction between the probe tip and bond pad, as well as the probe scrub of the probe tip on the bond pad, are important parameters that impact the failure of the ILD under the bond pad. The modeling results found that the small OT, flat probe tip, and large tip diameter may induce a small stress in the device under bond pad. There is a trade off between the probe scrub and the contact friction coefficient. Larger contact friction coefficient between probe pin and bond pad normally induces greater stress; however, it makes the probe scrub shorter, which may reduce the stress in the ILD. The trade-off point and optimization solution differ for different wafer bond pad structure and probe test system used.

The DOE probe test result with different ILD/metal thickness is correlated well with the related modeling solution. The modeling results further reveal that probe test can induce the local tensile first principal stress in ILD layer. Once the first principal stress (due to local bending or tensile) exceeds the ILD tensile yield strength, it induces the ILD crack. This is a root cause of the ILD failure in probe test.

Comparing probe test and wire bonding modeling with the same bond pad structure shown in Figure 14.13 shows that, for a normal wire bonding and a normal probe test under bond pad structure, the damage to the ILD layers between wire bonding and probe test is in the same level. However, the results vary for different bond pad structure, probe test, and wire bonding process. This might be observed by simulating different bond pad structures with different probe test and wire bonding process.

References

Chiang, C.L. and Hurley, D.T. (1998) Dynamics of backside wafer level microprobing, *IEEE 98CH36173 36th Annual International Reliability Physics Symposium,* Reno, Nevada.

Hartfield, C.D., Broz, J.J. and Moose, T.M. (2003) Mechanical and electrical characterization of an IC bond pad stack using a novel in-situ methodology, *Proceedings of 29th International Symposium for Testing and Failure Analysis, Santa Clara, CA.*

Hotchkiss, G. *et al.* (2001) Effects of probe damage on wire bonding integrity, *Proceedings of 51th Electronic Components and Technology Conference,* Las Vegas, May, 1175–1180.

King, J.A. (1998) *Materials Handbook for Hybrid Microelectronics*, Artech House, Boston, 1998.

Liu, Y., Desbiens, D., Irving, S. *et al.* (2005) Probe test failure analysis of bond pad over active structure by modeling and experiment, *Proceedings of 55th Electronic Components and Technology Conference,* Lake Buena Vista, Florida, May, 861–866.

Sauter, W. *et al.* (2003) Problems with wire bonding on probe marks and possible solutions, *Proceedings of 53th Electronic Components and Technology Conference,* New Orleans, May, 1350–1358.

Schmadlak, I. and Torsten, H. (2007) Simulation of wafer probing process considering probe dynamics, *Proceedings of EuroSimE 2007,* London, 278–282.

Timoshenko, S.P. (1951) *Theory of Elasticity*, McGraw-Hill, New York.

14

Power and Thermal Cycling, Solder Joint Fatigue Life

This chapter discusses the reliability issues of the electronic package in power cycling (PRCL), thermal cycling (TMCL), and solder joint fatigue life prediction. A nonlinear material model for die attach solder material is introduced, and an advanced finite element methodology is developed to target the coupled thermal-mechanical problem. For power cycling, a transient thermal analysis is presented first, which will show the thermal propagation as the power turns on and off, and the resulting temperature gradients. Then, cyclic transient temperature loads are applied to the model for nonlinear cycle stress analysis. For thermal cycling, the cycle temperature is directly applied to the model. A detail study regarding the solder joint stress loops and the fatigue with and without the underfill is conducted. The discussion and correlation with experimental data and actual process induced cracking are presented. This chapter also discusses the impact of process on the reliability of the package, such as the impact of the die attach process of power package, clean, and non-clean process for flip-chip solder joints.

14.1 Die Attach Process and Material Relations

A typical eutectic die attach process is shown in Figure 14.1.

A cavity collect head picks the die with metallization. A leadframe is held on the surface of heater block. The soft solder wire is then placed on the leadframe. The temperature of the heater block ramps up to the melting temperature of the solder wire that makes the wire fuse. Finally the cavity collect head moves the die to the melt solder in the leadframe. The die attach process, if in an improper assembly operation, might induce different bond line thicknesses or make the die tilt. Figure 14.2 shows a worst case of a tilted die example due to the die attach process and the 3D model (hiding the EMC) of a typical TO220 package.

In order to better describe the behavior of die attach material under a cyclic thermal load, the Anand model that includes a sinh function model and an evolution equation is introduced in Chapter 9, Section 9.3.1(Brown *et al.*, 1989; Maniatty *et al.*, 2001).

14.2 Power Cycling Modeling and Discussion

A simplified block diagram of the PRCL circuitry is shown in Figure 14.3. The device is powered by an external DC power supply. A resistor (R_L) is connected in series with the DUT. A control circuit monitors certain predefined parameters and actively participates in the power cycling process. These predefined parameters are the heating power, on/off time durations, and reference voltage (V_{ref}) which are manually set before running the test.

Modeling and Simulation for Microelectronic Packaging and Integration: Manufacturing, Reliability and Testing, Second Edition. Sheng Liu and Yong Liu.

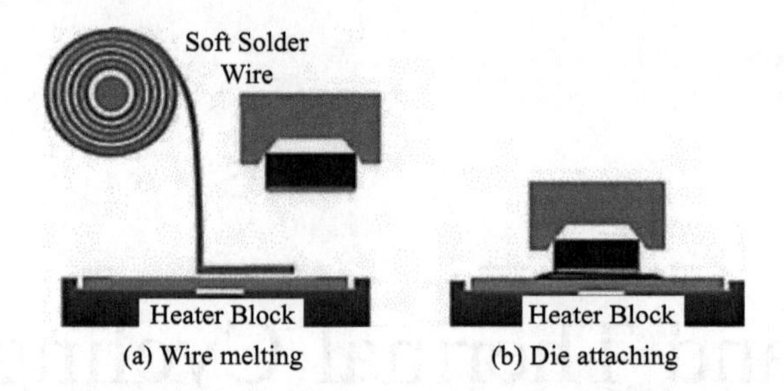

(a) Wire melting (b) Die attaching

Figure 14.1 Eutectic solder die attach process

Figure 14.4(a) shows the power cycling due to the power on 2 minutes and off 2 minutes. This will be the input for the package modeling. Because the heat source is caused by the current going though the die, which will induce the gradient of temperature distribution by on/off load condition in the package, which is different from the thermal cycle where the temperature will be applied as an uniform loading. Figure 14.4(b) shows test temperature cycle. Figure 14.5 shows the 3D model and mesh of the TO220 package. The power is applied in such a way that the junction temperature of the package in PRCL will be about 125 °C (Mid-STD-7500 Method 1037.2).

Figure 14.6 gives the temperature distribution of a typical PRCL, it shows the transient temperature change in a cycle from the thermal model. These transient temperature results in all nodes of the model being saved in a database which then is used as the temperature load to be applied to the thermal-mechanical

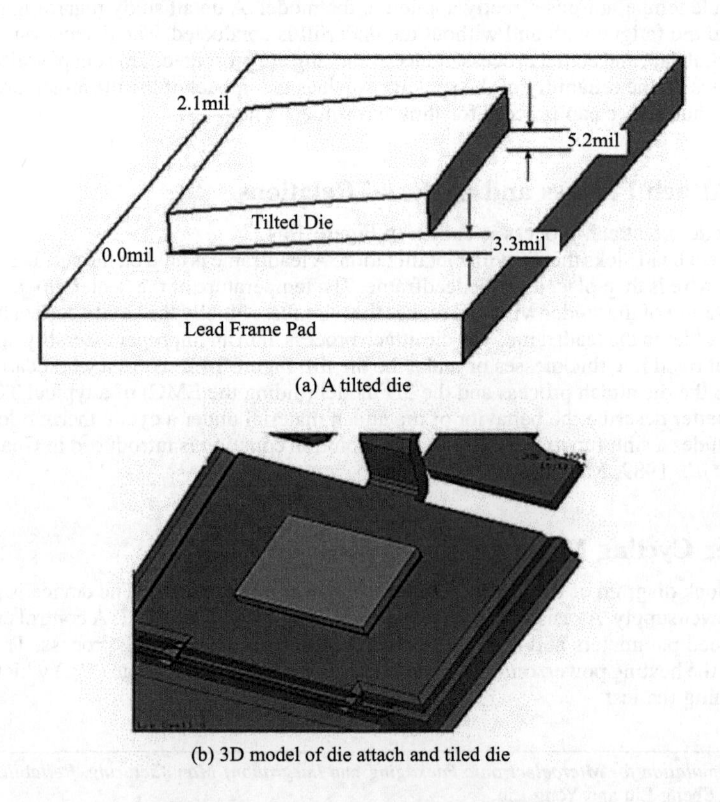

(a) A tilted die

(b) 3D model of die attach and tiled die

Figure 14.2 Tilted die due to die attach process

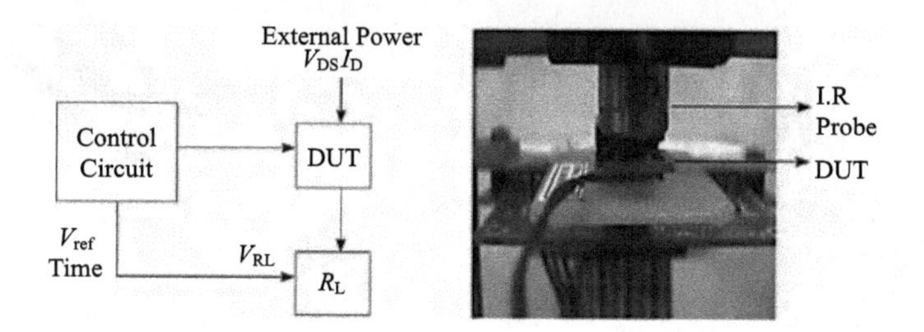

Figure 14.3 Simplified block diagram of PRCL circuitry

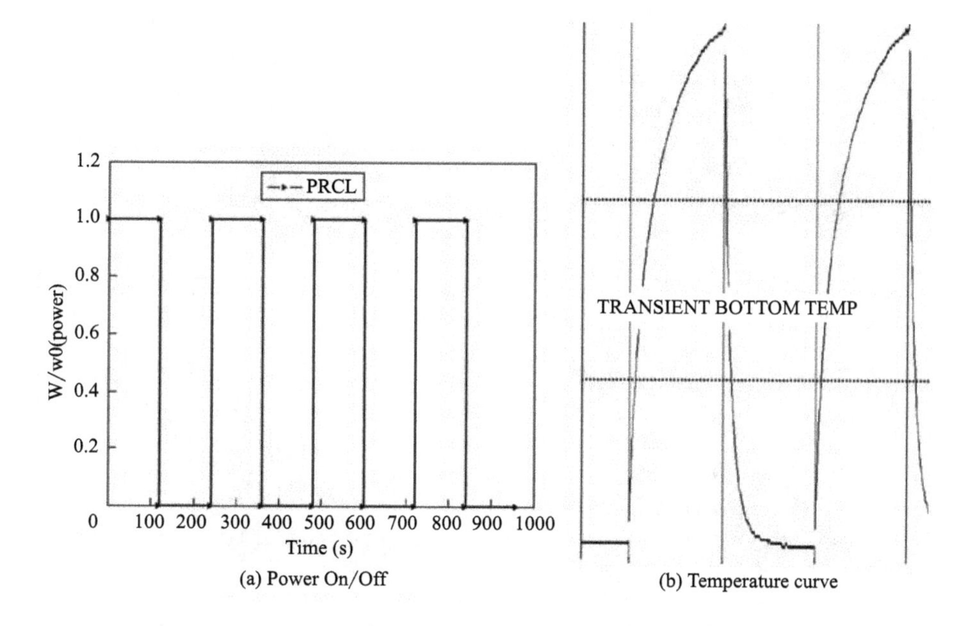

(a) Power On/Off (b) Temperature curve

Figure 14.4 Power cycles (2 minutes ON and 2 minutes OFF)

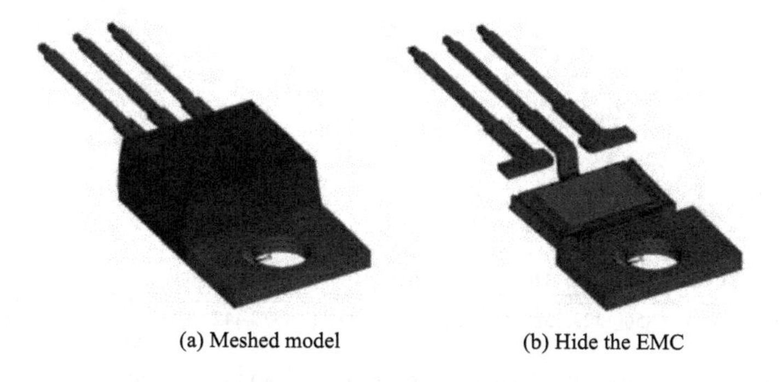

(a) Meshed model (b) Hide the EMC

Figure 14.5 TO 220 model and mesh

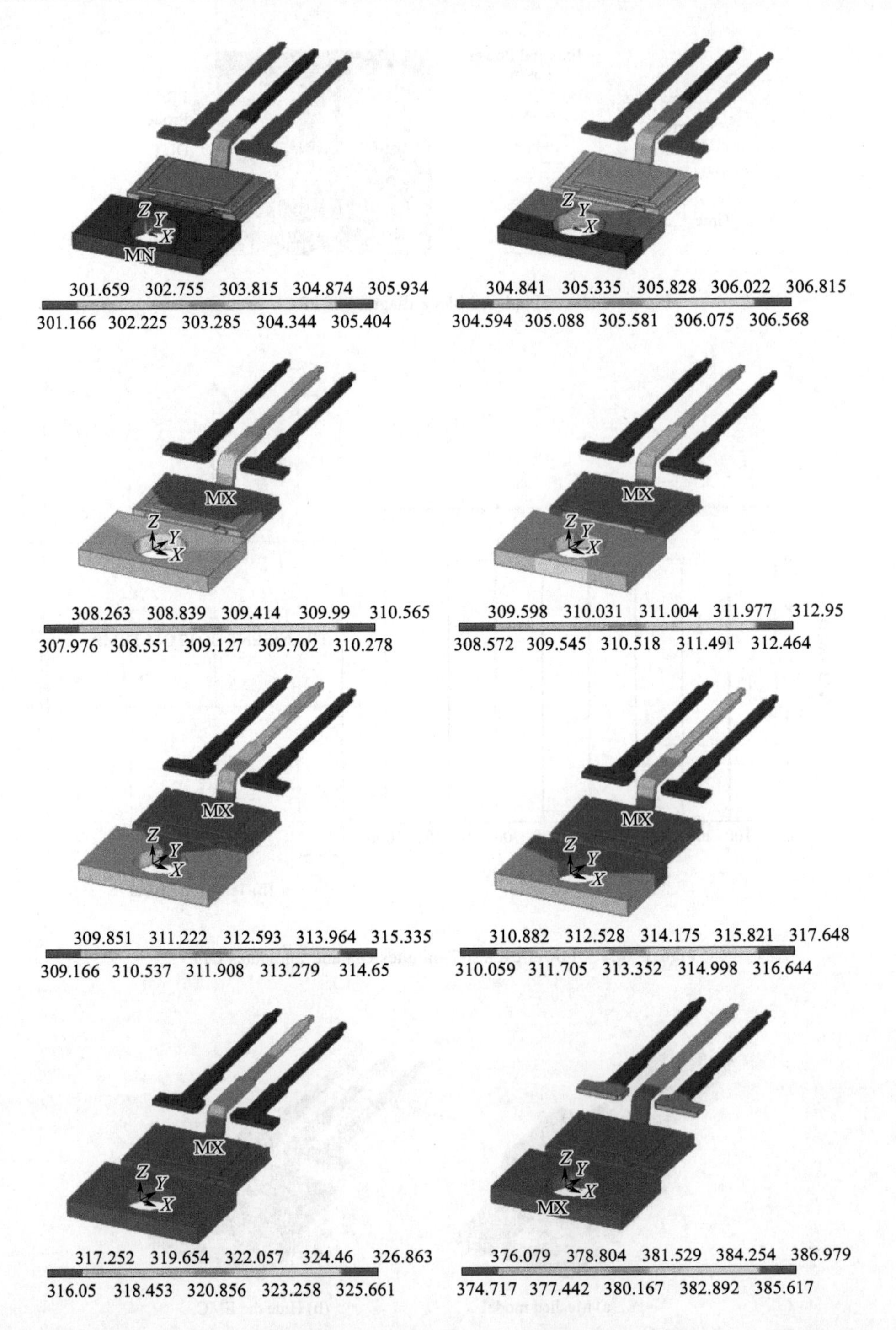

Figure 14.6 Transient temperature distribution in PRCL with die size 175 mil × 240 mil × 8 mil, unit in K *(Color version of this figure is available online)*

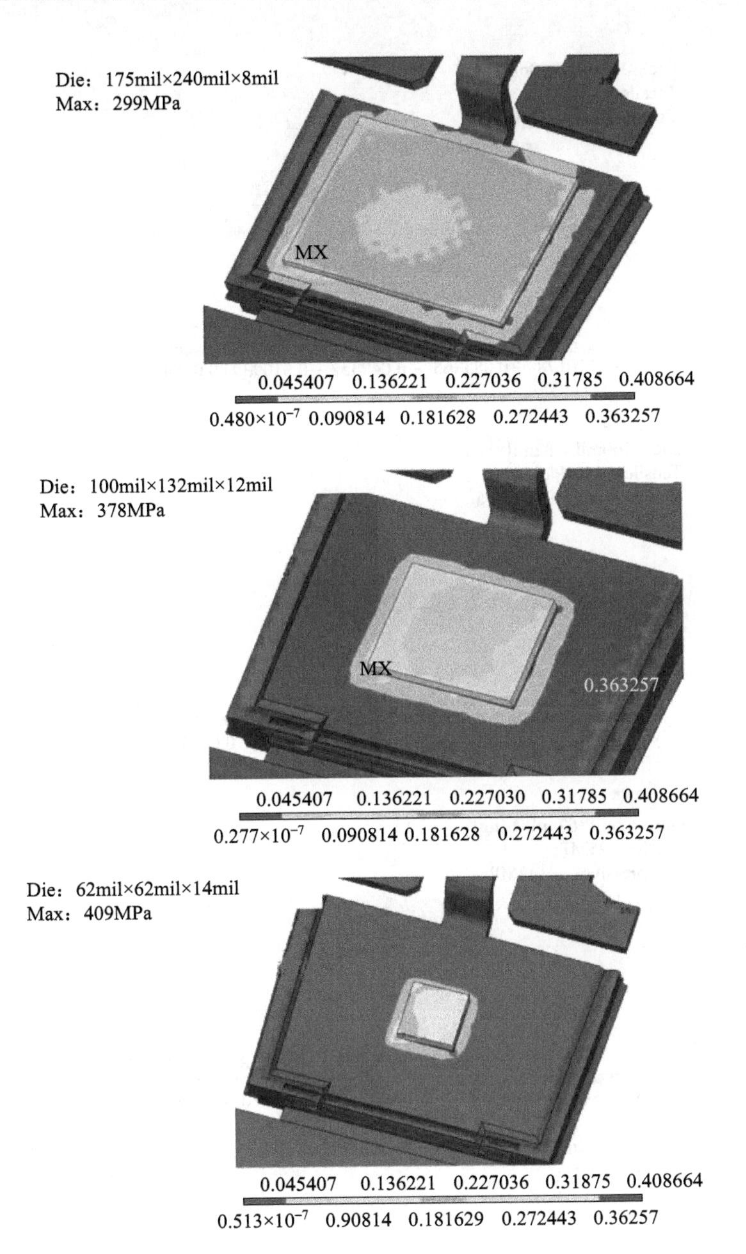

Figure 14.7 von-Mises stresses in three different dies when power is off for two minutes during PRCL *(Color version of this figure is available online)*

model for PRCL stress simulation. Four cycles are used in the simulation. The PRCL stress results are shown in Figures 14.7 to 14.10 and Tables 14.1 and 14.2. Figure 14.7 shows the von-Mises stress distribution inside the package (hiding the EMC) when power is off for two minutes, the maximum stress appears at the smallest case with largest die thickness. Figure 14.8 shows the first principal stress with different die size and thicknesses.

From Figure 14.8 it may be seen that the max stress locates at the point of the die that has contacted the pad due to tilt. Table 14.1 gives the comparison of first and third principal stress, maximum shear stress, and

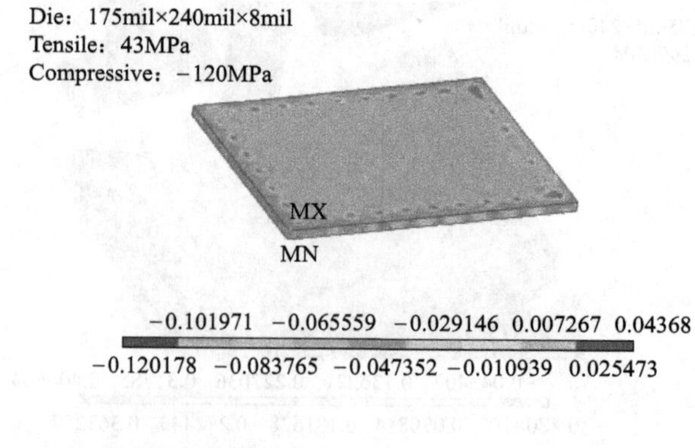

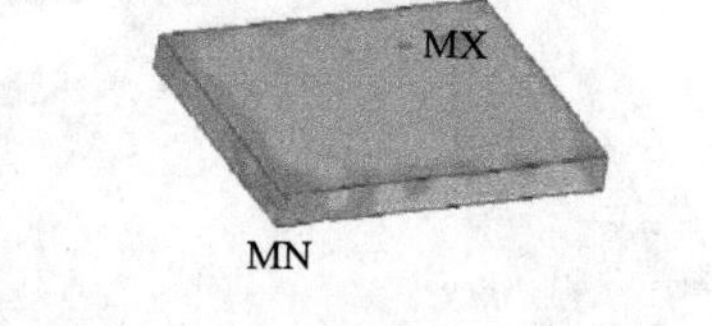

Figure 14.8 First principal stresses in three different dies when power is off for two minutes in PRCL *(Color version of this figure is available online)*

Table 14.1 Maximum die stress results when power is off for two minutes in PRCL. Reference temperature $T_g = 150\,°C$; T: Tensile stress; C: Compressive stress

Die Size	S_1 (MPa)	S_3 (MPa)	S_{xy} Max Shear Stress(MPa)	von-Mises Stress (MPa)
Die1: 175 mil × 240 mil × 8 mil	43(T) −120 (C)	−427 (C)	+94 −80	299
Die2: 100 mil × 132 mil × 12 mil	29 (T) −106 (C)	−538 (C)	+64 −110	378
Die 3: 62 mil × 62 mil × 14 mil	25(T) −142 (C)	−600 (C)	+46 −136	409

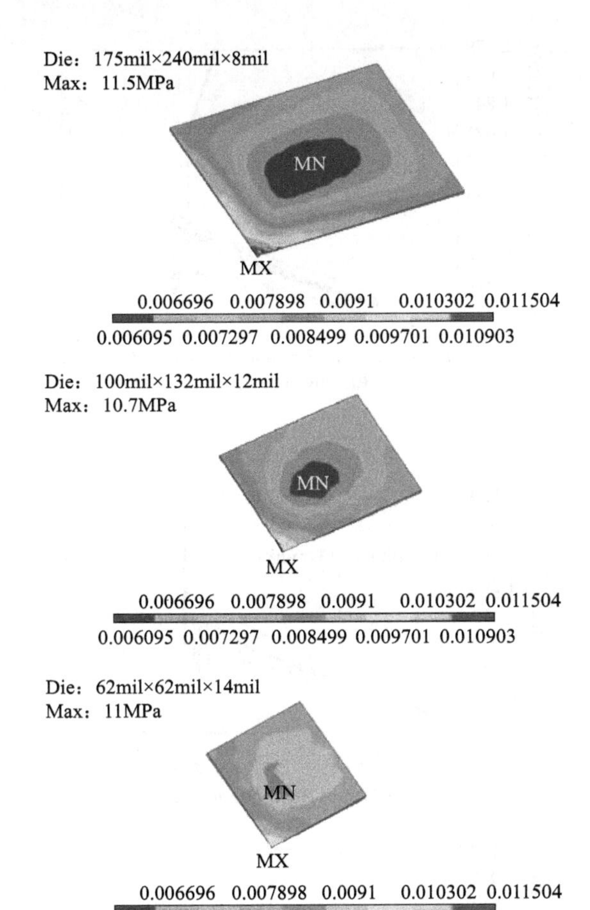

Figure 14.9 von-Mises stresses in die attach with three different dies when power is off for two minutes in PRCL *(Color version of this figure is available online)*

von-Mises stress. It shows that most of the die is in a compressive stress status, the greatest von-Mises stress, 3rd principal stress, and shear stress appear at the smallest size die with the biggest die thickness. Figures 14.8 to 14.9 show the von-Mises stress distribution of the die attach and the loops of effective stress-strain. Table 14.2 shows the comparison of effective strain, von-Mises stress, and the plastic strain energy density. The results show that the largest effective strain and stress appear in the die attach material for the largest die size. It locates at the point that the die has contacted the pad (BLT = 0.0) due to a tilted die. The tensile strength of die attach material 92.5Pb5Sn2.5Ag is about 30 MPa and failure strain is 0.4 based on vendor's test data. Although the die attach von-Mises stress is lower than the tensile strength, after many cycles if the effective strain is accumulated to a value that is higher than the failure strain, the die attach will fracture.

Table 14.2 Maximum die attach stress, strain and energy density when power is off for minutes in PRCL

Die Size	Max Effective Strain	Max Plastic Strain Energy Density (MPa)	von-Mises Stress (MPa)
Die1: 175 mil × 240 mil × 8 mil	0.0489	2.1	11.5
Die2: 100 mil × 132 mil × 12 mil	0.0439	2.2	10.7
Die3: 62 mil × 62 mil × 14 mil	0.0355	1.7	11.07

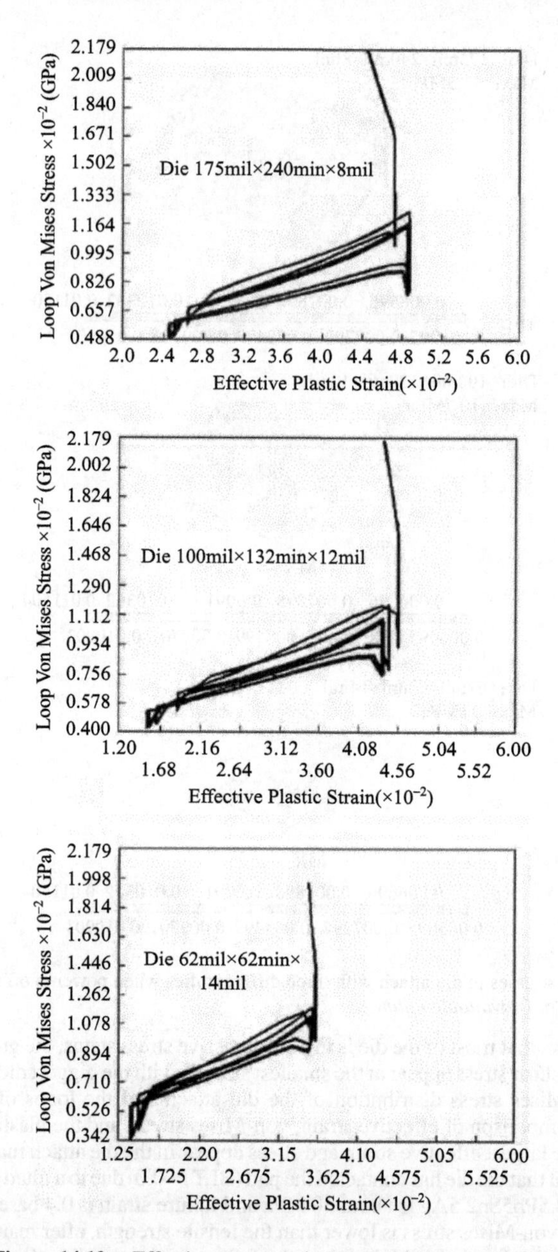

Figure 14.10 Effective stress-strain loops of die attach in PRCL

14.3 Thermal Cycling Modeling and Discussion

Thermal cycling is different from the power cycling. In thermal cycling, the thermal load temperature is considered as a uniform load and is applied to every part of the package, so there is no temperature gradient generated during thermal cycling. Figure 14.11 shows the temperature load cycles versus time.

The TMCL stress results are shown in Figures 14.12 to 14.16, Tables 14.3 to 14.5. Figure 14.12 shows the first principal stress with different die size and thickness. The results show that the maximum stress is

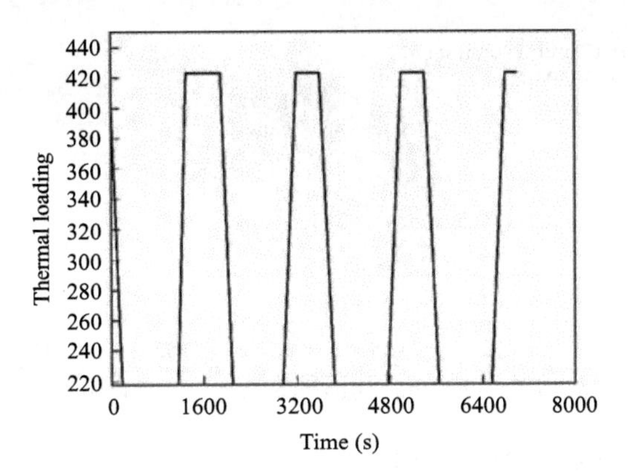

Figure 14.11 Temperature cycle. (−55 °C to 150 °C)

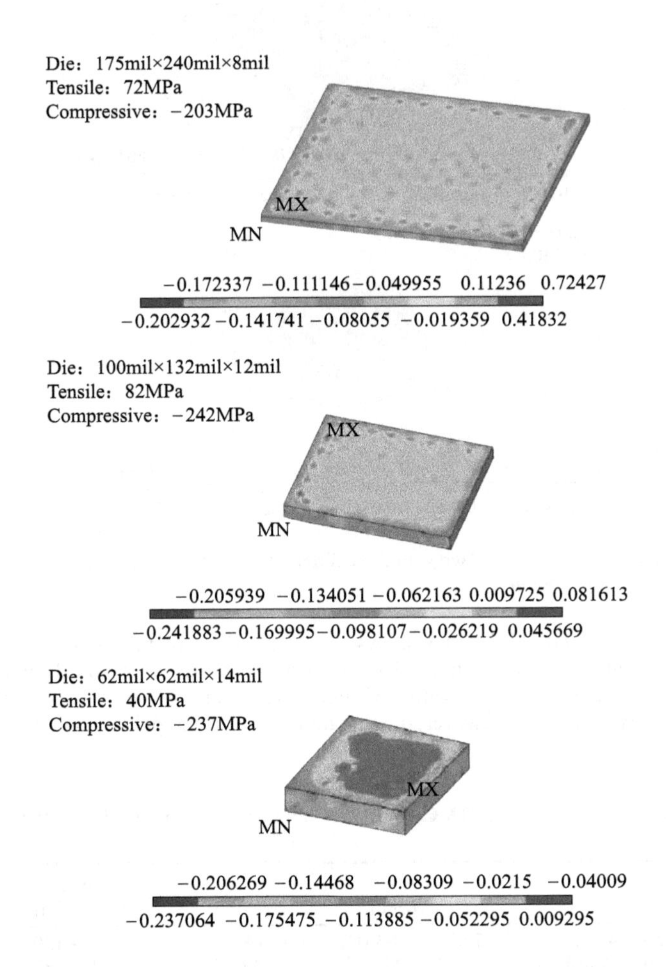

Figure 14.12 First principal stresses in three different dies when temperature ramps to −55 °C; T: Tensile; C: Compressive *(Color version of this figure is available online)*

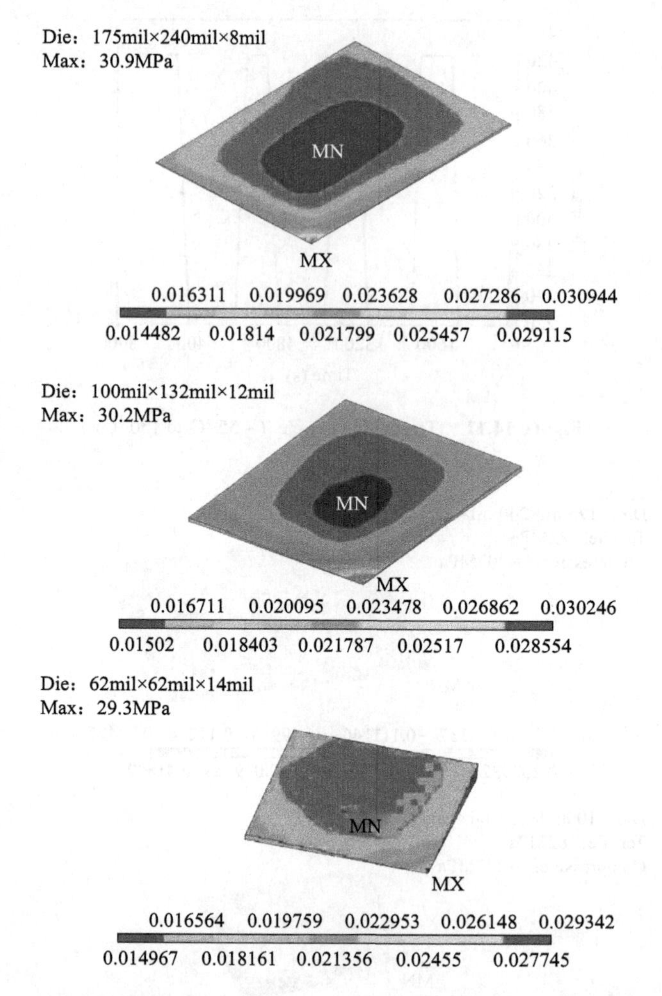

Figure 14.13 von-Mises stresses in die attach with three different dies when temperature ramps to −55 °C *(Color version of this figure is available online)*

located at the point BLT = 0.0. Table 14.3 shows the comparison of first and third principal stress, maximum shear stress, and von-Mises stress. The greatest von-Mises stress, third principal stress, and shear stress appear at the smallest size die with the biggest die thickness. This case is similar to the PRCL when power is off for two minutes. However, the maximum third principal stress in TMCL is 1004 MPa

Table 14.3 Maximum die stress results in TMCL at −55 °C reference temperature $T_g = 150$ °C; T: Tensile stress; C: Compressive stress

Die Size	S_1 (MPa)	S_3 (MPa)	S_{xy} Max Shear Stress (MPa)	von-Mises Stress (MPa)
Die 1: 175 mil × 240 mil × 8 mil	72(T) −203 (C)	−708 (C)	+157 −130	497
Die 2: 100 mil × 132 mil × 12 mil	82 (T) −242 (C)	−947 (C)	+173 −187	646
Die 3: 62 mil × 62 mil × 14 mil	40(T) −237 (C)	−1004 (C)	+77 −226	683

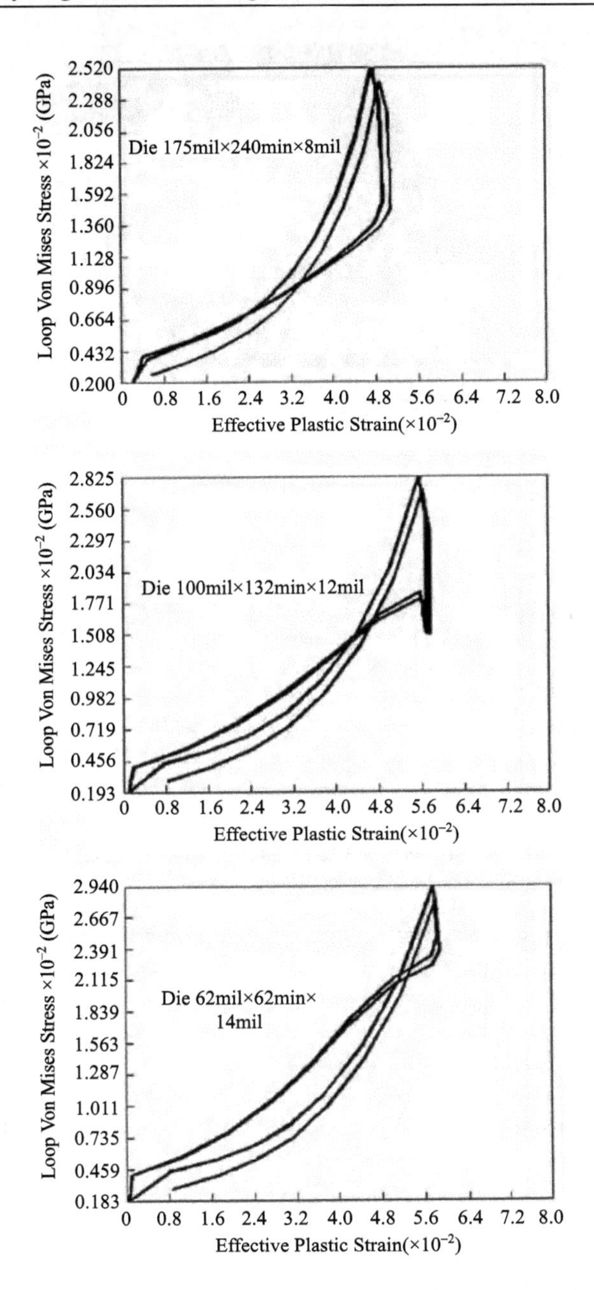

Figure 14.14 Effective stress-strain loops of die attach in TMCL

that is about 1.67 times greater than PRCL and is lower than the compressive strength of silicon (about 1100 MPa, see Silicon Properties, Engineering Design Research Laboratory 2005). Figures 14.13 and 14.14 show the von-Mises stress distribution of die attach and the loops of effective stress-strain in TMCL. Table 14.3 shows the comparison of effective strain, von-Mises stress, and the plastic strain energy density when temperature ramps to −55 °C. The results show that max von-Mises stress appears at

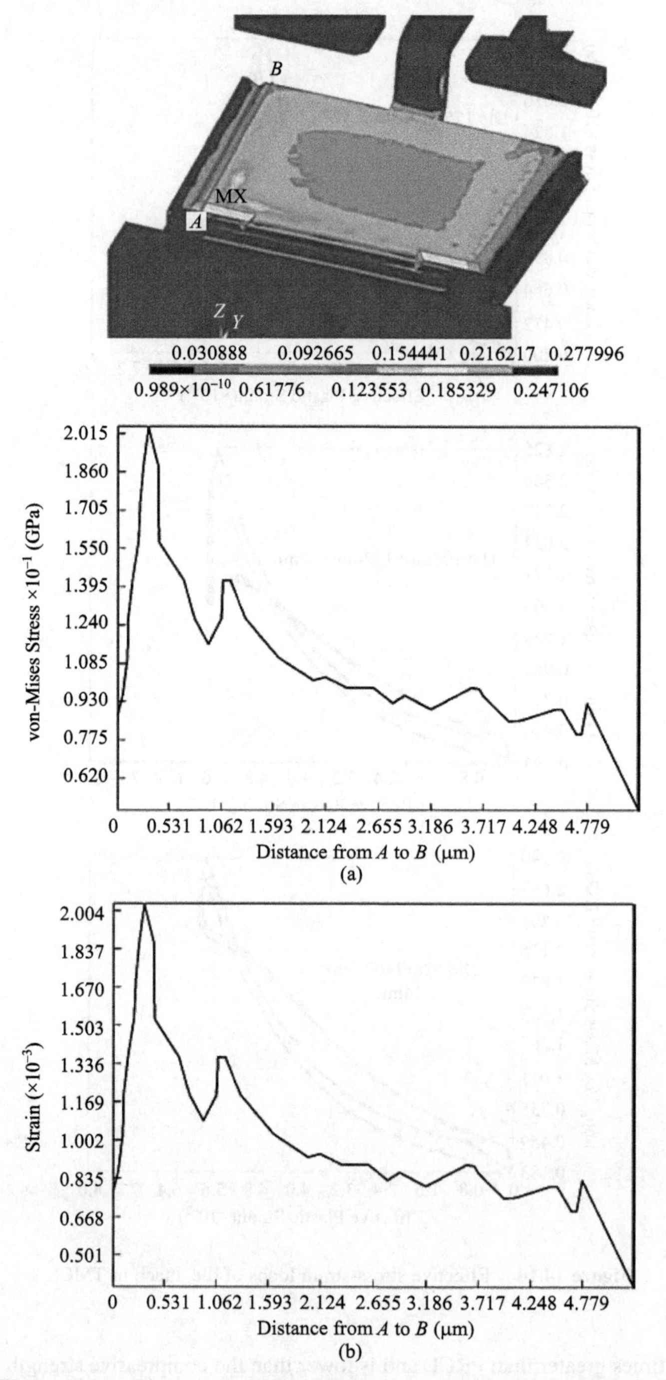

Figure 14.15 Stress concentration induced at pad due to BLT = 0.0 in TMCL

Table 14.4 Maximum die attach stress, strain and energy density in TMCL at −55 °C. Tensile strength: 30 MPa; Failure strain: 0.4

Die Size	Max Effective Strain	Max Plastic Strain Energy Density (MPa)	von-Mises Stress (MPa)
Die 1: 175 mil × 240 mil × 8 mil	0.077	5.6	30.9
Die 2: 100 mil × 132 mil × 12 mil	0.064	5.1	30.2
Die 3: 62 mil × 62 mil × 14 mil	0.058	4.6	29.3

Table 14.5 Maximum die stress results with different BLT in TMCL at −55 °C reference temperature $T_g = 150\,°C$; T: Tensile stress; C: Compressive stress; Die size: 100 mil × 132 mil × 12 mil without die tilt

BLT (mil)	S_1 (MPa)	S_3 (MPa)	S_{xy} Max Shear Stress (MPa)	von-Mises Stress (MPa)
0.0	13 (T) −184 (C)	−717 (C)	+123 −142	475
2.1	21 (T) −109 (C)	−606 (C)	+113 −156	452
3.3	38 (T) −104 (C)	−613 (C)	+114 −160	457

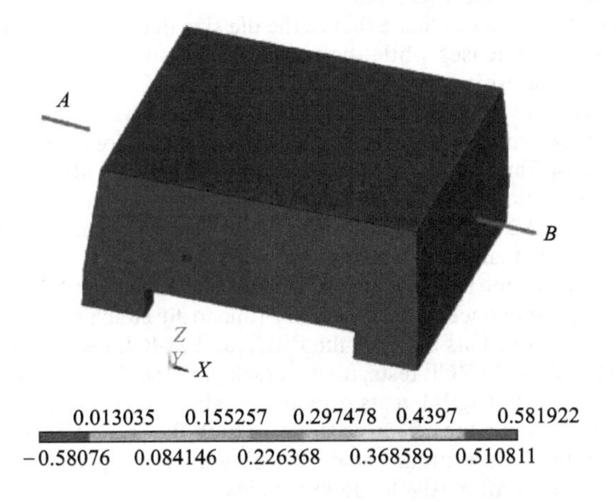

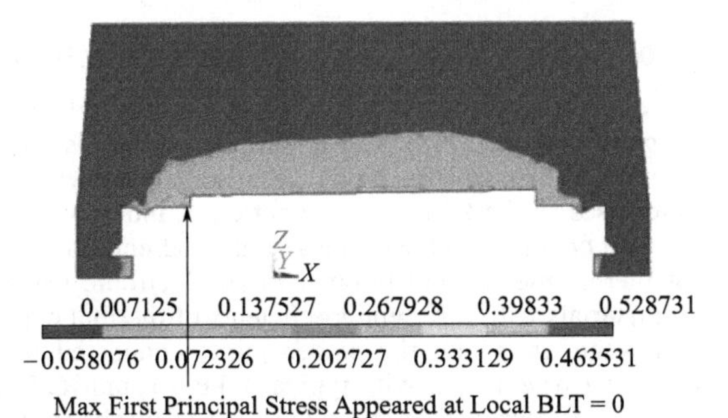

Figure 14.16 First principal stress of EMC in TMCL

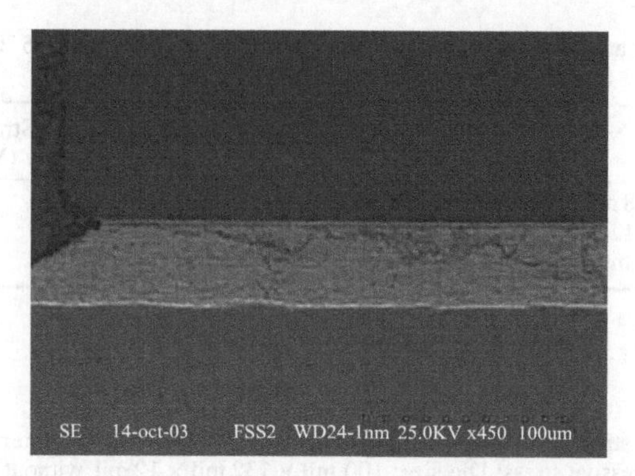

Figure 14.17 Die attach fracture found after 10000 PRCLs

the point of BLT = 0.0. The max die attach von-Mises stress in TMCL is about 2.68 times greater than PRCL. All of the die attach von-Mises stresses in three die cases have reached the die attach tensile strength, 30 MPa. However, the three max effective strains of die attach are much lower than the material failure strain (0.4 based on vendor's test data).

The results in Table 14.4 further indicate that as the die size decreases and die thickness increases, the max principal stress on die increases while the max effective strain reduces.

Figure 14.15(b) show the stress impact of a tilted die on pad and EMC; it can be seen that the stress peaks sharply at around BLT = 0.0 as well as highest principal stress in EMC. Table 14.5 gives the max stresses of a die(100 mil×132 mil×12 mil)with different die attach thickness for an ideal die attach process with die tilt. By comparing Table 14.5 and Table 14.3 (die 2), it may be seen that all of the stresses in Table 14.5 are less than the stress results in die 2 of Table 14.3. The maximum first principal stress S_1 of Table 14.5 (with BLT = 2.1 mil) is less than 2.2 times than maximum S_1 in Table 14.3, and the third principal stress S_3 of Table 14.5 is less 1.56 times than S_3 in Table 14.3.

Die crack is a big concern for the product due to the tilted die in PRCL and TMCL. Based on the modeling results, we have confidence that the tilted die (due to die attach process) will not induce die crack even in the worst die tilted case. This matches the PRCL and TMCL test results from our manufacturing site that showed after PRCL and PRCL tests, no die crack is found. The von-Mises stress of die attach in TMCL has reached its yield strength but its effective strain is still much lower than its failure strain. However, after many cycles once the effective strain of both PRCL and TMCL is accumulated to some value that is higher than the failure strain, the die attach will show appear fracture. Figure 14.17 shows such fracture in the die attach after 10000 power cycles.

14.4 Methodology of Solder Joint Fatigue Life Prediction

The long-term reliability of an electronic system is often directly related to the lifetime of solder interconnects in the package (Liu *et al.*, 2008). The solder interconnect is no longer simply an electrical conductor but is also the structural material that holds the package together. The importance of reliability of solder interconnects increases as the trend in the electronics industry moves toward surface mount technology, smaller joints, and finer pitch. One of the serious challenges for solder joint reliability is the failure that is a result of thermo-mechanical fatigue. In an electronic package, the solder is usually constrained between two materials with different coefficients of thermal expansion (CTE). The typical example is a flip-chip package, in which the solder joints connected by silicon chip (with CTE 2.4–2.8 ppm/°C) and substrate (for example with typical CTE 16 ppm/°C in plane and 60 ppm/°C out of the plane for the FR-4 substrate). During thermal cycling caused by either ambient temperature changes or by heat dissipation from the integrated circuit devices in the package, the solder joints must accommodate displacement due to a mismatch of coefficients of thermal expansion (CTE) between these two materials. The thermally induced displacement results in a complex stress and strain distribution in the solder joints. Fatigue cracks generally initiate at the bonded edges and may propagate along the interface or curve into the solder, leading to full failure of the solder joint.

Therefore, the approaches to thermal fatigue life prediction are extremely important for the reliability assessment of the solder joint. The most commonly used model to describe low-cycle fatigue of solders is Coffin–Manson relationship. It is an inelastic strain amplitude based method. The simple equation is given by Equations 4.1 to 4.3 in Chapter 4, Section 4.2.1.

When metals are cycled at a homologous temperature above 0.5, dwell time, strain rate, and temperature significantly affect lifetime. The reason is that in this temperature range, which for solders can extend below room temperature, creep occurs during cycling. In order to reflect the creep-fatigue interaction such as the effect of cycle frequency, the Coffin–Manson equation can be modified to include a term that accounts for the cycling frequency.

In high-temperature, low-cycle fatigue, either a slower strain rate or a dwell introduced increases creep deformation that reduces lifetime. Because both prolong the cyclic period, it is customary to describe lifetime as decreasing with decreasing frequency. Several experiments (Ren *et al.*, 1998, 2000) have found as exponential relationship between lifetime with frequency or dwell, in both isothermal and thermal cycling. The earliest proposed C–M modification came in 1969 (Norris and Landzberg, 1969). Based on mechanical test data on bulk solders, the following expression was proposed:

$$N_{\mathrm{f}} = (A/\varepsilon_{\mathrm{p}})^{1.9}(f^{1/3})\mathrm{e}^{0.0123/kT} \tag{14.1}$$

where f is the cyclic frequency; T is maximum temperature in degrees Kelvin; and k is Boltzmann's constant; ε_{p} the plastic strain amplitude, A material constant.

Based on evidence (Wild, 1974; Fox *et al.*, 1985) that the strain exponent in Equation 14.1 diverges significantly from 1.9 when solder is cycled very slowly, an alternative C-M adaptation has been proposed (Engelmaier, 1983) wherein the strain exponent is adjusted to take into account frequency and temperature effects. Curve fitting to mechanical cycle data of (Wild, 1974) for eutectic lead-tin solder, it was found that:

$$N_{\mathrm{f}} = \left(A/\varepsilon_{\mathrm{p}}\right)^{1/0.442+0.0006T-0.0174\ \ln(1+f)} \tag{14.2}$$

where, T is the mean temperature of the cycle.

Equations 14.1 to 14.2 are very widely utilized in the technical community. The popularity of these models traces to their simplicity and their easy compatibility with finite element modeling (FEM).

The energy method has long been known as an attractive alternative to the Coffin–Manson adaptations for solder interconnections. It incorporates solder relaxation and takes into account the structural influence of the entire package. Correlation with experimental data supports the basic precept of cumulative energy. A similar fashion of Coffin–Manson equation in terms of inelastic hysteresis energy density is also used in energy approach, which can be written as:

$$N_{\mathrm{f}} = C^{\mathrm{E}}\left(W^{\mathrm{I}}\right)^{-\delta} \tag{14.3}$$

where δ and CE are the material constants. As Coffin–Manson equation, it is also easy to be correlated with finite element modeling (FEM).

In this section, the Coffin–Manson equation is selected to predict the fatigue life coupled with the finite element analysis.

14.5 Fatigue Life Prediction of a Stack Die Flip-Chip on Silicon (FSBGA)

The stack die Flip-chip-on-Silicon BGA (FSBGA) Packages, either underfilled (Variation 3) or non-underfilled (Variation 1) in Chapter 12, Section 12.2, are selected to perform the analysis approach of fatigue life prediction. Since the non-woven substrate material is not recommended for the FSBGA

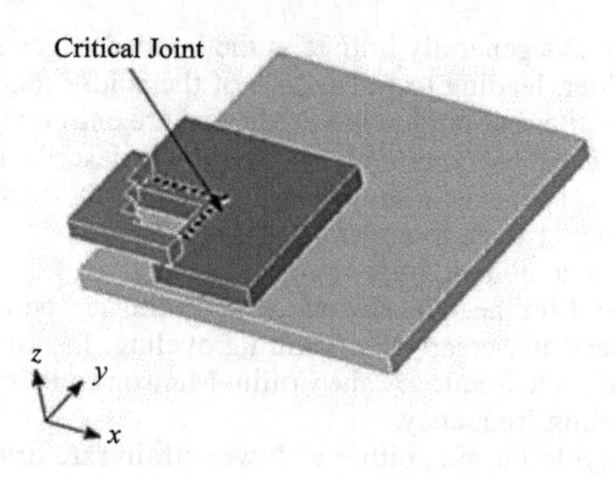

Figure 14.18 Critical solder joint location

architecture, here, both BGA substrate and motherboard are assumed to be FR-4 material. The study focuses on the critical solder joint, which is in the flip-solder joints and located in the corner of the silicon substrate as shown in Figure 14.18. In order to investigate the mesh dependence, the FEM model with fine mesh around the critical solder joint is generated. Figures 14.19 and 14.20 illustrate the local fine FEM mesh and the local FEM mesh used in the previous structure study, respectively. Both models are analyzed under thermal load from 23 °C to 125 °C. Comparison of the results is given in Table 14.6. The difference of maximum von-Mises stress is only 0.4%. It seems that there is no significant difference between the results from both meshes. That means the results are relatively independent in mesh. In succeeding section, all analyses are conducted by using the fine mesh shown in Figure 14.7.

To consider the nonlinear interaction during thermal cyclic loading, the strain rate dependent plasticity model in the previous section, is employed for both solder joints. The temperature profile used for thermal cyclic loading is shown in Figure 14.21, which is −40 °C to 125 °C, single zone chamber, 15 minute ramps and dwells, 1 hour cycle. This is one of the thermal fatigue test procedures used for FSBGA package. The first temperature raise step (Step 1: room temperature to 125 °C) and two complete cycles (Step 2 to Step 9) are executed in the FEM simulation for underfilled FSBGA Package. The first five steps (Step 1 to Step 5) are executed for a non-underfilled FSBGA Package.

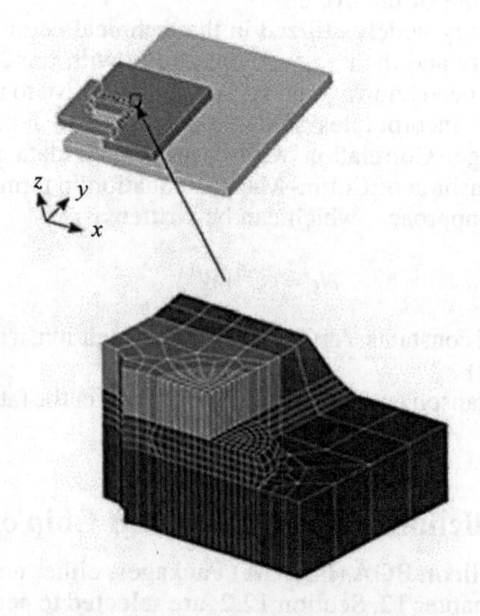

Figure 14.19 Local fine FEM mesh

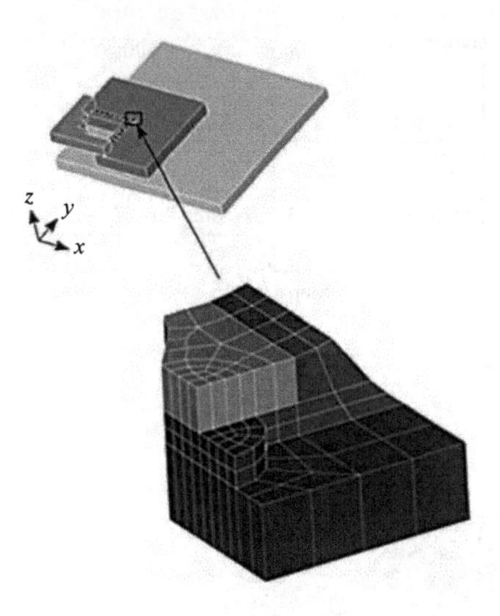

Figure 14.20 Local FEM mesh used in previous structure study

The responses of the solder interconnect for underfilled FSBGA Packages are shown in Figures 14.22 and 14.23 by the contours of accumulated equivalent inelastic strain on the critical solder joint at different cyclic loading steps. At the end of Step 1, the maximum equivalent inelastic strain (PEEQ) is along the diagonal plane at the lower right-hand corner as shown in Figure 14.22. But after one cycle, the maximum equivalent inelastic strain (PEEQ) shifts to the upper left-hand corner, location *A* as indicated in Figure 14.23, then keeps at the same location afterwards, Figure 14.24. Therefore, location *A* is considered as the critical point of the solder joint for underfilled FSBGA package in terms of fatigue

Table 14.6 Comparison of the results from different meshes

	Fine Mesh	Coarse Mesh	Difference
Max. von-Mises (MPa)	19.06	18.99	0.4%
Max. Equivalent Plastic Strain	6.108×10^{-3}	5.330×10^{-3}	12.7%

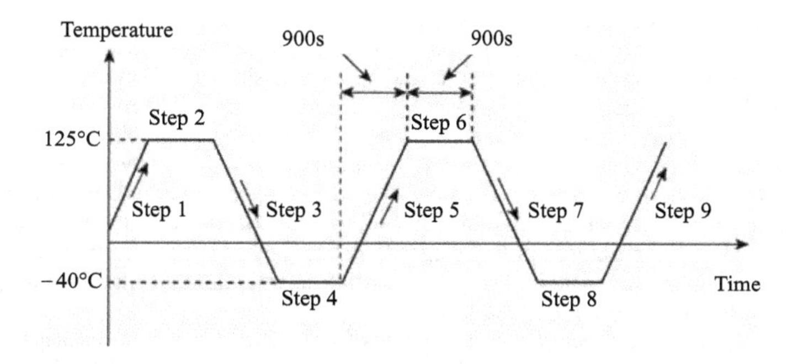

Figure 14.21 Temperature profile used for thermal cyclic loading

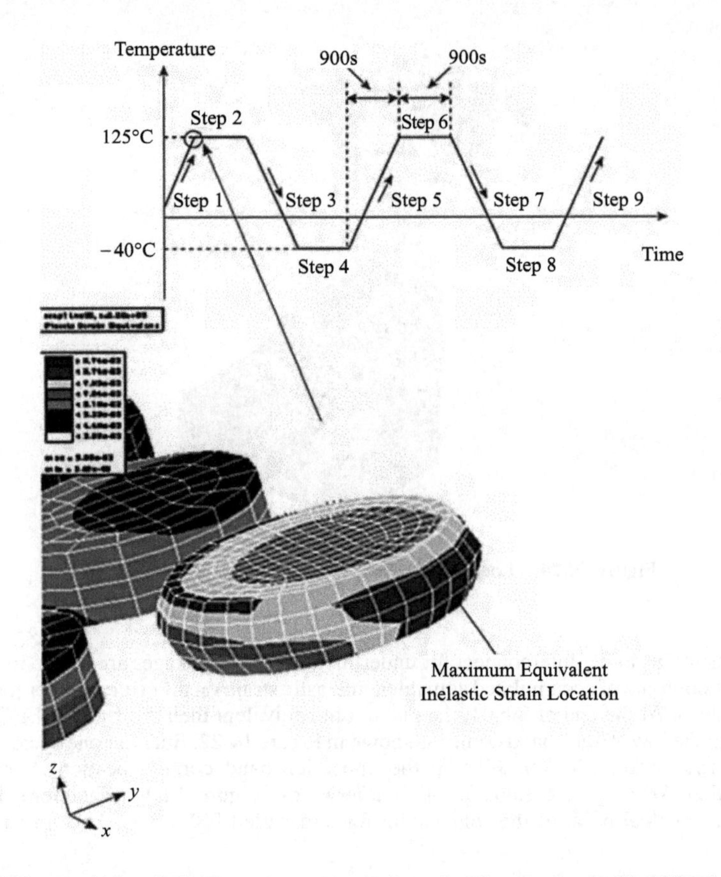

Figure 14.22 Equivalent inelastic strain contour at the end of Step 1 of underfilled FSBGA packages

life. Figure 14.25 presents the contour of equivalent inelastic strain on the critical solder joint for non-underfilled FSBGA package at the end of Step 5. The critical point of the solder joint, where the maximum equivalent inelastic strain occurs, is at location *B*. It is different from that of the underfilled FSBGA package due to different constrain around the solder joint. Thus, solder joint damage should initiate at these locations.

Figure 14.26 presents the curve of shear stress-shear strain in the critical solder joint at the location *B* where the maximum equivalent inelastic strain occurs during thermal cyclic loading for non-underfilled FSBGA package. The hysteresis curve is very similar to nonlinear characterization measured experimentally by solder joint specimens subject to thermal cycling as shown in Figure 14.27, which is taken from (Lau and Pao, 1997). During the heating cycle (a–b), inelastic strain accumulates, but because of the softening and creeping of the solder with increasing temperature, stress quickly reaches its maximum value, then decreases with the increase of strain. On the cooling region (c–d), the solder hardens when the temperature decreases, thus the characteristic negative stiffening is produced. Both time-independent plastic flow and time-dependent stress relaxation as well as creep occur in these two regions. During the high temperature and low temperature holding stages (b–c and d–a), it is obvious that the time-dependent stress relaxation or creep occurs. The inelastic shear strain is accumulated mainly during temperature ramp stages, that is, heating and cooling stages.

Since the deformation mode of solder joints changes from shear dominated deformation in the non-underfilled FSBGA package to a complicated 3D deformation state, the shape of the hysteresis curve for

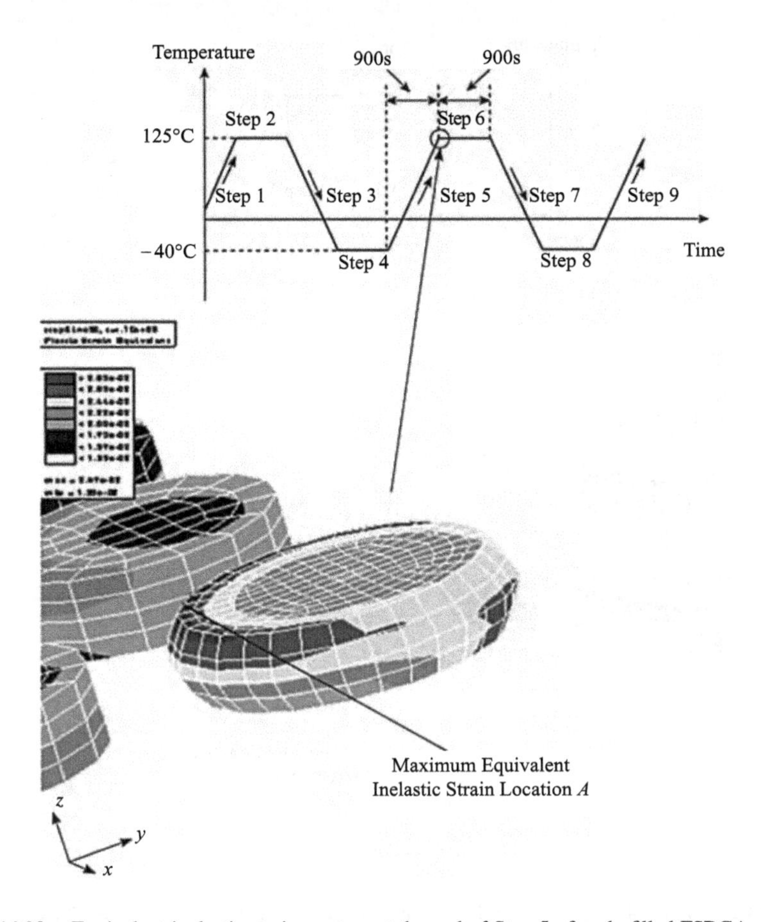

Figure 14.23 Equivalent inelastic strain contour at the end of Step 5 of underfilled FSBGA packages

the underfilled FSBGA package shows a small difference from that of the non-underfilled FSBGA package. However, it is noted that there still exists the time-dependent stress relaxation or creep. Furthermore, the stress/strain response tends to be stable rapidly. After the first heating step, the stress/strain curves of the succeeding two cycles are very alike.

The maximum equivalent inelastic strains in solder interconnection for the underfilled FSBGA package and non-underfilled FSBGA package are listed in Table 14.7. Then, the inelastic strain ranges are able to be calculated. The typical characterization of accumulated equivalent inelastic strain with temperature is shown in Figure 14.28, where the method used to calculate the equivalent inelastic strain range during temperature cycling is also indicated. It should be noted that the inelastic strain is accumulated twice per cycle.

Once the equivalent inelastic strain range is determined, thermal fatigue life can be estimated by Coffin–Manson relation. The results are given in Table 14.8. Since the inelastic strain range in the solder joint of underfill FSBGA under thermal cyclic loading is considerably reduced due to the underfill enhancement mechanisms, the thermal fatigue life is increased tremendously. It is one order of magnitude higher than that of non-underfilled FSBGA. This phenomenon is well correlated with the experiment results.

Therefore, the finite element analysis strongly supports the FSBGA package design in terms of reliability. It is recommended to use the underfill to be able to effectively prolong the fatigue life.

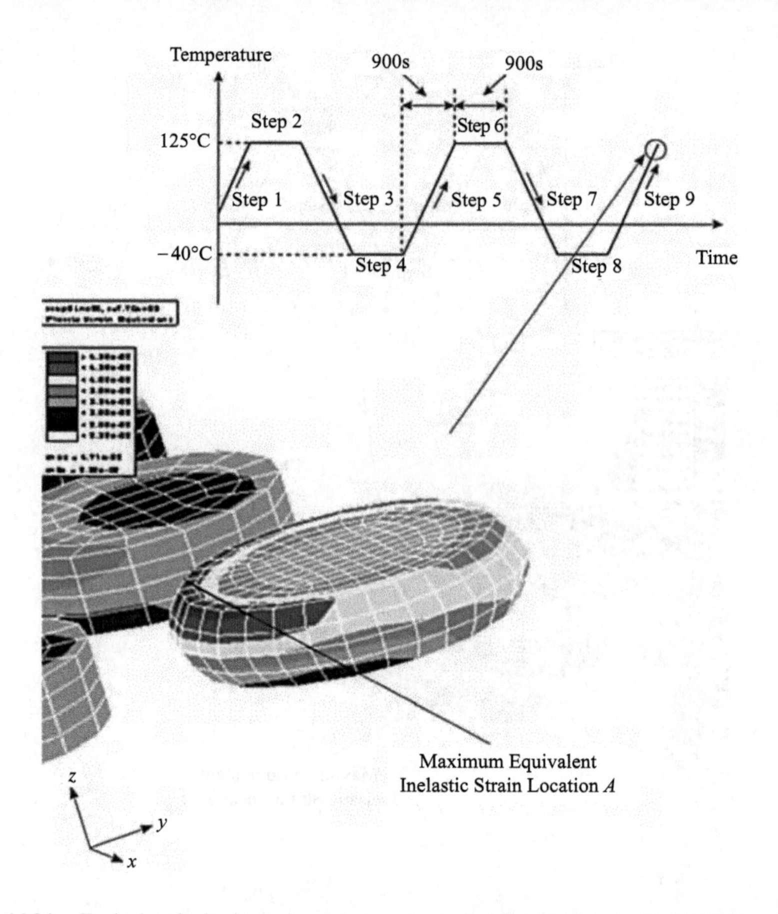

Figure 14.24 Equivalent inelastic strain contour at the end of Step 9 of underfilled FSBGA packages

Table 14.7 Maximum equivalent inelastic strains

	Maximum Equivalent Inelastic Strain	
Time	Underfilled FSBGA Location *A*	Non-underfilled FSBGA Location *B*
End of Step 1	8.648×10^{-3}	3.203×10^{-2}
End of Step 5	2.883×10^{-2}	1.086×10^{-1}
End of Step 9	4.719×10^{-2}	N/A

Table 14.8 Fatigue life of FSBGA packages

	Underfilled FSBGA Package	Non-underfilled FSBGA Package
Inelastic Strain Range	0.0101	0.0383
Fatigue Life (cycles)	3599	264

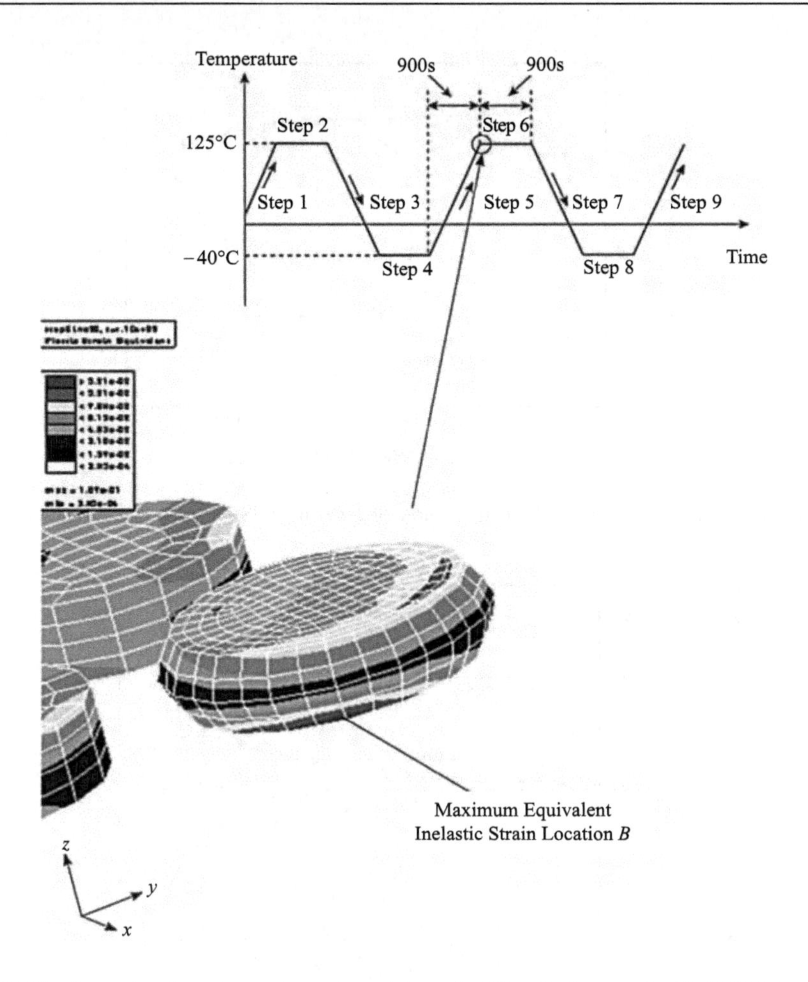

Figure 14.25 Equivalent inelastic strain contour at the end of Step 5 of non-underfilled FSBGA packages

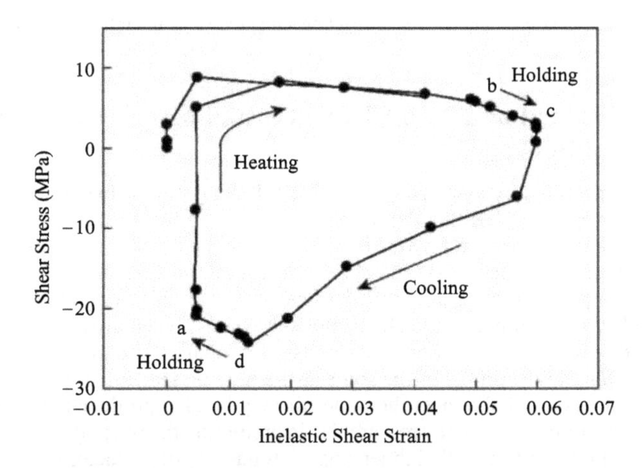

Figure 14.26 Shear stress-shear strain curve of non-underfilled FSBGA package at location *B*

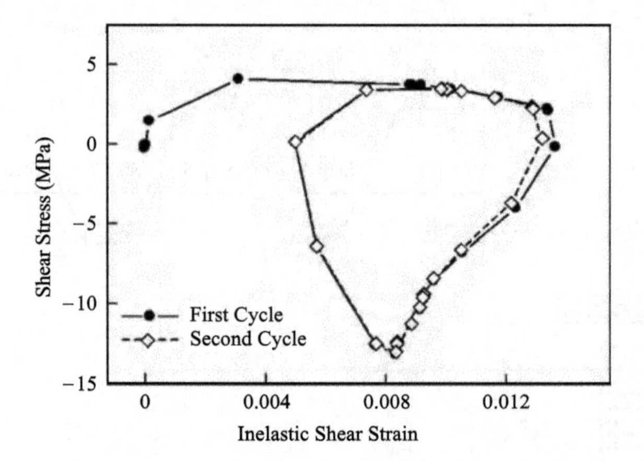

Figure 14.27 Shear stress-shear strain curve of underfilled FSBGA package at location *A*

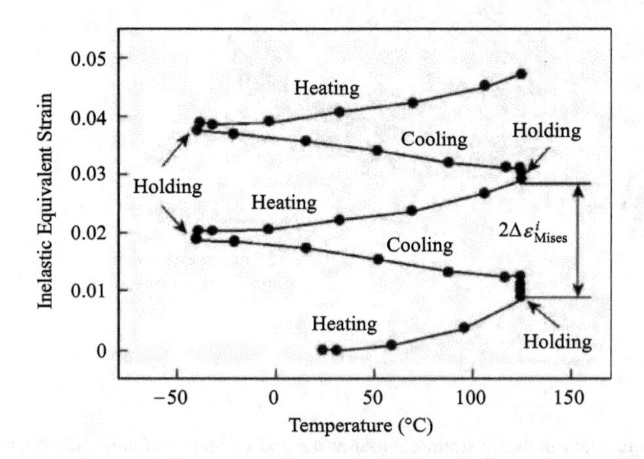

Figure 14.28 Equivalent inelastic strain via thermal cycling temperature for underfilled FSBGA package at location *A*

14.6 Effect of Cleaned and Non-Cleaned Situations on the Reliability of Flip-Chip Packages

Most flip-chips are underfilled by dispensing the underfill along one or two adjacent sides of the chip. Capillary flow then drives the underfill under the chip to fill the chip/substrate gap. However, it is hard to find perfectly filled layers for any real world samples without cracks, impurities, voids, inclusions, delaminations, and other defects in the corners of the solder joints and the underfill. Many factors can play a role in causing the imperfections of underfill. For instance, even if through the washing process after solder reflow the underfill sometimes cannot be completely filled in the corners of the solder balls because contamination like flux residues subsist on those parts (non-cleaned situation) which are difficult to be cleaned. Thus, the voids in the corners of the solder balls may be generated or the delaminations along the interfaces where the solder balls meet the underfill may be also produced. Nonetheless, a common practice which is often employed by many researchers in predicting the warpage, the stresses (strains), and the fatigue life of the solder joints directly influencing the reliability of packages is to assume perfect layers of generic underfill materials (cleaned situation). The effect of process-induced defects like the contamination of the flux residues has not been extensively studied and understood. Consequently, there is the concern that the imperfection (non-cleaned situation) of real underfill layers caused by the contamination of the flux residues in the corners of the solder joints may decrease mechanical stability, lose mechanical integrity, and shorten the life time of flip-chip packages to a level well below the predictions of the known simulation. Therefore, the increased demand for low cost, fine pitch, high performance, and high reliability of flip-chip packages with the advent of new materials and processes requires a better understanding of the effect of cleaned and non-cleaned situations on the reliability of flip-chip packages.

Therefore, the focus of the work in this section is to understand and quantify the impact of the non-cleaned situation caused by the flux residues in the corners of the solder joints on the thermomechanical reliability of the flip-chip solder system. It is our intention to understand whether the imperfection of the underfill in the corners of the solder joints accelerates the failure of the flip-chip solder interconnects.

14.6.1 Finite Element Models for the Clean and Non-Clean Cases

Assume the flip-chip package specimens satisfy the symmetric condition. Therefore, it can be simplified by a 2D symmetric FE model. The specimen is idealized to a finite element mesh shown in Figure 14.29. More fine mesh is used.

In order to simulate the non-cleaned situation caused by the flux residues in the corners of the solder joints, three non-cleaned FE models are considered. These include (1) gap model indicating that there are some gaps along the interfaces between the solder joints and the underfill where the underfill cannot reach the solder joints due to the existence of the flux residues (Figure 14.30(a)); (2) void model indicating that there are some voids in the corners of the solder joints (Figure 14.30(b)); and (3) contact model indicating that though the underfill can reach the solder joints, they are not bonded together and can slide along their interfaces arbitrarily (Figure 14.30(c)).

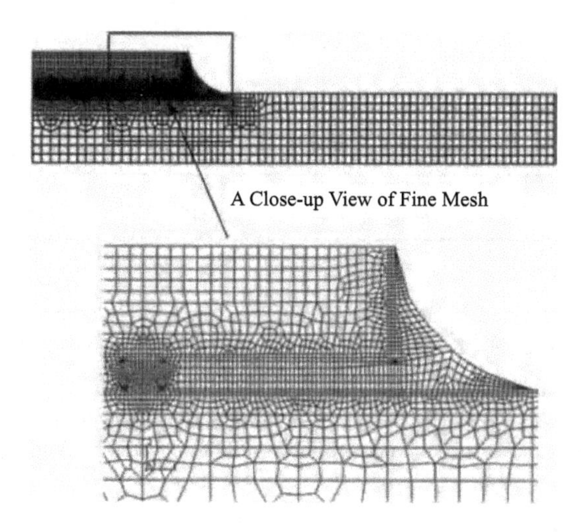

Figure 14.29 Finite element mesh for flip-chip package

14.6.2 Model Evaluation

In order to evaluate the finite element model, the results from the finite element analysis are compared to those from the test that is conducted by the real-time moiré interferometry. The flip-chip package is subjected to a temperature rise thermal load from room temperature 23 °C to 115 °C. The steady state deformed configuration of the non-cleaned flip-chip sample is selected during the test as the benchmark. The displacements at the prescribed locations, *C* and *D*, on the specimen obtained from the test and the finite element analysis are tabulated in Table 14.9. It is found that the relative errors between the

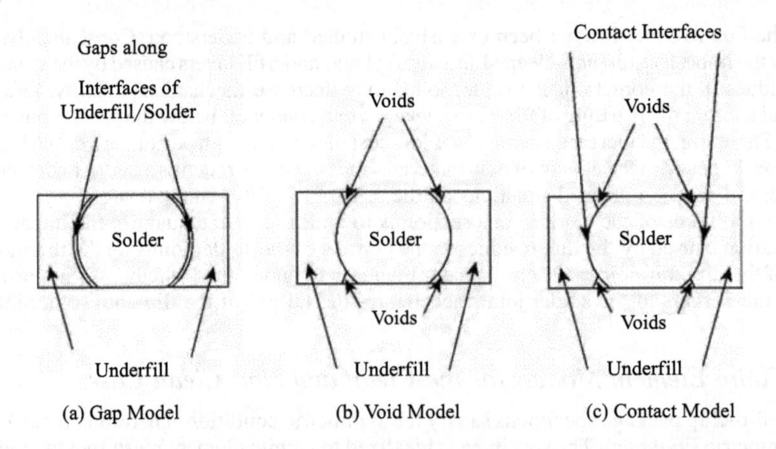

Figure 14.30 Three non-cleaned FE models in solder joints

Table 14.9 Comparisons of the displacement results between FEA and moiré test (μm)

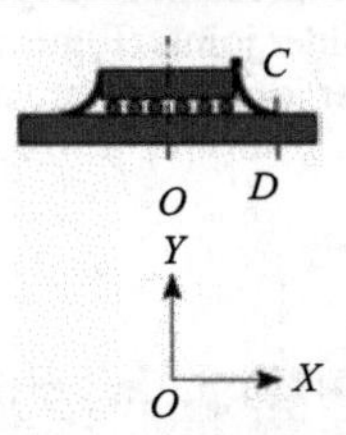

	Output Point	Reference Point	FEA Results	Test Results	Error (%)
U	*D*	*O*	6.908	5.838	18.3
V	*C*	*O*	9.046	10.842	16.5

displacements at the prescribed locations obtained from the finite element analysis and the test are generally less than 18%. A good agreement between the results of finite element analysis and those from test is obtained.

Furthermore, the displacement contours simulated by the finite element method and the fringe patterns captured by the moiré interferometry technique both in the x and y directions are shown in Figures 14.31 and 14.32. It is found that these displacement contours on the flip-chip package specimen obtained from

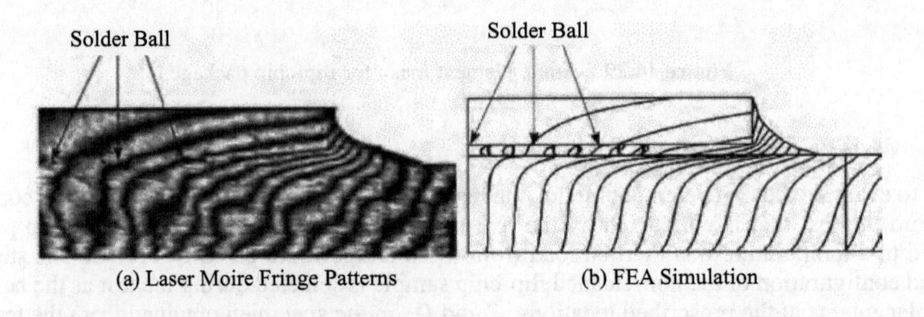

Figure 14.31 U-displacement fringe patterns and FEA simulation of cleaned flip-chip sample

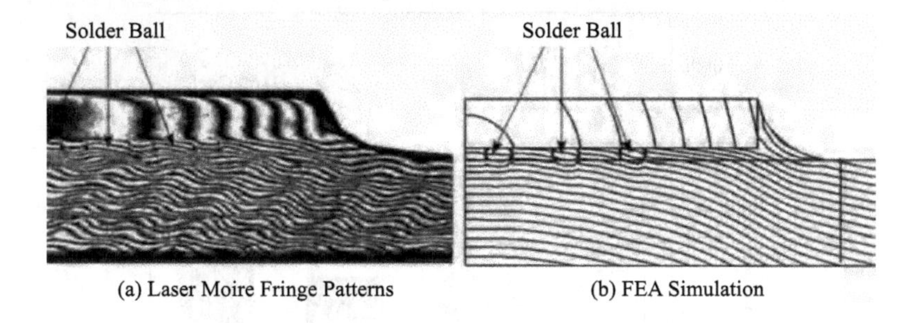

Figure 14.32 V-displacement fringe patterns and FEA simulation of cleaned flip-chip sample

the moiré interferometry show much of the same patterns compared with those modeled by the finite element method.

It is concluded from the above displacement comparison that the finite element model well represents the behavior of the flip-chip package specimen under the given test conditions.

14.6.3 Reliability Study for the Solder Joints

Since the solder joints in flip-chip packages play a key role as important interconnects, the reliability study will focus on the results of the solder joints between the cleaned FE model and the non-cleaned FE models.

Table 14.10 lists the inelastic strain ranges in the corners of the outmost solder joint obtained by using the cleaned model and the non-cleaned models under the thermal cyclic load from −40 °C to 125 °C. The ramping rate of the thermal cyclic load is 11 °C per minute. The temperature profile of the thermal cyclic load considered is plotted in Figure 14.33. It is noted that the inelastic strain ranges in the corners of the outmost solder joint obtained from the non-cleaned models are much higher than those obtained from the cleaned model. While the fatigue life of the solder joints in the flip-chip packages is directly determined by their inelastic strain ranges, as described in the well-known Coffin–Manson equation which has been

Table 14.10 Comparison of inelastic strain ranges between cleaned model and non-cleaned models

	Point *E*	Point *F*	Point *G*	Point *H*
Cleaning Model	0.0051	0.0084	0.0085	0.0108
Gap Model	0.1106	0.1157	0.0719	0.0514
Void Model	0.0148	0.0119	0.0285	0.0253
Contact Model	0.1059	0.1071	0.0689	0.0463

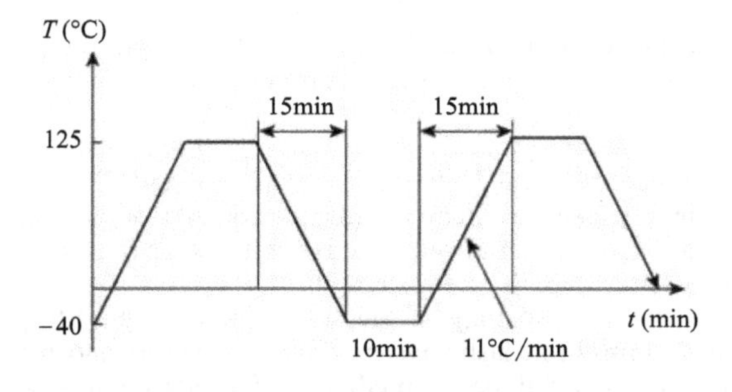

Figure 14.33 Temperature profile of thermal cyclic load

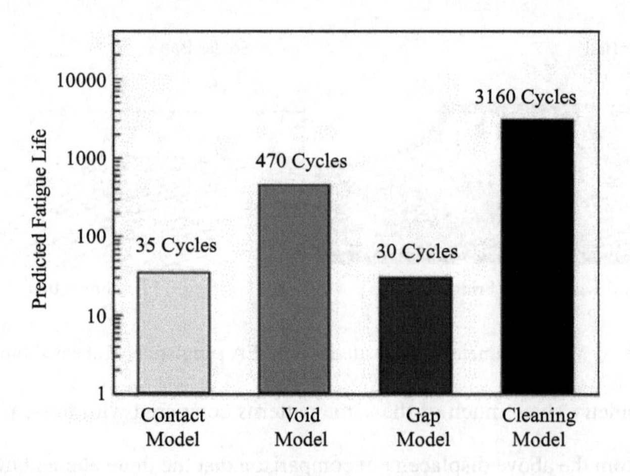

Figure 14.34 Predicted of fatigue life of the outmost solder joint by cleaned model and non-cleaned models (cycles)

widely used for fatigue life prediction of many solders. The simple equations, are given in Chapter 4, Section 4.2.1. Generally speaking, the larger the inelastic strain ranges, the shorter the fatigue life.

The fatigue life in the corners of the outmost solder ball predicted by Equation 14.2 using the cleaned FE model and the non-cleaned FE models is illustrated in Figure 14.34. It is obvious that the fatigue life in the corners of the outmost solder joint predicted by using the cleaned FEA models is much higher than that predicted by using the non-cleaned FE models. For example, the fatigue life of cleaned FE models is more than one hundred times higher than that of the non-cleaned gap FE model. Thus the non-cleaned situation decreases the mechanical stability and shortens the lifetime of flip-chip packages to a level well below the predictions of the cleaned simulation. Since it is hard to find perfect filled layers for any real world samples, the cleaned FE model yields an overly conservative fatigue life prediction for the outmost solder ball. Therefore, in order to improve the reliability of the solder interconnects, it is crucial to remove the flux residues in the corners of the solder joints,through a washing process or other effective method, to avoid the process-induced gaps or delaminations along the interfaces between the solder joints and the underfill.

14.7 Chapter Summary

This chapter develops the FEA simulation framework for PRCL and TMCL modeling that considers the impact of the die attach process. The work shown in the chapter is part of our modeling efforts for manufacturing quality and reliability. Major conclusions obtained based on our modeling are as follows.

The impact of the tilted die to the stress in PRCL and TMCL is significant. That has induced first principal stress 2.2 times greater and third principal stress 1.56 times greater than the case without die tilt. The tilted die also has induced the highest EMC and pad stress. Therefore, controlling the die attach process to get uniform die attach thickness without die tilt is important for reducing stress in the PRCL and TMCL.

The maximum stresses in the TMCL of this chapter are higher than that of PRCL. This is due to the temperature range in the PRCL and TMCL. In the current PRCL and TMCL conditions, the principal stresses in the die are not higher than the silicon failure criterion even at the worst die tilt case. However, for eutectic die attach material, the accumulated effective strain is the root cause of fractures after many cycles even if the stress is lower than the yield strength.

This chapter also gives the methodology of the solder joint fatigue life prediction based on the Coffin–Manson equation in terms of inelastic hysteresis energy density and experimental test data. The application of the methodology to a fatigue life prediction of a stack die flip-chip on silicon substrate is investigated and discussed with finite element analysis. The result recommended is that using the underfill will enable us to effectively prolong the fatigue life. The effect of cleaned and non-cleaned situations on the reliability of flip-chip packages is studied as well. The result disclosed that the fatigue life of cleaned FE models is more than one hundred times higher than that of non-cleaned gap FE model. Thus the non-cleaned situation decreases the mechanical stability and shortens the lifetime of flip-chip packages.

References

Brown, S.B., Kim, K.H. and Anand, L. (1989) An internal variable constitutive model for hot working of metals, *International Journal of Plasticity* **5**:95–130.

Darveaux, R. and Mawer, A. (1995) Thermal and power cycling limits of plastic ball grid array (PBGA) assemblies, *Proceedings of Surface Mount International*, 315–326.

Engelmaier, W. (1983) Fatigue life of leadless chip carrier solder joints during power cycling, *IEEE Transactions on Components, Hybrids and Manufacturing Technology*, CHMT-6(3): 232–237.

Fox, L.R., Sofia, J.W. and Shine, M.C. (1985) Investigation of solder fatigue acceleration factors, *IEEE Transactions on Components, Hybrids and Manufacturing Technology*, CHMT-4(2): 275–285.

Ham, S., Cho, M. and Lee, S. (2000) Thermal deformation of CSP assembly during temperature cycling and power cycling, *2000 International Symposium on Electronic Materials and Packaging, IEEE.*

Lau, J.H. and Pao, Y-H. (1997) *Solder Joint Reliability of BGA, CSP, Flip-chip, and Fine Pitch SMT Assemblies*, McGraw-Hill, New York.

Liu, Y. and Irving, S. (2003) Power cycling simulation of an IC Package: considering electromigration and thermal-mechanical failure, *Proceedings of 53th Electronic Components and Technology Conference*, New Orleans, Louisiana, May.

Liu, Y., Irving, S., Luk, T. and Kinzer, D. (2008) Trends of power electronic packaging and modeling, *Electronic Packaging Technology Conference 2008*, Singapore.

Maniatty, A.M., Liu, Y., Ottmar, K. *et al.* (2001) Stabilized finite element method for viscoplastic flow: Formulation and a simple progressive solution strategy, *Computer Methods in Applied Mechanics and Engineering* **190**:4609–4625.

Mid-STD-7500 Method 1037.2, Intermittent operation life.

Myllykoski, P., Reinikainen, T. and Rodgers, B. (2002) Power cycling simulation of BGA assemblies, *EuroSimE* **5**:95–130.

Norris, K.C. and Landzberg, A.H. (1969) Reliability of Controlled Collapse Interconnections, *IBM Journal of Research and Development* **13**(3):266–271.

Ren, W., Qian, Z. and Liu, S. (1998) Thermo-mechanical creep of two solder alloys, *48th Electronic Components and Technology Conference*, Seattle, Washington, 1431–1437.

Ren, W. (2000) *Thermo-mechanical Properties of Packaging Materials and Their Applications to Reliability Evaluation for Electronic Packages*, PhD Dissertation, Wayne State University.

Silicon Properties, Engineering Design Research Laboratory, California Institute of Technology (2005).

Syed, A. (2001) Predicting solder joint reliability for thermal, power and bend cycle within 25% accuracy, *Proceedings of 50th Electronic Components and Technology Conference*, Orlando, Florida, May.

Wild, R.N. (1974) Some fatigue properties of solder and solder joints, *Proceedings of NEPCON*, 105–117.

15

Passivation Crack Avoidance

A typical microelectronic device contains a silicon die in a package encapsulated with an epoxy. This polymer and various inorganic materials on the die, such as metal interconnects and ceramic passivation films, have dissimilar coefficients of thermal expansion (CTEs). When such a device is subjected to a change in temperature, the mismatch in the CTEs deforms the materials. In particular, as temperature cycles, the plastic deformation in a metal interconnect may accumulate incrementally, a phenomenon known as ratcheting plastic deformation. Ratcheting in the metal film may induce large stresses in an overlaying ceramic film, causing cracks to initiate and grow stably cycle by cycle. In this research, such ratcheting-induced stable cracking (RISC) is studied using a simplified three-layer model. We describe conditions under which ratcheting will occur in the metal layer, predict the number of cycles for the crack to initiate in the ceramic film, and discuss strategies to avoid RISC in design.

15.1 Ratcheting-Induced Stable Cracking: A Synopsis

D-PAK is an electronic package widely used in the automobile and power industry. Its structure is representative of many electronic packages that integrate diverse materials, as illustrated in Figure 15.1. On a silicon wafer, layers of polysilicon gates and dielectrics are deposited, which in turn are covered by metal interconnect, and then passivated by a SiN film. The wafer is then cut into dies, each of which is then mounted on a copper pad and encapsulated with an epoxy molding compound (EMC) at 175 °C.

Due to a mismatch between the coefficients of thermal expansion (CTEs), such an electronic package may exhibit various modes of failure in a thermal-cycling test. As an example, Figure 15.2 shows the micrographs of a D-PAK after 1000 cycles between − 65 °C and 150 °C. Cracks occur near the edges in the SiN film over wide aluminum pads, but rarely in the SiN film over narrow aluminum rings. These cracks typically initiate after some number of cycles, and grow stably cycle by cycle (Huang *et al.*, 2000; Huang *et al.*, 2001; Huang *et al.*, 2002; Zhang *et al.*, 2004). This failure mode is known as ratcheting-induced stable cracking (RISC).

This study studies RISC using a simplified three-layer model. We will elaborate the failure mechanism, describe conditions under which ratcheting will occur in the metal layer, predict the number of cycles for the crack to initiate in the ceramic film, and discuss strategies to avoid RISC in design.

This section describes qualitatively the process of RISC, leaving quantitative aspects to later sections. Figure 15.3 illustrates the structure to be studied as a model of the D-PAK. The passivation film, thickness h, lies on the metal underlayer, thickness H, which in turn lies on the semi-infinite substrate. The passivation film is elastic, the underlayer is elastic-perfectly-plastic, and the substrate is also elastic. Both a blanket metal film and a metal stripe are considered here. The gates and buried pillars

Modeling and Simulation for Microelectronic Packaging and Integration: Manufacturing, Reliability and Testing, Second Edition. Sheng Liu and Yong Liu.

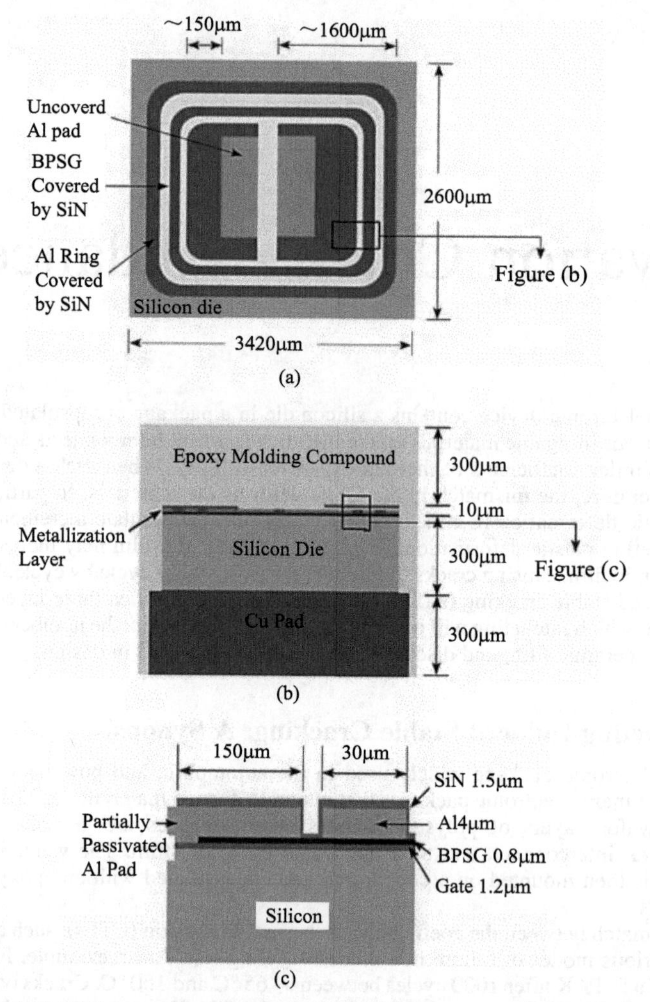

Figure 15.1 Schematics of a D-PAK structure. (a) Top view; (b) cross section; (c) magnified view of the interconnect structure of the lower right corner

surrounding gate (BPSG) are just a thin layer above silicon, and do not significantly affect the deformation of aluminum and passivation films. Consequently, the gates and BPSG are treated as a part of the substrate.

The EMC has a larger CTE than the silicon die. Upon cooling from the curing temperature, if the EMC and the silicon die were separated, the EMC would contract more than the silicon. However, this relative motion is constrained by the bonding between the EMC and the die, so that shear stresses develop on the EMC/die interface, concentrating at the edges and corners of the die. During a thermal-cycling test, the high-end temperature (150 °C) is lower than the curing temperature (175 °C). Consequently, at each corner or edge of the die, the magnitude of the shear stress varies with the temperature, but the direction of the shear stress is always biased towards the center of the die. We denote the shear stress on the EMC/die interface by τ_0, as indicated in Figure 15.3.

This interfacial shear stress τ_0 is partly sustained by the passivation film as a membrane stress σ, and partly transmitted to the metal film as a shear stress τ_m (Huang *et al.*, 2000; Huang *et al.*, 2001; Huang *et al.*, 2002). The shear stress τ_m in the metal, being limited by the yield strength of the EMC, may or may

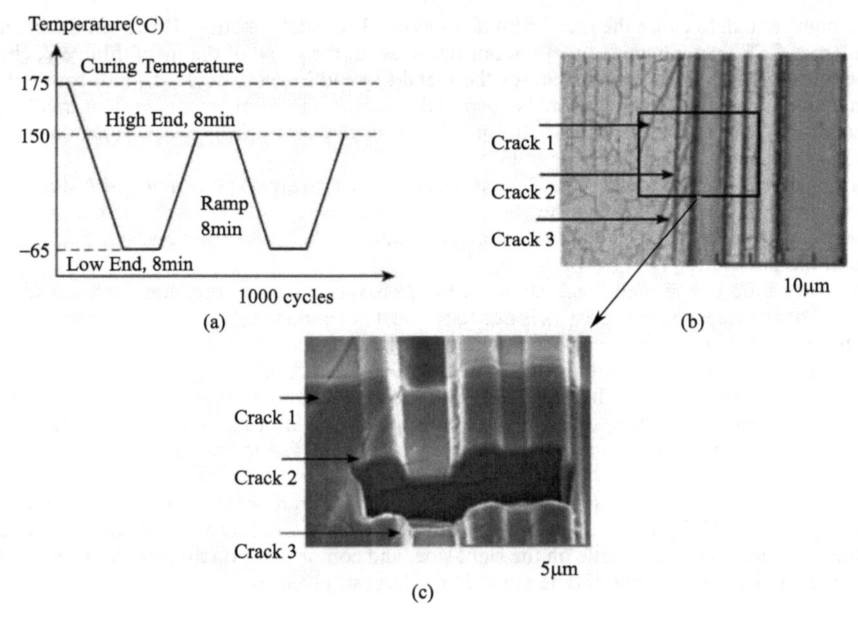

Figure 15.2 (a) Temperature history in a thermal-cycling test; (b) optical micrograph of a fractured die surface after 1000 temperature cycles; (c) SEM micrograph with a FIB cut

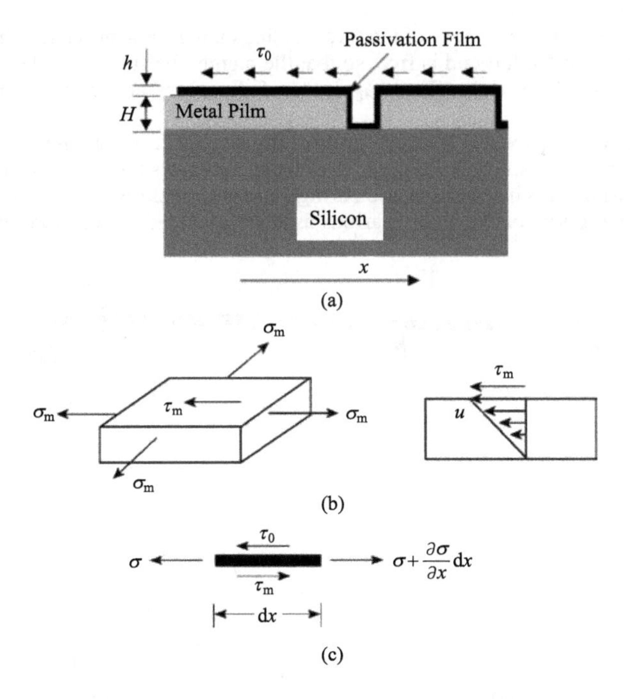

Figure 15.3 (a) A three-layer model of D-PAK; (b) the stress state in the metal film and the shear deformation; (c) the equilibrium of an infinitesimal element of the passivation film

not be high enough to cause the metal film to undergo plastic deformation. However, the metal film also has a larger CTE than silicon, so that the normal stress in the plane of the metal film, σ_m, changes with temperature. If this in-plane stress causes the metal to yield in every cycle, τ_m will cause plastic shear strain in the same direction as τ_m (that is, toward the center of the die), no matter how small τ_m is. Under this condition, the plastic shear strain in the metal will accumulate cycle by cycle, a phenomenon known as ratcheting plastic deformation (Bree, 1967).

As it ratchets toward the center, the metal film carries the overlaying passivation film along. However, if the passivation film is bonded with the die along an edge, the edge will be anchored and move negligibly. Consequently, the ratcheting metal will elastically deform the passivation and build up the membrane stress in the passivation.

The magnitude of the membrane stress in the passivation can be very high. This is understood as follows. The interfacial shear stress τ_0 is due to the mismatch in the CTEs of the epoxy and the silicon, so that the magnitude of τ_0 will not decay with the cycles. Recall that the interfacial shear stress τ_0 is partly sustained by the passivation film as the membrane stress σ, and partly transmitted to the metal film as the shear stress τ_m. As the metal film ratchets, the membrane stress in the passivation builds up, but the magnitude of the shear stress in the metal τ_m decays. After many cycles, τ_m will vanish, and τ_0 is fully sustained by the membrane stress in the passivation. The structure is said to have reached the steady state.

Figure 15.4 plots the steady-state distribution of the membrane stress in the passivation film for two kinds of geometry. In Figure 15.4(a), both ends of the passivation film are anchored to the silicon die, so that the membrane stress is tensile on the right side, and compressive on the left. At the two edges of the film, the magnitude of the membrane stress is the largest, given by:

$$\sigma_{ss} = \frac{\tau_0 L}{h}, \text{ at } x = L \tag{15.1}$$

This equation is obtained by the balance of the forces acting on the passivation film. In Figure 15.4(b), the right end is anchored, but the left end is free, so that the membrane stress vanishes on the left end, and linearly builds up toward the right. The magnitude of the stress at the right end is still, given by Equation 15.1.

Equation 15.1 shows that, even if the magnitude of the interfacial shear stress τ_0 is modest, the large aspect ratio *L/h* can greatly amplify the magnitude of the membrane stress. This is known as the shear-lag effect. When the tensile stress in the passivation is high enough, cracks will initiate and grow stably cycle by cycle. The following sections consider various aspects discussed above more quantitatively.

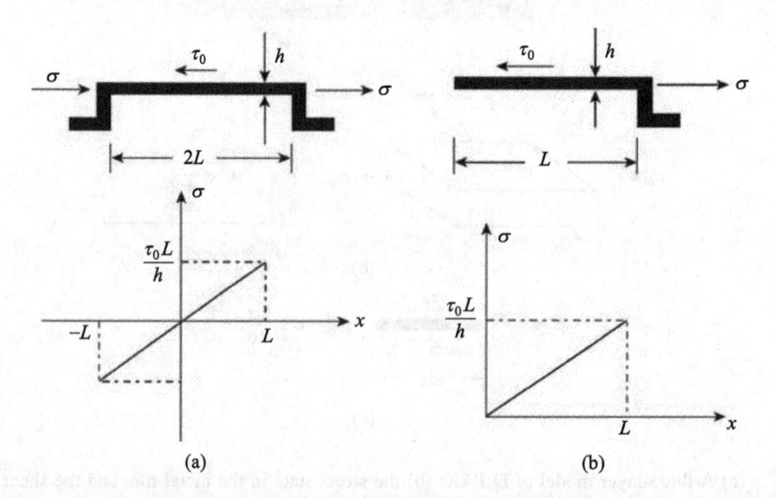

Figure 15.4 Stress distribution in steady state for both fully passivated and partially passivated SiN films. (a) Two ends are fixed; (b) the right end is fixed, but the left is free

15.2 Ratcheting in Metal Films

To better understand ratcheting plastic deformation, we next consider a blanket metal film on a silicon substrate, subject to cyclic temperature between T_L and T_H. For the time being, we assume that the shear stress τ_m is uniform in the metal and invariant with the temperature. The metal film is taken to be elastic-perfectly-plastic, with the CTE α_m, Young's modulus E_m, Poisson's ratio ν_m, and the temperature-independent yield strength Y_m. The substrate is elastic with CTE α_s. As shown in Figure 15.3(b), the metal film is under biaxial in-plane stress σ_m due to the temperature change, and a shear stress τ_m. The metal film is elastic when the stresses satisfy the von-Mises yield condition:

$$\sigma_m^2 + 3\tau_m^2 < Y_m^2 \tag{15.2}$$

As the temperature cycles between T_L and T_H, the normal stress σ_m also cycles (Figure 15.5). When the metal is plastic, the normal stress is fixed at constant levels:

$$\sigma_m = \pm\sqrt{Y_m^2 - 3\tau_m^2} \tag{15.3}$$

When the metal is elastic, however, the normal stress changes linearly with the temperature, having a slope $E_m(\alpha_m - \alpha_s)/(1 - \nu_m)$. It is evident from Figure 15.5, to ensure that the metal film remains elastic after the first cycle, the following condition must be satisfied (Huang *et al.*, 2001):

$$\left[\frac{E_m(\alpha_m - \alpha_s)(T_H - T_L)}{2(1 - \nu_m)Y_m}\right]^2 + 3\left(\frac{\tau_m}{Y_m}\right)^2 < 1 \tag{15.4}$$

Equation 15.4 is known as the shakedown condition. Under this condition the metal film may undergo plastic deformation in part of the first cycle, but will remain elastic in subsequent cycles.

Figure 15.6 shows a plane spanned by the normalized shear stress, τ_m/y_m, and the normalized temperature range, $E_m(\alpha_m - \alpha_s)(T_H - T_L)/(1 - \nu_m)Y_m$.

The plane is divided into four regimes: shakedown, plastic collapse, cyclic plastic deformation, and ratcheting. Such a diagram is called a Bree diagram (Bree, 1967). Equation 15.4 corresponds to part of an ellipse, inside which lies the shakedown regime. As a separate consideration, when $\tau_m/Y_m \geqslant 1/\sqrt{3}$, the metal film will deform plastically under the shear stress alone without the aid of the temperature change. In the other limit, when $\tau_m = 0$, the metal film undergoes cyclic plastic deformation if the temperature range is sufficiently large. The regime bounded within the above three boundaries is the ratcheting regime, where the metal film yields every cycle, but the amount of plastic shear strain is finite, accumulating cycle by cycle.

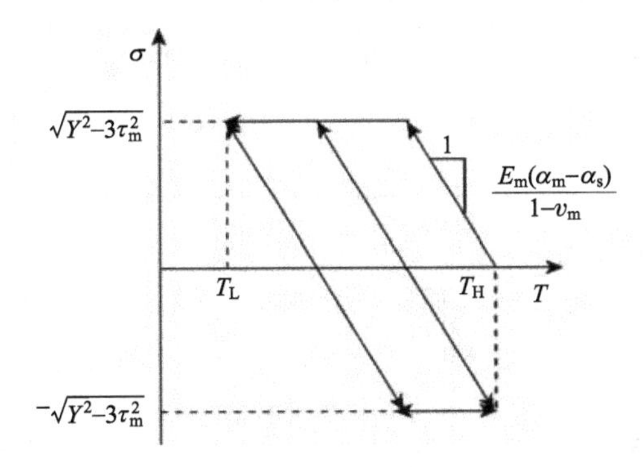

Figure 15.5 Biaxial stress in metal film as a function of the temperature

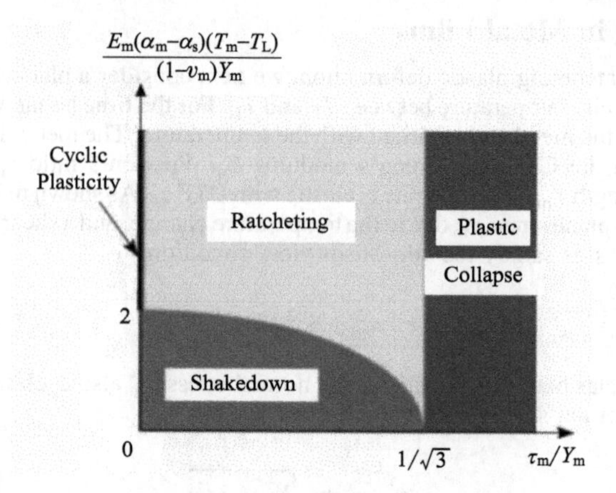

Figure 15.6 A Bree diagram for the elastic and perfectly plastic metal film

We next compare Al and Cu interconnects. As indicated by Table 15.1, the yield strength of EMC is comparable to that of Al, or even larger. Note that the interfacial shear stress τ_0 is linearly proportional to the temperature change when the EMC is elastic, and is limited by the shear yield strength of the EMC when the EMC is plastic. Hence, in this design of D-PAK, if the temperature range is large, τ_0 by itself can cause the Al film to deform plastically.

Figure 15.6 shows a Bree diagram for the elastic and perfectly plastic metal film. The plane is divided into four regimes: plastic collapse, shakedown, ratcheting, and cyclic plastic deformation.

To avoid plastic collapse, the design has to be modified to decrease the yield strength of the molding compound. Also notice that Al has a large CTE, such that the value of $E_m(\alpha_m - \alpha_s)(T_H - T_L)/(1 - \nu_m)Y_m = 4.29 > 2$. Consequently, inter-facial shear stress τ_0 of any magnitude can cause ratcheting. Indeed, ratcheting is commonly reported for aluminum interconnects (see References cited in Huang, *et al.*, 2000; Huang *et al.*, 2001; Huang *et al.*, 2002).

By contrast, ratcheting has not been reported in Cu interconnects. Using the values in Table 16.1, we find for Cu $E_m(\alpha_m - \alpha_s)(T_H - T_L)/Y_m(1 - \nu_m) = 1.77$. Using the yield strength of EMC as an estimate for τ_0, we find that $\tau_0/Y_m = 0.29$. These two estimates correspond to a point just above the ellipse in Figure 15.6, so that the Cu film will ratchet for the beginning cycles. However, the shear stress in Cu, τ_m, will relax from its initial value τ_0 cycle by cycle, so that the Cu film will approach the shakedown regime after a certain number of cycles. If the passivation film does not crack before the Cu film shakes down, it will never crack in the future cycles.

We next consider the plastic strain per cycle. When $E_m(\alpha_m - \alpha_s)(T_H - T_L)/(1 - \nu_m)Y_m > 2$, and the shear stress τ_m is much smaller than its yield strength Y_m, the plastic shear strain rate γ^p per cycle versus the shear stress τ_m can be approximated by a linear relation (Huang *et al.*, 2002):

Table 15.1 Materials properties

Materials	Young's Modulus E (GPa)	Poisson's Ratio ν	CTE α ($10^{-6}K^{-1}$)	Yield Strength Y (MPa)	Toughness K_{Ic} (MPa·m$^{1/2}$)
EMC	17	0.3	24	98–150	
BPSG	64	0.17	2.78		
Silicon Die	162.5	0.22	2.8		
Polygate	162.5	0.22	2.8		
SiN	275	0.24	4		5.8–8.5
Al	70.34	0.346	23.2	110	
Cu	130	0.33	17	345–310	

$$\frac{\partial \gamma^{p}}{\partial N}=\frac{\tau_{m}}{\eta} \tag{15.5}$$

where N is the number of cycles, and η, the linear ratcheting coefficient, is given as follows:

$$\eta=\frac{E_{m}}{12(1-\nu_{m})}\left[\frac{E_{m}(\alpha_{m}-\alpha_{s})(T_{H}-T_{L})}{(1-\nu_{m})Y_{m}}-2\right]^{-1} \tag{15.6}$$

for example, if $\tau_{m}/Y_{m}<0.2$, then the error of the linear approximation is less than 10%.

In the temperature range specified in Figure 15.2(a), all the materials in the structure do not creep, so dwelling time at high end and low end do not affect the analysis. In practice, it takes some time for the structure to reach a state of uniform temperature during the testing, but in the follow analysis, we neglect the non-uniformity in temperature. Consequently, various types of loading program, linear or sinusoidal, dwelling or not, will lead to the same accumulative effects over many cycles.

15.3 Cracking in Passivation Films

Let $\sigma(x,N)$ the membrane stress, and $u(x,N)$ the displacement. For simplicity, the interfacial shear stress τ_0 is assumed to be a constant, but the shear stress in the metal is taken to be a function $\tau_{m}(x,N)$. As shown in Figure 15.3(c), the equilibrium of an infinitesimal element of the passivation film requires that $\partial\sigma/\partial x+(\tau_{m}-\tau_{0})/h=0$. Hooke's law relates the membrane stress to the membrane strain, $\sigma=\bar{E}_{p}\partial u/\partial x$, where $\bar{E}_{p}=E_{p}/\left(1-\nu_{p}^{2}\right)$, and E_p and ν_p are Young's modulus and Poisson's ratio of passivation film. Moreover, as shown in Figure 15.3(b), the plastic shear strain relates to the displacement as $\gamma^{p}=u/H$. These relations, combining with Equation 15.5, give the governing equation for $u(x,N)$:

$$\frac{\partial u}{\partial N}=D\frac{\partial^{2}u}{\partial x^{2}}-\frac{\tau_{0}H}{\eta} \tag{15.7}$$

where $D=hH\bar{E}_{p}/\eta$. Equation 15.7 has the same form as the diffusion equation, and D can be interpreted as an effective diffusivity.

The characteristic number of cycles to reach the steady state can be estimated by:

$$N_{0}=\frac{L^{2}}{D}=\frac{L^{2}}{hH}\frac{\eta}{\bar{E}_{p}} \tag{15.8}$$

The complete solution of Equation 15.7 can be written as a series with the initial condition of $u=0$ at $N=0$. Because the first term in the series dominates when $N>0.05N_0$, the maximum tensile stress build-up on the right edge can be approximated by the following:

$$\sigma\approx\frac{\tau_{0}L}{h}\left[1-\frac{8}{\pi^{2}}\exp\left(-\frac{\pi^{2}}{4}\frac{N}{N_{0}}\right)\right] \tag{15.9}$$

Note that in this section we have neglected the residual stress in SiN due to the deposition processes.

Now let us consider an initial crack, of length $2a$, in the blanket passivation film, as shown in Figure 15.7. As the aluminum layer ratchets, the membrane stress in the passivation film builds up globally over the length scale L. However, locally near the crack, over the length scale a, the membrane stress relaxes. We can identify two distinct cycle scales.

The number of cycles to attain the steady state over the width of the passivation L scales as L^2/D. The number of cycles to relax near the crack scales as a^2/D. Assuming, $L\gg a$, after many cycles, the

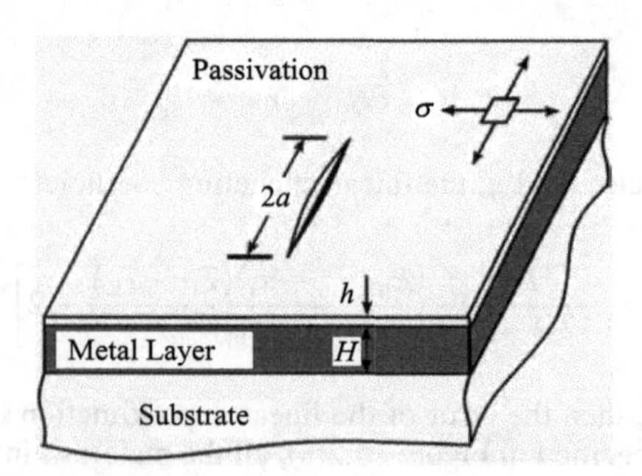

Figure 15.7 A schematic of an initially finite crack, of length $2a$, in the passivation film

crack behaves like a Griffith crack (Xia *et al*, 2000; Huang *et al.*, 2002; Liang *et al.*, 2003, 2004; Suo *et al.*, 2003). When the membrane stress is σ, the stress intensity factor is given by:

$$K = \sigma\sqrt{\pi a} \tag{15.10}$$

If this stress intensity factor is below the fracture toughness K_c of the passivation, that is, $\sigma\sqrt{\pi a} < K_c$, the crack does not grow. Consequently, the facture strength of the passivation σ_c is given by $\sigma_c = K_c/\sqrt{\pi a}$. In particular, if the fracture strength exceeds the steady state stress at the edge of the passivation, given by Equation 15.1, the crack will never grow. This condition is given by:

$$\sigma_c > \tau_0 L/h \tag{15.11}$$

When Equation (15.11) is violated, however, the crack will begin to grow after a certain number of cycles, denoted by N_i. Using Equation 15.12, we obtain:

$$\frac{N_i}{N_0} = -\frac{4}{\pi^2}\ln\left[\frac{\pi^2}{8}\left(1 - \frac{\sigma_c h}{\tau_0 L}\right)\right] \tag{15.12}$$

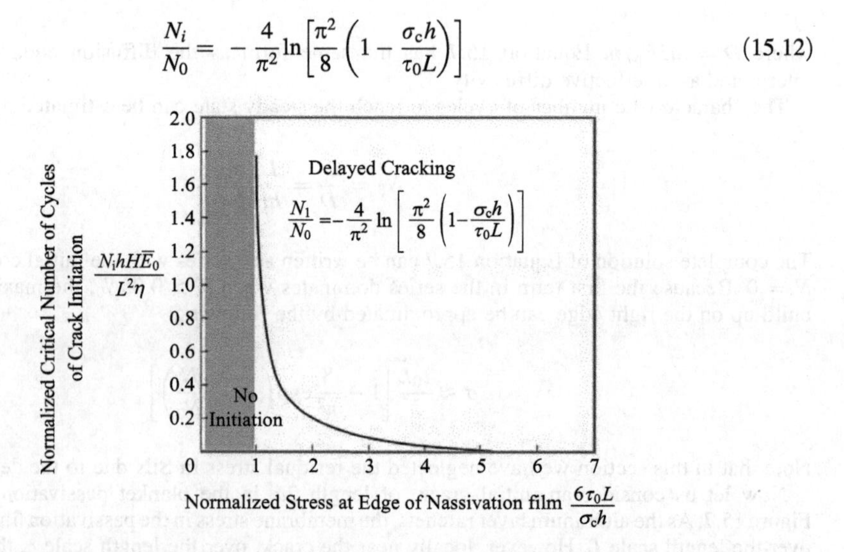

Figure 15.8 Normalized critical number of cycles to initiate the crack versus the normalized steady state stress level. If the fracture strength exceeds the steady-state stress, the crack remains stationary; otherwise, the crack initiates and grows after delayed cycles

This equation is valid when N_i is sufficiently large. Figure 15.8 plots the normalized number of cycles to crack initiation versus the normalized stress level in steady state. The whole plane is divided into two regimes: no crack initiation and delayed cracking as Equation 15.12.

Using the values in Table 15.1 and Figure 15.1, and the shear stress $\tau_0 = 50$MPa, the stress at steady state is $\sigma_{ss} = 5$GPa at the edge of the middle blanket film, and the characteristic number of cycles is about $N_0 = 50$. From Table 15.1, if we use 5.8 MPa·m$^{1/2}$ as the fracture toughness of SiN, and assume the largest initial crack length is comparable to the thickness, 1.5 m, then the critical stress to initiate the crack would be $\sigma_c = 4.73$GPa. Hence, from Figure 15.8 and Equation 15.12, it takes at least 55 cycles for crack initiation. In the narrow stripe, $\sigma_{ss} = 500$MPa at the edge, so that cracking is less likely to happen. The above conclusion can be verified by experimental pictures in Figure 15.2.

In the case of Figure 15.10(a), if the crack size is comparable or larger than the width of the stripe, then the stress intensity factor is determined by $K \sim \sigma_{ss}\sqrt{L}$, and so there exists a critical width of the stripe (Liang *et al.*, 2003):

$$L_c \sim \left(\frac{K_c h}{\tau_0}\right)^{2/3} \tag{15.13}$$

below which the crack never grows no matter how long the crack is. Based on this equation, and putting the values in, we can estimate that the critical width is 50 m for fully passivated stripe.

15.4 Design Modifications

From the above study, we may consider the following design modifications to avoid passivation cracking.

(i) Buffer Layer

If we add a compliant and soft buffer layer between the EMC and the silicon die, the shear stress τ_0 can reduce significantly, and so does the stress level in the steady state. Correspondingly, the lower bound of critical number of cycles of crack initiation, N_i, can increase by several orders of magnitude. It is even possible to prevent crack initiation. The comparisons are plotted in Figures 15.9 and 15.10.

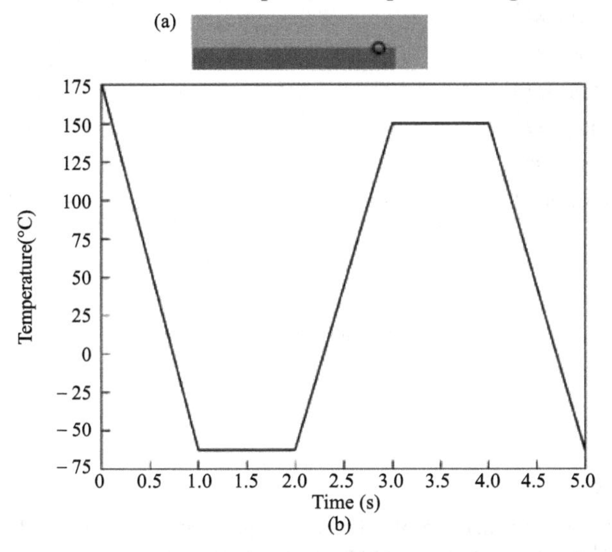

Figure 15.9 (a) A schematic of upper right quarter of the die package. (b) The temperature loading profile with unit °C. (c) The interfacial shear stress evolution around the corner as shown in (a) by circle, unit Pa. (d) Interfacial shear stress along the interface at $T = -65$ °C with unit Pa, and the inset shows the contour plot of FEA calculation

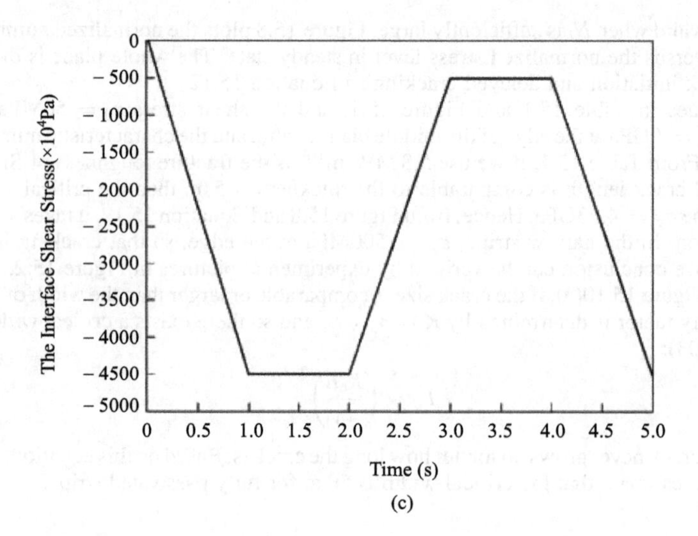

(c)

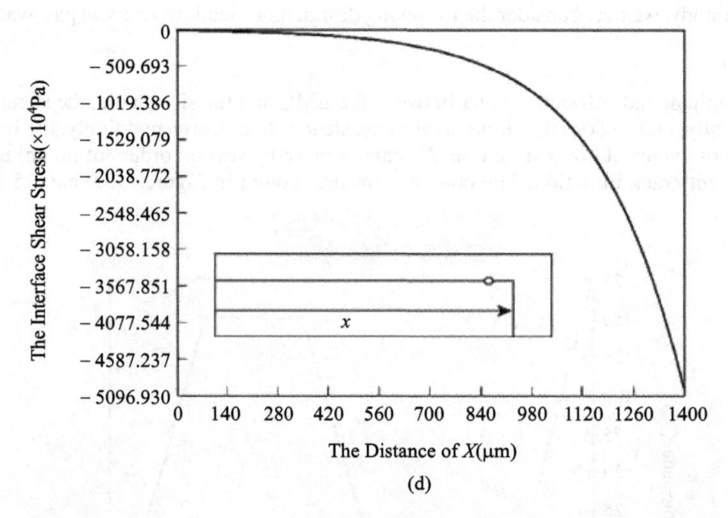

(d)

Figure 15.9 *(Continued)*

(ii) Small Passivation Width

The reduction of width and the increase of thickness of passivation can also reduce the stress level. The change of large or long passivation film to small or narrow pads is a practical method, for example, introducing slots into the passivation film. In Figure 15.11, a lower right corner is simulated by ANSYS. The original design of width 150 μm, the design with slots, and the design of smaller width 50 μm are shown. The stress level at steady-state is significantly reduced in the latter two cases.

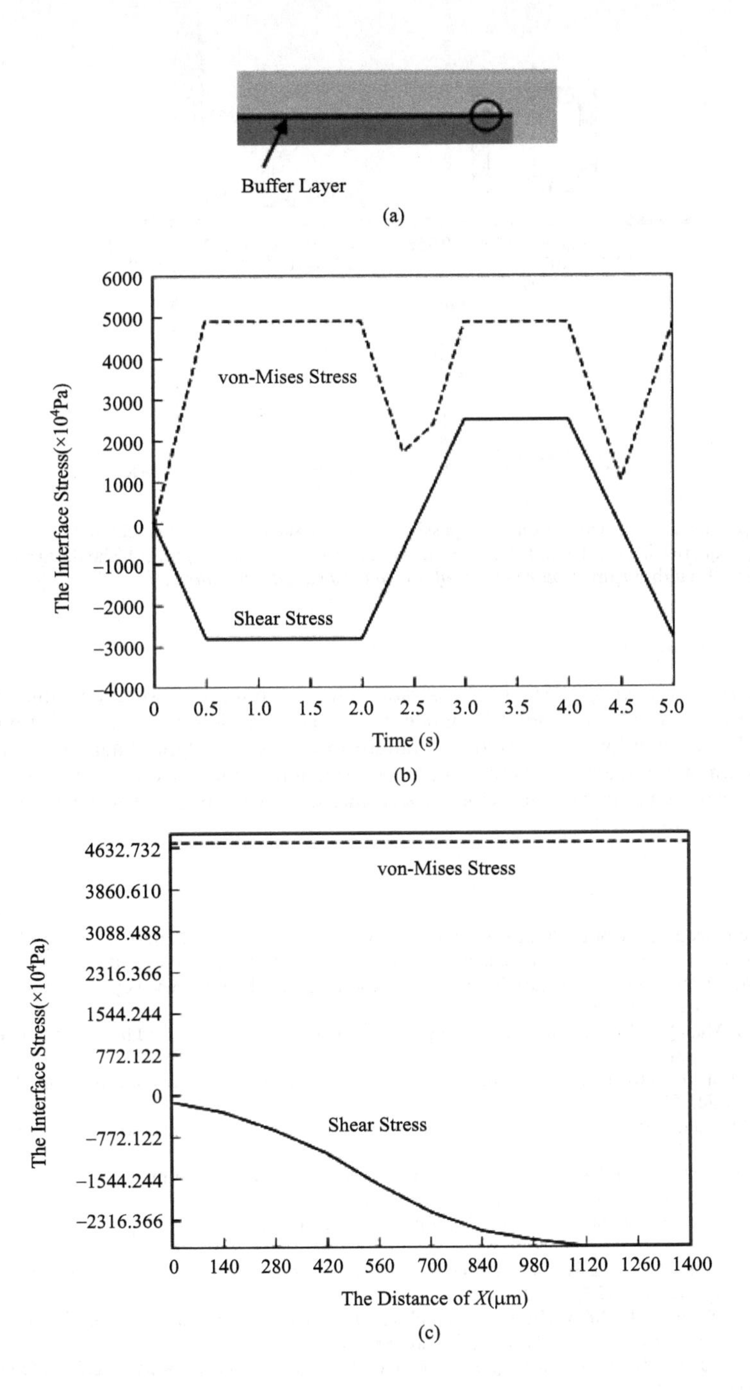

Figure 15.10 (a) A schematic of upper right quarter of the die package with buffer layer, under the same loading condition as in Figure 15.9(b). (b) The evolution of interfacial shear stress and von-Mises stress around the corner as shown in (a) by circle, unit Pa. (c) Interfacial shear stress and von-Mises stress along the interface at $T = -65\,°C$ with unit Pa

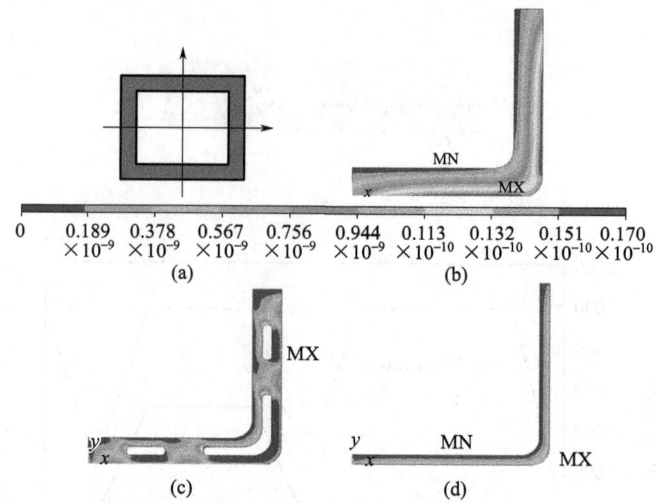

Figure 15.11 First principal stress distribution of the passivation film in steady state. (a) A schematic of passivation, due to symmetry only a quarter is calculated. (b) The original design with width 150 μm. (c) The design with slots. (d) The design with small width 50 μm (*Color version of this figure is available online*)

15.5 Chapter Summary

Ratcheting-induced stable cracking (RISC) in a passivation film is a commonly observed failure mode in microelectronic devices. This chapter uses a 1D model to estimate the lifetime in terms of temperature cycles, and to establish conditions that avoid this failure mode. A Bree-type diagram is plotted to ascertain if a metal film will undergo ratcheting plastic deformation. Under the condition that the metal does ratchet, we estimate the number of cycles for cracks to initiate. Several possible design modifications are discussed for the IC chips in design.

References

Bree, J. (1967) Elastic-plastic behavior of thin tubes subjected to internal pressure and intermittent high-heat fluxes with application to fast-nuclear-reactor fuel elements, *Journal of Strain Analysis* **2**:226–238.

Huang, M., Suo, Z., Ma, Q. and Fujimoto, H. (2000) Thin film cracking and ratcheting caused by temperature cycling, *Journal of Materials Research* **15**(6):1239–1242.

Huang, M., Suo, Z. and Ma, Q. (2001) Metal film crawling in interconnect structures caused by cyclic temperatures, *Acta Materialia* **49**:3039–3049.

Huang, M., Suo, Z. and Ma, Q. (2002) Plastic ratcheting induced cracks in thin film structures, *Journal of the Mechanics and Physics of Solids* **50**:1079–1098.

Huang, R., Prévost, J.H. and Suo, Z. (2002) Loss on constraint on fracture in thin film structures due to creep, *Acta Materialia* **50**:4137–1448.

Liang, J., Huang, R., Prévost, J.H. and Suo, Z. (2003) Evolving crack patterns in thin films with the extended finite element method, *International Journal of Solids Structures* **40**:2343–2354.

Liang, J., Huang, R., Prévost, J.H. and Suo, Z. (2003) Thin film cracking modulated by underlayer creep, *Experimental Mechanics* **43**:269–279.

Liang, J., Zhang, Z., Prévost, J.H. and Suo, Z. (2004) Time-dependent crack behavior in an integrated structure, *International Journal of Fracture* **125**:335–348.

Suo, Z., Prévost, J.H., and Liang, J. (2003) Kinetics of crack initiation and growth in organic-containing integrated structures, *Journal of the Mechanics and Physics of Solids* **51**:2169–2190.

Zhang, Z., Prévost, J.H. and Suo, Z. (2004) Cracking in interconnects induced by thermal ratcheting, *MRS fall meeting*, Boston, MA, December 2004.

Zhang, Z., Suo, Z., Liu, Y. and Irving, S. (2006) Methodology for avoidance of ratcheting-induced stable cracking (RISC) in microelectronic devices, *Electronic Components and Technology Conference 2006* San Jose, CA.

Xia, Z.C. and Hutchinson, J.W. (2000) crack patterns in thin films, *Journal of the Mechanics and Physics of Solids* **48**:1107–1131.

16

Drop Test

With the development of the information industry, portable electronics has been used in all the aspects of our human lives, from the cell phone, laptop notebooks, micro displays, to medical insulin pump, hearing aid, and pacemakers. It is well known that portable electronics products such as cellular phones, notebooks, pagers, two-way radios, and insulin pumps, are susceptible to accidental drop impact which can cause various damage modes such as interconnect breakage, shield housing fracture, Liquid Crystal Display (LCD) cracking, solder joint breaking, battery separation in cellular phones, display failure, solution leaking in some medical devices, and possible cracking/debonding along interfaces. Customers need not only various functions, but also need these functions to be reliable.

Industry is also pushing to form various consortiums and suggest different standards. For instance, in a recent JEDEC Standard, called "Board Level Drop Test Method of Components for Handheld Electronic Products" (JESD22-B111), it is important to conduct drop tests for handheld devices such as cameras, calculators, cell phones, pagers, palm size PCs, and the Personal Computer Memory Card. A few leading companies even have more strict in-house standards to win the increasing intensive competitions world wide.

International Association (PCMCIA) cards, smart cards, mobile phones, personal digital assistants (PDAs), and other electronic products that can be conveniently stored in a pocket and used while held in the user's hand. Short duration, high acceleration, soft interconnects between the component and the board make the material experience various loading rates. Miniaturized features of the interfaces and damage initiation sites make the measurement a very challenging task. Complexity of devices, board, process induced warpage and other imperfections, material, thickness and possible brittleness of interfaces, surface finish, interconnect material and standoff height, and component size, also make the impact modeling a challenging task.

16.1 Controlled Pulse Drop Test

There are three major types of methods used for the drop test evaluations: (1) free fall product level, (2) free fall board level, and (3) controlled pulse drop at board level (Xie *et al.*, 2003). In this study the third type "controlled pulse drop at board level" is utilized to investigate the reliability performance of an advanced molded leaded package (MLP) following the JEDEC standard (JEDEC Standard JESD22-B111 2003).

Due to the drop/impact event, the printed wiring board assembly inside the handheld product vibrates to cause a flexural motion of the board. The bending of the board induces severe stress in solder joints and other interconnects under the high level of G forces. This may not only cause mechanical failures, but also create electrical failures. Although there are many factors that affect the impact performance of the package, solder joints are generally recognized as the weakest part. The solder joint crack can occur either on the package side or the board side. The failure is a function of the combination factors of the board

Modeling and Simulation for Microelectronic Packaging and Integration: Manufacturing, Reliability and Testing, Second Edition. Sheng Liu and Yong Liu.

design, construction, material, thickness, surface finish, interconnect material, and standoff height, as well as component size.

Board level drop test experiment is an effective method to characterize the solder joint drop reliability performance under the guideline of JEDEC standards. However, the cost is very high and the test takes a long time to complete, which extends the design cycle. In addition, it is difficult to measure the dynamic strain/stress response in some parts due to the restriction with the mount-of-measurement devices (Groothuis *et al.*, 2005). In order to reduce cost and shorten the concept-to-market time, FEA simulation is widely adopted as an efficient alternative of the characterization tests. Early drop test simulation work was reported by Wu (1998, 2000), in which an explicit FE code LS-DYNA® was used to simulate a drop test at both component and product levels. Recently a novel dynamic nonlinear implicit algorithm with Input-G method is developed (Luan *et al.*, 2004, 2005). This Input-G method is useful for those only equipped with implicit-solver type of FE code.

In this section the implicit Input-G method is adopted to do the board level drop test simulation of the MLP package. A parametric study on solder joint height and MLP package thickness is conducted. The peeling stress and first principal stress of the solder joints are checked and compared for all the cases.

16.1.1 Simulation Methods

The drop test board based on the JEDEC Standard JESD22-B111 is shown in Figure 16.1, with a dimension of 132 mm × 77 mm × 1.0 mm, accommodated 15 components of same type in a three row by five column format. The MLP package has a dimension of 3 mm × 3 mm × 0.55 mm, which is mounted on the test board with 2.5 mil thick solder joint. According to the experiments, the central packages suffer the most drop impact damage. Therefore the packages at location U3 and U8 are thoroughly investigated. In order to simplify the model and reduce the total number of elements and nodes to save computation time, the packages at other locations are assumed to be a block with equivalent mass as the detailed packages. The FEA model of the whole system is shown in Figure 16.2, with a detailed package model at location U8, and all the simplified packages at other locations. The corresponding FEA mesh is shown is Figure 16.3. When the package at location U3 is studied, it will be modeled in all details and with all the other packages simplified.

Under a high-speed impact test, the solder material will become stronger and stiffer, and the yielding stress may increase three to four times compared with the static tensile test value. In this study, all the materials are assumed to be elastic, which reduces the accuracy of the calculations especially when the stresses near the edges of the solder joints are studied. However, it is believed that this does not affect the results in an unacceptable way when used for comparative purposes, but it does not allow us to consider the absolute values predictive. Since the greatest benefit is generally derived from comparative work, the method is very useful.

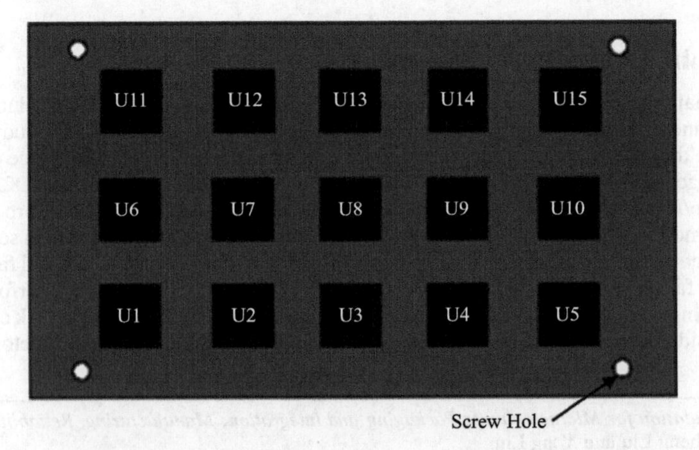

Figure 16.1 JEDEC drop test board and component locations

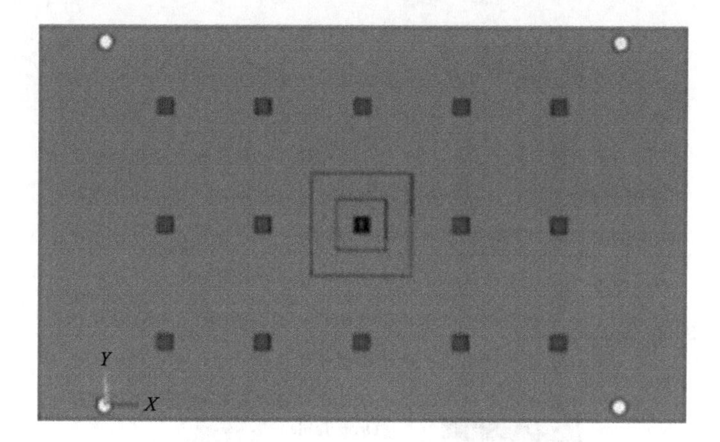

Figure 16.2 FEA model of the test board with 15 packages

The dynamic response of a package depends on many factors such as impact contact surface condition, fixture, screw mount and arrangement, and and so on. Adjusting all the factors to make the simulation results match the test results is usually very difficult. In order to save computational cost, an innovative approach using a predefined acceleration called "Input-G" method was proposed by Tee *et al.* (2003). This method simplified the simulation implementation. The acceleration or "G curve" is input as the load to the test board through the connector of package and fixture, which indirectly considers the complex effect of the contact condition and other factors (Groothuis *et al.*, 2005). The Input-G method is an efficient tool for DOE study or optimization of package/PCB design parameters for enhanced drop test performance (Figure 16.4). The first half-sine *G*-curve with a peak acceleration of 1500 *G* and duration of 0.5 ms is used as the loading of all the simulations in the study (Figure 16.5).

The implicit FE code ANSYS® is used to do the dynamic board level drop test simulations. The acceleration profile cannot be input directly into the implicit solver. Instead, the profile of

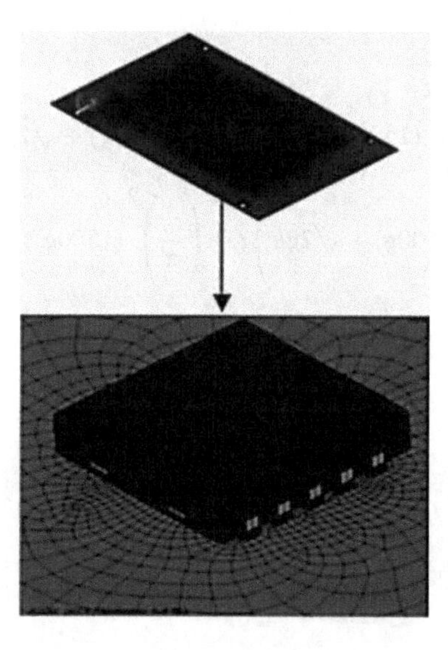

Figure 16.3 FEA mesh of the test board with 15 packages

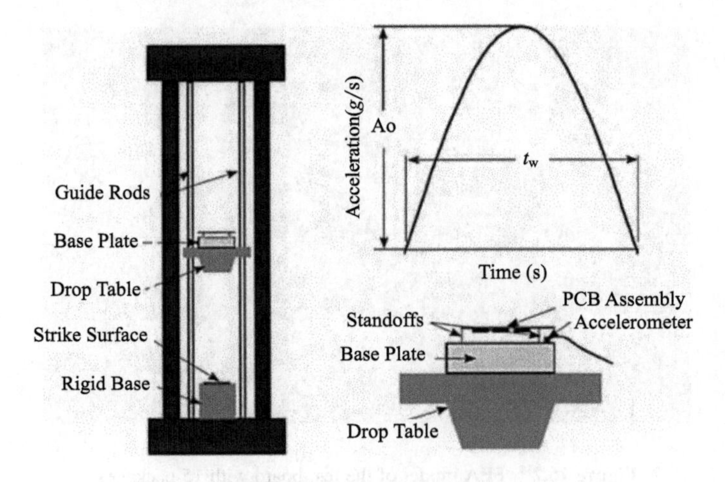

Figure 16.4 Drop test system

time-dependent displacement can be integrated from the impact pulse (to velocity profile first) and then directly applied to the four support screws as boundary conditions of implicit Input-G model (Luan *et al.*, 2004, 2005).

$$\{M\}[\ddot{u}_1] = \{C\}[\dot{u}_1] + \{K\}[u_1] = 0$$
$$\text{with initial conditions} \tag{16.1}$$
$$[u_1]|_{t=0} = 0$$
$$[\dot{u}_1]|_{t=0} = \sqrt{2gh}$$

and boundary condition

$$[u_1]|_{\text{athole}} = \left\{ \begin{array}{ll} -\left(\frac{t_w}{\pi}\right)^2 (1500g)\sin\frac{\pi t}{t_w} + \left(\frac{t_w}{\pi}(1500g) + \sqrt{2gh}\right)t, & t \leqslant t_w \\ \left(2\frac{t_w}{\pi}(1500g) + \sqrt{2gh}\right)t - \left(\frac{t_w}{\pi}\right)^2 (1500g)s, & t \leqslant t_w \end{array} \right\} \tag{16.2}$$

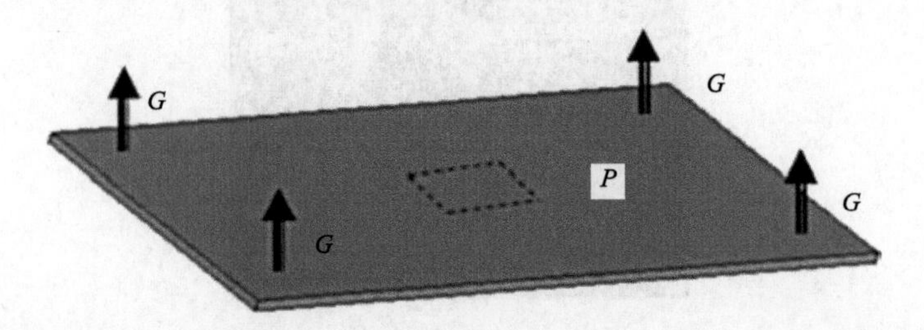

Figure 16.5 Input-G method

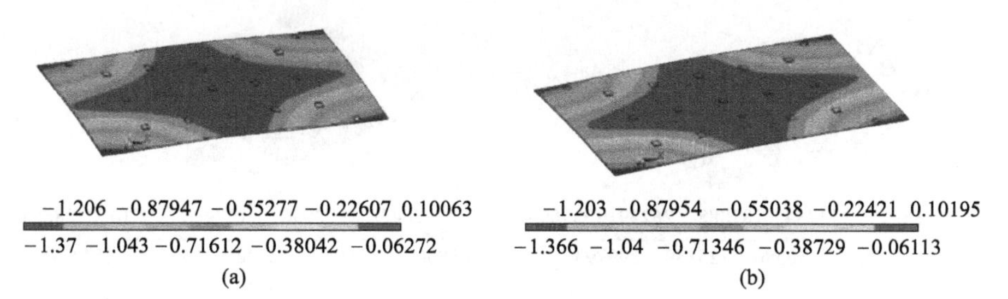

Figure 16.6 Deformation of the test board. (a) U8 location package; (b) U3 location package *(Color version of this figure is available online)*

16.1.2 Simulation Results

As mentioned above in Section 16.1.1, the simplified packages are assumed to be blocks with equivalent mass as the detailed packages. Through modal analysis of the whole system, it is found that the Young's modulus of the simplified packages has very little impact on the system's natural frequencies. Therefore, it is assumed that the simplified packages have the same Young's modulus as the epoxy molding compound material.

The MLP packages at two different locations are simulated based on the above information. One is focused on the U8 location package (with other packages simplified as blocks), the other is focused on the U3 location one (with other packages simplified as blocks). The deformation of the test board under impact at the end of the impact pulse at 0.5 ms is shown in Figure 16.6 for both cases. It can be seen that the deformation of the test board matches well for both cases. This indicates that the detailed package at U8 location or U3 location does not affect the performance of the whole test board. However, by this method, the detailed information at specific locations such as the solder joints performance can be obtained with less effort, avoiding using the global-local modeling technique.

The peeling stress (Sz) distribution of the solder joints for both cases is illustrated in Figure 16.7. The critical solder joint locates at the outer corner.The maximum peeling stress of solder joints at U8 location is a little bit larger than that at U3 location. Another observation is that the peeling stress of solder joints at U8 location reaches the maximum value at 0.825ms, while the peeling stress of solder joints at U3 location reaches the maximum value at 0.65ms. Figure 16.8 shows the first principal stress (S1) of the solder joints for both cases. It has the same trend as the peeling stress for different locations at U8 and U3. The time when the first principal stress reaches the maximum value is also consistent with that for the peeling stress.

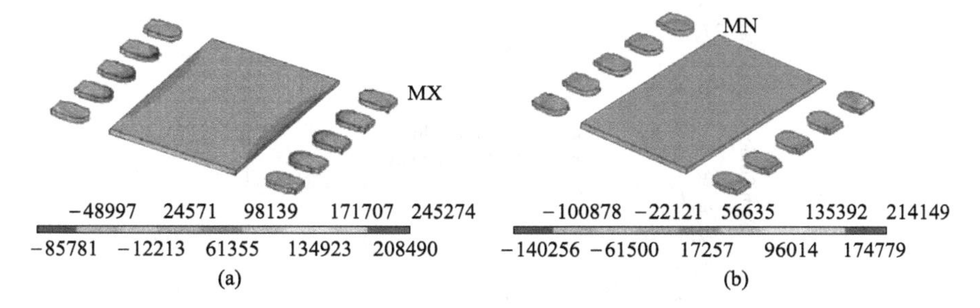

Figure 16.7 Peeling stress of the solder joints. (a) U8 location package; (b) U3 location package *(Color version of this figure is available online)*

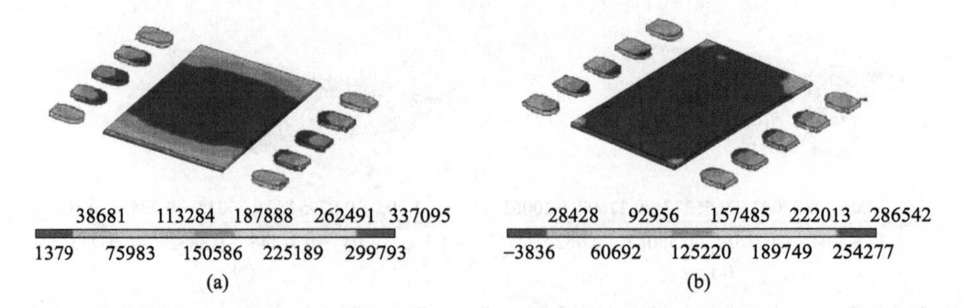

Figure 16.8 First principal stress of the solder joint. (a) U8 location; (b) U3 location *(Color version of this figure is available online)*

16.1.3 Parametric Study

The above loading condition and simulation method are followed to perform the parametric study under the dynamic drop impact. The effects of solder joint height and package thickness are investigated. Two solder joint heights, 2.5 mil and 4.0 mil, and three package thicknesses, 0.55 mm, 0.4 mm, 0.8 mm are considered. The parametric studies are focused on the package at U8 location.

The simulation results of the parametric study are summarized in Table 16.1. The peeling stress and first principal stress of solder joints reach the maximum value at 0.695 ms for the package with thickness of 0.8 mm, while the time is 0.825 ms for the package with 0.4 and 0.55 mm thickness. From the previous section, the time is 0.65 ms for the 0.55 mm thick package at U3 location. This indicates that the time when the stress of solder joints reaches its maximum value depends not only on the package information but also on the package location at the test board.

Table 16.1 Simulation results of parametric study

Solder Joint Height (mil)	Package Thickness (mm)	Time of Peak Stress (ms)	S_z in Solder Joints (MPa)	S_1 in Solder Joints (MPa)
2.5	0.8	0.695	281	382
	0.55	0.825	245	337
	0.4	0.825	224	312
4.0	0.8	0.695	300	396
	0.55	0.825	272	370
	0.4	0.825	253	348

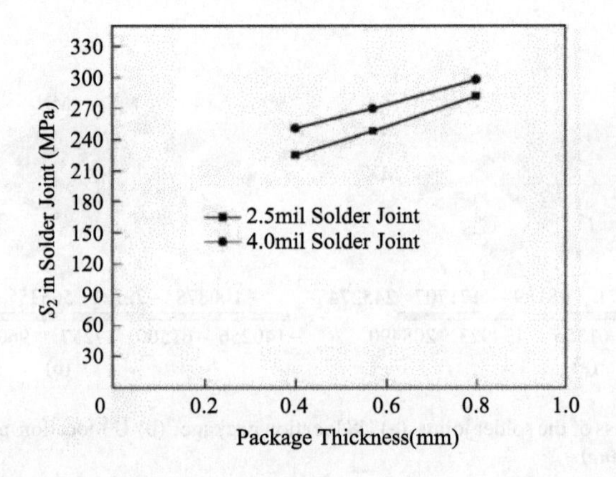

Figure 16.9 Maximum peeling stress in solder joints with respect to different package thicknesses

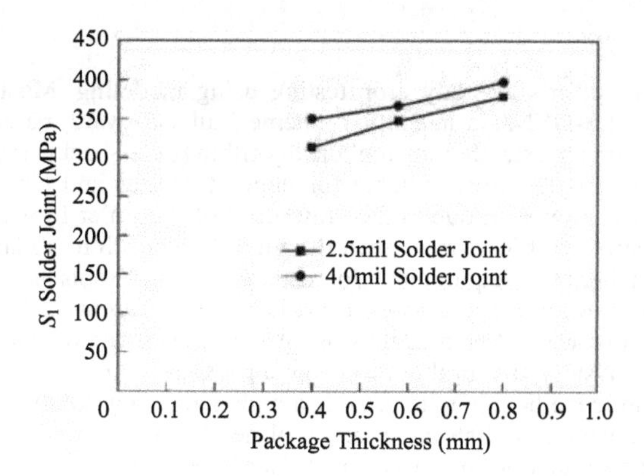

Figure 16.10 Maximum first principal stress in solder joints with respect to different package thicknesses

The effects of package thickness on the stresses of solder joints are illustrated in Figures 16.9 and 16.10. Both the maximum peeling stress and first principal stress increase steadily with the increment of package thickness for the same solder joints height. This means that thinner packages have better drop performance as compared with the thicker ones.

Figures 16.11 and 16.12 shows the effects of solder joint height on the stresses of solder joints under the dynamic impact. Higher solder joints produce larger peeling stress and first principal stress for the same packages thickness. This indicates that lower solder joints may increase the solder joints drop reliability under the dynamic impact.

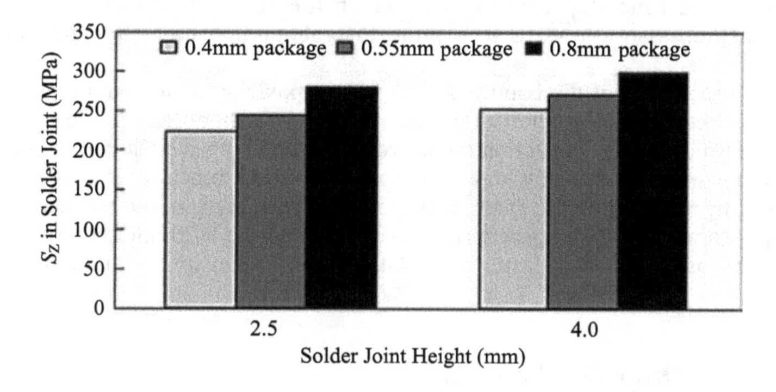

Figure 16.11 Maximum peeling stress in solder joints

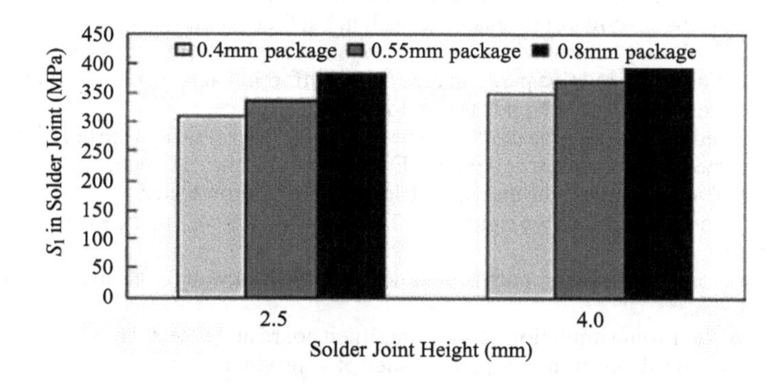

Figure 16.12 Maximum first principal stress in solder joints

16.2 Free Drop

Currently, there are many efforts to study drop testing using modeling. Most of the models utilize LS-DYNA FEA software. LS-DYNA is an explicit scheme (Euler-forward) based FEM code which has very high efficiency and is much faster than an implicit algorithm (Euler-backward), because of the lack of a local implicit-iteration. It is therefore suitable for large structures and high impact problems like automobiles and aerospace. However, due to the drawbacks of the Euler-forward scheme, the explicit scheme may be less accurate and less stable than the implicit algorithm. Today, as portable products become smaller and smaller, the requirement for accurate design tools becomes greater due to a decreasing available engineering margin of error. This section presents a fully nonlinear, transient dynamic implicit algorithm for free drop modeling using a normal commercial finite element framework such as ANSYS's multi-physics, structural or mechanical package. Our goals were two-fold: the first to develop an implicit dynamic methodology to provide a higher accuracy solution; secondly to maximize utilization of our current license pool by increasing tool flexibility.

The advantages of using simulation have been noted elsewhere (Tee *et al.*, 2003) and include lower cost and faster turn time. Through the advantages mentioned by Tee *et al.*, simulation enables techniques that are too costly to be performed by empirical methods. This includes the ability to optimize a package design under multiple requirements, for example we find that conditions that are optimal for drop test survival may not be optimal for thermal cycling, or even if examining drop test alone, it may not be practical to empirically look at multiple combinations of materials, drop angle, and package configuration. Using simulation we can quickly and affordably find the conditions that provide sufficient survivability under different sets of conditions.

In this new method, two models with dynamic contact pair settings are developed: one utilizes the full drop height as the free drop initial condition, and another model utilizes the position right before impact with an equivalent gap and corresponding velocity. Drop duration is defined in four phases: before drop impact, right before impact, right after impact, and well after impact for capturing stress wave propagation with time step control rules which are set up using the IC package's natural frequency. The method also allows us to capture the rebounding unit after first drop, or through multiple impacts.

Finally, drop test modeling of the Fairchild 6 lead Micropak™ mounted on a typical board and end product case is studied by this implicit drop test modeling methodology. Different package design parameters such as solder joint connection type, over mold and chip weight parameters are discussed. The package is compared to a device with an equivalent electrical function, in a flip-chip design using solder balls. Modeling results show the dynamic responses of acceleration, velocity and displacement, the stress wave propagation from the impact surface of the case to the PCB and then from the PCB to the package component, as well as the plastic impact strain energy density distribution in the whole drop impact process.

16.2.1 Simulated Drop Test Procedure

The simulated drop test model used here included the full package under test, we did not use a reduced model, such as a quarter model. The test board and equipment case used however, are simplified. As an IC manufacturer we know the specifics of the package design, but there is little point in using a specific board or case design, as this will vary between differing end customer applications. The simulation as performed will allow us to understand the relative performance of the different package options and will allow us to make recommendations regarding the package and its application.

In the work presented here we assume the case to be made of ABS plastic and the board is connected into the case. The case is the size of a typical cell phone. Further, we assume the impact velocity is not affected by air drag or other friction forces, for example, friction from a drop tester.

We assume that the contact surface is a rigid body. This corresponds well to the worst case actual surface like a concrete floor.

The entire energy loss during the impact is assumed to disperse through the stress wave, and thermal effects are not included.

Although we chose to do our simulation using a simplified board and case, there are no technical reasons that the test could not be done on a complete model of a product. While in the past, compute time restrictions may have limited the complexity, currently available hardware and parallel solvers can reduce runtimes to an acceptable duration.

The code written is designed to let the user choose a drop distance, an initial velocity or both. If the user chooses to include the free drop in the simulation it has been coded to use a very large timestep during the uneventful intial portion of the fall, prior to contact, so that there is minimal effect on the overall running time for the solution.

The timestep that will be used during the contact solution needs to be calculated so that we are able to capture the peak of the stress wave as it passes through the package on the board, or any other interesting features of the package.

We can calculate the size of the timestep required based on an analytical estimate for a simple case. The stiffness can be calculated as:

$$K = \frac{A \times M}{L} \tag{16.3}$$

where A is the characteristic area, L the characteristic length, M is the effective modulus, and K is the stiffness. The frequency of the system will be given by:

$$f = \sqrt{\frac{K}{m}} \tag{16.4}$$

where m is the effective mass. Finally we can estimate the timestep as:

$$\Delta t = \frac{1}{d \times f} \tag{16.5}$$

where d is the number of divisions of a single period and this may be selected to be a number between 50 and 100. The time it takes the stress wave to transit the system will be divided into d timesteps, providing sufficient resolution to capture the action.

16.2.2 Modeling Results and Discussion

Three package designs were modeled. Two were LGA type packages, differing only in the material of the substrate, one using ceramic (LGA-C) and the other BT (LGA-B). The third design was a wafer level flip-chip package that uses solder balls. The circuit function of the die was the same for all packages, although the specific die layout differed between the LGA packages and the flip-chip type. Table 16.2 summarizes pertinent information about the packages.

Figure 16.13 shows a sequence of selected frames for the flip-chip package. The initial two frames show the product package with the target and a closeup, at the first timestep past initial contact, both showing the von-Mises stress contours of the product package. From the sequence of frames that are shown one can see the stress propagates through the package in the manner expected. In the plots below and all plots in this section, the stress units are kPa.

Table 16.2 Key package design parameters

	LGA-B	LGA-C	Flip
Joint Height (μm)	65	65	110
Weight above Joint (mg)	1.62	2.33	1.054
Substrate Density (mg/mm^3)	1.6	3.83	NA
Substrate Type	BT	Cerm.	NA

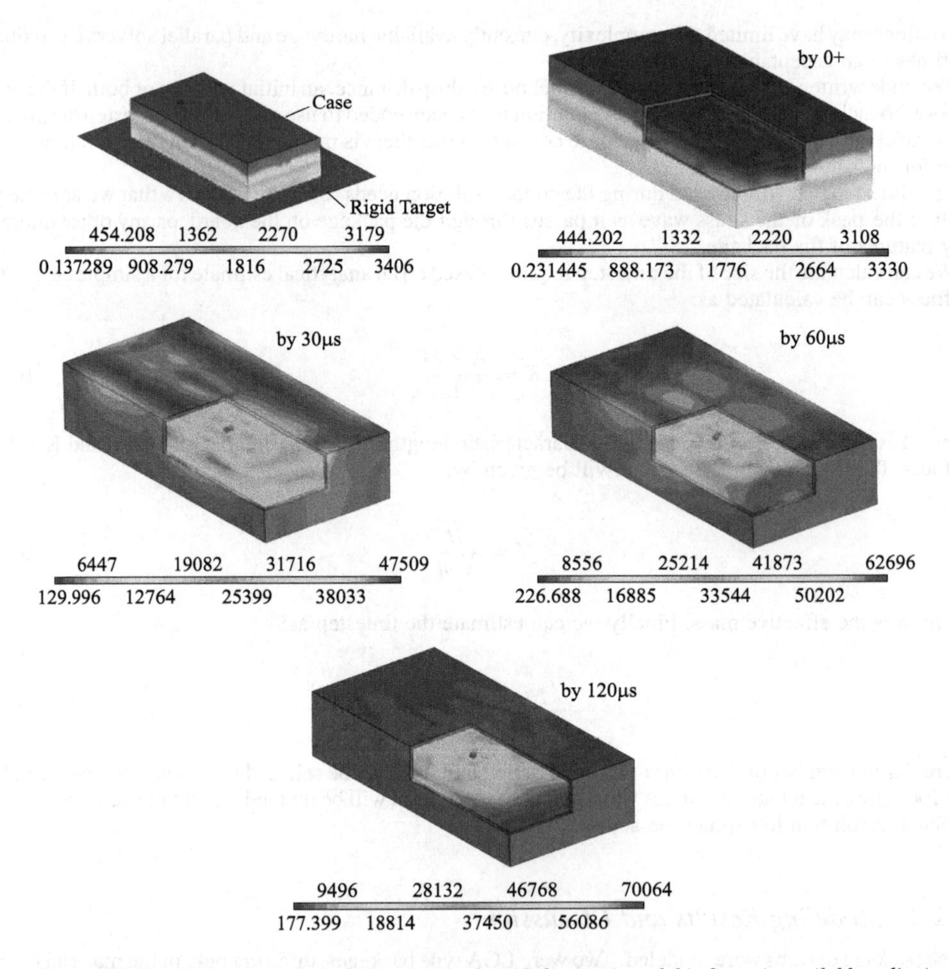

Figure 16.13 Stress wave view of cut-away section *(Color version of this figure is available online)*

Figure 16.14 shows a similar sequence for the board by itself with the case and the package removed. Again this shows the expected propagation of the stress wave. The initial frame is at the first timestep after contact, not exactly at $t = 0$.

The wave patterns of the stress in the board is a result of the transient vibration of the board, after the case strikes the surface. This can further be seen in Figure 16.15. Figure 16.15 shows the acceleration as experienced by a point on one of the corner solder joints. There is a large initial acceleration from the initial contact, and then ringing for a time of approximately 90 μs.

Some error is inherent in the calculation of the acceleration above. Using the implicit method in the current version of ANSYS the velocity and acceleration are only available from estimating the first and second derivatives in time of the displacement. If one is using an explicit method such as the ANSYS LS-DYNA module the velocity and accelerations are available within the simulator.

Figure 16.16 shows the maximum von-Mises stress in the LGA-C package over time. The time of impact is clearly delineated as is expected and the vibration of the board is clearly seen in the ringing of the stress. The ability to capture this level of detail is a good indicator that our selected timestep is sufficiently small to capture the relevant information. The von-Mises stress graphs for other two packages (LGA-B and Flip-chip) are similar except for magnitude.

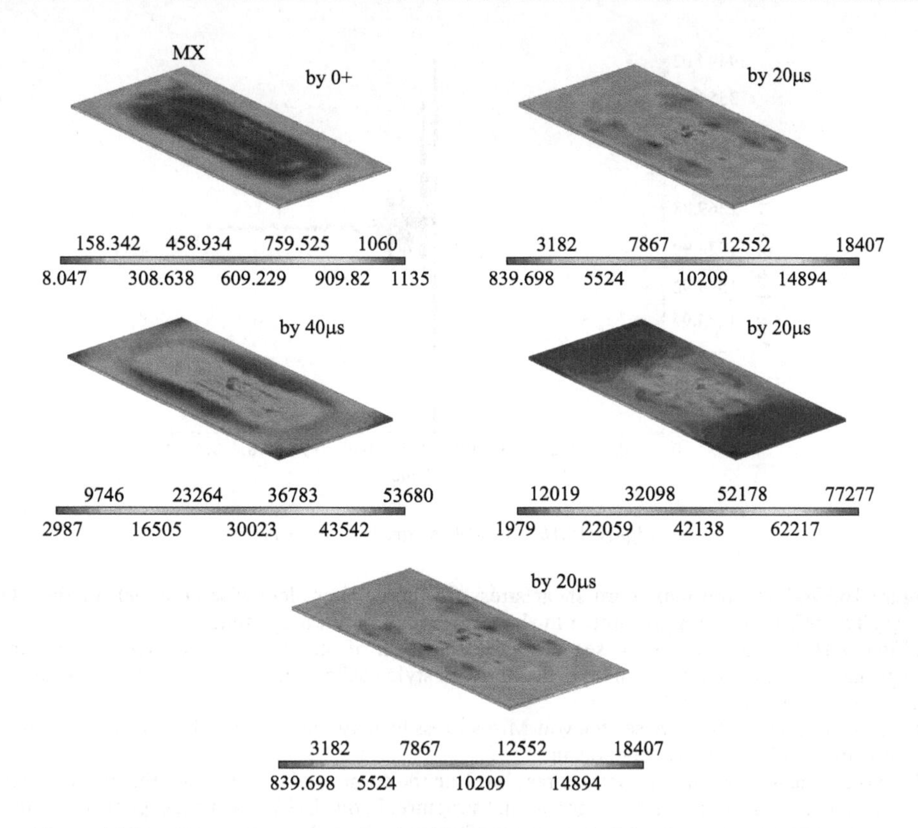

Figure 16.14 Stress wave propagation in PCB *(Color version of this figure is available online)*

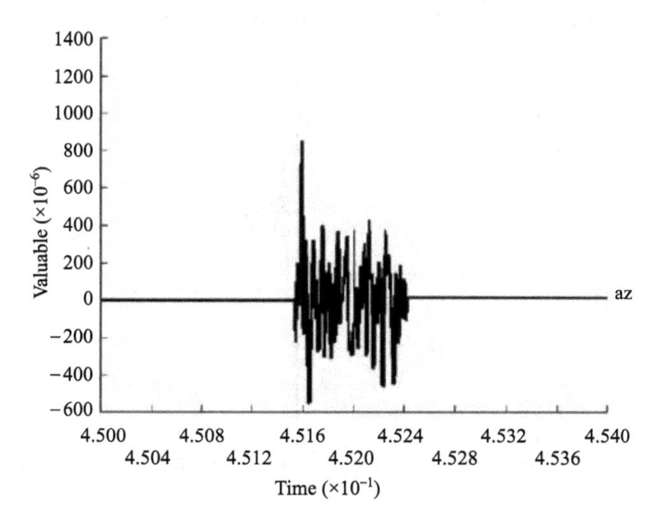

Figure 16.15 Acceleration a solder ball

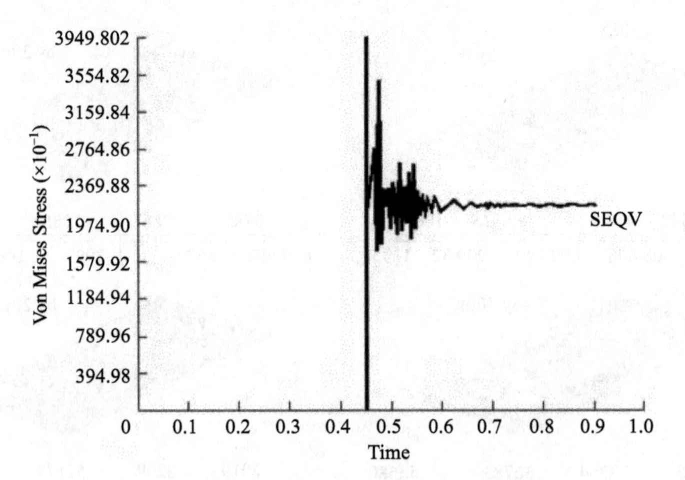

Figure 16.16 von-Mises stress at $t=33\ \mu s$

Figure 16.17 shows the maximum shear stress (in the counter-clockwise contour) for the LGA-C package. The other packages are similar in shape, although not in magnitude.

In Figures 16.18 and 16.19 we can see the increased level of stress in the heavier, stiffer, ceramic based package, Although not shown here, the flip-chip style package has a much greater stress level than the BT based package (LGA-B).

In Figures 16.20 and 16.21 we see the von-Mises stress levels at the solder pad level which is where the greatest stress levels in the package occur.

There is a noticable asymmetry of the stress levels in the corner pads. This shows that it is important to capture much of the detail of the package internal structure. In this LGA based package there are internal metal lines that run at an angle across the package. The level of detail was relatively easy to capture using the large number of element shapes available in ANSYS, and would have been more difficult to capture with the limited set of element shapes in a tool such as LS-DYNA.

Since the internal routing can affect the stress results it suggests that it may be possible to further optimize the package so that the stress is more evenly spread. This may allow us to raise the stress in two corner pads, while lowering it in the other two, which should be beneficial.

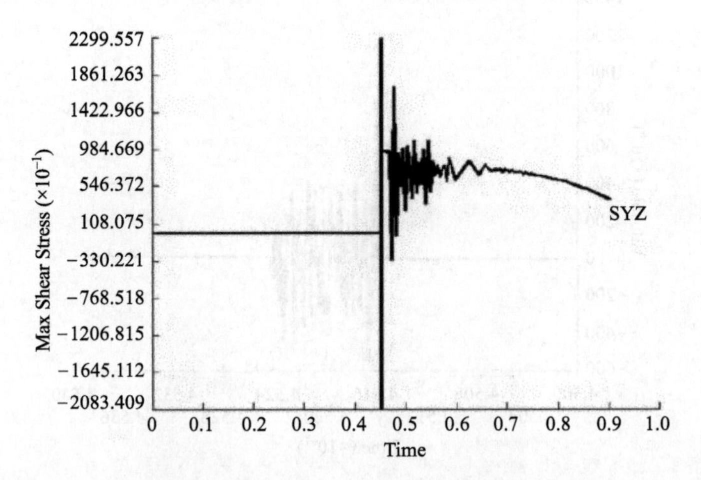

Figure 16.17 Maximum shear stress (LGAC)

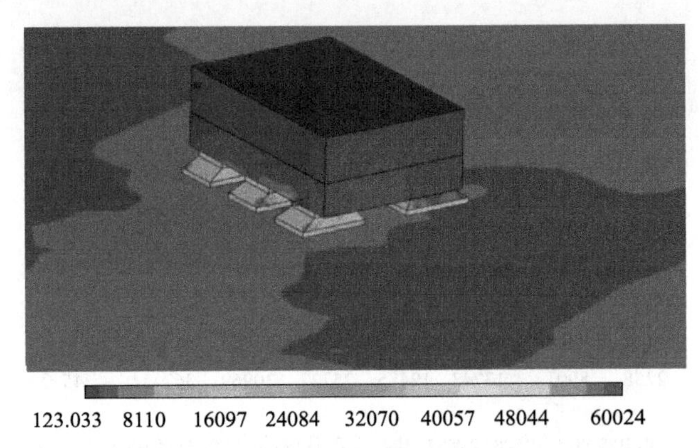

Figure 16.18 Peak von-Mises stress in the LGA-B *(Color version of this figure is available online)*

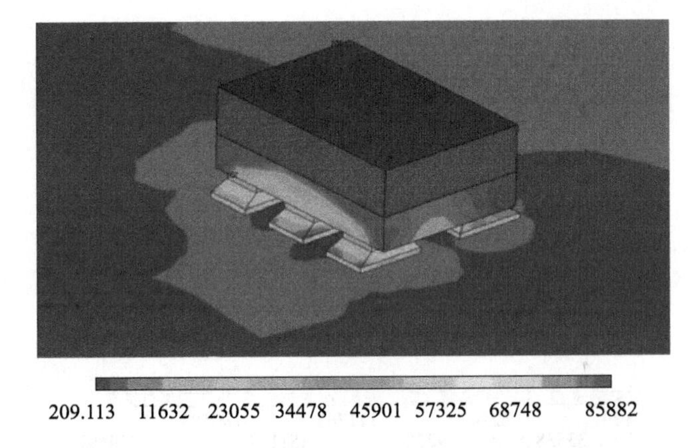

Figure 16.19 Peak von-Mises stress in the LGA-C *(Color version of this figure is available online)*

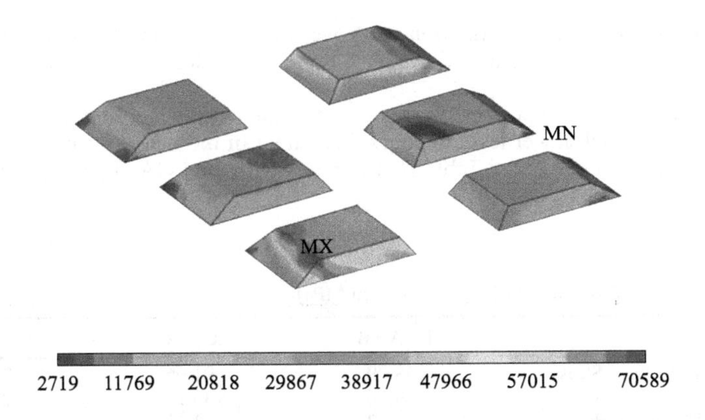

Figure 16.20 Representative von-Mises stress *(Color version of this figure is available online)*

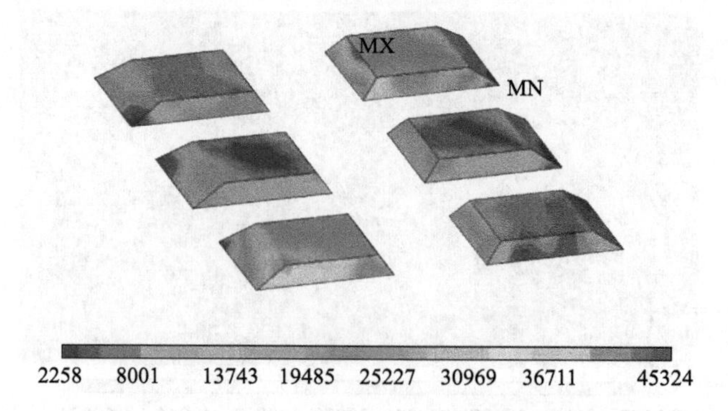

Figure 16.21 Representative von-Mises stress in the LGA-B package pads *(Color version of this figure is available online)*

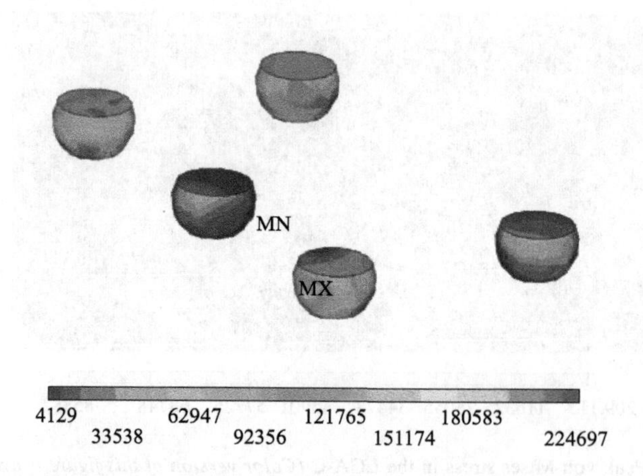

Figure 16.22 Representative von-Mises *(Color version of this figure is available online)*

Figure 16.22 shows the same information as above, but for the flip-chip type package. A major structural difference is evident here; the flip-chip uses five solder balls, not six pads as the LGA packages do.

Table 16.3 summarizes the important stresses found in the packages. From the data we can see the LGA-B package has a much lower level of stress than either of the other two packages.

The differences are summarized in Table 16.4 as ratios with the LGA-B package being normalized to unity.

Table 16.3 Maximum package stresses (in MPa)

	LGA - B	LGA - C	Flip-chip
Max von-Mises Stress	38.79	91.96	292.84
Max Shear Stress	19.16	39.13	127.34
Max Plastic Work	1.13	3.52	11.37

Table 16.4 Package stress ratios

	LGA-B	LGA-C	Flip-chip
Max von-Mises Stress	1	2.37	7.55
Max Shear Stress	1	2.04	6.65
Max Plastic Work	1	3.12	10.06

16.3 Portable Electronic Devices Drop Test and Simulation

In the sections, an insulin pump will be explained for the drop impact test and the simulation models of both the insulin pump and cellphone will be presented.

Preliminary data are collected out of an insulin pump on a free drop tester. Information on the acceleration, impact force, and local strains by strain gauges is obtained. The effects of the drop orientation and height are also discussed.The information is helpful to the dynamic design of an insulin pump against possible plastic deformation on the plastic case, needles, container for the insulin, electronic board, and interfaces. In addition, the drop impact simulation modeling results of the insulin pump are compared with the some experimental results.

For the cellphone modeling, the material rates dependent constitute modeling and nonlinear contact mechanics based modeling are also considered in the drop simulation with various impact orientations.

16.3.1 Test Set-Up

A unique drop tester is equipped with a drop control mechanism that is designed to control drop orientation and gain test results with a high degree of repeatability. A data acquisition and processing system are developed for this setup to measure impact force, acceleration, and strains. The system was developed in 1996 for Motorola by the first author's group in collaboration with Wisdom Technology Inc., now part of FineMEMS Inc. The same system is installed at Huazhong University of Science and Technology. Figure 16.23 shows a schematic illustration of the drop tester. In the picture, there is a drop frame that is arranged to slide along two guide cables. The cables are used to maintain the drop angle. Considerable effort was expended to insure that the apparatus would contribute negligible deviation from true free fall physics for any attached specimen. A specimen is hung at the frame with three strings. The string hanging is an ideal flexible fixture, allowing easy adjustment of drop orientation. The strings also isolate vibrations in the frame from the test specimen. The drop frame is attached to the stationary portion release of the consumer product from a desired height above the impact surface repeatedly. Specimen release is governed by computer control. After the moving frame is released and dropped, it will be stopped by rubber blocks. These rubber blocks are fixed at a lower position on the main frame, such that the test specimen hits the barrier as a free-fall dropped body.

The signal acquisition and processing system include sensors, signal amplifier, data acquisition module, anti-alias filter, A/D converter, and a computer for data display and post-processing. A load cell is installed under the barrier plate. The barrier plate is made of steel, or aluminum, or tile, or hard wood. Low mass accelerometers and strain gauges are installed inside the tested specimen. A program running in LabView environment has been developed to control the system and to process the test data. The parameter settings,such as dynamic range scales of sensors and amplifiers, data acquisition periods, and trigger time, are input to the computer. A trigger icon is toggled to release the electromagnet and to trigger the data acquisition system. The collected test data are displayed as time history curves and frequency spectra in various display windows. Then, the data are saved for further post-processing. Fundamental quantities such as impact force, acceleration, and strains are obtained. It is therefore essential to come up with some unique drop tester which can overcome the above mentioned difficulties. At the same time, the obtained data, when they are repeatable, can be used for the impact modeling by nonlinear finite element method.

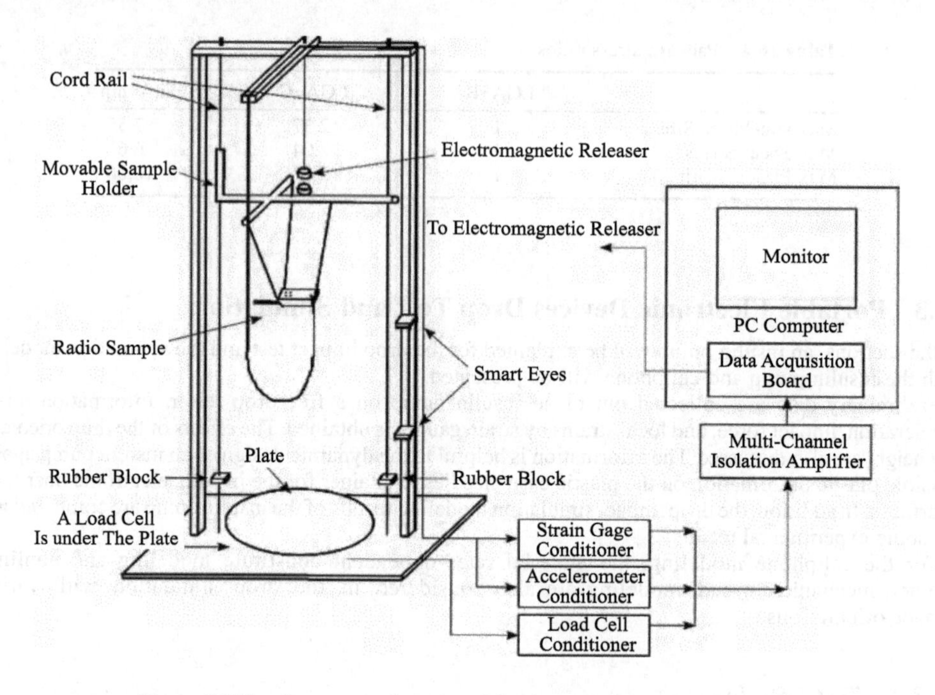

Figure 16.23 Structure and picture of the drop tester for portable device

16.3.2 Modeling and Simulation

Solid modeling tool ProE, and nonlinear finite element codes LS-DYNA are used for the modeling and simulation. Figure 16.24 shows the complex FEM models of the insulin pump and cellphone, and the most non-negative components are include in the models for the precise simulation modeling. Contact is modeled by master-slave. Shell element and 3D elements are all used. Global/local modeling is used for both global warpage/dynamic deformation and local stress/strain analysis. Plastic materials are assured to be rate dependent. For the comparison, as shown in the Figure 16.25, various drop impact orientations are conducted during the numerical simulation modeling.

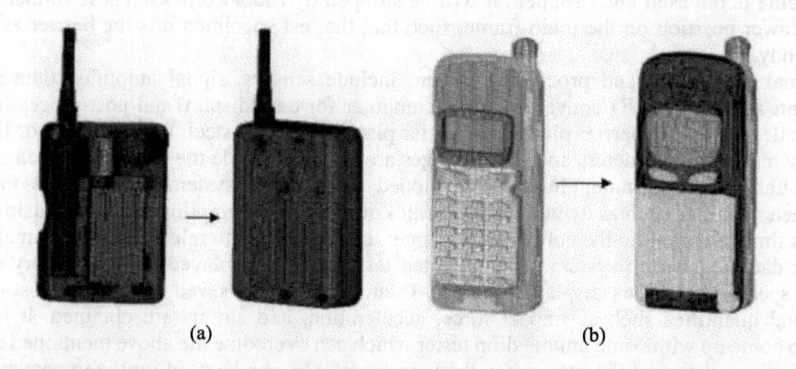

Figure 16.24 FEM model of two portable electronic devices. (a) Insulin pump; (b) cellphone

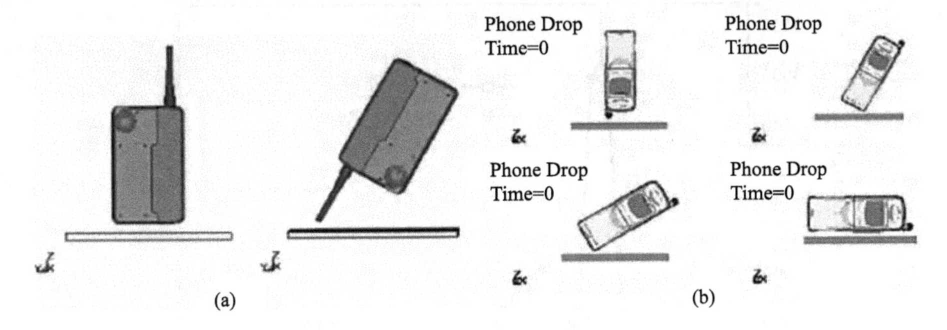

Figure 16.25 Picture of various impact orientations for simulation. (a) Insulin pump; (b) cellphone

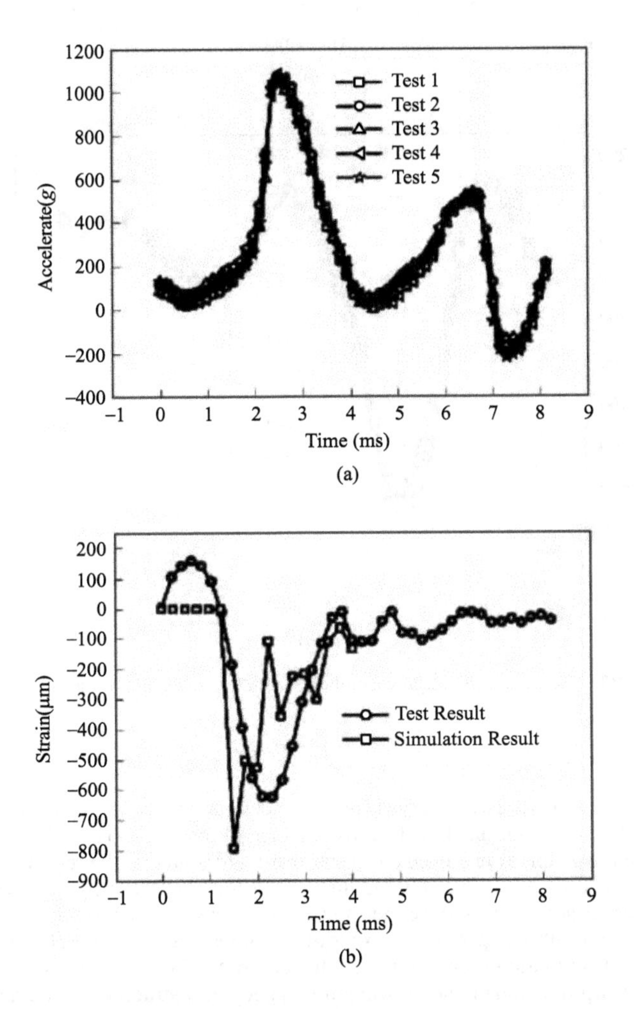

Figure 16.26 Impact under 750 mm height and impact direction of 90 degrees. (a) Acceleration; (b) impact strain

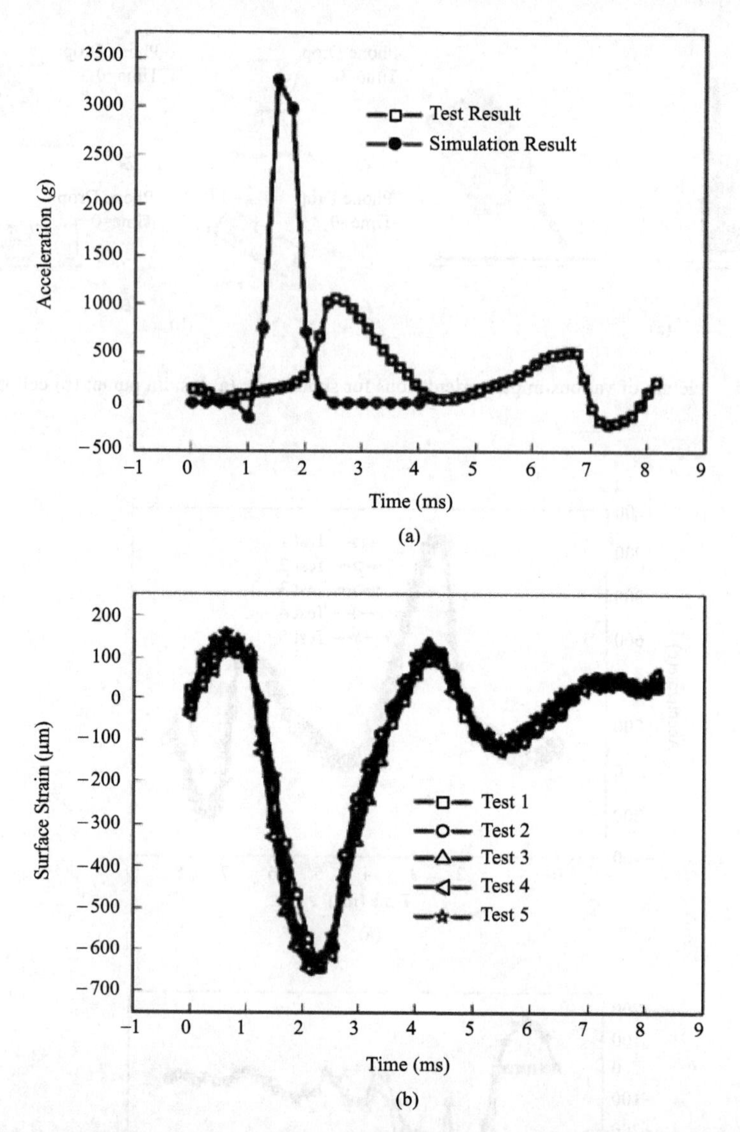

Figure 16.27 Test and drop impact simulation results. (a) Acceleration; (b) impact strain

16.3.3 Results

Figure 16.26 presents the first testing curves for the acceleration, and surface strain responses for an insulin pump with one an orientation of 90 degrees. The height for the drop is 750 mm. It shows that fairly high acceleration and large force are induced. It is interesting to note that the response is very repeatable, demonstrating that our drop tester is a reliable instrument for portable medical devices.

The same drop impact situations of the experiment test are implied to the simulation models of insulin pump. Drop simulation results compared with experimental impact response are shown in the Figure 16.27. Date points of test and simulation seem not to match well, that is for the un-modeled simulation models, such as the suitable simplification on the insulin pump's geometry structure. However, the trends of both can be found to be consistent.

16.4 Embedded Ultrathin Sensor Chip Drop Test and Simulation

Embedded packaging integrating both active and passive components into a single high density substrate, which can provide an effective solution to the system on package, has been paid more and more attention due to the demands of the high density, miniaturization, multifunction, high performance, and low cost of the semiconductor industry(Topper *et al.*, 1997; Cho *et al.*, 2002, 2004; Khan *et al.*, 2013; Lee *et al.*, 2012). The reliability of the embedded devices in the organic substrate packaging is one of the major concerns during its applications. A novel thermosetting resin with low elastic modulus, low dielectric constant, and low dielectric loss for the embedded passives and actives integral substrates was developed in (Uwada *et al.*, 2004). The results show that the embedded passives and actives integral substrates have good reliabilities during the reflow process and thermal cycling test. Finite element model of the embedded chip into the substrate was built in reference (Khong *et al.*, 2008) to study the effects of chip thickness, chip location, chip shape, solder bump, and material properties on the stress level under the thermal loading from 150℃ to 25℃. The heat dissipation and thermal stress of the active embedded chips in the organic substrate were investigated through the numerical simulation in reference (Guo *et al.*, 2011). The drop impact reliability of the high density embedded substrate is also one of the key issues for its mobile applications. However, there is still lack of references that could be found on drop impact reliability assessment for the embedded chip on the organic substrate packaging.

In this chapter, we applied the stress senor chip combined with the numerical simulation to study the drop impact reliability of embedded ultrathin chip in the organic substrate. The stress sensor based on the silicon piezoresistive effects was used for monitoring the stresses under the drop impact loading. The stress and strain behaviors of the embedded ultrathin chip and solder bumps were also investigated dynamic explicit finite element simulation with the input-G method. The effects of material properties and structural parameters on the stress and strain responses of the embedded ultrathin sensor chip and solder bump were investigated.

16.4.1 Stress Sensor and Embedded Package

The developed silicon piezoresistive stress sensors were placed orthogonally for the measurement of the parallel and perpendicular components of the applied stress, as shown in Figure 16.28. Four sensor rosettes were fabricated at different locations on a 5mm×5mm chip, which has 36 I/Os with 0.5mm in pitch. The manufacturing processes and calibration approach for the silicon piezoresistive stress sensor have been described in references (Zhang *et al.*,2009, 2012; Kumar *et al.*, 2011).

The 50μm thickness stress sensor chip embedded into a JEDEC drop impact test board is shown in Figure 16.29. Fine stress sensor chips were embedded on the PCB, which was located at U2, U4, U8, U12, and U14, respectively. In addition, the other locations on PCB were the dummy chips. The assembly and encapsulation processes are as follows. First, the stress sensor chips were bonded onto the copper pads inside the cavities of the printed circuit board (PCB) using the flip-chip bonder (FC150). The cavities were fabricated by the polyimide (PI) coating on the PCB surface. Then, it was reflowed with the conventional lead-free solder reflow temperature profile. The assembled stress sensor chip on the PCB is shown in Figure 16.29(a). After that a conventional underfill material was used to fill the cavity and the gap between the sensor chip and substrate, as shown in Figure 16.29 (b). The SnAgCu solder bumps with 200μm in diameter and 40μm in standoff were used as the interconnections. The structural specifications of the embedded stress sensor chip in the drop impact test board are listed in Table 16.5. Figure 16.30 shows the cross section of the embedded ultrathin stress sensor chip in the PCB test board after the underfill material curing at 150℃. It shows that the ultrathin stress sensor chip was fully covered by the underfill material.

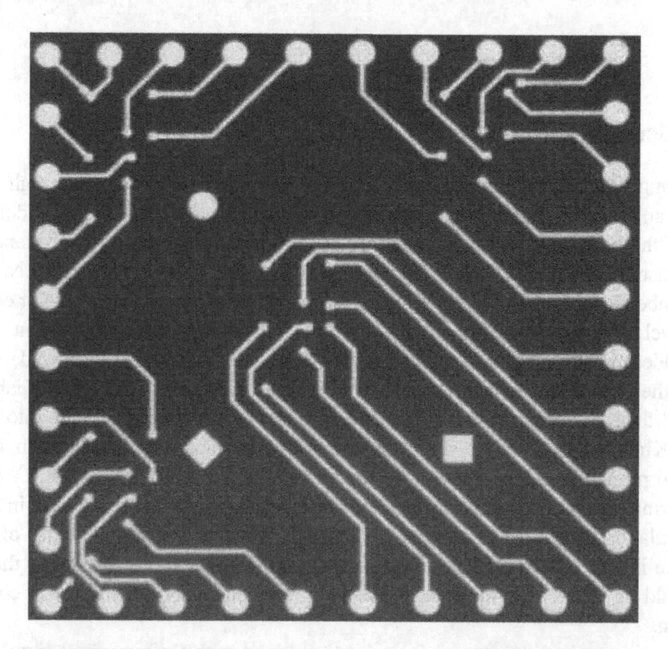

Figure 16.28 Optical picture of stress sensor based on the piezoresistive effects

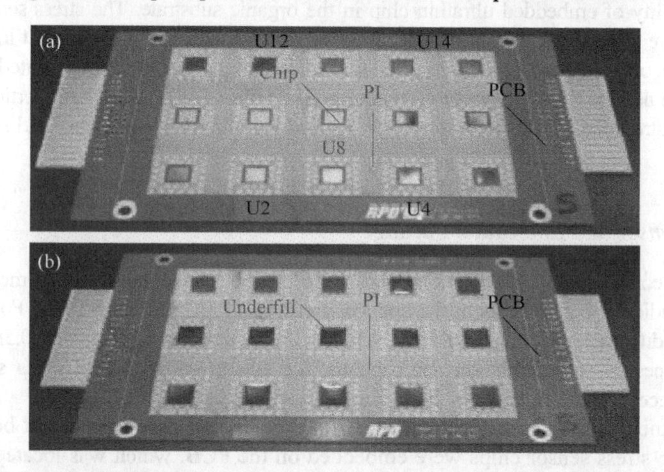

Figure 16.29 Drop impact test board with ultrathin stress senor chips. (a) After the flip-chip bonding and (b) after the embedded process with underfill material

Table 16.5 Structural details of the embedded ultrathin stress sensor chip in drop impact board

Component	Dimensions
Stress Sensor Chip	5mm × 5mm × 0.05mm
Test Board	132mm × 77mm × 1mm
Solder Bump Diameter/Pitch	200μm/500μm
Solder Bump Standoff	40μm

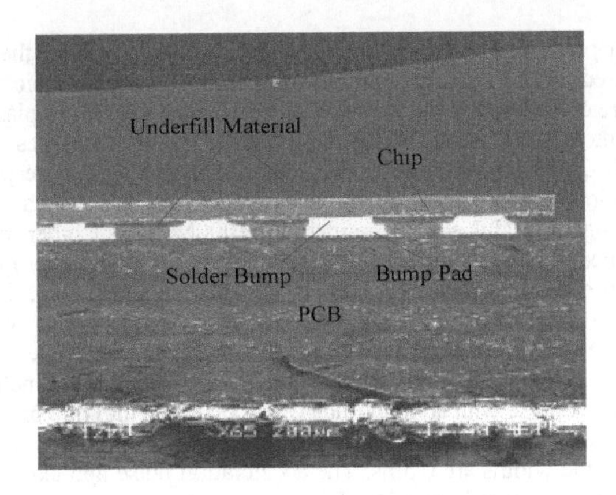

Figure16.30 SEM picture of the cross-section of the embedded ultrathin stress sensor chip in the drop impact test board

16.4.2 Drop Impact FEM Modeling and Validation

In order to further understand the stress and strain behaviors under the drop impact loading for the reliability design of the embedded substrate packaging, numerical simulation is an effective tool. In this chapter, a quarter 3D finite element model of the embedded stress sensor in the drop test broad was established, as shown in Figure 16.31. The embedded ultrathin sensor chip package at the center of PCB was built up in detail according to the geometry dimensions described in Table 16.5. And, the embedded sensor chip packages at the other positions on the drop impact test board were simplified as an equivalent structure. Then, coarse elements can be meshed on these packages in order to reduce the dynamic simulation task of the drop impact.

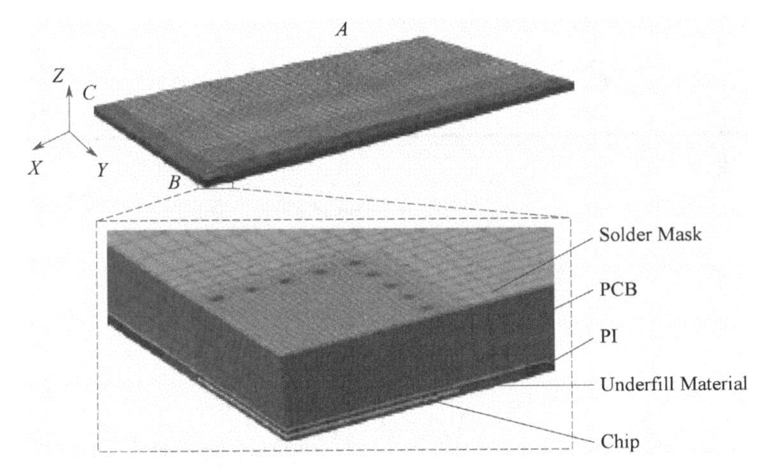

Figure 16.31 Finite element model of drop impact test for the stress sensor chip embedded in the PCB board

The drop impact tests of the embedded ultrathin sensor chip in the organic board were carried out according to the JEDEC standard JESD22-B111 under the loading of a half sine shock pulse with the amplitude of 1500*g* and pulse width of 0.5ms. The input-G method which directly applies the acceleration loading on the bolt holes of the PCB was demonstrated to be an effective way to simulate the drop impact test (Luan *et al.*, 2011; Liu, 2010; Liu *et al.*, 2010; Dhiman *et al.*, 2008; Che *et al.*, 2007; Yeh *et al.*, 2006). In our simulation model, 1500*g* acceleration within 0.5ms was directly applied onto the four bolt-holes on the PCB. In addition, a velocity of −4.77m/s along the *Z*-direction was applied to the whole test board as the initial condition. The symmetry boundary conditions were applied on the cutoff edge of the quarter model. The dynamic explicit nonlinear procedure of the commercial software was used for the drop impact simulation. The element C3D8 was used for the dynamic explicit simulation. The anisotropic material properties of the drop impact test board were considered in the model. The rate-dependent elastic-plastic material properties of solder bumps were taken into consideration (Wong *et al.*, 2008). The yield stresses of the SAC 305 under the strain rates 0.5, 1, 50, 100, 200, and 300 are 73, 77, 109, 124, and 130MPa, respectively. In addition, the copper has the elastic-plastic behavior. The yield stress of copper was assumed as 120MPa. Bilinear plastic model was used and the tangent modulus was assumed as 11000MPa. The other materials in the embedded package model were assumed to have the linear elastic properties. The material properties used for the model are listed in Table 16.6. In order to validate the developed drop impact model, the stress behaviors at the center of sensor chip from the numerical simulation results were compared with those from the drop experimental results. The AVEX shock tester instrumented with an accelerometer was used for drop impact test. The height and pressure of the AVEX System was varied until the accelerometer attached onto the drop table detected a half sine shock pulse with an amplitude of 1500*g* and width of 0.5ms. The acceleration pulses from different drop impact test boards are shown in Figure 16.32. It can be seen that the acceleration pulse amplitudes are ~1500*g* and the widths are 0.5ms. The acceleration pulse applied at the bolt-holes of the test board in the simulation is also shown in Figure 16.32.

Table 16.6 Material properties used for drop impact simulation model

Materials	Density (kg/m^3)	Elastic Modulus (GPa)	Poisson's Ratio	Yield Stress (MPa)
Si	2329	131	0.28	
Cu	8950	117	0.35	120
PI	2200	4	0.35	
PCB	2000	25/*x*, *y* 10/*z*	0.11/*x*, *y* 0.39/*z*	
Underfill	1900	6		
Solder Mask	1300	2.4		
SnAgCu(305)	7390	40		Rate Dependent

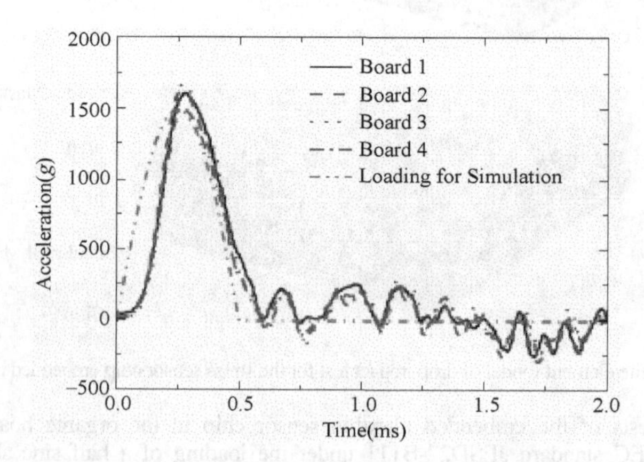

Figure 16.32 Acceleration pulses of drop impact tests of different boards and the accelerations pulse applied in the numerical simulations (*Color version online*)

The displacement evolutions of the different positions (position *A* is located at the bolt-hole, position *B* is located at the center of test board, and position *C* is located at edge of test board) of the drop test board are shown in Figure 16.33. It can be seen that the deflection of the center and edge of test board is cycling under the drop impact loading. The deflection at the bolt-hole position becomes constant after 0.5ms.

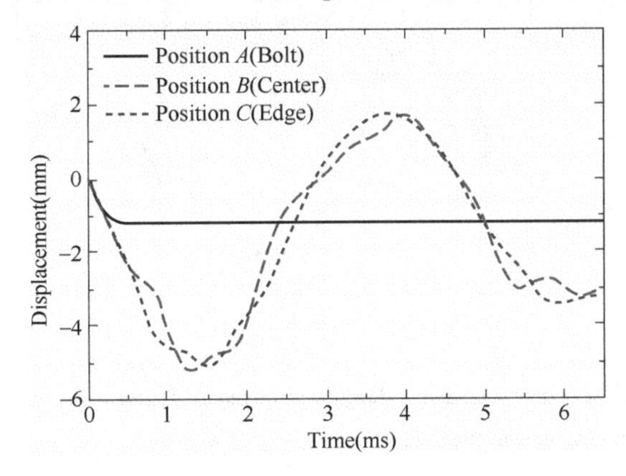

Figure 16.33 Displacement evolutions of the different positions on the PCB during the drop impact test

During the drop impact tests, the signals from the stress sensor at the center of embedded chip were monitored. The stress monitoring results during the drop impact tests are shown in Figure 16.34. It can be found that the normal stress σ_{11} along the *X*-direction is much higher than the normal stress σ_{22} along the *Y*-direction, which is due to that the deflection along the longitude direction of the drop test board is higher than that in width direction. The maximum normal stress σ_{11} at the center of embedded chip occurs during the first deflection cycle. The normal stresses reduce with time due to the dumping effects of the drop impact system.

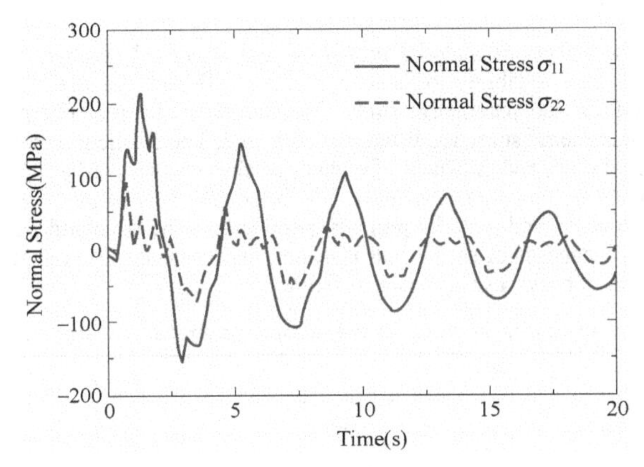

Figure 16.34 Evolutions of normal stresses in *X*-direction and *Y*-direction at the center of embedded stress sensor chip during the drop impact test

The normal stresses σ_{11} at the center of embedded ultrathin stress sensor chip from both experiment and simulation are compared in Fig. 16.35. The maximum normal stress σ_{11} from the simulation is ~276MPa and the maximum normal stresses σ_{11} from the experimental test board 2 and 4 are ~251 and 217MPa, respectively. The difference between the experimental stress curves of test board 2 and 4 might be due to the defects such as small voids and delamination induced by the fabrication processes. Therefore, by compared of the maximum normal stress σ_{11} from simulation result with experimental results of board 2 and 4, the discrepancies were found to be within 9.96% and 27.1%, respectively.

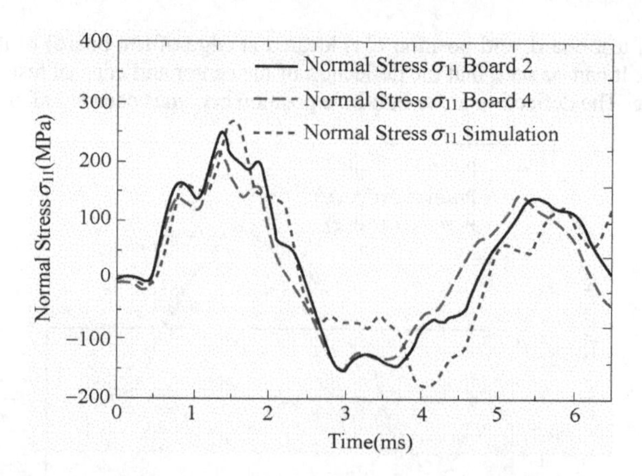

Figure 16.35 Evolutions of the normal stresses in *X*-direction at the center of stress sensor chip embedded with underfill from the experiment and numerical simulation results

The normal stress σ_{11} evolution curve from the simulation is fitted the experiment results quite well before 2.5ms. However, there are some discrepancies from 2.5ms to 4ms. The sources of the discrepancies may be due to the damping factor used in the simulation model is not the same as the experiment. The rate-dependent material properties of the other materials, such as copper, underfill, PCB may also contribute to the discrepancies. Based on this drop impact model, the parametric studies can be conducted to optimize the material selections and structural designs for the embedded ultrathin thin device in the organic substrate packaging.

16.4.3 Parametric Study

In order to investigate the effects of material properties and structural parameters on the stress level of embedded ultrathin chip and reliability of solder bumps, different material properties and structural parameters were selected for the parametric study. The range of the material properties and structural parameters used for the parametric study are listed in Table 16.7. The elastic modulus of underfill, thickness of PCB, standoff of solder bump, and thickness of embedded chip was selected as the factors. Three levels of each factor were considered in the parametric study. The basic case was defined for the parametric study in order to vary different structural and material property parameters. The structural parameters of basic case : thickness of PCB is 1mm, standoff of solder ball is 40μm, and thickness of embedded chip is 50μm. The elastic modulus of underfill of basic case is chosen as 6GPa.

Table 16.7 Factors and levels of the material properties and structural parameters for the parametric study

Factors	Levels		
Elastic Modulus of Underfill	1GPa	6GPa	15GPa
Thickness of PCB	0.5mm	1mm	2mm
Standoff of Solder Bump	40μm	80μm	120μm
Thickness of Embedded Chip	50μm	100μm	200μm

Under the drop impact loading two issues must be considered for the embedded ultrathin device in the organic substrate. The first one is the stress level on the embedded ultrathin chip, which might cause the cracking failure. The second one is the reliability of solder bumps. The plastic strain might be induced under the drop impact loading, which can cause the fracture failure.

Figure 16.36 shows the normal stress σ_{11} distribution of the embedded stress sensor chip after 1.5ms of drop impact test. It can be found that the maximum normal stress σ_{11} locates at the edge of chip surface opposite to the stress sensor side. The maximum normal stress σ_{11} at the chip edge of 316MPa is higher than that at the center of chip surface on the stress sensor side which is 276MPa.

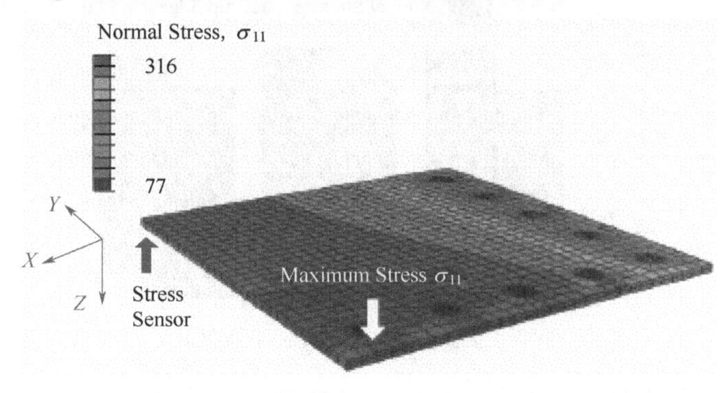

Figure 16.36 Distribution of normal stress in *X*-direction of embedded stress sensor chip after 1.5ms of drop impact test (*Color version online*)

In order to reduce the complication of simulation task, only 6.5ms step time was applied to the simulation model. Figure 16.37 shows the equivalent plastic strain distribution after 6.5ms of drop impact test of solder bumps in the package embedded with underfill. It can be found that the critical solder bump located at the diagonal corner of the solder bumps array. The maximum plastic strain happens to the chip side of the critical solder bump. The maximum equivalent plastic strain of the critical solder bump after 6.5ms drop impact test is 1.311%.

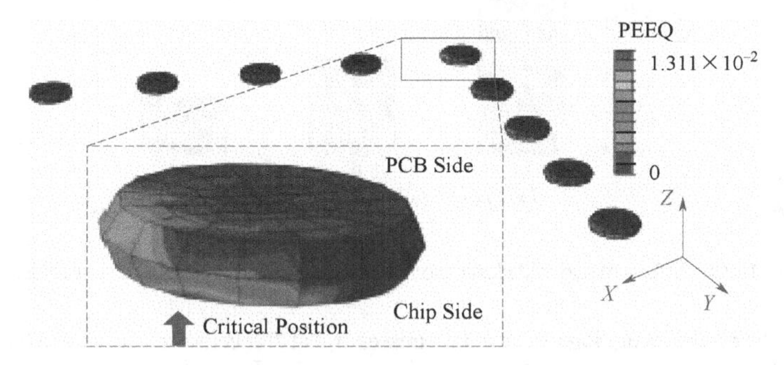

Figure 16.37 Equivalent plastic strain distribution of the solder bumps embedded with underfill after 6.5ms of drop impact test (*Color version online*)

In order to make the comparison and optimization for the parametric study, the maximum normal stresses σ_{11} at the center and edge of the embedded stress sensor chip and the maximum equivalent plastic strain of solder bump after 6.5ms drop impact test were selected as the indexes.

The effects of elastic modulus of underfill on the stress level of embedded thin chip and the equivalent plastic strain of solder bump are shown in Figures 16.38 and 16.39, respectively. It can be found that the effects of elastic modulus on the stress level of embedded chip are limited. The maximum equivalent plastic strain of the solder bump decreases from 4.53% to 0.79% when the elastic modulus of underfill increases from 1GPa to 15GPa. The underfill with higher elastic modulus provides a good protection to solder bump under the drop impact loading. However, the elastic modulus of underfill does not affect much on the stress level of embedded ultrathin chip. Therefore, in order to increase the reliability of solder bump, underfill with higher elastic modulus is recommended.

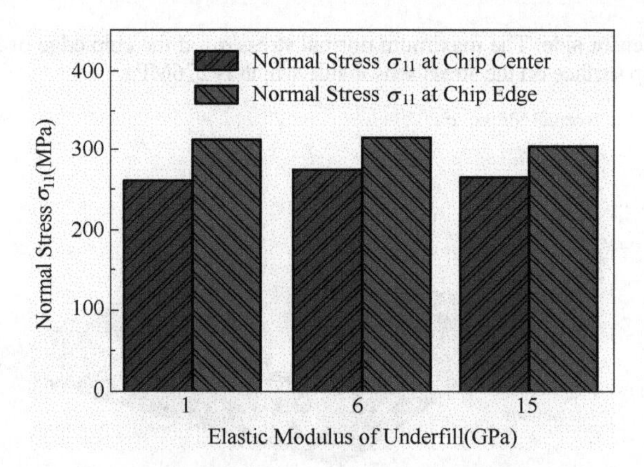

Figure 16.38 Effects of underfill elastic modulus on the normal stress in *X*-direction at the center and edge of embedded stress sensor chip

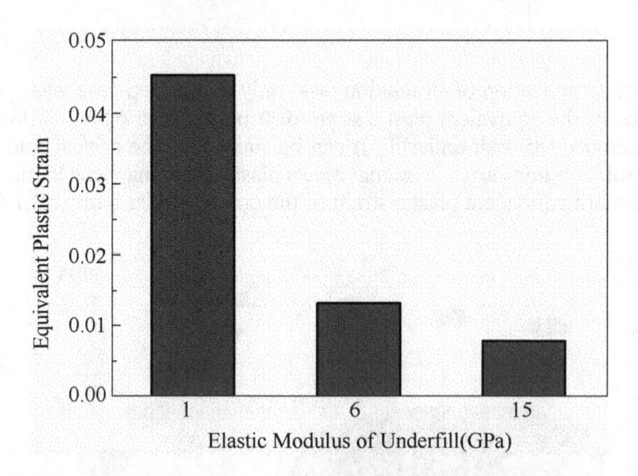

Figure 16.39 Effects of underfill elastic modulus on the equivalent plastic strain of solder bump

Figure 16.40 shows the evolutions of normal stresses σ_{11} at the center of stress sensor chip embedded without underfill from the experimental and simulation results. It can be seen that the simulated stress evolution is consistent with the experimental result, which also provides a validation of the simulation model. It also indicates that the normal stress σ_{11} of 85MPa at the center of stress sensor chip embedded without underfill is much lower than that embedded with underfill which is 251MPa (experimental results).

Figure 16.41 shows the normal stress σ_{11} distribution of embedded stress sensor chip of the case embedded without underfill after 1.5ms of drop impact test. It can be found that the maximum normal stress σ_{11} occurs at the diagonal corner edge of chip surface on the stress sensor side, which is different to the case with underfill, as shown in Figure 16.36. The locations of the maximum normal stress σ_{11} happening on the embedded ultrathin chip are different of the cases embedded with and without underfill.

Figures 16.41 and 16.42 show the effects of underfill on the stress level of embedded chip and the equivalent plastic strain of solder bump (simulation results). It can be found that the maximum normal stress σ_{11} at the center of embedded chip reduces dramatically from 276MPa to 83MPa when no underfill was used for embedding, which is consistent to the stress sensor monitoring result. However, the maximum normal stress σ_{11} located at the edge of embedded chip increases from 316MPa to 362MPa. As shown in Figure 16.43, the maximum equivalent plastic strain of solder bump increases dramatically from 1.31% to 22.7%. Therefore, underfill is required for enhancing the reliability of solder bumps of the embedded packaging under the drop impact loading.

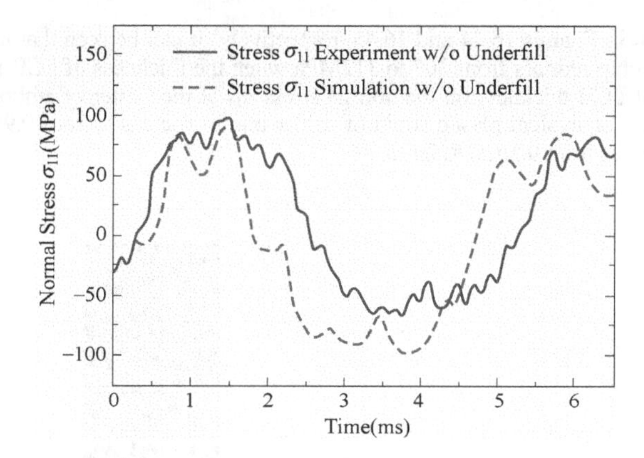

Figure 16.40 Evolutions of normal stresses in *X*-direction at the center of stress sensor chip embedded without underfill from the experiment and simulation results

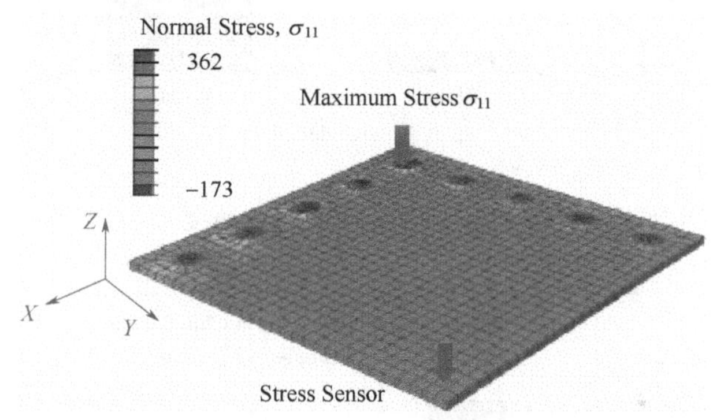

Figure 16.41 Distribution of normal stress in *X*-direction of embedded stress sensor chip after 1.5ms of drop impact test of the case embedded without underfill (*Color version online*)

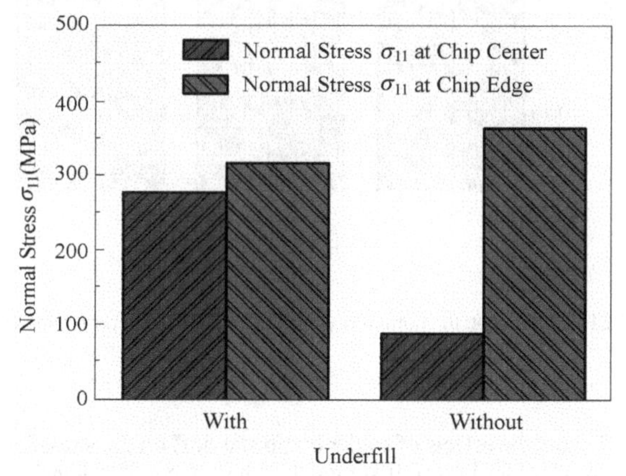

Figure 16.42 Effects of underfill on the normal stress in *X*-direction at the center and edge of embedded stress sensor chip

The effects of PCB thickness on the stress level of embedded chip and the equivalent plastic strain of solder bump are shown in Figures 16.44 and 16.45, respectively. It can be seen that the normal stress σ_{11} at the edge of embedded chip reduces from 367 to 312MPa when the thickness of PCB increases from 0.5mm to 2mm. The effects of PCB thickness on the normal stress σ_{11} at the center of embedded chip are limited. However, the maximum equivalent plastic strain of solder bumps increases from 0.9% to 1.88% when the thickness of PCB increases from 0.5mm to 2mm.

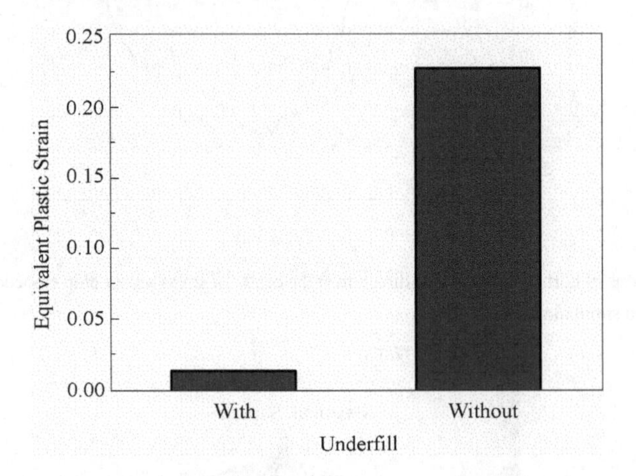

Figure 16.43 Effects of underfill on the equivalent plastic strain of solder bump

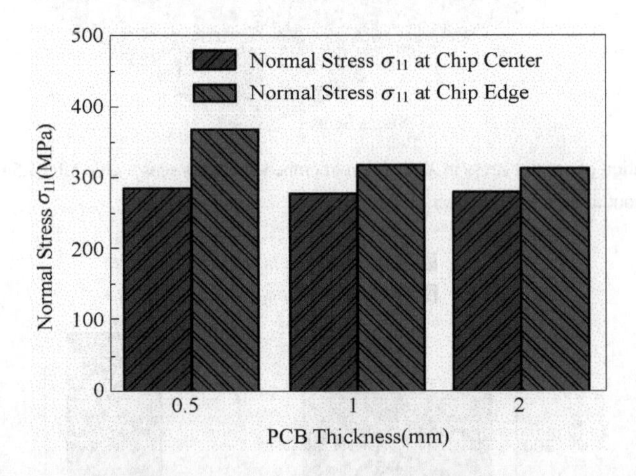

Figure 16.44 Effects of PCB thickness on the normal stress in *X*-direction at the center and edge of embedded stress sensor chip

Figures 16.46 and 16.47 show the effects of solder bump standoff on the stress level of embedded chip and the equivalent plastic strain of solder bump. It shows that the effects of solder bump standoff on the stress level of embedded chip are limited. The normal stress σ_{11} at the edge of embedded stress sensor chip increase slightly from 316MPa to 339MPa with increasing of solder bump standoff from 40μm to 120μm. However, the maximum equivalent plastic strain of solder bump decreases from 1.31% to 0.89% when the solder bump standoff increases from 40μm to 120μm.

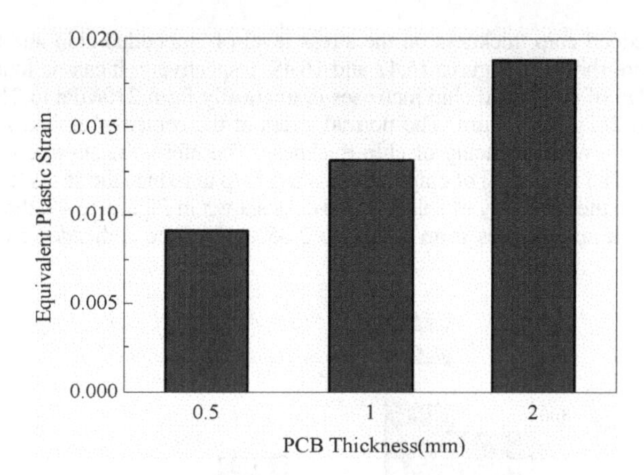

Figure 16.45 Effects of PCB thickness on the equivalent plastic strain of solder bump

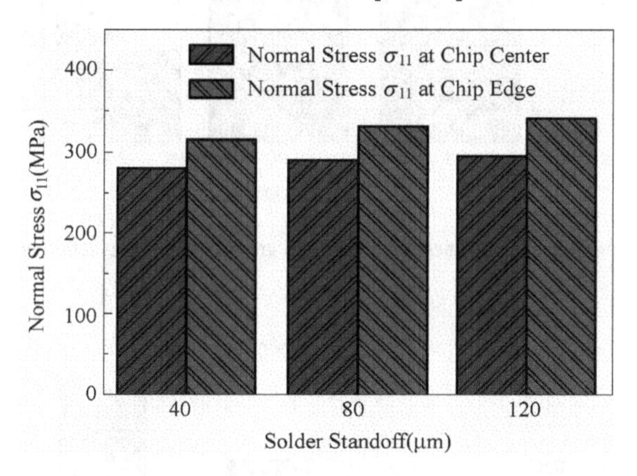

Figure 16.46 Effects of solder bump standoff on the normal stress in *X*-direction at the center and edge of embedded stress sensor chip

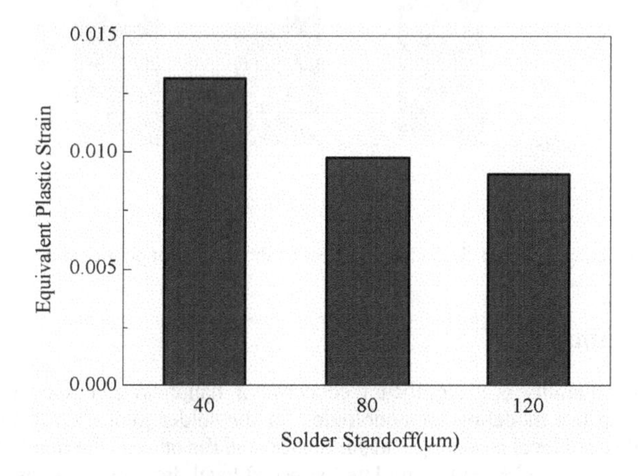

Figure 16.47 Effects of solder bump standoff on the equivalent plastic strain of solder bump

The effects of embedded chip thickness on the stress level of embedded chip and the equivalent plastic strain of solder bump are shown in Figures 16.48 and 16.49, respectively. It can be found that the maximum normal stress at the edge of embedded chip increases dramatically from 210MPa to 316MPa when the chip thickness reduces from 200μm to 50μm. The normal stress at the center of embedded chip also increases from 144MPa to 276MPa when reducing of chip thickness. Therefore, the stress control of the embedded ultrathin chip is critical. The increasing of chip thickness can help to reduce the stress level of embedded chip. However, it is harmful to the reliability of solder bumps. As shown in Figure 16.49, the maximum equivalent plastic strain of solder bump increases from 1.31% to 2.66% when the embedded chip thickness increases from 50μm to 200μm.

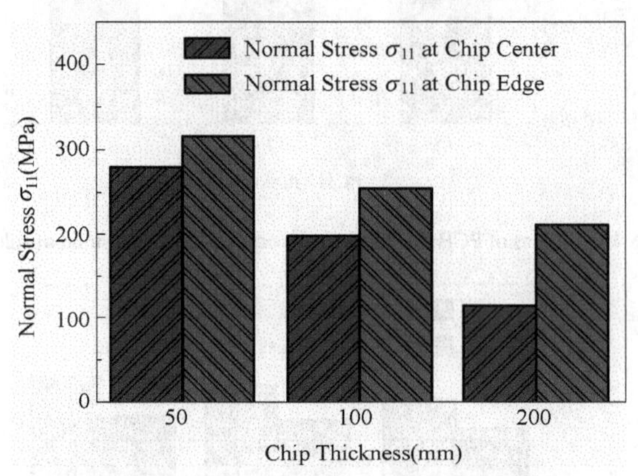

Figure 16.48 Effects of embedded chip thickness on the normal stress in *X*-direction at the center and edge of embedded stress sensor chip

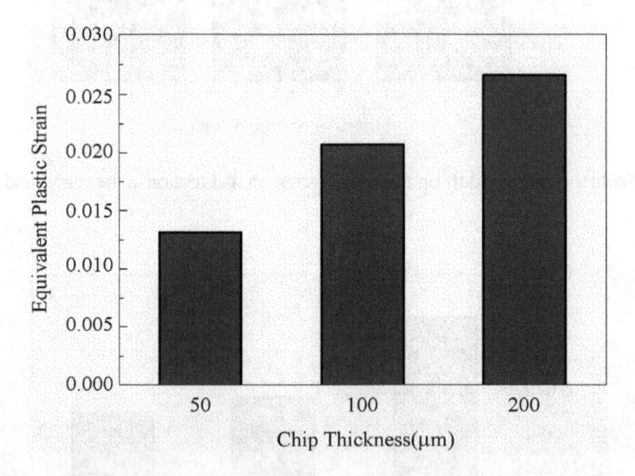

Figure 16.49 Effects of embedded chip thickness on the equivalent plastic strain of solder bump

16.5 Chapter Summary

The solder joints drop reliability is a significant concern for handheld electronic products. This chapter presented two major drop test modeling methodologies for the solder joint drop test reliability. One is the controlled pulse drop at board level modeling and simulation and the other is the free fall board level drop test. The implicit Input-G method is adopted to simulate the board level drop test of an advanced molded leaded package by using the commercial finite element code for the controlled pulse drop test with JEDEC standard. The JEDEC standard test board accommodates 15 packages in a 3 row by 5 column format. Parametric studies on solder joints height and MLP package thickness are conducted in the board level drop test simulations. The peeling stress and first principal stress of the solder joints are examined and compared. Simulation results show that thinner packages have better drop performance compared with the thicker ones, and lower solder joints may increase the solder joints drop reliability under the dynamic impact. The methodology of the full free drop test modeling is developed in the chapter after the controlled pulse drop test simulation section. Two models with dynamic contact pair settings are developed: one utilizes the full drop height as the free drop initial condition, and another model utilizes the position right before impact with an equivalent gap and corresponding velocity. Drop duration is defined in four phases: before drop impact, right before impact, right after impact, and well after impact for capturing stress wave propagation with time step control rules which are set up using the IC package's natural frequency. The method also allows us to capture the rebounding unit after first drop, or through multiple impacts. Finally the full free drop test simulation for a cell phone with various impact orientations is investigated. The results of free drop test and simulation are comparison and discussion. In the last part of this chapter, the embedded test chip has been demonstrated to be a powerful tool when coupled finite element method for design for testability and design for manufacturability for the optimum packaging design.

References

Groothuis, S., Chen, C. and Kovacevic, R. (2005) Parametric investigation of dynamic behaviour of FBGA solder joints in board-Level drop simulation, *Proceedings of Electronic Components and Technology Conference 2005* Lake Buena Vista FL, 499–503.

JEDEC, Standard JESD22-B111 (2003) Board Level Drop Test Method of Components for Handheld Electronic Products.

Kujala, A. *et al.* Nokia Transition to Pb-free manufacturing using LGA technology, *Electronic Components and Technology Conference 2002* San Diego CA.

Luan, J.E. and Tee, T.Y. (2004) Novel board level drop test simulation using implicit transient analysis with input-G method, *Proceedings of the 6th Electronic Packaging Technology Conference*, Singapore, 671–677.

Luan, J.E. and Tee, T.Y. (2005) Drop impact simulation using implicit input-G method, *5th ANSYS Asian Conference*, Singapore.

Tee, T.Y., Ng, H.S., Lim, C.T., Pek, E. and Zhong, Z. (2003) Board level drop test and simulation of TFBGA packages for telecommunication applications, *Electronic Componants and Technology Conference*, New Orleans, LA, 121–129.

Wong, E.H., Lim, K.M., Lee, N., Seah, S. and Wang, H.C. (2002) Drop impact test - mechanics and physics of failure, *Journal of Electronics Packaging Technology Conference 4th*, December, Issues 10–12, 327–333.

Wu, J., Song, G., Yeh, C. and Wyatt, K. (1998) Drop/impact simulation and test validation of telecommunication products, *Proceedings of Inter Society Conference on Thermal Phenomena*, ITHERM, 330–336.

Wu, J. (2000) Global and local coupling analysis for small components in drop simulation, *Proceedings of the 6th International LS-DYNA Users Conference*, Detroit, 11:17–11:26.

Xie, D., Arra, M., Yi, S. and Rooney, D. (2003) Solder joint behavior of area array packages in board level drop for handheld devices, *Proceedings of 53rd Electronic Components and Technology Conference*, New Orleans, Louisiana, May, 130–134.

17

Electromigration

This chapter studies the numerical simulation method for electromigration in IC device and solder joint in a package under the combination of high current density, thermal load, and mechanical load. The three dimensional electromigration finite element model for IC device/interconnects and solder joint reliability is developed and tested. Numerical experiments are carried out to obtain the electrical, thermal, and stress fields with the migration failure under high current density loads. The direct coupled analysis and indirect coupled analysis that include electrical, thermal, and stress fields are investigated and discussed. The viscoplastic Anand constitutive material model with both SnPb and SnAgCu lead-free solder materials is considered in the research. An IC device is studied to show the modeling methodology and the comparison with previous test data. A global CSP package with PCB is modeled using relative coarse elements. In order to reduce the computational costs and to improve the calculation accuracy, a submodel with a refined mesh is constructed. The submodel technique is studied in a direct and indirect coupled multiple fields. The comparison of voids generation through numerical example in this chapter and previous experimental result is given.

17.1 Basic Migration Formulation and Algorithm

Based on Black's equation (Black, 1967), the atomic flux due to electromigration, thermomigration and stress migration in a conductor can be expressed as following (Dalleau *et al.*, 2001):

$$\vec{J}_{\mathrm{Em}} = \frac{ND}{kT} Z^{*} e\rho \vec{j} \tag{17.1a}$$

$$\vec{J}_{\mathrm{Th}} = -\frac{ND}{kT} Q^{*} \frac{\nabla T}{T} \tag{17.1b}$$

$$\vec{J}_{\mathrm{S}} = -\frac{ND}{kT} \Omega \nabla \sigma_{\mathrm{m}} \tag{17.1c}$$

$$\vec{J}_{\mathrm{Tol}} = \vec{J}_{\mathrm{Em}} + \vec{J}_{\mathrm{Th}} + \vec{J}_{\mathrm{S}} \tag{17.1d}$$

where N is the atomic concentration; k is Boltzmann's constant; e is the electronic charge; Z^{*} is the effective charge which is determined experimentally; T is the absolute temperature, ρ is the resistivity which is calculated as $\rho = \rho_0(1+\alpha(T-T_0))$, α is the temperature coefficient of the metallic material; D is the diffusivity, $D = D_0 \exp\left(-\frac{E_a}{kT}\right)$, E_a is the activation energy, D_0 is the thermally activated diffusion

Modeling and Simulation for Microelectronic Packaging and Integration: Manufacturing, Reliability and Testing, Second Edition. Sheng Liu and Yong Liu.

coefficient; $\vec{j}$ is the current density vector; Q^* is the heat of transport; Ω is the atomic volume; and σ_m is the local hydrostatic stress.

From Equation 17.1a, the atomic flux divergence due to the electromigration can be expressed as:

$$\mathrm{div}(J_{Em}) = \left(\frac{E_a}{kT^2} - \frac{1}{T} + \alpha\frac{\rho_0}{\rho}\right)\cdot \vec{J}_{Em}\cdot \nabla T + \frac{1}{N}\vec{J}_{Em}\cdot \nabla N \tag{17.2a}$$

The atomic flux divergence related to the thermomigration can be written as:

$$\mathrm{div}(\vec{J}_{Th}) = \left(\frac{E_a}{kT^2} - \frac{3}{T} + \alpha\frac{\rho_0}{\rho}\right)\cdot \vec{J}_{Th}\cdot \nabla T + \frac{NQ^*D_0}{3k^3T^3}j^2\rho^2e^2\exp\left(-\frac{E_a}{kT}\right) + \frac{1}{N}\cdot \vec{J}_{Th}\cdot \nabla N \tag{17.2b}$$

From Equation 17.1c, the atomic flux divergence due to stress migration can be derived as:

$$\mathrm{div}(\vec{J}_S) = -\frac{N\Omega D}{kT}\left\{\left(\frac{E_a}{kT^2} - \frac{1}{T}\right)\nabla\sigma_m\cdot\nabla T + \mathrm{div}\nabla\sigma_m\right\} - \frac{\Omega D}{kT}\nabla\sigma_m\cdot\nabla N$$

The coupling of temperature effects on the expansion of a material is then defined by Hooke's law:

$$\varepsilon_{xx} = \frac{1}{E}\left[\sigma_{xx} - \nu(\sigma_{yy} + \sigma_{zz})\right] + \alpha_l(T - T_0)$$

$$\varepsilon_{yy} = \frac{1}{E}\left[\sigma_{yy} - \nu(\sigma_{xx} + \sigma_{zz})\right] + \alpha_l(T - T_0)$$

$$\varepsilon_{zz} = \frac{1}{E}\left[\sigma_{zz} - \nu(\sigma_{zz} + \sigma_{xx})\right] + \alpha_l(T - T_0)$$

where E is Young modulus, ν is the Poisson ratio, αl is the expansion coefficient of the metallization. The divergence of the hydrostatic stress gradient can then be expressed as:

$$\mathrm{div}\nabla\sigma_m = -\frac{2E}{3(\nu-1)}\alpha_l\mathrm{div}\nabla T$$

The final formula of the atomic flux divergence due to stress migration can be represented as:

$$\begin{aligned}\mathrm{div}(\vec{J}_S) = &\left(\frac{E_a}{kT^2} - \frac{1}{T}\right)\cdot\vec{J}_S\cdot\nabla T + \frac{1}{N}\vec{J}_S\cdot\nabla N + \\ &\frac{2EN\Omega D_0\alpha_l}{3(1-\nu)kT}\exp\left(-\frac{E_a}{kT}\right)\left[\frac{j^2\rho^2e^2}{3k^2T} + \left(\frac{1}{T} - \alpha\frac{\rho_0}{\rho}\right)\nabla^2T\right]\end{aligned} \tag{17.2c}$$

Without considering the effect of the divergence item (related to gradient of ∇N) of atomic density, the divergences of total atomic flux for electronic migration, thermal migration, and stress migration can be expressed as:

$$
\begin{aligned}
\operatorname{div}(\vec{J}_{\mathrm{Tol}}) = {} & \left(\frac{E_{\mathrm{a}}}{kT^2} - \frac{1}{T} + \alpha\frac{\rho_0}{\rho}\right)\cdot\vec{J}_{\mathrm{Em}}\cdot\nabla T + \left(\frac{E_{\mathrm{a}}}{kT^2} - \frac{3}{T} + \alpha\frac{\rho_0}{\rho}\right)\cdot\vec{J}_{\mathrm{Th}}\cdot\nabla T + \\
& \frac{NQ^{*}D_0}{3k^3T^3}j^2\rho^2e^2\exp\left(-\frac{E_{\mathrm{a}}}{kT}\right) + \left(\frac{E_{\mathrm{a}}}{kT^2} - \frac{1}{T}\right)\cdot\vec{J}_{\mathrm{S}}\cdot\nabla T + \\
& \frac{2EN\Omega D_0\alpha_1}{3(1-\nu)kT}\exp\left(-\frac{E_{\mathrm{a}}}{kT}\right)\left(\frac{1}{T} - \alpha\frac{\rho_0}{\rho}\right)\nabla^2T - \frac{2EN\Omega D_0\alpha_1}{3(1-\nu)kT}\exp\left(-\frac{E_{\mathrm{a}}}{kT}\right)\frac{j^2\rho^2e^2}{3k^2T}
\end{aligned}
\tag{17.2d}
$$

The divergences of atomic flux can be further expressed as:

$$
\operatorname{div}(\vec{J}_{\mathrm{Tol}}) = N\cdot F(\vec{j}, T, \sigma_{\mathrm{m}}, E_{\mathrm{a}}, D_0, E, \cdots) \tag{17.3}
$$

The time dependent evolution of the local atomic concentration is given as:

$$
\operatorname{div}(\vec{J}_{\mathrm{Tol}}) + \frac{\partial N}{\partial t} = 0 \tag{17.4}
$$

From Equation 17.3, the divergence value is in proportion to the atomic concentration N and to the function F in which different physical parameters are included. So, Equation 17.4 is equivalent to the following Equation 17.5:

$$
NF + \frac{\partial N}{\partial t} = 0 \tag{17.5}
$$

Integrating Equation 17.5, it can be obtained that:

$$
\int_{N_i}^{N_{i+1}} \frac{\mathrm{d}N}{N} = \int_{t_i}^{t_{i+1}} -F\mathrm{d}t \tag{17.6}
$$

Due to the explicit Euler algorithm, Equation 17.6 can be simplified as:

$$
N_{i+1} = N_i\mathrm{e}^{-F_i\Delta t_i} \tag{17.7}
$$

where N_i and F_i are the atomic concentration and the value of F in ith step, respectively. Equation 17.7 is the incremental formula of the atomic concentration. From Equations 17.1 to 17.3 and Equation 17.7 the local time-dependent iterative scheme has been developed, which is different from previous work.

In this chapter, both the direct coupled analysis and indirect coupled analysis are studied. The directly coupled method and its flow chart have been proposed in our previous papers (Liang *et al.*, 2006). Figure 17.1 shows the algorithm of indirectly coupled analysis. Submodeling is a finite element technique that one can use to obtain more accurate results in a particular region of a model. In the thermalelectric submodel solution procedure, the degree of freedom (DOF) variables (temperature and voltage) at the nodes along the cut boundaries by interpolating results from the gobal model are applied on the cut boundaries of the submodels. In the structure submodel solution producer, the DOFs are temperature and displacement to be transferred from the gobal model to the submodel. In Figure 17.1, the initial atomic density N_0 is set to be 1. This will not lose the generality for the void generation algorithm because the void criterion is based on the ratio of N/N_0. In addition, if people are interested in the actual atomic flux divergence, they can easily obtain it by multiplying the actual initial atomic density N_0 by the current value of simulated atomic flux divergence.

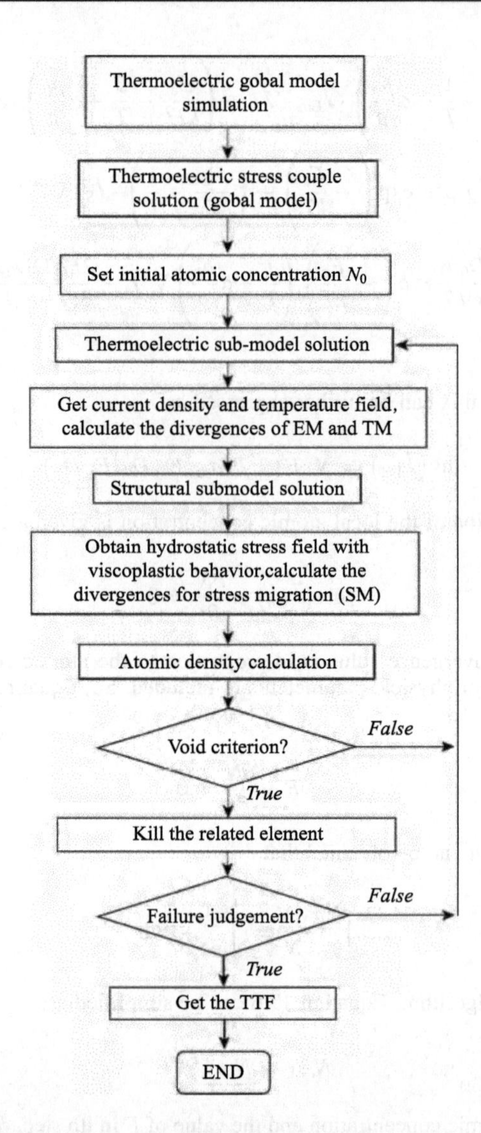

Figure 17.1 Flow chart of indirect coupled analysis

In structure analysis, the solder bump is meshed with the visco107 element for the ANAND model. But the visco107 element in ANSYS® dose not have the element birth and death feature. To achieve the element death effect, ANSYS® does not actually remove "killed" elements. Instead, it deactivates them by changing the element material attribute from an ANAND model to an elastic model with 10×10^{-6} of the elastic module and the density of the source model. This actually multiplies the stiffness by a severe reduction factor. Here, with the previously presented comsiderations extracted from the experimental analysis and the obsevations (Dalleau *et al.*, 2002), due to the exponential form taken by Equation 17.7, the criterion is that the material of an element in the model is considered to be empty when its atomic concentration reaches 10% of the initial concentration. So, once the atomic concentration value of an element has reached the above criterion, the element will be killed.

In the indirect coupled algorithm flow chart, finding the solution of divergence of electric, thermal, and stress migration is a core part of the methodology. Following is the iterative algorithm.

Let i step be the ith iterative step of Equation 17.7 from initial time T_0 to failure time T_f.

1. Resume electric-thermal couple submodel and solve;
2. Compute the divergences for electronic migration and thermal migration:
   2.1 Create an element table for temperature, current density, and temperature gradient
   2.2 Element table operation: $\vec{j} \cdot \nabla T, \nabla^2 T, j^2$
   2.3 For elem $\in$ [1, $e$max], where emax is the size of element table
      (a) Get $T, \vec{j} \cdot \nabla T$, $\nabla^2 T$ and j^2 of every element
      (b) AT(elem) = T, AJDT(elem) = $j \cdot \nabla T$, ADT2(elem)= $\nabla^2 T$, AJS2(elem) = j^2, ATG (elem)= ∇T
      (c) Compute div$(\vec{J}_{\mathrm{Em}})$ and div$(\vec{J}_{\mathrm{th}})$
   2.4 Output all the parameters and array to an external file
   2.5 Save and exit
3. Resume the structural submodel, which the thermal boundary condition comes from as the solution of electric-thermal couple analysis above and solve;
4. Compute the divergences for stress migration at the ith step:
   4.1 Resume all the parameters and array from the saved external file
   4.2 For elem $\in$ [1, emax], build a local stress reconstruction:
      (a) Get $T, \vec{j} \cdot \nabla T$, $\nabla^2 T$ and j^2 of every element
      (b) Get the local hydrostatic stress σ_m^k of every node inside the current element
      (c) Create the shape function array ψ_k
      (d) $\sigma_m = \sum\limits_{k=1}^{n} \sigma_m^k \psi_k, \frac{\partial \sigma_m}{\partial x_k} = \sum\limits_{k=1}^{n} \sigma_m^k \frac{\partial \psi_k}{\partial x_k}$
      (e) Compute div$(\vec{J}_{\mathrm{S}})$ at the i th step
      (f) Find the total div$(\vec{J}_{\mathrm{Tol}})$ = div$(\vec{J}_{\mathrm{EM}})$ + div$(\vec{J}_{\mathrm{Th}})$ + div$(\vec{J}_{\mathrm{S}})$ at i th step. The function F_i of Equation 17.3 can be obtained, which can used for iterative Equation 17.7 to get the atomic density $N_i + 1$
5. Repeat the above procedure for next step.

17.2 Electromigration Examples from IC Device and Package

This section will give two application examples, one is an IC device and another one is a flip-chip CSP package. Direct couple analysis methodology is applied to the IC device and both direct and indirect couple methodologies are applied to the solder joint inter-connection of the CSP package.

17.2.1 A Sweat Structure

The three-dimensional finite element model for a sweat structure (from Dalleau *et al.*, 2002) is shown in Figure 17.2. Here due to symmetry, only a quarter of the structure was modeled. The symmetry from this simplification can reduce the number of elements and save computing time. The linear regular hexahedral elements were map-meshed, which can also save computing time and improve the accuracy.

This section focuses only on high load condition from 12×10^6 A/cm^2 up to 24×10^6 A/cm^2.Different values of loading are taken within this range in order to study the influence of the electrical charge on the lifetime of the metallization structure. The structure is considered to be stress-free at room temperature for the simulations. The boundary conditions are applied the same as the reference (Dalleau *et al.*, 2002) for a bench work comparison. At the top metal fan out end 10 μm section, apply the current load $I = 0.123$ A and at the center metal 1 um section, apply the voltage $V = 0$ (Figure 17.2).

Figure 17.3 shows the temperature distribution and current density distribution under current load of 0.123A. Figure 17.4 shows the atomic density distribution of the sweat structure, which indicates that failure firstly happens near the middle of the sweat structure. The result was consistent with the picture observed in the experiment (Dalleau *et al.*, 2002), as shown in Figure 17.4(b). Furthermore, according to

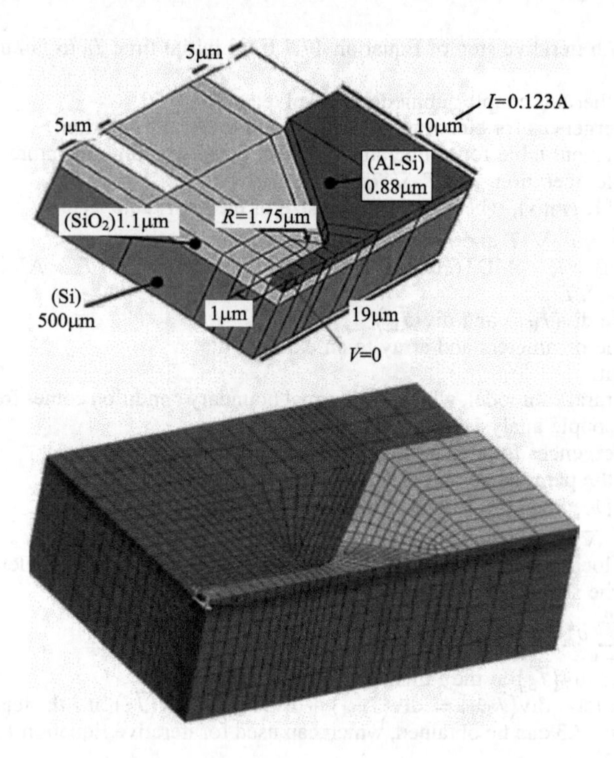

Figure 17.2 A 3D sweat structure model and mesh

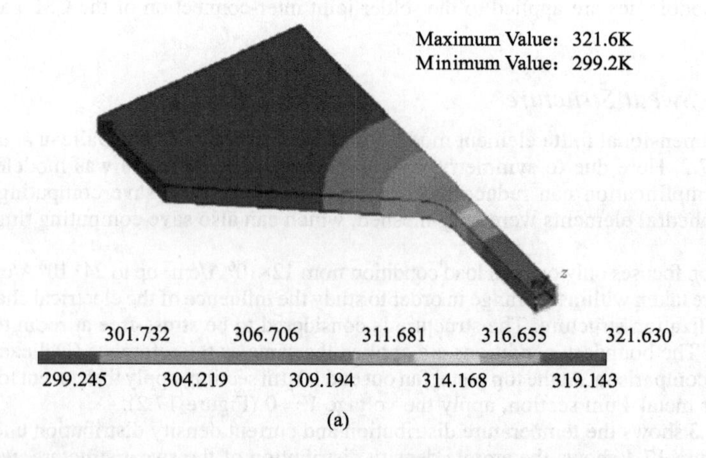

Figure 17.3 Temperature and current density distributions with current load 0.123 A. (a) Temperature distribution; (b) current density distribution *(Color version of this figure is available online)*

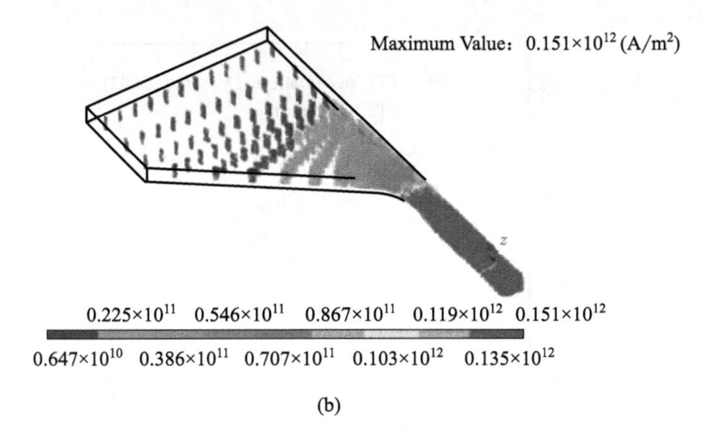

(b)

Figure 17.3 (*Continued*)

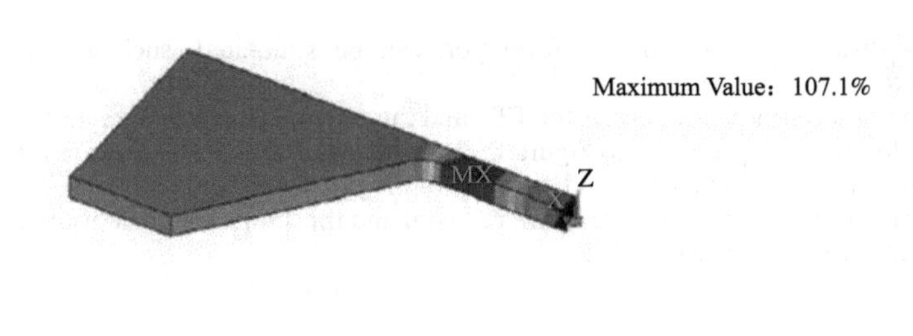

(a)

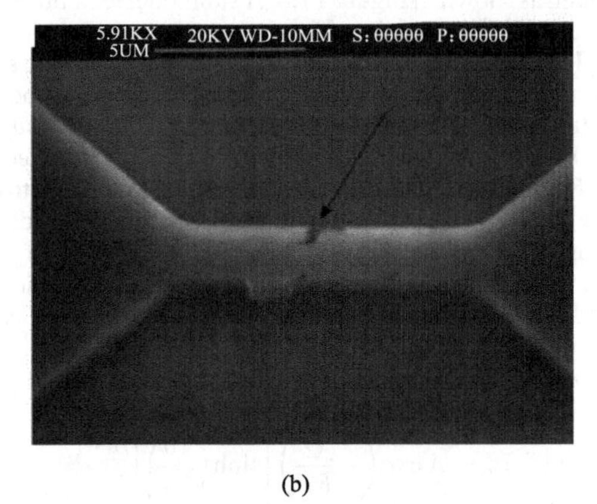

(b)

Figure 17.4 Void comparison between and simulation and test. (a) Relative atomic density distribution (dark color indicates very small value: void) with $N_0 = 1$; (b) void generation (Dallleau *et al.*, 2002)

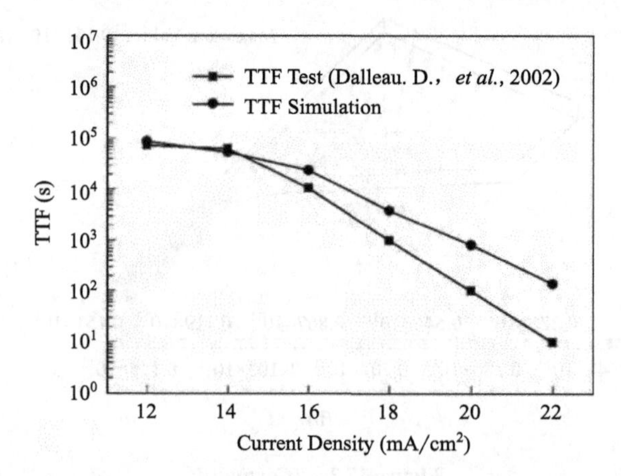

Figure 17.5 Time to failure (TTF) comparison between the test data (Dalleau *et al.*, 2002) and our algorithm

atomic density distribution, the hillock formation can be simulated, such as the red location in Figure 17.4(a).

A comparison of the simulation results for TTF and the previous experimental test results obtained in (Dalleau *et al.*, 2002) is also presented in Figure 17.5. The simulated TTF life data are very consistent with the experimental data. In considering all the assumptions (Al diffusivity, effective charge number, vacancy formation criterion, and so on) that were taken and the simplifications of the models made, the simulation results seem quite reasonable.

17.2.2 A Flip-Chip CSP with Solder Bumps

Gee *et al.* (2005) have performed the electromigration test for lead-free material 95.5Sn4.0Ag0.5Cu and lead material 37Pb63Sn solder joints in a CSP package. Components used in the system include: silicon chip, under bump metallurgy (UBM), aluminum trace, copper trace, and solder bumps. To correlate our simulation methodology with the experimental data. The CSP package in reference (Gee *et al.*, 2005) is modeled, which has 36 bumps with 500 μm pitch. The exterior 20 solder bumps are assumed to connect with each other in a daisy chain as shown in Figure 17.6. A submodel technique in ANSYS is introduced to get a better response of the electronic migration. The global thermal-electric coupled field model uses a Solid-69 element and the global stress model uses a Visco-107 element for solder bumps and Solid-45 element for the remaining parts of the model. The global structure is modeled using relative coarse elements in the first step. In the second step, a refined thermal-electric coupled field submodel and a refined stress submodel with UBM (Al/NI(V)/Cu) layer are then constructed as shown in Figure 17.7. Gee *et al.* (2005) confirmed that the UBM has greater resistance to the electromigration, and the UBM/Solder interface (solder side) seems more likely to have electromigration voids. Therefore, in this section, a very fine mesh is set along the interface.

The material constitutive model of the solder bump is based on an Anand model. The Anand model unifies the creep and rate-independent plastic behaviour of the solder by making use of a flow equation and an evolution equation. The basic equations are listed as follows.

Viscoplastic flow equation:

$$\dot{\varepsilon}_{\mathrm{p}} = A\exp\left(-\frac{Q}{RT}\right)\left[\sinh\left(\xi\frac{\sigma}{s}\right)\right]^{1/m} \tag{17.8}$$

where $\dot{\varepsilon}_{\mathrm{p}}$ is the inelastic strain rate, A is the pre-exponential factor, Q is the activation energy, R is gas constant, T is absolute temperature, ξ is the multiplier of stress, σ is the equivalent stress, and m is strain rate sensitivity.

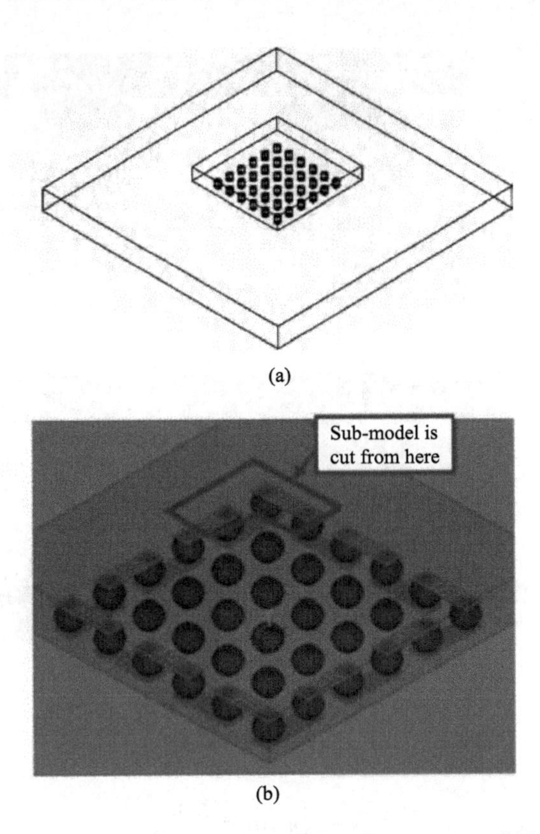

Figure 17.6 A CSP package model. (a) A CSP package structure; (b) local view

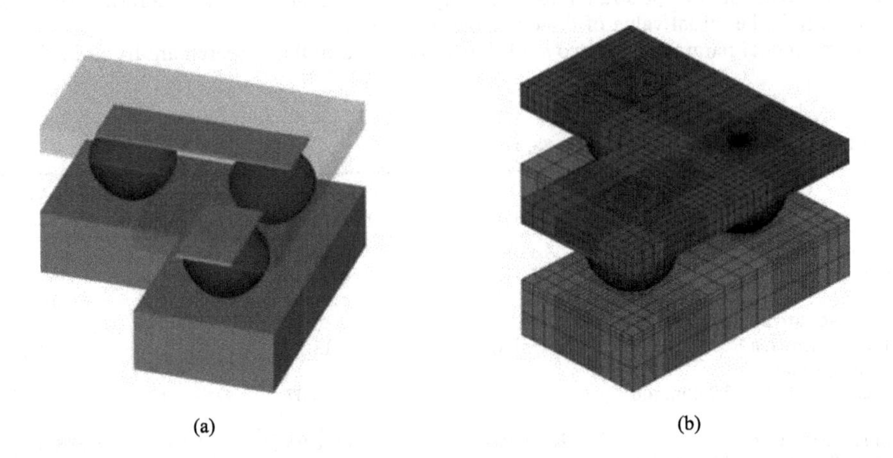

Figure 17.7 Thermal-electric submodel and stress submodel. (a) Solid submodel; (b) submodel mesh; (c) front view of the submodel with a fine mesh on corner solder bump; (d) local view of the UBM structure

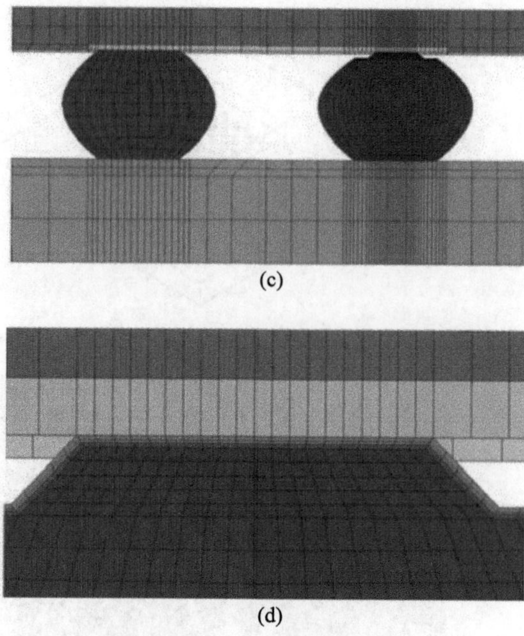

(c)

(d)

Figure 17.7 *(Continued)*

Evolution equation:

$$s = \left\{ h_0(|B|)^a \frac{B}{|B|} \right\} \dot{\varepsilon}_p \tag{17.9}$$

where $B = 1 - \frac{s}{s^*}$ and $s^* = \hat{s}\left[\frac{\dot{\varepsilon}_p}{A}\exp\left(\frac{Q}{RT}\right)\right]^n$, h_0 is the hardening/softening constant, a is the strain rate sensitivity of hardening/softening. The quantity s^* represents a saturation value of deformation resistance s. $\hat{s}$ is a coefficient, and n is the strain rate sensitivity for the saturation value of deformation resistance, respectively. s_0 is the initial value of the deformation resistances.

The Anand model parameters of two solder materials used in this research are listed in Table 17.1.

Table 17.1 ANAND model parameters for SnAgCu and SnPb

Description	Symbol	63Sn37Pb (Darveaux, 2002)	95.5Sn4.0Ag0.5Cu (Wang *et al.*, 2007)
Pre-exponential Factor	A (1/s)	4×10^6	325
Activation Energy	Q/R (K)	9400	10561
Stress Multiplier	ξ	1.5	10
Strain Rate Sensitivity of Stress	m	0.303	0.32
Coef. for Deformation Resistance Saturation Value	$\hat{s}$ (MPa)	13.79	42.1
Strain Rate Sensitivity of Saturation Value	n	0.07	0.02
Hardening Coefficient	h_0 (MPa)	1378.95	800000
Strain Rate Sensitivity of Hardening Coefficient	a	1.3	2.57
Initial Value of s	s_0 (MPa)	12.41	20

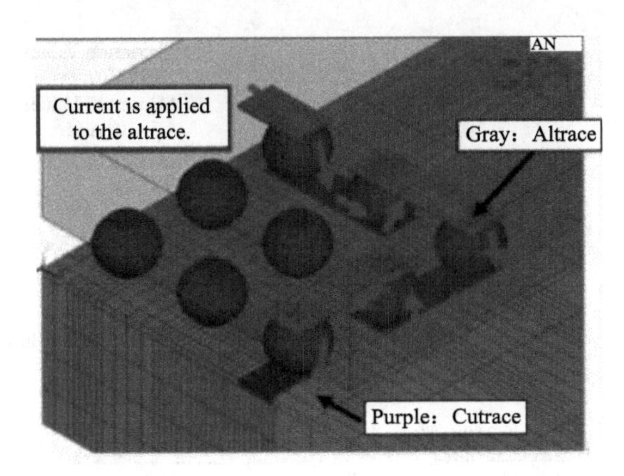

Figure 17.8 Current flow direction in a global model

The electromigration parameters of both SnPb and SnAgCu solder bumps are selected from previous references (Choi *et al.*, 2002; Tu, 2003; Liang *et al.*, 2005; Liang *et al.*, 2006; Nah *et al.*, 2006; Lai *et al.*, 2006; Yue *et al.*, 2005; Gee *et al.*, 2005). For SnPb material: activation energy E_a=0.8eV; effective charge number $Z^*=-33$; self-diffusion-coefficient $D_0=0.016\,m^2/s$; heat of transport $Q^*=0.0094$ eV; initial electrical resistivity $R_0=15.5\times10^{-8}\,\Omega\cdot m$; temperature coefficient resistance $\alpha=3.0\times10^{-3}/K$; atomic volume $\Omega=2.48\times10^{-29}m^3/atom$. For SnAgCu material: activation energy $E_a=0.8$ eV; effective charge number $Z^*=-23$; self-diffusion- coefficient $D_0=0.027\,m^2/s$; heat of transport $Q^*=0.0094$ eV; initial electrical resistivity $R_0=13.3\times10^{-8}\,\Omega\cdot m$; temperature coefficient resistance $\alpha=2.8\times10^{-3}/K$; atom volume $\Omega=2.71\times10^{-29}m^3/atom$.

Figure 17.8 shows the current flow direction in a global model. The free convection boundary condition is applied with 20 W/($m^2\cdot$°C) film coefficient and 50 °C bulk temperature.

In this section, four topics are discussed: (1) the impact of direct coupling and indirect coupling analysis; (2) impact of different current loadings; (3) impact of different solder material (SnPb versus SnAgCu); (4) void generation.

(i) The impact of direct couple analysis and indirect couple analysis

In the simulation the solder bump model/element is considered to be elastic in the direct couple analysis while it is considered to be viscoplastic in the indirect coupling analysis, because there is no directly coupled element which can be used for non-linear analysis.

Figure 17.9 shows the current density distribution in the submodel and solder bump with 1.7 A. From Figure 17.10 it can be seen that the current density at the corner is approximately one order of magnitude higher than the average current density. It agrees the conclusion of reference (Black, 1967). Figure 17.10 shows the temperature gradient distribution in the submodel and bump with 1.7 A. The temperature distribution agrees with the test results very well (Gee *et al.*, 2005). Due to conductivity performance, the maximum temperature gradient value is 17046 K/m (relative small), which would induce small divergences of thermomigration.

Figure 17.11 shows the hydrostatic stress distribution in the submodel with direct coupling analysis (elastic) and indirect coupling analysis (viscoplastic solder) under the current loading 1.7 A. From Figure 17.11, it can be observed that the maximum hydrostatic stress obtained by indirect coupling analysis with a viscoplastic solder bump is less than one third of the value obtained by the direct coupling method with an elastic solder bump.

Figure 17.12 shows the total atomic flux divergence contour due to electromigration, thermomigration, and stress migration in both the elastic solder bump (direct couple method) and viscoplastic SnPb solder bump (indirect couple method) with current loading 1.7 A. From Figure 17.12, it can be seen that the significantly different divergence distributions from the direct coupling method and the indirect coupling method, and the maximum total divergence value of the direct method is larger than that of the indirect method.

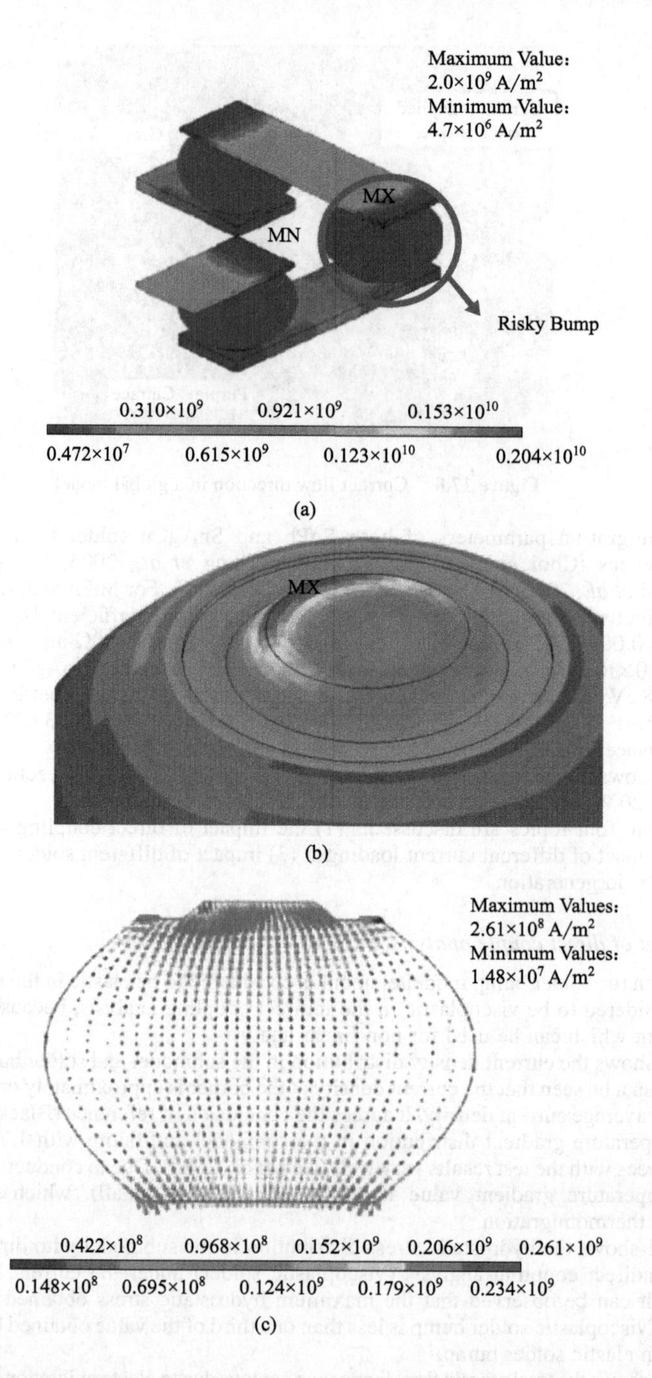

Figure 17.9 Current density distribution with current load 1.7 A. (a) Current density distribution for trace-and-bump; (b) local current density in corner risky bump; (c) current density vector contour for risky bump *(Color version of this figure is available online)*

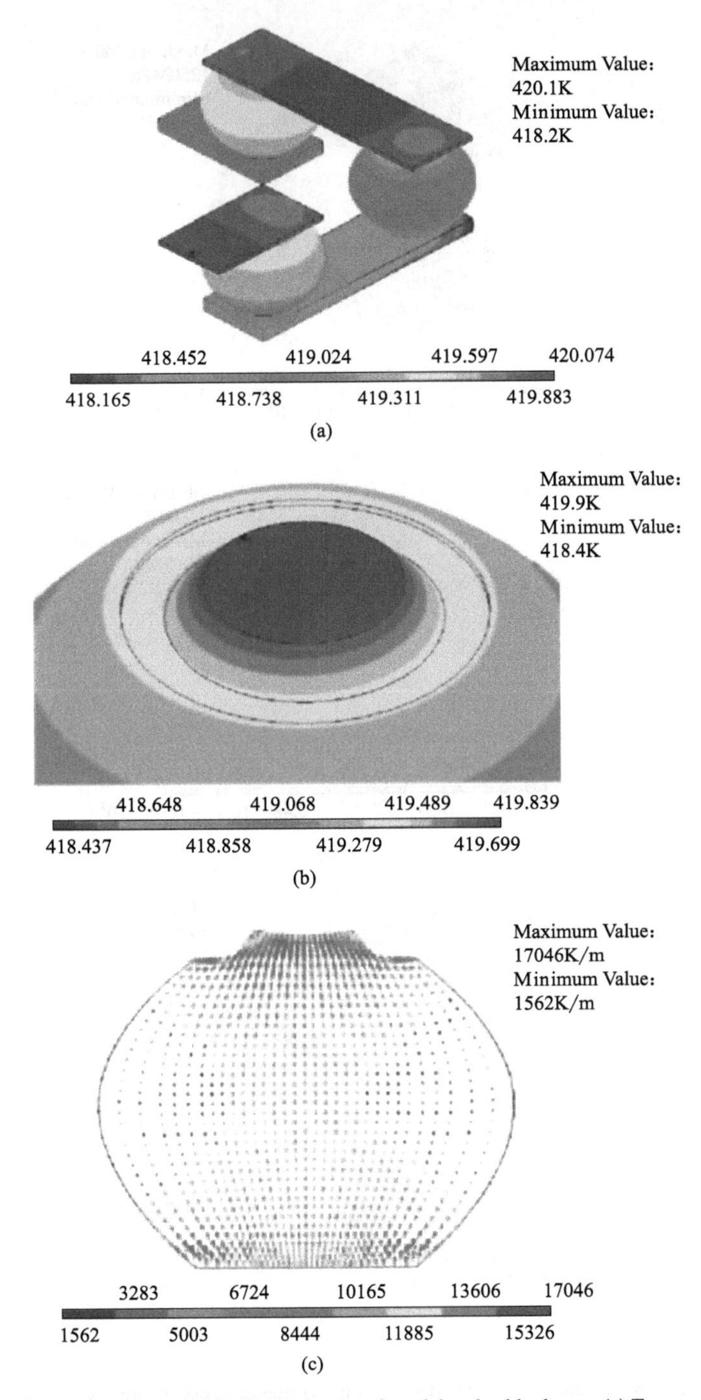

Figure 17.10 Temperature and its gradient distribution in submodel and solder bump. (a) Temperature distribution for trace-and-bump; (b) temperature distribution for corner bump; (c) temperature gradient vector contour for risky bump *(Color version of this figure is available online)*

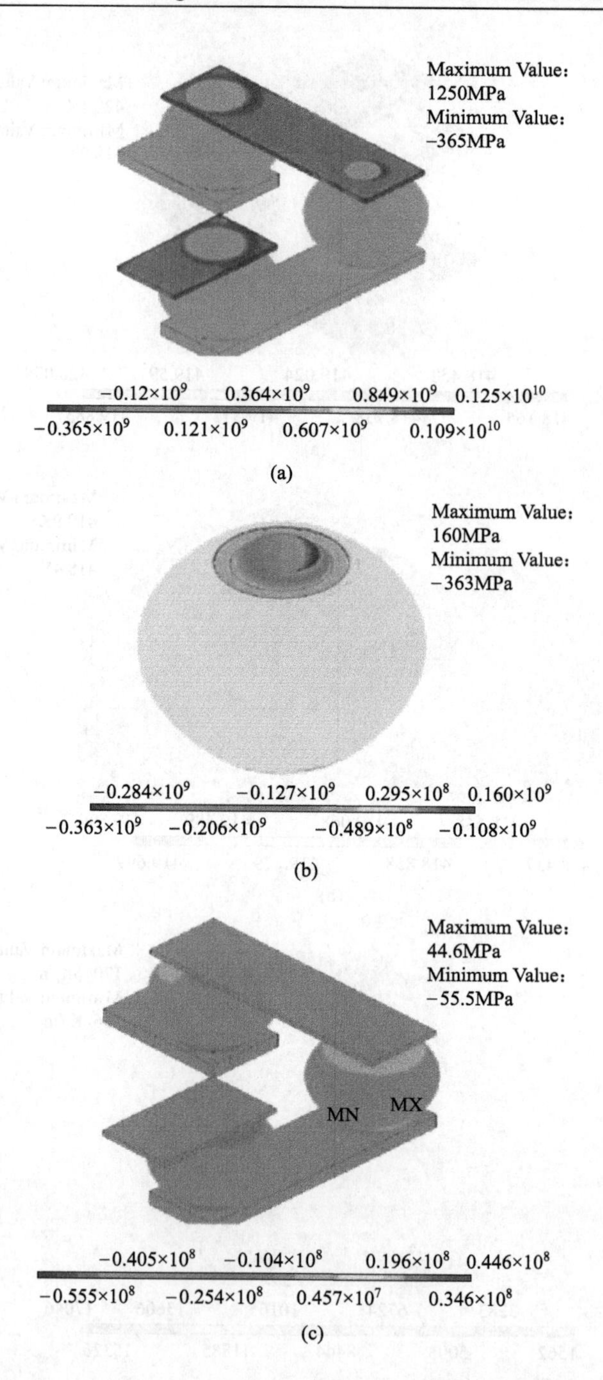

Figure 17.11 Hydrostatic stress (pressure) distribution. (a) Elastic stress of submodel by direct couple method; (b) elastic stress in the corner solder bump; (c) viscoplastic stress of submodel with indirect couple analysis; (d) viscoplastic stress in the corner solder bump *(Color version of this figure is available online)*

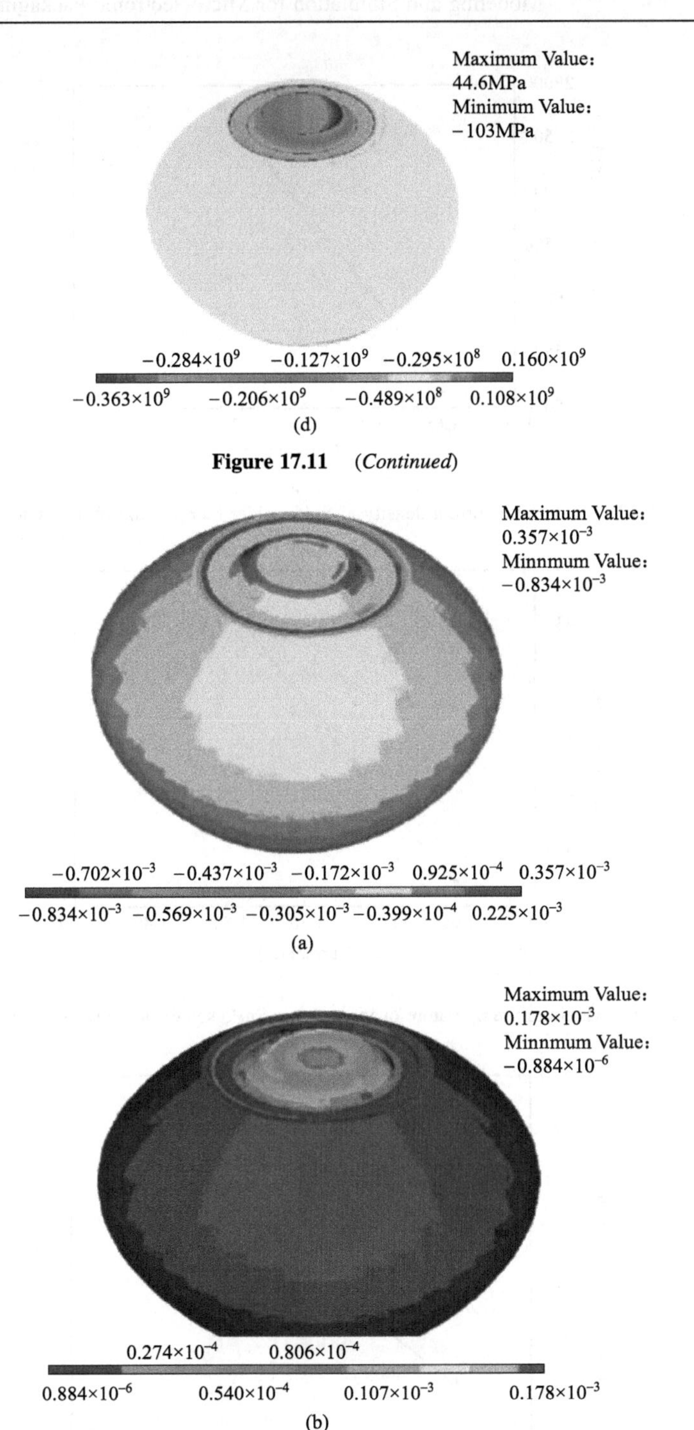

(d)

Figure 17.11 (*Continued*)

(a)

(b)

Figure 17.12 Total atomic flux divergence due to electromigration, thermomigration and stress migration with direct and indirect couple methods. (a) Elastic solder bump with direct couple method; (b) viscoplastic solder bump with indirect coupled method *(Color version of this figure is available online)*

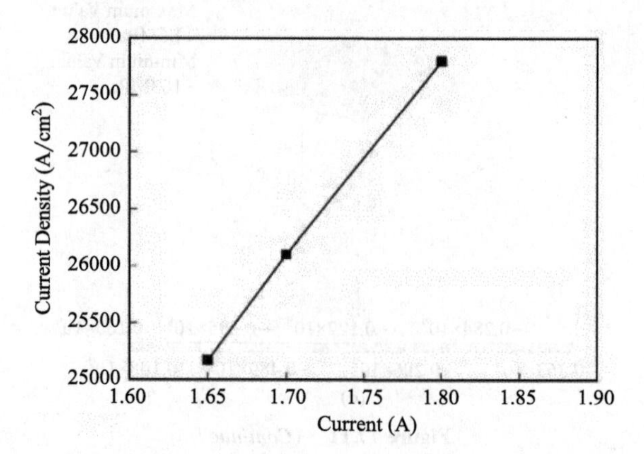

Figure 17.13 Maximum current density of SnPb solder bump with different current loads

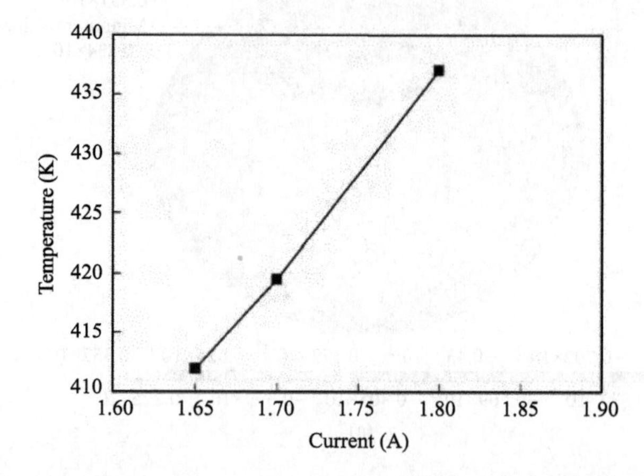

Figure 17.14 Maximum temperature of viscoplastic SnPb solder bump with different current

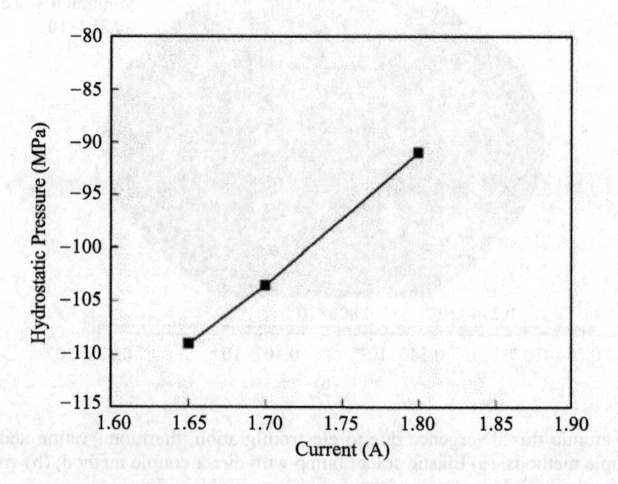

Figure 17.15 Maximum hydrostatic stress of viscoplastic SnPb solder bump with different current

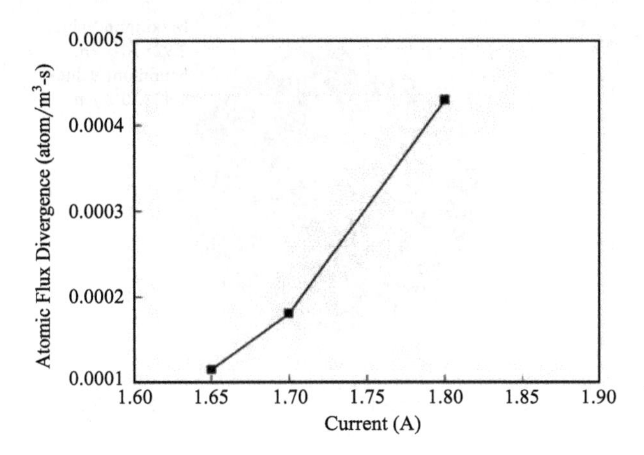

Figure 17.16 Maximum total atomic divergence of viscoplastic SnPb solder bump with different current

(ii) Impact of different current loadings

The results of Figures 17.13 to 17.16 give the solution of a corner SnPb joint under the indirect method with different current loads. Figure 17.13 shows the maximum current density of the corner solder bump with different current loadings 1.65 A, 1.7 A, and 1.8 A. Figure 17.14 shows the maximum temperature of the corner solder bump subjected to current loads 1.65 A, 1.7 A, and 1.8 A. Figure 17.15 shows the maximum hydrostatic stress in the corner solder bump with current 1.65 A, 1.7 A, and 1.8 A. From these figures, it can be observed that the current density, the temperature, and the hydrostatic pressure linearly increase as the current load increases. Figure 17.16 shows the total atomic flux divergence in the corner solder joint varies with different current 1.65 A, 1.7 A, and 1.8 A. The result indicates that as the current increases the total divergence increases rapidly.

(iii) Impact of different materials SnPb (63Sn37Pb) versus SnAgCu (95.5Sn4.0Ag0.5Cu)

Figure 17.17 shows the current density distribution in viscoplastic SnPb and SnAgCu corner solder bump with 1.7A. Figure 17.18 shows the temperature distribution in viscoplastic SnPb and SnAgCu solder bump with 1.7 A. Figure 17.19 shows the hydrostatic stress distribution in viscoplastic SnPb and SnAgCu corner solder bump with 1.7 A. Figure 17.20 shows the total atomic flux divergence contour due to electromigration, thermomigration, and stress migration in viscoplastic SnPb and SnAgCu corner solder bump with 1.7 A. Table 17.2 lists the maximum divergences due to electromigration, thermomigration,

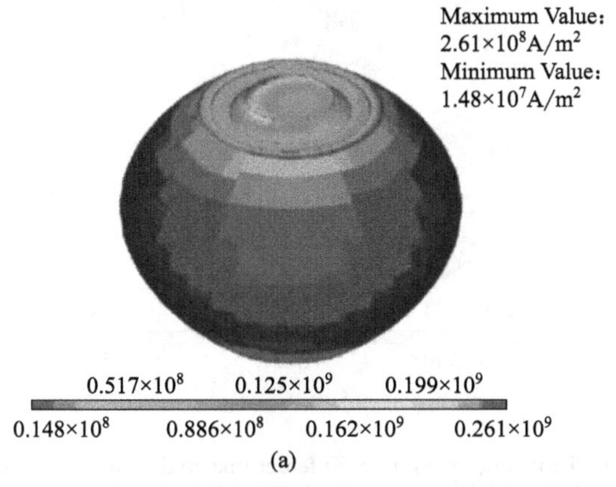

Figure 17.17 Current density distribution with two different materials. (a) SnPb solder bump; (b) SnAgCu solder bump *(Color version of this figure is available online)*

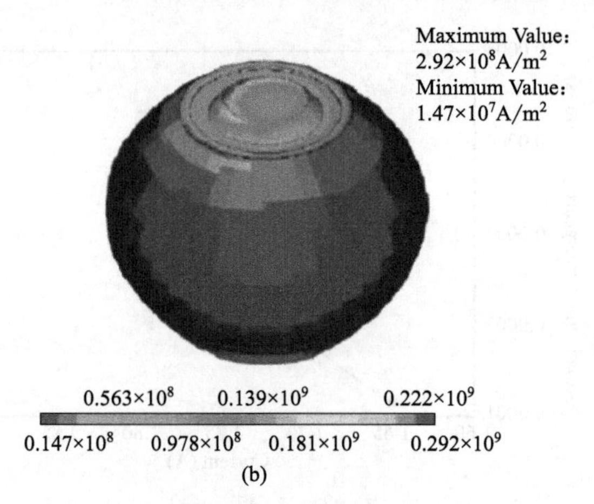

Figure 17.17 (*Continued*)

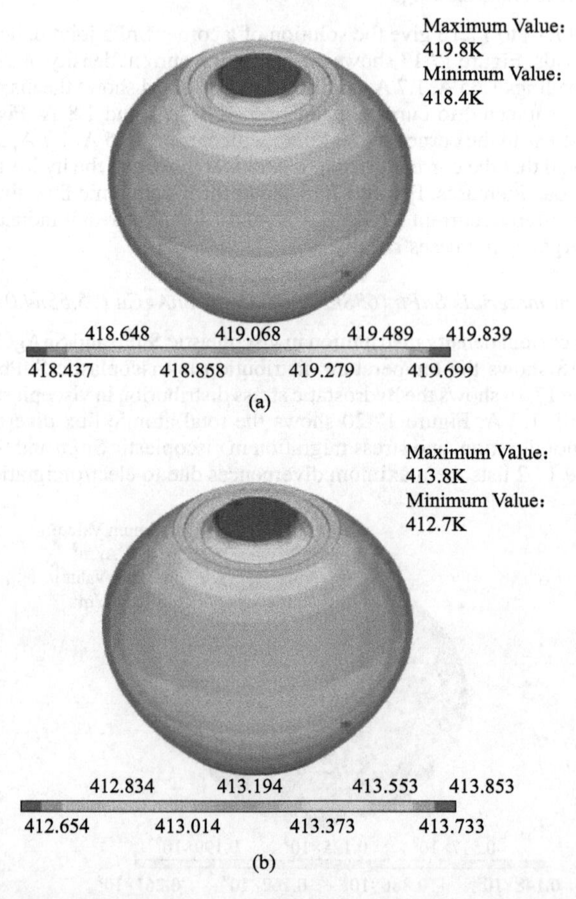

Figure 17.18 Temperature distribution with two different materials. (a) Viscoplastic SnPb solder bump; (b) viscoplastic SnAgCu solder bump *(Color version of this figure is available online)*

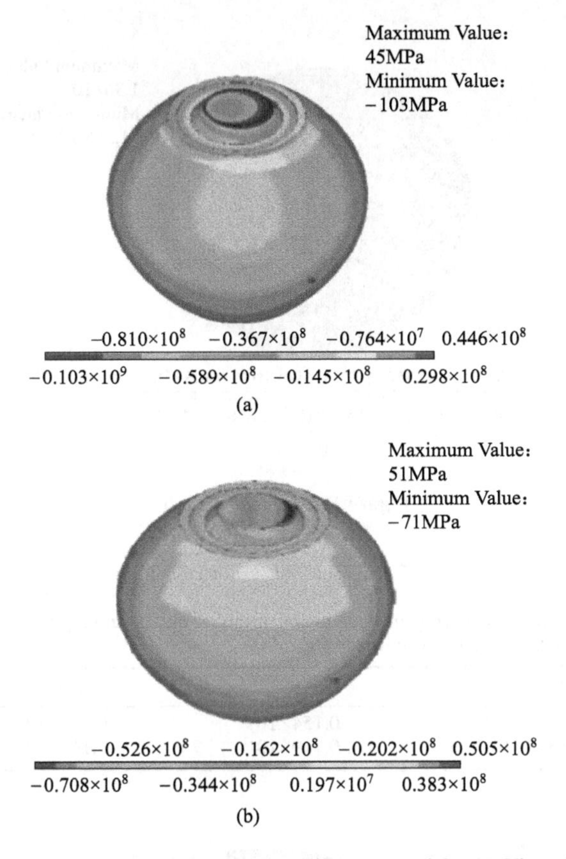

Figure 17.19 Hydrostatic stress distribution with two different materials. (a) Viscoplastic SnPb solder bump; (b) viscoplastic SnAgCu solder bump *(Color version of this figure is available online.)*

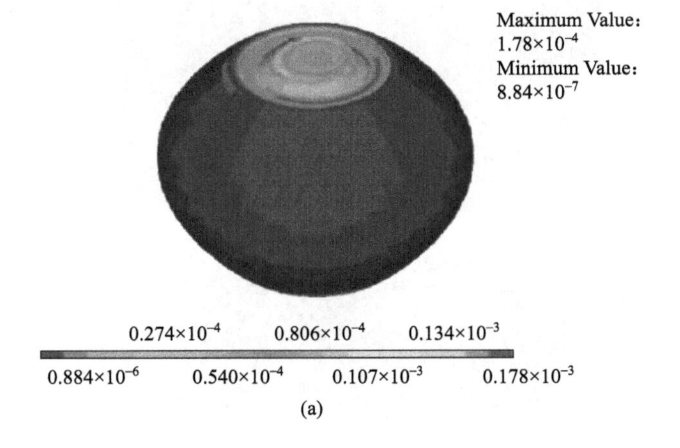

Figure 17.20 Total atomic flux divergence with different materials. (a) Viscoplastic SnPb solder bump; (b) viscoplastic SnAgCu solder bump *(Color version of this figure is available online)*

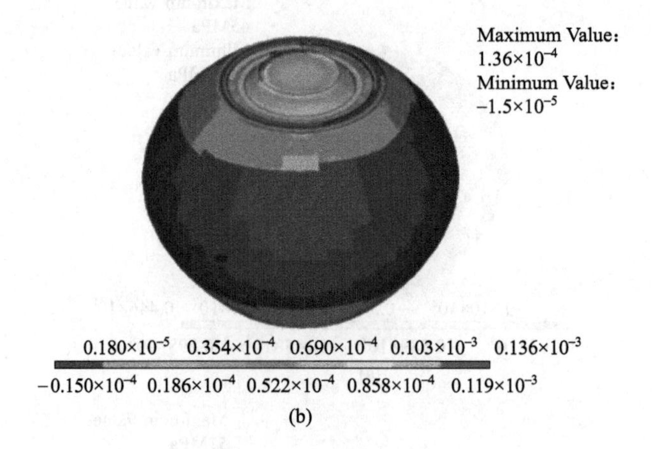

Figure 17.20 (*Continued*)

Table 17.2 Maximum atomic flux divergence due to electromigration, thermomigration and stress migration with two different solder materials with N0 = 1. unit: atoms/(m^3 s)

Material	Div J_{th}	Div J_{em}	Div J_s	Div J_{tot}
PbSn	0.758×10^{-6}	0.154×10^{-3}	0.237×10^{-4}	0.178×10^{-3}
SnAgCu	0.841×10^{-6}	0.119×10^{-3}	0.162×10^{-4}	0.136×10^{-3}

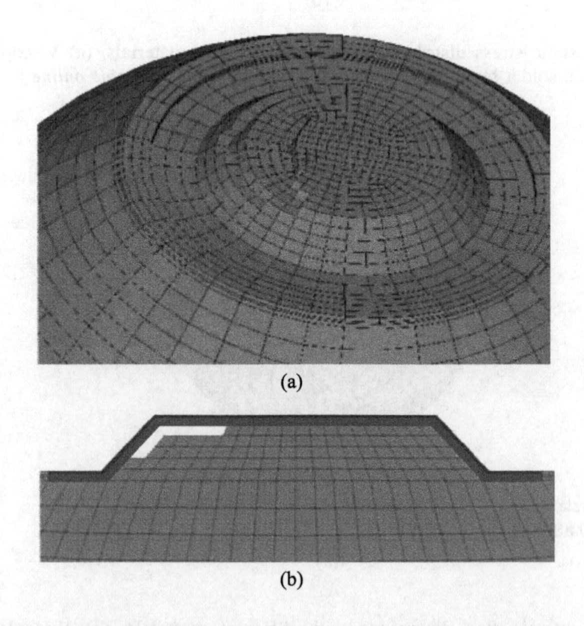

Figure 17.21 Void formation of SnPb solder bump at 414 hours. (a) Void formation; (b) cross-section of void formation with UBM; (c) current density contour *(Color version of this figure is available online)*

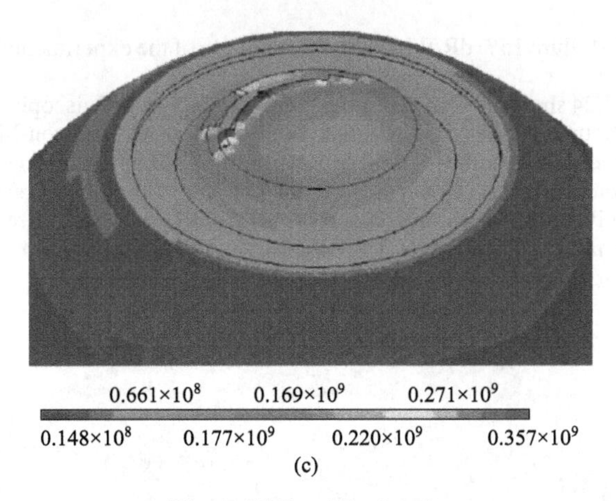

Figure 17.21 *(Continued)*

and stress migration in viscoplastic SnPb and SnAgCu corner solder bump. It can be seen that for the same temperature and current densities, the SnAgCu solder seems slightly more resistant to electromigration failure than the PbSn solder. In both materials, the effect of electromigration is dominate, next is the stress mugration and the thermal migration is the smallest.

(iv) Void formation

Figure 17.21 shows the void formation along the interface of the solder and UBM in a SnPb corner solder bump at 414 hours under 1.7 A. Figure 17.22 shows that the time to failure by simulation is 817.5

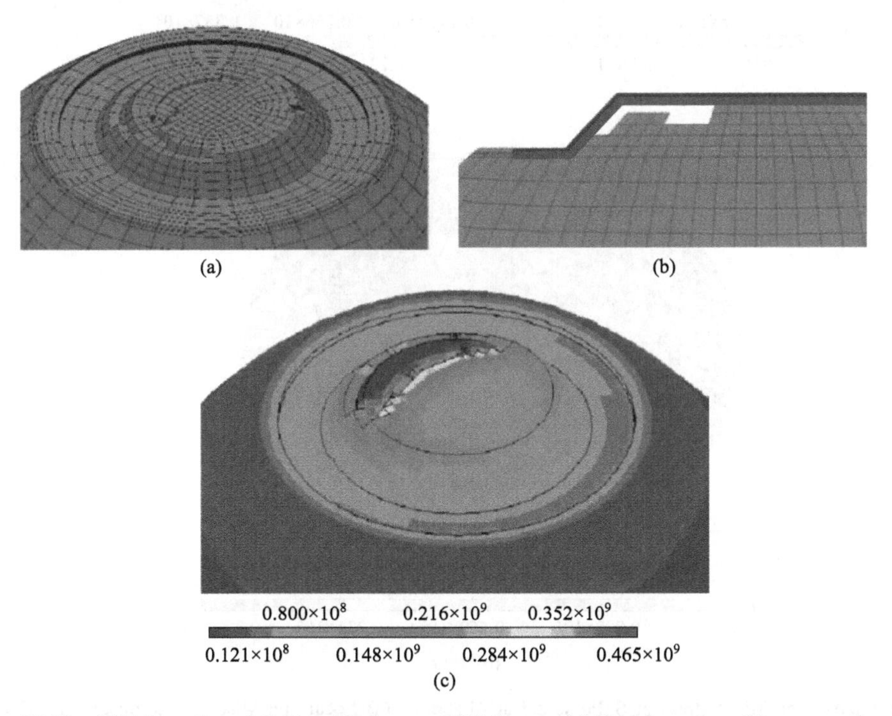

Figure 17.22 Void formation of SnPb solder bump at 817.5 hours. (a) Void formation; (b) cross-section of void formation with UBM; (c) current density contour *(Color version of this figure is available online)*

hours, while the time to failure 15% dR/R in PbSn solder bump of the experimental result is 500–1500 h in (Gee *et al.*, 2005).

Figures 17.23 and 17.24 show the von-Mises stress contour pictures of viscoplastic SnPb solder bump at 414 hours and 817.5 hours. The results show that as the void grows, the von-Mises stress relaxes.

Figure 17.25 gives the time to failure of void generation at 1074.6 hours and the void shape comparison with the cross section picture in test (Gee *et al.*, 2005) which shows similar void shapes between modeling and test. Figure 17.26 gives von-Mises stress distribution at TTF. The modeling results have disclosed that the pb-free material SnAgCu has a longer TTF life (about 24%) than SnPb material.

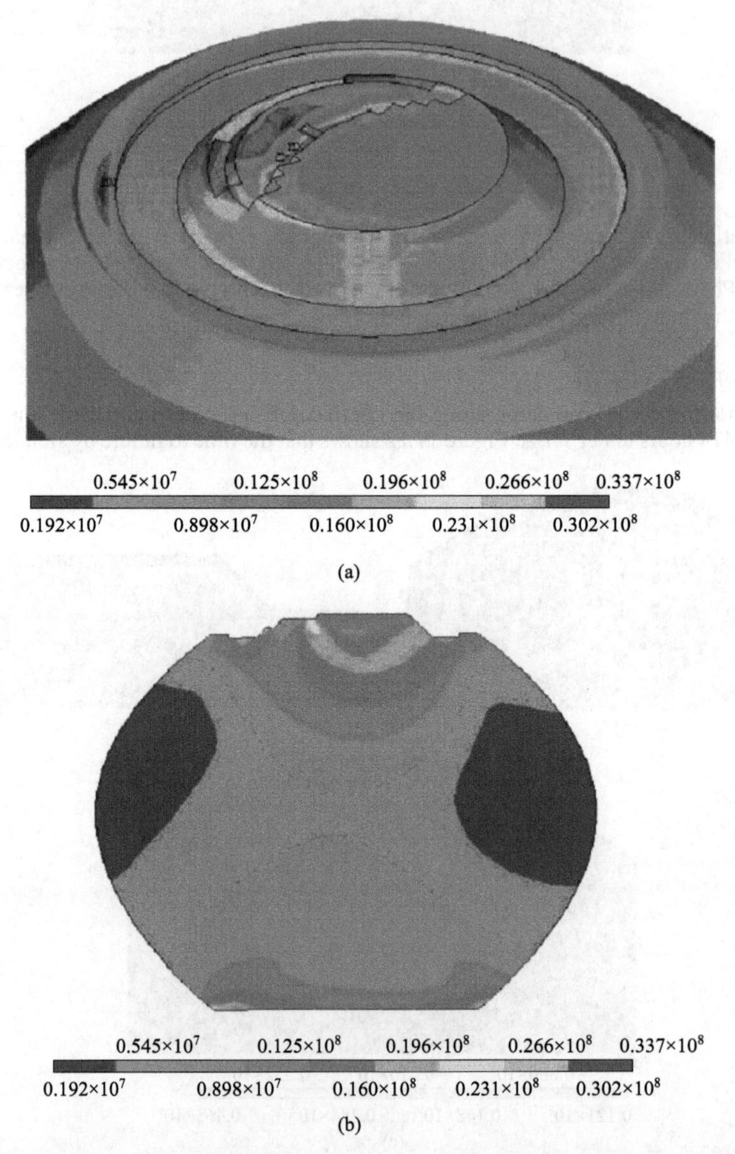

Figure 17.23 von-Mises stress of SnPb solder at 414 hours. (a) Local von-Mises stress, maximum: 33.7 MPa; (b) cross-section of the corner solder bump *(Color version of this figure is available online)*

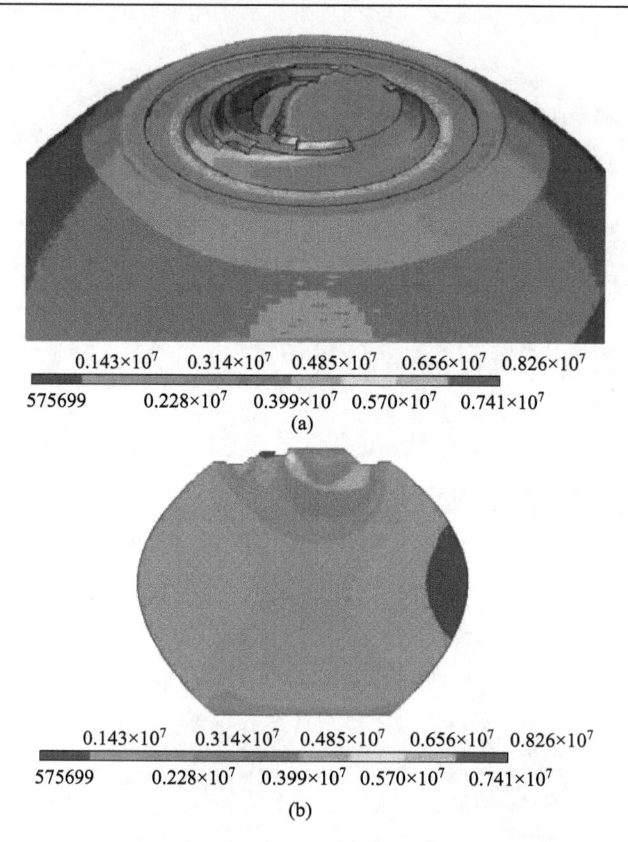

Figure 17.24 von-Mises stress of SnPb solder at 817.5 hours. (a) Local von-Mises stress, maximum 8.26 MPa; (b) cross-section of the corner solder bump *(Color version of this figure is available online)*

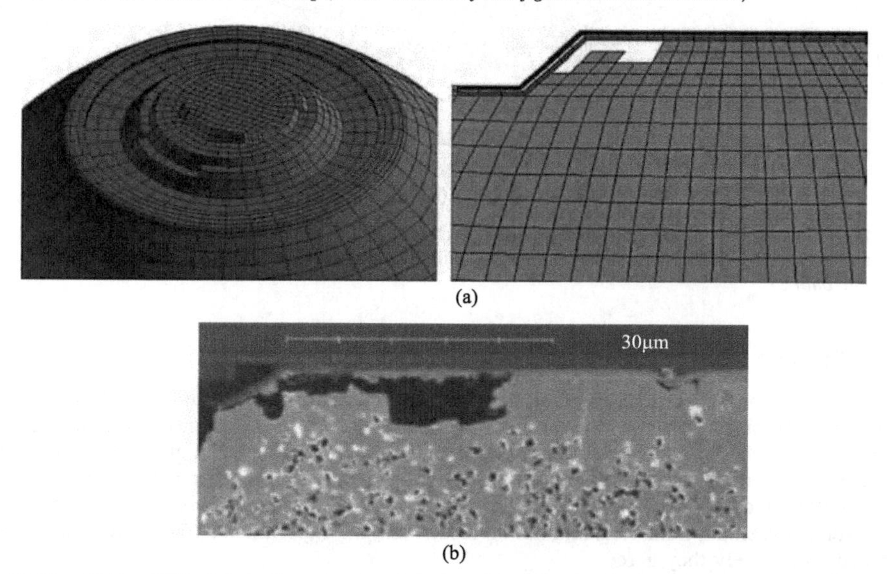

Figure 17.25 Void at TTF of SnAgCu and comparison with test data. (a) Void at TTF of SnAgCu solder material. Its TTF is 1074.6 hours; (b) void at TTF 1475 hours in SnAgCu electromigration test (Gee *et al.*, 2005)

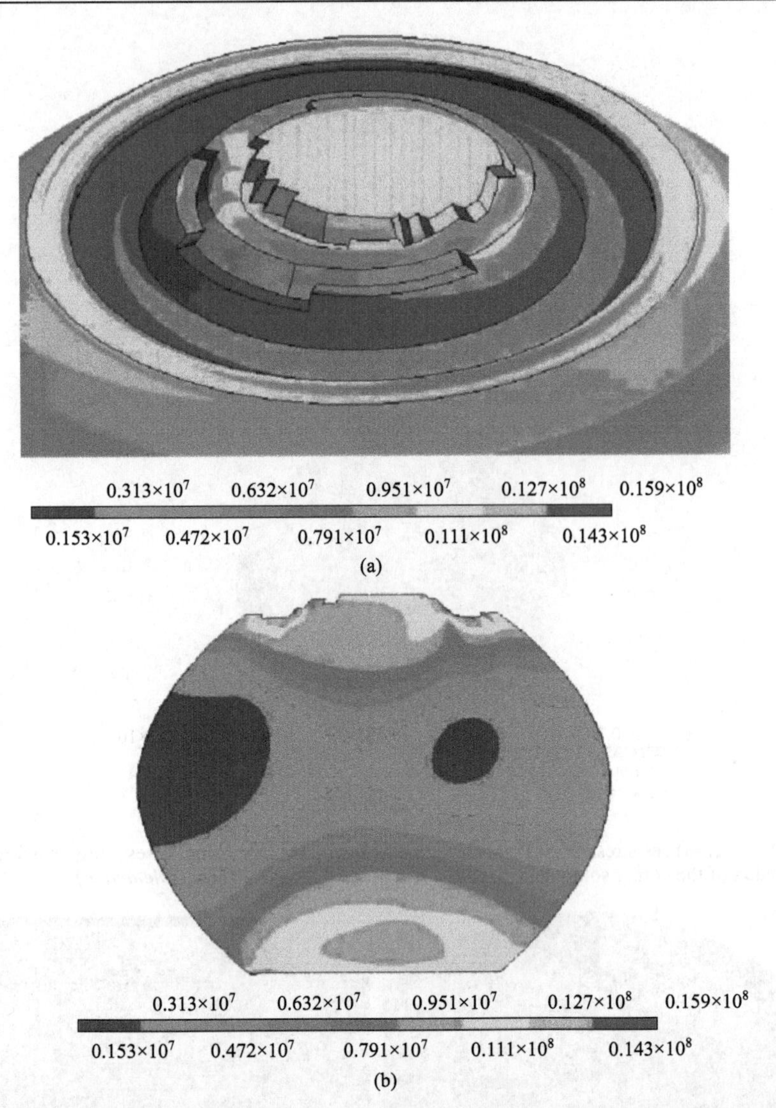

Figure 17.26 von-Mises stress of Pb-free material SnAgCu at TTF 1074.6 h. (a) von-Mises stress at TTF of SnAgCu, maximum 15.9MPa; (b) cross section of von-Mises stress at TTF of SnAgCu *(Color version of this figure is available online)*

17.3 Chapter Summary

The three-dimensional electrical, thermal, and stress direct and indirect coupled analyses for electromigration simulation are presented to examine and quantify the effects of current crowding, joule heating, and stress in an IC device and solder joint of a package. A submodeling technique is developed to better simulate the electromigration, thermal migration, and stress migration. The simulation has shown the 3D void generation and TTF for both the device and solder joint of an IC package in electromigration. Comparison between our numerical simulation results and the experimental results from previous work has shown how closely they agree.

The simulation has disclosed that the direct coupling method seems to give a higher stress and atomic flux divergence than an indirect couple method; this is because the element in an ANSYS direct coupling analysis only has linear properties, while the indirect method may make full use of the material's non-linear properties. The simulated results also found that Pb-free material SnAgCu seems to have less migration failure as compared to SnPb material.

Since the ultimate goal of this chapter is to develop a damage mechanics model for both IC interconnects and solder joints under high current density stressing, it is important to investigate the material mechanical property degradation due to the IMC and dissolution of UBM and solder alloy during the migration process. Also, to get the right electromigration parameters for a Pb-free material requires a large amount of experimental work. These will be electromigration future research directions.

References

Black, J.R. (1967) *Proceedings of 6th Annual Reliability Physics Symp.*, IEEE, New York, 148–159.

Blech, I.A. (1976) Electromigration in thin aluminum films on tianium nitride, *Journal of Applied Physics* **47**(4):1203–1208.

Choi, W.J., Yeh, E.C.C. and Tu, K.N. (2002) Electromigration of flip-chip solder bump on Cu/Ni(V)/AI thin film under bump metallization, *Electronic Components and Technology Conference*, San Diego, CA, 1201–1205.

Dalleau D. and Weide-Zaage, K. (2001) Three-dimensional voids simulation in chip-level metallization structures: a contribution to reliability evaluation, *Microelectronics Reliability* **41**:1625–1630.

Dalleau, D. and Weide-Zaage, K. (2002) 3D time-depending simulation of void formation in a SWEAT metallization structure, *Proceedings of EuroSimE*, Paris, France, April.

Darveaux, R. (2002) Effect of simulation methodology on solder joint crack growth correlation and fatigue life prediction, *ASME Journal of Electronic Package* **124**:147–152.

Gee, S., Kelkar, N., Huang, J. and Tu, K.N. (2005) Lead-free and PbSn bump electromigration testing, *Proceedings of InterPACK 2005, IPACK 2005–73417*, San Francisco, California, July 17–22.

James, R.B. (1969) Electromigration failure modes in aluminum metallization for semiconductor devices, *Proceedings of the IEEE* **57**(9):1587–1594.

Joseph, A.K. (1998) *Material Handbook for Hybrid Microelectronics*, Artech House, Boston.

Lai, Y.S., Chen, K.M., Kao, C.L. *et al.* (2006) Electromigration of Sn-37Pb and Sn-3Ag-1.5Cu/Sn-3Ag-0.5Cu composite flip-chip solder bumps with Ti/Ni(V)/Cu under bump metallurgy, *Microelectronics Reliability*.

Liang, S.W., Shao, T.L. and Chen, C. (2005) 3D simulation on current density distribution in flip-chip solder joints with thick Cu UBM under current stressing, *Electronic Components and Technology Conference*, Lake Buena, Vista, FL, 1416–1420.

Liang, S.W., Chang, Y.W., Shao, T.L., Chen, C. and Tu, K.N. (2006) Effect of three-dimensional current and temperature distributions on void formation and propagation in filip-chip solder joints, *Applied Physics Letter* **89**(2):021117.

Liang, L.H., Xu, Y.J. and Liu, Y. (2006) Electro-migration study in solder joint and interconnects of IC packages, *Proceedings of EuroSIME 2006* April, Como/Italy, 464–470.

Liang, L.H. and Liu, Y. (2006) Reliability study in solder joint under electromigration thermal-mechanical load, *International Conference on Electronics Packaging Technology*, 2006, Shanghai, China, 861.

Liu, Y. and Scott, I. (2003) Power cycling simulation of an IC package: considering electromigration, and thermal-mechanical failure, *Electronic Components and Technology Conference 2003* New Orleans, LA, 415–421.

Nah, J.W., Ren, F., Tu, K.N. *et al.* (2006) Electro migration in Pb-free flip-chip solder joints on flexible substrates, *Applied Physics Letter* **99**(2):023520.

Rinne, G. (2004) Electromigration in SnPb and Pb-Free solder bumps, *IEEE Electronic Components and Technology Conference 2004* Las Vegas, NE, 974–978.

Tu, K.N. (2003) Recent advances on electromigration in every-large-scale-integration of interconnects, *Journal of Applied Physics* **94**(9):5451–5473.

Wang, Q. *et al.* (2007) Experimental determination and modification of the anand model constants for 95.5Sn4.0Ag0.5Cu, *Eurosime 2007*, London, UK, April.

Weide-Zaage, K., Dalleau, D. and Yu, X.B. (2003) Static and dynamic analysis of failure locations and void formation in interconnects due to various migration mechanisms, *Materials Science in Semiconductor Prcessing* **6**:85–92.

Yue, H., Basaran, C., Hopkins, D. and Lin, M. (2005) Modeling deformation in microelectronics BGA solder joints under high current density, part I: simulation and testing, *IEEE Electronic Components and Technology Conference 2005*, Lake Buena, Vista, FL, 1437–1444.

18

Popcorning in Plastic Packages

Plastic IC packaging subjected to moisture is vulnerable to delamination and cracking during the solder reflow process, both a classical yield issue in plastic packaging and other emerging packaging technology. During the early stage of packaging development, no matter if it is plastic packaging for an IC chip, or LED chip, some partial delamination or voids are always present along part of the interface. Modeling the response of the delaminated plastic package is therefore an essential step to fundamentally understand failure mechanisms and mechanics involved in the cooling, moisture absorption, and solder reflow process. A nonlinear finite element model is developed for predicting the deformation, stress, and fracture behavior of delaminated plastic packages induced by mechanical and hygro-thermal loads. The model consists of a sequentially coupled hygro-thermo-mechanical analysis considering moisture absorption, evaporation, and interface contact and fracture analysis. A Lagrange Multiplier method is utilized to model the delamination interface condition. A general contact model is adopted which can handle complex contact conditions, such as arbitrary slippage and discontinuous curvature. A thermal contact is utilized to model the contact along delamination surfaces. Mixed mode fracture modes are elaborated. The model is verified by comparing existing analytical results with the predictions in the case of pulsed heating of the IC chip. Packaging responses and failure mechanisms due to encapsulation cooling, moisture absorption and evaporation, wave soldering, and interfacial moisture pressure loading were investigated, with consideration of temperature-dependent material properties change around the glass transition temperature for both die attach and molding compound. Packaging deformation and stress, crack tip driving force, doming of the delamination, and delamination growth stability as a function of time are discussed.

18.1 Statement of Problem

Consider a simplified packaging subjected to a hygro-thermo-mechanical load as shown in Figure 18.1(a). Axisymmetric FE models, with or without thermal contact resistance or mechanical contact at the delaminated interface, were utilized in the modeling. Molding compound thickness above the IC chip and below the leadframe may vary. The model is shown in Figure 18.1(b). Physically, only one point is supported at the center of the package. The temperature cycle is shown in Figure 18.2. The first cooling stage is used to represent cooling from the encapsulation, followed by a moisture absorption process of duration with long enough time of 10^8s at room temperature. The moisture content at the saturation level is assumed to be 0.3%. The third stage is a solder reflow process.

In the stage, the process temperature is the preheating to 150 °C for 30s, and holding at this temperature for 60s, then ramping further to a peak temperature of 220 °C for 20s, then cooling to room temperature for 90s, finally holding at room temperature for 100s. When the temperature is higher than the boiling temperature of the water, the moisture in the molding compound and die attach is assumed to evaporate.

Modeling and Simulation for Microelectronic Packaging and Integration: Manufacturing, Reliability and Testing, Second Edition. Sheng Liu and Yong Liu.

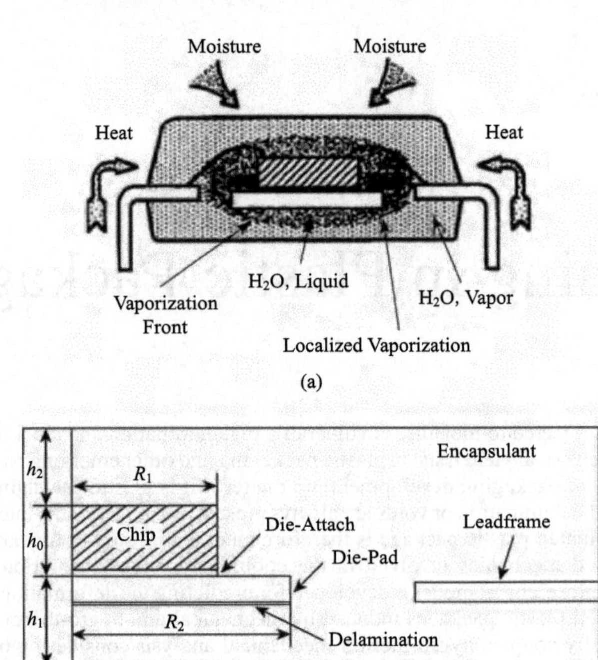

Figure 18.1 (a) A simplified packaging subjected to a hygro-thermo-mechanical load; (b) A half model for the case of symmetric delamination

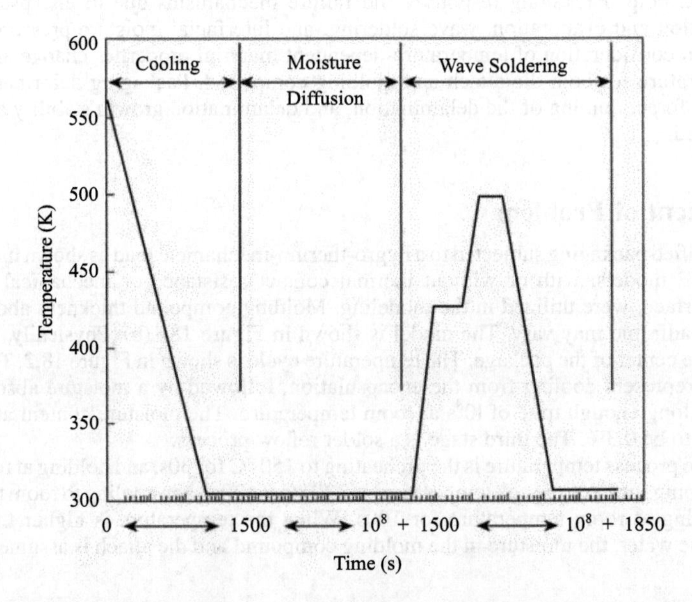

Figure 18.2 Temperature cycle for cooling, moisture absorption, and wave soldering processes (Reproduced with permission from D. Dalleau and K. Weide-Zaage, "3D time-depending simulation of void formation in a SWEAT metallization structure," *Proceedings of EuroSimE*, April 2002. © 2002 IEEE.)

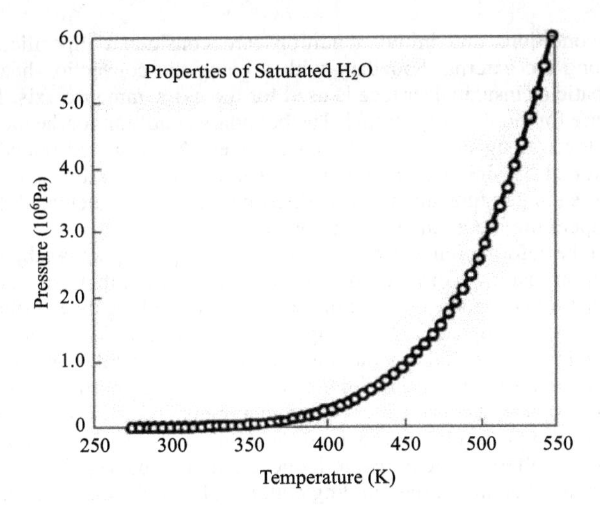

Figure 18.3 A relationship for interface moisture induced pressure versus temperature

Pressure versus temperature at the delaminated interface full of moisture is assumed to follow the relationship shown in Figure 18.3.

In this investigation, it was desired to determine:

(1) The effect of the delamination on the hydro-thermal-mechanical response in cooling, moisture absorption, and wave soldering induced moisture evaporation;
(2) Fracture modes and mechanisms in terms of the strain energy release rate distribution at the crack tip;
(3) The effect of variations of material properties; and
(4) Stability of the crack growth.

18.2 Analysis

It is assumed that the deformation and induced damage do not generate internal heat in the three processes considered, a sequential coupling among the temperature, moisture, and deformation is adopted in the modeling. The current modeling, therefore, consists of a heat-transfer analysis, a moisture-diffusion analysis, a stress analysis, and a fracture analysis. In the heat-transfer analysis, both heat conduction in the packaging materials and heat convection between the packaging and the environment are considered. The moisture diffusion is analogous to the heat conduction equations with the appropriate replacements of the material constants and boundary conditions.

In the following, brief presentations for these sub-models are given to facilitate the discussion of the predicted results.

The heat diffusion equation can be expressed in the cylindrical coordinates (r, ϕ) as:

$$\frac{1}{r}\frac{\partial}{\partial}\left(kr\frac{\partial T}{\partial R}\right)+\frac{1}{r^2}\frac{\partial}{\partial}\left(k\frac{\partial T}{\partial \phi}\right)+q_g=\rho c_p\frac{\partial T}{\partial t} \tag{18.1}$$

where k is the thermal conductivity (W/(m·°C)), T is the temperature (in K or °C, and $T(K)=T(°C)+273.15$), c_p is the specific heat at constant pressure (J/(kg·°C)), ρ is the mass density (kg/m^3), and q_g is the rate of energy generation per unit volume (W/m^3). If the thermal conductivity is a constant independent of position or temperature, thermal diffusivity $D_T = k/\rho c_p$, can be defined. This constant is important for moisture absorption, as the associated equation is the same as the heat diffusion equation expressed here.

Relevant boundary conditions and initial conditions are considered. Specifically, heat convection boundary is used along the external boundary with h being the convection heat transfer coefficient (W/(m·°C)). An adiabatic or insulated surface is used for the axi-symmetric axis. Initial temperature is the cooling temperature for molding compound. The boundary condition for the moisture diffusion is the prescribed moisture level along the same external surfaces. An insulated boundary condition in the symmetric axis is assumed. The initial condition for moisture content is assumed to be zero. Therefore, by the current analysis, the temperature and moisture distributions can be calculated. It has to be noted that the moisture and temperature are assumed not to be coupled.

It is understood that the deformation may not be large when the packaging is subjected to the considered processing and operation loading. However, due to the consideration that progressive damage may be induced and various material and force nonlinearity such as contact along the interface are considered, the finite deformation theory is adopted for handling any local large deformation induced. Incremental modeling in stress and strain calculation is naturally needed and the relevant theories can be found in literature and various manuals for commercial finite element codes. Important quantities here are various strain increments. For instance, the mechanical strain increment is $\Delta\varepsilon^{M} = \Delta\varepsilon^{T} - \Delta\varepsilon^{th} - \Delta\varepsilon^{C}$, where the superscript T, th and C designate total, thermal, and moisture induced strain, respectively.

It is noted here that the effect of the moisture induced strain is the same as the temperature induced strain. When the moisture evaporates, the molding compound and die attach material will have a larger swelling coefficient β_t. Interface moisture pressure burst will be created during the moisture evaporation.

Both thermal contact and mechanical contact are considered along the concerned interface. Thermal contact resistance is used to model the element heat conduction equations between two dissimilar materials via a resistance across a gap. We are aware that the thermal contact resistance is a function of the physical properties of the contact material, the surface conditions and finish, the contact pressure, and the presence of a fluid or vacuum in the gap between the surfaces. Such a contact resistance is numerically taken into consideration in those elements across the material interface (Mix *et al.*, 1993; Shapiro, 1986). In order to handle mechanical contact of the delamination, a Lagrange multiplier method is employed to implement contact constraints (Shapiro, 1986). The proposed contact fracture algorithms have been applied to study problems such as fiber matrix debonding and cracking of metal matrix composites, powder compaction and sintering, debonded thin film substrate systems, toughened composite delaminations, ceramic/conductive epoxy/glass systems, plastic packaging, and resin/copper systems (Shapiro, 1986; Nguyen *et al.*, 1993; Liu *et al.*, 1994). Details are omitted here for the brevity of the presentation.

The initiation of delamination growth (onset of delamination growth) was predicted based on interfacial linear elastic fracture mechanics. The well-known virtual crack closure technique by Rybicki *et al.* (1977) served as the basis for the mines Mode I, Mode II, and the total strain energy release rates (G_{I}, G_{II}, G_{III}, and phase angle $\tan\psi = \sqrt{G_{\mathrm{II}}/G_{\mathrm{I}}}$ from the energy required to close the delamination over a small area. In terms of the finite element framework, it has been demonstrated that very accurate results for the energy release rates can be obtained as long as finite element mesh is reasonable, even with a fairly coarse mesh (Shapiro, 1986; Liu *et al.*, 1994; Rybicki *et al.*, 1977). The demonstrated examples covered isotropic, orthotropic, anisotropic, long fiber reinforced composites, and woven composites by the first author (Shapiro, 1986) and many others.

Once the total energy release rate and the fracture toughness are obtained, the onset of crack growth is predicted when $G(\psi_r) = G_c(\psi_r)$ is satisfied, where mode mixity angle ψ_r is defined with r indicating that ψ_r is a slowly changing function with respect to a length T. With the given mode mixity ψ_r the interface fracture toughness G_C is defined and it is a property of the bimaterial interface. The above criteria were recently developed and called the mixed-mode fracture mechanics theory, which has been summarized in some recent review articles (Hutchinson and Suo, 1992; Shih *et al.*, 1991). However, in the framework of the current implementation, the T dependence of G and G_C are eliminated, which is much easier for industrial laboratories in terms of both fracture toughness testing and numerical modeling of interfacial crack growth.

Finite element model can be formulated by using standard techniques (Marsden *et al.*, 1981; Bathe, 1982) Temperature, moisture, deformations, and stresses inside the specimens are calculated. Four-node axisymmetric elements are used. The four-point Gauss integration scheme is used for all the modeling work. Contact elements which allow arbitrary sliding and opening are used for the interface crack modeling. The convergence has been checked for all the problems and part of the verification will be presented.

18.3 Results and Comparisons

Axisymmetric and plane strain models are used for cases to be discussed in the coming sections, as they represent approximately the 3D rectangular shape of a typical plastic IC package in this study (Mix *et al.*, 1993). The dimensions used in the axisymmetric model are also shown in Figure 18.4, which is the same as that used by Mix *et al.* (1993).

The finite element model consisted of 1689 nodes and 1617 elements. A typical finite element mesh is shown in Figure 18.4. The mesh is believed to provide a good balance between the accuracy and cost due to its unique transition zones linking fine mesh to coarse mesh while maintaining good aspect ratios of elements used. The initial temperature is assumed to be the room temperature (303 K) or the curing temperature in the encapsulation (468 K). The stress-free temperature is assumed to be the same as the initial temperature. T_g is assumed to be 443 K for the molding compound. The assumption of room temperature as the stress-free temperature is valid for molding compounds with fully relaxed stress state. The residual stress is considered if the curing temperature is the stress-free temperature and it is assumed that the residual stresses cannot be relaxed in the molding compound and die attach. Here the die attach also assumes a value of 388 K in T_g.The curing temperature for the die attach is 468 K.

The boundary condition is such that the entire external surface of the package participated in the convective heat transfer, with a heat transfer coefficient of 50 W/(m^2·°C) to a surrounding applied temperature as shown in Figure 18.5. This value was used in the investigation by Mix *et al.* (1993) for a plastic IC packaging subjected to a pulsed chip heating. Other two heat transfer coefficients, 5 W/(m^2·°C) and 15 W/(m^2·°C), were also evaluated but the results showed similar trend. Thermal contact resistance for the delaminated interface is assumed to be 1 °C/W at the considered delaminated interface for most simulations. Other values were also evaluated. The material properties are shown in Table 18.1.

18.3.1 Behavior of a Delaminated Package due to Pulsed Heating-Verification

For this specific example, the package is assumed to be stress-free at room temperature, which represents a package that has been stored for a long time under standard conditions (Mix *et al.*, 1993). It has to be pointed out that a local fine mesh along the encapsulant/die interface and the active chip surface for internal heat generation. Figure 18.5 shows the temperatures along the axis of the symmetry on both upper and lower surfaces of a fully delaminated interface between the chip and the molding compound with a pulsed power input for chip heating. In the finite element modeling, a thin layer of the elements are assumed to generate the power shown in the figure. It should be pointed out that the prediction here coincides with what was reported by Mix *et al.* (1993). It is also observed that the crack tip is initially in sliding mode and then in opening mode in steady state. This is also confirmed by observing a fully closed

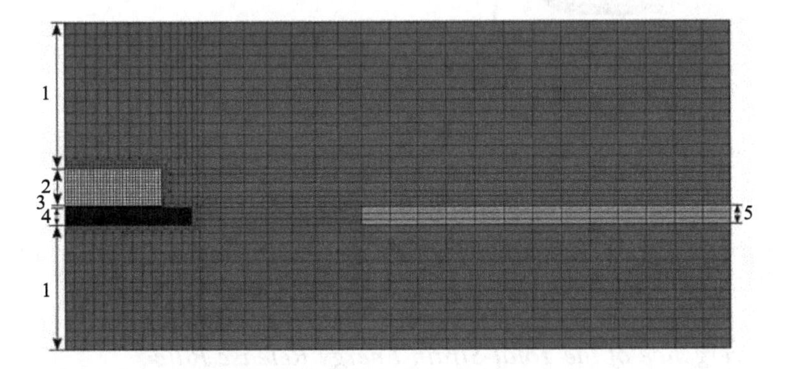

Figure 18.4 A typical finite element mesh used for the investigation (Reproduced with permission from D. Dalleau and K. Weide-Zaage, "3D time-depending simulation of void formation in a SWEAT metallization structure," *Proceedings of EuroSimE*, April 2002. © 2002 IEEE.)

Table 18.1 Material properties of the sample problem

	Die	D/A Water	D/A Steam	Die-Pad	L/F	EMC Water	EMC Steam
Young's Modulus (GPa)	165.5	8.96 ($T<T_1$) 2.48 ($T>T_2$)	8.96 ($T<T_1$) 2.48 ($T>T_2$)	119.3	119.3	12.753 ($T<T_1$) 2.55 ($T>T_2$)	12.753 ($T<T_1$) 2.55 ($T>T_2$)
CTE(10^{-6}/°C)	2.3	41 ($T<T_1$) 15 ($T>T_2$)	41 ($T<T_1$) 15 ($T>T_2$)	16.9	16.9	14 ($T<T_1$) 28 ($T>T_2$)	14 ($T<T_1$) 28 ($T>T_2$)
Poisson Ratio	0.25	0.25	0.25	0.34	0.34	0.25	0.25
Thermal Conductivity [W/(m·°C)]	153	4.5	4.5	196.6	196.6	0.67	0.67
Density(kg/m^3)	2330	3800	3800	8920	8920	1820	1820
Specific Heat [J/(kg·°C)]	703	703	703	400	400	1884	1884
CME(1/M)	0	0.6	1.2	0	0	0.6	1.2
D_M(10^{-13}m^2/s)	0	2.28	N/A	0	0	2.28	N/A

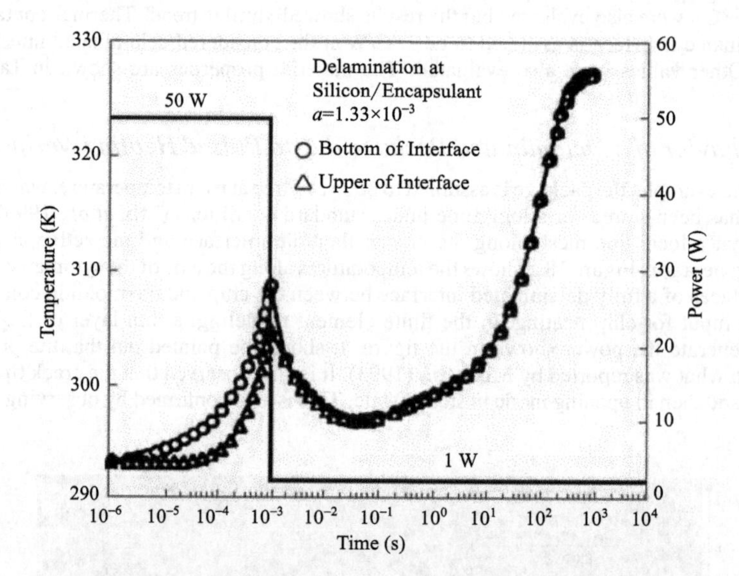

Figure 18.5 Temperatures along the axis of the symmetry on both upper and lower surfaces of a fully delaminated interface between chip and molding compound due to a pulsed power input for chip heating

delamination interface during the heating and a fully opened delamination interface in the steady state. Details are not presented here due to the space limitation.

18.3.2 Convergence of the Total Strain Energy Release Rate

In terms of common practice of industrial laboratories, two values are used for both Young's modulus and *CTE*, before and after T_g. Of course, true nonlinear relationships can be implemented in the finite element code, as has been done in a recent work (Liu *et al.*, 1994) for conductive epoxy modeling.

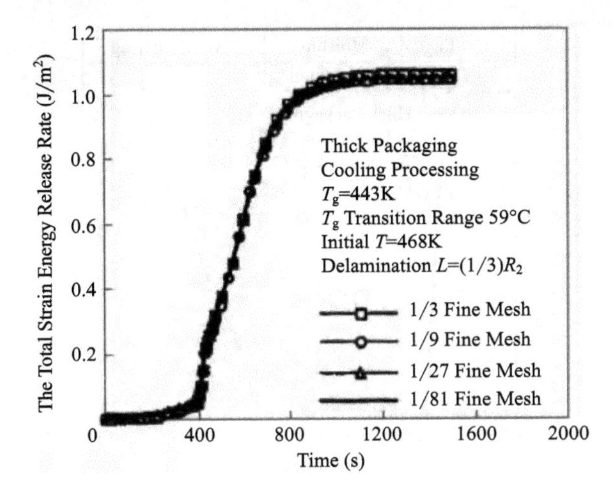

Figure 18.6 Crack tip driving forces for a thick packaging with various meshes in the transient cooling process: a convergence study

However, this information is not available currently for this research. Parameters typical of the molding compounds and die attach materials are used instead. In the following, three delamination sizes and two ratios of CTE's (two and five) before and after T_g are presented. The transition range for T_g is 50 °C. It is assumed that the ratio for Young's modulus is 5.0 and only a center delamination is considered for axisymmetric models.

Several refined meshes along the delamination line have been used to check the convergence of the total strain energy release rate for a delamination of one-third the pad length. Four different meshes are used with the finest mesh being 1/81 the size of the original mesh in terms of the element size at the crack tip. Actually the finest mesh has 2361 nodes and 2361 elements, which significantly enhanced the local mesh without increasing much the total nodes and elements. The results are shown in Figure 18.6 for a thick packaging in the transient cooling process. It is shown in the figure that the crack tip driving force is relatively small for the packaging considered. The crack tip driving force is relatively small during the initial cooling before the glass temperature, followed by a fast increase after the T_g and finally reaches a constant value at room temperature. More physical explanation will be given in the following section. If the interface has a weak or contaminated bond line, the delamination may grow.

18.3.3 Effect of Delamination Size and Various Processes for a Thick Package

First, a temperature evolution of the crack tip is predicted as shown in Figure 18.7. Following the applied temperature cycle as shown in Figure 18.2, the crack tip temperature first decreased to room temperature gradually due to the encapsulation cooling and then staggered for a while when the package is held at room temperature, then gradually increased to a peak temperature.

It has to be pointed out here that a log-scale is used due to the different time used for three processes, where the moisture diffusion has the longest time. Figure 18.8 presents the strain energy release rate, the phase angle, and the gap in the center of the delamination as a function of time for three different crack sizes. The longest crack is slightly smaller than the pad size as the total strain energy release rate is not convergent due to the complicated distribution of corner materials.

It is interesting to observe the evolution of the three quantities in three typical processes: encapsulation cooling, room temperature moisture absorption to simulate a storage condition, and solder reflow. Gap distribution provides the information of the crack opening displacement at the center of the delamination. The phase angle provides the relative contribution of opening and sliding at the crack tip. The strain energy release rate is the crack tip driving force.

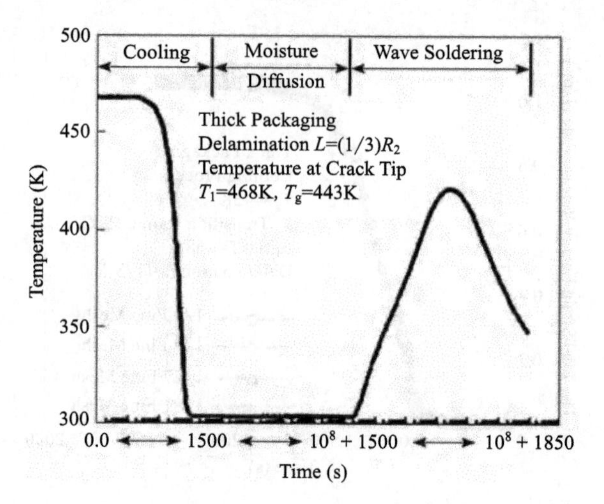

Figure 18.7 Crack tip temperature evolution in three processes: cooling, moisture diffusion, and wave soldering

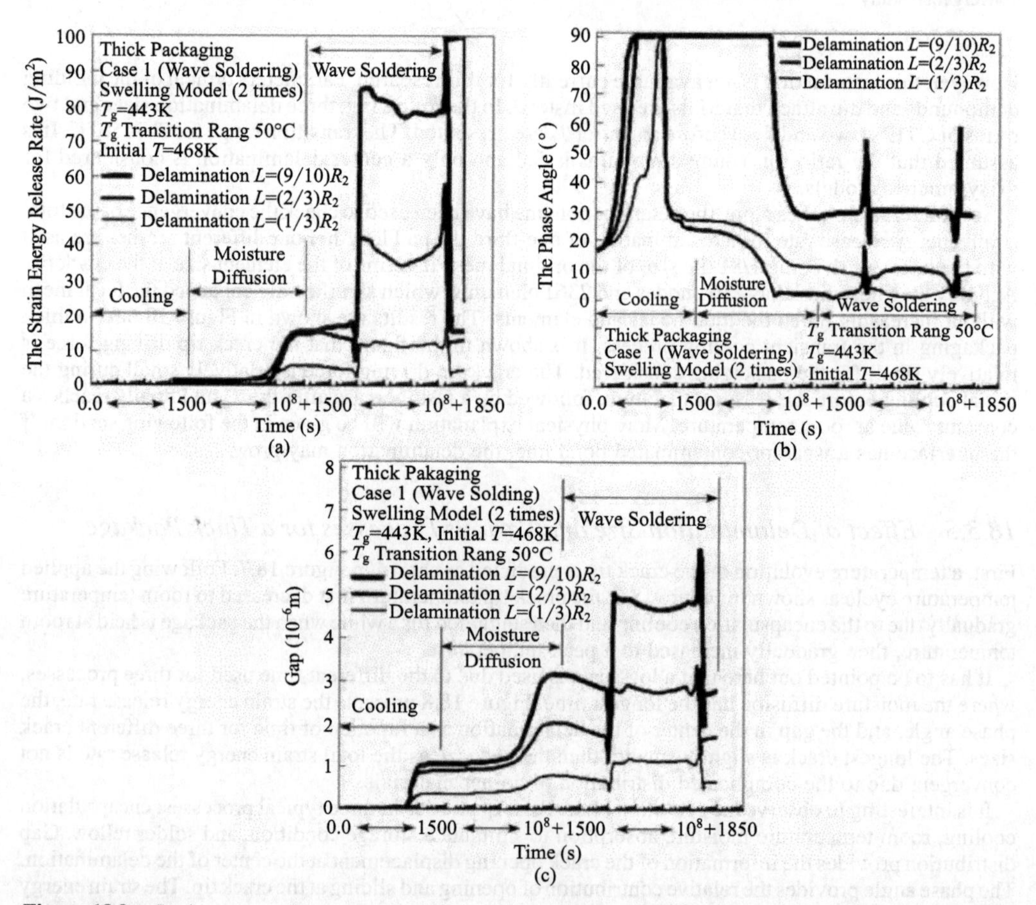

Figure 18.8 Strain energy release rate, the phase angle and the crack center gap as a function of time for three different crack sizes

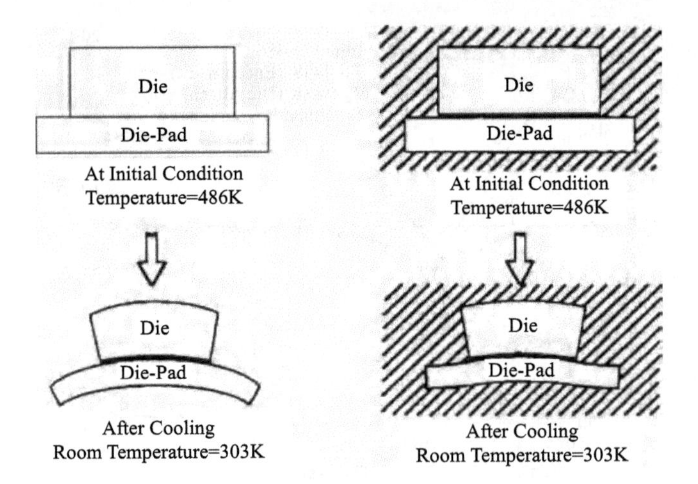

Figure 18.9 A representative deformation pattern for the die pad system bonded by die attach due to the encapsulation cooling

(1) *Encapsulation Cooling*: If there is no molding compound, the die/pad system bonded by die attach may deform in the way shown in Figure 18.8(a). When the molding compound cools and compacts the die/pad assembly, a typical deformation pattern may be shown in Figure 18.9. A deformed mesh at room temperature is shown in Figure 18.10. It is interesting to observe the double curvature induced for the die-pad during the cooling, as this is important for analyzing cracks with different crack sizes. For instance, when the crack is a long one, extending to the second curvature, it is likely to have a sliding (closed) crack tip, as to be demonstrated later. Two factors contribute to this pattern: difficulty for the die to deform in cooling which resists the contraction of molding compound from the right top portion; and compression due to the molding compound upon the right portion of the die-pad which does not have the resistance from the die.

During the initial stage of the cooling, a transient cooling front approaches the molding compound from the right top and bottom corners on the external surface. This global bending tends to open the crack tip but close the center of the crack. Two important temperatures need special attention: T_g^M for the molding compound and T_g^D for the die attach. Figure 18.11(a) and (b) present temperature at the crack tip, peeling and shearing stresses at the Gauss point closest to the crack tip, the strain energy release rate as a function of time. It is noted that the crack tip driving force for three sizes of delamination are presented in Figure 18.11(b). It is recalled that there is a 50 °C temperature transition around T_g^M (443 K) for the molding compound and 10 °C around T_g^D (388 K) for the die attach material. When the temperature drops to 417 K, the peeling stress is zero, indicating a sliding mode and zero gap, which can be clearly shown in Figure 18.12.

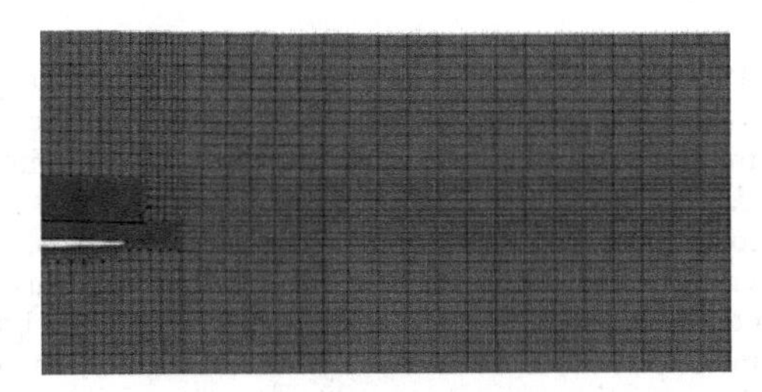

Figure 18.10 A deformed mesh at room temperature after the encapsulation cooling

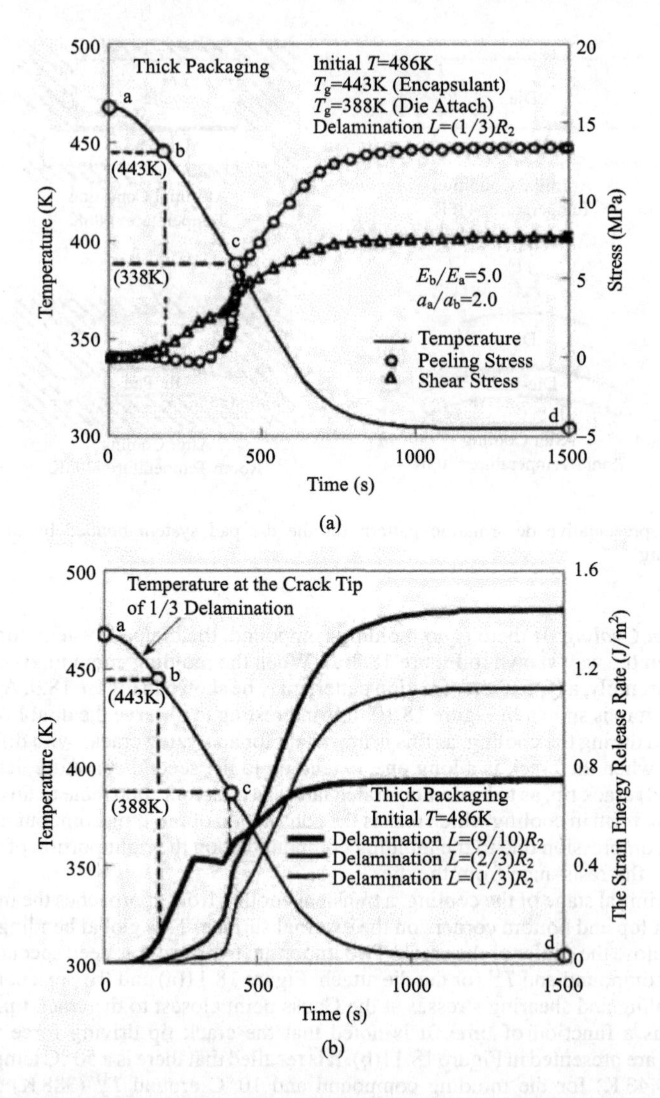

Figure 18.11 Evolution of crack tip temperature, stress, driving force. (a) Peeling and shearing stresses at the crack tip as a function of time in the cooling process; (b) the crack tip driving force

As shown in Figure 18.11(b), when the temperature is above the T_g^D range, the die pad support due to the die attach is relatively weak, resulting in a lower stress state at the crack tip. It should be pointed that all the points for stress output in this figure are the closest Gauss points near those corners inside the molding compound. This mechanism, coupled with lower Young's modulus and small temperature drop can help explain why the phase angles have lower numbers, gaps are small or zero, and have extremely small crack tip driving force. When the cooling front inside the molding compound continues to approach to the crack tip in the packaging from both the bottom and right side of the package, the longer crack tip tends to be closed by the contraction due to the molding compound below the pad earlier than the shorter crack due to its short path to the cold lead frame and the pad edge, creating a phase angle of 90° (sliding contact). For other two shorter cracks, crack tips tend to be

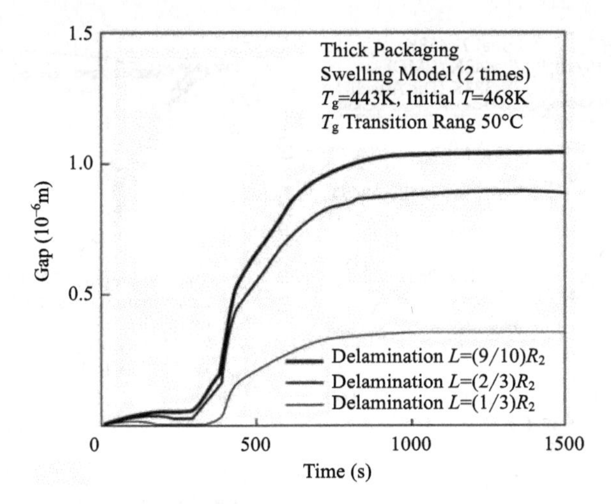

Figure 18.12 Evolution of crack center gap for three delamination cracks

closed based on the same mechanism. However, if the crack is small, the crack tip may close rather late, as shown in Figure 18.11(b).

When the temperature drops around the T_g^D range, the die attach provides stiffer support to the die pad, creating a sharp increase in peeling stress and a steady increase in shearing stress around the crack tip, resulting in a sharp increase in the crack tip driving force. When the temperature in the packaging is relatively uniform, the crack tip driving force, the center gap, and phase angle tend to remain a steady value.

Figure 18.13(a)–(c) present the normal stress in z direction, peeling and shearing stresses for the long crack as a function of time. The stresses are enhanced when T cools down below T_g^M range, and experience significant variations around T_g^D. For instance, the decrease in σ_x results from the hardening of the die-attach, reducing the bending at point B and C. As compared to the one-third short crack case, both the peeling and shearing stresses are much lower, which could explain the decrease of the crack tip driving force as shown in Figure 18.13(a). The physics behind the crack driving force drop when the molding compound hardens, which is due to the reduced deformation and stress. When the die attach gets hardened, the stresses and crack tip driving force increases but with a reduced rate. It is also observed that the long crack tip is in sliding mode after molding compound hardens, while for those two shorter cracks considered, the opening mode dominates the crack tip driving force. From a geometric point of view, when crack extends beyond the pad with curvature change, the crack tip tends to be in sliding mode. If the interface is relatively brittle and fracture toughness is low, the interfacial crack may grow unstable first and then stable when the crack extends to some size. Of course, certain caution must be exercised here as the fracture toughness is not available for the concerned interface.

(2) *Moisture Absorption*: In the moisture absorption process, the swelling moisture absorption tends to open the crack, resulting in an increasing gap in the center of the crack and the shift of the crack tip mode from sliding to opening mode. Even for the longest crack, the crack tip tends to open up gradually in the final stage of moisture saturation. It is clear that there is a critical level of moisture absorption beyond which the crack tip driving force and doming of the delamination will have a sharp increase. This sharp increase in crack tip driving force is due to the sharp increase of the moisture content at the crack tip and the molding compound, as clearly demonstrated by Figure 18.14(a). It is observed here that the moisture content increases fast in the initial moisture stage up to the saturation level. When the moisture is fully saturated, the crack tip driving force and the delamination gap also reach a saturation level. At saturated level, the crack tip is a mixed mode and dominated by the opening mode (Mode I). However, for the long crack, before the moisture diffusion front reaches the

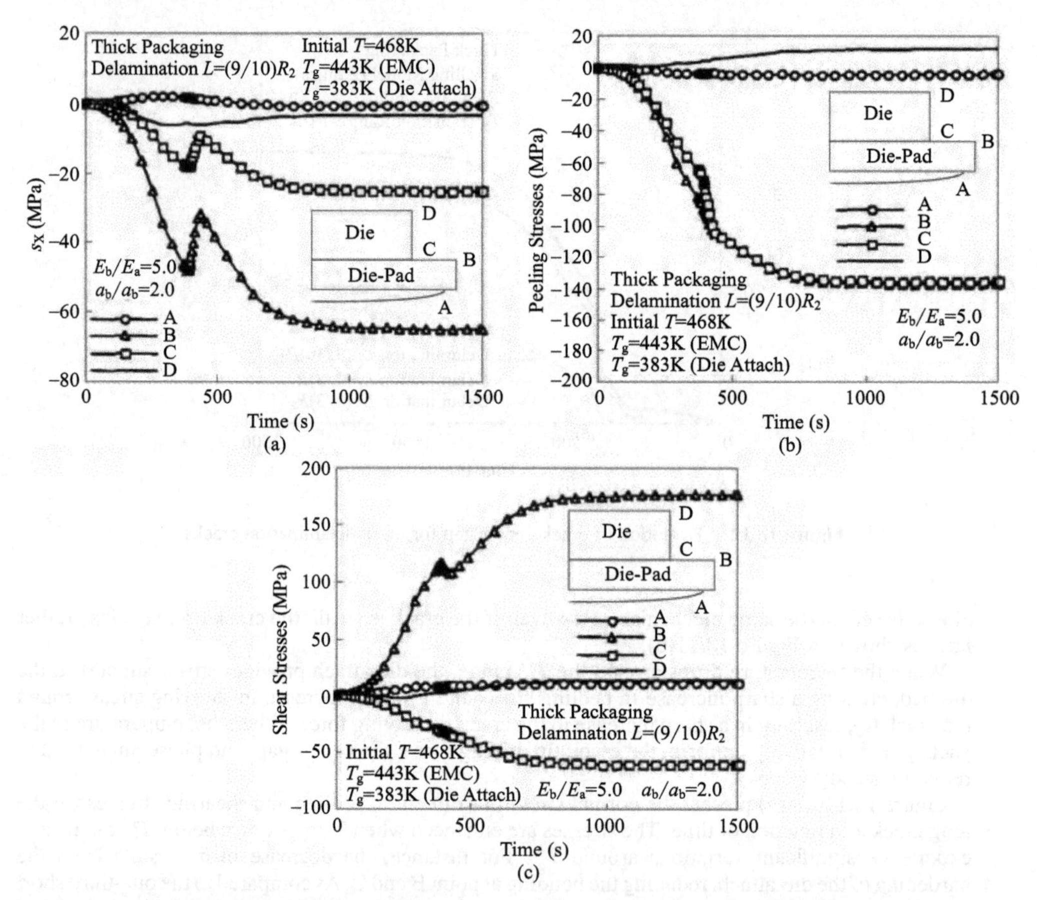

Figure 18.13 Normal stress in x-direction, peeling and shearing stresses at several corner Gauss point function of time. (a) σ_x; (b) peeling stress; (c) shearing stress

interface zone, the crack tip is initially closed and in sliding mode. However, when the moisture front is close enough to the interface zone, the molding compound expands sufficiently to open the interface and crack tip significantly. It is observed that the energy release rate is in the range of typical fracture toughness for the molding compound die pad interface. In general, representative values of fracture energy can range from 1–10 J/m^2 for SiO_2, 30–50 J/m^2 for Al_2O_3, about 200 J/m^2 for epoxy, about 50 J/m^2 for Au/Al_2O_3 interface, about 10 J/m^2 for Cu/SiO_2 interface (Suo, 1993). For molding compound, the typical Mode I fracture toughness in terms of the critical stress intensity factor K_{IC} can range from 0.5 Pa·m$^{1/2}$ at 215 °C (Sawada, 1994) to 3 MPa·m$^{1/2}$ at room temperature for fracture resistant molding compounds. If the relationship $G = (1 - \nu^2)K_2/E$ is used and let E= 12 GPa, the G_{IC} is 17 J/m^2 at 215 °C. Currently, no fracture toughness data is available for the concerned interface. It is well known that the molding compound and leadframe interface is weak and some researchers assumed only 5% of the K_{IC} value of the EMC were used for the interface fracture toughness, very small fracture toughness for the interface. If the processing condition is not controlled well or inappropriate surface adhesion enhancement surface treatment is exercised, the delamination may grow either during the cooling, or moisture absorption, or at the saturation. The energy release rate is observed to increase with the increasing crack size when the crack is not close to the pad edge, and the crack has a smaller driving force for a crack closer to the pad edge. It may be concluded that if the delamination grows, the growth can be unstable initially and then stable when the crack approaches the corner of the pad.

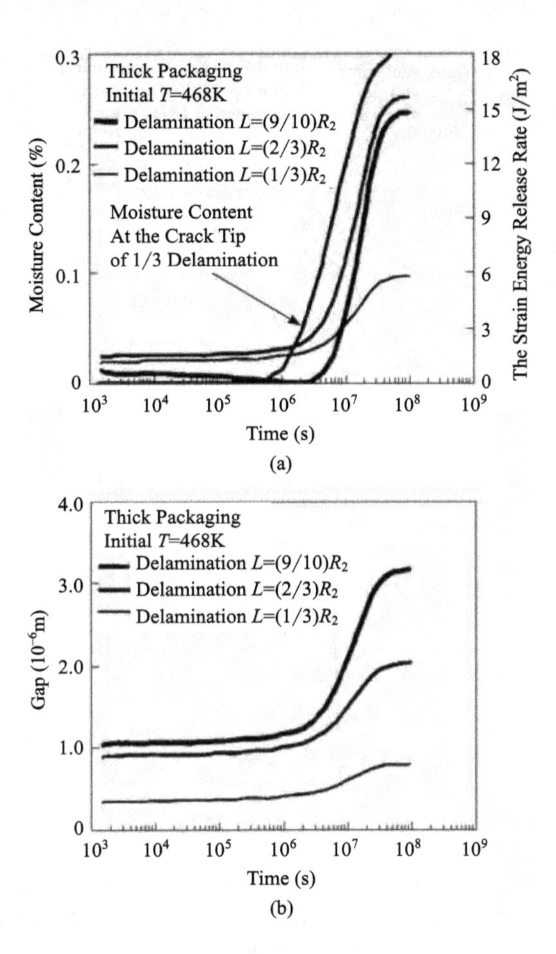

Figure 18.14 (a) Evolution of crack tip driving forces versus moisture content for one third crack; (b) crack center gaps

More realistic fracture toughness data as a function of mode mixity is required to give a more reasonable prediction of crack growth for the moisture absorption process.

(3) *Solder Reflow*: In the solder reflow process, the evolution of the relevant quantities is complex due to the involvement of many variables. Figure 18.15(a) and (b) present a relationship between the crack tip temperature and the crack tip driving force for a short crack. Various relevant temperatures are shown in the upper figure to help explain the heat transfer and deformation mechanisms as follows. Figure 18.16 presents evolution of several stress components for a short crack at the crack tip, bottom corner of the die pad, and the upper corner of the die attach (shown as A, B, and D in Figure 18.16). During the first phase of the solder reflow process, when the temperature is below the boiling temperature of the moisture (100 °C or 373 K, point b), the heating of the leadframe tends to counteract the moisture absorption as the moisture evaporation is not significant. Both the thermal expansion of the molding compound near the right corners in both the x- and y-directions and the leadframe/pad expansion create a local stretching and global bending due to the relatively thick packaging considered. This deformation mechanism tends to reduce the center gap. If the crack is short, the crack tip opening displacement tends to be smaller, resulting in a decreasing driving force. However, for a longer crack extending beyond the double-curvature point, the crack may be actually opened up, resulting in an increase in the crack tip driving force.

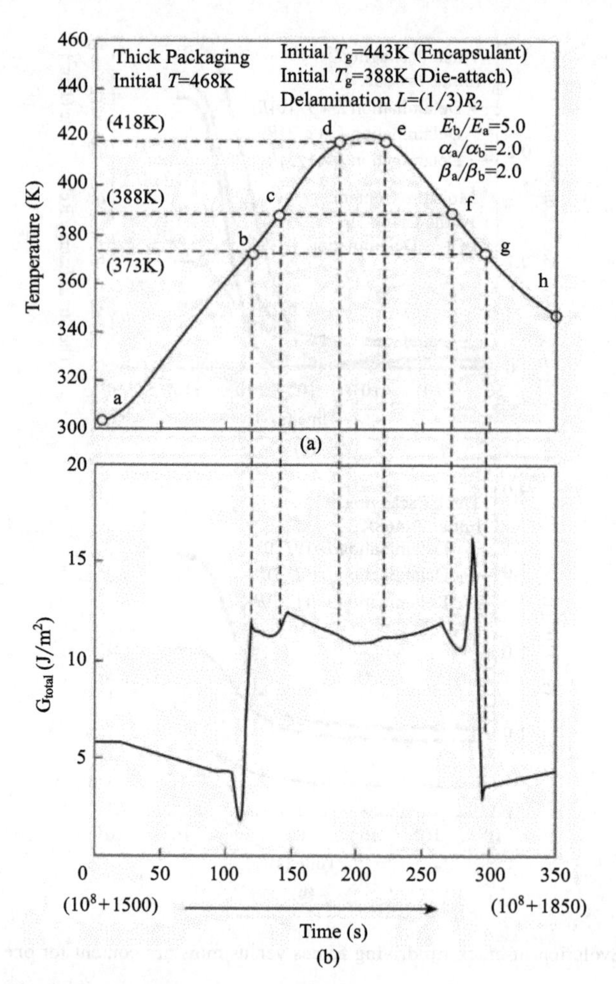

Figure 18.15 A relationship between the crack tip temperature and the crack tip driving force for a short crack as functions of time during a solder reflow process

When the moving evaporation front reaches an influence zone around the interface, covering the whole external surface, there exists a competing mechanism between the vapor force which is directly pulling the interface zone and the bending induced force that is pushing the influence zone. If the bending induced pushing upon the influence zone dominates, the center gap and crack tip displacement tend to decrease, resulting in a reduction in the crack tip driving force and gap, increasing in the phase angle.

When the molding compound is sufficiently swept by the vaporization front, the swelling (pulling) effect tends to be dominating, although the temperature along the interface crack zone may have not yet reached boiling temperature. There exists a critical influence zone, beyond which the interface begins to open up when the interface zone is all heated up to the boiling temperature. This is the instant that the crack tip diving force and center gap (doming) reach their peak values, and the fracture is opening mode dominated. Here again, if the interface is weak, the delamination will grow and it is likely in an unstable manner, resulting in the cracking within the molding compound.

When the temperature continues to increase up to the T: range (point c with 10 °C variation), die attach material tends to get softened, reducing local stress level at the crack tip and therefore reducing the crack tip driving force. Here the thermal expansion for the molding compound and leadframe/pad also contribute to the deformation in addition to the material softening.

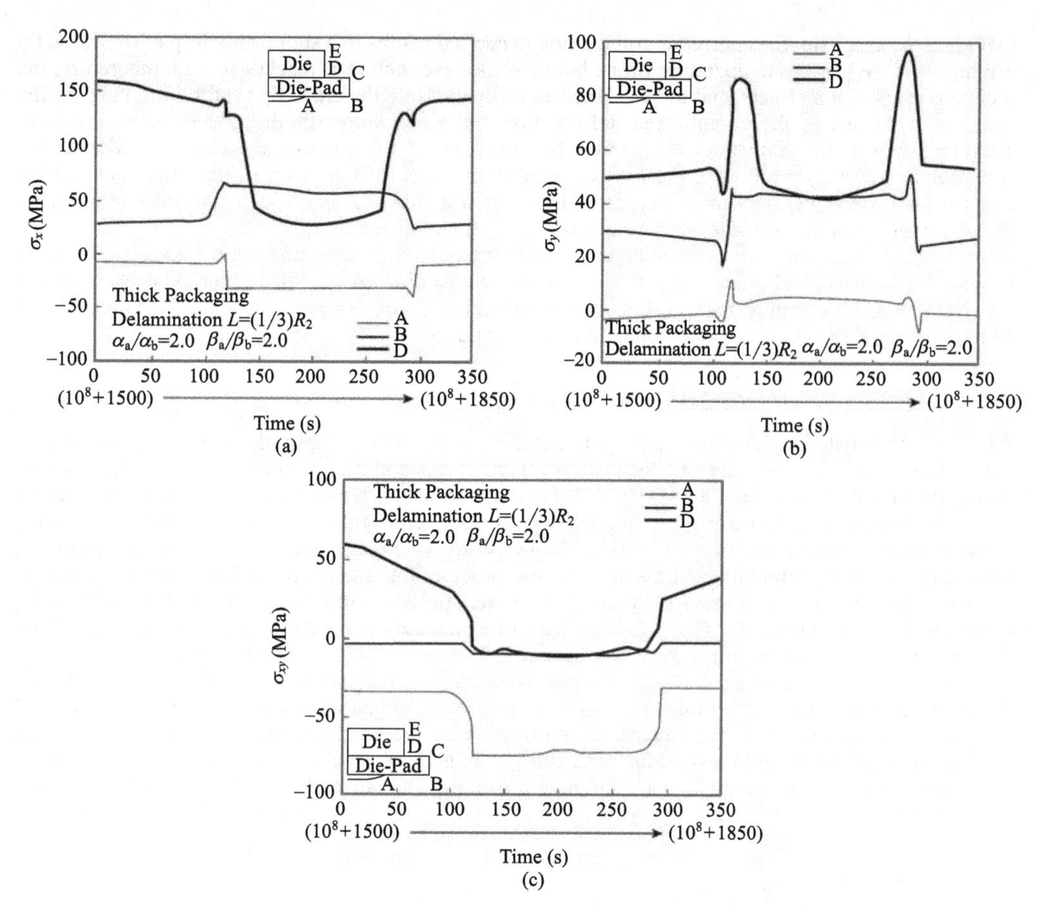

Figure 18.16 Evolution of several stress components for a short crack at the crack tip, bottom corner of die pad, and the upper corner of the die attach

For one-third delamination, the driving force tends to decrease around T_g^D and the phase angle does increase slightly and center gap drops. After the point c range, the thermal deformation of the leadframe/pad and the thermal expansion of the molding contribute to an increase of the crack tip driving force, a decrease in the phase angle, and an increase in the gap. When the temperature front approaches to the Ti' range of the molding compound (50 °C), the molding compound gets softened significantly, continuing to decrease the local stress distribution, resulting in the decrease in the crack tip driving force. This process stops when the peak temperature is reached.

For the same one-third crack, when the temperature begins to drop from its peak until it reaches the T_g^M range for molding compound material, the reverse process begins, resulting in a steady increase of the crack tip driving force and the phase angle and a decrease of the crack center gap due to the hardening of the molding compound. When the temperature reaches T_g^D range of die attach, a decrease in deformation due to the hardening of the die attach material causes a decrease in the crack tip driving force and phase angle and center gap. Most parts of the molding compound have been cooled down below the boiling temperature, even though the crack tip and its neighboring molding compound are still in the boiling temperature. When the cooling front approaches the so-called critical influence zone around the interface crack, the crack tip driving force increases sharply. When the cooling front moves into this influence zone but with cooling down the crack tip below the boiling temperature, the dominating global contraction reduces the crack tip driving force and the center gap. When the cooling front sweeps the whole crack including the crack tip, the enormous compaction generated results in a significant drop of the crack tip driving force. When the temperature drops below some level before it reaches room temperature, the moisture evaporation effects tend to be relaxed extensively. Here the crack tip is still dominated by the opening mode due to the complicated deformation and various contributing mechanisms involved. Therefore, when the temperature continues to drop toward the room temperature, similar to the encapsulation cooling process, there exists a gradual increase of the crack tip driving force until it reaches a saturated level at room temperature. Of course, it should be pointed out that the crack tip driving force is larger than the one after the encapsulation cooling.

For larger delaminations (two-thirds and nine-tenths crack), the deformation mechanisms are similar to the one-third crack case. For the nine tenths crack, the decrease in the crack tip driving force due to the die attach softening is large due to the fact that more molding compound is heated above the T_g of the molding compound, resulting in more reduction of the stress induced.

18.3.4 *Effect of Moisture Expansion Coefficient*

As demonstrated above, the moisture evaporation is very important in controlling the crack tip driving force. Therefore, we would like to look at its effect. In the following, three values of the moisture evaporation coefficient were selected. The interface vapor pressure is also assumed when the interface is heated up to the boiling temperature. Figure 18.17 presents the crack tip driving force as a function of time. If one does not consider the increase in the p, the moisture evaporation coefficient, the crack tip driving force curve is fairly smooth. In the solder reflow process, the crack tip driving force continues to decrease up to the peak temperature due to the both die attach and molding compound softening at T_g range for these two materials. The crack driving force increases again during the cooling stage where the molding compound and die attach get hardened. The gap shows the similar trend to the crack tip driving force. The phase angle has a slight change, more toward the opening mode. Overall the crack tip driving force is lower than the moisture absorption process. The phase angle tends to be fully open and the gap is getting smaller. If we increase A ten times after the temperature evaporation the crack tip driving force can reach up to several hundred J/m^2 (Figure 18.18), which is believed to be much larger than the typical fracture toughness data at high temperature. Of course, the interface fracture toughness could be fairly large which can be 1000 J/m^2 for Mode I, such as for copper foil/resin interface for

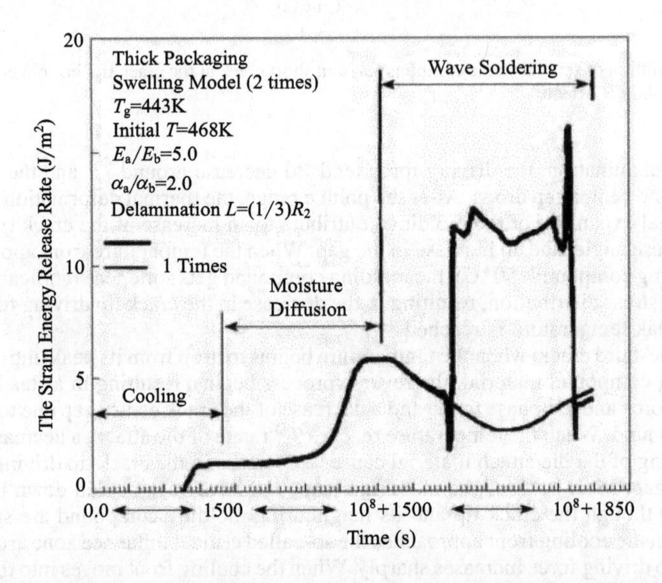

Figure 18.17 Crack tip driving force as a function of time in the solder reflow process for two different moisture expansion coefficients

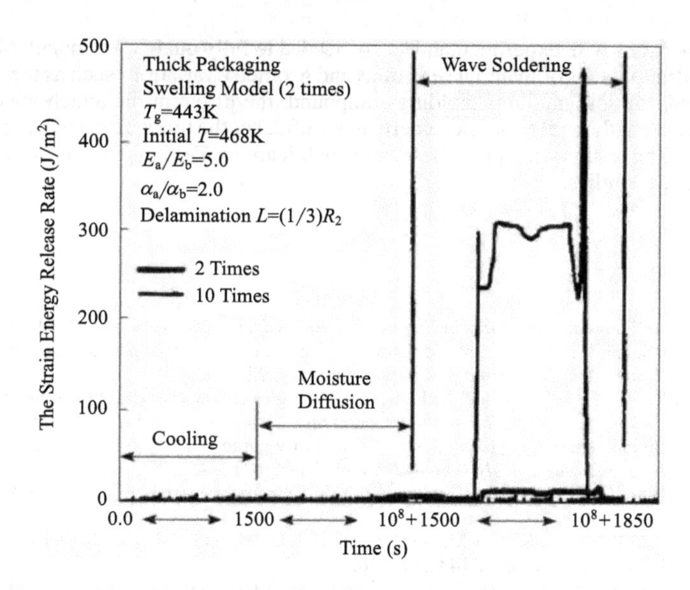

Figure 18.18 Crack tip driving force as function of time in the solder reflow process for two different moisture expansion coefficients

printed wiring board (PWB) (Liu *et al.*, 1994). The evolution of the crack tip driving force is similar to the case when β equals 2. It is shown that the moisture expansion coefficient β for bulk molding compound and die attach has a significant effort on the plastic IC packaging delamination failure. Also, it is observed that interface vapor pressure has a smaller effect. From a material point of view, it is essential to develop molding compounds with less moisture absorption and smaller moisture expansion coefficient. It has to be pointed out that the β values used here are all assumed ones, similar to the ones for graphite fiber reinforced composites. Realistic data should be measured for the typical molding compounds used in the industry.

18.4 Chapter Summary

As a first effort, transient models for temperature and moisture absorption and evaporation have been constructed to investigate the mechanisms and mechanics involved for a delaminated plastic IC package subjected to encapsulation cooling, room temperature storage, and solder reflow process.

Major conclusions are:

(1) For pulsed heating, there is significant difference between the transient and steady state in terms of temperature distribution, crack closure, and fracture mode mixity.
(2) The delamination could grow in either a stable or an unstable manner if a weak bond between leadframe and molding compound exists in different processes considered.
(3) Moving the cooling front, moisture diffusion front, and the moisture evaporation front are found to be important for controlling the deformation and fracture behaviors.
(4) Moisture expansion coefficient is an important parameter for controlling package fracture.
(5) Material property changes at T_g for both die attach and molding compound are important for the crack tip driving force, mode mixity, and the crack center gap.
(6) Transient efforts are important in the popcorning model.
(7) The interface fracture toughness is essential for both the reliability enhancement of the package.
(8) The bulk moisture expansion within the molding compound is the dominating parameter for the cracking of the thick package studied.

However, comparisons with experimental data are needed to fully qualify the model. More parametric study will be devoted to issues of material selections and geometry variations such as the thickness of the molding compound, the low modulus molding compound, the different die attach materials, molding compound with different thermal expansion coefficients, effect of the thermal and mechanical contact for different packages. The results will appear in separate publications. Design guidelines and curves will be provided in the future studies.

References

Bathe, K.J. (1982) *Finite Element Procedures in Engineering Analysis*, Englewood Cliffs, Prentice-Hall, New Jersey.

Fukuzawa, I., Ishiguro, S. and Nanbu, S. (1985) Moisture resistance degradation of plastic LSI's by reflow soldering, *Reliability Physics Symposium, 23rd Annual*, Orlando, Florida, 192–197.

Ganesan, G.S. and Berg, H.M. (1993) Model and analysis for solder reflow cracking phenomenon in SMT plastic packages, *Electronic Components and Technology Conference*, Orlando, FL, USA, ECTC, 653–660.

Hattori, T., Sakata, S. and Watanabe, T. (1988) A stress singularity parameter approach for evaluating adhesive and fretting strength, *Advances in Adhesively Bounded Joints*, Chicago, IL, ASME.

Hutchinson, J.W. and Suo, Z. (1992) Mixed mode cracking in layered materials, *Advance in Applied Mechanics* **29**:63–191.

Kawai, S., Nishimura, A., Hattori, T., Kitano, M., and Shimuizu, T. (1991) Reliability evaluation of electronic devices, *Material Science Engineering, Volume A* **143**:247–256.

Liu, S. (1992) Damage mechanics of cross-ply laminates resulting from transverse concentrated loads, PhD dissertation, Mechanical Engineering Department, Stanford University, CA, August 1992.

Liu, S. (1993) Debonding and cracking of microlaminates due to mechanical and hydro-thermal loads for plastic packaging, *Proceedings of the ASME Winter Annual Meeting*, USA, 1–11.

Liu, S., Kutlu, Z. and Chang, F.K. (1993) Matrix cracking and delamination propagation in laminated composites subjected to quasistatic transverse concentrated loading, *Journal of Composite Material* **27**(5):436–470.

Liu, S., Mei, Y.H., Zhou, S.G. and Zhu, J.S. (1994) Effect of processing induced defects on reliability of plated through holes in PWB, *Proceedings of 27th International Symposium on Microelectronics*, Boston, MA, USA, 576-581.

Liu S., Zhu, J.S., Mei, Y.H., Hu, J.M. and Pao, Y.H. (1994) Fracture behaviors of surface notched ceramic conductive epoxy glass specimens subjected to thermo-mechanical loading, *The ASME WAM, Symposium on Materials and Mechanics in Electronic Packaging for the 21st Century: Quality through Robust Design and Manufacturing*, Chicago, 167–184.

Liu, S., Mei, Y.H. and Wu, T.Y. (1994) Bimaterial interfacial crack growth as a function of mode-mixity in resin/copper interface, *The 26th ASME WAM, Symposium on Materials and Mechanics in Electronic Packaging for the 21st Century: Quality through Robust Design and Manufacturing*, Chicago, 101–111.

Liu, S. (1994) Quasi impact damage initiation and growth of thick-section and toughened composite materials, *International Journal of Solids and Structures* **31**(22):3079–3098.

Liu, S. and Zhu, J.S. (1994) Micromechanics of high temperature composites containing cracking and debonding, *The 26th International SAMPE Technical Conference*, Atlanta, USA.

Liu, S. and Zhu, J.S. (1994) Elastic-plastic finite element models for micro indentation of Al/Si and Si/Al thin film substrate systems with debonding, *The 7th International SAMPE Electronic Materials and Processes Conference*, Parsippany, NJ, 21–23.

Liu, S. and Mei, Y.H. (1994) Effects of voids and their interactions on SMT solder joint reliability, *Journal of Soldering and Surface Mount Technology* **6**(3):21–28.

Marsden, J.E. and Hughes, T.J.R. (1981) *Mathematical Foundations of Elasticity*, Academic Press, New York.

Mix, D.E., Bar-Cohen, A. and Tamma, K.K. (1993) Influence of chip/encapsulant delamination on the thermo-structural behavior of a thermally-pulsed plastic IC package, *Proceedings of the ASME International Electronics Packaging Conference, Binghamton*, NY, USA, 55–79.

Nguyen, L.T., Lo, R.H.Y. and Belani, J.G. (1993), Molding compound trends in a denser packaging world I. Technology evolution, *IEEE International Electronic Manufacture Technology Symposium*, Kanazawa, Japan, 204–212.

Nguyen, L.T., Lo, R.H.Y., Chen, A.S., Takiar, H. and Belani, J.G. (1993) Molding compound trends in a denser packaging world II. Qualification tests and reliability concerns, *IEEE Transactions on Reliability* **42**(4):518–535.

Nishimura, A., Tatemichi, A., Miura, H., and Sakamoto, T. (1987) Life estimation of IC plastic under temperature cycling based on fracture mechanics, *Electronic Components and Technology Conference*, Washington, DC, USA, ECTC, 477–483.

Oizumi, S., Ito, S. and Suzuki, H. (1987) Analysis of reflow soldering by finite element method, Nitto Technical Report, Special Number: Semiconductor Encapsulant.

Rybicki, E.F. and Konninen, M.F. (1977) A Finite element calculation of stress intensity factors by a modified crack closure integral, *Engineering Fracture Mechanics* **9**(4):931–938.

Sawada, K., Nakazawa, T., Kawamura, N., Matsumoto, K., Hiruta, Y. and Sudo, T. (1994) Simplified and practical estimation of package cracking during reflow soldering process, *Proceeding of International Reliability Physics Symposium*, San Jose, CA, USA, 114–119.

Segelken, J.M. and Lustiger, A. (1992) Acoustic microscopy-A research tool for reliability improvements through failure mode analysis of materials interaction, *ASME Winter Annual Meeting*, Anaheim, CA, 1–17.

Shapiro, A.B. (1986) TOPAZ2D-A two dimensional finite element code for heat transfer analysis, electrostatic and magetostatic problems, *User Manual, UCID-20824, Lawrence Livermore National Laboratory.*

Shih, C.F., Asaro, R.J. and Dowd, N.P.O. (1991) Elastic-plastic analysis of cracks on bimaterial interfaces: Part III-large-scale yielding, *Journal Apply Mechanical* **113**:450463.

Simon, B.R., Yuan, Y., Umaretiya, J.R., Prince, J.L. and Staszak Z.J. (1989) Parametric study of a VLSI plastic package using locally refined finite element models, *IEEE Semiconductor Thermal and Temperature Measurement Symposium*, San Diego, CA, USA, 52–58.

Simon, B.R., Yuan, Y., Umaretiya, J.R., Bavirisetty, R. and Prince, J.L. (1990) Thermal and mechanical finite element analysis of a VLSI package including spatially varying thermal contact resistance, *IEEE Semiconductor Thermal and Temperature Measurement Symposium*, Phoenix, AZ, USA, 74–81.

Sun, C.T. and Manoharan, M.G. (1989) Strain energy release rates of an interfacial crack between two orthotropic solids, *Journal of Composite Materials* **23**(5):460–478.

Suo, Z. (1993) Cracking and debonding of microlaminates, *Journal of Vacuum Science Technology* **11**(4):1367–1372.

19

Modeling and Simulation of Power Electronic Modules

High power electronic products such as insulated gate bipolar transistor (IGBT) are becoming more and more popular in the area of power equipment, for instance, traction motor control, hybrid vehicles, uninterruptible power supplies (UPS), servomechanisms, welders, alternative energy sources such as solar energy, and wind energy （Liu, 2012）. As the requirements of higher power and smaller size are increasingly demanded, power density of high power electronic devices increases rapidly. At the same time, performance of these devices, and reliability such as the lifetime are closely related to the junction temperature of the power diode or transistor, a challenging for all the semiconductor modules. According to the statistics, nearly sixty percent of electronics failures are caused by temperature rise and the failure rate nearly doubles if operating temperature increases by every 10℃ (Tummala *et al.*, 1989.). Thus, if the heat generated from the chips could not be dissipated away as quickly and effectively as possible from the devices, the failure rate would rise rapidly. Conventionally, the industry utilized heat sink made by aluminum or copper on which module was mounted by either thermal grease or thermal pad at the interface. The heat sink could be an air-cooled, plate-fin heat sink or a liquid-cooled cold plate (Lee *et al.*, 2000). However, there are challenges for such cooling system, including high interface thermal resistances brought by multilayer materials, their possible non-perfect interfaces, thermal stresses caused by coefficient of thermal expansion (CTE) mismatch, non-uniform temperature distribution across devices, and higher system cost. If these problems could not be resolved, the chips would be over burned by the high junction temperature or cracking induced by large mechanical stress, and as a result the failure rate of module would rise rapidly. These defects may affect long term durability and the damage tolerance of those modules may be a critical challenge, as moisture may degrade material interfaces in terms of mechanical strength, which may induce more failures similar to those encountered in low power plastic modules, for instance, popcorning phenomenon (Mei *et al.*, 1995; Wang *et al.*,1998). It can be perceived that with high voltage, the failure can be more dramatic, as often seen in the field of railway customer service centers.

19.1 Structure Analysis of Power Electronics with Microchannel Coolers

A 1200V/75A IGBT module was investigated. The module layout and internal schematic diagram ignored the bonded wires were shown in Figure 19.1. In this module, six IGBT chips (size of each chip was 9×9 mm^2), six rectifier diode chips (size of each chip was 7×7 mm^2), several freewheeling diode (FWD) chips (size of each chip was 3×6 mm^2), and one IGBT chip for chopping wave (size of chip was 6×6 mm^2) were attached by solder on the DBC (direct bonding copper, diffusion bonded sandwich structure of copper, ceramic and copper) substrates. Then the DBC substrates were bonded to the copper baseplate. After the bonding process, thick aluminum wires were bonded on the chips for the use of electronic interconnects. Finally, the module was filled with silicone gel and encased within epoxy molding compound (EMC) in order to protect the module from moisture, vibration or short circuit.

Modeling and Simulation for Microelectronic Packaging and Integration: Manufacturing, Reliability and Testing, Second Edition. Sheng Liu and Yong Liu.

The power losses of an IGBT module mainly were generated from the inner IGBT chips and FWD chips. On one hand, IGBT is not a perfect switch. When it switches on, the breakover voltage would generate the conduction loss and switching losses. On the other hand, FWD chips also have two types of losses, conduction loss stemming from forward conductive process and switching loss from reverse recovery process. Power loss of IGBT module is as large as 550W for this investigated module so that proper cooling method should be chosen to make sure the operating life and enhance the reliability of modules. Traditionally, heat dissipating structure of IGBT module is to attach the module on a finned heat sink by thermal grease, as shown in Figure 19.2. However, the cooling efficiency of this method is low due to the existence of multiple thermal interfaces, and the whole device is bulky, thus, to optimize the cooling design, microchannel was developed in baseplate instead of attaching the whole module onto a huge heat sink.

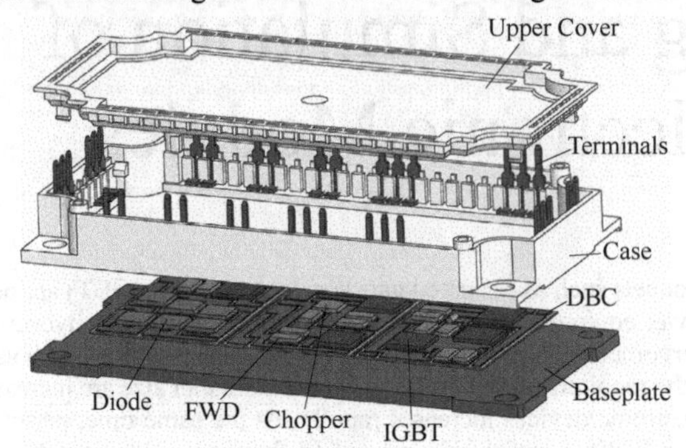

Figure 19.1 IGBT module pattern and internal schematic diagram

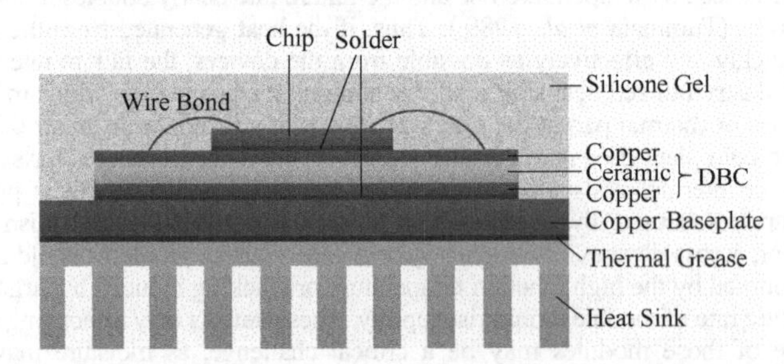

Figure 19.2 Traditional heat dissipating structure of IGBT module

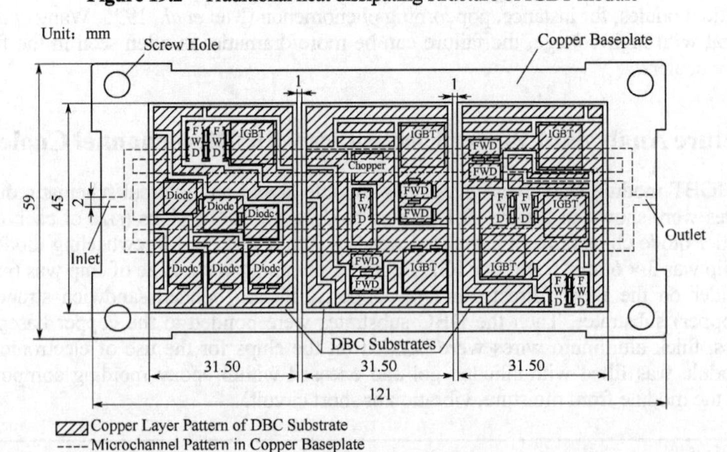

Figure 19.3 Design configuration of copper microchannel baseplate

The design configuration and the real scanning acoustic microscopy (SAM) image of copper microchannel baseplate are shown in Figures 19.3 and 19.4 respectively. To manufacture a proper microchannel, firstly, it should be designed to enable the micropump to be integrated in the module. This micropump consisted of motor, rotor, rotor cover, cover plate, base and two protrusive tubes that is able to deliver high flow rate under closed-loop control (shown in Figure 19.5). The nominal voltage of the micropump is 24V, nominal power is 10W, pressure head is about 10m and its flow rate is 2400mL/min. The leak rate of the micropump is a key parameter due to the high hermeticity requirement, which is 108Pa·m^3/s. The mass of the pump is about 45g, diameter is 24mm and the total length is 38mm, the tube length stretched out is 14.5mm. Compared to the size of the power electronics devices, this pump is tiny and light, and can be directly integrated into the power electronics devices. Next, through-channel appearing to be the white channel in Figure 19.4 was manufactured by laser drilling method, and then the undesired segments were closed by casting. Finally, the simple real product picture of integrated IGBT module with microchannel cooler baseplate and micropump was given in Figure 19.5.

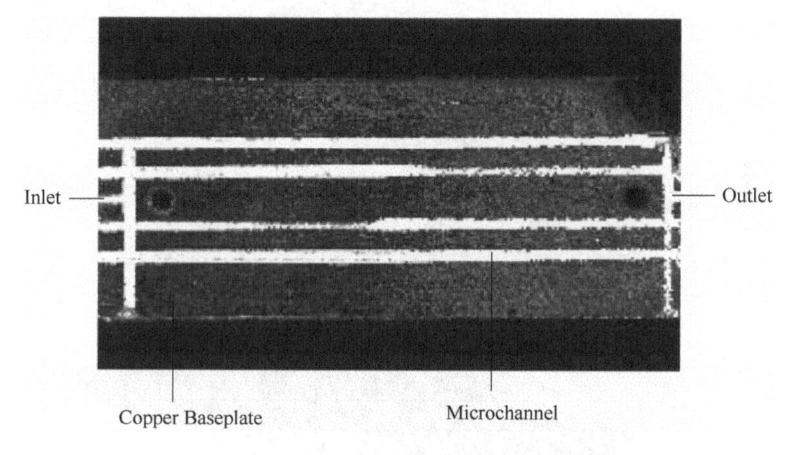

Figure 19.4 SAM image of copper microchannel baseplate

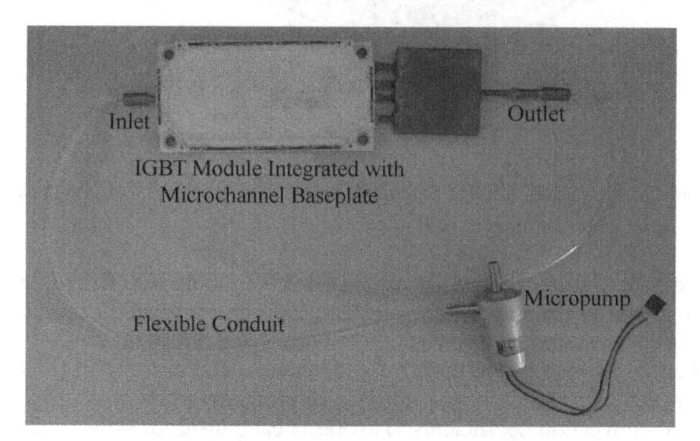

Figure 19.5 Integrated IGBT module with micropump

19.2 Thermal Simulation of IGBT Module on Copper Microchannel Baseplate

To investigate the cooling performance of the module attached on the copper microchannelled baseplate, the COMSOL finite element simulation software was chosen to conduct the heat dissipation simulation. Moreover, considering the fluid performance in the microchannel, we used finite-volume based computational fluid dynamics software package, ANSYS FLUENT, to verify the analytical method of the calculation of convection coefficient of the baseplate integrated with microchannel. The main geometry and finite element mesh for IGBT module integrated with copper microchannelled baseplate were shown in Figure 19.6, in which the mesh was reasonably set based on the mesh dependence study. The size of finite element mesh ranged from 1mm to 2mm. The orientations of the coordinate axes were indicated by the arrows. Power losses of chips are set to be the heat source of this FEM analysis. Here, power loss of each IGBT chip was 60W, power loss of each FWD chip was 6W, power loss of the chopper was 58W, and power loss of each rectifier diode was 10W (Xu, *et al.*, 2002). There were six IGBT chips, twelve FWD chips, one chopper chip, and six rectifier diode chips in total. All the power losses generated from chips internal were applied as loads in volume.

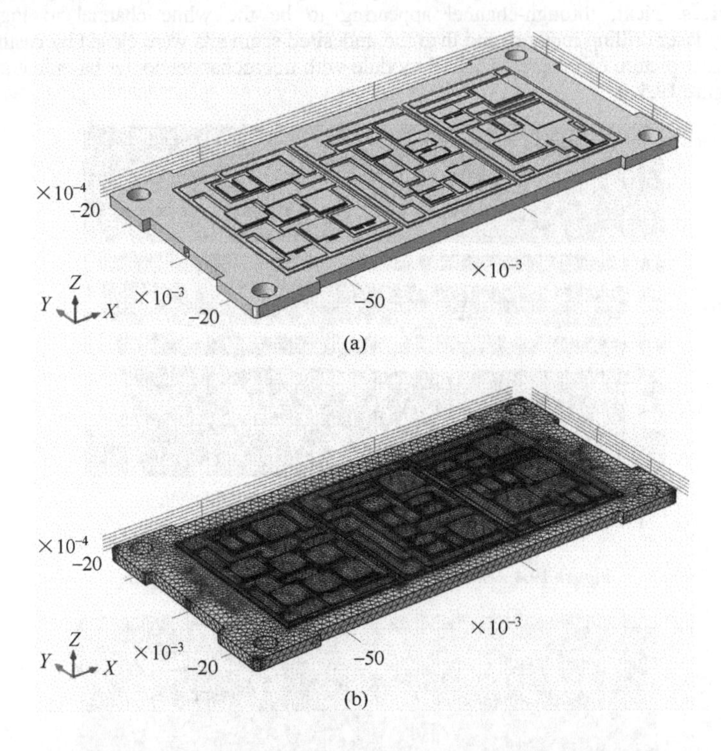

Figure 19.6 Main model and FEA mesh of IGBT module. (a) Main model geometry; (b) finite element mesh of main model

To model the heat transfer in the rectangular channel and the whole IGBT module, we assumed that the fluid was incompressible and its properties kept constant. Heat transfer path is mainly from the top chips to the bottom baseplate. We also assume the convection boundary conditions inside the microchannel walls and the bottom surface of baseplate exposed to the air. As the thermal conductivity of silicone gel covered on the chips and wire was only 0.8W/(m·K), it was assumed to be insulating in order to simplify the calculation. We would use the calculated convection coefficients to express heat transfer of flowing fluid to take into account the fluid coupling effect (Chen *et al.*, 2012). Material properties and thickness of each layer of the model were listed in Table 19.1.

Table 19.1 Material properties used for thermo-mechanical modeling of module

Material	Density (kg/m³)	Specific Heat (J/(kg·℃))	Themal Conductivity (W/(m·K))	Young's Modulus (GPa)	Poisson Ratio	CTE (ppm/K)	Yield Strength (MPa)	Thickness (mm)
Copper (Baseplate)	8950	385	385	110	0.34	17.4	121	3

(continued)

Material	Density (kg/m³)	Specific Heat (J/(kg·℃))	Themal Conductivity (W/(m·K))	Young's Modulus (GPa)	Poisson Ratio	CTE (ppm/K)	Yield Strength (MPa)	Thickness (mm)
Die	2329	−73℃: 557 27℃: 713 127℃: 785	124	112.4	0.28	2.6		0.2
Copper (DBC)	8950	385	385	110	0.34	17.4	121	0.3
Ceramic (96%Al_2O_3)	3800	880	25	300	0.22	6.4		0.38
Solder (SAC305)	7390	217	57	−50℃: 57.3 25℃: 57.3 125℃: 57.3	0.35	−50℃: 57.3 25℃: 57.3 125℃: 57.3	−50℃: 57.3 25℃: 57.3 125℃: 57.3	0.15

In order to simulate the thermal performance, the convection coefficient h between the channel walls and water should be estimated primarily. In this section, analytical and numerical methods were both applied to investigate the module. Researchers had proposed many correlations to calculate the single-phase fluid of constant physical properties in the past years. In 1971, Webb (Webb, 1971) tested out the correlations by experiment and finally he recommended Petukhov-Kirillov correlation and pointed out that it was most accurate when $0.5 \leqslant Pr \leqslant 2000$ and $10^4 \leqslant Re \leqslant 5\times10^6$. In this section, for the investigated situation, the calculated Prandtl number $Pr = 6.8174$, and Reynolds number $Re = 19960$, both satisfied the requirements. Thus we chose Petukhov-Kirillov turbulent flow correlation formula to calculate the heat transfer, in which the Nusselt number Nu was given by [Petukhov *et al.*, 1958]

$$Nu = \frac{(f/2)Re \times Pr}{1.07 + 12.7\sqrt{f/2}(Pr^{2/3} - 1)} \tag{19.1}$$

where f is the friction factor given by $f = (1.58 \ln Re - 3.28)^{-2}$, Re stands for the internal-flow Reynolds number described by $Re = vD\rho/\mu$, and Pr represents the Prandtl number and $Pr = \mu c_p / k$.

Here, v is the linear flow velocity, k is the coolant thermal conductivity, ρ is the coolant density, μ and c_p are respectively the coolant viscosity and heat capacity. Besides, D stands for the hydraulic diameter, which equals to four times the channel area divided by perimeter, and here D is set to be 2mm.

And then h would be achieved by

$$h = \frac{Nu}{D} \tag{19.2}$$

Thus, we assume the coolant is water at 20℃ with 10m/s flow velocity which was within the pressure drop range of the operational micropump in the channels, the value of h was calculated to be 2.0688W/(cm^2·K). To verify the analytical model, we also applied CFD simulation as shown in Figure 19.2 for traditional heat dissipating structure of IGBT module, based on commercial software, ANSYS FLUENT. For the investigated single-phased turbulent flow, standard k-e model, double precision solver and SIMPLE algorithm were applied to solve the pressure-velocity coupling equations (Lee *et al.*, 2006; Sakanova *et al.*, 2014; Yin *et al.*, 2013). The boundary conditions are listed below: At the inlet, velocity normal component $v = 10$m/s, temperature $T = 20$℃. At the outlet, the flow was assumed pressure outlet. For computational region, constant heat flux was assumed as 5W/cm^2 at the top, and the other boundaries were assumed to be insulated. The calculated heat transfer coefficient distribution of microchannel with flowing water is shown in Figure 19.7. Then we calculated the volume average of heat transfer coefficient on water and got the final result of 1.9618W/(cm^2·K), which coincided well with the analytical result, $h = 2.0688$W/(cm^2·K).

To clearly reveal the advantage of microchannel baseplate, we compared it with the traditional copper baseplate. What called for special attention is that in practical application, the traditional module would be attached on a heat sink to satisfy the thermal need. But for the comparison, we got rid of the heat sink to make all other conditions the same. Simulation results of temperature distribution are shown in Figures 19.8(a) and (b). We could find that the highest temperature of module with traditional copper baseplate without any cooling methods already reached to about 28.00℃ (in only air natural convection), while the hottest point of module with copper microchannel baseplate would appear in the diode chip and was only about 80.43℃, a significant drop.

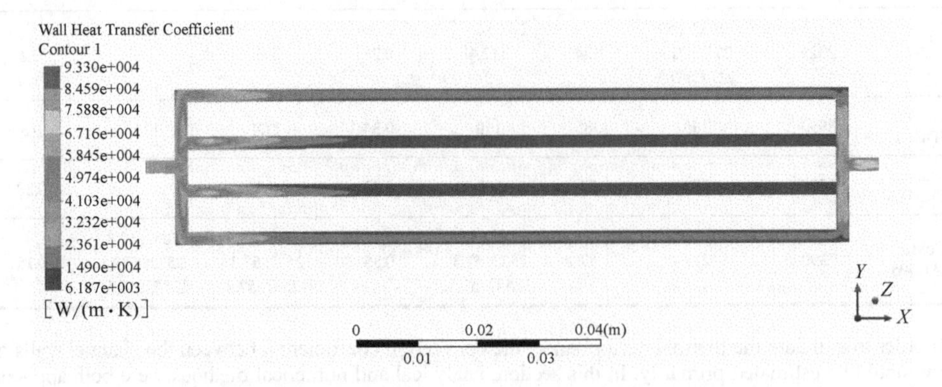

Figure 19.7 Heat transfer coefficient distribution of microchannel

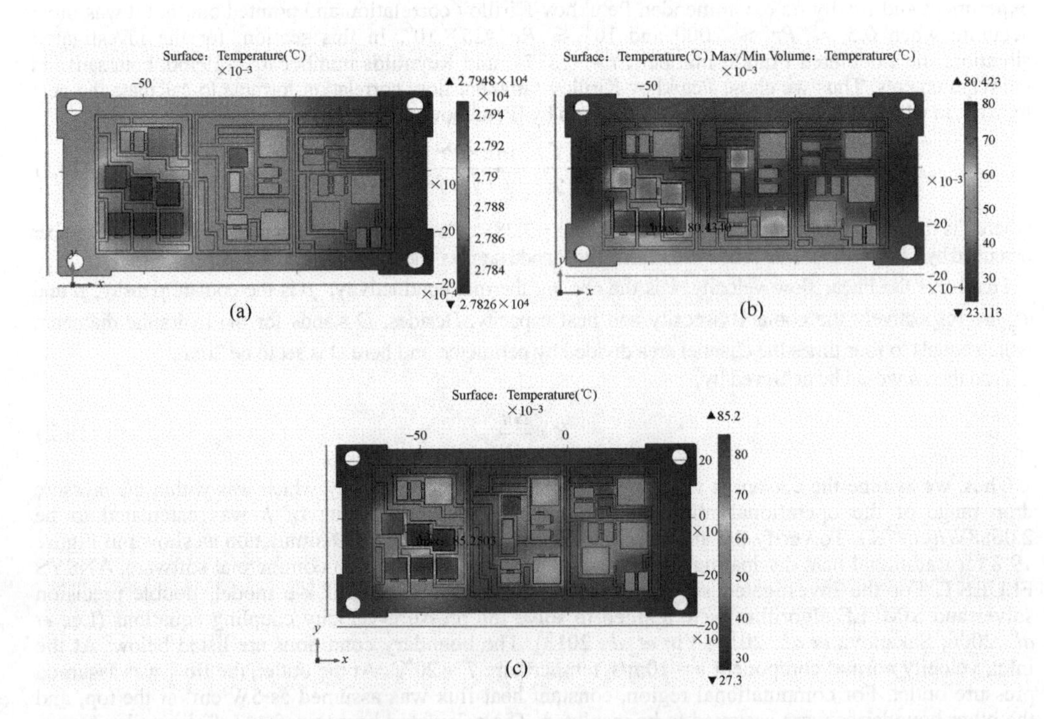

Figure 19.8 Temperature distribution comparison between two kinds of modules. (a) Temperature distribution of IGBT module with traditional copper baseplate (no microchannel); (b) temperature distribution of IGBT module with copper microchannel baseplate; (c) temperature distribution of IGBT module with traditional copper baseplate (fin heat sink)

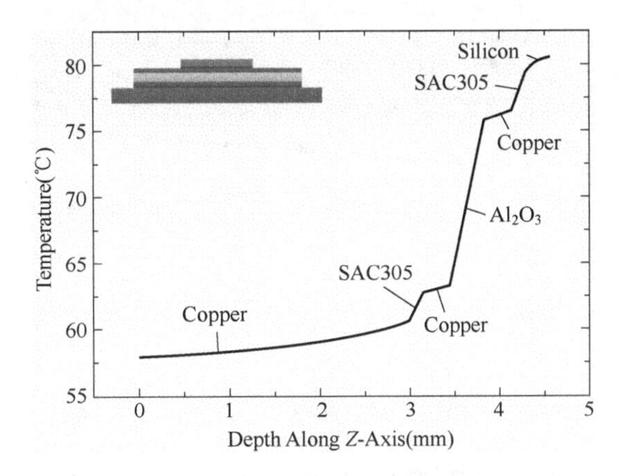

Figure 19.9 Temperature profile through the package

However, the simulation without heat sink for traditional IGBT modules have no practical meaning. Luo *et al.* (Luo *et al.*, 2009) had proposed a design and optimization method used for the engineering design of horizontally-located plate fin heat sink applied in electronic cooling systems. Based on that theory, traditional fin heat sink could be specially designed so that the temperature could satisfy the industry needs. If this kind of module need to show the same thermal performance as the one with microchannel, the average heat transfer coefficient of the fin heat sink rolled on the bottom surface of copper baseplate should be as high as 7000W/(cm^2·K). The simulation results in practical applications could be seen in Figure 19.8(c), the highest temperature was about 85.2℃. Nevertheless, the fin heat sink with such high heat transfer coefficient was difficult to heat transfer coefficient and the volume of this kind of heat sink was so huge compared to the module itself. What is more, the cost would rise due to the usage of metal heat sink.

Furthermore, the impact of thermal resistance of each layer of module was also studied. A plot of the vertical (*Z*-axis) temperature distribution at the hottest location (center of the diode die at *X* and *Y*) of module with copper microchannel baseplate is shown in Figure 19.9. From the plot, we can find that largest rate of temperature gradient appeared in the ceramic layer in the middle of the DBC, nearly 14℃ temperature drop across this layer. It can be also seen that slope of temperature variation was related to the material thermal conductivity of each layer. The temperature of the microchannel embedded in the copper baseplate located on the left hand edge of the curve was only 58.0℃. The whole bulk temperature increased slightly, only about 2.7℃ across the baseplate. Except for the design of microchannel, the thermal management of the module could be also optimized by choosing material with higher thermal conductivity, such as AlN [thermal conductivity j = 319W/(m·K)(Watari *et al.*, 2003)] instead of Al_2O_3 [j = 25W/(m·K)], or nanosilver paste [j = 240W/(m·K) (Chen *et al.*, 2012)] instead of SAC305 (j = 57W/(m·K). But changing material would increase costs, we need to do a tradeoff between the cost and the performance if it is really needed.

One alternative approach was to adjust the thickness of ceramic layer. The relationship between thickness of ceramic layer and the highest temperature of IGBT module was shown in Figure 19.10. It could be seen that for Al_2O_3 ceramic and AlN ceramic, the relationships had different change trend. When thickness of ceramic layer increased, if the module used Al_2O_3 ceramic, the highest temperature (appearing on IGBT chip) would increase; however, if the module used AlN ceramic, it would decrease. It was because that the thermal conductivity of AlN was more than ten times larger than Al_2O_3 ceramic, almost comparable to copper [j = 385W/(m·K), thus AlN ceramic would facilitate the ability of heat dissipation, but Al_2O_3 ceramic would make against the thermal performance. Considering the industry cost and requirement of thermal performance, we could utilize thinner AlN ceramic layer to replace thicker Al_2O_3 ceramic layer in order to achieve desired performance.

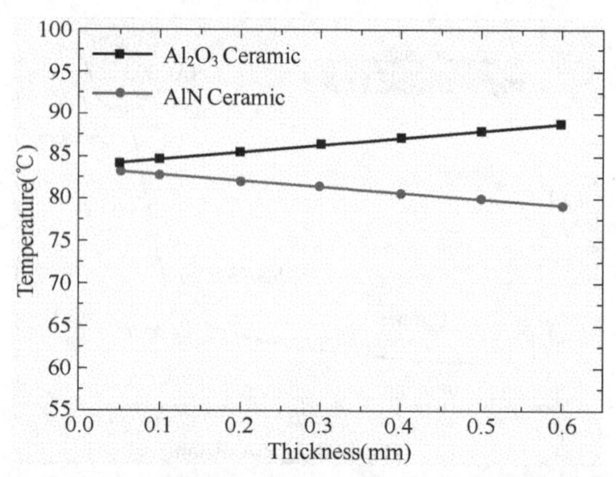

Figure 19.10 Relationship between thickness of ceramic layer and highest temperature of IGBT module

19.3 Residual Stress Analysis of IGBT Module on Copper Microchannel Baseplate

Even though the IGBT module was not on working and no heat dissipation generated, residual stress would still exist in the package due to the CTE mismatch of different material during the assembly process such as reflow soldering process, in which the temperature would change between the solidification temperature (if SAC305 solder was used, 220℃) and the room temperature. During the reflow operation, on account of the asymmetric structure of the package, the temperature distribution in the package is nonuniform, as well as the CTE difference among the die, solder, copper and ceramic, therefore the residual stress was created. The reflow profile of SAC305 was given in Figure 19.11, in which the peak temperature is 236℃, 20℃ higher than the solidification temperature, and the duration of whole process is 280s. Nonlinear mechanical behavior was considered for the FEA calculation, because the soft die attach solder material and copper would present the plastic performance, in this case, we chose Anand viscoplasticity rate-dependent behavior model for the soft solder and Chaboche nonlinear kinematic hardening model for the copper and the reasons are described as follows. The detailed properties of these two models were respectively given in Tables 19.2 and 19.3. Relationship between temperature and yield stress of copper was additionally described in Figure 19.12. Besides, since we assumed that plastic stress-strain behavior of copper was rate-independent (Zhu, *et al.*, 2007), creep and other long-duration effects would be ignored for copper. As we desire to calculate the stress distribution right after the rapid cooling or heating process and judge whether the dies would crack or not at the most severe condition, therefore, neglecting creep of copper which tend to reduce the package stresses over time for a given strain was reasonable (Calame *et al.*, 2005).

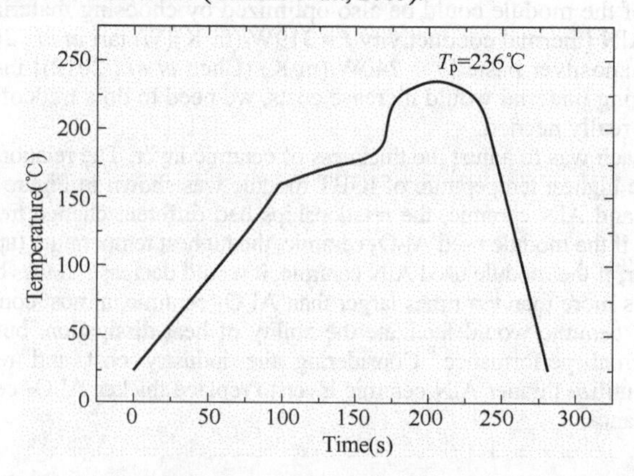

Figure 19.11 Reflow profile of solder SAC305

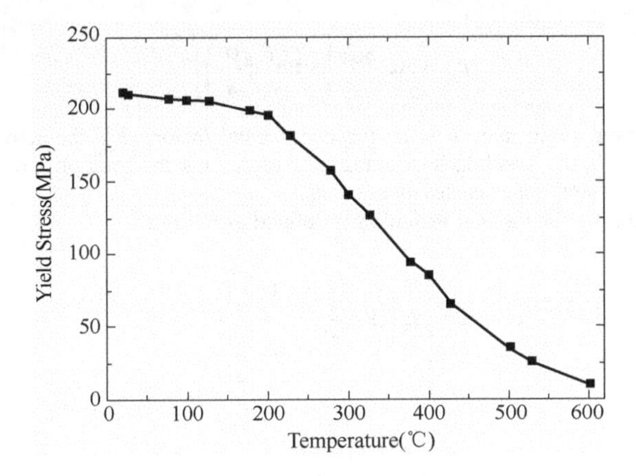

Figure 19.12 Relationship between temperature and yield stress of copper

Table 19.2 Parameters in Anand viscoplasticity constitutive model of solder

Constant	Quantity	SAC305
s_0	Initial Value of Deformation Resistance	2.15MPa
Q/R	Activation Energy/Boltzmann's Constant	9970K
A	Pre-exponential Factor	17.994 1/s
ξ	Multiplier of Stress	0.35
m	Strain Rate Sensitivity of Stress	0.153
h_0	Hardening/Softening Constant	1525.9MPa
$\hat{s}$	Coefficient for Deformation Resistance Saturation Value	2.536 MPa
n	Strain Rate Sensitivity of Deformation Resistance	0.028
a	Strain Rate Sensitivity of Hardening/Softening	1.69

Table 19.3 Parameters in Chaboche constitutive model of copper (Chaboche,1989)

Temperature (℃)	C_1 (MPa)	γ_1	C_2 (MPa)	γ_2
20	54041	962	721	1.1
50	52880	1000	700	1.1
150	45760	1100	600	1.1

On one hand, Anand constitutive model was chosen to describe the material properties of the solder because it might accurately reflect the temperature sensitivity, strain hardening and dynamic recovery and better reflected the creep behavior of softer die attach solder. Anand model was a single-scalar internal variable model for large, isotropic, viscoplastic deformations formulated by Anand (Anand *et al.*, 1985) and further developed by Brown *et al.* (Brown *et al.*, 1989). Anand model includes a flow equation and an evolution equation.

The flow equation is shown below

$$\dot{\varepsilon}_{\text{in}} = A e^{-Q/RT}\left[\sinh\left(\xi\frac{\sigma}{s}\right)\right]^{1/m} \tag{19.3}$$

where $\dot{\varepsilon}_{\text{in}}$ is the inelastic strain rate, A is the pre-exponential factor, Q is the activation energy, R is the Boltzmann's constant, T is the absolute temperature, ξ represents the multiplier of stress, s is equivalent stress and m stands for the strain rate sensitivity of stress.

The evolution equation for the internal variable s is defined as follows

$$\dot{s} = \left[h_0\left(\left|1-\frac{s}{s^*}\right|\right)^a \frac{1-\frac{s}{s^*}}{\left|1-\frac{s}{s^*}\right|}\right]\dot{\varepsilon}_{\text{in}} \tag{19.4}$$

$$s^* = \hat{s}\left[\frac{\dot{\varepsilon}_{\text{in}}}{A}e^{Q/RT}\right]^n \tag{19.5}$$

where s^* is saturation stress, h_0 represents the hardening/softening constant, a is the strain rate sensitivity of hardening/softening, $\hat{s}$ is a coefficient for deformation resistance saturation value, and n stands for the strain rate sensitivity of deformation resistance.

On the other hand, as the temperature of IGBT module changed with the operating process, the plastic behavior of copper was described by temperature-dependent Chaboche nonlinear kinematic hardening model because it took temperature effect into account. The yield function for Chaboche nonlinear kinematic hardening rule is shown below (Chaboche,1989):

$$F = \sqrt{\frac{3}{2}(\{s\}-\{\alpha\})^{\mathrm{T}}[M](\{s\}-\{\alpha\})} - Y = 0 \tag{19.6}$$

where $\{s\}$ represents the deviator stress, $\{\alpha\}$ is the back stress and Y refers to the yield stress of the material.

The back stress $\{\alpha\}$ for the Chaboche model is as follows:

$$\{\alpha\} = \sum_{i=1}^{n}\{\alpha_i\} \tag{19.7}$$

$$\{\Delta\alpha\}_i = \frac{2}{3}C_i\{\Delta\varepsilon_{\text{pl}}\} - \gamma_i\{\alpha_i\}\Delta\hat{\varepsilon}_{\text{pl}} + \frac{1}{C_i}\frac{\mathrm{d}C_i}{\mathrm{d}\theta}\Delta\theta\{\alpha\} \tag{19.8}$$

where $\hat{\varepsilon}_{\text{pl}}$ is the accumulated plastic strain, θ is the temperature, C_i and γ_i respectively represent the Chaboche material parameters, C_i is initial hardening modulus and γ_i controls the decreasing rate of hardening modulus with increasing plastic strain, and i is the number of kinematic models, here $i = 1, 2$. In Equation 19.8, the first term represents the hardening modulus, the second term represents the nonlinear effect.

The same model as the thermal simulation part was used to analyze the mechanical performance. Part of the mesh was shown in Figure 19.13. During the cooling down process, the copper would shrink more than the Si chip because of the CTE mismatch, and thus copper would bend into a convex shape, and die would suffer from tensile stress.

The displacement of the IGBT module when the reflow soldering process finished was shown in Figure 19.14. It could be seen that the module would deform and the maximum displacement of the module was 160.21μm. In order to verify the validity of FEA results, three-coordinate measuring machine was used to test the warpage of the copper baseplate and warpage of sample 1 was 164.2μm while sample 2 was 166.0μm. The experimental schematic diagram was given in Figure 19.15, in which 25 points were measured precisely in order to project out the warpage of whole baseplate. The trends of deformation change keep the same between simulation results and experimental test results, and the difference rate was about 3.0%. We thought it was acceptable, and the simulation was valid.

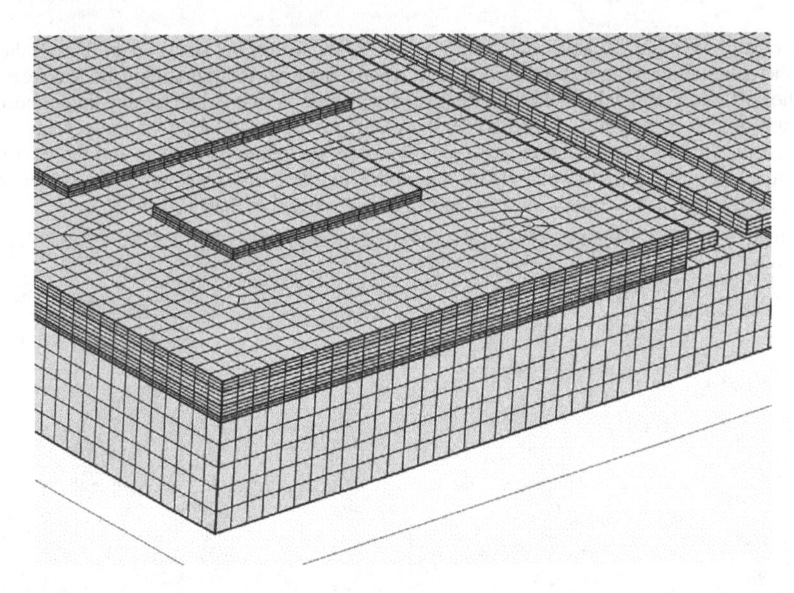

Figure 19.13 Part of the mesh of the investigated module for thermal-mechanical analysis Time=280s Surface: Total displacement(μm)

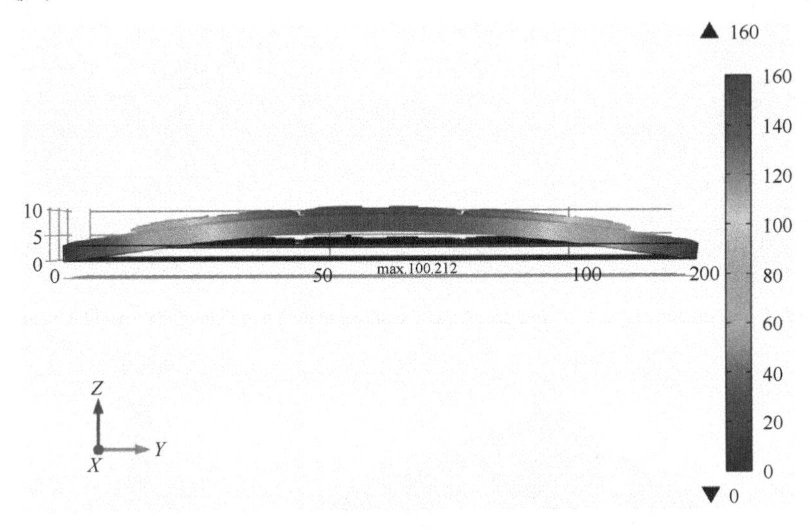

Figure 19.14 Displacement distribution when reflow process finished

Moreover, the detailed stress contour of each component of IGBT module was shown in Figures 19.16 to 19.18. From Figure 19.16, it could be found that the maximum von-Mises stress distribution of copper baseplate when reflow process finished was 142MPa, at which the copper would be in yield. In DBC substrate as shown in Figure 19.17, the maximum von-Mises stress of copper layer was 186MPa appearing at the edge of bottom copper layer which would contact the soft solder. The ceramic layer would be bent and the maximum first principal stress was 43MPa while the third principal stress was 87.5MPa, both were lower than the fracture strength of Al_2O_3, which was as high as 300MPa. It could be clearly seen that the stress concentrated where the chips were plated. For the silicon chips, we analyzed in details the principal stress distribution in Figure 19.18, from which it could be seen that the maximum third principal stress on the top surface with the value of 275MPa was the most dangerous part, however, it was still below the tensile strength of silicon die, which was typically 380MPa (Zhao *et al.*, 2008), so that crack would not occur even though the dies were bent due to the residual stress, but the high value in tensile stress in the die should be a concern. We also focused on the soft solder and calculated the effective creep strain of two solder layers, the highest strain was 0.13, appearing at the edge point of the solder between baseplate and bottom copper layer, while the maximum strain of the other solder layer was 0.08. We drew the relationship between the maximum strain and the area of solder in Figure 19.19 and clearly found that the strain increased when the solder area became larger and the change curve was nearly linear when the area was smaller than $100mm^2$.

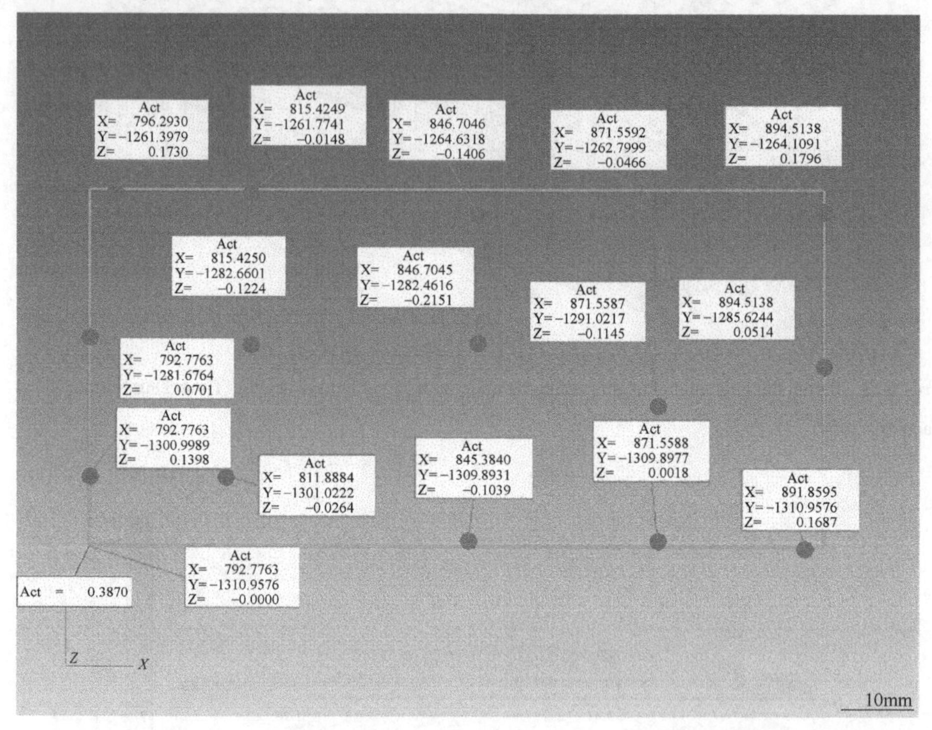

Figure 19.15 Schematic diagram of three-coordinate measuring experiment Time=280s von-Mises-stress(MPa)

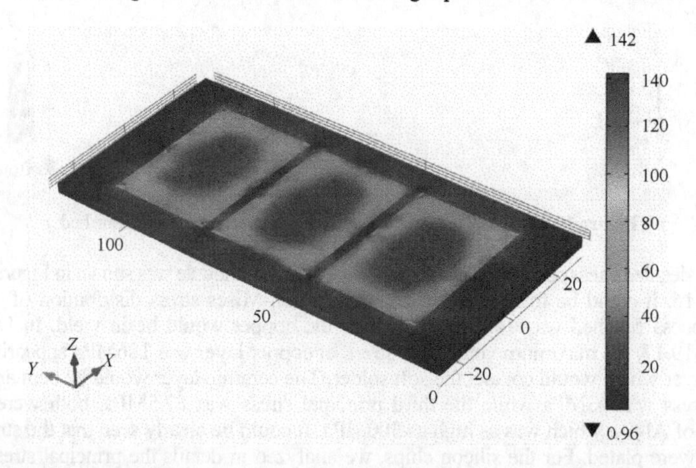

Figure 19.16 von-Mises stress distribution of copper baseplate when reflow process finished

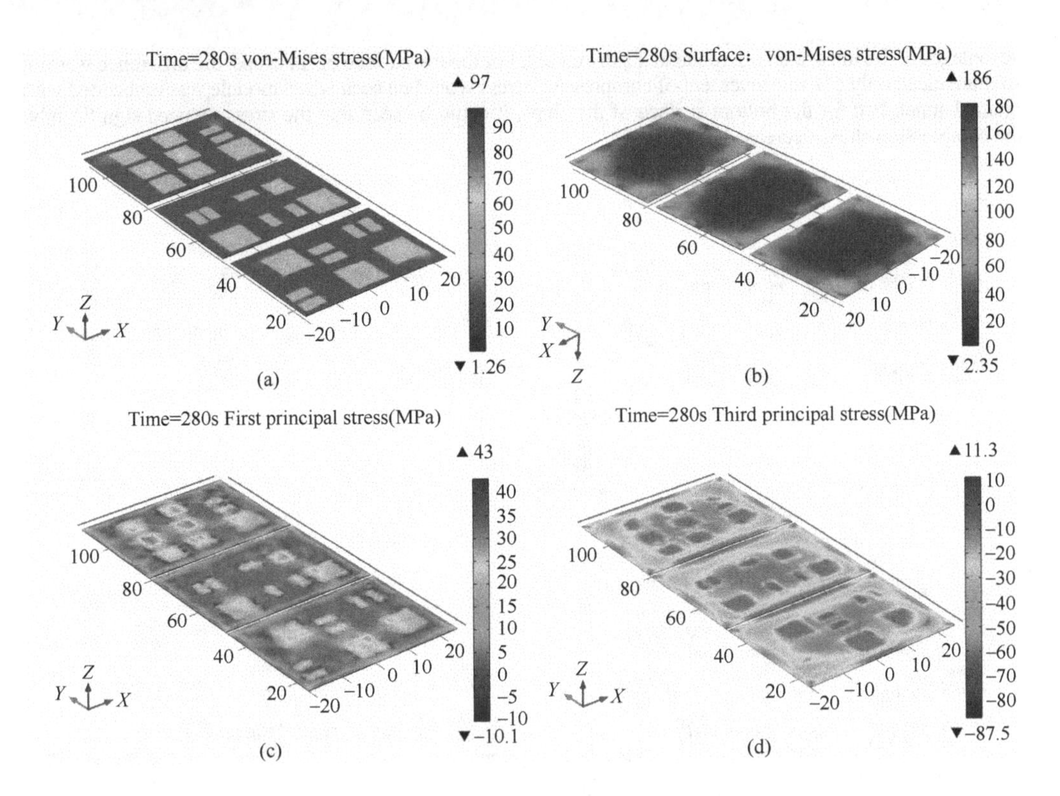

Figure 19.17 Stress distribution of DBC substrate when reflow process finished. (a) von-Mises stress on the top copper layer ; (b) von-Mises stress on the bottom copper layer ; (c) first principal stress of ceramic layer ; (d) third principal stress of ceramic layer

Operating stress analysis of IGBT module on copper microchannel baseplate was done. Considering the influence of residual stress calculated in this section, the module was power on and the power losses were shown in Section 19.1. The maximum displacement of operating module was 3.41μm, which was given in Figure 19.20. In addition, three-coordinate measuring machine was used to measure the warpage and result showed that sample 1 was 18μm and sample 2 was 12μm. The deformation trend from reflowing process to operating process of FEA results and experimental test results kept the same. Although in FEA method, the residual stress due to reflow process would cause deformation of 160.21μm, it could be compensated for by the heat generated from operating chips and then the final warpage would decrease sharply, and nearly became planar. In other words, the module would be bent in opposite direction and therefore the module itself would be planar. The tensile stress (the first principal stress) and the compressive stress (the third principal stress) on the bottom and top of the die when module was operating are respectively shown in Figures 19.21 and 19.22. On the bottom of the die where soft solder was attached on, the trend of tensile stress and compression stress were almost the same. The peak tensile stress was 70.1MPa and compression stress reached 57.3MPa. On the top of the die, because there was no solder to release the strain and Si chip was bent, tensile stress was nearly zero, while compression stress increased along the *Y*-axis and reached to the peak at the center of the die. The peak compression stress reached 188.2MPa, which was much larger than the bottom of the die. All of these stresses were below failure levels and die would be crack-free. In contrast, we also simulated the traditional IGBT module without microchannel and focused on the stress of chips. The comparison results were shown in Table 19.4. From the table we could found that although the displacement of the whole module increased when the module was embedded with microchannel, stress level changed little. The hollow section of the baseplate would cause larger deformation of the whole module and residual stress on the chips during reflowing process, but when the modules were working and water flew in the channel, the advantages of the integrated microchannel showed out. For the top surface of the chips, the difference was not so dramatical, only 5% improvement of compression stress would be seen when module was embedded with microchannel. But for the bottom surface of the chips, it could be seen that the stress reduced significantly, the compression stress decreased 35%.

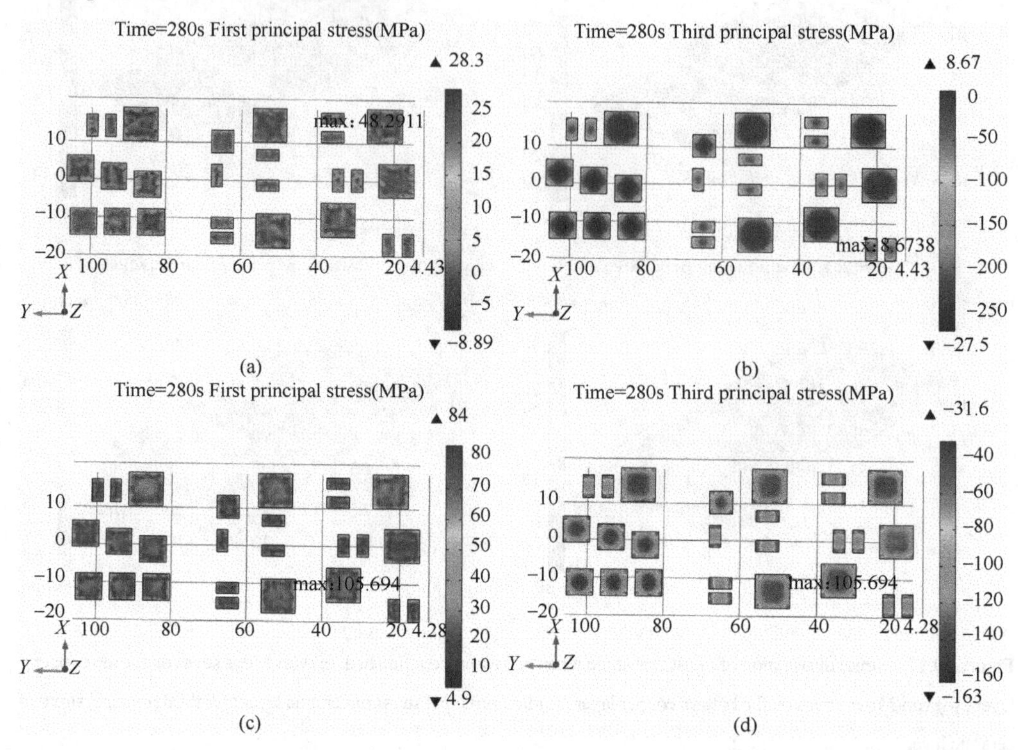

Figure 19.18 Principal stress distribution of chips when reflow process finished. (a) First principal stress on top surface; (b) third principal stress on top surface; (c) first principal stress on bottom surface; (d) third principal stress on bottom surface

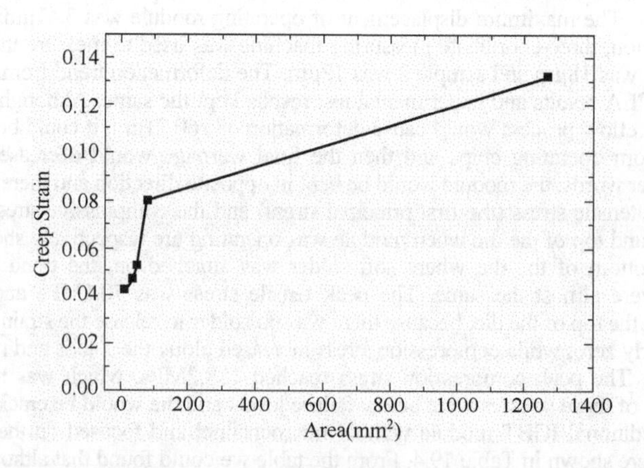

Figure 19.19 Relationship between the maximum strain and the area of solder

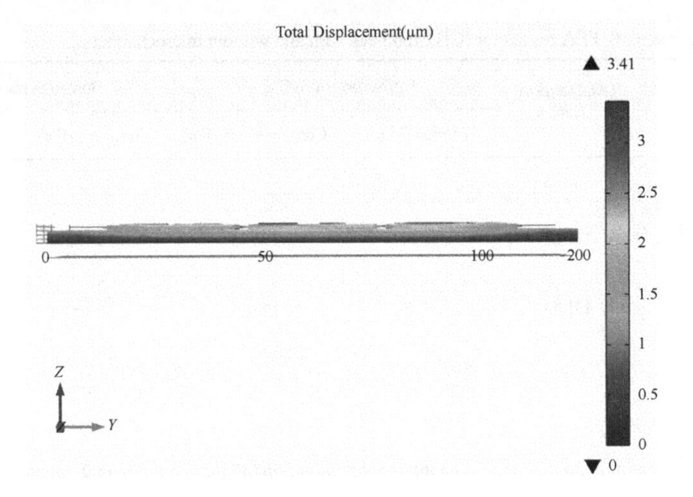

Figure 19.20 Displacement distribution of operating module

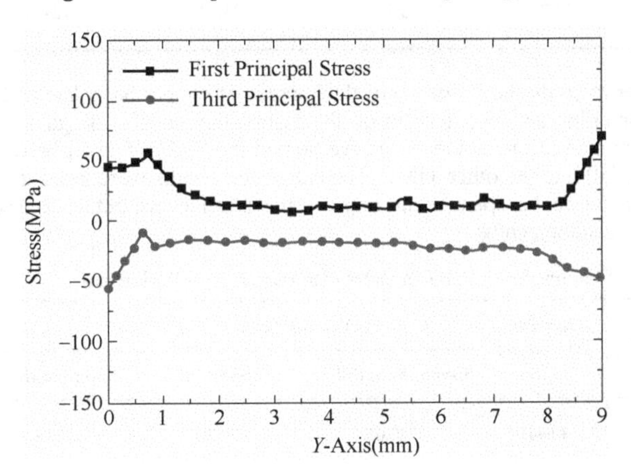

Figure 19.21 Principal stress along *Y*-axis at bottom of chip during operating

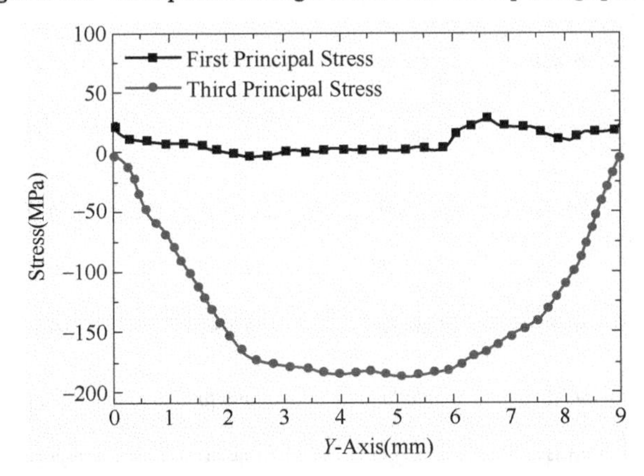

Figure 19.22 Principal stress along *Y*-axis at top of chip during operating

Table 19.4 Comparison of FEA results for IGBT modules with and without microchannel

Condition	Displacement (μm)	Top Surface of Si Chips		Bottom Surface of Si Chips	
		Tension (MPa)	Compression (MPa)	Tension (MPa)	Compression (MPa)
Residual Stress					
IGBT Module with Microchannel	160.21	48.29	276.75	105.71	175.30
IGBT Module without Microchannel	131.85	51.26	265.01	93.08	180.93
Operating Stress					
IGBT Module with Microchannel	3.41	64	268.88	66.27	91.66
IGBT Module without Microchannel	13.7	73.86	282.8	123.36	123.87

Furthermore, in order to study the influence of thicknesses of copper baseplate and solder layer thickness on the stresses of silicon chips, and also to optimize the design, the model was again analyzed with parameter sweep. To make a difference of two solder layer, we named the SAC305 layer between Si and DBC TIM1 (thermal interface material) and the other TIM2. The calculated results were listed in Table 19.5. It could be clearly seen that thicken copper baseplate could improve the stress levels, but thicken solder layer would help to reduce the stress on the silicon chips.

Table 19.5 Summary of FEA results for IGBT modules with various layer thicknesses

Condition	Thickness	Junction Temperature (℃)	Si Residual Stress		Si Operation Stress	
			Compression (MPa)	Tension (MPa)	Tension (MPa)	Compression (MPa)
Copper baseplate	2	80.441	230.66	70.35	74.95	288.05
(TIM1 = 0.15mm,	2.5	80.430	254.32	80.20	71.50	276.32
TIM2 = 0.15mm)	3	80.423	276.75	105.70	66.27	268.88
TIM1 (copper = 3mm,	0.10	80.407	288.52		71.12	279.23
TIM2 = 0.15mm)	0.15	80.423	276.75		66.27	268.88
	0.2	80.439	265.28	268.88	60.28	244.12
TIM2 (copper = 3mm,	0.10	80.403	284.66	110.95	69.63	277.67
TIM1 = 0.15mm)	0.15	80.423	276.75	105.70	66.27	268.88
	0.2	80.443	267.85	98.65	63.53	250.12

19.4 Optimization for Warpage and Residual Stress Due to Reflow Process in IGBT Modules

The warpage of the IGBT module not only impacts on the subsequent package processes, but also increases the gap between the copper substrate and the heat sink, which may result in drastic change of the system thermal resistance as shown in Figure 19.23. Besides, warpage of the entire module can cause an extra mechanical stress in IGBT modules when the IGBT modules are mounted on a heat sink by bolts.

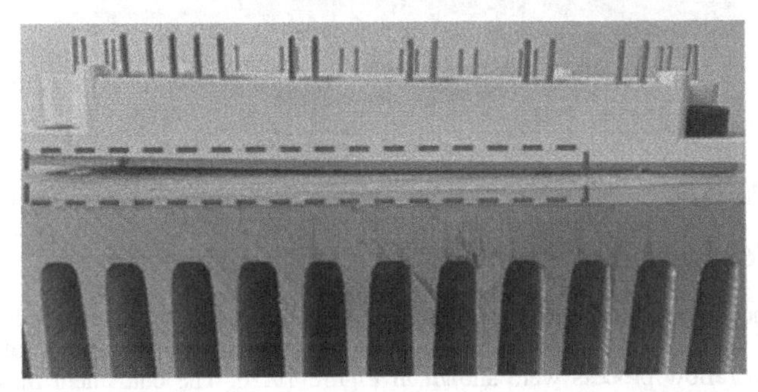

Figure 19.23 Assembly structure of IGBT module mounted on a heat sink

In order to simplify the finite element model, a 1/4 symmetry model was adopted and small edges and through-holes of the copper substrate were consequently neglected. The model of the DBC plates with and without copper layer patterns were both modeled to investigate the influence of the copper layer patterns on warpage and stress developed in reflow process, respectively as shown in Figures 19.24 and 19.25. In order to minimize aspect ratio, each assembly component, apart from the copper substrate layer, was meshed with only three elements through the thickness. The reflow process was modeled via a coupled thermo-mechanical analysis. Convection was also considered in this analysis because nitrogen was used as protection gas to avoid oxidation during the reflow process. Thermal loads, as shown in Figure 19.11, were applied to the bottom surface of copper substrate.

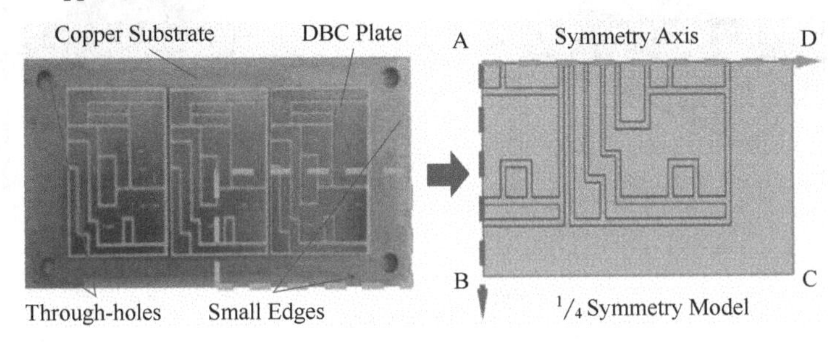

Figure 19.24 Simplified quarter symmetry model of IGBT assembly and patterns of the top copper layer

The finite element model of the reflow process was computed in two steps. Step 1 was heating process from room temperature to peak temperature and then back to solidus temperature, and step 2 was cooling process from solidus temperature to room temperature. In the heating process, owing to the significantly low stiffness properties of the paste-state and liquid solder, solder elements contributed few to the accumulation of thermal and plastic strains, therefore the solid properties of solder elements were ''killed'' and Young's modulus of solid solder was multiplied by a coefficient of 1E−6. In step 2 the solid properties of solder elements were activated and adopted their natural stiffness properties, and the initial temperature and stress and strain values of all elements were obtained from the result of step 1. This method was successfully applied for the first time to a flip-chip packaging process in [Wang *et al.*, 1998], which we called it the sequential process mechanics.

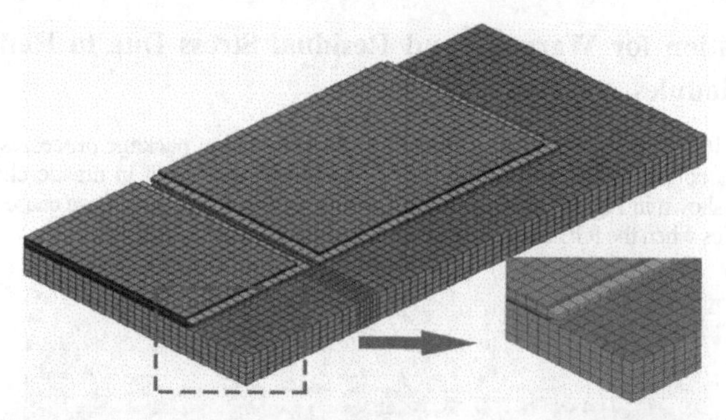

Figure 19.25 Finite element model without copper layer patterns of DBC plate

19.4.1 Effects of Copper Layer Patterns of DBC on Warpage and Stress

The DBC plates and copper substrate were assumed to be stress free and perfectly flat initially. The simulation results of the model with and without copper layer patterns on distribution of residual stress and final warpage due to reflow process were shown in Figure 19.26. The equivalent maximum von- Mises stresses of models with and without copper patterns both appeared in alumina layers due to the CTE mismatch between alumina and copper and the sandwich structure of DBC plate. Furthermore, the maximum residual stress in IGBT assembly increased from 82.3MPa to 96.9MPa when copper layer patterns of the DBC plates were concerned in the model, but this increase was very limited. Besides, it can be also seen obviously from Figure 19.26 that the copper patterns of DBC plates had little perceptible impact on residual stresses in solder layers and final warpage in IGBT assembly. Consequently, we concluded that the copper layer patterns of DBC plates had little influence on the warpage and residual stress so that they were not considered in the subsequent research in this study.

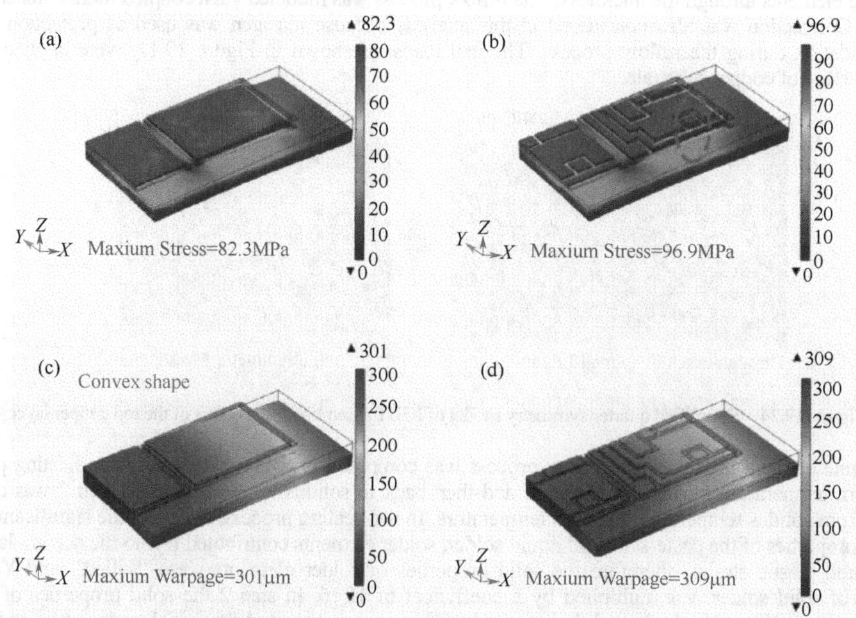

Figure 19.26 von-Mises stresses and warpage developed in IGBT assembly after reflow. (a) and (b) revealed residual stresses in module; (c) and (d) represented the final warpage in module; (e) and (f) showed residual stresses in top solder layer

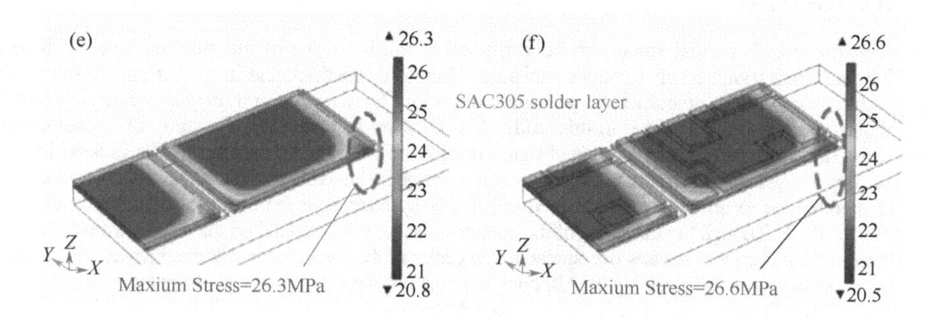

Figure 19.26 (Continued)

19.4.2 Effects of the Arrangement of DBC Plates on Warpage and Residual Stress in IGBT Modules

Residual stress and warpage developed in reflow process not only were induced by the CTE mismatch of packaging materials but also depended on the structure of IGBT assembly. Finite element models with different arrangement distance between DBC plates were simulated to reveal the effect of the arrangement of DBC plates on warpage and residual stress. The simulation results were shown in Figure 19.27. Another focus of this study was the stress in the lead-free solder between the DBC plates and copper substrate. Obviously as the distance between DBC plates increased, maximum residual stresses in IGBT module and in solder layers changed slightly, and the maximum difference value (*D*-value) of maximum stress in module and in solder layers were only 2.9MPa and 0.7MPa, respectively, which indicated that the distance between DBC plates had small influence on residual stress due to reflow process. But with the increase of the distance between DBC plates, the final maximum warpage of IGBT module decreased significantly from 301μm to 265μm. Therefore, the warpage of the IGBT module due to reflow process could be improved by increasing the arrangement distance between DBC plates without changing the residual stress in IGBT module.

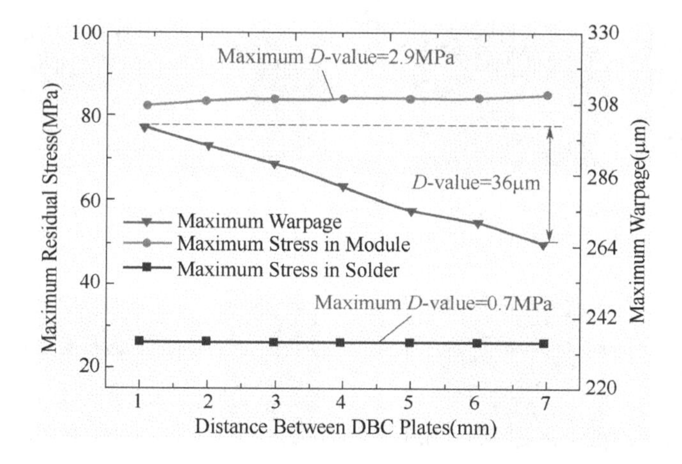

Figure 19.27 Effects of the arrangement of DBC plates on warpage and residual stress due to reflow process

19.4.3 Effects of the Thickness of Packaging Components on Warpage and Residual Stress in IGBT Modules

Almost all warpage and thermal stress can be attributed to the influence of the alumina layer with nearly double CTE mismatch mounted on a copper substrate. Therefore, the thickness in copper and alumina of the packaging components as key parameters, which are important for warpage and residual stress developed in IGBT modules, were also investigated in this study. A simulation-based parametric study of copper substrate thickness was conducted by examining a broad range from 2mm to 6mm. Simulation results indicated that the thickness of copper substrate had significant influences on warpage and maximum residual stress in module, and the warpage in module due to reflow can be reduced effectively by increasing the thickness of copper substrate as shown in Figure 19.28. Though the thickness of the copper substrate affected little on the maximum residual stress in solder layer, the high residual stresses overspread to the center of the solder layer with the increase of the copper substrate thickness as shown in Figure 19.29. Although the maximum residual stress in solder layers was below the strength point 42MPa, the large area of high stress could potentially affect long term reliability of IGBT module.

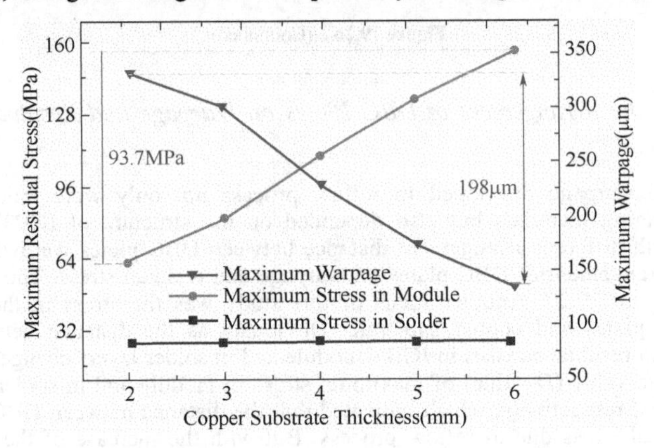

Figure 19.28 Effects of the copper substrate thickness on warpage and residual stress due to reflow process

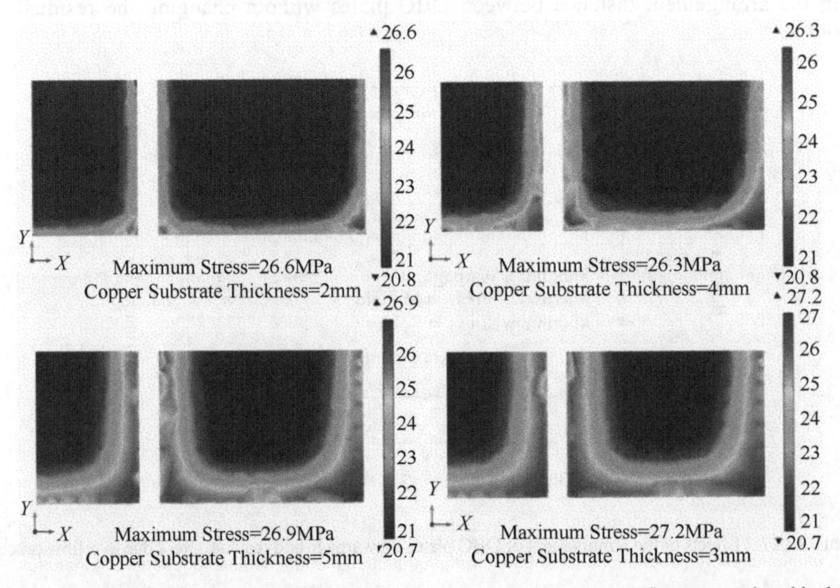

Figure 19.29 Effects of the copper substrate thickness on residual stress due to reflow process in solder layer

The similar analysis was performed by considering the thickness of the solder layer, alumina layer and copper layer of DBC plate, and effects of the thickness of packaging components on warpage and residual due to reflow process in IGBT modules were taken for comparison as shown in Figure 19.30. It can be seen that the thickness variation of packaging components had small impact on maximum stress in solder layers. As the thickness of solder layers increased from 0.05μm to 0.45μm, the maximum stress in module and solder stress and the warpage changed little. The maximum residual stresses in IGBT module, which always appeared in alumina layers, decreased sharply with the increase of the thickness of copper layers in DBC plates, but increased significantly with the decrease of the thickness of alumina layers. Those fore-mentioned behaviors were induced by the high level mismatch in CTE between copper and alumina and the sandwich structure of DBC plate. Besides, the maximum warpage in IGBT module different from maximum residual stress in solder layer was highly sensitive to variation in thickness of alumina layer and copper layer of DBC plate. All those analyses indicated that the warpage and residual stress due to reflow process in IGBT modules could be reduced by decreasing the thickness of alumina layers and increasing the thickness of copper layers in DBC plates and the arrangement distance between DBC plates. Moreover, it was noted that when the thickness of copper substrate increased, the warpage would reduce significantly, though meanwhile a substantial increase of maximum residual stress would be induced in module, which may be a potential danger for long term reliability of IGBT modules.

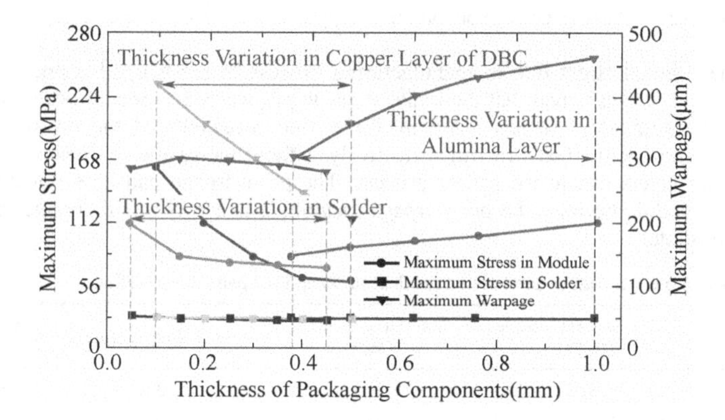

Figure 19.30 Effects of the thickness of packaging components on warpage and residual stress due to reflow process in IGBT modules

19.4.4 Effects of Pre-warped Copper Substrate on Warpage and Stress in IGBT Modules

The fore-mentioned studies on structure parameters of IGBT modules can improve the residual stress and warpage developed in reflow process, but this improvement is limited considering the electrical properties and manufacture costs. However, the warpage of IGBT module, more than 50μm, cannot be acceptable in some special applications in industry. Furthermore, a convex IGBT module which is difficult to be flattened in assembly processes is also not acceptable in industry. In order to get flatter IGBT modules after reflow process and avoid convex shape of the copper substrate, the concave shape copper substrate pre-warped by stamping, casting or other mechanical methods can be used to compensate for the warpage due to reflow process. Another two models with different pre-warped copper substrates were established to study the influence of the pre-warpage magnitude of the copper substrate on the deformation and residual stress of IGBT package module. Copper substrate pre-warped in width direction and length direction was created to be symmetrical for the center point A of the substrate, as shown in Figure 19.31. The center line of the center cross section of copper substrate both in width and length directions was a circular arc. In follow-up studies, the z-coordinate value of point C relative to point A would be used to represent the warpage of copper substrate. The pre-warpage magnitude of two models was 310μm and 387μm, respectively.

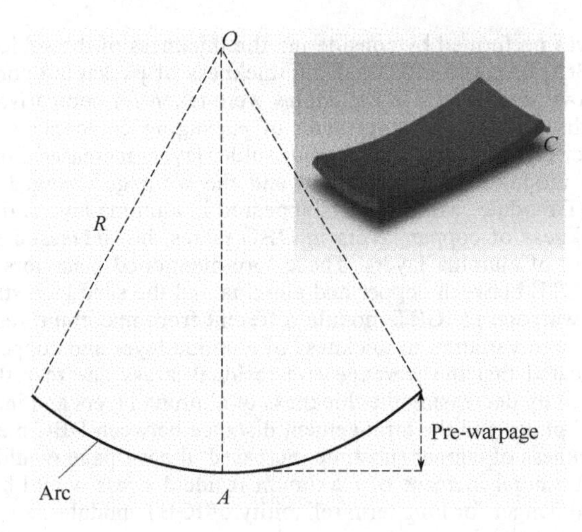

Figure 19.31 Cross section contour profile of copper substrate in width and length direction

In this study the residual stress due to manufacturing process in pre-warped copper substrate was not considered, therefore, it was assumed that the initial stress in pre-warped copper substrate was zero in finite element models. The simulation results listed in Table 19.6 revealed that the pre-warped method can compensate for the warpage of IGBT module effectively without increasing the residual stresses in IGBT module and in solder layers due to the reflow process. The pre-warpage had little impact on deformation developed in IGBT module because the pre-warpage magnitude is a very small change relative to a 3mm thickness copper substrate.

Table 19.6 Influence of pre-warpage magnitude on residual stress and warpage due to reflow

Pre-warpage (μm)	Max. Stress in Module (MPa)	Max. Stress in Solder (MPa)	Max. Warpage (μm)	Variation in Warpage (μm)
0	82.3	21.8	−301	301
310	82.7	23.7	11	299
387	82	23.3	87	300

The comparison of the copper substrate shape was represented by the *z*-coordinate value of line *AD* in Figure 19.32. Both two pre-warped copper substrates were still regular concave shape after reflow process, but when the copper substrate was pre-warped to 310μm before reflow, the final maximum warpage of the copper substrate in *AD* direction was no more than 5μm, which means that the pre-warping substrate is an effective and safe method to reduce warpage due to reflow process without inducing any other reliability problem.

19.4.5 Experiment

Three pre-warpage copper substrates with pre-warpage magnitude of 0μm, 310μm and 387μm were selected for the verification experiment in this study. Three samples were soldered in the vacuum reflow oven VADU100 and the transient temperatures subjected to copper substrate were measured and recorded by means of thermocouples as shown in Figure 19.31. The initial and final warpages were recorded by measuring the out-of-plane deformation at the four corners and the 24 isomeric points on line *AD* of the copper substrate, by means of a three coordinate measuring machine with accuracy of 1μm. The warpage measurements were referenced to the center of the copper substrate, as shown in Figure 19.33. Vertical displacement of the assembly was defined as the difference in elevation between reference point and point *C*.

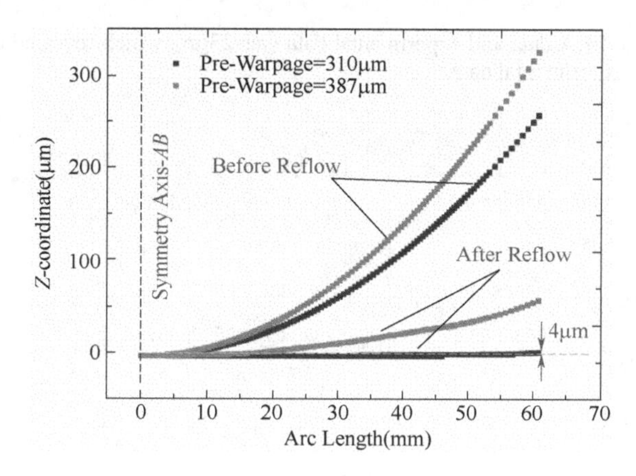

Figure 19.32 Warpage in IGBT module along the *A-D* path after utilizing the proposed pre-bending method before and after reflow process

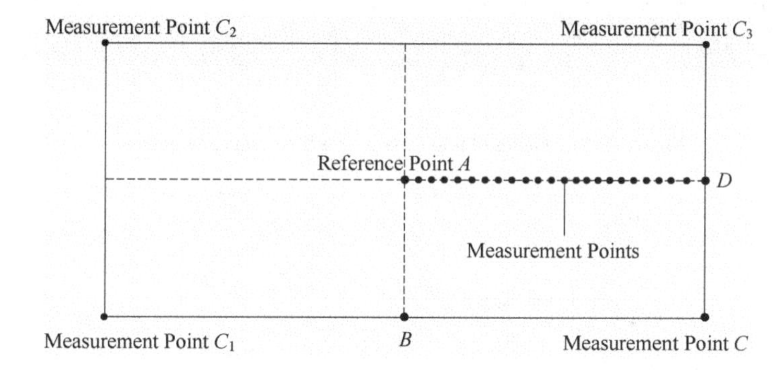

Figure 19.33 Warpage measurement points referenced to the center of the copper substrate

In order to reduce research variables, three samples were soldered at the same time in the vacuum oven, where the temperature difference can be controlled within ±2°C on the heating plate. The properties of three samples and the initial deformation of as-received copper substrates were listed in Table 19.7. It was noted that the as-received copper substrates were not perfectly symmetric due to error in the copper substrate fabrication process, but the difference of the initial deformation at the four corners of the copper substrate was very small. Therefore, the copper substrate can be seen as symmetry and the vertical displacement of point *C* still was utilized to characterize the overall deformation of the copper substrate.

Table 19.7 Properties of three samples

Sample number	Pre-warpage (μm)	Point *C* (μm)	Point C_1 (μm)	Point C_2 (μm)	Point C_3 (μm)
1	0	0	−2	4	−3
2	310	310	315	311	311
3	387	387	387	389	383

The final shape of the pre-warped copper substrate after soldering was shown in Figure 19.34. The final warpage of assemblies measured by coordinate measuring machine and the results of finite element analysis were taken for comparison as shown in Figure 19.35. It was obvious that two samples were concave shape and one sample was convex shape after reflow process. The maximum difference of *z*-coordinate of the points in arc *AD* between FEA data and experimental data was 23μm, which revealed that the finite analysis results agreed well with experimental data.

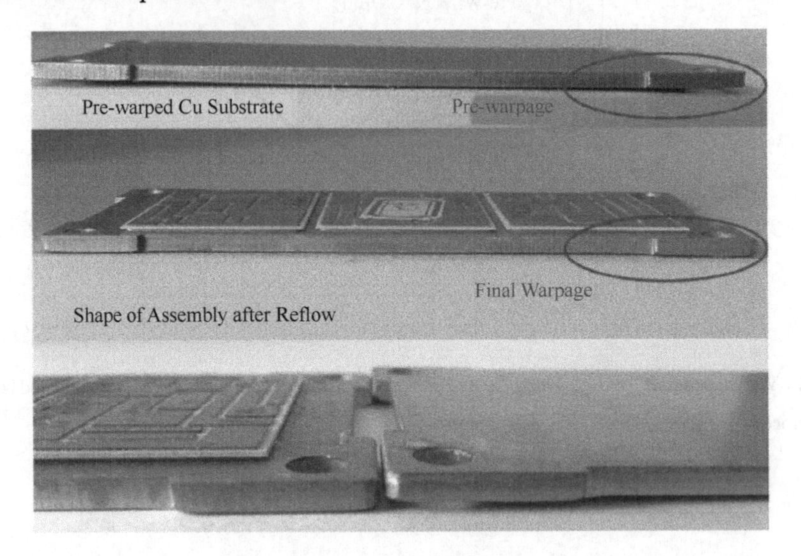

Figure 19.34 Photos of IGBT assemblies before and after reflow

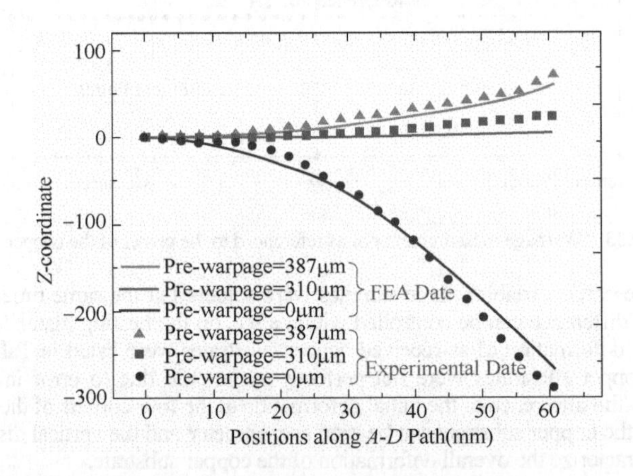

Figure 19.35 Comparison of experimental and FEA results on final warpage of IGBT module along *A-D* path

19.5 An Optimal Structural Design to Improve the Reliability of Al_2O_3-DBC Substrates under Thermal Cycling

A diagram of the DBC substrate fabrication process is schematically shown in Figure 19.36. The alumina (Al_2O_3) and thin copper laminates are assembled together. After heating up to 1065℃ under oxygen atmosphere, the oxidation reaction occurs on the interface between alumina and copper laminates. Strong bonding is achieved by diffusion reaction process and subsequent cooling process. On the other hand, the process flow is shown in Figure 19.37 for mass production in industry. A large substrate is fabricated at first. Single sized DBC substrate is then arranged in an array based on the substrate design. Designed copper pattern is further formed by an etching process. Finally, small pieces are obtained by laser cutting process and mechanical breaking method through punched holes.

The DBC substrates are subjected to thermo-mechanical fatigue during fabrication processes and power cycling under operation. The thermal-mechanical stress resulted from the thermal expansion (CTE) mismatch of the DBC substrate materials. The dominant failure modes of DBC substrates are identified as ceramic substrate cracking and interfacial delamination. Material defects can induce micro or/and macro cracks in the ceramic substrate. However, fatigue cracks in copper can be initiated from the geometric singularity sites such as edges and the copper/ceramic interface, which will determine the thermal fatigue life of the DBC substrate (Pietranico *et al.*, 2009; Dupont *et al.*, 2006).

Plastic deformation in the copper layer can significantly impact the interface cohesion strength and cause the delamination of copper-ceramic interfaces (Wei *et al.*, 1997). Based on our experiments more than fifty percent of DBC substrate failures were caused by the delamination after forty thermal cycles from −55℃ to +150℃. When the peak temperature increased to 250℃, the average thermal fatigue lives of Al_2O_3-DBC decreased to approximately 20 cycles (Dong *et al.*, 2010). The upper copper layer was delaminated at the edges (Lutz *et al.*, 2008). To improve the reliability and lifetime of DBC substrates under high temperature and harsh environment, a few methods have been proposed based on the failure mode and fatigue mechanism. One method is to replace the metallization layer from copper to aluminum, *i.e.*, direct bonding aluminum (DBA) substrate. The fatigue life of 1500 cycles from −55℃ to 250℃ without any delamination of the aluminum was reported. Instead, the surface would be roughened (Lei *et al.*, 2009). Alternative method is to use AlN as ceramic substrate. The thermal conductivity and mechanical strength are improved. However, the tradeoff is the substrate cost. The other solution is proposed to optimize the design of DBC structures, which is investigated in this section.

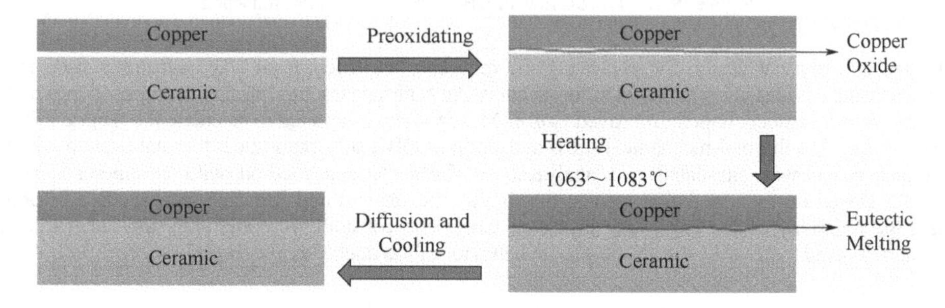

Figure 19.36 Schematic diagram of the fabrication processes of the DBC substrate

According to investigation by Pietranico *et al.* (Pietranico *et al.*, 2009), the elastic-plastic behavior of the copper layer plays an important role. It has been observed that the cracks initiate from geometric singularities created by the chemical etching of the upper copper layer. Therefore, copper layer pattern design should be a key factor to improve the fatigue life of the DBC substrate. There have been several works to optimize the DBC structure. Dupont (Dupont *et al.*, 2006) has studied the effects of copper layer thickness of ceramic substrates on the reliability under high temperature cycling. It was found that reducing copper layer thickness can considerably increase the number of cycles before failure. The dimples set in the edge of copper layer could act as a local reduction of the metallization in order to increase the solder lifetime, reported by Schulz-Harder and Dupont (Schulz-Harder *et al.*, 2003; Dupont *et al.*, 2006). The edge tail length of copper layer also has a significant effect on the thermal crack under high temperature cycling (Dong *et al.*, 2009). However, the approaches discussed above have some disadvantages. Firstly, when the thickness of copper layer reduces, the through-current capability of DBC substrate will also decrease. Secondly, etching dimples in the metallization of DBC substrate will increase the difficulty and cost of fabrication. Thirdly, the patterns on the copper layer are achieved by etching process, thus the designed edge tail can be hardly realized.

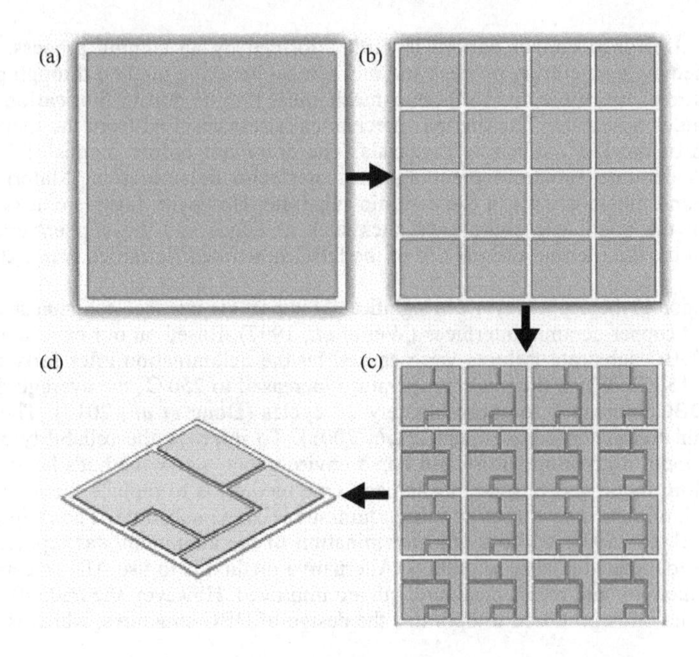

Figure 19.37 Process flow of DBC substrates in mass production

The structure optimal design for improving the reliability and lifetime of DBC substrates used in high power electronic devices is investigated in this section. An optimal structural design is proposed in which the copper layers were ladder shaped. Improved Coffin-Mason law was applied to calculate the fatigue life under thermal cycling. The thermal-mechanical stress and strain of DBC substrates under thermal cycling with wide temperature range was simulated and analyzed by finite element method with commercial software COMSOL. Based on the simulation results, the fatigue mechanism and life time were studied in order to evaluate the reliability of the improved substrates. Chaboche constitutive model with non-linear kinematic hardening was used to describe the elastic-plastic behavior of the ductile copper layer.

19.5.1 Failure Mechanisms of DBC Substrate

It is known that common failure modes of DBC substrate used in power electronic modules include delamination of the copper layer from ceramic and cracking within the ceramic layer (McCluskey *et al.*, 2012). The root cause derives from the inherent defects of ceramic layer and voids at the interface between copper and ceramic, as shown in Figure 19.38. The interfacial voids are formed during fabrication processes of the DBC substrate. On the other hand, failure of substrate was caused by the coefficient of thermal expansion (CTE) mismatch between the copper (CTE = 19.5ppm/K) and ceramic (alumina: 6.8ppm/K$\leqslant$ CTE$\leqslant$9ppm/K, aluminum nitride: 4.3ppm/K$\leqslant$CTE$\leqslant$6.2ppm/K).When the substrate suffers from temperature variation such as temperature cycling and power cycling, the accumulated stress/strain from the CTE mismatch between the ceramic and copper can lead to interfacial delamination and ceramic crack. Scan acoustic microscope (SAM) method is used to inspect the interfacial delamination of DBC substrate non-destructively. The SAM images of the interface between upper copper layer and ceramic layer before and after thermal cycling as well as the corresponding optical photograph are shown in Figure 19.39. It can be found that the delamination all initiate from the outside edge of each copper pattern.

The inherent defects of ceramic and voids caused by manufacturing technique can be solved by the approach of the improvement of processing technology. However, the delaminations and cracks under thermal or power cycling caused by CTE mismatch were physical nature of material, which cannot be eliminated but only be relieved in order to extend the lifetime of DBC substrate without changing material types.

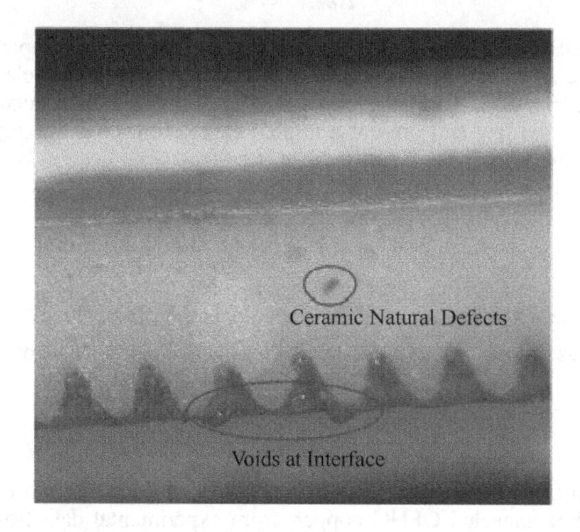

Figure 19.38 Inherent defects of ceramic layer and voids at interface of DBC substrate

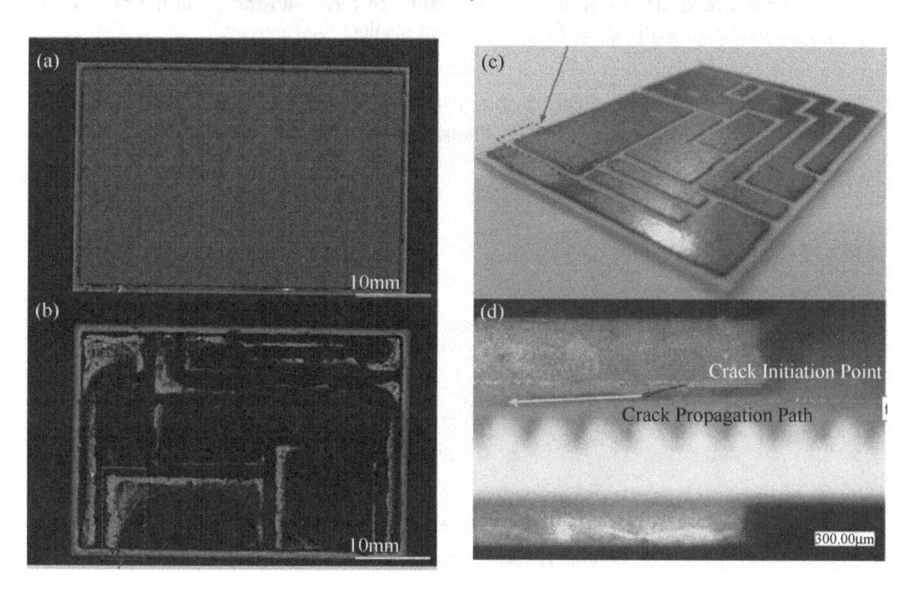

Figure 19.39 Interfacial delamination of DBC substrate. (a) SAM image of the interface between upper copper layer and ceramic layer before thermal cycling; (b) after thermal cycling; (c) optical photograph of DBC substrate after thermal cycling; (d) cross-section view: crack initiated from copper edge and propagated along the directions of the arrows

As the copper foil suffers from periodic plastic load during thermal cycling test, the plastic strain will accumulate and finally cause the copper to damage failure. If cycle numbers are below 104 cycles, plastic damage is the root cause of the failure mechanism, this phenomenon is often called low cycle fatigue (LCF), the cyclic life of which can be predicted by the well-known Coffin-Manson law (Coffin, 1962) based on the plastic strain amplitude: where C and k are material constants.

$$\Delta\varepsilon_{p}N^{k}=C \tag{19.9}$$

However, the life of DBC substrate under thermal cycling with wide temperature range is less than hundred cycles. This type of failure phenomenon is called extremely low cycle fatigue (ELCF). In this situation, the Coffin-Manson law cannot fit well according to Shimada's experiments (Shimada *et al.*, 1987; Xue *et al.*, 2008) modified the law and present a unified expression for extremely low cycle fatigue. The expression was shown below:

$$N=\frac{1}{2}\frac{e^{\lambda}-1}{e^{\lambda\left(\frac{\varepsilon_{d}}{\varepsilon_{f}}\right)^{m}}-1} \tag{19.10}$$

where N is fatigue life; m and λ are material constants; ε_f is the plastic strain when the copper fractured; ε_d is equivalent plastic distortion, ε_1, ε_2, ε_3 are three principal components of the plastic strain tensor:

$$\varepsilon_{d}=\sqrt{\frac{2}{3}\left(\varepsilon_{1}^{2}+\varepsilon_{2}^{2}+\varepsilon_{3}^{2}\right)} \tag{19.11}$$

The schematic diagram of the equivalent plastic strain and distortion under a periodic plastic loading is shown in Figure 19.40. For annealed OFHC copper, from experimental data from Coffin and Tavernelli (Chaboche *et al.*, 1989), the constants $m = 1.56$, $\lambda = 1.55$, and the tensile ultimate stress is 202 MPa. Thus, plastic strain of copper would directly influence the lifetime of DBC substrate, and the fatigue life can be predicted using the calculated plastic strain from simulation results.

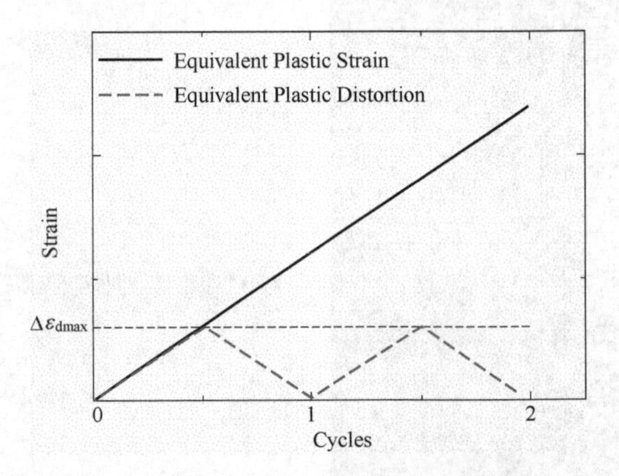

Figure 19.40 Schematic diagram of the equivalent plastic strain and distortion under a periodic plastic loading

19.5.2 Optimal Structure Design of DBC Substrate

(i) Optimal Structure Design

DBC substrate is a multilayered structure which consists of a ceramic layer with thin sheet of copper bonded on both sides of the ceramic. We modified the copper foil structure and designed it to be ladder shape, as shown in Figure 19.41. We etched the copper foil to form a ladder based on the traditional substrate to decentralize the stress on copper. The copper foil contacting with the ceramic was the first copper ladder, and the above one on which the chips would be soldered was the second copper ladder. In this study, the area of ceramic layer was 30 × 30mm^2. The thickness of ceramic h_0 was set as 0.38mm. The thicknesses of first and second copper ladder were defined as h_1 and h_2 respectively, and the sum of h_1 and h_2 was 0.3mm. The lengths of edge tail of first and second copper ladders were set as w_1 and w_2 respectively.

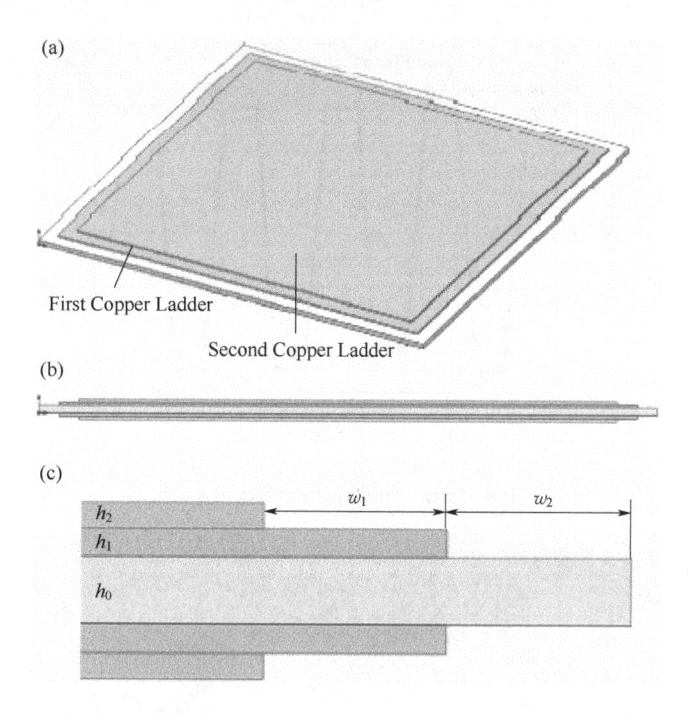

Figure 19.41 Ladder shaped DBC substrate. (a) Global view; (b) cross-section view; (c) partial enlarged detail

(ii) Thermal Cycling Test

According to standard of JESD22-A104D, the thermal cycling test was designed as below: the temperature ranges from −55℃ to 150℃, 1h a cycle and the dwell time was 15min. The temperature ramp rate was less than 15℃/min in order to get rid of the influence of temperature gradient. Both measured temperature profile and set temperature profile of thermal cycling are shown in Figure 19.42. The DBC substrates were placed in the Espec EGNZ12-6NA, a thermal cycling chamber with the temperature uniformity of ±2℃. As the ceramic in the middle of the DBC substrate is electric insulating, the DBC can be regarded as a plate capacitor, the value of capacitance is directly related to the area of copper layer and the distance between copper layers. The formula of capacitance value of plate capacitor can be simply expressed as $C=S\varepsilon/d$, where S represents for the area of copper layer, ε is the dielectric coefficient of Al_2O_3, and d is the distance between parallel copper layers. If the DBC substrate failure occurs, there are interfacial delaminations or cracks, the copper will separate from the ceramic, the distance between copper layers will consequently increase, thus the value of capacitance will decrease. Therefore, based on the simple theory, we designed an effective qualitative test to estimate whether the substrate fails. After several cycles, the DBC substrates were taken out of the chamber and connected to the digital multimeter to measure the capacitance value of each DBC substrate, as shown in Figure 19.43. The capacitance of flawless DBC substrate is 180pF. If the value decreased, it is the proof of the existence of failure. Ideally, when the distance increase from original 0.38mm to 0.39mm, the value of capacitance will decrease by 2.6%. Considering the slight deformations of copper layers and the measurement error, we define that when the value reduced by 5%, it is believed that the DBC fails. Thermal cycling tests were conducted on both traditional Al_2O_3-DBC substrates (Group 1) and ladder shaped Al_2O_3-DBC substrates with h_1 = 0.15 mm, w_1=w_2=2mm (Group 2). Each group took ten samples for testing. The experimental results of number of cycles when substrate failure are shown in Figure 19.44. It can be found that the average number of cycles of DBC substrate in Group 1 was only around 41, while the value of Group 2 was about 122. The ladder shape structure was demonstrated to be effective to prolong the lifetime of DBC.

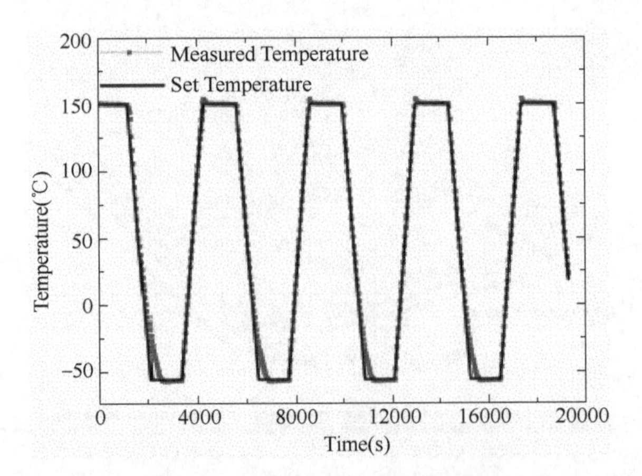

Figure 19.42 Profile of thermal cycling test

Figure 19.43 Measurement of the capacitance of DBC substrate for the failure evaluation

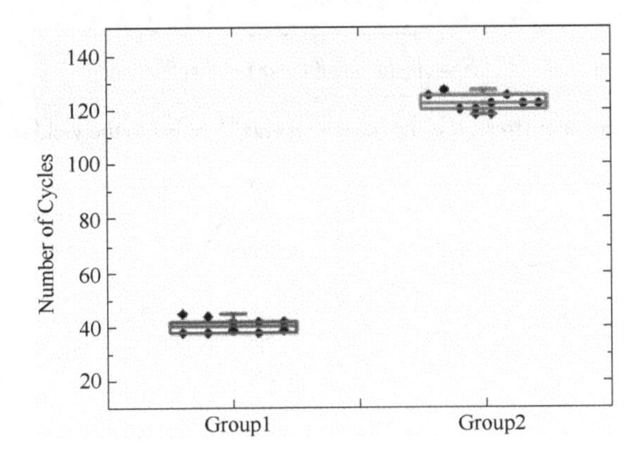

Figure 19.44 Fatigue lifetime of Al_2O_3-DBC substrates under thermal cycling test

(iii) Finite Element Model under Thermal Cycling

As the DBC substrate was square and symmetric in the FE analysis, in order to simplify the calculation amount, quarter model was used for the simulation. The general view of the quarter FE model for ladder shaped DBC is shown in Figure 19.45. It is known that there is no temperature gradient in DBC substrate under thermal cycling condition, so that we ignore the heat transfer of DBC substrate in the chamber and directly apply temperature as shown in Figure 19.41 on the whole body of DBC substrate. The displacement of center point *A* was prescribed at *Z*-axis direction to ensure the free deformation caused by the CTE mismatch. All of elements were hexahedral, ratio of element length was less than 10, which was reasonable for finite element analysis. The complete mesh consisted of 9440 domain elements, 6064 boundary elements, and 936 edge elements.

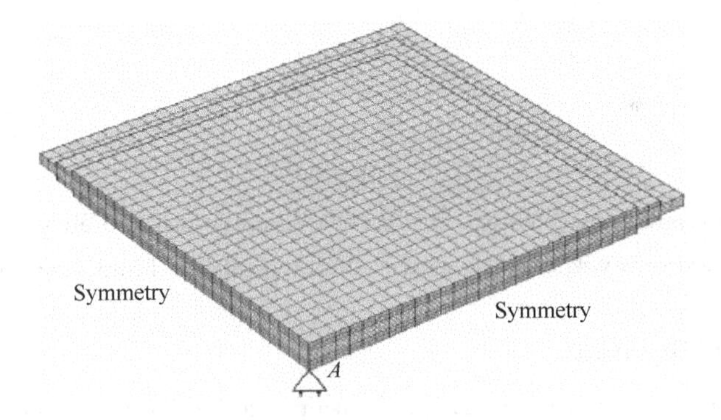

Figure 19.45 Quarter FE model and mesh of the ladder shaped DBC substrate

In this simulation, the material of ceramic was Al_2O_3, the copper was oxygen-free high-conductivity (OFHC) copper. The basic material properties are shown in Table 19.8. As the temperature range was wide, temperature dependency of the thermal and mechanical properties of OFHC copper are given in Figure 19.46. The elastic-plastic behavior of copper was described by Chaboche model with non-linear kinematic hardening, which took temperature effect into account and fitted cyclic thermal loading well. The yield function for Chaboche nonlinear kinematic hardening rule is shown below (Chaboche 1989):

Table 19.8 Material properties of DBC substrate at room temperature

Material	Young's Modulus (GPa)	Poisson Ratio	CTE (ppm/K)	Yield Strength(MPa)
OFHC Copper	110	0.34	17.4	121
96% Al_2O_3	300	0.22	6.4	

$$F = \sqrt{\frac{3}{2}(s-\alpha)(s-\alpha)} - Y = 0 \tag{19.12}$$

where s represents the deviator stress, α is the back stress and Y refers to the yield stress of the material.

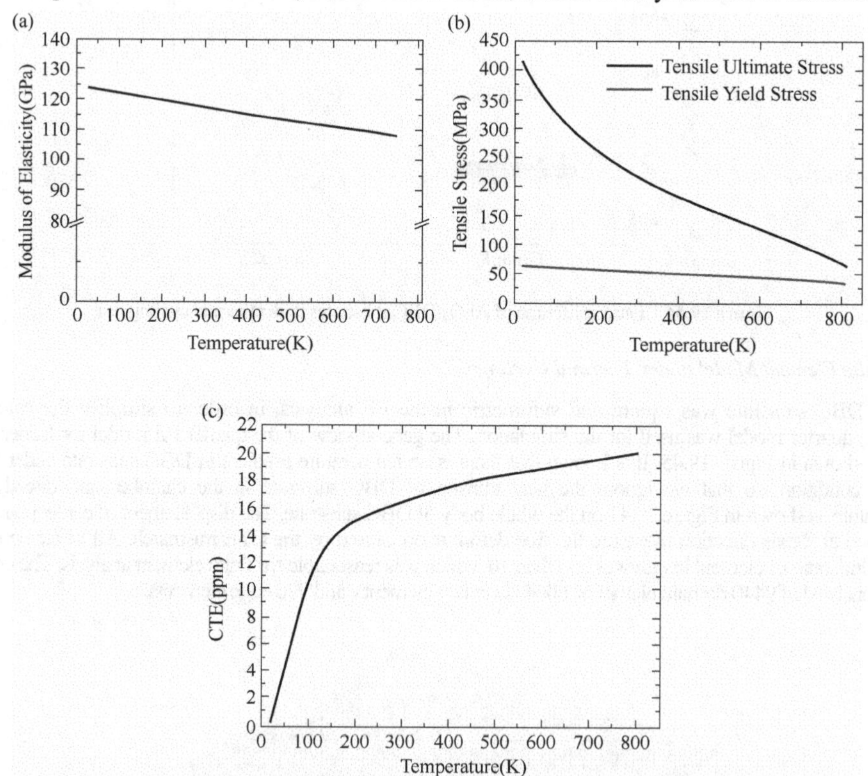

Figure 19.46 Material properties varied with temperature for OFHC copper. (a) Modulus of elasticity (Esposito *et al.*, 1975); (b) tensile ultimate stress and tensile yield stress (Esposito *et al.*, 1975); (c) coefficient of thermal expansion (Hahn *et al.*, 1970)

19.5.3 Results and Discussion

To increase the optimization efficiency, the design of experimental (DoE) numerical simulations were used to determine which parameter played the key role for the optimal structure. With regard to the traditional DBC structure, h_1 = 0.3mm and h_2 = 0. The value of h_1, w_1 and w_2 as variables were analyzed to study their influences to the reliability of the DBC substrate.

Firstly, the influence of h_1 to the reliability of DBC substrate was studied when the values of w_1 and w_2 were both set as 1mm. Figure 19.47 shows the distribution of von-Mises stress at the end of thermal cycling of parametric study of h_1 from 0.05mm to 0.3mm. The figures illustrate that the maximum stress appeared at the edge of the copper. It accords with the phenomenon that the delamination initiated from the boundary of the copper foil. Moreover, with the increase of the value of h_1, the maximum von-Mises stress transfers from the edge of second copper ladder to the edge of first copper ladder which directly bonded on the ceramic layer. Besides, from Figure 19.47(a) to (f), it can be found that the maximum stresses of ladder shaped DBC with h_1 = 0.05mm and traditional DBC with h_1 = 0.3mm are 60.1248MPa and 59.9281MPa, respectively. When we adjust the thickness of copper ladder and set h_1 as 0.15 mm, the stress decreases to 58.5459MPa, as shown in Figure 19.47(c).

As the copper is ductile material, the plastic strain would accumulate in the first and second copper ladder during thermal cycling test. When changing the value of h_1, it can be found that the maximum plastic strain accumulation will locate at the corner point of second copper ladder if h_1 is lower than 0.15 mm, otherwise it will locate at the corner of first copper ladder. Figure 19.48 shows the equivalent plastic strain during thermal cycling at points where maximum plastic strains of first and second copper ladder respectively locate when w_1=w_2= 1mm and h_1= 0.15mm. In this case, the maximum strains of first and second copper ladders separately locate on the corner points of the Cu-Cu interface and the Cu-Al_2O_3 interface. The maximum equivalent plastic distortion closely related to the fatigue life can be calculated from the simulation results based on Equation19.11.

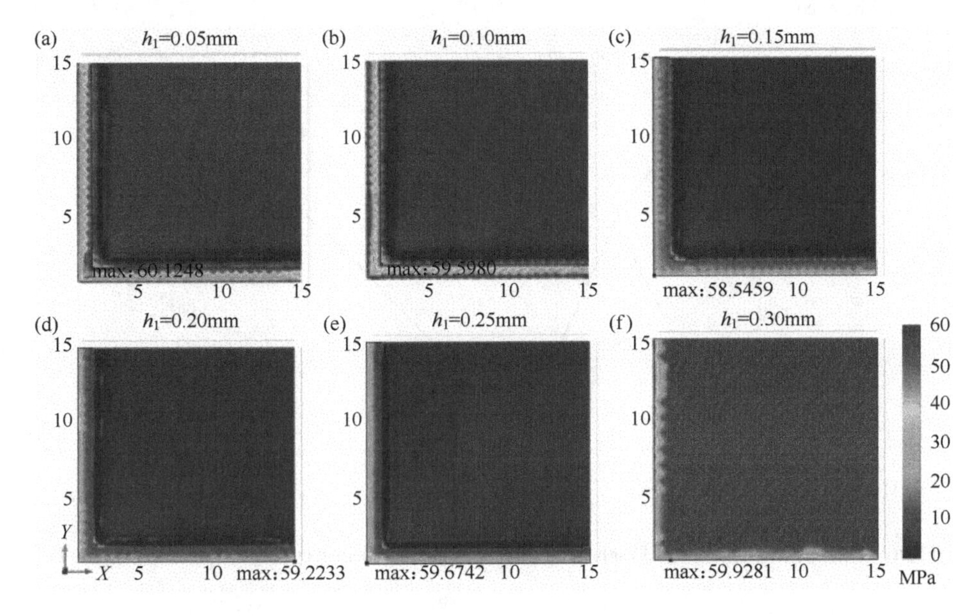

Figure 19.47 Distributions of von-Mises stress at the end of thermal cycling of parametric study of h_1. (a) h_1=0.05mm; (b) h_1=0.10mm; (c) h_1=0.15mm; (d) h_1=0.20mm; (e) h_1=0.25mm; (f) h_1 = 0.30mm

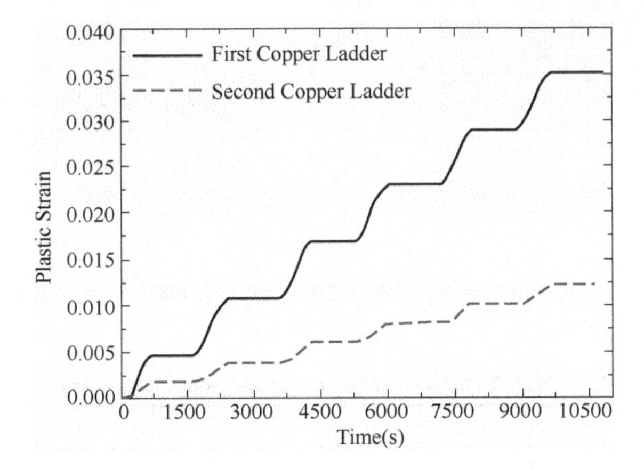

Figure 19.48 Equivalent plastic strain during thermal cycling of the first and second copper ladder (h_1 = 0.15mm)

The maximum plastic distortion ε_d of the copper and the warpage of the DBC substrate at the end of the thermal cycling are shown in Figure 19.49. It can be seen that the warpage increases with the increase of h_1, but the influence is slight. When the value of h_1 increases from 0.05mm to 0.30mm, the warpage of the substrate only increases by 4.2%. Besides, the equivalent plastic distortion ε_d decreases sharply when h_1 increases from 0.05 mm to 0.15mm, but then increases when h_1 increases from 0.15mm to 0.30mm. Through the whole diagram, it is evident that the influence of the thickness of the first copper ladder is significant to the plastic distortion. To obtain the longest lifetime of cycles calculated from Equation. 19.10, the plastic distortion should be as small as possible, thus the best optimized thickness of first copper ladder should be set as 0.15mm when the total thickness of copper foil is 0.30mm. In other words, the thicknesses of each copper ladder should be equal.

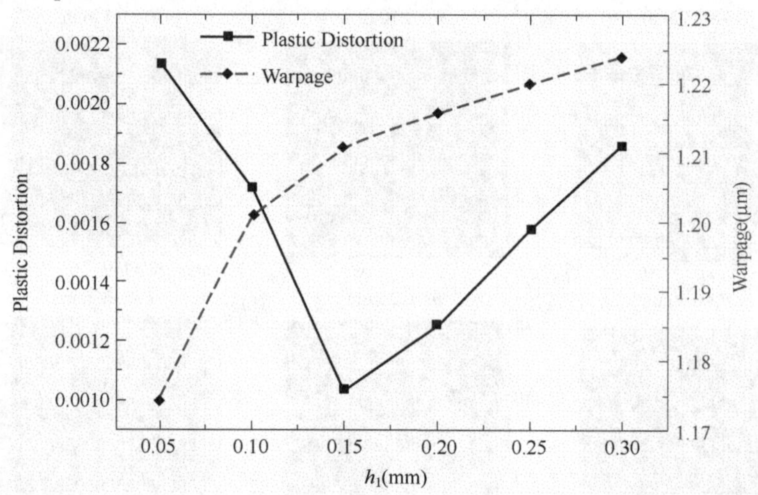

Figure 19.49 Plastic distortion and warpage of DBC substrate after thermal cycling with various values of h_1

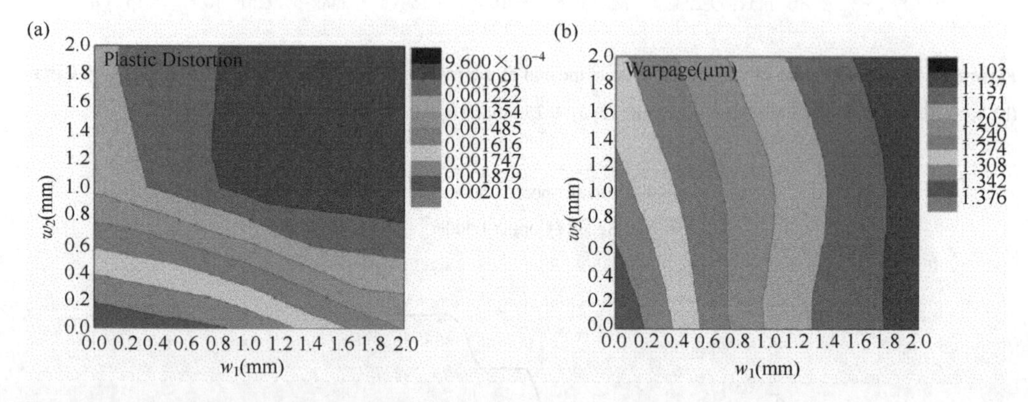

Figure 19.50 Plastic distortion and warpage of DBC substrate after thermal cycling with various values of w_1 and w_2. (a) Plastic distortion; (b) warpage

Table 19.9 Fatigue life in DoE scheme for the simulation with various geometric parameters.

Group	h_1(mm)	w_1(mm)	w_2(mm)	N_f
1#	0.05	1	1	38.4815
	0.10			54.0779
	0.15			119.6389
	0.20			88.4497
	0.25			61.8661
	0.30			47.7581

(*continued*)

Group	h_1(mm)	w_1(mm)	w_2(mm)	N_f
2#	0.15	0	0	42.2113
		1		47.7581
		2		68.6080
3#	0.15	0	1	82.1937
		1		119.6389
		2		134.0133
4#	0.15	0	2	88.4497
		1		117.8525
		2		135.6734

When the value of h_1 was fixed as 0.15mm, the values of w_1 and w_2 were parametric designed to reach the smallest plastic distortion. The results of plastic distortion and warpage of parametric design combined both w_1 and w_2 are shown in Figure 19.50(a) and (b) respectively, where the FE model swept the w_1 and w_2 from 0mm to 2mm at the step of 0.5mm. When the value of w_2 was set to be zero, it represented for the traditional DBC substrate. It can be found obviously that the longer the edge tail, the smaller the plastic distortion and warpage. This finding is also applicable to traditional DBC substrate. Moreover, from Figure 19.50(a), it can also be observed that the influence of w_2 on the plastic distortion is more significant than w_1. To the contrary, the influence of w_1 on the warpage is greater than w_2 as shown in Figure 19.50(b). Although the longer edge tails w_1 and w_2 can prolong the lifetime of DBC substrate, the effective area utilization of copper should also be concerned to satisfy the requirement for chip attachment. We should make a tradeoff between them.

The fatigue life in DoE scheme for the simulation with various geometric parameters calculated from Equation. 19.10 are listed in Table 19.9. The fatigue life under the thermal cycling test that ranged from −55℃ to 150℃ of traditional DBC substrate is only about 48 cycles. However, when applying the ladder shape structure and adjusting the thickness and edge tail length of each copper ladder, the fatigue life of DBC substrate can increase to more than 100 cycles, and will reach to the peak lifetime of about 136 cycles when the parameter of h_1, w_1, w_2 are respectively set as 0.15mm, 2mm and 2mm. These results can be proven by the experimental test results shown in Figure 19.44.

19.6 Chapter Summary

The thermo-mechanical behaviors of power electronics modules were studied using the FEA method in order to better understand the cooling effect of copper microchannel and mechanisms of stress origin. The pre-warping method has been verified to be a safe and effective method to reduce warpage due to reflow process without increasing the residual stress in IGBT module in this study. The results of the parametric analyses showed that warping of the IGBT power module can be effectively controlled and that satisfactory co-planarity can be achieved after packaging assembly. The analysis results indicated that the pre-warpage magnitude had little influence on deformation magnitude and residual stress developed in IGBT module due to reflow, which meant that there must be an optimal value of pre-warpage magnitude for a packaging structure. Improved Coffin-Manson law was applied to calculate the extremely low cycle fatigue life of DBC substrate. The simulation results of DBC fatigue life were verified by experiments.

References

Addagarla, A., Siva, N. (2012) Finite element analysis of flip-chip on board (FCOB) assembly during reflow soldering process, *Soldering and Surface Mount Technology*, 2492-99.

Anand, L. (1985) Constitutive equations for hot-working of metals, *International Journal of Plasticity*, 1(3):213-31.

Brown, S. B, Kim, K. H, Anand, L. (1989) An internal variable constitutive model for hot working of metals, *International Journal of Plasticity*, 5(2):95-130.

Brown, S. B., Kim, K. H., Anand, L. (1989) Internal variable constitutive model for hot working of metals, *International Journal of Plasticity*, 5 95-130.

Calame, J. P, Myers R. E, Wood, F. N, Binari, S.C. (2005) Simulations of direct-die-attached microchannel coolers for the thermal management of GaN-on-SiC microwave amplifiers, *IEEE Transactions on Components, Packaging and Manufacturing*

Technology, 28(4):797-809.

Chaboche, J. L. (1989) Constitutive equations for cyclic plasticity and cyclic viscoplasticity, *International Journal of Plasticity*, 5(3):247-302.

Chen G, Han D, Mei Y-H, Cao X, Wang T, Chen X, *et al.* (2012) Transient thermal performance of IGBT power modules attached by low-temperature sintered nanosilver, *IEEE Transactions on Device and Materials Reliability*, 12(1):124-32.

Chen, Z., Zhang, Q., Wang, K., Chen, M., Liu, S. (2012) Fluid-solid coupling thermomechanical analysis of high power LED package during thermal shock testing, *Microelectronics Reliability*, 52(8):1726-34.

Ciappa, M. (2002) Selected failure mechanisms of modern power modules, *Microelectronics Reliability*, vol. 42: 653-667.

Coffin Jr., L. (1962). Low Cycle Fatigue—A Review, 1962.

Dong, G. Chen, X. Zhang X., Ngo, K.D.T., Lu, G.Q. (2010) Thermal fatigue behaviour of Al_2O_3-DBC substrates under high temperature cyclic loading, *Soldering and Surface Mount Technology*, 22(2) 43-48.

Dong, G., Lei, G., Chen, X., Ngo, K., Lu, G. Q. (2009) Edge tail length effect on reliability of DBC substrates under thermal cycling, *Soldering and Surface Mount Technology*, 21(3):10-5.

Dong, G., Lei, G., Chen, X., Ngo, K., Lu, G.Q. (2009) Edge tail length effect on reliability of DBC substrates under thermal cycling, *Soldering and Surface Mount Technology*, 21, 10-15.

Dunne, F. P. E. (1992) Hayhurst, D. R., Continuum damage based constitutive equations for copper under high temperature creep and cyclic plasticity, *Proceedings of the Royal Society of London, Series A (Mathematical and Physical Sciences)*, UK, 1992, pp. 545-566.

Dupont, L., Khatir, Z., Lefebvre, S., Bontemps, S. (2006) Effects of metallization thickness of ceramic substrates on the reliability of power assemblies under high temperature cycling, *Microelectronics Reliability*, *46*, 1766-1771.

Dupont, L., Khatir, Z., Lefebvre, S., Bontemps, S. (2006) Effects of metallization thickness of ceramic substrates on the reliability of power assemblies under high temperature cycling, *Microelectronics Reliability*, 46 (9-11) 1766-1771.

Esposito, J. J., Zabora, R. F. (1975) Thrust chamber life prediction, Volume 1: Mechanical and Physical Properties of High Performance Rocket Nozzle Materials [Final Report].

Hahn, T. A. (1970) Thermal expansion of copper from 20 to 800 K—standard reference material 736,Journal of Applied Physics, 41 (13) 5096-5101.

Herkommer, D., Punch, J., Reid, M. (2013) Constitutive modeling of joint-scale SAC305 solder shear samples, *IEEE Transactions on Components Packaging and Manufacturing Technology*, 3 :275-281.

Hung, Thang B, Van Chuc N, Van Trinh P, Thi Thanh Tam N, Ngoc Minh P. (2011) Thermal dissipation media for high power electronic devices using a carbon nanotube-based composite, *Advances in Natural Sciences: Nanoscience and Nanotechnology*, 2(2):025002-6.

Incropera, F. P, Lavine, A. S, DeWitt, D. P. (2011) Fundamentals of heat and mass transfer, *John Wiley & Sons Incorporated.*

Knight, R. W., Hall, D. J., Goodling, J. S., Jaeger, R. C. (1992) Heat sink optimization with application to microchannels, *IEEE Transactions on Components Hybrids & Manufacturing Technology*, 15(5):832-42.

Lee, C. C., Kao, K. S., Lin, L., Chang, J. Y., Leu, F. J., Lu, Y. L., and Chang, T. C. (2014) Investigation of pre-bending substrate design in packaging assembly of an IGBT power module, *Microelectronic Engineering*, Vol. 120, pp. 106-113.

Lee, P. S., Garimella, S. V. (2006) Thermally developing flow and heat transfer in rectangular microchannels of different aspect ratios, *International Journal of Heat and Mass Transfer*, 49(17-18):3060-7.

Lee, T.Y. (2000) Design optimization of an integrated liquid-cooled IGBT power module using CFD technique, *IEEE Transactions on Components, Packaging and Manufacturing Technology*, 23(1):55-60.

Lei, T. G., Calata, J. N., Ngo, K. D. T., Lu, G. Q. (2009) Effects of large-temperature cycling range on direct bond aluminum substrate, *IEEE Transactions on Device and Materials Reliability*, 9 (4) 563-568.

Lhommeau, T., Perpiñà, X., Martin, C., Meuret, R., Mermet-Guyennet, M., Karama, M. (2007) Thermal fatigue effects on the temperature distribution inside IGBT modules for zone engine aeronautical applications, *Microelectronics Reliability*, 47, 1779-1783.

Liu, S., Liu, Y. (2011) Modeling and Simulation for Microelectronic Packaging Assembly: Manufacturing, Reliability and Testing, *John Wiley & Sons.*

Liu, S., Luo, X. (2011) LED Packaging for Lighting Applications: Design, Manufacturing, and Testing, *John Wiley & Sons.*

Liu, Y. (2012) Power electronic packaging: design, assembly process, Reliability and modeling, *Springer.*

Luo, X., Liu, S. (2007) A microjet array cooling system for thermal management of high brightness LEDs, *IEEE Transactions on Advanced Packaging*, 30(3):475-84.

Luo, X., Xiong, W., Cheng, T., Liu, S. (2009) Design and optimization of horizontally located plate fin heat sink for high power LED street lamps, *In: 2009 IEEE 59th electronic components and technology conference*, p. 1-4.

Lutz, J. Herrmann, T., Feller, M., Bayerer, R., Licht, T., Amro, R. (2008) Power cycling induced failure mechanisms in the viewpoint of rough temperature environment, *Proceedings of the 5th International Conference on Integrated Power Electronic Systems*,

2008, pp. 55-58.
McCluskey, P. (2012) Reliability of power electronics under thermal loading, *7th International Conference on Integrated Power Electronics Systems*, 2012, 1-8.
Miller, T. (1991) Strength and fatigue of dispersion-strengthened copper, *Journal of Nuclear Materials*, 179-181:263-6. Part 1(0).
Motalab, M., Cai, Z., Suhling, J. C. and Lall, P. (2012) Determination of Anand constants for SAC solders using stress-strain or creep data, *13th Inter Society Conference on Thermal and Thermomechanical Phenomena in Electronic Systems, pp.* 910-922.
Mysore, K, Subbarayan, G, Gupta, V, Zhang, R. (2009) Constitutive and aging behavior of Sn3.0Ag0.5Cu solder alloy, *IEEE Transactions on Electron Packaging Manufacturing*, 32(4):221-32.
Mysore, K., Subbarayan, G., Gupta, V., Zhang, R. (2009) Constitutive and aging behavior of Sn3.0Ag0.5Cu solder alloy, *IEEE Transactions on Components Packaging and Manufacturing Technology*, 32, 221-232.
Petukhov, B., Kirillov, V. (1958) The problem of heat exchange in the turbulent flow of liquids in tubes, *Teploenergetika* 4, (4):63-8.
Pietranico, S., Pommier, S., Lefebvre, S., Khatir, Z., Bontemps, S. (2009) Characterisation of power modules ceramic substrates for reliability aspects, *Microelectronics Reliability*, 49 (9-11) 1260-1266.
Pietranico, S., Pommier, S., Lefebvre, S., Pattofatto, S. (2009) Thermal fatigue and failure of electronic power device substrates, *International Journal of Fatigue*, 31 (11-12) 1911-1920.
Popova, N., Schaeffer, C., Avenas, Y., Kapelski, G. (2006) Fabrication and thermal performance of a thin flat heat pipe with innovative sintered copper wick structure, In: *Conference record of the 2006 IEEE industry applications conference, Forty-First Ias Annual Meeting*, p. 791-6.
Rodriguez, P., Fusaro, J. M. (1997) Integration of heat exchanger with high current hybrid power module, In: *Proceedings of The International Technical Conference & Exhibition on Packaging & Integration of Electronic and Photonic Microsystems 1997*, vol. 2, 15-19 p. 2173-8.
S. Liu, Y. Mei. (1995) Behavior of delaminated plastic IC packages subjected to encapsulation cooling, moisture absorption, and wave soldering, *IEEE Transactions on Components, Packaging and Manufacturing Technology Part A*, 18, 634-645.
Sakanova, A., Yin, S., Zhao, J., Wu, J. M., Leong, K. C. (2014) Optimization and comparison of double-layer and double-side micro-channel heat sinks with nanofluid for power electronics cooling, *Applied Thermal Engineering*, 65(1-2):124-34.
Sanchez, J. (1999) State of the art and trends in power integration, 20-29.
Sanchez, J., Bourennane, A., Breil, M., Austin, P., Brunet, M., Laur, J. P. (2007) Evolution of the classical functional integration towards a 3D heterogeneous functional integration, *14th International Conference on Mixed Design of Integrated Circuits and Systems* , pp. 23-34.
Schulz-Harder, J. (2003) Advantages and new development of direct bonded copper substrates, *Microelectronics Reliability*, 43 (3) 359-365.
Schulz-Harder, J., Dezord, J. B., Schaeffer, C., Avenas, Y., Puig, O., Rogg, A., Exel, K., Utz-Kistner, A. (2006) DBC (Direct Bond Copper) substrate with integrated flat heat pipe, *In: Semiconductor thermal measurement and management symposium, 2006 IEEE twenty-second annual IEEE*, p. 152-6.
Shammas, N. Y. (2005) The role of semiconductor devices in high power circuit breaker applications, *WSEAS Transactions on Circuits and Systems*, 4826.
Shimada, K., Komotori, J., Shimizu, M. (1987) The applicability of the Manson-Coffin law and Miner's law to extremely low cycle fatigue, *Transactions of the Japan Society of Mechanical Engineering A*, 53 (491) 1178-1185.
Smet, V., Forest, F., Huselstein, J., Richardeau, F., Khatir, Z., Lefebvre, S., Berkani, M. (2011) Ageing and failure modes of IGBT modules in high-temperature power cycling, *IEEE Transactions on Industrial Electronics*, 58 (10) 4931-4941.
Squiller, D., Greve, H., Mengotti, E., McCluskey, F. P. (2014) Physics-of-failure assessment methodology for power electronic systems, *Microelectronics Reliability*, 54 (9-10) 1680-1685.
Tummala, R. R., Rymaszewski, E. J. (1989) Microelectronics packaging handbook, *Van Nostrand Reinhold Company*.
Ume, I. C., Martin, T. (1997) Finite element analysis of PWB warpage due to cured solder mask sensitivity analysis, *IEEE Transactions on Components, Packaging and Manufacturing Technology Part A*, 20, 307-317.
Wang, J., Qian, Z., Zou, D., Liu, S. (1998) *Electronic Components & Amp; Technology Conference, 1998. 48th IEEE 1998*, pp. 1438-1445.
Wang, J., Zou, D., Qian, Z., Ren, W., Liu, S. (1998) Effect of manufacturing induced defects on the reliability of flip-chip packages, *In: Proceedings of the 1998 ASME international mechanical engineering congress and exposition*, vol. 24, November 15, 1998-November 20 p. 35-42.
Watari, K, Nakano, H, Tsugoshi, T, Nagaoka, T, Urabe, K., Cao, S, *et al.* (2003) Microstructure and thermal conductivity of AlN ceramic with eliminated grain boundary phase, *Advanced ceramics and composites. Zurich-Uetikon: Trans Tech Publications Ltd.*, p. 361-4.
Webb, R. L. (1971) A critical evaluation of analytical solutions and Reynolds analogy equations for heat and mass transfer in smooth tubes, *Waerme und Stoffuebertrag*, 4:197-204.

Wei, Y., Hutchinson, J.W. (1997) Nonlinear delamination mechanics for thin films, *Journal of the Mechanics and Physics of Solids*, 45 (7) 1137-1159.

Wen, T., Ku, S. (2007) Efficient evaluation of substrate warpage by finite element method and factorial design analysis, *Proceedings of the 57th Electronic Component & Technology Conference*, Reno, Nevada, 1754-1759.

Xu, D., Lu, H., Huang L. (2002) Power loss and junction temperature analysis of power semiconductor devices, *IEEE Transactions on Industry Applications*, 3, 8(5):1426-31.

Xue, L. (2008) A unified expression for low cycle fatigue and extremely low cycle fatigue and its implication for monotonic loading, *International Journal of Fatigue*, 30 (10-11) 1691-1698.

Ye, H., Lin, M., Basaran, C. (2002) Failure modes and FEM analysis of power electronic packaging, *Finite Elements in Analysis and Design*, 38601, 612.

Yi Jiun C, TaiFa Y. (2009) Thermal stress and heat transfer characteristics of a Cu/diamond/Cu heat spreading device, *Diamond and Related Materials*,18(2-3):283-6.

Yin, S., Tseng, K. J., Zhao, J. (2013) Design of AlN-based micro-channel heat sink in direct bond copper for power electronics packaging, *Applied Thermal Engineering*, 5 2(1):120-9.

Yu, X., Zhang, L., Zhou, E., Feng, Q. (2011) Heat transfer of an IGBT module integrated with a vapor chamber, *Journal of Electronic Packaging*, 133(1):011008.

Zhang, J., Ding, H., Baldwin, D. F., Ume, I. C. (2003) Characterization of in-process substrate warpage of underfilled flip chip assembly, *Electronics Manufacturing Technology Symposium, 2003. IEMT 2003. IEEE/CPMT/SEMI 28th International 2003*, pp. 291-297.

Zhao, J., Tellkamp, J., Gupta, V., Edwards, D. (2008) Experimental evaluations of the strength of silicon die by 3-point-bend versus ball-on-ring tests, *In: 11th Intersociety conference on thermal and thermomechanical phenomena in electronic systems*, p. 687-94.

Zhu, F., Zhang, H., Guan, R., Liu, S. (2007) Effects of temperature and strain rate on mechanical property of Sn96.5Ag3Cu0.5, *Journal of Alloys and Compounds*, 438(1-2):100-5.

20

Analytical Models for Thermal Resistances in Electronics Packaging

In electronics packaging, packaging thermal resistance or junction temperatures of dies are the most-used indexes for evaluating the thermal management performance of packaging. Generally, the electronic devices are mounted on a printed circuit board. Heat generated by chips or dies conducts through packaging and then transfers onto print circuit board. As dies and packaging usually are much smaller than the substrate or print circuit board where they are located on, the heat dissipation processes can be treated as that heat flux from a portion surface conducts into a larger plate. For these applications, the thermal resistances of print circuit board or heat spreader where chips are bonded on commonly take significant impact on thermal characterization of electronic device or packaging. Therefore, it is critical to find methods to calculate thermal resistance of small heat sources on a larger plate with various boundary conditions.

Thermal spreading resistance takes the majority part of the thermal resistances when heat conducts from small area into a large plate, therefore, most of the calculations in describing such a heat transfer case were based on the concept of thermal spreading resistance. Kennedy (Kennedy, 1960) studied thermal spreading resistance of uniform heat flux source on a finite cylinder and obtained analytical solutions for a wide range of geometrical parameters with different boundary conditions. Kadambi and Abuaf (Kadambi *et al.*, 1985) presented analytical solutions for three-dimensional steady-state and transient thermal conduction with a uniform heat flux on the top surface and a convective heat transfer boundary condition at the bottom. John and Krane (John, 1991) obtained the temperature field in the same rectangular geometries as in reference (Kadambi *et al.*, 1985) and calculated thermal resistance by using the temperature field. The boundary conditions were based on constant and uniform temperature. Yovanovich (Yovanovich, 1980) and Negus *et al.* (Nugus *et al*, 1984) found the solution for the thermal spreading resistance of a single centered heat source on cylinder disk with a heat transfer coefficient at the lower surface. Lee *et al.* and Song *et al.* (Lee *et al.*, 1995; Song *et al.*, 1994) analyzed the thermal constriction and spreading resistance in a plate with a uniform heat flux region on one surface and a third kind thermal boundary condition over the other surface and obtained the approximate equation of the thermal constriction and spreading resistance based on the analytical solutions of Yovanovich (Yovanovich, 1980). Yovanovich *et al.* did series of studies on thermal spreading resistance for various geometrics (Yovanovich, 1976, 1998, 1999). Muzychka *et al.* (Muzychka, *et al.*, 2003, 2006) presented analytical solutions of thermal spreading resistance for rectangular flux channels and discussed the influence of geometric and edge cooling. In the models presented in references (Yovanovich, 1976, 1998, 1999; Muzychka, 2003, 2006), the lower surface of the bodies is convective cooling while the other boundaries are considered to be adiabatic. Since most of Yovanovich's thermal spreading resistance formulas were based on mean source temperature, Ellison (Ellison, 2003) put forward the solution for the maximum thermal spreading resistance in rectangular geometry with one surface convective cooling. Moreover, Ellison provided extensive graphical results which are easily used for engineers.

20.1 Resistances Eccentric Heat Source on Rectangular Plate with Convective Cooling at Upper and Lower Surfaces

Figure 20.1 illustrates an eccentric heat source on isotropic plate with upper and lower surfaces cooling. Here,

Modeling and Simulation for Microelectronic Packaging and Integration: Manufacturing, Reliability and Testing, Second Edition. Sheng Liu and Yong Liu.

an eccentric heat source with uniform heat flux is located on an isotropic plate. The upper surface and lower surface of the plate are convectively cooled with heat transfer coefficient ht and hl respectively. Sides of the plate are adiabatic. In steady-state situation, the governing equation for describing the heat transfer in Figure 20. 1 is the Laplace's equation:

$$\nabla^2 T = \frac{\partial^2 T}{\partial x^2} + \frac{\partial^2 T}{\partial y^2} + \frac{\partial^2 T}{\partial z^2} = 0 \tag{20.1}$$

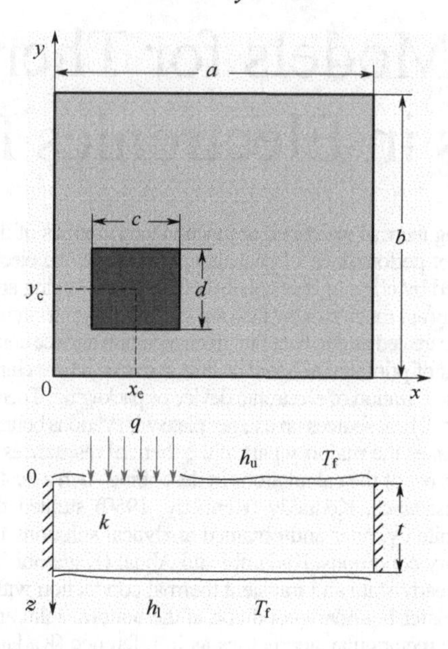

Figure 20.1 Eccentric heat source on isotropic plate with upper and lower surfaces cooling

where T is the temperature field of the plate at steady-state and it is a function of space as $T(x, y, z)$. Boundary conditions for the system are listed as follows:

$$\left.\frac{\partial T}{\partial z}\right|_{z=0} = -\frac{q}{k}, (x, y) \in A_s \tag{20.2}$$

$$\left.\frac{\partial T}{\partial z}\right|_{z=0} = -\frac{h_u}{k}[T(x, y, 0) - T_f], (x, y) \in (A - A_s) \tag{20.3}$$

$$\left.\frac{\partial T}{\partial z}\right|_{z=t} = -\frac{h_l}{k}[T(x, y, t) - T_f], (x, y) \in A \tag{20.4}$$

$$\left.\frac{\partial T}{\partial x}\right|_{x=0} = 0, \left.\frac{\partial T}{\partial x}\right|_{x=a} = 0 \tag{20.5}$$

$$\left.\frac{\partial T}{\partial y}\right|_{y=0} = 0, \left.\frac{\partial T}{\partial y}\right|_{y=b} = 0 \tag{20.6}$$

A_s is heat source located area. A is the plate's area.

Usually, the separation of variables method is applied to solve partial differential Equation 20.1 with boundary conditions Equations 20.2~20.6. The solution is assumed to have the form $\theta(x, y, z) =$

$X(x)\cdot Y(y)\cdot Z(z)$. Here $\theta(x,y,z)$ is the temperature excess and it is defined as $T(x,y,z)-T_f$. Equation 20.1 and boundary conditions Equations 20.5, 20.6 constitute an eigenvalue problem. Therefore, $\theta(x,y,z)$ can be expressed as

$$\theta(x,y,z)=A_0+B_0z+\sum_{m=1}^{\infty}\cos(\lambda x)[A_1\cosh(\lambda z)+B_1\sinh(\lambda z)]+$$
$$\sum_{n=1}^{\infty}\cos(\delta y)[A_1\cosh(\delta z)+B_1\sinh(\delta z)]+ \tag{20.7}$$
$$\sum_{m=1}^{\infty}\times\sum_{n=1}^{\infty}\cos(\lambda x)\cos(\delta y)[A_3\cosh(\beta z)+B_3\sinh(\beta z)]$$

where $\lambda=m\pi/a$, $\delta=n\pi/b$, $\beta=\sqrt{\lambda^2+\delta^2}$, A_i and B_i (i= 0, 1, 2, 3) are the Fourier coefficients. If h_u is zero, taking Fourier expansion at boundary z=0 can obtain coefficients A_i, B_i and finally find the analytical solution as Equation 20.7. This is the common way to obtain analytical solutions. As h_u is positive, it becomes very difficult to get A_i and B_i by taking Fourier expansion at boundary z=0. The problem will change into a Laplacian problem with mixed boundary conditions. Read (Read, 2007) once proposed an analytic series method for some of those problems with specific mixed boundary conditions as both Dirichlet and Neumann conditions that appeared on one surface. However, this method is still not suitable for solving the mixed boundary condition problem presented in Equations 20.1-20.6. Up till now there are no analytical solutions for the mixed boundary conditions problem described as Equations 20.1-20.6. In the following sections, thermal resistance network method will be used to analyze this problem and resistance network to calculate thermal resistance of the plate with upper and lower surfaces cooling will be presented.

20.1.1 Network Model

Model Description and Establishment

For a plate with upper and lower surfaces cooling, heat flux from heat sources spreads into the plate and then transfers to ambient through upper and lower surfaces respectively. Figure 20.2 shows the heat flow lines of the system. In the analysis, it is assumed that portion of heat transferring into ambient by the upper surface spreads in an upper layer with thickness of t_u. Similarly, the rest heat, which dissipates into ambient through the lower surface, spreads in a lower layer with thickness t_l.

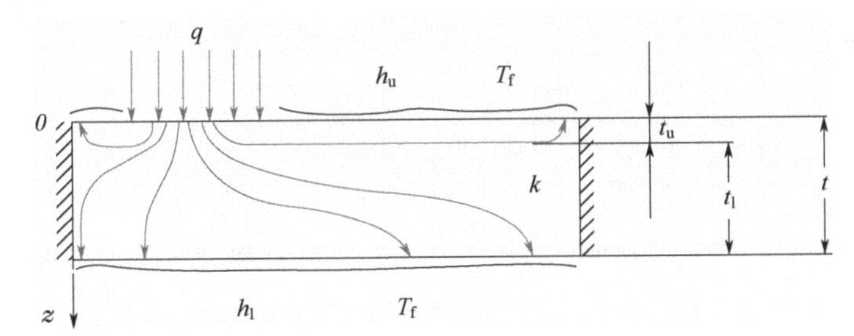

Figure 20.2 Heat flows of the system

Thicknesses of upper and lower spreading layers are determined by relative values of the heat transfer coefficients at upper and lower surfaces. In cases that h_u is very small compared to h_l, most of heat will transfer to ambient through the lower surface. As a result, the lower spreading layer will take up a majority of the whole thickness of the plate. Contrarily, if h_l is very small compared to h_u, most of heat will transfer to ambient through the upper surface and the lower spreading layer will take up a majority part of the whole plate. Therefore, t_u and t_l are defined as

$$t_u=\frac{h_u}{h_u+h_l}t \tag{20.8}$$

$$t_1 = \frac{h_1}{h_u + h_1}t \tag{20.9}$$

Since the plate with both surfaces cooling is divided into upper and lower spreading layers, thermal characterization of both layers should be analyzed firstly. Thermal characterization of the lower spreading layer can be modeled based on the expressions presented in reference (Muzychka *et al.*, 2003), where rectangular eccentric heat source is on rectangular plate with only lower surface convective cooling. Thermal characterization of the upper spreading layer will be analyzed by analog with the case of only lower surface cooling in the following.

Figure 20.3 shows the schematic of eccentric heat source on plate with lower surface convective cooling and upper surface convective cooling respectively. Before conducting thermal analysis of the plate, two simulations were done by software COMSOL. In the first simulation, the plate's upper surface except heat source area was loaded a heat transfer coefficient 10W/(m^2·K) while the other surfaces, including sides and lower surfaces, were set as adiabatic. In the second simulation, the lower surface was given a heat transfer coefficient 10W/(m^2·K) while the other surfaces were adiabatic. Thermal conductivity of the plate was 5W/(m·K) and the ambient temperature was 25℃ in these two simulations. Heat flux in the heat source area was 1667W/m^2. Heat flow lines in the plate obtained by the two simulations are illustrated in Figure 20.4. We define the heat source block as a cuboid whose upper surface is exactly the area of heat source. From Figure 20.4, it can be found that heat conducts through the cuboid to the lower surface first and then spread over the plate, finally transfers to ambient from upper surface because of its convective cooling condition. For plate with lower surface cooling, some of heat flow to the lower surface directly in the cuboid and transferred to ambient through lower surface, while the rest heat conducts through parts of the cuboid and then spreads over the plate, finally dissipates to the ambient.

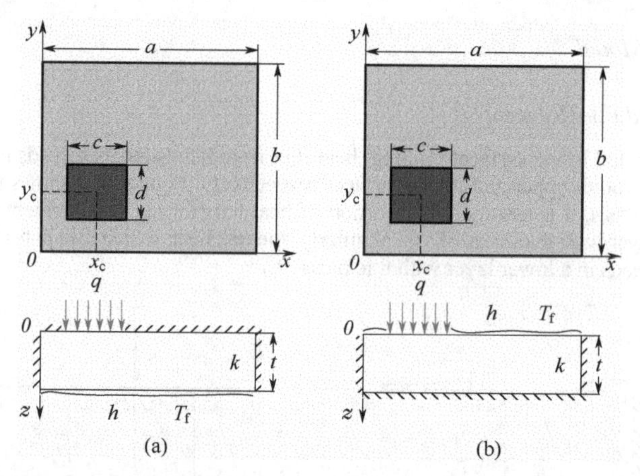

Figure 20.3 Schematic of eccentric heat source on plate. (a) Lower surface convective cooling; (b) upper surface convective cooling

According to the heat flow lines shown in Figure 20.4, we can establish thermal resistances networks for plates presented in Figure 20.3, which are shown in Figure 20.5. $\overline{T}_s$ is mean temperature of the heat source. T_f is ambient temperature. R_s' is defined as thermal sub-spreading resistance. Since the plate is usually thin in electronics packaging, the thermal spreading process happened in the two simulation cases are nearly the same. As a result, R_s' in the two network models in Figure 20.5 are treated to be equal. R_{1D_cu} is one-dimensional thermal resistance of the cuboid and it is given as $t/(kcd)$. R_{h1} is the convective film resistance at lower surface of the cuboid while R_{h2} is the one at the rest area of lower surface when the plate is only with lower surface cooling. R_h is convective film resistance at upper surface when the plate is only with upper surface cooling. Because the upper surface of the cuboid is exactly the area of heat flux, R_{h2} equals R_h . r is a constant whose value is in the range 0-1 and it is given as 1/4 in the model. Based on the thermal network model shown in Figure 20.5(a), the total thermal resistance of the system with lower surface cooling is expressed as

$$R_{\mathrm{to\_1}} = rR_{\mathrm{1D\_cu}} + \frac{(R_s' + R_{\mathrm{h2}})[(1-r)R_{\mathrm{1D\_cu}} + R_{\mathrm{h1}}]}{R_s' + R_{\mathrm{h2}} + (1-r)R_{\mathrm{1D\_cu}} + R_{\mathrm{h1}}} \tag{20.10}$$

$$R_{\mathrm{h1}} = \frac{1}{hcd}, R_{\mathrm{h2}} = \frac{1}{h(ab-cd)} \tag{20.11}$$

For plate with lower surface cooling, analytical solutions have been found in reference (Muzychka *et al.*, 2003) to calculate the total thermal resistance and it is given by

$$R_{\mathrm{to\_1}} = \frac{t}{kab} + R_s + \frac{1}{hab} \tag{20.12}$$

where R_s is the thermal spreading resistance of the system and it is given as

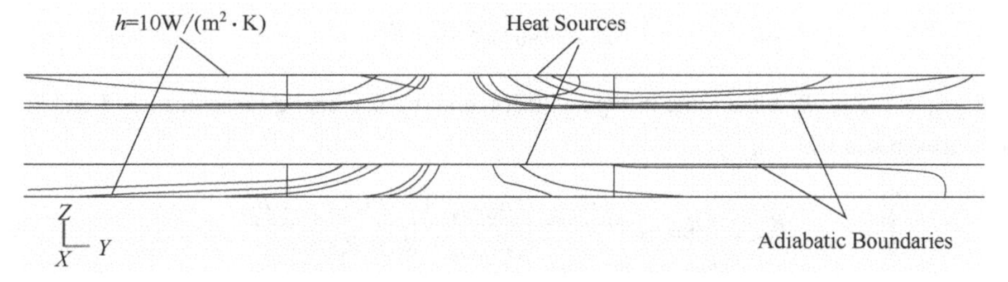

Figure 20.4 Heat flow lines in the plate with different cooling boundaries

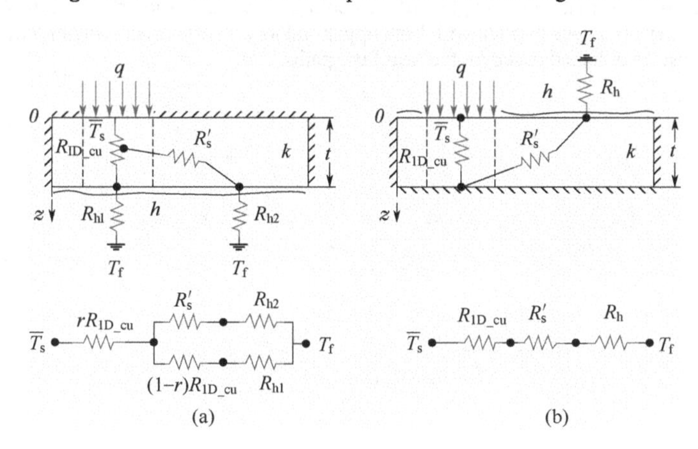

Figure 20.5 Thermal network models for cases shown in Figure 20.3

$$R_s = \frac{8}{abc^2k}\sum_{m=1}^{\infty}\frac{\cos^2(\lambda x_c)\sin^2\left(\frac{c}{2}\lambda\right)}{\lambda^3\phi(\lambda)} + \frac{8}{abd^2k}\sum_{n=1}^{\infty}\frac{\cos^2(\delta y_c)\sin^2\left(\frac{d}{2}\delta\right)}{\delta^3\phi(\delta)} + \frac{64}{abc^2d^2k}\sum_{m=1}^{\infty}\times\sum_{n=1}^{\infty}\frac{\cos^2(\lambda x_c)\sin^2\left(\frac{c}{2}\lambda\right)\cos^2(\delta y_c)\sin^2\left(\frac{d}{2}\delta\right)}{\beta\lambda^2\delta^2\phi(\beta)} \tag{20.13}$$

With $$\phi(\zeta)=\frac{\zeta\sinh(t\zeta)+h/k\cosh(t\zeta)}{\zeta\cosh(t\zeta)+h/k\sinh(t\zeta)} \tag{20.14}$$

where $\lambda = m\pi/a$, $\delta = n\pi/b$, $\beta=\sqrt{\lambda^2+\delta^2}$, and ζ is replaced by λ, δ, β, accordingly. Thus R_s' in Figure 20.5 can be obtained as follows:

$$R_s'=\frac{(R_{to\_1}-rR_{1D\_cu})[(1-r)R_{1D\_cu}+R_{h1}]}{R_{1D\_cu}+R_{h1}-R_{tol}}-R_{h2} \tag{20.15}$$

For plate with upper surface cooling, the total thermal resistance of the plate is expressed as Equation 20.16 according to thermal resistances network shown in Figure 20.5(b).

$$R_{to\_u}=R_{1D\_cu}+R_s'+R_h \tag{20.16}$$

As R_{h2} equals R_h, R_s' is the same in Figure 20.5(a) and (b), combining Equations 20.15 and 20.16 can get the final expression for R_{to_u},

$$R_{to\_u}=R_{1D\_cu}+\frac{(R_{to\_1}-rR_{1D\_cu})[(1-r)R_{1D\_cu}+R_{h1}]}{R_{1D\_cu}+R_{h1}-R_{tol}} \tag{20.17}$$

As stated previously, heat transfer processes in system with both upper and lower surfaces cooling are combination of the ones in plate with single surface cooling, like plates shown in Figure 20.3. Figure 20.6 shows the heat transfer processes in plate with both upper and lower surfaces cooling and thermal resistances network model is also established based on the heat flow paths.

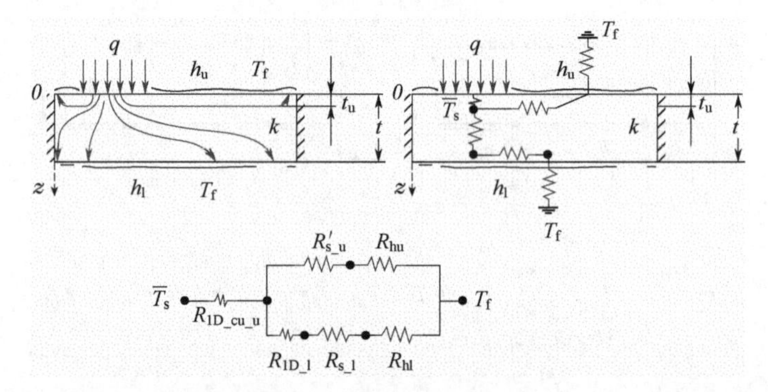

Figure 20.6 Heat transfer process and thermal resistances network of plate with upper and lower surfaces cooling

Previous analysis and results of thermal characterization of single surface cooling plate are applied here to calculate thermal resistance in the network shown in Figure 20.6. R_{1D_cu} is one-dimensional thermal conductive resistance of the cuboid with thickness t_u and it is given as $t_u/(kcd)$. R'_{s_u} is thermal sub-spreading resistance of spreading layer. According to Equation. 20.15, R'_{s_u} is given by

$$R'_{s\_u}=\frac{(R_{to\_1\_u}-rR_{1D\_cu\_u})[(1-r)R_{1D\_cu\_u}+R_{hu1}]}{R_{1D\_cu\_u}+R_{hu1}-R_{to\_1\_u}}-R_{hu} \tag{20.18}$$

where $R_{to\_1\_u}$ is total thermal resistance of plate with thickness t_u and h_u at its lower surface, R_{hu1} is $1/(h_u cd)$ and R_{hu} is $1/[h_u(ab-cd)]$. Equations 20.12-20.14 are applied to calculate $R_{to\_1\_u}$ by replacing h and t with h_u and t_u, respectively.

R_{1D_1} is one-dimensional thermal conductive resistance of lower spreading layer and is given as

$$R_{1D\_1} = \frac{t_1}{kab} \tag{20.19}$$

R_{s_1} is thermal spreading resistance of the lower spreading layer. Equations 20.13 and 20.14 are used to calculate R_{s_1} by replacing h and t with h_1 and t_1, respectively. R_{hl} is convective film resistance at lower surface of the plate and it is given by

$$R_{hl} = \frac{1}{h_1 ab} \tag{20.20}$$

The total thermal resistance of resistances network shown in Figure 20. 6 is

$$R_{to} = R_{1D\_cu\_u} + \frac{(R_s' + R_{hu})(R_{1D\_1} + R_{s\_1} + R_{hl})}{R_s' + R_{hu} + R_{1D\_1} + R_{s\_1} + R_{hl}} \tag{20.21}$$

As a result, mean temperature of heat source predicted by the network model is expressed as

$$\overline{T}_s = qabR_{to} + T_f \tag{20.22}$$

20.1.2 Comparisons and Discussion

To validate the present network model, random application cases were used to check. In the verification cases, the dimensions of the plate are 150mm × 100mm × 2mm. The size of the heat source is 30mm by 20mm and its centroid is (100, 60) mm. The heat flux is 1667W/m^2. Thermal conductivity of the plate and heat transfer coefficients at upper and lower surfaces are variable parameters in the calculation. To prove and compare the modeling results, simulations with the same parameters were also done by software COMSOL to provide a reference. Since all the heat transfer processes in the present paper is based on thermal conduction, good simulation result is very reliable for comparison.

As different mesh structures in simulations may lead to different results, the mesh structure was verified to ensure accuracy and avoid unacceptable errors before the computation. The mesh verification was done by comparison between simulation results and analytical solutions presented in reference (Muzychka *et al.*, 2003). In the mesh verification simulations, ht was set as 0W/(m^2·K) to meet boundary conditions stated in reference (Muzychka *et al.*, 2003). All other parameters in the simulation were the same as those in the analytical solution. For this case, the plate was meshed by free mesh. Predefined mesh sizes was set as normal and maximum mesh element size of the small cuboid is given by 0.001m. Using this mesh structure, simulations were done under variable conditions. It is found that mean temperatures of the heat source obtained by simulations and analytical solution are nearly the same while thermal conductivity of the plate k and heat transfer coefficient at lower surface h_1 change in wide ranges. The comparison demonstrates that the mesh structure is good enough to offer reliable simulation results.

The data comparisons between the network model and simulations are presented in Figures 20.7-20.9. As shown in these Figures, the mean temperatures of heat source predicted by the network model are very close to the ones obtained by simulations at the same conditions. If we defined relative error between network model and simulation as

$$E_r = \frac{\overline{T}_s - \overline{T}_{s\_si}}{\overline{T}_{s\_si} - T_f} \times 100\% \tag{20.23}$$

where $\overline{T}_s$ is mean temperature of heat source predicted by the network model and $\overline{T}_{s\_si}$ is the one obtained by simulation, the maximum relative error is 3.47%.

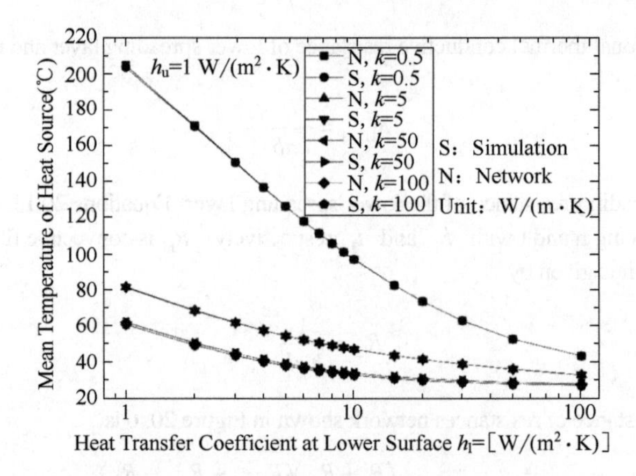

Figure 20.7 Data comparisons among network model and simulations as h_u 1 W/(m²·K)

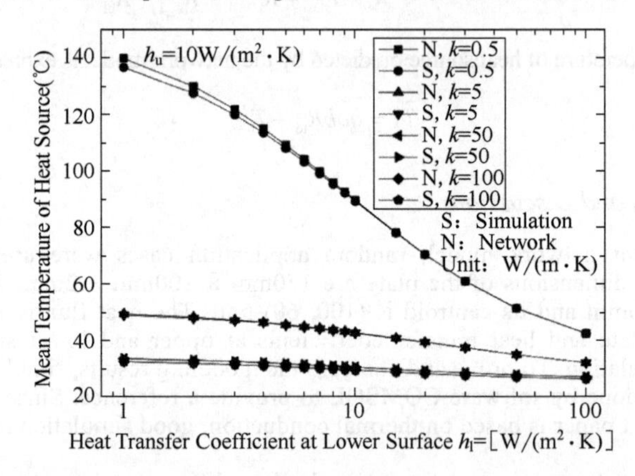

Figure 20.8 Data comparisons among network model and simulations as h_u 10W/(m²·K)

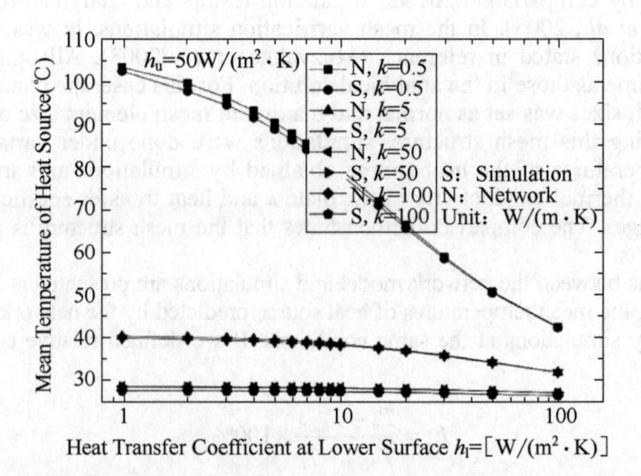

Figure 20.9 Data comparisons among network model and simulations as h_u 50W/(m²·K)

20.2 Thermal Resistance Model for Calculating Mean Die Temperature of A Typical BGA Packaging

Ball Grid Array (BGA) packaging is one of the most commonly used packaging types applied for microelectronic devices. To guarantee the reliability and longevity of BGA productions, it is necessary and important to analyze thermal characterization of BGA packaging, especially the die temperature of BGA packaging. In the semiconductor industry, numerical simulations and experiments are the mostly used methods to obtain the die temperature of a chip packaging. These methods commonly cost much time and resources. To reduce the time of obtaining die temperature and improve the efficiency of microelectronic device design, it is of significance to establish an analytical thermal model for analyzing and calculating die temperature of chip packaging.

Traditionally, thermal resistance is defined as:

$$R_{\mathrm{th}} = \frac{T - T_{\mathrm{ref}}}{Q} \tag{20.24}$$

where T is some critical temperature (usually the junction temperature), T_{ref} is reference temperature, and Q is usually the steady-state heat dissipation of the component. Since chip packaging usually contains multilayers, the structure is complex and the ambient conditions are variable, there is no such analytical single thermal resistance model can be used to calculate the die temperature until today. Bar-Cohen *et al.* (Bar-Cohen *et al.*, 1992) firstly proposed a thermal resistance network model to calculate junction temperatures. The network topology is star-shaped and only under isothermal condition can it predict the junction temperature accurately. Europe (Rosten *et al.*,1997) carried out a 3-year European collaborative project, named DELPHI, whose goal was to find solutions to predict the operating temperatures of critical electronic parts at the component-, board- and system-level. In the project DEPHI, Lasance *et al.* (Lasance *et al.*,1995) established an improved star-shaped compact thermal resistance network model which was boundary condition independent. Surface-to-surface resistors were added to the improved star-shaped network for better representing realistic characterization of heat transfer. In all of the 38 kinds of different conditions mentioned in reference (Lasance *et al.*,1995), the improved model can predict the junction temperature accurately, the error is only 1%-2%. Aranyosi *et al.* (Aranyosi *et al.*,2000) used the thermal model presented in reference (Lasance *et al.*,1995) to analyze conduction cooled electronic applications. Two general network topologies, incorporating both simple star-shaped network mentioned in reference. (Bar-Cohen *et al.*, 1992) and more complex, shunted network, were developed. They found that optimized star-shaped compact thermal model predicted the junction temperature accurately. Considering the complexity of multi-resistor thermal network models like the one in reference (Lasance *et al.*,1995) and unavailability of the required information about the packaging internal structure and materials for end-users, Tal and Nabi (Tal *et al.*, 2001) put forward an analytic method for converting standardized IC-packaging thermal resistances, junction to ambient resistance (R_{ja}) and junction to case resistance (R_{jc}), into a two-resistor thermal model which contains a junction to top resistance (R_{jt}) and a junction to board resistance (R_{jb}). With the help of the method named PERIMA, one can evaluate the two resistances in the two-resistor model using an analytic algorithm. Although lacking of the information about packaging, this method can evaluate resistances in a reasonable range. Garcia and Chiu (Garcia *et al.*,2005) used two-resistor model to analyze multiple-die and multi-chip packaging. They concluded that chip-scale packaging was suitable for two-resistor thermal model because of small thermal gradients and low thermal resistance between the die stack.

The die in packaging is commonly smaller than the substrate or heat spreader, which the die is located on. In addition, the area of PCB, where packaging is bonded on, is much more than the one of packaging. Therefore, there exists spreading resistance as heat flows from die into heat spreader or from packaging to PCB. The spreading resistance was obtained by means of the definition proposed by Mikic and Rohsenow (Mikic *et al.*,1966):

$$R_{\mathrm{s}} = \frac{\overline{T}_{\mathrm{source}} - \overline{T}_{\mathrm{contact\_plane}}}{Q} \tag{20.25}$$

The mean temperature of the heat source area was obtained from:

$$\overline{T}_{\text{source}} = \frac{1}{A_{\text{source}}} \int_{A_{\text{source}}} T_{\text{source}} \mathrm{d}A_{\text{source}} \tag{20.26}$$

And the mean temperature of the contact plane was obtained from:

$$\overline{T}_{\text{contact\_plane}} = \frac{1}{A_{\text{contact\_plane}}} \int_{A_{\text{contact\_plane}}} T_{\text{contact\_plane}} \mathrm{d}A_{\text{contact\_plane}} \tag{20.27}$$

Spreading resistance is much more than one-dimension conduction resistance. Thus spreading resistance is the major thermal resistance when heat transfers from die to substrate or heat spreader, as well as when heat dissipates from packaging to PCB in steady state. Yovanovich *et al.* did series of researches on spreading resistance and obtained many analytical solutions for calculating spreading resistances in different situations. Yovanovich *et al.* (Yovanovich *et al.*, 1999) analyzed spreading resistance of isoflux rectangular and strips on compound flux channels and obtained general spreading resistance expression. They also studied spreading resistance in compound and orthotropic systems and also found analytical solutions (Muzychka *et al.*, 2001, 2004). In references. (Muzychka *et al.*, 2003, 2004), Yovanovich *et al.* conducted research on spreading resistance in rectangular flux channel and obtained spreading resistance expressions based on different boundary conditions. Moreover, they discussed the influence of geometry and edge cooling on spreading resistance.

There are many other methods used for evaluating junction temperatures besides thermal resistance network models. One method uses the superposition method. Lall *et al.* (Lall *et al.*,1995) presented superposition method for evaluating junction temperatures of multichip modules. Although this method needs few experiments, it is very accurate for a model with equal die size and symmetric die location. Zahn (Zahn, 1998) presented the linear superposition theory on the non-linear matrix multiplier. This theory can be applied to a small multiple-output device working over a wide range of operating power at a natural convection steady state environment only. Another method is RSM(response surface method). As explained by Roux, RSM is a method for constructing global approximations of a system's behavior based on the results calculated at various points in the design space. Zahn (Zahn, 2000) also discussed the RSM for the thermal characterization of multiple heat source packaging. Accurate results can be obtained using RSM, but it requires many experiments to establish the response surface model. In addition, the number of experiments increases exponentially according to the number of heat sources. Im *et al.* (Im, *et al.*, 2004) proposed an approach that uses both the linear superposition and RSM to overcome the disadvantages of each method. This composed method can calculate device junction temperature accurately and it requires much fewer experiments than RSM, especially in case of lots of heat sources.

In this section, an analytical thermal resistance network model for calculating the mean die temperature of typical plastic BGA packaging is presented. Unlike other thermal network models mentioned above, the model proposed in this section does not need analysis tools. It also does not needs any experiments to evaluate mean die temperature. All of the resistances in the network have analytical solutions. Therefore, the mean die temperature can be calculated by programming based on the analytical solutions of thermal resistance in the network. The mean die temperatures of a typical plastic BGA packaging which are predicted by this model are close to the ones obtained by numerical simulations at various thermal conditions. This model is accurate enough for the semiconductor industry. This model can also be used to guide the thermal management design of plastic BGA packaging.

20.2.1 Model Development

As shown in Figure 20.10, a typical plastic BGA packaging contains mold compound, die, substrate, epoxy, solder balls and Au wire which is used to connect die and substrate. In steady-state, the heat generated by die dissipates to ambient mainly in two paths: (1)parts of heat flows from die to mold compound and then transfers to ambient from top surface and the edge of mold compound; (2)the rest heat flows from die to substrate and then separates into two parts: one transfers to ambient directly from the area where is not covered by solder balls at the bottom surface of substrate, the other one conducts to solder balls and spreads into PCB, and then transfers to ambient from top and bottom surfaces of PCB respectively. According to the two paths of heat dissipation from die to ambient, the thermal resistance network is established as shown in Figure 20. 11.

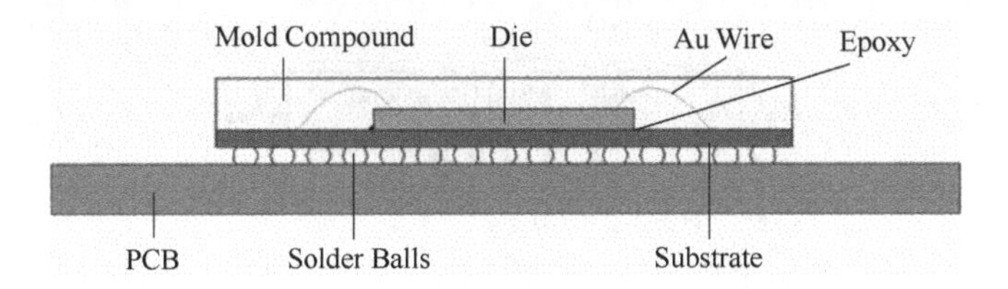

Figure 20.10 Cross section of a typical BGA packaging bonded on PCB

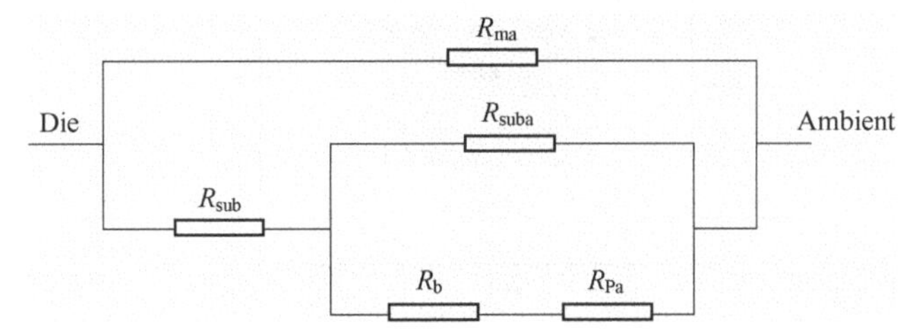

Figure 20.11 Thermal resistance network of the typical BGA packaging shown in Figure 20.10

R_{ma} includes the resistances of mold compound and the resistance between surfaces of mold compound and ambient. R_{sub} is spreading resistance and one-dimensional conductivity resistance of substrate. R_{suba} is resistance between exposed bottom surface of substrate and ambient. R_b is resistance of solder balls. R_{Pa} includes the resistance of PCB and the resistance between PCB surfaces and ambient. The detailed analytical solution of every thermal resistance in Figure 20.11 will be discussed in the following parts respectively.

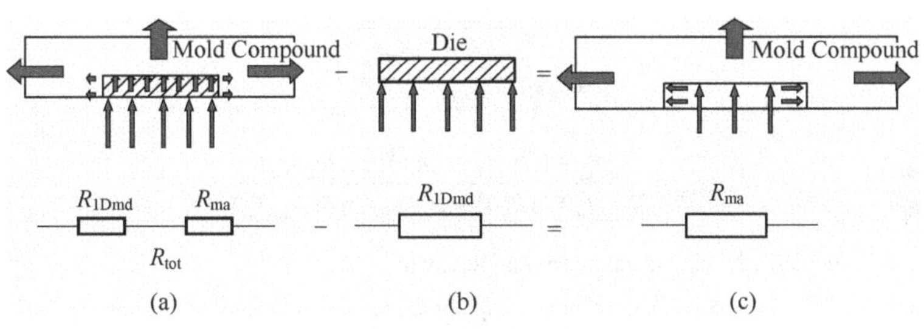

Figure 20.12 Schematic diagram for calculating R_{ma}. (a) A rectangular flux channel of mold compound with die; (b) mold compound with die size; (c) actual flux channel ofmold compound

(i)Calculation of R_{ma}

The actual heat transfer in top part of the packaging is that heat generated by die spreads into mold compound and then dissipates to ambient as shown in Figure 20.12(c). To calculate R_{ma}, as shown in Figure 20.12(a), it is supposed that heat flows from the bottom of die and dissipates through the die, this heat transfer process is the same as actual situation. Therefore, as shown in Figure 20.12, R_{ma} can be calculated by the following expression, $R_{ma} = R_{tot} - R_{1Dmd}$, where R_{tot} is the total thermal resistance as heat flows from the bottom of die to ambient, R_{1Dmd} is one-dimensional thermal conduction resistance of die size mold compound.

Muzychka *et al.* (Muzychka *et al.*, 2003) obtained a total thermal resistance expression for calculating total thermal resistance of rectangular flux channels with edge cooling. As the present case shown in Figure 20.12(a) is as same as that in reference. (Muzychka *et al.*,2003), thus R_{tot} can be calculated using the expression obtained by Muzychka *et al.* in reference. (Muzychka *et al.*,2003). The total resistance R_{tot} is calculated from the following general expression according to the notions in Figure 20.13.

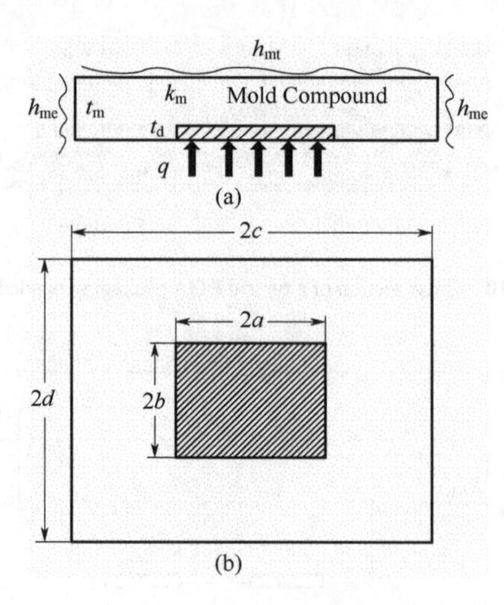

Figure 20.13 Rectangular flux channel with edge cooling and relative dimensions. (a) Section view; (b) bottom view

$$R_{tot} = \frac{cd}{k_m a^2 b^2} \sum_{m=1}^{\infty} \sum_{n=1}^{\infty} \frac{\sin^2(\delta_{xm} a / c) \sin^2(\delta_{yn} b / d) \phi_{mn}}{\delta_{xm} \delta_{yn} \beta_{mm} [\sin(2\delta_{xm}) / 2 + \delta_{xm}][\sin(2\delta_{yn}) / 2 + \delta_{yn}]} \tag{20.28}$$

$$\delta_{xm} \tan(\delta_{xm}) = \frac{h_{me} c}{k_m}, \delta_{yn} \tan(\delta_{yn}) = \frac{h_{me} d}{k_m} \tag{20.29}$$

$$\beta_{mm} = \sqrt{\left(\frac{\delta_{xm}}{c}\right)^2 + \left(\frac{\delta_{ym}}{d}\right)^2} \tag{20.30}$$

$$\phi_{mn} = \frac{\beta_{mm} t_m + [(h_{mt} t_m) / (k_m)] \tan h(\beta_{mm} t_m)}{[(h_{mt} t_m) / (k_m)] + \beta_{mm} t_m \tan h(\beta_{mm} t_m)} \tag{20.31}$$

As shown in Figure 20.12, R_{ma} is calculated as follows:

$$R_{ma} = R_{tot} - R_{1Dmd} \tag{20.32}$$

where

$$R_{1Dmd} = \frac{t_d}{k_m 4ab} \tag{20.33}$$

Therefore, R_{ma} can be obtained by substituting Equations 20.28 and 20.33 into Equation 20.32.

(ii) Calculation of R_{suba}

R_{suba} is the thermal resistance between the bottom surface of substrate and ambient. Here the bottom substrate is not covered by solder balls. R_{suba} is given by:

$$R_{suba} = \frac{1}{h_{subbo} A_{subbo}} \tag{20.34}$$

where

$$A_{subbo} = 4cd - \frac{\pi D_1^2}{4} N \tag{20.35}$$

where N is the number of solder balls and D_1 will be given in next section

(iii) Calculation of R_b

In the model, a solder ball is assumed to be equivalent to a truncated cone as shown in Figure 20.14. Thermal resistance of the truncated cone is given by:

$$R_{b1} = \frac{4L}{\pi k_b D_1 D_2} \tag{20.36}$$

where D_1 and D_2 are diameters of circles at the top and bottom surfaces of truncated cone respectively, L is the height of truncated cone.

Heat flows through solder balls in parallel. Therefore, the total thermal resistance of solder balls is:

$$R_b = \frac{R_{b1}}{N} \tag{20.37}$$

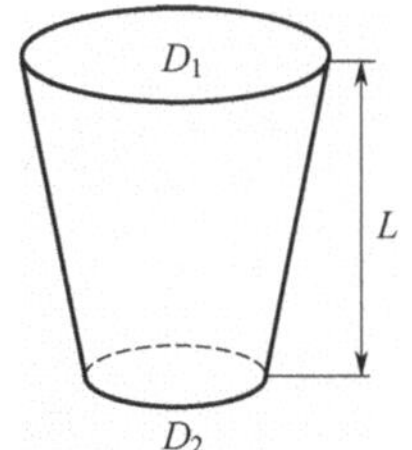

Figure 20.14 Solder ball equivalent

(iv) Calculation of R_{Pa}

Ball array of a typical BGA packaging is shown in Figure 20.15(a). Generally, array of balls at bottom of BGA packaging consists of two parts.Some balls are arranged at the central area of the bottom. The other balls are arrayed around the central ones. Since solder balls are arranged to be close to each other, the areas where solder balls are bonded on can be simplified to heat sources as shaded parts shown in Figure 20.15(b). The shaded parts in Figure 20.15(b) are the areas where solder balls are located on PCB.

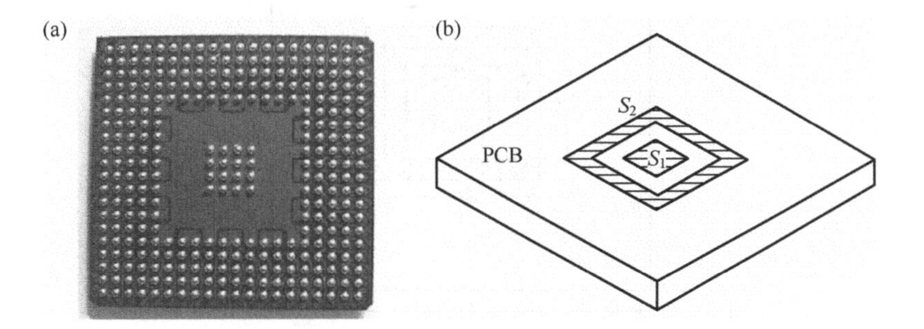

Figure 20.15 (a) Solder ball array of a typical plastic BGA packaging; (b) the areas where solder balls located on PCB (S_1 and S_2)

Heat dissipation paths through PCB are presented as follows. Heat coming from packaging flows into S_1 and S_2 as shown in Figure 20.16(a), then it conducts and spreads onto PCB and finally transfers to ambient by

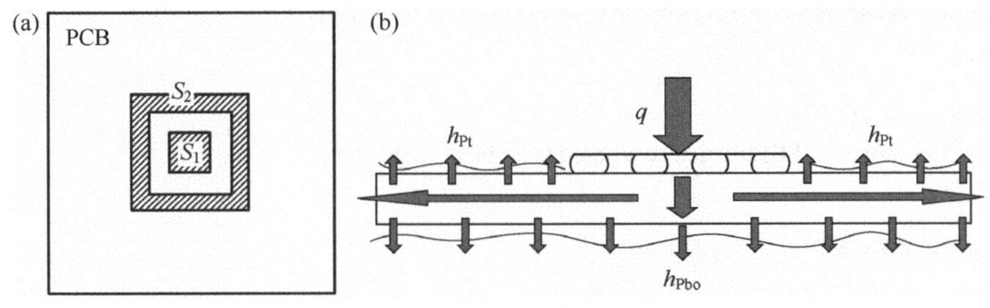

Figure 20.16 (a) Simplified heat sources on PCB; (b) heat dissipation paths through PCB

convection from the top and bottom surfaces of PCB as shown in Figure 20.16(b), respectively. According to the heat dissipation paths through PCB, the thermal resistance network for calculating R_{Pa} is established as Figure 20.17. Since the PCB is very thin compared to its width and length, it is assumed that the heat flows from the solder balls only spreads along the horizontal directions and then dissipates to the ambient form top and bottom surfaces of the PCB. Thus the resistances in Figure 20.8 are connected in a parallel way.

In Figure 20.17, R_{Ps} is spreading resistance of PCB. R_{1DP} is one dimensional thermal conduction resistance of PCB. R_{hPt} is thermal resistance between top of PCB, which is exposed to ambient. R_{hPbo} is the thermal resistance between PCB bottom part and ambient.

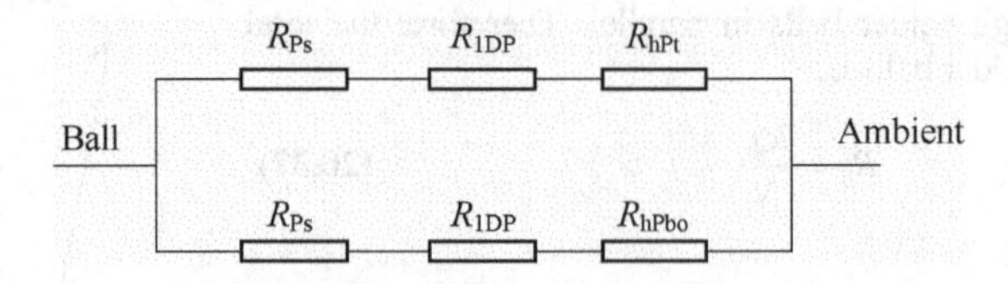

Figure 20.17 Thermal resistances network used to calculate R_{Pa}

Dimensions of heat sources S_1, S_2 and PCB are shown in Figure 20.18(a). Thermal parameters of PCB and boundary conditions are shown in Figure 20.18(b).

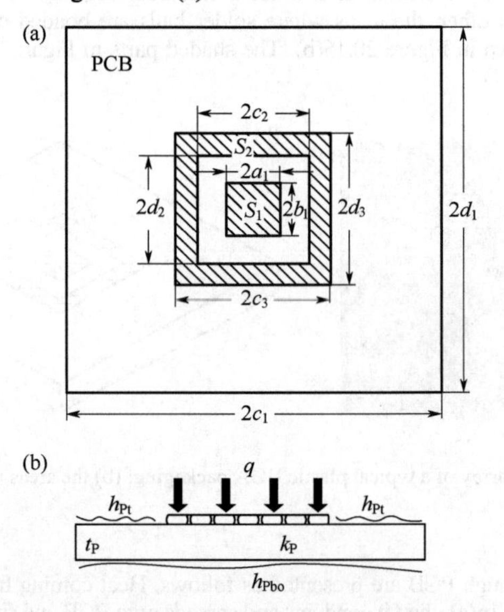

Figure 20.18 (a) Dimensions of heat sources S_1, S_2 and PCB; (b) thermal parameters of PCB and boundary conditions

(1) Calculation of R_{hPt}

The area of top surface of PCB, which is exposed to ambient, is given by:

$$A_{Pt} = 4c_1d_1 - 4cd \tag{20.38}$$

R_{hPt} is expressed as:

$$R_{hPt} = \frac{1}{h_{Pt}A_{Pt}} \tag{20.39}$$

(2) Calculation of R_{hPbo}

The area of PCB's bottom surface is given by:

$$A_{\mathrm{Pbo}} = 4c_1 d_1 \tag{20.40}$$

R_{hPbo} is expressed as:

$$R_{\mathrm{hPbo}} = \frac{1}{h_{\mathrm{Pbo}} A_{\mathrm{Pbo}}} \tag{20.41}$$

(3) Calculation of R_{Ps} and R_{1DP}

For isotropic plate with single heat source on its surface, the total thermal resistance is defined as:

$$R_{\mathrm{total}} = \frac{\overline{T}_{\mathrm{source}} - T_{\mathrm{a}}}{Q} \tag{20.42}$$

where $\overline{T}_{\mathrm{source}}$ is mean sources temperature given by:

$$\overline{T}_{\mathrm{source}} = \frac{1}{A_{\mathrm{source}}} \iint_{A_{\mathrm{source}}} T(x,y,0) \mathrm{d}A_{\mathrm{source}} \tag{20.43}$$

where A_{source} is the source area, $(x, y, 0)$ is coordinate of the top surface and 0 is z direction component.

Muzychka *et al.* (Muzychka *et al.*, 2003) obtained the general solution for temperature distribution of isotropic plate, on which single rectangular heat source is located, with top and edge boundaries adiabatic. For multiple sources, the temperature distribution of isotropic plate is obtained by superposition (Muzychka *et al.*, 2003).

Heat sources S_1 and S_2 are separated into five rectangular heat sources as shown in Figure 20.18(a) for applying the solution obtained in reference. (Muzychka *et al.*, 2003). So the temperature distribution on top surface of PCB can be achieved by superposition as:

$$T(x,y,0) - T_{\mathrm{a}} = \sum_{i=1}^{5} \theta_i(x,y,0) \tag{20.44}$$

where θ_i is the temperature rise for each heat source by itself. In the model, the PCB is supposed to be mounted on a heat sink. Heat transfer rate at bottom surface of the PCB is much more than the one at top surface. As a result, the equivalent heat transfer coefficient at bottom surface of the PCB will be much more than the one at top surface where nature convection happens. Thus the top surface of the PCB can be considered to be adiabatic so that R_{Ps} and R_{1DP} are assumed to be only relevant with heat transfer coefficient at bottom surface of the PCB h_{Pbo}. For those cases that heat transfer coefficient at top surface of PCB is close to or even larger than the one at bottom surface, R_{Ps} and R_{1DP} should be calculated in another method, which is represented in reference. (Luo *et al.*, 2009). As the thickness of PCB is much less than its width and length, edges of PCB are also treated as adiabatic. Therefore, θ_i can be obtained using the general solution for temperature distribution of isotropic plate with adiabatic conditions at boundaries. θ_i is given by:

$$\theta_i(x,y,0) = Q^i \left[A_0^i + \sum_{m=1}^{\infty} A_m^i \cos(\lambda x) + \sum_{n=1}^{\infty} A_n^i \cos(\delta y) + \sum_{m=1}^{\infty}\sum_{n=1}^{\infty} A_{mn}^i \cos(\lambda x)\cos(\delta y) \right] \tag{20.45}$$

Having obtained temperature distribution of top surface of PCB, the mean temperature rise of heat source S_1 and S_2 can be calculated by integrating Equation 20.24 over the regions S_1 and S_2. However, it is complex to calculate the mean temperature rise because of the special shape of S_2. To simplify the calculation, the mean temperature rise of rectangular region consisted by S_1, S_2 and the area between them substitutes the mean temperature rise of heat sources S_1 and S_2 in the present model. As temperature of this area is close to S_1 and S_2, this simplified treatment will not bring unacceptable error. The mean temperature rise of the substitute rectangular area is given by:

$$\overline{\theta} = \sum_{i=1}^{5} \frac{1}{A_{\text{substitute}}} \iint_{A_{\text{substitute}}} \theta_i(x,y,0)\mathrm{d}A_{\text{substitute}} = \sum_{i=1}^{5} \overline{\theta_i} \tag{20.46}$$

where $A_{\text{substitute}} = 4c_3 d_3$ and

$$\begin{aligned}\overline{\theta_i} = Q^i \Bigg[& A_0^i + 2\sum_{m=1}^{\infty} A_m^i \frac{\cos(\lambda x_{\text{BGA}})\sin(\lambda c_3)}{2\lambda c_3} + \\ & 2\sum_{n=1}^{\infty} A_n^i \frac{\cos(\delta y_{\text{BGA}})\sin(\delta d_3)}{2\delta d_3} + \\ & 4\sum_{m=1}^{\infty}\sum_{n=1}^{\infty} A_{mn}^i \frac{\cos(\lambda x_{\text{BGA}})\sin(\lambda c_3)\cos(\delta y_{\text{BGA}})\sin(\delta d_3)}{4\lambda\delta c_3 d_3} \Bigg]\end{aligned} \tag{20.47}$$

where Q^i is the heat flow of ith heat source, (x_{BGA} , y_{BGA}) is substitute rectangular centroid and:

$$A_0^i = \frac{1}{4c_1 d_1}\left(\frac{t_{\text{p}}}{k_{\text{p}}} + \frac{1}{h_{\text{Pbo}}}\right) \tag{20.48}$$

$$A_m^i = \frac{\cos(\lambda x_i)\sin[(1/2)\lambda L_i]}{c_1 d_1 L_i k_{\text{p}} \lambda^2 \phi(\lambda)} \tag{20.49}$$

$$A_n^i = \frac{\cos(\delta y_i)\sin[(1/2)\delta W_i]}{c_1 d_1 W_i k_{\text{p}} \delta^2 \phi(\delta)} \tag{20.50}$$

$$A_{mn}^i = \frac{\cos(\lambda x_i)\sin[(1/2)\lambda L_i]\cos(\delta y_i)\sin[(1/2)\delta W_i]}{c_1 d_1 L_i W k_{\text{p}} \beta\delta\lambda\phi(\beta)} \tag{20.51}$$

where $\lambda = m\pi / 2c_1$, $\delta = n\pi / 2d_1$, $\beta = \sqrt{\lambda^2 + \delta^2}$ and

$$\phi(\xi) = \frac{\xi \sin h(\xi t_{\text{p}}) + h_{\text{Pbo}} / k_{\text{p}} \cos h(\xi t_{\text{p}})}{\xi \cos h(\xi t_{\text{p}}) + h_{\text{Pbo}} / k_{\text{p}} \sin h(\xi t_{\text{p}})} \tag{20.52}$$

ξ is replaced by λ , δ , or β , accordingly. (x_i, y_i) is ith source centroid. L_i and W_i are the length and width of ith source respectively. These parameters are given by:

$$x_1 = c_1 - \frac{c_2 + c_3}{2}, x_2 = c_1 + \frac{c_2 + c_3}{2} \tag{20.53}$$

$$x_3, x_4, x_5 = c_1 \tag{20.54}$$

$$y_1, y_2, y_5 = d_1 \tag{20.55}$$

$$y_3 = d_1 - \frac{d_2 + d_3}{2}, y_4 = d_1 + \frac{d_2 + d_3}{2} \tag{20.56}$$

$$L_1, L_2 = c_3 - c_2, W_1, W_2 = 2d_2 \tag{20.57}$$

$$L_3, L_4 = 2c_3, W_3, W_4 = d_3 - d_2 \tag{20.58}$$

$$L_5 = 2a_2, W_5 = 2b_2 \tag{20.59}$$

As $\overline{\theta}_i$ has been obtained, the total thermal resistance is:

$$R_{\text{total}} = \frac{\overline{\theta}}{Q} = R_{\text{Ps}} + R_{\text{1DP}} + R_{\text{hPbo}} \tag{20.60}$$

where Q is heat flow rate of heat source S_1 and S_2. It is assumed that heat flux of S_1 and S_2 is constant q. Therefore, Q_i and Q are given by:

$$Q_i = L_i W_i q, Q = q\sum_{i=1}^{5} L_i W_i \tag{20.61}$$

R_{Ps} and R_{1DP} are obtained by summary of Equations 20.46, 20.47, 20.60 and 20.61 as follows:

$$\begin{aligned} R_{\text{Ps}} + R_{\text{1DP}} = \sum_{i=1}^{5} \frac{L_i W_i}{\sum_{i=1}^{5} L_i W_i} \Bigg[& A_0^i + 2\sum_{m=1}^{\infty} A_m^i \frac{\cos(\lambda x_{\text{BGA}})\sin(\lambda c_3)}{2\lambda c_3} + \\ & 2\sum_{n=1}^{\infty} A_n^i \frac{\cos(\delta y_{\text{BGA}})\sin(\delta d_3)}{2\delta d_3} + \\ & 4\sum_{m=1}^{\infty}\sum_{n=1}^{\infty} A_{mn}^i \frac{\cos(\lambda x_{\text{BGA}})\sin(\lambda c_3)\cos(\delta y_{\text{BGA}})\sin(\delta d_3)}{4\lambda\delta c_3 d_3} \Bigg] - R_{\text{hPbo}} \end{aligned} \tag{20.62}$$

(4) Summary of R_{Pa} Calculation

As all resistances in the network shown in Figure 20.17 have been calculated, R_{Pa} is expressed as:

$$R_{\text{Pa}} = \frac{1}{1/(R_{\text{Ps}} + R_{\text{1DP}} + R_{\text{hPbo}}) + 1/(R_{\text{Ps}} + R_{\text{1DP}} + R_{\text{hPt}})} \tag{20.63}$$

(v) Calculation of R_{sub}

Muzychka *et al.* obtained a spreading resistance expression for calculating spreading resistance of isotropic plate with central heat source and edge adiabatic. In present case, the substrate is thin compared to its length and width. Most of heat coming from die conducts to solder balls and transfers to ambient by convection at the bottom surface of the substrate. Therefore, the edge of substrate can be considered to be adiabatic. And die is located at central of substrate. R_{subs} can therefore be calculated by the spreading resistance expression obtained in reference (Muzychka *et al.*, 2003). Because there exists both conduction and convection at the bottom surface of substrate, the expression mentioned above cannot be used to calculate R_{subs} directly. To obtain R_{subs}, an equivalent heat transfer coefficient which is equivalent to actual heat transfer at bottom surface of the substrate should be found. It is supposed that heat transfers from substrate to ambient by convection with heat transfer coefficient h_{equ}. The total heat flowing from substrate is provided by:

$$q_{\text{suba}} = h_{\text{equ}} A_{\text{subb}} (\overline{T}_{\text{subb}} - T_{\text{a}}) \tag{20.64}$$

Actually, according to thermal resistances network as shown in Figure 20.11, the total heat flowing from substrate is:

$$q_{\text{suba}} = \frac{\overline{T}_{\text{subb}} - T_{\text{a}}}{1/[(1/R_{\text{suba}}) + 1/(R_{\text{b}} + R_{\text{Pa}})]} \tag{20.65}$$

Eliminating q_{suba} in Equations 20.64 and 20.65, h_{equ} is obtained as:

$$h_{\text{equ}} = \frac{(1/R_{\text{suba}}) + 1/(R_{\text{b}} + R_{\text{Pa}})}{A_{\text{subb}}} \tag{20.66}$$

As h_{equ} has been found, the spreading resistance of substrate R_{subs} is obtained from the following general expression which shows the explicit relationships with the geometric and thermal parameters of the system according to the notations in Figure 20. 19:

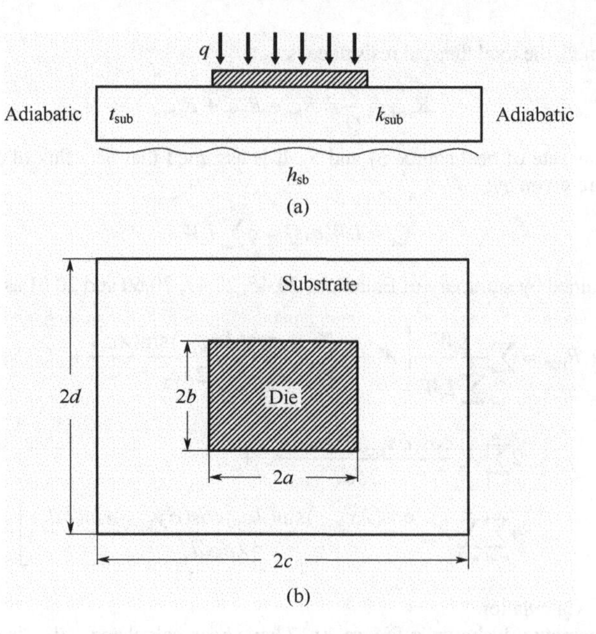

Figure 20.19 Substrate with die located in center. (a) Section view; (b) top view

$$R_{subs} = \frac{1}{2a^2cdk_{sub}}\sum_{m=1}^{\infty}\frac{\sin^2(a\delta_m)}{\delta_m^3}\varphi(\delta_m) + \frac{1}{2b^2cdk_{sub}}\sum_{n=1}^{\infty}\frac{\sin^2(b\lambda_n)}{\lambda_n^3}\varphi(\lambda_n) + \frac{1}{a^2b^2cdk_{sub}}\sum_{m=1}^{\infty}\sum_{n=1}^{\infty}\frac{\sin^2(a\delta_m)\sin^2(b\lambda_n)}{\delta_m^2\lambda_m^2\beta_{mn}}\varphi(\beta_{mn}) \tag{20.67}$$

where

$$\delta_m = m\pi / c, \lambda_n = n\pi / d \tag{20.68}$$

$$\beta_{mn} = \sqrt{\delta_m^2 + \lambda_n^2} \tag{20.69}$$

$$\phi(\xi) = \frac{(e^{2\xi t_{sub}} + 1)\xi - (1 - e^{2\xi t_{sub}})h_{equ} / k_{sub}}{(e^{2\xi t_{sub}} - 1)\xi + (1 + e^{2\xi t_{sub}})h_{equ} / k_{sub}} \tag{20.70}$$

And R_{sub} is given by:

$$R_{sub} = R_{subs} + R_{1Dsub} \tag{20.71}$$

where

$$R_{1Dsub} = \frac{t_{sub}}{k_{sub}4cd} \tag{20.72}$$

Therefore, R_{sub} can be obtained by substituting Equations 20.67 and 20.72 into Equation 20.71.

(vi) Calculation of R_{to}

Based on the above equations, the total thermal resistance between die and ambient is given by:

$$R_{to} = \frac{1}{(1/R_{ma}) + (1/(R_{sub} + (1/((1/R_{suba}) + (1/(R_b + R_{Pa}))))))} \tag{20.73}$$

(vii) Prediction of Mean Die Temperature $\overline{T}_d$

Generally, the total thermal resistance between die and ambient is defined as:

$$R_{to} = \frac{\overline{T}_d - T_a}{Q} \tag{20.74}$$

where $\overline{T}_d$ is the mean temperature of die, Q is the power input into die. Therefore, $\overline{T}_d$ is obtained as:

$$\overline{T}_d = R_{to}Q + T_a \tag{20.75}$$

20.2.2 *Analysis and Calculation*

To demonstrate the feasibility of the present model, the mean die temperatures of system containing a typical plastic BGA packaging and PCB under three series of thermal conditions were predicted by the proposed model respectively. Simulations to obtain the mean die temperatures of the system under the same thermal conditions as the prediction model were done by commercial software COMSOL. The accuracy of the presented model is proven by comparing the simulation data with the predicted ones.

Table 20.1 Dimensions of the system

	Parameter	Symbol	Dimension
Component			
Mold Compound	Length(mm)	$2c$	23
	Width(mm)	$2d$	23
	Thickness(mm)	t_m	8
Die	Length(mm)	$2a$	8
	Width(mm)	$2b$	0.25
	Thickness(mm)	t_d	23
Substrate	Length(mm)	$2c$	23
	Width(mm)	$2d$	23
	Thickness(mm)	t_{sub}	0.67
Solder Balls	Thickness(mm)	L	0.46
	Diameter (mm)		0.70
	Cone Diameter (mm)	D_1, D_2	0.52
	Ball Number	N	233
	Ball Array in Center		5×5
	Pitch (mm)		1.27
PCB	Length(mm)	$2c_1$	76
	Width(mm)	$2d_1$	76
	Thickness(mm)	t_p	1
Simplified Structure			

(continued)

	Parameter	Symbol	Dimension
Heat Source S_1	Length(mm)	$2a_1$	5.78
	Width(mm)	$2b_1$	5.78
Heat Source S_2	Length 1(mm)	$2c_2$	12
	Width 1(mm)	$2d_2$	12
	Length 2(mm)	$2c_3$	21
	Width 2(mm)	$2d_3$	21

Table 20.2 Invariable thermal parameters in three series of thermal conditions

Thermal Parameter	Symbol	Quantity
Heat Transfer Coefficient at Top of Mold Compound, W/(m²·K)	h_{mt}	5
Heat Transfer Coefficient at Edge of Mold Compound, W/(m²·K)	h_{me}	5
Heat Transfer Coefficient at Bottom of Substrate Exposed to Ambient, W/(m²·K)	h_{subbo}	1
Heat Transfer Coefficient at Top of PCB, W/(m²·K)	h_{Pt}	5
Thermal Conductivity of Mold Compound, W/(m·K)	k_m	0.2
Thermal Conductivity of Solder Balls, W/(m·K)	k_b	20
Ambient Temperature (K)	T_a	293.15
Power Input Die (W)	Q	5

Table 20.3 Three series of thermal conditions.

Thermal Conditions Series	k_{sub} ,W/(m·K)	k_p ,W/(m·K)	h_{Pbo} ,W/(m·K)
I	1–10	5	500
II	5	1–10	500
III	5	5	100–1000

In the first series, thermal conductivity of substrate k_{sub} is changed, other parameters are constant. In the second series, thermal conductivity of PCB k_p is the only variable thermal parameter. In the last series, heat transfer coefficient at the bottom surface of PCB h_{Pbo} is variable while other thermal parameters remain to be constants. Besides k_{sub}, k_p and h_{Pbo}, other thermal parameters and dimensions of plastic BGA packaging and PCB are selected based on commonly used ones in semi-conductor industry. The dimensions of the system are the same in all three series and they are given in Table 20.1. The invariable thermal parameters in the three series of thermal conditions are shown in Table 20.2. The three series of thermal conditions are presented in Table 20.3.

In series Ⅰ, thermal conductivity of substrate k_{sub} is changed from 1 to 10W/(m·K) and the step is 1W/(m·K). In series II, thermal conductivity of PCB k_p is variable from 1 to 10W/(m·K) and the step is 1W/(m·K) too. In series III heat transfer coefficient at bottom surface of PCB h_{Pbo} is varied from 100 to 1000W/(m²·K) and the step is 100W/(m²·K). In each series, only one thermal parameter is variable while the other ones are constant as shown in Table 20.3.

MATLAB was employed for the present calculation in three series of thermal conditions given above. In Equation 20.28, 100 terms were used in double summation. In other equations which are used to calculate spreading resistance or temperature distribution 100 terms were used in each single summation and every double summation consists of 10000 terms.

COMSOL was employed for simulations. All of the simulations were done in the same three series of thermal conditions as the ones used in the proposed model for calculating the mean die temperature. To make simulations close to the actual heat transfer processes, heat transfer coefficient at edge of substrate and PCB were added on. Since it is nature convection from edge of substrate and PCB, the heat transfer coefficients were given as 5W/(m^2·K).

20.2.3 Results and Discussions

Tables 20.4–20.6 present the data obtained by the proposed model and simulations. And the difference between mean die temperatures calculated by the presented model and the ones obtained by simulations, errors are also shown in the tables. The error is defined as:

$$\text{Error} = \frac{\left(\overline{T}_{\text{dmodel}} - T_{\text{a}}\right) - \left(\overline{T}_{\text{dsimulation}} - T_{\text{a}}\right)}{\overline{T}_{\text{dsimulation}} - T_{\text{a}}} \times 100\%$$

Table 20.4 Comparison of data obtained by the present model and simulations in thermal conditions series Ⅰ

k_{sub} ,W/(m·K)	$\overline{T}_{\text{dmodel}}$ (K)	$\overline{T}_{\text{dsimulation}}$ (K)	$\overline{T}_{\text{dmodel}} - \overline{T}_{\text{dsimulation}}$ (K)	Error (%)
1	441.818	469.499	-27.681	-15.697
2	407.508	415.694	-8.186	-6.680
3	391.782	394.189	-2.407	-2.382
4	381.733	381.767	-0.034	-0.038
5	374.425	373.356	1.069	1.333
6	368.738	367.142	1.596	2.157
7	364.126	362.294	1.832	2.650
8	360.279	358.369	1.910	2.929
9	357.005	355.105	1.900	3.067
10	354.174	352.335	1.839	3.107

Table 20.5 Comparison of data obtained by the present model and simulations in thermal conditions series Ⅱ

k_{p} ,W/(m·K)	$\overline{T}_{\text{dmodel}}$ (K)	$\overline{T}_{\text{dsimulation}}$ (K)	$\overline{T}_{\text{dmodel}} - \overline{T}_{\text{dsimulation}}$ (K)	Error (%)
1	396.895	410.226	-13.331	-11.387
2	385.242	391.036	-5.794	-5.919
3	380.051	382.559	-2.508	-2.805
4	376.797	377.226	-0.429	-0.510
5	374.425	373.356	1.069	1.333
6	372.546	370.323	2.223	2.881
7	370.98	367.831	3.149	4.217
8	369.63	365.719	3.911	5.389
9	368.439	363.889	4.550	6.432
10	367.369	362.277	5.092	7.366

Table 20.6 Comparison of data obtained by the present model and simulations in thermal conditions series Ⅲ

h_{Pbo} ,W/(m²·K)	$\overline{T}_{dmodel}$ (K)	$\overline{T}_{dsimulation}$ (K)	$\overline{T}_{dmodel} - \overline{T}_{dsimulation}$ (K)	Error (%)
100	430.694	419.257	11.437	9.069
200	404.625	396.259	8.366	8.114
300	390.452	385.070	5.382	5.855
400	381.136	378.154	2.982	3.508
500	374.425	373.356	1.069	1.333
600	369.31	369.789	−0.479	−0.625
700	365.261	367.010	−1.749	−2.368
800	361.963	364.774	−2.811	−3.925
900	359.217	362.927	−3.710	−5.317
1000	356.893	361.371	−4.478	−6.564

From Tables 20.4-20.6, it is found that in most cases the error is less than ±10%. The maximum error is 15.697% in thermal conditions series Ⅰ k_{sub}, 1W/(m·K). Those thermal conditions with error more than ±10% are not commonly used in typical BGA packaging. Therefore, the present model can be used to predict mean die temperature of typical BGA packaging with good accuracy. As shown in Tables 20.4 and 20.6, mean die temperature decreases considerably as k_{sub} and h_{Pbo} increases. These trends give a guide that k_{sub} and h_{Pbo} are two key factors which take significance impact on mean die temperature of typical BGA packaging. Therefore, the present model provides an optimization method to choose materials and boundary conditions used in typical plastic BGA packaging for good thermal management.

20.3 Chapter Summary

A thermal resistances network model to calculate thermal resistance of eccentric heat source on rectangular plate with upper and lower surface cooling was established based on heat transfer processes in the plate. A random application case was used verification check of the network model. Simulations were also conducted to offer a reference. The data comparisons show that the network model is able to calculate thermal resistance of eccentric heat source on rectangular plate with upper and lower surface cooling accurately. An analytical thermal resistance network model for predicting the mean die temperature of a typical plastic BGA packaging is presented. The proposed model was applied to predict the mean die temperature of a typical BGA packaging under three series of thermal conditions. Comparing the data obtained by the presented model and simulations, it is found that the proposed model predicted the mean die temperature in good accuracy. As each resistance in the network has an analytical solution, the proposed model can help find what the key factors influencing mean die temperature in a typical plastic BGA packaging are. An optimization method to typical plastic BGA design for good thermal management has been provided. The extensions to other packages are under way.

References

Aranyosi, A., Ortega, A., Griffin, R.A., West, S., and Edwards, D. (2000) Compact thermal models of packages used in conduction cooled applications, *IEEE Transactions on Components and Packaging Technology*, Vol. 23, No. 3, 470-480.

Bar-Cohen, A., Elperin, T., Eliasi, R., Bar-Cohen, A., Elperin, T. and Eliasi, R. (1989) Q_{jc} characterization of chip packages-justification, limitations, and future, *IEEE Transactions on Components, Hybrids, and Manufacturing Technology* Vol.12(4), 724-731.

Ellison, G. N. (2003) Maximum thermal spreading resistance for rectangular sources and plates with nonunity aspect ratios, *IEEE Transactions on Components and Packaging Technologies* 26 (No. 2).

Garcia, E., Chiu, C. (2005) Two-resistor compact modeling for multiple die and multi-chip packages, *Proceedings of* 21*th*

IEEE Semiconductor Thermal Measurement and Management Symposium, San Jose, CA, 2005.

Hein, V.L., Lenzi, V.D. (1969) Thermal analysis of substrates and integrated circuits, *Proc. Electronic Components and Technology Conference*, 166-177.

Im, Y., Kwon, H., Kim, H., Kim, T., Cho, T., Oh, Y. (2004) Methodology for accurate junction temperature estimation of SIP (system in package), *Proceedings of 20th IEEE Semiconductor Thermal Measurement and Management Symposium*, San Jose, CA.2004.

John, M., Krane, M. (1991) Constriction resistance in rectangular bodies, *Transactions of the ASME* 113 392-396.

Kabir, H., Ortega, A. (1998) A new model for substrate heat spreading to two convective heat sinks: application to the BGA package, *Proceedings of 17th IEEE Semiconductor Thermal Measurement and Management Symposium*, San Diego, CA, 24-30.

Kadambi, V., Abuaf, N. (1985) An analysis of the thermal response of power chip packages, *IEEE Transactions on Electron Devices* 32 (No. 6).

Kennedy, D. P. (1960) Spreading resistance in cylindrical semiconductor devices, Journal of Applied Physics 311490-1497.

Lall, B. S., Guenin, B. M., Molnar, R. J. (1995) Methodology for thermal evaluation of multichip modules, *Proceedings of 11th IEEE Semiconductor Thermal Measurement and Management Symposium* 758-764.

Lasance C. J. M., Rosten, H. I., Parry, J. D. (1997) The world of thermal characterization according to DELPHI-Part II: experimental and numerical methods, *IEEE Transactions on Components, Packaging, and Manufacturing Technology*, Part A 20(4):392-398.

Lasance, C. J. M., Vinke, H., Rosten, H., Weiner, K. L. (1995) A novel approach for the thermal characterization of electronic parts, *Proceedings of 11th Semiconductor Thermal Measurement and Management Symposium*, San Jose, CA, 1-9.

Lee, S., Song, S., Au, V., Moran, K.P. (1995) Constriction/spreading resistance model for electronic packaging, *Proceeding of the 4th ASME/JSME Thermal Engineering Joint Conference*, 199-206.

Luo, X. B., Mao, Z. M., Liu, S. (2009) A thermal model for calculating thermal resistance of eccentric heat source on rectangular plate with convective cooling existing at upper and lower surfaces, *Proceeding of 10th International Conference on Electronic Packaging Technology & High Density*, Beijing, China,294-298.

Mikic, B. B., Rohsenow, W. M. (1966) The effect of surface roughness and waviness upon the overall thermal contact resistance, Heat Transfer Lab. Rept. 4542-41, MIT, Cambridge, MA.

Muzychka, Y. S. (2006) Influence coefficient method for calculating discrete heat source temperature on finite convectively cooled substrates, *IEEE Transactions on Components and Packaging Technologies* 29 (No.3) 636-643.

Muzychka, Y. S., Culham, J. R., Yovanovich, M. M. (2003) Thermal spreading resistances of eccentric heat sources on rectangular flux channels, *Journal of Electronic Packaging* 125 178-185.

Muzychka, Y. S., Yovanovich, M. M., Culham, J. R. (2006) Influence of geometry and edge cooling on thermal spreading resistance, *Journal of Thermophysics and Heat Transfer* 20 (No. 2) 247-255.

Muzychka, Y.S., Culham, J.R., Yovanovich, M.M. (2003) Thermal spreading resistances in rectangular flux channels part I - geometric equivalences, *36th AIAA Thermophysics Conference* Orlando, Florida AIAA 2003-4187.

Nugus, K. J., Yovanovich, M. M. (1984) Constriction resistance of circular flux tubes with mixed boundary conditions by linear superposition of Neumann solutions, *Proceeding ASME 22nd Heat Transfer Conference, Niagara Falls*, NY.

Read, W.W. (2007) An analytic series method for Laplacian problems with mix boundary conditions, *Journal of Computational and Applied Mathematics* 209 22-32.

Rosten, H. I., Lasance, C. J. M., and Parry, J. D. The world of thermal characterization according to DELPHI -part I: Background to Delphi, *IEEE Transactions on Components, Packaging, and Manufacturing Technology*, Part A, 20(4):384-391.

Song, S., Lee, S., Au, V. (1994) Closed form equation for thermal constriction/spreading resistances with variable resistance boundary condition, *Proceeding of the 1994 IEPS Conference*, 111-121.

Tal, Y., Nabi, A. (2001) A simple analytic method for converting standardized IC-package thermal resistances (Q_{ja}, Q_{jc}) into a two-resistor model (Q_{jb}, Q_{jt}), *Proceedings of 17th IEEE Semiconductor Thermal Measurement and Management Symposium*, San Jose, CA, 2001.

Yovanovich, M. M. (1976) General thermal constriction parameter for annular contacts on circular flux tubes, *AIAA Journal* 14 (No. 6) 822-824.

Yovanovich, M. M. (1980) General solution of constriction resistance within a compound disk, *Progress in Astronautics and Aeronautics: Heat Transfer, Thermal Control, and Heat Pipes*. MIT Press, Cambridge, MA, 47-62.

Yovanovich, M. M., Culham, J. R., Teertstra, P. M. (1998) Analytical modeling of spreading resistance in flux tubes, half spaces, and compound disks, *IEEE Transactions on Components, Packaging, and Manufacturing Technology -Part A* 21 168-176.

Yovanovich, M. M., Muzychka, Y.S., Culham, J. R. (1999) Spreading resistance of isoflux rectangles and strips on compound flux channels, *Journal of Thermophysics and Heat Transfer* 13 (No. 4) 495-500.
Zahn, B. A. (1998) Using design of experiment simulation responses to predict thermal performance limits of the heatsink small outline package (HSOP) considering both die bond and heatsink solder voiding, *Proceedings of 11th IEEE Semiconductor Thermal Measurement and Management Symposium*, San Jose, CA.
Zahn, B. A. (2000) Steady state thermal characterization and junction temperature estimation of multichip module packages using the response surface method, *IEEE Transactions on Components and Packaging Technologies* 23(1):33-39.

21

3D Through Silicon Via (TSV) Packaging

With increasing demands of functionality and portability for electronic devices, the proliferation of 3D chip stacking technique has occurred within the last several years or so in the semiconductor industry (Knickerbocker *et al.*, 2008; Lau, 2011). This rapid development may be attributed to its advantages including high bandwidth, low latency, low power dissipation, and small form factor, which greatly enhance the performance of integrated circuit systems and enrich the applications in modules based on 3D through silicon via (TSV) packaging gradually become the key components for the success of the Internet of Things (IoT) proposed by International Business Machines Corporation and Internet of Everything by Cisco Systems.

21.1 A New Prewetting Process of TSV Electroplating for 3D Integration

This high aspect ratio of TSV can bring challenge for the full filling of the electrolyte into the vias, because for blind vias plating in micro scale, as the existence of surface tension of the electrolyte and the air in the blind vias, the electrolyte is hard to completely flow in and fully fill the blind vias, which may then result in voids at the bottom of the blind vias and limits the filled aspect ratio and quality (Zhang *et al.*, 2015; Wei *et al.*,2012).

Therefore, several pre-wetting methods, such as an ultrasonic bath and vacuum pumping, are applied before electroplating. Yet performing an ultrasonic bath takes a significant amount of time and may destroy the copper seed layer; furthermore, in this condition, the ultrasonic bath cannot work well and the aspect ratio of TSV cannot be plated, the best aspect ratio at present by this method is 5:1 to 7:1, such as by Jian Cai team from Tsinghua University (Wei *et al.*, 2012). The effect of prewetting is good by bathing the wafer in vacuum, but by vacuum pumping technology, as the vacuum degree can limit the aspect ratio of filled TSV, very high vacuum chambers must be generated or maintained, which can take long time of process and run up the cost for high performance vacuum pump is needed. In addition, the limited volume of the chamber can limit the size and quantity of the wafers processed per time, which will further run up the cost of TSV filling process for mass production. Therefore, to solve the problem, in this work, the infiltration process of TSV is studied, and a new 3-step process of ethanol-based prewetting is developed to fill the high aspect ratio TSV. By this method, high vacuum chamber is no longer needed, and very simple equipment can be designed to achieve high aspect ratio TSV filling, which has a low cost and can be adapted to industrial standards. In this prewetting process, at present, void-free filling can be achieved in TSVs with aspect ratio of 10:1.

21.1.1 Modeling and Simulation

(i) 3-Step Prewetting Process

As schematically shown in Figure 21.1 (a) and (b), firstly, inserting slowly the TSV samples vertically into the ethanol, then they can be fully filled with ethanol for its good wettability. Secondly, the samples are

Modeling and Simulation for Microelectronic Packaging and Integration: Manufacturing, Reliability and Testing, Second Edition. Sheng Liu and Yong Liu.

infiltrated in deionized water. Since ethanol is miscible with water at any ratio, the ethanol in the TSVs can be replaced by deionized water, as schematically shown in Figure 21.1 (c). Finally, the samples are infiltrated in the copper electrolyte and filled with solutions, as schematically shown in Figure 21.1 (d).

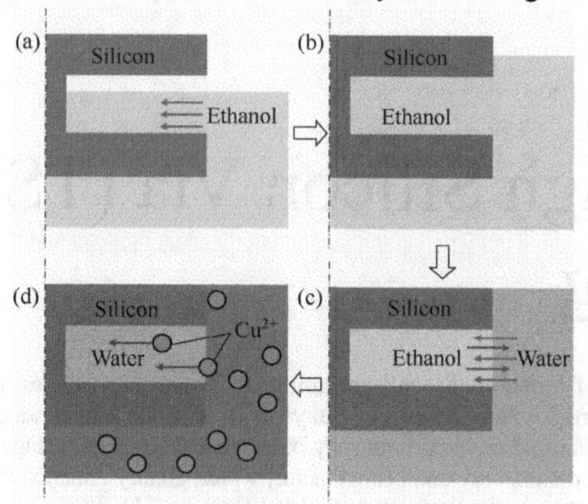

Figure 21.1 Schematic diagram of 3-step prewetting process

(ii) Model Setup

To study and the effect of the second step of the 3-step prewetting process, a simulation model is setup using COMSOL Multi-physics software. As schematically shown in Figure 21.2, for this pre-wetting process, the wafer is needed to vertically insert into the plating solution slowly. So the boundary conditions of the simulation model could be set up as shown in Figure 21.3. The whole modeling area is filled with air at the beginning, as the

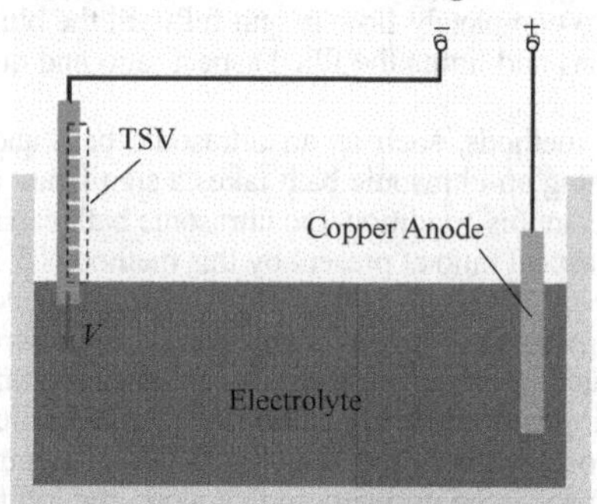

Figure 21.2 Schematic of the regular pre infiltration process

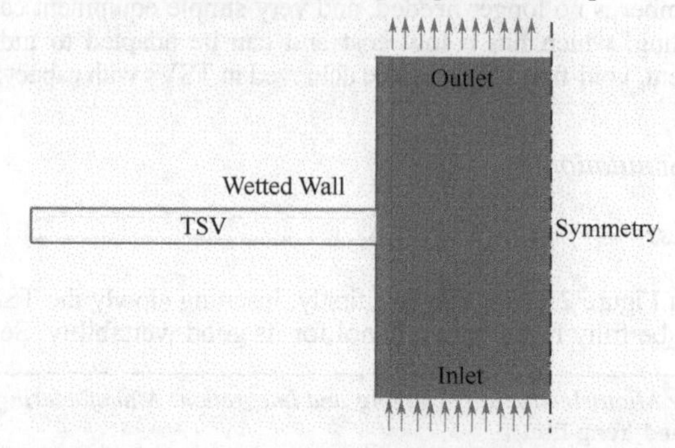

Figure 21.3 Boundary conditions of the simulation model

wafer is inserted vertically into the copper electrolyte, the solution can be regarded as flowing in from the bottom inlet and flowing out from the top outlet. The wafer surface and the TSV wall can be seen as wetted wall boundary and the right side border can be considering as symmetry boundary. Besides, gravity is loaded in the whole modeling area. Then, Laminar two-phase flow and phase field method of COMSOL Multi-physics, is utilized to solve the model.

(iii) Simulation Results

Finite element models of different contact angles (including 90°, 60°, and 45°) of infiltrating liquid and wetted wall are established and solved by COMSOL Multi-physics 5.2 software from COMSOL Inc.

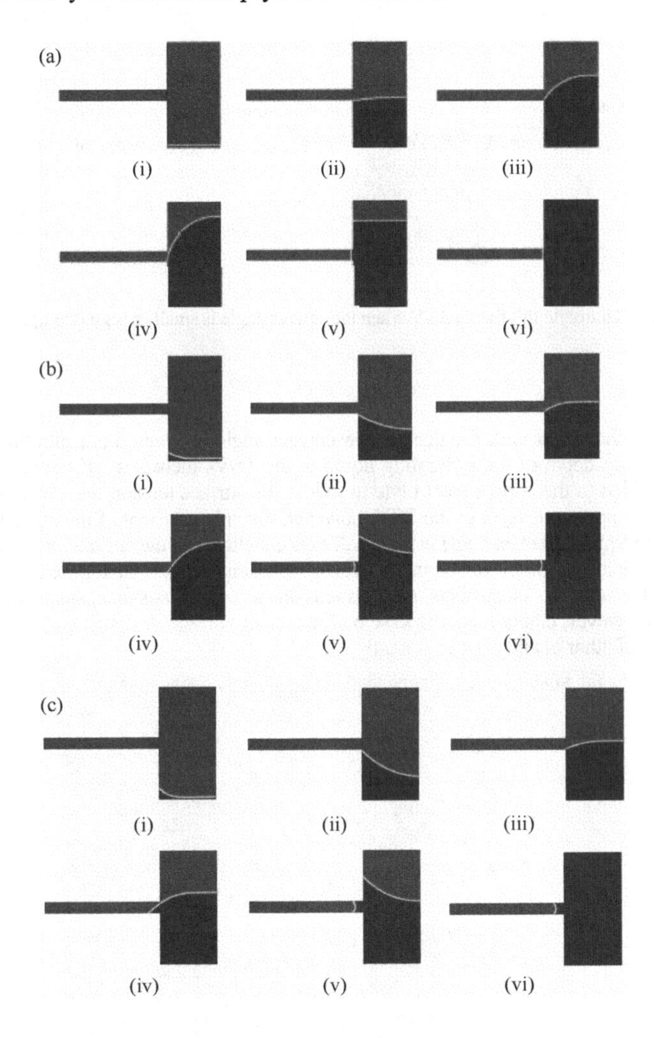

Figure 21.4 Simulation results of different contact angle. (a) 90°; (b) 60°; (c) 45°

The simulation results of infiltration process are shown in Figure 21.4 (a), (b) and (c), which represent cases of contact angle 90°, 60° and 45°, respectively. Firstly, as the wafer inserting into the plating solution, the electrolyte flows up along the wafer surface with the contact angle. Then the electrolyte reaches and stopped at the TSV orifice while the other parts of the electrolyte continue to flow upward. When the angle of the liquid surface and the TSV wall is equal to the contact angle, the electrolyte starts to flow into the TSV until the liquid surface reaches near the upper wall of TSV [when the contact angle is 90°, the electrolyte will not flow into the TSV, as shown in Figure 21.4 (a)]. Then the electrolyte flows over the TSV and a chamber is formed due to the surface tension of the liquid. It can be observed from the figures that with the decrease of contact angle, the depth of infiltration will be gradually increase. And it can be inferred that if the contact angle is small enough, the electrolyte will flow to the bottom of the TSV, as shown in Figure 21.5, then the liquid flows back along the upper wall of TSV, as shown in Figure 21.4 (b), in this case, the whole TSV can be filled with the plating solution. When the aspect ratio of the TSV is 10:1 and the aperture is small enough, the contact angle can be a value which is a little larger than 5.7° (arctan 0.1). And theoretically, if the contact angel is 0°, TSV of aspect ratio much greater than 10 can be infiltrated.

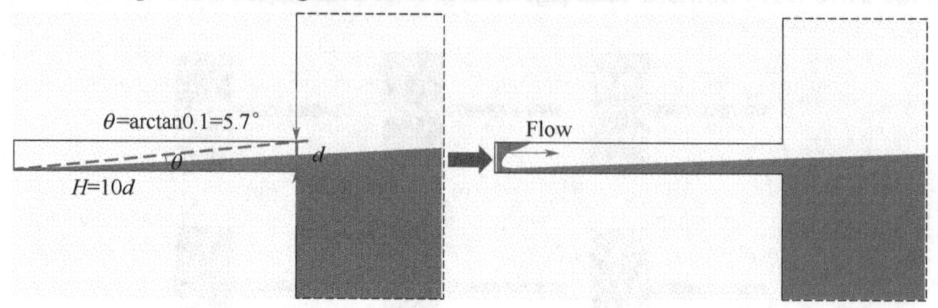

Figure 21.5 Electrolyte full fill the TSV when the contact angle is small enough (the upper TSV wall)

21.1.2 Experiments

The simulation results show that with the decrease in contact angle between the infiltrating liquid and copper seed layer, the infiltration depth of the infiltrating liquid in the TSVs increases. Of course, it is known that an active agent can be added to the copper electrolyte to reduce the surface tension and contact angle between the infiltrating liquid and copper seed layer in the TSV, however, the enhancement of the infiltration property is not sufficient because the required contact angle is too small as is concluded in the simulation results.

Ethanol easily disperses on many different materials, including copper. In Figure 21.6(a), liquid droplets are formed when the copper electrolyte is dropped on a copper surface, and the contact angle is approximately 60°. However, ethanol can disperse on the copper surface, according to Figure 21.6(b) and (c), *i.e.*, the contact angle of ethanol and copper is nearly 0°.

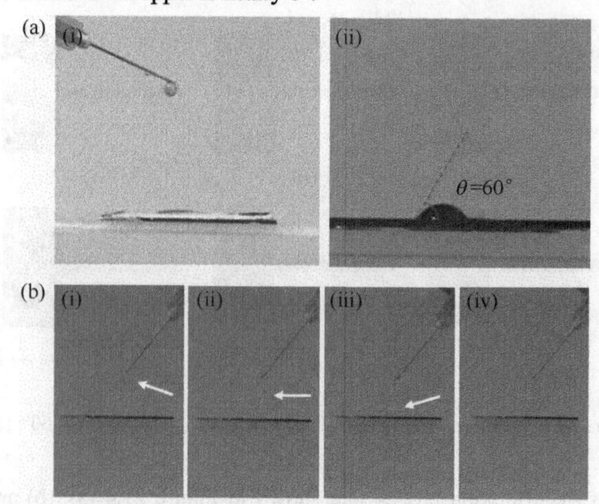

Figure 21.6 Infiltation tests of copper surface by copper electrolyte and ethanol. (a) Liguid beads formed by the copper electrolyte; (b) spreading of ethanol from the side view; (c) spreading of ethanol from the top view

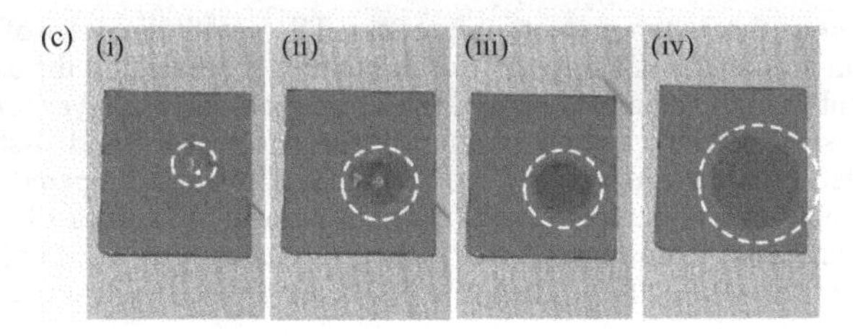

Figure 21.6 (Continaed)

To study the infiltration process of ethanol and the TSV copper wall, a batch of samples were fabricated. First, a channel array with a width of 300μm and length of 3mm was etched onto silicon wafers using deep reactive ion etching (DRIE), surface technology systems (STS), and the Bosch process (Ranganathan *et al.*, 2011; Fursenko *et al.*, 2015; Singulani *et al.*, 2013). Second, a 500nm-thick copper layer was sputtered to form the seed layer, and the wafers were diced into 5mm×5mm chips. Finally, the surfaces of the channels were bonded with transparent silica glass (10μm-thick PDMS coated on the surface of the glass). The infiltration details were obtained from the transparent silica glass using a camera system, as schematically shown in Figure 21.7. When the infiltrating liquid was slowly injected into the water tank through the micro pipe and the liquid surface slowly rose over the etched channels, the infiltration process was recorded by the camera. The physical photo of the system is shown in Figure 21.8.

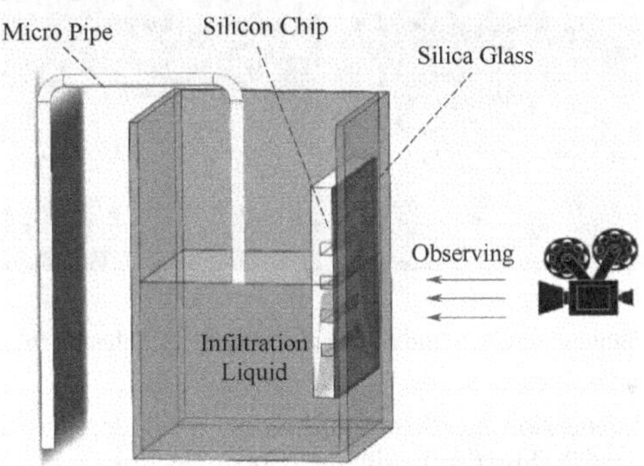

Figure 21.7 Observation of the infiltration process

Figure 21.8 Camera system built up for observing the infiltration process in the samples

To confirm the effect of the 3-step prewetting process, TSVs with diameter of 20μm and depths of 200μm were etched on silicon chips using the Bosch etching process, and the aspect ratio was 10:1. A sputtering equipment at high vacuum was used to ensure that the seed layer fully covered the TSV. The prewetting processes of the vacuum pumping, ultrasonic bath, and 3-step prewetting were conducted. Then, a plating table (SYM-WB from Shanghai Xinyang Company) was used to conduct the processes to compare the effect of these wetting methods. Additives of 3360 from Shanghai Xinyang Company were used to form void free filling (Vereecken *et al.*, 2005; Broekmann *et al.*, 2011; Tsai *et al.*, 2010; Shi *et al.*, 2013). Then, the TSVs were diced to observe the filling state from the section.

21.1.3 Results and Discussions

To clearly observe the infiltration status of ethanol, a red pigment needs to be added. Figure 21.9 (a) shows that in the infiltration process, the color of the blind vias became red from the top to the bottom, which implies that the blind vias were fully filled with the stained ethanol. For the copper electrolyte, as shown in Figure 21.9 (b), only a small section at the opening of the blind vias was filled with the blue liquid, which implies that the solution did not fill the blind vias. Thus, the TSV cannot be fully filled without the prewetting process.

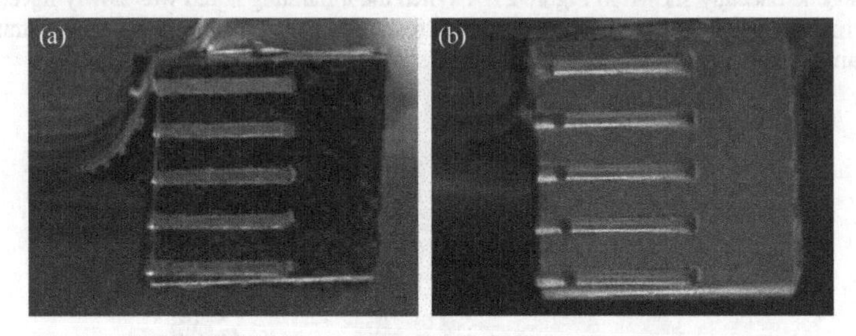

Figure 21.9 Infiltration results of the copper electrolyte. (a) Elthanol; (b) copper electrolyte

To further verify the effectiveness of the ethanol infiltration, the 3-step prewetting process was used to fill blind vias with a 20μm width and 200μm depth with the copper electrolyte. For comparison, the samples used in the prewetting processes of vacuum pumping and an ultrasonic bath and samples without the prewetting process were plated. The results are shown in Figure 21.10. Figure 21.10 (a) shows that without the prewetting process, the blind vias were not filled with copper; only bumps remained on the chip surface from the small amount of solution that entered the vias. As shown in Figure 21.10 (b), for the blind vias of this size,

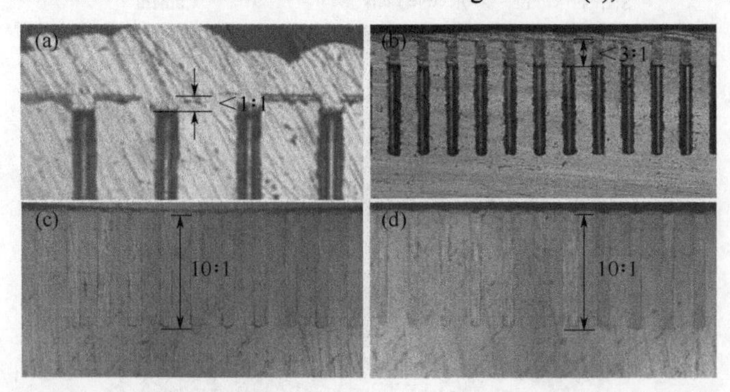

Figure 21.10 Electroplating results of different prewetting processes. (a) Without prewetting process, filled aspect ratio <1:1; (b) ultrasonic bath, filled aspect ratio < 3:1; (c) vaccum pumping, filled aspect ratio = 10:1; (d) 3-step prewetting process, filled aspect ratio = 10:1

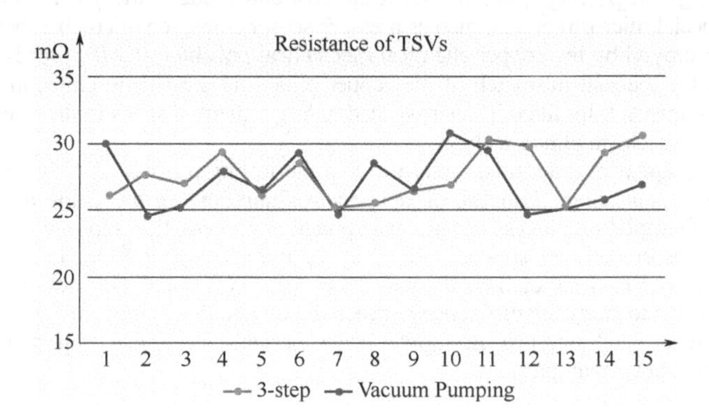

Figure 21.11 15 points measured resistance of TSVs by 3-step prewetting process and vacuum pumping process

the ultrasonic bath prewetting process can result in a copper filling with an aspect ratio of approximately 3:1 to 4:1. As shown in Figure 21.10 (c) and (d), for both vacuum pumping and the 3-step prewetting process, the blind vias with an aspect ratio of 10:1 can be fully filled with copper. To confirm the electrical conductivity of the TSVs fabricated by the 3-step prewetting process, the resistance of the 20μm× 200μm TSVs fabricated by this prewetting method and vacuum pumping process are measured, the result is shown in Figure 21.11, 15 points are measured and it is found that the resistance of the 20μm × 200μm TSVs fabricated by the two prewetting processes are almost equal, which are both within the scope of 25mΩ to 30mΩ.

Another important factor of the 3-step pre-wetting process which is needed to pay attention is the method of the wafer that is put into the ethanol. Although the speed of the liquid that flows into the TSV is very fast (from the observing of the high speed camera of the channels of 300μm × 3000μm, the needed flowing time is about tens of milliseconds, and for 20μm×200μm blind vias, the time is even much shorter), the wafer must be insert vertically into the ethanol slowly to offer enough time for the ethanol to flow to the bottom of the blind vias, or otherwise, the ethanol will seal the blind vias ahead of time and make air exist in the bottom of the blind vias, which will then cause void in the bottom of blind vias after plating.

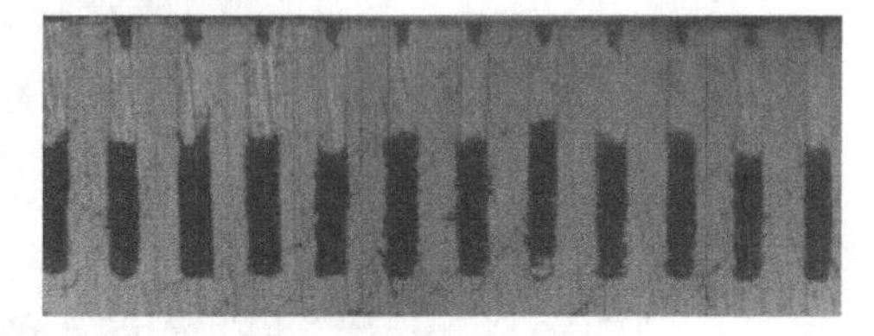

Figure 21.12 Void formed in the bottom of the blind vias by too fast inserting to the ethanol in 3-step prewetting process

And for higher aspect ratio of blind vias, the flowing time needed is longer and the inserting speed is needed to be slower. The experimental results show that, for 20μm × 200μm blind vias, a vertical inserting speed of less than about 5mm/s is OK for fully filling, yet for inserting speed of more than 5mm/s, unstable dimensions of void may occur in the TSV plating results, such as shown in Figure 21.12, which utilized a 8mm/s speed inserting into the ethanol.

21.2 Study of Annular Copper-Filled TSVs of Sensor and Interposer Chips for 3D Integration

In the conventional through silicon via (TSV) filling process for 3D integration, TSV is often void-free filled with copper by the electroplating process (Shi *et al.*, 2013; Jin *et al.*, 2013, 2016). Yet for the 3D integration of microelectromechanical systems (MEMS) sensor chips, the TSV plating process is not compatible with the chip fabricating process, and a rather complex process flow is needed to integrate MEMS sensor chips and interposers with TSV, including the TSV etching [deep reactive-ion etching (DRIE)], insulating layer formation [plasma-enhanced chemical vapor deposition (PECVD) or thermal oxidation], barrier layer and seed layer sputtering, copper filling (electroplating), double-face copper polishing, and surface circuit fabrication (sputtering and peeling); thus, it raised the cost and reduced the yield. Moreover, for MEMS sensor chips with special microstructures, such as pressure sensor chips, the microchambers with ultrathin top caps that may be destroyed by the copper chemical mechanical polishing (CMP) process needed after TSV plating. Besides, as the thermal mismatch of the copper pillar and the silicon chips, in thermal cycles and thermal shock environment, fully filled TSVs may lead to larger thermal stress in the interface of copper and silicon and may cause cracks in ultrathin chips.

Therefore, in this chapter, in order to simplify the chip fabricating process, TSV is annularly filled by the copper only utilizing sputtering. Samples of temperature-humidity sensor chips are fabricated by the conventional TSV electroplating process and annular sputtering process, the chips are thinned to 100μm, and 10μm-diameter TSV is etched and annularly filled by 2μm-thick copper from double side of the wafer. Finite-element analysis (FEA) simulation, thermal cycle test (TCT), and electrical conductivity property measuring are conducted to study the difference between the two kinds of chips. In the results, the TSV chips fabricated by the annular sputtering process show a better thermal-mechanical property and a slightly worse but enough electrical conductivity property.

21.2.1 Experiments

The scheme structure of a typical 3D MEMS sensor chips integration is shown in Figure 21.13. It contains temperature-humidity sensors and a pressure sensor with TSVs connected to an interposer. Then, the interposer is connected to application-specific integrated circuit (ASIC) and substrate by TSVs. Temperature and humidity sensors are fabricated by the conventional TSV electroplating process and annular sputtering process.

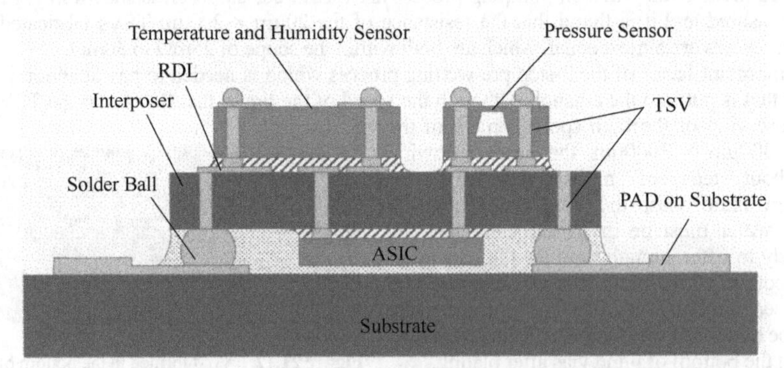

Figure 21.13 Schematic of the 3D MEMS sensor chips integration

The flow diagram of the conventional TSV filling process is shown in Figure 21.14(a); the approach contains more than 17 steps. Some steps are compressed to one step.

Step Ⅰ: 10μm blind via arrays are etched in the pad areas of the wafer by the DRIE process (surface technology systems); using the Bosch process, the etched depth is 120μm.

Step Ⅱ: The wafer is thinned and polished to 100μm and makes the blind vias into TSVs by an integrated wafer thinning process developed in (Zhou *et al.*, 2011; Li *et al.*, 2011, 2012). Then, 2μm SiO_2 is deposited onto the double-wafer surface and TSV walls to form the insulation layer.

Step Ⅲ: As the wafer thickness is too thin to support the photoresist spin coating, a carrier wafer is needed for temporary bonding of the device wafer to offer extra mechanical strength. Besides, the carrier can prevent the photoresist from sucking through TSV into the vacuum chuck when spin coating.

Step Ⅳ: Photoresist is spin coated onto the device wafer and lithographed to expose the pads needed to deposit the barrier layer and adhesive layer. After developing, the device wafer is debonded from the device wafer.

Step Ⅴ: 100nm TaN and 50nm Cr are sputtered onto the frontside surface to form the barrier layer and the adhesive layer.

Step Ⅵ: By ultrasonic bath in stripping liquid, the photoresist is removed, and the metal layer on it is peeled off.

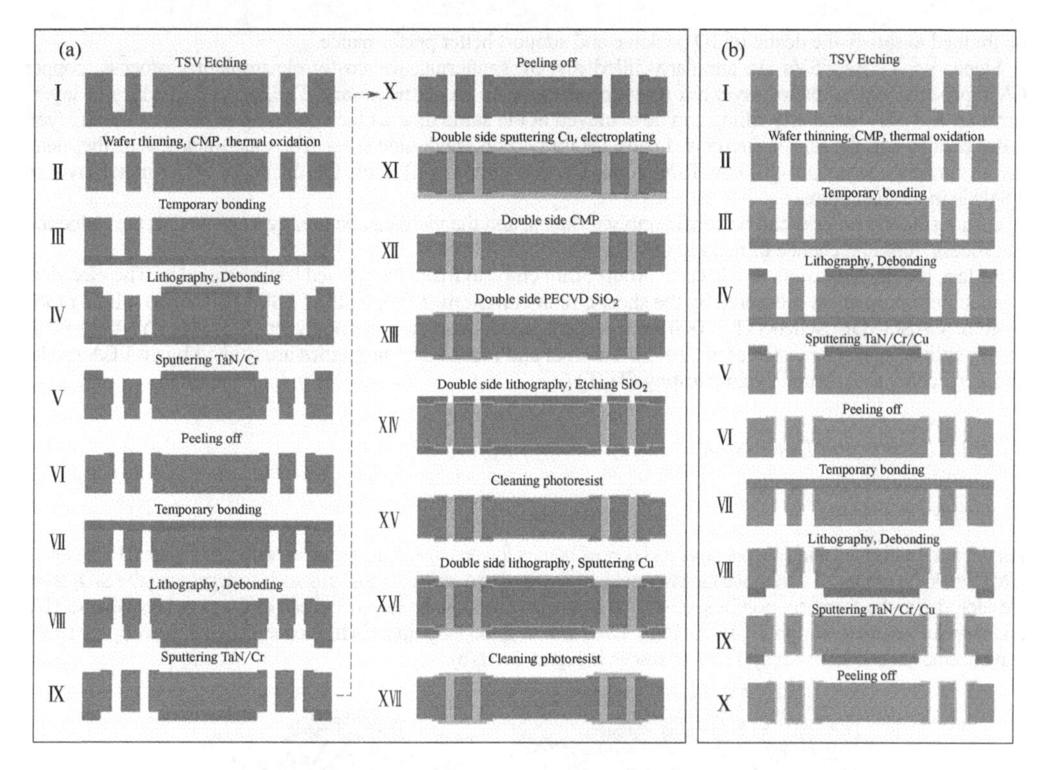

Figure 21.14 Flow diagram of temperature-humidity sensor chip fabricating process. (a) Conventional filling process; (b) annular filling process

Steps Ⅶ-Ⅹ: The same process as steps III-VI is conducted at the backside surface to form the barrier layer and the adhesive layer. As for the conventional magnetron sputtering equipment, the sputtered metal layer in the deeper area of TSV is much thinner than the top area, so double-side sputtering is needed to form the uniform thick barrier layer and adhesive layer.

Step Ⅺ: A 2-μm-thick copper seed layer is sputtered onto double surface of the device wafer, and TSV is filled by a double-side electroplating method presented developed in reference (Shi *et al.*, 2013). By adding appropriate additives, void-free filling can be achieved.

Step Ⅻ: CMP is conducted at the frontside and backside surfaces to remove the thick layer of copper and bumps formed in electroplating process. This is a depressing process that will also remove the SiO_2 insulating layer on the wafer surface after removing copper, and as the unflatness of the wafer can be more than 5μm, the removal of the SiO_2 is nearly unavoidable if the copper layer is completely removed.

Step XIII : As a result of the above, SiO_2 is needed to be deposited on to double-wafer surfaces by PECVD to reform the insulation layer.

Steps XIV-XV: To expose TSV from the SiO_2 layer, another lithographing process is needed to etch the SiO_2 covering TSVs.

Steps XVI-XVII: After the lengthy and complex TSV fabricating process, the circuit layer at the frontside and the bonding pad at the backside are finally fabricated by the sputtering and peeling-off processes.

As described above, due to the poor compatibility of sensor chips fabricating process and conventional TSV fabricating process, a rather long and complex process flow is needed to fabricate just these simple chips, if for sensor chips with more complex structures such as pressure sensors, it is nearly unimaginable how complex the fabricating process would be, not to mention the yield and cost.

To solve the problem, a simplified as well as low-cost annular copper sputtering process is implemented to fabricated temperature-humidity sensor chips with equal or better quality and reliably. The flow diagram of the process is shown as in Figure 21. 14(b); the approach contains about ten steps.

Steps Ⅰ-Ⅳ: These steps are all the same as the conventional process because the chip thickness needs to be thinned to satisfy the desire of 3D package and support better performance.

Steps V-X: As TSVs are annularly filled just by sputtering, the costly electroplating process, copper CMP process, and insulation layer reforming process are no longer necessary. The barrier and adhesive layers forming and TSV annularly filling can be achieved at the same time as the sputtering process of metal layer. After peeling off, the sensor circuit and pads are formed on the double sides. The total thickness of the metal layer on each side is 1μm (100nm TaN, 50nm Cr, and 850nm Cu); thus, the thickness of the metal layer in TSVs is more than 1μm.

As a result, the process can be considerably reduced, and the yield can be greatly improved as the reduction of process steps and chance of making mistakes.

At last, the device wafers are diced to 3mm×3mm chips to make the needed testing samples. The electrical conductivity property is measured by the short-circuit current measurements of TSVs performed with a probe system (CASCADE SUMMTI 11000) and a semiconductor parameter analyzer (KEITHLEY 4200-SCS) that is utilized in reference (Li *et al.*, 2015). Thermal and mechanical properties are studied by an FEA model and verified by the thermal cycling testing (TCT).

21.2.2 Results and Discussion

(i) Sample Observation

As the non-uniformed thick copper layer and copper bumps formed by electroplating as shown in Figure 21.15, copper CMP has to be processed to remove the layer, which may lead to a series of problems such as destroying the SiO_2 layer. Besides, due to the chemistry polishing effect, the copper in TSVs can be etched and causes the steps between the TSV and chip surface, as shown in Figure 21.16(a). Even worse, if too long-time CMP is conducted, all the copper in TSV can be removed and leave empty TSVs, as shown in Figure 21.16(b).

Figure 21.15 Optical microscopes images of the thick copper layer and copper bumps

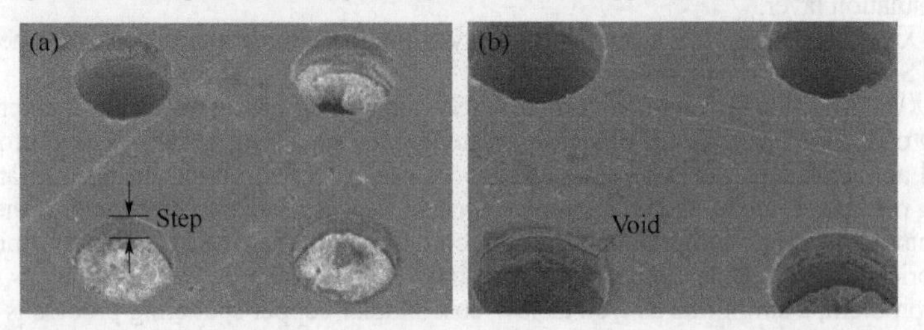

Figure 21.16 Scanning electron microscope images of TSVs after CMP. (a) Steps between TSV and chip surface; (b) empty TSVs

Therefore, the conclusion can be drawn that bonding pads with fully filled TSVs would not offer larger bonding area and better bonding strength than annularly filled TSV.

The optical microscope images of temperature-humidity sensor chips fabricated by annularly filled TSV are shown in Figure 21.17. In order to offer good electrical conductivity, a 7 × 7 TSV array is formed in each 150μm × 150μm pad, and the filling ratio can be calculated as

$$\lambda=1-\frac{\pi(r-t)^2 h}{\pi r^2 h}=1-\frac{(r-1)^2}{r^2} \tag{21.1}$$

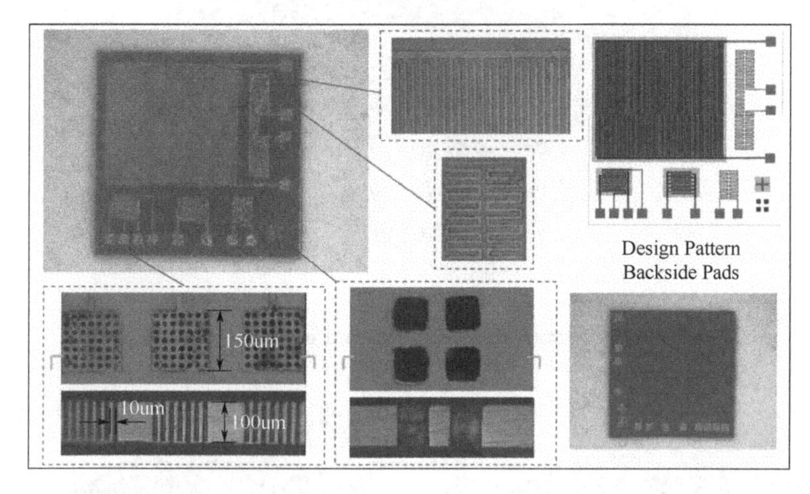

Figure 21.17 Optical microscope images of temperature-humidity sensor chips fabricated by annularly filled TSV

where λ represents the filling ratio, r represents the TSV radius (5μm), t represents the thickness of the metal layer (1–2μm), and h represents the thickness of the chip (100μm).

The value of the filling ratio is at least 0.36. When multiplying the value with the amount of TSVs (7×7 = 49), it can be drawn that the conductivity of the annularly filled TSV array of each pad is equal to a pad with 17.64 fully filled TSVs, which is good enough for electrical interconnection. In fact, in this case, it is the thin circuit layer itself instead of TSVs that limits the electrical conductivity.

(ii) Thermal-Mechanical Reliability

To compare the thermal-mechanical reliability of fully filled TSV chips and annularly filled chips, a thermal-cycle FEA model is built up by commercial computer software COMSOL to investigate the stress distribution in copper TSV and silicon substrate under high temperature (125℃) and low temperature (−45℃); the stress-free temperature is the room temperature (25℃). The temperature of the thermal cycling is from −45℃ to 125℃, following the sequence of 5 min ramping to hot, 30 min dwelling at hot, 5 min ramping to cold, and 30 min dwelling at cold. The intercepted pad area mesh model of annularly filled sample is shown in Figure 21.18, as the copper pad is symmetrical in the *x*-, *y*-, and *z*-directions, 1/8 symmetry simplified model is built up; thus, the dimension of the model is 100μm×100μm×50μm as is labeled, and the aperture and pitch of TSV are 10 and 20μm, which is consistent with the samples. Models of different annularly filled copper thicknesses, 1μm and 2μm, and fully filled TSV are built up.

The simulation results of high temperature are shown in Figure 21.19. It can be found that the maximum von-Mises stress in silicon substrate is located at the TSV edge near the chip surface, as shown in the dashed box. This is because the deformation of the silicon caused by the expansion of copper near the orifice area of TSV is the largest in TSV, as the coefficient of thermal expansion (CTE) of copper is much bigger than silicon. The maximum von-Mises stress in annularly filled copper TSV is located at the inside surface of the copper layer, as shown in the right-side dashed box in Figure 21.19(c). This is because the inside wall of the copper layer in TSV received the max extrusion deformation from the silicon substrate. The simulation results of low temperature are shown in Figure 21.20. It can be found that the von-Mises stress distribution is similar to high temperature, which is caused by the similar factor, the maximum stress in silicon substrate is located at the TSV edge near the chip surface, and the maximum von Mises stress in annularly filled copper TSV is located at the inside surface of the copper layer.

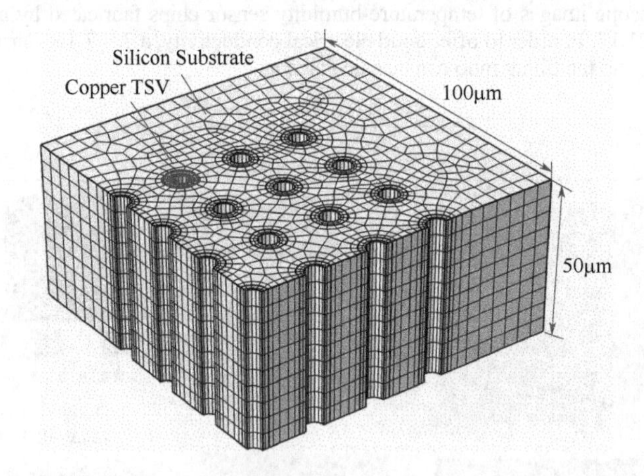

Figure 21.18 Intercepted pad area mesh model of annularly filled sample

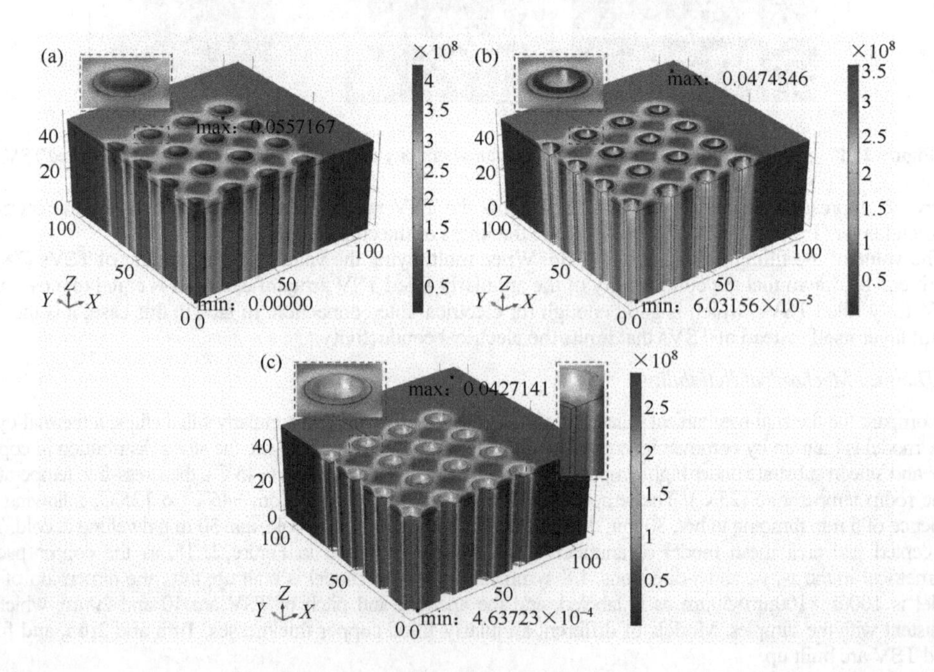

Figure 21.19 von-Mises stress (Pa) and deformation (μm) under high temperature (125°C). (a) Fully filled; (b) 2μm-thick annular filled; (c) 1μm-thick annular filled

The data of maximum von-Mises stress in silicon substrate and copper TSV under high temperature (125℃) and low temperature (−45℃) are shown in Figure 21.21. Under high temperatuire (125℃), when TSV is fully filled, the maximum stress in silicon substrate can reach up to 316.71MPa, which is a very dangerous value for silicon material, as the fracture strength of silicon is about 350MPa (different kinds of silicon wafer have different strengths) (Liu *et al.*, 2010; Zhou *et al.*, 2015).

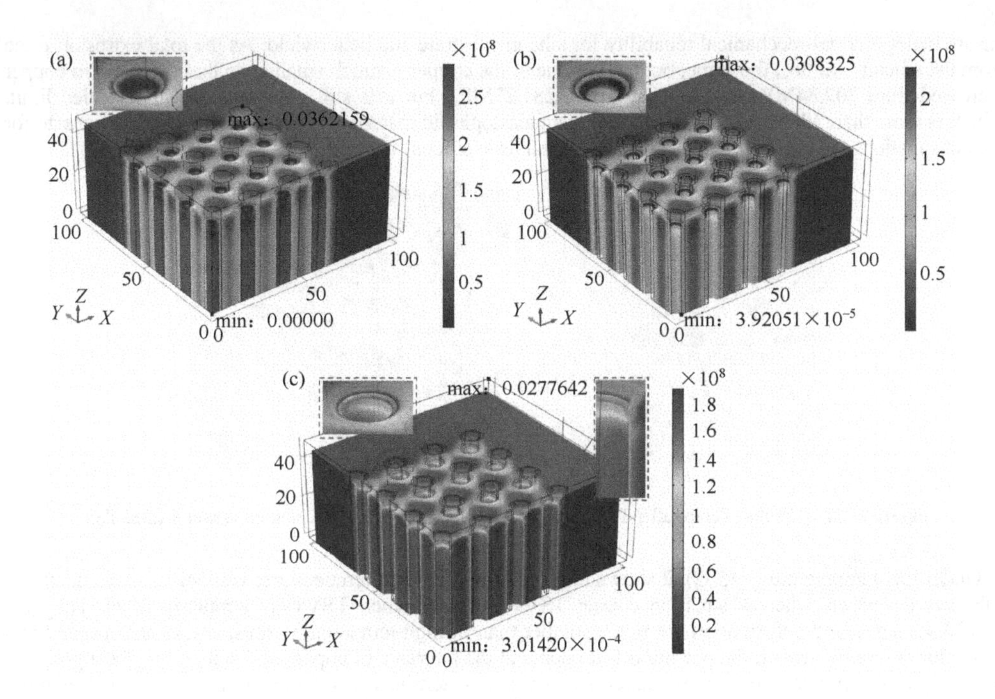

Figure 21.20 von-Mises stress (Pa) distribution and deformation (μm) under low temperature (−45℃). (a) Fully filled; (b) 2μm-thick annular filled; (c) 1μm-thick annular filled

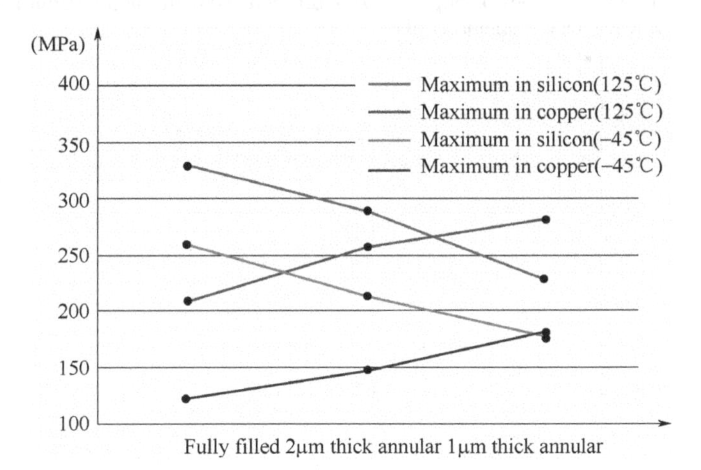

Figure 21.21 Maximum von-Mises stress in silicon and copper under high

Moreover, if microcracks or microdefects exist in the area, stress concentration will cause much greater stress and lead to chip fragmentation, as shown in Figure 21.22(a), which is a cracked fully filled TSV sample obtained in TCT. When TSV is annularly filled by 2μm-thick copper, as the thickness of copper is thinner, the total thermal expansion force applied to silicon is reduced, so the maximum von-Mises stress in silicon substrate is reduced significantly to 286.92MPa, and continually reduced to 229.13MPa when the copper thickness is reduced to 1μm, which is a much safer value and can ensure better thermal-mechanical reliability for silicon substrate and better yield. As the total extrusion force from the silicon is almost the same, but the volume of the copper is much smaller, so that the stress in copper increased from 207.84MPa to 255.24MPa and 281.27MPa, but it is still not more than the yielded limit, which is more than 300MPa; besides, copper is an elastoplastic material; its plastic stage can compensate for the deformation, and cracks are not so easy to appear as in silicon substrate.

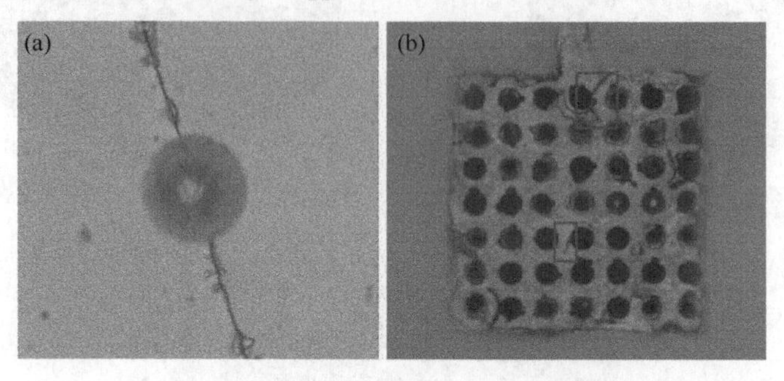

Figure 21.22 Cracks of chips after TCT. (a) Crack in fully filled TSV; (b) crack in annularly filled TSV

Under low temperature (−45°C), it shows the similar rule as high temperature, von-Mises stress in silicon substrate decreases, whereas stress in copper TSVs increases when TSVs are annularly filled. The max von-Mises stress and deformation are much smaller than in high temperature; thus, cracks are not the main factor for failure. However, the peeling off of copper at the interface of copper and silicon may be a potential reliability problem (because CTE of copper is larger, and its tensile stress in the interface), and the maximum von-Mises stress that occurs near the interface of the three models is recorded as 258.76MPa (full), 214.45MPa (2μm), and 178.64MPa (1μm), which means that the annularly filled sample is better.

As shown in Figure 21.23, the maximum deformation of silicon under high temperature (125℃) and low temperature (−45℃) is reduced from 0.056μm to 0.047μm and 0.042μm, and 0.036μm to 0.031μm and 0.028μm, respectively, when TSV is annularly filled.

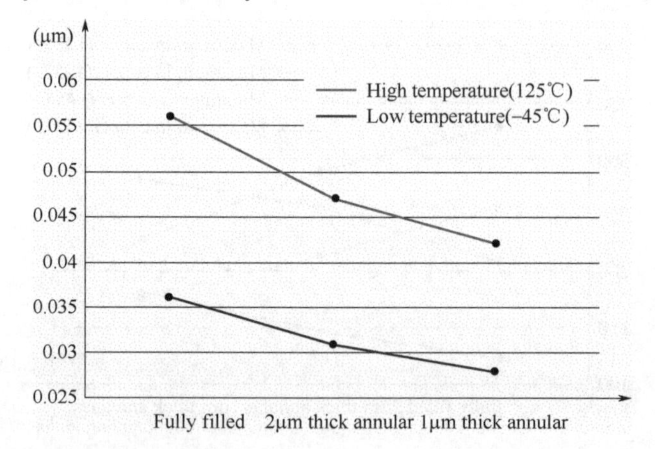

Figure 21.23 Maximum deformation under high temperature (125℃) and low temperature (−45℃)

Then, in TCT, 100 of both the two kinds of samples (fully filled and 1μm thick annularly filled) are placed in a thermal chamber. The temperature of the thermal cycling is from −45℃ to 125℃, following the sequence of 5min ramping to hot, 30min dwelling at hot, 5min ramping to cold, and 30min dwelling at cold. The samples are examined after 1000 cycles to detect the cracks in chips. After TCT, cracks appeared in seven of the fully filled samples, yet for annularly filled samples, the number of cracked chips is reduced down to two, which confirmed the simulation results. Besides, as shown in Figure 21.22(a), cracks in chips of fully filled TSVs are long cracks along the crystal direction, yet for chips of annularly filled TSVs, all of the cracks in TSV are small and are in random directions, as shown in Figure 21.22(b). This is because the stress in TSV is not strong enough to lead the cracks to further expand alone the crystal direction.

The Weibull distribution was a probability distribution function proposed by the Swedish scientist Weibull in 1951. The Weibull distribution is widely applicable for reliability analysis method. In this paper, ten chips, respectively, of fully filled TSV and 1μm-thick annularly filled TSV are put into TCT box until all of them are failed. The temperature of the thermal cycling is from −45°C to 125°C, following the sequence of 5 min ramping to hot, 5 min dwelling at hot, 5 min ramping to cold, and 5 min dwelling at cold. The chips are checked every 100 cycles, and based on the data, the three parameter Weibull distribution is calculated in MATLAB, and the results are shown in Figure 21.24. From Figure 21.24, it can be found that the chips of 1μm-thick annularly filled TSV have better reliability, which is consistent with the simulation results.

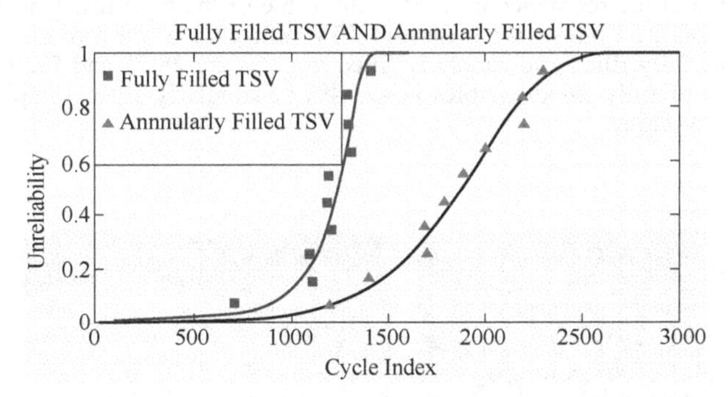

Figure 21.24 Three-parameter Weibull distribution curve of fully filled and annularly filled TSVs under TCT

(iii) Electrical Conductivity

For the non-uniformed thickness of copper after sputtering, the deeper, the thinner (Raynal *et al.*, 2009; Ledain *et al.*, 2008). For double-side sputtering by PVD, as shown in Figure 21.25(a), in an ideal situation, the thickness of the seed layer in all the areas of TSV is equal, but in the actual situation of PVD sputtering, as shown in Figure 21.25(b), in deeper area of the TSV, the copper layer will be much thinner. Therefore, by double-side sputtering, the copper seed layer in TSV is thinner than ideal, which leads to the resistance of TSVs in the fabricated samples that are larger than the ideal value. The gap of the resistance will be different because of different dimensions of TSV, different PVD equipments, and different processes. Therefore, the electrical conductivity challenges could be brought in.

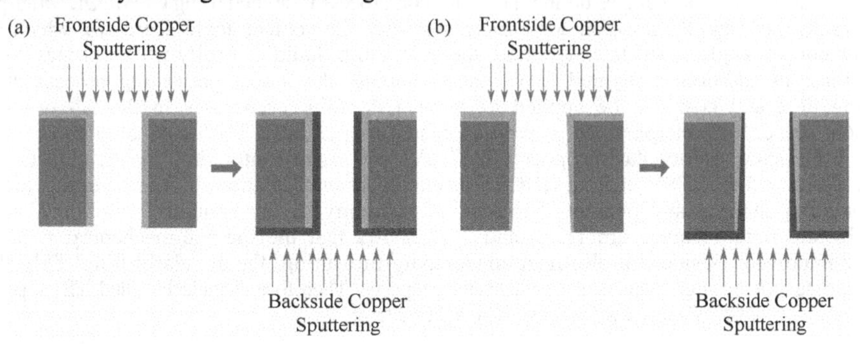

Figure 21.25 Non-uniformity of copper seed layer by PVD sputtering. (a) Copper seed layer under ideal situation; (b) actual copper seed layer by sputtering

Table 21.1 Measured TSV resistance of the two kinds of samples(mΩ)

No.	1	2	3	4	5	Average
Full	27.98	26.74	28.35	27.83	30.64	28.31
Annular	58.69	57.77	59.64	61.48	58.93	59.30

To verify the electrical conductivity of annularly filled TSVs, the fully filled and 2-μm-thick annularly filled chips are bonded to interposer chips with Cu-Sn pad alloying (Li, *et al.*, 2012). (A 1μm-thick Sn was evaporated onto the 2μm copper pads of the interposers.) Then, a probe system (CASCADE SUMMTI 11000) and a semiconductor parameter analyzer (KEITHLEY 4200-SCS) mentioned before are utilized to measure the resistance, as shown in Figure 21.26. The resistance of the circuit on the interposer and temperature and humidity sensor chips that link the probes and TSVs are measured, and the resistance of TSV is the value of the total resistance that subtracts the circuit resistance. Table 21.1 shows the measured TSV resistance of the two kinds of samples. The average values of the fully filled and annularly filled samples are 28.31 and 59.30mΩ, respectively. The TSVs resistance of fully filled samples is 47.74% of annularly filled samples, which is good enough for most sensor chips.

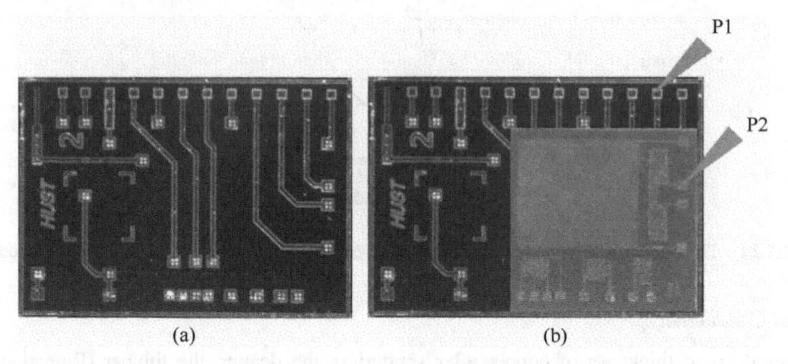

Figure 21.26 Measured structure of bonded chips. (a) Interposer chip; (b) interposer chip with temperature and humidity sensor chips

21.3 Chapter Summary

The effect of different contact angles during the infiltration process is studied using COMSOL Multi-physics, and the results show that the invasion depth increases when the contact angle decreases, which implies that if the contact angle is sufficiently small, the infiltrating liquid can fully fill blind vias with high aspect ratios. In addition, compared to vacuum pumping, the 3-step prewetting process with less processing time and cost can be applied to some TSV filling processes in the lab or industrial production. sensor and interposer chips with annularly copper filled TSVs and fully filled TSVs are fabricated. Utilizing the annularly copper filled TSVs can significantly simplify the chip fabricating process, which includes TSV etching (DRIE), insulating layer formation, barrier layer sputtering and copper layer sputtering, and peeling. The thermal reliability of the annularly filled TSV structure studied by the finite-element simulation and TCT shows that the thermal-mechanical reliability is markedly improved. Besides, in electrical conductivity measuring, the annularly filled TSV shows a slightly worse but enough electrical conductivity property. Therefore, annularly filled TSVs process is better than fully filled TSVs process for sensor chips fabricating.

References

Broekmann, P., *et al.* (2011) Classification of suppressor additives based on synergistic and antagonistic ensemble effects,

Electrochimica Acta, vol. 56, no. 13, pp. 4724-4734.

Bryan, B., *et al.* (2006) Die stacking (3D) microarchitecture, in *Proc. The 39th Annual IEEE/ACM International Symposium on Microarchitecture*, 469-479.

Dong, X., Wu, X., Sun, Xie, G. Y., Li, H., and Chen, Y. (2008) Circuit and microarchitecture evaluation of 3D stacking magnetic RAM (MRAM) as a universal memory replacement, in *Proc. IEEE 5th ACM Design Automation Conference*, Jun. 2008, pp. 554-559.

Fursenko, O., Bauer, J., Marschmeyer, S., and Stoll, H. P. (2015) Through silicon via profile metrology of Bosch etching process based on spectroscopic reflectometry, *Microelectronic Engineering*, vol. 139, pp. 70-75.

Jin, S., Seo, S., Park, S., and Yoo, B. (2016) Through-silicon-via (TSV) filling by electrodeposition with pulse-reverse current, *Microelectronic Engineering*, vol. 156, pp. 15-18, Apr. 2016.

Jin, S., Wang, G., and Yoo, B. (2013) Through-silicon-via (TSV) filling by electrodeposition of Cu with pulse current at ultra-short duty cycle, *Journal of The Electrochemical Society.*, vol. 160, no. 12, pp. D3300-D3305, 2013.

Kgil, T., *et al.* (2006) PicoServer: Using 3D stacking technology to enable a compact energy efficient chip multiprocessor, in *Proc. 12th International Conference on Architectural Support for Programming Languages and Operating Systems*, pp. 117-128, Nov. 2006.

Knickerbocker, J. U., *et al.* (2008) Three-dimensional silicon integration, *IBM Journal of Research and Development*, vol. 52, no. 6, 553-569, Nov.

Kosemura, D., and Wolf, I. D. (2015) Three-dimensional micro-Raman spectroscopy mapping of stress induced in Si by Cu-filled through-Si vias, *Applied Physics Letters*, vol. 106, no. 19, Aug. 2015, Art. no.191901.

Lau, J. H. (2011) Evolution, challenge, and outlook of TSV, 3D IC integration and 3D silicon integration, in *International Symposium on Advanced Packaging Materials: Processes, Properties and Interface*, 462-488.

Lau, J. H. (2011) Overview and outlook of through-silicon via (TSV) and 3D integrations, *Microelectronics International* vol. 28, no. 2, pp. 8-22.

Ledain, S., *et al.* (2008) An evaluation of electrografted copper seed layers for enhanced metallization of deep TSV structures, in *Proc. International Interconnect Technology Conference (IITC)*, Burlingame, CA, USA, 159-161.

Li, C. Zhou, S., Chen, R., Peng, T., Wang, X., and Liu, S. (2012) Integrated wafer thinning process with TSV electroplating for 3D stacking, in *Proc. 13th International Conference on Electronic Packaging Technology & High Density Packaging*, Guilin, China, 945-948.

Li, C., Wang, X., Chen, M., Zhou, S., Lv, Y., and Liu, S. (2013) Novel design and reliability assessment of a 3D DRAM stacking based on Cu-Sn microbump bonding and TSV interconnection technology, in *Proc. IEEE 63rd Electronic Components and Technology Conference*, May 2013, pp. 1861-1865.

Li, C., Wang, X., Song, S. and Liu, S. (2015) 21-layer 3-D chip stacking based on Cu-Sn bump bonding, *IEEE Transactions on Components, Packaging and Manufacturing Technology* vol. 5, no. 5, pp. 627-635.

Li, C., Zhou, S. Hu, C., Wang, X., and Liu, S. (2011) A novel design of handling system for silicon wafer thinning, in *Proc. 12th International Conference on Electronic Packaging Technology & High Density Packaging*, Shanghai, China, 1-4.

Liu, S., Liu, Y. (2010) *Modeling and Simulation for Microelectronic Packaging Assembly: Manufacturing, Reliability and Testing*. Hoboken, NJ, USA: Wiley, pp. 443-474.

Ranganathan, N., Lee, D. Y., Youhe, L., Lo, G. Q., Prasad, K., and Pey, K. L. (2011) Influence of Bosch Etch Process on Electrical Isolation of TSV Structures, *IEEE Transactions on Components, Packaging and Manufacturing Technology*, vol. 1, no. 10,1497-1507.

Raynal, F., *et al.* (2009) Electrografted seed layers for metallization of deep TSV structures, in *Proc. 59th Electronic Components and Technology Conference*, San Diego, CA, USA, 1147-1152.

Ryu, S. K., *et al.* (2012) Characterization of thermal stresses in through-silicon vias for three-dimensional interconnects by bending beam technique, *Applied Physics Letters*, vol. 100, no. 4, Jan. 2012, Art. no. 041901.

S. Zhou, C. Liu, X. Wang, X. Luo, and S. Liu. (2011) Integrated process for silicon wafer thinning, in *Proc. IEEE 61st Electronic Components and Technology Conference*, Orlando, FL, USA, 1811-1814.

Shi, S., Wang, X., Xu, C., Yuan, J., Fang, J., and Liu, S. (2013) Simulation and fabrication of two Cu TSV electroplating methods for wafer-level 3D integrated circuits packaging, *Sensors and Actuators A: Physical*, vol. 203, 52-61.

Singulani, A. P., Ceric, H., and Selberherr, S. (2013) Stress evolution in the metal layers of TSVs with Bosch scallops, *Microelectronics Reliability*, vol. 53, nos. 9-11, 1602-1605.

Tsai, H. C., Chang, Y.C., and Wu, P.W. (2010) Rapid galvanostatic determination on levelers for superfilling in Cu electroplating, *Electrochemical and Solid-State Letters*, vol. 13, no. 2, D7-D10.

Vereecken, P. M., Binstead, R. A., Deligianni, H., and Andricacos, P. C. (2005) The chemistry of additives in damascene copper plating, *IBM Journal of Research and Development*, vol. 49, no. 1, 3-18.

Wei, T., *et al.* (2012) Copper filling process for small diameter, high aspect ratio Through Silicon Via (TSV), in *Proc. IEEE*

13th International Conference on Electronic Packaging Technology & High Density Packaging, 483-487.

Zhang, J., Luo, W., Li, Y., Gao, L., and Li, M. (2015) Wetting process of copper filling in through silicon vias, *Applied Surface Science*, vol. 359, 736-741.

Zhou, L., *et al.* (2015) Fracture strength of silicon wafer after different wafer treatment methods, in *Proc. 16th International Conference on Electronic Packaging Technology & High Density Packaging*, Changsha, China, 871-874.

Zhu, Q. S., Toda, A., Zhang, Y., Itoh, T., and Maeda, R. (2014) Void-Free Copper Filling of Through Silicon Via by Periodic Pulse Reverse Electrodeposition, *Journal of The Electrochemical Society*, vol. 161, no. 5, pp. D263-D268, 2014.

Part IV

Modern Modeling and Simulation Methodologies: Application to Nano Packaging

22

Classical Molecular Dynamics

Molecular dynamics method has been developed by Alder, B. J., and Wainwright, T. E., from the 1950s (Alder *et al.*, 1957). In the 1970s, the molecular dynamics method experienced rapid development and wide applications. It has become an important approach to computer modeling and simulation (Binder *et al.*, 2004). The molecular dynamics method can handle a lot of issues, such as interface issues (Slawomir *et al.*, 1996; Watanabe *et al.*, 2004), fluid transport properties, materials, surface melting (Lutsko *et al.*, 1989; Wang *et al.*, 1997), which can not only calculate the equilibrium systems but can also analyze and calculate the non-equilibrium processes(Kirkpatriek, 1984, 1983). These advantages make molecular dynamics very useful in physics, chemistry, and materials sciences and have attracted considerable attention.

The molecular dynamics method calculates and determines system configuration changes by intrinsic kinetics. In this method, the equations of motion are established for a group of molecules, and then the coordinates and momentum of individual molecules are obtained by the numerical solution of the system equation of motion, finally a static or dynamic nature of a many-body system is calculated by statistical mechanics, and thus the macro-system properties are obtained.

Due to the calculation capacity restriction of current computers, the simulated system is much smaller than the size of the actual system, which means that the applications of the molecular dynamics method have two basic limitations: one is the time, the other is the size limit. The calculated results can be close to the actual situation by building a reasonable calculation model. In this chapter, the MD is applied to some fundamental processes invarious low dimensional material systems and thin film systems potentially applicable to emerging systems in engineering and nanoelectronics.

22.1 General Description of Molecular Dynamics Method

The molecular dynamics simulation method is based on Newton's second law for the equation of motion, $F = ma$, where F is the force exerted on the particle, m is its mass of the particle, and a is its acceleration applied on the particle. From the value of the force applied on each atom, it is possible to determine the acceleration of each atom in the simulation system.

The basic principles of classical molecular dynamics are as follows. For a system with N atoms, the system state is described by the N atoms of the position $\{r_i\}$ and velocity $\{v_i\}$. Energy of the system is expressed as $H(\{r_i, v_i\})$, $V(\{r_i\})$ is the interaction potential energy function between atoms, the system equations of the motion are defined as:

$$H = \frac{1}{2}\sum_{i=1}^{N} m_i v_i^2 + V(\{r_i\}) \tag{22.1}$$

Modeling and Simulation for Microelectronic Packaging and Integration: Manufacturing, Reliability and Testing, Second Edition. Sheng Liu and Yong Liu.

$$m_i \frac{d^2}{dt^2} r_i = -\frac{\partial}{\partial r_i} V(\{r_i\}) \tag{22.2}$$

For a particle i, the surrounding particles j was applied with a force on it, if we describe the force by interaction potential energy, then the potential energy of particle I is as follows:

$$u(r_i) = \sum_j u(r_{ij}) \tag{22.3}$$

The forces applied on particle i induced by surrounding particles are as follows:

$$\vec{f}_i = -\nabla u(r_i) \tag{22.4}$$

According to Newton's law, the position and velocity of particle in the next moment are as follows:

$$m\vec{a}_i = m\frac{\partial \vec{v}_i}{\partial t} = m\frac{\partial^2 \vec{r}_i}{\partial t^2} = \vec{f}_i \tag{22.5}$$

After a period of movement time with a certain velocity $\vec{v}_i$, we can obtain a new particle distribution. By solving the above equation we can obtain the trajectories of atoms in the space.

22.2 Mechanism of Carbon Nanotube Welding onto the Metal

In this section, we use classical molecular dynamics for studying the welding process of carbon nanotube onto the metal surface. The relationship between surface melting and the contact length of the carbon nanotube and the metal is investigated. The results indicate that melting firstly occurs on the surface of the metal as the temperature is rising. While the contact length increases with the melting propagating from the open surface to the interior of the metal. In addition, the wetting property plays an important role during the welding of the CNT onto the metal. We use these results in order to illustrate the possibility of welding the CNT onto the metal under controllable local energy.

22.2.1 Computational Methodology

For molecular dynamics simulation of welding CNT onto the metal surface, the setup of the simulations performed in the present work is shown in Figure 22.1, where a 3.566 nm long (5, 5) open ended carbon nanotube with 350 atoms forms an interface on the free surface of the nickel (Ni) slab with a set of 6200 atoms distributed in 31 layers with 200 atoms per layer. Periodic boundary conditions are imposed in x and y directions, and free boundary conditions in the z direction are imposed. Our simulations are based on the massively parallel LAMMPS code (Arcidiacono *et al.*, 2005).

The interactions for Ni–Ni are modeled by the embedded atomic method (EAM) models which have been successfully used in modeling various bulk metals (Hoover, 1986), while the C–Ni interactions are modeled with Lennard–Jones 12–6 potential defined by:

$$u_{ij} = 4\varepsilon_{ij}\left[\left(\frac{\sigma_{ij}}{r_{ij}}\right)^{12} - \left(\frac{\sigma_{ij}}{r_{ij}}\right)^{6}\right] \tag{22.6}$$

where r_{ij} denotes the scalar distance between sites i and j, σ_{ij} and ε_{ij} are the interaction parameters. The parameters of σ_{ij} and ε_{ij} are obtained from the literature (Hoover, 1985) ($\sigma_{\text{C-NI}} = 2.45$Å, $\varepsilon_{\text{C-NI}} = 0.456$ eV). A cut-off distance of 9Å is used for Lennard–Jones forces. The carbon nanotube is assumed to be rigid for computational efficiency. This means that for each time step, the total force and torque on the carbon nanotube is computed and the coordinates and velocities of the atoms in the carbon nanotube are updated so that they move as a rigid body. Accounting for such interactions in test runs cannot materially alter the results while dramatically increasing the computational time to reach equilibrium. In addition, to avoid fluctuation of the CNT in the x and y directions, the forces on the CNT of these two components are omitted from the present simulation.

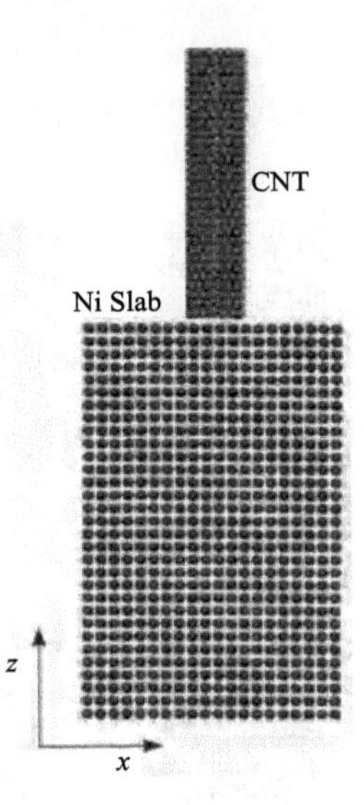

Figure 22.1 Schematic diagram of the simulated system

For each temperature, the constant NPT (N is the number of the atoms, P is pressure; T is temperature) integration is performed for a bulk Ni system to calculate the lattice constants at desired temperatures using a Nose–Hoover temperature thermostat (Zhang *et al.*, 2000) and Nose–Hoover pressure barostat (Chen *et al.*, 2005) and setting the bulk pressure to zero. And then the NI surface and CNT systems are created with these lattice constants corresponding to the temperatures. The number of layers in each computational cell is chosen such that the top and the bottom surfaces are beyond the interaction (Fujishima *et al.*, 2003). Under the conditions of constant volume and constant temperature (NVT), these Ni–CNT systems are equilibrated to the desired temperatures using a Nose–Hoover temperature thermostat (Zhang *et al.*, 2000) with a time step of 1 fs for 200 ps. Afterwards, the systems are left undisturbed in 300 ps constant-energy simulations during which the coordinates and velocities of the atoms are stored for analyses of structural and dynamical properties of the systems.

22.2.2 *Results and Discussion*

Figure 22.2 shows results for the contact length of the CNT and the metal at different temperatures in the range from 1400 K to 1550 K. Here the contact length represents the length of the CNT embedded into the metal. At a lower temperature of 1400 K or below, the metal can preserve its crystalline and there is no melting behavior. Hence, the CNT cannot be embedded into the metal and therefore there is a weak contact between the CNT and the metal. While at the temperature of 1450 K, some disordered atoms begin to appear on the surface of the metal, the CNT and metal atoms get closer. Beyond this temperature the configuration of the first layer becomes completely disordered and the second layer structure keeps ordered partly, which indicates that surface melting occurs and a liquid layer comes into being. As a result, the metal atoms begin to wet the CNT surface and part of the CNT is embedded into the metal. As the temperature rises, the liquid layer spreads inward, which means that more metal atoms are able to wet the CNT surface. Consequently, with increasing temperature the contact length increases until most of the CNT atoms are embedded into the metal at the temperature of 1550 K.

In the light of these facts, we conclude that in order to weld the CNT onto the metal, two conditions should be satisfied. First, the metal should be melting. From our simulation, the melting process can be divided into two stages, that is, surface melting and melting as the temperature increases. It is indicated that the welding can occur during both stages depending on the energy exerted on the metal. The obtained results are similar with previous experimental ones which indicate that the contact resistance decreases

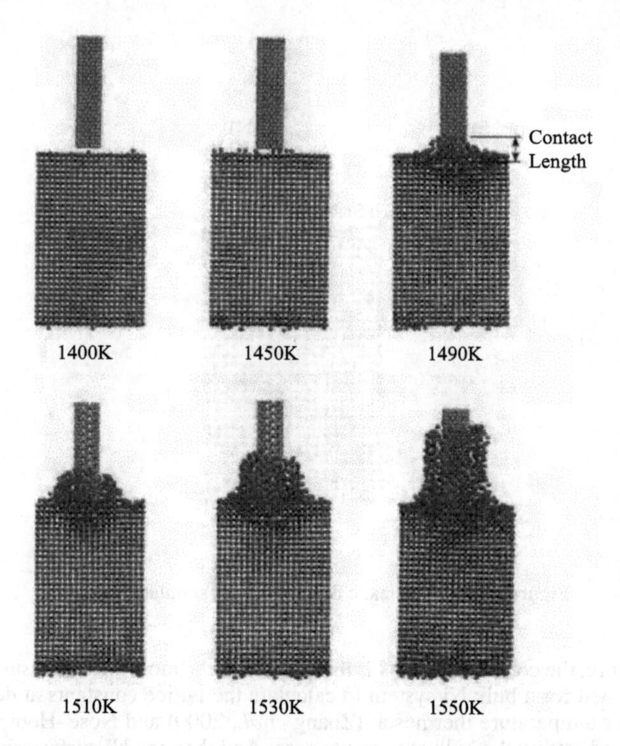

Figure 22.2 Schematic of snapshots for configurations of Ni-CNT system at different temperature

more dramatically with larger electric or ultrasonic power. In addition, the metal should wet the CNT surface. In the wetting case, the CNT can be easily embedded into the metal. Otherwise, it is difficult to weld the CNT onto the metal even though it is at a higher temperature.

In order to study the surface melting behavior and the relation between the order at the surface and the temperature effects, the layer structure factor is defined by (Fujishima *et al.*, 2003):

$$S_l(q) = \left| \left\langle \frac{1}{N_l} \sum_{i \in l} e^{-i\vec{q}.\vec{R}} \right\rangle \right| \tag{22.7}$$

where $S_l(q)$ is a measurement of the order in the nuclear positions along q in layer l, N is the number of the atoms in the layer l, and $\langle\rangle$ denotes the ensemble averages, q is the two dimensional reciprocal lattice vector. The unit of $S_l(q)$ indicates a perfect order solid and being zero of $S_l(q)$ refers to the surface melting. The calculated three uppermost layer structure factors in the equilibrium state as a function of temperature for metal surface are shown in Figure 22.3. As shown in Figure 22.3, with the increasing of the temperature, the outermost layer atoms lose their crystalline order gradually as $S_l(q)$ decreasing until $S_l(q)$ becomes zero which indicates total melting of this layer. The inner layers exhibit the same trends but a higher temperature is needed. During this process, the disordered atoms begin to wet the CNT surface. As shown in Figure 22.2, at higher temperature, the contact length is larger than that at lower temperature due to more atoms becoming disordered at this temperature. Therefore, the contact length is primarily determined by the number of the disordered atoms which result from the surface melting. It is also noted that full melting of the metal is not necessary for nano-welding, as the surface melting can provide enough disordered atoms for wetting the CNT surface. That is why previous experiments can obtain better contact without destroying the electrodes. Especially for local point annealing and ultrasonic method, the energy only heats the metal surface which induces high temperature onto the metal surface and therefore the surface melting is enhanced.

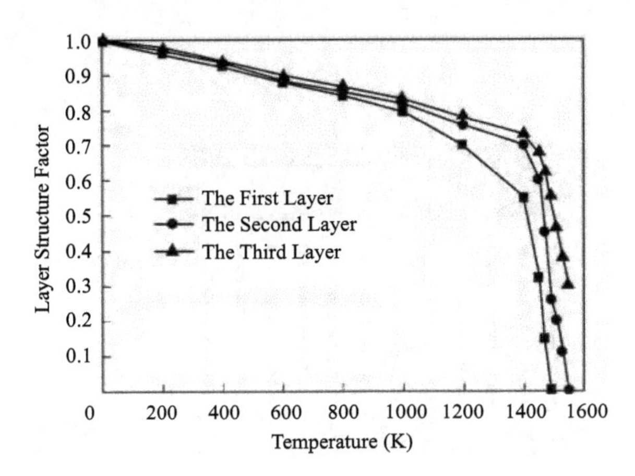

Figure 22.3 Layer structure factor as a function of the temperature for the uppermost three layers of the Ni slab

In addition, the wetting property of the metal atoms to the CNT surface is another factor which determines the welding process. According to Equation 22.7, we can change the wetting ability of the metal atoms to the CNT surface by varying the relative energy parameter $\varepsilon_{\text{C-metal}}$ (larger $\varepsilon_{\text{C-metal}}$ means good affinity of metal atoms to the CNT surface). Figure 13.7 presents data obtained by molecular dynamic calculation. As seen from Figure 13.7, the contact length decreases as the relative energy parameter becomes small. Since the small energy parameter indicates that metal atoms will be difficult to stick to the CNT surface even at high temperature, larger contact length cannot be obtained during welding the CNT onto the metal which results in weak contact bonding. A previous experimental study has demonstrated that the metal with excellent wetting property such as Ti, Ni, and Pd can form continuous or quasi-continuous metal–CNT contact, while the metal with bad wetting property such as Au, Al, and Fe only form isolated discrete particles on the CNT 19. These results indicate that the metal with better wetting ability to the CNT surface can obtain larger contact length during the welding. Therefore, during the welding process of the CNT onto the metal, the type of the metal should be selected properly or an effective method is utilized to enhance the wetting property of the metal atoms to the CNT surface.

Several different methods of bonding between CNTs and metallic substrates have been developed such as low temperature solder method (Lan *et al.*, 2008; Kumar *et al.*, 2006), conductive polymer composites method and thermocompression bonding. In all these methods, the main point is coating the substrate with conductive materials which is capable of welding the CNTs onto the substrate at low temperature in a melting state (Lan *et al.*, 2008; Kumar *et al.*, 2006; Jiang *et al.*, 2007) or solid state. For solid-state thermocompression bonding as shown in Figure 22.4, CNTs are coated by the metal followed by bonding the CNT–metal structure to the metallic substrates. This process allows contacting surfaces to be joined under pressure at much lower temperature via interdiffusion of metal atoms across the face of the joint, during which unexpected phases at the bond interface are not involved and expensive solders, fluxes, or shielding gases are not needed (Johnson *et al.*, 2009). Unlike the conventional solid-state diffusion bonding process, the nanoscale structures at the CNT–metal interfaces may strongly affect the bonding mechanism. In this case, the interplay between mechanical and thermal diffusion process needs further understanding. Computer simulations may provide powerful methods to investigate the structural and dynamical properties of interfaces which otherwise are very difficult to study by means of experiments. Modeling of this process can be used to identify the mechanism of CNTs bonding which may be used to optimize the process window to obtain reliable contact bonding.

First, the forming process of Ni–CNT contact interface on the CNT tip is studied. Initially, CNT forms an interface on the free surface of the nickel (Ni) cluster with a set of 1099 atoms as shown in Figure 22.5. Free boundary conditions are imposed in all directions. Under the conditions of constant volume and constant temperature (NVT), this Ni–CNT system is equilibrated to the desired temperatures using a Nose–Hoover temperature thermostat (Hoover, 1986) with a time step of 0.1 fs for 500 ps.

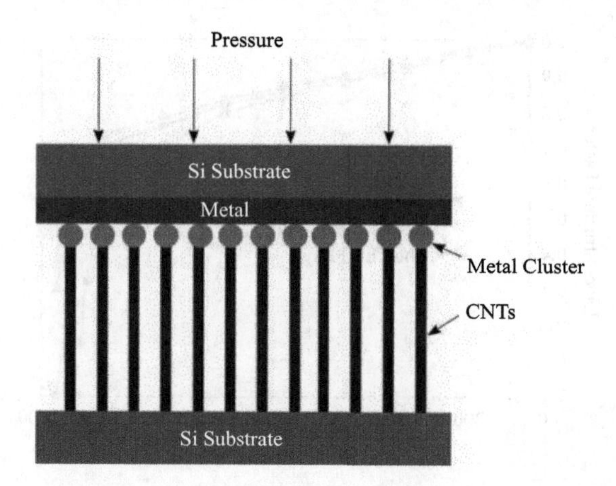

Figure 22.4 Schematic for concept of thermocompression bonding of carbon nanotubes to metallic substrates

Next, the obtained Ni–CNT system forms an interface on the free surface of Al slab with a set of 11625 atoms. Periodic boundary conditions are imposed in transverse directions, and free boundary condition in the z direction is imposed. For each atomic model, the constant NPT (N is the number of the atoms, P is pressure, and T is temperature) integration is performed for a bulk Al system to calculate the lattice constants at desired temperatures using a Nose–Hoover temperature thermostat and Nose–Hoover pressure barostat (Hoover, 1985) and setting the bulk pressure to zero. And then the Al surface and CNT systems are created with these lattice constants corresponding to the temperatures. The number of layers in each computational cell are chosen such that the top and the bottom surfaces are beyond the interaction (Yang *et al.*, 2007). Under the conditions of constant volume and constant temperature (NVT), these CNT–Ni–Al systems are equilibrated to the desired temperatures using a Nose–Hoover temperature thermostat (Brenner *et al.*, 2002) with a time step of 0.1 fs for 1000 ps. The external pressure is applied in the z direction by applying an additional external force on CNT atoms in such a way that every C atom experiences the same force.

Figure 22.5 shows the results of the Ni–CNT interfacial structure obtained by placing a nickel cluster cut from the bulk nickel face centered cubic (fcc) structure and equilibrating the system at temperatures between 1000 and 1050 K. The initial Ni cluster geometry is fcc as shown in Figure 22.5. After equilibration at a constant temperature, the cluster becomes more spherical. As the temperature increases, the metal atoms in the cluster begin to wet the CNT surface. At higher temperature levels, the larger contact length can be observed.

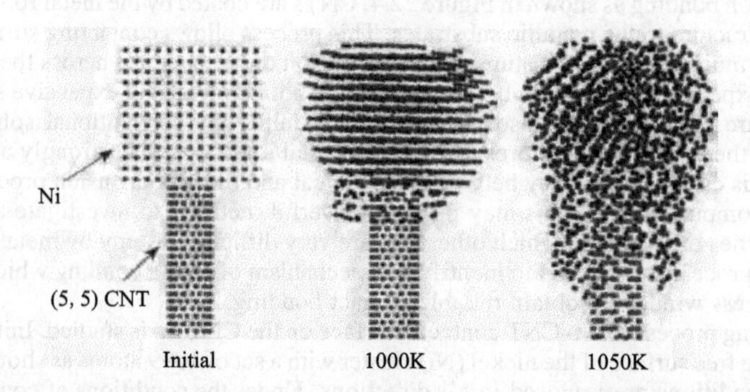

Figure 22.5 Schematic of snapshots for time averaged configurations of Ni–(5, 5) CNT system at different temperature levels

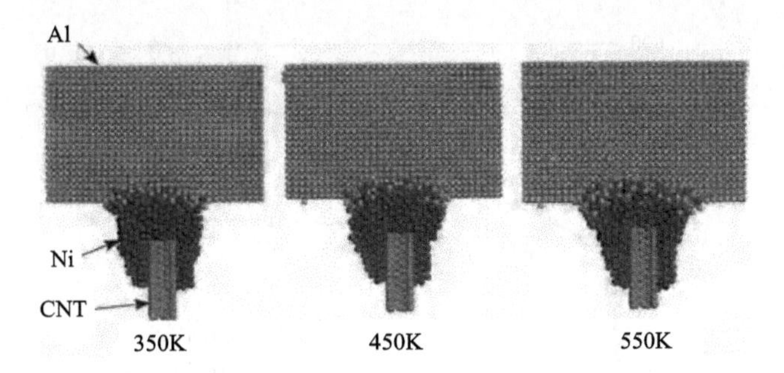

Figure 22.6 Schematic of snapshots for time averaged configurations of a cross-section of CNT–Ni–Al system at different temperature levels after 500 ps

Due to the large ratio of surface to bulk atoms of nanoscale metal cluster, the high surface energy will drive the cluster to become spherical. In addition, the small cluster has a much lower melting point than the bulk material. Particularly, the surface melting behavior will dominate the melting process which induces metal atoms on the surface of the cluster to become disordered at a much lower temperature. With the increasing of the temperature, the melting propagates from the surface to the inner part. Consequently, more metal atoms on the surface of the cluster become disordered, and these disordered atoms can wet the CNT surface to realize coalescence of CNT and the metal cluster. This process is critical for the bonding. For one thing, the nanoscale metal cluster has a high surface energy which can enhance the diffusion bonding of the metal atoms to the substrate and therefore low temperature bonding is achievable. In addition, to achieve higher bonding strength, the contact length between the metal cluster and CNT should be large enough to avoid debonding of CNT from the cluster.

Figure 22.6 shows the configuration of a cross-section of the structure after 500 ps of diffusion at different temperatures with a constant pressure of 1 MPa. At 350 K, the interfacial diffusion between the Ni cluster and Al slab occurs, however, only a small number of Al atoms have diffused into the Ni side. While at 450 K, more Al atoms have diffused into the Ni side. When the temperature is higher than 450 K, a more significant diffusion of Al atoms into Ni cluster is seen. To characterize the interfacial region during diffusion bonding, we count the number of alloyed Al atoms which indicate the quantity of Al–NI alloy bonds formation induced by the diffusion. Figure 22.7 shows the number of alloyed Al atoms as a function of time at different temperatures. At a lower temperature of 350 K, the number of alloyed Al atoms increases at the first 20 ps and the diffusion becomes more slowly thereafter until there is no further increase after 100 ps. At a higher temperature of 450 K and 550 K, the diffusion experiences a similar process, and the difference is that the diffusion is more rapid and the number of alloyed Al atoms can increase continuously without saturation over the duration of calculation.

The results indicate that the diffusion between the Ni cluster and Al slab can occur at a low temperature for CNT–Ni–Ai system. As the temperature increases, the diffusion becomes more rapid. Unlike the diffusion between two bulk materials, the nanoscale cluster provides a high energy surface on which the atoms are very active to form an alloy with the contacted bulk metal material even at much lower temperature. Initially, these surface atoms coalesce with Al atoms and form Ni–Al alloy at the contact interface. As a result, the number of alloyed Al atoms increases rapidly at the initial stage as shown in Figure 22.7. Afterwards, as the increase becomes slower, this process may be determined by the diffusion mechanism driven by thermal movement. Higher temperature will fascinate forming vacancies by breaking the bonds and driving the metal atoms into these vacancies.

Figure 22.8 presents the number of alloyed Al atoms as a function of time at different pressures with a constant temperature of 450 K. When there is no external pressure applied on the CNT, the number of alloyed Al atoms increases rapidly at the initial 30 ps and the diffusion becomes more slowly thereafter until there is no further increase after 100 ps. As the system is under higher pressure of 1 MPa and 2 MPa, the diffusion rate increases as the pressure increases and the saturation will continue for a longer time at larger external pressures. Unlike the effects of the temperature, the results show that the pressure will facilitate the initial interfacial diffusion process. The diffusion rate at the initial stage under higher pressures is higher than that without external pressure.

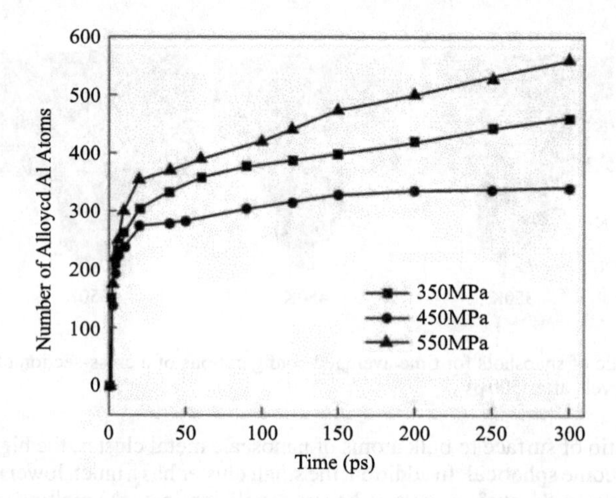

Figure 22.7 Number of alloyed Al atoms versus time under different temperature

Like the effects of the temperature, the pressure is another driving force of the diffusion. On one hand, the pressure can reduce the distance between the Ni cluster and Al slab, making more Ni and Al atoms get closer at the contact interface and therefore facilitating the initial contact bonding. From an experimental point of view, the external pressure can help more CNTs with different length get close contact with the objective substrate, which will result in larger bonding strength and more transport channels of electrons and heat. In addition, during the diffusion process, the stress induced by the external pressure will be a driving force to break the metal bonds and form vacancies, making it easier for interdiffusion (Mishin, 2004). Therefore, the pressure is an effective factor to enhance sufficient interfacial diffusion of the CNT–Ni–Al system. However, higher pressure may result in deformation, even buckling of CNTs, which indicates that we cannot use too high a pressure in the experiments (Dong *et al.*, 2007).

In order to examine the mechanical properties and the failure modes during the debonding process, tensile loading is applied at 300 K by fixing one end of the Al slab and moving three layers of C atoms in the z direction as shown in Figure 22.9. To effect the tensile deformation, the displacement of the boundary atoms is controlled by time steps. The simulations may be viewed as

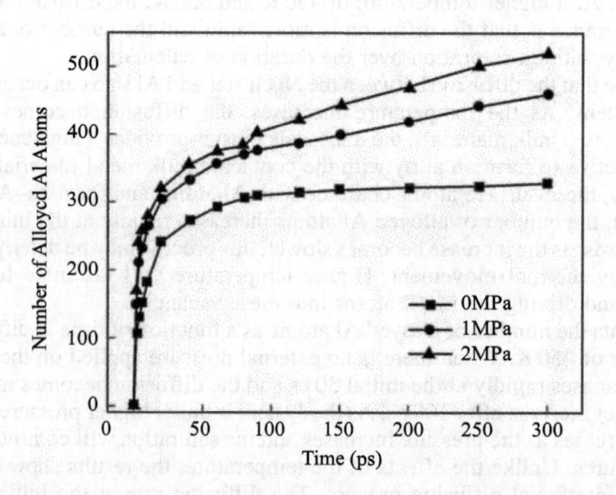

Figure 22.8 Number of alloyed Al atoms versus time under different pressure

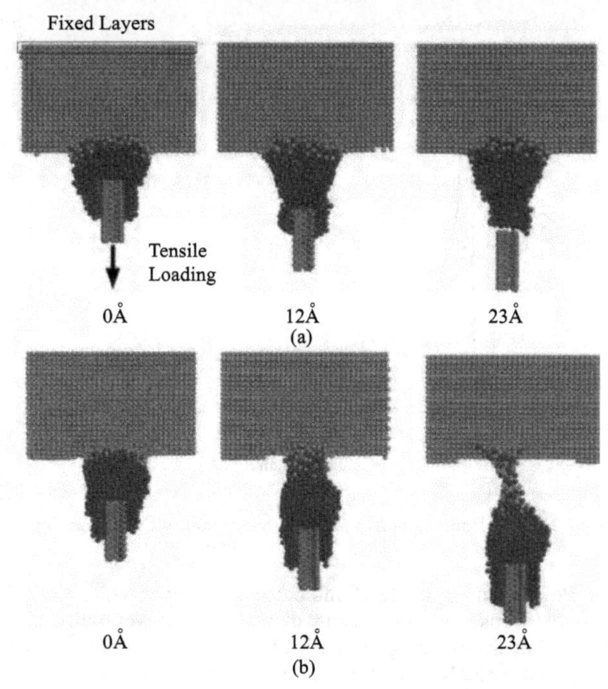

Figure 22.9 Configurations of a cross-section of CNT–Ni–Al system under different displacements. (a) Debonding process of the sample bonding under 450 K, 1MPa; (b) debonding process of the sample bonding under 350 K, 0MPa (1Å=10^{-10}m)

displacement-controlled experiments. Each loading increment corresponds to a displacement of 0.05 Å and is followed by a period of 2 ps of equilibration at constant displacement.

Figure 22.9 shows a series of images from simulations of debonding processes. The initial configuration of the debonding simulation is obtained from the previous bonding model. For Figure 22.9(a), the bonding is achieved at 450 K with pressure of 1 MPa, while for Figure 22.9(b), the bonding is achieved at 350 K with no pressure applied on the system. The results show that with the tensile loading applied on the boundary atoms the Ni cluster is elongated and the contact length between the CNT and the cluster decreases, and debonding occurs when the displacement is large enough. There are two potential debonding positions, one is at the Ni–Al interface which is determined by the diffusion bonding strength and the other is at the CNT–Ni interface which is determined by the coalescence strength between the CNT and the nickel cluster. The debonding position of the samples under tensile loading depends on the competition of these two interface strengths. For the example shown in Figure 22.9(a), the diffusion bonding is under high temperature and pressure which induces strong interfacial bonding strength between the Ni cluster and Al slab. As a result, the debonding occurs at the CNT and Ni cluster interface. This behavior indicates that the strength of CNT–metal interface is an important factor for bonding of CNTs to metallic substrates as well as the strength of metal–metal interface. Selecting metal having excellent adhesion properties to the CNT surface and forming coalescence structure between CNT and the metal cluster at high temperature to obtain larger contact length can be helpful to enhance the strength of CNT–metal interface. For the example shown in Figure 22.9(b), the bonding is under low temperature and there is no pressure applied which results in weak interfacial bonding strength between the metal cluster and the substrate, therefore, the failure firstly occurs at the Ni–Al interface.

To further understand the mechanical properties of the samples during the debonding process, we measure the tensile force (F) during the simulation by summing the forces exerted on boundary atoms by other atoms. The corresponding force-displacement curves of the debonding process with different bonding conditions are shown in Figure 22.10, where the force data are based on averaging over 1000 time steps. Initially, the force increases with increasing displacement which can be viewed as an elastic

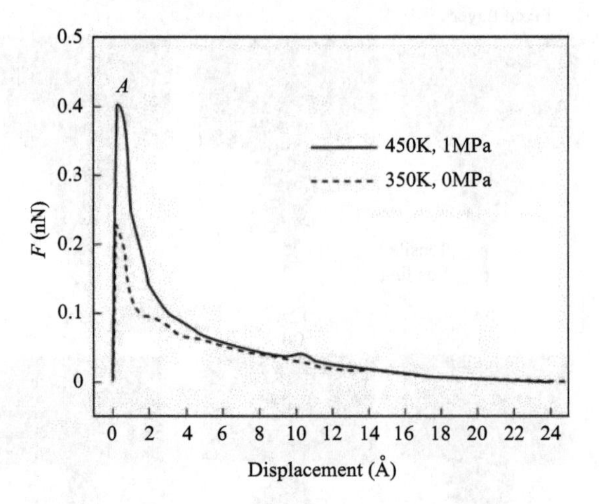

Figure 22.10 Force (*F*)–displacement curve during the debonding process

deformation behavior. Point A in Figure 22.10 indicates the point at which the debonding process starts. After point A, the force drops with increasing displacement, eventually going to very near zero. In addition, it is shown that the bonding strength of the sample bonding under 450 K and 1 MPa is larger than that bonding under 350 K and 0 MPa. This behavior may be attributed to different debonding positions. The former is determined by the CNT–Ni interface strength due to full interface contact between the metal cluster and the metallic substrate during the diffusion bonding. While the latter is determined by the Ni–Al interface strength due to the weak interface contact.

22.3 Applications of Car–Parrinello Molecular Dynamics

Car–Parrinello molecular dynamics (CPMD) simulations based on density functional theory (DFT) were carried out to determine the initial growth process of gallium nitride on the surface of single-walled carbon nanotubes (SWCNTs). The theoretical results reveal that the nitrogenadsorbed surface was more easily formed than the galliumadsorbed surface after the impinging of atoms on the pristine surface of SWCNTs. A gallium atom and a nitrogen atom form a bond with a length of approximately 1.9 Å. The results also demonstrate that the growth on metallic SWCNTs is more stable than that on semiconducting ones.

22.3.1 Car–Parrinello Simulation of Initial Growth Stage of Gallium Nitride on Carbon Nanotube

The first-principles calculations were performed using Car–Parrinello molecular dynamics (CPMD) method (Car *et al.*, 1985), which has been successfully used for the study of molecular adsorption (Langel, 1996; Mischler *et al.*, 2005). The CPMD code was an implementation of density functional theory (DFT) in the Kohn–Sham (KS) formulation and of the Car–Parrinello scheme. In the present work, the interactions of the ion cores with the valence electrons were described by Trouiller–Martins norm-conserving pseudo-potentials (Goedecker *et al.*, 1996), and nonlocal core corrections (NLCC) were included for the gallium species. The electronic wavefunctions were expanded in a plane wave basis set with an energy cutoff of 100 Rydberg. The exchange functional was given by Becke (1988) and correlation energy expression by Lee–Yang–Parr (LYP) (Lee *et al.*, 1988) respectively. The CPMD parallel code was used.

A super cell box model consisting of a single SWCNT, an atomic layer of adatoms, and 20 Å of vacuum region above the adsorption surface was used for calculations. We chose the zigzag (10, 0) and armchair (5, 5) SWCNTs with diameters $d_{(10,0)} = 0.783$ nm and $d_{(5,5)} = 0.678$ nm, as representative of semi-conducting and metallic SWCNTs, respectively. Several possible adsorption sites for clean surfaces and subsequent adatom surfaces have been considered. Periodic boundary conditions were used, while the periodicity of adatom was identical to both zigzag (10, 0) and armchair (5, 5) carbon nanotubes. In our lattice super cell, the distance between the centers of two adjacent nanotubes was 20 Å, which was sufficient to avoid the in-plane interactions between SWCNTs in adjacent unit cells. In order to minimize the interaction along the carbon nanotube axial direction between atoms adsorbed on SWCNTs in adjacent super cells, we employed several repeated carbon layers to simulate the (10, 0) and (5, 5) SWCNTs, resulting in 80 and 60 carbon atoms in the super cell unit, respectively. Our simulations were started from different initial conditions of nitrogen and gallium atoms coverage on the carbon nanotube surface. Four possible adsorption ways of gallium and nitrogen atoms on SWCNT are considered: (A) top of the carbon atom, (B) top of the bridge center of the C–C bond, (C) top of the center of the carbon hexagon ring, (D) top of the bridge center of another C–C bond. In each adsorption way, two possible adsorption sites with different distances above the carbon nanotube surface are considered. Full wavefunction optimizations are then performed. The total energies and the optimized structures are obtained for each adsorption way.

The adsorption energy of adatom on a SWCNT is defined as follows:

$$E_{\text{adsorb}} = E_{\text{SWCNT+adatom}} - E_{\text{SWCNT}} - E_{\text{adatom}} \tag{22.8}$$

where $E_{\text{SWCNT+adatom}}$, E_{SWCNT} and E_{adatom} denote the total energy of the adatom adsorbed SWCNT, the pristine SWCNT, and the adatom, respectively.

The chemical potential of gallium and nitrogen atoms in vapor phase is given by the equation (Kangawa *et al.*, 2007):

$$\mu = -k_{\mathrm{B}} T \ln \left[\frac{g k_{\mathrm{B}} T}{p} \left(\frac{2\pi m k_{\mathrm{B}} T}{h^2} \right)^{\frac{3}{2}} \right] \tag{22.9}$$

where k_{B} is the Boltzmann constant; T is the growth temperature; g is the degree of degeneracy of electron energy level; p is the equivalent pressure of the particle; m is the mass of one particle, and h is the Planck's constant.

The chemical potentials are calculated on the assumption that the adsorptions were under typical metal organic chemical vapor deposition (MOCVD) growth conditions. The temperature is 1300 K, and the pressure is set as atmospheric pressure.

With the approach described above, the present study is focused on the adsorption behavior of atomic gallium and nitrogen on the ideal SWCNTs surface firstly. Figure 22.11(a) and (b) show a schematic drawing of the (10, 0) and (5, 5) SWCNTs surface, while the adsorption way denoted by the capital letters A–D, respectively. The subscript number 1 and 2 represented two different sites of the same adsorption way, as number 1 denotes the distance between adsorb atom and carbon nanotube surface is 1.5 Å, while 2.0 Å for number 2. Corresponding adsorption energies are summarized in Figure 22.11(c), (d), (e), and (f). By comparing the adsorbtion properties between different adsorption sites on (10, 0) and (5, 5) carbon nanotubes, we suggested that the most energetic sites for nitrogen atom adsorption were the C_1 site for both (10, 0) SWCNTs and (5, 5) SWCNTs. The corresponding absolute value of nitrogen adsorption energies are larger than that of the chemical potential of nitrogen in the vapor phase, which means the C_1 site is the favorable adsorption site among all the considered sites. As shown by the schematic, the adsorption effect of nitrogen on subscript 2 sites becomes smaller than that on the subscript 1 site. On the other hand, the most energetic site for gallium atom adsorption is the C_2 site for both (10, 0) SWCNTs and (5, 5) SWCNTs. However, we noticed that some adsorption energies of gallium are positive, which indicates these are an exothermic reaction and less stable, namely, these adsorb configurations will be almost impossible to exist under typical gallium nitride growth conditions. Therefore, the formation of a gallium closest surface is a negligible probability. These results implied that a nitrogen-adsorbed surface is easily formed after the impinging of atoms on the initial surface of SWCNTs. In other words, gallium nitride growth may firstly occur in the case of a thin nitride layer formed on the surfaces of carbon nanotube.

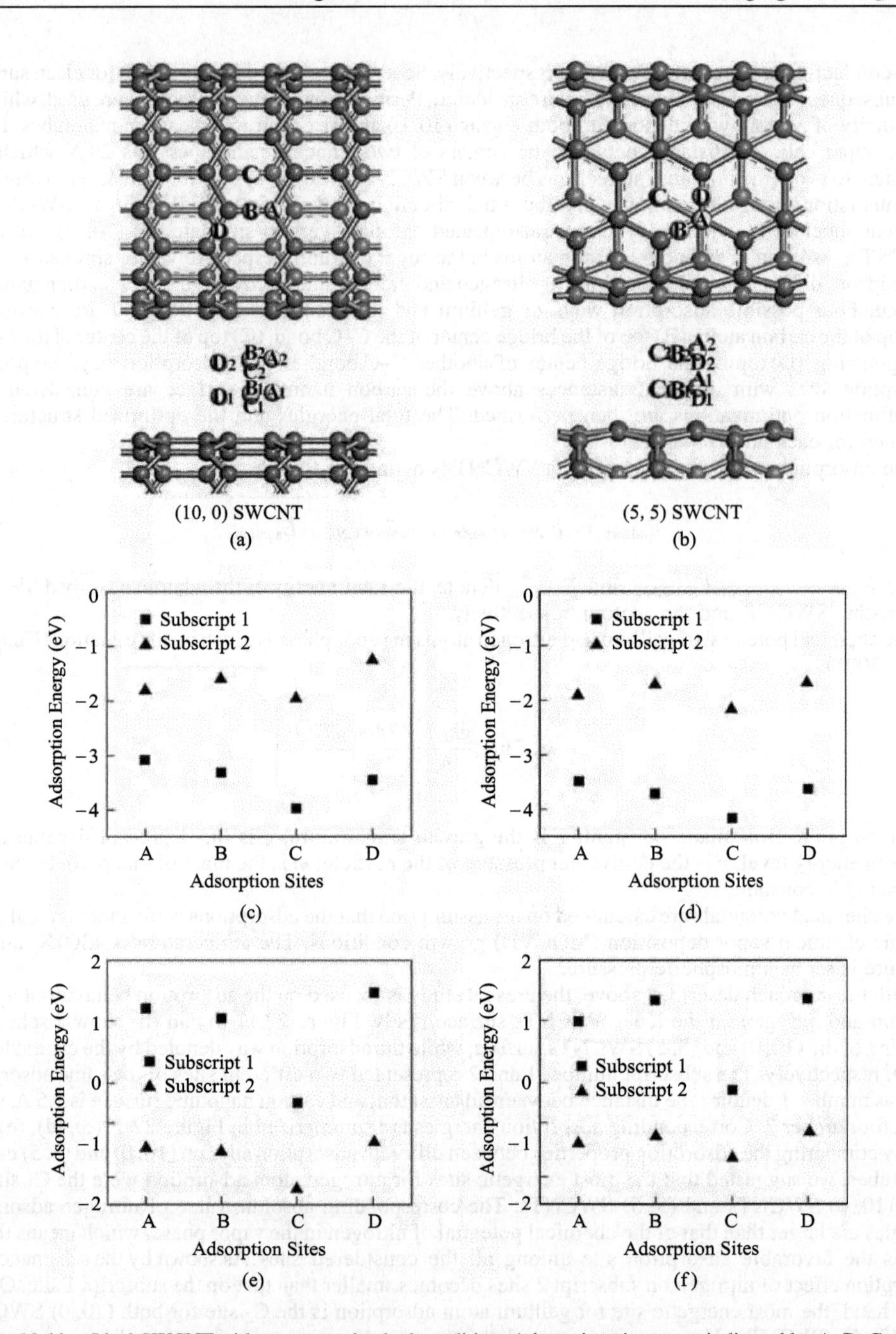

Figure 22.11 Ideal SWCNT without atom adsorbed condition. Adsorption sites were indicated by A, B, C, and D with subscript number 1 and 2. Top and side view of (10, 0)(a) and (5, 5)(b) SWCNT. The adsorption energies of each site for nitrogen (c) (d) and gallium (e) (f) atom on (10, 0) and (5, 5) SWCNT, respectively

It is shown that the adsorption energy of nitrogen adatoms on SWCNTs depended on the chirality of SWCNTs. Under the same adsorption way, the calculated adsorption energies on the (5, 5) SWCNT is slightly larger by 0.1–0.6 eV than that on the (10, 0) SWCNT. It indicated that interactions between metallic SWCNTs and nitrogen adatoms are a bit stronger than that between their semiconducting counterparts and adatom with similar diameter. That is, nitrogen atoms adsorbtion on metallic SWCNTs will be more stable than on semiconducting ones. The larger polarizability of metallic SWCNTs is the reason for the more induced charge (Krupke *et al.*, 2003). However, on part of instances, gallium atom adsorptions showed a totally opposite trend, which revealed the metallic atom would be less stable on metallic SWCNTs.

Subsequently, we investigated the adsorption stages of the second layer on the optimal nitrogen-adsorbed SWCNT surface. Figure 22.12(a) and (b) shows a schematic drawing of the nitrogen-adsorbed SWCNT surface with the adsorption sites denoted by the capital letters A′–D′, respectively. The distance between the nitrogen adsorbed surface and the second adatom site is 1.5 Å. The adsorption energies on each site are shown in Figure 22.12 (c) and (d) , respectively. If a nitrogen atom adsorbs near the precursors nitrogen adatom on the surface, nitrogen atoms would be desorbed as a nitrogen gas molecule instead to form a nitrogen-dimer, as a result, a nitrogen-dimer was unstable on the surface (Kangawa *et al.*, 2007). At the optimal adsorption sites C′ for (10, 0) SWCNT and D′ for (5, 5) SWCNT, the absolute values of gallium adsorption energy become larger than that of the chemical potential of gallium in the vapor phase, while the gallium atom could adsorb on the nitrogen-adsorbed surface. The gallium atom bonded to a substrate nitrogen atom, with a bond length of approximately 1.9 Å, that is, for (10, 0) SWCNT case, the distance between C and the closest nitrogen atom was 1.91 Å, which is close to the distance between the closest gallium and nitrogen atom along *c* axis in wurtzite gallium nitride. From a crystallography point of view, under common ambient conditions, gallium nitride crystallizes in the

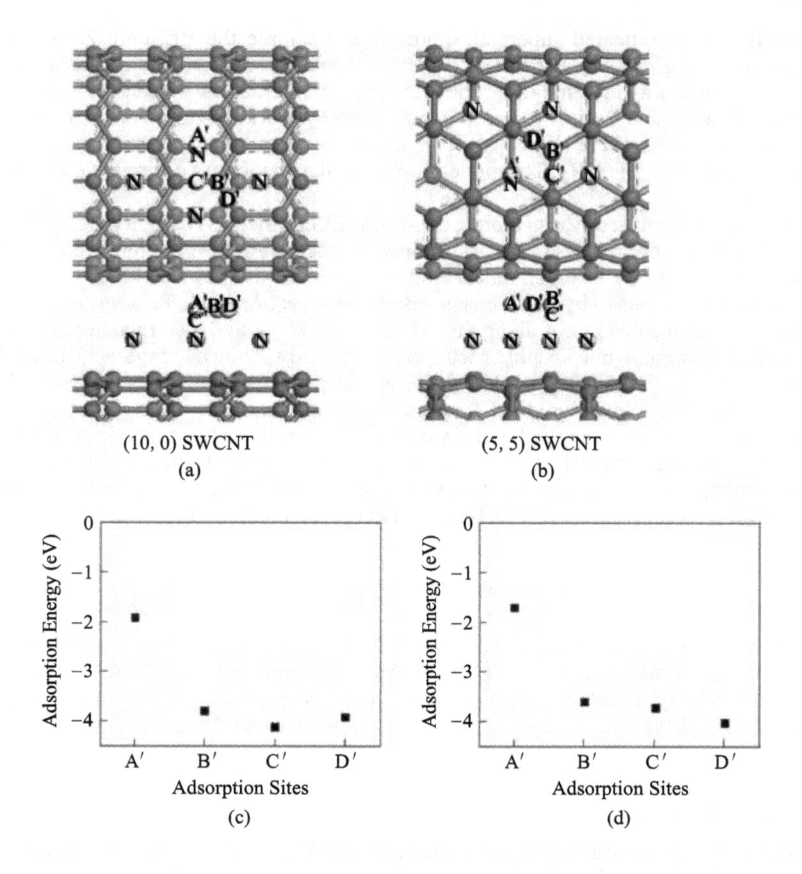

Figure 22.12 Adatom SWCNT with nitrogen atom adsorbed condition. Adsorption sites were indicated by A′, B′, C′, and D′. Top and side view of (a)(10, 0) and (b)(5, 5) SWCNT. Corresponding adsorption energies of gallium atom on each site for (c)(10, 0) and (d)(5, 5) SWCNT, respectively

hexagonal wurtzite structure rather than metastable zinc-blende structure (Munoz *et al.*, 1993). We could speculate that the gallium nitride nanowire will form a wurtzite-like structure.
Conclusions:
We determined the initial growth mechanisms of gallium nitride on SWCNTs by theoretical analyses based on Car–Parrinello molecular dynamics calculations. The results indicated that a nitrogen adsorption occurs on the initial surface under typical growth conditions, followed by gallium adsorption on the nitrogen adatom surface. Moreover, the most favorable site for adsorption of adatom is suggested to be on top of the center of the carbon hexagon ring (C site). Theoretically, the nitrogen adsorbed onto (10, 0) SWCNT is less stable than adsorbed onto (5, 5) SWCNT, namely, the growth on metallic SWCNTs will be more easily than that on semiconducting counterparts. The results implied that the nitrogen coverage condition of the surface may have profound influence on the subsequent gallium nitride growth, while gallium could adsorb on the nitrogen-adsorbed surface. However, further studies are needed to elucidate the details of the entire process.

22.3.2 *Effects of Mechanical Deformation on Outer Surface Reactivity of Carbon Nanotubes*

(*i*) *Methods*

We used a periodically repeated supercell approach and sample the Brillonin Zone with a single k–point, and our supercell was large enough to avoid spurious interaction between periodic replicas of an oxygen atom. For atoms in simulation, the electron–ion interaction was described by Tronullier–Martins nonlocal norm-conserving pseudopotentials. The generalized gradient corrected exchange functional Becke–Lee–Yang–Par (BLYP) (Gill *et al.*, 1992) was adopted, and the plane wave basis set had a kinetic energy cutoff of 60 Ry. The models are relaxed by geometry optimization with conjugate-gradient scheme.

In order to illustrate the interaction of atomic oxygen with deformed SWCNT, a semiconducting (10, 0) single wall carbon nanotube (SWCNT) with 140 atoms is modeled. An oxygen atom is placed at a distance of 1.5 Å above the center of a hexagon on each adsorption sites, as shown in Figure 22.13.

The axial strains are obtained by axial tension (Buongiorno *et al.*, 1998; Takanori *et al.*, 2002) which are applied by shifting the end atoms along the axis by small steps, and then relaxing the model via the conjugate-gradient method, while keeping each end constrained as shown in Figure 22.13(a). The strains of 2.5%, 5%, 7.5%, 10%, 12.5%, and 15% are applied and computed respectively with the same pseudopotentials and generalized gradient corrected exchange function.

The bending angle of 30° is applied and six different adsorption sites that denote different strains are considered. To obtain relax structures under bending and to calculate atomic adsorption properties, the atoms of two rings at each end are fixed, as shown in Figure 22.13(b), and then the model is relaxed via the conjugate-gradient method. The binding energy of atomic oxygen on different system is calculated as:

$$E_b = E(\text{oxygen}) + E(\text{CNT}) - E(\text{oxygen} + \text{CNT}) \tag{22.10}$$

where E (oxygen) is the energy of oxygen atom, E (CNT) is the energy of the isolated CNT system and E (oxygen + CNT) is the energy of the whole system including the adsorbed oxygen atom. All calculations of total energy at certain strain use the same supercell and k-points. A positive E_b corresponded to an adsorption process.

(*ii*) *Results and Discussion*

First we study the adsorption of atomic oxygen on a pristine SWCNT. The results show that atomic oxygen is very reactive and the adsorption takes place with no barrier to form epoxide-like structure, as shown in Figure 22.14, the adsorption energy is 2.12 eV. The bond lengths of C–C and C–O are 1.573 Å and 1.465 Å respectively, and the oxygen atom preferentially attacks "twisted C–C bonds" shown in Figure 22.14,

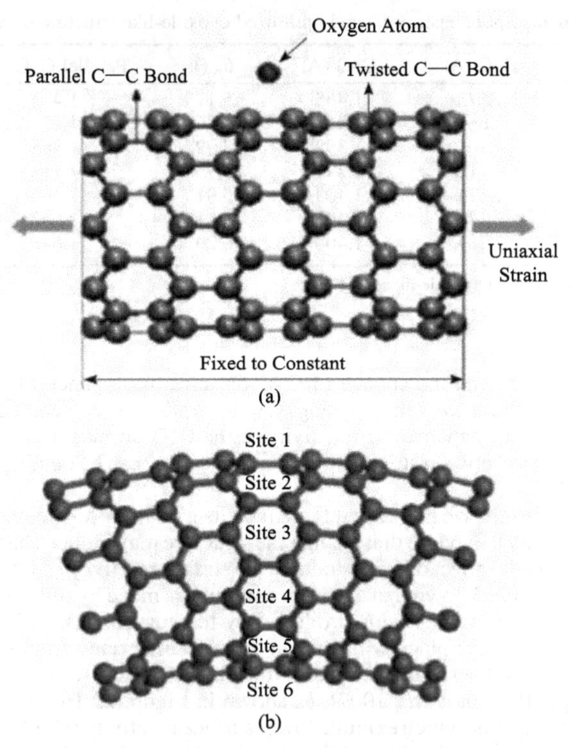

Figure 22.13 Schematic for computational model: (a) axial strains of 2.5%, 5%, 7.5%, 10%, 12.5%, and 15% were applied respectively to introduce deformation on CNTs and; (b) a bending angle of 30° was applied to introduce bending deformation and six different adsorption sites denoted different strains on CNTs surface

because they have a larger curvature and more reactive. The results are in agreement with previous study (Liu *et al.*, 2007).

Next, the structure is considered under various axial strains, with the adsorption energy E_b and epoxide-like structure parameters for the optimized structures listed in Table 22.1. It is shown that the adsorption energy increases with increasing axial strain, the C–C bonds elongate and are apt to break with the

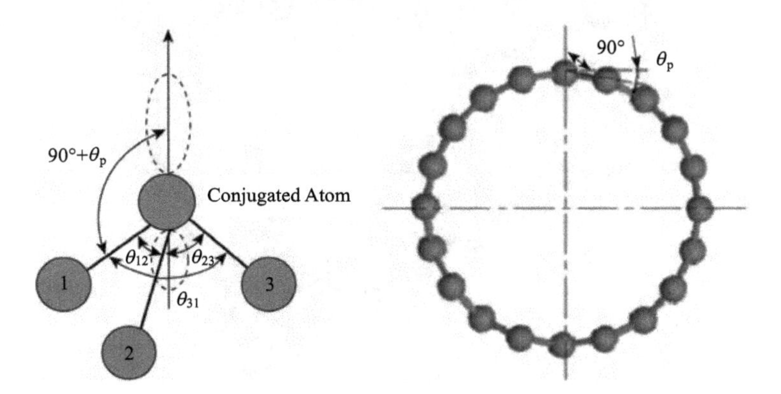

Figure 22.14 Schematics for the θ_p angle. Three angles made by a conjugated atom and its three adjacent atoms are θ_{12}, θ_{23}, and θ_{31}. For CNTs, the θ_p angle falls between 0° and 19.5°

Table 22.1 Calculated adsorption energy E_b and optimized epoxide-like structure parameters

Strain	E_b (eV)	C–C (Å)	C–O (Å)	θ_P ()	Parallel C–C (Å)	Twisted C–C (Å)
0.000	2.12	1.573	1.465	5.11	1.434	1.422
0.025	2.25	1.666	1.442	5.36	1.456	1.435
0.050	2.55	2.101	1.398	5.59	1.485	1.440
0.075	2.76	2.112	1.399	5.75	1.518	1.442
0.100	3.03	2.174	1.403	5.91	1.566	1.443
0.125	3.29	2.179	1.405	6.11	1.642	1.444
0.150	3.57	2.185	1.408	6.29	1.676	1.445

C–C and C–O denote bond length of epoxide-like structure.

adsorption of oxygen atom, and the change of C–O distance has a general trend towards more stable equilibrium under larger strains. All these changes of adsorption energy and CNT structure parameters indicate that the axial strains enhance the reactivity of the CNT surface to adsorb the oxygen atom: an approximate linear relation between strain and adsorption energy can be obtained from the calculation as plotted in Figure 22.15.

From our simulation results we notice that there is a close relationship between the surface reactivity and the CNT structure; any behaviors that change the structure may induce change of surface reactivity. Therefore, we can explain the effect of axial strain on the surface reactivity of CNTs from the point of view of structure deformation. It is known that a carbon nanotube is made by rolling up a graphene sheet into a cylinder (Iijima, 1991), and the strains produced by the curved surface induce σ-π hybridization. An additional reaction can take place with carbon atoms transformed from sp2 to sp3 hybridization. The reactivity is correlated with the extent of this hybridization transformation denoted by the pyramidalization angle θ_P (Liu *et al.*, 2007), as shown in Figure 22.16. For $\theta_P = 0°$, there is the sp2 hybridization as a flat graphene which exhibits lower surface reactivity (Goldsmith *et al.*, 2007), and for $\theta_P = 19.5°$, there is the sp3 hybridization as CH_4. The θ_P value for the carbon atom on a carbon nanotube falls between these two values, and both the curvature and the reactivity increase with the θ_P value

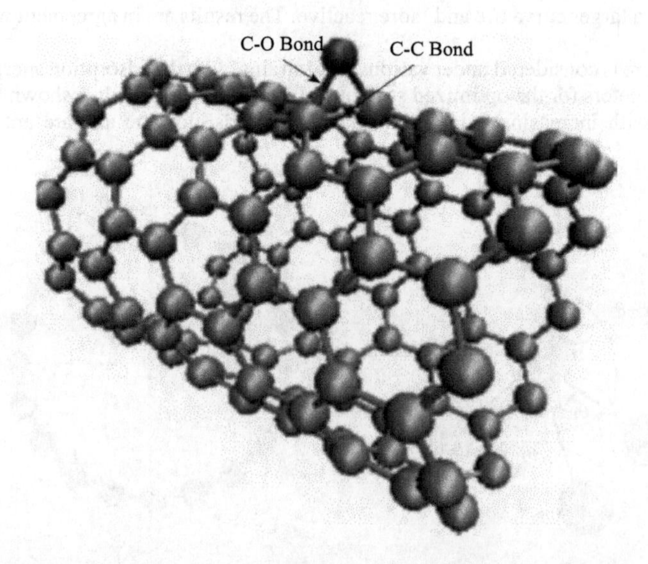

Figure 22.15 Structure of epoxide-like product after adsorption of atomic oxygen on the surface of CNT under tension strain

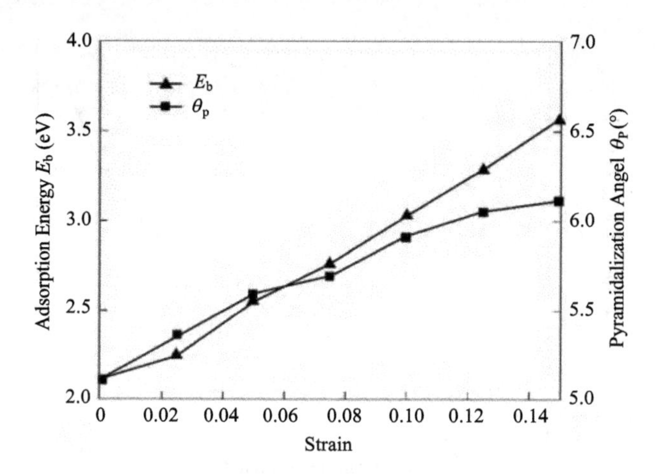

Figure 22.16 Variation of adsorption energy E_b and pyramidalization angle θ_P against strain. Both E_b and θ_P increase with the increase intension strains

(Niyogi *et al.*, 2002). For the (*n*, 0) tubes in our calculation, the pyramidalization angle θ_P can be calculated by the Equation:

$$\begin{bmatrix} x_1 \\ y_1 \\ z_1 \end{bmatrix} = \begin{bmatrix} -1 \\ 0 \\ 0 \end{bmatrix} \tag{22.11}$$

$$\begin{bmatrix} x_2 \\ y_2 \\ z_2 \end{bmatrix} = \begin{bmatrix} -\cos\theta_{12} \\ \sin\theta_{12} \\ 0 \end{bmatrix} \tag{22.12}$$

$$\begin{bmatrix} x_3 \\ y_3 \\ z_3 \end{bmatrix} = \begin{bmatrix} -\cos\theta_{31} \\ \dfrac{\cos\theta_{23} - \cos\theta_{12}\cos\theta_{31}}{\sin\theta_{12}} \\ \sqrt{1-\cos^2\theta_{31} - \left(\dfrac{\cos\theta_{23} - \cos\theta_{12}\cos\theta_{31}}{\sin\theta_{12}}\right)^2} \end{bmatrix} \tag{22.13}$$

and:

$$\cos(\theta_p + 90°) = \frac{x_1 y_2 z_3}{\sqrt{(y_2 z_3)^2 + [z_3(x_2 - x_1)]^2 + [y_3(x_2 - x_1) - y_2(x_3 - x_1)]^2}} \tag{22.14}$$

as reported by Haddon (2001). The results of pyramidalization angle θ_P at different adsorption sites are shown in Table 22.1.

In a previous report (Zhang *et al.*, 2004), an approximate linear relation between adsorption energy and the pyramidalization angle has been fitted by calculating various pristine carbon nanotubes of different diameters to illustrate the crucial effects of the pyramidalization angle on the surface reactivity of

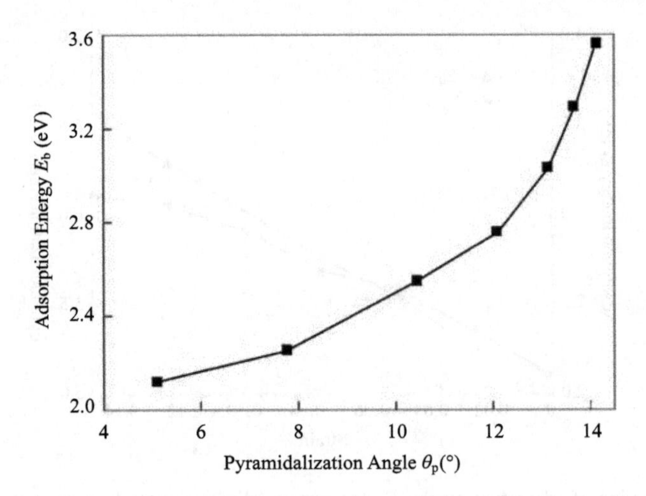

Figure 22.17 Binding energy E_b plotted versus pyramidalization angle θ_p. Binding energy E_b increases with the increase in pyramidalization angle θ_p

pristine CNT. However, in our study, an approximate linear relationship between adsorption energy and the pyramidalization angle can only be obtained in strains that are below 10%, as the strain increases, the pyramidalization angle alters little but the adsorption energy alters a lot, as plotted in Figure 22.16. It is clear that the pyramidalization angle is not the only factor that affects the surface reactivity of the CNT.

It has been reported that less energy is needed to change the bond angle than the bond length of CNT under small deformation, as evidenced from MD simulation (Wu *et al.*, 2004) and the finite element method (Najib *et al.*, 2007). Therefore, the CNT structure is apt to deform by changing the bond angle as shown in Figure 22.16 rather than bond length under low strains. And the axial strain makes the diameter that determines the pyramidalization angle of the CNT structure decrease with an increase of axial strain, nevertheless the hexagon structure of CNT under strain is close to that of pristine CNT for little change of bond length.Therefore, the approximate linear relation between adsorption energy and pyramidalization angle is obtained in low axial strains as plotted in Figure22.17, and pyramidalization angle plays a crucial role in surface reactivity of CNT in which the hexagon structure is close to that of pristine CNT. Furthermore, there is an approximate linear relation between the strain and pyramidalization angle in low strains as shown in Figure 22.17.

On the other hand, with the increase of strain, elongation of "parallel C–C bond" will be more obvious, as shown in Table 22.1, and the CNT structure is not an approximate hexagon structure as pristine CNT, the results are in agreement with the theoretical study on structural mechanics of zigzag carbon nanotubes (Najib *et al.*, 2007). It is clear that elongation of "parallel C–C bonds" shown in Table 22.1 will change the electron structure greatly under larger strains, and the electron structure has been reported as an important factor affects surface reactivity (Zhang *et al.*, 2004). In this case, the hybridization is complex and cannot be denoted by the pyramidalization angle, and the elongation of C–C bonds induces high surface potential energy that increases the surface reactivity of CNT. Here, we can also obtain an approximate linear relation between the strain and the adsorption energy in high strains.

Finally, the structure was considered under a bending angle of 30°, with the adsorption energy E_b at each site for the optimized structures listed in Table 22.2. The structure of an epoxide-like product after adsorption of atomic oxygen is shown in Figure 22.18. It is shown that the application of the bending angle to a carbon nanotube induces bond elongation or shortening at different locations, and the adsorption energy at each site is different. The adsorption energy levels at site_1, site_2, and site_3 which are under tension strains are larger than those on a pristine SWCNT, while the adsorption energy levels at site_4, site_5, and site_6 which are under compression strain is smaller than those on a pristine SWCNT.

The bending deformation of SWCNT which can be illustrated as a classic elastic beam by generalized Hooke's law and elasticity theory, produces different strains on the surface of SWCNT. For sites 1–3,

Table 22.2 Calculated adsorption energy E_b and pyramidalization angle θ_p

Sites	E_b (eV)	θ_p (°)
1	3.59	5.6141
2	3.51	5.5294
3	2.65	5.2780
4	2.09	5.0307
5	1.89	4.8861
6	1.81	4.8023

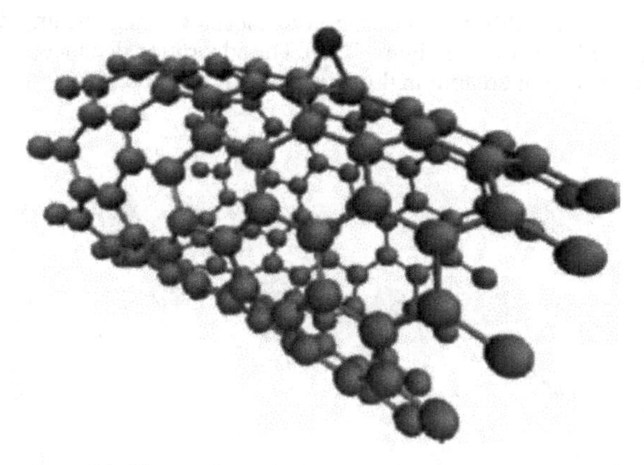

Figure 22.18 Structure of epoxide-like product after adsorption of atomic oxygen on the surface of CNT under bending deformation

the tension strains induce larger pyramidalization angles than pristine SWCNT, the pyramidalization angle increases on increasing the strain, as a result, site 1 has the largest pyramidalization angle. In contrast, for sites 4–6, the compression strain induces smaller pyramidalization angles, and the pyramidalization angle decreases with the compression strain increasing, hence site_6 has the smallest pyramidalization angle. Since the pyramidalization angle is directly related to reactivity, it indicates that the six sites in our study have different surface reactivities, and the oxygen atom will tend to be adsorbed on site_1, which has the highest surface reactivity. Therefore, the bending deformation of CNT offers a method to control the pattern of adsorption.

22.3.3 Adsorption Configuration of Magnesium on Wurtzite Gallium Nitride Surface Using First-Principles Calculations

(i) Computation Details

The first-principles calculations are performed using the Car–Parrinello molecular dynamics (CPMD) method, which has been successfully used for the study of atom and molecular adsorptions (Langel, 1996; Mischler *et al.*, 2005; Song *et al.*, 2009; Yan *et al.*, 2009). The density functional theory is in widespread use in the adsorption process of molecular and atom (Yuan *et al.*, 2009; Martins *et al.*, 2004).The CPMD code is an implementation of the density functional theory in the Kohn–Sham (KS) formulation and of the Car–Parrinello scheme. In the present work, the interactions of the ion cores with the valence electrons are described by Trouiller–Martins norm-conserving pseudo-potentials (Goedecker *et al.*, 1996), and non-local core corrections (NLCC) are included for the gallium species. The electronic wave-functions are expanded in a plane wave basis set with an energy cutoff of 80 Ry. The exchange functional was given by (Becke, 1988) and correlation energy expression by (Becke, 1988; Lee *et al.*, 1994). The CPMD parallel code, version 3.11.1, is used. Wavefunction optimizations are taken in CPMD to obtain the total energies of the simulation systems.

A super-cell slab model consisting of two layers {0001} basal plane wurtzite gallium nitride with 16 atoms in total and a 5 nm vacuum region above the adsorption surface is used for calculations. A magnesium adsorbed atom is placed on the surface. Ga-terminated surface and N-terminated surface are modeled, respectively. Periodic boundary conditions are used. Magnesium adatoms are placed on the pristine Ga- and N-terminated surfaces, six possible adsorption sites for initial surfaces have been considered, namely, (a) top of the first-layer atom; (b) top of the bridge center of two adjacent first-layer atoms; (c) top of the center of the first-layer triangle ring; (d) another top of the bridge center of two adjacent first-layer atoms; (e) top of the bridge center of first-layer atom and second-layer atom; (f) top of the second-layer atom. A schematic model of magnesium adsorption sites is shown in Figure 22.19.

In each adsorption site, three different adsorption distances of magnesium adatom above the polar surface plane are considered, as shown in Figure 22.20. The adsorption distances are defined as the nearest distance between the magnesium adatom and gallium nitride surface plane.

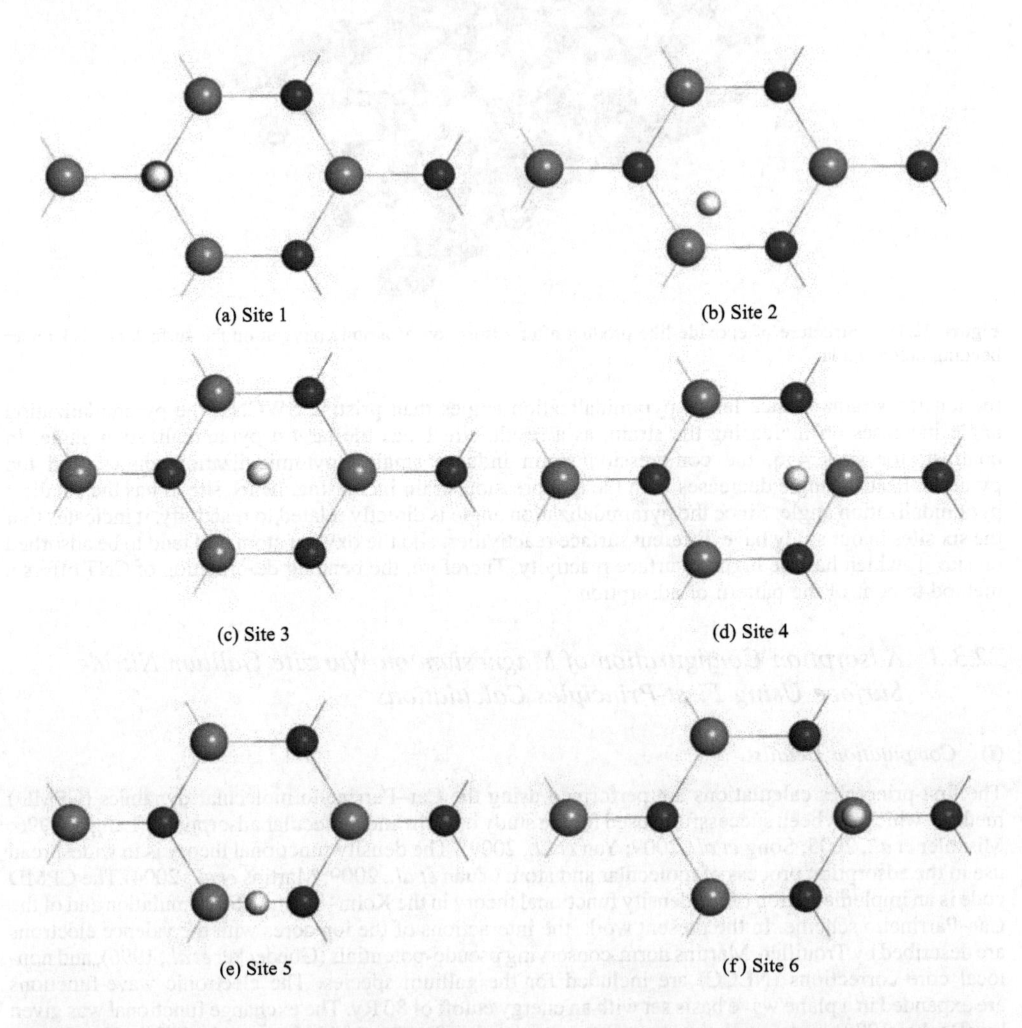

Figure 22.19 Magnesium adsorption sites with gallium-terminated or nitrogen-terminated gallium nitride surface. Black spheres are surface-terminated layer atom, large grey spheres are second layer atom, and small white spheres are magnesium adatom

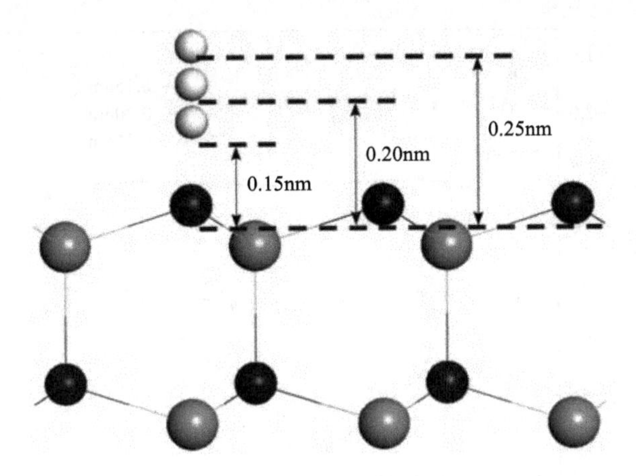

Figure 22.20 Slab model for the Ga-terminated and N-terminated surfaces. Three considered adsorption distances above the surface atom plane are shown. Black spheres are surface-terminated layer atom, large grey spheres are second layer atom and small white spheres are magnesium adatom

The adsorption energy of magnesium adatom on gallium nitride surface is defined as follows:

$$E_{\text{adsorb}} = E_{\text{Mg+GaN}} - E_{\text{Mg}} - E_{\text{GaN}} \tag{22.15}$$

where $E_{\text{Mg+GaN}}$ is the total energy of the gallium nitride surface with the adsorbed magnesium atom, E_{GaN} stands for the total energy of the pristine wurtzite gallium nitride surface without the adatom, and E_{Mg} is the total energy of magnesium atom. We optimized the lattice parameters for GaN in the wurtzite structure, which is the most stable energy state. The total energies were calculated for each adsorption condition.

(ii) Results and Discussion

With the approach described above, we plot the adsorption energy of magnesium on the Ga-terminated and N-terminated surfaces as a function of adsorption sites. The results are shown in Figure 22.21. The present study firstly focuses on the adsorption behavior of atomic magnesium on the N-terminated gallium nitride surface, while the corresponding adsorption energies are summarized in Figure 22.22(a). On the N-terminated surface, it is noticed that in all the sites of the adsorption distances of 0.25 nm, the adsorption energies are similar, and the average absolute value is 2.75 eV, while in all the sites of the adsorption distances of 0.2 nm, the adsorption energies are also similar, the average absolute value of adsorption energy is 3.75 eV.At sites 1 and 6, the adsorption energies decrease when magnesium atom is too close or too far from the N-surface, and the best adsorption distance is 0.2 nm. At sites 2–5, the adsorption energies increase as the adsorption distance decreases. The adsorption is more stable when the magnesium atom is much closer to the N-surface. It is suggested that site 3 is the preferred site on the condition that the adsorption distance is 1.5 nm, which indicates that the magnesium atom preferentially adsorbs at the triangle ring hollow site. Subsequently, we investigate the adsorption stages of Ga-terminated surface. The general tendency at site 1 is that the adsorption energy becomes positive as the adsorption distances decrease, due to a strong repulsive interaction between the adatom at site 1 and corresponding surface atoms, and a similar tendency is shown for site 6. The absolute value of adsorption energies decrease as the adsorption distance decreases at sites 2–5. The reduction of adsorption energies suggests that these adsorbed configurations will be unlikely to occur under typical gallium nitride p-type doping growth conditions. It is also remarkable that the magnesium adsorption energies are much more stable among sites in which the adsorption distances are 0.2 nm and 0.25 nm than those in which the adsorption distance is 0.15 nm. Compared with the absolute value of adsorption energies on

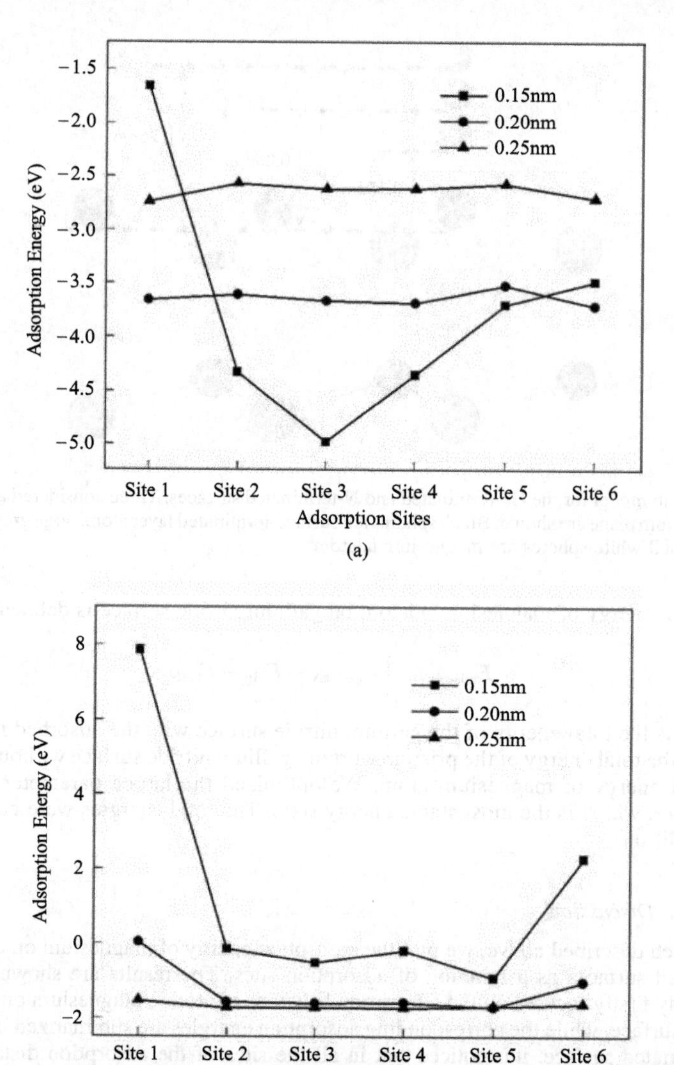

Figure 22.21 Adsorption energy as functions of adsorption sites. (a) N-terminated surface; (b) Ga-terminated surface

Ga-terminated surface and that on N-terminated surface, we conclude that the magnesium atom is strongly apt to adsorb on N-terminated surfaces rather than on Ga-terminated surfaces.

The adsorption energies as a function of adsorption distance are summarized in Figure 22.22. It is noticed that the discrepancy of adsorption energies at each site increase as the adsorption distance decreases, which means the diffuse potential barrier of magnesium adatom increases as the adsorption distance decreases. As a result, the surface mobility of magnesium becomes lower while the adatom is closer to the gallium nitride surface. This tendency is similar for both the Ga-terminated surface and the N-terminated surface. Furthermore, the surface diffusivity of magnesium atom on the N-terminated surface is much lower than that on the Ga-terminated surface due to both the larger average adsorption

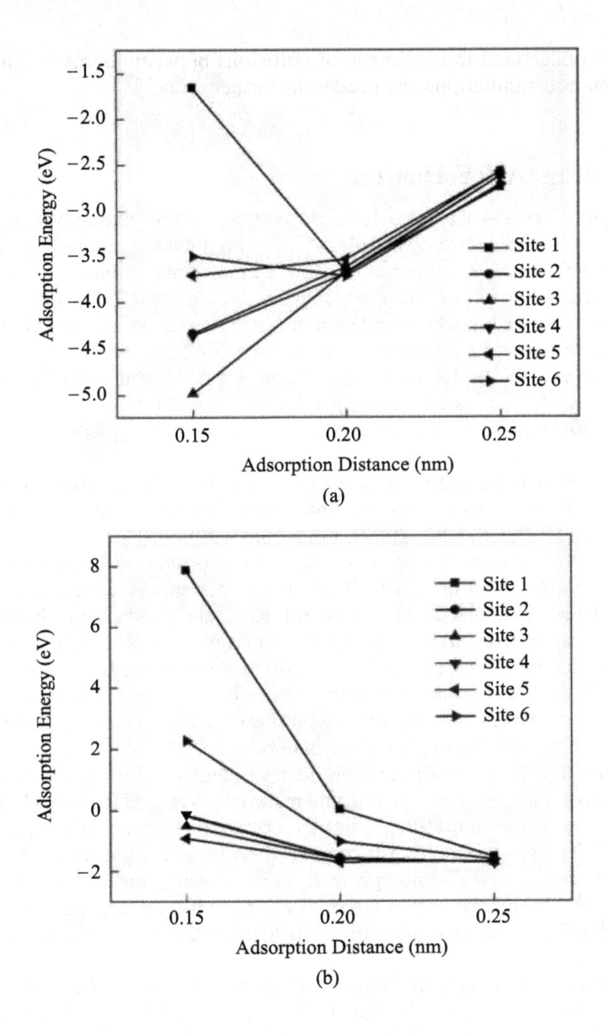

Figure 22.22 Adsorption energy as functions of adsorption distance. (a) N-terminated surface; (b) Ga-terminated surface

energies and the lower adsorption distance on the N-terminated surface than that on Ga-terminated surface. It implies that a higher probability for magnesium adatoms to diffuse on Ga-terminated surface and even finally replace gallium atoms. The results indicate that the p-type doping on the Ga-terminated surface will be better distributed than that on the N-terminated surface.

It has been reported that high quality epitaxial gallium nitride films deposited by MOCVD on c-plane sapphire substrates grown in the (0001) direction with Ga-terminated surfaces, while molecular beam epitaxy (MBE) growth commonly occurs in the (000–1) direction, yielding an N-terminated film (Smith *et al.*, 1997; Kazimirov *et al.*, 1998). Besides, clean N-terminated gallium nitride (0001) surfaces are found to be thermodynamically unstable (Smith *et al.*, 1999). Our calculation results show that the MOCVD system is better for p-type doping processes of gallium nitride.

In metal-organic chemical vapor deposition (MOCVD) condition, there are many collisions between atoms and molecules. However, the mean free-path length of gas molecules in MOCVD condition is about 102 Å orders of magnitude, which is significantly larger than the adsorption distance in our models. Therefore, the simulation results could be generally applicable for MOCVD growth condition.

Moreover, in order to understand the influence of collisions between atoms and molecules on adatom adsorption, detail dynamic simulations are needed in further work.

22.4 Nano-Welding by RF Heating

Owing to their unique electrical and structural properties, carbon nanotubes (CNTs) are promising candidate for use in nanoscale electronics, with potential applications, including field effect transistors (Tans *et al.*, 1998), novel sensor (Baughman *et al.*, 2002) and interconnect (Jiang *et al.*, 2007). While in actual device applications, CNTs must be contacted with metallic structures. However, simply bringing CNTs into the contact cannot be relied upon to generate secure and stable bonds, which may cause weak mechanical performance and high contact resistances between CNTs and electrodes (Chen, *et al.*, 2005; Heinze *et al.*, 2002). These disadvantages prevent further practical application of CNT-based devices. The key challenges for creating reliable contact bonding between the CNTs and the metallic structures are forming mechanically strong bonds and obtaining larger contact length (Andriotis *et al.*, 2008).

In order to improve the contact bonding between CNTs and metallic structures, many welding methods have been proposed including focused on nanoink deposition (Dockendorf *et al.*, 2007), local Joule heating (Dong *et al.*, 2007; Peng *et al.*, 2009), ultrasonic welding (Chen *et al.*, 2007) and local point annealing (Woo *et al.*, 2007). These experimental studies focus on exploring the methods to coat metals on CNT-electrodes interface (Dockendorf *et al.*, 2007) and create energy which can be utilized to melt or soften the surface of the metallic electrodes and therefore induce good contact bonding. However, these methods are not likely to realize large scale reliable bonding of CNTs onto metal electrode without destruction of other structures. They can only deal with a limited number of interfaces, one at a time. In addition, using the ultrasonic tips may destroy the electrodes. In order to realize reproducible large scale fabrication of CNT-based devices, the welding technology should be able to provide batch fabrication capability, which cannot be achieved by existing methods.

In this research, the high frequency induction heating method which is a simple yet cost-effective technology for producing a temperature rise in some materials very rapidly is used to obtain high thermal energy on metal electrodes within short time. This method provides sufficient energy to melt the surface of the metal and weld CNTs onto the melting layers. Furthermore, this non-contact and selective heating allows us to bond CNTs onto electrodes in large-scale without heating the whole structures. In addition, we employ multi-physics and molecular dynamics simulations to investigate mechanism of welding CNTs onto metallic electrodes. The results can be utilized to identify the factors that affect the welding process and optimize the process.

For the purpose of the experiment, the single wall carbon nanotubes (SWCNTs) suspensions were prepared by sonicating the raw materials in dimethylformamide (DMF) for 8 hours at a power level of 80 watts and subsequently centrifuged at 16000 rpm for 1 hour, and the upper 10% of the supernatant was then carefully decanted. The resulting CNTs suspension had a concentration of 10 mg/L approximately. The suspension was then ultrasonically treated for 2 hours to sufficiently disperse the SWCNTs. The arrays of parallel Cr/Ni/Au electrodes whose Cr, Ni, Au layers were 0.5 nm, 1 μm, and 500 nm thick were prepared using a standard optical lithographically on thermally oxidized silicon substrates have a width of 50 μm, and a length of 100 μm. The distance between the source and the drain electrodes was approximately 5 μm. The dielectrophoresis (DEP) deposition of aligned SWCNTs was achieved by driving the electrodes at a frequency of 3 MHz and a peak-to-peak voltage of 20 V for 30 s during exposure to SWCNTs suspensions. And then the substrate consisting of several electrode pairs with aligned SWCNTs network was obtained. The welding of CNTs onto electrodes was achieved by the induction heating system. As shown in Figure 22.23, an induction heating system consists of a high frequency power supply, a cooling system, an induction coil, and work pieces. A high frequency power supply drives the induction coil, generating a magnetic field. Meanwhile, the eddy current is induced in the work piece as it is placed inside the magnetic field. This causes heat to be generated by the work piece due to the eddy current. According to the induction heating mechanism, a higher temperature will be generated at the parts with higher conductivity and permeability. Therefore, the heating process is non-contact and selective. In this study, a radio-frequency (RF) coil was at power levels in the 600 W range with a frequency of 13.56 MHz, and the heating time was 10 s.

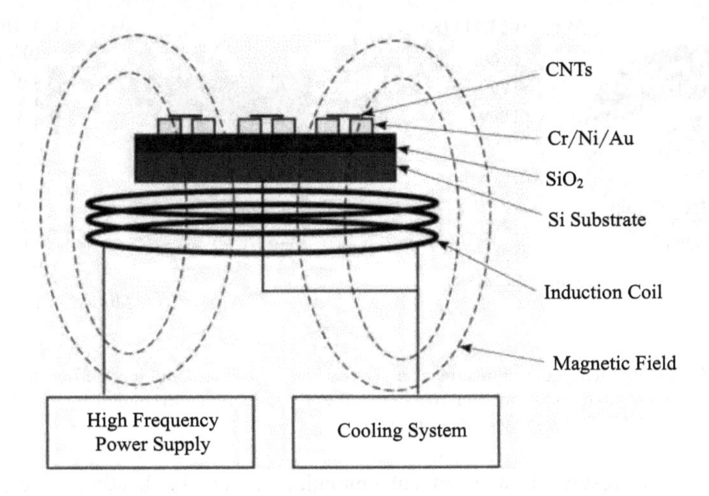

Figure 22.23 Schematic illustration for induction heating

Figure 22.24 shows the morphology of the SWCNTs on the electrode before and after induction heating process.Before the welding,one end of the CNTs can be observed lying on the electrode surface.While after induction heating, the CNTs on the electrode surface are covered with metal. Furthermore, we cannot observe an obvious change of the electrode morphology. The most important thing is that the CNTs on the electrodes are not burned during heating. These results indicate that the induction heating selective occurs on the metallic electrode and makes the surface of the electrode melt which can wet the CNT surface to form CNT-metal hybrid structure.

The temperature increase ΔT of the structure due to the induction heating of time t is expressed as (Yang *et al.*, 2005):

$$\Delta T \propto CA_s \frac{\mu_r^2}{\rho} f^2 H^2 t \tag{22.16}$$

where C and A_s are determined by the geometry of the structure, frequency f and magnetic intensity H are determined by the power supply and the coil, relative permeability μ_r and electrical resistivity ρ are material properties of the structures. According to Equation 22.16, the temperature increase will prefer to occur on a structure with high relative permeability and low electrical resistivity. Therefore, the metallic electrodes with much lower electrical resistivity will be selectively heated. On the other hand, the Ni layer with high relative permeability can also increase the heating efficiency.

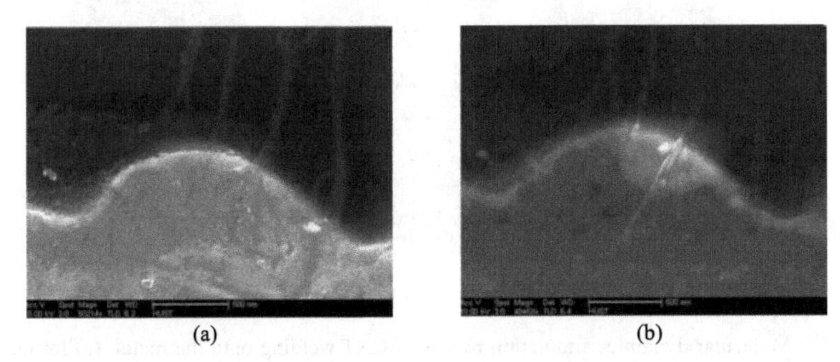

Figure 22.24 Scanning electron microscopy images of CNTs welding onto metal electrodes. (a) Before the high frequency inducing heating process; (b) after the high frequency inducing heating process

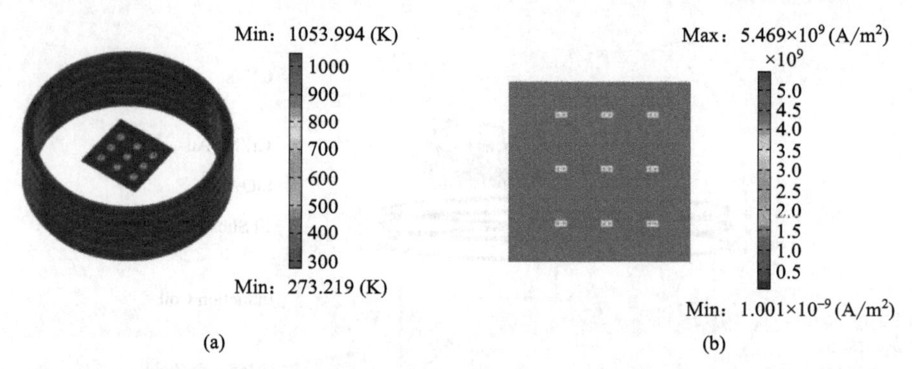

Figure 22.25 Finite element simulation results of induction heating carried out by coupling electromagnetism and heat transfer. (a) Temperature distribution after 10 s of heating; (b) induction current density distribution after 10 s of heating (*Color version of this figure is available online.*)

Figure 22.25 presents results of finite element simulation of induction heating process carried out by coupling electromagnetism and heat transfer. The calculated temperature distribution after 10 s heating shown in Figure 22.25(a) indicates that temperature on the electrodes is much higher than the substrate due to selective heating. As shown in Figure 22.25(b), the induction current is focused on the surface and edge of the electrodes, which results in a quick temperature increase in these particular regions. On the other hand, the heat transfer may induce temperature increase of the substrate, so that the substrate should be cooled during the heating process.

To further understand the welding mechanism, we perform molecular dynamics to investigate the interface configuration. the setup of the simulations performed in the present work is shown in Figure 22.26(a), where a 3.908 nm long (5, 5) open ended carbon nanotube with 380 atoms forms interface on the free surface of the gold (Au) slab with a set of 5534 atoms. Periodic boundary conditions are imposed in x and y direction, and free boundary conditions in the z direction are imposed. Our simulations are based on the massively parallel LAMMPS code (Plimpton, 1995). The interactions for Au–Au are modeled by embedded atomic method (EAM) models (Foiles *et al.*, 1986), while the C–Au interactions are modeled with Lennard–Jones 12–6 potential (Arcidiacono *et al.*, 2005) ($\sigma_{\text{c-NI}}$ = 2.29943 Å, $\varepsilon_{\text{c-Ni}}$ = 0.01273 eV). A cut-off distance of 9 Å is used for Lennard–Jones forces. The constant NPT integration is performed to the desired temperatures using a Nose–Hoover temperature thermostat (Hoover, 1986) and Nose–Hoover pressure barostat (Hoover, 1985) and setting the bulk pressure to zero. And this Au-CNT system is equilibrated to the desired temperature using a Nose–Hoover temperature thermostat (Hoover, 1986) with a time step of 0.1 fs for 500 ps.

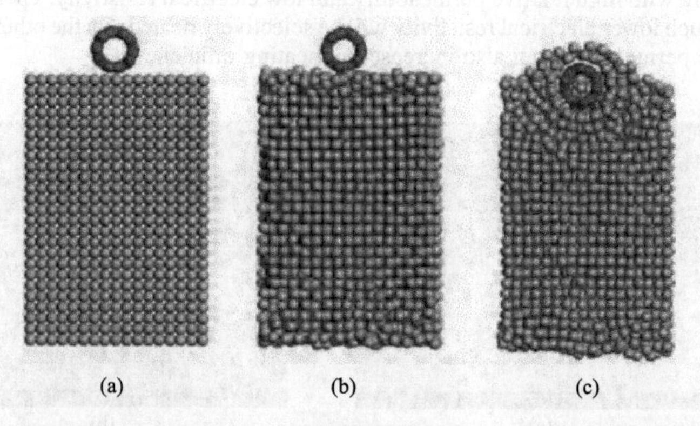

Figure 22.26 Molecular dynamics simulation results of CNT welding onto the metal. (a) Initial configuration of Au-CNT system; (b) time averaged configuration of Au–CNT system at 1050 K after 500 ps; (c) time averaged configuration of a cross-section of Au–CNT system with surface charges at 1050 K after 500 ps

Figure 22.26(b) presents the time averaged Au–CNT configuration at a temperature of 1050 K which is lower than the bulk melting point of Au. We cannot observe the welding behavior even though surface melting occurs on the electrodes. The result indicates that heating the metal structure to the bulk melting point is not necessary for nano welding as the surface melting can occur in lower temperature. In addition, Au has poor wetting properties to the CNT surface which has been reported by previous study (Zhang *et al.*, 2000). This condition restricts the choice of electrodes, in fact, it exclude most metals (Zhang *et al.*, 2000). However, we can observe Au coating on the CNT continuous in the experiment as shown in Figure 22.24. We contribute this behavior to electrowetting. Since the surface charges may reduce the surface tension of the melting metal layers and enhance attraction between metal atoms and the CNT surface, which may strongly influence the behavior of the dynamical properties of the interfaces and the wetting properties of the metal atoms (Chen *et al.*, 2005). For induction heating, the induction eddy current exist in both in plane and out of plane, and the surface charges are induced where the eddy current flow is restricted at the interface in the conductor and the air regions (Fujishima *et al.*, 2005). The exact distribution of the charge at the interface is determined by both the induction eddy current and the microscopic state. Particularly, for the nano welding process, the evolution of the interface state will inevitably lead to charge rearrangements which cannot be handled by empirical approaches. In our simulation model, we use constant charge distribution to describe the effects of interfaces charges on nano welding, which may be different from the real experimental conditions. Although a simplification of the charge distribution, this model contains the salient features of charged interface and should give a reasonable description of the effect of surface charges on the welding process.

In the case with charges, we assume that the negative charges with value of 6 e are evenly distributed on the first layer of the metal atoms, and the charge densities are presented by exact charge quantity per surface site. To keep the simulation system in electrical neutrality, the uniform positive charges with the same amount of charges as that of negative ones distributed on the tube. In this case, Coulombic pairwise interaction is added to the Lennard–Jones 12–6 potential. As shown in Figure 22.26(c), the welding of CNT onto the metal can be observed under consideration of surface charges. This electrowetting behavior indicates that the induction heating method can be used to weld CNT onto the metals with poor wetting properties to the CNT, which provides more choices for metallic electrodes.

Figure 22.27 shows the representative current-voltage (I–V) characteristic curves of electrode–CNTs–electrode configuration before and after the high frequency induction heating welding process. Before the bonding process the two-terminal resistances measured between the two electrodes are high, at the level of 1 MΩ. However, the two-terminal resistance can be significantly lowered after high frequency induction heating process, at the level of 10 KΩ. The mechanism of this reduction is based on the fact that the molten metal wets the surface of CNTs at the contact areas and the stable bonding between carbon and metal atoms are formed, the enlarged contact area also enhance the electron transporting

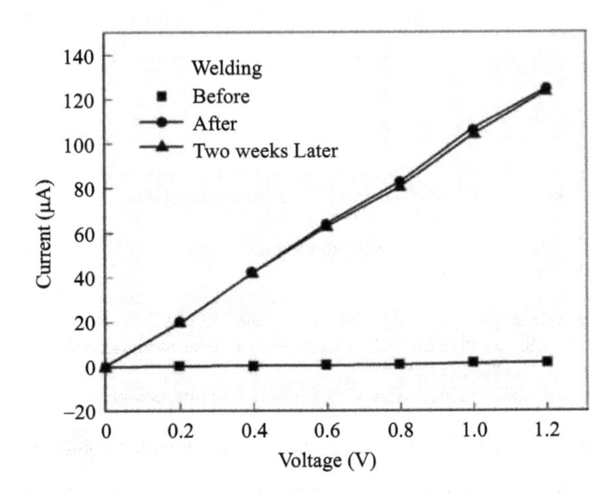

Figure 22.27 I–V measurements before and after high frequency induction heating process

through the contact interface. In addition, the heating can desorb some foreign molecules absorbed in the interface region, resulting in a decrease of the contact resistance.

The I–V characteristics of the assembled SWCNTs indicate that the optimum contact between aligned CNTs and Cr/Ni/Au electrode is realized by high frequency induction heating process. The results also suggest that the reduction of contact resistance achieved by high frequency induction heating is irreversible, which can be demonstrated by measuring the same samples after two weeks of initial measurements. There are no obvious changes in the contact resistance, as shown in Figure 22.27. Therefore, we can conclude that the localized induction heating can create reliable and larger contact bonding area between the CNTs and the metallic structures.

In conclusion, we have demonstrated that welding of CNTs onto metallic electrodes can be achieved by the high frequency induction heating on the basis of its non-contact and selective heating. The experiments have obtained optimum contact between CNTs and metal electrodes. We note that the reduction of contact resistance achieved by high frequency induction heating is irreversible. In particular, there is no obvious destruction of the electrodes and the CNTs during the heating. The process is promising for large-scale and wafer-level CNT-based devices fabrication. By performing multiphysics and molecular dynamics study, we investigate the mechanism of the welding process. The results indicate that heating will selectively occur on the metallic structure with high electronic conductivity and relative permeability. In addition, the atomic study indicates that the surface charge induced by the induction potential plays an important role in nano welding which can improve wetting between metal atoms and the CNT surface. These behaviors provide more choices of metallic electrodes rather than using metals with excellent wetting properties to the CNT.

22.5 Chapter Summary

We have carried out molecular dynamics simulation to investigate the contact configuration modification at CNT-metal interface during the nanowelding. By modeling the welding of the CNTs with various diameters onto the metal surface at different temperatures, we find that both side contact and core filling contact structures are formed at the interfaces. The results indicate that the metal atoms prefer filling into the core to migrating along the outer surface of the CNT, which results in larger contact length in the core than that on the outer walls. In addition, the filling of metal atoms can occur at much lower temperature than the melting point due to the local melting behaviors. We also note that the entrance of the metal atoms into the core may strongly enhance the local melting at the interfaces, which results in a larger contact length between the metal and the CNTs with larger diameters. These findings raise the possibility of welding the CNTs onto the metal at low temperature and forming CNT-metal hybrid structures at the interface to enhance the electrical and thermal conductivity. From point of view of experiment, the welding can be achieved by exerting heat onto the CNT-metal interface such as joule heating.

References

Alder, B.J. and Wainwrihgt T.E. (1957) Phase transition for a hard sphere system, *Journal of Chemical Physics* **27**:1208.

Andriotis, A., Menon, M. and Gibson, H. (2008) Realistic nanotube-metal contact configuration for molecular electronics applications, *IEEE Sensors Journal* **8**:910–913.

Antonio, T. and de Leeuw, N.H. (2006) Ab initio molecular dynamics study of 45S5 bioactive silicate glass, *Journal of Physical Chemistry B* **110**:25810–25816.

Arcidiacono, S., Walther, J.H., Poulikakos, D., Passerone, D. and Koumoutsakos, P. (2005) Solidification of gold nanoparticles in carbon nanotubes, *Physical Review Letters* **94**:105502.

Banerjee, S., Naha, S. and Puri, I.K. (2008) Molecular simulation of the carbon nanotube growth mode during catalytic synthesis, *Applied Physics Letter* **92**:233121.

Baughman, R.H., Zakhidov, A.A. and de Heer, W.A. (2002) Carbon nanotubes–the route toward applications, *Science* **297**(5582):787–792.

Becke, A.D. (1988) Density-functional exchange-energy approximation with correct asymptotic behavior, *Physical Review A* **38**:3098.

Binder, K., Horbaeh, J., Kob, W., Paul, W. and Varnik, F. (2004) Moleeular dynamics Simulations, *Journal of Physics: Condensed Matter* **16**:S429–S453.

Brenner, D.W., Shenderova, O.A., Harrison, J.A., Stuart, S.J., Ni, B. and Sinnott, S.B. (2002) A second-generation reactive empirical bond order (REBO) potential energy expression for hydrocarbons, *Journal of Physics: Condensed Matter* **14**:783–802.

Buongiorno, M., Nardelli, B.I., Bernholc, Y.J. (1998) Mechanism of strain release in carbon nanotubes, *Physics Review B* **57**:4277.

Burghard, M. (2005) Electric and vibrational properties of chemically modified single walled carbon nanotubes, *Surface Science Reports* **58**:1.

Car, R. and Parrinello, M. (1985) Unified approach for molecular dynamics and density-functional theory, *Physics Review Letter* **55**:2471–2474.

Chang, H. and Lee, J.D. (2001) Adsorption of NH_3 and NO_2 molecules on carbon nanotubes, *Applied Physics Letter* **79**:3863–3865.

Chen, C.C., Yeh, C.C., Liang, C.H. *et al.* (2001) Preparation and characterization of carbon nanotubes encapsulated GaN nanowires, *Journal of Physics and Chemistry of Solids* **62**:1577–1586.

Chen, J.Y., Kutana, A., Collier, C.P. and Giapis, K.P. (2005) Electrowetting in carbon nanotubes, *Science* **310**:1480–1483.

Chen, Z., Appenzeller, J., Knoch, J., Lin, Y.M. and Avouris, P. (2005) The role of metal-nanotube contact in the performance of carbon nanotube field-effect transistors, *Nano Letters* **5**(7):1497–1502.

Chen, C.X., Liu, L., Lu, Y., Kong, E.S., Zhang, Y., Sheng, X. and Ding, H. (2007), A method for creating reliable and low-resistance contacts between carbon nanotubes and microelectrodes, *Carbon* **45**:436–442.

Daw, B. (1983) Semiempirical, quantum mechanical calculation of hydrogen embrittlement in metals, *Physical Review Letter* **50**:1285–1288.

Dockendorf, C.P.R., Steinlin, M., Poulikakos, D. and Choi, T.Y. (2007) Individual carbon nanotube soldering with gold nanoink deposition, *Applied Physics Letter* **90**(19):193116.

Dong, Z.H., Xue C.S., Zhuang, H.Z. *et, al.* (2005) Synthesis of three kinds of GaN nanowires through Ga_2O_3 films' reaction with ammonia, *Physica E* **27**:32–37.

Dong, L.F., Youkey, S., Bush, J. and Jiao, J. (2007) Effects of local Joule heating on the reduction of contact resistance between carbon nanotubes and metal electrodes, *Journal of Applied Physics* **7101**:024320.

Enrico, T., Ivano, T. and Ursula, R. (2007) Trajectory surface hopping within linear response time-dependent density-functional, *Theory Physical Review Letter* **98**:023001-1-023001-4.

Foiles, S.M., Baskes M.I. and Daw, M. S. (1986) Embedded-atom-method functions for the fcc metals Cu, Ag, Au, Ni, Pd, Pt, and their alloys, *Physical Review B* **33**:7983–7991.

Fujishima, Y. and Wakao, S. (2003) Surface charge analysis in eddy current problems, *Electrical Engineering in Japan* **145**:59–66.

Giannozzi, P., Car, R. and Scoles, G. (2003) Oxygen adsorption on graphite and nanotubes, *Journal of Chemical Physics* **118**:1006.

Gill, P.M.W., Johnson, B.G., Pople, J.A. and Frisch, M.J. (1992) A standard grid for density functional calculations, *Chemical Physics Letters* **197**:499.

Goedecker, S., Teter, M. and Hutter, J. (1996) Separable dual-space gaussian pseudopotentials, *Physics Review B* **54**:1703–1710.

Goldsmith, B.R., Coroneus, J.G., Khalap, V.R., Kane, A.A., Weiss, G.A. and Collins, P.G. (2007) Conductance-controlled point functionalization of single-walled carbon nanotubes, *Science* **315**:77.

Gulseren, O., Yildirim, T. and Ciraci, S. (2001) Tunable adsorption on carbon nanotubes, *Physical Review Letter* **87**:116802.

Haddon, R.C. (2001) Comment on the relationship of the pyramidalization angle at a conjugated carbon atom to the sigma bond angles, *Journal of Physical Chemistry A* **105**:4164.

Han, W.Q., Fan, S.S., Li, Q.Q. *et al.* (1997) Synthesis of gallium nitride nanorods through a carbon nanotube–confined reaction, *Science* **277**:1287–1289.

Heinze, S., Tersoff, J., Martel, R., Derycke, V., Appenzeller, J. and Avouris, P.H. (2002) Carbon nanotubes as schottky barrier transistors, *Physical Review Letters* **89**:106801.

Hoover, W.G. (1985) Canonical dynamics: Equilibrium phase-space distributions, *Physics Review A* **31**:1695–1697.

Hoover, W.G. (1986) Constant-pressure equations of motion, *Physics Review A* **34**:2499–2500.

Hwang, H.J., Byun, K.R. and Kang, J.W. (2004) Carbon nanotubes as nanopipette: modellingand simulations, *Physica E* **23**:208–216.

Iijima, S. (1991) Helical of graphitic carbon, *Nature* **354**:56.

Jang, J.H., Herrero, A.M., Gila, B., Abemathy, C. and Craciun, V. (2008) Study of defects evolution in GaN layers grown by metal-organic chemical vapor deposition, *Journal of Applied Physics* **103**:063514.

Jiang, H., Zhu, L., Moon, K. and Wong, C.P. (2007), Low temperature carbon nanotube film transfer via conductive polymer composites, *Nanotechnology* **18**:125203.

Johnson, R.D., Bahr, D.F., Richards, C.D., Richards, R.F., McClain, D., Green J. and Jiao, J. (2009) Thermocompression bonding of vertically aligned carbon nanotube turfs to metalized substrates, *Nanotechnology* **20**:065703.

Kangawa, Y., Matsuo, Y., Akiyama, T. *et al.* (2007) Ab initio-based approach on initial growth kinetics of GaN on GaN (001), *Journal Crystal Growth* **301**:75–78.
Kazimirov, A., Scherb, G., Zegenhagen, J., Lee, T.L., Bedzyk, M.J., Kelly, M.K., Angerer, H. and Ambacher, O. (1998) A structure study of the electroless desposition of Au on Si(111):H, *Applied Physics Letter* **84**:1703.
Kirkpatriek, S., Gelatt Jr., C.D. and Vecchi, M.P. (1983) Optimization by simulated annealing, *Science* **220**:671–680.
Kirkpatriek, S. (1984) Optimization by simulated annealing: quantitative studies, *Journal of Statistical Physics* **34**:975–986.
Krupke, R., Hennrich, F., Lohneysen, H. *et al.* (2003) Separation of metallic from semiconducting single-walled carbon nanotubes, *Science* **301**:344–347.
Kühne, T.D., Krack, M., Mohamed, F.R. and Parrinello, M. (2007) Efficient and accurate car-parrinello-like approach to born-oppenheimer molecular dynamics, *Physical Review Letter* **98**:066401.
Kumar, A., Pushparaj, V.L., Kar, S., Nalamasu, O., Ajayan, P.M. and Baskaran, R. (2006) Contact transfer of aligned carbon nanotube arrays onto conducting substrates, *Applied Physics Letter* **89**:163120.
Kumar, V.S., Kumar, J., Srivastava, R.K. *et al.* (2008) Growth and characterization of gallium nitride nanocrystals on carbon nanotubes, *Journal of Crystal Growth* **310**:2260–2263.
Lan, C., Srisungsitthisunti, P., Amama, P.B., Fisher, T.S., Xu, X. and Reifenberger, R.G. (2008) Measurement of metal/carbon nanotube contact resistance by adjusting contact length using laser ablation, *Nanotechnology* **19**:125703.
Langel, W., (1996) Car-Parrinello simulation of NH3 adsorbed on the MgO(100) surface, *Chemical Physics Letters* **259**(1–2):7–14.
Lee, C., Yang, W. and Parr, R.G. (1988) Development of the Colle-Salvetti correlation-energy formula into a functional of the electron density, *Physical Review B* **37**:785.
Li, J., Ye, Q., Cassell, A. *et al.* (2003) Bottom-up approach for carbon nanotube interconnects, *Applied Physics Letter* **82**:2491–2493.
Lutsko, J., Wolf, F.D., Phillpot S.R. *et al.* (1989) Molecular-dynamics study of lattice-defect-nucleated melting in silicon, *Physical Review B* **40**(5):2841.
Marco, P., Marcella, I., Gianni, C., Michele, P. and Vincenzo, S. (2006) Lithium hydroxide phase transition under high pressure: an Ab initio molecular dynamics study, *Chem Phys Chem* **7**:141–147.
Margulis, Vl.A. and Muryumin, E.E. (2007) Atomic oxygen chemisorption on the sidewall of zigzag single-walled carbon nanotubes, *Physical Review B* **75**:035429-1–12.
Martins, J.B.L., Longo, E., Salmon, O.D.R., Espinoza, V.A.A. and Taft, C.A. (2004) First principles study of the polar O-terminated ZnO surface, *Chemical Physics Letter* **400**:481.
Mathias, R., Robert, B., Thomas, H. and Gotthard, S. (2007) Car–Parrinello treatment for an approximate density-functional theory method, *The Journal of Chemical Physics* **126**:124103.
Mischler, C., Horbach, J., Kob, W. *et al.* (2005) Water adsorption on amorphous silica surfaces: a Car-Parinello simulation study, *Journal of Physics: Condensed Matter* **17**:4005–4013.
Mishin, Y. (2004), Atomistic modeling of the γ and γ'-phases of the Ni-Al system, *Acta Materials* **52**:1451.
Munoz, A. and Kunc, K. (1991) High-Pressure phase of gallium nitride, *Physics Review B* **44**:10372–10373.
Munoz, A. and Kunc, K. (1993) First-principles calculation of the structural, electronic, and vibrational properties of gallium nitride and aluminum nitride, *Physical Review B* **48**:7897–7902.
Najib, A. and Kasti, Z. (2007) Zigzag carbon nanotubes molecular/structural mechanics and the finite element method, *International Journal of Solids and Structures* **44**:6914.
Nakamura, S., Mukai, T. and Senoh, M. (1994) Candela-class high-brightness InGaN/AlGaN blue-light-emitting diodes, *Applied Physics Letter* **64**:1687.
Nemec, N., Tománek, D. and Cuniberti, G. (2006) Contact dependence of carrier injection in carbon nanotubes: an Ab initio study, *Physical Review Letter* **96**:076802.
Nigra, P. and Freeman, D.L. (2004) On the encapsulation of nickel clusters by molecular nitrogen, *Journal of Chemical Physics* **121**:475.
Niyogi, S., Hamon, M.A. and Hu, H. (2002) Chemistry of single-walled carbon nanotubes, *Accounts of Chemical Research* **35**:1105.
Orazio, F., Simona, Q., Giovanna, V., Ettore, F., Aldo, G. and Gloria, T. (2002) High-pressure behavior of bikitaite: An integrated theoretical and experimental approach, *American Mineralogist* **87**:1415–1425.
Park, N. and Hong, S. (2005) Electronic structure calculations of metal-nanotube contacts with or without sites Eb (eV) θP (°) oxygen adsorption, *Physics Review B* **72**:045408.
Pearton, S.J. and Ren, F. (2000) GaN electronics, *Advanced Materials* **12**:1571.
Peng, Y., Cullis, T. and Inkson, B. (2009) Bottom-up nanoconstruction by the welding of individual metallic nanoobjects using nanoscale solder, *Nano Letter* **9**(1):91–96.
Pietro, B., Simona, Q., Uartieri, A. Sani, and Giovanna, V. (2002) High-pressure deformation mechanism in scolecite: A combined computational experimental, *Study American Mineralogist* **87**:1194–1206.

Ponce, F.A., Bour, D.P., Young, W.T., Saunders, M. and Steeds J.W. (1996) Determination of lattice polarity for growth of GaN bulk single crystals and epitaxial layers, *Applied Physics Letter* **69**:337.

Plimpton, S.J. (1995) Fast parallel algorithms for short-range molecular dynamics, *Journal of Computational Physics* **117**:1–119.

Sergei, I. and Gregory, A. (2005) Voth Ab initio molecular-dynamics simulation of aqueous proton salvation and transport revisited, *The Journal of Chemical Physics* **123**:044505.

Slawomir, B. and Stephen H. (1996) Garofalini, molecular dynamics study of silica-alumina interfaces, *Journal of Physics Chemistry* **100**:2201–2205.

Smith, A.R., Feenstra, R.M., Greve, D. W., Neugebauer, J., and Northrup, J. E. (1997), Reconstructions of the GaN (0001) surface, *Physical Review Letter*, **79**:3934.

Smith, A. R., Feenstra, R. M., Greve, D.W., Shin, M.S., Skowronski, M., Neugebauer, J. and Northrup, J.E. (1999) GaN (0001) surface structures studied using scanning tunneling microscopy and first-principles total energy calculations, *Surface Science* **423**:70.

Song, X.H., Liu, S., Yan, H. and Gan, Z.Y. (2009) First-principles study on effects of mechanical deformation on outer surface reactivity of carbon nanotubes, *Physica E* **41**:626.

Stuart, T. and Harrison, J.A. (2000) A reactive potential for hydrocarbons with intermolecular interactions, *Journal of Chemical Physics* **112**:6472.

Sun. Y.P., Fu. K.F., Lin. Y. and Huang, W.J. (2002) Functionalized carbon nanotubes properties and applications, *Accounts of Chemical Research* **35**:1096.

Sun, Y., Zhu, L, Jiang, H., Lu, J., Wang, W. and Wong, C.P. (2008) A paradigm of carbon nanotube interconnects in microelectronic packaging, *Journal of Electronic Materials* **37**:1691–1697.

Takanori, I. and Kazume, N. (2002) First principles calculations for electronic band structure of single-walled carbon nanotube under uniaxial strain, *Surface Science* **514**:222.

Taniyasu, Y. and Yoshikawa, A. (2001) In-situ monitoring of surface stoichiometry and growth kinetics study of GaN (0001) in MOVPE by spectroscopic ellipsometry, *Journal of Electronic Materials* **30**:1402–1407.

Tans, S.J., Verschueren A.R.M. and Dekker, C. (1998) Room-temperature transistor based on a single carbon nanotube, *Nature* **393**:49–52.

Ulbricht, H., Moos, G. and Hertel, T. (2002) Physisorption of molecular oxygen on single-wall carbon nanotube bundles and graphite, *Physics Review B* **66**:075404.

Wang, J., Li, J., Yip, S., Phillpot, S. *et al.*, (1997) Mechanical instabilities of homogeneous crystals, *Physica, A* **240**:396.

Watanabe, T., Tatsumura, K., and Ohdomari, I. (2004) SiO_2/Si interface structure and its formation studied by largr-scale molecular dynamics simulation, *Applied Surface Science* **237**:125–133.

Woo, Y., Duesberg, G.S. and Roth, S. (2007) Reduced contact resistance between an individual single-walled carbon nanotube and a metal electrode by a local point annealing, *Nanotechnology* **18**:095203.

Wu, J., Zang, J., Larade, B., Guo, H. and Gong, X.G. (2004) Computational design of carbon nanotube electromechanical pressure sensors, *Physical Review B* **69**:153406.

Xu, J. and Fisher, T.S. (2006) Enhancement of thermal interface materials with carbon nanotube arrays, *International Journal of Heat and Mass Transfer* **49**:1658–1666.

Xue, C.S., Wu, Y.X. and Zhuang, H.Z. *et al.* (2005) Growth and characterization of high-quality GaN nanowires by ammonification technique, *Physica E* **20**:179–181.

Yang, H., Wu, M.C. and Fang, W. (2005) Localized induction heating solder bonding for wafer level MEMS packaging, *Journal of Micromechanics Microengineering* **15**:394–399.

Yan, H., Gan, Z., Song, X., Chen, Z., Xu, J. and Liu, S. (2009) Adsorption configuration of magnesium on wurtzite gallium nitride surface using first-principles calculations, *Physica B: Condensed Matter* **404**(20):594–3597.

Yang, J., Hu, W. and Xiao, S. (2007) Surface melting of close-packed Mg (0001), *Solid State Communications* **143**:545–549.

Yuan, Q.Z. and Zhao, Y.P. (2009) Hydroelectric voltage generation based on water-filled single-walled carbon nanotubes, *Journal of American Chemical Society* **131**:6374–6376.

Zhang, Y., Franklin, N.W., Chen, R.J., and Dai, H. (2000) Metal coating on suspended carbon nanotubes and its implication to metal-tube interaction, *Chemical Physics Letters* **331**:35–41.

Zhang, Y.F. and Liu, Z.F. (2004) Oxidation of zigzag carbon nanotubes by singlet O2: dependence on the tube diameter and the electronic structure, *Journal of Physical Chemistry B* **108**:11435.

Zhao, J.J., Buldum, A., Han, J. *et al.* (2002) Gas molecule adsorption in carbon nanotubes and nanotube bundles, *Nanotechnology* **13**:195–200.

23

Aluminum Nitride Deposition

Bestowed with properties of wide band gap (6.2eV), high thermal conductivity [3.3W/(K·cm)], high electrical resistivity (10^{13}V·cm) and piezoelectric effect and high strength, AlN is widely used in the area of photoelectric devices, piezoelectric devices, electronic packaging, microelectronic devices and hard coating especially in the field of photoelectric devices, it is extensively used for fabrication of the deep-ultraviolet light-emitting diodes (UV-LEDs) (Hirayama *et al.*, 2009; Subramani *et al.*, 2014; Yan *et al.*, 2015).

However, because of the high cost of AlN substrate and its limited performance, growing AlN films on a foreign substrate is still the main method. Sapphire (Potì, 2006), SiC (Lin *et al.*, 2010) and Si (Mino *et al.*, 2012) are the cardinal substrates. Nevertheless, the lattice mismatch with those materials cannot be ignored, which results in high threading dislocation density, and it is still difficult to reduce threading dislocation density to 105cm^2 (Imura *et al.*, 2007; Dong *et al.*, 2014; Lee *et al.*, 2005; Paduano *et al.*, 2003; Bai *et al.*, 2006); therefore it is hard to improve the performance of devices (Pernot *et al.*, 2010; Shatalov *et al.*, 2012; Hirayama *et al.*, 2002) Growing AlN films on AlN substrate with zero lattice mismatch is beyond doubt the best way; however, commercially available AlN substrate is still immature. Thus, the technique for inducing homoepitaxial AlN growth needs to be further exploited.

Moreover, the growth mechanisms concerning homoepitaxial growth of c-plane AlN are still not enough. A lot of scholars only studied homoepitaxial growth of c-plane AlN from experimental aspects, and there is still a long way to go, although some progress has been made (Kinoshita *et al.*, 2013; Bryan, *et al.*, 2013; Bryan *et al.*, 2016). Although c-plane is a critical plane for growing AlN (Funato *et al.*, 2012), there are a few studies concerning the corresponding mechanism of polar c-plane AlN. Similar theoretical studies have been reported on the deposition of GaN (Zhou *et al.*, 2006) and non-polar AlN (Leathersich *et al.*, 2013). Atomic-scale simulation of the growth and its dependency on various deposition conditions will facilitate effective understanding of the c-plane AlN growth process. The atomic assembly mechanism, crystallinity and defects can be visualized directly by molecular dynamics (MD) simulations instead of experiments. The growth of AlN on c-plane AlN is studied by the large-scale atomic/molecular massively parallel simulator (Plimpton *et al.*, 1995) with the Tersoff potential (Tungare *et al.*, 2011). At a N: Al flux ratio of 2.0, the growth temperature is varied from 1000K to 2000K with an increment of 200K. Then the N: Al flux ratio is varied from 0.8 to 2.8 with an increment of 0.4 at 1800K. The growth rate, crystallinity and stoichiometry under different conditions are discussed in detail. In addition, the reaction mechanism between substrate atoms and adatoms is revealed. Furthermore, the stresses of deposited AlN films are investigated.

23.1 Study Effects of Temperature and N: Al Flux Ratio on Deposited AlN

23.1.1 Model and Methods

The three-dimensional model of the AlN substrate is built, as shown in Figure 23.1. *x*, *y* and *z* represent the directions of , $<1\bar{1}00>$ $<11\bar{2}0>$ and <0001> correspondingly. The substrate filled with 10800 atoms (5400 Al atoms and 5400 N atoms) has a dimension of 46.65<Å, 80.8002Å and 29.868Å, accordingly. The red atoms represent the Al atoms, while the blue atoms represent the N atoms. The atoms on the top layers are

Modeling and Simulation for Microelectronic Packaging and Integration: Manufacturing, Reliability and Testing, Second Edition. Sheng Liu and Yong Liu.

N atoms. The substrate is divided into three groups: two pairs of closely spaced Al, and N planes along the z direction at the bottom are fixed to prevent the moving of the substrate due to the hitting of Al and N atoms during the deposition, which is called the fix group. The middle eight pairs of atoms make up the thermal control group, where the canonical (NVT) ensemble is used to perform time integration on Nose-Hoover style non-Hamiltonian equations of motion to update the positions and velocities of the atoms to get the prescribed substrate temperature (Leathersich *et al.*, 2013). The uppermost two pairs of atoms form the free group in which the atoms are entirely free to interact and transmit energy with the deposited atoms. Periodic boundary conditions are applied to the directions of x and y in order to create a seemingly infinite boundary in these dimensions, while the free boundary condition is applied to the z direction to enable the deposition of Al and N atoms towards the substrate surface. The growth of AlN is simulated by periodically injecting an Al atom or a N atom from a 25-fold-lattice height towards the substrate surface, and the x and y coordinates of injected atoms are allocated randomly. The incident angle was 0° (perpendicular to the substrate) (Cao *et al.*, 2013), which is indicated by the black arrow in Figure 23.1. The injected atoms were all assigned a kinetic energy of 0.17eV. Such thermalized fluxes are analogous to those of a molecular beam epitaxy or a high pressure sputtering process. The three-body Tersoff potential is applied to describe the atomic interactions between atoms, and its reliability and application have already been proved (Leathersich *et al.*, 2013; Tungare *et al.*, 2011; Xiang *et al.*, 2017).

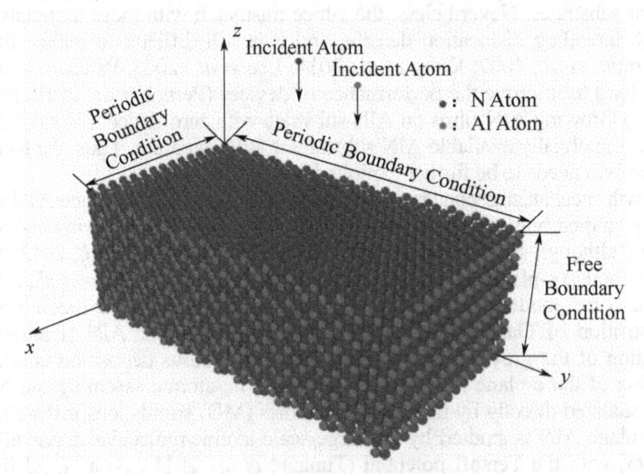

Figure 23.1 Model of the substrate; the red atoms represent the Al atoms, the blue atoms represent the N atoms (*Color version online*)

The time interval of injected Al atoms is kept as a constant of 5ps per atom, while that of N atoms is varied to get a range of N: Al flux ratios, ranging from 0.8 to 2.8. The total number of injected Al atoms is 4000, while the injected N atoms are varied according to the N: Al ratio in each simulation. The deposition process takes 20000ps. After the deposition, a relaxation process of 10000ps is taken to equilibrate the system. The visualization of the model is realized by OVITO (Stukowski *et al.*, 2009).

The stress of the substrate and deposited films is calculated. Because stress is closely correlated to crystal quality, atom stress can be described by Basinski *et al.* as follows (Basinski *et al.*, 1971):

$$\sigma = \frac{1}{V_i}\left(m_i v_i + \frac{1}{2}\sum_{i \neq j}^{N} \boldsymbol{r}_{ij} \otimes \boldsymbol{f}_{ij}\right) \tag{23.1}$$

$$\sigma_{mn}^{\text{avg}} = \frac{1}{N}\sum_{i=1}^{N} \sigma_{mn}^{i} \tag{23.2}$$

In Equation 23.1, V_i represents the volume of the atoms, m_i represents the mass of atom i, v_i represents the velocity of atom i, r_{ij} denotes the distances between atom i and the atoms around it, f_{ij}

denotes the inter-atomic forces between them, and $\otimes$ represents the tensor product between the two vectors.

In Equation 23.2, σ_{mn}^{i} represents the stress of the *i*th atom, *m* and *n* are indices of the stress tensor, and *N* is the number of atoms in the substrate or the deposited film.

Here we calculate the average normal stress σ_{zz}^{avg} and the average mean biaxial stress: $(\sigma_{xx}^{\text{avg}}+\sigma_{yy}^{\text{avg}})/2$. They are selected because the first one is one order of magnitude higher than the shear stress in the deposition case, and the latter is related to the residual stress of the subject, which are interrelated to their structure composition (Basinski *et al.*, 1971; Stukowski *et al.*, 2012).

To quantitatively compare the atomic structure of deposited films, dislocation analysis (DXA) (Stukowski *et al.*, 2012) is adopted, in which the local environment of each atom is analyzed to identify atoms that form a perfect crystal lattice. Atomic structure identification is based on the common neighbor analysis method, which can identify the hexagonal diamond, cubic diamond and other structures. The DXA ignores the chemical atom types, thus the hexagonal diamond (cubic diamond) structure could be treated as wurtzite (zinc blende).

23.1.2 Results and Discussion

(i) Effect of Temperature

Temperature is a critical factor for growth of AlN, and high temperature is favorable for growing AlN (Ambacher *et al.*, 1998). However, it is still a great challenge to enhance the temperature of the reactor chamber higher than 2000K experimentally. Hence the temperature is changed from 1000K to 2000K in order to discuss the effect of growth temperature, and the N: Al flux ratio is set to be 2.0. The growth results under different temperatures are shown in Figure 23.2. The deposition process is divided into four stages by time and the total number of deposited atoms at all stages are calculated as shown in Figure 23.2(a)–(d). We find that as the temperature increases from 1000K to 2000K, the total number of finally deposited atoms decreases. For a certain stage, the total number of deposited atoms decreases as the temperature increases despite the fact that the injected atoms are all the same. The number of deposited atoms indicates the growth rate of AlN films. It is obvious that the growth rate decreases at all

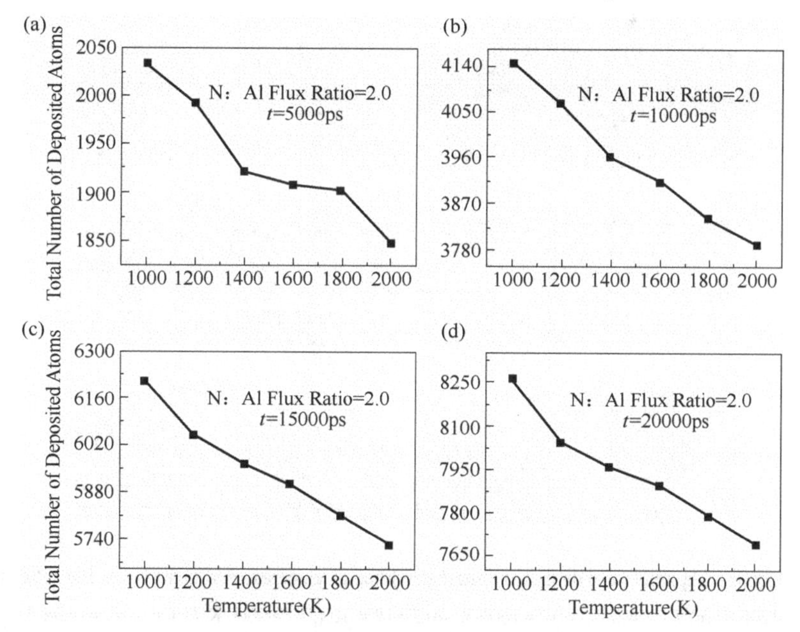

Figure 23.2 Total number of atoms during and after deposition under different deposition temperatures

stages as the temperature increases from 1000K to 2000K. This is owing to the improved desorption effect when the temperature increases (Choi *et al.*, 2007). During the deposition, some of the injected atoms adsorb on the surface, while other atoms desorb from the surface owing to thermal fluctuation. The injected Al atoms are mainly consumed to form AlN, while the injected N atoms are easier to desorb when the temperature increases. Increasing the temperature increases the probability of desorption, thus reducing the growth rate.

Crystal lattice perfection is essential for the AlN film. Deviation from the ideal lattice configuration can change the coordinates of atoms and overstretch bonds, which result in changed electron populations in the valence and conduction bands. In addition, the existence of defects and even polytypism seriously affects the electrical and optical properties of AlN (Nepal *et al.*, 2004). Thus, the perfect lattice, a stoichiometric composition and low defect concentration are expected to produce a high-quality AlN film.

To explore the lattice arrangement and defects, the atomic-scale structures of deposited AlN films under different growth temperatures are obtained as shown in Figure 23.3. The light blue atoms indicate the N atoms in the substrate, dark blue atoms indicate the Al atoms in substrate, red atoms indicate the injected N atoms and green atoms indicate the injected Al atoms. For a perfect lattice, a hexagonal mesh structure can be observed when seeing along the *x*-axis, as shown by the substrate atoms. The atomic-scale structure of the deposited AlN film at low temperature (1000K) contains a high population of defects and even some amorphousness, as shown in Figure 23.3(a). As the temperature increases from 1000K to 1800K, the deposited films exhibit much better developed crystalline features. The deposited AlN film shows regular epitaxial planes and contains a low defect concentration at 1800K, as shown in Figure 23.3(e). The crystalline feature gets saturated from 1800K to 2000K. To quantitatively compare the atomic structure of deposited films under different temperatures, Figure 23.4 shows the structure component of AlN films with a N: Al flux ratio of 2.0. Three types of structures: wurtzite, zinc blende and other structure are counted. Among them, wurtzite is the ideal structure of the AlN film. Figure 23.4 shows an increased component of wurtzite as the growth temperature increases from 1000K to 1800K, which indicates the improved crystallinity as the temperature increases from 1000K to 1800K. It is obvious that high temperature is good for getting better crystallinity. With the diffusion barrier of 1.17eV, it is difficult for Al adatoms to diffuse and adsorb (Bryan *et al.*, 2016; Choi *et al.*, 2007). Increasing the temperature increases the mobility of Al adatoms, promoting better crystallinity of deposited films. The effect of temperature on crystalline quality has been experimentally explored. Claudel *et al.* (Claudel *et al.*, 2011) found that the crystalline quality of epitaxial AlN layers grown on AlN templates increased with increasing deposition temperature from 1400℃ to 1500℃, which is clearly consistent with our simulation.

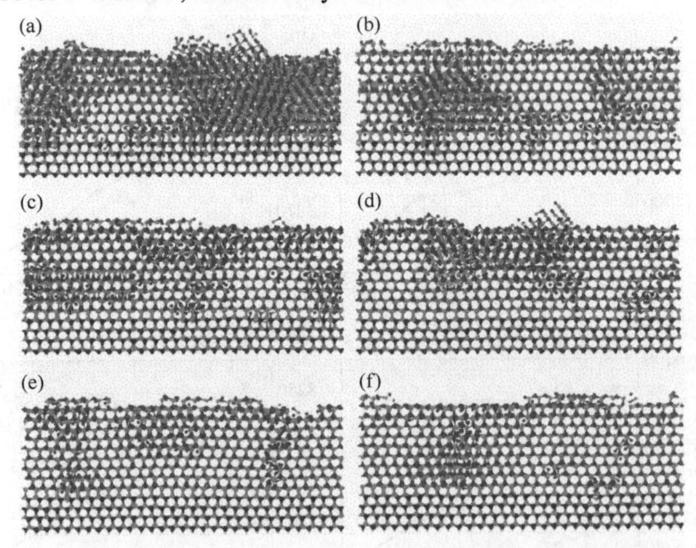

Figure 23.3 Atomic-scale structure of deposited films under different temperatures with the N : Al flux ratio of 2.0. (a) T= 1000 K; (b) T=1200 K; (c) T= 1400 K; (d) T=1600 K; (e) T=1800 K; (f) T=2000 K(*Color version online*)

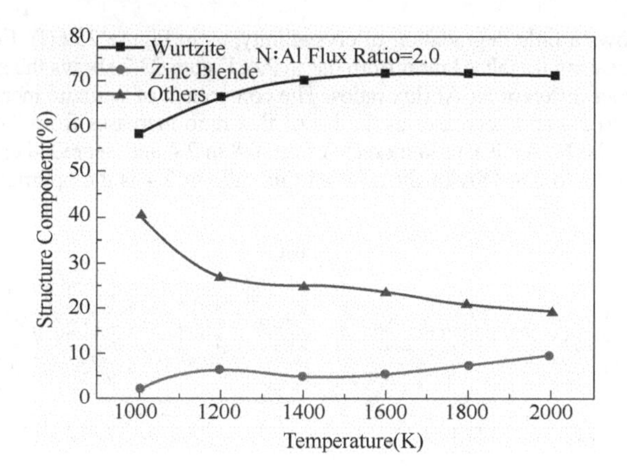

Figure 23.4 Structure component of the AlN film under different temperatures with a N : Al flux ratio of 2.0

The ideal stoichiometry of the AlN film is 1, which means the perfect N fraction in AlN is 50%. Greater than or less than 50% means excess of N or Al atoms, correspondingly, both of which cause defects in deposited AlN film. Here, the N fractions of deposited AlN films with a N: Al flux ratio of 2.0 under different temperatures are calculated in Figure 22.32. It is clear that as the temperature increases, the N fraction decreases. A N fraction much greater than 50% appears at 1000K and 1200K; the related defect clusters are shown as insets in Figure 23.5. It is approximately 50% from 1800K to 2000K. This indicates that a higher N: Al flux ratio is needed to get an ideal stoichiometric ratio as the temperature increases. Combining Figures 23.3 and 23.4, we can clearly see that the AlN film at 1800K has both a relatively good crystallinity and N fraction, which indicates the connection between good crystallinity and a near 50% N fraction.

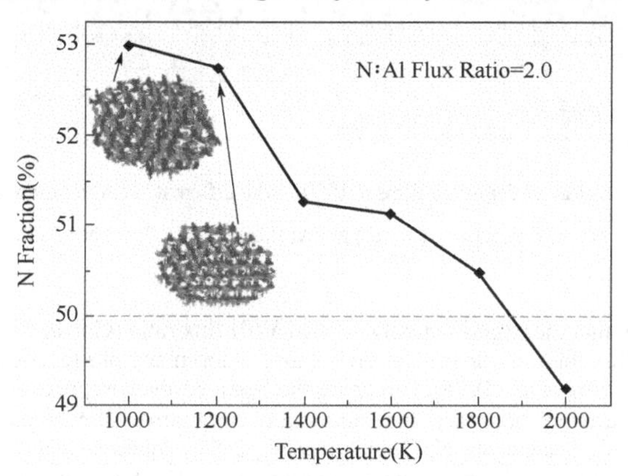

Figure 23.5 N fraction of deposited films with a N : Al flux ratio of 2.0 under different deposition temperatures ranging from 1000K to 2000K, with a increment of 200 K

(ii) Effect of N : Al Flux Ratio

As we know from the above, the N: Al flux ratio is an important factor influencing crystallinity as well as stoichiometry. Here, the N: Al flux ratio of deposited AlN at 1800K is further optimized. The atomic scale structures of the deposited AlN films at 1800 K are shown in Figure 23.6(a)–(f) with six different N: Al ratios (0.8, 1.2, 1.6, 2.0, 2.4 and 2.8). It can be observed in Figure 23.6(a) and (b) that at a low N: Al flux ratio (0.8 and 1.2), the atoms are arranged in great disorder. As the N: Al flux ratio increases to 1.6, there is still some disorder but the overall film crystallinity is improved. When the N: Al flux ratio is further increased to 2.0 and 2.4, the crystallinity of the film is further improved. Little disorder and few defects remain. At the N: Al flux ratio of 2.8, the film shows a little degradation of crystallinity, as in Figure 23.6(f). Epitaxial film growth is clearly promoted by increasing the N: Al ratio from 0.8 to 2.4. Figure 23.7 shows the structure component of AlN films at 1800 K under different N: Al flux ratios. The component of wurtzite increases as the N: Al flux ratio increases from 0.8 to 2.4 and decreases as the N: Al flux ratio increases from 2.4 to 2.8, indicating the improved crystallinity as the N: Al flux ratio increases from 0.8 to 2.4 and degraded crystallinity as the N: Al flux ratio increases from 2.4 to 2.8. Obviously, a N: Al flux ratio of 2.4 is the optimal flux ratio for the AlN film grown at 1800K.

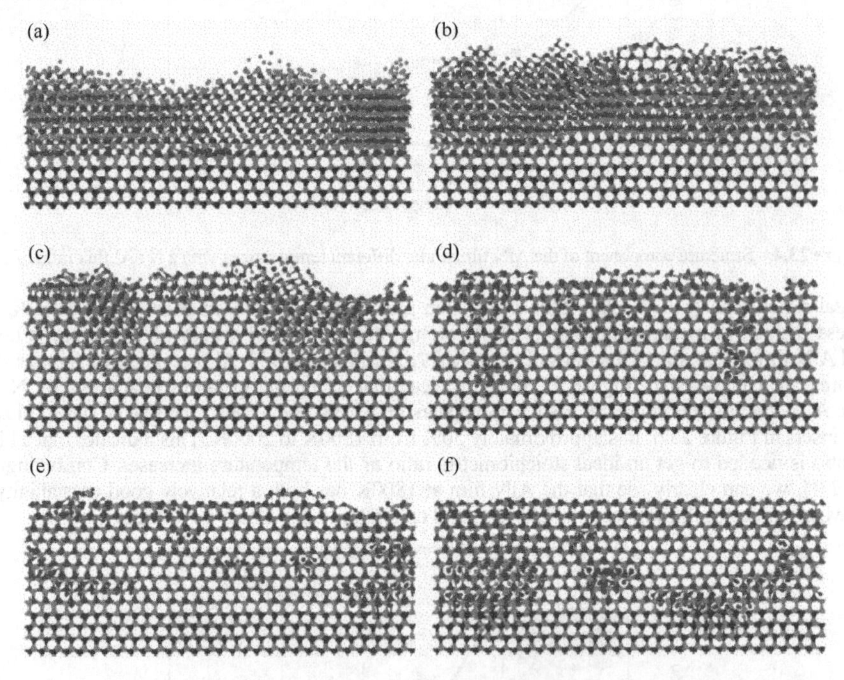

Figure 23.6 Atomic-scale structure of deposited films at 1800K with different N : Al flux ratios. (a) N : Al flux ratio = 0.8; (b) N : Al flux ratio = 1.2; (c) N : Al flux ratio = 1.6; (d) N : Al flux ratio = 2.0; (e) N : Al flux ratio = 2.4; (f) N : Al flux ratio = 2.8 (*Color version online*)

It should be pointed out that the experimentally applied V/III flux ratio (Guo *et al.*, 2017; Imura *et al.*, 2006) is not equal to the N: Al flux ratio in our simulation because of too many pre-reactions between the Al source and the N source (Mihopoulos *et al.*, 1998). But using the beam equivalent pressure ratio to represent the N: Al flux ratio can approximately approach our simulation conditions. The N plasma that dominates the reaction could be observed. Kaneko *et al.* (Kaneko *et al.*, 2016) found that the emission spectrum of N plasma is dominated by a series of sharp atomic N emission peaks (N*) during the growth of AlN. Thus, the N : Al flux ratio could be denoted by N* : Al. They found that the high Al : N ratios (1.67 and 1.18) leads to bad AlN crystal quality, which indicates that a lower Al : N ratio (lower than 1.18), in other words, a higher N : Al ratio (higher than 1/1.18, ~0.85 is needed to get good crystal quality. Meanwhile, our simulation also indicates that a N : Al flux ratio higher than 1 is needed to get good crystallinity under different temperatures. The trends in the experiment and our simulation are consistent.

The N fraction of deposited AlN at 1800K is explored as a function of N: Al flux ratio in Figure 23.7. The deposited film has a N fraction below 50% when the N: Al flux ratio increases from 0.8 to 1.6 and becomes approximately 50% at the N: Al flux ratios of 2.0 and 2.4. Sub-stoichiometric surfaces are formed when the N: Al flux ratio is below 2.0. As the N: Al flux ratio increases to 2.8, the N fraction of the deposited film becomes greater than 50%. Less or greater than 50%, both cause defects and degraded crystallinity. The related defects cluster at the N: Al flux ratios of 0.8 and 1.2 are shown as insets in Figure 23.8. This could be definitely proved by Figures 23.6 and 23.7.

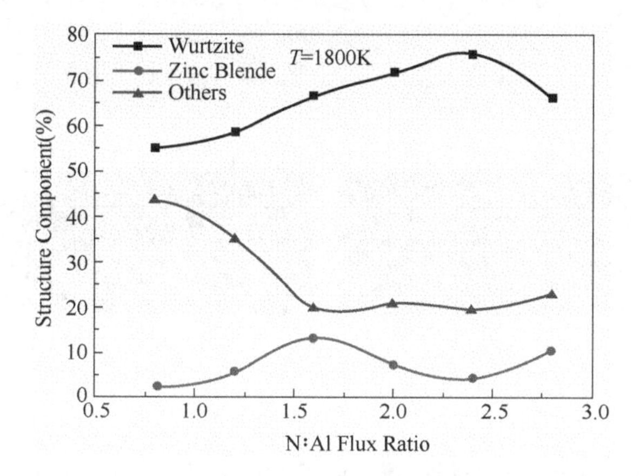

Figure 23.7 Structure component of the AlN film under different temperatures with a N : Al flux ratio of 2.0

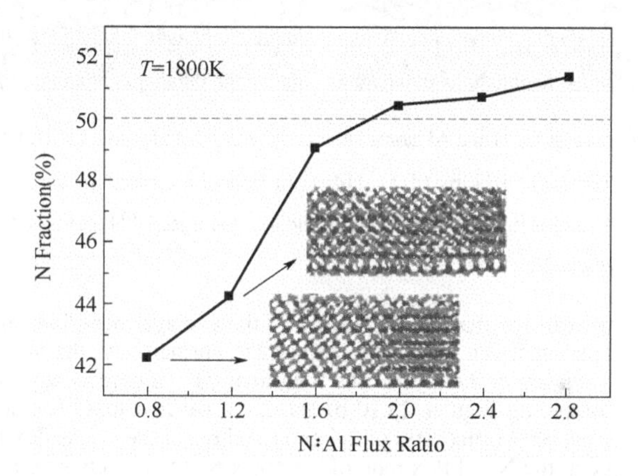

Figure 23.8 N fraction of deposited films at 1800K with different N: Al flux ratios from 0.8 to 2.8, with an increment of 0.4

Figures 23.5 and 23.8 show a phenomenon that the N fraction decreases as the temperature increases and increases as the N: Al flux ratio increases. The ideal N: Al flux ratio at 1800K is 2.4 (greater than 1). Here, time-resolved images of the interaction between injected Al and N atoms with substrate atoms on the surface are shown in Figure 23.9(a) − (d). There are two N atoms and one Al atom moving towards the substrate in Figure 23.9(a), then one N and one Al atom impact with atoms on the substrate surface correspondingly in Figure 23.9(b). In Figure 23.9(c), the impacting Al atom is bonded to the atoms on the substrate surface, while the impacting N atom rebounds and moves away. Figure 23.9(d) shows another rebounded N atom and a new impacting Al atom. The different mechanism of Al and N atoms interacting with substrate atoms means that the number of injected N atoms is always more than the number of injected Al atoms and the number of injected N atoms is always greater than deposited N atoms. Film with good crystallinity requires a near stoichiometry. Thus, the AlN film is always grown with a N: Al ratio greater than 1.

The crystal quality of the film is closely related to the stress in it. Therefore, the stress of the films is discussed. Here we discuss the average normal stress and the average mean biaxial stress of the films at 1800K with the N: Al flux ratios of 2.0, 2.4 and 2.8. The deposited films are divided into 26 layers along the z-axis while ignoring the fixed atoms at the bottom of the substrate. The height of each layer is 2Å . Then, the average normal stress and the average mean biaxial stress of each layer are calculated by Equations 23.1 and 23.2, respectively.

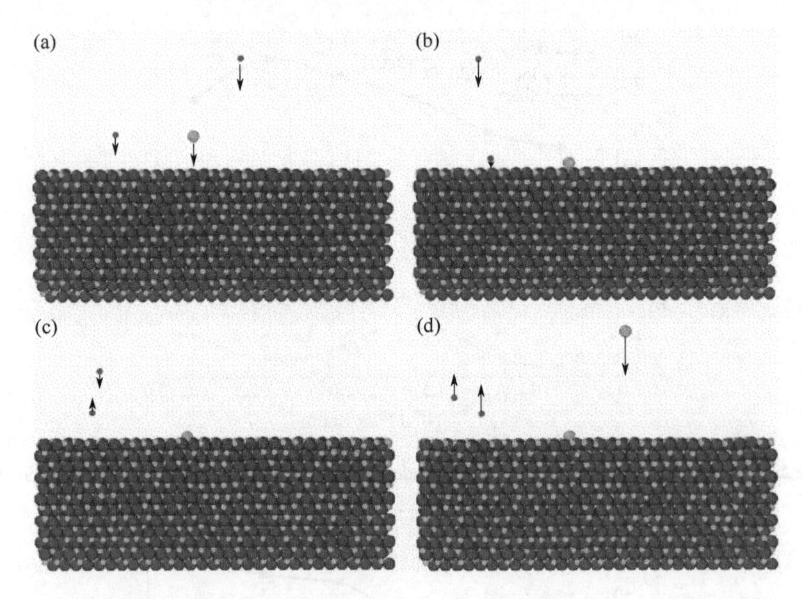

Figure 23.9 Time-resolved process for N and Al atoms interacting with substrate atoms at 1800K with a N: Al flux ratio of 2.0. (a) t = 111ps; (b) t = 112ps; (c) t = 113ps; (d) t = 114ps (the light blue atoms indicate the N atoms in the substrate, dark blue atoms indicate the Al atoms in substrate, red atoms indicate the injected N atoms and green atoms indicate the injected Al atoms) (*Color version online*)

From Figure 23.10(a), we can see that the fluctuation of the average mean biaxial stress at a N: Al flux ratio of 2.4 is minimal and maximal at a N: Al flux ratio of 2.0. Furthermore, the biggest fluctuation range at the N: Al flux ratio of 2.0 appears at the layers where the majority of defects appears, as shown in Figure 23.6(d). Similarly, when comparing Figures 23.10(b) and 23.6, we find that the fluctuation of the average normal stress is minimal at the N: Al flux ratio of 2.4 and maximal at the N: Al flux ratio of 2.8. Moreover, the biggest fluctuation range at the N: Al flux ratio of 2.8 appears at the layers where the majority of defects appears, as shown in Figure 23.6(f). Taken together, the defects and stresses are interrelated. The defects appear where stresses occur. Furthermore, good crystallinity at 1800K is obtained when the fluctuation of both the average mean biaxial stress and the average normal stress are minimal.

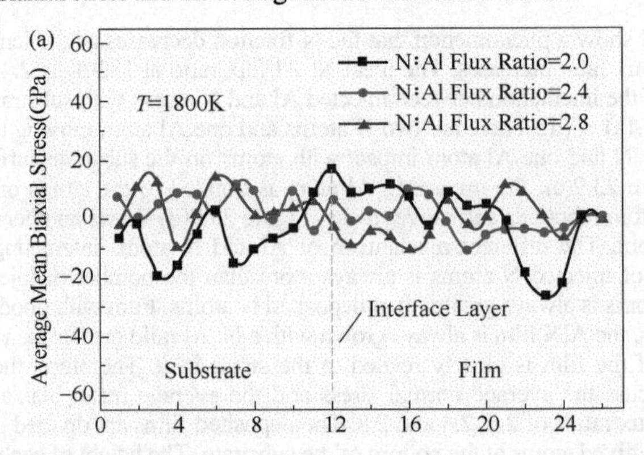

Figure 23.10 (a) Average mean biaxial stress and (b) the average normal stress of the films after deposition at 1800K with different N: Al flux ratios (2.0, 2.4 and 2.8)

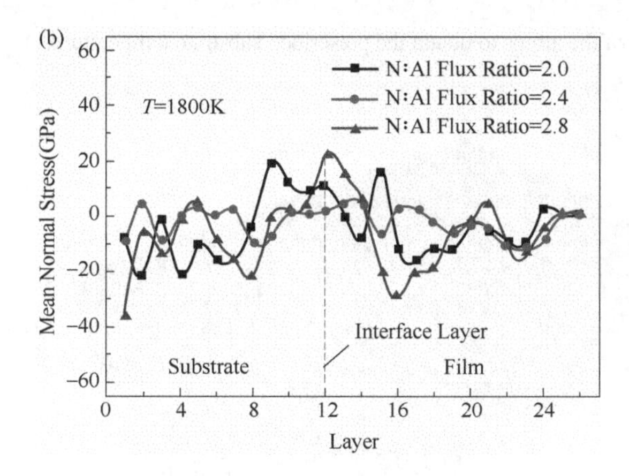

Figure 23.10 (Continued)

23.2 AlN Deposition on GaN Substrate

The Al-Ga-N SW potential (Zhou *et al.*, 2013) is adopted to describe the deposition of AlN on GaN substrate. The related parameters are shown in Table 23.1 below. Although there are many potentials independently describing the properties of AlN (Tungare *et al.*, 2011; Vashishta *et al.*, 2011) or GaN (Nord *et al.*, 2003; Béré 2006; Zhou *et al.*, 2015), there are few potentials simultaneously describing the properties of AlN and GaN. The Al-Ga-N SW potential can well describe the lattice constant, cohesive energy, bulk modulus and thermal expansion coefficient of AlN and GaN. And it captures the lowest energies for the equilibrium wurtzite phase of both AlN and GaN.

Table 23.1 SW potential parameters

Parameter	Ga-Ga	N-N	Al-Al	Ga-N	Ga-Al	N-Al
ε (eV)	1.2000	1.2000	0.5650	2.1700	0.5223	2.2614
σ (Å)	2.1000	1.3000	2.6674	1.6950	2.7322	1.7103
α	1.60	1.80	1.55	1.80	1.55	1.80
λ	32.5	32.5	0.0	32.5	0.0	40.5
γ	1.2	1.2	1.2	1.2	1.2	1.2
A	7.9170	7.9170	17.8118	7.9170	17.8118	7.9170
B	0.72	0.72	0.72	0.72	0.72	0.72

The 3D model of GaN substrate is shown in Figure 23.11. To make the crystallographic vectors properly represent a-, m-, and c-plane surfaces, the crystallographic vectors $\boldsymbol{a}=a(1,0,0)$, $\boldsymbol{b}=a(0,\sqrt{3},0)$, and $\boldsymbol{c}=a(0,0,c/a)$ with eight atoms per unit cell are introduced. x, y and z axis represents a, m and c direction correspondingly. Periodic boundary conditions are applied to the x and y directions in order to create a seemingly infinite boundary in these dimensions. Meanwhile, the free boundary condition is applied to the z direction to enable the deposition of Al and N atoms towards the GaN substrate surface. The GaN substrate filled with 10800 atoms has a dimension of 47.79, 82.78, and 29.17 nm in x, y, and z directions, respectively. It is divided into three groups: two pairs of atoms at the bottom of the substrate make up the fixed group to prevent movement of the substrate due to the deposition of atoms. Two pairs of atoms on the top of the substrate constitute the free group, in which the atoms are entirely free to interact and transmit energy with the deposited atoms. The rest of the atoms in the middle of GaN substrate act as the thermal control group, where the NVT ensemble is used to perform time integration on Nose-Hoover style non-Hamiltonian equations of motion to update the positions and velocities of the atoms to obtain the prescribed substrate temperature.

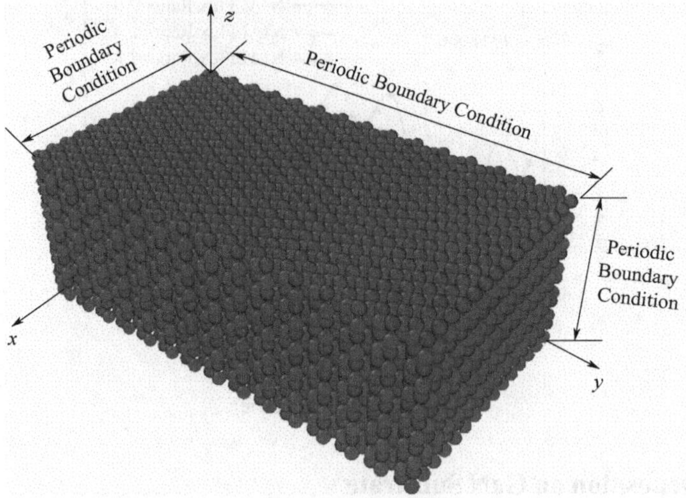

Figure 23.11 3D model of GaN substrate

The growth of AlN is simulated by periodically injecting a single (either aluminum or nitrogen) atom towards the GaN substrate from a 30-fold lattice height toward the AlN substrate surface with an incident angle of 0° (perpendicular to the substrate). The injected atoms are all assigned a kinetic energy of 0.17eV (Zhou *et al.*, 2006). The time interval of injected Al atoms is kept as a constant of 5ps per atom while that of N atoms is varied according to the N: Al flux ratio. It should be noted that the deposition rate in MD simulation is several orders of magnitude higher than realistic values, which is needed to overcome the computational expense so that sufficient materials can be grown within reasonable computing time. Because of the fast deposition rate, the adatoms are buried by new arrivals before they sufficiently diffused and relaxed. The effects of accelerated growth rate are partly compensated by using elevated growth temperature (Gruber *et al.*, 2017; Chu *et al.*, 2018).

23.2.1 Analysis Methods

The visualization of GaN substrate and deposited AlN film is realized by the open visualization tool (OVITO) (Stukowski *et al.*, 2009). Identify Diamond Structure (IDS) (Maras *et al.*, 2016) is adopted to identify the atoms arranged in hexagonal (wurtzite) or cubic diamond (zinc blende) lattice. To identify the central atom, the second nearest neighbors are taken into account to distinguish between hexagonal and cubic diamond structures. This method involves the characterization of the geometrical arrangement of the second nearest neighbors and can be used to visualize and identify the stacking faults and dislocations. In order to display more clearly the defect region, the local lattice structures are indicated by color type and numerical type. The specific classification is shown in the discussion about atomic structure distribution and component below.

The surface roughness of deposited AlN films under different conditions is denoted by root-mean-square (RMS) roughness *Ra*. The formula is shown below:

$$Ra = \sqrt{\frac{\sum_{1}^{N}(Z_i - \bar{Z})^2}{N}} \tag{23.3}$$

In this formula, Z_i represents the Z coordinate of the surface atom i, $\bar{Z}$ represents the average Z coordinate of all the surface atoms, and N denotes the total number of surface atoms of the deposited AlN films.

23.2.2 *Results and Discussion*

(i) Effect of Substrate Termination Surface

For GaN substrate, either surfaces can be N terminated or Ga terminated. The polar planes for GaN in the (0001) direction can be cleaved in two ways: combination of Ga-terminated (0001) and N-terminated (000-1); combination of Ga-terminated (000-1) and N-terminated (0001) plane. However, the formation energy of Ga-terminated (0001) and N-terminated (000-1) are lower than that of Ga-terminated (000-1) and N-terminated (0001) plane (Jindal *et al.*, 2009, 2010), which suggests that Ga-terminated (0001) and N-terminated (000-1) surfaces are more stable. Thus, (0001) Ga-terminated GaN and (000-1) N-terminated GaN are applied to explore the effect of the substrate surface on the growth of AlN film.

The number of injected Al atoms is 4000 and the injected Al: N flux ratio is set to be 1:1. Figure 23.12 shows the surface morphology of deposited AlN film on two GaN surfaces. It can be clearly seen that the deposited AlN film shows terrible surface morphology when growing on (000-1) N-terminated GaN surface. There are islands with irregular shape on deposited AlN surfaces. When growing on (0001) Ga-terminated GaN surface, the deposited AlN films display comparatively smooth surfaces. The atoms are piled up layer by layer and no obvious island is observed.

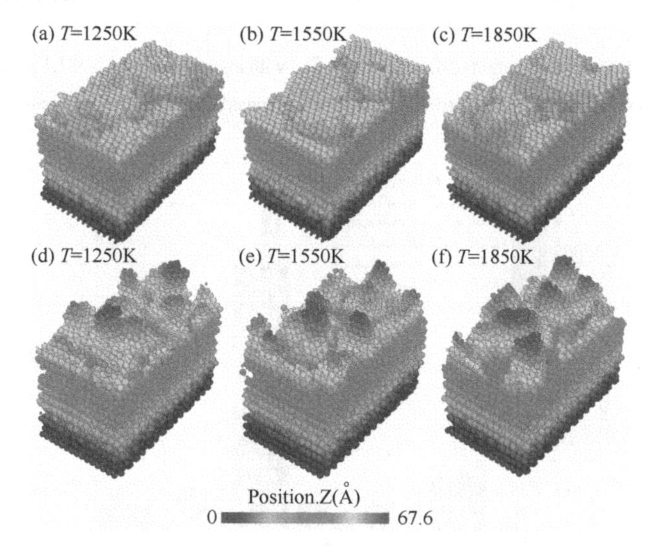

Figure 23.12 Surface morphology of deposited AlN film with injected Al:N ratio of 1:1 at 1250K, 1550K and 1850K on: (a)–(c) (0001) Ga-terminated GaN; (d) – (f) (000−1) N-terminated GaN (*Color version online*)

The atomic structure distribution of deposited AlN film on two GaN surfaces is shown in Figure 23.13. The ideal lattice structure is hexagonal diamond structure (wurtzite). The structure distribution of deposited AlN film on two GaN surfaces can be clearly observed. To quantitatively calculate the component of all types of structure in Figure 23.13, the structural component of deposited AlN film is shown in Figure 23.14. The color classification in Figure 23.13 is one-to-one correspondence with the numeric classification of abscissa in Figure 23.14. The specific correspondence is shown in Figure 23.14. It can be seen in Figures 23.13 and 23.14 that the component of wurtzite structure of deposited AlN increases as the temperature increases from 1250K to 1850K on two GaN surfaces. However, the component of wurtzite structure of deposited AlN film on (0001) Ga-terminated GaN surface is much higher than that on (000-1) *N*–terminated GaN surface. Meanwhile, the component of zinc blende and other structure on (0001) Ga-terminated GaN surface is lower than that on (000-1) N-terminated GaN surface under all temperatures, indicating that better crystallinity of deposited AlN film is achieved on (0001) Ga-terminated GaN surface than on (000-1) N-terminated GaN surface.

Figure 23.13 Atomic structure distribution of deposited AlN film with injected Al:N ratio of 1:1 at 1250K, 1550K and 1850K on: (a)–(c) (0001) Ga-terminated GaN; (d)–(f) (000-1) N-terminated GaN (*Color version online*)

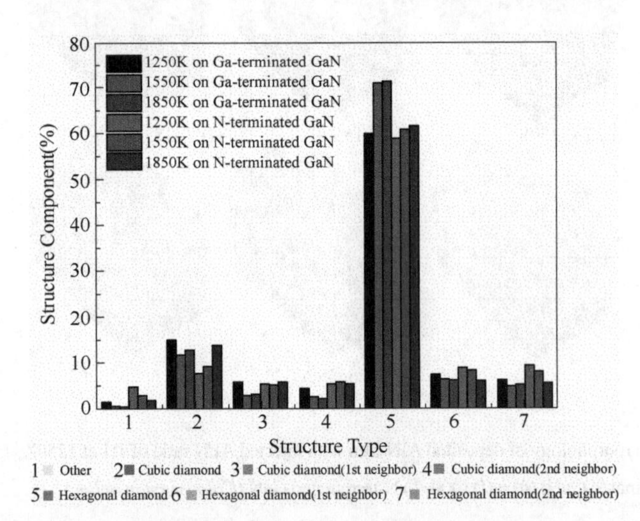

Figure 23.14 Atomic structure type and component of deposited AlN film with injected Al: N ratio of 1:1 at 1250K, 1550K and 1850 K on two GaN surfaces (*Color version online*)

The adatom diffusion is considered as a key parameter controlling the material quality and the surface morphology during the epitaxial growth process (Jindal *et al.*, 2009). It has been reported that the diffusion barriers for Al, Ga and N adatoms on (0001) Al-terminated AlN surface are 1.17, 0.99 and 0.9eV, respectively. Meanwhile, the diffusion barriers for Al, Ga and N adatoms on (000-1) N-terminated AlN surface are 2.28, 1.96 and 2.62eV, respectively (Jindal *et al.*, 2009). Similar study has also proven that the diffusion barriers for Ga and N atoms on (0001) Ga-terminated GaN surface are 0.32 and 1.03eV while that are 0.85 and 2.4 eV on (000-1) N-terminated GaN surface (Jindal *et al.*, 2010). In addition, similar tendency that Al and N adatoms have lower diffusion barriers on (0001) Ga-terminated GaN surface than on (000-1) N-terminated GaN surface are obtained by using the SW potential. These results all showed that the metal adatom and N adatom both have lower diffusion barriers on (0001) metal-terminated surface than on (000-1) N-terminated surface, indicating that the Al and N adatoms are more mobile on (0001) Ga-terminated GaN surface than on (000-1) N-terminated GaN surface. Thus, better film crystallinity and surface morphology are obtained on (0001) Ga-terminated GaN surface. In addition, increasing the growth temperature increases the surface kinetics of Al and N adatoms, enhancing the mobility of Al and N adatoms. Thus, it is easier for Al and N adatoms to find the ideal lattice site to form AlN lattice, resulting in better surface morphology and crystallinity.

From the above analysis, it can be observed that AlN film shows better crystal quality on (0001) Ga-terminated GaN substrate than on (000-1) N-terminated GaN substrate.

Therefore, our following discussions only focus on the growth of AlN film on (0001) Ga-terminated GaN substrate.

(ii) Effect of Temperature and N: Al Flux Ratio

(1) Surface Morphology

To further investigate the effect of temperature and N:Al flux ratio on growth of AlN film, the growth temperature and N:Al flux ratio are varied. The number of injected Al atom is 8000 and the number of injected N atoms is varied according to the corresponding N:Al flux ratio.

Figures 23.15–23.17 show the surface morphology of AlN films under different temperatures at N: Al flux ratio of 0.8, 1.0, and 1.2, respectively. The results show that AlN films achieve smoother surface as the temperature increases under all N: Al flux ratios. Increasing the growth temperature does effectively increase the surface morphology of deposited AlN films.

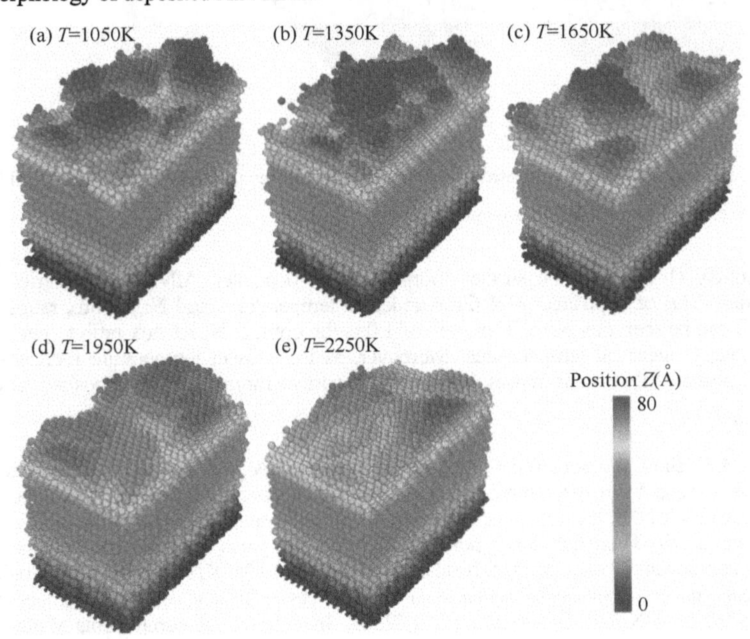

Figure 23.15 Surface morphology of deposited AlN film under different temperatures with injected N: Al flux ratio of 0.8 on (0001) Ga-terminated GaN surface (*Color version online*)

In addition, it can be identified that N:Al flux is another sensitive factor affecting the surface morphology of deposited AlN film. At N:Al flux ratio of 0.8, the deposited AlN film shows relatively rough surface at low temperature (1050K and 1350K) and displays improvements as the temperature increases in Figure 23.15. Compared with the surface morphology of deposited AlN films at N:Al flux ratio 0.8, the surface morphology of AlN films shows the overall improvements under all temperatures at N:Al flux ratio of 1.0 as shown in Figure 23.16. When the N:Al flux ratio is further increased to 1.2, the surface morphology of deposited AlN films show overall degradations under all temperatures in Figure 23.17.

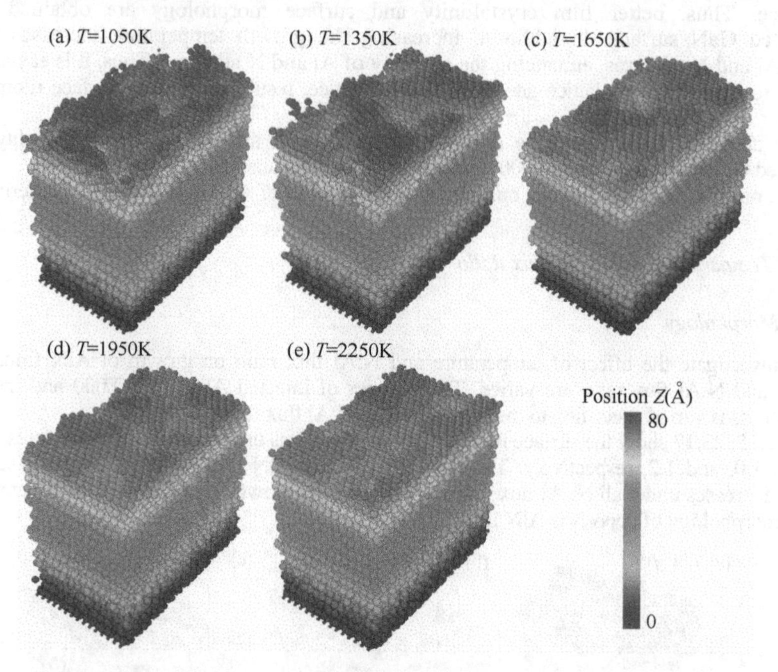

Figure 23.16 Surface morphology of deposited AlN film under different temperatures with injected N: Al flux ratio of 1.0 on (0001) Ga-terminated GaN surface (*Color version online*)

To quantitatively characterize the surface morphology of deposited AlN film, numerical value of the surface roughness (*Ra*) of deposited AlN films under all temperatures and N: Al flux ratios are shown in Figure 23.18. It can be seen that N: Al flux ratio of 1.0 is the optimal N: Al flux ratio to get relatively good surface morphology under all temperatures. Moreover, as the growth temperature increases, the surface roughness of deposited AlN film decreases, improving the surface morphology of deposited AlN film.

(2) Crystallinity

Figures 23.19-23.21 show the structure component of deposited AlN under different temperatures at N:Al flux ratio of 0.8, 1.0 and 1.2 respectively. The classifications of structure type are the same as Figure 23.14. At the N: Al flux ratio of 0.8, the deposited AlN film contains relatively low composition of wurtzite AlN as shown in Figure 23.19. There are large portions of other structures, zinc blende and its related defect structures. The component of wurtzite AlN reaches a maximum at 1950K. As the N:Al flux ratio is increased from 0.8 to 1.0, the composition of wurtzite AlN improves distinctly under all temperatures and the component of wurtzite AlN reaches a maximum at 2250K. In addition, the components of other structure are very low under all temperatures. When the N:Al flux ratio is further increased to 1.2, the component of wurtzite AlN decreases obviously at 2250K while those are less obvious at other temperatures. Meanwhile, the component of other structure increases distinctly at low temperatures (1050K and 1350K) as the N: Al flux ratio increases from 1.0 to 1.2.

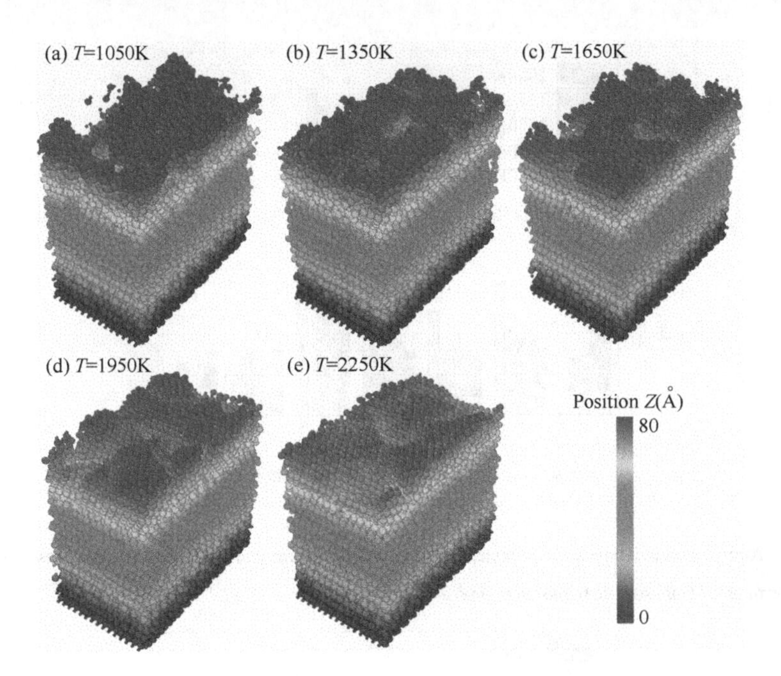

Figure 23.17 Surface morphology of deposited AlN film under different temperatures with injected N: Al flux ratio of 1.2 on (0001) Ga-terminated GaN surface (*Color version online*)

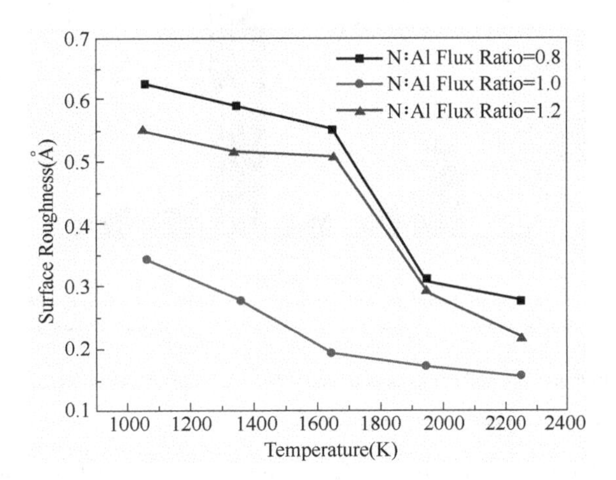

Figure 23.18 RMS roughness of deposited AlN film under different temperatures and injected N: Al flux ratios on (0001) Ga-terminated GaN surface

Taken together, temperature and N: Al flux ratio are two critical factors synergistically affecting the crystallinity of deposited AlN film. Increasing the growth temperature increases the crystallinity of deposited AlN film at N: Al flux ratio of 1.0. Moreover, increasing the N: Al flux ratio from 0.8 to 1.0 definitely improves the crystallinity of deposited AlN film under all temperatures.

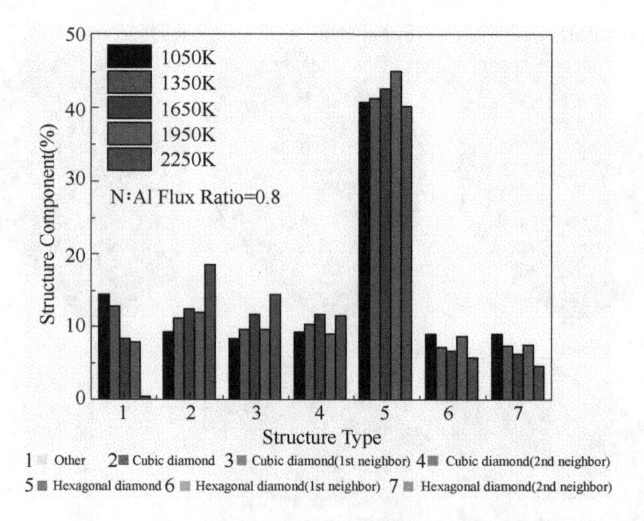

Figure 23.19 Atomic structure type and component of deposited AlN films under different temperatures at N:Al flux ratio of 0.8 on Ga-terminated GaN surface (*Color version online*)

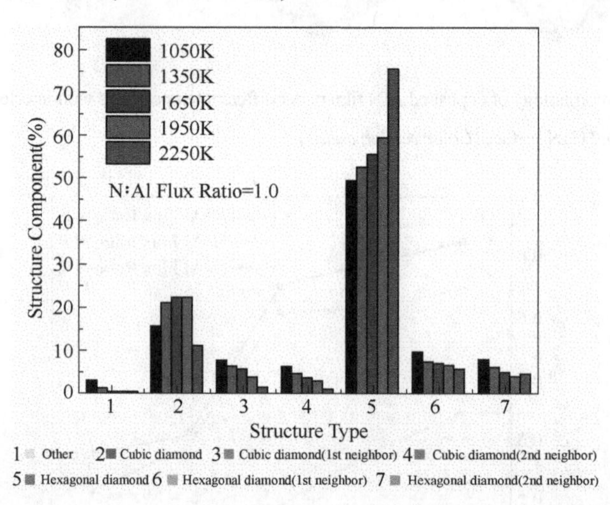

Figure 23.20 Atomic structure type and component of deposited AlN films under different temperatures at N: Al flux ratio of 1.0 on Ga-terminated GaN surface (*Color version online*)

Combined with the above discussion about surface morphology and crystallinity, we find that the surface morphology and crystallinity are both closely related to temperature and N:Al flux ratio. Increasing the growth temperature improves the surface kinetics of adsorbed Al and N adatoms, which facilitates the adatoms to find their ideal lattice sites, thus improving the crystal quality of deposited AlN film. In addition, the injected N: Al flux ratio directly affects the balance of stoichiometry. For a balanced stoichiometry, the N fraction of deposited AlN should be 50%. It has been proved in our previous study that film with good crystallinity is connected with a near stoichiometry (Zhang *et al.*, 2018). Film with the non-stoichiometric ratio is more likely to have vacancies, interstitial atom and even other defects. Thus, the N fraction of deposited film greater or less than 50% both lead to more defects and worse crystal quality. To further verify this, the N fraction of deposited AlN film is shown in Figure 23.22. It can be clearly seen that a near 50% N fraction appears when the injected N:Al flux ratio is 1.0 under all temperatures. N fraction greater and less than 50% appear at injected N: Al flux ratio of 1.2 and 0.8 correspondingly. Moreover, the comparatively better crystal quality of deposited AlN film is obtained at N: Al flux ratio of 1.0 other than 0.8 and 1.2. This could explain why the relatively good surface morphology and crystallinity appears at N: Al flux ratio of 1.0. It should be noted that the N/Al ratio in our simulation is different from that in most practical experiments (Chen *et al.*, 2018; Nemoz *et al.*, 2018; Boichot *et al.*, 2013). In our simulation, the reaction sources are Al and N atoms. The N/Al ratio is simply denoted by the N atoms/Al atoms in our simulations. By contrast, in practical AlN growth in MOCVD(Chen *et al.*, 2018), MBE (Nemoz *et al.*, 2018) and HVPE (Boichot *et al.*, 2013), the reaction sources are usually chemical compounds: TMAl and NH3 in MOCVD, the III elements and NH3 in MBE, $AlCl_3$ and NH_3 in HVPE, in which the N/Al ratio is denoted by N-sources/Al-sources. There are many pre-reactions between those chemical compounds while this does not exist in our simulation. Therefore, it is not strange that the practical N concentration is much higher than that in our simulation.

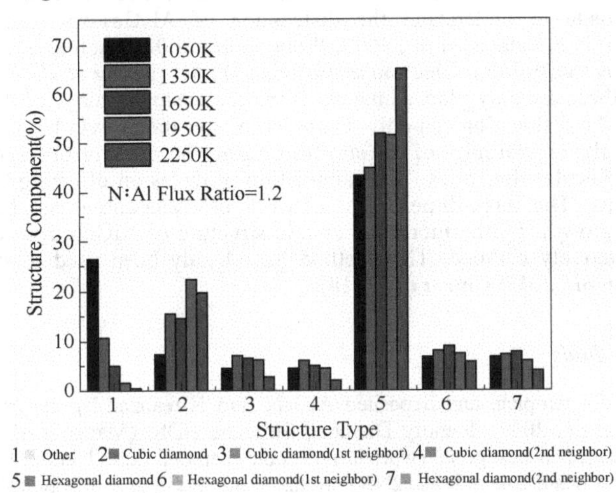

Figure 23.21 Atomic structure type and component of deposited AlN films under different temperatures at N: Al flux ratio of 1.2 on Ga-terminated GaN surface (*Color version online*)

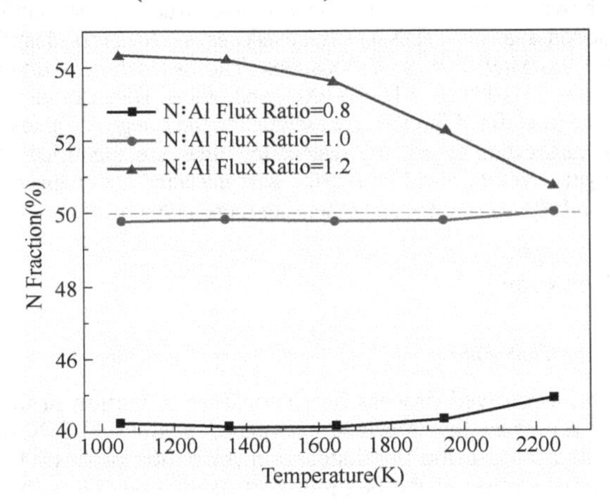

Figure 23.22 N fraction of deposited AlN films under different temperatures and N: Al flux ratios on Ga-terminated GaN surfaces

23.3 Atomic Simulation of AlGaN Film Deposition on AlN Template

AlGaN alloys are currently of great interest for the realization of deep-ultraviolet (UV) optical devices like light-emitting diodes (LEDs) because the direct transition energy can be adjusted between 6.2eV (AlN) light-emitting devices is limited and worsens as the wavelength is reduced and the aluminum (Al) mole fraction is increased owing to a reduction of electron and hole concentrations, low injection efficiencies, and a high density of crystalline defects (Lochner *et al.*, 2013). In particular, high dislocation density has potential influence on the formation of mid-gap states, which can act as non-radiative recombination centres in AlGaN-based active devices thus lowering the emission efficiency (Kida *et al.*, 2003; Hirayama *et al.*, 2015; Massabuau *et al.*, 2017; Besendfer *et al.*, 2019). It is therefore important to understand the properties of dislocations in Al*x*Ga1-*x*N. Tremendous efforts have been made to understand the dislocation of Al*x*Ga1-*x*N film on AlN template (Massabuau *et al.*, 2017; Follstaedt *et al.*, 2009; Wang *et al.*, 2002; Kaun *et al.*, 2012). Dislocations are usually detected by transmission electron microscope (TEM) (Leung *et al.*, 2014; Raghavan *et al.*, 2012).Nevertheless, these are only planar images. Three-dimensional image of dislocation cannot be observed. Moreover, the dislocation can only be observed after the growth but not during the growth process. In addition, the crystallinity of AlGaN film cannot be directly observed or quantitatively counted. Notably, molecular dynamics (MD) simulation provides an effective theoretical means to handle these problems. The three-dimensional network of dislocations can be directly observed during and after the growth. Furthermore, the atomic structure of AlGaN film can also be directly observed and quantitatively counted. This method has already been used to study the growth of InGaN film (Gruber *et al.*, 2017; Chu *et al.*, 2018).

23.3.1 Analysis Methods

The visualization of AlN template and deposited AlGaN film is realized by the open visualization tool (OVITO) (Stukowski *et al.*, 2009). Identify Diamond Structure (IDS) (Maras *et al.*, 2016) is adopted to identify the atoms arranged in hexagonal (wurtzite) or cubic diamond (zinc blende) lattice. To identify the central atom, the second nearest neighbors (2NN) are taken into account to distinguish between hexagonal and cubic diamond structures. This method involves the characterization of the geometrical arrangement of the second nearest neighbors and can be used to visualize and identify the stacking faults and dislocations. In order to display more clearly the defect region, the local lattice structures are indicated by color type. The specific classification is shown in the discussion about atomic structure distribution below. Moreover, we used the dislocation extraction algorithm (DXA) (Stukowski *et al.*, 2012) to identify dislocation patterns in crystals and determine the indexing in these dislocations. The dislocation is divided into five types: b = 1/3<1-210>, b = <0001>, b = <1-100>, b = 1/3<1-100>, and others. Dislocations with a burgers vector that does not belong to any of the predefined families are assigned to the category 'others'.

To further display the dislocation types, the dislocation lines are identified and indicated by colors: dislocation lines with burgers vectors b = 1/3<1-210> is represented by green, b = <0001> by blue, b = <1-100> by pink, b = 1/3<1-100> by orange, and others are represented by red.

23.3.2 Results and Discussion

(i) Dislocation in Deposited AlGaN Film

Figure 23.23 shows the time-resolved process for dislocation formation in deposited AlGaN film at 2000K. The dotted lines represent the interface between deposited AlGaN film and AlN template. Six snapshots are presented to describe the dislocation formation process. It can be seen in Figure 23.23(a) that there is no dislocation at the beginning of deposition. Then the dislocation appears at the interface between AlGaN film and AlN template at 15000ps in Figure 23.23(b). Over time, the dislocations continue to grow in Figure 23.23 (c)–(f). Moreover, there are three types of dislocation denoted by burgers vector in deposited AlGaN film: b = 1/3<1-100>, b = 1/3<1-210>, and other dislocations. Among the three types of dislocation, b = 1/3<1-100> is the most common one. It is known that the larger the burgers vector, the higher the elastic energy of the crystal caused by dislocation (Hull *et al.*, 2011). In other words, the smaller the burgers vector, the easier the corresponding dislocations will occur. That is why b = 1/3<1-100> is the most common dislocation at 2000K.

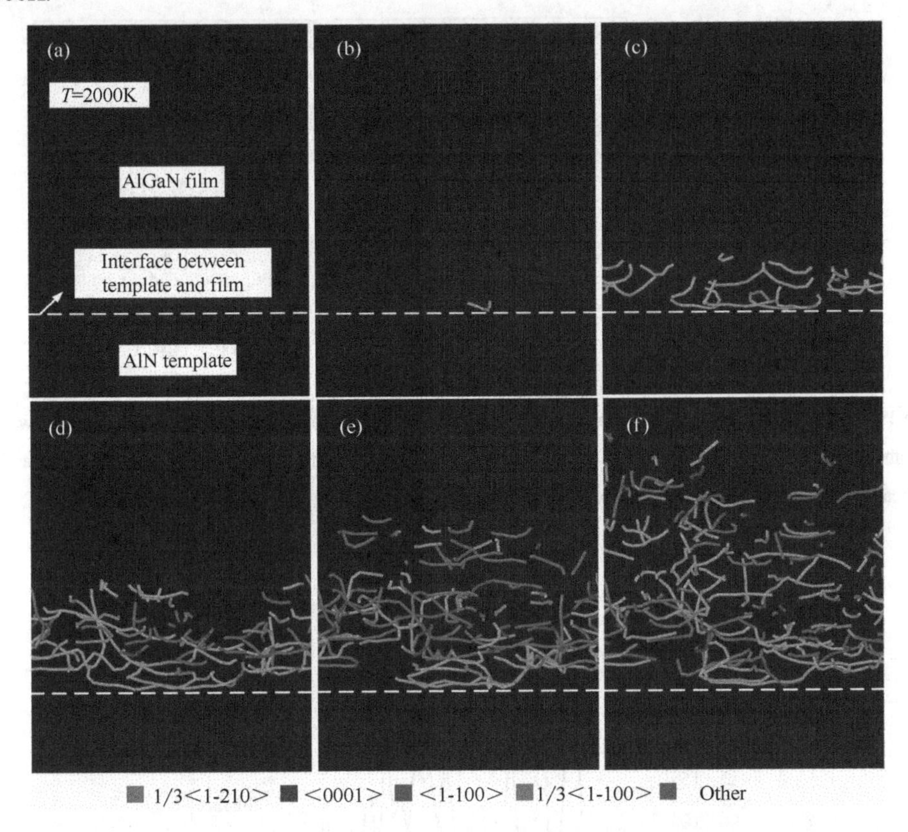

Figure 23.23 Time-resolved process for dislocation formation in deposited AlGaN at 2000K. (a) t = 0ps; (b) t = 15000ps; (c) t = 30000ps; (d) t = 45000ps; (e) t = 60000ps; (f) t = 75000ps (*Color version online*)

Moreover, it also can be seen in Figure 23.23 that significant dislocations develop during the deposition of AlGaN film. A comparison of the dislocation plan views is shown in Figure 23.24. Figures 23.24 (b), (c) show the dislocation of deposited AlGaN film within the bulk and at the AlGaN film/AlN template interface respectively. It can be seen that dislocations primarily occur at the interface between AlGaN film and AlN template. This is understandable because the lattice mismatch exits between AlGaN film and AlN template, which inevitably leads to atom mismatch and involve dislocation (Hull *et al.*, 2011). Furthermore, it can be clearly seen in Figures 23.24(b), (c) that dislocation in the bulk of AlGaN film is less than that in the interface between AlGaN film and AlN template. This indicates that increasing the thickness of deposited AlGaN film to a certain extent is beneficial to reducing dislocation. As the AlGaN film grows thicker, the lattice mismatch effect between the substrate and the film decreases. Thus the dislocation density decrease as the film grows thicker. It should be noted that, some dislocations seem to come out of the side of the supercell. This is because the periodic boundary conditions are applied to the x and y directions, which means the box is periodic, so that particles and dislocations interact across the boundary, and they can exit at one end of the box and re-enter the other end.

To further quantify the observations in Figures 23.23 and 23.24, dislocation densities are calculated as functions of layer thickness along the growth direction, and the results are shown in Figure 23.25. It can be seen that dislocation densities in deposited AlGaN films have a pronounced peak near the interface. The dislocation density decreases with the increase of AlGaN film thickness. Similarly, AlGaN film with a certain thickness is often grown on AlN to filter dislocations in the process of experimentally growing LED (Nishida *et al.*, 2004; Ponce *et al.*, 1994; Li *et al.*, 2018).

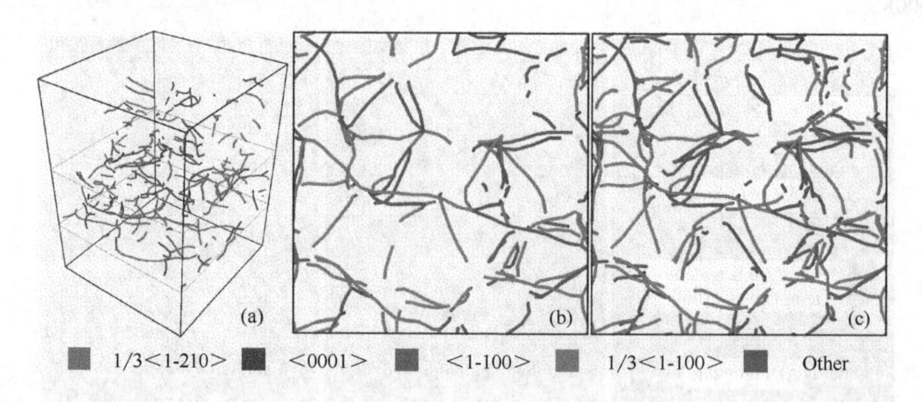

Figure 23.24 V isualization of dislocation networks in deposited AlGaN film at 2000K. (a) A perspective overview; (b) a slice within the bulk of the film; (c) a slice at the AlGaN film/AlN template interface. Location of cross-sections (b) and (c) are indicated in (a) by two planes respectively (*Color version online*)

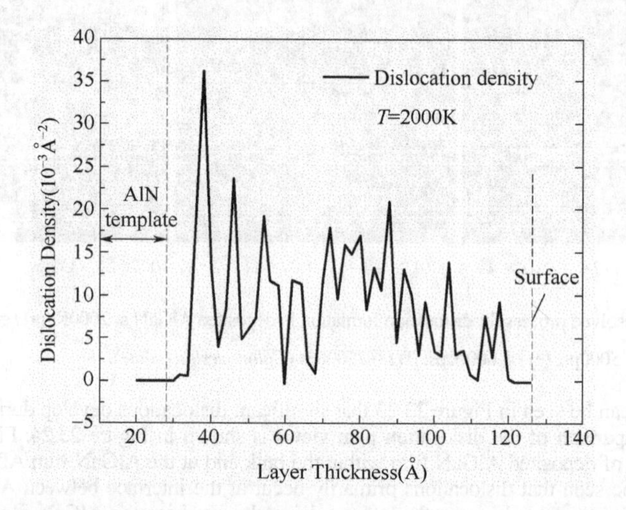

Figure 23.25 Dislocation densities in deposited AlGaN films as a function of the layer thickness along growth direction (The original AlN template surface is at 28Å)

Figures 23.26 and 23.27 show the perspective overview and side view of deposited AlGaN films under different temperatures respectively. From Figures 23.26 and 23.27, we can see that dislocations are densely distributed at 1600K. As the temperature increases from 1600K to 2400K, the dislocation distribution becomes sparse. Moreover, Figure 23.28 quantitatively counts the total dislocation length of deposited AlGaN films under different temperatures. From Figures 23.26−23.28 it can be seen that the dislocation decreases with the temperature increases from 1600K to 2400K, indicating that the dislocation are probably kinetically trapped during deposition. Increasing the growth temperature reduces dislocation densities.

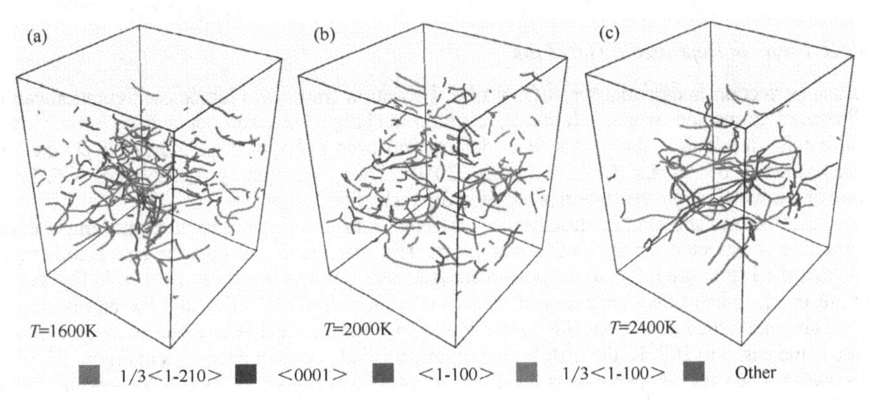

Figure 23.26 Perspective overview deposited AlGaN films under different temperatures. (a) $T = 1600K$; (b) $T = 2000K$; (c) $T = 2400K$ (*Color version online*)

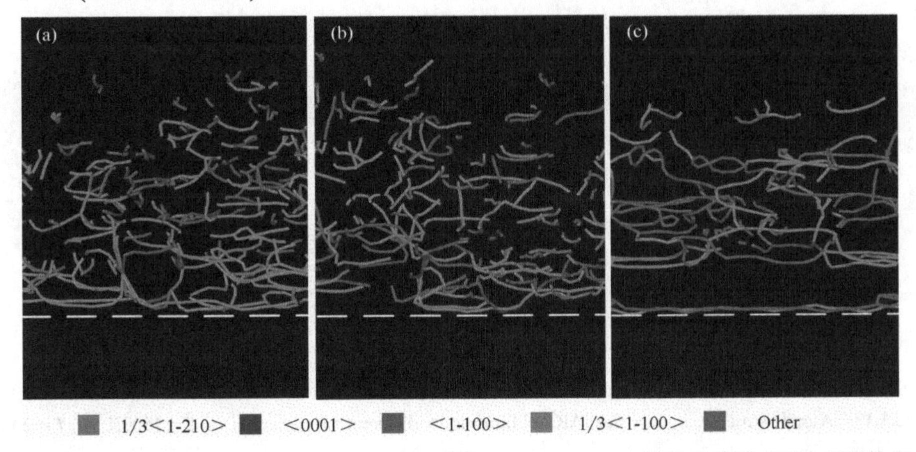

Figure 23.27 Side view of deposited AlGaN films under different temperatures. (a) $T = 1600K$; (b) $T = 2000K$; (c) $T = 2400K$ (*Color version online*)

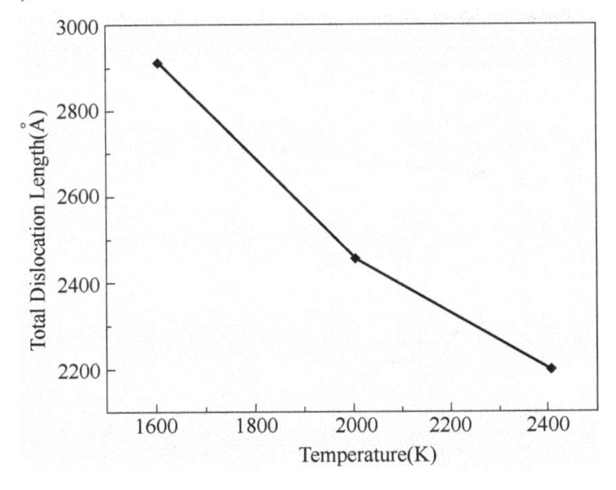

Figure 23.28 Total dislocation length of deposited AlGaN films under different temperatures

(ii) Atomic Structure in Deposited AlGaN Film

Crystal lattice perfection is essential for AlGaN film. Deviation from ideal lattice configuration can change the coordinates of atoms and overstretch bonds, resulting in changed electron populations in the valence and conduction bands. In addition, the existence of defects and even polytypism seriously affects the electrical and optical properties of AlGaN (Makaram *et al.*, 2010). Thus, high-quality AlGaN film with low lattice deformation and low defect concentration is required for such applications.

Figure 23.29 shows the atomic structure in deposited AlGaN films under different temperatures. The wurtzite structure is expected for the ideal AlGaN film. The zinc blende denotes the zinc blende inclusions. And the unidentified structure refers to the unwanted defects or polytypism. It can be seen in Figure 23.29 (a) that there are much unidentified structures in deposited AlGaN film at 1600K and the atoms are arranged irregularly. There are lattice distortions due to the deviation from the ideal lattice sites of some atoms. As the temperature is increased to 2000K, the distribution of unidentified structure decreases in Figure 23.29 (b). It is further decreased when the temperature is increased to 2400K in Figure 23.29 (c). Meanwhile, the overall atomic arrangement is more orderly as the temperature increases. The regular pattern of the deposited atoms can be more easily observed.

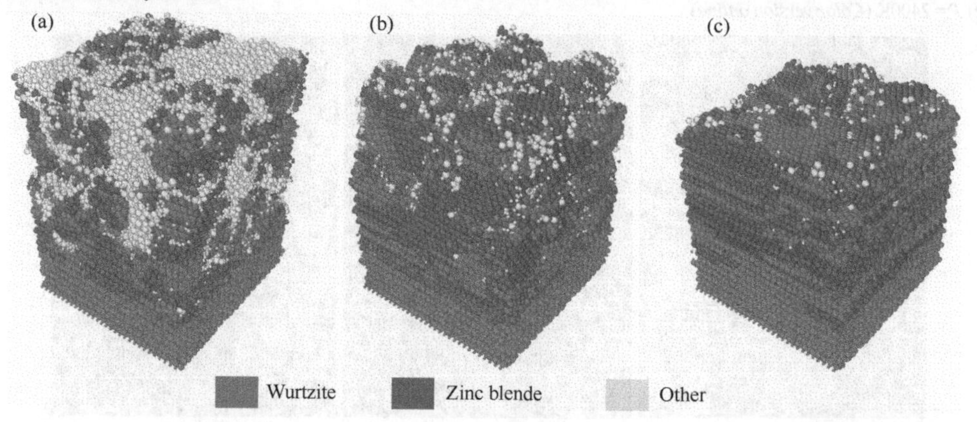

Figure 23.29 Atomic structure in deposited AlGaN films under different temperatures. (a) $T=1600$K; (b) $T=2000$K; (c) $T=2400$K (*Color version online*)

To quantitatively count the structural component of deposited AlGaN film, the structure component of deposited AlGaN films as a function of temperature is shown in Figure 23.30. It can be clearly seen in Figure 23.30

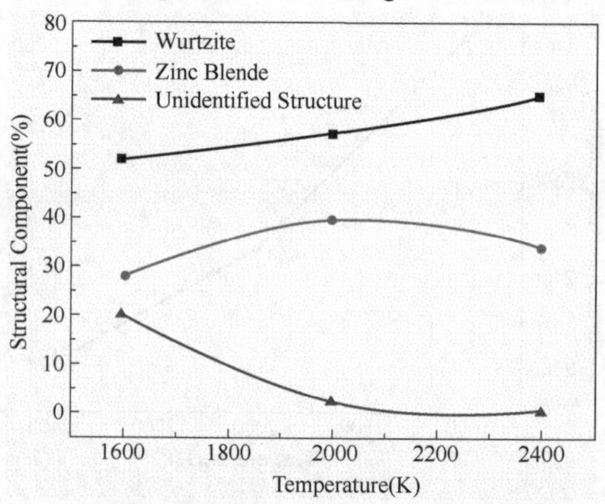

Figure 23.30 Structure component of deposited AlGaN films as a function of temperature

that the component of unidentified structure decreases as the temperature increases while that of wurtzite structure increases with the temperature increasing. Moreover, there are considerable portion of zinc blende structure under all temperatures. This is due to the stacking faults in deposited AlGaN film. However, this may be exaggerated because the energy differences between wurtzite and zinc blende are both small for AlN and GaN (Serrano *et al.*, 2000), and the SW potential ensures the lowest energies for the equilibriumwurtzite and zinc blende phase for both AlN and GaN. Overall, the crystallinity of deposited AlGaN is improved as the temperature increases. This is consistent with the dislocation discussion above. The mobility of atoms increases with increasing growth temperature. It is easier for adatoms to find ideal lattice points at higher temperature. Thus the dislocation and other defects can be effectively reduced and the crystal quality could be improved.

23.4 Chapter Summary

We study the atomic growth mechanism of AlN grown on c-plane AlN by MD simulations. The effects of temperature and N:Al flux ratio are investigated in detail. We have studied the effect of substrate surface on the deposition of AlGaN film on two AlN surfaces: (0001) Al-terminated and 0001 N-terminated AlN surface. Growth temperatures and injected Al:Ga ratios were varied to comprehensively explore the effects. The surface morphology and atomic structure of deposited AlGaN film were discussed in detail. Increasing the growth temperature can also effectively reduce the dislocation in deposited AlGaN film. Furthermore, the crystallinity of deposited AlGaN film could be improved by increasing the growth temperature. This is consistent with the dislocation discussion. The mobility of adatoms increases as the growth temperature increases. So it is easier for adatoms to find their ideal lattice points at higher temperature. Thus the dislocation and other defects can be effectively reduced and the crystal quality of deposited AlGaN film could be improved.

References

Allerman, A. A., Crawford, M. H., Fischer, A. J., Bogart, K. H. A., Lee, S. R., Follstaedt, D.M., Provencio, P.P., and Koleske, D.D. (2004) Growth and design of deep-UV (240 290 nm) light emitting diodes using AlGaN alloys, *Journal of Crystal Growth*, 272, 227.

Ambacher, O. (1998) Growth and applications of Group III-nitrides, *Journal of Physics D: Applied Physics* -31, 2653-2710.

Bai, J, Dudley, M, Sun, W. H, Wang, H. M, Khan, M. A. (2006) Reduction of threading dislocation densities in AlN/sapphire epilayers driven by growth mode modification, *Applied Physics Letters*, 88, 051903.

Basinski, Z. S., Duesbery, M. S., Taylor R. (1971) Influence of shear stress on screw dislocations in a model sodium lattice, *Canadian Journal of Physics*, 49, 2160-2180.

Béré, A., and Serra, A. (2006) On the atomic structures, mobility and interactions of extended defects in GaN: dislocations, tilt and twin boundaries, *Philosophical Magazine*, 86(15), 2159-2192.

Besendfer, S., Meissner, E., Lesnik, A., Friedrich, J., Dadgar, A., and Erlbacher, T. (2019) Methodology for the investigation of threading dislocations as a source of vertical leakage in AlGaN/GaN-HEMT hetero structures for power devices, *Journal of Applied Physics*, 125, 095704.

Boichot, R., Coudurier, N., Mercier, F., Lay, S., Crisci, A., Coindeau, S., Claudel, A., Blanquet, E., and Pons, M. (2013) Epitaxial growth of AlN on c-plane sapphire by high temperature hydride vapor phase epitaxy: Influence of the gas phase N/Al ratio and low temperature protective layer, *Surface and Coatings Technology* 237, 118.

Bryan, I., Bryan, Z., Mita, S., Rice, A., Tweedie, J., Collazo, R., Sitar, Z. (2016) Surface kinetics in AlN growth: a universal model for the control of surface morphology in III-nitrides, *Journal of Crystal Growth*, 438, 81-89.

Bryan, I., *et al.* (2013) Strain relaxation by pitting in AlN thin films deposited by metalorganic chemical vapor deposition, *Applied Physics Letters*, 102, 061602.

Cao, Y., Zhang, J., Wu, C., Yu, F. (2013) Effect of incident angle on thin film growth: a molecular dynamics simulation study, *Thin Solid Films*, 544, 496-499.

Chen, Y., Zhang, Z., Jiang, H., Li, Z., Miao, G., and Song, H. (2018) The optimized growth of AlN templates for back-illuminated AlGaN-based solar-blind ultraviolet photodetectors by MOCVD, *Journal of Materials Chemistry C*, 6, 4936.

Choi, S. (2007) Spectroscopic ellipsometry of group-III adatom kinetics on III-nitride semiconductor surfaces, PhD thesis, *Duke University*, Durham, NC, USA.

Chu, K., Gruber, J., Zhou, X.W., Jones, R.E., Lee, S.R., and Tucker, G.J. (2018) Molecular dynamics studies of InGaN growth on nonpolar (1120) GaN surfaces, *Physical Review Materials*, 2, 013402.

Chu, K., Gruber, J., Zhou, X.W., Jones, R.E., Lee, S.R., and Tucker, G.J. (2018) Molecular dynamics studies of InGaN growth on nonpolar GaN surfaces, *Physical Review Materials*, 2 (1), 013402.

Claudel, A., *et al.* (2011) Investigation on AlN epitaxial growth and related etching phenomenon at high temperature using high temperature chemical vapor deposition process, *Journal of Crystal Growth*, 335, 17-24.

Dong, P *et al.*, (2014) AlGaN-based deep ultraviolet light-emitting diodes grown on nano-patterned sapphire substrates with significant improvement in internal quantum efficiency, *Journal of Crystal Growth* ,395, 9-13.

Follstaedt, D. M. Lee, S. R., Allerman, A. A., and Floro, J. A. (2009) Strain relaxation in AlGaN multilayer structures by inclined dislocations, *Journal of Applied Physics*, 105, 083507.

Funato, M., Matsuda, K., Banal, R. G., Ishii, R, Kawakami, Y. (2012) Homoepitaxy and photoluminescence properties of (0001) AlN, *Applied Physics Express*, 5, 082001.

Gruber, J., Zhou, X.W., Jones, R.E., Lee, S.R., and Tucker, G.J. (2017) Molecular dynamics studies of defect formation during heteroepitaxial growth of InGaN alloys on (0001) GaN surfaces, *Journal of Applied Physics*, 121, 195301.

Guo, Y., Fang, Y., Yin, J., Zhang, Z., Wang, B., Li, J., Lu, W., Feng, Z. (2017) Improved structural quality of AlN grown on sapphire by 3D/2D alternation growth, *Journal of Crystal Growth*, 464, 119-122.

Hirayama, H, Fujikawa, S, Noguchi, N, Norimatsu J, Takano T, Tsubaki K, Kamata N. (2009) 222-282 nm AlGaN and InAlGaN-based deep-UV LEDs fabricated on high-quality AlN on sapphire, *Physica Status Solidi (a)*, Vol. 206, 1176-1182.

Hirayama, H., Enomoto, Y., Kinoshita, A., Hirata, A., and Aoyagi, Y. (2002) Efficient 230-280 nm emission from high-Al-content AlGaN-based multiquantum wells, *Applied Physics Letters*, 80, 37.

Hirayama, H., Fujikawa, S., and Kamata, N. (2015) Recent Progress in AlGaN - Based Deep - UV LEDs, *Electronics and Communications in Japan.*, 98, 1.

Hirayama, H., Kinoshita, A., Yamabi, T., Enomoto, Y., Hirata, A., Araki, T., Nanishi, Y., Aoyagi, Y. (2002) Marked enhancement of 320-360 nm ultraviolet emission in quaternary InxAlyGa1-xyNInxAlyGa1-x-yN with In-segregation effect, *Applied Physics Letters*, 80, 207-209.

Hull, D., and Bacon, D.J. (2011) *Introduction to Dislocations*, Elsevier, University of Liverpool.

Hull, D., and Bacon, D.J. (2011) *Introduction to Dislocations*, 5th ed. Department of Engineering, Materials Science and Engineering, University of Liverpool.

Imura, M. *et al.* (2007) Dislocations in AlN epilayers grown on sapphire substrate by high-temperature metal-organic vapor phase epitaxy, *Japanese Journal of Applied Physics*, 46, 1458-1462.

Imura, M., *et al.* (2006) High-temperature metalorganic vapor phase epitaxial growth of AlN on sapphire by multi transition growth mode method varying V/III ratio, *Japanese Journal of Applied Physics*, 45, 8639-8643.

Jindal, V., and Shahedipour-Sandvik, F. (2009) Density functional theoretical study of surface structure and adatom kinetics for wurtzite AlN, *Journal of Applied Physics*, 105, 084902.

Jindal, V., and Shahedipour-Sandvik, F. (2009) Density functional theoretical study of surface structure and adatom kinetics for wurtzite AlN, *Journal of Applied Physics*, 105,084902.

Jindal, V., and Shahedipour-Sandvik, F. (2010) Computational and experimental studies on the growth of nonpolar surfaces of gallium nitride, *Journal of Applied Physics*, 107, 054907.

Kaneko, M., Kimoto, T., Suda, J. (2016) Strong impact of the initial III/V ratio on the crystalline quality of an AlN layer grown by rf-plasma-assisted molecular-beam epitaxy, *Applied Physics Express*, 9, 025502.

Kaun, S.W. Wong, M.H. Mishra, U.K., and Speck, J.S. (2012) Correlation between threading dislocation density and sheet resistance of AlGaN/AlN/GaN heterostructures grown by plasma-assisted molecular beam epitaxy, *Applied Physics Letters*, 100, 262102.

Kida, Y., Shibata, T., Miyake, H., and Hiramatsu, K. (2003) Metalorganic vapor phase epitaxy growth and study of stress in AlGaN using epitaxial AlN as underlying Layer, *Japanese Journal of Applied Physics*, 42, L572.

Kinoshita, T., *et al.* (2013) Performance and reliability of deep-ultraviolet light-emitting diodes fabricated on AlN substrates prepared by hydride vapor phase epitaxy, *Applied Physics Express*, 6, 092103.

Leathersich, J., Suvarna, P., Tungare, M., (Shadi) Shahedipour-Sandvik, F. (2013) Homoepitaxial growth of non-polar AlN crystals using molecular dynamics simulations, *Surface Science*, 617,36-41.

Lee, S. R., West, A. M., Allerman, A A., Waldrip K E, Follstaedt, D. M., Provencio, P. P, Koleske, D,D, Abernathy C. R. (2005) Effect of threading dislocations on the Bragg peak widths of GaN, AlGaN, and AlN heterolayers, *Applied Physics Letters*, 86, 241904.

Leung, B., Han, J., and Sun, Q. (2014) Strain relaxation and dislocation reduction in AlGaN step - graded buffer for crack - free GaN on Si (111), *Physica Status Solidi C*, 11, 437-441.

Li, D., Jiang, K., Sun, X., and Guo, C. (2018) AlGaN photonics: recent advances in materials and ultraviolet devices,

Advances in Optics and Photonics, 10, 43-110.

Lin C. M., Lien, W. C, Felmetsger V. V., Hopcroft, M. A., Senesky DG, Pisano A. P. (2010) AlN thin films grown on epitaxial 3C-SiC (100) for piezoelectric resonant devices, *Applied Physic Letters*, 97, 658-663.

Lochner, Z., Kao, T. T, Liu Y-S, Li X-H, Satter M. M, Shen S-C, Yoder PD, Ryou J. H, Dupuis RD, Wei Y, Xie H, Fischer A, Ponce FA. (2013) Deep-ultraviolet lasing at 243 nm from Photo-pumped AlGaN/AlN heterostructure on AlN substrate, *Applied Physics Letters*, 102:101110.

Luo, W., Li, L., Li, Z., Yang, Q., Zhang, D., Dong, X., Peng, D., Pan, L., Li, C., Liu, B., and Zhong, R. (2017) Influence of the nucleation layer morphology on the structural property of AlN films grown on c-plane sapphire by MOCVD, *Journal of Alloys and Compounds*, 697, 262.

Luo, X, Hu, R, Liu, S, Wang, K. (2016) Heat and fluid flow in high-power LED packaging and applications, *Progress in Energy and Combustion Science*, Vol. 56, 1-32.

Makaram, P. J. Joh, J.A., Alamo, del., Palacios, T., and Thompson, C.V. (2010) Evolution of structural defects associated with electrical degradation in AlGaN/GaN high electron mobility transistors, *Applied Physics Letters*, 96, 233509.

Maras, E., Trushin, O., Stukowski, A., Ala-Nissila, T., and Jonsson, H. (2016) Global transition path search for dislocation formation in Ge on Si (001), *Computer Physics Communications*, 205, 13.

Massabuau, F. C. P., Rhode, S. L., Horton, M. K., Hanlon, T. J. O' Kovács, A., Zielinski, M.S., Kappers, M.J. Dunin-Borkowski, R.E., Humphreys, C.J., and Oliver, R. A. (2017) Dislocations in AlGaN: Core structure, atom segregation, and optical properties, *Nano Letter*, 17, 4846.

Mihopoulos, T. G., Gupta, V., Jensen, K. F. (1998) A reaction-transport model for AlGaN MOVPE growth, *Journal of Crystal Growth* ,195, 733-739.

Mino, T., Hirayama, H., Takano, T., Tsubaki, K, Sugiyama, M. (2012) Characteristics of epitaxial lateral overgrowth AlN templates on (111) Si substrates for AlGaN deep-UV LEDs fabricated on different direction stripe patterns, *Physica Status Solidi C*, 9, 802-805.

Nemoz, M., Dagher, R., Matta, S., Michon, A., Vennegues, P. (2017) Dislocation densities reduction in MBE-grown AlN thin films by high-temperature annealing, *Journal of Crystal Growth*, 461, 10-15.

Nepal, N., Nam, K. B., Nakarmi, M. L., Lin, J. Y., Jiang, H. X., Zavada, J. M., Wilson, R. G. (2004) Optical properties of the nitrogen vacancy in AlN epilayers, *Applied Physic Letters*, 84, 1090-1092.

Nishida, T., Makimoto, T., Saito, H., and Ban, T. (2004) AlGaN-based ultraviolet light-emitting diodes grown on bulk AlN substrates, *Applied Physics Letters*, 84, 1002.

Nord, J., Albe, K., Erhart, P., and Nordlund, K. (2003) Modelling of compound semiconductors: analytical bond-order potential for gallium, nitrogen and gallium nitride, *Journal of Physics: Condensed Matter*, 15, 5649.

Paduano, Q. S., Drehman, A. J., Weyburne, D. W., Kozlowski, J., Serafinczuk, J, Jasinski, J, Liliental-Weber Z. (2003) X-ray characterization of high quality AlN epitaxial layers: effect of growth condition on layer structural properties, *Physica Status Solidi C*, 0, 2014-2018.

Peng, Y, Li, R, Cheng, H, Chen, Z, Li, H, Chen, M. (2017) Facile preparation of patterned phosphorin-glass with excellent luminous properties through screen-printing for high-power white light-emitting diodes, *Journal of Alloys and Compounds*, Vol. 396, 279-284.

Pernot, C., *et al.* (2010) Improved efficiency of 255-280 nm AlGaN-based light-emitting diodes, *Applied Physics Express*, 3, 061004.

Plimpton, S. J. (1995) Fast parallel algorithms for short-range molecular dynamics, *Journal of Computational Physics*, 117, 1-20.

Ponce, F.A., Major, J.S., Plano, W.E., and Welch, D.F. (1994) Crystalline structure of AlGaN epitaxy on sapphire using AlN buffer layers, *Applied Physics Letters*, 65, 2302.

Potì B, Tagliente MA, Passaseo A. (2006) High quality MOCVD GaN film grown on sapphire substrates using HT-AlN buffer layer, *Journal of Non-Crystalline Solids*, 352, 2332-2334.

Raghavan, S., Manning, I.C., Weng, X., and Redwing, J.M. (2012) Dislocation bending and tensile stress generation in GaN and AlGaN films, *Journal of Crystal Growth*, 359, 35-42.

Serrano, J., Rubio, A., Hernández, E., Muaz, A., and Mujica, A. (2000) Theoretical study of the relative stability of structural phases in group-III nitrides at high pressures, *Physical Review B*, 62, 16612.

Shatalov, M., *et al.* (2012) AlGaN deep-ultraviolet light-emitting diodes with external quantum efficiency above 10%, *Applied Physics Express*, 5, 082101.

Stukowski, A. (2009) Visualization and analysis of atomistic simulation data with OVITO-the open visualization tool, *Modelling and Simulation in Materials Science and Engineering*, 18, 015012.

Stukowski, A. (2009) Visualization and analysis of atomistic simulation data with OVITO-the Open Visualization Tool, *Modelling and Simulation in Materials Science and Engineering*, 18 (1), 18, 015012.

Stukowski, A., Bulatov, V. V., Arsenlis, A. (2012) Automated identification and indexing of dislocations in crystal interfaces, *Modelling and Simulation in Materials Science and Engineering*, 20, 085007.

Stukowski, A., Bulatov, V.V., and Arsenlis, A. (2012) Automated identification and indexing of dislocations in crystal interfaces, *Modelling and Simulation in Materials Science and Engineering*, 20, 085007.

Subramani S, Devarajan M. (2014) Influence of AlN thin film as thermal interface material on thermal and optical properties of high-power LED, *IEEE Transactions on Device and Materials Reliability*, Vol. 14, 30-34.

Susilo, N., Hagedorn, S., Jaeger, D., Miyake, H., Zeimer, U., Reich, C., Neuschulz, B., Sulmoni, L., Guttmann, M., Mehnke, F., Kuhn, C., Wernicke, T., Weyers, M., and Kneissl, M. (2018) AlGaN-based deep UV LEDs grown on sputtered and high temperature annealed AlN/sapphire, *Applied Physics Letters*, 112(4), 041110.

Tonisch, K., Cimalla, V., Foerster, C, Romanus, H., Ambacher, O., Dontsov, D. (2006) Piezoelectric properties of polycrystalline AlN thin films for MEMS application, *Sensors and Actuators A: Physical*, Vol. 132, 658-663.

Touzi, C., Omnès, F., Jani, B. E., and Gibart, P. (2005) LP MOVPE growth and characterization of high Al content AlxGa1−xN epilayers, *Journal of Crystal Growth*, 279, 31.

Tran, AT, Wunnicke, O, Pandraud, G, Nguyen, MD, Schellevis, H, Sarro, PM. (2013) Slender piezoelectric cantilevers of high quality AlN layers sputtered on Ti thin film for MEMS actuators, *Sensors and Actuators A: Physical*, Vol. 202, 118-123.

Tungare, M., Shi, Y., Tripathi, N., Suvarna, P., and Shahedipour-Sandvik, F.S. (2011) A Tersoff - based interatomic potential for wurtzite AlN, *Physica Status Solidi (a)*, 208, 1569.

Tungare, M., Shi, Y., Tripathi, N., Suvarna, P., and Shahedipour-Sandvik, F.S. (2011) A Tersoff - based interatomic potential for wurtzite AlN, *Physica Status Solidi* A, 208, 1569.

Tungare, M., Shi, Y., Tripathi, N., Suvarna, P., Shahedipour-Sandvik, F. S. (2011) A Tersoff-based interatomic potential for Wurtzite AlN, *Physica status solidi (a)*, 208, 1569-1572.

Vashishta, P., Kalia, R. K., Nakano, A., and Rino, J.P. (2011) Interaction potential for aluminum nitride: A molecular dynamics study of mechanical and thermal properties of crystalline and amorphous aluminum nitride, *Journal of Applied Physics*, 109, 033514.

Vashishta, P., Kalia, R.K., Nakano, A., and Rino, J.P. (2011) Interaction potential for aluminum nitride: A molecular dynamics study of mechanical and thermal properties of crystalline and amorphous aluminum nitride, *Journal of Applied Physics*, 109, 033514.

Wang, H.M., Zhang, J.P. Chen, C.Q., Fareed, Q., Yang, J.W., and Khan, M.A. (2002) AlN/AlGaN superlattices as dislocation filter for low-threading-dislocation thick AlGaN layers on sapphire, *Applied Physics Letters*, 81, 604.

Xiang, H., Li, H., Peng, X. (2017) Comparison of different interatomic potentials for MD simulations of AlN, *Computational Materials Science*, 140, 113-120.

Xiang, HG, Li, HT, Fu, T, Zhao, YB, Huang, C, Zhang, G, Peng, XH. (2017) Molecular dynamics simulation of AlN thin films under nanoindentation, *Ceramics International*, Vol. 43, 4068-4075.

Yan J, Wang J, Zhang Y, Cong P, Sun L, Tian Y, Zhao C, Li J. (2015) AlGaN-based deep-ultraviolet light-emitting diodes grown on high-quality AlN template using MOVPE, *Journal of Crystal Growth*, 414, 254-257.

Zhang, L., Yan, H., Zhu, G. Liu, S., Gan, Z., and Zhang, Z. (2018) Effect of substrate surface on deposition of AlGaN: a molecular dynamics simulation, *Crystals*, 8, 279.

Zhang, L.B., H. Yan, G. Zhu, S. Liu, and Z.Y. Gan. (2018) R. Soc.Open. Sci. 5, 180629.

Zhang, Y. F., Liu, Z. F. (2004) Oxidation of zigzag carbon nanotubes by singlet O_2: dependence on the tube diameter and the electronic structure, *Journal of Physical Chemistry*, B Vol.108: 11435.

Zhou, X. W., Murdick D. A, Gillespie, B, Wadley, H. N. G. (2006) Atomic assembly during GaN film growth: molecular dynamics simulations, *Physical Review B*, 73, 1079.

Zhou, X.W., Foster, M.E., Jones, R., Yang, P., Fan, H., and Doty, F.P. (2015) A modified stillinger-weber potential for TlBr, and its polymorphic extension, *Journal of Materials Science Research*, 4, 15.

Zhou, X.W., Jones, R. E., Kimmer, C.J., Duda, J.C., and Hopkins, P.E. (2013) Relationship of thermal boundary conductance to structure from an analytical model plus molecular dynamics simulations, *Physical Review B*, 87, 094303.

Zhou, X.W., Jones, R.E., Duda, J.C., and Hopkins, P.E. (2013) Molecular dynamics studies of material property effects on thermal boundary conductance, *Physical Chemistry Chemical Physics*, 15, 11078.

Zhou. X. W., D.A. Murdick, B. Gillespie, and H.N.G. Wadley. (2006) Atomic assembly during GaN film growth: Molecular dynamics simulations, *Physical Review B* ,73, 045337.

24

Mechanical Properties of AlN and Graphene

Aluminum nitride (AlN) is a wide-bandgap (6.01−6.05eV at room temperature) semiconductor material, with potential applications as the core of various key technologies including MIS (metal-insulator- semiconductor) devices (Adam *et al.*, 2001). AlN is widely used in mobile phone radio frequency filters (Loebl *et al.*, 2004) and ultra-violet solid state light sources (Taniyasu *et al.*, 2006). In spite of remarkable progress in the growth of III-nitride films and device fabrication technologies, some basic properties of AlN remain poorly studied, especially those related to monitoring techniques for mass production and reliability testing techniques for micro scale devices. In focus of these problems, lattice mismatch and thermal mismatch between film and substrate (Liu *et al.*, 2014; Bryan *et al.*, 2013) cause large strain, which leads to high dislocation density or even cracking (Raghavan *et al.*, 2011). Directly measuring engineering strain states of these micro scale semiconductors is one of the key techniques required nowadays. Since its first exfoliation from graphite crystals in 2004 (Novoselov *et al.*, 2004), graphene, with its distinctive band structure and fascinating combination of extraordinary properties, has attracted enormous attention in scientific and industrial communities because of its potential for various applications in graphene-based composites (Huang *et al.*, 2012), nano-devices (Jayasekera *et al.*, 2007), energy storage materials (Zhu *et al.*, 2014), and so on. As the strongest ever measured (Lee *et al.*, 2008), gas-impermeable (Bunch *et al.*, 2008), chemically and thermally stable (Zhou *et al.*, 2013; Margine *et al.*, 2008), and atomically thick material (Novoselov *et al.*, 2004), graphene is an excellent candidate as a protective coating or solid lubricant for enhancing surfaces, reducing adhesion, friction, and wear when coated on various surfaces so as to extend the longevity of graphene-based devices (Yu *et al.*, 2014; Su *et al.*, 2014; Berman *et al.*, 2014; Kim *et al.*, 2011; Peng *et al.*, 2015).

24.1 Mechanical Properties of AlN with Raman Verification

Raman scattering is widely used as a convenient and nondestructive strain detection method. Raman active phonons are straightforward signatures of bonds to sample strain fields, phase mixing, and interface morphologies. The quality of film crystallization and its in-situ evolution can also be characterized by Raman scattering (Kuball, 2011). Theoretically, Raman frequencies are in general calculated by density functional perturbation theory (DFPT) (also termed as linear response method) (Baroni *et al.*, 2001).

There are extensive studies of mechanical constants (Wagner *et al.*, 2002; Wright 1997; Łepkowski, 2015; Pandey *et al.*,2014) and phonon deformation potentials (Goni *et al.*, 2001; Callsen *et al.*, 2014; Sarua *et al.*, 2002; Gleize *et al.*, 2003; Kallel *et al.*, 2014) by first-principles calculations. Most results do not agree with experimentally detected Raman frequencies. Moreover, most of the investigators neglect to report whether or not their results are consistent with Raman frequencies. In fact, the phonon spectrum from Raman scattering is one of the fundamental indicators for crystal dynamics (Born *et al.*,1954), which in turn works as a bridge between mechanical constants and Raman frequencies. Specially, recent publications of Łepkowski (Łepkowski, 2015) and Callsen *et al.* (Callsen *et al.*, 2014) reveal that some elastic stiffness constants are sensitive to Raman frequency shifts and calculation methods. The main difficulties might be the high accuracy of elastic constants calculations. While the main idea of elastic constants determination (Wagner *et

Modeling and Simulation for Microelectronic Packaging and Integration: Manufacturing, Reliability and Testing, Second Edition. Sheng Liu and Yong Liu.

al., 2002; Wright, 1997) is by setting a finite value as stress/strain and then calculating corresponding strain/stress after re-minimizing the total energy, the two calculations have some intrinsic relationship all based on total energy which is mainly comprised of potential energy and kinetic energy. Considering the Born-Oppenheimer approximation (temperature →K), the kinetic energy should be zero. This implies that the kinetic energy error is a strong function of E_{cut} (scaling as E_{cut}^{-4}) (Mattsson *et al.*, 2005). Moreover, this small error in the value of E_{cut} will further lead to greater errors in the elastic constants and Raman spectra calculations due to the derivative operation of total energy. As a result, a quantitative prediction of Raman frequencies from first-principles calculations agreeing well with experiments is still lacking but desirable for further implication of Raman based measurement, such as, film growth monitoring and stress detection.

24.1.1 Methodology

The first-principles calculations have been done using CASTEP (Cambridge Sequential Total Energy Package) (Segall *et al.*, 2002), by which the DFT scheme within the GGA approximation is carried out. The well-established Perdew-Wang (PW91) version of the GGA approximation is adopted and norm-conserving pseudopotentials are used to describe electron-ion interactions.

The plane-wave energy cut-off, E_{cut} is 120Ry (1632eV) and Monkhorst-Pack grids of 11 × 11 × 11 are used for Brillouin Zone (BZ) sampling (FFT grid density is 45 × 45× 72). The geometry optimizations are achieved while the maximum force on every atom is smaller than 3.0×10^{-7} eV/Å; Additionally, norm-conserving pseudopotentials are adopted for considering the weak crystal symmetry and inhomogeneous electron gas distribution of AlN. Raman frequencies are calculated by a hybrid finite displacement/density functional perturbation theory (DFPT) (Baroni *et al.*, 2001; Gonze, 1997).

In addition to the GGA-PW91 calculations above, calculations based on the local density approximation (LDA) are also performed. The convergence thresholds used in the geometry optimizations and all other settings are kept as above.

There are mainly three variables related to stress/strain and polarization analyses, the biaxial modulus $\bar{C}$, bulk modulus B and Poisson ratio γ_{zx}, which are calculated by (Wagner *et al.*, 2002; Goni *et al.*, 2001)

$$\begin{cases} \bar{C} = C_{11} + C_{12} - 2\dfrac{C_{13}^2}{C_{33}} \\ \sigma_{xx} = \bar{C}\varepsilon_{xx} \\ B = \dfrac{(C_{11} + C_{12})C_{33} - 2C_{13}^2}{C_{11} + C_{12} + 2C_{33} - 4C_{13}} \\ \gamma_{zx} = \dfrac{C_{13}}{C_{11} + C_{12}} \end{cases} \tag{24.1}$$

where $\bar{C}$ stands for elastic constant, B stands for bulk modulus, γ_{zx} stands for Poisson ratio, σ and ε stand for stress and strain tensors.

E_{cut} is not usually known in advance for given calculation. Therefore, trying increased values of E_{cut} and inspecting the change of total energy to be small enough is one of the two important convergence tests, see Figure 24.1.

BZ sampling is realized by the Monkhorst-Pack method (Monkhorst *et al.*, 1976), which is an important bridge to treat the electronic structure for periodic systems in that it is able to take advantage of translational symmetry through Bloch's theorem with a finite number of k-points rather than full integration over the BZ (Mattsson *et al.*, 2005). Therefore, the approximation with a finite number of k-points should also be verified by convergence tests (see Figure 24.2).

It is worth mentioning that there is a balanced input for total energy minimization. As shown in Figures 24.1 and 24.2, we attempt to explore balanced E_{cut} and k-points for total energy minimization. The slopes of the total energy changing in Figure 24.1 (4.738×10^{-6}) and Figure 24.2($\times 10^{-6}$) are more or less in the same order of magnitude. At the same time, curvilinear trends in figures show an obvious difference. Therefore, well designed fitting functions are also shown in both figures to extrapolate the slope between total energy and E_{cut} or k-points. In fact, the four total energy related items, E_{cut}, k-points, SCF tolerance, and convergence thresholds in energy minimization, should be selected carefully with balanced control for a good final precision but with less computational cost.

24.1.2 Results and Analysis

Wurtzite-type AlN belongs to the $C_{6v}^4(P6_3mc)$ space group (Arguello *et al.*, 1969), with two formula units and four atoms per primitive cell. The primitive cell information used as original input is from ICSD#41482 (Xu *et al.*, 1993). After geometry optimizations, the lattice parameters are $a = 3.073$Å, $c = 4.948$Å, and $u = 0.381$, and $a = 3.064$Å, $c = 4.925$Å, and $u = 0.381$ for GGA and LDA calculations, respectively.

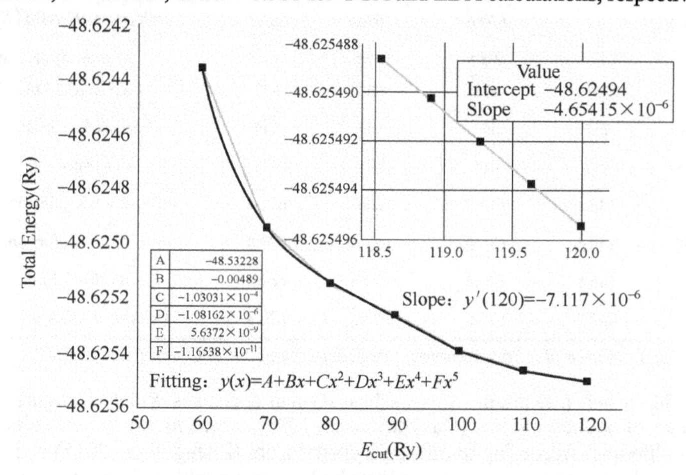

Figure 24.1 Convergence test of E_{cut}

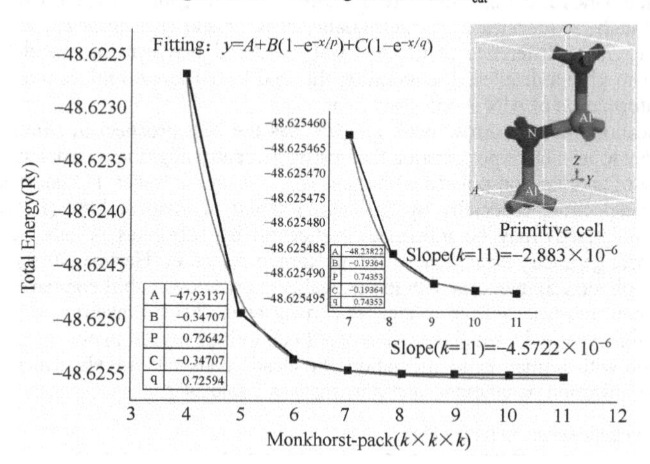

Figure 24.2 Convergence test of *k*-point grid

There are six optical modes: $\Gamma = A_1 + E_1 + 2E_2 + 2B_1$. A_1 and E_1 are both infrared and Raman active [polarized along *z* (c-lattice) optical axis and in the basal (*x*,*y*) plane respectively]. At the same time, both of them will split into LO and TO frequencies. Short-range atomic forces and long-range Coulomb fields are responsible for $A_1 - E_1$ and LO-TO splitting respectively (Demangeot *et al.*, 1996). This provides an insight for the calculation methods selection (*e.g.*, LDA or GGA). Additionally, E_2 (E_2^{low} and E_2^{high} none polarized) modes are Raman active and $2B_1$ are silent modes without meaning in practical stress detection.

We summarize our results of Raman frequencies and compared to experiments in Table 24.1.

Table 24.1 Comparison with Raman experiments (unit: cm^{-1})

Refs.	E_2^{low}	A_1 (TO)	E_2^{high}	E_1 (TO)	More Info
Zheng *et al.*, 2015		611	657	671	Bulk crystal, FWHM $\left(E_2^{\text{high}}\right) < 3.8$ cm^{-1}

(continued)

Refs.	E_2^{low}	A_1 (TO)	E_2^{high}	E_1 (TO)	More Info
Zhuang *et al.*, 2004		609.6	656	669.3	Bulk crystal, FWHM $(E_2^{high}) \approx 6.6$ cm^{-1}
Goni *et al.*, 2001	247.5	608.5	655.5	669.3	Bulk crystal, (whisker)
Goni *et al.*, 2001	241	618	667	677	Calculated, LDA, E_{cut} =75 Ry
Manjón *et al.*, 2008	237	594	636	649	Calculated, GGA-PBE, E_{cut} =60 Ry
Davydov *et al.*, 1998	252	615	667	673	Calculated, E_{cut} unknown
Davydov *et al.*, 1998	248.4	613.8	660	673.4	$T = 6$ K, film on α - Al_2O_3
Davydov *et al.*, 1998	248.6	611.0	657.4	670.8	$T = 6$ K, film on α - Al_2O_3
LDA	243.6	629.2	672.3	684.2	Calculated, LDA E_{cut} =120 Ry
GGA	247.3	615.6	656.5	670.5	Calculated, GGA=PW91 E_{cut} =120 Ry

According to reference (Davydov *et al.*, 1998), frequencies change little between 6K (0K, ground state) and 300K

FWHM (full width at half maximum) line-width in Raman spectra is widely recognized as an inversely proportional indicator of phonon lifetime (Davydov *et al.*, 1998), which in turn denotes higher crystal quality with longer phonon lifetime. According to this, bulk crystal from (Zheng *et al.*, 2015) is the best one among these carefully selected references (according to their bulk crystal quality). Thus, it is reasonable to take its frequencies as a benchmark for comparison here. As can be seen in Table 24.1, our calculated results show excellent agreement with frequencies from high quality bulk crystal (Zheng *et al.*, 2015). The only flaw appeared in A_1 (TO) with a difference of $615.6 - 611 = 4.6\text{cm}^1$ between the calculated and benchmark frequencies. In order to give a detailed discussion on this and later Raman implications on stress detection, some space group information of AlN is necessary here.

In practical implication, strong narrow peak in E_2^{high} has the best prospect in strain/stress detection for nitride compounds due to its lack of polarization that resists geometry alignment errors in testing (Zheng *et al.*, 2015). (The intensity of another non-polarized phonon, E_2^{low} is much lower.) Other polarized phonons are easy to influence in detection, especially by the angle/rotation of incident light (Bergman *et al.*, 1999). Therefore, A_1(TO) (polarized) may be influenced (reflected) by substrates or other crystal faces due to deviation from the ideal geometry arrangement during Raman detection (Harima, 2002). Another unverified trend we found is the phonon frequencies with lower quality show a lower shift compared with higher quality bulk crystal, see Raman frequencies from references (Zheng *et al.*, 2015; Zhuang *et al.*, 2004) in Table 24.1 for example. This could explain the small deviation of A_1(TO) of bulk crystal in our calculations.

After verification with Raman experimentation, the elastic constants are also calculated with the same mode and energy minimization parameters, as summarized in Table 24.2.

Table 24.2 Calculated elastic constants (unit: GPa)

Refs.	C_{11}	C_{12}	C_{13}	C_{33}	C_{44}	$\bar{C}$	B	γ_{zx}
LDA	397	144	115	372	115	469.9	211.9	0.2126
LDA	401	143	112	369	118	476.0	210.6	0.2059
GGA	361	130	93	339	107	440.0	187.1	0.1894
Expt.	411	149	99	389	125	509.6	210.4	0.1768
Expt.	390±15	145±20	106±20	398±20	105±10	478.5	210.0	0.1981
Expt.	140.5	148.4	98.9	384.3	124	508.0	209.5	0.1770
LDA	411.5	136.9	101.8	402.5	123	496.9	211.4	0.1856
GGA	408.4	124.0	90.7	401.2	125.4	491.4	202.8	0.1704

The elastic constants predicted in references. (Łepkowski, 2015; Pandey *et al.*, 2014) (see Table 24.2) both based on LDA show a good agreement with each other, but not experiment. On the contrary, although the Raman frequencies from our LDA calculations (in Table 24.1) have considerable difference with experimental data—especially the reliable experimental phonon E_2^{high} the elastic constants agree well with experiments (Table 24.2), even better than GGA from which the structure is able to rebuild Raman frequencies very well in Table 24.1. It is widely recognized that lattice parameters will increase if GGA is adopted as the exchange-correlation functional in geometry optimizations. However, the lattice constants decrease in our calculation compared with experimental lattice constants. At the same time, Raman frequencies calculated with the same exchange-correlation functional and optimized structure compare well with experiment. This makes it worthy to explore mechanisms for additional compounds in terms of pseudopotential design and geometry optimization algorithm improvements.

It is worth noting that though Raman frequencies calculated by LDA are not as acceptable as GGA-PW91 using norm-conserving pseudopotentials, the difference in calculated elastic constants is not so obvious compared with experiment ones as shown in Table 24.2, considering difficulty in evaluation of experiment error. Theoretically, Raman frequency and elastic constant calculations are based on second and first order derivative operations of the total energy respectively. In this viewpoint, elastic constants calculated by GGA-PW91 are more reasonable.

Here the phonon pressure coefficients are considered, which are also necessary in Raman stress/strain detection. Under the framework of Hooke's law, phonon pressure coefficients are defined (Callsen *et al.*, 2014) as the proportion between Raman frequency shift $\Delta\omega_\lambda$ and strain/stress (λ is indicator of the selected phonon. As mentioned above, E_2^{high} is the first choice in practical application):

$$\Delta\omega_\lambda = 2a\varepsilon_{xx} + b\varepsilon_{zz} = 2\tilde{a}\sigma_{xx} + \tilde{b}\sigma_{zz} \tag{24.2}$$

where ε and σ are the strain and stress tensors respectively. The constants a, b, $\tilde{a}$, and $\tilde{b}$ are the coefficients. There is a reasonable assumption for common AlN thin films on different substrates: $\varepsilon_{xx} = \varepsilon_{yy}$ and $\sigma_{xx} = \sigma_{yy}$.

The idea for determining these coefficients is to impose a known stress/strain condition to the studied system, followed with total energy minimization, and then calculate the corresponding frequency shift $\Delta\omega$ to determine the constants by Equation 24.2 (Wagner *et al.*, 2002; Callsen *et al.*, 2014). For example (Callsen *et al.*, 2014):

$$\begin{cases} \sigma_{xx} = \sigma_{yy} = 0, \sigma_{zz} \neq 0 \Rightarrow \Delta\omega \sim \tilde{b} \ \text{uniaxial-pressure} \\ \sigma_{xx} = \sigma_{yy} = \sigma_{zz} \neq 0 \Rightarrow \Delta\omega \sim 2\tilde{a} + \tilde{b} \ \text{hydrostatic-pressure} \end{cases}$$

A similar process for a and b can be found in reference (Wagner *et al.*, 2002). However, what we want to emphasize is that, if the uniaxial-pressure condition is imposed on the model, the cell will not retain its symmetry or else it would be unstable, which means a computational cost greater than the high symmetry crystal calculation but yielding lower precision (the hydrostatic-pressure condition could be imposed without destroying crystal symmetry). Results from experiment under uniaxial-pressure conditions maybe more dependable and could be used to ascertain $\tilde{b}$. Thus, we just calculate the value of $2\tilde{a} + \tilde{b}$ as shown in Table 24.3.

Table 24.3 Comparison with Raman experiments (unit: cm^{-1})

Pressure	E_2^{low}	A_1(TO)	E_2^{high}	E_1(TO)	Peak- E_2^{high} (cm^{-1})
−5.0	0.0651	5.0441	6.1916	5.6571	625.5646
−1.5	0.0427	5.0133	6.1130	5.5973	647.3532
−0.5	0.0364	5.0005	6.0613	5.5817	653.4770
0.5	0.0307	4.9966	6.0720	5.5673	659.5587
1.5	0.0257	4.9910	6.0536	5.5525	665.6031
5.0	0.0081	4.9632	5.9902	5.5050	686.4735
Average	0.0348	5.0015	6.0853	5.5768	
Goni *et al.*, 2001	0.12±0.05	4.4±0.1	4.99±0.3	4.55±0.03	Expt. $2\tilde{a}+\tilde{b}$
Callsen *et al.*, 2014	0.75±0.04	1.46±0.02	1.66±0.02	1.20±0.02	Expt. $\tilde{b}$
Manjón *et al.*, 2008	0.07±0.02	4.35±0.03	5.40±0.04	5.33±0.04	Expt. $2\tilde{a}+\tilde{b}$

(1) Pressure > 0 means being compressed.

(2) Considering that other phonons are easy to be influenced in detection, we should pay more attention on E_2^{high}.

(3) We have evaluated AlN films with different substrates appeared in recent years [including sapphire, SiC, Si (001) and ourselves experiment on Si (111)], stress in film on Si (111) substrate is the largest, more or less than 2.0GPa, thus, we did not calculate more higher pressure status.

For the first time in the literature, we calculated $2\tilde{a} + \tilde{b}$ under different pressures (similar studies in the literature only give one coefficient as constant). Correspondingly, we find a new trend never reported before: $2\tilde{a}+\tilde{b}$ is decreasing slightly according to the pressure, as shown in Figure 24.3. Commonly, this coefficient is considered a constant with pressure. Therefore, we also list peaks of E_2^{high} in the right highlighted column for further correction. For example, if a detected peak of E_2^{high} is 662cm^{-1}, a suitable coefficient can be found between 6.0720 and 6.0536 by interpolation.

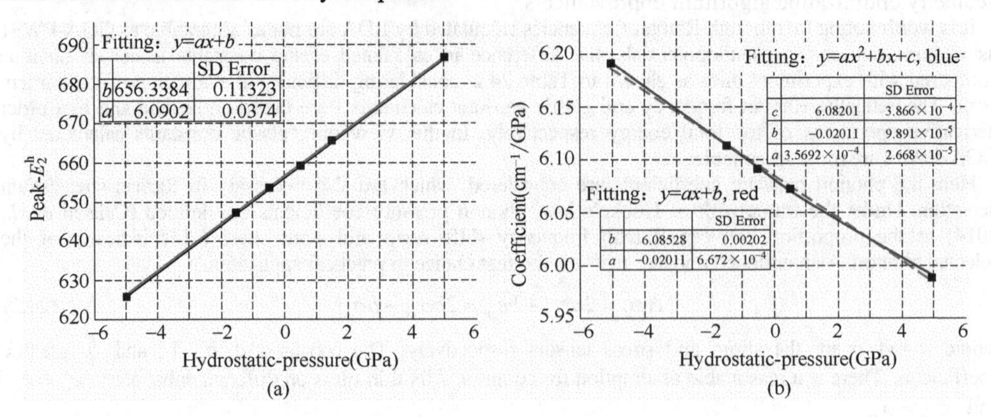

Figure 24.3 Curves of hydrostatic-pressure between Peak of E_2^{high} and $2\tilde{a}+\tilde{b}$ *(Color version online)*

24.2 Stress Evolution in AlN and GaN Grown on Si(111): Experiments and Theoretical Modeling

Gallium nitride (GaN) and aluminum nitride (AlN) have extensive potentials in new generation broad band semiconductor material exploration due to their perfect optical and electrical properties (Hadi *et al.*, 2014). Widely recognized hot topics include light-emitting diodes, ultraviolet (UV) radiation sensors, high-electron mobility transistors (HEMTs) (Del Alamo *et al.*, 2009) and etc. Though these exciting applications have been reported for years, some basic problems remain poorly studied, especially the crystal quality control during hetero-epitaxy (Falub *et al.*, 2012), lattice and thermal mismatch alleviation (Del Alamo *et al.*, 2009; Scholz 2012), and crack prevention (Wei *et al.*, 2011; Meneghesso *et al.*, 2008). These problems present major limitations in terms of LED device yield, performance, and cost (Leach *et al.*, 2014). One of the most widely identified parameters is the warping (a popular statement is bow or wafer bow) which is commonly utilized as an important and convenient indicator of quality in production. In fact, Leach *et al.* (Leach *et al.*, 2014) have analyzed and summarized the 2012 SSL Manufacturing R&D Roadmap and pointed out that wafer bow is directly responsible for at least 6 of the total 24 problem areas and the FLAAT (flat layers at all temperatures) technology of template is related to each of these issues. Moreover, problems are even more remarkable aiming at films grown on silicon wafers as emphasized in the Roadmap of 2014.

Wafer bow is intrinsically caused by thermal mismatch between the film and substrate, which manifests itself more obviously with high temperature variation like the films grown by metal-organic chemical vapor deposition (MOCVD). There are indeed a lot of efforts (Zhang *et al.*, 2014; Hirayama *et al.*, 2015) that attempt to control wafer bow to be minimum which is strongly related to the stress and strain induced mechanically, such as, strain relaxation in our previous work using step-graded buffer layers (Leung *et al.*, 2014) and Leach's work using FLAAT technology with a polycrystalline GaN film on the backside of the substrate for force balance between both sides (Leach *et al.*, 2014). As mentioned in references (Leach *et al.*, 2014; Hirayama *et al.*, 2015) and Roadmap of 2014, utilizing a thicker substrate is another necessary way to compensate for large bow effect specially for a thicker film required in high performance and powerful devices. In fact, lattice power company recently supplies a product with a much thicker substrate [the thickness of film and 2-inch. Si(111) substrate are 5 and 1500 μm respectively] aimed for higher performance and better reliability, which implies a novel investigation trend in this field. In practice, the gradient or flatness of film surface which is particularly detrimental to LED performance (Leach *et al.*, 2014) should be controlled carefully for fewer defects such as hexagonal pits caused by stress observed in (Leung *et al.*, 2014). At the same time, the stress state of film is strongly related to the reliability of metal line interconnection in devices and the percentage of dislocations in post thermal process (Campbell, 1996). For example, the deformation of solder pad (such as p-electrode or n-electrode which is connected on the surface of film as a component in devices) is readily influenced by the stress in film because the deformations between the film and solder pad are much easier to misfit under cyclic workloads coupled with the residual stress in film. Thermal-expansion and cold-contraction further compiling with the deformation caused by the residual stress will also greatly increase the complexity of treatment in packaging (Wang *et al.*, 2015). On the other hand, factories spend a lot of time and energy on several times of quenching process (post thermal process) as important manufacture steps so as to release the residual stress in films and detect killer defects at an early stage, which in turn optimize process yields (Campbell, 1996) (see also the Roadmap of 2014).

The problems mentioned above have already attracted tremendous research interests in literature especially on experimental investigations for more than 30 years. However, it still lacks of enough numerical models for residual stress forecasting in the complicated process flows. In view of this, the aim here is at constructing a numerical model for films grown on Si(111) substrate. The model is of practical significance for stress state prediction especially considering that the experiments would exhaust money and time greatly (van Driel *et al.*, 2011). The main theme of this work is to prepare good samples and analyze Raman frequency shifts to ascertain the residual stress values which are adopted for model verification. The temperature dependent stress evolutions and the impact factors are discussed in depth together with model building and wafer bow inspecting as intuitionistic indicators to reveal the mechanisms.

24.2.1 Sample Preparation and Material Characteristics

Wurtzite-type epitaxial structures are grown by metal-organic chemical vapor phase deposition (MOCVD). Trimethylgallium, trimethylaluminum and ammonia are used as precursors for gallium, aluminum, and nitrogen respectively, with hydrogen as the carrier gas. 2-inch. silicon(111) wafers are loaded into the reactor in a N_2 glovebox and the temperature is raised to 950℃ for thermal cleaning and removal of native oxide. Al-preseeding is performed after substrate preparation and a thickness of 200nm AlN epitaxial layer is grown at 1150℃, then, followed with a thickness of 500nm GaN growth at the same temperature. The V/III ratio is always controlled to be at about 1100 with a pressure of 100 mbar during growth. We prepare three types of samples and the growth conditions for AlN samples are the same but without GaN growth further. [Sample-1: GaN ~500nm, AlN ~200nm, Si(111) ~700μm; Sample-2: AlN ~200nm, Si(111) ~1500μm; Sample-3: AlN ~200nm, Si(111) ~700μm.]

There are different formulas to calculate thermal expansion coefficient (TEC), which lead to corresponding inconformity in results, *e.g.*, difference in selection of reference temperature (mainly including room and 0K temperatures) and datum length. Considering anisotropy, we use the format about the TEC α below to resist the errors from datum length selection.

$$\alpha = \frac{\mathrm{d}L(t)}{L(t)\mathrm{d}t} \tag{24.3}$$

where $L(t)$ corresponds to a-lattice and t is an indicator of temperature. In this way, we fit TECs directly from temperature dependent lattice variations and resample them together for comparison, as shown in Figure 24.4. The original temperature dependent lattice datas are from reference (Reeber *et al.*, 1996, 2000; Slack *et al.*, 1975). Another thermal stress related variable is temperature dependent Young's modulus. Temperature dependent experimental elastic constants can be found in reference (Reeber *et al.*, 2001; Landolt-Bornstein 1979; Kluge *et al.*, 1986) and have been fitted here (experimental temperature range AlN : T∈[0, 1350]K; GaN : T∈[0, 900] K; Si : T∈[0, 1164] K).

$$\text{AlN}:\begin{cases} C_{11} = 412.00 - 4.1524\times10^{-3}T - 5.9638\times10^{-6}T^2 \\ C_{12} = 149.08 - 1.6913\times10^{-3}T - 2.7353\times10^{-6}T^2 \\ C_{13} = 99.611 - 2.4723\times10^{-3}T - 1.2923\times10^{-6}T^2 \\ C_{33} = 386.20 - 6.3559\times10^{-3}T - 4.5396\times10^{-6}T^2 \\ C_{44} = 124.19 - 0.7009\times10^{-3}T - 0.5724\times10^{-6}T^2 \end{cases} \tag{24.4}$$

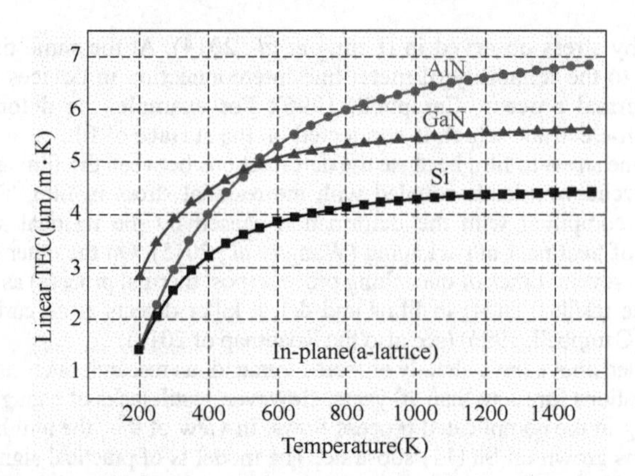

Figure 24.4 Curves of linear TEC (in plane, corresponding to a-lattice)

$$\text{GaN:}\begin{cases} C_{11}=376.77-5.7006\times10^{-3}T-1.2454\times10^{-6}T^2 \\ C_{12}=142.56-2.4076\times10^{-3}T-6.1198\times10^{-6}T^2 \\ C_{13}=99.33-3.3499\times10^{-3}T-3.2915\times10^{-6}T^2 \\ C_{33}=387.74-9.1662\times10^{-3}T-1.1020\times10^{-6}T^2 \\ C_{44}=98.57-0.6854\times10^{-3}T-1.1182\times10^{-6}T^2 \end{cases} \tag{24.5}$$

$$\text{Si:}\begin{cases} C_{11}=170.06-1.634\times10^{-2}T \\ C_{12}=65.59-0.569\times10^{-2}T \\ C_{44}=81.66-0.717\times10^{-2}T \end{cases} \tag{24.6}$$

Only the silicon is fitted linearly due to its weak higher order fitting coefficients. Moreover, considering the anisotropic properties of crystal, we deduce the Young's modulus, Poisson's ratio, and Shear modulus (only suitable for hexagonal crystals):

$$\text{AlN and GaN:}\begin{cases} E_x=E_x=\dfrac{C_{11}^2C_{33}-2C_{13}^2C_{11}-C_{12}^2C_{33}+2C_{13}^2C_{12}}{C_{11}C_{33}-C_{13}^2} \\ \gamma_{12}=\dfrac{C_{12}C_{33}-C_{13}^2}{C_{11}C_{33}-C_{13}^2} \\ G_{xy}=\dfrac{C_{11}-C_{12}}{2}, G_{zx}=G_{yz}=C_{44} \end{cases} \tag{24.7}$$

Commonly, there are differences between crystallographic coordinates and mechanical coordinates to characterize mechanical properties. After this transformation, values of Equation 24.4 are enough to express the plane stress appearing in films and $G_{zx}=G_{yz}$ are required for modeling transverse shear deformations in a shell given stress component $\sigma_z=0$.

For Si(111) wafers, reference (Brantley 1973) has already investigated special formulas for considering anisotropy in {111} plane. In order to cover the high temperature part, the empirical formula $E=E_0-BT\exp\left(-\frac{T_0}{T}\right)$ (E_0 is the Young's modulus at 0K. *B* and T_0 are decided by fitting, without clear physical meanings) is widely recognized to realize high temperature trend extrapolation. However, just as shown in reference (Bruls *et al.*, 2001), the trend for high temperature part is nearly the same as polynomial fitting. Thus, we directly utilize the formulas above for trend extrapolation to cover the high temperature part. In this way, the comparison of in-plane Young's modulus is shown in Figure 24.5.

In brief, AlN and GaN possess an anisotropic performance for TEC, while silicon is isotropic material; for Young's modulus, they are all significant anisotropic materials [*e.g.*, for Young's modulus of silicon, the maximum is 188GPa and the minimum is just 130GPa within different crystal directions (Hopcroft *et al.*, 2010)]. Comparing with Figures 24.4 and 24.5, AlN presents a larger deformation potential due to its higher magnitudes both in TEC and Young's modulus at high temperatures. At the same time, GaN presents a high temperature softening compared with AlN. To this end, a temperature dependent anisotropic model is expected to estimate the above relationships as the main problems behind in film crystal growth.

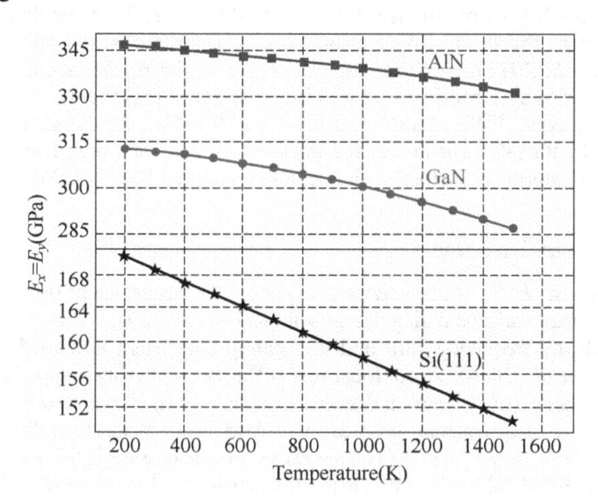

Figure 24.5 Temperature dependent Young's modulus of AlN, GaN, and Si(111) (in plane)

24.2.2 *Stress Characterization*

(i) Stress Relationship Analysis

The X-ray diffraction patterns have been shown in Figure 24.6. According to Bragg law, the small

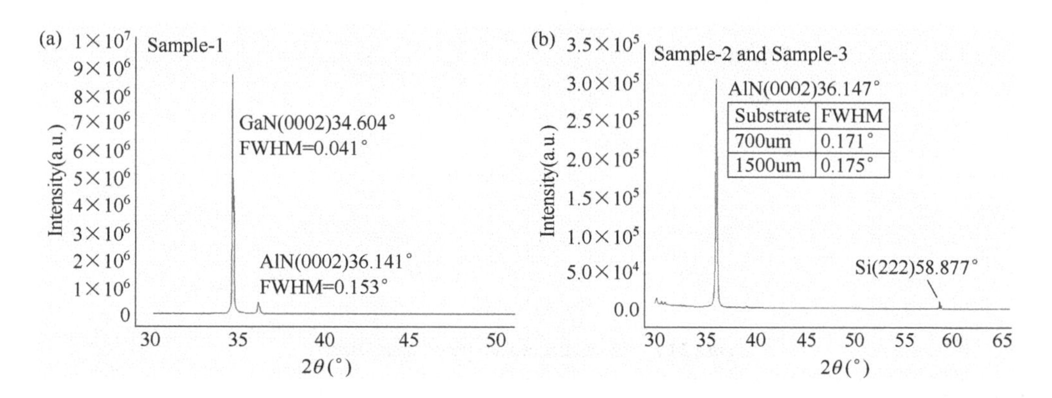

Figure 24.6 X-ray diffraction patterns of GaN and AlN films. (a) X-ray pattern of Sample-1; (b) X-ray patterns of Sample-1 and Sample-2

values of FWHM (full width at half maximum) imply a high crystal quality (Moram *et al.*, 2009). In addition, the FWHM of AlN as an interlayer of Sample-1 is improved after GaN growth (0.153° < 0.171° means a less crystal quality degradation caused by the residual stress). This may be explained as crystal structure protection action from the GaN layer to the AlN interlayer during warping caused by temperatures, *e.g.*, the existence of GaN may lead to a more uniform force distribution in surface of AlN during warping. At the same time, both AlN and GaN films will shrink more than silicon, thus the films impose a compressive stress to the silicon substrate which in turn yields a bowl-like appearance. In comparison with Sample-2 and Sample-3, the overall effect is beneficial to the interlayer AlN crystal quality to result in a lower FWHM value. This mechanism should also be a good indicator to optimize the thickness relationships between films and substrate aimed at a crack free GaN layer (Wei *et al.*, 2011). In short, the couplings of the thicknesses, temperature dependent thermal expansions and Young's modulus should be considered together.

Furthermore, the nearly equal FWHM values, 0.171° and 0.175°, for the same thickness of AlN films but different thickness of substrates, mean that the thickness of substrate only has a weak influence on the crystal films in c-lattice direction. Therefore, a shell element adopted later in FEM simulations is reasonable for actual situations.

(ii) Stress Ascertain by Raman Scattering

The Raman frequency shift of E_2^{high} is often adopted to identify the magnitude of stress in III-nitride compounds, even in situ monitoring the stress variation during film growth (Kuball *et al.*, 2001). Theoretically, the stress in crystal is a linear relationship with the frequency shift and the scaling coefficient is named as phonon deformation potential (PDP) (Callsen *et al.*, 2014). A confocal micro-Raman scattering setup (LabRAM HR800, Laser length: $\lambda = (532 \pm \Delta\lambda)$ nm, $\Delta\lambda \leqslant 0.65$nm; diameter of laser: ~1μm) is used to accomplish the experiments at room temperature. We choose backscattering geometry with laser beam incident on (0001) surface. According to reference (Arguello *et al.*, 1969), only E_2 and A_1(TO) are allowed to the detector in this mode. The Raman frequency peaks of E_2^{high} and their FWHM values fitted by Gaussian function have been marked in Figure 24.7. Based on the theory above, the evaluated stress values have been arranged in Table 24.4, followed with an explanation.

The bold values of AlN in Table 24.4 are based on single crystals (stress free), specially, (Zheng *et al.*, 2015) is selected due to its smaller FWHM value of E_2^{high} [(Zheng *et al.*, 2015) <3.8 cm^{-1}; (Zhuang *et al.*, 2004)≈6.6 cm^{-1}]. Considering the smaller stress in GaN film and the difficulty in determining a better stress free phonon peak, the stress precision in GaN film is not estimated further and two sets of stress range are listed based on (Kisielowski *et al.*, 1996; Zhang *et al.*, 2013) respectively.

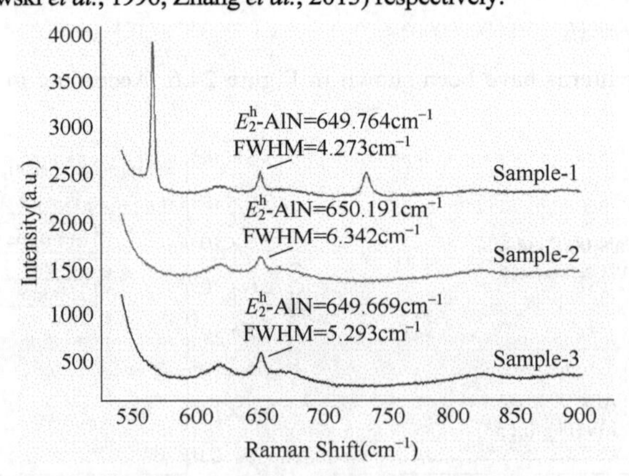

Figure 24.7 Raman frequencies (resolution: 0.4cm^{-1})

Table 24.4 Stress evaluation

	E_2^{high} (cm^{-1})	$-2a$(cm^{-1}/GPa)	Stress(GPa)
AlN	657	3.74, 3.39	Sample-1: 1.386–2.412, 1.935
	656	4.5, 3.0	Sample-2: 1.386–2.270, 1.821

(continued)

	E_2^{high} (cm^{-1})	-2a(cm^{-1}/GPa)	Stress(GPa)
			Sample-3: 1.407 to 2.444, 1.960
GaN	566.2	2.77, 2 .4, 1.7, 2.43	Sample-1: 0.095 to 0.345
	565	2.7, 2.9, 4.47, 6.2	Sample-1: -0.361 to -0.099

24.2.3 Simulations

(i) Theory in Simulation

According to thermal stress mathematical model, the thermal stress can be expressed through α by:

$$\sigma_{\rm th} = \frac{E_{\rm f}}{1-\gamma_{\rm f}}\int_{T_1}^{T_2}(\alpha_{\rm s}-\alpha_{\rm f})\,{\rm d}T \tag{24.8}$$

where subscripts f and s stand for film and substrate respectively; E and γ denote Young's modulus and Poisson's ratio. The formula intrinsically expresses an integration from film growth temperature T_1 (1425K) to room temperature T_2 (300K). At the same time, it can also be expressed with Stoney equation deduced specially for films on Si(111) substrate (Nix, 1989):

$$\sigma_{\rm f} = \frac{6C_{44}(C_{11}+2C_{12})}{C_{11}+2C_{12}+4C_{44}}\times\frac{h_{\rm s}^2}{6h_f}\times\frac{1}{R} \tag{24.9}$$

where h denotes the thickness, and C_{ij} represent elastic constants of silicon prepared in Equation 24.5. R is the radius of curvature which has been recognized as an important indicator of film quality. In practice, a more general formula is used to predict R:

$$R = \left|\frac{(1+y'^2)^{3/2}}{y''}\right| \tag{24.10}$$

where y indicates the off-plane displacement (out of plane displacement), and y' and y'' are its first and second derivatives.

(ii) Model Information and Scaling Verification

The square element size in simulations is about 0.75mm as shown in Figure 24.8. The simulation

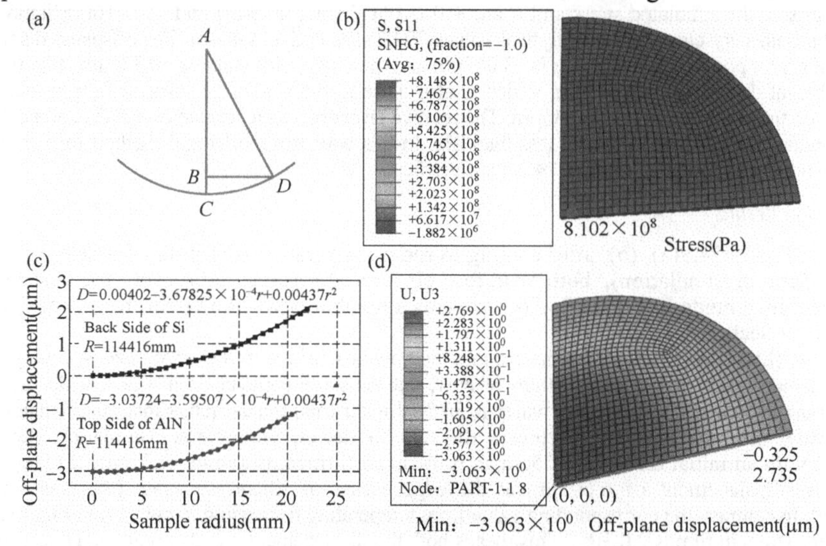

Figure 24.8 Scaling verification. (a) Schematic of radius of curvature; (b) stress on surface of film.; (c) off-plane displacement along radius; (d) off-plane displacement distribution (*Color version online*)

along film thickness is realized by 9 uniform points integration (in Abaqus CAE software, this item is titled as Thickness Integration Points for input) and 15 points for silicon substrate with a size decreasing ratio of 5 towards the film surface (in Abaqus CAE software, this item is titled as Sizing Controls of Bias Ratio for element meshing). SC8R and C3D8R elements are adopted for film and substrate respectively.

Additionally, symmetry is considered and only one quarter of the wafer is constructed in simulation model.

Temperature range in simulation is from 1425K to 300K. Both films and substrate are considered in plane stress state and thus the silicon is considered as isotropic but using {111} plane parameters, such as Young's modulus and Poisson's ratio. The parameters for shell element have been prepared in Section 24.2.1 and the model is constructed compatible with the same coordinates to consider anisotropy. The center point on back sides of silicon is fixed in z direction as boundary condition and the coordinate origin. The units in simulations are Pa and μm.

Scale issue is the main doubtful point considering the thin film in a nanometer scale in thickness and the substrate in micrometer scale. This is verified by assuming that the properties (of Sample-3) are constant values at 600K without varying with temperatures (600K is a random selection so as to artificially calculate the thermal stress by Equation 24), and their values have been listed in Table 24.5. Considering both Equations 24.8 and 24.9 expressing film stress, analytical results can be solved by equaling them with temperatures at T_1 = 1425K and T_2 = 300K. In this way, the analytical values are obtained as, σ_{th} = 794.3MPa, R = 114.444m.

Table 24.5 Variables used in scaling verification (600 K)

Si(111) ~ substrate				AlN~film			
E_s (GPa)	γ_s	TEC_s (μm/m·K)	h_s (μm)	E_s (GPa)	γ_s	TEC_s (μm/m·K)	h_s (μm)
164.2	0.262	3.888	700	343.4	0.319	5.320	0.2

The first part in Equation 24.7 is 222.62GPa based on Equation 24.6 and the diameter of sample is 5cm.

The corresponding simulation results have been shown in Figure 24.8 with a schematic drawing illustrating the relationship between off-plane displacements (out of plane displacement) on edge point and radius of curvature R in Figure 24.8(a), BD = 25mm, AC = AD = R=114.444m, according to the Pythagorean theorem, the off-plane displacement of edge point D is BC≈2.731μm.

In comparison, the simulated stress and R are 810.2MPa [Figure 24.8(b)] and 114.416m [Figure 24.8(c)] respectively, both very close to the analytical values 794.3MPa and 114.444m. For off-plane displacement, the value of center point on film surface is −3.063μm and the edge point value is −0.325μm [Figure 24.8(d)]. Thus, the height difference is 2.738μm, which is also close to the analytical value, 2.731μm and the edge point value on the corner of silicon, 2.735μm. The relative errors between simulation results and analytic ones, including radius of curvature R, are all less than 2%. In this way, the model and method for variable value preparation presented above in Section. 24.2.1 are verified.

(iii) Simulation Results

As shown in Figure 24.9(a), (b), after cooling to room temperature (the initial diameter of sample at 1450K is 5cm in simulation), both manifest obvious shrinkage, moreover, the distribution of displacement magnitude in Figure 24.9(e) presents a notable bowl-like effect that is consistent with the analysis in Section. 24.2.2.

Leach *et al.* (Leach *et al.*, 2014) emphasize that the bow is sensitive to the wafer size, specially, 15μm of GaN on a 2-inch. wafer might cause 90μm of bow, but the same epilayer on a 4-inch wafer might cause 170μm of bow even though the 4-inch wafer is 50 % thicker than the 2-inch. substrate. Considering this and the wafer size at high temperature is difficult to detect, the inverse simulation (temperature increasing) with an initial diameter of 5cm at 300K is performed as shown in Figure 24.9(c), yielding an off-plane displacement of 6.602μm at the edge corner of silicon substrate (the stress in film is −1.923GPa). In contrast, the corresponding value from temperature decreasing simulation in Figure 24.9(a) is 6.209μm (the stress in film is 2.0GPa). This means both the displacement on edge point and stress in film are sensitive to the wafer size.

Therefore, the simulated film stresses are prepared with temperature increasing simulations for higher precision in Table 24.6. Theoretically, the film stress value from this inverse simulation is of the same magnitude but converting its sign. Others, stresses in silicon section [such as Figure 24.9(d)] and off-plane displacements are prepared as conservative estimations by temperature decreasing simulations for directly understanding of the mechanical mechanisms between layers.

Table 24.6 Stresses in AlN layers (GPa)

Sample	Stress(simulated)	Stress(Raman)	FWHM(cm^{-1})
Sample-1	1.919	1.386–2.412, 19.35	4.723
Sample-2	1.925	1.291–2.270, 1.821	6.342
Sample-3	1.923	1.407–2.444, 1.960	5.293

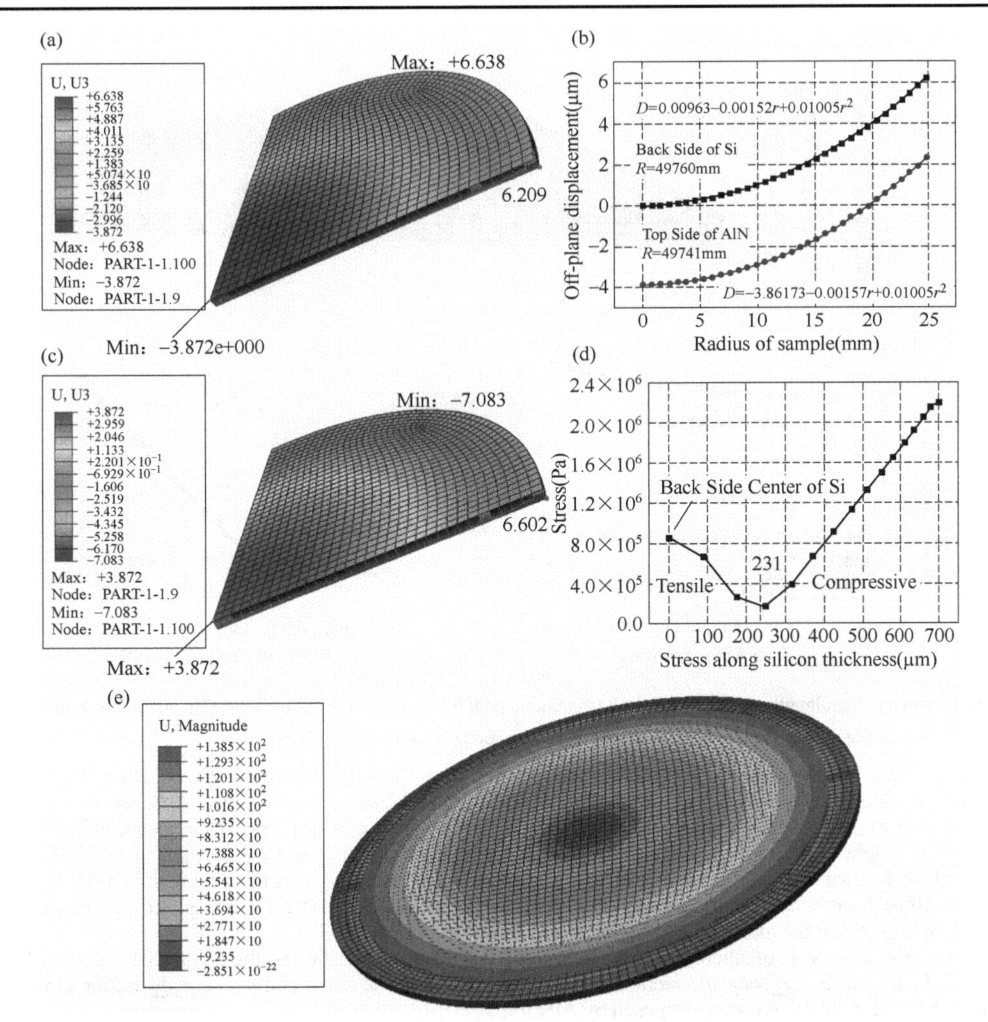

Figure 24.9 imulated results of sample-3 (thickness: AlN ~200nm, Si ~700μm). (a) off-plane displacement; (b) off-plane displacement along radius; (c) off-plane displacement (temperature increasing); (d) stress in silicon section; (e) whole show of displacement magnitude (deformation scale factor: 20) (*Color version online*)

Comparing with the stresses detected from Raman scattering, the simulated stresses are all in acceptable error range, especially very close to the stresses based on Single crystals. The small contradiction on Sample-2 (substrate thickness: 1500μm) can be explained with crystal quality degradation indicated by FWHM. The thicker substrate leads to a stronger resistance to deform as shown in Figure 24.10(a) below, and the structure of film crystal is weakened as indicated by FWHM.

Considering Figures 24.9, 24.10, and Table 24.6, it is verified that a thicker substrate is beneficial to the flatness of film surface in growth, and the film stress is slightly influenced. In fact, the film stresses with different substrate thicknesses should be equal as implied by Equation 24.6 in which there is no parameter related to the size of substrate at all. Furthermore, the two samples grown with the same growth conditions should also have a fixed relationship between their radiuses of curvatures implied by Equation 24.7:

$$\frac{h_{s1}^2}{h_{f1}R_1}=\frac{h_{s2}^2}{h_{f2}R_2} \tag{24.11}$$

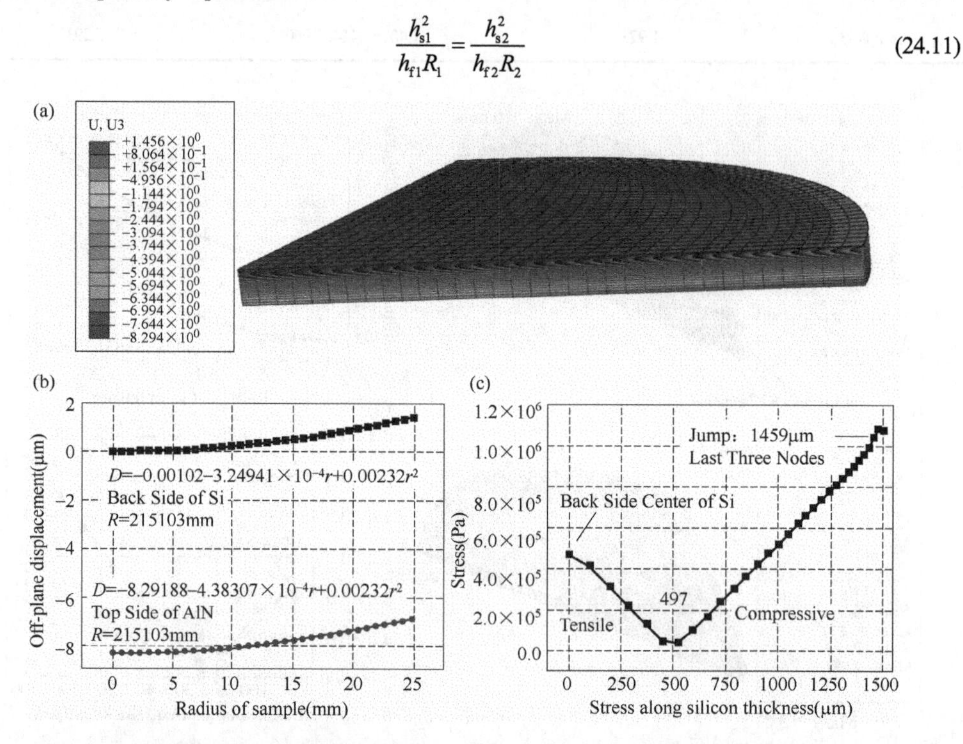

Figure 24.10 Simulated results of Sample-2 (thickness: AlN ~200 nm, Si ~1500 μm). (a) Off-plane displacement; (b) off-plane displacement along radius; (c) stress in silicon section (*Color version online*)

Given the substrate thickness, h_{s1} = 0.7mm and h_{s2} = 1.5mm (h_{f1} = h_{f2} = 200nm), this equation is satisfied well in simulation results of Figures 24.9(b) and 24.10(b). The relative error is less than 6.2%, which is also a verification of our model constructed in Section. 24.2.1. At the same time, this formula should give an important insight to design films. The simulated results of Sample-1 have been shown in Figure 24.11, and the main difference is the higher off-plane displacement on the edge points as comparing with Figures 24.9(b) and 24.10(b). In order to give an explanation of this, we further summarize the Figures 24.9(d), 24.10(c), and 24.11(b) to ascertain the mechanical mechanisms as shown in Figure 24.12.

Firstly, the stress state of silicon substrate in Figure 24.12 has been verified by the Figures 24.9(d), 24.10(c), and 24.11(b), that is, the backside section in silicon is bearing tensile stress and the top side section close to film is bearing compressive stress imposed by AlN film.

Secondly, both films are considered as shells because of their less thicknesses compared with the thickness of substrate. At the same time, both materials of films will shrink more than silicon substrate, as clarified in Section. 24.2.2. Therefore, the AlN interlayer is just working as a bridge to transfer deformation from substrate to GaN layer, that is to say, both films work together to resist the deformation from substrate (the bending stiffness of film is not considered in a shell element at all).

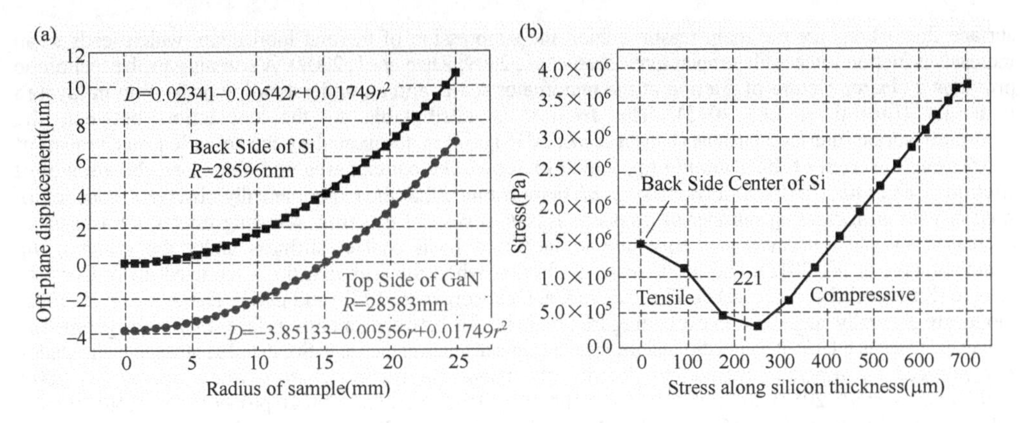

Figure 24.11 Simulated results of Sample-1 (thickness: GaN~500nm, AlN ~200nm, Si~700μm). (a) Off-plane displacement along radius; (b) Stress in silicon section

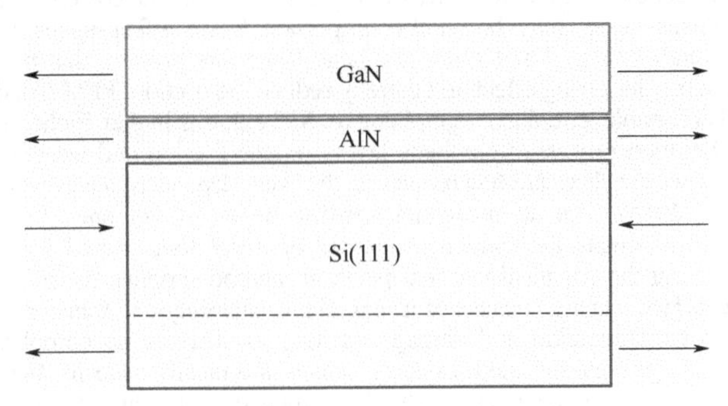

Figure 24.12 Stress state

Finally, according to the equilibrium sections in Figure 24.9(d) (231μm), Figure 24.10(c) (497μm), and Figure 24.11(b) (221μm), and the off-plane displacements in Figures. 24.9(b), 24.10(b), and 24.11(a), the bow is sensitive to the magnitude of stress in GaN, as demonstrated in Figure 24.11(a) with higher off-plane displacements on the edge points. The reason is that the total force increases with its thickness though the magnitude of stress is lower than AlN. This is particularly important or difficult for a thicker film grown on thinner substrate, which is consistent with common sense in production.

24.3 Molecular Distinctive Nanofriction of Graphene Coated Copper Foil

The microscale and nanoscale friction characteristics of micro-electro-mechanical-systems (MEMS) and nano-electro-mechanical-systems (NEMS) devices play an important role in determining performance, reliability, and lifetime of the overall system (Kim *et al.*, 2007; Bhushan *et al.*, 2007). As such, controlling and minimizing friction and wear-related mechanical failures remains one of the greatest challenges for contemporary moving assembles. Nanoscale friction on graphene exhibits different behaviors from its 3D counterpart (Mo *et al.*, 2009; Lee *et al.*, 2010) because of its large surface-to-volume ratio, super-high mechanical stiffness, and strength (Lee *et al.*, 2008); all are highly desirable for wear protection. Most experimental studies have reported ultralow friction (Lee *et al.*, 2009), or even superlubricity (Feng *et al.*, 2013) sometimes accompanied by wear reduction (Berman *et al.*, 2013). The coefficient of friction for SiO_2 reduces from 0.68 to 0.12-0.22 after coating it with monolayer graphene (Kim *et al.*, 2011). Suspended graphene exhibits distinctive behavior in which nanoscale friction first increases and then decreases with increasing loads (Ye *et al.*, 2014). Deng *et al.* reported an adhesion-dependent negative friction coefficient on chemically modified graphite at the nanoscale (Deng *et al.*, 2012). Zhang *et al.* experimentally studied the nanoscale friction characteristics of graphene exfoliated onto weakly adherent silica substrates, and found that surface fluctuations are the main reason behind the suppression of thermal lubrication, which leads to an increase in friction force with temperature (Zhao *et al.*, 2009; Zhao *et al.*, 2007). A scanning probe technique provides a clearer picture of friction at the nanometer scale, and an understanding of friction in layered materials (Liechti *et al.*, 2015). The friction is attributable to the interaction between the incommensurate interface lattices (Koren *et al.*, 2015), and is dominated by the so-called mechanism of "puckering" in front of the scanning tip, which increases the contact area and, therefore, the amount of friction (Lee *et al.*, 2010). Smolyanitsky *et al.* demonstrated that the experimentally observed reduction of friction with an increasing number of graphene layers in case of a narrow scanning tip can be a result of decreased sample deformation energy due to increased local contact stiffness under the scanning tip (Smolyanitsky *et al.*, 2012), and an increase of friction when the scanning tip is retracted away from the sample (Smolyanitsky *et al.*, 2012). The JKR model (Johnson *et al.*, 1971) and DMT (Derjaguin *et al.*, 1975) model are generally successful in explaining friction behavior between smooth surfaces of various geometries, however, there is a lack of atomistic insight into the tip-sample interactions. So, detailed atomistic simulation has emerged as important method to describe the nanoscale friction characteristics (Mo *et al.*, 2009; Smolyanitsky *et al.*, 2012). In brief, the factors including load (Ye *et al.*, 2014), properties and roughness of substrate (Paolicelli *et al.*, 2015), scratch velocity (Gnecco *et al.*, 2000; Li *et al.*, 2011), adhesion strength between contact surfaces (Deng *et al.*, 2012), tip sizes (Smolyanitsky *et al.*, 2012) and temperature (Zhang *et al.*, 2015) significantly affect the friction properties of graphene or grapheme coated materials.

The number of graphene layers may be another important factor affecting the friction behaviors and characteristics of the overall system. Once more graphene layers are present, the interlayer van der Waals interaction can minimize the puckering effect and thereby reduce the friction. FFM (friction force microscopy) measurements found that monolayer epitaxial graphene on SiC exhibits higher friction than bilayer graphene (Filleter *et al.*, 2009). Solution processed graphene layers reduce friction and wear on steel surfaces in air (Berman *et al.*, 2013). However, there are few reports on the layer dependent nanofriction of graphene coated systems (Egberts *et al.*, 2014). The atomic characteristics are as of yet unexplored. Furthermore, it is especially interesting to investigate the friction properties of metal foils coated with graphene. Graphene grown on copper foils using the chemical vapor deposition method supplies the perfect test sample which shows great merit as a surface coating because of its excellent scalability and transferability (Li *et al.*, 2009). In addition, there is no contamination and damage existing as there is no complex transfer process of graphene, such as PMMA (polymethyl methacrylate), which is typically used for the transfer process, that may have a significant effect on the results. The fundamental understanding of the atomistic process governing the friction behaviors and characteristics of multi-layer graphene/copper systems is still lacking.

In this chapter, we employ a faithful atomistic modeling to elucidate the friction behavior of graphene coated copper films, as well as the characteristics. The graphene/Cu system is selected because graphene on Cu(111) has a higher quality than that grown on other crystal orientations of copper (Gao *et al.*, 2010). We conduct atomic nano-indentation and scratching simulations of monolayer, bilayer, and trilayer graphene supported by a single crystal Cu(111) substrate. We have found that the friction reduces as the number of graphene layers increases. The corresponding behaviors and characteristics are explored in atomistic detail and excellent agreement with experiment is achieved.

24.3.1 Modeling and Method

Our molecular dynamics (MD) calculations in this study are performed using the Large-scale Atomic/Molecular Massively Parallel Simulator (LAMMPS) (Plimpton, 1995). We have used a rigid hemispherical diamond tip with a 2.5 nm radius to indent graphene layers grown on a Cu(111) substrate as shown in Figure 24.13. Our model is smaller than that used in recent nano-indentation experiments, which has a size of 15.37nm, 16.24nm, 14.81nm. However, our model dimension sizes are sufficiently large to simulate a semi-infinite boundary since the Cu(111) substrate block is cubic with a side length of around 5 times the tip radius (Klemenz *et al.*, 2014) and the stress is localized and not impacted by the boundaries. In addition, the periodic boundaries are also employed in both lateral directions to avoid spurious edge effects. The choice above represents a compromise that minimizes finite-size effects while keeping the simulations affordable.

The quality of the MD simulation results significantly depends on the accuracy of the potential function used. The adaptive intermolecular reactive empirical bond order (AIREBO) potential (Stuart *et al.*, 2000) is used to describe the interatomic interaction for the tip and graphene, because it has been shown to accurately capture the bond-bond interaction between carbon and hydrogen atoms as well as bond breaking and bond reforming. Therefore, the AIREBO potential is also reliable for the studies of friction on graphene. The interaction cutoff distance in the switching function of the AIREBO potential must be selected carefully to avoid the known spurious post-hardening behavior arising from an improper cutoff distance. We set the cutoff distance to be 2.0Å as suggested in reference (Pastewka *et al.*, 2008). Interatomic forces within the Cu(111) substrate are derived from an embedded atom method potential (Bai *et al.*, 2010). The interactions between tip and graphene, tip and copper, and graphene and copper are described by the van der Waals interaction as a 12-6 Lennard-Jones (LJ) potential. The parameters of the LJ potentials in our model are as follows: $\varepsilon_{C-C}=2.86$ meV, $\sigma_{C-C}=0.347$ nm, $\varepsilon_{C-Cu}=11.7$ meV, $\sigma_{C-Cu}=0.300$nm (Huang *et al.*, 2015), which actually indicates the adhesion strength between the tip, graphene, and substrate, a key factor in determining the frictional characteristics of the system.

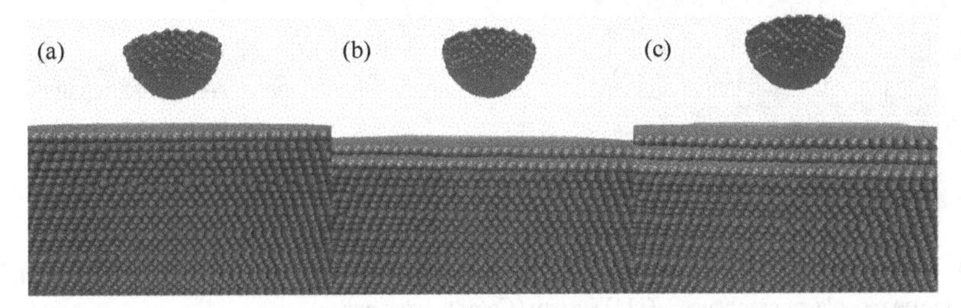

Figure 24.13 Model of nano-indentation and scratch on (a) monolayer graphene/Cu(111), (b) bilayer graphene/Cu(111), and (c) trilayer graphene/Cu(111) (*Color version online*)

The time increment of simulations is fixed at 1.0fs, and a Langevin thermostat is applied to the whole system to maintain a stable temperature of 300K. Firstly, the system is relaxed to minimum energy state for 15ps. Then, the tip moves down and penetrates into the graphene/Cu(111) system at the speed of 0.05nm/ps while the bottom copper atoms with thickness of 0.55nm are fixed. Finally, the tip scratches laterally at a speed of 0.015nm/ps while the normal (F_N) and lateral forces (F_L) acting on the rigid tip are recorded, which gives us much information of the scratch process of the tip on monolayer, bilayer, and trilayer graphene/Cu(111). Using the three procedures above, we can study the nanoscale friction behaviors and characteristics of different layers of graphene/Cu(111) systems by gradually changing the nano-indentation depth. In order to assess graphene's protective potential, analogous indentation and scratch simulations with a bare Cu(111) substrate are also performed.

24.3.2 Results and Discussion

(i) Scratching

A qualitatively distinct scratch behavior for a 2.06nm nano-indentation depth on a monolayer graphene/Cu(111) system is presented in Figure 24.14. After a nano-indentation depth of 2.06nm, the tip scratch laterally [Figure 24.13(a)] above the surface, and the F_L and F_N on it are shown in Figure 24.14(b). Initially, F_L increases while the F_N decreases with fluctuations, and both of them become stable after scratching about 2.40nm laterally. The phenomenon is also present in that of tip scratching laterally on graphene/Pt system (Klemenz *et al.*, 2014) because of periodic lattices of graphene and the underlying substrate. We consider the values after $L=2.40$nm as $F_L=(142.57\pm8.56)$ nN, $F_N=(494.45\pm14.45)$nN.

Similarly, the F_L and F_N at different nano-indentation depths are calculated at the state before the rupture of graphene, as shown in Figure 24.15. We choose some critical value point and insert its corresponding state of nano-indentation and scratching laterally. The state after the rupture of graphene is shown in the bottom right.

From Figure 24.15, we can see that both F_L and F_N increase as the nano-indentation depth increases before the rupture of graphene. However, the relationship is not linear which is in contrast to classical friction laws.

At small nano-indentation depths, monolayer graphene/Cu(111) mainly deforms elastically under the sliding of the tip. The lateral force is small and shows a modulation of 0.25nm lattice constant of graphene , a behavior reminiscent of stick-slip which requires dragging springs (Dong *et al.*, 2013). The characteristics at very small nano-indentation depths are consistent with a previous study (Klemenz *et al.*, 2014) and experiments on SiO_2 (Kim *et al.*, 2011). The lateral force F_L always fluctuates around 0nN before F_N increases to 74.21nN. These results indicate that when the system deforms in the elastic regime, there can be very low and even negligible friction for sliding on the tip, which is more pronounced for graphene-covered Cu than for the bare Cu.

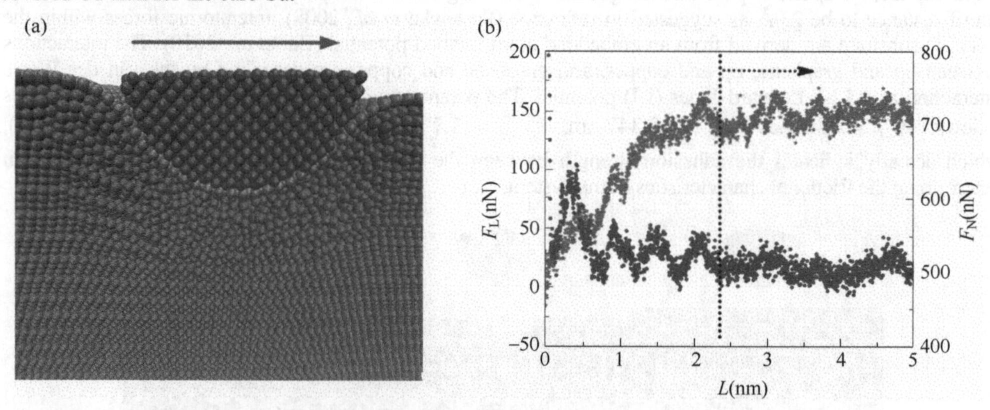

Figure 24.14 (a) Cross section of nanoscratch at 2.06nm normal nano-indentation depth; (b) its corresponding lateral force and normal force on monolayer graphene/Cu(111) system (*Color version online*)

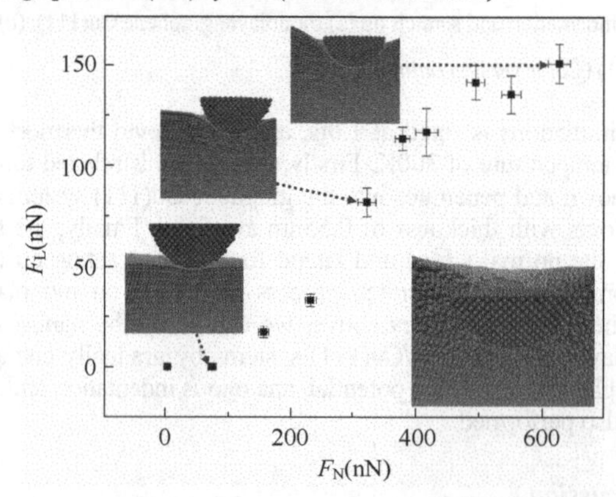

Figure 24.15 F_L-F_N relationship of monolayer graphene/Cu(111) with inserted figures of critical nano-indentation (*Color version online*)

When the penetration depth re aches deep enough to induce plastic deformation in the Cu(111) substrate, F_L becomes finite as shown in Figure 24.14. As the tip has penetrated into the substrate deep enough to induce fracture of monolayer graphene, the graphene ruptures further during the scratch and the lateral force increases suddenly and significantly while the normal force drops down to that of bare Cu(111), which is in agreement with that of graphene/Pt system (Klemenz *et al.*, 2014). This jump means that graphene has little or no effect on indentation and scratch behaviors, indicating that graphene layers wear out and lose their effectiveness on the mechanical behaviors of the Cu(111) substrate. Similar to the monolayer graphene/Cu(111) system above, the F_L–F_N relationship of bilayer and trilayer graphene/Cu(111) systems are shown in Figures 24.15 and 24.16, respectively.

(ii) F_L–F_N and μ–F_N

In order to quantitatively evaluate how different the friction characteristics of the three graphene/Cu(111) systems and the protective effect of graphene, their F_L–F_N relationships are shown in Figure 24.16(a) considering bare Cu(111). The coefficient of friction is defined as $\mu = F_L/F_N$, and the corresponding μ-F_N relationship is shown in Figure 24.16(b). It can be seen from Figure 24.16 that the lateral force changes nonlinearly with relatively high normal force, which are different from the results at significantly lower normal loads as shown in Figure 24.17 predicting sub-linear dependence of lateral force on the normal load. The nonlinearity comes from the large plastic deformation of underlying copper substrate at relatively high normal loads.

The graphene layers significantly change the friction characteristics of bare Cu(111), and the monolayer, bilayer, and trilayer graphene/Cu(111) also show very different characteristics. Furthermore, the friction force of Cu(111) covered with more layers is reduced. Compared to bare Cu(111), graphene coated Cu(111) can hold larger normal force. As to bare Cu(111), the normal force remains stable while the lateral force monotonically increases with the increase of nano-indentation depth, which is consistent with that of bare Pt substrate (Klemenz *et al.*, 2014).

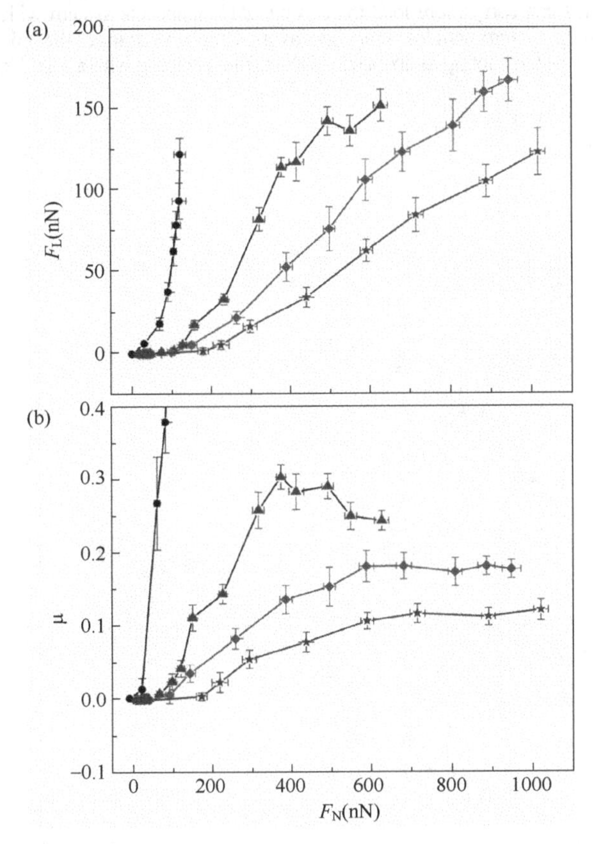

Figure 24.16 Normal load dependence of (a) lateral force, (b) coefficient of friction for bare Cu(111), mono-, bi-, and tri-layer graphene/Cu(111)

—●—Bare Cu(111); —▲—Monolayer Graphence/Cu(111); —◆—Bilayer Graphene/Cu(111); —★—Trilayer Graphene/Cu(111)

Furthermore, the F_L-F_N curves of graphene/Cu(111) systems mainly consist of four regimes (Klemenz *et al.*, 2014): the elastic regime, the small plastic regime, the large plastic regime, and the regime after rupture of graphene layers. In the first regime, the lateral force remains small and relatively stable while the normal force increases, which means negligible friction and super lubricity. This elastic regime holds until the normal force reaches 74.22nN, 98.97nN, and 192.30nN for the monolayer, bilayer, and trilayer graphene/Cu(111) systems, respectively. In the next regime, the lateral force increases monotonically rather than linearly with the normal load ranging from 74.22nN to 378.36nN for the monolayer graphene/Cu(111) system [from 98.97nN to 589.26nN for the bilayer graphene/Cu(111) system, and from 179.50nN to 715.43nN for the trilayer graphene/Cu(111) system]. In the large plastic regime, the lateral force increases almost linearly with the normal force from 589.26nN to 946.93nN for the bilayer graphene/Cu(111) system [from 715.43 nN to 1018.56nN for the trilayer graphene/Cu(111) system]. However, the lateral force increases in fluctuations with the normal force from 378.36nN to 627.82nN for the monolayer graphene/Cu(111) system because the tip interacts with the Cu(111) substrate more strongly than that in the other two systems. Finally, the normal force drops down to 112.70nN while the lateral force increases significantly and steadily with the nano-indentation depth, which means graphene layers crack and rupture, and then lose their protective effects. In the last regime, the scratch characteristics are the same as that of bare Cu(111), which is not shown in Figure 24.16(a).

In order to reveal the physical mechanism of the characteristics observed above, we obtain the relationship between the lateral force, the normal force and the nano-indentation depth as shown in Figure 24.17.

From Figure 24.17(a), we find that the normal force strongly differs between bare Cu(111), monolayer graphene/Cu(111), bilayer graphene/Cu(111) and trilayer graphene/Cu(111) systems. For a given h, graphene/Cu(111) systems can carry more load than bare Cu(111), and the system with more graphene layers can carry more load than the system with less graphene layers. Graphene changes the indentation hardness of the overall graphene/Cu(111) system, acting as an elastic high-stiffness coating with a high Young's modulus.

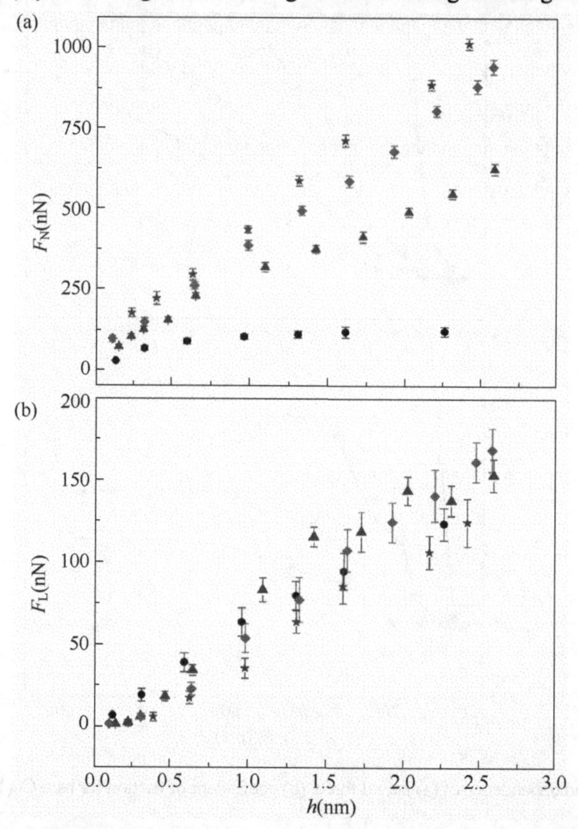

Figure 24.17 Nano-indentation depth dependence of (a) normal force, (b) lateral force for bare Cu(111), mono-, bi-, and tri-layer graphene/Cu(111)

• Bare Cu(111); ▲ Monolayer Graphene/Cu(111); ◆ Bilayer Graphene/Cu(111); ⋆ Trilayer Graphene/Cu(111)

In contrast to the normal force, the lateral force mainly depends on the nano-indentation depth and changes less when the bare Cu(111) is covered with graphene, as well as with increasing number of layers, as shown in Figure 24.17(b). Graphene does not change the work that is necessary to scratch the copper substrate.

At the same normal load, the indentation depth decreases from bare Cu(111), monolayer graphene/Cu(111), bilayer graphene/Cu(111) to trilayer graphene/Cu(111) which indicates that it is the indentation depth reduction responsible for the reduction of friction.

The characteristics of graphene/Cu(111) above come from that graphene should only change the indentation hardness of the overall graphene/Cu(111) system but not the work that is necessary to scratch it, which are in excellent agreement with that of graphene coated Pt substrate (Klemenz *et al.*, 2014) indicating that our results are reasonable. What's more, different number of graphene layers shows different nanofriction characteristics that is also in agreement with simulations and experiments of suspended graphene and supported graphene by other substrates (Kim *et al.*, 2011; Filleter *et al.*, 2009).

The coefficients of friction of the bare Cu(111), monolayer, bilayer, and trilayer graphene/Cu(111) systems are a function of the normal force, as shown in Figure 24.16(b). For the graphene/Cu(111) systems, the nanofriction curve can be divided into four below 0.01. This is a sliding regime. The coefficient of friction monotonically increases to 0.30 with normal forces ranging from 74.22nN to 378.36nN for the monolayer graphene/Cu(111) system [to 0.18 from 98.97nN to 589.26nN for the bilayer graphene/Cu(111) system, and to 0.11 from 192.30nN to 715.43nN for the trilayer graphene/Cu(111) system], which means a large amount of elastic and little plastic deformation. In part III, the coefficient of friction drops down to a stable 0.24 ranging from 378.36nN to 627.82nN for the monolayer graphene/Cu(111) system while it holds stably at about 0.18 ranging from 589.26nN to 946.93nN for the bilayer graphene/Cu(111) system, and about 0.11 ranging from 715.43nN to 1018.56nN for the trilayer graphene/Cu(111) system, indicating a large amount of plastic and little elastic deformation. In the part IV, the rupture of graphene layers occurs. The friction coefficient increases beyond 1.00 and increasing monotonically with the nano-indentation depth thereafter.

The same behavior is observed in bare Cu(111). However, the coefficient of friction of bare Cu(111) significantly increases with greater nano-indentation depth while the normal force holds stably around 112.70nN, which is not shown in Figure 24.16(b).

The results above indicate that the existence of graphene layers enhances the load carrying capacity, extends the low friction range, and effectively decreases the frictional force of Cu(111). In addition, the bilayer graphene reduces friction more than the monolayer graphene while the trilayer graphene reduces friction more than the bilayer graphene.

The normal force is large enough to stabilize the friction coefficient before the rupture of the graphene layers, and the relationship between the friction coefficient and the number of layers of graphene (N) is shown in Table 24.7.

Table 24.7 Relationship of friction coefficient and the number of graphene layers of graphene/Cu(111) system

N	0	1	2	3
μ	>1.0	0.24–0.30	0.18	0.11

Table 24.7 shows the distinctive nanofriction characteristics of monolayer, bilayer, trilayer graphene/Cu(111), and bare copper foil. Such a relationship can be used to determine the existence, and the number of layers of graphene, which suggests a routine to characterize the number of graphene layers on a substrate, as an alternative approach to Raman spectroscopy.

There are few studies about the friction coefficient of graphene coated copper foil. As the friction coefficient of SiO_2 substrate reduces from 0.68 to 0.12-0.22 after coating it with graphene (Kim *et al.*, 2011), our results about the friction coefficients of different graphene layers are reasonable. The differences are caused by the different surface interaction forces between the underlying substrate and graphene.

There are differences in the nanoscale friction characteristics of monolayer graphene/Cu(111), bilayer graphene/Cu(111), trilayer graphene/Cu(111), and the bare Cu(111). Graphene, due to its ultra-strong mechanical properties, significantly improves the normal load bearing capacity of a copper substrate through effectively constraining dislocation motion and providing increased resistance against dislocation propagation across the graphene-metal interface (Kim *et al.*, 2013).

24.4 Chapter Summary

We successfully explored modes and parameters to rebuild Raman frequencies detected from a perfect single crystal, and then, use this Raman verified mode to investigate properties of AlN, specifically aimed at future stress/strain and polarization analyses of AlN films or micro semiconductor devices by Raman detection. We investigate the nanoscale frictional behaviors and characteristics of mono-, bi-, and tri-layer graphene/Cu(111) systems using molecular dynamics simulations. By using the technique of MD, we published a paper (Wang *et al.*, 2016) in *Computational Materials Science* about the atomic details of the distinguished frictional behaviors of different layered graphene on Cu substrate in February 2016. Li *et al.* (Li *et al.*, 2016) published in *Nature* in November the same year about the atomic details of such frictional behaviors of graphene on silicon by molecular dynamics simulations (MD). Both works reproduced the experimental friction behaviors of different layered graphene systems.

References

Adam, T., Kolodzey, J., Swann, C. P., *et al.* (2001) The electrical properties of MIS capacitors with ALN gate dielectrics, *Applied Surface Science*, 175 428-435.

Arguello, C. A., Rousseau, D. L., Porto, S. P. S. (1969) First-order Raman effect in wurtzite-type crystals, *Physical Review*, 181(3), 1351.

Arguello, C. A., Rousseau, D. L., Porto, S.P.D.S. (1969) First-order Raman effect in wurtzite-type crystals, *Physical Review*, 181 (3) 1351.

Bai, X. M., Voter, A. F., Hoagland, R. G., Nastasi, M., Uberuaga, B. P. (2010) Efficient annealing of radiation damage near grain boundaries via interstitial emission, *Science*, 327 (5973) 1631-1634.

Baroni, S., De Gironcoli, S., Dal Corso, A., *et al.* (2001) Phonons and related crystal properties from density-functional perturbation theory, *Reviews of Modern Physics*, 73 (2) 515.

Bergman, L., Dutta, M., Balkas, C., *et al.* (1999) Raman analysis of the E1and A1 quasi-longitudinal optical and quasi-transverse optical modes in wurtzite AlN, *Journal of Applied Physics*, 85 (7) 3535-3539.

Berman, D., Erdemir, A., Sumant, A. V. (2013) Few layer graphene to reduce wear and friction on sliding steel surfaces, *Carbon*, 54, 454-459.

Berman, D., Erdemir, A., Sumant, A. V. (2014) Graphene: a new emerging lubricant, *Materials Today*, 17 (1) 31-42.

Bhushan, B. (2007) Nanotribology and nanomechanics of MEMS/NEMS and BioMEMS/BioNEMS materials and devices, *Microelectronic Engineering*, 84 (3) 387-412.

Born, M., Huang, K. (1954) Dynamical theory of crystal lattices.

Brantley, W. A. (1973) Calculated elastic constants for stress problems associated with semiconductor devices, *Journal of Applied Physics*, 44(1), 534-535.

Bruls, R. J., Hintzen, H. T., De With, G., *et al.* (2001) The temperature dependence of the Young's modulus of $MgSiN_2$, AlN and Si3N4, *Journal of the European Ceramic Society*, 21(3), 263-268.

Bryan, I., Rice, A., Hussey, L., Bryan, Z., *et al.* (2013) Strain relaxation by pitting in AlN thin films deposited by metalorganic chemical vapor deposition, *Applied Physics Letters*,102 (6) 061602.

Bunch, J. S., Verbridge, S. S., Alden, J. S., van der Zande, A. M., Parpia, J. M., Craighead, H. G., McEuen, P.L. (2008) Impermeable atomic membranes from graphene sheets, *Nano Letters*, 8 (8) 2458-2462.

Callsen, G., Reparaz, J. S., Wagner, M. R., *et al.* (2011) Phonon deformation potentials in wurtzite GaN and ZnO determined by uniaxial pressure dependent Raman measurements, *Applied Physics Letters*, 98(6), 061906.

Callsen, G., Wagner, M. R., Reparaz, J. S., *et al.* (2014) Phonon pressure coefficients and deformation potentials of wurtzite AlN determined by uniaxial pressure-dependent Raman measurements, *Physical Review B*, 90 (20) 205206.

Callsen, G., Wagner, M. R., Reparaz, J. S., *et al.* (2014) Phonon pressure coefficients and deformation potentials of wurtzite AlN determined by uniaxial pressure-dependent Raman measurements, *Physical Review B*, 90(20), 205206.

Campbell, S. A. (1996) The Science and Engineering of Microelectronic Fabrication, 2nd ed. Oxford University Press, New York.

Davydov, V. Y., Kitaev, Y. E., Goncharuk, I. N., *et al.* (1998) Phonon dispersion and Raman scattering in hexagonal GaN and AlN, *Physical Review B*, 58 (19) 12899.

Del Alamo, J. A., Joh, J. (2009) GaN HEMT reliability, *Microelectronics-Reliability*, 49(9), 1200-1206.

Demangeot, F., Frandon, J., Renucci, M. A., *et al.* (1998) Raman study of resonance effects in $Ga_{1-x}Al_xN$ solid solutions, *Materials Research Society Internet Journal of Nitride Semiconductor Research*, 3, e52.

Deng, Z., Klimov, N. N., Solares, S. D., Li, T., Xu, H., Cannara, R. J. (2012) Nanoscale interfacial friction and adhesion on supported versus suspended monolayer and multilayer graphene, *Langmuir*, 29 (1) 235-243.

Deng, Z., Smolyanitsky, A., Li, Q., Feng, X. Q., Cannara, R. J. (2012) Adhesion-dependent negative friction coefficient on chemically modified graphite at the nanoscale, *Nature Materials*, 11 (12) 1032-1037.

Derjaguin, B. V., Muller, V. M,. Toporov, Y. P. (1975) Effect of contact deformations on the adhesion of particles, *Journal of Colloid and Interface Science*, 53 (2) 314-326.

Dong, Y., Li, Q., Martini, A. (2013) Molecular dynamics simulation of atomic friction: A review and guide, *Journal of Vacuum Science & Technology*, 31 (3) 030801.

Falub, C. V., von Ka¨nel, H., Isa, F., *et al.* (2012) Scaling hetero-epitaxy from layers to three-dimensional crystals, *Science* 335(6074), 1330-1334.

Feng, X., Kwon, S., Park, J. Y., Salmeron, M. (2013) Superlubric sliding of graphene nanoflakes on graphene, *ACS Nano*, 7 (2) 1718-1724.

Filleter, T., McChesney, J. L., Bostwick, A., Rotenberg, E., Emtsev, K. V., Seyller, T., Bennewitz, R. (2009) Friction and dissipation in epitaxial graphene films, *Physical Review Letter*,102 ,086102.

Gao, L., Guest, J. R., Guisinger, N. P. (2010) Epitaxial graphene on Cu(111), *Nano Letter*, 10 (9) 3512-3516.

Gleize, J., Renucci, M. A., Frandon, J., Bellet-Amalric, E., *et al.* (2003) Phonon deformation potentials of wurtzite AlN, *Journal of Applied Physics*, 93 (4) 2065-2068.

Gleize, J., Renucci, M. A., Frandon, J., *et al.* (2003) Phonon deformation potentials of wurtzite AlN, *Journal of Applied Physics*, 93(4), 2065-2068.

Gnecco, E., Bennewitz, R., Gyalog, T., Loppacher, C., Bammerlin, M., Meyer, E., Güntherodt, H. J. (2000) Self-organized patterning of an insulator-on-metal system by surface faceting and selective growth: NaCl/Cu (211), *Physical Review Letters*, 84 (6) 1172-1175.

Goni, A. R., Siegle, H., Syassen, K. (2001) Effect of pressure on optical phonon modes and transverse effective charges in GaN and AlN, *Physical Review B*, 64 (3) 035205.

Gonze, X. (1997) Dynamical matrices, Born effective charges, dielectric permittivity tensors, and interatomic force constants from density-functional perturbation theory, *Physical Review B*, 55 (16) 10337.

Hadi, W. A., Shur, M. S., O'Leary, S. K. (2014) Steady-state and transient electron transport within the wide energy gap compound semiconductors gallium nitride and zinc oxide: an updated and critical review, *Journal of Materials Science: Materials in Electronics*, 25(11), 4675-4713.

Harima, H. (2002) Properties of GaN and related compounds studied by means of Raman scattering, *Journal of Physics: Condensed Matter*, 14 (38) R967.

Hirayama, H., Fujikawa, S., Kamata, N. (2015) Recent progress in AlGaN-Based deep-UV LEDs, *Electronics and Communications in Japan*, 98(5), 1-8.

Hopcroft, M., Nix, W. D., Kenny, T. W. (2010) What is the Young's Modulus of Silicon?, *Journal of Microelectromechanical Systems*, 19(2), 229-238.

Huang, H., Tang, X., Chen, F., Yang, Y., Liu, J., Li, H., Chen, D. (2015) Radiation damage resistance and interface stability of copper-graphene nanolayered composite, *Journal of Nuclear Materials*, 460, 16-22.

Huang, X. Qi, X., Boey, F., Zhang, H. (2012) Graphene-based composites, *Chemical Society Reviews* 41 (2) 666-686.

Jayasekera, T., Mintmire, J. W. (2007) Transport in multiterminal graphene nanodevices, *Nanotechnology* 18 (42) 424033.

Johnson, K. L., Kendall, K., Roberts, A. D. (1971) Surface energy and the contact of elastic solids, *Proceedings of the Royal Society A*, 324 (1558) 301-313.

Kallel, T., Dammak, M., Wang, J., *et al.* (2014) Raman characterization and stress analysis of AlN: Er^{3+} epilayers grown on sapphire and silicon substrates, *Materials Science and Engineering: B*, 187 46-52.

Kim, K. S., Asay, D. B., Dugger, M.T. (2007) Nanotribology and MEMS, *Nano Today* ,2 (5) 22-29.

Kim, K. S., Lee, H. J., Lee, C., Lee, S. K., Jang, H., Ahn, J. H., Lee, H. J. (2011) Chemical vapor deposition-grown graphene: the thinnest solid lubricant, *ACS Nano*, 5 (6) 5107-5114.

Kim, Y., Lee, J., Yeom, M. S., Shin, J. W., Kim, H., Cui, Y., Han, S. M. (2013) Strengthening effect of single-atomic-layer graphene in metal-graphene nanolayered composites, *Nature Communications*, 4: 3114.

Kisielowski, C., Kru¨ger, J., Ruvimov, S., *et al.* (1996) Strain-related phenomena in GaN thin films, *Physical Review B*, 54(24), 17745.

Klemenz, A., Pastewka, L., Balakrishna, S. G., Caron, A., Bennewitz, R., Moseler, M. (2014) Atomic scale mechanisms of friction reduction and wear protection by graphene, *Nano Letter*, 14(12): 7145−7152.

Kluge, M. D., Ray, J. R., Rahman, A. (1986) Molecular dynamic calculation of elastic constants of silicon, *The Journal of Chemical Physics*, 85(7), 4028-4031.

Koren, E., Lörtscher, E., Rawlings, C., Knoll, A.W., Duerig, U. (2015) Adhesion and friction in mesoscopic graphite contacts, *Science*, 348 (6235) 679-683.

Kuball, M. (2001) Raman spectroscopy of GaN, AlGaN and AlN for process and growth monitoring/control, *Surface and Interface Analysis*, 31(10), 987-999.

Kuball, M. (2011) Raman spectroscopy of GaN, AlGaN and AlN for process and growth monitoring/control, *Surface and Interface Analysis*, 31 (10) 987-999.

Landolt-Bornstein, H. (1979) in Crystal and Solid State Physics, ed. By K.-H. Hellwege (Springer, Berlin, 1979), p. 116.

Leach, J. H., Shishkin, Y., Udwary, K., *et al.* (2014) Large-area bow-free n+ GaN Templates by HVPE for LEDs SPIE OPTO, *International Society for Optics and Photonics*, pp. 898602-1-898602-13.

Lee, C., Li, Q., Kalb, W., Liu, X. Z., Berger, H., Carpick, R.W., Hone, J. (2010) Frictional characteristics of atomically thin sheets, *Science*, 328(5974) 76-80.

Lee, C., Wei, X., Kysar, J. W., Hone, J. (2008) Measurement of the elastic properties and intrinsic strength of monolayer graphene, *Science* 321 (5887) 385-388.

Lee, H., Lee, N., Seo, Y., Eom, J., Lee, S. (2009) Comparison of frictional forces on graphene and graphite, *Nanotechnology*, 20 (32) 325701.

Łepkowski, S. P. (2015) Inapplicability of Martin transformation to elastic constants of zinc-blende and wurtzite group-III nitride alloys, *Journal of Applied Physics*, 117 (10) 105703.

Leung, B., Han, J., Sun, Q. (2014) Strain relaxation and dislocation reduction in AlGaN step-graded buffer for crack-free GaN on Si(111) , *Physica Status Solidi*, 11(3-4), 437-441.

Li, Q., Dong, Y., Perez, D., Martini, A., Carpick, R.W. (2011) Speed dependence of atomic stick-slip friction in optimally matched experiments and molecular dynamics simulations, *Physical Review Letter*, 106 (12) 126101.

Li, S. Z., Li, Q. Y., Carpick, R. W., *et al.* (2016) The evolving quality of frictional contact with graphene, *Nature*, 539:541-545.

Li, X., Cai, W., An, J., Kim, S., Nah, J., Yang, D., Ruoff, R. S. (2009) Large-area synthesis of high-quality and uniform graphene films on copper foils, *Science*, 324 (5932) 1312-1314.

Liechti, K. M. (2015) Understanding friction in layered materials, *Science*, 348 (6235) 632-633.

Liu, P., De Sarkar, A., Ahuja, R. (2014) Shear strain induced indirect to direct transition in band gap in AlN monolayer nanosheet, *Computational Materials Science*, 86 206-210.

Loebl, H. P., Metzmacher, C., Milsom, R. F., *et al.* (2004) RF bulk acoustic wave resonators and filters, *Journal of Electroceramics*, 12 (1-2) 109-118.

Lu, J. Y., Wang, Z. J., Deng, D. M., *et al.* (2010) Determining phonon deformation potentials of hexagonal GaN with stress modulation, *Journal of Applied Physics*, 108(12), 123520.

Manjón, F. J., Errandonea, D., Romero, A. H., *et al.* (2008) Lattice dynamics of wurtzite and rocksalt AlN under high pressure: Effect of compression on the crystal anisotropy of wurtzite-type semiconductors, *Physical Review B*, 77 (20) 205204.

Margine, E. R., Bocquet, M. L., Blase, X. (2008) Thermal stability of graphene and nanotube covalent functionalization, *Nano Letters*, 8 (10) 3315-3319.

Mattsson, A. E., Schultz, P. A., Desjarlais, M. P., *et al.* (2005) Designing meaningful density functional theory calculations in materials science—a primer, *Modelling and Simulation in Materials Science and Engineering*,13 (1) R1.

McNeil, L. E., Grimsditch, M., French, R. H. (1993) Vibrational spectroscopy of aluminum nitride, *American Ceramic Society*, 76: 1132.

Meneghesso, G., Verzellesi, G., Danesin, F., *et al.* (2008) Reliability of GaN high-electron-mobility transistors: state of the art and perspectives, *IEEE Transactions on Device & Materials Reliability*, 8(2), 332-343.

Monkhorst, H. J., Pack, J. D. Special points for Brillouin-zone integrations, *Physical Review B*, 13 (12) (1976) 5188.

Moram, M. A., Vickers, M. E. (2009) X-ray diffraction of III-nitrides, *Reports on Progress in Physics*, 72(3), 036502

Nix, W. D. (1989) Mechanical properties of thin films, *Metallurgical and Materials Transactions A*, 20(11), 2217-2245.

Novoselov, K. S., Geim, A. K., Morozov, S.V., Jiang, D., Zhang, Y., Dubonos, S.V., Grigorieva, I. V., Firsov, A. A. (2004) Electric field effect in atomically thin carbon films, *Science* 306 (5696) 666-669.

Pandey, B. P., Kumar, V., Proupin, E. M. (2014) Elastic constants and Debye temperature of wz-AlN and wz-GaN semiconductors under high pressure from first-principles, *Pramana* 83 (3) 413-425.

Paolicelli, G., Tripathi, M., Corradini, V., Candini, A., Valeri, S. (2015) Nanoscale frictional behavior of graphene on SiO_2 and Ni (111) substrates, *Nanotechnology*, 26 (5) 055703.

Pastewka, L., Pou, P., Pérez, R., Gumbsch, P., Moseler, M. (2008) Describing bond-breaking processes by reactive potentials: Importance of an environment-dependent interaction range, *Physical Review B*, 78, (16) 161402.

Peng, Y., Wang, Z., Zou, K. (2015) Friction and wear properties of different types of graphene nanosheets as effective solid lubricants, *Langmuir*, 31 (28) 7782-7791.

Philip Egberts, Han, G. H., Liu, X. Z., Johnson, A. C., Carpick, R.W. Frictional behavior of atomically thin sheets: hexagonal-shaped graphene islands grown on copper by chemical vapor deposition, *ACS Nano*, 8 (5) (2014) 5010-5021.

Plimpton, S. (1995) Fast parallel algorithms for short-range molecular dynamics, *Journal of Computational Physics*, 117 (1) 1-19.

Polian, A., Grimsditch, M., Grzegory, I. (1996) Elastic constants of gallium nitride, *Journal of Applied Physics*, 79 (6) 3343-3344.

Raghavan, S., Redwing, J. M. (2011) *Group III-A Nitrides on Si: Stress and Microstructural Evolution*, CRC Press, Boca Raton.

Reeber, R. R., Wang, K. (1996) Thermal expansion and lattice parameters of group IV semiconductors, *Materials Chemistry and Physics*, 46(2), 259-264.

Reeber, R. R., Wang, K. (2000) Lattice parameters and thermal expansion of GaN, *Journal of Material Research*, 15(01), 40-44.

Reeber, R. R., Wang, K. (2001) High temperature elastic constant prediction of some group III-nitrides, *Materials Research Society Internet Journal of Nitride Semiconductor Research* 6: e3.

Reeber, R. R., Wang, K. (2001) High temperature elastic constant prediction of some group III-nitrides, *MRS Internet Journal of Nitride Semiconductor Research* 6, 3.

Sarua, A., Kuball, M., Van Nostrand, J. E. (2002) Deformation potentials of the E_2(high) phonon mode of AlN, *Applied Physics Letters*, 81 (8) 1426-1428.

Scholz, F. (2012) Semipolar GaN grown on foreign substrates: a review, *Semiconductor Science and Technology*, 27(2), 024002.

Segall, M. D., Lindan, P. J. D., Probert, M. J., *et al.* (2002) First-principles simulation: ideas, illustrations and the CASTEP code, *Journal of Physics: Condensed Matter*, 14 2717-2744.

Slack, G. A., Bartram, S. F. (1975) Thermal expansion of some diamondlike crystals, *Journal of Applied Physics*, 46(1), 89-98.

Smolyanitsky, A., Killgore, J. P. (2012) Anomalous friction in suspended graphene, *Physical Review B*, 86 (12) 125432.

Smolyanitsky, A., Killgore, J. P., Tewary, V. K. (2012) Effect of elastic deformation on frictional properties of few-layer graphene, *Physical Review B*, 85 (3) 035412.

Stuart, S. J., Tutein, A. B., Harrison, J. A. (2000) A reactive potential for hydrocarbons with intermolecular interactions, *The Journal of Chemical Physics*, 112 (14) 6472-6486.

Su, Y., Kravets, V. G., Wong, S. L., Waters, J., Geim, A. K., Nair, R. R. (2014) Impermeable barrier films and protective coatings based on reduced graphene oxide, *Nature Communications*, 5: 4843.

Taniyasu, Y., Kasu, M., Makimoto, T. (2006) An aluminium nitride light-emitting diode with a wavelength of 210 nanometres, *Nature*, 441 (7091) 325-328.

Van Driel, W. D., Yuan, C. A., Koh, S., *et al.* (2011) LED system reliability *12th International Conference on Thermal, Mechanical and Multi-Physics Simulation and Experiments in Microelectronics and Microsystems (EuroSimE)*, IEEE, pp. 1-5.

Wagner, J. M., Bechstedt, F. (2002) Properties of strained wurtzite GaN and AlN: Ab initio studies, *Physical Review B*, 66 (11) 115202.

Wang, L., Xu, C., Zhang, W., *et al.* (2015) Investigation of thermal-mechanical stress and chip-packaging-interaction issues in low-k chips, *16th International Conference on Electronic Packaging Technology (ICEPT) IEEE*, pp. 627-630.

Wang, W. H., Peng, Q., Dai, Y. Q., *et al.* (2016) Distinctive nanofriction of graphene coated copper foil, *Computational Materials Science*, 117:406-411.

Wei, M., Wang, X.. Pan, X., *et al.* (2011) Effect of high temperature AlGaN buffer thickness on GaN epilayer grown on Si(111) substrates, *Journal of Materials Science: Materials in Electronics*, 22(8), 1028-1032.

Wright, A. F. (1997) Elastic properties of zinc-blende and wurtzite AlN, GaN, and InN, *Journal of Applied Physics*, 82 (6) 2833-2839.

Xu, Y. N., Ching, W. Y. (1993) Electronic, optical, and structural properties of some wurtzite crystals, *Physical Review B*, 48 (7) 4335.

Y. Mo, K.T. Turner, I. Szlufarska .(2009) Friction laws at the nanoscale, *Nature*, 457 (7233) 1116-1119.

Ye, Z., Martini, A. (2014) Atomistic simulation of the load dependence of nanoscale friction on suspended and supported graphene, *Langmuir*, 30 (49) 14707-14711.

Yu, X., Tao, J., Shen, Y., Liang, G., Liu, T., Zhang, Y., Wang, Q. J. (2014) A metal-dielectric-graphene sandwich for surface enhanced Raman spectroscopy, *Nanoscale*, 6 9925-9929.

Zhang, B., Liu, Y. (2014) A review of GaN-based optoelectronic devices on silicon substrate, *Chinese Science Bulletin*, 59(12), 1251-1275.

Zhang, X. H., Zhao, C. L., Han, J. C., *et al.* (2013) Observation of symmetrically decay of A1 (longitudinal optical) mode in free-standing GaN bulk single crystal from Li_3N flux method, *Applied Physics Letters*, 102(1), 011916.

Zhang, Y., Dong, M., Gueye, B., Ni, Z., Wang, Y. (2015) Temperature effects on the friction characteristics of graphene, *Applied Physics Letters*, 107 (1) 011601.

Zhao, X., Hamilton, M., Sawyer, W. G., Perry, S. S. (2007) Thermally activated friction, Tribology Letters, 27 (1) 113-117.

Zhao, X., Phillpot, S. R., Sawyer, W. G., Sinnott, S. B., Perry, S. S. (2009) Transition from thermal to athermal friction under cryogenic conditions, *Physical Review Letters*, 102 (18) 186102.

Zheng, W., Zheng, R., Huang, F. *et al.* (2015) Raman tensor of AlN bulk single crystal, *Photonics Research*, 3 (2) 38-43.

Zheng, W., Zheng, R., Huang, F., *et al.* (2015) Raman tensor of AlN bulk single crystal, *Photonics Research*, 3(2), 38-43.

Zhou, S., Bongiorno, A. (2013) Origin of the chemical and kinetic stability of graphene oxide, *Scientific reports*, 3 2484.

Zhu, J. X., Yang, D., Yin, Z., Yan, Q., Zhang, H. (2014) Graphene and grapheme-based materials for energy storage applications, *Small*, 10 (17) 3480-3498.

Zhuang, D., Edgar, J. H., Liu, B., *et al.* (2004) Bulk AlN crystal growth by direct heating of the source using microwaves, *Journal of Crystal Growth*, 262(1), 168-174.

Zhuang, D., Edgar, J. H., Liu, B., Huey, H. E., *et al.* (2004) Bulk AlN crystal growth by direct heating of the source using microwaves, *Journal of Crystal Growth*, 262 (1) 168-174.

Appendix Conversion Tables and Constants

Table A1 General conversion factors

Quantity	U.S. unit	SI equivalent
Length	1in	=0.025400m
	1mil	$=0.025400\times10^{-3}$m
Area	$1in^2$	$=0.64516\times10^{-3}m^2$
Volume	$1in^3$	$=0.016387\times10^{-3}m^3$
Temperature	Temp.(°F)	=9/5×Temp.(°C)+32

Table A2 Gonversion factors for stress analysis

Quantity	U.S. unit	SI equivalent
Force	1lbf	=4.4482N
	1dyn	$=1\times10^{-5}$N
Pressure, Stress	1psi	=6894.8Pa
	1msi	$=6894.8\times10^{6}$Pa

Table A3 Important constants

Constant	U.S. unit	SI unit
Absolute zero	−459.67°F	−273.15°C
Acceleration of gravity	$32.174ft/s^2$	$9.8066m/s^2$
Atmospheric pressure	14.694psi	0.10132×10^{6}Pa
Stefan-Boltzmann constant	$0.1714\times10^{-8}Btu/(h\cdot ft^2\cdot °R^4)$ where Temp.(R)=Temp.(°F)+459.67	$5.669\times10^{-8}W/(m^2\cdot K^4)$ where Temp.(K)=Temp.(°C)+273.15